Recent Advances in Active Control of Sound and Vibration

Proceedings of the Conference on

Recent Advances in Active Control of Sound and Vibration

General Chairman: C. R. Fuller
Technical Chairman: R. A. Burdisso
Program Editors: C. A. Rogers, C. R. Fuller

Virginia Polytechnic Institute
and State University
Blacksburg, Virginia

15–17 April, 1991

Sponsored by:

NASA Langley Research Center
Office of Naval Research
U.S. Army Aerostructures Directorate

TECHNOMIC
PUBLISHING CO., INC.
LANCASTER · BASEL

Recent Advances in Active Control of Sound and Vibration
a **TECHNOMIC**® publication

Published in the Western Hemisphere by
Technomic Publishing Company, Inc.
851 New Holland Avenue
Box 3535
Lancaster, Pennsylvania 17604 U.S.A.

Distributed in the Rest of the World by
Technomic Publishing AG

Printed in the United States of America
10 9 8 7 6 5 4 3 2 1

Main entry under title:
Recent Advances in Active Control of Sound and Vibration

A Technomic Publishing Company book
Bibliography: p.

Library of Congress Card No. 91-65229
ISBN No. 87762-839-4

Contents

Preface

The last decade has witnessed a dramatic rise in research and development activity in the area of active control of sound and vibration. Although the potentiality of success appears high it is well to remember that the present research represents the third "wave" in the field. Perhaps the first applications were stimulated by the work of Paul Leug who in 1934 demonstrated that sound could be electronically "cancelled" with additional active acoustic control sources. The next peak in active control was centered around the classic work of Jessel, Swinbanks, and others who in the early 1970s derived many of the basic system equations to model closed loop behavior and established preliminary understanding of the physics. The present day research is of course due to advances in computer chips which enable the "real time" implementation of digital signal processing and control algorithms. These approaches have allowed the extremely fast implementation of system ID approaches, stimulating the use of adaptive algorithms and, as in the work of Widrow, reducing the need for accurate system models.

Associated with the present research activity is the development of advanced actuators and sensors. The area of smart or intelligent material systems which contain highly integrated actuator, sensor and control capabilities has matured a great deal over the past few years. The fundamental objectives of intelligent material systems of decreasing mass and energy needs or requirements to facilitate adaptive and/or multifunctioning active structure has been demonstrated most successfully in vibration and acoustic control and has inspired multidisciplinary research programs devoted to synthesizing new actuator and sensor material systems. The fusion of various scientific and technological disciplines to achieve optimal material system designs for active control is evident in the papers presented herein. We are perhaps on the threshold of a new era of material systems synthesis whereby design for static performance is certainly necessary but no longer sufficient.

The key to the successful application of active control may lie in the understanding of the physical mechanisms inherent in the control approaches. Thus much work has also been concentrated on studying the open and closed loop behavior of the physical system, with a view to efficient control approaches. Another important goal is to identify areas that need further research and development as well as promising new directions of research.

The main aim of the conference is to bring together researchers and engineers in the above disciplines. To some degree research in the key areas has happened independently. Thus one goal of the conference is to create synergism among the above research topics as all control approaches involve an integrated use of each component. To this end the conference will focus on recent advances in the component areas, as well as new applications of the integrated approaches.

The future for active control looks very promising as indicated by the richness and diversity of the papers presented at this conference. However, there are a number of critical problems to be addressed and many of these are key to the use of active control in realistic situations. Perhaps it is well to remember the fluctuating fortunes of active control in the past and it is with some restraint that applications in active control should be pursued to ensure that the present period reaches its deserved maturity and potential.

The conference organizers gratefully acknowledge the sponsorship of this meeting by NASA Langley Research Center, the Office of Naval Research, and the U.S. Army Aerostructures Directorate, and the participation of the following professional organizations, the American Institute of Aeronautics and Astronautics, the American Helicopter Society, the Acoustical Society of America, and the Institute of Noise Control Engineering.

C. R. Fuller
C. A. Rogers

Actuators

PIEZOELECTRIC ACTUATOR MODELS FOR ACTIVE SOUND AND VIBRATION CONTROL OF CYLINDERS

Harold C. Lester,[1] **Sylvie Lefebvre**[2]

ABSTRACT

Bending and in-plane piezoelectric actuator models, adapted from flat plate theory, are developed for controlling sound radiation and transmission associated with vibrating cylinders. The actuator models are evaluated in the low frequency regime (20 – 500) Hz. where active control is a viable concept for reducing vibration and noise levels. Coupling between the cylinder and interior acoustic cavity is examined by studying the modal spectra, particularly for the low-order cylinder modes which couple efficiently with the cavity at low frequencies. Over the frequency range investigated, the in-plane piezoelectric actuator model produces a more favorable distribution of low-order modes necessary for efficient interior noise control.

NOMENCLATURE

A_{mn}	modal coefficients for f_x
a	radius of cylinder
C_f	speed of sound in air
C_L	axial phase speed in cylinder
$\widehat{C}_{m'm}$	modal coupling coefficients - see eqs. (16)
E	modulus of elasticity of cylinder
F_c	piezoelectric actuator complex amplitude, in-plane model
F_{mn}	modal coefficients for f_x
f_x, f_θ, f_n	external forces acting on cylinder
G_{mn}	weighting coefficients, see eq. (6)
h	cylinder thickness
$H(x), H(\theta)$	unit step functions
J_n	complex Bessel function of order n

[1]Structural Acoustics Branch, Acoustics Division, NASA Langley Research Center, Hampton, VA 23665-5225, U.S.A.
[2]Department of Mechanical Engineering, Virginia Polytechnic Institute and State University, Blacksburg, VA 24061-0238, U.S.A.

k	free space wavenumber, $k = \omega/C_f$
L	length of cylinder
M_c	piezoelectric actuator complex amplitude, bending model
m	axial wavenumber
m_x, m_θ	external moments acting on cylinder
n	circumferential wavenumber
P_{mn}	modal coefficients for acoustic pressure inside cylinder
p	acoustic pressure in cylinder
T_{mn}	modal coefficients for f_θ
t	time
U_{mn}	modal amplitudes for u cylinder displacement
u, v, w	axial, circumferential, and normal displacements of cylinder
V_{mn}	modal amplitudes for v cylinder displacement
W_{mn}	modal amplitudes for w cylinder displacement
$\widehat{W}_{mn}$	modal coefficients for w - see eqs. (14)-(16)
x, r, θ	cylindrical coordinates
x_c, θ_c	piezoelectric actuator center coordinates
y	transformed coordinate, $y = a\theta$
α	cylinder stiffness parameter
α_m	radial wavenumber - see eqs. (12)
β	cylinder thickness parameter
$\Delta x, \Delta\theta$	dimensions of piezoelectric actuator
$\delta(x), \delta(\theta)$	Dirac delta functions
ε_m	equals 2 if m=0, equals 1 otherwise
ν	Poisson's ratio
ρ_f	density of air
ρ_s	density of cylinder
Ω	nondimensional frequency, $\Omega = \omega a/C_L$
ω	steady state circular frequency

Subscripts

m	denotes axial mode component
n	denotes circumferential mode component
p	denotes piezoelectric actuator property

Superscripts

c	denotes $\cos(n\theta)$ component
s	denotes $\sin(n\theta)$ component

INTRODUCTION

There has been a growing interest recently in the use of piezoelectric actuators for controlling structural vibration and the attendant transmission and radiation of sound. These materials are capable of applying direct force inputs to a structure and are much lighter and more compact than conventional shaker type actuators. Most of the recent applications of piezoelectric actuators have been concerned with controlling vibration levels and power flow in beams and plates[1 – 4]. Similar research devoted to using piezoelectric materials to control sound and vibration associated with shell structures has been primarily based on layered shell theory[5, 6]. That is, the analytical models assume that the piezoelectric material comprises a total, distinct layer of the shell configuration. It is of interest, however, to examine the coupling effects of piezoelectric materials of small patch areas[7], similar to beam and plate applications, when installed on vibrating shell type structures.

In this paper, the bending and in-plane piezoelectric actuator models developed for plates[2, 3] are adapted to a simply-supported cylindrical shell. The bending model simulates the effect of two out-of-phase piezoelectric actuators attached on opposite sides of the cylinder wall. By driving the two actuators out-of-phase, a normal stress distribution is produced which varies linearly through the shell's thickness. This approximates a state of pure bending about the middle surface. The essence of the modelling involves replacing the two piezoelectric actuators by a uniform, line moment distribution applied along the perimeter of the patch area. With the in-plane piezoelectric actuator model, the adjacent two actuators are driven in-phase creating an extensional deformation of the middle surface. The two piezoelectric actuators are therefore replaced by a uniform, in-plane line force distribution applied along the patch perimeter. The amplitude of the line moment or line force distribution acting on the cylinder's middle surface is a function of the piezoelectric input voltage.

The bending and in-plane piezoelectric actuator models are evaluated in the low frequency regime (under 500 Hz) where active control has proven to be a viable concept for reducing vibration levels and interior and radiated noise. The study will therefore be particularly concerned with the coupling of the piezoelectric actuator models with the low-order axial and circumferential cylinder modes. At low frequencies, it is these modes that most readily couple with the interior acoustic cavity modes. This coupling will be demonstrated by comparing computed cylinder and acoustic cavity responses as well as mode amplitude spectra for each type of piezoelectric actuator model. It will be established that the bending model couples to the cylinder in a fundamentally different manner than does the in-plane model. This produces some significant differences in the excitation of low-order cylinder and cavity modes. In addition, some limited calculations are presented to indicate how low-order mode coupling can be improved by using arrays of piezoelectric elements.

ANALYSIS

In the following sections, the time harmonic response equations of a simply-supported, elastic cylinder are formulated for two types of piezoelectric actuator models: (1) a bending model and (2) an in-plane model. The coordinate details for the two models are shown in figure 1, where an actuator model is shown centered at coordinates (x_c, θ_c) of a uniform cylinder of radius a, length L, and thickness h. Within the context of the present paper, both types of piezoelectric actuator models consist of two, congruent, rectangular sections of piezoelectric material bonded to opposite sides of the cylinder wall. Thus, one actuator model consists of two equal sections of piezoelectric material: one bonded on the inside of the cylinder and the other opposite the first, but bonded on the outside cylinder wall. With the bending model, the two opposite sections of piezoelectric material are driven out-of-phase, whereas with the in-plane model, the sections are driven in-phase. The differences between these two piezoelectric actuator models, particularly with regards to the type of cylinder response induced, will be elaborated upon further in the following sections.

It should be noted that the cylinder is assumed to possess simply-supported boundary conditions at the ends $x = 0$ and $x = L$ and the loading coefficients will be defined in terms of the natural vibration mode shapes for these boundary conditions. However, the modal approach can be easily generalized to other boundary conditions. It is further assumed that the cylinder is closed by rigid (acoustically hard) endcaps, thus forming a closed cylindrical cavity. Of particular concern in the low frequency regime will be how well these two piezoelectric actuator models excite low-order axial and circumferential modal responses in the cylinder. These modes are known to couple efficiently with the interior acoustic cavity.

In the following sections, the analyses for both the bending and in-plane piezoelectric actuator models are summarized. This will involve defining the respective loading distributions and determining the modal loading coefficients for the appropriate cylinder response equations. Finally, the analysis for the acoustic cavity response is outlined. All response variables and inputs are assumed to vary harmonically with time. An $e^{j\omega t}$ time dependence is assumed throughout the paper.

Cylinder Response For Bending Piezoelectric Actuator Model

As mentioned previously, the actuator models used in this paper are adapted from flat plate applications, where the bending (out-of-plane) and in-plane deformations are uncoupled. Therefore, for a flat plate application, driving the opposite two sections of piezoelectric material (comprising a single actuator model) out-of-phase will produce an axial stress distribution in the plate which varies linearly through the thickness. This produces the classical out-of-plane bending displacement situation. For this piezoelectric loading condition, it can be shown[3] that the effect of the actuator (acting on the plate) can be replaced by a line moment distribution acting around the perimeter of the piezoelectric patch area. This concept is illustrated in figure 2a. The piezoelectric actuator is located at coordinates (x_c, θ_c) and the patch dimensions are given by Δx and $a\Delta\theta$, where a is the radius of the cylinder.

It should be noted that the bending (out-of-plane) and in-plane cylinder displacements are coupled by curvature effects (eqs. A-1). Hence, any in-plane loading (f_x or f_θ) and/or out-of-plane loading (f_n) will produce both in-plane deformations (u and v) as well as out-of-plane displacements (w). Curvature effects are not accounted for in the piezoelectric actuator loading model, since the models are adapted from flat plate theory. It is therefore assumed that, if the patch dimensions are kept small relative to the cylinder radius, the error due to neglecting curvature effects in the actuator loading will remain small and that the piezoelectric actuator models developed herein will be valid for studying the low-order cylinder/cavity coupling effects.

The equations governing the motion of a cylinder are summarized in the Appendix (see eqs. A-1). For the bending piezoelectric actuator model, the in-plane loads are nulled, that is, $f_x = f_\theta = 0$. In the absence of a directly applied normal pressure, the loading provided by the line moment distribution shown in figure 2a can be developed as follows[8, 9]:

$$f_n(x,\theta) = \frac{\partial m_\theta}{\partial x} + \frac{1}{a}\frac{\partial m_x}{\partial \theta} \tag{1}$$

where:

$$m_x = F_c[H(x - x_1) - H(x - x_2)][\delta(y - y_2) - \delta(y - y_1)] \tag{2a}$$

$$m_\theta = F_c[H(y - y_1) - H(y - y_2)][\delta(x - x_2) - \delta(x - x_1)] \tag{2b}$$

Here $H(x)$ is the unit step function and $\delta(x)$ is the Dirac delta function. In equations (2), the change of variables $y = a\theta$ has been used along with the following definitions:

$$x_1 = x_c - \frac{\Delta x}{2} \tag{3a}$$

$$x_2 = x_c + \frac{\Delta x}{2} \tag{3b}$$

$$y_1 = y_c - \frac{\Delta y}{2} \tag{3c}$$

$$y_2 = y_c + \frac{\Delta y}{2} \tag{3d}$$

In equations (2) and (3), the actuator dimensions are specified by Δx and $\Delta\theta$ and the patch is centered at coordinates (x_c, θ_c) as indicated in figure 2. The parameter F_c is a complex constant proportional to the piezoelectric input voltage and dielectric properties.

The normal loading as expressed by equations (1) - (3) is now represented in terms of a modal series in which the basis functions are the mode shapes for the radial, $w(x, \theta)$, cylinder displacement (see eq. A-4c). Therefore,

$$f_n(x, \theta) = \sum_{n=0} \sum_{m=1} \sin\left(\frac{m\pi x}{L}\right) \left[F^c_{mn} \cos(n\theta) + F^s_{mn} \sin(n\theta)\right] \tag{4}$$

Using standard modal expansion techniques, it can be established that:

$$F^c_{mn} = \frac{8F_c}{L^2} \sin\left(\frac{m\pi\Delta x}{2L}\right) \sin\left(\frac{m\pi x_c}{L}\right) \begin{cases} \frac{m}{4}\Delta\theta & \text{if } n = 0 \\ G_{mn} \sin\left(\frac{n\Delta\theta}{2}\right) \cos(n\theta_c) & \text{otherwise} \end{cases} \tag{5a}$$

$$F^s_{mn} = \frac{8F_c}{L^2} \sin\left(\frac{m\pi\Delta x}{2L}\right) \sin\left(\frac{m\pi x_c}{L}\right) \begin{cases} 0 & \text{if } n = 0 \\ G_{mn} \sin\left(\frac{n\Delta\theta}{2}\right) \sin(n\theta_c) & \text{otherwise} \end{cases} \tag{5b}$$

The weighting coefficients G_{mn} is identified as follows:

$$G_{mn} = \left(\frac{m}{n} + \frac{nL^2}{m\pi^2 a^2}\right) \tag{6}$$

This coefficient, which also appears in the bending model for a flat plate[3], plays an important role with regard to how the low-order cylinder modes are excited by the bending piezoelectric actuator model. The weighting coefficient, G_{mn}, will be examined in more detail when results are presented and discussed.

Once the loading coefficients are known, the displacement response of the cylinder can be determined[8, 9]. Only the radial displacement, $w(x, \theta)$, is of interest here, since it is this motion that couples with the cylindrical cavity modes and thereby radiates sound into the cylinder's interior. This displacement is given by equation (A-3c) which is repeated here as follows:

$$w(x, \theta) = \sum_{n=0} \sum_{m=1} \sin\left(\frac{m\pi x}{L}\right) \left[W^c_{mn} \cos(n\theta) + W^s_{mn} \sin(n\theta)\right] \tag{7}$$

where the W^c_{mn} and W^s_{mn} modal coefficients for the cylinder displacement can be determined on a mode by mode basis in terms of the F^c_{mn} and F^s_{mn} modal coefficients for the loading $f_n(x, \theta)$.

Response Equations For In-Plane Piezoelectric Actuator Model

For the in-plane actuator model, the two opposite sections of piezoelectric material are driven in-phase. This model, when utilized on a flat plate[3], will produce a uniform stress distribution through the plate's thickness. In addition, only in-plane displacements will be produced, since the in-plane and bending (out-of-plane) plate displacements are uncoupled. Here, with the cylinder adaptation of this actuator plate model concept, in-plane deformations will produce, through the intrinsic curvature (coupling) effects of shell theory, bending (out-of-plane) displacements. Later, when results are presented and discussed, it will be seen that this coupling allows the in-plane piezoelectric actuator model to effectively excite significant levels of low-order mode participation in the out-of-plane cylinder displacement $w(x, \theta)$.

In order to simulate in-plane actuator loading, the effect of the piezoelectric material on the cylinder is replaced by a line force distribution acting on the perimeter of the patch area[3] as illustrated in figure 2b. Referring to equations (A-1), $f_n = 0$, since there are no external pressure loads acting on the cylinder, and the actuator loading can be expressed as

$$f_x = M_c[H(y - y_1) - H(y - y_2)][\delta(x - x_2) - \delta(x - x_1)] \tag{8a}$$

$$f_\theta = M_c[H(x - x_1) - H(x - x_2)][\delta(y - y_2) - \delta(y - y_1)] \tag{8b}$$

where, as before, $y = a\theta$ (see also eqs. 3).

Following the same procedure as was used for the bending piezoelectric actuator model, these loadings are representative as a modal series, that is,

$$f_x(x, \theta) = \sum_{n=0}\sum_{m=0} \cos\left(\frac{m\pi x}{L}\right)[A^c_{mn}\cos(n\theta) + A^s_{mn}\sin(n\theta)] \tag{9a}$$

$$f_\theta(x, \theta) = \sum_{n=0}\sum_{m=1} \sin\left(\frac{m\pi x}{L}\right)[T^c_{mn}\cos(n\theta) + T^s_{mn}\sin(n\theta)] \tag{9b}$$

The loading coefficients can be determined by standard modal expansion techniques and for $f_x(x, \theta)$ are as follows:

$$A^c_{mn} = \frac{-8M_c}{\pi L}\sin\left(\frac{m\pi\Delta x}{2L}\right)\sin\left(\frac{m\pi x_c}{L}\right)\begin{cases}\frac{\Delta\theta}{4} & \text{if } n = 0\\ \frac{1}{n}\sin\left(\frac{n\Delta\theta}{2}\right)\cos(n\theta_c) & \text{otherwise}\end{cases} \tag{10a}$$

$$A^s_{mn} = \frac{-8M_c}{\pi L}\sin\left(\frac{m\pi\Delta x}{2L}\right)\sin\left(\frac{m\pi x_c}{L}\right)\begin{cases}0 & \text{if } n = 0\\ \frac{1}{n}\sin\left(\frac{n\Delta\theta}{2}\right)\sin(n\theta_c) & \text{otherwise}\end{cases} \tag{10b}$$

Similarly, the modal loading coefficients for $f_\theta(x,\theta)$ are:

$$T^c_{mn} = \frac{-8M_c}{\pi^2 a}\sin(\frac{m\pi\Delta x}{2L})\sin(\frac{m\pi x_c}{L})\begin{cases} 0 & \text{if } n=0 \\ \frac{1}{m}\sin(\frac{n\Delta\theta}{2})\sin(n\theta_c) & \text{otherwise} \end{cases} \quad (11a)$$

$$T^s_{mn} = \frac{8M_c}{\pi^2 a}\sin(\frac{m\pi\Delta x}{2L})\sin(\frac{m\pi x_c}{L})\begin{cases} 0 & \text{if } n=0 \\ \frac{1}{m}\sin(\frac{n\Delta\theta}{2})\cos(n\theta_c) & \text{otherwise} \end{cases} \quad (11b)$$

The cylinder modal coefficients W^c_{mn} and W^s_{mn} appearing in equation (7) can be determined on a mode by mode basis in terms of the modal loading coefficients (eqs. 10 and 11).

Acoustic Response Inside Vibrating Cylinder

This section summarizes the equations defining the acoustic response produced inside a cylindrical cavity with rigid endcaps and vibrating sidewalls. The acoustic response inside the cylindrical cavity is coupled to the cylinder response through the momentum boundary condition evaluated on the inside shell wall. As will be established in the following, since the cylinder and cavity responses are expressed in terms of different modal basis functions, this will lead to the definition of a modal coupling coefficient.

The acoustic pressure response inside the vibrating cylinder is governed by Helmholtz's equation. Since the cylinder is assumed to be closed by hard end caps, it is convenient to expand the interior solution in terms of the natural vibration mode shapes of the cylindrical cavity as follows[10]:

$$\begin{aligned} p(x,r,\theta) = & \sum_{n=0}\sum_{m=0} P^c_{mn} J_n(\alpha_m r)\cos\left(\frac{m\pi x}{L}\right)\cos(n\theta) \\ & + \sum_{n=1}\sum_{m=0} P^s_{mn} J_n(\alpha_m r)\cos\left(\frac{m\pi x}{L}\right)\sin(n\theta) \end{aligned} \quad (12a)$$

$$\alpha^2_m = k^2 - \left(\frac{m\pi}{L}\right)^2 \quad (12b)$$

The modal coefficients P^c_{mn} and P^s_{mn} are determined from

$$\rho_f\omega^2 w(x,\theta) = \frac{\partial p(x,a,\theta)}{\partial r} \quad (13)$$

where the derivative of $p(x,r,\theta)$ is evaluated on the inside wall of the cylinder $(r=a)$.

The problem in applying this boundary condition is that the displacement response of the cylinder is represented with $\sin\left(\frac{m\pi x}{L}\right)$ basis functions (see eq. 7), whereas the acoustic pressure response (eqs. 12a) inside the vibrating cylinder is described in terms of $\cos\left(\frac{m\pi x}{L}\right)$ functions. These basis functions are the respective mode shapes for the two regimes. Therefore, before applying equation (13), it is necessary to expand the cylinder displacement $w(x,\theta)$ in terms of the $\cos\left(\frac{m\pi x}{L}\right)$ functions as follows:

$$w(x,\theta) = \sum_{n=0}\sum_{m=0} \cos\left(\frac{m\pi x}{L}\right)\left[\widehat{W}^c_{mn}\cos(n\theta) + \widehat{W}^s_{mn}\sin(n\theta)\right] \tag{14}$$

The modal amplitudes $\widehat{W}^c_{mn}$ and $\widehat{W}^s_{mn}$ appearing in equation (14) can therefore be determined from the W^c_{mn} and W^s_{mn} amplitudes in equation (7). The relationship is

$$\widehat{W}^c_{mn} = \sum_{m'=1} \widehat{C}_{m'm} W^c_{m'n} \tag{15a}$$

$$\widehat{W}^s_{mn} = \sum_{m'=1} \widehat{C}_{m'm} W^s_{m'n} \tag{15b}$$

where the $\widehat{C}_{m'm}$ modal coupling coefficient is defined as follows:

$$\widehat{C}_{m'm} = \begin{cases} \Delta^{(-)}_{m'm} + \Delta^{(+)}_{m'm} & \text{if } m' \neq m \\ 0 & \text{otherwise} \end{cases} \tag{16a}$$

$$\Delta^{(\pm)}_{m'm} = \frac{1}{\pi\varepsilon_m}\left[\frac{1-\cos\left(m' \pm m\right)\pi}{m' \pm m}\right] \tag{16b}$$

The parameter $\varepsilon_m = 2$ if $m = 0$, otherwise $\varepsilon_m = 1$. It should be emphasized again that this mode matching procedure couples the cylinder and acoustic cavity responses through the respective axial wave numbers for each domain. Equations (15) delineate this axial wave number coupling. With regards to the circumferential wave number, the cylinder and cavity responses can be determined on a mode by mode basis for the same value of n.

The pressure coefficients appearing in equation (12) can thus be calculated from

$$P^c_{mn} = P_{mn}(\omega)\,\widehat{W}^c_{mn} \tag{17a}$$

$$P^s_{mn} = P_{mn}(\omega)\,\widehat{W}^s_{mn} \tag{17b}$$

where the modal transfer function $P_{mn}(\omega)$ for the interior pressure is defined as follows:

$$P_{mn}(\omega) = \frac{\rho_f \omega^2}{\alpha_m J'_n(\alpha_m a)} \tag{18}$$

Summarizing, the modal transfer function $P_{mn}(\omega)$ relate the acoustic cavity mode amplitudes to the amplitudes of the $\cos\left(\frac{m\pi x}{L}\right)$ cylinder displacement terms as indicated in equations (17). This change in basis functions is required in order to couple the cylinder and cavity responses. The $\cos\left(\frac{m\pi x}{L}\right)$ amplitudes are related to the amplitudes of the cylinder natural vibration modes, that is, the $\sin\left(\frac{m\pi x}{L}\right)$ basis functions, through the coupling coefficient $\widehat{C}_{m'm}$ as indicated by equations (15) and (16).

DISCUSSION OF RESULTS

Typical response calculations are presented and interpreted for the bending and in-plane piezoelectric actuators models developed in the previous sections. These results will consist of both cylinder and interior cavity acoustic responses as well as the associated modal spectra.

A simply-supported, aluminum cylinder having hard end-caps is used for these calculations. It should be noted that the aluminum cylinder has the overall dimensions of $a = 0.832m$, $L = 3.65m$, and $h = 0.0017m$. Values of Young's modulus of $E = 7.1(10)^{10}$ N/m^2 and $E_p = 6.5(10)^{10}$ N/m^2 were used for the cylinder and piezoelectric materials, respectively. Poisson's ratio for both materials was assumed to be $\nu = \nu_p = 0.30$. Unless noted otherwise, the actuator configurations consist of two small, rectangular patches of piezoelectric material bonded to opposite sides of the cylinder wall. The patches measured 0.0381m (1.5 in.) by 0.0635m (2.5 in.). The patches were orientated with the longer dimension running circumferentially around the cylinder. Hence, in the equations for the piezo loading coefficients, $\Delta x = 0.0381$m and the angle subtended by the actuator models circumferentially is $\Delta\theta = 4.34°$. The actuators were always located at $x = \frac{L}{2}$. For the single piezoelectric actuator model cases, the actuators were located circumferentially at $\theta = 0°$. For the multiple actuator cases, symmetrical distributions about $\theta = 0°$ were used and the reader is referred to the figure captions for nominal changes in the size and number of actuators used for a particular calculation. Structural damping was accounted for in the cylinder equations by adding a small imaginary part $\eta_s C_L$ to the phase speed C_L. In a similar manner, a small imaginary part $\eta_f C_f$ was added to the speed of sound C_f to account for dissipative effects in air. Values of $\eta_s = 0.100$ and $\eta_f = 0.050$ were arbitrarily chosen as providing representative damping levels.

Cylinder Responses and Modal Spectra

Cylinder radial displacement responses, produced by the bending and in-plane actuator models located at $x_c = L/2$ and $\theta_c = 0.0$, are shown in figure 3. The nondimensional driving frequency is $\Omega = 0.194$, which corresponds to 200.0 Hz for the geometry under consideration. This is very close to the resonant frequency of the ($m = 0, n = 2$, lowest-order radial) acoustic cavity mode which is about 198.9 Hertz. This cavity mode is significant since it would most likely be excited by any external loading field, for example, a propeller noise field, acting on the cylinder. Since the cylinder is excited symmetrically about its center, only odd ($m = 1, 3, ...$) $sin(\frac{m\pi x}{L})$ cylinder modes and even ($m = 0, 2, ...$) $cos(\frac{m\pi x}{L})$ cavity modes are excited. In a similar manner, because the excitation is symmetric about the $\theta = 0$ axis, all $sin(n\theta)$ modal coefficients are identically zero. Each bending and in-plane data set graphed in figure 3 is normalized to its own respective maximum. Of interest here is not the absolute levels produced by the two different piezoelectric actuator models, but rather, the modal character of the respective response distributions.

From the results presented in part (a) of figure 3, the axial distributions look quite similar. That is, the contributions of axial mode order m is about the same. However, with reference to part (b) of figure 3, which shows the circumferential distributions at mid-cylinder, examination of the mode amplitude spectra to follow will establish that the cylinder response induced by the in-plane actuator model contains contributions of low-order circumferential modes, which are not excited by the bending actuator model. This fundamental difference, between the responses induced by the bending and in-plane actuator models, will be better delineated by the examination of the respective modal spectra which will be presented and discussed shortly.

Before introducing modal spectra, it is of interest to examine the interior acoustic cavity pressure response excited by the cylinder wall displacements presented in figure 3. The interior acoustic wall pressures are presented in figure 4, where again each respective response data set is normalized to its own maximum. In part (a) of figure 4, the in-plane case shows greater low-order axial mode participation as compared to the bending piezoelectric actuator model result. The circumferential distribution for the in-plane actuator model (fig. 4b) shows significant low-order circumferential mode participation.

Modal spectra for the cylinder and cavity responses shown in figures 3 and 4 are presented in figure 5 for the bending actuator model. Amplitudes in decibels are indicated by the grey scale, which ranges from black (0 dB) for the maximum level to white (−21 dB) for the minimum level by three decibel increments. The modal spectra W^c_{mn} for the cylinder displacement shows a distinct lack of low-order circumferential (that is, n) modes. In fact, the dominant modes are the ($m = 1, 3$ and,$5, n = 18$) modes, which have resonant frequencies of about $\Omega = 0.190$ (195.9 Hz.), $\Omega = 0.193$ (198.9 Hz.), and $\Omega = 0.200$ (206.8 Hz.), respectively. This emphasizes one of the complications of shell models such as the simply-supported cylinder used for these calculations. That is, the abundance of many modes (of different mode orders), but with nearly the same resonance frequencies[8, 9]. For example, the ($m = 3, n = 4$) cylinder mode has a resonant frequency of about $\Omega = 0.206$ (212.9 Hz), yet this low-order mode was not excited by the bending piezoelectric actuator model. The bending piezoelectric actuator model therefore appears to have the tendency to select, or to couple more efficiently with, the high-order cylinder modes.

Thus, it is not surprising that in part (b) of figure 5, which shows the modal spectra for the interior acoustic pressure response P^c_{mn}, the interior cavity response for the bending piezoelectric actuator model shows a lack of significant low-order circumferential mode participation. Thus, even though the cylinder is excited very close to the resonant frequency of the $(m = 0, n = 2)$ acoustic cavity mode, there is so little input from the cylinder in this mode that the primary interior acoustic response is forced by the bending actuator model to respond in the high-order circumferential modes. It should be recalled that the cavity response is described in terms of $\cos(\frac{m\pi x}{L})$ modes and for symmetric loading about $x = \frac{L}{2}$ only even $(m = 0, 2, ...)$ are excited.

This difference in low-order mode participation is important when considered from an active noise control point of view. For example, at 200.0 Hz a primary cavity response would most likely include a significant $(m = 0, n = 2)$ modal response. In order to reduce this contribution in the interior noise field, a single bending actuator would be highly ineffective, since this type of piezoelectric element does not produce adequate modal content at the required level.

Contrast the results of figure 5 with the cylinder displacement spectra W^c_{mn} shown in part (a) of figure 6 for the in-plane actuator model. Here low-order circumferential modal response is readily apparent. For example, the $(m = 3, n = 4)$ cylinder mode is only 6.1 dB down from the peak $(m = 3, n = 18)$ mode. Also, it should be pointed out that the $(m = 1, n = 2)$ cylinder mode is 19.6 dB down and is just out of the range of values covered by the grey scale and is therefore not shown on part (a) of figure 6. Although the amplitude of the $(m = 1, n = 2)$ cylinder mode is quite small, it produces, as seen in part (b) of figure 6, a dominant response in the $(m = 0, n = 2)$ acoustic cavity mode. Therefore, the in-plane piezoelectric actuator model couples much more efficiently with the low-order interior cavity modes and would thus be much better suited for noise control applications. It should be noted that the respective spectra for the bending and in-plane cases (figs. 5 and 6) are normalized to the respective maximum for each data set.

The reason for this difference in the manner in which the two piezoelectric actuator models excite the cylinder can be traced to a fundamental difference between the loading coefficients for the bending model (eqs. 5) and the loading coefficients for the in-plane model (eqs. 10 and 11). In the case of the former, a weighting coefficient G_{mn} has been identified (eq. 6). The coefficients for the in-plane model involve $\frac{1}{n}$

and $\frac{1}{m}$ factors which tend to accentuate the participation of the modes with small wave numbers, in either spatial dimension. The behavior of G_{mn} is considerably more complicated. These weighting coefficients are presented in spectral form in figure 7. It is immediately obvious that, (depending on the length to radius ratio of the cylinder), G_{mn} assumes reduced values for small mode orders and also when m and n have nearly the same value. Thus, the nature of the weighting coefficients G_{mn} is to accentuate the participation of the cylinder modes where the ratio of the mode orders is large. Hence, the single bending moment actuator model is not as effective as the in-plane model in exciting low-order cylinder responses. In support of these predictions, it should be noted that for cylinders deforming in low-order circumferential modes, most of the elastic strain energy is associated with stretching[11 – 13]. It therefore appears that direct in-plane force inputs can provide efficient coupling with the low-order cylinder and cavity modes.

The cylinder and cavity responses presented in figures 5 and 6 were generated by a single, small (1.5 by 2.5 in.) bending or in-plane actuator model located at $x = \frac{L}{2}$ and $\theta = 0°$. Recall that these actuator models consist of two sections of piezoelectric material bonded to opposite sides of the cylinder wall and driven out-of-phase or in-phase. It is of interest to determine the effect of either making these actuator models longer or using a number of smaller elements, particularly in the azimuthal direction. By making the effective circumferential length of the piezoelectric material longer, it should be possible to enhance coupling with the low-order circumferential modes.

In figures 8 - 11, cylinder and cavity responses are presented for the situation when the piezoelectric material extends from $-45.0°$ to $45.0°$ but remains the same width of $\Delta x = 0.0831$m (1.5 in.). The motivation for choosing this length comes from equations (5), (10), and (11). Previously it has been established that the $(m = 0, n = 2)$ acoustic cavity mode is near resonance at 200.0 Hertz. In these equations $\sin(\frac{n\Delta\theta}{2}) = 1.0$ when $n = 2$ and $\Delta\theta = \frac{\pi}{2}$. Therefore, all calculations to follow are for the case when $\Delta\theta = \frac{\pi}{2}$ with the actuator centered at $\theta = 0°$. In order to minimize any error in neglecting curvature effects associated with actuator loading, it was decided to use eighteen actuator segments ($N_c = 18$) with each piezoelectric section subtending a 5.0° arc, that is, $\Delta\theta = 5.0°$ as stated in the figure titles. Each of the eighteen actuator segments is therefore about 0.0731m (2.88 in.) long.

Cylinder displacement and acoustic cavity spectra are presented in figure 8 for the bending piezoelectric actuator model. Unlike the results for the single actuator segment discussed earlier (fig. 5), this particular actuator configuration produces a small contribution (not shown) for the $(m = 1, n = 2)$ cylinder mode. This contribution is about 37.9 dB down from the peak ($m = 3, n = 18$) mode and is out of the grey scale range used in figure 8. Although the cylinder response in the ($m = 1, n = 2$) mode is extremely small, it drives an appreciable response in the ($m = 0, n = 2$) cavity mode as shown in part (b) of figure 8. It should be recalled that this acoustic cavity mode is very close to resonance and the response level produced is no doubt very sensitive to the amount of acoustic damping assumed in the cavity model. For the in-plane actuator model, the spectra is shown in figure 9 and, as before (fig. 6), there is significant response produced in the ($m = 1, n = 2$) cylinder mode. In this instance, the ($m = 0, n = 2$) cavity mode, as shown in part (b) of figure 9, is dominant with just a slight amount of the ($m = 2, n = 1, 2$) modes being present. Thus, by making the circumferential length of the piezoelectric actuator model optimal with respect to the $n = 2$ cavity mode, the in-plane actuator model still couples more favorably with the low-order modes than does the bending actuator model.

In figures 10 and 11, the situation is similar to that reported in figures 8 and 9 except here the actuator amplitudes are weighted with a $\cos(2\theta)$ variation. The cylinder displacement spectra shown in figure 10 for the bending piezoelectric actuator model is similar to that of figure 8, although there is a more significant excitation of the low-order modes. This likewise produces high levels of excitation of the low-order cavity modes as shown in part (b) of figure 10. Results for the in-plane actuator model are shown in figure 11. Although there is appreciable excitation of the ($m = 3, n = 4$) cylinder mode, the dominant low-order response in the cavity remains in the ($m = 0, n = 2$) mode. It should be noted that the ($m = 3, n = 4$) cylinder mode has a resonant frequency of about ($\Omega = 0.206$, 212.9 Hz.). Here again, the in-plane actuator model retains its ability to couple efficiently with the low-order cylinder and acoustic cavity modes. Additionally, for the in-plane actuator case, the $\cos(2\theta)$ weighting results in significantly less structural response in the high-order cylinder modes (fig. 11a). This perhaps would have possible implications from a material fatigue life consideration.

CONCLUDING REMARKS

In this paper, two piezoelectric actuator models were evaluated for controlling the noise produced inside a vibrating cylindrical shell. These models were adapted from flat plate theory. Each actuator model consisted of two small patches of piezoelectric material bonded to opposite sides of the cylinder wall. For the bending model, the opposite actuators were driven out-of-phase. To approximate this type of loading on the cylinder, the two patches of piezoelectric material were approximated by a uniform, line moment distribution applied around the perimeter of the patch area. For the in-plane model, the two opposite actuators were driven in-phase. The shell loading for this situation was approximated by a uniform, line force distribution acting around the perimeter of the patch area.

Cylinder and interior acoustic pressure responses and modal spectra indicate that the in-plane piezoelectric actuator model couples more efficiently with the low-order cylinder modes than does the bending model. It is the low-order cylinder modes which couple efficiently with the interior cavity in the lower frequency regime (20 – 500 Hz.). Hence, it is the low-order modes which must be controlled in order to provide reduced noise levels inside the vibrating cylindrical shell. The bending actuator model lacked significant low-order mode participation and would probably not be as effective as the in-plane actuator model in an active noise control system. A limited number of calculations were made for distributions of piezoelectric actuators configured to couple with the low-order cylinder modes. Again, the in-plane model produced significantly higher low-order mode participation in comparison to the bending model. Furthermore, by properly weighting the actuator responses, the excitation of high-order cylinder modes was considerably reduced.

Finally, with regards to future work, it should be noted that the piezoelectric actuator models evaluated in this paper are probably valid only for small patch dimensions (relative to the cylinder radius). This is because the models were adapted from similar flat plate models. Therefore, any curvature effects in formulating the out-of-plane (bending) and in-plane loadings were neglected. What is needed is a cylinder (shell) type actuator model. By conforming with the curvature properties of the shell, such an actuator model would include the curvature effects both in the formulation of the loading terms as well as the coupling effects intrinsically included in the shell equations. Additionally, the range of applicability for the in-plane and bending piezoelectric actuator models studied in this paper needs to be established by comparison with experimental data[14] and/or the calculations produced by more exact shell type actuator models.

REFERENCES

1. Crawley, E. F. and Luis, Javier de: "Use of Piezoelectric Actuators as Elements of Intelligent Structures", *AIAA Journal*, Vol. 25, No. 10, October, 1987, pp. 1373-1385.

2. Clark, R. L. Jr.; Fuller, C. R.; and Wicks, A.: "Characterization of Multiple Piezoelectric Actuators for Structural Excitation", *Journal of Acoustical Society of America*, to be published, 1991.

3. Dimitriadis, E. K.; Fuller, C. R.; and Rogers, C. A.: "Piezoelectric Actuators for Distributed Noise and Vibration Excitation of Thin Plates", Proceeding of the 8th ASME Conference on Failure Prevention, Reliability, and Stress Analysis, 1989, pp. 223-233.

4. Fuller, C. R.; Gibbs, G. P.; and Silcox, R. J.: "Simultaneous Active Control of Flexural and Extensional Waves in Beams", *Journal of Intelligent Material Systems and Structures*, vol. 1, No. 2, April 1990.

5. Tzou, H. S. and Gadre, M.: "Theoretical Analysis of a Multi-Layered Thin Shell Coupled With a Piezoelectric Actuator for Distributed Vibration Controls", *Journal of Sound and Vibration*, 132(3), 1989, pp. 433-450.

6. Jia, J. and Rogers, C. A.: "Formulation of a Laminated Shell Theory Incorporating Embedded Distributed Actuators", *The American Society of Mechanical Engineers*, Reprinted From AD-Vol. 15, Adaptive Structures, Editor: B. K. Wada, Book No. H00533, 1989.

7. Fuller, C. R.; Snyder, S. D.; Hansen, C. H.; and Silcox, R. J.: "Active Control of Interior Noise in Model Aircraft Fuselages Using Piezoceramic Actuators", The AIAA 13th Aeroacoustics Conference, Paper 90-3922, Oct. 22-24, 1990.

8. Leissa, A. W.: Vibration of Shells, NASA SP-288, 1973.

9. Kraus, H.: *Thin Elastic Shells*, John Wiley and Sons, Inc., New York, 1967.

10. Junger, M. C. and Feit, D.: *Sound, Structures, and Their Interaction*, 2nd ed., The MIT Press, Cambridge, 1986.

11. Smith, P. W. Jr.: "Phase Velocities and Displacements of Free Waves in a Thin Cylindrical Shell", *Journal of the Acoustical Society of America*, Vol. 27, No. 6, November, 1955, pp. 1065-1072.

12. Blevins, R. D.: "Formulas for Natural Frequency and Mode Shape", Van Nostrand Reinhold Co., NY.

13. Fahy, F. J.: "Sound and Structural Vibration: Radiation, Transmission and Response", Academic Press, New York, 1985.

14. Lefebvre, S.: Masters Thesis, Virginia Polytechnic Institute and State University, to be published, 1991.

APPENDIX - CYLINDER EQUATIONS

In this appendix, the Donnell-Mushtari cylinder equations are summarized. These equations neglect transverse shear and rotary inertia effects and represent the simplest class of thin shell equations available for a cylinder. The reader is referred to the literature[8, 9] for more exact treatments.

Let $u(x,\theta,t)$, $v(x,\theta,t)$ and $w(x,\theta,t)$ represent the axial, circumferential and radial displacements, respectively, of the middle surface of the shell. These cylinder displacements are functions of time t and position on the middle surface as given by (x,θ). In terms of these three displacements, the Donnell-Mushtari cylinder equations can be written as follows:

$$\left(\frac{1}{C_L^2}\right)\frac{\partial^2 u}{\partial t^2} - \frac{\partial^2 u}{\partial x^2} - \left(\frac{1-\nu}{2a^2}\right)\frac{\partial^2 u}{\partial \theta^2} - \left(\frac{1+\nu}{2a}\right)\frac{\partial^2 v}{\partial x\,\partial \theta} - \left(\frac{\nu}{a}\right)\frac{\partial w}{\partial x} = \left(\frac{1-\nu^2}{Eh}\right) f_x \qquad (A-1a)$$

$$\left(\frac{1}{C_L^2}\right)\frac{\partial^2 v}{\partial t^2} - \left(\frac{1+\nu}{2a}\right)\frac{\partial^2 u}{\partial x\,\partial \theta} - \left(\frac{1-\nu}{2}\right)\frac{\partial^2 v}{\partial x^2} - \left(\frac{1}{a^2}\right)\frac{\partial^2 v}{\partial \theta^2} - \left(\frac{1}{a^2}\right)\frac{\partial w}{\partial \theta} = \left(\frac{1-\nu^2}{Eh}\right) f_\theta \qquad (A-1b)$$

$$\left(\frac{1}{C_L^2}\right)\frac{\partial^2 w}{\partial t^2} + \left(\frac{\nu}{a}\right)\frac{\partial u}{\partial x} + \left(\frac{1}{a^2}\right)\frac{\partial v}{\partial \theta} + \left(\frac{1}{a^2}\right) w + \left(\frac{h^2}{12}\right)\nabla^4 w = -\left(\frac{1-\nu^2}{Eh}\right) f_n \qquad (A-1c)$$

In equations (1), a is the radius of the middle surface of the cylinder and h is the shell wall thickness. The material properties are: the mass density ρ, Young's modulus E, and Poisson's ratio ν. These parameters define the following two constants fundamental for thin shell analysis:

$$\beta^2 = \frac{h^2}{12a^2} \qquad (A-2a)$$

$$C_L = \left[\frac{E}{\rho_s\left(1-\nu^2\right)}\right]^{\frac{1}{2}} \qquad (A-2b)$$

The parameter β is related to the direct effects of flexural bending in the shell wall, as contrasted to transverse shear effects which are neglected in the Donnell-Mushtari theory. The other defining parameter of thin shell theory is C_L which is the phase speed of axial propagating waves in the cylinder wall.

The effect of the external load field is accounted for by the distributions $f_x(x,\theta,t)$, $f_\theta(x,\theta,t)$, and $f_n(x,\theta,t)$. These loadings are assumed to act on the middle surface and are expressed in units of force per unit area. The in-plane loadings are given by f_x and f_θ. The pressure loading f_n acts normal to the cylinder wall. The negative sign on this loading component reflects the usual sign convention that an external pressure loading on a shell is positive, which is in opposite sense to a positive radial displacement $w(x, \theta)$.

A harmonic $e^{j\omega t}$ time variation is now assumed for all the displacements and external forces f_x, f_θ, and f_n. Therefore, in the following, the time variable t is suppressed in accordance with this convention. For a harmonic time dependence, the displacements for a simply supported cylinder can most conveniently be expressed in terms of the natural vibration mode shapes as follows :

$$u(x, \theta) = \sum_{n=0}\sum_{m=0} \cos\left(\frac{m\pi x}{L}\right)\left[U^c_{mn}\cos(n\theta) + U^s_{mn}\sin(n\theta)\right] \qquad (A-3a)$$

$$v(x, \theta) = \sum_{n=0}\sum_{m=1} \sin\left(\frac{m\pi x}{L}\right)\left[V^c_{mn}\cos(n\theta) + V^s_{mn}\sin(n\theta)\right] \qquad (A-3b)$$

$$w(x, \theta) = \sum_{n=0}\sum_{m=1} \sin\left(\frac{m\pi x}{L}\right)\left[W^c_{mn}\cos(n\theta) + W^s_{mn}\sin(n\theta)\right] \qquad (A-3c)$$

In a similar manner, the external loading can also be expanded in terms of the natural vibration modes of the cylinder. This allows the direct coupling of the load field with the cylinder displacement field so that the modal coefficients appearing in equations (A-3) can be determined on a mode by mode basis. Hence, in the interval ($0 \le x \le L, 0 \le \theta \le 2\pi$), the loading can be expressed as:

$$f_x(x, \theta) = \sum_{n=0}\sum_{m=0} \cos\left(\frac{m\pi x}{L}\right)\left[A^c_{mn}\cos(n\theta) + A^s_{mn}\sin(n\theta)\right] \qquad (A-4a)$$

$$f_\theta(x, \theta) = \sum_{n=0}\sum_{m=1} \sin\left(\frac{m\pi x}{L}\right)\left[T^c_{mn}\cos(n\theta) + T^s_{mn}\sin(n\theta)\right] \qquad (A-4b)$$

$$f_n(x, \theta) = \sum_{n=0}\sum_{m=1} \sin\left(\frac{m\pi x}{L}\right)\left[F^c_{mn}\cos(n\theta) + F^s_{mn}\sin(n\theta)\right] \qquad (A-4c)$$

As shown in many standard references[8,9], the cylinder modal displacement coefficients W^c_{mn}, W^s_{mn}, U^c_{mn}, U^s_{mn}, V^c_{mn}, and V^s_{mn} can be determined on a mode by mode basis by substituting equations (A-3) and (A-4) into equations (A-1).

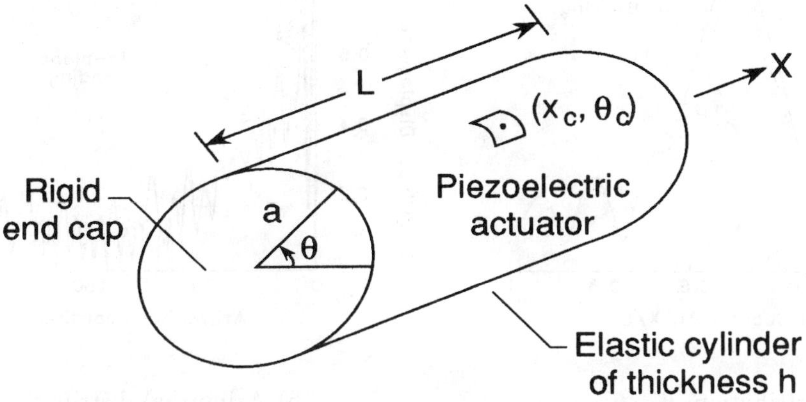

Figure 1. Cylinder and piezoelectric actuator coordinate geometry.

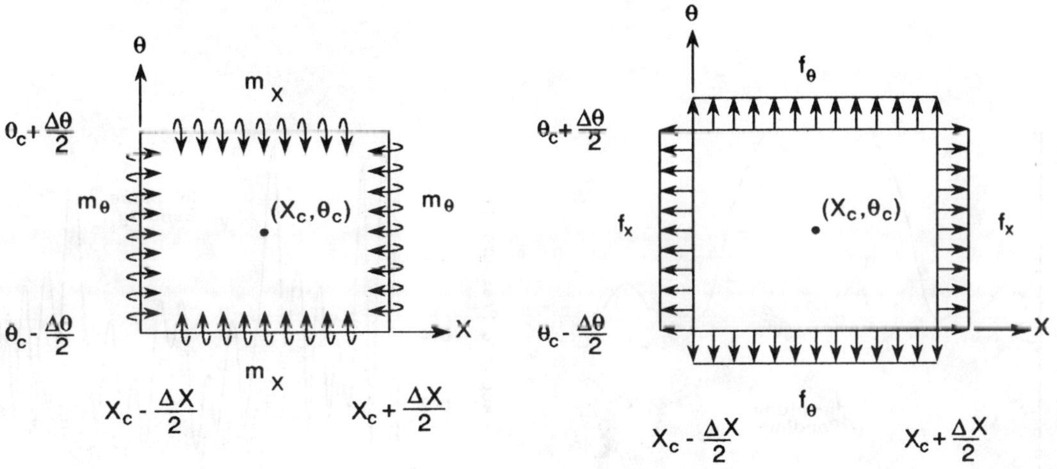

(a) Bending model (b) In-plane model

Figure 2. Piezoelectric actuator models used on cylinder.

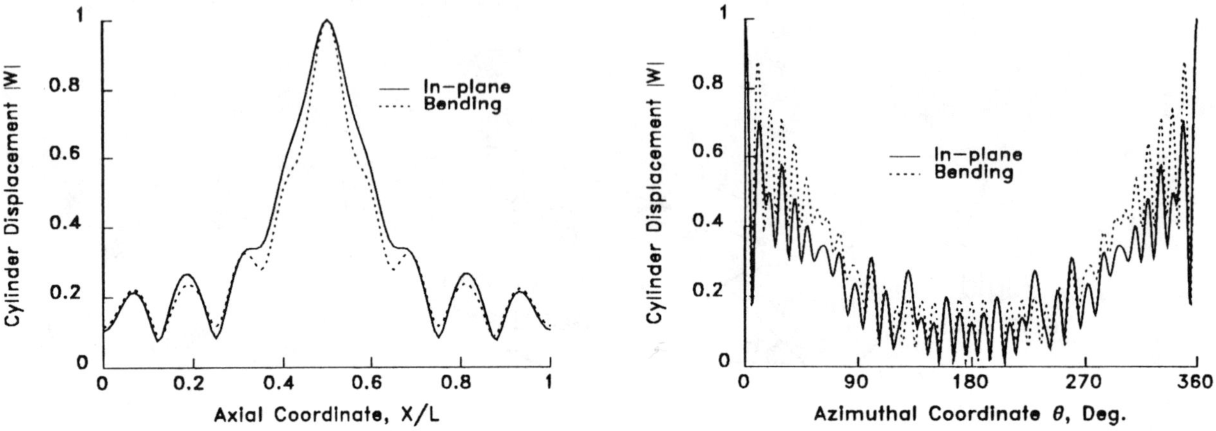

(a) Axial distribution, $\theta = 0°$

(b) Azimuthal distribution, $x = \frac{L}{2}$

Figure 3. Cylinder displacement responses induced by piezoelectric actuator models, $N_c = 1$, $V_c = 1.0$, $x_c = \frac{L}{2}$, $\theta_c = 0°$, $\Delta x = 0.0381$m (1.5 in.), $\Delta\theta = 4.34°$ (2.5 in.), $\Omega = 0.194$ (200.0 Hz).

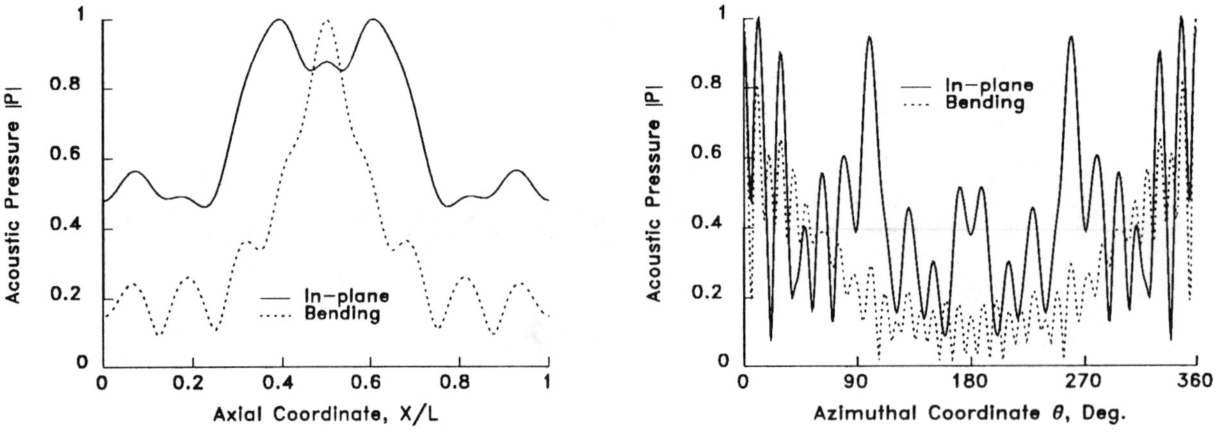

(a) Axial distribution, $\theta = 0°$

(b) Azimuthal distribution, $x = \frac{L}{2}$

Figure 4. Cavity acoustic pressure responses induced by piezoelectric actuator models, $N_c = 1$, $V_c = 1.0$, $x_c = \frac{L}{2}$, $\theta_c = 0°$, $\Delta x = 0.0381$m (1.5 in.), $\Delta\theta = 4.34°$ (2.5 in.), $\Omega = 0.194$ (200.0 Hz).

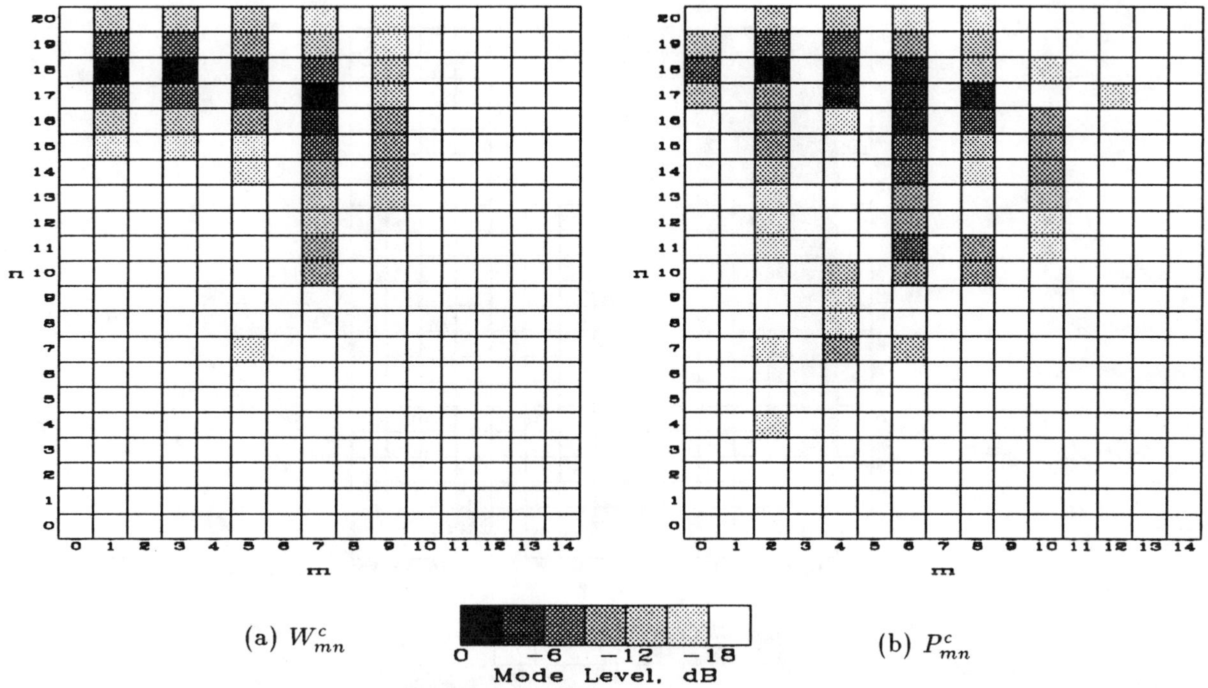

(a) W^c_{mn} (b) P^c_{mn}

Figure 5. Modal spectra for bending piezoelectric actuator model, $N_c = 1$, $V_c = 1.0$, $x_c = \frac{L}{2}$, $\theta_c = 0°$, $\Delta x = 0.0381$m (1.5 in.), $\Delta\theta = 4.34°$ (2.5 in.), $\Omega = 0.194$ (200.0 Hz).

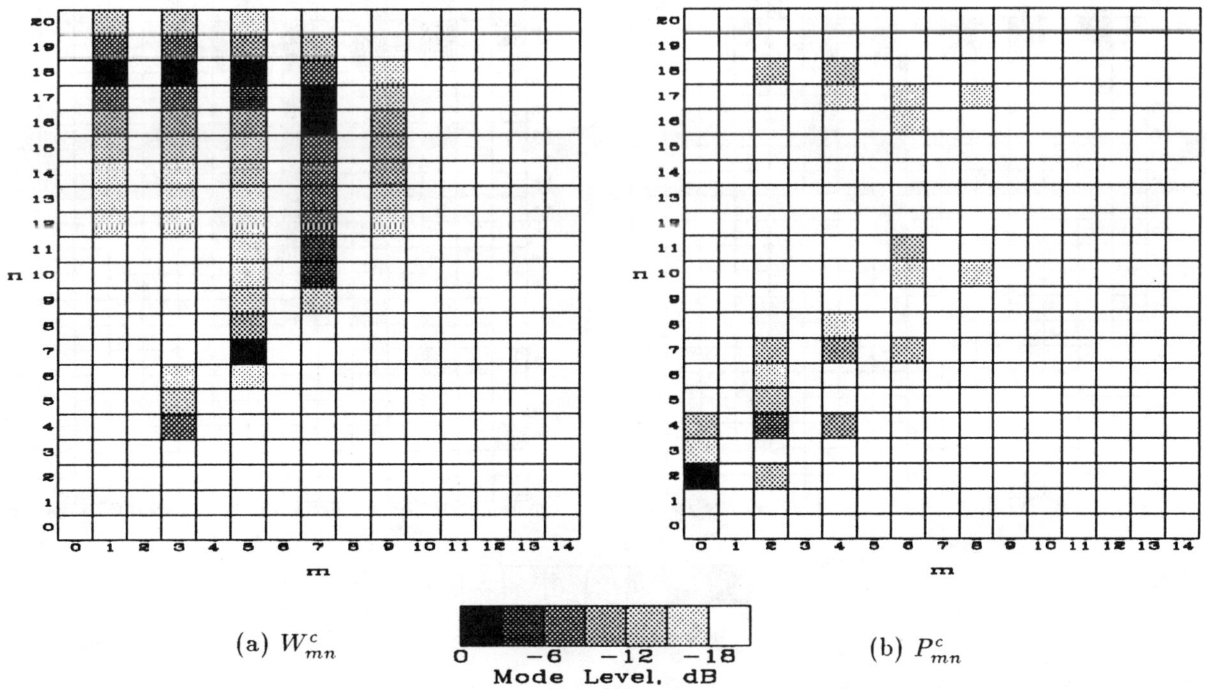

(a) W^c_{mn} (b) P^c_{mn}

Figure 6. Modal spectra for in-plane piezoelectric actuator model, $N_c = 1$, $V_c = 1.0$, $x_c = \frac{L}{2}$, $\theta_c = 0°$, $\Delta x = 0.0381$m (1.5 in.), $\Delta\theta = 4.34°$ (2.5 in.), $\Omega = 0.194$ (200.0 Hz).

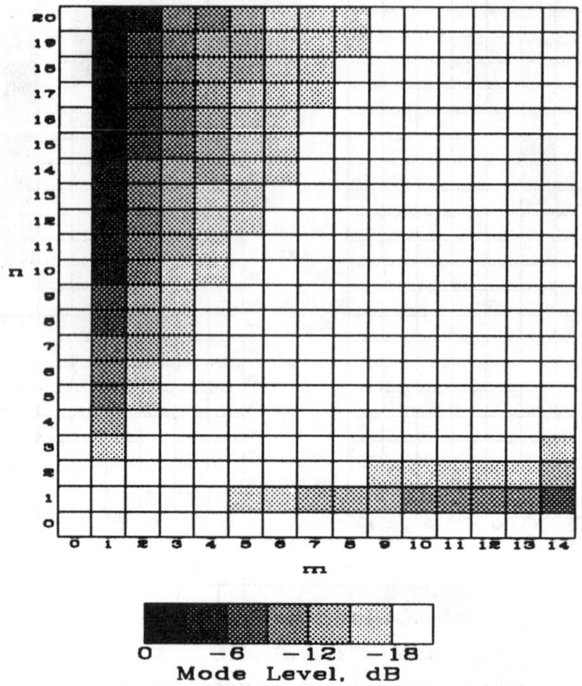

Figure 7. Weighting coefficient G_{mn}.

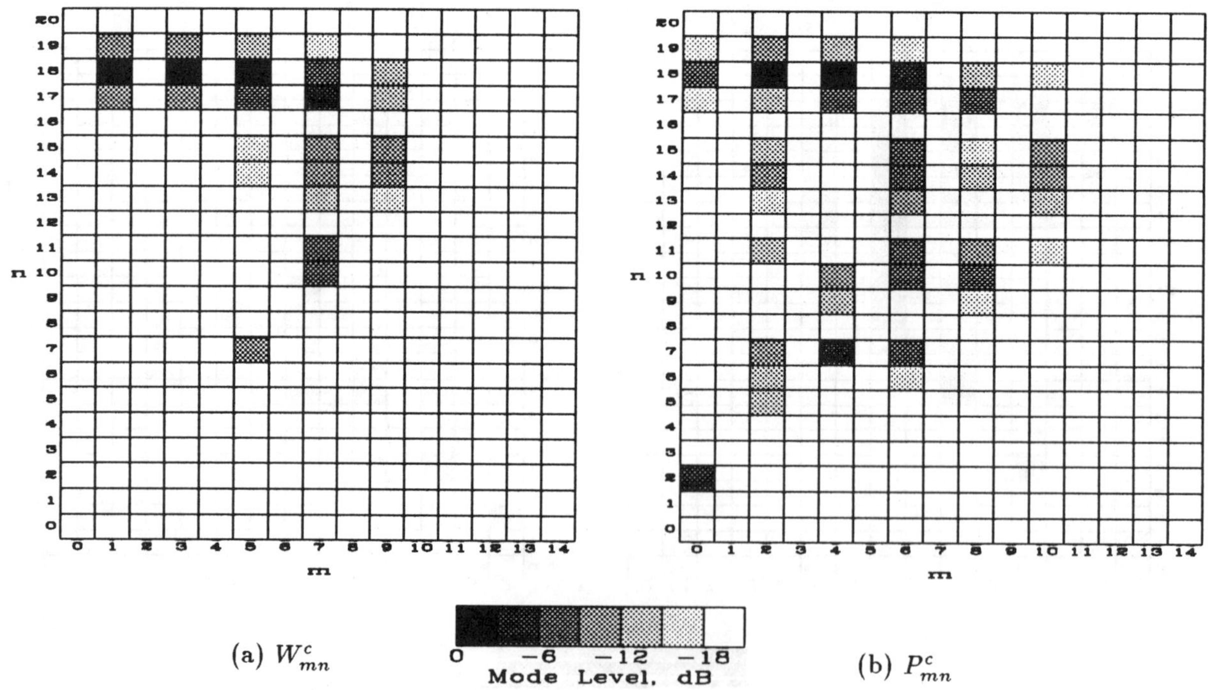

(a) W^c_{mn} (b) P^c_{mn}

Figure 8. Modal spectra for multiple bending piezoelectric actuators, $N_c = 18$, $V_c = 1.0$, $x_c = \frac{L}{2}$, $\theta_c = -42.5°$, $-37.5°$,...,$37.5°$,$42.5°$, $\Delta x = 0.0381$m (1.5 in.), $\Delta\theta = 5.0°$ (2.9 in.), $\Omega = 0.194$ (200.0 Hz).

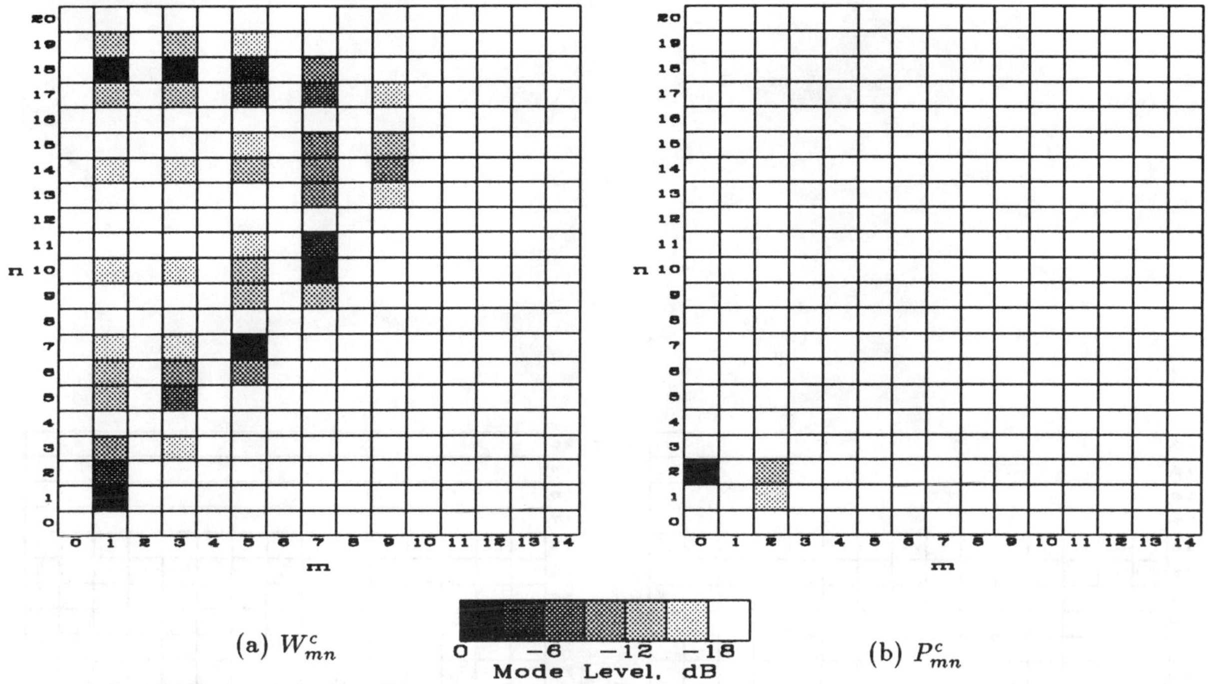

(a) W^c_{mn} (b) P^c_{mn}

Figure 9. Modal spectra for multiple in-plane piezoelectric actuators, $N_c = 18$, $V_c = 1.0$, $x_c = \frac{L}{2}$, $\theta_c = -42.5°, -37.5°, ..., 37.5°, 42.5°$, $\Delta x = 0.0381$m (1.5 in.), $\Delta\theta = 5.0°$ (2.9 in.), $\Omega = 0.194$ (200.0 Hz).

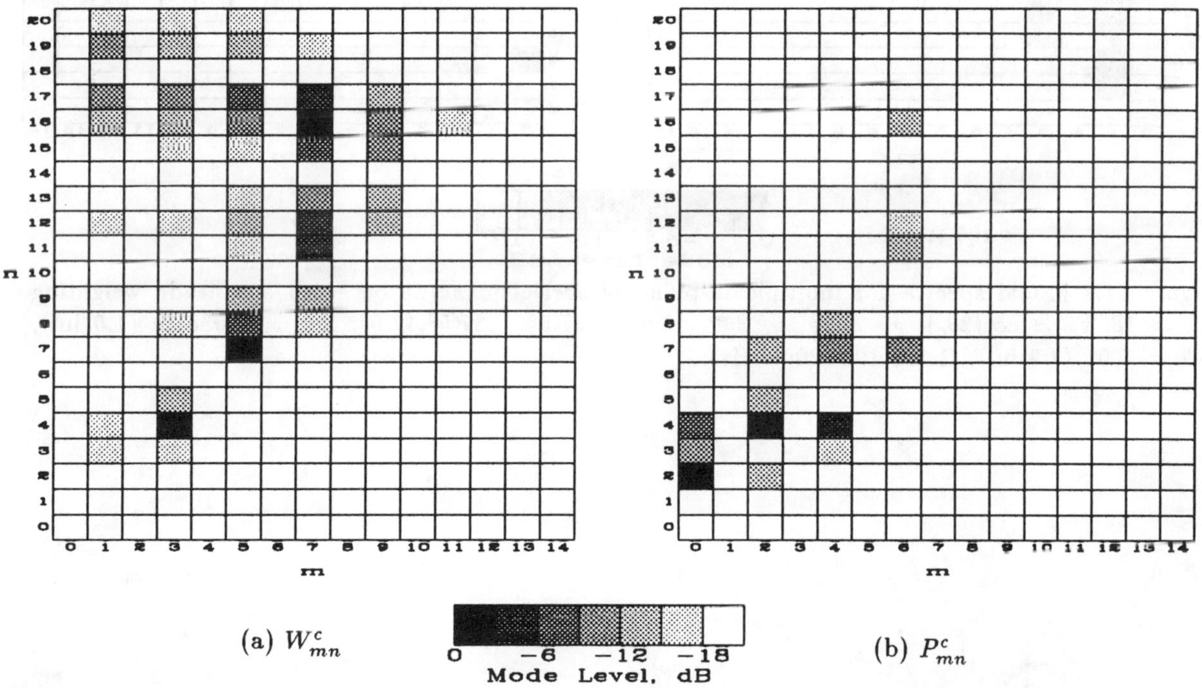

(a) W^c_{mn} (b) P^c_{mn}

Figure 10. Modal spectra for multiple bending piezoelectric actuators with amplitude weighting, $N_c = 18$, $V_c = \cos(2\theta_c)$, $x_c = \frac{L}{2}$, $\theta_c = -42.5°, -37.5°, ..., 37.5°, 42.5°$, $\Delta x = 0.0381$m (1.5 in.), $\Delta\theta = 5.0°$ (2.9 in.), $\Omega = 0.194$ (200.0 Hz).

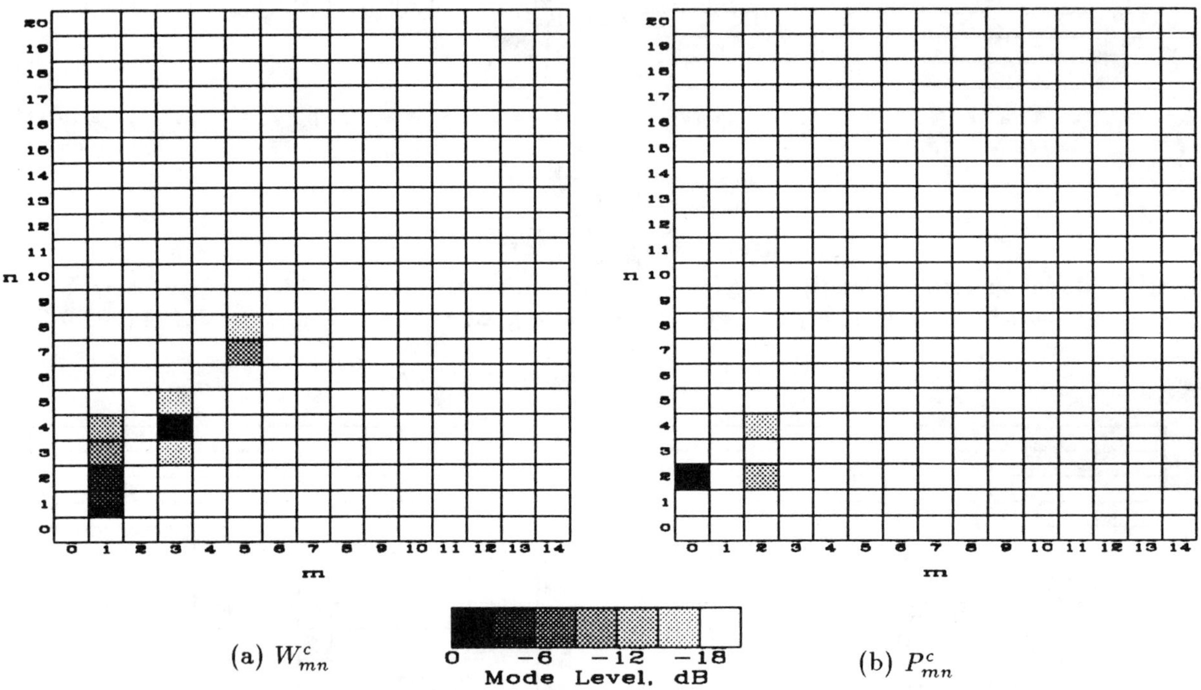

(a) W^c_{mn} (b) P^c_{mn}

Figure 11. Modal spectra for multiple in-plane piezoelectric actuators with amplitude weighting, $N_c = 18$, $V_c = \cos(2\theta_c)$, $x_c = \frac{L}{2}$, $\theta_c = -42.5^\circ, -37.5^\circ, \ldots, 37.5^\circ, 42.5^\circ$, $\Delta x = 0.0381$m (1.5 in.), $\Delta\theta = 5.0^\circ$ (2.9 in.), $\Omega = 0.194$ (200.0 Hz).

ACTIVE VIBRATION CONTROL OF THIN CYLINDRICAL SHELLS USING PIEZO-ELECTRIC ACTUATORS

Venkata R. Sonti,[1] J. D. Jones[1]

ABSTRACT

A simplified analytical model is developed to study the modal coupling behavior of piezo-actuators surfaced-bonded to a finite circular cylindrical shell. The primary input to the model is generated by a line force while the secondary control input is due to a single piezo-actuator patch modelled as a uniform moment over the surface of the patch. Direct comparisons are made between the performance of the piezo-actuator control patch and a point control force for both on- and off-resonance cases. Methods for selectively tailoring the modal coupling behavior are discussed. Results suggest that small patches couple more selectively into higher order modes whereas the selectively of larger patches varies strongly with the patch length and location.

NOMENCLATURE

INTRODUCTION

Active control of sound transmission into enclosed flexible cylinders has been an ongoing research area of interest for several years because of its implications in reducing interior cabin noise in turboprop aircraft. Previous work [1,2] has shown that low frequency cabin noise is directly coupled to fuselage vibration. One novel active noise control (ANC) strategy involves the use of vibration actuators to directly control the shell response and consequently reduce sound transmission into the cabin. Early analytical and experimental work [3,4,5] has demonstrated the use of point actuators to reduce strongly-coupled structural modes of the shell thereby reducing the sound transmission into the enclosed cavity. However, because point actuators are "spectrally-white" in a spatial sense, a significant fraction of the control energy was spilled into neighboring uncontrolled modes of the structure. Since these modes are often poorly coupled to the acoustic cavity, the acoustic control performance is often not compromised. Nevertheless, the increase in the shell response can lead to accelerated fatigue problems in the structure.

Recently, the use of piezo-actuators has drawn much attention because of the ability to vary the geometry, location and material properties of the actuators in order to specify (*apriori*) the desired modal coupling. This characteristic has been successfully demonstrated for simple geometries such as beams and plates [6,7]. One novel concept [8] involves the development of a composite structure with modal piezoelectric sensors and actuators capable of detecting and controlling select modes without spillover. Although the notion of modal actuators and sensors

[1]Ray W. Herrick Laboratories, School of Mechanical Engineering, Purdue University, West Lafayette, IN 47907, U.S.A.

is not new, practical methods for achieving this goal (namely using piezoelectric materials) have only recently been forthcoming.

Recent experimental studies [9] have successfully demonstrated the use of piezo-ceramic actuators for controling sound transmission into cylindrical cavities. Under acoustic and structural resonance conditions, global attenuations of the interior sound field of approximately 10 dB have been demonstrated. However, for the acoustic resonance case, a higher modal density in the shell resulted in significant increases in the shell vibration. Thus, a better fundamental understanding of the modal coupling behavior of piezo-actuators is needed in order to retain the acoustic performance while minimizing control spillover in the shell.

In the current work, a simplified analytical model is used to study the modal coupling behavior of a piezo-actuator patch surfaced-bonded to a finite circular cylindrical shell. The piezo-actuator is modelled as a uniform moment over the surface of the patch. Methods for selectively tailoring the modal coupling behavior for active noise and vibration control purposes are presented. Implications to controlling sound transmission into enclosed flexible cylinders with minimum increase in the shell vibration are discussed.

MODEL DESCRIPTION

An analytical model is developed which represents the dynamic response characteristics of a finite simply-supported cylindrical shell in-vacuo. The assumed dimensions of the steel shell include a radius (a) of 0.75 m, a length (L) of 3 m and a thickness (h) of 2 mm. The response of the finite cylinder is characterized by the classical thin shell theory of Love [10] which neglects the effects of transverse shear deformation and rotary inertia. The primary forcing function is assumed to be a line force extending over $\theta = \pm 15$ deg. and located at $x = 0.5$ as shown in Figure 1. This forcing function was chosen because it provides a simple representation of the distributed character of the structureborne engine vibrations carried along a wing spar or engine pylon attachment. Since the primary objective of his work is to study the fundamental coupling behavior of piezo-actuators, there is little need to more exactly model the forcing function more accurately.

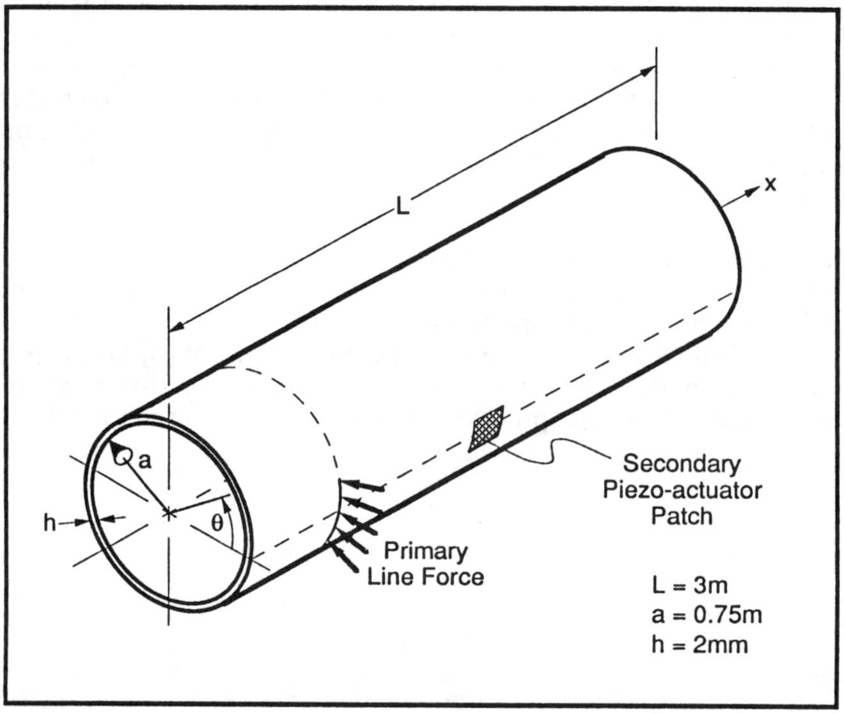

Figure 1. Shell geometry and coordinate system

The control input is assumed to be due to either a point force or a single piezo-actuator patch surfaced-bonded to the surface of the shell. The influence of the patch is modeled as a uniform moment over the surface of the patch. Although the shell curvature inherently causes some coupling between the transverse and in-plane vibration (in the direction of curvature), this effect is neglected in the current study for simplicity. The influence of the curvature is to create an anisotropic piezo-actuator/substructure coupling. However, if the radius of the shell is large with respect to the patch size, this effect will be minor. The passive (mass loading and stiffness) influence of the piezo-actuators is also neglected, since only relatively small piezo-actuator patches are considered in this investigation.

ANALYSIS

The study of active vibration control using the piezo-actuator patches requires a knowledge of the modal character of the shell due to the primary and secondary forcing functions. To this end, general structural response functions due to both the primary line force and the secondary point control force and uniform control moment are presented. Minimization of the shell vibration is performed based on a quadratic optimization [11] of the shell response at a single point error sensor location. In some cases, a modal sensor (capable of detecting only one select structural mode) was used for optimization.

RESPONSE TO A LINE FORCE

The general separable solution for the transverse displacement of a simply-supported cylindrical shell is given by

$$U_3(x,\theta,t)=\sum_{m=1}^{\infty}\sum_{n=0}^{\infty}\eta_{mn}(t)u_{3mn}(x,\theta) \tag{1}$$

where u_{3mn} is the mode shape function given by

$$u_{3mn}(x,0)=\sin(\frac{m\pi x}{L})\cos(n(\theta-\phi)) \tag{2}$$

and η_{mn} is the modal participation factor obtained by solving the harmonic oscillator equation [Soedel].

$$\ddot{\eta}_{mn}+2\xi_{mn}\dot{\eta}_{mn}+\omega_{mn}^2\eta_{mn}=F_{mn} \tag{3}$$

In equations (1) - (3) the indices m and n refer to the axial and circumferential mode numbers, respectively. The subscript 3 refers to transverse (radial) direction while ϕ represents the relative angular location of the forcing function with respect to the coordinate system. In equation (3), x_{mn} and ω_{mn} are the modal damping coefficients and the natural frequencies of the system and F_{mn} is the effective forcing function. For all cases considered, a small level of damping was introduced in order to ensure convergence of the solution.

For the case of a line force between $\theta = \pm 15$ deg. and located along $x = 0.5$, the effective forcing function (F_{mn}) is given by

$$F_{mn}=\frac{1}{\rho hN_{mn}}\int_0^{2\pi}\int_0^{L}P_3U_{3mn}\delta(x-x_1)\{h(\theta-\theta_1)-h(\theta-\theta_2)\}a\,dx\,d\theta \tag{4}$$

In Equation (4), P_3 is the magnitude of the line force [in N/m], ρ is the shell density, δ represents a dirac delta function, h is a unit step function and N_{mn} is a normalization factor given by

$$N_{mn}=\int_0^{2\pi}\int_0^{L}\{\sin(\frac{m\Pi x}{L})\cos(n(\theta-\phi))\}^2\, a\, dx\, d\theta \tag{5}$$

Equations (1)-(5) can be used to obtain the modal solution of the transverse shell vibration as given below.

$$U_{3mn}(x,\theta)=\sum_{m=1}^{\infty}\sum_{n=0}^{\infty}[\frac{F_{mn\phi=0}\cos(n\theta)+F_{mn\phi=\frac{\pi}{2n}}\sin(n\theta)}{\omega_{mn}^2\sqrt{[1-(\frac{\omega}{\omega_{mn}})^2]^2+[2\xi_{mn}\frac{\omega}{\omega_{mn}}]^2}}]\sin(\frac{m\pi x}{L}) \tag{6}$$

where there will be a special case for n = 0 (not shown).

RESPONSE TO A POINT FORCE

The procedure for deriving the response characteristics due to a point force is identical to that of the line force except that the forcing function (F_{mn}) differs. For a point force applied at (x_1,θ_1), F_{mn} is given by

$$F_{mn}=\frac{1}{\rho h N_{mn}}\int_0^{2\pi}\int_0^{L}P_3\sin(\frac{m\pi x}{L})\cos(n(\theta-\phi)) \tag{7}$$

where P_3 is the magnitude of the point force [in N/m^2]. A detailed derivation of the point force response is given in reference [10], thus only the final result is given here.

$$U_{3mn}(x,\theta)=\frac{2P_3}{\rho h a L\pi}\sum_{m=1}^{\infty}\sum_{n=0}^{\infty}\frac{\sin(\frac{m\pi x_1}{L})\sin(\frac{m\pi x}{L})\cos(n(\theta-\theta_1))}{\varepsilon_n\omega_{mn}^2\sqrt{[1-(\frac{\omega}{\omega_{mn}})^2]^2+4\xi_{mn}^2(\frac{\omega}{\omega_{mn}})^2}} \tag{8}$$

RESPONSE TO A UNIFORM MOMENT

The effective forcing function of the piezo-actuator patch is modeled as a uniform moment over the surface of the patch. As in the previous cases, the harmonic oscillator equation (3) must be solved. For an applied moment on the shell surface, the effective forcing function (F_{mn}) is given by

$$F_{mn}=\frac{1}{\rho h N_{mn}}\int_0^{2\pi}\int_0^{L}u_{3mn}(x,\theta)\{\frac{\partial M_x a}{\partial x}+\frac{\partial M_\theta}{\partial \theta}\}\, dx\, d\theta \tag{9}$$

where M_x and M_θ are the effective moments in x and θ directions, respectively.

In most investigations of piezo-actuators, a uniform moment distribution on the surface of the patch is assumed in both x and θ directions. This will not be exactly true for the cylinder problem because the curvature will cause the effective moment to be different in the two directions. However, if the patch size is small compared to the radius of the shell, then this effect will be negligible. Thus, in this investigation a uniform moment over the patch is assumed and is given by

$$M = \overline{M}\{h(x-x_1)-h(x-x_2)\}\{h(\theta-\theta_1)-h(\theta-\theta_2)\} \tag{10}$$

The variables $x_1, x_2, \theta_1, \theta_2$ represent the boundaries of the piezo-actuator patch. and $\overline{M}$ is the uniform amplitude of the moment [in N-m/m]. The moment equation then can be differentiated in x and θ and substituted into equation (9) which can subsequently be integrated to formulate the applied effective forcing function (F_{mn}). This procedure is relatively straight forward and hence again only the final result is presented.

$$F_{mn}=-\left(\frac{\overline{M}am\pi}{\rho h N_{mn} Ln}+\frac{\overline{M}Ln}{a\rho h m\pi N_{mn}}\right)\{\cos(\frac{m\pi x_1}{L})-\cos(m\pi x_2)\}^{*}$$

$$\{\sin(n(\theta_2-\phi))-\sin(n(\theta_1-\phi))\} \tag{11}$$

for n not equal to 0. Similarly, for n equal to zero it is given by

$$Fmn=\frac{-\overline{M}am\pi}{\rho h N_{mn} L}\{\cos(\frac{m\pi x_1}{L})-\cos(\frac{m\pi x_2}{L})\}\{\theta_2-\theta_1\} \tag{12}$$

Equations (11) and (12) may be substituted into the harmonic oscillator equation (3) to obtain the solution to the modal participation factor.The final transverse response of the shell has the following form

$$U_{3mn}(x,\theta)=\sum_{m=1}^{\infty}\sum_{n=0}^{\infty}\left[\frac{F_{mn\phi=0}\cos(n\theta)+F_{mn\phi=\frac{\pi}{2n}}\sin(n\theta)}{\omega_{mn}^2\sqrt{[1-(\frac{\omega}{\omega_{mn}})^2]^2+[2\xi_{mn}\frac{\omega}{\omega_{mn}}]^2}}\right]\sin(\frac{m\pi x}{L}) \tag{13}$$

RESULTS AND DISCUSSION

In this section, active vibration control of the simply-supported cylindrical shell model is presented for both an on-resonance and an off-resonance case. For the on-resonance case, a driving frequency of 55.7 Hz was chosen because it represents an isolated resonance of the shell (see Figure 2). In contrast, an off-resonance frequency of 135.3 Hz was chosen because it is in close proximity to several natural frequencies of the shell. Structural modes of both high and low circumferential order are excited. This represents a classic case in the shell where the lower order modes will likely be well-coupled to the interior acoustic cavity, while the higher order modes will be selectively filtered and thus have little contribution to the interior acoustic field. For both cases, the control performance of a point control force is compared with that of the piezo-actuator patch. Implications to sound transmission into the enclosed cavity are discussed.

ON-RESONANCE CASE

As discussed in the Model Description section, the primary input to the shell model is a line force over θ = ± 15 deg. and located at x = 0.5 as shown in Figure 3(a). The primary force is driven at 55.7 Hz which corresponds to the natural frequency of the (m=1, n=8) mode of the shell. For both cases presented a single point error sensor, located at over θ = ± 5 deg and 1.48 m ≤ x ≤ 1.52 m, was used to optimize the control actuator input.

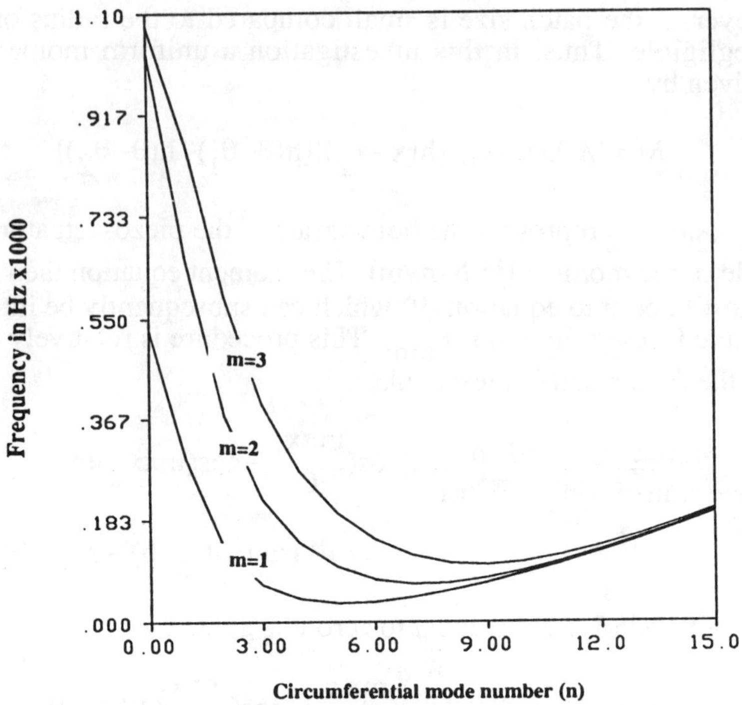

Figure 2. Cylinder natural frequencies for m=1,2,3 and n=0,15

Point Control Force

A point control force is positioned at $\theta = 0^o$ and x=1.5 m, which corresponds to the axial centreline of the shell as shown in Figure 3(a). Figures 3(b)-3(d) show the uncontrolled and controlled modal amplitudes of the shell model for the n=0,15 circumferential modes.with axial order m=1, 2 and 3, respectively. The shell response is decomposed into symmetric A_{mn} and asymmetric B_{mn} modes with a cos(nθ) and a sin(nθ) distribution, respectively. Because of the symmetric location of both the primary and the secondary point actuators, the B_{mn} modes are not excited and thus are omitted from the results presented here.

As expected for the uncontrolled case, the (1,8) mode is much more strongly excited than other adjacent modes. Similarly, the action of the point control force reduces this mode by approximately 38.7 dB with little control spillover. Interestingly, the other m=1 modes are also reduced slightly because they are all in-phase with the dominant (1,8) mode. Conversely, the m=3 modes increase slightly because they are out-of-phase with the (1,8) mode. Nevertheless, this spillover is negligible because the natural frequencies of the m=3 modes are far removed from the driving frequency resulting in a low modal participation factor.

Uniform Control Moment

The piezo-actuator control patch is modeled as a uniform moment over $\theta = \pm 5$ deg and 1.48 m $\leq$ x $\leq$ 1.52 m, as shown in Figure 4(a). This represents a patch size of approximately 13 cm by 4 cm with the longer dimension in the θ direction. The specified patch size is roughly typical for commercially-available piezo-ceramic actuators. Figures 4(b)-4(d) show the controlled and uncontrolled modal amplitudes of the shell. As in the point control force case, the (1,8) mode is largely reduced (by 46.6 dB), however, the overall vibration in the shell is actually slightly higher for the uniform control moment than for the point control force. Similar results are obtained as long as the patch remains entirely within the node lines of the (1,8) mode. As with the point control force, all of the m = 1 modes are reduced while the m = 3 modes experience some slight spillover. The control action of the piezo-actuator patch is maximized when it is centrally located within an anti-nodal region. In essence, the patch behaviour resembles that of a point force placed at the patch centre. That is, its maximum coupling occurs when the centre of the patch coincides with the antinode of the dominant mode and no coupling occurs when the centre of the patch coincides with the node. Thus, for a structural resonance case, the control performance of the piezo-actuator is only a weak function of the position except when a node line is crossed.

The two on-resonance cases illustrated above show that control of a structural resonance is relatively simple to achieve when isolated from other natural modes of the system. For such a case, the type, size, and location of the actuator does not generally have a significant influence on the control performance of the ANC system (assuming that the actuator can couple into the dominant mode of the shell).

OFF-RESONANCE CASE

In order to better examine the distinguishing features of the point control force versus the piezo-actuator patch, a more realistic off-resonance case was examined. For this case, the driving

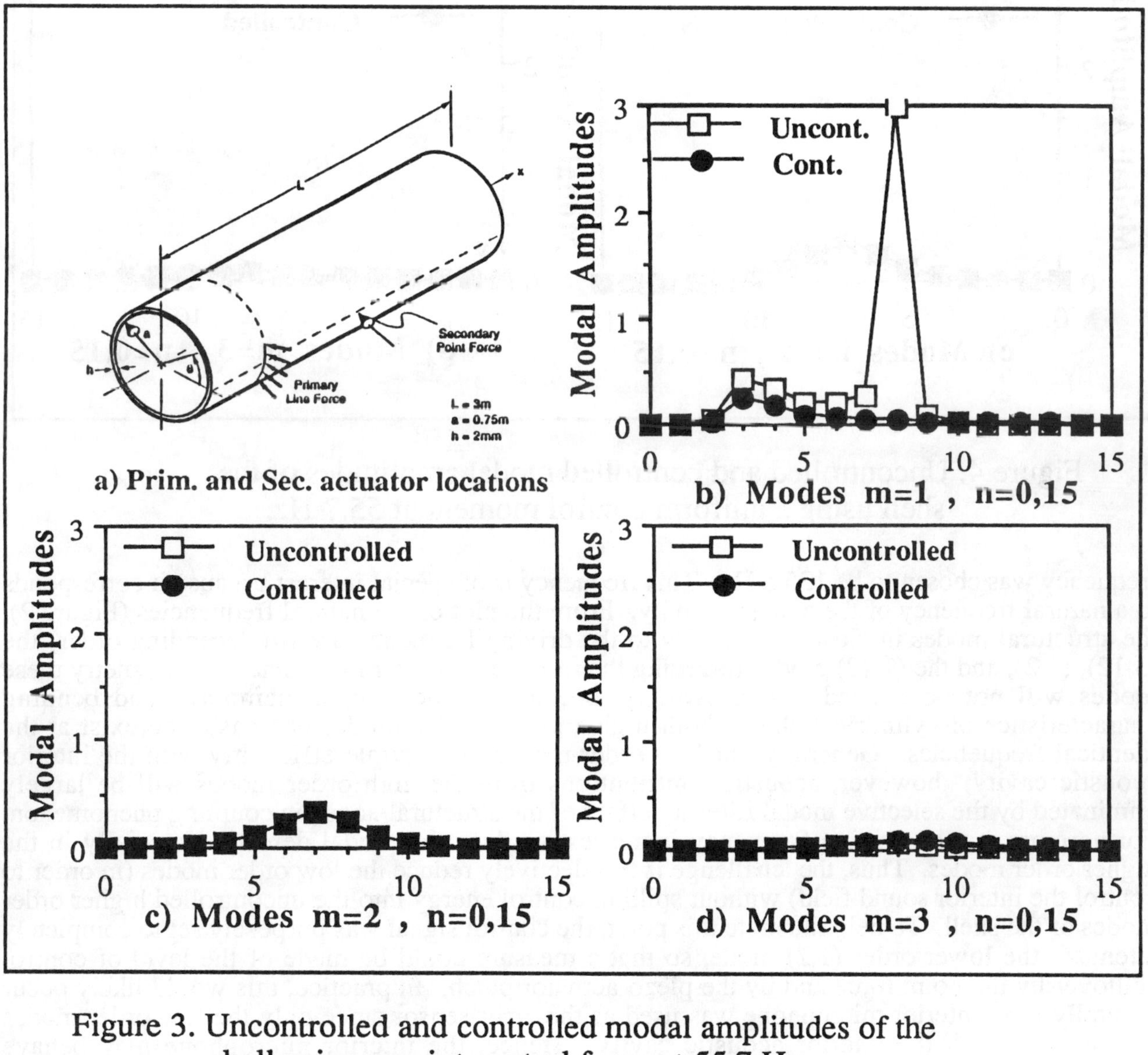

Figure 3. Uncontrolled and controlled modal amplitudes of the shell using a point control force at 55.7 Hz.

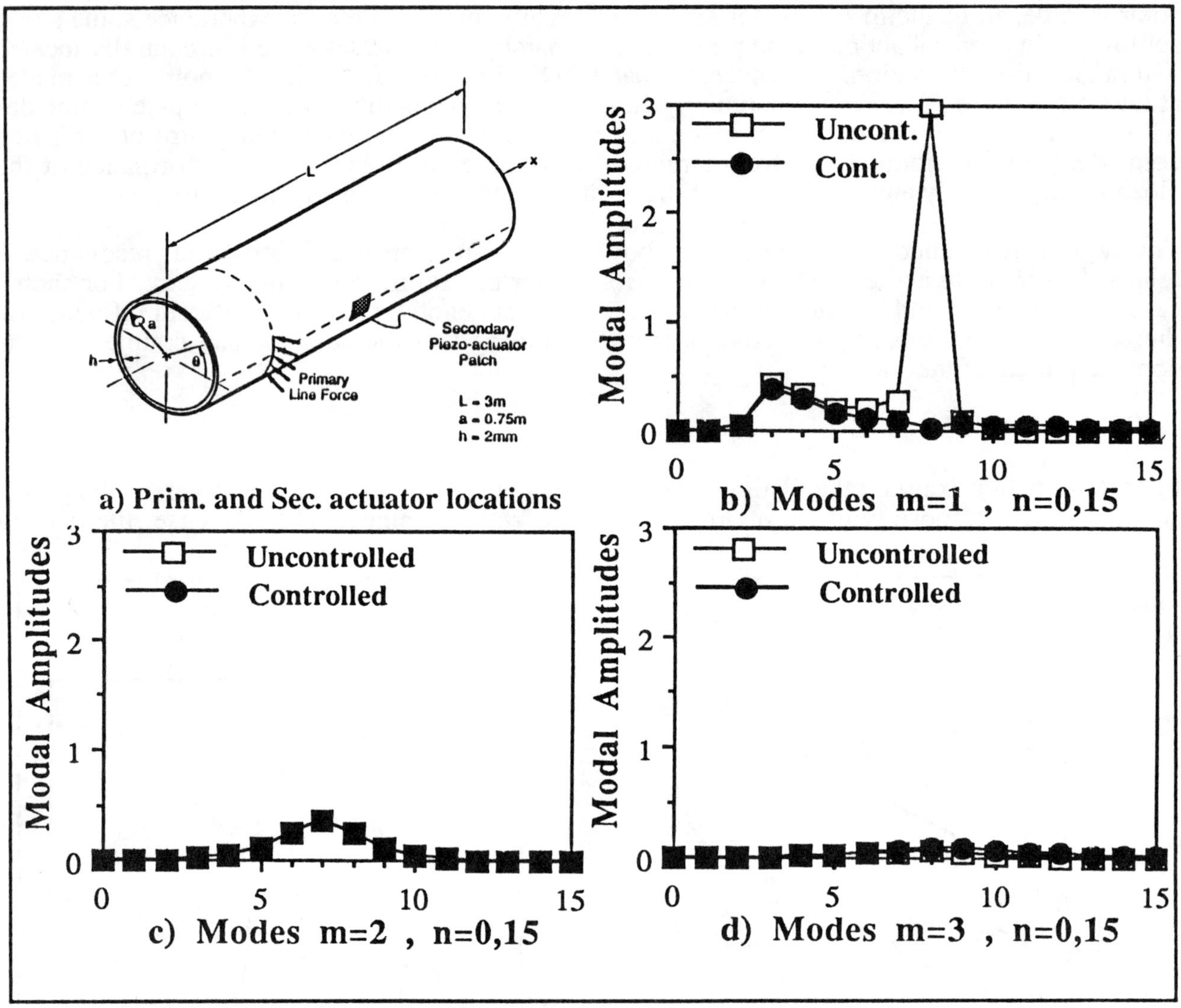

Figure 4. Uncontrolled and controlled modal amplitudes of the shell using a uniform control moment at 55.7 Hz.

frequency was chosen to be 135.5 Hz. This frequency is of special interest because it corresponds to a natural frequency of the acoustic cavity. From the plot of the natural frequencies (Figure 2), the structural modes in closest proximity to the driving frequency are (in decending order) the (3,12), (1,2), and the (1,12) modes (omitting the m = 2 and m= 4 modes since by symmetry these modes will not be excited). Interestingly, because of the unique membrane and bending characteristics of cylindrical shells, both high and low order modes can easily co-exist at the identical frequencies. Generally, the low order modes will couple efficiently with the interior acoustic cavity; however, acoustic contributions from the high order modes will be largely eliminated by the selective modal filtering effect of the structural/acoustic coupling phenomenon. Furthermore, as the driving frequency increases, so does the modal density, especially in the higher order modes. Thus, the challenge is to selectively reduce the low order modes (in order to control the interior sound field) without spilling control energy into the uncontrolled higher order modes of the shell. To help illustrate this point, the control signal was purposely set to completely attenuate the lower order (1,2) mode, so that a measure could be made of the level of control spillover by the point force and by the piezo-actuator patch. In practice, this would likely occur naturally if an interior microphone was used as the error sensor since only the low order modes couple well with the interior acoustic cavity. Hence, the interior microphone may behave something like a modal sensor and direct the controller to eliminate a single mode regardless of the consequences in the other modes.

Point Force Control

As with the on-resonance case, the point control force is positioned at $\theta = 0^o$ and x=1.5 m, as shown in Figure 5(a). Figures 5(b)-5(d) show the uncontrolled and controlled modal amplitudes of the shell model. The distributed character of the primary line force selectively couples into the (1,2), (2,4), (3,6) and the (3,12) modes with about equal amplitudes. Application of the point control force reduces the (1,2) mode as designed (with a 60 dB reduction), but consequently induces a significant increase in the vibration level of the uncontrolled (3,12) mode. This occurs because of the close proximity of the natural frequency of the (3,12) mode to the driving frequency (see Figure 2) results in an extremely high modal participation factor. This result is in close agreement with other similar experimental and analytical results using point actuators [3,4,5] and

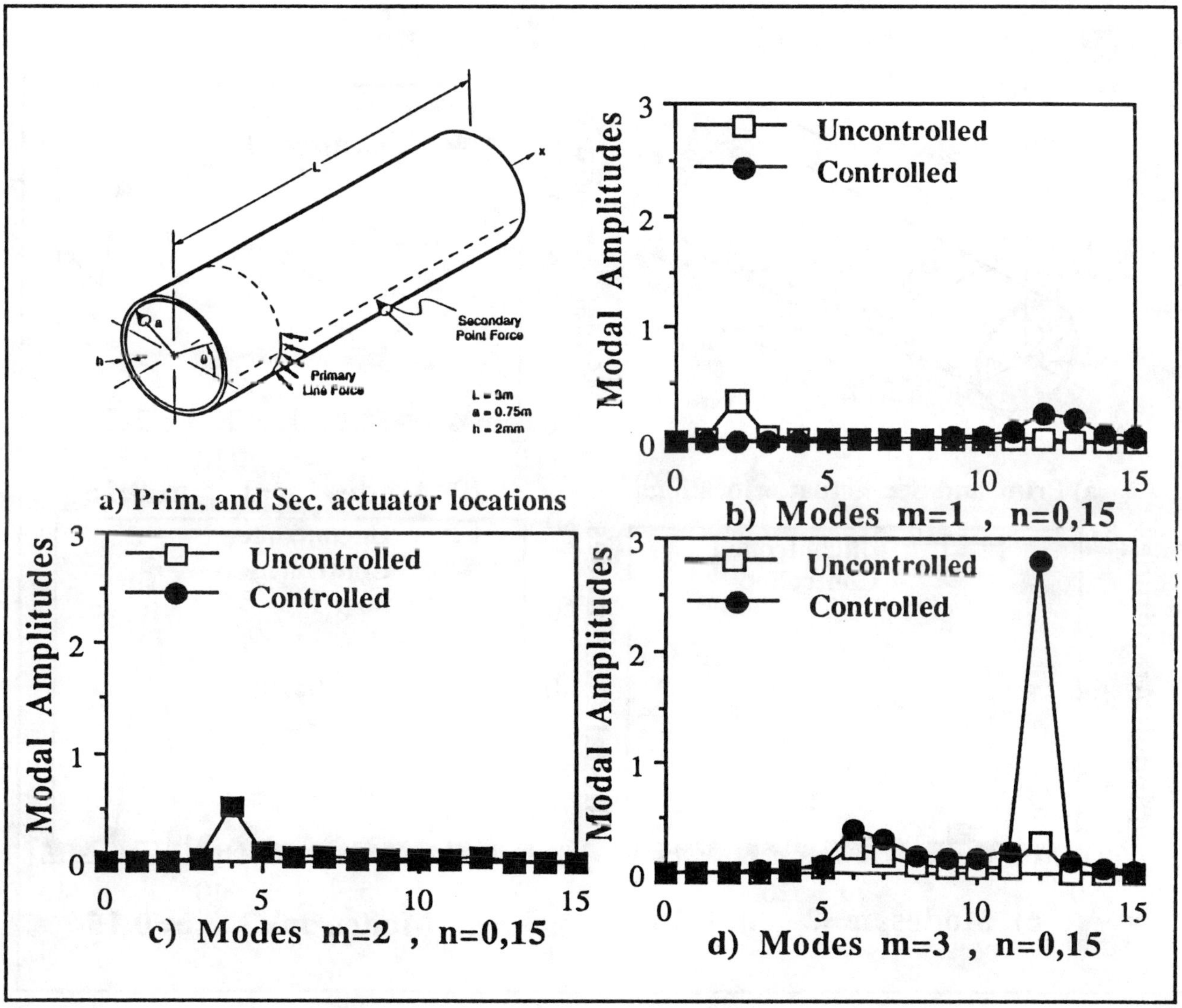

Figure 5. Uncontrolled and controlled modal amplitudes of the shell using a point control force at 135.3 Hz.

thus is not unexpected. Control spillover in the shell also occurs to a lesser degree in several other adjacent modes with lower modal participation factors.

Uniform Moment Control

The piezo-actuator control patch is again distributed over $\theta = \pm 5$ deg and $1.48 \text{ m} \leq x \leq 1.52$ m, as shown in Figure 6(a). Figures 6(b)-6(d) show the controlled and uncontrolled modal amplitudes of the shell. Although its difficult to see on Figures 6(b)-6(d) (because of scaling differences), the distributed character of the primary line force again selectively couples into the (1,2), (2,4), (3,6) and the (3,12) structural modes as before. Application of the piezo-actuator control patch reduces the (1,2) mode as designed (with a 60 dB reduction), but consequently creates a dramatic increase in the vibration level of the uncontrolled (1,12), (3,12) and the (3,14) modes. This result is quite surprising since intuitively one might expect for the piezo-actuator patch to give less control spillover considering its distributed character. However, the reason for

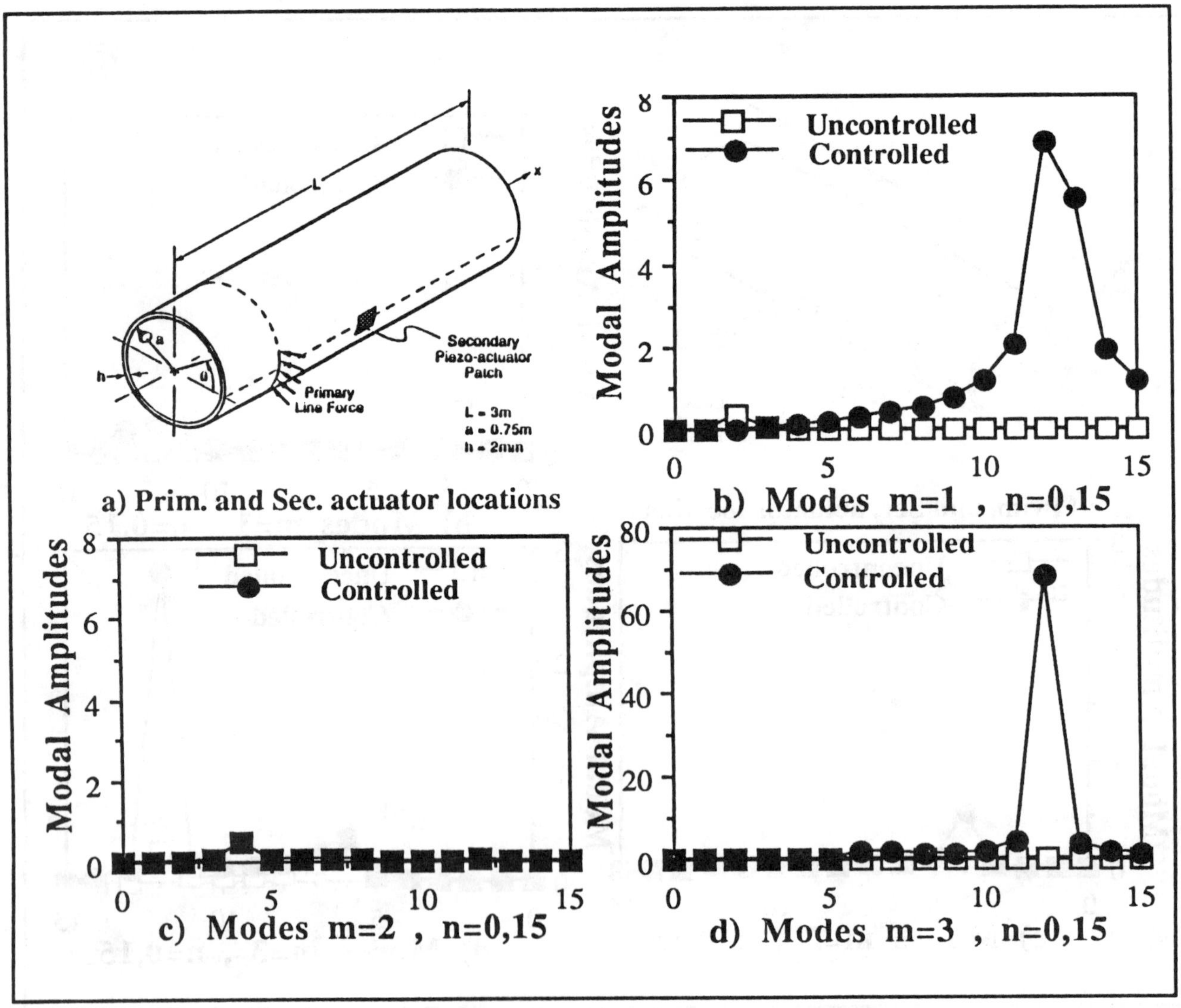

Figure 6. Uncontrolled and controlled modal amplitudes of the shell using a uniform control moment at 135.3 Hz.

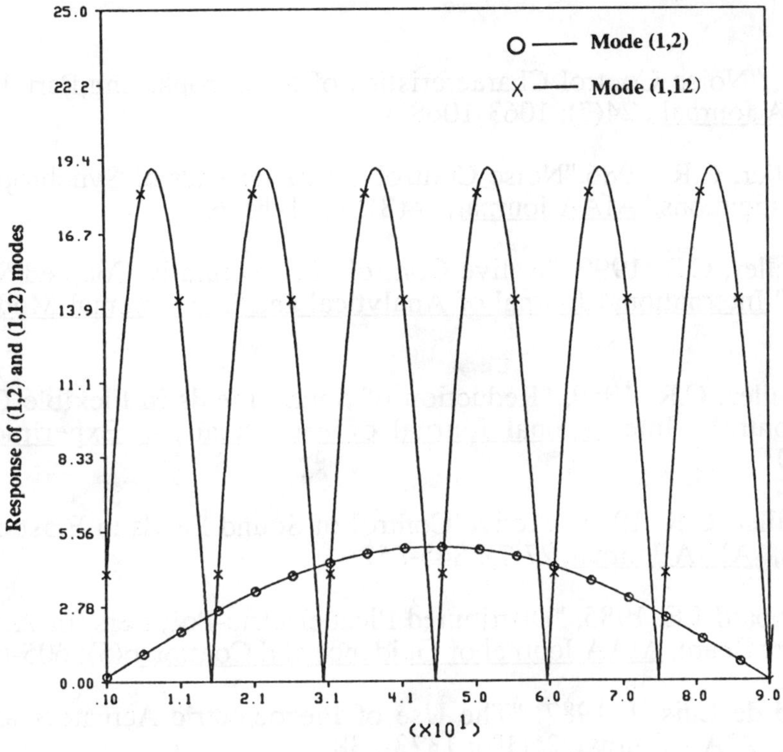

Figure 7. Variation of modal response with angular patch length

this unexpected result is due to the size of the patch size incomparison the spatial wavelength of the mode. A small patch size will be able to couple into higher order modes well but not lower order modes. Consequently if the piezo-actuator patch is too small it can cause even more spillover than the point force. This is because the point force tends to distribute its energy evenly to all the modes where as a small piezo-actuator patch willselectively emphasize the higher order modes more than lower the order modes.

CONCLUSIONS

A simplified analytical model was developed for the purpose of studying the modal coupling behavior of piezo-actuators to a finite circular cylindrical shell. Comparisons between point control forces and piezo-actuators showed little difference for a structural resonance condition. However, off the structural resonance small piezo-actuators selectively coupled better with higher order modes and can potentially induce more control spillover than point actuators. To advoid this problem larger piezo-actuator patches will generally provide better coupling with lower order modes.

ACKNOWLEDGEMENTS

The authors gratefully acknowledge the support of this work by the National Science Foundation under NSF Grant MSM-8810384. The authors would also like to thank Mr. Sung Jin Kim for his valuable contributions to this work.

REFERENCES

1. Fuller, C.R. 1986. "Noise Control Characteristics of Synchrophasing,Part 1: Analytical Investigations," AIAA Journal , 24(7): 1063-1068.

2. Jones, J.D.and Fuller, C.R. 1986."Noise Control Characteristics of Synchrophasing, Part 2: Experimental Investigations,' AIAA Journal, 24(8): 1271-1276.

3. Jones, J.D. and Fuller, C.R. 1990. "Active Control of Structurally-Coupled Sound Fields in Elastic Cylinders," International Journal of Analytical and Experimental Modal Analysis, 5(3): 123-140.

4. Jones, J.D. and Fuller, C.R. 1989. "Reduction of Sound Fields in Flexible Cylinders by Active Vibration Control," International Journal of Analytical and Experimental Modal Analysis, 4(2): 45-50.

5. Jones, J.D. and Fuller, C.R. 1987. "Active Control of Sound Fields in Elastic Cylinders by Multi-Control Forces," AIAA Journal, 27(7): 845-852.

6. Bailey ,T. and Hubbard J.E. 1985, "Distributed Piezoelectric-Polymers for Active Vibration Control of a Cantilever Beam, AIAA Journal of Guidance and Control, 6(5): 605-611.

7. Crawley, E.F. and de Luis, J. 1987. "The Use of Piezoelectric Actuators as Elements of Intelligent Structures," AIAA Journal, 25(10): 1373-1385.

8. Lee, C.K. and Moon, F.C. 1990. "Modal Sensors/Actuators," Journal of Applied Mechanics, 57(6): 434-441.

9. Fuller, C.R., Snyder, S., Hansen, C and Silcox, R. 1990."Active Control of Interior in Model Aircraft Fuselages Using Piezoceramic Actuators," AIAA Paper 90-3922, AIAA 13th Aeroacoustics Conference, Tallahassee,FL.

10. Soedel,W. 1981. Vibrations of Shells and Plates, Marcel Dekker, Inc., pp.199-228 and 248-260.

11. Nelson, P.A., Curtis, A.R.D., Elliott, S.J. and Bullmore, A.J. "The Minimum Power Output of Free Field Point Sources And the Active Control Of Sound," Journal of Sound and Vibration, 116(3): 397-414.

A STUDY OF THE RESPONSE OF A SIMPLY SUPPORTED BEAM TO EXCITATION BY A PIEZOELECTRIC ACTUATOR

Jie Pan,[1,2] Colin H. Hansen,[1] Scott D. Snyder[1]

ABSTRACT

This paper presents a dynamic model of the response of a simply supported beam to excitation by actuators made using piezoelectric ceramics glued to the beam surface. The dynamic extensional strain on the beam surface is calculated by including the free stress conditions at the piezoelectric actuator boundaries and by considering the dynamic coupling between the actuator and the structure. The results obtained from this analysis are compared with those from past studies, which were based on the assumption of static coupling between the structure and the actuator. The analysis shows that the strain field (and so the beam displacement) close to the actuator is significantly different from that obtained by using static stress analysis results, and that to obtain an accurate description of the response, the coupled structural and piezoceramic actuator system must be modelled dynamically.

INTRODUCTION

Piezoelectric ceramics[1, 2, 3] show great promise as actuators for the control of structural vibration and noise radiation. The structural vibration (for beams or plates) due to the excitation of a piezoceramic actuator has been modelled by Crawley and de Luis[1] and Dimitriadis *et al.*[2, 3]. In particular, Crawley and de Luis presented a comprehensive static model for a piezoelectric actuator and a substructural coupled beam. In their model, the relationship between static structural strains, both in the structure and in the actuator, and the applied voltage cross the actuator thickness was clearly demonstrated. These static results were then used to predict the dynamical behavior of the coupled system. The structural dynamic response thus obtained is approximate and needs to be qualified by direct dynamic modelling of the coupled system. Dimitriadis *et al.* used a similar static approach and extended the one dimensional beam vibration problem into a two dimensional plate vibration problem. In their approach, the stress free condition on the piezoelectric actuator boundaries was not satisfied. For the case of shape memory alloy actuators embedded in a composite material, Liang and Rogers[4] showed that their finite element analysis did not accurately estimate strain values at the free edge, although away from the edge the strains were accurately calculated and relatively constant. Although their results were obtained from the shape memory alloy actuators, they were used to justify the assumption used by Dimitriadis *et al.*. However, the stress distribution obtained by Liang and Rogers was also for the static condition. To estimate the validity and

[1]Department of Mechanical Engineering, University of Adelaide, South Australia, 5001, Australia.
[2]Current address: Department of Mechanical Engineering, The University of Western Australia, Nedlands 6009, Western Australia.

the accurate frequency and spatial range for these approximations, it is necessary to investigate the free boundary effect for piezoelectric actuators, and it is necessary to investigate how this effect varies with the operating frequency.

In this article, a dynamic model is developed for a simply supported elastic beam with a piezoelectric actuator glued to each of its upper and lower surfaces. In this model the free stress boundary conditions characterizing the finite size piezoelectric actuator are also taken into account. The results obtained here are compared with those obtained by using a static analysis and by assuming the stress distribution of an infinite piezoelectric element.

SYSTEM DYNAMIC MODELLING

Figure 1 shows a Bernoulli-Euler beam with an identical piezoelectric actuator mounted on each of two opposite sides of the beam. The actuators are driven by a pair of electrical fields with the same amplitude and in opposite phase. They are perfectly bonded on to the beam, which allows the assumption of strain continuity at the bonding interface. The beam is divided into three sections as shown in Fig. 1, and the dynamic behavior of each section will be analysed separately.

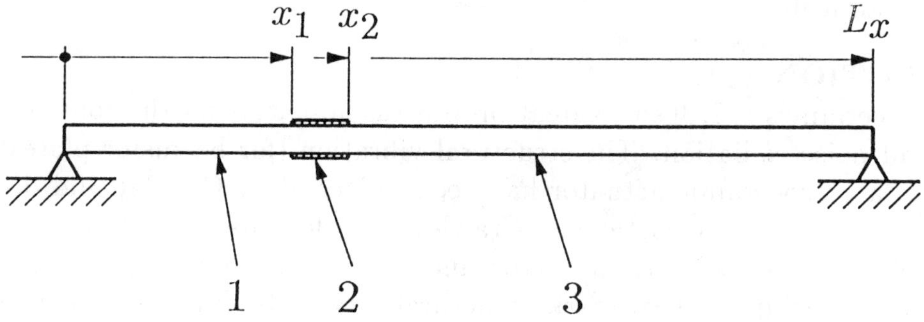

Fig. 1: A simply supported beam with piezoelectric actuators.

A finite element of length dx in the second section $x_1 < x < x_2$ is shown in Fig. 2(a). Because of the thinness of the piezoelectric elements, the longitudinal stress σ_{pe} in the element is assumed to be uniform in the Z direction. For the longitudinal motion of the top piezoelectric elements, equating the external forces to the element inertia force gives:

$$\frac{\partial \sigma_{pe}}{\partial x} t_c - \tau = \rho_{pe} t_c \frac{\partial^2 u_{pe}}{\partial t^2} \tag{1}$$

where t_c and ρ_{pe} are the piezoelectric element thickness and density respectively, τ is the shear stress on the interface surface, and u_{pe} is the longitudinal displacement of the element in the x direction. The strain ϵ_{pe} of piezoelectric elements is defined by:

$$\epsilon_{pe} = \frac{\partial u_{pe}}{\partial x} \tag{2}$$

and the tensile stress σ_{pe} in the piezoelectric element is related to the pure longitudinal strain ϵ_{pe} in the element and the piezoelectric strain[1] $d_{31}V/t_c = \Lambda$ by:

$$\sigma_{pe} = E_{pe}(\epsilon_{pe} - \Lambda) \; , \tag{3}$$

where E_{pe} is the Young's modulus of the piezoelectric elements, V is the voltage applied across the element thickness and d_{31} is the piezoelectric constant. Substituting Eqs. (2) and (3) into Eq. (1) and differentiating with respect to x gives:

$$E_{pe}t_c\frac{\partial^2 \epsilon_{pe}}{\partial x^2} - \frac{\partial \tau}{\partial x} = \rho_{pe}t_c\frac{\partial^2 \epsilon_{pe}}{\partial t^2} . \tag{4}$$

Figure 2(b) shows the isolated beam element. Because the mass loading of the piezoelectric element on the beam element is negligible, the vertical force equilibrium gives:

$$\frac{\partial T}{\partial x} = \rho_p t_b b\frac{\partial^2 w}{\partial t^2} . \tag{5}$$

where ρ_p and t_b are respectively the density and thickness of the beam, T is the vertical shear force on the beam cross section, b is the width of the beam and w is the vertical displacement of the beam element. If the effects of the transverse shear and rotary inertia of the element are ignored, (the assumptions of the Bernoulli-Euler beam theory), the moment equilibrium about an axis normal to the $X-Y$ plane and passing through the center of the element gives:

$$-\frac{\partial M}{\partial x} + T - \tau t_b b = 0 , \tag{6}$$

where M is the moment force about the neutral plane generated by the distributed normal stress $\sigma_p\frac{2z}{t_b}$ (where z is the local coordinate in the beam thickness (or y) direction, and is zero at the beam neutral axis) and is specified as:

$$M = b\int_{-0.5t_b}^{0.5t_b}\sigma_p\frac{2z^2}{t_b}dz = \frac{t_b^2 b}{6}\sigma_p . \tag{7}$$

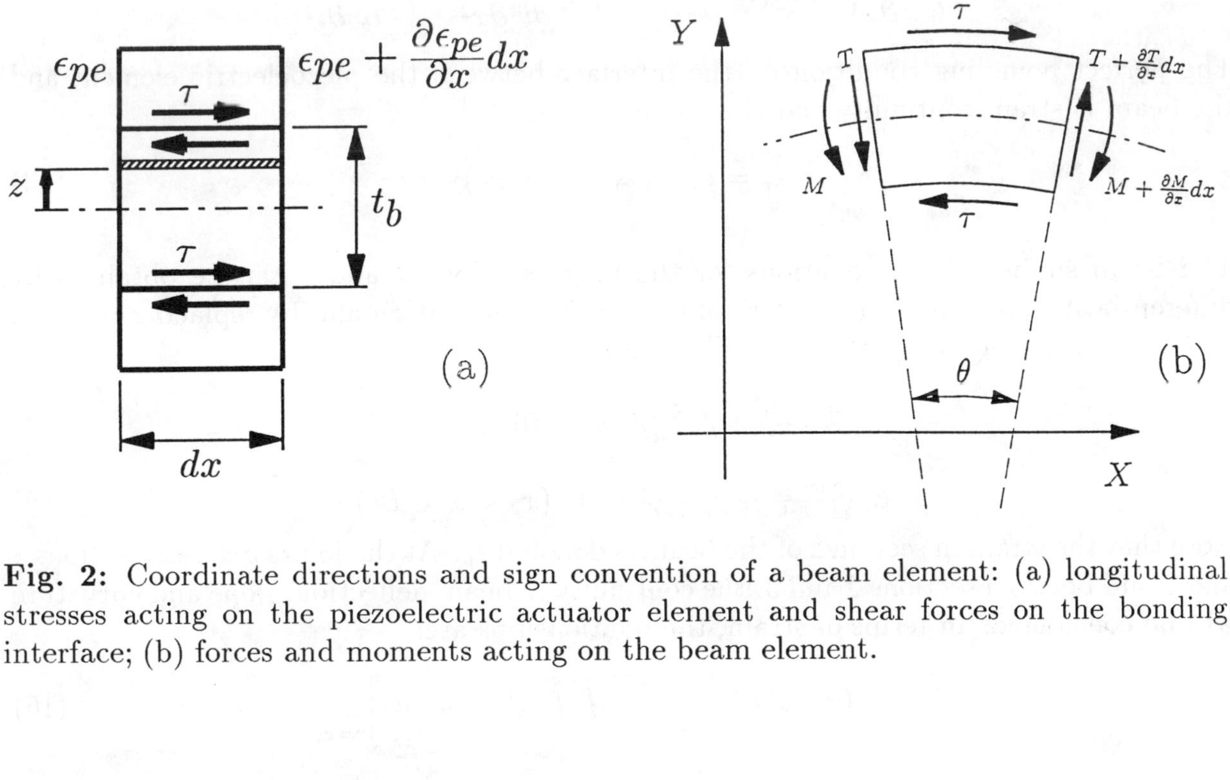

Fig. 2: Coordinate directions and sign convention of a beam element: (a) longitudinal stresses acting on the piezoelectric actuator element and shear forces on the bonding interface; (b) forces and moments acting on the beam element.

σ_p is the beam surface tensile stress and is related to the beam surface strain ϵ_p by:

$$\sigma_p = E_p \epsilon_p \ . \tag{8}$$

To include the damping effect, the complex Young's moduli E_p and E_{pe} are used in the analysis.

Using Eqs. (5)-(8), T, M and σ_p can be eliminated and the resultant equation is:

$$\frac{t_b E_p}{6} \frac{\partial^2 \epsilon_p}{\partial x^2} + \frac{\partial \tau}{\partial x} = \rho_p \frac{\partial^2 w}{\partial t^2} \ . \tag{9}$$

The angular displacement θ of the cross section plane relative to that perpendicular to the neutral beam plane can be calculated by[4]:

$$\theta = \frac{\partial w}{\partial x} = -\frac{2u_p}{t_b} \tag{10}$$

where u_p is the beam surface longitudinal displacement, which is related to the beam surface strain by $\epsilon_p = \frac{\partial u_p}{\partial x}$. Therefore, using Eq. (10), the beam vertical displacement is related the beam surface strain by:

$$\frac{\partial^2 w}{\partial x^2} = -\frac{2}{t_b} \epsilon_p \tag{11}$$

By adding Eqs. (4) and (9), the terms including τ disappear, and differentiating both sides of the resultant equation twice with respect to x (so that Eq. (11) can be used), the following equation can be obtained:

$$\frac{t_b E_p}{6} \frac{\partial^4 \epsilon_p}{\partial x^4} + E_{pe} t_c \frac{\partial^4 \epsilon_{pe}}{\partial x^4} = \rho_{pe} t_c \frac{\partial^4 \epsilon_{pe}}{\partial t^2 \partial x^2} - \rho_p \frac{2}{t_b} \frac{\partial^2 \epsilon_p}{\partial t^2} \ . \tag{12}$$

The perfect bounding condition on the interface between the piezoelectric element and the beam is strain continuity and this requires that:

$$\epsilon_p = \epsilon_{pe} \quad (x_1 < x < x_2) \tag{13}$$

The beam surface strain equations for the beam sections 1 and 3 can be obtained by differentiating both sides of the Bernoulli-Euler beam equation and by replacing w by ϵ_{p1} and ϵ_{p3} respectively using Eq. (11). They are:

$$E_p \frac{t_b^2}{12} \frac{\partial^4 \epsilon_{p1}}{\partial x^4} + \rho_p \frac{\partial^2 \epsilon_{p1}}{\partial t^2} = 0 \quad (0 < x < x_1) \tag{14}$$

$$E_p \frac{t_b^2}{12} \frac{\partial^4 \epsilon_{p3}}{\partial x^4} + \rho_p \frac{\partial^2 \epsilon_{p3}}{\partial t^2} = 0 \quad (x_2 < x < L_x) \tag{15}$$

Note that the strain in section 2 of the beam is denoted ϵ_p. At the joints between sections 1 and 2 and between sections 2 and 3, the continuity in beam deflection, slope and curvature will be considered. In terms of strain, the relationships are:

$$\int\int \epsilon_{p1}(x,t) dx dx \Big|_{x=x_1} = \int\int \epsilon_p(x,t) dx dx \Big|_{x=x_1} \ , \tag{16}$$

$$\int \epsilon_{p1}(x,t)dx\Big|_{x=x_1} = \int \epsilon_p(x,t)dx\Big|_{x=x_1}, \epsilon_{p1}(x_1,t) = \epsilon_p(x_1,t) , \tag{17}$$

and

$$\int\int \epsilon_{p3}(x,t)dxdx\Big|_{x=x_2} = \int\int \epsilon_p(x,t)dxdx\Big|_{x=x_2} , \tag{18}$$

$$\int \epsilon_{p3}(x,t)dx\Big|_{x=x_2} = \int \epsilon_p(x,t)dx\Big|_{x=x_2}, \epsilon_{p3}(x_2,t) = \epsilon_p(x_2,t) . \tag{19}$$

The free stress condition for the piezoelectric element at the location $x = x_1$ and $x = x_2$ constrains the strain values at these two locations. This constraint ($\sigma_{pe}|_{x=x_1,x_2} = 0$) leads to the following boundary conditions.

$$\epsilon_{p1} = \epsilon_p = \Lambda \quad (x = x_1) \tag{20}$$

and

$$\epsilon_{p3} = \epsilon_p = \Lambda \quad (x = x_2) \tag{21}$$

Equations (12)-(15) and the joint conditions (Eqs. (16)-(21)) as well as the boundary conditions at the two ends of the beam ($x = 0$ and $x = L_x$) form a boundary value problem. The dynamic strain response to the excitation by the applied voltage term Λ can be determined from the solution of this boundary value problem.

SYSTEM EQUATIONS

In the solution of the system equations, simply supported boundary conditions are assumed for the two ends of the beam. For single frequency excitation (angular frequency ω) of the piezoelectric actuator, Eqs. (12),(14) and (15) become fourth order ordinary homogenous differential equations. The solutions for ϵ_{p1}, ϵ_p and ϵ_{p3} are respectively:

$$\epsilon_{p1}(x,\omega) = a_1 e^{jk_1 x} + a_2 e^{-jk_1 x} + a_3 e^{k_1 x} + a_4 e^{-k_1 x} , \tag{22}$$

$$\epsilon_p(x,\omega) = a_5 e^{jk_2^i x} + a_6 e^{-jk_2^i x} + a_7 e^{k_2^r x} + a_8 e^{-k_2^r x} , \tag{23}$$

and

$$\epsilon_{p3}(x,\omega) = a_9 e^{jk_1 x} + a_{10} e^{-jk_1 x} + a_{11} e^{k_1 x} + a_{12} e^{-k_1 x} , \tag{24}$$

where the wavenumbers k_1, k_2^i and k_2^r are respectively

$$k_1 = (\frac{12\rho_p \omega^2}{E_p t_b^2})^{1/4} , \tag{25}$$

$$k_2^i = \left[\frac{\sqrt{(\rho_{pe} t_c \omega^2)^2 + 8\rho_p \omega^2 (E_p t_b/6 + E_{pe} t_c)/t_b} + \rho_{pe} t_c \omega^2}{2(E_p t_b/6 + E_{pe} t_c)}\right]^{1/2} , \tag{26}$$

and

$$k_2^r = \left[\frac{\sqrt{(\rho_{pe} t_c \omega^2)^2 + 8\rho_p \omega^2 (E_p t_b/6 + E_{pe} t_c)/t_b} - \rho_{pe} t_c \omega^2}{2(E_p t_b/6 + E_{pe} t_c)}\right]^{1/2} . \tag{27}$$

Physically, the exponential terms $e^{-jk_1 x}$ and $e^{-jk_2^i x}$ in Eqs. (22)-(24) with the time factor $e^{j\omega t}$ used in this analysis represent the right travelling waves in the beam. Similarly, the terms $e^{jk_1 x}$ and $e^{jk_2^i x}$ correspond to the left travelling waves.

The twelve unknown coefficients $[A] = [a_1, a_2 ... a_{12}]^T$ in Eqs. (22)-(24) are determined by the joint conditions and the simply supported boundary conditions at the beam ends. The resultant 12 algebraic equations are presented in the Appendix, and the matrix expression of the equations can be written as:

$$[\alpha][A] = [\bar{\Lambda}] \tag{28}$$

$[\bar{\Lambda}]$ is a 12×1 dimensional matrix:

$$[\bar{\Lambda}] = [0, 0, 0, 0, \Lambda, \Lambda, 0, 0, \Lambda, \Lambda, 0, 0]^T \tag{29}$$

For a given input Λ and excitation frequency, $[A]$ can be calculated by Eq. (28). The dynamic strain distributions can be then obtained by using Eqs. (22)-(24).

RESULTS AND DISCUSSION

Results from numerical evaluation of the theory outlined in the previous section are compared here with results of some recent experimental work and also with the predictions reported in references 1 and 6. The dimensions of the simply supported steel beam (the same as used in reference 6), are length, $L_x = 380mm$, width, $b = 40mm$ and thickness $t_b = 2mm$. The piezoelectric patches (PZTG -1195) are 38.0 mm in length, $40mm$ in width and $0.2mm$ in thickness. The center of the patches are located at $x = 98.0mm$. The parameters of the beam and piezoelectric materials used in the calculation are listed in Table 1.

Table 1: Material parameters used in calculation

ρ_p	7800 (kg/m^3)	ρ_{pe}	7275 (kg/m^3)
E_p	21.6×10^{10} (N/m^2)	E_{pe}	6.3 $\times 10^{10}$ (N/m^2)
t_b	0.002 (m)	t_c	0.0002 (m)
		d_{31}	1.9 $\times 10^{-10}$ (m/V)

The numerical results obtained here are in terms of the beam surface strain. To compare these results with those obtained by others[1, 6], the beam vertical displacement has been converted into beam surface strain by Eq. (11). The converted results are expressed as:

$$\epsilon_p = \bar{A}_o \sum_{n=1}^{\infty} \frac{n^3}{(\omega_n^2 - \omega^2)} [\cos \frac{n\pi x_1}{L_x} - \cos \frac{n\pi x_2}{L_x}] \sin \frac{n\pi x}{L_x} \tag{30}$$

where $\bar{A}_o$ is a constant and x_1 and x_2 are the locations of the piezoelectric actuator edges. The nth resonance frequency of the beam is determined by[6]:

$$\omega_n^2 = \frac{E_p t_b^2}{12\rho_p} (\frac{\pi n}{L_x})^4 . \tag{31}$$

Figures 3 and 4 show the beam response to a cyclic voltage $V = 50.0e^{j\omega t} Volts$ applied to a single piezoelectric actuator bonded to the beam. The strain response (dB re 1 strain) shown in Fig. 3 is calculated at $x = 340mm$, which is far away from the piezoelectric actuator located at $x = 98.0mm$. Both the dynamic analysis outlined here and the static beam/actuator coupling based analysis of references 1 and 6 show similar predictions for the beam resonance frequencies and the general shape of the beam strain response. The higher frequency resonance peaks are higher in frequency for the dynamic model than for the static model. This indicates the dominant influence of the piezoelectric actuator stiffness on the beam high frequency response. The response shown in Fig. 4 is calculated at $x = 100mm$, which is right at the position where the piezoelectric actuator is located. At this position, the response predicted by the two models differs significantly, when the driving frequency is off resonance. The dynamic model indicates that the strain at the actuator location is mainly controlled by the external voltage and the beam resonance effect is only observable when the driving frequency is at a beam resonance frequency.

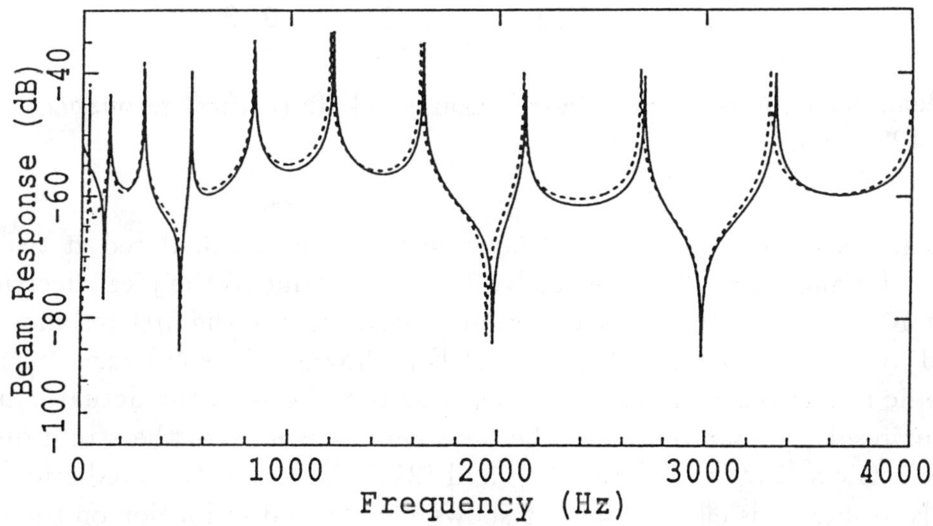

Fig. 3: Far field beam longitudinal strain response. Calculated at $x = 340$ mm; (——) = by Eqs. (28), (22), (23) and (24), (- - - -) by Eq. (30).

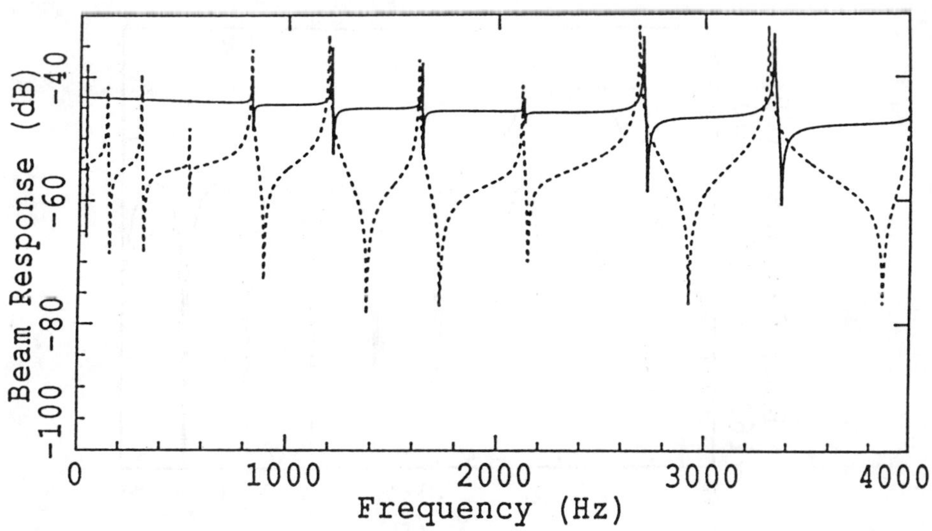

Fig. 4: Near field beam longitudinal strain response. Calculated at $x = 100$ mm; (——) = by Eqs. (28), (22), (23) and (24), (- - - -) by Eq. (30).

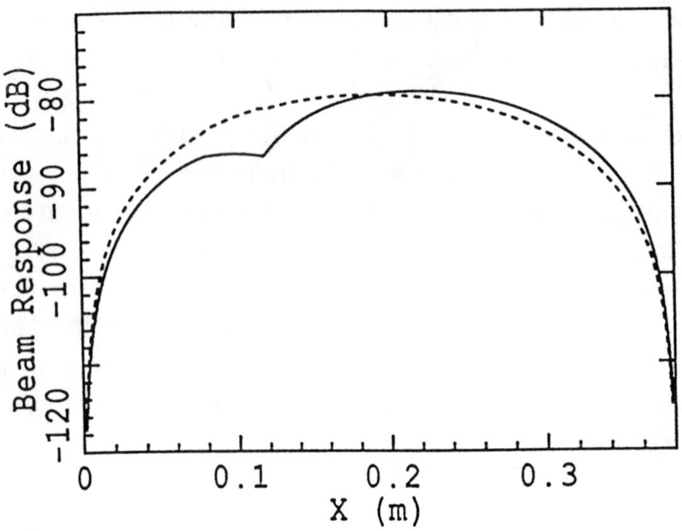

Fig. 5: Beam longitudinal strain distribution at 33 Hz (at first resonance); (——) = by Eqs. (28), (22)-(24), (- - - -) by Eq. (30).

Figures 5 and 6 show the distributed beam surface strain calculated at two driving frequencies 31 Hz and 3000 Hz respectively. The central line of the piezoelectric actuator is located at $x = 98.0mm$. In Fig. 5, the strain distribution of the first resonant beam mode calculated by each of the two different models is shown. The solid curve calculated using the dynamic model indicates that the strain at both ends of the actuator has the same value. This result is consistent with the physical description of the strain distribution in the piezoelectric actuator (see Eqs. (20) and (21)). However, the prediction of the static model fails to show this effect. Figure 6 shows the strain distribution on the beam surface at an off resonance frequency. Away from the actuator, the same strain distributions in the beam are predicted by both models; however, a large difference can be observed close the actuator.

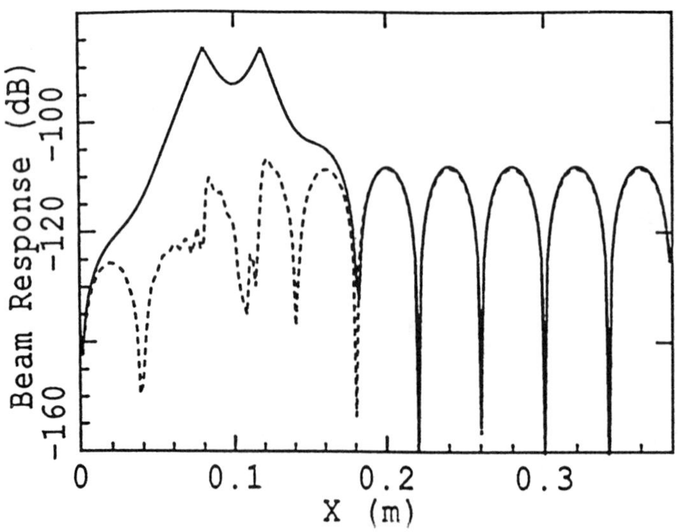

Fig. 6: Beam longitudinal strain distribution at 3000 Hz (off resonance); (——) = by Eqs. (28), (22)-(24), (- - - -) by Eq. (30).

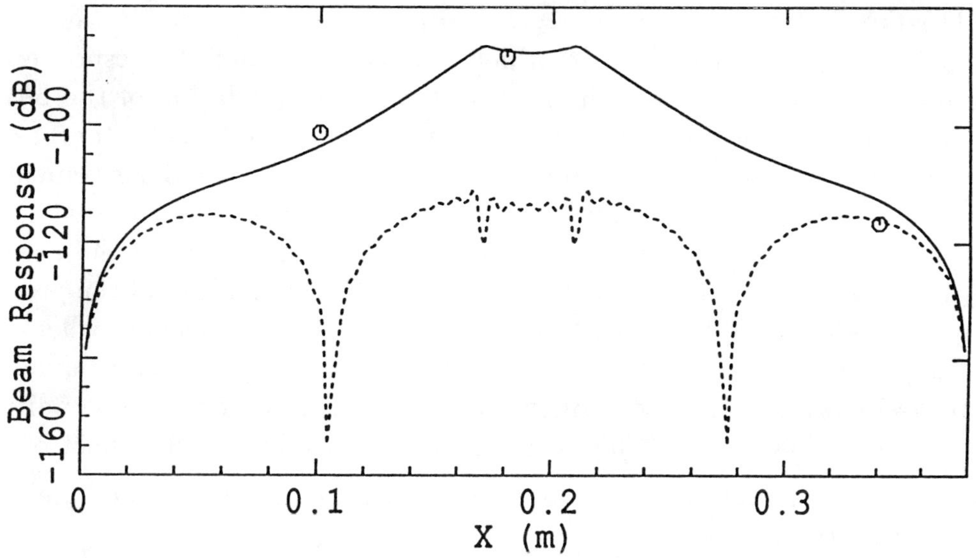

Fig. 7: Beam longitudinal strain distribution at 410 Hz and the actuator is located at beam center; (——) = by Eqs. (28), (22)-(24), (- - - -) = by Eq. (30) and (o) = measured.

To obtain the experimental data, a simply supported beam with piezo-electric actuators was made. The beam has the same dimensions as that used for the calculation. The piezo-electric actuator was located at the middle of the beam, and the beam surface strain was measured at three locations using strain gauges. The measured beam surface strain distributions at 410 Hz and 600 Hz are presented in Figs. 7 and 8, together with the theoretical predictions made using the dynamic theory in this paper and the static theory used in early work by others[1, 6]. This preliminary experimental results indicate that the dynamic theory accurately predicts the surface strain in the beam at locations close to the actuator as well as at locations far away from it.

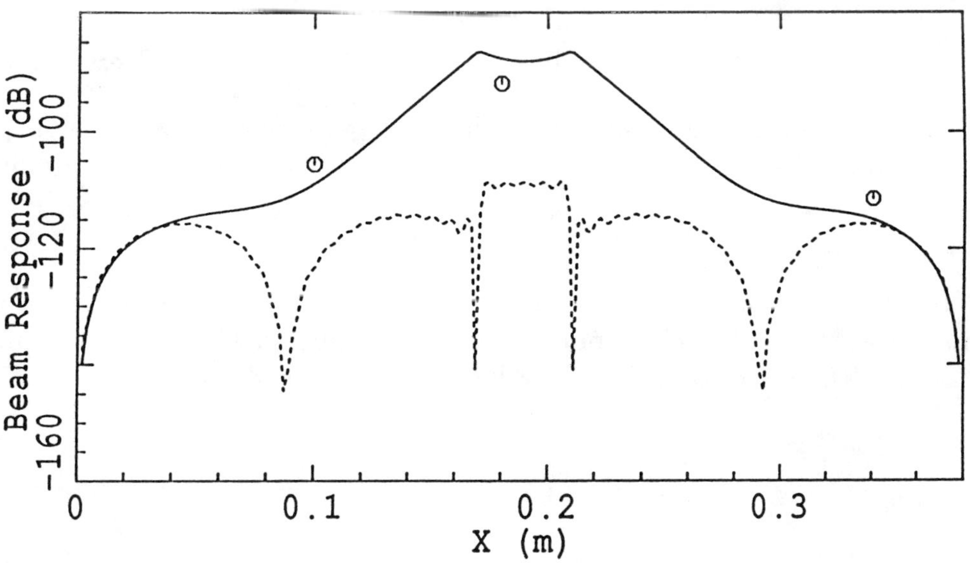

Fig. 8: Beam longitudinal strain distribution at 600 Hz and the actuator is located at beam center; (——) = by Eqs. (28), (22)-(24), (- - - -) = by Eq. (30) and (O) = measured.

CONCLUSIONS

A dynamic model has been developed which is able to predict the response of a beam driven by piezoelectric actuators. The model allows the calculation of the beam surface strain at any location of the beam, (including locations close to the actuators) as a function of the voltage applied to the actuators. The results are in good agreement with the results of static coupling based analyses, developed by previous authors at locations on the beam which are far away from the actuators. However, close to the actuators, the results of the dynamic analysis shown here demonstrate the inability of the previous static analyses to accurately predict the surface strain of the beam, especially when the driving frequency does not correspond to a beam resonance frequency. The model presented here has resolved the problems introduced by the stress free boundary conditions of the piezoelectric actuator and the problems introduced by excluding the effect of the inertia forces. However, more experimental work is needed to verify this model adequately, and this work is currently underway.

REFERENCES

1. Crawley, E.F. and J. de Luis, 1987. "Use of piezoelectric actuators as elements of intelligent structures". AIAA Journal, **25**: 1373-1385.

2. Dimitriadis, E.K., C.R. Fuller and C.A. Rogers, 1989. "Piezoelectric actuators for distributed noise and vibration excitation of thin plates". 8th ASME Conference on Failure, Prevention, Reliability and Stress Analysis, Montreal.

3. Dimitriadis, E.K. and C.R. Fuller, 1989. "Investigation on active control of sound transmission through elastic plates using piezoelectric actuators". AIAA conference paper, AIAA-89-1062.

4. Liang. C. and C.A. Rogers, 1989. "Behavior of shape memory alloy actuators embedded in composites". Proceedings of the 1989 Composites Conference, Beijing, China, August.

5. Graff, K. F. 1975. "Wave motion in elastic solids". Clarendon Press, Oxford.

6. Clark, Jr.,R. L., C. R. Fuller and A. Wicks, "Characterization of multiple piezoelectric actuators for structural excitation". J. Acoust. Soc. Am. to be published

ACKNOWLEDGMENTS

Financial support for this work from the Australian Defence Science and Technology Organization, Material Research Laboratories is gratefully acknowledged. Authors are gratefully acknowledge Mr. J. Rivory for his assistance in obtaining the experimental data.

APPENDIX

The components of the matrix $[\alpha]$ are:

$$\begin{aligned}
\alpha_1 = &\quad [1,1,1,1,0,0,0,0,0,0,0,0] \\
\alpha_2 = &\quad [-1,-1,1,1,0,0,0,0,0,0,0,0] \\
\alpha_3 = &\quad [0,0,0,0,0,0,0,0,e^{jk_1L_x},e^{-jk_1L_x},e^{k_1L_x},e^{-k_1L_x}] \\
\alpha_4 = &\quad [0,0,0,0,0,0,0,0,-e^{jk_1L_x},-e^{-jk_1L_x},e^{k_1L_x},e^{-k_1L_x}] \\
\alpha_5 = &\quad [e^{jk_1x_1},e^{-jk_1x_1},e^{k_1x_1},e^{-k_1x_1},0,0,0,0,0,0,0,0] \\
\alpha_6 = &\quad [0,0,0,0,e^{jk_2^ix_1},e^{-jk_2^ix_1},e^{k_2^rx_1},e^{-k_2^rx_1},0,0,0,0] \\
\alpha_7 = &\quad [\frac{-je^{jk_1x_1}}{k_1},\frac{je^{-jk_1x_1}}{k_1},\frac{e^{k_1x_1}}{k_1},\frac{e^{-k_1x_1}}{k_1},\frac{je^{jk_2^ix_1}}{k_2^i},\frac{-je^{-jk_2^ix_1}}{k_2^i},\frac{-e^{k_2^rx_1}}{k_2^r},\frac{-e^{-k_2^rx_1}}{k_2^r},0,0,0,0] \\
\alpha_8 = &\quad [\frac{-e^{jk_1x_1}}{k_1^2},\frac{-e^{-jk_1x_1}}{k_1^2},\frac{e^{k_1x_1}}{k_1^2},\frac{e^{-k_1x_1}}{k_1^2},\frac{e^{jk_2^ix_1}}{(k_2^i)^2},\frac{e^{-jk_2^ix_1}}{(k_2^i)^2},\frac{-e^{k_2^rx_1}}{(k_2^r)^2},\frac{-e^{-k_2^rx_1}}{(k_2^r)^2},0,0,0,0] \\
\alpha_9 = &\quad [0,0,0,0,e^{jk_2^ix_2},e^{-jk_2^ix_2},e^{k_2^rx_2},e^{-k_2^rx_2},0,0,0,0] \\
\alpha_{10} = &\quad [0,0,0,0,0,0,0,0,e^{jk_1x_2},e^{-jk_1x_2},e^{k_1x_2},e^{-k_1x_2}] \\
\alpha_{11} = &\quad [0,0,0,0,\frac{je^{jk_2^ix_2}}{k_2^i},\frac{-je^{-jk_2^ix_2}}{k_2^i},\frac{-e^{k_2^rx_2}}{k_2^r},\frac{-e^{-k_2^rx_2}}{k_2^r},\frac{-je^{jk_1x_2}}{k_1},\frac{je^{-jk_1x_2}}{k_1},\frac{e^{k_1x_2}}{k_1},\frac{e^{-k_1x_2}}{k_1}] \\
\alpha_{12} = &\quad [0,0,0,0,\frac{e^{jk_2^ix_2}}{(k_2^i)^2},\frac{e^{-jk_2^ix_2}}{(k_2^i)^2},\frac{-e^{k_2^rx_2}}{(k_2^r)^2},\frac{-e^{-k_2^rx_2}}{(k_2^r)^2},\frac{-e^{jk_1x_2}}{k_1^2},\frac{-e^{-jk_1x_2}}{k_1^2},\frac{e^{k_1x_2}}{k_1^2},\frac{e^{-k_1x_2}}{k_1^2}]
\end{aligned}$$

WAVE PROPAGATION IN PIEZOELECTRIC LAYERED MEDIA WITH SOME APPLICATIONS

B. Honein,[1] A. M. B. Braga,[2] P. Barbone,[1] G. Herrmann[1]

ABSTRACT

In this paper we introduce a systematic methodology to investigate wave propagation in piezoelectric layered media. It is based on a matrix formalism and the concept of the surface impedance tensor which relates the components of particle displacement and the normal component of the electric displacement along a surface to the electric potential and the components of traction acting along the same surface. Once the surface impedance tensor for a single layer is calculated, a simple recursive algorithm allows the evaluation of the surface impedance tensor for any number of layers. As an example, the surface impedance tensor is used to find the dispersion curves for a bilaminated piezoelectric plate. Also the dispersion curves for the subsonic interfacial waves when the plate is in contact with a non-conducting acoustic fluid is investigated. Floquet theory is also applied to study wave propagation when the layered medium is periodic.

INTRODUCTION

The purpose of the present contribution is to advance and implement a systematic study of harmonic wave propagation in piezoelectric layered media.

Piezoelectric materials have been widely used for a long time as electromechanical transducers, such as ultrasonic generators, filters, sensors, and actuators, to mention only a few. In recent years, composite materials, consisting of combinations of two or more different piezoelectric and non-piezoelectric material phases, have been designed to meet specific technical needs. Such composites permit the tailoring of special properties, unavailable in homogeneous phases, and therefore they are becoming increasingly important in diverse areas of modern technology.

Due to the inherent complexity of such structures, as a result of the great variety of possible geometries and related type of waves, only scant information regarding these problems has been available.

An attractive formulation to study wave propagation in piezoelectric composite materials appears to be the eight-dimensional state vector formalism proposed originally by Kraut [1]. At the heart of this formulation is the well known fact that a system of ordinary differential equations can be transformed into a first-order system. More

[1]Division of Applied Mechanics, Stanford University, Stanford, CA 94305-4040, U.S.A.

[2]Department of Mechanical Engineering, Pontifícia Universidade Católica do Rio de Janeiro, Rio de Janeiro, CEP 22453, Brazil.

specifically the piezoelectric equations can be regarded as a system of second-order differential equations with respect to the variable whose axis is perpendicular to phase-interfaces. This system can be transformed into a first-order one by introducing new dependent variables which, however, are not chosen arbitrarily but are formed of the additional quantities which have to be continuous across phase-interfaces. When specialized to harmonic waves, the above formulation leads to a homogeneous first-order system with piecewise constant coefficients whose solution is simply the product of the solution for each phase.

This matrix formulation has been applied in one form or another to problems in isotropic and anisotropic elasticity by Stroh [2], [3], Gilbert and Backus [4], Ingebrigtsen and Tonning [5], Barnett and Lothe [6], [7], [8], Chadwick and Smith [9], Gilbert [10], Braga and Herrmann [11], [12] and others.

A faithful application of the Kraut formulation leads, as its counterpart in isotropic and anisotropic elasticity, to inherent numerical problems since both growing and decaying exponentials have to be evaluated, a problem known in numerical analysis as the exponential dichotomy [13]. Many schemes to circumvent this problem have been devised. An efficient and numerically stable algorithm appears to be the one based on the evaluation of the surface impedance tensor, for a single layer. A recursive algorithm is then employed to evaluate the surface impedance tensor for any number of layers. The surface impedance tensor has successfully been applied in the study of surface waves by some of the aforementioned investigators. Recursive algorithms to study the reflection and refraction of elastic waves in isotropic stratified materials have been used by Kennett [14], [15] and Hager and Rostamian [16]. Braga and Herrmann [11], [12] applied similar techniques to investigate wave propagation in anisotropic elastic layered media.

In this contribution we extend the above-mentioned results to piezoelectric materials. In particular, the Kraut formulation is derived in a way which allows extension to curvilinear coordinates. It was shown in an earlier paper [17] that the arrangement of the components within the (column) state vector is of significance in simplifying the analysis. In some special but important cases, the overall problem can be simplified by exploiting material symmetries and thus reducing the order of the eigenvalue problem by a factor of two.

It was also shown that our formulation for the anti-plane and plane strain cases is mathematically equivalent to the plane strain and three- dimensional cases, respectively, for anisotropic elasticity which was discussed extensively by Braga and Herrmann [11], [12]. We then indicate how Floquet theory can be applied in the study of wave propagation in periodically layered media. We show that in the case when the axis perpendicular to phase interfaces is an axis of reflection symmetry, Floquet wave numbers can be obtained by solving a quartic equation whose coefficients involve invariants of the advance mapping matrix for a unit cell.

The impedance surface tensors for a single layer are introduced and a straightforward method is applied to arrive at the impedance tensors for a laminated plate. This method allows more flexibility to deal with different types of boundary conditions. The interaction of a layered piezoelectric solid with an overlying acoustic fluid is investigated. Particular attention is paid to the subsonic interfacial waves of the Scholte-Gogoladze type. Numerical calculations performed for a number of piezoelectric laminates are presented.

I. STATE VECTOR FORMULATION

The governing field equations of piezoelectricity [18], in the quasi-static approximation, are

$$\text{div } \boldsymbol{\sigma} = \rho \frac{\partial^2 \mathbf{u}}{\partial^2 t}, \tag{1}$$

and

$$\text{div } \mathbf{D} = 0, \tag{2}$$

where ρ is the material density and the stress tensor $\boldsymbol{\sigma}$ is given in terms of the displacement gradient $\nabla \mathbf{u}$ and the electric potential ϕ by

$$\boldsymbol{\sigma} = \mathbf{C} : \nabla \mathbf{u} + \nabla \phi \cdot \mathbf{e}, \tag{3}$$

while the electric displacement $\mathbf{D}$ is given by

$$\mathbf{D} = \mathbf{e} : \nabla \mathbf{u} - \boldsymbol{\epsilon} \cdot \nabla \phi. \tag{4}$$

In (3) and (4), $\mathbf{C}$ designates the elastic constants tensor and $\mathbf{e}$ and $\boldsymbol{\epsilon}$ are the piezoelectric and dielectric constants tensors, respectively.

We consider an orthonormal basis $(\mathbf{m_1}, \mathbf{m_2}, \mathbf{n})$ of coordinates, refered to also as (x, y, z) and by subscripts $(1, 2, 3)$, which do not necessarily coincide with the crystal axes X, Y, and Z. We define $\mathbf{t}$ as the traction across planes normal to $\mathbf{n}$, i.e.,

$$\mathbf{t} = \boldsymbol{\sigma} \mathbf{n} \quad \text{and} \quad D_n = \mathbf{D} \cdot \mathbf{n}. \tag{5}$$

We can show that the governing equations (1)-(4) can be written in the form

$$\partial_n \boldsymbol{\xi} = \mathcal{L}\ \boldsymbol{\xi}, \tag{6}$$

where $\boldsymbol{\xi} = (\mathbf{u},\ iD_n,\ i\mathbf{t},\ \phi)^T$ and

$$\mathcal{L} = \begin{bmatrix} \mathcal{S}_1 & \mathcal{S}_2 \\ \mathcal{S}_3 & {\mathcal{S}_1}^T \end{bmatrix}, \tag{7}$$

where

$$\mathcal{S}_1 = \begin{bmatrix} {\mathcal{S}_1}^{11} & {\mathcal{S}_1}^{12} \\ {\mathcal{S}_1}^{21} & {\mathcal{S}_1}^{22} \end{bmatrix}, \tag{8}$$

with

$${\mathcal{S}_1}^{11} = -\mathbf{Q}^*(\mathbf{n})^{-1}\mathbf{R}^*(\mathbf{n}, \mathbf{m}_\alpha)\partial \mathbf{m}_\alpha, \tag{9}$$

$${\mathcal{S}_1}^{12} = -i\mathbf{Q}^*(\mathbf{n})^{-1}\frac{\mathbf{S}(\mathbf{n})}{\epsilon(\mathbf{n}, \mathbf{n})}, \tag{10}$$

$${\mathcal{S}_1}^{21} = -i\left[\mathbf{P}^*(\mathbf{m}_\beta, \mathbf{m}_\alpha) - \mathbf{P}^*(\mathbf{n}, \mathbf{m}_\alpha) \cdot \mathbf{Q}^*(\mathbf{n})^{-1}\mathbf{R}^*(\mathbf{n}, \mathbf{m}_\beta)\right] \partial \mathbf{m}_\alpha \partial \mathbf{m}_\beta, \tag{11}$$

and

$${\mathcal{S}_1}^{22} = \frac{-1}{\epsilon(\mathbf{n}, \mathbf{n})}\left[\mathbf{P}^*(\mathbf{n}, \mathbf{m}_\alpha) \cdot \mathbf{Q}^*(\mathbf{n})^{-1}\mathbf{S}(\mathbf{n}) + \epsilon(\mathbf{n}, \mathbf{m}_\alpha)\right] \partial \mathbf{m}_\alpha, \tag{12}$$

and

$$\mathcal{S}_2 = \begin{bmatrix} {\mathcal{S}_2}^{11} & {\mathcal{S}_2}^{12} \\ {\mathcal{S}_2}^{21} & {\mathcal{S}_2}^{22} \end{bmatrix}, \tag{13}$$

with

$${\mathcal{S}_2}^{11} = -i\mathbf{Q}^*(\mathbf{n})^{-1}, \tag{14}$$

$$\mathcal{S}_2{}^{12} = -\mathbf{Q}^*(\mathbf{n})^{-1}\mathbf{P}^*(\mathbf{n},\mathbf{m}_\alpha)\partial\mathbf{m}_\alpha, \tag{15}$$

$$\mathcal{S}_2{}^{21} = -\mathbf{P}^*(\mathbf{n},\mathbf{m}_\alpha)\cdot\mathbf{Q}^*(\mathbf{n})^{-1}\partial\mathbf{m}_\alpha, \tag{16}$$

and

$$\mathcal{S}_2{}^{22} = i\left[\begin{array}{c}\mathbf{P}^*(\mathbf{n},\mathbf{m}_\alpha)\cdot\mathbf{Q}^*(\mathbf{n})^{-1}\mathbf{P}^*(\mathbf{n},\mathbf{m}_\beta) + \epsilon(\mathbf{m}_\alpha,\mathbf{m}_\beta)\\ -\frac{\epsilon(\mathbf{n},\mathbf{m}_\alpha)\epsilon(\mathbf{n},\mathbf{m}_\beta)}{\epsilon(\mathbf{n},\mathbf{n})}\end{array}\right]\partial\mathbf{m}_\alpha\partial\mathbf{m}_\beta, \tag{17}$$

and

$$\mathcal{S}_3 = \begin{bmatrix}\mathcal{S}_3{}^{11} & \mathcal{S}_3{}^{12}\\ \mathcal{S}_3{}^{21} & \mathcal{S}_3{}^{22}\end{bmatrix}, \tag{18}$$

with

$$\mathcal{S}_3{}^{11} = i\left\{\rho\frac{\partial^2}{\partial t^2} - \left[\mathbf{R}^*(\mathbf{m}_\alpha,\mathbf{m}_\beta) - \mathbf{R}^*(\mathbf{n},\mathbf{m}_\alpha)^T\mathbf{Q}^*(\mathbf{n})^{-1}\mathbf{R}^*(\mathbf{n},\mathbf{m}_\beta)\right]\partial\mathbf{m}_\alpha\partial\mathbf{m}_\beta\right\}, \tag{19}$$

$$\mathcal{S}_3{}^{12} = \frac{1}{\epsilon(\mathbf{n},\mathbf{n})}\left[\mathbf{P}(\mathbf{m}_\alpha,\mathbf{n}) - \mathbf{R}^*(\mathbf{n},\mathbf{m}_\alpha)^T\mathbf{Q}^*(\mathbf{n})^{-1}\mathbf{S}(\mathbf{n})\right]\partial\mathbf{m}_\alpha, \tag{20}$$

$$\mathcal{S}_3{}^{21} = \frac{1}{\epsilon(\mathbf{n},\mathbf{n})}\left[\mathbf{P}(\mathbf{m}_\alpha,\mathbf{n}) - \mathbf{S}(\mathbf{n})\cdot\mathbf{Q}^*(\mathbf{n})^{-1}\mathbf{R}^*(\mathbf{n},\mathbf{m}_\alpha)\right]\partial\mathbf{m}_\alpha, \tag{21}$$

and

$$\mathcal{S}_3{}^{22} = \frac{-i}{\epsilon(\mathbf{n},\mathbf{n})}\left[\mathbf{S}(\mathbf{n})\cdot\mathbf{Q}^*(\mathbf{n})^{-1}\frac{\mathbf{S}(\mathbf{n})}{\epsilon(\mathbf{n},\mathbf{n})} - 1\right]. \tag{22}$$

The tensor $\mathbf{R}^*(\mathbf{a},\mathbf{b})$ and vector $\mathbf{P}^*(\mathbf{a},\mathbf{b})$ are defined by

$$\mathbf{R}^*(\mathbf{a},\mathbf{b}) = \mathbf{R}(\mathbf{a},\mathbf{b}) + \frac{\mathbf{P}(\mathbf{a},\mathbf{n})\otimes\mathbf{P}(\mathbf{b},\mathbf{n})}{\epsilon(\mathbf{n},\mathbf{n})}, \tag{23}$$

$$\mathbf{P}^*(\mathbf{a},\mathbf{b}) = \mathbf{P}(\mathbf{a},\mathbf{b}) - \frac{\mathbf{P}(\mathbf{a},\mathbf{n})\epsilon(\mathbf{n},\mathbf{b})}{\epsilon(\mathbf{n},\mathbf{n})}, \tag{24}$$

$\mathbf{R}$ and $\mathbf{P}$ being the following tensor and vector, respectively,

$$\mathbf{R}(\mathbf{a},\mathbf{b})\mathbf{v} = \mathbf{C} : [\mathbf{v}\otimes\mathbf{b}]\mathbf{a}, \tag{25}$$

$$\mathbf{P}(\mathbf{a},\mathbf{b}) = \mathbf{b}\cdot\mathbf{e}\cdot\mathbf{a}, \tag{26}$$

Here

$$\mathbf{Q}^*(\mathbf{n}) = \mathbf{R}^*(\mathbf{n},\mathbf{n}) \quad \text{and} \quad \mathbf{S}(\mathbf{n}) = \mathbf{P}(\mathbf{n},\mathbf{n}). \tag{27}$$

In the remainder of this paper we will concentrate on plane harmonic waves propagating in the x-z plane. Hence we consider the following expressions

$$\mathbf{u}(x,z,t) = \overline{\mathbf{u}}(z)e^{i(\omega t - k_x x)}, \tag{28}$$

$$D_z(x,z,t) = \overline{D}_z(z)e^{i(\omega t - k_x x)}, \tag{29}$$

$$\mathbf{t}(x,z,t) = \overline{\mathbf{t}}(z)e^{i(\omega t - k_x x)}, \tag{30}$$

$$\phi(x,z,t) = \overline{\phi}(z)e^{i(\omega t - k_x x)}. \tag{31}$$

Then equation (6) takes the form

$$\frac{d}{dz}\overline{\boldsymbol{\xi}} = -i\mathbf{N}\overline{\boldsymbol{\xi}}, \tag{32}$$

where we have defined

$$\overline{\boldsymbol{\xi}} = (\overline{\mathbf{u}},\ i\overline{D}_z,\ i\overline{\mathbf{t}},\ \overline{\phi})^T, \tag{33}$$

and $\mathbf{N}$ is the matrix

$$\mathbf{N} = \begin{bmatrix} \mathbf{S}_1 & \mathbf{S}_2 \\ \mathbf{S}_3 & \mathbf{S}_1{}^T \end{bmatrix}, \tag{34}$$

where

$$\mathbf{S}_1 = \begin{bmatrix} -k_x\mathbf{Q}^*(3)^{-1}\mathbf{R}^*(3,1) & \mathbf{Q}^*(3)^{-1}\frac{\mathbf{s}(3)}{\epsilon(3,3)} \\ -k_x^2\begin{pmatrix} \mathbf{P}^*(1,1) \\ -\mathbf{P}^*(3,1)\cdot\mathbf{Q}^*(3)^{-1}\mathbf{R}^*(3,1)\end{pmatrix} & \frac{-k_x}{\epsilon(3,3)}\begin{pmatrix}\mathbf{P}^*(3,1)\cdot\mathbf{Q}^*(3)^{-1}\mathbf{s}(3) \\ +\epsilon(3,1)\end{pmatrix} \end{bmatrix}, \tag{35}$$

$$\mathbf{S}_2 = \begin{bmatrix} \mathbf{Q}^*(3)^{-1} & -k_x\mathbf{Q}^*(3)^{-1}\mathbf{P}^*(3,1) \\ -k_x\mathbf{P}^*(3,1)\cdot\mathbf{Q}^*(3)^{-1} & k_x^2\begin{pmatrix}\mathbf{P}^*(3,1)\cdot\mathbf{Q}^*(3)^{-1}\mathbf{P}^*(3,1) \\ +\epsilon(1,1)-\frac{\epsilon(3,1)\epsilon(3,1)}{\epsilon(3,3)}\end{pmatrix} \end{bmatrix}, \tag{36}$$

$$\mathbf{S}_3 = \begin{bmatrix} \rho\omega^2\mathbf{I} - k_x^2\begin{pmatrix}\mathbf{R}^*(1,1) \\ -\mathbf{R}^*(3,1)^T\mathbf{Q}^*(3)^{-1}\mathbf{R}^*(3,1)\end{pmatrix} & \frac{k_x}{\epsilon(3,3)}\begin{pmatrix}\mathbf{P}(1,3) \\ -\mathbf{R}^*(3,1)^T\mathbf{Q}^*(3)^{-1}\mathbf{s}(3)\end{pmatrix} \\ \frac{k_x}{\epsilon(3,3)}\begin{pmatrix}\mathbf{P}(1,3) \\ -\mathbf{s}(3)\cdot\mathbf{Q}^*(3)^{-1}\mathbf{R}^*(3,1)\end{pmatrix} & \frac{1}{\epsilon(3,3)}\left(\mathbf{s}(3)\cdot\mathbf{Q}^*(3)^{-1}\frac{\mathbf{s}(3)}{\epsilon(3,3)} - 1\right) \end{bmatrix}. \tag{37}$$

The matrix $\mathbf{N}$ has some interesting properties which, however, will not be explored here. The general solution of equation (6) for a homogeneous layer can be written immediately as

$$\overline{\boldsymbol{\xi}}(z) = e^{-i\mathbf{N}z}\overline{\boldsymbol{\xi}}(0). \tag{38}$$

The matrix $\mathbf{M}(z) = e^{-i\mathbf{N}z}$ is known in the geophysics literature as the propagator matrix.

II. PERIODICALLY PIEZOELECTRIC LAYERED MEDIA

In this section, we consider the case of an infinite periodically layered medium. The unit cell is formed by n layers of piezoelectric (or non-piezoelectric) materials with different properties. The thicknesses are $h_1, h_2, \cdots, h_n$, respectively (see Fig. 1).

We denote by

$$\mathbf{M}_j = e^{-i\mathbf{N}_j h_j} \qquad (j = 1, \cdots, n) \tag{39}$$

the propagator matrix corresponding to the jth layer where $\mathbf{N}_j$ is the fundamental piezoelectric matrix of material j. Then the period advance mapping will be given by

$$\Omega = \mathbf{M}_n\mathbf{M}_{n-1}\cdots\mathbf{M}_1. \tag{40}$$

The eigenvalues of Ω are the (Floquet) multipliers.

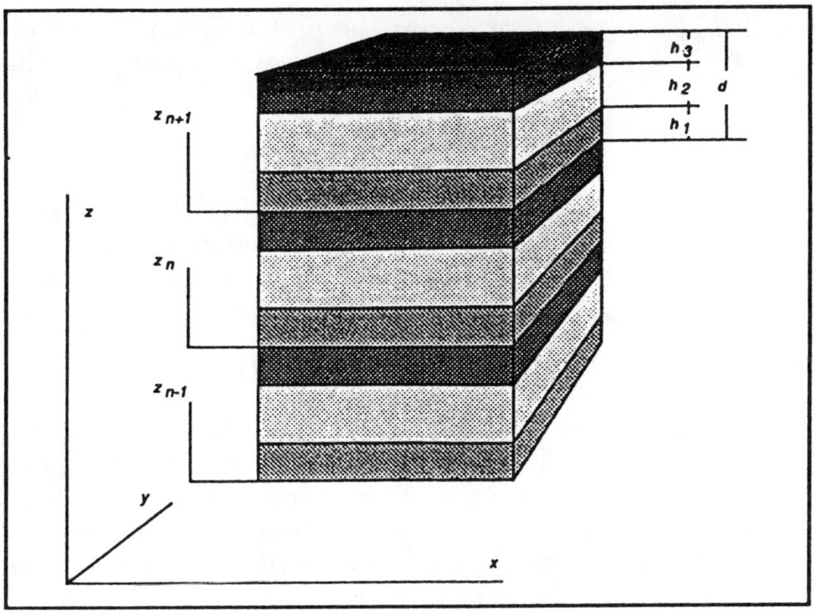

Figure 1. Three-phase periodically layered media

Let $\mathbf{R} = \frac{1}{id}\log \boldsymbol{\Omega}$, where $d = h_1 + \cdots + h_n$ denotes the total thickness of the unit cell. A well known result of Floquet theory states that

$$\overline{\xi}(z) = \mathbf{F}(z)e^{i\mathbf{R}z}\overline{\xi}(z_0), \tag{41}$$

where z_0 is any reference point and $\mathbf{F}$ is a d-periodic matrix. The eigenvalues of $\mathbf{R}$ are the Floquet wave numbers. These are related to the Floquet multipliers by

$$\kappa = \frac{1}{id}\log \tau = \frac{1}{id}(\log|\tau| + i\arg \tau), \tag{42}$$

where κ are the eigenvalues of $\boldsymbol{\Omega}$ and arg τ designates the multi-valued argument of τ.

The dispersion spectrum will be divided into passing or stopping bands depending whether the imaginary part of κ is zero or not. In other words the passing bands are the regions of the (k_x, ω) plane where $|\tau| = 1$ while the stopping bands are the regions where $|\tau| \neq 1$.

In general the solution to the eigenvalue problem involves an 8×8 matrix. However, for the case when a reflective symmetry is present with respect to the z-axis, we can reduce the eigenvalue problem by a factor of two.

Indeed in this case, the Floquet multipliers for waves propagating in the positive z-direction must coincide with the Floquet multipliers for waves propagating in the negative z-direction. This implies that if τ is an eigenvalue of $\boldsymbol{\Omega}$ so is τ^{-1}, therefore the characteristic equation reduces to

$$\tau^8 - I_1(\tau^7 + \tau) + I_2(\tau^6 + \tau^2) - I_3(\tau^5 + \tau^3) + I_4\tau^4 = 0, \tag{43}$$

where $I_i = I_i(\omega, k_x)$ $(i = 1, \cdots, 4)$ are the first four invariants of $\boldsymbol{\Omega}$. Effecting the change of variable $\chi = \tau + \tau^{-1}$, equation (43) can be written in the form

$$\chi^4 - I_1\chi^3 + (I_2 - 4)\chi^2 + (3I_1 - I_3)\chi - 2I_2 + I_4 + 2 = 0. \tag{44}$$

This quartic equation has a closed-form solution. After χ is obtained, τ is calculated by solving the quadratic equation. A detailed discussion can now be carried out regarding the boundaries separating the passing bands regions from the stopping bands regions as was done in [11].

III. THE SURFACE IMPEDANCE TENSORS

As mentioned earlier, a numerical difficulty arises when writing down the solution for many layers as the product of the solutions given by equation (38) for a single layer. This is attributed to the inherent stiffness present in the matrix $M = e^{-i\mathbf{N}z}$ due to the fact that the eigenvalues of $\mathbf{N}$ are associated with decaying as well as growing waves. This difficulty can be overcome by the introduction of the concept of the surface impedance tensors relating the vector $\mathbf{V} = (i\mathbf{t},\ \phi)^T$ to the vector $\mathbf{U} = (\mathbf{u},\ iD_3)^T$ at the plane normal to the z-axis. After the surface impedance tensor for a single layer is evaluated, a simple recursive algorithm which is numerically stable can be written down to evaluate the surface impedance tensor for many layers. The definitions of $\mathbf{U}$ and $\mathbf{V}$ can be modified to accommodate different types of boundary conditions. The definition of the surface impedance tensor has to be modified accordingly. For example, if one is interested in problems where the traction $\mathbf{t}$ and the electric displacement D_3 vanish along the boundary then $\mathbf{U}$ should be defined as $\mathbf{U} = (\mathbf{u},\ \phi)^T$ while $\mathbf{V}$ must be defined as $\mathbf{V} = (i\mathbf{t},\ iD_3)^T$. There are simple relations among the different impedance tensors which can be easily employed to treat problems with different types of boundary conditions along the bottom and the top surfaces. We define the surface impedance tensor $\mathbf{G}$ by $\mathbf{V} = \mathbf{G}\ \mathbf{U}$.

DERIVATION OF THE SURFACE IMPEDANCE TENSOR

We consider the situation depicted in Fig. 2 of a single layer formed of a homogeneous piezoelectric material. We assume that the bottom surface is characterized by the surface impedance $\mathbf{G_0}$. Our aim is to evaluate the surface impedance $\mathbf{G_1}$ at the top of the surface. We designate by

$$\begin{bmatrix} \mathbf{A}_1 & \mathbf{A}_2 \\ \mathbf{L}_1 & \mathbf{L}_2 \end{bmatrix}$$

the matrix of eigenvectors of $\mathbf{N}$ arranged in such a manner that the first four eigenvalues of $\mathbf{N}$ correspond to decaying waves in the positive z-direction. Then the solution of equation (38) can be written as

$$\begin{pmatrix} \mathbf{U}(z) \\ \mathbf{V}(z) \end{pmatrix} = \begin{bmatrix} \mathbf{A}_1 & \mathbf{A}_2 \\ \mathbf{L}_1 & \mathbf{L}_2 \end{bmatrix} \begin{bmatrix} \boldsymbol{\Phi}_1(z) & 0 \\ 0 & \boldsymbol{\Phi}_2(z) \end{bmatrix} \begin{pmatrix} \mathbf{c}_1 \\ \mathbf{c}_2 \end{pmatrix}, \tag{45}$$

where

$$\boldsymbol{\Phi}_1 = \mathrm{diag}(e^{-i\lambda_1 z},\ e^{-i\lambda_2 z},\ e^{-i\lambda_3 z},\ e^{-i\lambda_4 z}), \tag{46}$$

and

$$\boldsymbol{\Phi}_2 = \mathrm{diag}(e^{-i\lambda_5 z},\ e^{-i\lambda_6 z},\ e^{-i\lambda_7 z},\ e^{-i\lambda_8 z}), \tag{47}$$

λ_i being the eigenvalues of $\mathbf{N}$.

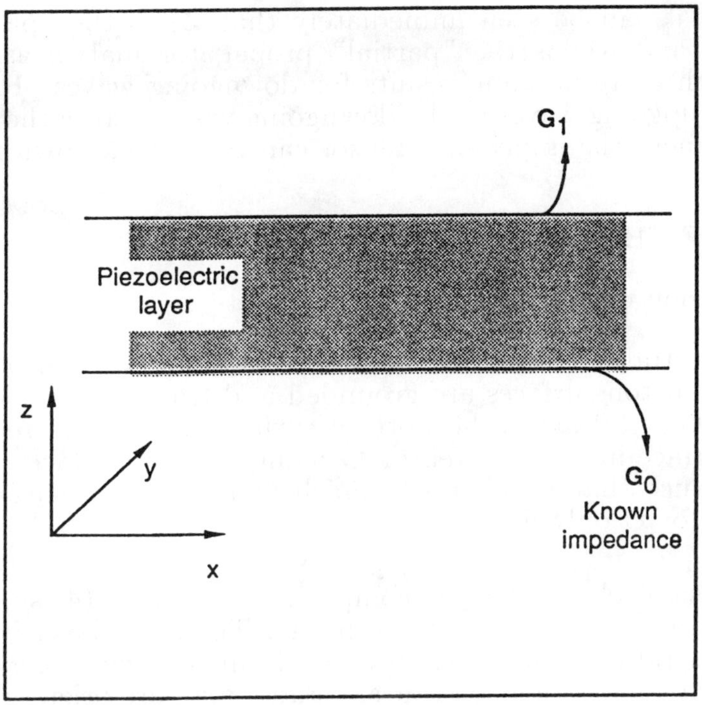

Figure 2. Single piezoelectric layer.

From equation (45), we deduce that

$$\mathbf{U}(z) = \mathbf{A}_1\boldsymbol{\Phi}_1(z)\mathbf{c}_1 + \mathbf{A}_2\boldsymbol{\Phi}_2(z)\mathbf{c}_2, \tag{48}$$

and

$$\mathbf{V}(z) = \mathbf{L}_1\boldsymbol{\Phi}_1(z)\mathbf{c}_1 + \mathbf{L}_2\boldsymbol{\Phi}_2(z)\mathbf{c}_2. \tag{49}$$

For $z = 0$, we have

$$\mathbf{U}(0) = \mathbf{A}_1\mathbf{c}_1 + \mathbf{A}_2\mathbf{c}_2, \tag{50}$$

and

$$\mathbf{V}(0) = \mathbf{L}_1\mathbf{c}_1 + \mathbf{L}_2\mathbf{c}_2. \tag{51}$$

The relation $\mathbf{V}(0) = \mathbf{G}_0\ \mathbf{U}(0)$ serves to eliminate one of the constants $\mathbf{c}_1$ or $\mathbf{c}_2$ in favor of the other in equations (48) and (49). This leads to the result that

$$\mathbf{V}(z) = \mathbf{G}(z)\ \mathbf{U}(z), \tag{52}$$

where

$$\mathbf{G} = (\mathbf{Z}_1\mathbf{H} + \mathbf{Z}_2)(\mathbf{I} + \mathbf{H})^{-1}, \tag{53}$$

and

$$\mathbf{Z}_1 = \mathbf{L}_1\ \mathbf{A}_1{}^{-1} \quad \text{and} \quad \mathbf{Z}_2 = \mathbf{L}_2\ \mathbf{A}_2{}^{-1}, \tag{54}$$

$$\mathbf{H} = \mathbf{M}_1\ \mathbf{R}_0\ \mathbf{M}_2{}^{-1}, \tag{55}$$

$$\mathbf{R}_0 = (\mathbf{Z}_1 - \mathbf{G}_0)^{-1}\ (\mathbf{G}_0 - \mathbf{Z}_2), \tag{56}$$

$$\mathbf{M}_1 = \mathbf{A}_1\boldsymbol{\Phi}_1\mathbf{A}_1{}^{-1} \quad \text{and} \quad \mathbf{M}_2 = \mathbf{A}_2\boldsymbol{\Phi}_2\mathbf{A}_2{}^{-1}. \tag{57}$$

By setting $\mathbf{c}_2 = 0$ it can be seen immediately that $\mathbf{Z}_1$ is the "partial" (or "local") impedance tensor, and $\mathbf{M}_1$ is the "partial" propagator matrix associated with the upgoing waves, with corresponding results for downgoing waves. $\mathbf{R}_0$ is the reflection tensor relating the upgoing waves to the downgoing waves across the surface $z = 0$. By repeating the argument the impedance tensor can be written down for any number of layers.

APPLICATION OF THE SURFACE IMPEDANCE TENSOR

Example 1. Dispersion curves for a bilaminated plate

Suppose we wish to study the wave propagation in a laminated piezoelectric plate (Fig. 3) whose bottom and top surfaces are grounded and traction-free. We set the surface impedance tensor $\mathbf{G}_0 = 0$ along the bottom surface. Using the recursive algorithm, we evaluate the surface impedance tensor $\mathbf{G}$ along the top surface. From the relation $\mathbf{V} = \mathbf{G}\ \mathbf{U}$ we deduce that $\mathbf{G}\ \mathbf{U}$ must vanish along the top surface. We obtain a nontrivial solution for $\mathbf{U}$ only if

$$\det \mathbf{G} = 0, \tag{58}$$

which is the dispersion relation for the composite plate. Fig. (4) shows the dispersion curves for a laminated plate formed of two layers. The lower layer is formed of PZT-2 with 80% thickness ratio. The upper layer is formed of zinc oxide. The solid line represents the dispersion curve when the x,y, and z axes are oriented along the crystal axes X, Y, and Z, respectively, in both materials. The dotted line corresponds to the case when $x = X$, $y = \frac{\sqrt{3}}{2}Y + \frac{1}{2}Z$, and $z = -\frac{1}{2}Y + \frac{\sqrt{3}}{2}Z$ in PZT-2. The reference speed is taken to be $\bar{c} = \sqrt{C_{YZYZ}/\rho}$ of PZT-2.

Example 2. Fluid/piezoelectric solid interfacial subsonic waves

Let an acoustic fluid of density ρ_f and speed of sound c_f occupy the semi-infinite

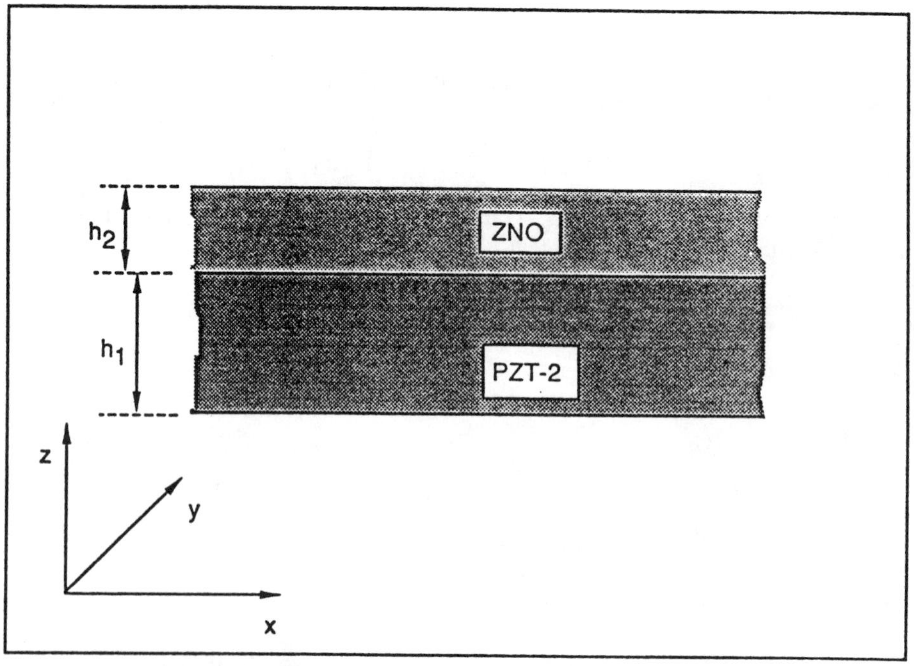

Figure 3. Bilaminated piezoelectric plate.

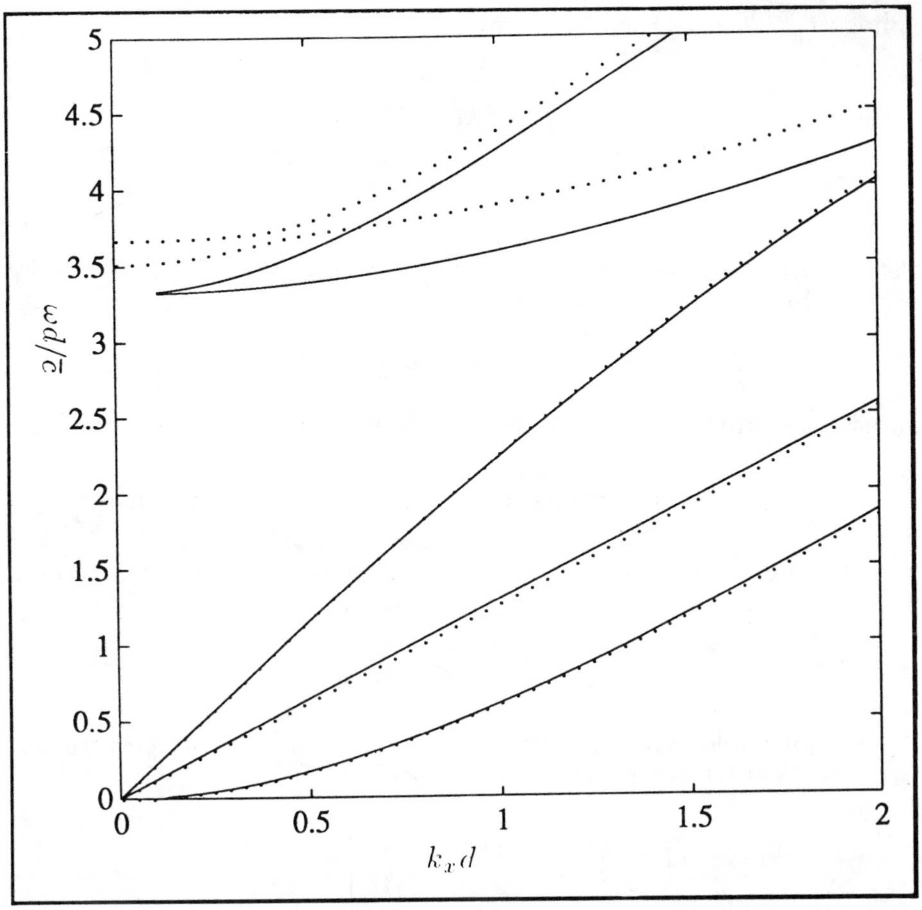

Figure 4. Dispersion curves for bilaminated piezoelectric plate. d is the plate thickness, $\bar{c}$ Is a reference speed. The solid line corresponds to coincidence of crystal axes X, Y, Z, with plate axes x, y, z. The dotted line corresponds to a rotation of π/6 about the x-axis in PZT-2.

region $z > 0$. This fluid half-space is in contact at $z = 0$ with the plane surface of a piezoelectric material. We consider plane harmonic motions in the fluid of the type

$$\mathbf{v}(x, z, t) = \overline{\mathbf{v}}(z)e^{i(\omega t - k_x x)}, \tag{59}$$

and

$$p(x, z, t) = \overline{p}(z)e^{i(\omega t - k_x x)}, \tag{60}$$

where $\mathbf{v}$ and p are, respectively, the particle velocity and pressure fields. The solid interface is characterized by the piezoelectric impedance tensor $\mathbf{G}$. The unit vector $\mathbf{n}$, parallel to the Cartesian z-axis, is normal to the plane interface.

In the absence of interior sources, the motion in the fluid half-space is described, in addition to equations (59) and (60), by equations

$$\overline{\mathbf{v}}(z) \cdot \mathbf{n} = e^{-i\theta z}\overline{\mathbf{v}}(0) \cdot \mathbf{n} \qquad \text{and} \qquad \overline{p}(z) = Z_f \overline{\mathbf{v}}(z) \cdot \mathbf{n}, \tag{61}$$

where the phase $\theta = \theta(\omega, k_x)$ is chosen as

$$\theta(\omega, k_x) = \begin{cases} -i(k_x^2 - \frac{\omega^2}{c_f^2})^{\frac{1}{2}}, & \frac{\omega}{c_f} < k_x \\ (\frac{\omega^2}{c_f^2} - k_x^2)^{\frac{1}{2}}, & \frac{\omega}{c_f} \geq k_x \end{cases} \tag{62}$$

in order to satisfy the radiation condition. In equation (61), $Z_f = Z_f(\omega, k_x)$ is the fluid impedance given by

$$Z_f(\omega, k_x) = \frac{\rho_f\, \omega}{\theta(\omega, k_x)}. \tag{63}$$

Along the fluid/solid interface we impose the following continuity conditions:

$$\overline{\mathbf{v}}(0) \cdot \mathbf{n} = i\omega\overline{\mathbf{u}}(0) \cdot \mathbf{n} \qquad \text{and} \qquad \overline{\mathbf{t}}(0) = -\overline{p}(0)\mathbf{n}, \tag{64}$$

where $\overline{\mathbf{u}}$ and $\overline{\mathbf{t}}$, which are, respectively , the displacement and traction vectors along the solid surface, are related by the surface impedance tensor, i.e.,

$$\overline{\mathbf{t}} = i\omega\mathbf{G}_m\overline{\mathbf{u}}, \tag{65}$$

where $\mathbf{G}_m$ is the piezoelectrically stiffened mechanical impedance tensor. To evaluate $\mathbf{G}_m$, we partition $\mathbf{G}$ into submatrices and write

$$\begin{pmatrix} i\overline{\mathbf{t}} \\ \overline{\phi} \end{pmatrix} = \begin{bmatrix} -\omega\mathbf{Y} & \omega\mathbf{g}_1 \\ \mathbf{g}_2^T & \mathcal{G} \end{bmatrix} \begin{pmatrix} \overline{\mathbf{u}} \\ i\overline{D} \end{pmatrix}, \tag{66}$$

leading to

$$i\overline{\mathbf{t}} = -\omega\mathbf{Y}\overline{\mathbf{u}} + i\omega\mathbf{g}_1\overline{D}, \tag{67}$$

$$\overline{\phi} = \mathbf{g}_2 \cdot \overline{\mathbf{u}} + i\mathcal{G}\overline{D}. \tag{68}$$

Suppose the electric potential is prescribed along the surface and given by, say, $\overline{\phi} = 0$. Then by eliminating $\overline{D}$ from equations (67) and (68) we conclude that

$$\overline{\mathbf{t}} = i\omega(\mathbf{Y} + \frac{1}{\mathcal{G}}\mathbf{g}_1 \otimes \mathbf{g}_2)\overline{\mathbf{u}}. \tag{69}$$

Hence the piezoelectrically stiffened mechanical impedance tensor is given by

$$\mathbf{G}_m = (\mathbf{Y} + \frac{1}{\mathcal{G}}\mathbf{g}_1 \otimes \mathbf{g}_2). \tag{70}$$

By combining equation $(61)_2$ with expressions (64) and (65), and solving for the normal component of the partial velocity at the interface, we obtain

$$(Z_f + Z_s)\overline{\mathbf{v}}(0) \cdot \mathbf{n} = 0, \tag{71}$$

where $Z_s = Z_s(\omega, k_x)$ is the normal piezoelectrically stiffened mechanical impedance tensor, given by

$$Z_s(\omega, k_x) = (\mathbf{n} \cdot \mathbf{G}_m{}^{-1}\mathbf{n})^{-1}. \tag{72}$$

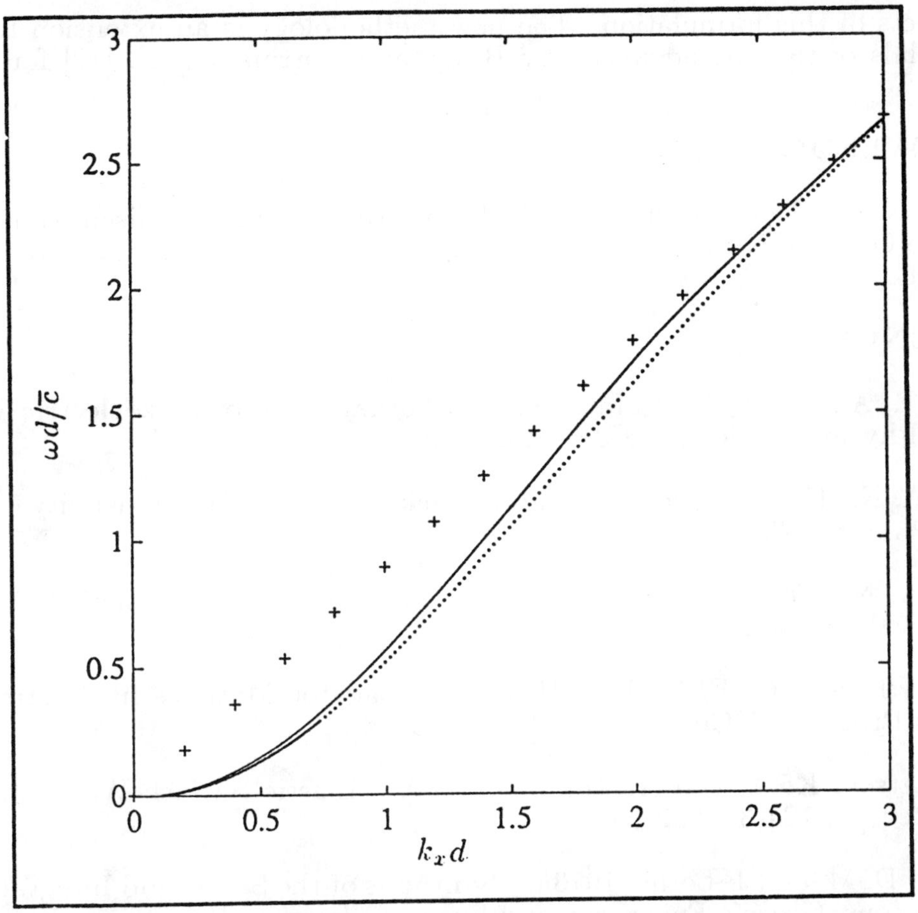

Figure 5. Dispersion curves for subsonic interfacial waves at the fluid/bilaminated plate interface. The solid line corresponds to coincidence of crystal axes X, Y, Z with plate axes x, y, z. The dotted line is for x=Z, y=X, z=Y in PZT-2. The '+' line is the sonic line.

Equation (71) admits non-trivial solutions only if

$$Z_s(\omega, k_x) + Z_f(\omega, k_x) = 0. \tag{73}$$

Hence, expression (73) is the dispersion equation for the propagation of free harmonic waves along the fluid/solid interface. Fig. (5) shows the dispersion curve for the subsonic interfacial wave propagating along the surface of the bilaminated plate treated in example 1 when its upper surface is in contact with water. The solid line corresponds to the case when the x, y, and z axes are aligned with the crystal axes X, Y, and Z, respectively, in both materials. The dotted line represents the case when $x = Z$, $y = X$, and $z = Y$ in PZT-2.

V. CONCLUSIONS

In this paper we presented a systematic methodology to study wave propagation in piezoelectric layered media. Lack of space has prevented us to fully explore in detail some aspects of this formulation. The new methodology is an extension to piezoelectric materials of the one advanced by Braga and Herrmann [11], [12] for anisotropic elasticity.

ACKNOWLEDGEMENTS

The authors would like to thank Prof. B. A. Auld for helpful discussions. The support of the Office of Naval Research through a grant to Stanford University is greatly appreciated.

REFERENCES

1. Kraut, E. A., 1969. "New Mathematical Formulation for Piezoelectric Wave Propagation." Physical Review, 188(3): 1450-1455.

2. Stroh, A. N., 1958. "Dislocations and Cracks in Anisotropic Elasticity." Philosophical Magazine, 3: 625-646.

3. Stroh, A. N., 1962. "Steady State Problems in Anisotropic Elasticity." Journal of Mathematical Physics, 41: 77-103.

4. Gilbert, F. and G. E. Backus, 1966. "Propagator Matrices in Elastic Wave and Vibrations Problems." Geophysics, 31(2): 326-332.

5. Ingebrigtsen, K. A. and A. Tonning, 1969. "Elastic Surface Waves in Crystals." Physical Review, 184(3): 942-951.

6. Barnett, D. M. and J. Lothe, 1973a. "Synthesis of the Sextic and Integral Formalism for Dislocations, Green's Functions, and Surface Waves in Anisotropic Elastic Solids." Physica Norvegica, 7: 13-19.

7. Barnett, D. M. and J. Lothe, 1974. "Consideration of the Existence of Surface Wave (Rayleigh Wave) Solutions in Anisotropic Elastic Crystals." J. Physics F: Metal Physics, 4: 671-686.

8. Barnett, D. M. and J. Lothe, 1985. "Free Surface (Rayleigh) Waves in Anisotropic Elastic Half-Spaces: The Surface Impedance Method." Proceedings of the Royal Society, London A, 402: 135-152.

9. Chadwick, P. and G. D. Smith, 1977. "Foundations of the Theory of Surface Waves in Anisotropic Elastic Materials." Adv. Appl. Mech., 17: 303-376.

10. Gilbert, K. E., 1983. "A Propagator Matrix Method for Periodically Stratified Media." J. Acoust. Soc. Amer., 73: 137-142.

11. Braga, A. M. B. and G. Herrmann, 1988. "Plane Waves in Anisotropic Layered Composites." in Wave Propagation in Structural Composites, A. K. Mal and T. C. T. Ting, ed. New York: ASME-AMD Vol. 90, pp. 81-98.

12. Braga, A. M. B. and G. Herrmann, 1990. "Free Waves at a Fluid/Layered - Composite Interface," in Proceedings of the IUTAM Symposium on Elastic Wave Propagation and Ultrasonic Nondestructive Evaluation, S. K. Datta, J. D. Achenbach, and Y. S. D. Rajapakse, ed. Amsterdam: Elsevier.

13. Chin, R. C. Y., G. W. Hedstrom, and L. Thigpen, 1984. "Matrix Methods in Synthetic Seismograms." Geophys. J. R. Astr. Soc., 77: 483-502.

14. Kennett, B. L. N., 1974. "Reflections, Rays and Reverberations." Bull. Seism. Soc. of Am., 64: 1685-1696.

15. Kennett, B. L. N., 1983. Seismic Wave Propagation in Stratified media, London: Cambridge University Press.

16. Hager, W. W. and R. Rostamian, 1988. "Reflection and Refraction of Elastic Waves for Stratified Materials." Wave Motion, 10: 333-348.

17. Honein, B., A.M.B. Braga, P. Barbone, G. Herrmann, 1991. "Wave Propagation in Piezoelectric Layered Media," in Proceedings of the 2nd ISME Meeting, International Symposium on the Application of Electromagnetic Forces, Sendai, Japan, J. Tani, ed. Amsterdam: Elsevier.

18. Auld, B. A., 1990. Acoustic Fields and Waves in Solids, Vol. 1, Florida: R. E. Krieger Publishing Company.

TRANSFER FUNCTION ANALYSIS FOR A FLEXIBLE BEAM WITH PIEZOELECTRIC FILM ACTUATORS AND SENSORS

Thomas E. Alberts,[1] Joseph A. Colvin[1]

ABSTRACT

In this paper, some fundamental relationships for beams incorporating piezoelectric film actuators and sensors are examined. The differential equation of motion for a beam with piezoelectric film bonded to both sides is used to develop Laplace domain transfer function models of the system. These transfer functions are exact Laplace domain representations of the system equations of motion. The transfer functions are cast into a closed rational form using Maclaurin series expansions representing a specific number of modes. In this form, the transfer functions lend themselves to classical control analysis. It is shown that the transfer function relating a voltage applied to a full coverage actuating layer, to the voltage induced in a full coverage sensing layer on the opposite beam face, behaves like a classic colocated system with alternating poles and zeros and accordingly the system is easy to stabilize with low order compensation. In contrast to this result, it is shown that in spite of the effective colocation of actuator and sensor in the case of the transfer function from actuating voltage to tip position, the desirable alternating pole/zero pattern is not exhibited due to incompatibility of actuating and sensing signals. This result is verified experimentally.

1. INTRODUCTION

Piezoelectric films are under investigation as actuators and sensors for vibration control of flexible space-based structures. These films are thin semicrystalline polymers that are made of poly-vinylidene fluoride (PVDF) and have an electrical conducting layer on both sides. PVDF film strains when exposed to a voltage and, conversely, produces a voltage when strained. This property exists because of the permanent dipole nature of the film which is induced by exposing it to a high electric field during manufacture. For a detailed explanation of the piezoelectric nature of PVDF film see [1] and [2]. Film layers bonded to the surface of flexible structure members can act as either control actuators or sensors depending upon whether voltage is applied or measured.

[1]Department of Mechanical Engineering and Mechanics, Old Dominion University, Norfolk, VA 23529-0247, U.S.A.

A unique characteristic of PVDF film actuators and sensors is that they are spatially distributed over the surface the member which is being controlled. This is in contrast to discrete actuators and sensors more customarily employed in the control of structures. Due to the spatially distributed nature of piezoelectric films, it is not entirely clear from the outset whether or not actuator/sensor pairs are colocated. As such, it is not known whether closed loop systems will benefit from certain highly desirable properties associated with colocated actuator/sensor pairs or whether, for example, a given actuator/sensor pair leads to non-minimum phase zeros typically associated with non-colocation in flexible structures. In this paper a distributed parameter model of a beam-film system is considered. The system investigated is a cantilever beam with piezoelectric film bonded to both sides. The differential equation of motion for this system is transformed to Laplace domain transfer functions after application of the appropriate boundary conditions. Transfer functions relating actuator voltage to tip position and relating actuator voltage to the voltage produced at a piezoelectric sensor are developed. Because the film distribution is spatially invariant, the effect of the actuating layer is equivalent to a moment applied to the free end of the beam and the sensing layer senses the angular displacement at the free end of the beam. Root locus plots are employed as well as analytical and experimental frequency response plots for model verification and analysis.

2. PVDF FILM ELECTROMECHANICAL RELATIONS

The governing equations of motion for beams incorporating PVDF film layers have been documented in [3] and [4]. These results are adapted here to a cantilever beam with a uniform distribution of PVDF film bonded to both sides. Consider the three layer composite beam illustrated in Figure 1.

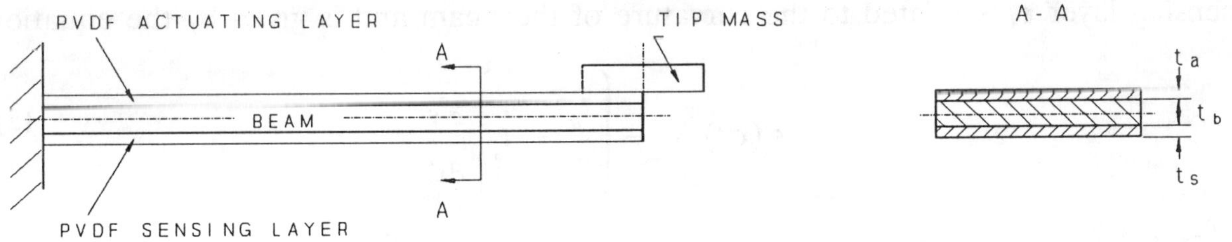

Figure 1. Schematic of Experimental Beam with PVDF Layers

The subscripts *a*, *s* and *b* indicate the actuating layer, sensing layer and beam respectively. The effect of applying a voltage to a free layer of PVDF film is to induce a strain ϵ_a in that layer is given by

$$\epsilon_a(x,t) = V(x,t)\,\frac{d_{31}}{t_a} \tag{1}$$

where $V(x,t)$ is the applied voltage and d_{31} is the piezoelectric strain constant. The value of d_{31} and all other piezoelectric constants used in this paper are as given in [5] and are listed in Table 1. The PVDF actuating layer induces a longitudinal strain in the beam surface. Assuming perfect bonding between the film and beam, this longitudinal strain ϵ_b is found by considering a force equilibrium at the interface of the two layers and is given by

$$\epsilon_b(x) = -\frac{E_a t_a}{E_b t_b + E_a t_a}\epsilon_a(x) \tag{2}$$

where E_b and E_a are the Young's modulus of the beam and the actuator respectively. The electrically induced strain in the PVDF actuating layer also exerts a force on the beam, acting through a distance from the neutral axis of the beam to the neutral axis of the actuating layer. The moment $M(x,t)$ applied by the actuating layer changes with the geometry of the composite beam and is given for this case as

$$M(x,t) = cV(x,t) \tag{3}$$

where c is a geometric constant and is given by the equation

$$c = \frac{2E_b E_a W d_{31}(t_b^2 + t_b t_a)}{E_b t_b + 2E_a t_a} \tag{4}$$

and W is the width of both the beam and the PVDF layer.

A similar analysis must be done for the sensing layer. The strain developed in the sensing layer ϵ_s is related to the curvature of the beam and is given by the equation:

$$\epsilon_s(x,t) = -\left(\frac{t_b}{2} + t_a\right)\frac{\partial^2 y}{\partial x^2} \tag{5}$$

The applied strain at a given point x on the beam gives rise to a charge $q(x,t)$ on the sensing layer surface which is given by the equation:

$$q(x,t) = \left(\frac{k_{31}^2}{g_{31}}\right)\epsilon_s(x,t) \tag{6}$$

Here k_{31} is the piezoelectric electromechanical coupling constant and g_{31} is the piezoelectric stress constant. The total charge developed on the surface of the sensing layer is the spatial integral of the charge over the length of the beam. The total charge $Q(t)$ developed on the sensing layer is given by the equation:

where $V(x,t)$ is the applied voltage and d_{31} is the piezoelectric strain constant. The value of d_{31} and all other piezoelectric constants used in this paper are as given in [5] and are listed in Table 1. The PVDF actuating layer induces a longitudinal strain in the beam surface. Assuming perfect bonding between the film and beam, this longitudinal strain ϵ_b is found by considering a force equilibrium at the interface of the two layers and is given by

$$\epsilon_b(x) = -\frac{E_a t_a}{E_b t_b + E_a t_a}\epsilon_a(x) \tag{2}$$

where E_b and E_a are the Young's modulus of the beam and the actuator respectively. The electrically induced strain in the PVDF actuating layer also exerts a force on the beam, acting through a distance from the neutral axis of the beam to the neutral axis of the actuating layer. The moment $M(x,t)$ applied by the actuating layer changes with the geometry of the composite beam and is given for this case as

$$M(x,t) = cV(x,t) \tag{3}$$

where c is a geometric constant and is given by the equation

$$c = \frac{2E_b E_a W d_{31}(t_b^2 + t_b t_a)}{E_b t_b + 2E_a t_a} \tag{4}$$

and W is the width of both the beam and the PVDF layer.

A similar analysis must be done for the sensing layer. The strain developed in the sensing layer ϵ_s is related to the curvature of the beam and is given by the equation:

$$\epsilon_s(x,t) = -\left(\frac{t_b}{2} + t_a\right)\frac{\partial^2 y}{\partial x^2} \tag{5}$$

The applied strain at a given point x on the beam gives rise to a charge $q(x,t)$ on the sensing layer surface which is given by the equation:

$$q(x,t) = \left(\frac{k_{31}^2}{g_{31}}\right)\epsilon_s(x,t) \tag{6}$$

Here k_{31} is the piezoelectric electromechanical coupling constant and g_{31} is the piezoelectric stress constant. The total charge developed on the surface of the sensing layer is the spatial integral of the charge over the length of the beam. The total charge $Q(t)$ developed on the sensing layer is given by the equation:

The differential equation and boundary conditions given above describe a system where the voltage V varies with both position and time. For this to be true the PVDF film actuating layer must have a spatially varying distribution. In the case considered here the actuating layer covers the entire beam surface and thus the spatial distribution is constant with respect to x. Thus, for this case, the partial derivative of voltage with respect to the spatial variable x is zero and it is clear from equations (9) and (10) that the differential equation reduces to the standard Bernoulli-Euler beam equation. The boundary conditions reduce to those associated with a cantilever beam with a moment applied to the free end.

4. SOLUTION OF BERNOULLI-EULER EQUATION

The Bernoulli-Euler equation is transformed to an ordinary differential equation by eliminating the partial derivatives with respect to time by Laplace transformation. The Laplace transformed Bernoulli-Euler equation and boundary conditions are given by

$$
\begin{aligned}
&\frac{d^4 y(x,s)}{dx^4} + \frac{\rho A}{EI} s^2\, y(x,s) = 0\\
&y(0,s) = \frac{dy(0,s)}{dx} = 0\\
&EI\frac{d^2 y(L,s)}{dx^2} = -I_T s^2 \frac{dy(L,s)}{dx} + cV(s)\\
&EI\frac{d^3 y(L,s)}{dx^3} = M_T s^2 y(L,s)
\end{aligned}
\tag{11}
$$

To simplify the notation, a dimensionless frequency operator and position variable are defined respectively as follows:

$$
\lambda^4 = -\frac{\rho A L^4}{EI} s^2 \qquad and \qquad \chi = \frac{x}{L} \tag{12}
$$

The Bernoulli-Euler equation can be expressed in terms of these variables as

$$
\frac{d^4 y(\chi,s)}{d\chi^4} - \lambda^4 y(\chi,s) = 0 \tag{13}
$$

This equation is a linear homogeneous ordinary differential equation and can be solved by the method of undetermined coefficients. The general solution to this differential equation can be written as

$$y(\chi,s) = A_1 \cos\lambda\chi + A_2 \sin\lambda\chi + A_3 \cosh\lambda\chi + A_4 \sinh\lambda\chi \tag{14}$$

where A_1, A_2, A_3 and A_4 are coefficients to be determined from the boundary conditions. Application of the boundary conditions reveals that the constants satisfy $A_1 = -A_3$, $A_2 = -A_4$ and the relationship:

$$\begin{bmatrix} A_1 \\ A_2 \end{bmatrix} = \left[M(\lambda)\right]^{-1} \begin{bmatrix} \dfrac{cV(t)}{EI} \\ 0 \end{bmatrix} \tag{15}$$

The matrix $[M(\lambda)]$ is given by

$$\begin{bmatrix} -\dfrac{\lambda^2}{L^2}(c_\lambda + ch_\lambda) + \dfrac{I_T\lambda^5}{\rho AL^5}(s_\lambda + sh_\lambda) & -\dfrac{\lambda^2}{L^2}(s_\lambda + sh_\lambda) - \dfrac{I_T\lambda^5}{\rho AL^5}(c_\lambda - ch_\lambda) \\ \dfrac{\lambda^3}{L^3}(s_\lambda - sh_\lambda) + \dfrac{M_T\lambda^4}{\rho AL^4}(c_\lambda - ch_\lambda) & -\dfrac{\lambda^3}{L^3}(c_\lambda + ch_\lambda) + \dfrac{M_T\lambda^4}{\rho AL^4}(s_\lambda - sh_\lambda) \end{bmatrix} \tag{16}$$

$$c_\lambda = \cos\lambda,\ s_\lambda = \sin\lambda,\ ch_\lambda = \cosh\lambda,\ sh_\lambda = \sinh\lambda$$

Note that $[M(\lambda)]$ is the characteristic matrix for this system and that $\det[M(\lambda)] = 0$ is the characteristic equation. Equations (14), (15) and (16) are combined to give:

$$A_1 = -A_3 = \left[-(c_\lambda + ch_\lambda) + \frac{M_T\lambda}{\rho AL}(s_\lambda - sh_\lambda)\right]\frac{cV(s)}{2EI\Delta}$$

$$A_2 = -A_4 = \left[-(s_\lambda - sh_\lambda) - \frac{M_T\lambda}{\rho AL}(c_\lambda - ch_\lambda)\right]\frac{cV(s)}{2EI\Delta} \tag{17}$$

$$\Delta = \left(\frac{\lambda}{L}\right)^6 \frac{I_T M_T}{(\rho A)^2}(1 - c_\lambda ch_\lambda) - \left(\frac{\lambda}{L}\right)^5 \frac{I_T}{\rho A}(c_\lambda sh_\lambda + s_\lambda ch_\lambda)$$
$$+ \left(\frac{\lambda}{L}\right)^3 \frac{M_T}{\rho A}(c_\lambda sh_\lambda - s_\lambda ch_\lambda) + \left(\frac{\lambda}{L}\right)^2 (1 + c_\lambda ch_\lambda)$$

Equations (14) and (17) represent the exact Laplace domain solution of the differential equation chosen to model this system. The equation shows the relationship between the actuating voltage and the deflection of the cantilever beam at any point x along its length.

5. TRANSFER FUNCTIONS

The deflection equation can be used to develop transfer functions relating actuating layer voltage to tip position and actuating layer voltage to sensing layer voltage The transfer functions obtained from equations (14) and (17) are in terms of transcendental functions and must be rationalized for a specific number of modes of vibration. The transfer function between actuating layer voltage and tip position is written directly from equations (14) and (17) and is given by

$$\frac{y}{V}(\lambda) = \frac{c}{\Delta\, EI}\sin\lambda \sinh\lambda \tag{18}$$

The transfer function relating the actuating layer voltage to the sensing layer voltage is determined with the use of equation (8) which relates sensing layer voltage to the slope of the beam's displacement. Substituting the first derivative of the deflection equation into equation (8) yields the transfer function

$$\frac{V_s}{V_a}(\lambda) = -\frac{k_{31}^2 c\lambda}{WL^2CEIg_{31}\Delta}\left(\frac{t_b}{2}+t_a\right)\left[(s_\lambda ch_\lambda + c_\lambda sh_\lambda) + \frac{M_T\lambda}{\rho AL}(c_\lambda ch_\lambda - 1)\right] \tag{19}$$

The transfer functions of equations (18) and (19) are exact relationships containing transcendental terms. These functions can be rationalized as shown by Schmitz [6], for a specific number of modes, by expanding each numerator and denominator into a Maclaurin series of the form

$$f(\lambda) = \frac{\lambda^p}{p!}\frac{d^pf(0)}{d\lambda^p}(1 + k_1\lambda^4 + k_2\lambda^8 + \dots) \tag{20}$$

or in the equivalent product form

$$f(\lambda) = \frac{\lambda^p}{p!}\frac{d^pf(0)}{d\lambda^p}\prod_{i=1}^{N}\left[1 - \left(\frac{\lambda}{\lambda_i}\right)^4\right] \tag{21}$$

The coefficient in front of this equation is obtained by differentiating each transcendental numerator and denominator until a non-zero constant is obtained when the derivative is evaluated at zero. The p in equation (21) is the index of that first non-zero derivative and the N corresponds to the number of modes to be retained in the rationalized representation. To obtain the Laplace domain transfer functions, equation (12) is used to make the substitution

$$\lambda_i^4 = -\frac{\rho A L^4}{EI} s_i^2 \tag{22}$$

Rationalization of equation (18) as described above, yields the Laplace domain transfer function relating tip position to actuating voltage for the first four modes of this system.

$$\frac{y}{V}(s) = \frac{cL^2}{2EI} \prod_{i=1}^{4} \left[\frac{1 + \left(\frac{s}{s_{ni}}\right)^2}{1 + \left(\frac{s}{s_{di}}\right)^2} \right] \tag{23}$$

The subscripts ni and di designate to the ith root of the transcendental numerator and denominator respectively. Similarly the Laplace domain transfer relating actuating layer voltage to sensing layer voltage is obtained by rationalizing equation (19).

$$\frac{V_s}{V_a}(s) = \frac{k_{31} c}{WLCg_{31} EI} \left(\frac{t_b}{2} + t_s \right) \prod_{i=1}^{4} \left[\frac{1 + \left(\frac{s}{s_{ni}}\right)^2}{1 + \left(\frac{s}{s_{di}}\right)^2} \right] \tag{24}$$

6. PASSIVITY

Gervarter [7] and Martin [8] have shown that beam-like structures in torsion or bending, always have transfer functions with poles and zeros that alternate along the imaginary axis providing that the actuator sensor pair are compatable and colocated. This property is highly desirable in flexible structures under feedback control because it allows the system to be adequately stabilized using simple low order compensation such as PD control or rate feedback. Recently, Wang and Vidyasagar [9] have pointed out the connection between this property and the passivity property of classical network theory. According to network theory [10], certain transfer functions posses the property of passivity which guarantees that such a system can be stabilized using simple low order compensation such as PD control. Formally, the passivity theorem [11] states that a system with a passive transfer function can be stabilized by any finite gain strictly passive compensator. In simple terms the passivity property means that the phase of the transfer function always lies between +90 degrees and -90 degrees in the right half s-plane. Rational transfer functions which have a single pole or zero at the origin and additionally only simple poles and zeros which alternate along the imaginary axis, are formally known as reactive [10]. Reactive transfer functions are known to be passive. Hence, any flexible structure with compatible colocated actuators and sensors exhibit the desirable passivity property to within a time derivative. In the next section the transfer functions derived above are examined from this perspective.

7. RESULTS

The transfer functions of equations (23) and (24) will be analyzed here to evaluate the characteristic dynamics of the systems under consideration. The physical constants used in the analysis are listed in Table 1. The values of s_{ni} and s_{di} are listed in Table 2. First we consider Equation (24), relating actuating layer voltage to sensing layer voltage for the cantilever beam with a tip mass and tip inertia. The analytical frequency response and root locus (P-control) plots of this system are given in Figure 2.

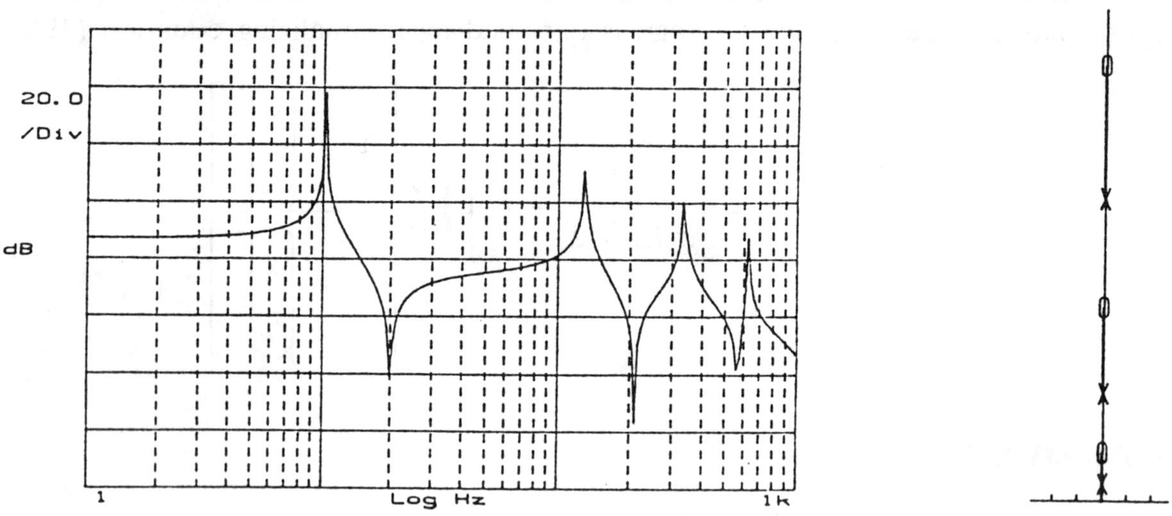

Figure 2. Bode Magnitude Plot and P-Control Root Locus for "Voltage to Voltage" System

The figures illustrate that this system is characterized by alternating poles and zeros on the imaginary axis and hence the system, once differentiated, is reactive. In this example the actuator and sensor signals are compatible because, as we have shown, the actuator effectively applies a moment to the end of the beam and the sensor is in effect sensing the angle of the end of the beam.

Now, let us consider the transfer function relating actuating layer voltage to beam tip position. Two cases will be considered, first a very small payload will be considered, and in the second case a larger tip payload is added which has correspondingly larger tip mass and inertia. The values of s_{ni} and s_{di} for these cases are also given in Table 1. The analytical transfer functions for these two cases were experimentally verified using a Hewlett-Packard 3526A two channel dynamic signal analyzer. A random noise signal generated by the analyzer and amplified by a KEPCO BOP-1000M amplifier was input to the piezoelectric actuating layer. The system response was measured with an accelerometer fixed at the tip of the beam. Figures 3 and 4 show the analytical and experimental frequency response plots for the cases with a light tip mass and a heavy tip mass respectively.

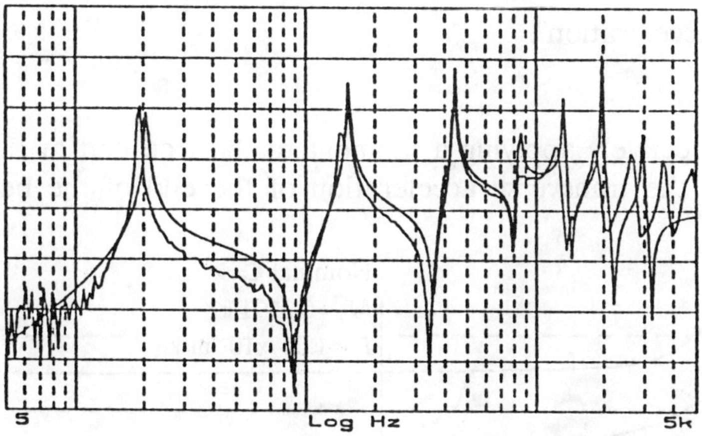

Figure 3. Analytical Versus Experimental Bode Magnitude Plot for Small Payload, "Voltage to Tip Acceleration".

Inspection of Figures 3 and 4 shows that initially the poles and zeros alternate along the imaginary axis, as one might expect for a structure with colocated actuators and sensors. However, at some point determined by the payload, the alternating pattern is interrupted by a pair of poles occurring in sequence, without a zero in between. This phenomenon occurs between the 4th and 5th pole for the small tip mass case and between the 2nd and 3rd pole for the heavy tip mass case. As payload mass is increased, the unwanted sequential pair of poles occurs at lower frequencies. Figure 5 shows the dimensionless frequencies λ corresponding to the first seven poles and zeros as they vary with the dimensionless tip mass to beam mass ratio. Tip inertia I_T is varied from the nominal "large mass" case according to the 5/3 power of the mass scaling factor to provide proper dynamic scaling of the tip inertia's influence. Note that the zeros are invariant to tip mass while the poles undergo frequency compression which forces some pole pairs to lie between successive zeros. In theory, the zero payload case has alternating poles and zeros for all frequencies. The non-alternating pole-zero pattern can be explained by the fact that even though the actuator and sensor are effectively colocated, their signals are not compatible. The

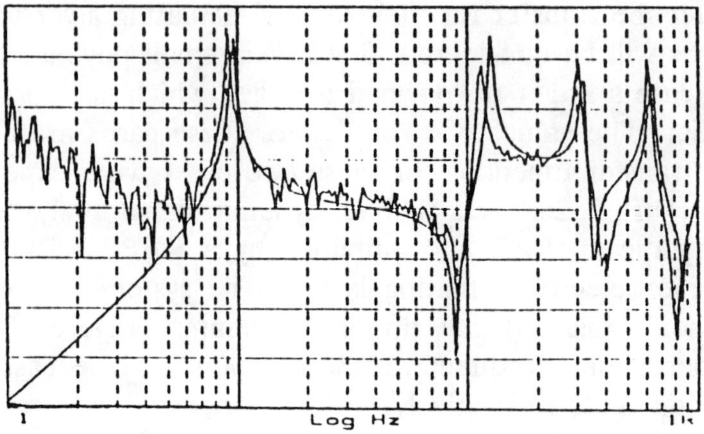

Figure 4 Analytical Versus Experimental Bode Magnitude Plot for Large Payload, "Voltage to Tip Acceleration".

actuating layer is effectively applying a moment to the free end of the beam and the sensor is measuring the transverse acceleration of the end of the beam.

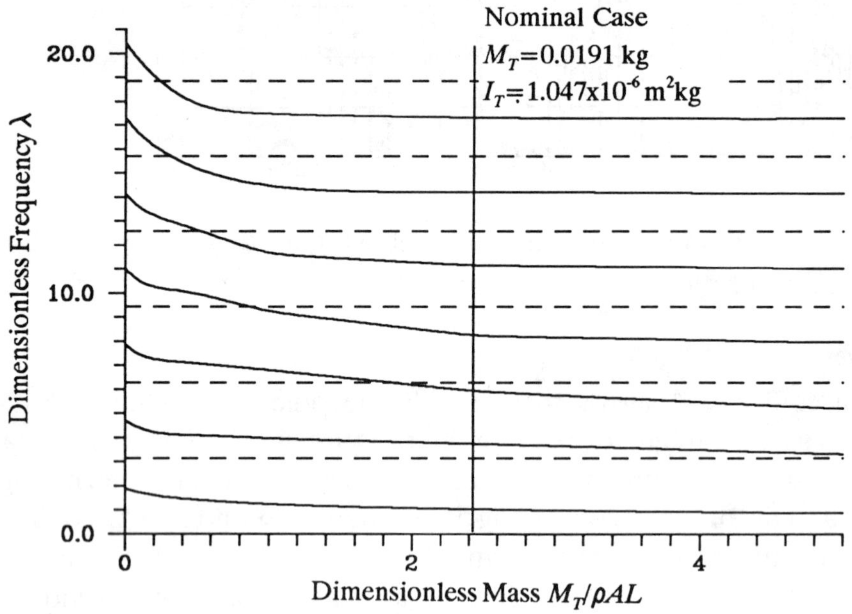

Figure 5. Variation of Dimensionless Poles and Zeros with Tip Mass for Equation (18).

The non-alternation of the poles and zeros exhibited by the voltage to tip position transfer function has stability implications on the closed loop system. In particular, this system is fundamentally more difficult to stabilize than one with alternating poles and zeros. Consider the root locus for the case with a heavy tip mass and position feedback. The closed-loop system for this case will be unstable for low order compensation as illustrated by the root locus of Figure 6.

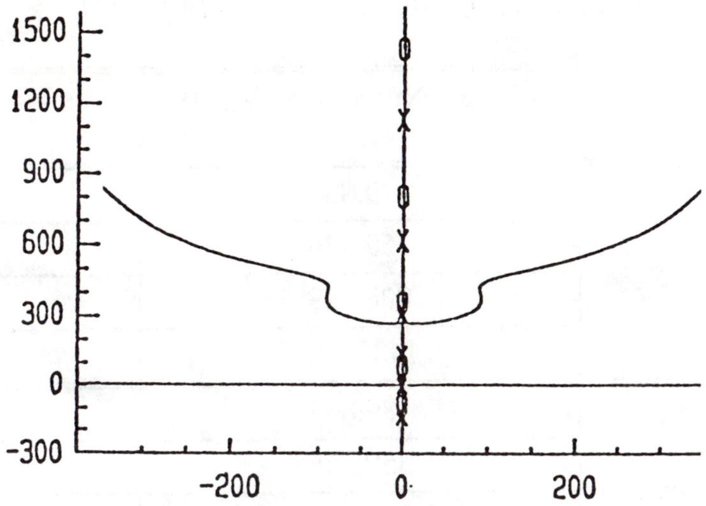

Figure 6. P-Control Root Locus for "Voltage to Tip Position" Transfer Function.

Note that the instability occurs in between the two poles located together and is a result of the non-alternating pole-zero pattern. This situation can be avoided by using a feedback signal which is compatible with the moment being applied by the actuating layer, for example the tip angle measurement sensed by the piezoelectric film layer in the example represented by equation (24).

8. CONCLUSION

The transfer functions relating actuating layer voltage to sensing layer voltage and relating actuating layer voltage to tip position have been derived for spatially invariant PVDF film bonded to the surface of a cantilever beam. The transfer function relating voltage to tip position has been experimentally verified and was shown to be an accurate model of the system. It has been shown that despite the effective colocated nature of the actuator and sensor for this system it does not exhibit the desirable alternating pole-zero pattern because the actuator signal is not compatible with the sensor signal. Neither of the systems considered exhibits the non-minimum phase zeros associated with non-colocation of actuators and sensors.

9. ACKNOWLEDGEMENT

The work presented in this paper is supported by National Science Foundation grant ENG-8811633.

10. APPENDIX

Table 1. Transfer Function Roots for Equation (23)

	s-Plane Numerator Roots (Hz)	s-Plane Denominator Roots (Hz)
No Tip Mass	89.837	32.004
	359.346	200.566
	808.529	561.580
	1433.855	1100.580
Small Tip Mass $M_T = 0.00267$ kg $I_T = 1.36\text{x}10^{-7}\ m^2kg$	89.837	20.689
	359.346	154.864
	808.592	447.754
	1433.855	850.650
Large Tip Mass $M_T = 0.0191$ kg $I_T = 1.047\text{x}10^{-6}\ m_2kg$	89.837	9.613
	359.346	127.183
	808.529	321.448
	1433.855	619.915

Table 2 Physical Parameters of the Experimental System

Piezoelectric material properties and geometrical constants		Beam constants	Beam/Film system constants
$d_{31}=23\times10^{-12}$ m/v	$I_a=2.138\times10^{-13}m^4$	$I_b=9.8154x10^{-13}m^4$	$EI=0.0750$N-m^2
$g_{31}=216\times10^{-3}$(v/m)/(n/m^2)	$E_a=2\times10^9$ N/m^2	$E_b=76\times10^9$N/m^2	$\rho A=0.0549$ kg/m
$k_{31}=0.12$	$t_a=52\times10^{-6}$ m	$t_b=8.12\times10^{-4}$ m	$W=0.022$ m
$C=379\times10^{-12}$F/cm^2	$c=1.667\text{x}10^{-6}$Nm/v		$L=0.144$ m

11. REFERENCES

1. Lee, C.K., "Piezoelectric Laminates for Torsional and Bending Modal Control: Theory and Experiment", Ph.D. Dissertation, Cornell University, 1987.

2. Cudney, H.H., "Distributed Structural Control Using Multilayered Piezoelectric Actuators", Ph.D Dissertation, SUNY Buffalo, NY., 1989.

3. Bailey, T., and Hubbard, J.E. "Distributed Piezoelectric Polymer Active Vibration Control of a Cantilever Beam", *J. Guidance, Control and Dynamics*, Vol. 8(5), pp. 605-611.

4. Miller, S.E., and Hubbard, J.E. "Observability of a Bernoulli-Euler Beam Using PVF_2 as a Distributed Sensor," C.S. Draper Laboratory, Cambridge, MA., 1987.

5. Penwalt Corp., "KYNAR PIEZO FILM Technical Manual", Valley Forge, PA., 1987.

6. Schmitz, E., "Experiments on the End-Point Position Control of A Very Flexible One-Link Manipulator", Ph.D Dissertation, Stanford University, Stanford CA., 1985.

7. Gevarter, W.B., "Basic Relations for Control of Flexible Vehicles", *AIAA Journal*, Vol. 8, April 1970, pp.666-672.

8. Martin, G.D., "On thc Control of Flexible Mechanical Systems", Ph.D. Dissertation, Stanford University, Stanford CA., 1978.

9. Wang, D. and Vidyasagar, M., "Passive Control of a Single Flexible Link", *Proc. 1990 IEEE International Conference on Robotics and Automation*, May 13-18, 1990, pp.1432-1437.

10. Balabanian, N. and Bickart, T.A., *Electrical Network Theory*, John Wiley & Sons, Inc., 1969.

11. Desoer, C.A. and Vidyasagar, M., *Feedback Systems: Input-Output Properties*, Academic Press, 1975.

OPTIMIZATION OF PIEZO-ACTUATOR/SUBSTRUCTURE COUPLING FOR ACTIVE NOISE AND VIBRATION CONTROL

Sung J. Kim,[1] **James D. Jones**[1]

ABSTRACT

The optimization of piezo-actuator/substructure coupling for active noise and vibration control is investigated. The composite structure of interest consists of an elastic substructure with surface-bonded piezo-actuator layers which cover the entire substructure. A generalized formulation based on Love's equations of motion for the composite structure is derived and used to estimate the optimal thickness of the piezo-actuator layers. Modified equations are developed for the degenerate case of small piezo-actuator patches and are shown to give identical results. Comparisons between piezo-ceramic (PZT) and piezo-film (PVDF) actuators are also presented.

INTRODUCTION

Low frequency structural noise and vibration is a persistent problem in a variety of lightweight flexible structures such as aircraft, rotorcraft, space structures and automobiles. One promising method of reduction is active noise and vibration control using piezo-actuators. Piezo-actuators have attracted significant attention in recent years because of their distributed character of actuation which allows them to be tailored to selectively reduce structural modes with little control spillover. Piezo-actuators also have several other inherent advantages over conventional actuators including being inexpensive, space efficient, lightweight and easily shaped and bonded-to (or embeded-in) a variety surfaces. Furthermore, in contrast to point actuators, piezo-actuators do not require a back reaction.

Early studies [1,2,3] demonstrated the use of piezo-actuators as active dampers in reducing the free-vibration decay time of one-dimensional beams. Subsequent investigations [4,5,6] focused on development of static analytical models for embeded and surface-bonded piezo-actuator patches for the purpose of providing a better fundamental understanding of the piezo-actuator/substructure coupling. The simplified static models are effective for providing fundamental insights and design

[1]Ray W. Herrick Laboratories, School of Mechanical Engineering, Purdue University, West Lafayette, IN 47909, U.S.A.

guidelines; however, these models neglect the passive (local mass loading and stiffening) influence of the piezo-actuators on the dynamic behavior of the composite structure.

More recent investigations [7,8] have adopted more rigorous generalized mathematical formulations which treat the structure as a composite shell composed of alternate active (piezo-actuator) and passive layers (elastic substructure or bonding layer). These generalized formulations can be used to account for the passive (mass loading and stiffening) influence of the surface-bonded piezo-actuators on the dynamic behavior of the composite structure.

In the current work, a generalized approach to the optimal design of a piezo-actuator for active vibration control of a thin elastic shell is presented. Love's equations of motion of the composite structure are developed with the use of linear piezoelectric constitutive equations [8]. An index function which represents the effectiveness of the piezo-actuator/substructure coupling is developed to determine the optimal thickness of a piezo-actuator layer which fully covers the substructure. Furthermore, for the degenerate case of the substructure with smaller piezo-actuator patches, a modified equation of motion is developed under the assumption of negligible piezo-actuator mass loading and stiffening effects. The optimal thickness of a piezo-actuator under a constant electric field which maximizes the effective bending moment is estimated for both piezo-ceramic (PZT) and piezo-film (PVDF) actuators. These results lend insight into the optimal design of piezo-actuators for active noise and vibration control.

EQUATIONS OF MOTION FOR THE COMPOSITE STRUCTURE

The composite structure of interest in this study is in general assumed to consist of several active (piezo-actuator) and non-active (bonding, substructure) layers (see Figure 1), which are sufficiently thin to consider the entire composite structure as thin. The piezo-ceramic (PZT) actuators are assumed to be ferroelectric ceramics (polarized in the α_3-direction) and have the symmetry of a hexagonal crystal in class C_{6v}=6mm [9]. In contrast, the piezo-film actuators are assumed to be ferroelectric polyvinylidene fluoride polymers (PVDF) (also polarized in the normal α_3-direction) with the crystalline form of antipolar Form II [10]. All layer materials are assumed to be elastically isotropic.

In order to identify the electro-mechanical influence of the piezo-actuator, Tzou and Gadre [7] assumed that resultant in-plane displacements of a layer can be divided into two components: an elastic displacement due to in-plane external forces and an electro-mechanical displacement induced by the piezoelectric forces. However, this approach results in discontinuities of the displacement at the active/non-active layer interfaces because electro-mechanical displacements exist in the active (piezoelectric) layers but not in the adjacent non-active (bonding) layers. In contrast, Lee [8] incorporated the piezoelectric term into a simplified linear stress-strain constitutive equation. Because this approach guarantees strain continuity across the thickness of the composite structure, Lee's method of incorporating the piezoelectric effect is used in this investigation.

For i-th layer, the stress-strain relationships can be expressed as

$$\sigma_{1i} = \frac{E_i}{1-\nu_i^2}[\varepsilon_{1i}+\nu_i\varepsilon_{2i}-(1+\nu_i)\Lambda_i] \tag{1}$$

$$\sigma_{2i} = \frac{E_i}{1-\nu_i^2}[\varepsilon_{2i}+\nu_i\varepsilon_{1i}-(1+\nu_i)\Lambda_i] \tag{2}$$

where $\mathbf{E_i}$ and ν_i are the Young's modulus and Poisson's ratio of the i-th layer and subscripts **1** and **2** refer to the two in-plane directions (α_1 and α_2). The parameter Λ_i is the free piezoelectric strain which is defined as

$$\Lambda_i = \frac{V_i}{h_i}d_{31i} \tag{3}$$

where $\mathbf{V_i}$ is the applied voltage, $\mathbf{h_i}$ is the thickness of i-th layer and $\mathbf{d_{31i}}$ is the piezoelectric strain constant. Note that if the i-th layer is non-active, the free piezoelectric strain (Λ_i) is zero.

Love's simplifications (which assume a thin shell with negligible rotary inertia and transverse shear strain) and Hamilton's principle can be applied to derive the equations of motion for each layer. Since a detailed derivation is shown in references [7,11], only the final equations are presented in this study.

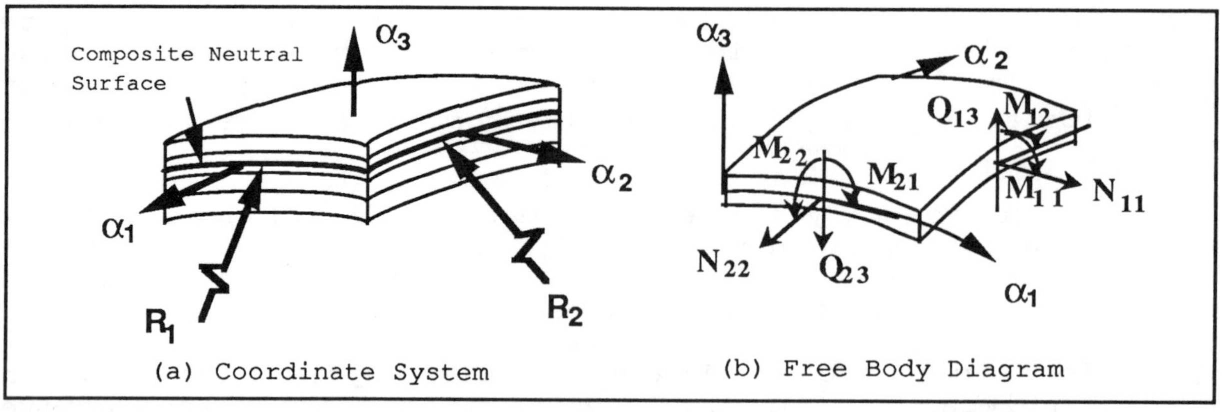

Figure 1. Coordinate System and Free Body Diagram of Composite Structure.

For the entire composite structure, the Love's equation of motion in the normal direction (α_3) is given by

$$-\frac{\partial}{\partial\alpha_1}(Q_{13}A_2) - \frac{\partial}{\partial\alpha_2}(Q_{23}A_1) + A_1A_2\left[\frac{N_{11}}{R_1} + \frac{N_{22}}{R_2}\right] + A_1A_2\rho h\ddot{u}_3 = A_1A_2q_3 \tag{3}$$

where $\mathbf{u_3}$ is the normal displacement, $\mathbf{q_3}$ is the applied normal force, ρ is the shell density, $\mathbf{Q_{13}}$ and $\mathbf{Q_{23}}$ are the composite transverse shear forces and $\mathbf{N_{11}}$ and $\mathbf{N_{22}}$ are the composite normal forces and $\mathbf{R_1}$ and $\mathbf{R_2}$ are the radii of curvature for the two in-plane directions (see Figure 1 (a) and (b)). The variables $\mathbf{A_1}$ and $\mathbf{A_2}$ are the Lame parameters, which depend on the geometry of the shell. The remaining companion equations for the entire composite structure are given below:

$$\frac{\partial}{\partial\alpha_1}(M_{11}A_2) + \frac{\partial}{\partial\alpha_2}(M_{12}A_1) + M_{12}\frac{\partial A_1}{\partial\alpha_2} - M_{22}\frac{\partial A_2}{\partial\alpha_1} - Q_{13}\,A_1A_2 = 0 \tag{4}$$

$$\frac{\partial}{\partial\alpha_1}(M_{12}A_2) + \frac{\partial}{\partial\alpha_2}(M_{22}A_1) + M_{12}\frac{\partial A_2}{\partial\alpha_1} - M_{11}\frac{\partial A_1}{\partial\alpha_2} - Q_{23}\,A_1A_2 = 0 \tag{5}$$

where, $$M_{mn} = \sum_{i=1}^{n} M_{mni}\,, \tag{6}$$

$$N_{mn} = \sum_{i=1}^{n} N_{mni}\,, \tag{7}$$

$$q_3 = \sum_{i=1}^{n} q_{3i}\,, \tag{8}$$

$$\rho h = \sum_{i=1}^{n} \rho_i h_i\,, \tag{9}$$

$$Q_{k3} = \sum_{i=1}^{n} Q_{k3i} \tag{10}$$

where $\mathbf{M_{11}}$, $\mathbf{M_{12}}$ and $\mathbf{M_{22}}$ are the resultant moment reactions of the composite structure and the subscript i represents the i-th layer. The composite shear forces are related to the composite moments by Equations (4)-(10). Using Equations (3)-(10), the normal equation of motion for the entire composite structure can then be expressed explicitly as a differential equation in the unknown normal displacement $\mathbf{u_3}$.

RESULTS AND DISCUSSION

In this section, the generalized composite shell formulation presented in Equations (3)-(10) is used to investigate the influence of select piezo-actuator design parameters on the effective moment applied to the elastic substructure. The ultimate objective of this investigation is to provide a better fundamental understanding of the piezo-actuator/substructure coupling in order to optimize the design of piezo-actuators for active noise and vibration control purposes. In the current study, two design areas are addressed: the optimal thickness of piezo-actuators and the use of piezo-ceramic (PZT) versus piezo-film (PVDF) actuators.

OPTIMAL THICKNESS OF PIEZO-ACTUATORS

In designing piezo-actuators, one parameter of critical importance is the actuator thickness. Early studies suggested that under a constant applied voltage the thinner the piezo-actuator, the better. This common misconception is derived from equation (3) which indicates that for a constant applied voltage ($\mathbf{V}$) the smaller the piezo-actuator thickness ($\mathbf{t_z}$), the larger the free piezoelectric strain (Λ). However, in practice, piezoelectric materials have a maximum allowable electric field strength ($\mathbf{V/t_z}$) which if exceeded will cause the material to lose its piezoelectric properties over time [12]. For most commercially-available piezo-ceramic materials, the maximum allowable electric field strength for continuous usage is less than 200(V/mm). Thus, in optimizing the piezo-actuator thickness, the electric field strength ($\mathbf{V/t_z}$) should be held constant rather than the applied voltage ($\mathbf{V}$) to correctly represent the true nature of piezoelectric materials.

In a recent investigation [6], the authors demonstrated that under a constant electric field strength there exists an optimal piezo-actuator thickness which maximizes the effective bending moment applied to the substructure. The composite structure studied was an elastic plate with twin piezo-actuator patches symmetrically surface-bonded on the upper and lower surfaces of the plate. However, the model was based on a static analysis of the piezo-actuator which does not account for the passive (mass loading and stiffening) effects of the actuators on the plate. Recently, Lee [8] presented theoretical and experimental work concerning modal sensors and actuators which use piezo-actuator (PVDF) layers surface-bonded over the entire substructure [13]. In the investigation, the dynamic behavior of the composite structure (including the piezo-actuator layers) was considered. Thus, it is desirable to analyze the optimal thickness of the piezo-actuator layer using the more generalized approach of Lee [8].

In this section, the basic analytical expression for determining the optimal thickness of a piezo-actuator layer using Love's equations of motion for a composite shell is presented. It is assumed in the formulation that the piezo-actuator layer covers the entire substructure. The degenerate case of small piezo-actuator patches is also presented and compared with the previous static analysis [6]. In this case, the mass loading and stiffening effects of the piezo-actuators is neglected. These two cases are chosen for presentation because they represent limiting extremes of the piezo-actuator influence and because they can be handled in a closed form fashion.

Plate With Piezo-Actuator Layers

Consider an elastic plate with surface-bonded piezo-actuator layers, as shown in Figure 2. Assume that there are no applied external normal forces (q_3=0). Also, assume that there are no in-plane forces and consequently the in-plane (membrane) strains are zero ($\varepsilon^o_{11i} = \varepsilon^o_{22i} = 0$). The normal equation of motion for the composite structure then can be expressed as

$$D_c \nabla^4 u_3 + \nabla^2 m_{pz} + m_c \ddot{u}_3 = 0 \tag{11}$$

where D_c, m_{pz} and m_c are the flexural rigidity of the composite structure, the effective bending moment induced by the piezo-actuator, and the surface mass of the composite structure, respectively. These parameters are defined as follows:

$$D_c = \frac{E_p}{3(1-\nu_p^2)}(\delta_{t1}^3 - \delta_{b1}^3) + \frac{2E_b}{3(1-\nu_b^2)}(\delta_{t2}^3 - \delta_{b2}^3) + \frac{2E_p}{3(1-\nu_p^2)}(\delta_{t3}^3 - \delta_{b3}^3) \tag{12}$$

$$m_{pz} = \frac{E_z}{1+\nu_z}(\delta_{t3}^2 - \delta_{b3}^2)\Lambda \tag{13}$$

$$m_c = \rho_p h_p + 2\rho_b h_b + 2\rho_z h_z. \tag{14}$$

Here, E_p, E_b and E_z are the Young's modulus of the plate, the bonding layer and piezo-actuator layer, respectively. The subscripts '**1**,**2** and **3**' refer to the elastic substructure, the bonding layer and the piezo-actuator layer, respectively, as shown in Figure 1. Similarly, the δ parameters represent the distances from the neutral surface of the composite structure to the top and bottom of the various composite layers as listed in Figure 1.

Knowing the resultant displacements of the composite structure are proportional to the ratio of the effective bending moment to

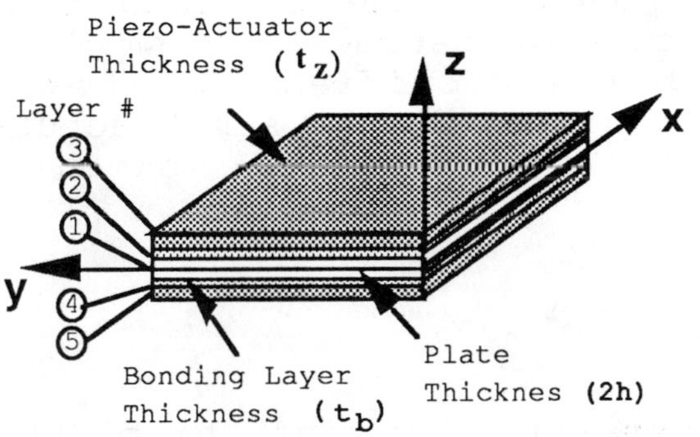

$\delta_{t1} = h$
$\delta_{b1} = -h$
$\delta_{t2} = -\delta_{b4} = h + t_b$
$\delta_{b2} = -\delta_{t4} = \delta_{t1}$
$\delta_{t3} = -\delta_{b5} = h + t_b + t_z$
$\delta_{b3} = -\delta_{t5} = \delta_{t2}$

Figure 2. Plate with Surface-Bonded Piezo-Actuator Layers.

the flexural rigidity of the composite structure ($\mathbf{m_{pz}/D_c}$), the index for the determination of the optimal thickness is given by:

$$m_{opt} = \frac{m_{pz}}{D_c}. \tag{15}$$

Note that if the material properties (including the piezoelectric properties of the active layer) are uniform, then the second term of Equation (11) becomes zero. Therefore, there is no apparent exciting forces applied to the composite structure by the piezo-actuator layer. However, in this case the effect of the piezo-actuator layer appears in the boundary conditions in the form of line-moments. The boundary conditions are not specified in this study but are shown in reference [8].

From Equations (12) and (13), it is easy to see that both $\mathbf{m_{pz}}$ and $\mathbf{D_c}$ increase without bound as the piezo-actuator thickness increases. However, the flexural rigidity ($\mathbf{D_c}$) increases more rapidly than the effective moment ($\mathbf{m_{pz}}$). Thus, the ratio ($\mathbf{m_{pz}/D_c}$) approaches zero in the limit as the thickness of the piezo-actuator ($\mathbf{t_z}$) goes to infinity. As $\mathbf{t_z}$ approaches zero, the flexural rigidity approaches a finite value while the effective moment approaches zero. Thus, the ratio ($\mathbf{m_{pz}/D_c}$) again approaches zero in the limit. These two limiting cases suggest that an optimal piezo-actuator thickness exists which maximizes the piezo-actuator/substructure coupling.

The behavior of this index function along with the piezo-actuator thickness will be compared with the effective bending moment induced by small piezo-actuator patches in the following section to discuss the optimal thickness for both cases.

Plate with Small Piezo-Actuator Patches

In contrast to PVDF, piezo-ceramic (PZT) actuators are difficult to manufacture in sheets large enough to fully cover a substructure. Therefore, it is worthwhile to investigate the optimal thickness of piezo-actuator patches which are small compared with the in-plane dimensions of the substructure. It is assumed that the piezo-actuator patches are small enough so that the passive (mass loading and stiffening) effects have a negligible influence on the dynamics of the substructure. Given this assumption, the governing equation of motion (Equation (11)) can be transformed into

$$D_p \nabla^4 u_3 + \nabla^2 \frac{m_{pz}}{D_c} D_p + \rho_p h_p \ddot{u}_3 = 0 \tag{16}$$

where $\mathbf{D_p}$ represents the flexural rigidity of the plate as given by the first term of Equation (12).

The argument of the second term in the above equation represents the effective bending moment applied to the plate by the piezo-actuator patches:

$$m_x = m_y = m_{pz}\frac{D_p}{D_c}. \tag{17}$$

Therefore, it is assumed that small piezo-actuator patches apply a bending moment to the substructure within the actuator boundaries without changing the dynamic characteristics of the plate. However, the magnitude of the effective bending moment depends on the material properties, the actuator thicknesses and the applied electric field strength as shown in Equation (17).

It is worthwhile to compare Equations (15) and (17) which represent the degree of piezo-actuator/substructure coupling for the piezo-actuator layer and small piezo-actuator patches, respectively. The only difference between the two expressions is the presence of the plate flexural rigidity (**D_p**) in Equation (17). However, for a given substructure, the thickness and material properties are fixed resulting in a constant flexural rigidity (i.e., **D_p**=constant). Therefore, Equations (15) and (17) will show the same functional behavior for the thickness and material property variations of the piezo-actuator. In other words, the optimal thicknesses which maximize the piezo-actuator/substructure coupling is identical for both the piezo-actuator layers and the small piezo-actuator patches. Since Equations (15) and (17) are functionally similar, the design issues of this study will be illustrated in terms of the effective bending moment of Equation (17).

<u>Comparison of Current Analysis with Previous Study [6]</u>

When the plate has small double-sided piezo-actuator patches, the entire structure is identical to that of reference [6]. Thus, it is worthwhile to compare the current results with those of reference [6]. For this structure, the effective bending moment of Equation (17) can be simplified and cast into a non-dimensional form with the aid of Figure 1 as follows:

$$m_x = m_y = \frac{\rho_z(2+2\rho_b+\rho_z)}{1+\alpha\rho_b(3+3\rho_b+\rho_b^2)+\beta\rho_z(3+3\rho_b^2+\rho_z^2+6\rho_b+3\rho_z+3\rho_b\rho_z)}h^2\gamma\Lambda \tag{18}$$

$$\text{where,}\quad \rho_b = \frac{t_b}{h} \quad \text{and} \quad \rho_z = \frac{t_z}{h}, \tag{19}$$

$$\text{and}\ \alpha = \frac{1-\nu_p^2}{1-\nu_b^2}\frac{E_b}{E_p},\ \beta = \frac{1-\nu_p^2}{1-\nu_z^2}\frac{E_z}{E_p},\ \gamma = \frac{E_z}{1-\nu_z}. \tag{20}$$

In Equations (18)-(20), the geometrical and material properties of the bonding layer and the piezo-actuator are non-dimensionalized by those of the plate. Thus, the parameters ρ_b

and ρ_z represent the non-dimensional thickness ratios of the bonding layer and piezo-actuator, respectively. Similarly, the variables α and β represent non-dimensional bonding layer and piezo-actuator material property ratios as shown in Equations (19) and (20), while γ is the piezo-actuator stress-strain ratio.

In contrast, Kim and Jones [6] developed the effective moment from a static analysis of the composite structure as follows:

$$m_x = m_y = \frac{\rho_z(2+2\rho_b+\rho_z)}{1+\alpha^*\rho_b(3+3\rho_b+\rho_b^2)+\beta^*\rho_z(3+3\rho_b^2+\rho_z^2+6\rho_b+3\rho_z+3\rho_b\rho_z)} h^2\gamma\Lambda \qquad (21)$$

$$\text{where,} \quad \alpha^* = \frac{1-\nu_p}{1-\nu_b}\frac{E_b}{E_p}, \quad \beta^* = \frac{1-\nu_p}{1-\nu_z}\frac{E_z}{E_p}, \quad \gamma = \frac{E_z}{1-\nu_z}. \qquad (22)$$

where α^* and β^* represent slightly modified forms of the bonding layer and piezo-actuator material property ratios. Since the basic assumptions of both studies are identical (i.e., thin composite structure with linear strain variation across the composite thickness), both results might be expected to be equivalent. However, when the force and moment equilibrium conditions are applied in the static analysis case [6], the two-dimensional effects are omitted because the two orthorgonal directions are treated independently instead of simultaneously. Thus, there is a difference of squaring of Poisson's ratio between the material property ratios α^* and β^* in Equation (22) and α and β in Equation (20).

A comparison of the effective moment induced by a double-sided piezo-ceramic actuator (assuming a bonding layer thickness of zero) from the current and previous study [6], is shown in Figure 3. The applied electric field strength is assumed to be 100 [V/mm], which approaches the maximum allowable field strength for most piezo-ceramic materials in continuous use. In this figure, the effective moment is given as a line moment with units of [N-m/m] and is a function of the non-dimensional piezo-actuator thickness (ρ_z). Since the only difference between the two expressions is the squared Poisson's ratio in Equation (20) and since the Poisson's ratios for each layer are almost identical, the two expressions give approximately equivalent results and thus the differences are undetectable in Figure 3.

Optimal Thickness of the Piezo-Actuator

In Figure 3, it is shown that the effective bending moment reaches the maximum value when the piezo-actuator thickness approaches the half plate thickness, ρ_z~1 (for the steel plate) and the quarter plate thickness, ρ_z~0.5 (for the aluminum plate). This illustrates that the optimal piezo-actuator thickness depends on the substructure material properties, especially Young's modulus ($\mathbf{E_p}$) [6]. Similarly, the optimal

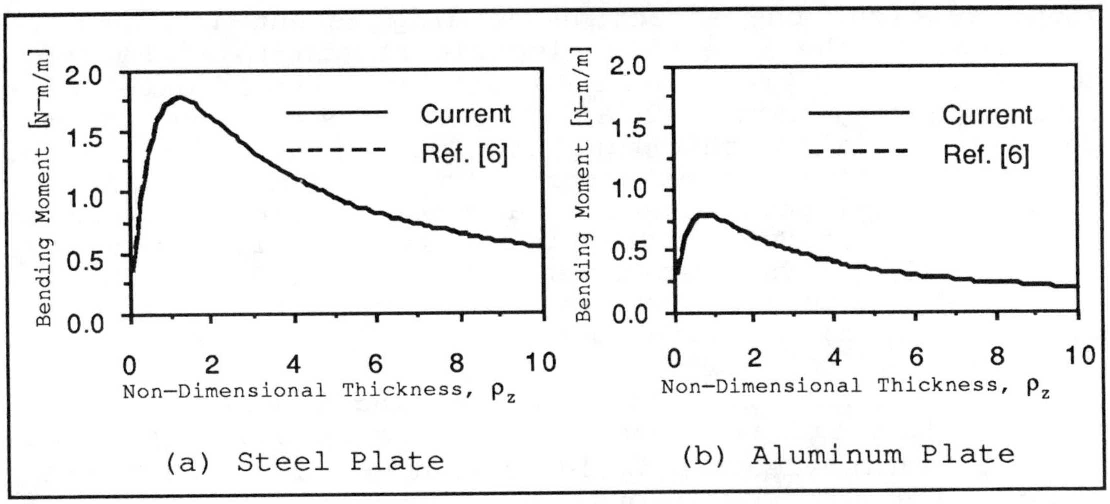

Figure 3. Comparison of the Effective Bending Moment Induced by a Pair of Surface-Bonded Piezo-Actuators from the Current and Previous Study [6].

piezo-actuator thickness is also strongly dependent on Young's modulus of the piezo-actuator ($\mathbf{E_z}$), giving smaller optimal thicknesses with stiffer piezo-actuators. In previous work [6], it was also shown that the effective bending moment induced by the piezo-actuators increases with a diminishing bonding layer thickness without significant effect on the optimal piezo-actuator thickness. Therefore, the influence of the bonding layer thickness is neglected in this study by setting its thickness to zero (ρ_b=0). Since other minor effects on the optimal thickness are well documented in authors' previous study [6], the other important design parameter, the use of piezo-ceramic (PZT) versus PVDF is presented in the following section.

USE OF PIEZO-CERAMICS (PZT) VERSUS PIEZO-FILM (PVDF) ACTUATORS

In recent years, PVDF has gathered much attention as a new piezoelectric transducer for many sensor and actuator applications since it is inexpensive, compliant, available in large sheets and easy shaped into complex configurations. Naturally, its application to active noise and vibration control is an increasingly important topic. While it is generally believed that the PZT actuators are superior to PVDF actuators, there has been little attempt to compare quantitatively the performance differences between the two actuator types. Since the two materials have quite different material properties, a number of factors should be considered for direct comparison.

As for the degree of piezoelectricity, PVDF has almost order of magnitude lower piezoelectric strain constant ($\mathbf{d_{31}}$) than PZT. However, the maximum allowable electric field strength of PVDF is almost hundred times greater than that of PZT. Therefore, the maximum free piezoelectric strain (Λ) for PVDF actuators is expected to be ten times greater than PZT actuators. In the previous section, the effective bending moment is shown to be proportional to the free piezoelectric strain (Λ). However, it is not necessarily true that PVDF actuators would give a higher effective bending moment than PZT actuators because the moment expression is also the function of the elastic material properties of the piezo-actuator. Crawley and de Luis [4] and Kim and Jones [6] demonstrated that the stiffer piezo-actuators give larger effective bending moments. Usually, the Young's modulus of PVDF is hundred times smaller than that of the PZT. Therefore, PZT actuators compensate the small free piezoelectric strain (Λ) with high stiffness.

Figures 4 and 5 show the effective bending moment as a function of the non-dimensional piezo-actuator thickness (ρ_z) using the maximum allowable electric field strengths for both PZT and PVDF actuators (assuming continuous use). The assumed values of the PZT and PVDF material properties include piezoelectric strain constants ($\mathbf{d_{31}}$) of $200*10^{-12}$ and $20*10^{-12}$ (m/V), applied electric field strengths ($\mathbf{V/t_z}$) of 100 and 10000 (V/mm) and Young's moduli ($\mathbf{E_p}$) of $6.5*10^{11}$ and $0.02*10^{11}$ (N/m^2). The Poisson's ratios are assumed to be 0.3 for both actuator types.

Surprisingly, the maximum bending moment for the PVDF actuators is approximately four times that estimated for the PZT actuators. However, the optimal thickness of the PVDF actuators necessary to maximize the applied bending moment is approximately five times that of the PZT actuators. For a typical substructure thickness of 1 mm, this would require the PVDF actuator to be approximately 4-5 mm thick. Since most commercially-available PVDF materials are very thin (generally less than 0.2 mm), current PVDF actuators are ineffective.

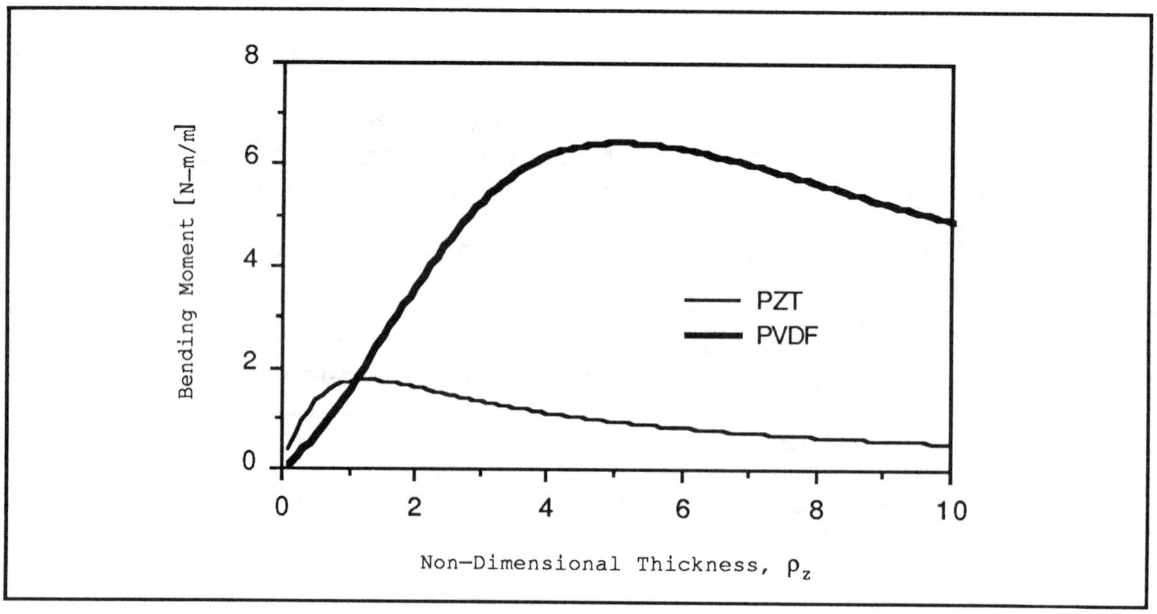

Figure 4. Comparison of PZT Actuator with PVDF Actuator with a Steel Substructure.

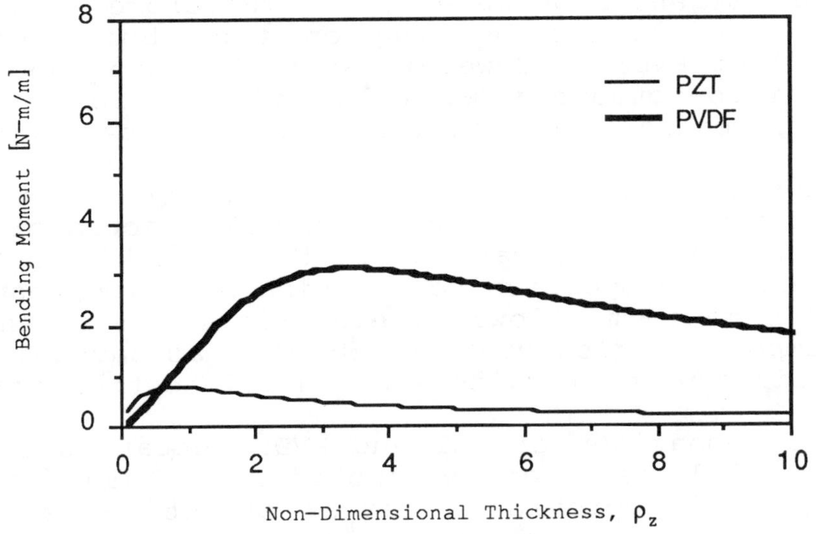

Figure 5. Comparison of PZT Actuator with PVDF Actuator with an Aluminum Substructure.

In contrast, since ultra thin PZT actuators are difficult to manufacture because of excessive brittleness, PZT actuators within typical optimal thicknesses are readily-available. These factors seem to have contributed to the success of PZT actuators over PVDF.

Interestingly, even near the PZT optimal thickness, the effective bending moment induced by a PVDF actuator appears to be approximately equal to that of an equivalent PZT actuator. However, most PVDF has substantially higher mechanical damping properties than PZT. In a previous study [6], the authors demonstrated that the effective bending moment decreases with increasing damping. Consequently, the effective bending moment of PVDF shown in Figures 4 and 5 likely overestimates the true applied moment. Thus, the PZT actuators will generally out perform the equivalent PVDF actuators, especially for thin actuators. Also, at the same thickness, PVDF requires hundred times higher voltage which approaches 10000 (V) for a 1 (mm) thick PVDF and thus presents additional safety issues.

Piezoelectric transducers have two primary loss mechanisms: the dielectric loss factor and the mechanical loss factor. Dielectric loss is specified by the loss tangent value (tan δ) in the manufacturer's data sheet whereas mechanical loss is specified by the quality factor (Q) at one of the resonant frequencies. Of these losses, it is generally believed that dielectric losses are more significant in practical applications. While, it is difficult to establish which type of piezo-actuator (PVDF or PZT) has the lower dielectric losses (primarily because these losses vary depending on the manufacturer), the dielectric losses appear to be approximately the same for PZT and PVDF materials. Therefore, the dielectric loss factor has little bearing on the choice of actuator material (PZT or PVDF). However, special care should be taken to choose piezo-actuators with low dielectric constants for improved performance, especially for high power applications.

Another important factor to be considered is the thermal stability of the piezo-actuator. Since PVDF actuators have high mechanical damping, heat can be built up quickly and may cause the piezoelectric properties to diminish overtime, especially if used near the maximum allowable electric field strength. This may be an important disadvantage for PVDF actuators, especially for applications which require substantial control energy.

An objective comparison of PZT and PVDF actuators reveals that neither material is clearly superior in all cases. PZT actuators generally provide a larger effective bending moment for thin actuators but cannot be easily manufactured in large sheets or formed into complex geometries. Thus, PZT is more suitable for small piezo-actuator patches. In contrast, PVDF actuators are readily available in large compliant sheets (suitable for molding to irregular structures) but are not commercially-available in thicknesses generally needed to maximize the effective moment. Furthermore, the low electro-mechanical coupling and higher mechanical damping factors further reduce the effectiveness of the PVDF actuators. Nevertheless, PVDF is suitable for larger piezo-actuator layers, especially where irregular geometries are encountered. Thus, the final choice of actuator materials is highly dependent on the specific objectives of the application.

CONCLUSIONS

Love's equations of motion for the composite shell with piezo-actuator layers are derived. A modified equation of motion for the degenerate case of small piezo-actuator patches is given. An index function defined as the ratio of the bending moment to the flexural rigidity of the composite structure is used to determine the optimal thickness of the piezo-actuator layers and small piezo-actuator patches. It is shown that both piezo-actuator layers and small piezo-actuator patches have identical optimal thicknesses.

Direct comparisons of PZT and PVDF actuators indicate that neither material exhibits clear superiority. PZT actuators are more suitable as small patches whereas PVDF actuators serve better in large layers and especially for irregular shapes. The final choice of actuator material is highly dependent on the specific objectives of the application.

ACKNOWLEDGEMENTS

The authors gratefully acknowledge the support of this work by the National Science Foundation under NSF grant MSS-8957191.

The authors also appreciate GoldStar Co., Ltd., in Korea, whose funding supports the doctoral studies of Mr. Sung Jin Kim.

REFERENCES

1. Bailey, T. and Hubbard, J. E., 1985, "Distributed Piezoelectric-Polymer Active Vibration Control of a Cantilever Beam", AIAA Journal of Guidance and Control, 6(5):605-611.

2. Fansen, J. L, and Chen, J. C., 1986, "Structural Control by the Use of Piezoelectric Active Members", Proceedings of NASA/DOD Control-Structures Interaction Conference, NASA CP-2447, Part II.

3. Baz, A. and Poh, S., 1987, "Optimum Vibration Control of Flexible Beams by Piezoelectric Actuators", NASA CR-180209.

4. Crawley, E. F., and de Luis, J., "Use of Piezoelectric Actuators as Elements of Intelligent Structures", AIAA Journal, 25(10):1373-1385.

5. Dimitriadis, E. K., and Fuller, C. R., 1989, "Piezoelectric Actuators for Noise and Vibration Control of Thin Plates", 12th ASME Conference on Mechanical Vibration and Noise, Montreal.

6. Kim, S. J. and Jones, J. D., 1990, "Optimal Design of Piezo-Actuators for Active and Vibration Control", AIAA-90-3925, AIAA 13th Aeroacoustics Conference, Tallahassee.

7. Tzou, H. S. and Gadre, M., 1989, "Theorctical Analysis of a Multi-Layered Thin Shell Coupled with Piezoelectric Shell Actuators for Distributed Vibration Controls", J. Sound and Vibration, 132(3):433-450.

8. Lee, C. K., 1990, "Theory of Laminated Piezoelectric Plates for the Design of Distributed Sensors/Actuators. Part I: Governing Equations and Reciprocal Relationships", J. Acoust. Soc. Am., 87(3):1144-1158.

9. Jaffe, B., Cook, W. R. and Jaffe, H., 1971, Piezoelectric Ceramics, Academic Press, pp. 1-20.

10. KYNAR PIEZO FILM Technical Manual, 1987, Pennwalt Corp., Valley Forge, PA 19482, pp. 5-7.

11. Soedel, W.,1981, Vibrations of Shells and Plates, Marcel Dekker, Inc., pp. 1-66.

12. Randeraat, J and Setterington, R. E., 1974, Piezoelectric Ceramics, Mullard Limited, pp. 15-16.

13. Lee, C. K. and Moon, F. C.,1990, "Modal Sensors/Actuators", Trans. ASME, J. Appl. Mechanics., v. 57:434-441

Fluid Phenomena

ADAPTIVE ACTIVE CONTROL OF INSTABILITIES

Guy Billoud,[1] Marie Annick Galland,[1] Can Huynh Huu,[2] Sébastien Candel[2]

ABSTRACT

This paper reports the results of an experimental investigation in which adaptive methods are applied to various situations of active control of flow instabilities : combustion instabilities, flow-excited cavity oscillations, compressor surge. The present work draws especially attention to the adaptive controller behaviour and capacities.The main results are as follows: i) The adaptive controller manages to automatically find its optimal characteristics in order to cancel the instability in the three experiments under consideration ii) It is able to cope with time-dependent changes in the operating conditions such as flow velocity sweeps for the oscillating cavity and air and propane mass flow rate variations in the combustion instability experiment iii) In the case of the Helmholtz cavity resonator, the algorithm has also exhibited capacity to recover effective control even when initially set such that artificial closed-loop instabilities induced by the control system occurred.

INTRODUCTION

Attention has since many years been paid to the possibility of using active control techniques for stabilising self-sustained oscillations. This phenomenon of self-sustained instabilities, also referred to as natural instabilities, or simply instabilities, results from an unstable coupling between an excitation and the corresponding response of the system. Several previous studies have shown that active control techniques introducing an additional coupling with very simple electronic circuitry (referred to from now on as "controllers") lead to a stabilization of the system. This feature has been demonstrated in many different situations, such as combustion instabilities [1 - 3], flow-excited cavity oscillations [4 - 6], and compressor surge [7], [8].

In the experiments cited above, however, artificial instabilities may arise due to closed loop feedback induced by the controller. Moreover, the simple controllers implemented may fail when operating under varying flow conditions. The purpose of this study is therefore to introduce adaptive principles to the control strategy to provide : i) automatic optimal control, ii) accomodation of variable flows, iii) elimination of artificial instabilities.

Three experimental situations of active control are considered in this paper : compressor surge (section 2), combustion instabilities (section 3) and self-sustained pressure oscillations in a flow-excited cavity (section 4). In addition to the analysis of the control

[1]Fluid Dynamics and Acoustics Laboratory, CNRS URA 263, Ecole Centrale de Lyon, BP 163, 69131 Ecully Cedex, France.
[2]EM2C Laboratory, CNRS, Ecole Centrale de Paris, 92295 Chatenay Malabry Cedex, France.

efficiency in both steady and varying operating conditions, special attention will be paid to the algorithm dynamics during its operation, and to the principle of operation of the control system.

1.GENERAL DESCRIPTION OF THE ADAPTIVE METHOD

The algorithm used to adapt the controller parameters is based on the idea of energy minimization, classically referred to as LMS. In the present study, the controller used is a recursive digital filter (IIR filter). Hence the algorithm task is to minimise the energy of a global-response sensor signal through an optimization of the coefficients of this IIR filter. Since energy is intimately related to the strength of the instability, this would appear to be a logical strategy for achieving control.

1.1. ALGORITHM DESCRIPTION

The input-output law of an IIR filter has the general form:

$$y_n = \sum_{i=0}^{N-1} a_i x_{n-i} - \sum_{j=1}^{N} b_j y_{n-j} \quad \text{or} \quad y_n = {}^{T}A\, X^{(n)} - {}^{T}B Y^{(n)}$$

$$\text{with } {}^{T}A = (a_0, a_1, \ldots, a_{N-1});\; {}^{T}B = (b_1, b_2, \ldots, b_n)$$

$${}^{T}X^{(n)} = (x_n, x_{n-1}, \ldots, x_{n-N+1});\; {}^{T}Y^{(n)} = (y_{n-1}, y_{n-2}, \ldots, y_{n-N})$$

where x_n and y_n are the nth samples of the filter input and output, respectively,

a_i and b_i are the filter coefficients.

The basic idea is to chose the coefficients a_i and b_i such that the controller minimizes flow instabilities. Note that for time-dependent flows, a_i and b_i are functions of time. Hence an iterative approach is necessary to update the coefficents. Widrow[9] introduced it for transverse (or FIR filters and Feintuch [10] extended it to IIR filters. Its iteration scheme is as follows:

$$A^{(n+1)} = A^{(n)} - k\, \nabla_A\, E(\varepsilon_n^2)$$

$$B^{(n+1)} = B^{(n)} - k\, \nabla_B\, E(\varepsilon_n^2) \qquad (1)$$

where $E(\varepsilon_n^2)$ is the energy, k is a positive constant designated as the convergence factor.

In the case of active control, the LMS-gradient algorithm is used to minimize the interference of two physical fluctuations. The first being the natural fluctuation and the second being the fluctuation indirectly generated by the active control system, so that the algorithm structure includes an additional unit which represents the cancelling fluctuations source, referred to as auxiliary path.The algorithm form must take this additional path into account. Considering the case of an IIR filter, and adopting appropriate estimates of the involved gradients, it can be shown that the algorithm takes the form :

$$\begin{cases} A^{(n+1)} = A^{(n)} + 2\,k\,\varepsilon_n\,U^{(n)} \\ B^{(n+1)} = B^{(n)} + 2\,k\,\varepsilon_n\,V^{(n)} \end{cases} ; \begin{cases} V^{(n)} = (v_n, v_{n-1}, \dots v_{n-N+1}) \\ U^{(n)} = (u_n, u_{n-1}, \dots u_{n-N+1}) \\ u_n = F^T X^{(n)}, v_n = F^T Y^{(n)} \end{cases} \quad (2)$$

where F is the weight vector of a filter expected to ensure convergence and stability (see the block diagram in Fig. 1 in which the box D represents the auxiliary path).

This last type of LMS-gradient algorithm for IIR filters has been successfully applied to active sound cancellation experiments (Erikssonn [11], Billoud [12]).

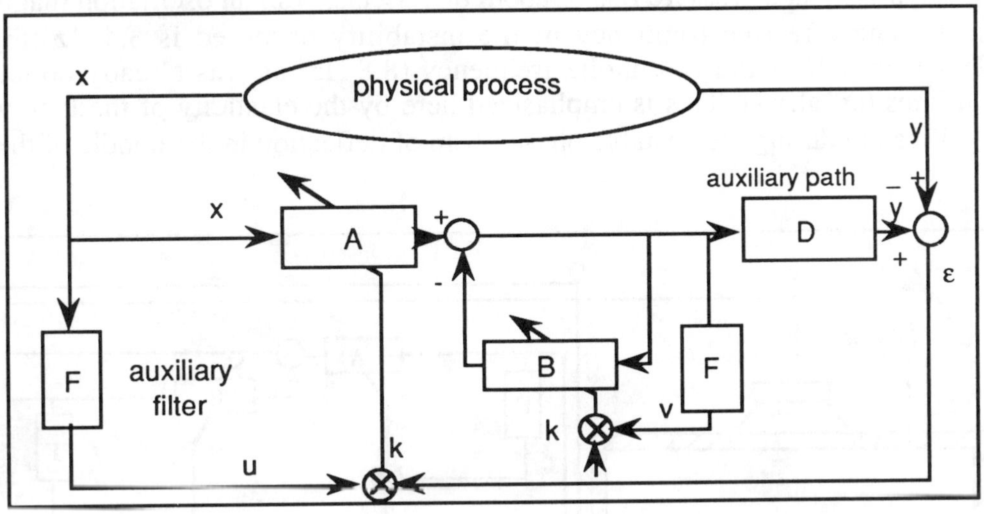

Figure 1: Adaptive algorithm block diagram

1.2. CONVERGENCE PROBLEM

The control algorithm convergence has been studied in detail by Morgan [13], when optimizing a FIR filter (weight vector A) in the presence of an auxiliary path modelled by a second FIR filter (weight vector D). Morgan has shown that the convergence is guaranteed as long as the eigenvalues of $[I - 2kE\,(X^{(n)}\,FD^TX^{(n)})]$ stay inside the unit circle. Consequently, it is sufficient but not necessary that F equals D as is sometimes implicitely admitted. In practice, however, insofar as D has to be determined , it is simpler to take the correcting filter F equal to D, as has been done for the experiments presented below.

2.COMPRESSOR SURGE CONTROL

The compressor surge phenomenon has been studied and described by Epstein (et al.)[14] and Emmons (et al.)[15]. When a compressor connected to a plenum operates in the region of low mass flow rates and high pressure rises, it can suddenly switch to an unstable operation. This instability is due to a coupling between the unsteady flow excitation (e.g. turbulence, blade wakes) and the Helmholtz resonance of the plenum.

Active control techniques are relevant to the avoidance of such unstable coupling. Their feasibility has been theoretically and experimentally shown by Epstein (et al.)[14] and Ffowcs Williams (et al.)[7]. Experimental studies were first conducted with a simple phase shifter as a controller. More recently, a real scale test was conducted on a gas turbine engine[16]. The controller design has been in that case more intricated and a fixed-response broadband filter had to be used.

2.1. EXPERIMENTAL FACILITY

In order to study the feasibility of adaptive control in compressor surge, a simple set-up has been built, comprised of an axial rotary blower mounted on a 0,7 m^3 tank made of plastic (Fig. 2). A throttle serves to control the location of the operating point. The active control actuator is a commercial electrodynamic loudspeaker, 40 cm in diameter. The instability of this system is typically at a pressure rise of about 88 Pascals, with an oscillation magnitude of 30 Pascals. The effective frequency of the instability observed is 3.4 Hz (Fig. 4), which is less than half of the Helmholtz frequency (8.8 Hz) as was already noticed by Ffowcs Williams (et al.)[7]. This is emphasized here by the elasticity of the tank which vibrates significantly during the unstable phase (2 cm of deflection in the middle of the tank walls).

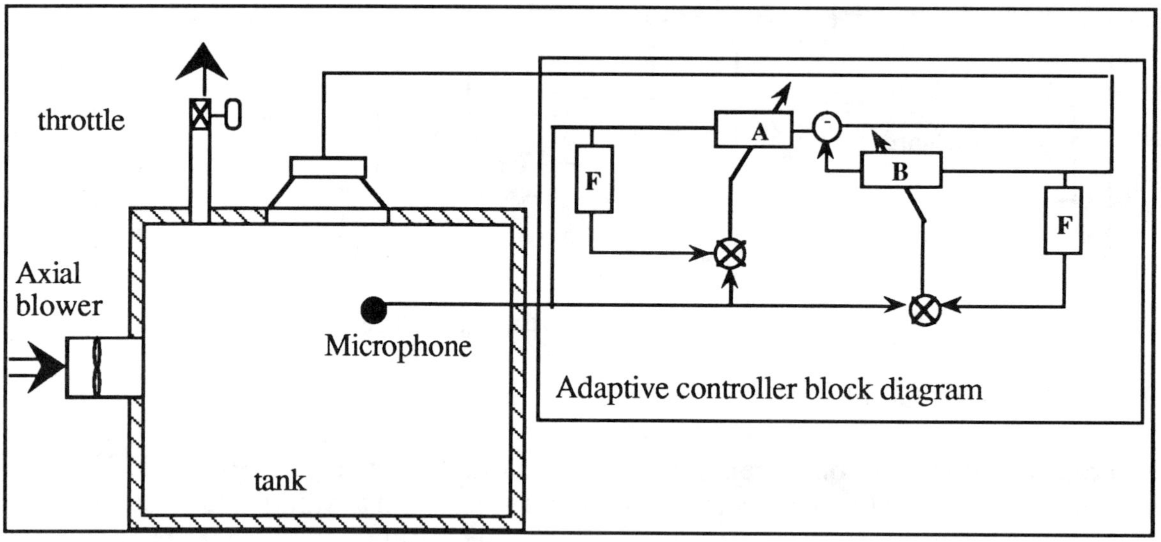

Figure 2: Surge control test facility

2.2. ADAPTIVE ACTIVE CONTROL PERFORMANCE

Experimental Procedure

Adaptivity is introduced by using a programmable digital filter as a controller in which the algorithm described in section 1 is programmed. As seen in section 11, a secondary filter F is necessary to ensure the algorithm convergence.

Accordingly, the first step of the procedure is to determine the modelling filter D of the auxiliary path, hereafter referred to as identification procedure. This is usually done by feeding the control actuator with band-limited random noise and implementing a signal-fitting algorithm with the blower off as shown in figure 3.

The following step is then the actual implementation of the whole adaptive active control strategy. The schematization of this implementation is depicted in figure 2. It differs from the classical implementation in ANC systems insofar as the same signal, here the pressure fluctuation signal, is used as both the input of the controller and the error signal to be minimized. The only information that can be obtained on the instability is indeed given by

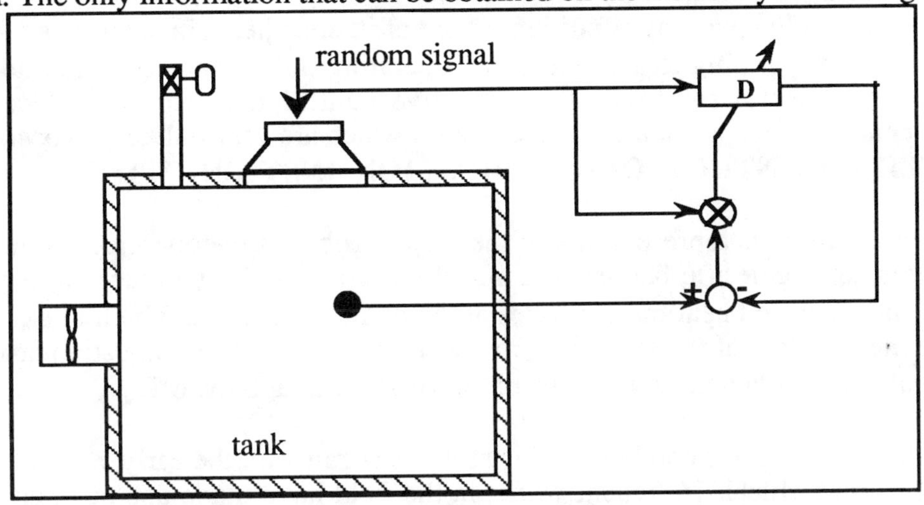

Figure 3: Experimental identification of D

the pressure fluctuation which is uniform in magnitude and phase allover the tank. Hence there is no way to detect two separate input-output signals as supposed in the algorithm fundamentals. This raises a question on the validity of using the algorithm. The posititive experimental results, however , partially yields an answer(see below). Further analysis will be given in the case of a flow-excited cavity (see section 4).

Experimental Results

The experiment has been conducted with an IIR filter with N = 15. After the onset of instability , the controller was switched on with all coefficients set to a null value. The algorithm then succeeded in finding a parameter set such that the compressor surge

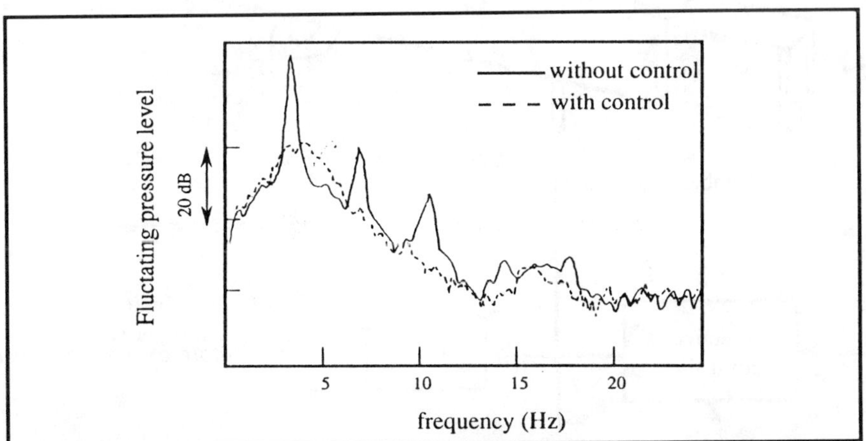

Figure 4: Inside-tank fluctuating pressure spectra

vanished. This can be seen by comparing the fluctuating pressure spectra with and without control (Fig. 4). The active control effect is also subjectively observed because the flow changes from a pulsating to a regular velocity .This stabilization proves the capacity of the adaptive method to automatically find a satisfactory configuration for the controller under steady conditions.

The study of unsteady operating conditions is not of interest here since there is a single type of resonance corresponding to a single given operating point. The controller adaptivity is thus much more useful for finding an effective control than for the self-adapting to significant changes in the instability characteristics which are very unlikely to occur.

3. ADAPTIVE CONTROL OF COMBUSTION INSTABILITIES

Combustion instabilities represent one of the major problems encountered while designing rocket motors, jet engine afterburners and industrial furnaces. They induce large mechanical vibrations and enhanced heat transfer rates at the combustor walls, which in extreme cases can lead to the total loss of the system. In many circumstances the combustion process is not able to sustain large oscillations which result in partial or total blow off.

Active methods to suppress combustion instabilities began with the early work of Tsien[17], Crocco [18]and Marble[19]. Recent works in the field have made use of fixed-parameter controllers which amplify and delay a signal taken from the combustor, and then feed it into an actuator mounted on the propellant feed system to achieve control [1-3].

3.1. EXPERIMENTAL CONFIGURATION

Experimental Facility

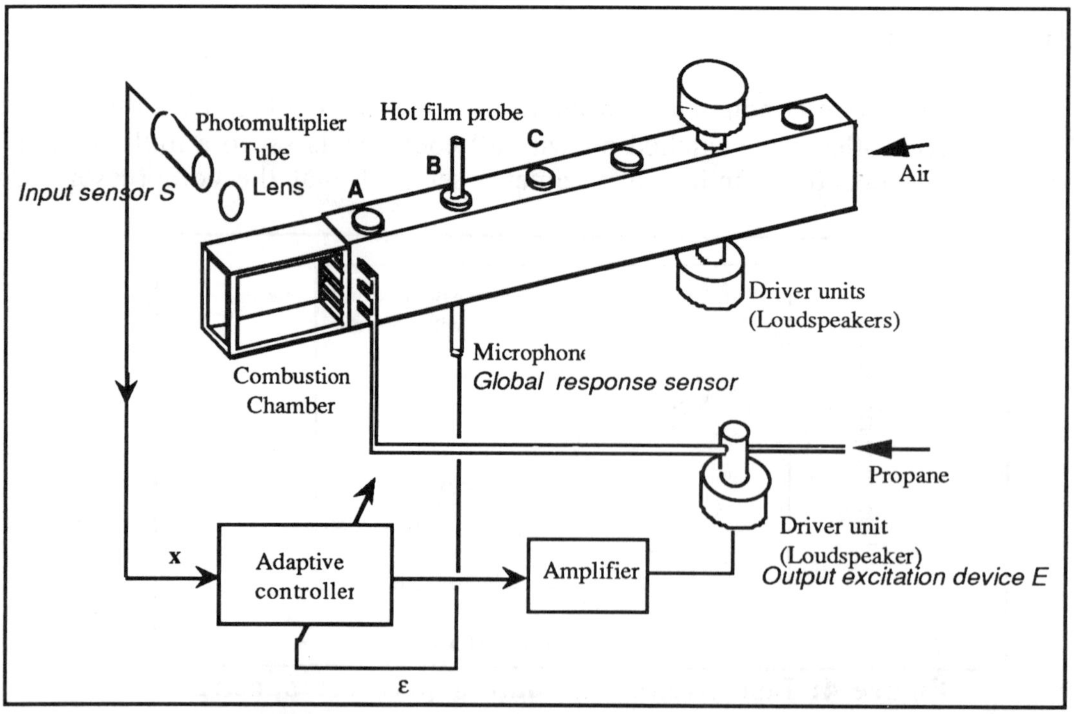

Figure 5: Combustion instabilities test facility

The turbulent combustor used in this study is a 250 kW non premixed device sketched in Figure 5. Air is supplied by a long duct on which pressure transducers and hot wire probes can be installed. Propane is fed into the combustion chamber through six narrow rectangular slots arranged to provide three pairs of flame. The side walls are equipped with quartz windows which enable optical access. A photomultiplier is used to detect the global emission from C_2 radicals which is related to heat release in the combustor.

Combustor Operating Regimes

A convenient way to characterise the combustion chamber is to analyse its stability chart (Fig 6). Unstable regimes characterized by frequencies of 180, 210, and 280 Hz are observed in zones 2, 3, 5. Partial extinction of one pair of flames occurs in zone 4 because of a powerful low frequency pulsation at 100 Hz. Unless one provides a quick shift in air flow rate over this region, region 5 would be unreachable. We also noted that a slight shift in the frequency characterizing an unstable regime is observed when propellant flow rates are changed. The simultaneous extinction of two pairs of flame is observed in zone 6. Hence stable combustion only occurs in region 1. This situation is even worsened by the presence of a partial extinction zone at low air flow and high propane flow rates.

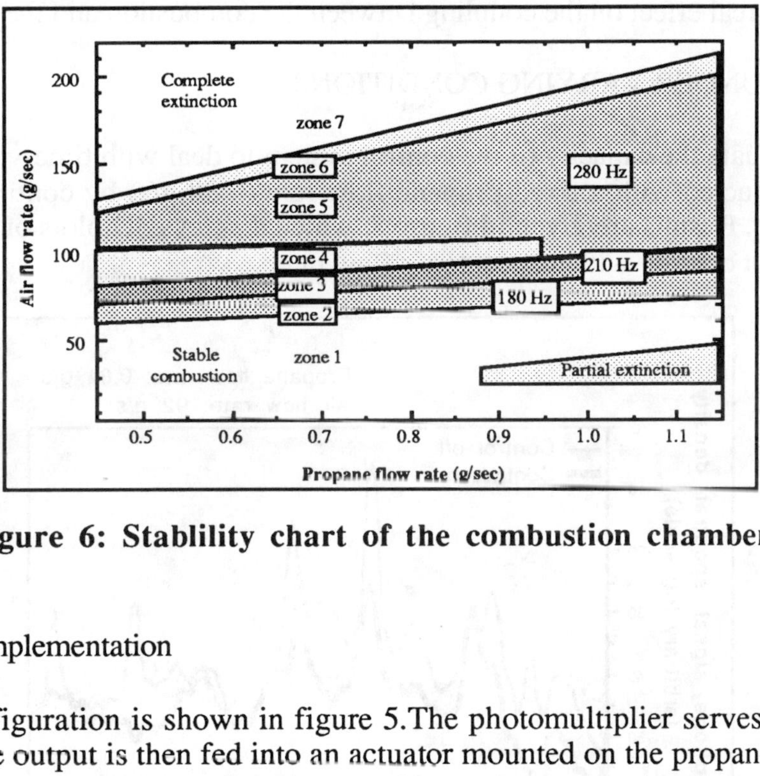

Figure 6: Stablility chart of the combustion chamber

Active Control Implementation

The control configuration is shown in figure 5.The photomultiplier serves as input to the controller, whose output is then fed into an actuator mounted on the propane feed line. The global response is collected by a microphone and is used to update the filter coefficients.

Experimental Procedure

The procedure is similar to that used in the preceeding situation.

The identification procedure in this case consists of exciting the actuator with a sine wave at a frequency close to the one of the natural mode of the system and taking the global response delivered by the pressure transducer. This excitation is required for an identification under conditions as close as possible to the actual operating conditions. This determination of D would seem suspect because the global response signal is strongly perturbated by the natural flame oscillations. Nevertheless, in general the estimate of D was accurate enough to obtain positive results. The identification problem is still not entirely solved and is currently being studied.

When the identification procedure has been completed, the test can be conducted by setting the combustor in an unstable regime and turning the controller on with nulled initial values.

3.2 RESULTS UNDER STEADY CONDITIONS

It is first interesting to see if the various observed unstable modes are controlable. Figure 7 gives the pressure power spectral density with and without control for one of the three unstable modes. It shows that this adaptive control method is able to stabilize the combustor operation (Additionnal work has shown the same for the two other modes). When the stabilization is achieved, the controller output is always at a low level. This kind of behavior differs from ANC systems, where the actuator input signal never decreases. In the ANC configuration the controller never acts on the primary source of noise. Hence the power needed by the actuator is always high. In contrast, in the combustion instability case the controller has a real effect on the coupling between the combustion and the acoustics.

3.3 RESULTS UNDER VARYING CONDITIONS

In order to evaluate the capacity of the control system to deal with time-varying conditions, tests were conducted with a fixed propane flow rate (0.88 g/s) by continuously changing the air flow rate. Results are given in figure 8, which displays 3D plots of spectral densities with and without control with respect to air flow rate.

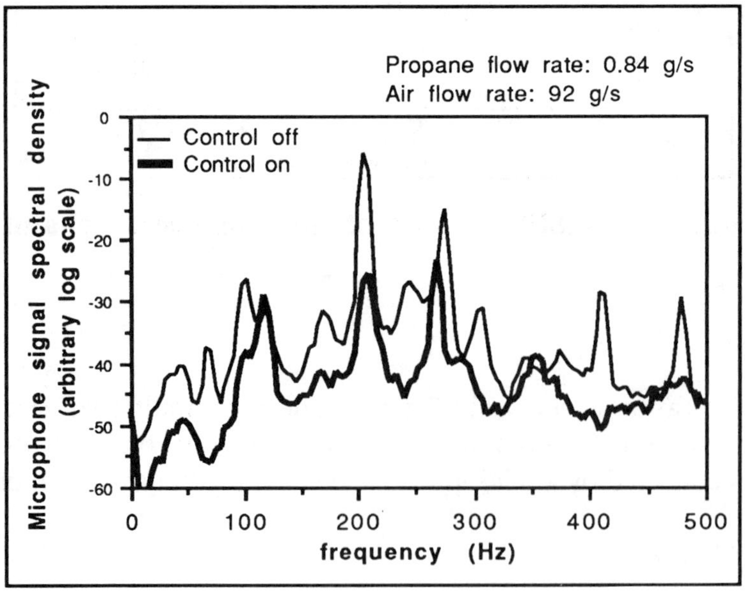

Figure 7

The system appears able to control the different modes appearing while increasing the air flow rate. Further, when the controller was turned on just after ignition, the combustion chamber operation was stable even inside the partial extinction region (zone4, figure 6).

3.4 SYSTEM VERSATILITY

Other transducer configurations were tried, exchanging the photomultiplier and the microphone roles or using a single signal for both functions. All these configurations led to an effective system which is due to the very similar behavior of the two sensor signals.

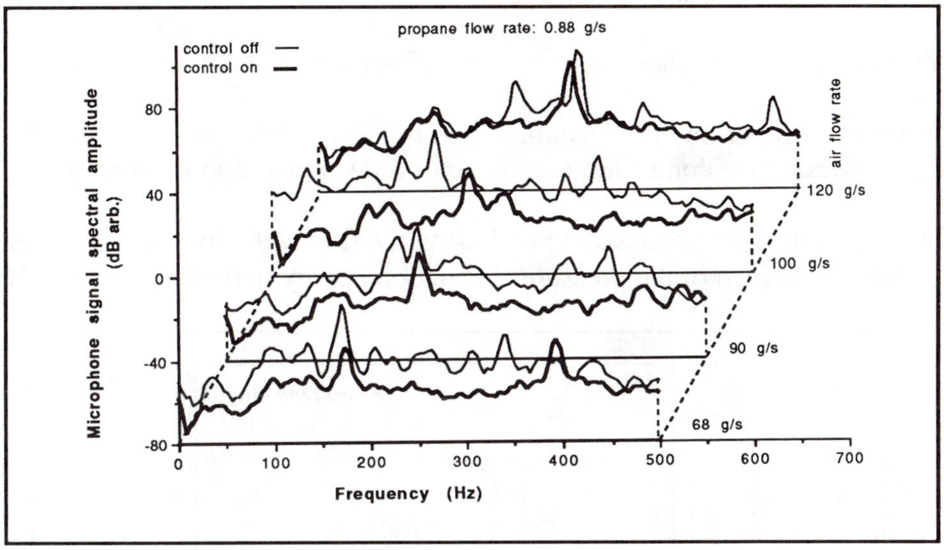

Figure 8: Acoustic pressure spectra with and without control during a change in the air flow rate

4. ADAPTIVE CONTROL OF SELF-SUSTAINED PRESSURE OSCILLATIONS IN A FLOW-EXCITED CAVITY

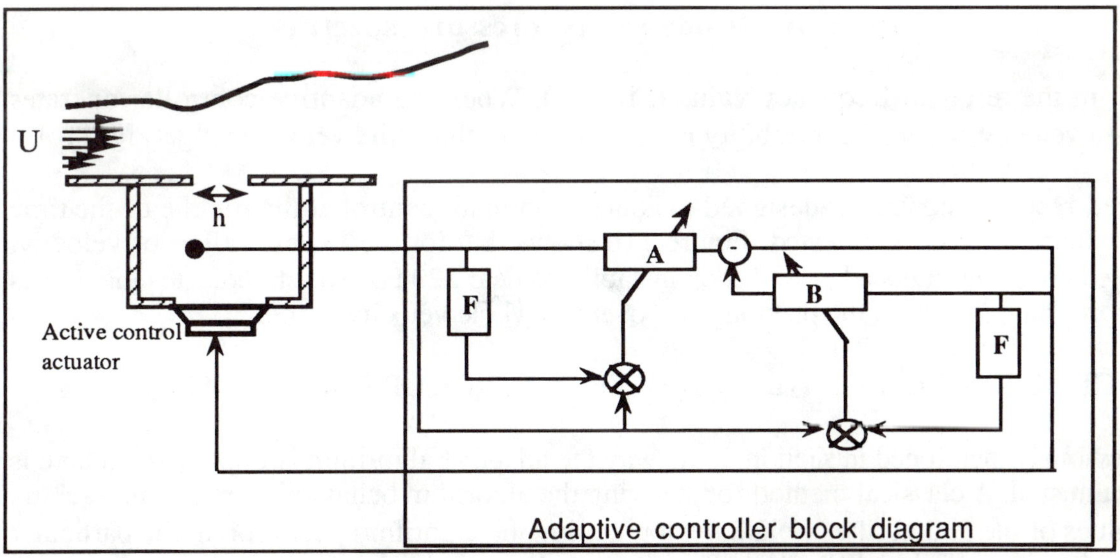

Figure 9: Experimental set-up

When a cavity is excited by a shear flow, a coupling between the natural shear layer instabilities and acoustic modes of the cavity occurs(see Elder[6] and Möser[5]). This coupling may become unstable, resulting in a strong sound radiation. The principle of active stabilization for these types of instabilities has been previously studied by Sunyach & Ffowcs Williams [4], and Möser[5].

4.1. ADAPTIVE CONTROL PERFORMANCE

The experimental procedure is exactly the same as the one used in §1 (fig. 9)

During a steady operation, the instability frequency observed is about 180 Hz. When switched on, the adaptive controller quickly achieved stabilization (Figure 10).

Unsteady conditions were obtained by increasing the flow velocity regularly in time. A time-frequency grey-scale map of the pressure fluctuation energy in the cavity therefore exhibits

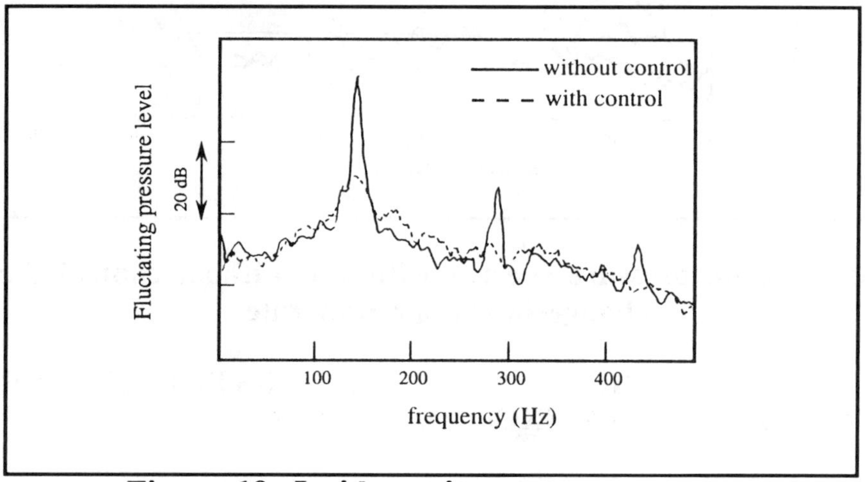

Figure 10: Inside-cavity pressure spectra

a shift in the resonant frequency value (Fig 11a). When the adaptive controller operates during a velocity sweep, the instability is cancelled over the entire velocity range(Fig. 11c).

A fixed-response controller designed to achieve optimal control at the middle of the time axis in figure 11 has been tested. Figure 11b shows that for the lowest values of velocity, the fixed controller induced closed-loop instabilities near 220 Hz, which indicates that it was not appropriate to control the phenomenon over the whole velocity range.

4.2. DETAILED STUDY OF THE ADAPTIVE CONTROL OPERATION

As previously mentioned in section 2, the way the adaptive algorithm is implemented here is rather unusual. A classical method for studying the algorithm behaviour consists in tracking the values of the digital filter weight vectors during the algorithm operation in the particular case of a two-coefficient FIR filter. This feature allows indeed the weight vectors tracks to be represented by curves in a (a1,a2) plane, a1 and a2 being the filter coefficients.This method has been applied in two situations: i)active control of a sinusoidal sound(fig. 12a), ii) in the present case (fig. 12b). Results are plotted on a grey-scale background representing the achieved attenuation. On each figure, we display the tracks for different intial values of a1 and a2 (starting point marked with a circle). The major difference between figures 12a

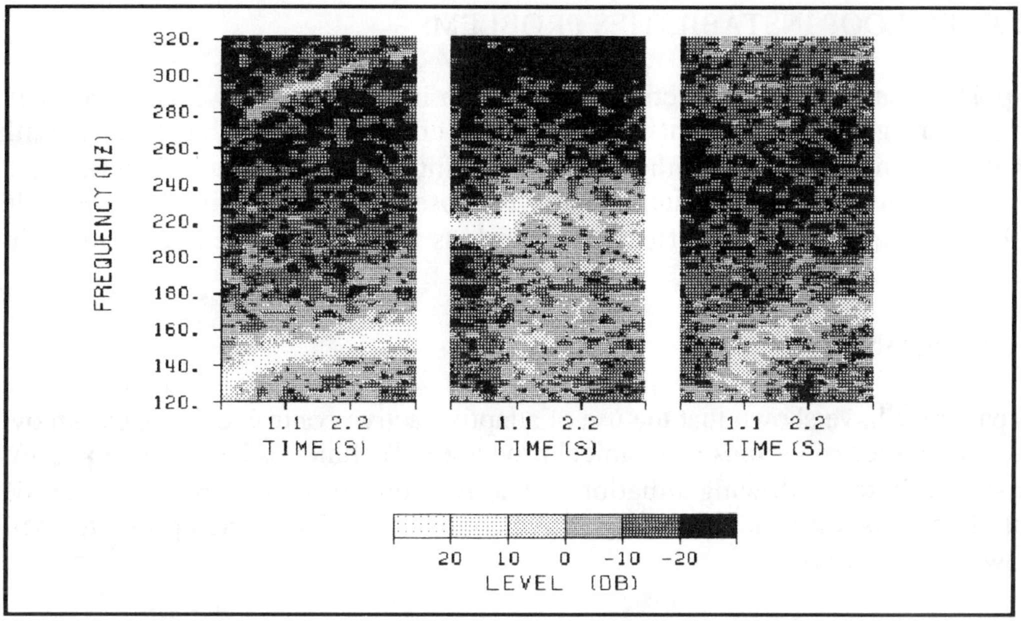

Figure 11: Time-frequency plot of the pressure energy (a)without control (left), (b)with a fixed controller (center), (c) with adaptive control (right)

and 12b lies in the pattern of the attenuation. A single minimum appears in the case of anti-noise, which corresponds to the best interference between the sound and anti-sound waves. Whatever starting point is chosen, all the trajectories converge to this focal point. In the present case of active control of self-sustained oscillations, no single focal point exists. A wide region of minimum values is observed. This suggests that the active control of instabilities consists of modifying the cavity response rather than creating an interference. The trajectories in 12b converge to the closest minimum value to the starting point, approximately following a steepest-slope line, which indicates that the algorithm behaves as if it were a true gradient algorithm.

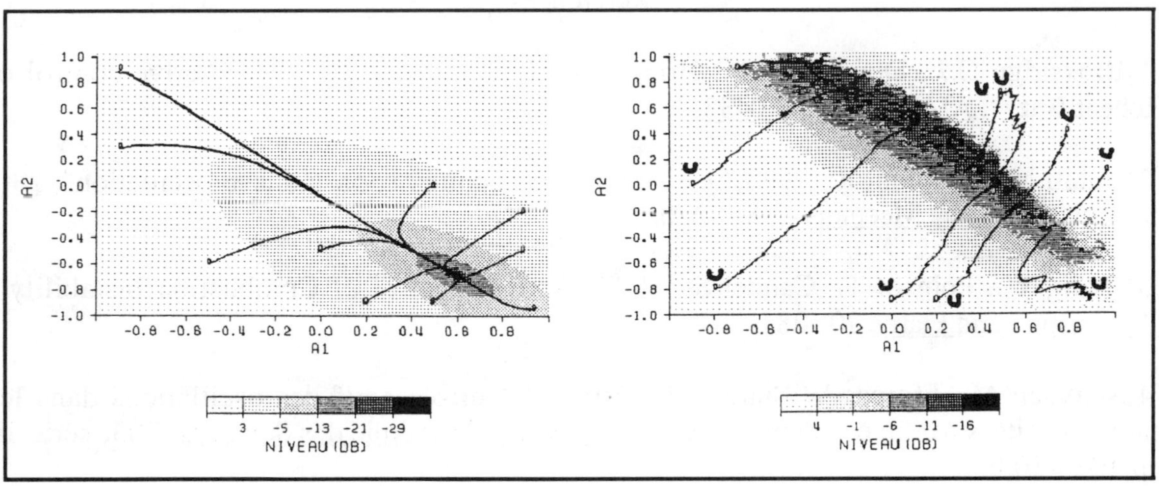

Figure 12: Weights tracks: (a) case of ANC (b) present case

4.3. CLOSED-LOOP INSTABILITIES PROBLEM:

As pointed out in the introduction, closed-loop instabilities may arise in active control systems. During the experiments, the adaptive controller was able to deal with these artificial instabilities. Indeed, with the starting points marked with a "U" in figure 12b, the initial coefficient values led systematically to closed-loop instabilities. Nevertheless, the algorithm managed to recover effective control, as is shown by the track pattern in figure 11b.

CONCLUSIONS

In this paper we have shown that the use of adaptive active control techniques can overcome natural instabilities of various mechanical processes. Techniques have been experimentally demonstrated in the following situations: i) active control of compressor surge, ii) active control of combustion instabilities, iii) active control of self sustained pressure oscillations in a flow-excited cavity.

The adaptive method developed for the present study has exhibited the following properties:

.It is able to automatically find an optimal control configuration to cancel the unwanted instability.

.It manages to cope with continuous changes in the operating conditions - even when the corresponding unstable coupling switches from one mode to another, as seen in the combustion experiment.

.It has also shown capacity to recover effective control when initially set to a configuration such that artificial instabilities occurred (in the case of the flow-excited cavity).

The promising results obtained during these first experiments lead us to think that this principle of adaptive active control of instabilities could be applied to real-scale experiments on industrial systems.

References

[1]Bloxsidge G.J., Dowling A.P., Hooper N., Langhorne P.J., 1987,"Active control of Reheat Buzz", AIAA Paper 87-0433.

[2]Gulati M. and Mani R., 1990,"Active control of combustion induced oscillation", 28th Aerospace Science Meeting A. 90-0270.

[3] Lang W., Poinsot T., Candel S., 1987,"Active Control of Combustion Instability", Combustion and Flame 70, 281.

[4]Sunyach M., Ffowcs Williams J.E., 1986, "Contrôle actif des oscillations dans les cavités excitées par un écoulement", Comptes rendus Académie des Sciences, 303, série II, pp 1085-1088.

[5]Möser M., "Aktive Kontroller einfacher, selbsterregter Resonatoren", Acustica, 69, 1989, pp 175-184.

[6]Elder, S.A., "Self-excited depth-mode resonance for a wall-mounted cavity in turbulent flow". J.A.S.A. 64, 1978, pp 877-890.

[7]Ffowcs Williams, J.E., Huang,X.Y., Active stabilisation of compressor surge, J. Fluid. Mech. 204, 1989, pp 245-262.

[8]Galland M.A.,Billoud G., Sunyach M.,"Adaptive Active Control of natural flow instabilities", Workshop on applications of active control to engineering systems, University of Cambridge, 1990.

[9]Widrow B., Adaptive Filters, Aspects of network and system theory, R.E. Kalman and N. De Claris, Eds, New York: Holt, Rinehart and Winston, 1970.

[10]Feintuch P.L., "An adaptive recursive filter", Proceedings of the IEEE, Vol 64, pp 1622-1624, nov. 1976.

[11]Eriksson L.J., Allie M.C., Greiner R.A., "The selection and application of an IIR Adaptive filter for use in active sound attenuation", IEEE trans. on ASSP, Vol 35, n° 4, Apr 87.

[12]Billoud G., Galland, M.A., Sunyach M.,"The use of time algorithms for the realization of an active sound-attenuator", Proceedings of ICASSP, IEEE 1989 International conference on Acoustics, Speech, and Signal Processing, Glasgow, 1989.

[13]Morgan D., "Analysis of multiple correlation cancellation loops with a filter in the auxiliary path", IEEE trans. on ASSP, Vol ASSP-28, n°4, pp 454-467, 1980.

[14]Epstein A., Ffowcs Williams J.E., Greitzer E.M., "Active suppression of compressor instabilities", AIAA 10th Aeroacoustics Conference, Paper AIAA-86-1994, 1986.

[15]Emmons H.W., Pearson C.E., Grant H.P., "Compressor surge and stall propagation", Transactions of the ASME, vol. 77, 455-469.

[16]Ffowcs williams J.E., Graham W.R., "An engine demonstration of active control of surge", Journal of Fluid Mechanics, 1990.

[17]Tsien H.S.,"Servo-stabilization of combustion in rocket motors", American Rocket Society Journal 22, pp 256-263, 1952.

[18]Crocco L., Cheng S.L., "Theory of combustion instability in liquid proppellant rocket motors", Butterworths Science Publication, AGARDOGRAPH N 8, 1956.

[19]Marble F.E., Cox D.W.J., "Servo-stabilization of low frequency oscillation s in liquid rocket motors", Journal of Appl. Math. Phys. 6, pp 1, 1955.

OPTIMAL CONTROL OF THERMAL DISTURBANCES IN A SHEAR FLOW

P. L. Chow,[1] **Lucio Maestrello**[2]

ABSTRACT

A problem of suppressing the thermally excited periodic disturbances in a shear flow is studied analytically. By the principle of optimal control, the temperature fluctuation and its adjoint state are shown to satisfy a system of coupled differential equations. The optimality system is solvable by a Fourier transform and the solution yields the optimal control of the surface heating. For the case of a uniform or slowly varying velocity profile, the solutions are found in a simple closed form, if the disturbance is a surface wave propagating downstream. The results show that the optimal control is most effective in reducing the thermal disturbances with small wave numbers, and the optimal heat input along the wall is proportional to the amplitude of the disturbance but inversely, to the wave number and the control cost parameter.

I. INTRODUCTION

The paper is concerned with an analytical study on the optimal boundary control of an externally excited temperature fluctuation in a parallel flow, relevant to a flow control problem by surface heating. The main objective is to ascertain, via a simplified model, the effectiveness of the thermal boundary control in reducing the temperature fluctuation. Even though the model is not realistic, the results of this study could provide a qualitative assessment of what can be achieved by means of the active surface-heating control. The paper is organized as follows. In Section II the optimal boundary-control problem for the governing equation is formulated. By minimizing the objective function, the optimality system for the problem is derived in Section III. It is shown in Section IV that, by the transform method, the optimality system is reducible to a one-dimensional problem in the transverse direction to the wall. The transformed system can then be solved to give the solution by the inverse transform. In Section V the problem is solved explicitly for some special cases, where the disturbances are generated by a surface wave and the shear velocity is uniform or slowly varying in the transverse direction. Also, some numerical results are obtained. The major conclusion of this study is that the effectiveness of the surface-heating control

[1]Wayne State University, Detroit, MI 48202, U.S.A.
[2]NASA Langley Research Center, Hampton, VA 23665, U.S.A.

depends critically on the wave number of the distrubance. The optimal control is most effective in the lower end of the wave number spectrum. This and other results are discussed in Section VI.

II. FORMULATION OF PROBLEM

In a two-dimensional parallel shear flow over a flat plate ($y = 0$), let the uniform temperature be perturbed by an external periodic heat source inside the fluid or on the wall. The temperature fluctuation θ satisfies the following nonhomogeneous convective heat equation:

$$\left(\frac{\partial}{\partial t} + U\frac{\partial}{\partial x}\right)\theta = \nu\left(\frac{\partial^2}{\partial x^2} + \frac{\partial^2}{\partial y^2}\right)\theta + f(x,y,t), \tag{1}$$

which is subject to the Neumann boundary condition

$$k_o\frac{\partial \theta}{\partial y}(x,o,t) = -q(x,t), \tag{2}$$

for $t > 0, -\infty < x < \infty$, and $y > 0$, where $\nu = k/c$ is the thermal diffusivity and f is the source of thermal disturbance; k_0 is the thermal diffusivity of the wall and the heat flux q is the control variable. Of particular physical interest is the case where the perturbation is oscillatory in time and localized in space. Thus the perturbing source function f is assumed to satisfy the mean-square integrability condition

$$\lim_{T\to\infty}\frac{1}{T}\int_0^T\int_{-\infty}^{\infty}\int_0^{\infty}|f(x,y,t)|^2 dt\ dx\ dy < \infty. \tag{3}$$

As a consequence, the heat flux q, the active control, is expected to have a similar property

$$\lim_{T\to\infty}\frac{1}{T}\int_0^T\int_{-\infty}^{\infty}|q(x,t)|^2 dt\ dx < \infty. \tag{4}$$

As an optimality criterion, the objective function J given below will be minimized,

$$J(q) = \lim_{T\to\infty}\frac{1}{T}\int_0^T\int_{-\infty}^{\infty}\{|\theta(x,0,t)|^2 + N|q(x,t)|^2\}dt\ dx \tag{5}$$

where the first term in the integrand yields the mean-square temperature fluctuation along the wall, while the second one is a measure of the mean-square control cost, where $N > 0$ is a cost parameter. For $N = 0$, the minimization of J gives the optimal control $\tilde{q}$ which is the best possible for the system to realize. On the other hand, for $N > 0$, the optimal solution $\tilde{q}$ will be to minimize the wall temperature fluctuation without excessive use of the control action. The choice of the number N depends subjectively on the relative importance assigned to two competing factors in the optimality criterion. Now the mathematical problem for the control of thermal disturbances can be formulated as follows: For a given heat source perturbation $f(x,y,t)$ satisfying the condition (3), find the optimal control $\tilde{q}(x,t)$, with the property (4), which minimizes the objective function (5) . Here the "state" θ is the solution of the system (1) and (2) subject to the zero initial condition. Note that, since the transient part of the solution to the system will be wiped out by the time average in (5), without loss of generality the initial condition may be taken to be zero.

III. THE OPTIMALITY SYSTEM

Since the governing equation is linear and the functional J is quadratic, the problem can be solved by the principle of superposition. Therefore it is sufficient to consider a time-harmonic perturbation of the form:

$$f(x,y,t) = g(x,y)e^{-i\omega t}. \tag{6}$$

Then it is possible to seek a time-harmonic solution

$$\theta(x,y,t) = \phi(x,y)e^{-i\omega t}, \tag{7}$$

and

$$q(x,t) = r(x)e^{-i\omega t} \tag{8}$$

A substitution of (6)–(8) into (1), (2), and (5) yields

$$L\phi = \nu\nabla^2\phi - U(y)\frac{\partial\phi}{\partial x} + i\omega\phi = -g(x,y) \tag{9}$$

$$k_o\frac{\partial\phi}{\partial y}(x,o) = -r(x), \tag{10}$$

$$J(r) = \int_{-\infty}^{\infty}\{|\phi(x,o)|^2 + N|r(x)|^2\}dx, \tag{11}$$

where ϕ vanishes as $|x| \to \infty$ or as $y \to \infty$. It is well known that the necessary condition for J being minimal is

$$\delta J(r) = 0, \tag{12}$$

where δJ means the variation of J with respect to r. Note that the quantities ϕ, r are complex. Denote their complex conjugates by $\phi^\star$ and $r^\star$ respectively. By introducing an adjoint state ψ to ϕ, one can derive the following optimality system [6]

$$L\phi = \nu\nabla^2\phi - U(y)\ \frac{\partial\phi}{\partial x} + i\omega\phi = -g(x,y), \tag{13}$$

$$L^\star\psi = \nu\nabla^2\psi + U(y)\ \frac{\partial\psi}{\partial x} - i\omega\psi = 0, \tag{14}$$

$$k_o\frac{\partial\phi}{\partial y}(x,o) - \frac{1}{N}\psi(x,o) = 0, \tag{15}$$

$$\phi(x,o) + k_o \frac{\partial \psi}{\partial y}(x,o) = 0. \tag{16}$$

It is seen that, in order to find the optimal control $\tilde{r}$, one must solve an extended system of coupled equations (13)–(16). Then the optimal solution is given by

$$r = \tilde{r}(x) = -\frac{1}{N}\psi(x,o) \tag{17}$$

and

$$\tilde{q}(x,t) = \tilde{r}(x)\, e^{-i\omega t} \tag{18}$$

gives the optimal control for the time-harmonic disturbance (6). In the multiple frequency case, the optimal control $\tilde{q}$ is a sum of the single-frequency solutions by superposition.

IV. SOLUTION OF THE OPTIMALITY SYSTEM

Clearly, by a Fourier transform in x, the optimality system (13)–(16) is reducible to a one-dimensional problem:

$$\nu\hat{\phi}'' - [\nu\lambda^2 + i(\lambda U + \omega)]\hat{\phi} = -\hat{g}(\lambda, y), \tag{19}$$

$$\nu\hat{\psi}'' - [\nu\lambda^2 - i(\lambda U + \omega)]\hat{\psi} = 0, \tag{20}$$

$$k_o\hat{\phi}'(\lambda,o) - \frac{1}{N}\hat{\psi}(\lambda,o) = 0, \tag{21}$$

$$k_o\hat{\psi}'(\lambda,o) + \hat{\phi}(\lambda,o) = 0, \tag{22}$$

where the Fourier transform $\hat{f}(\lambda)$ of a function $f(x)$ is defined as

$$\hat{f}(\lambda) = \int_{-\infty}^{\infty} f(x)e^{i\lambda x}dx.$$

Let $\hat{\phi}_p$ be a particular solution and $\hat{\phi}_1, \hat{\phi}_2$, two linearly independent complementary solutions of the equation (19) such that $|\hat{\phi}_2| \to \infty$ as $y \to \infty$. Then the complex conjugates $\hat{\phi}_1^\star$ and $\hat{\phi}_2^\star$ are linearly independent solutions of (20). Therefore the bounded solutions of (19) and (20) are given by

$$\hat{\phi}(\lambda,y) = \hat{\phi}_p(\lambda,y) - a\hat{\phi}_1(\lambda,y), \tag{23}$$

$$\hat{\psi}(\lambda,y) = b\hat{\phi}_1^\star(\lambda,y), \tag{24}$$

where, in view of the boundary conditions (21) and (22), the constants a and b are found to be

$$\begin{aligned} a(\lambda) &= [k_o^2(\hat{\phi}_1^\star)'\hat{\phi}_p' + \frac{1}{N}\hat{\phi}_1^\star\hat{\phi}_p](\lambda,o)/d(\lambda), \\ b(\lambda) &= k_o(\hat{\phi}_1\hat{\phi}_p' - \hat{\phi}_1'\hat{\phi}_p)(\lambda,o)/d(\lambda), \end{aligned}$$

where

$$d(\lambda) = (k_0^2|\hat{\phi}_1'|^2 + \frac{1}{N}|\hat{\phi}_1|)(\lambda, o).$$

In view of the equations (19) and (24), by an inverse Fourier transform, the optimal control law is obtained,

$$\tilde{r}(x) = -\frac{1}{2\pi N}\int_{-\infty}^{\infty} b(\lambda)\hat{\phi}_1(\lambda, o)e^{-i\lambda x}d\lambda. \tag{25}$$

Under the optimal control, the state of thermal fluctuation is given by the inverse transform of (23):

$$\phi(x,y) = \frac{1}{2\pi}\int_{-\infty}^{\infty} e^{-i\lambda x}[\hat{\phi}_p(\lambda, y) - a(\lambda)\hat{\phi}_1(\lambda, y)]d\lambda. \tag{26}$$

On the other hand, when there is no control ($r \equiv 0$), the temperature fluctuation $\phi_o(x,y)$ is given by

$$\phi_o(x,y) = \frac{1}{2\pi}\int_{-\infty}^{\infty} e^{-i\lambda x}[\hat{\phi}_p(\lambda, y) - a_o(\lambda)\hat{\phi}_1(\lambda, y)]d\lambda, \tag{27}$$

where

$$a_o(\lambda) = \hat{\phi}_p'(\lambda, o)/\hat{\phi}_1'(\lambda, o).$$

To assess the effectiveness of the control, we may introduce either the uniform reduction ratio

$$\varepsilon_o = \max \varepsilon(x) = \max.|\frac{\phi(x,o)}{\phi_o(x,o)}|, \tag{28}$$

or the (root) mean reduction ratio

$$\begin{aligned} \varepsilon_m &= \left\{\int_{-\infty}^{\infty} |\phi(x,o)|^2 dx \Big/ \int_{-\infty}^{\infty} |\phi_o(x,o)|^2 dx\right\}^{1/2} \\ &= \left\{\int_{-\infty}^{\infty} |\hat{\phi}(\lambda,o)|^2 d\lambda \Big/ \int_{-\infty}^{\infty} |\hat{\phi}_o(\lambda,o)|^2 d\lambda\right\}^{1/2}, \end{aligned} \tag{29}$$

by Parseval's equality in Fourier transform. To give a qualitative physical interpretation of our results, it is instructive to go through some examples in detail.

V. EXAMPLES

Two examples corresponding to a time-harmonic perturbation $f(x, y, t)$ of the form (6) will be considered. In particular the spatial distribution function g is assumed to be

$$g(x,y) = A \ \exp\{i\kappa x - \frac{1}{2}\alpha x^2 - \beta y\} \tag{30}$$

which represents a decaying surface-wave disturbance with amplitude A, the wave number κ, and the exponential decay parameters α and β. Then the Fourier transform of g in x is given by

$$\hat{g}(\lambda, y) = \hat{g}_o(\lambda)e^{-\beta y}, \tag{31}$$

where

$$\hat{g}_o(\lambda) = A\sqrt{\frac{2\pi}{\alpha}}e^{-\frac{1}{2\alpha}(\lambda+\kappa)^2}. \tag{32}$$

As the first example, let the velocity $U = U_o$ be a constant. Then the transformed optimality system (19)–(22) with g given by (30) can be solved analytically. The transformed optimal solutions (23) and (24) become

$$\hat{\phi}(\lambda, y) = h(\lambda)\left[e^{-\beta y} - \left(\frac{1 + Nk_0^2\beta\xi^*}{1 + Nk_0^2|\xi|^2}\right)e^{-\xi^* y}\right], \tag{33}$$

$$\hat{\psi}(\lambda, y) = h(\lambda)\left[\frac{Nk_o(\xi - \beta)}{1 + Nk_o^2|\xi|^2}\right]e^{-\xi^* y}, \tag{34}$$

where

$$h(\lambda) = \frac{\hat{g}_o(\lambda)}{\xi^2(\lambda) - \beta^2}, \tag{35}$$

and $\xi^*(\lambda)$ is the complex conjugate of $\xi(\lambda)$ defined by

$$\xi(\lambda) = \{\lambda^2 + i(\lambda U_o + \omega/\nu)\}^{1/2}, Re\{\xi(\lambda)\} > 0. \tag{36}$$

By the inverse Fourier transform (25), the optimal control law is found to be

$$\tilde{r}(x) = -\frac{k_o}{2\pi}\int_{-\infty}^{\infty}\frac{\hat{g}_o(\lambda)e^{-i\lambda x}}{(1 + Nk_o^2|\xi(\lambda)|^2)[\xi(\lambda) + \beta]}. \tag{37}$$

Along the wall, from (26) and (27), the controlled and the uncontrolled temperature fluctuations are given by

$$\phi(x, o) = \frac{Nk_o^2}{2\pi}\int_{-\infty}^{\infty}\frac{\xi^*(\lambda)\hat{g}_o(\lambda)}{[\xi(\lambda) + \beta](1 + Nk_o^2|\xi(\lambda)|^2)}e^{-i\lambda x}dx, \tag{38}$$

$$\phi_o(x, o) = \frac{1}{2\pi}\int_{-\infty}^{\infty}\frac{\hat{g}_o(\lambda)}{\xi(\lambda)[\xi(\lambda) + \beta]}e^{-i\lambda x}d\lambda. \tag{39}$$

One observes that, if N is large, then $\tilde{r}(x) \sim 0$ and $\phi(x, o) \sim \phi_o(x, o)$. That is, when the control is costly, no control action is taken so that the state is unchanged. While, if the control cost parameter N is small, we get

$$\tilde{r}(x) \sim \int_{-\infty}^{\infty}\frac{\hat{g}_o(\lambda)e^{-i\lambda x}}{\xi(\lambda) + \beta}d\lambda, \tag{40}$$

and $\phi(x, o) \sim 0$. Thus the unrestricted thermal control almost completely eliminates the temperature fluctuation along the wall. In fact, in this case, the control action is to cancel out the heat flux across the wall at any cost.

To simplify the results, let the decay factor α be small. In view of (32), as $\alpha \to 0, \hat{g}$ approaches a δ-function and the otpimal control (31) reduces to

$$\tilde{r}(x) \sim -\frac{Ak_o e^{i\kappa x}}{[\xi(-\kappa)+\beta](1+Nk_o^2|\xi(-\kappa)|^2)}. \tag{41}$$

The corresponding residual temperature fluctuation (27) becomes

$$\phi(x,o) \sim \frac{ANk_0^2\xi^*(-\kappa)e^{i\kappa x}}{[\xi(-\kappa)+\beta](1+Nk_o^2|\xi(-\kappa)|^2)}. \tag{42}$$

Recall, by noting (30), that

$$\xi(k) = \{k^2 + i(kU_o + \omega/\nu\}^{1/2}, Re\{\xi\} > 0, \tag{43}$$

which gives

$$|\xi(k)|^2 = \{k^4 + (kU_o + \omega)^2/\nu^2\}^{1/2}.$$

As it turns out, in this case, the uniform reduction ratio and the mean reduction ratio, defined by (28) and (29) respectively, coincide. In fact we have

$$\varepsilon = \frac{Nk_o^2|\xi(-\kappa)|^2}{1+Nk_o^2|\xi(-\kappa)|^2} < 1, \tag{44}$$

where $\varepsilon = \varepsilon_o = \varepsilon_m$.

From (39) one sees that, as $\alpha \to 0$, the excited disturbance is a surface plane wave. The otpimal response $\tilde{r}$ in (41) is also of the same wave form with a phase-shift arg.$\{\xi(-\kappa)+\beta\}$. Under the control, the surface temperature fluctuation is reduced to (42), which has the reduction factor ε given by (44). It shows clearly that ε increases from 0 to 1 as $\eta = Nk_o^2|\xi(-\kappa)|^2$ increases from 0 to ∞. For a fixed control cost parameter N, η is an increasing function of κ and ω, but a decreasing function of ν. Therefore the reduction ratio ε and, hence, the effectiveness of the thermal control decreases as the wave number κ or the frequency ω of the distrubance increase, while they increase as the thermal diffusivity ν of the fluid increases. These results are physically plausible and may provide some insights into a certain flow control problem by surface heating or cooling.

As the second example, assume the shear velocity U varying slowly from zero at the wall to the free stream velocity U_o at the height $y = \frac{1}{\delta}$, where δ is small. Then, by applying the multiple-scale or the WKB method [7], one can construct asymptotic solutions to the optimality system (19)–(22), by using δ as the small parameter. Without giving the derivation, it can be shown that the first-order asymptotic approximation yields the complementary and particular solutions to equation (19) as follows:

$$\hat{\phi}_{1,2}(\lambda, y) \sim a(\lambda, y)e^{\mp\Omega(\lambda,y)} \tag{45}$$

and

$$\hat{\phi}_p(\lambda, y) \sim [2\eta(\lambda, o)]^{-1}\{\hat{\phi}_1(\lambda, y)\int_0^y \hat{\phi}_2(\lambda, s)\hat{g}(\lambda, s)ds + \hat{\phi}_2(\lambda, y)\int_y^\infty \hat{\phi}_1(\lambda, s)\hat{g}(\lambda, s)ds\} \tag{46}$$

where

$$\eta(\lambda, y) = \{\lambda^2 + i\,\nu^{-1}[\lambda U(y) + \omega]\}^{1/2}, Re\eta > 0, \tag{47}$$

$$a\,(\lambda, y) = \left[\frac{\eta(\lambda, o)}{\eta(\lambda, y)}\right]^{1/2}, \tag{48}$$

and

$$\Omega(\lambda, y) = \int_0^y \eta(\lambda, s)ds.$$

For a small decay factor, $\alpha \approx 0$, by some tedious by straight-forward algebraic manipulations, one can obtain the otpimal control $\tilde{r}$ and the residual wall temperature fluctuation $\phi(x, o)$ in a simple form:

$$\tilde{r}(x) \sim -A\frac{k_o\gamma(-\kappa)e^{i\kappa x}}{(1 + Nk_0^2|\eta(-\kappa, o)|^2)}; \tag{49}$$

$$\phi(x, o) \sim \gamma(-\kappa)\left[\frac{k_o^2 N\eta^\star(-\kappa, o)}{1 + k_o^2 N|\eta(-\kappa, o)|^2}\right]e^{i\kappa x}, \tag{50}$$

where

$$\gamma(\lambda) = \int_0^\infty a(\lambda, \tau)e^{\,\Omega(\lambda,\tau)-\beta\tau}d\tau \tag{51}$$

From (42) it follows that the reduction ratio

$$\varepsilon = \frac{Nk_o^2|\eta(-\kappa, o)|^2}{1 + Nk_o^2|\eta(-\kappa, o)|^2}. \tag{52}$$

In contrast with the results (41)–(44) for a uniform shear profile, the above results (48)–(51) show that, for a slowly varying shear flow, the velocity variation introduces a shape factor γ as defined by (45) which modifies the control and the corresponding temperature fluctuation. In fact it is easy to verify that if $u \equiv U_o$, the results (48)–(51) reduce to the previous ones (41)–(44) with a uniform shear velocity, as they should. Therefore the previous physical interpretation of those results is still valid, at least qualitatively. Some numerical calculations have been performed when the shear velocity profile is uniform. The numerical results have contributed to the concluding remarks given in the next section.

VI. CONCLUDING REMARKS

In this paper, the optimal control of thermally excited oscillatory disturbances is studied analytically. By applying the principle of optimal control, the optimality system of coupled equations is derived for the state of temperature fluctuation and its adjoint state. The equations are solved analytically by the transform method. For constant and slowly varying velocity profiles, the analytical expressions for the optimal control heat flux across the boundary and the residual wall-temperature fluctuation are obtained in a closed form. These expressions are also evaluated numerically for a certain set of parameter values. Based on the the above results, several concluding remarks and discussions are given as follows.

1) It is well known that in the optimal control of distributed parameter system, the boundary control, in general, will not be able to cancel out the undesired thermal fluctuation completely. The results (44), on the reduction ratio ε show that an optimal control always results in a reduced fluctuation, depending on the excitation wave number κ. In fact, at small wave numbers and low control cost, a nearly complete cancellation is possible. However, as the wave number increases, the effectiveness of control decreases monotonically.

2) From the result (40), it is seen that the optimal heat input is proportional to the amplitude A of the disturbance and inversely proportional to the wave number κ and the control cost parameter N, with other parameters fixed. The amplitude max $.|\tilde{r}|$ seems to exhibit a single peak value as a function of κ or $a = Nk_o^2$. Also the phase of the control, depending on κ, is shifted in order to affect an optimal cancellation.

3) Even thought the optimality criterion is to minimize the wall-temperature fluctuation without excessive use of control, it is will actually lead to a reduction in thermal fluctuation inside the flow as well. This is due to the maximum principle for a parabolic equation which says the maximum temperature fluctuation inside the flow is always less than that on the boundary.

4) In these examples, only the case of time-periodic disturbances is considered, but the present method can be easily applied to the general case. However the computational problem becomes more complicated because the control $q(x,t)$ is no longer time-harmonic.

ACKNOWLEDGEMENTS

The authors wish to thank Mr. F.F. Badavi for computational support. The research of the first author (P.L. Chow) was supported by NASA Langley Research Center under Grant NAG 1-460.

REFERENCES

1. Schlichting, H., Boundary-Layer Theory, McGraw Hill, 1979.

2. Liepmann, H.W., Brown, G.L. and Nosenchuck, D.M., Control of laminar-instability waves using a new technique. J. Fluid Mech., 118 (1982), p. 187.

3. Liepmann, H.W. and Nosenchuck, D.M., Active control of laminar-trubulent transition, J. Fluid Mech., 118 (1982), p. 187.

4. Maestrello, L., Active transition fixing and control of boundary layer in air, AIAA paper 85-0564, March, 1985.

5. Maestrello, L. and Ting, L., Analysis of active control by surface heating, AIAA Journal, 23 (1985), p. 1038.

6. Lions, J.L., Optimal Control of Systems Governed by Partial Differential Equations, Springer Verlag, N.Y., 1971.

7. Cole, J.D., Perturbation Methods in Applied Mathematics, Blaisdell Pub. Co., Waltham, Mass. 1968.

AN ADAPTIVE CONTROL SYSTEM FOR ACTIVE ATTENUATION OF WIND TUNNEL NOISE

S. Anwar,[1] P. J. Hollis[2]

ABSTRACT

Active control of noise is particularly important when the unwanted noise is in the low frequency range. Such is the case with wind tunnel fan drive noise. This paper involves the design and simulation of an active noise control (ANC) system that can be adopted for use in a wind tunnel.

Since the wind tunnel is subjected to changes in air velocity, ambient temperature, and sometimes to some nonlinearities in its sound field, the acoustic plant cannot be modeled by a fixed transfer function. Therefore, an adaptive control system is designed to take care of any changes in system transfer function or input noise. A finite impulse response (FIR) filter, which is always stable, is used to model the plant. A recursive least squares (RLS) adaptive algorithm, which has robustness, is employed to adapt the filter coefficients. The active noise control system is simulated using a number of plant models, input processes and using least mean squares (LMS) as well as RLS algorithms. The RLS algorithm shows better stability, convergence and noise attenuation properties than the LMS algorithm for all types of inputs and plant models. The effect of the plant model on the simulated results is significant. Use of a pole-zero model for the acoustic plant shows a slower convergence rate of the residual noise when the ANC system is simulated using the RLS algorithm. It is found that an increase in filter length, although it increases the maximum attainable attenuation, also decreases the convergence rate. The filter length is selected by making a compromise between these two factors. The final design of the ANC system is simulated and it shows considerable attenuation with a reasonable rate of convergence.

1. INTRODUCTION

Unwanted sound or noise can be attenuated in two basic ways, either by passive control or by active control. The first method is prevalent in most noise control problems. The principal reasons for the wide use of passive noise controllers are their low cost and their efficiency in medium and high frequency ranges. However, for low frequency noise, passive controllers become bulky, expensive and relatively ineffective when compared with active controllers. In cases where the unwanted noise is of low frequency and of high amplitude, an active noise control system can give desired attenuation at reasonable cost. One particular case is wind tunnel fan drive noise. This paper involves the design and simulation of an active noise control (ANC) system that can be adopted for use in a wind tunnel.

[1]Dept. of Mechanical Engineering, University of Arizona, Tucson, AZ, U.S.A.

[2]Dept. of Mechanical Engineering, FAMU/FSU College of Engineering, Tallahassee, FL, U.S.A.

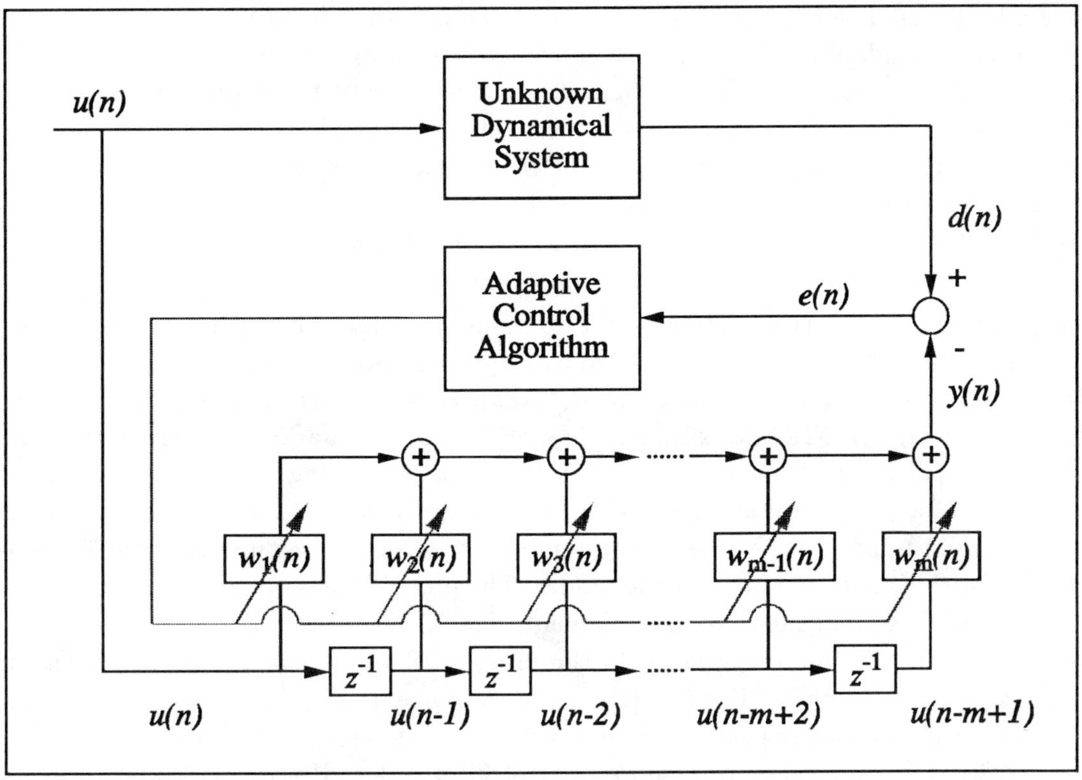

Figure 1. Adaptive control applied to a system identification problem.

Since the wind tunnel is subjected to changes in air velocity, ambient temperature, and sometimes to some nonlinearities in its sound field, the acoustic plant cannot be modeled by a fixed transfer function. Therefore, an adaptive control system is needed to take care of any changes in system transfer function or input noise.

1.1 SELECTION OF ADAPTIVE ALGORITHM FOR THE ANC SYSTEM

Adaptive control techniques are very useful in a system identification problem. In most practical problems, the system is usually an unknown plant or the system behavior changes with time. In these cases, it is necessary to model the system using the output of the actual system and the output of the model. Obviously, the weights of the model have to be changed according to the system behavior through some adaptive algorithm. Figure 1 shows an adaptive control system in a system identification framework illustrating its basic principle in terms of block diagrams. The figure shows an unknown dynamic system with a set of discrete-time measurements $d(n)$ defining variations of the output signal of the system in response to a known stationary signal $u(n)$ applied to the system input.

The system can be linear or nonlinear and the process can be stationary or time varying. For simplicity, let the system be linear and the process be stationary. In system identification, it is needed to develop a model for this system in the form of a transversal filter consisting of a set of delay line elements (z^{-1}) and a corresponding set of adjustable coefficients (w_i) as shown in the figure. At time n, the available signal consists of a set of samples $u(n)$, $u(n-1), \ldots, u(n-m+1)$. These samples are multiplied by a corresponding set of adjustable tap weights $w_1(i)$, $w_2(i)$, $\ldots$, $w_m(i)$, to produce an output signal $y(n)$. If the actual signal output of the unknown system is denoted by $d(n)$, then the estimation error $e(n)$, is given by the difference between $d(n)$ and $y(n)$. Usually $e(n)$ is nonzero indicating the deviation of the model from the unknown system. In order to minimize this deviation, the estimation error $e(n)$ is fed into an adaptive algorithm which controls the corrections applied to the individual tap weights in the transversal filter. Thus, the tap weights of the filter assume a new set of values for use on the next iteration and a new filter output along with a new value for the estimation error is produced at time n+1. The iteration process is continued for a number of times until estimation error $e(n)$ becomes sufficiently small in some statistical sense. If the process is time-varying, the adaptive algorithm not only adjusts the filter coefficients, but also continually tracks the statistical variations in the process.

A finite impulse response (FIR) filter is used to model the plant. FIR (finite impulse response) filters have certain advantages over IIR (infinite impulse response) filters. FIR filters are always stable even in a nonstationary environment. Adaptive IIR filters are computationally more involved than adaptive FIR filters and their stability is not guaranteed. Since the goal here is to use an adaptive filter which will be stable in all operating conditions and which will allow the use of a programmable controller to implement the adaptive filter, an FIR filter is used. Keeping in mind that the desired system model is an all-zero model, some suitable FIR adaptive algorithms are examined for the model. The first choice is obviously the LMS adaptive algorithm which is simple and easy to implement. The LMS algorithm can be modified to track the system with a little nonstationarity in the process. The LMS algorithm has been used for active noise control system by Burgess [1], Poole and Leventhall [2] and Eriksson [3]. Burgess, [1], did not consider acoustic feedback in his system and as a result, his simulation of the ANC system showed a fast convergence rate for a lower filter order. However, Eriksson, [3], showed that if the acoustic feedback is considered, the simulation of the system does not converge for a pseudo-random input. He attributed this to the fact that since acoustic feedback adds poles to the system, an IIR filter is needed in order for the algorithm to converge. It will be shown that the LMS algorithm converges for FIR filters in a Gaussian random process if a higher order filter is used with a very small convergence factor, although at a slow rate. In his final design, Eriksson [4] used an IIR filter with the RLMS adaptive algorithm for active sound attenuation in a duct. Poole and Leventhall, [2], used three LMS algorithms in different paths of the control network, including error path. However, their system did not produce satisfactory results for pure tones and band-limited noise. That was because their model was such that even though the error term was zero, one of the adaptation algorithms did not stop adapting the filter coefficient as pointed out by Eriksson. For random noise, however, the system was successful.

Considering all these facts, the LMS algorithm does not seem to be a robust one for the ANC system. The next adaptive algorithm that has been considered is the RLS algorithm which has the property of robustness and fast convergence rate, which come at the cost of computational complexity. Since RLS algorithms are deterministic and use all the past data of the process to adapt the filter coefficients, their convergence properties are much better than those of LMS algorithms. Also with a slight change in the forgetting factor of the algorithm, the robustness of the filter can be preserved in a nonstationary environment. Since the exponentially weighted RLS algorithm with U-D factorization is numerically better suited for implementation, it seems a good choice for the ANC system. A brief description of the LMS and the RLS algorithm is given in the appendix.

2. COMPUTER SIMULATION AND RESULTS

The ANC system is simulated to examine various algorithms and to examine the effects of

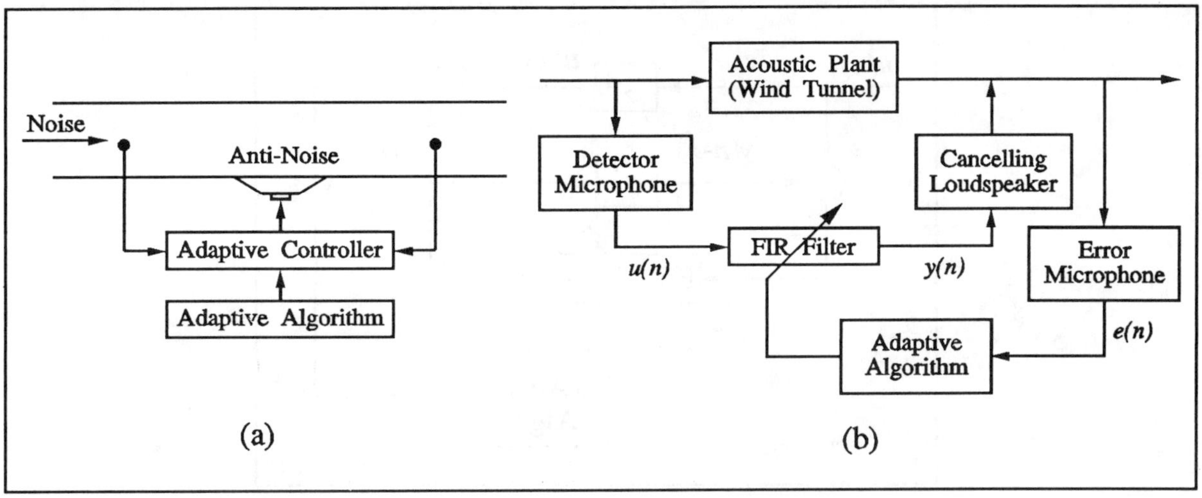

Figure 2. a) Basic structure of the hardware for simulation, b) Block diagram representation.

different design parameters on the system dynamics both in the time and frequency domains. The basic assumptions that are made about the structural hardware of the ANC system are as follows:

The ANC system consists of the following equipment:

1. one detector microphone preferably unidirectional,
2. one noise canceling loudspeaker,
3. one error microphone (unidirectional or omnidirectional),
4. a programmable controller or digital computer with A/D, D/A converters, and anti-aliasing or low-pass filters.

Other assumptions made about the system itself are:

1. Only plane waves travel through the duct making the system one dimensional.
2. There is no delay in the error path. Usually there will be some delay in the error path, unless the error microphone is placed right at the top of the canceling loudspeaker. However, for acoustical reasons, it is better to place the microphone at least some distance away from the canceling loudspeaker in the downstream direction.
3. The fundamental duct resonance frequency is higher than all operating frequencies. This is to ensure that the canceling loudspeaker produces only plane waves in the direction of the duct axis.
4. The acoustic plant is a pure delay plant. This assumption is reasonable if the noise terminates anechoically at the ends the duct. It was shown by Elliott and Nelson [5] that the acoustic plant can be represented as a delay plant for a duct. If the end reflections are low in amplitude compared to the actual noise itself, the system can be assumed to be a delay plant for all practical purposes.

The basic structure of the hardware of the ANC system that is going to be used in the simulation is shown in Fig. 2(a). The corresponding block diagram representation is shown in Fig. 2(b).

Three types of input process have been considered in the simulation.

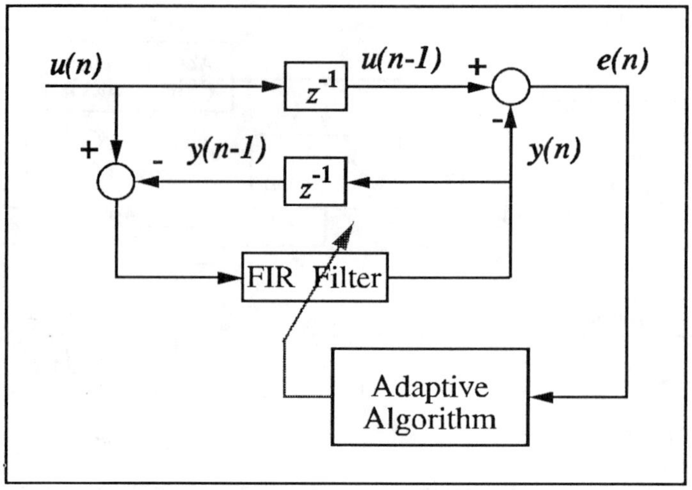

Figure 3. Simulation diagram of the ANC system with acoustic feedback.

1. Discrete frequency noise containing 3 or more sinusoids of different frequencies.
2. Gaussian random noise with zero mean and unity standard deviation which is usually known as white noise.
3. Discrete frequency noise masked with white noise which is of broadband nature.

2.1 ANC SYSTEM WITH ACOUSTIC FEEDBACK

The block diagram for the simulation of the ANC system with acoustic feedback is shown in Fig. 3. The acoustic plant is considered to be a single delay plant. The filter length was chosen to be 10. For the RLS algorithm, the forgetting factor λ was chosen to be 1, meaning that the adaptation process has infinite memory. The convergence factor μ of the LMS algorithm was chosen to be 0.025. A simulation program was run for all three types of input processes. The results of the simulation are explained below. The attenuation is defined by

$$\text{Attenuation, dB} = 20\log_{10} UF - 20\log_{10} EF \tag{1}$$

where

UF = Fourier Spectra of the input noise and,

EF = Fourier Spectra of the attenuated noise.

The normalized frequency is defined

$$\text{Normalized Frequency} = \frac{\text{Actual Frequency}}{\text{Sampling Frequency}} \tag{2}$$

Since the maximum frequency of an FFT spectra is governed by the Nyquist frequency, the maximum value of the normalized frequency will be $\frac{\text{Nyquist Frequency}}{\text{Sampling Frequency}} = 0.5$.

2.1.1 Discrete Frequency Noise

Three equally spaced frequencies were chosen as 100, 150, and 200 Hz which are in the low frequency range. The input noise was the summation of two sine waves (100 and 200 Hz) and one cosine wave (150 Hz). The amplitude of each of the sine and cosine functions was unity. Figure 4 (a) and (c) show the squared residual noise at different iterations for both algorithms. The convergence rate of the RLS algorithm is certainly faster than that of the LMS algorithm. As can be seen in Fig. 4(c), the squared residual noise diminishes to almost zero at around the 30th iteration giving complete attenuation, although there was an accumulation of error in the first few iterations. Desired attenuation was not achieved even after 128 iterations when the system was simulated with the LMS algorithm, Fig. 4(a). The adaptation rate of the filter coefficients was much slower for the LMS algorithm than that of RLS algorithm. Figure 4(b) and (d) show the frequency domain attenuation for the simulated ANC system. Attenuation occurs in all three discrete frequencies with a maximum of about 11 dB in the case of the system with the RLS adaptive filter as can be seen from Fig. 4(d). However, considerable intensification at certain other frequencies accompanies this attenuation. The use of a higher order filter may reduce this kind of intensification. When simulated with LMS algorithm, the maximum attenuation in the FFT spectra was about 31 dB, Fig. 4(b).

2.1.2 White Noise

For white noise input, the convergence properties are much better for the system with the RLS

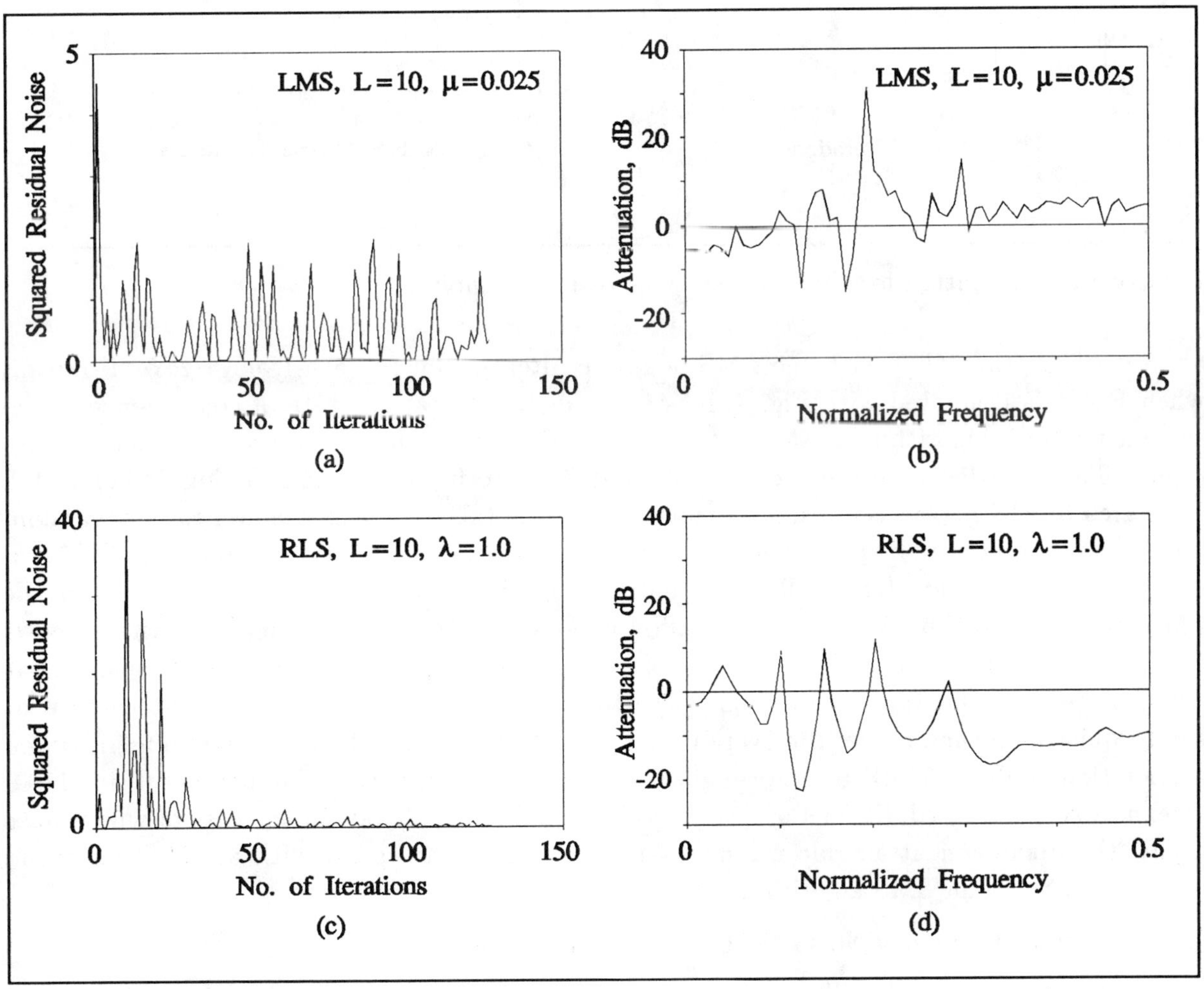

Figure 4. Attenuation for the ANC system with discrete frequency noise.

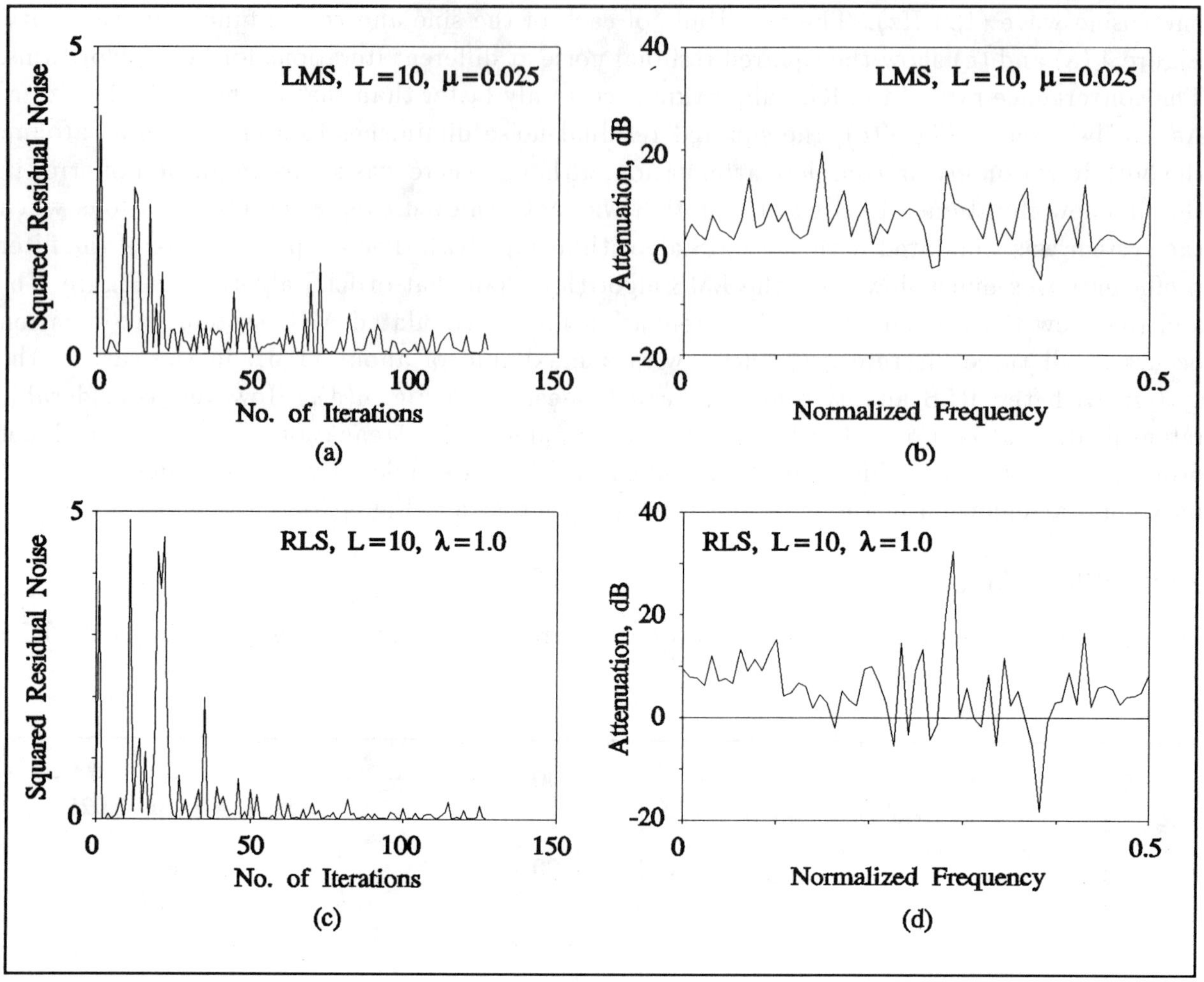

Figure 5. Attenuation for the ANC system with white noise.

adaptive filter than that with the LMS adaptive filter as can be seen from Fig. 5. However, when Fig. 5 is compared with Fig. 4, it is easily deduced that the rate of convergence of the system for both algorithms is slower for white noise input than for discrete frequency noise input. This may be attributed to the fact that discrete frequency noise is highly correlated compared to white noise as a result of which adaptation takes place at a slower rate for random inputs than for discrete frequency inputs. The desired level of squared residual noise is achieved at about 50th iteration for the RLS simulation, Fig. 5(c), whereas it is not obtained for the LMS simulation within the range of iteration covered in Fig. 5(a). The RLS algorithm shows considerable attenuation over a wide range of frequencies except for a few discrete frequencies at which some intensification occurs, Fig. 5(d). However, since these frequencies are located near the Nyquist frequency range, the overall attenuation figure is good. The maximum simulated attenuation is about 32 dB at a normalized frequency of about 0.3. The maximum simulated attenuation is about 20 dB for the LMS algorithm, Fig. 5(b). Thus, in the case of white noise input, the maximum attainable attenuation is more for the system with an RLS algorithm than for one with an LMS algorithm.

2.1.3 Discrete Frequency Noise Masked with White Noise

Probably this kind of input is the most practical case that is encountered in noise control problems. As stated earlier, the noise spectra of a wind tunnel contains certain discrete characteristic frequencies masked with a hum which is of random nature. Time domain simulation results are shown in Fig. 6. The squared residual noise diminishes quickly to near-zero values for the ANC system with RLS adaptive filter as can be seen from Fig. 6(c). The algorithm converges to the desired value in about 30 iterations. In the case of the system with the LMS algorithm, the system becomes unstable, Fig. 6(a). Instead of diminishing, the squared residual noise keeps on increasing in each iteration, making the system unstable. For this kind of noise, the robustness of RLS algorithm prevented the system from becoming unstable. However, the frequency domain simulation results are less bright than those of the previous two cases. As can be seen from Fig. 6(b) and (d) which shows that although considerable attenuation is obtained at a number of discrete frequencies, moderate intensification occurs in certain areas of the frequency domain, with a maximum of 20 dB. The maximum simulated attenuation was 25 dB.

2.2 EFFECT OF PLANT MODEL ON THE SIMULATION RESULTS

In all the simulations presented here, the system model of the acoustic plant, i.e. the duct carrying the noise, is assumed to be a single delay plant. This assumption is reasonable to some extent as long as there are no end reflections of the transmitted noise through the duct. This is particularly true for a closed loop wind tunnel which does not have any end. However, sudden

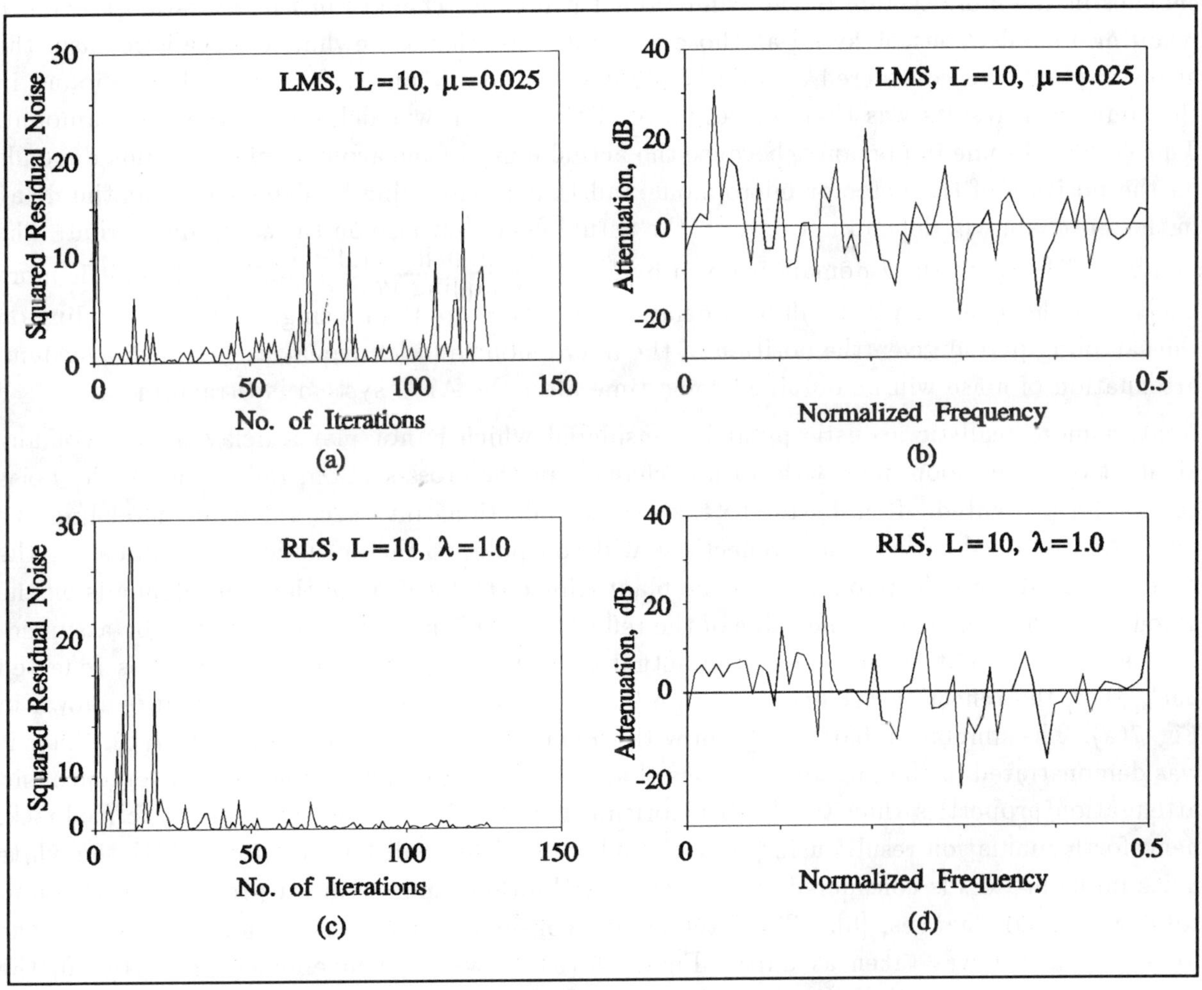

Figure 6. Attenuation for the ANC system with discrete frequency noise masked with white noise.

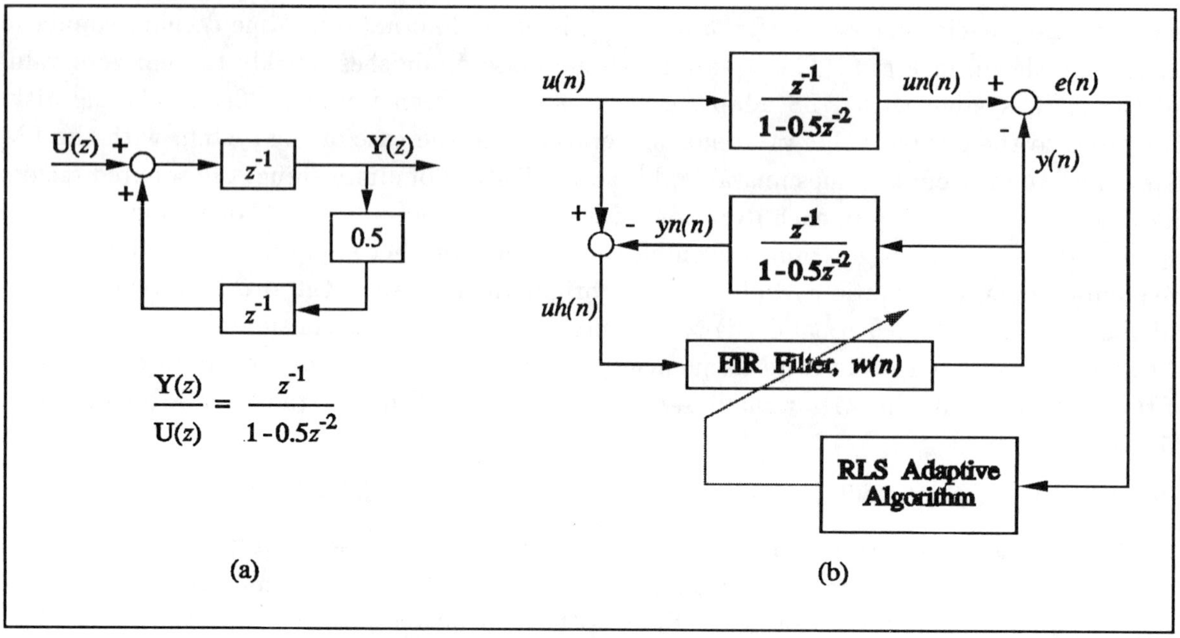

Figure 7. Modified acoustic plant and simulation diagram.

changes in the cross-section of the wind tunnel will cause changes in the acoustic impedance resulting in reflections of sound at those sections. For this somewhat less realistic case, the acoustic plant was considered as a delay plant with 5 delay units. The only change found in the simulation results was that the response of the system was delayed by the same amount. This case is of some importance, because the actual delay of the acoustic plant in time depends on the position of the detector microphone and that of canceling loudspeaker. But the delay in the z-domain will depend not only on the actual delay but also on the sampling period. The number of delays in the z-domain is given by $\delta = \frac{\text{Actual Delay in Time}}{\text{Sampling Period}}$. Therefore, it is more likely that the acoustic plant will be a delay plant with more than a single delay, depending on the sampling period given the position of the microphone and loudspeaker. For such a system, attenuation of noise will be obtained some time after the ANC system is turned on.

Next, a more realistic acoustic plant is considered which is not just a delay plant. For any air duct or closed loop duct with abrupt changes in the cross-section, reflections of the noise cannot be prevented. It is better to take these reflections into account when modeling the duct. As discussed earlier, these reflections will add poles to the acoustic plant indicating the need to consider a pole-zero model of the plant. Since the location of the poles depends on the amount of reflected noise, some value of the reflection coefficient of the duct has to be assumed. If it is assumed that 50 percent of the actual noise reaching the end of the duct is reflected back, then the reflection coefficient will be 0.5. This model of the acoustic plant is shown in Fig. 7(a). The simulation block diagram with the modified plant is shown in Fig. 7(b). Since it was demonstrated in the previous sections that the RLS algorithm has better convergence and attenuation properties than the LMS algorithm for an ANC system with acoustic feedback, henceforth simulation results using only the RLS algorithm will be presented. Only the white noise input process is considered, since results with other kinds of input processes were found to show similar changes, [6]. The filter length considered here was 5 and the value of the forgetting factor was taken as unity. Figure 8 (a) shows the convergence properties of the squared output noise. The rate of convergence is slower than that of the ANC system with single delay plant, Fig. 5(c). The reason for the slower convergence rate may be due to fact that the additional poles of the acoustic plant that are added to the total system cause the adaptation to take place at a slower pace which in turn reduces the rate of convergence of the ANC system. Figure 8(b) shows the frequency domain attenuation for this simulation. As can be seen, some intensification of noise occurs over the lower and upper frequency ranges. Considerable attenuation is achieved over the middle range of frequencies with a maximum of 15 dB.

2.3 EFFECT OF FILTER ORDER ON SYSTEM PERFORMANCE

In previous sections, it was demonstrated that the RLS algorithm is more stable than the LMS algorithm for an adaptive ANC system. The rate of convergence of the RLS algorithm was found to be faster than that of the LMS algorithm for the types of inputs considered. Frequency domain attenuation was better for the ANC system with the RLS adaptive filter than with the LMS adaptive filter in most cases, particularly in the case of discrete frequencies mixed with random noise. Considering all these factors, the RLS algorithm is adopted for the desired ANC system. Next it is necessary to determine the length of the filter. To see the effect of filter length on the convergence properties in the time domain and on the maximum attainable attenuation in the frequency domain of the ANC system, the simulation program was run for different filter lengths from 3 to 20. The results are shown in Fig. 9 for three types of noise inputs. The forgetting factor of the RLS algorithm was kept unity in all these cases. All the simulations done in this section assume the ANC system with modified acoustic plant as shown in Fig. 7. Figure 9(a) shows the variation of the rate of convergence of noise output with filter length L for discrete frequency noise. The rate of convergence has been defined by the inverse of the number of iterations needed for the squared system output noise averaged over 5 iterations to converge to a particular value δ. The value of δ was chosen to be 0.1 for all the cases presented here. Although the figure shows some abrupt changes, it can be concluded that the rate of convergence generally decreases with increasing filter length, i.e. more iterations are required for a desired convergence rate if a higher order filter is used.

This is likely to occur since the use of a higher order filter involves adaptation of a large number of filter coefficients. As a result, the overall adaptation rate will slow down because a longer period of time is needed for the algorithm to adapt the coefficients. Subsequently it results in a slower convergence rate. Figure 9(b) shows the change in maximum attainable noise

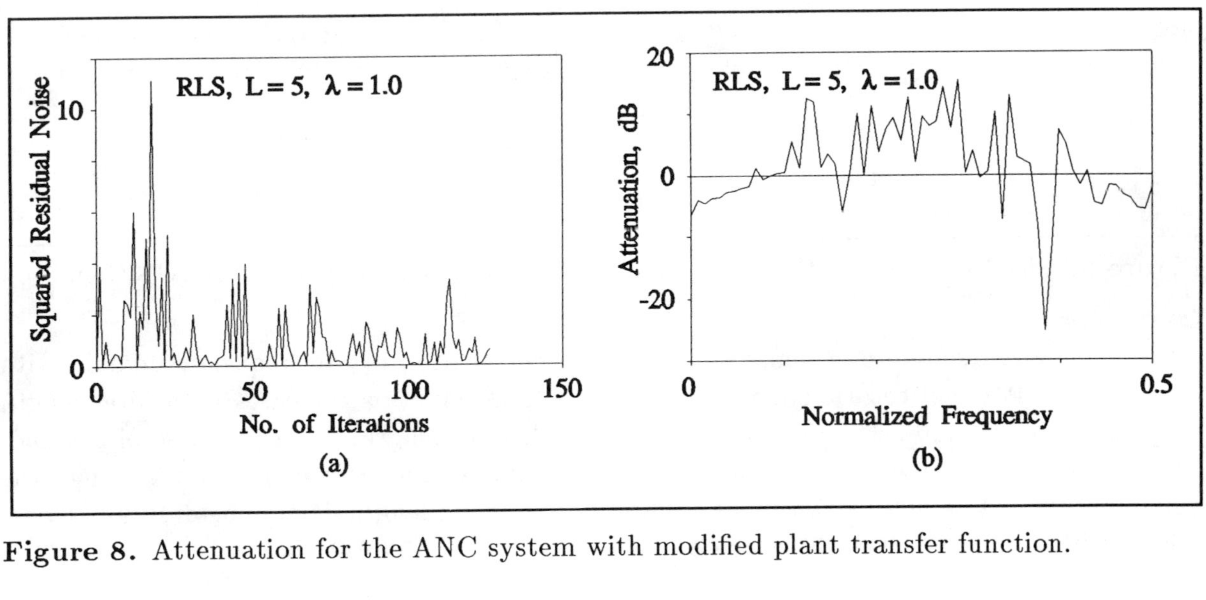

Figure 8. Attenuation for the ANC system with modified plant transfer function.

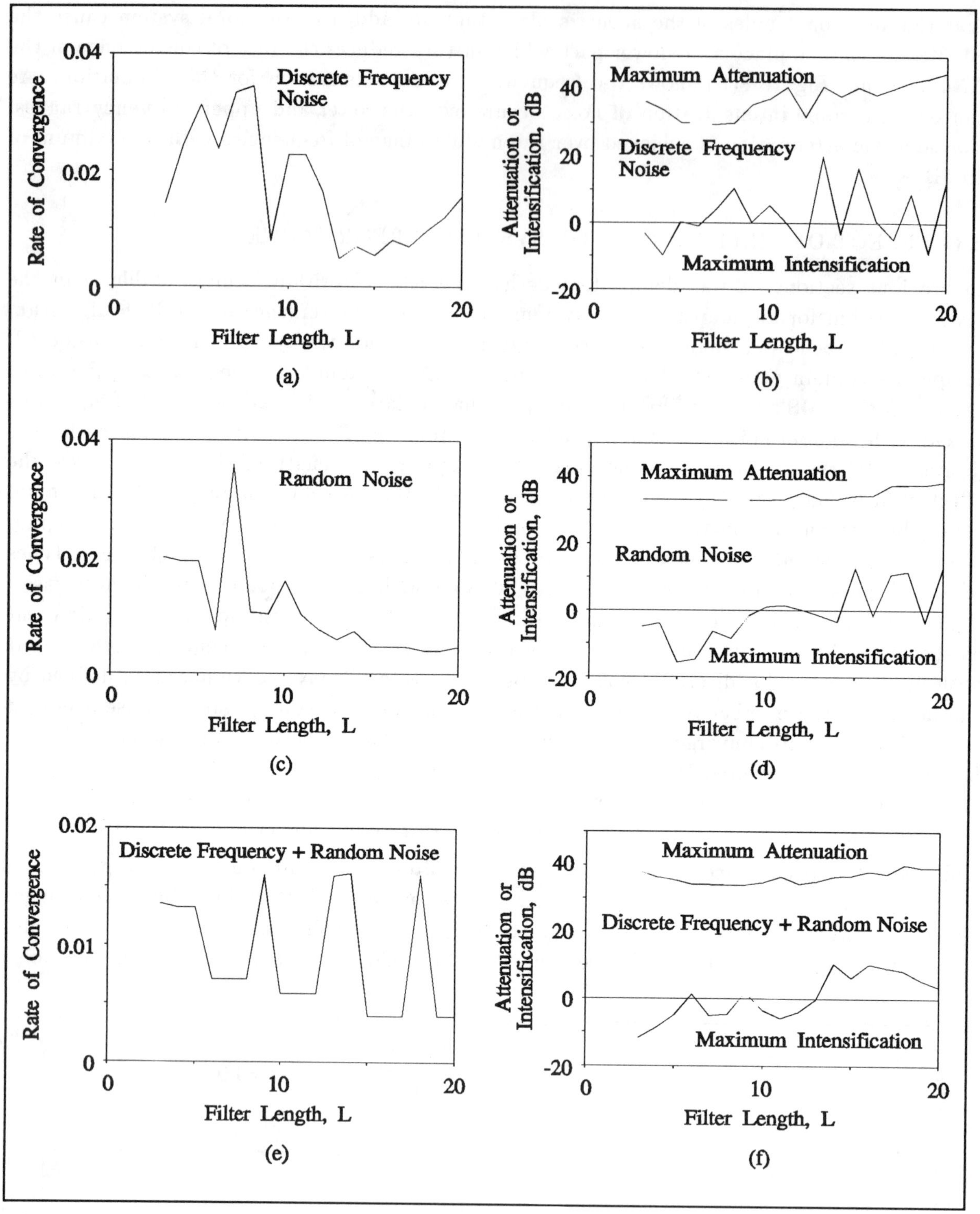

Figure 9. Effect of filter length on the convergence and attenuation of the ANC system.

attenuation and maximum accompanying noise intensification in the frequency domain with filter length. It is noticed that maximum attainable attenuation increases with filter length whereas maximum intensification varies greatly with filter length. Hence the use of a higher order filter may not intensify noise at all for some value of the filter length, L. Although the maximum attainable attenuation increases slowly with filter length, it may be desirable to use a higher order filter in some cases to achieve a particular attenuation.

Figure 9(c) shows the change in convergence rate with filter length for random noise input. In this case, higher order filters are again found to have a slower convergence rate. Similar results were obtained for frequency domain attenuation and intensification with varying filter length. Figure 9(e) shows variations of convergence rate with filter length L for combined discrete frequency and random noise. Once again the rate of convergence of the ANC system gets slower with the increase in filter length. Again, maximum attainable attenuation increases slowly with L.

Taking all three kinds of inputs into consideration, it was found that a filter order of 10 or less can give fast convergence as well as considerable attenuation without much intensification. A filter of 8 is finally chosen for this purpose. With the selection of the length of adaptive FIR filter, the initial design process of the ANC system comes to an end. To verify the performance of the designed system, it was simulated with filter length 8 and the results for combined noise (i.e. discrete frequency+random) input are shown in Fig. 10. As can be seen from both the time and frequency domain simulation results, desired attenuation is obtained with reasonable rate of convergence.

3. CONCLUSIONS

The need for an adaptive active control system was realized, since a wind tunnel is subjected to variations in air velocity, temperature, air turbulence, and some other factors such as non-linearity. Among many adaptive algorithms, the LMS algorithm, which is the simplest one, has been used in an ANC system by some researchers. However, Eriksson's results, [4], showed instability in the ANC system with an FIR filter for random noise input. The RLS algorithm, although computationally more complex than the LMS algorithm, has better convergence and stability properties. The ANC system in the present work was simulated by using an FIR filter with an RLS algorithm for different inputs and for a number of plant models. Simulation results for the ANC system with acoustic feedback showed better robustness when used with the RLS algorithm than with the LMS algorithm for all types of inputs. Overall attenuation in the frequency domain was better in the case of the RLS adaptive algorithm.

Simulation results with a more realistic plant model and an RLS adaptive algorithm showed a slower convergence rate than the case with an acoustic plant as a delay plant. However the

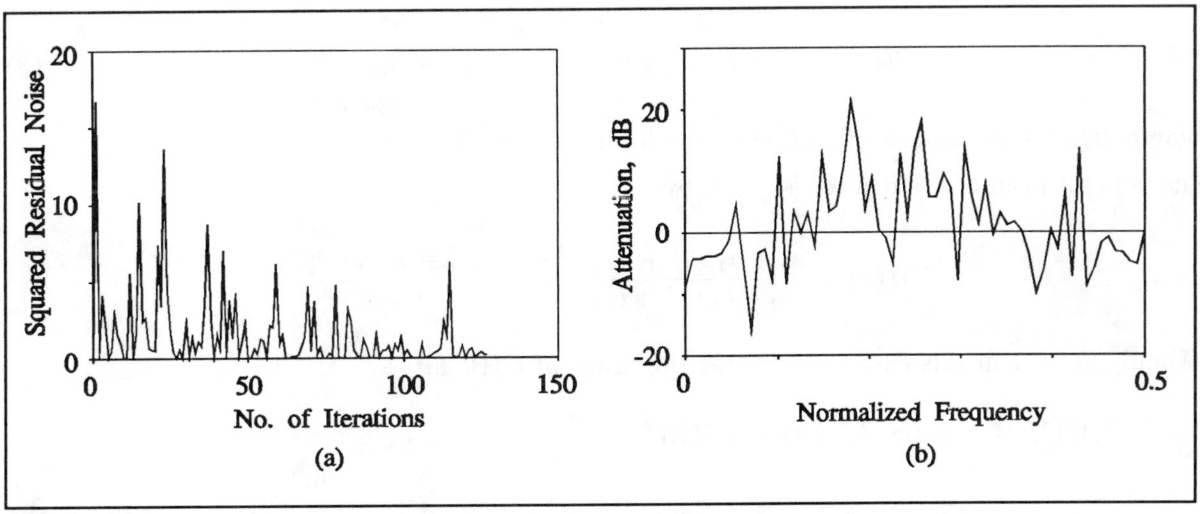

Figure 10. Simulation results for the final design of the ANC system.

attenuation properties were satisfactory for the simulated system with the modified plant. The effect of filter length on the rate of convergence and attenuation was studied. It was found that the rate of convergence decreases with increase in filter length whereas the attenuation increases with increase in filter length. A compromise between convergence rate and attenuation was made and finally the length of the FIR adaptive filter was selected as 8. The final design of the ANC system was simulated and found to have a reasonable convergence rate and good attenuation properties for an input containing discrete frequencies and random noise.

4. REFERENCES

1. Burgess, J.C., (1981). "Active adaptive sound control in a duct: A computer simulation." *J. Acoust. Soc. Am.*, 70(3): 715–726.

2. Poole, J.H.B., and Leventhall, H.G., (1976). "An experimental study of Swinbanks' method of active attenuation of sound in ducts." *J. Sound Vib.*, 49(2): 257–266.

3. Eriksson, L.J., et al, (1987). "The selection and application of an IIR filter for use in active sound attenuation." *IEEE Transactions on ASSP*, ASSP-35(4): 433–437.

4. Eriksson, L.J., (1985). "Active sound attenuation using adaptive digital signal processing techniques." *Ph.D. Thesis*, University of Wisconsin-Madison, Madison, Wisconsin.

5. Elliot, S.J., and Nelson, P.A., (1984). "Models for describing active noise control in ducts." *I.S.V.R. Technical Report No. 127.*

6. Sohel, A., (1990). "Design of an Adaptive Control System for Active Attenuation of Wind Tunnel Noise." MS thesis, Florida State University, Tallahassee, Florida.

APPENDIX A. MODELING OF THE PLANT

In a system identification problem, any unknown plant is usually modeled in z-domain either by an all-zero transfer function which has a finite impulse response (FIR) or by a pole-zero transfer function which has an infinite impulse response (IIR). The general form of an FIR filter in z-domain is given below:

$$H(z) = a_0 + a_1 z^{-1} + a_2 z^{-2} + \dots\dots\dots + a_m z^{-m} \tag{3}$$

where $H(z)$ is the transfer function of the filter having a length m.
The general form of a IIR filter is given by

$$H(z) = \frac{a_0 + a_1 z^{-1} + a_2 z^{-2} + \dots\dots\dots + a_m z^{-m}}{b_0 + b_1 z^{-1} + b_2 z^{-2} + \dots\dots\dots + b_n z^{-n}} \tag{4}$$

Usually, $b_0 = 1$ in this case. n indicates the order of a IIR filter.

APPENDIX B. RLS ALGORITHM

The final form of the standard RLS algorithm is presented below. The derivation of the following can be found in any standard text book on adaptive filters.

B.1 NOTATION

n = Number of iterations

$\mathbf{P}(n)$ = Inverse of the correlation matrix

δ = A small positive constant ¡¡ 1

$\mathbf{w}(n)$ = Tap weight vector

$\mathbf{k}(n)$ = Gain vector

$e(n)$ = A priori estimation error

λ = Forgetting factor

$\mathbf{u}(n)$ = Tap input vector at time n

$d(n)$ = Desired or unknown plant output

B.2 INITIALIZATION

$$\mathbf{P}(0) = \delta^{-1}\mathbf{I} \tag{5}$$

$$\mathbf{w}(0) = \mathbf{0} \tag{6}$$

B.3 ITERATIVE MAIN LOOP

$$\mathbf{k}(n) = \frac{\lambda^{-1}\mathbf{P}(n-1)\mathbf{u}(n)}{1 + \lambda^{-1}\mathbf{u}^T(n)\mathbf{P}(n-1)\mathbf{u}(n)} \tag{7}$$

$$e(n) = d(n) - \mathbf{w}^T(n-1)\mathbf{u}(n) \tag{8}$$

$$\mathbf{w}(n) = \mathbf{w}(n-1) + \mathbf{k}(n)e(n) \tag{9}$$

$$\mathbf{P}(n) = \lambda^{-1}\mathbf{P}(n-1) - \lambda^{-1}\mathbf{k}(n)\mathbf{u}^T(n)\mathbf{P}(n-1) \tag{10}$$

The computational cost of the RLS algorithm may be reduced by defining

$$\mathbf{x}(n) = \lambda^{-1}\mathbf{P}(n-1)\mathbf{u}(n) \tag{11}$$

and then by rewriting first and fourth lines of main loop as

$$\mathbf{k}(n) = [1 + \mathbf{u}^T(n)\mathbf{x}(n)]^{-1}\mathbf{x}(n) \tag{12}$$

$$\mathbf{P}(n) = \lambda^{-1}\mathbf{P}(n-1) - \mathbf{k}(n)\mathbf{x}^T(n) \tag{13}$$

Here, in the recursion for $\mathbf{P}(n)$, the Hermitian property of $\mathbf{P}(n-1)$ has been used.

APPENDIX C. LMS ALGORITHM

The final form of the LMS adaptive algorithm is given below.

C.1 PARAMETERS

M = Number of taps

μ = Convergence factor, where $0 < \mu < \dfrac{2}{\text{Total input power}}$ and usually, $\mu << 1$

C.2 INITIALIZATION

$$\mathbf{w}(0) = \mathbf{0} \tag{14}$$

C.3 DATA

a. Given:

$\mathbf{u}(n) = M \times 1$ tap input vector at time n

$d(n) =$ Desired response at time n

b. To be computed:

$\mathbf{w}(n+1) =$ Estimate of tap-weight vector at time $n+1$.

Computation for $n = 0, 1, 2, \ldots$

$$e(n) = d(n) - \mathbf{w}^T(n)\mathbf{u}(n) \tag{15}$$

$$\mathbf{w}(n+1) = \mathbf{w}(n) + \mu\mathbf{u}(n)e(n) \tag{16}$$

ACTIVE CONTROL OF SURGE IN A GAS TURBINE ENGINE

M. Harper[1]

ABSTRACT

We have carried out a series of experiments on a small, robust 60Hp engine with a centrifugal compressor, with the aim of investigating the potential for active control of surge. The work was motivated by a theoretical study [1] suggesting that surge might be initiated when the Helmholtz acoustic resonance mode of an engine becomes negatively damped, and that this mode should be amenable to linear active control.

We have discovered that the Helmholtz mode is not responsible for surge in our test engine, and that it is not sufficient to treat the engine as a linear system. Instead, surge appears to result from a compressor instability akin to the 'rotating stall' reported in axial compressors.

Nonetheless, a single-channel control system was able to suppress surge at the normal surge line. We present a simple nonlinear model to explain how this comes about.

1. INTRODUCTION

This work was motivated by a paper by Epstein, Ffowcs Williams and Greitzer [1] in which they presented a simple model of the unsteady dynamics of a gas turbine engine. They worked with the linearised equations of state of their model, and concluded that surge could be initiated when the model's Helmholtz mode of oscillation became negatively damped. This would occur when the slope of the compressor characteristic became positive at sufficiently small mass flow. They showed that by using linear single-channel feedback to affect unsteady plenum pressure, extra damping of the Helmholtz mode could be introduced, rendering their model engine stable down to lower mass flows.

[1]Topexpress Limited, Poseidon House, Castle Park, Cambridge, CB3 0RD, England.

The theory has been tested in the laboratory on turbochargers discharging into a plenum (representing the combustion chamber) with an outflow valve (representing the turbine in a real engine). Both Ffowcs Williams and Huang [2] and Pinsley et al [3] found that the control strategy did indeed extend the stable operating range down to significantly lower mass flows.

The next step was to apply the method to a working engine. Ffowcs Williams and Graham [4] selected a Rover IS/60 engine, a 60-HP unit with a centrifugal compressor (Figure 1). The engine had a mechanical droop governor. They instrumented the engine with four pressure sensors at the impeller intake (station 1), four at diffuser inlet (station 2) and four at diffuser delivery (station 3).

The most striking difference between conditions in the laboratory models and those in the Rover engine was the level of noise. While the plenum of a laboratory rig [2] showed very low levels of pressure fluctuation when not in surge, the station 3 signals in the Rover engine showed a standard deviation of $^1/_2$ psi over the frequency band 0 - 1kHz. We shall see that this is relevant to their different responses to control.

Ffowcs Williams and Graham used a narrowband control system to suppress surge induced by transient overfuelling at a normally-stable operating condition. They found, however, that it could not affect surge occurring at the normal surge line, and concluded that a broadband system was required. It is the results of using broadband control that we report here.

2. THE LINEAR ENGINE MODEL

Figure 2 shows a schematic of the simple lumped-parameter model of the compression system. We work with dimensionless variables, as in references [1], [2] and [3].

These are

ϕ_1 unsteady compressor mass flow

ϕ_2 unsteady throttle mass flow

ψ_p unsteady plenum pressure

$C(\phi_1)$	compressor pressure rise
$T(\phi_2)$	throttle pressure drop
ϕ_N	equivalent noisy mass flow from combustion
ϕ_C	control mass flow into plenum
B	Greitzer's B parameter

The linearised equations for the system are (c.f. [1], [2]):

$$\dot{\phi}_1 = B\left(C'\,\phi_1 - \psi_P\right) \quad (1)$$

$$\dot{\psi}_P = \frac{1}{B}\left(\phi_1 - \phi_2 + \phi_C + \phi_N\right) \quad (2)$$

$$\psi_P = T'\phi_2 \quad (3)$$

where apostrophes indicate differentiation with respect to the argument.

By imposing a control law of the form

$$\phi_C = F\left(\psi_P\right) \quad (4)$$

we can influence stability. The control law suggested in [1] is

$$\phi_C = z\,\psi_P \quad (5)$$

which results in the modified eigenvalues

$$s = \frac{1}{2}\left\{BC' - \frac{1}{BT'} + \frac{z}{B} \pm \left[\left(BC' - \frac{1}{BT'} + \frac{z}{B}\right)^2 + 4C'\left(\frac{1}{T'} - z\right) - 4\right]^{\frac{1}{2}}\right\} \quad (6)$$

where 's' is the Laplace transform variable.

Negative real values of z thus tend to reduce the real part of the eigenvalues. We shall call this 'control strategy I'.

However this is not the only control law that we can envisage; for example, an alternative strategy suggested by the form of equation (4) is

$$\phi_c (s - BC') = zs\psi_P \tag{7}$$

where z is again a real constant. This control law however has the disadvantage that it must be tuned to 'follow' the engine's operating point, since it involves the slope of the compressor characteristic. The resulting eigenvalues are given by

$$s = \frac{1}{2}\left\{BC' - \frac{1}{BT'} + \frac{z}{B} \pm \left[\left(BC' - \frac{1}{BT'} + \frac{z}{B}\right)^2 + 4\left(\frac{C'}{T'} - 1\right)\right]^{\frac{1}{2}}\right\} \tag{8}$$

so that once again, negative values of z tend to stabilise the system. We shall call this 'control strategy II'. The two strategies differ in their effects on the movement of the eigenvalues in the complex plane as feedback gain is increased. We therefore planned to try both strategies, since for the same z they will affect the pressure spectrum in different ways.

3. EXPERIMENTS

3.1 CONTROL SYSTEM

The feedback path used for control is shown in Figure 3. The digital filter was programmed to implement the required control law.

3.2 EFFECT OF THE ACTUATOR

The injection into the plenum of a mean flow of air through the actuator may be expected to affect the engine parameters at which surge occurs, particularly fuel flow. However, it made no difference to the position of the surge line on the compressor map, but <u>did</u> make a considerable difference to the spectrum of unsteady pressure in the plenum. This is evident in Figure 4, which shows the spectra just before surge, without (solid line) and with (broken line) the actuator attached but not driven. The Helmholtz mode between 40 and 80Hz, and a higher-order mode at 260Hz, are evidently strongly damped by the presence of the actuator. Although we have not ascertained exactly how this damping arises, the control of self-excited resonances by mechanical damping is a recognised phenomenon [5] and has been successfully applied to the Helmholtz mode of a turbocharger rig by Gysling et al [6].

We regard the fact that this strong damping does not affect the position of the surge line as strong evidence that the Helmholtz mode is not responsible for surge in our test engine.

3.3 EFFECTS OF CONTROL ON PLENUM PRESSURE SPECTRUM

We first tested our control system with the engine at medium load, well away from the surge line. We applied strategy I control with a dimensionless feedback gain (see section 2) of (-0.2). The effect on the plenum pressure spectrum can be seen in Figure 5: the solid line is the spectrum without control, the dotted is with control. Control produces a 5dB decrease at low frequencies; its effect diminishes with frequency until, above about 60Hz, it makes no discernible difference.

We next applied strategy II control. There was no significant effect below about 30Hz, and a maximum of 1dB suppression between 30 and 80Hz.

In both cases, these effects were accurately predicted by linear theory: away from the surge line, the engine responds linearly.

3.4 EFFECT ON THE SURGE LINE

We tested the effect of the filters by first bringing the engine to an operating point at a safe margin from the surge line; switching control on; and then very slowly increasing the engine load to reduce mass flow and so bring the engine to the surge line.

Trials were carried out over a period of days, during which the engine was repeatedly surged, with and without control. The results are summarised by figure 6, which shows corrected speed plotted against corrected mass flow at surge. We have used different symbols for surge with no control ('o'), strategy I control ('x'), and strategy II control ('*'). An estimate of the uncontrolled surge line is drawn in as a dashed line. Our expected uncertainty in measured mass flow is ±1%. In view of this, it would appear from figure 6 that strategy I control is moving it the wrong way (i.e. towards higher mass flow) by at least 1%.

Strategy II control produced different results. Designing with the steady-state value of (C') did not appear to affect the engine's behaviour at surge; designing with a value of 1% of this produced radically new behaviour. As the surge line was reached, instead of going into continuous surge the engine underwent what appeared to be single cycles of surge every few seconds; figure 7 shows a typical time-history of plenum pressure with surge cycle events spaced by about 2.5 seconds. On one occasion we continued to raise engine load after the onset of this behaviour; the single surge cycles appeared to become gradually more frequent. We did not continue the experiment to discover at how low a mass flow the surge process became continuous because of the risk of damage to the engine.

In figure 8 we show corresponding time histories of plenum pressure (lower trace), and transducer voltage drive (upper trace), during a 'single surge cycle' event. Note that a positive voltage tends to close the actuator air valve: figure 8 shows that the control system reacts to the fall in plenum pressure by actually increasing the flow of air through the actuator into the plenum, the reverse of the action that a conventional bleed system would take in order to move the engine away from surge.

3.5 BEHAVIOUR OF THE COMPRESSOR CLOSE TO SURGE

An oscillation can be detected on our station 2 (diffuser channel inlet) sensors at any engine load; its relative phase between different channels varies roughly as two degrees per degree of angular position around the diffuser. As load is increased, the oscillation grows to dominate over a background of noise of up to 1psi and rises in frequency from 160Hz at idle to about 186Hz at the surge line. Figure 9 shows this evolution as a growth in the coherence, near the resonance frequency, between the signals from pressure transducers in different diffuser channels. At low load (9(a)), coherence reaches a peak of 0.5 at 160Hz: the resonance no more than equals uncorrelated flow noise in time-average power. At medium load, the peak is about 0.62 at 170Hz. At high load, just before surge, it has reached 0.85 at a frequency of 186Hz. It is now almost six times the power of the noise. This rise in frequency accompanying a rapid rise in amplitude strongly suggests a resonance becoming progressively more lightly damped. The oscillation occurs in bursts, and the onset of surge is always immediately preceded by a rapid growth of the oscillation. This is persuasive evidence that surge is triggered by stall in the compressor.

3.6 RESPONSE TO ASYNCHRONOUS FORCING

We noted in section 3.2 that the reduction in unsteady plenum pressure brought about by the passive damping effect of our actuator did not affect the surge point. We also carried out experiments to determine whether increasing the broadband pressure fluctuation had any effect. We did this by applying large asynchronous drive signals to our actuator. We found that the engine could be surged well short of the normal surge line; the required forcing level decreased as the normal surge line was approached.

4. NONLINEAR ENGINE MODEL

It is evident from section 3 that the engine's behaviour is highly non-linear. We present here an extension of the analysis of the simple model of section 2 to include nonlinear effects, and show that it is capable of explaining the observed phenomena.

We simplify the model of section 2 by prescribing a fixed turbine mass flow. This is in any case a reasonable approximation, since the turbine characteristic in our real engine is extremely steep compared to the compressor. The equations of our simplified model are

$$\dot{\phi} = B\,(C\,(\phi) - \psi) \tag{9}$$

$$\dot{\psi} = \frac{1}{B}\,(\phi - \phi_T) \tag{10}$$

where ϕ is compressor mass flow, ψ is plenum pressure and ϕ_T is the fixed turbine mass flow. Note that we may combine these equations to obtain

$$\ddot{\phi}_\Delta - BC'\,\dot{\phi}_\Delta + \frac{1}{B}\,\phi_\Delta = 0 \tag{11}$$

where

$$\phi_\Delta = \phi - \phi_T$$

which describes an oscillator with nonlinear damping given by (-BC').

Figure 10 shows the two-dimensional phase space of ψ plotted against ϕ with the compressor characteristic $C(\phi)$ superimposed, and ϕ_T indicated by a vertical dashed line. There is an equilibrium point at $(\phi_T, C(\phi_T))$. The trajectories will tend to circulate anti-clockwise about the equilibrium point. The nature of the equilibrium point and the possible existence of limit cycles surrounding it, are governed by the exact form of $C(\phi)$. Let ϕ_0 be the local maximum of $C(\phi)$. When $\phi_T < \phi_0$ the equilibrium is unstable; when $\phi_T > \phi_0$ it is stable; when $\phi_T = \phi_0$ its behaviour is more complex, as we shall see.

We now wish to know whether the two-dimensional non-linear model can describe behaviour observed in the real engine:

1) the engine always surges abruptly, going directly into a very large limit cycle rather than experiencing the exponential growth of an initially linear oscillation in plenum pressure, as predicted in [1];

2) the engine can be caused to surge from a normally-stable equilibrium by sufficiently large asynchronous forcing; the level required decreasing as the surge line is approached.

3) the engine can be returned from surge to stable equilibrium by a simple, linear feedback system.

To show that the 2D model can indeed represent these phenomena, we will first describe an arrangement of attractors in phase space which can give these phenomena, and then discuss how this arrangement might be achieved, given certain conditions on the form of (C).

In figure 10, we show the phase space, with a linearly-stable operating point surrounded first by an unstable periodic orbit and outside that by a stable limit cycle. Trajectories inside the unstable orbit always lead to the stable operating point at $(\phi_T, C(\phi_T))$; all others lead to the stable orbit. Provided the radius of the unstable orbit exceeds the largest displacements from equilibrium caused by the engine's internal forcing, the engine will operate stably; large transients applied to the engine may however cause the engine state to cross the unstable orbit, and it will then move out to the stable orbit, i.e. it will surge. The requirement of sudden surge at the surge line can be met if, as ϕ_T is decreased towards ϕ_0, the radius of the unstable orbit decreases towards zero while the stable orbit remains large. In this case, at some point before ϕ_T reaches ϕ_0, the radius of the unstable orbit will fall below the amplitude of the engine's internal forcing. As soon as the engine crosses the unstable orbit it will again move out to the stable orbit: thus the model could account for abrupt surge at a point on the characteristic with a significantly negative slope.

This shrinking of the unstable orbit can also explain how the engine can response to control as a linear system when it is far from the surge line: the engine's response will remain approximately linear as long as it remains well within the basin of attraction of the stable operating point. According to our model, this is necessarily the case when the engine is far from surge.

We shall make use of two pre-existing results in non-linear dynamics. The first result is that a Hopf bifurcation, with bifurcation parameter ϕ_T, occurs at $\phi_T = \phi_0$. As the throttle setting passes through this value, the equilibrium point changes its stability and a periodic orbit, a limit cycle, is born. The limit cycle amplitude will increase from zero as ϕ_T departs from ϕ_0. This is shown rigorously elsewhere [7]. It is also shown there that which way round the above sequence happens (i.e. whether for increasing or decreasing ϕ_T) and whether the attractors are stable or unstable, depends on the sign of the third derivative of $C(\phi_0)$. In our case we wish an unstable operating point to turn stable, and an unstable limit cycle to appear around it, as ϕ_T increases through ϕ_0. This requires $C'''(\phi_0) > 0$.

The second result is an unpublished result due to Villari, and is demonstrated elsewhere [7]. It states that, for a given $C(\phi)$ and ϕ_T, if for some $\phi > 0$,

$$C(\phi_T - \phi) < C(\phi_T + \phi)$$

then a stable orbit can exist outside the unstable. This requires $C(\phi)$ to be steeper at $(\phi_0 - \phi)$ than at $(\phi_0 + \phi)$ for some range of ϕ; this is compatible with the previous requirement that $C'''(\phi_0) > 0$.

5. DISCUSSION

The most significant points to arise from this work are as follows:

- the irrelevance of the Helmholtz acoustic mode of the engine to surge;
- the highly nonlinear behaviour of the engine, and of the effect of demonstrated control;
- the importance of the very high levels of intrinsic noise in both the diffuser and the plenum;
- the importance of the unsteady behaviour of the compressor.

Our interpretation of the results is that surge is normally initiated, not by the Helmholtz oscillation but by stalling in the compressor. This stalling is itself provoked by noise excursions in the engine, even when the time-averaged mass-flow through the compressor is not low enough to stall it: this accounts for the frequent but short bursts of rotating stall seen in the diffuser at mass flows close to the surge line. It also accounts for the way in which, after surge has been suppressed by our control system, it is re-initiated after a short interval.

Because surge is initiated by compressor stall, our single-channel control system, which is connected only to the plenum, cannot prevent its inception. It can, however, suppress surge after a single cycle. This may well have some utility as a 'rapid recovery' system; however, we believe the natural next step is to develop a control system aimed at the compressor. We have started work on such a system, and hope shortly to attempt control of the 'rotating stall' mode of the compressor.

6. CONCLUSIONS

The surge point of the test engine appears to be controlled, not by the acoustic Helmholtz mode of the engine, but by a stall phenomenon in the centrifugal compressor akin to the 'rotating stall' seen in axial compressors.

The normal operating point co-exists with a surge limit cycle over a wide range of operating conditions. The engine may be moved from normal, stable operation into surge by forcing with an unsteady pressure signal into the plenum; it may be returned from the surge cycle to stable operation using simple feedback control of the unsteady plenum pressure. Such control operates effectively at the normal surge line; it may be useful for rapid recovery from surge induced by operating transients. However, it does not appear capable of moving the surge line to significantly lower mass flows.

We believe that the way forward is to address the stall instability in the compressor. Work is currently underway, both to model the mechanism of the linkage of surge and stall, and to implement feedback control of the compressor instability itself.

7. ACKNOWLEDGEMENTS

The support of the US Office of Naval Research is gratefully acknowledged: this work has been funded through grant number N00014-87-J-1214.

The support and technical guidance of Professor J E Ffowcs Williams, Principal Investigator under the ONR grant, is also gratefully acknowledged.

The test engine was maintained and run by Noel Penny Turbines, Coventry, UK. The work could not have progressed so quickly without the able assistance of Mr Hugh Freeman (project engineer) and Mr Malcolm Dredge (technician).

Finally, the author would like to acknowledge the help of Dr David Allwright of Topexpress in establishing the conditions under which the nonlinear model of section 4 would be valid.

8. REFERENCES

1. A H EPSTEIN, J E FFOWCS WILLIAMS and E M GREITZER. Active Suppression of Compressor Instabilities. AIAA 10th Aeroacoustic Conference, paper no. AIAA-86-1994.

2. J E FFOWCS WILLIAMS and X Y HUANG. Active Stabilisation of Compressor Surge. J. Fluid Mech. 204, 245-262 (1989).

3. J E PINSLEY, G R GUENETTE, E M GREITZER and A H EPSTEIN. Active Stabilisation of Centrifugal Compressor Surge. 35th ASME International Gas Turbine Conference, 1990. Paper No. 90-GT-123.

4. J E FFOWCS WILLIAMS and W R GRAHAM. An Engine Demonstration of Active Surge Control. 35th ASME International Gas Turbine Conference, 1990. Paper No. 90-GT-224.

5. K R ASFAR. Quenching of Self-Excited Vibrations. Trans. ASME. JVASRD 111, No. 2, April 1989.

6. D L GYSLING, J DUGUNDJI, E M GREITZER and A H EPSTEIN. Dynamic Control of Centrifugal Compressor Surge using Tailored Structures. 35th ASME International Gas Turbine Conference, 1990. Paper No. 90-GT-122.

7. J E FFOWCS WILLIAMS, M HARPER and D J ALLWRIGHT. Report No. TOP/ONR-N00014-87-J-1214/90/1 to the US Office of Naval Research on the Active Control of Surge in a Gas Turbine Engine, Phase III. August 1990. Appendices I and II.

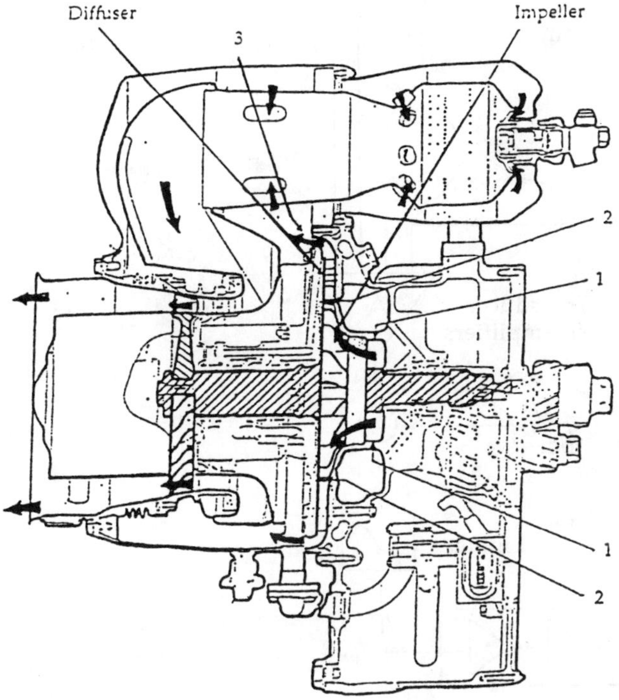

- Complete Set of Steady State Performance Measurements
- Pressure Measurements 1 - Four at 90° Spacing
  2 - Four at 90° Spacing
  3 - Four at 90° Spacing
- Casing Acceleration Measurements
- Sound Pressure Measurements at Engine Inlet and Outlet

Figure 1 Engine Sensors

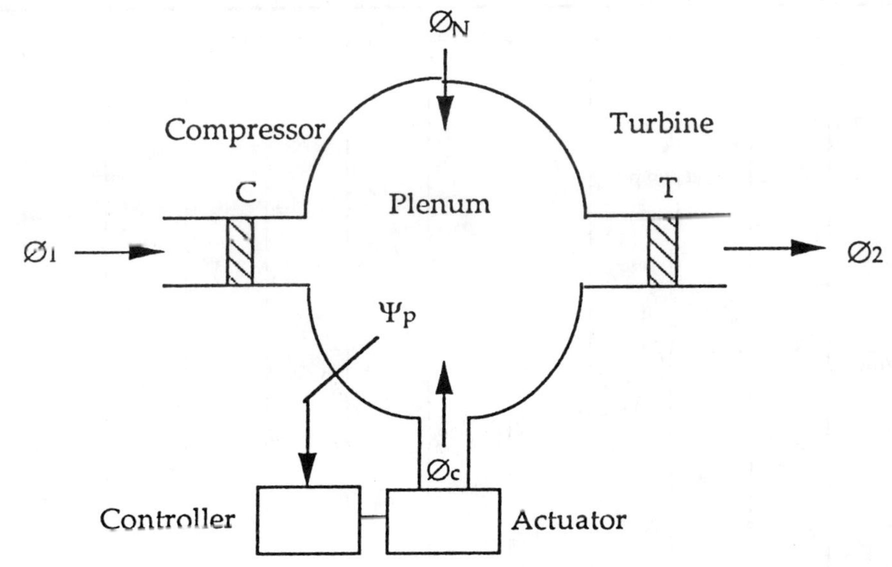

Figure 2 Lumped Parameter Engine Model

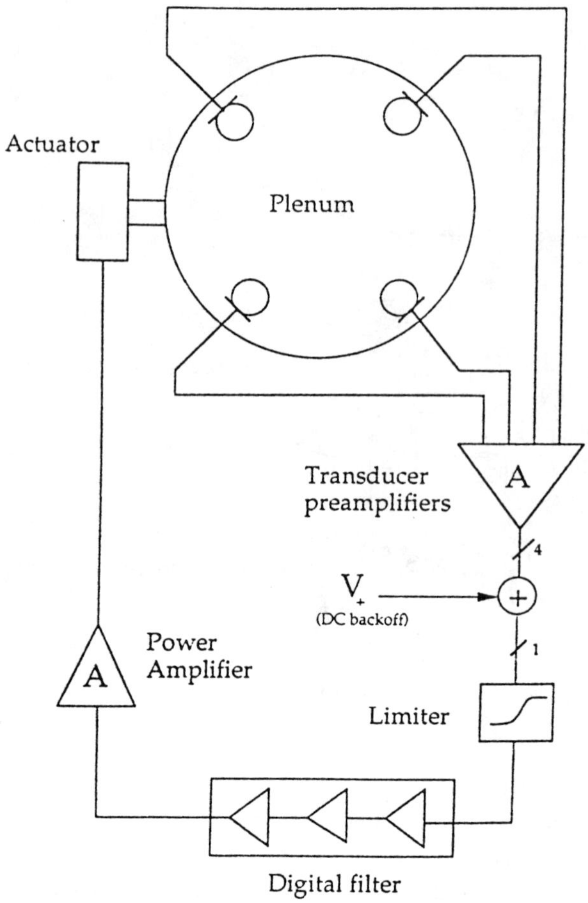

Figure 3 Block diagram of the feedback path

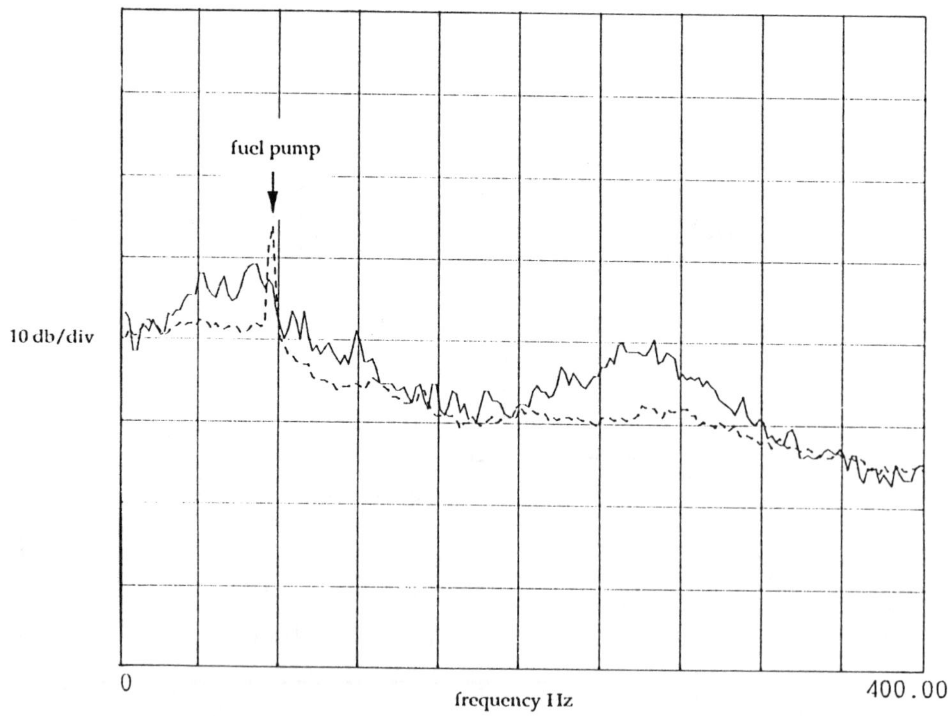

Figure 4 Plenum pressure spectrum at high load, without (solid line) and with (dashed line) the actuator fitted.

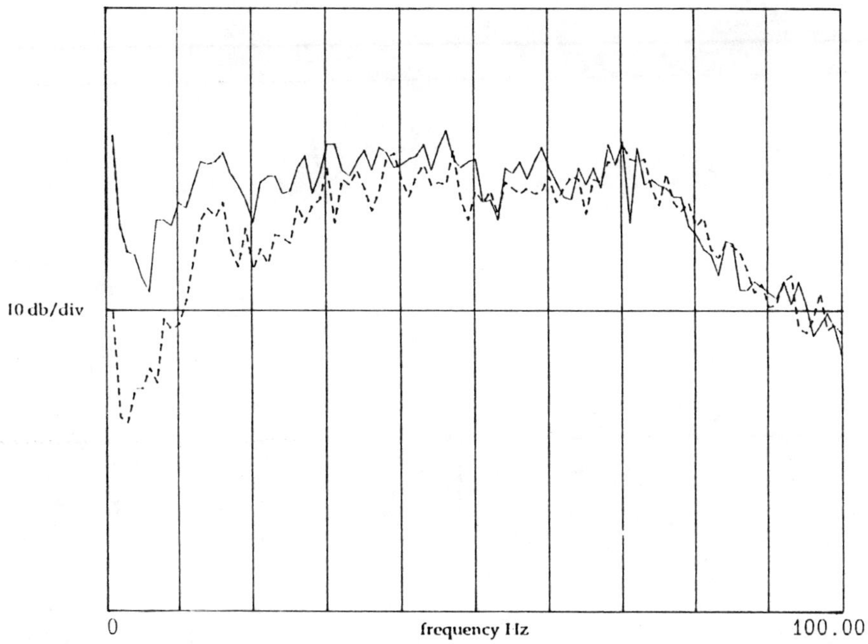

Figure 5 Effect of "Strategy I" control on plenum pressure spectrum.
Solid line: no control
dashed line: control on

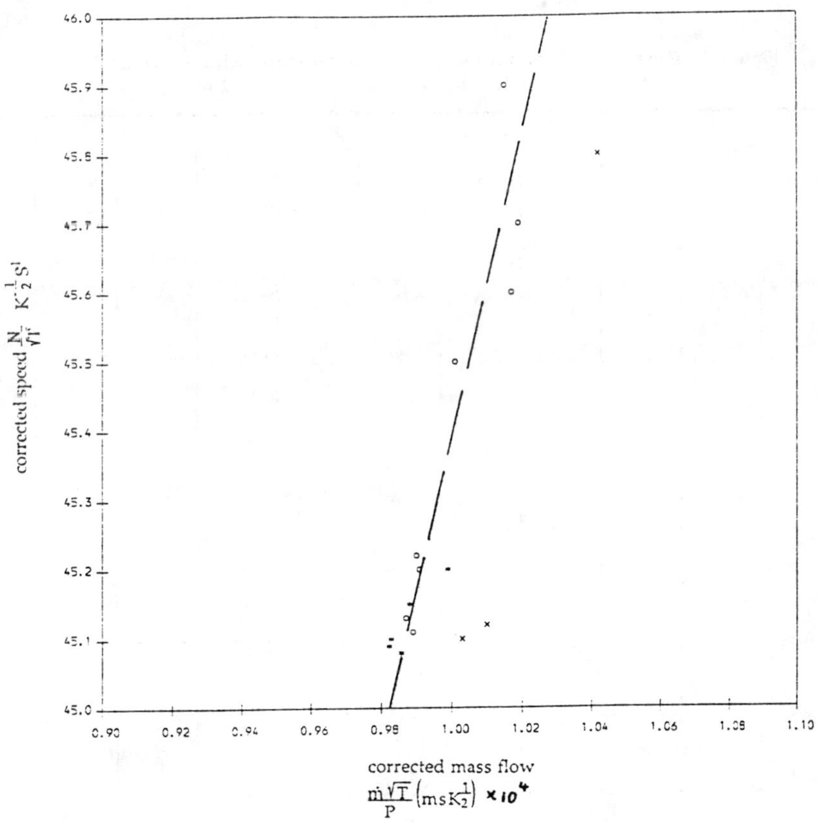

Figure 6 Effect of control on the surge line.
Symbols:
o no control
x "strategy I" control
* "strategy II" control
- - - normal surge line

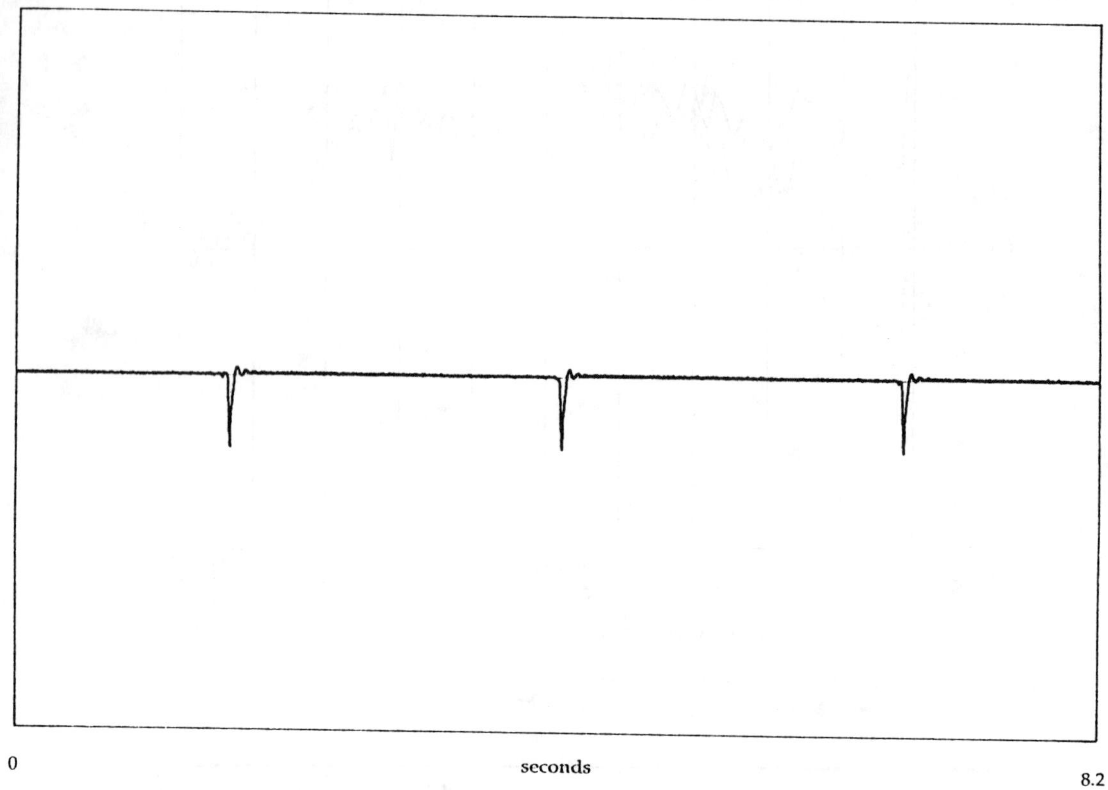

Figure 7 Time history of unsteady plenum pressure with "strategy II" control applied at the surge line. Time window 8.2 sec.

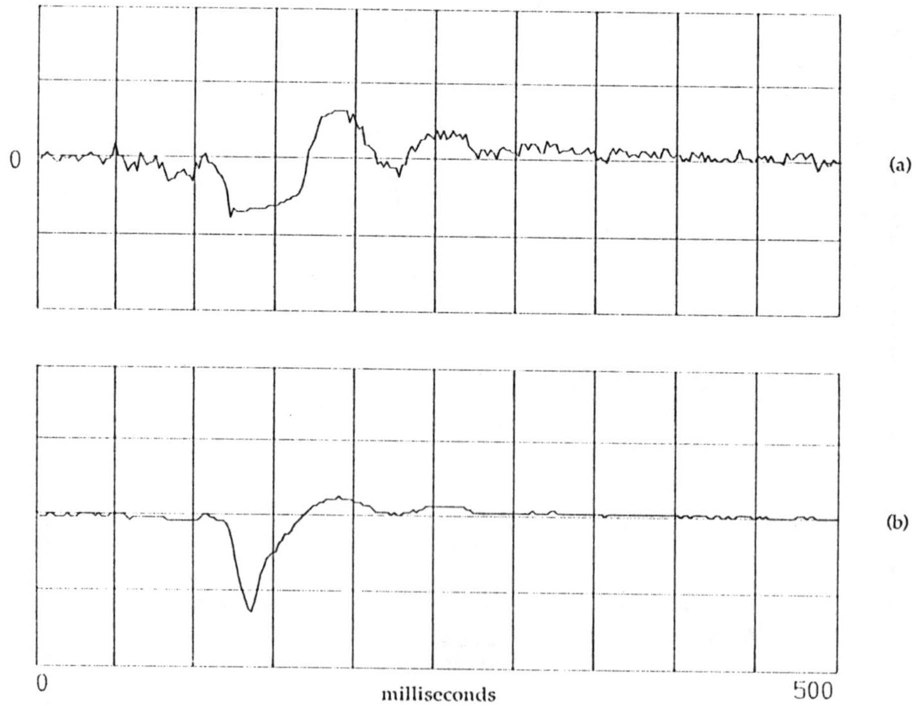

Figure 8 Time histories of actuator drive voltage and unsteady plenum pressure during a single-surge-cycle event.
(a) drive voltage
(b) pressure

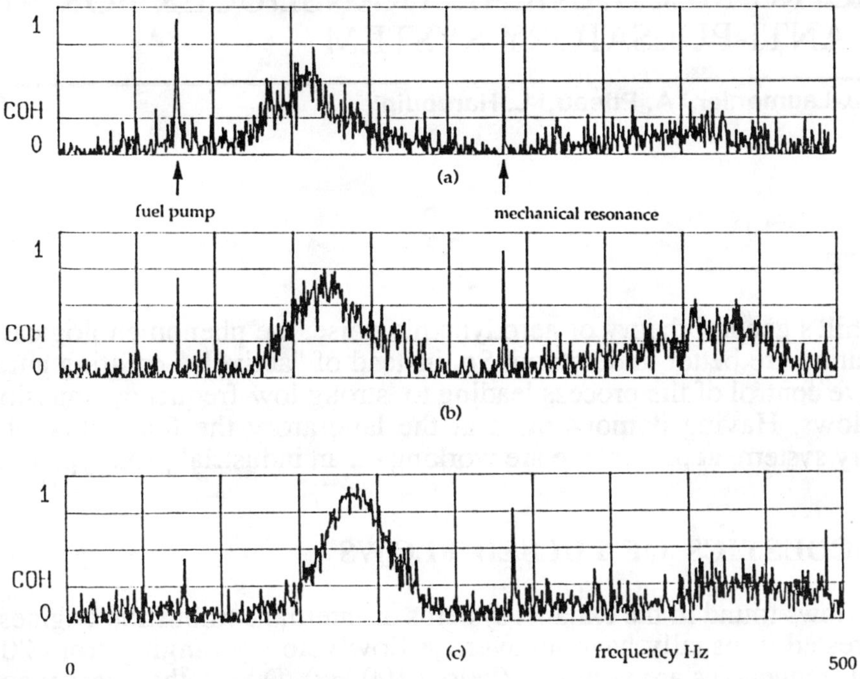

Figure 9 Coherence of pressure signals from two diffuser channels as a function of load:

(a) low load
(b) medium load
(c) high load

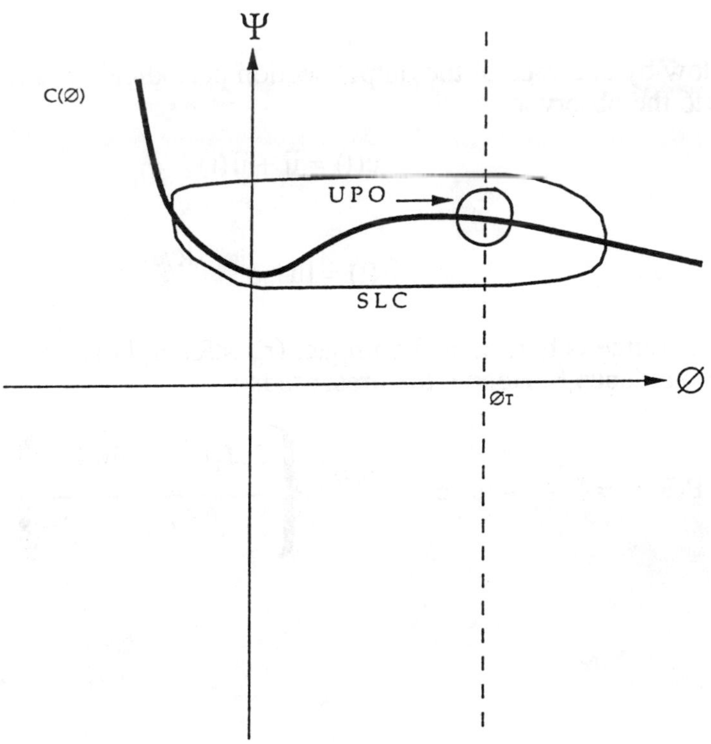

Figure 10 Proposed Structure for 2D model
SLC : Stable Limit Cycle
UPO : Unstable Periodic Orbit

AN ELECTRO-MECHANICAL TRANSDUCER FOR AN ACTIVE ANTI-PULSATORY SYSTEM

J. Tartarin,[1] J. Laumonier,[1] A. Piteau,[1] L. Hardouin[1]

ABSTRACT

Since Lighthill's general theory of aerodynamic noise, the phenomenological mechanisms of noise generation are better understood. So, instead of "actively" creating anti-noise, we have tried an active control of the process leading to strong low-frequency emission at the exhaust of pulsed flows. Having demonstrated at the laboratory the feasibility of such an active anti-pulsatory system, at present we are working on an industrial prototype to be tested soon.

I. AEROACOUSTICS OF PULSED FLOWS

The type of flows found in the exhaust systems of internal combustion engines or compressors we are interested in usually have an average flow velocity ranging from 20 to 60 m/s. The fundamental frequencies are very low (below 100 Hz). Due to the quasi-incompressibility of the fluid at this low Mach number, the pressure fluctuations are almost in phase everywhere in straight ducts and, strictly speaking, do not "acoustically" propagate. The exhaust mouth of incompressible pulsed flows can be identified with an acoustic monopole (of diameter r_0). At a distance R from this source, the acoustic pressure (P(R,t)) is proportional to the variation in time in the mass flow (Q_m (t - R/c)) :

$$P(R,t) = \frac{1}{4\pi R} \frac{\partial Q_m(t - R/c)}{\partial t} \quad (I.1)$$

Whenever mass flow by unit area of the output section periodically varies to pulsation ω, the following applies to the observer :

$$u(t) = \bar{u} + \tilde{u}(t) \quad (I.2)$$

with:

$$\tilde{u}(t) = |\tilde{u}|.\, e^{j\omega(t-R/c)} \quad (I.3)$$

Hypothetically, the source is baffled and compact (r_o<<R) and the observer is in the near field (R<<λ) under the incidence θ. Hence, the pressure is :

$$P(R,t) = \frac{j\rho\omega r_o^2}{2R} |u|\, e^{j\omega(t-R/c)} \left\{ \frac{2.J_1(\frac{\omega r_o}{c} \sin\theta)}{\frac{\omega r_o}{c} \sin\theta} \right\} \quad (I.4)$$

[1]Laboratoire d'Etudes Aerodynamiques (unité de Recherches associée au C.N.R.S. n° 191, Poitiers, France).

For low frequencies, the value of ($\omega r_o/c$) is small and the relative variation of Bessel's function tends towards unity, so we approximate the relation (I.4) :

$$P(r,t) = \frac{j\omega\rho r_o^2}{2R} |\tilde{u}| e^{j\omega(t-R/c)} \quad \text{(I.5)}$$

and its R.M.S value :

$$P_{ac} = \frac{\rho\omega r_o^2}{\sqrt{2}R} \cdot |\tilde{u}| \quad \text{(1.6)}$$

The acoustic energy radiated by a pulsed flow is related mainly to the pulsatory character of the flow. **This causal mechanism between Fluid Mechanics and Acoustics shows that reducing the flow modulation amplitude should attenuate the low frequency sound level**. However, the existence of walls and the turbulence of the flow constitute "dipolar" and "quadripolar" acoustic sources. Because of their poor efficiency, they will produce a background large band noise, below which the discrete spectrum lines linked to the pulsation cannot be attenuated.

II. PRINCIPLE OF THE ACTIVE CONTROL OF PULSED FLOWS

The regulation of a flow is usually provided by a device which modifies the "head loss" (e.g. gates, walves...). When the flow fluctuates periodically around a fixed mean value, the idea is to fix a unit which will control that modulation in real time to remove it. So, how should the command law of an inserted movable singularity be made to achieve this smoothing? What sort of singularity is the easiest to move and what actuator should be chosen?

Controlling a pulsed flow by a head loss varying instantly creates a physical system with a double non-stationary character. The first theoretical approach to modelise this system consists in decomposing the double unsteadiness into two non-stationary separated problems, provided they are considered as quasi-linear (in the sense that the causes and the effects are nearly proportional):

- a fixed singularity in a pulsed flow
- a periodically fluctuating head loss in a steady flow

Then, respecting the synchronicity of the fluctuations and the quasi-steadiness hypothesis, we consider the superposition of these two flow configurations as the physical model. The hypothesis of quasi-steadiness is confirmed in such flows when the time scale characterizing the fluctuations (the fundamental period T) is much greater than the one based upon duct diameter and flow velocity (D/u).

1) Fixed singularity

By definition, for a **steady flow** in a duct, the dimensionless mean head loss coefficient (λ) of a singularity (for instance a flap with a slight incidence (α) (figure 1)) is the ratio of the difference of static (upstream(1)-downstream(2)) pressures to dynamic pressure:

$$\overline{P_1} - \overline{P_2} = \lambda(\overline{\alpha}) \cdot \frac{1}{2}\rho\overline{u}^2 \quad \text{(II.1)}$$

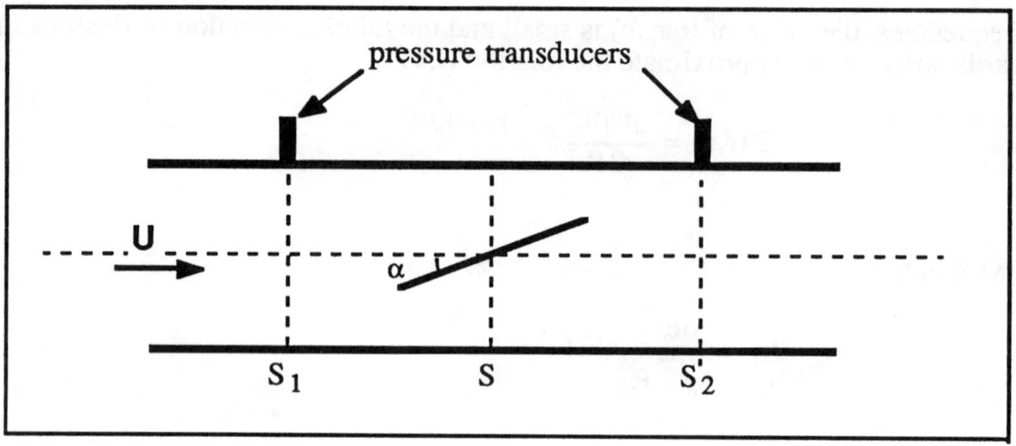

Figure 1.

In presence of **incompressible pulsed flow,** with variables separated as follows :

$$u(t) = \bar{u} + \tilde{u}(t) \text{ (with } |\tilde{u}| << \bar{u}) \ ; \ P(t) = \bar{P} + \tilde{P}(t) \ ; \ \rho(t) = \bar{\rho} \tag{II.2}$$

(II.1) leads to :

$$(\overline{P_1} + \tilde{P}_1(t)) - (\overline{P_2} + \tilde{P}_2(t)) = \lambda(\bar{\alpha}) \cdot \frac{1}{2} \rho (\bar{u} + \tilde{u}(t))^2 \tag{II.3}$$

It is easy to demonstrate theoretically and verify experimentally that the mean and periodic contributions of these variables are related as follows :

$$\begin{cases} \overline{P_1} - \overline{P_2} = \lambda(\bar{\alpha}) . \dfrac{\rho \bar{u}^2}{2} & \text{(II.4)} \\ \tilde{P}_1(t) - \tilde{P}_2(t) = \lambda(\bar{\alpha}) \rho \bar{u} \tilde{u}(t) & \text{(II.5)} \end{cases}$$

or :

$$\frac{\tilde{P}_1(t) - \tilde{P}_2(t)}{\overline{P_1} - \overline{P_2}} = 2 \frac{\tilde{u}(t)}{\bar{u}} \tag{II.6}$$

2) Fluctuating head loss in an initially steady flow

Let us consider a movable singularity such as a flap oscillating around a non-zero incidence ($\bar{\alpha}$), introduced in a flow which is steady far upstream from it (figure 2).

If the amplitudes of the periodic oscillations are not too large, the causes (periodical fluctuations (λ) in head loss) and the effects (local perturbations of the flow) are quasi-linear so that relation (II.1) leads to :

$$\overline{P_1} + \tilde{P}'_1(t) - (\overline{P_2} + \tilde{P}'_2(t)) = (\lambda(\bar{\alpha}) + \tilde{\lambda}(\bar{\alpha}, t)) \cdot \frac{1}{2} \rho (\bar{u} + \tilde{u}'(t))^2 \tag{II.7}$$

Separating mean and unsteady contributions and neglecting quite small quadratic terms in relation (II.7) leads to :

$$\begin{cases} \overline{P_1} - \overline{P_2} = \lambda(\overline{\alpha}) \cdot \dfrac{\rho \overline{u}^2}{2} & \text{(II . 8)} \\ \tilde{P}_1'(t) - \tilde{P}_2'(t) = \tilde{\lambda}(\alpha(t)) \cdot \dfrac{\rho \overline{u}^2}{2} + \lambda(\overline{\alpha}) \rho \overline{u} \tilde{u}'(t) & \text{(II . 9)} \end{cases}$$

or :

$$\frac{\tilde{P}_1'(t) - \tilde{P}_2'(t)}{\overline{P_1} - \overline{P_2}} = \frac{\tilde{\lambda}(\alpha(t))}{\lambda(\overline{\alpha})} + 2\frac{\tilde{u}'(t)}{\overline{u}} \qquad \text{(II.10)}$$

Some obvious facts about unsteady contributions are supported by our experiments [2] : fundamental periodicities are the same, quadratic terms are negligible, upstream and downstream pressures are in phase opposition...

3) Superposition

Assuming that the two previous flows have the same mean physical variables (velocity, pressure and temperature) and comparable limit conditions, the system which results from the superposition of the two phenomena has to be written (in compliance with the equations (II.3) and (II.6) and the same hypotheses) :

$$(\overline{P_1} + \tilde{P}_1(t) + \tilde{P}_1'(t)) - (\overline{P_2} + \tilde{P}_2(t) + \tilde{P}_2'(t)) = \overline{\lambda}\frac{\rho \overline{u}^2}{2} + \overline{\lambda}\rho\overline{u}\tilde{u}(t) + \overline{\lambda}\overline{u}\tilde{u}'(t) + \tilde{\lambda}\frac{\rho \overline{u}^2}{2} \qquad \text{(II.11)}$$

Separating mean and unsteady contributions, relation (II.8) becomes :

$$\begin{cases} \dfrac{(\tilde{P}_1(t) + \tilde{P}_1'(t)) - (\tilde{P}_2(t) + \tilde{P}_2'(t))}{\overline{P_1} - \overline{P_2}} = \dfrac{\tilde{\lambda}(t)}{\overline{\lambda}} + 2 \cdot \dfrac{\tilde{u}(t)}{\overline{u}} + 2 \cdot \dfrac{\tilde{u}'(t)}{\overline{u}} & \text{(II.12)} \\ \overline{P_1} - \overline{P_2} = \overline{\lambda} \cdot \dfrac{\rho \overline{u}^2}{2} & \text{(II.13)} \end{cases}$$

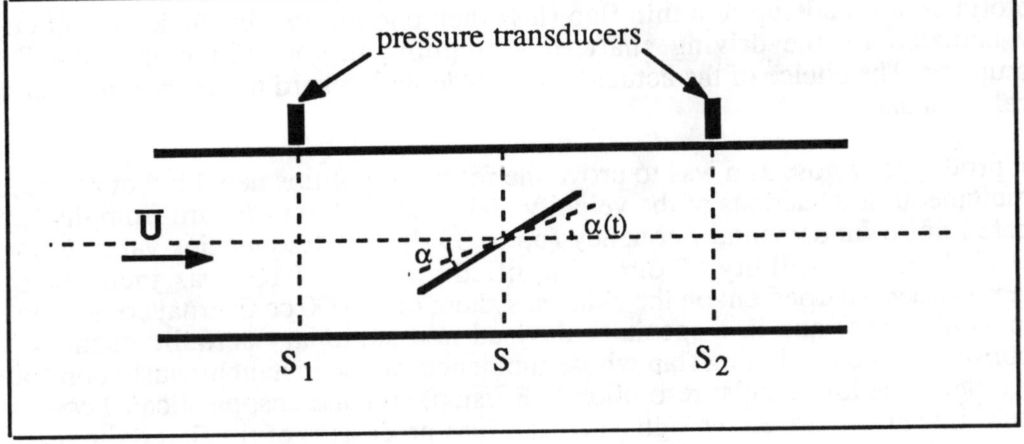

Figure 2.

4) Simplified model of the flow active control process

As the aim is to cancel the fluctuation in downstream flow, this results locally in an attenuation of associated pressure fluctuation ; i.e., ideally,by :

$$\tilde{P}_2(t) + \tilde{P}_2'(t) = 0 \qquad \text{(II.14)}$$

Then, the head loss coefficient is logically governed by the following temporal law :

$$\frac{\tilde{\lambda}(t)}{\bar{\lambda}} = \frac{\tilde{P}_1(t) + \tilde{P}_1'(t)}{\overline{P_1} - \overline{P_2}} - 2\left(\frac{\tilde{u}(t)}{\bar{u}} + \frac{\tilde{u}'(t)}{\bar{u}}\right) \qquad \text{(II.15)}$$

The reference quantities $\tilde{u}(t)/\bar{u}$ and $\tilde{u}'(t)/\bar{u}$ (which are, strictly speaking, not measurable) are ipso facto small, of the same order of magnitude and in phase opposition [1] : their sum is negligible. So, the temporal command law governing the relative variation of head loss coefficient ($\tilde{\lambda}$) which should lead to active fluctuations reduction is approximately :

$$\frac{\tilde{\lambda}(t)}{\bar{\lambda}} = \frac{\tilde{P}_1(t) + \tilde{P}_1'(t)}{\overline{P_1} - \overline{P_2}} \qquad \text{(II.16)}$$

We have shown and experimentally confirmed that, for optimal attenuation, $\tilde{P}_1(t)$ and $\tilde{P}'_1(t)$ are in phase ; the "counter-reaction" (the counter-pressure) caused by the movement of the device thus merely results in local amplification (of factor K). In this way the command law may be established from the upstream pressure observed before the actuation of the device by taking factor K into account ($1<K<2$).

Therefore, another form of relation (II.13) is :

$$\tilde{\lambda}(t) = \frac{K.\tilde{P}_1(t)}{\overline{P_{ref}}} \qquad \text{(II.17)}$$

with : $$P_{ref} = \rho\bar{u}^2/2 \qquad \text{(II.18)}$$

III. Introducing the prototypes

The prototypes are made up of a thin flap (less then one millimeter thick) in a circular duct directly actuated by the driving shaft of a stepping motor, which is controlled by a microcomputer. The choice of the actuator was made with regard to positioning accuracy and numerical command.

The first prototype, whose aim was to prove the feasibility of this new kind of active control, gave simultaneous attenuations of the velocity and pressure (downstream from the apparatus) and radiated noise fundamental frequency (up to 35 dB, for frequencies varying from 20 to 100 Hz) [1,2,3]. The validity of our "simplified" model (II.16) was then demonstrated through experiments carried out on the exhaust system of a 2000 cc internal combustion engine (without combustion) and in a specially devised non-stationary aeraulic channel [2]. The device consists of the oscillating flap whose incidence can be instantaneously controlled by a stepping motor. The low angular resolution (1.8 °/step) and the unsophisticated command law ("sawtooth" movement driven through an 8-bit micro-processor) of the first prototype led us to improve and automate the device [1.2.3].

1) Command law

The first automation exercise is to compute the flap position $\alpha(t)$ from head loss coefficient ($\lambda(t)$), because of nonlinearity of $\lambda(\alpha)$ (fig.3).

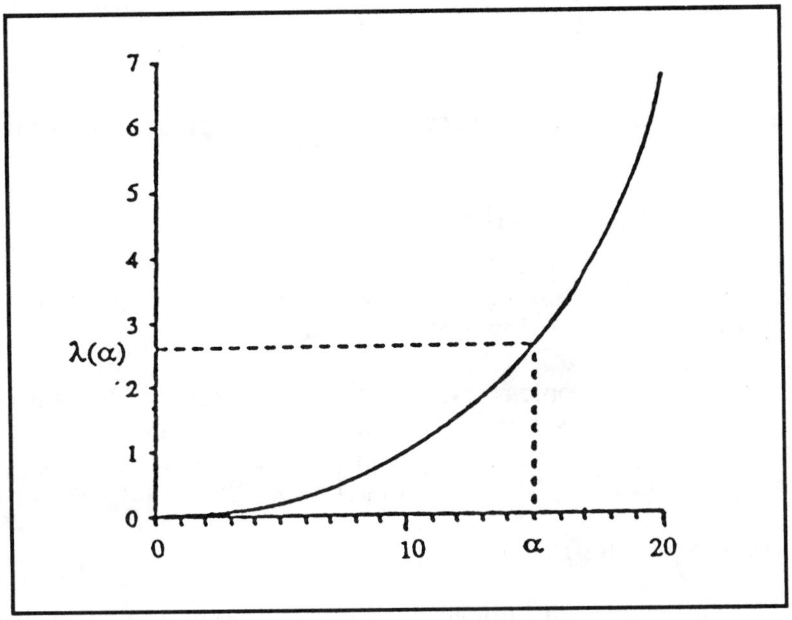

Figure 3.

Relation (II.16) or (II.17) shows we have to compute the coefficient's relative variation due to the flap's oscillations from the measurements of pressure fluctuations. There appear several pairs of corresponding relative positions $\tilde{\alpha}(t)/\overline{\alpha}$. Whether the mean incidence ($\overline{\alpha}$) is weak, and so is the mean head loss, the amplitude ($|\tilde{\alpha}(t)|$) is large and needs better dynamical performances (acceleration, etc.) of the actuator. A compromise solution has to be found.

We adopt a local linearization of the polynomial function $L(\alpha)$ as near as possible to $\lambda(\alpha)$. A finite-differences scheme (based upon Crank-Nicolson) leads to the discrete form of (II.17) :

$$K.\frac{P_{k+1} - P_k}{\Delta t} = \overline{P_{ref}} \cdot \left(\frac{d\lambda}{d\alpha}\right)_{\alpha_{k+\frac{1}{2}}} \cdot \left(\frac{\alpha_{k+1} - \alpha_k}{\Delta t}\right) \qquad \text{(III.1)}$$

and using derivative $L'(\alpha)$ to estimate $d\lambda/d\alpha$, we obtain the instantaneous position of the flap :

$$\alpha_{k+1} = \alpha_k + L'\left(\frac{\alpha_{k+1} + \alpha_k}{2}\right)^{-1} . K. \frac{P_{k+1} - P_k}{\overline{P_{ref}}} \qquad \text{(III.2)}$$

The iterative algorithm :

$$\alpha_{k+1}^{i+1} = \alpha_k + L'\left(\frac{\alpha_{k+1}^{i}+\alpha_k}{2}\right)^{-1}. K. \frac{P_{k+1} - P_k}{\overline{P_{ref}}} \qquad \text{(III.3)}$$

with initial value :

$$\alpha^1_{k+1} = \alpha_k \qquad \text{(III.4)}$$

and convergence criterion based upon the angular positioning resolution of the actuator (r_e) :

$$|\alpha_{k+1}^{i+1} - \alpha_{k+1}^{i}| \leq r_e \qquad \text{(III.5)}$$

allows us to compute the discrete command law from the conditional sampling of upstream pressure and the knowledge of the flow's characteristics.

The major advantage of dimensionless relation (III.3) is that it is to be quite "universal", being based upon similitude of flap's head loss coefficient (λ). It only requires to know the "dynamical" pressure (P_{ref}) of the flow (measured for instance by a "Pitot tube" or deduced from the volume of the engine's cylinders) and to choose the initial position of the flap (α_0).

2) Prototype definition

The previous explanations give an indication of what is needed to build up an **experimental** automated prototype:

- a detection signal from the conditional sampling and treatment of upstream pressure.

- an optical encoder, connected to one shaft of the pulsating engine, to generate the clock needed for conditional sampling and synchronisation of the actuator's movement.

- a multitask computer with intelligent boards for signals acquisition and treatment, frequency and events counting, command law computing and generation.

- an actuator able to reproduce the command law perfectly.

Figure 4 shows the synoptic of our second laboratory prototype (see next page).

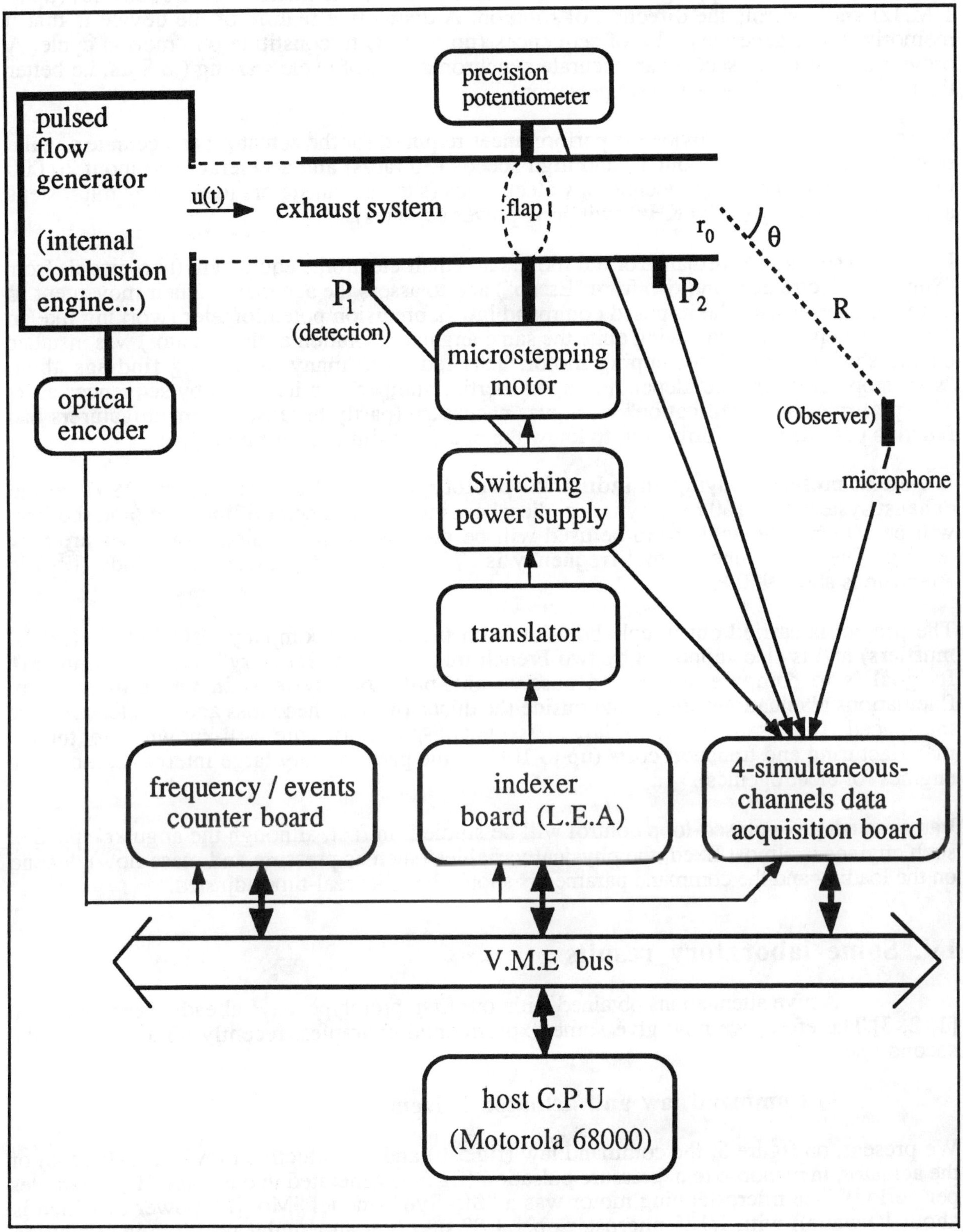
precision potentiometer
pulsed flow generator
(internal combustion engine...)
u(t)
exhaust system
flap
r_0
θ
R
P_1
(detection)
P_2
microstepping motor
optical encoder
(Observer)
microphone
Switching power supply
translator
frequency / events counter board
indexer board (L.E.A)
4-simultaneous-channels data acquisition board
V.M.E bus
host C.P.U (Motorola 68000)

Figure 4.

The **"Indexer"** board was specially conceived and realized at the laboratory to control the movement of micro-stepping motors. It generates pulses at the micro-steps frequencies (up to 1 MHz) and controls the direction of rotation. A distinctive feature of the device is that it memorizes and generates a lot of sequences (up to 128) to constitute one motion cycle. A programmable delay secures an accurate synchronization of cycle starting (to 8 µs, i.e better than 0.3° phase shift for a 100 Hz pulsation).

Microstepping motors provide the performances required for the actuator : an accurate angular positioning resolution (about 1') and high speeds (50 rad/s) and accelerations (about 20 000 rad/s^2). But they need high-technology electronics (such as translators to generate microsteps and high-frequency (≈100 KHz) switching power supplies).

In order to test the performance of our motors and their electronic equipment (two models from "Superior Electronics" and two from "Escap") and to assess the accuracy of their movement so as to compare it with the imposed command law, a precision potentiometer (working like an "analogical encoder" and having quite the same angular resolution as the actuator) was installed on the shaft of the microstepping motor. This led us to many interesting findings about "start-stop" performance, depending upon inertia, number of microsteps by sequences a.s.o. This preliminary "qualification" work was necessary (partly because the manufacturers had failed to provide any information) to know the operating limits of our protoypes.

We are at present studying an **industrial prototype** we will test next spring 1991 on the exhaust system of a 150 hp-6 cylinder -diesel engine. The diameter of the duct (and the flap) will be 10 cm : the actuator to be used will be one the microstepping motors we are now testing. The fixed fundamental frequency is 72 Hz and the pressure amplitude ($|\tilde{P}_2|$) to attenuate is about 900 Pa.

The project is carried out in collaboration with the French company "**BOET**" (industrial mufflers) and is also supported by two French ministries for "Industry" and "Environment". Its goal is to compare active and passive anti-pulsatory systems in terms of pressure fluctuations attenuations inside and ouside the ducts, of mean head loss and counter-pressure introduced. The disadvantages of large passive anti-pulsatories are well-known: size, tuning, manufacturing and transport costs (up to 10% of the price of very large internal combustion engines for electrogenics).

Later the adaptive closed-loop control will be studied: in fact, although the angular speed of such engines is almost fixed, the physical variables (such as pressure and mass flow) depend on the loading and the command parameters should be quite real-time adjusted.

IV. Some laboratory results

Active attenuations obtained with our first prototype have already been presented [1, 2, 3].Therefore, we now give some experimental examples, recently obtained with the second one.

1) Command law and actuator movement

We present, on figure 5, the command law (fig.5-b) and the effective movement (fig 5-c) of the actuator, in response to a pressure pulsation (fig 5-a) generated in our channel (32 samples per period). The microstepping motor was a "Slo-Syn" (model "Mo 61", power of which is about 10 watts)) with 1/125 microsteps per 1.8° step (i.e. an angular positioning resolution equal to 0'52").

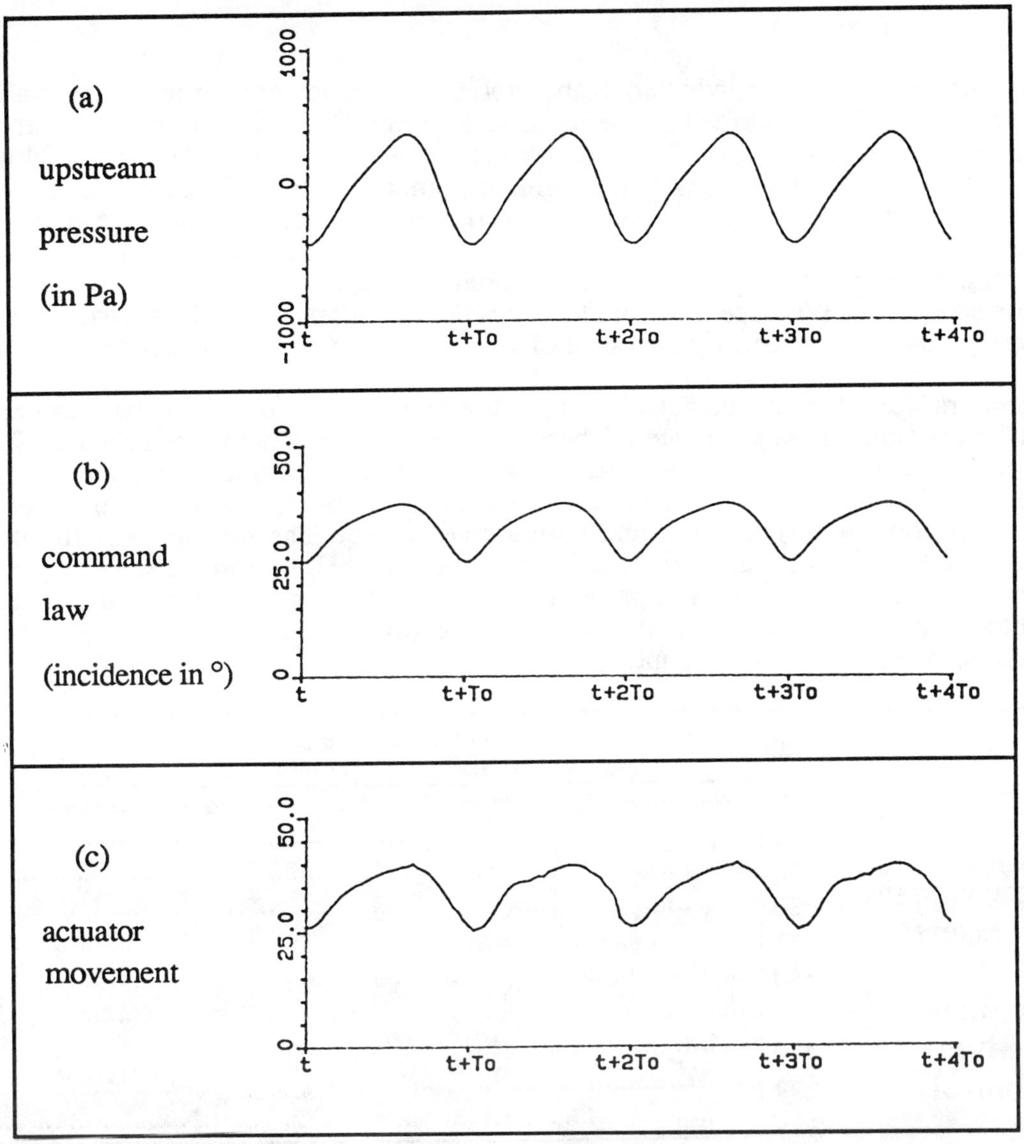

Figure 5 (a,b,c)

The physical parameters to compute the command law were :

$$N = 40\ \text{Hz} \ ; \ P_{ref} = 151\ \text{Pa} \ ; \ \bar{u} = 15.9\ \text{m/s} \ ; |\tilde{P}_1| = 300\ \text{Pa (R.M.S value)} \ ;$$

The period contains also 32 sequences and the actuator motion shows :

$$|\tilde{\alpha}| = 6.3° \ ; \ \bar{\alpha} = 32.5\ ° \ ; \ |\tilde{\lambda}| = 4.55 \ ; \ \bar{\lambda} = 5.28$$

The greatest velocity and acceleration were : 120.10^3 µpsteps/s and 47.10^6 µpsteps/s^2

A lot of tests, for fundamental frequencies ranging from 20 to 100 Hz and for pressure fluctuations up to 1000 Pa, showed the limits of such actuators, when the load has practically the same inertia than the rotor : the speed is limited to 200 10^3 µsteps/s (i.e. 50.2 rad/s) and the acceleration to 70 10^6 µpsteps/s^2 (i.e. 17 600 rad/s^2). For safety precaution these criteria are now included in the command law computation.

2) Active attenuations

The first attenuation tests conducted at the laboratory on our second prototype validate the discrete command law described by the iterative scheme (III.3). The indexer's programmable delay allows a precise study of the phase parameter in order to develop a closed-loop automation of the device. Besides, another major point of interest will be the determination of its impulsive response (or its transfer function) : our laboratory channel can be instantly transformed into a steady flow which allows us to study the various physical parameters (pressure, counter-pressure, velocity a.s.o) in relation, for instance, to a sinusoidally varying head loss coefficient. We hope therefore to gain a better understanding of physical phenomena in unsteady flows and to identify the range of validity of our "superposition" model.

By a temporal treatment of the detection pressure signal, we can choose the number and the range of harmonics to be attenuated. Therefore, the device is a sort of selective filter. For instance, if the fundamental and the third harmonic are the frequencies to reduce, we observed that the rest of the spectrum is indeed unmodified. Figures 6-a and 6-b show the spectrums of downstream pressure (P_2) without and with active control. The fundamental frequency (80 Hz) is reduced by about 20 dB and the third harmonic (240 Hz) disappears. The other discrete frequencies (50, 100... Hz) are not generated by our pulsating device, but are linked to the "unwanted" ventilator noise. For information : the flow and command law parameters are nearly the same as in the next example.

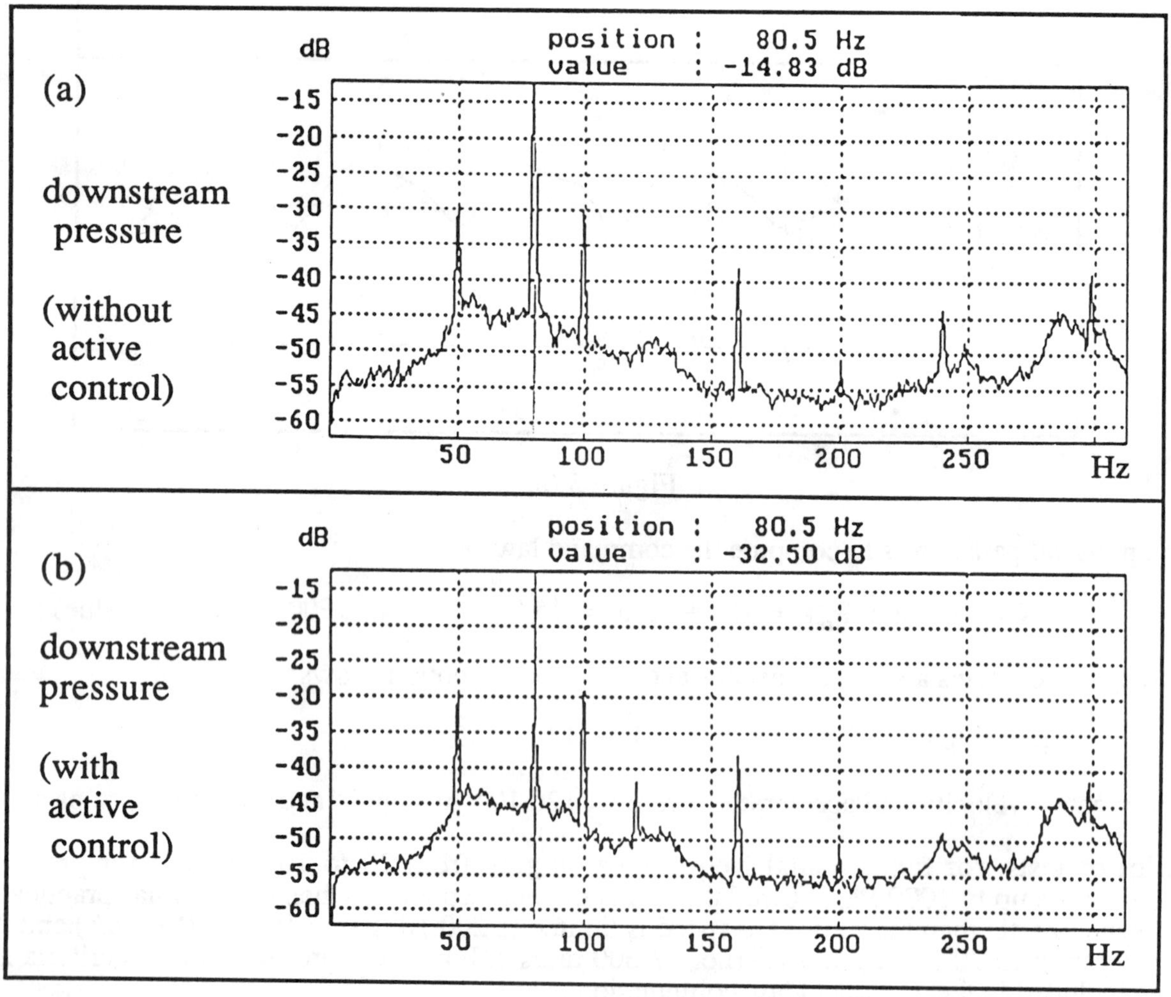

Figure 6 (a,b).

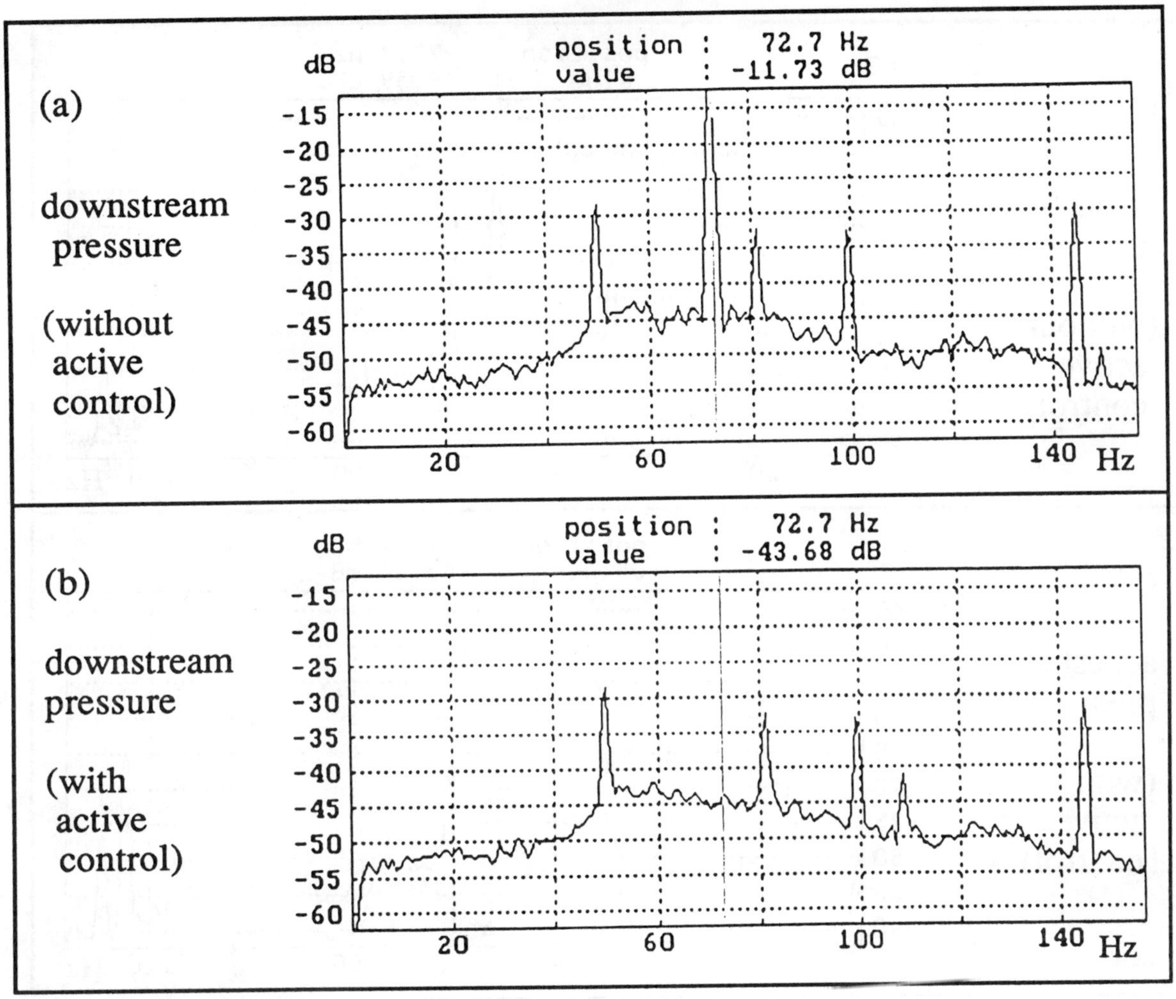

Figure 7.

For the second example (figures 7 and 8), the industrial flow conditions we shall meet are reproduced in our channel accurately (same dimensionless parameters and same fundamental frequency). As the fundamental (72 Hz) will be the most energetic, we first chose to reduce as much as possible the downstream pressure fluctuation (fig. 7-a and 7-b) to observe the resulting attenuation of the acoustic pressure near the exhaust mouth (fig 8-a and 8-b). The corresponding flow parameters are:

$$\bar{u} = 32 \text{ m/s} \;;\; P_{ref} = 600 \text{ Pa} \;;\; |\tilde{P}_2|_{(\text{without control})} = 840 \text{ Pa}$$

The command law leads to a mean incidence ($\bar{\alpha}$) of 38.5° and a movement amplitude ($|\tilde{\alpha}|$) of 2.37° for the flap, leading to a mean head loss coefficient (λ) of about 8 (which is undoubtedly less than the coefficient of a passive anti-pulsatory device).
We observe that a "complete" attenuation of the in-duct pressure fundamental (here 32 dB reduction) leads to 20 dB attenuation of the acoustic one. An explanation is the acoustic energy (convected by the flow or conducted by boundary vibrations) is "masked" by the larger non-stationary fluid mechanical energy. When the active control, based upon "Fluid Mechanics" laws, is "perfect", "Acoustics" is appearing. That's a real problem, encountered for instance in active anti-noise system, when the low-frequency pressure fluctuations measured by microphones are not linked to acoustic waves but to turbulences : the device regenerates noise instead of reducing it! (A searcher of our laboratory is now developing a method based upon signal treatment for "acoustic/turbulence" separation in flows).

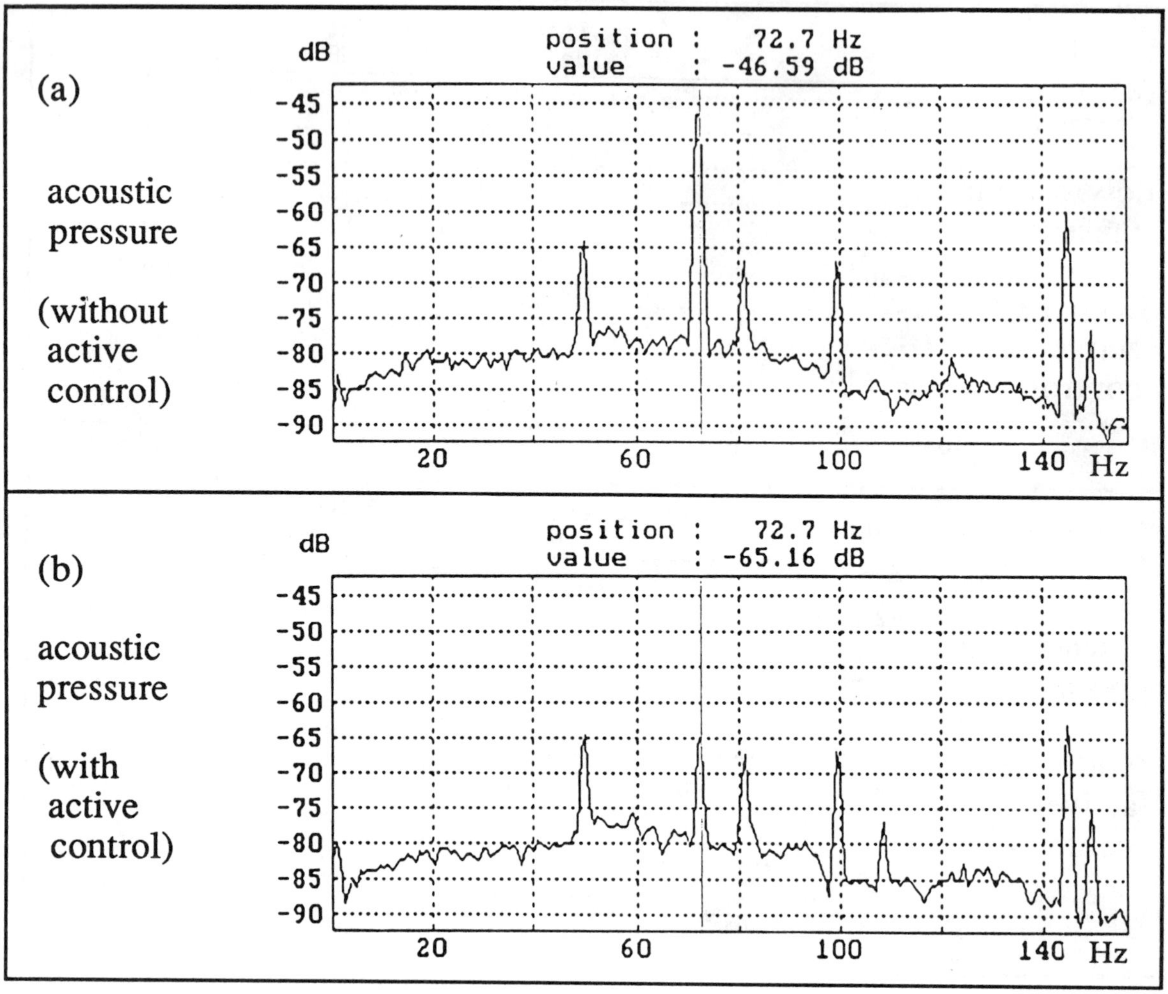

Figure 8.

Conclusion

If the industrial feasibility of our active control device is proved, we will have to study the closed-loop automation and the adaptive control. This should be facilitated through an understanding of physical unsteady phenomena and the knowledge of the "electro-mechanical" transducer's transfer function. Another question will be how to associate active anti-pulsatory devices (most appropriate for low frequency and discrete harmonic noise attenuation) with conventional "anti-noise systems"(using electro-acoustic transducers and most appropriate for medium and high frequencies) in order to design a "multipurpose" active muffler?

References

1. J.L Peube, J.Tartarin, M. de Coligny and J.Laumonier, 1988. "A new device for active attenuation of low-frequency fluctuations in flows". Proceedings Internoise 88, vol.2, pp. 1073-1076.

2. J. Laumonier, 1990. "Etude d'un dispositif de contrôle actif de fluctuations de débits pulsés utilisable comme silencieux". Thèse de Doctorat de l'Université de Poitiers, Janvier 1990.

3. J. Laumonier, J. Tartarin and J.L. Peube, 1990. "Active control of pulsed flows for low frequency noise attenuation". Proceedings Internoise 90

PASSIVE AND ACTIVE CONTROL OF WAVES ON PLATE STRUCTURES INDUCED BY CONVECTIVE LOADING

Lucio Maestrello[1]

ABSTRACT

Passive and active methods used experimentally to control boundary layer instability, plate vibration and sound generation are reviewed. Rubber wedges glued to the plate along its edges are an effective passive method to reduce plate vibration and acoustic radiation. The oscillation of the flexible plate generates Tollmien-Schlichting waves in the boundary layer, and an upstream acoustic source has been found to affect both the boundary layer transition location and the response of the flexible plate. Experiments show that localized surface heating at the leading edge delays the transition location downstream of the flexible plate, and that active vibration control reduces the displacement amplitude of the plate induced by the acoustic source.

1. INTRODUCTION

In this paper, methods of controlling a boundary layer and flexible plate interaction are studied. It is known that a flexible plate can excite instability waves in a laminar boundary layer much earlier than an identical flow over a rigid surface. The pressure fluctuations in the flow field can also excite plate oscillation, which in turn generates noise on the unwetted side of the plate.

The response of a structure to laminar or transitional boundary layers is different from the response to turbulent boundary layers. In a laminar boundary layer, the oscillation of the plate can be attributed to the starting flow in a tunnel or the oscillation of the airframe, and hence is a source of instability. In a turbulent boundary layer, it is the turbulence that forces the plate oscillation [1,2]. These experimental results on flow over a flexible structure suggest the need for controlling the structure's response, and the growth of the instability to prevent early transition as compared to a rigid structure.

To control the interior noise level of an aircraft, a passive technique was developed in the 60's. The control principle was to suppress waves created by a convective turbulent boundary layer flow reflecting from the upstream and downstream boundaries of the plate. Rubber wedges glued to the boundaries were used to modify the plate boundary conditions. Control was most effective at aerodynamic coincidence, when the speeds of the waves in the boundary layer and in the plate matched. Initially, experiments were conducted in laboratories, and later in flight [2-3].

Several techniques to control structural vibration and the resultant sound radiation by active means have been reported in the 80's [4-6]. However, control of flow transition over a flexible plate, and plate vibration forced by instability waves has been investigated only recently [7,8]. The flow control was accomplished by using a thin electrically heated

[1]NASA Langley Research Center, Hampton, VA 23665-5225, U.S.A.

wire embedded in a substrate at the leading edge of a flat plate. Localized heating at the leading edge alters the growth of the instabilities due to progressive cooling downstream, and therefore the flow stability is increased.

An upstream acoustic source was found to affect the flow at the leading edge as well as the response of the flexible plate, effects known as boundary layer receptivity. The acoustic disturbances dominate the response of the plate when the boundary layer is turbulent, and even when the boundary layer is laminarized by localized surface heating. Active vibration control was used to reduce the amplitude of the plate oscillation induced by the acoustic disturbance [8].

Application of passive control to waves in flow and structure is simple and very effective as compared to active control. Active techniques, when applied to a broad range of coupled waves like instability waves in flow and structure, require the implementation of a feedback system. Such a system is difficult to operate in a convective nonuniform flow, like the boundary layer. Also, a feedback rule for the control must be established such that the response is most effective.

2. COUPLING BETWEEN PLATE OSCILLATION AND INSTABILITY WAVES

In this section, experimental results on the excitation of instability waves by the oscillation of a flexible plate are presented. The plate is mounted on a cut out section of a rigid

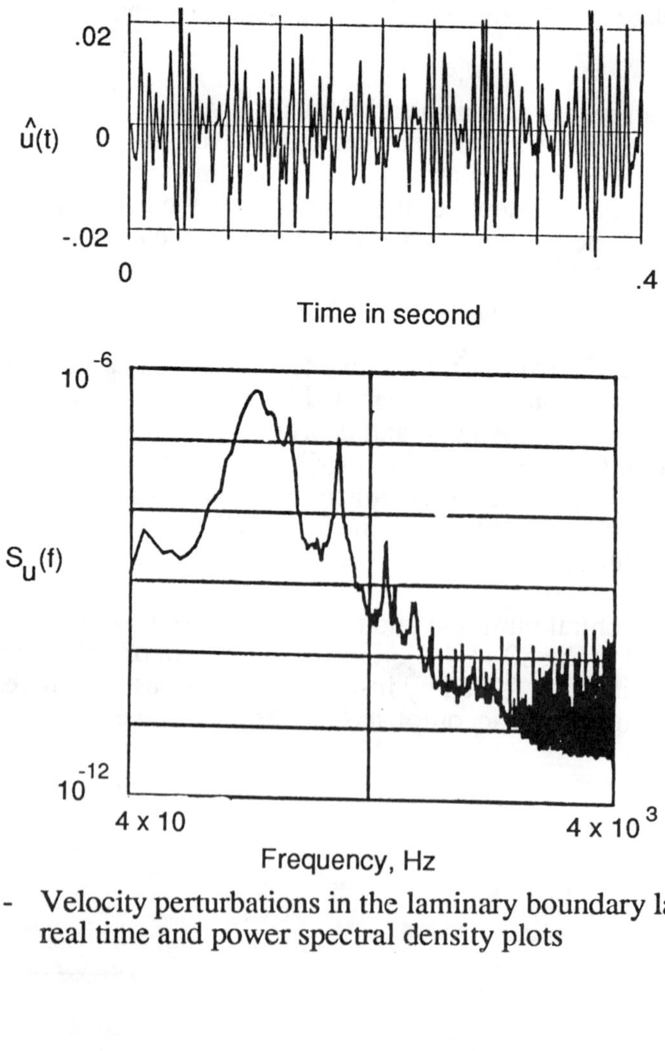

Figure 1 - Velocity perturbations in the laminary boundary layer above the flexible plate, real time and power spectral density plots

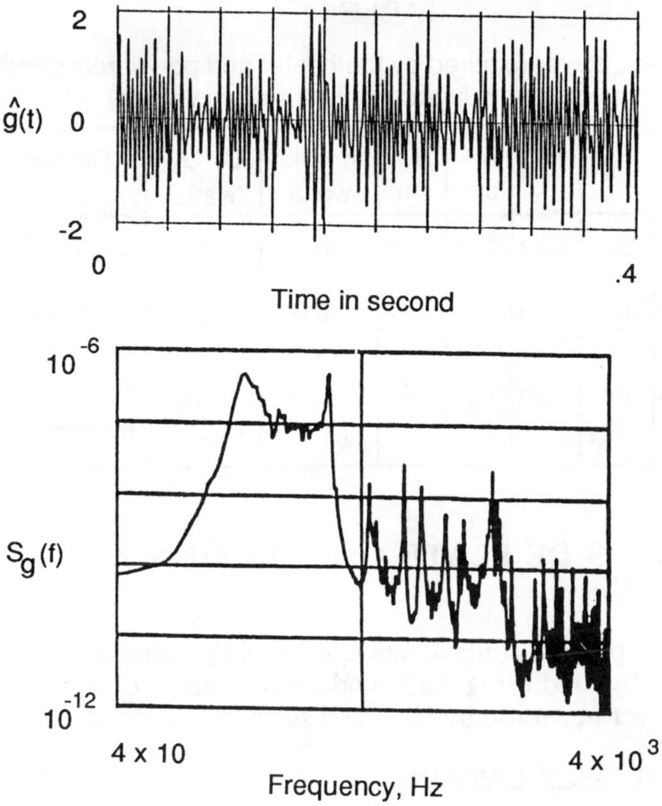

Figure 2 - Plate accelerations in the laminar boundary, real time and power spectral density plots

surface to simulate a panel on the wall of a fuselage. A 20.3 x 30.5 x 0.102 cm. flexible aluminum plate with clamped edges was tested, the same configuration described in [7,8].

Figure 1 shows the time history and power spectral density of the velocity perturbation $\hat{u}(t)$, $S_u(f)$, taken by a hot wire anemometer in the boundary layer above the plate, while Fig. 2 displays similar plots of the plate acceleration, $\hat{g}(t)$, $S_g(f)$, taken by an accelerometer mounted beneath the surface. The plate deflection is linear, i.e. very small compared to the thickness. Details of the boundary layer parameters measured above the 20.3 x 30.5 x 0.102 cm. plate are tabulated in the Table, where C_f is the skin friction coefficient; R_θ, Reynolds number (based on momentum thickness); H, shape factor; δ, boundary layer thickness; and θ, momentum thickness. As can be seen in the figures, the velocity fluctuations and the plate accelerations have similar behavior. These results are unique experimental evidence on the receptivity of a laminar boundary layer to plate vibration.

Wind tunnel results show that a laminar boundary layer remains undisturbed when the flexible plate is replaced by a rigid one. However, instability waves in the boundary layer over a flexible plate are observed as images of the plate oscillations. The waves grow rapidly with distance downstream. As a result, the motion of a flexible plate is identified as the source of instability in a laminar boundary layer. The plate oscillation can be attributed to the starting flow in a tunnel or to the oscillation of the airframe. Thus, it can be concluded that the flexible surface accelerates the transition process as compared to the rigid one.

TABLE

Mean flow parameters, power required by the heater and power reduction in the flow for different states of the boundary layer with free stream velocity U_e = 18.29 m/sec

State of the flow	C_f	R_θ	H	δ, cm.	θ, cm.	P = IE watts	D_f U_e watts	Remark
Laminar	.00085	756	2.43	0.48	.063	0	0	Natural state
Turbulent	.00470	1890	1.42	1.19	.147	0	0	Turbulent state induced by sound
Laminar	.00070	790	2.50	0.50	.058	129	54	Transition delayed by heating

3. PASSIVE CONTROLS OF PLATE OSCILLATION AND BOUNDARY LAYER TRANSITION

In this section, two types of passive control, one to control the plate oscillation and the resulting sound generation by modifying the boundary conditions of the plate, and the other to delay boundary layer transition by localized surface heating, are described.

3A. CONTROL OF PLATE OSCILLATION

Experiments on the control of structural response to turbulent boundary layers and sound radiation were conducted in a wind tunnel using a typical aircraft panel, and in flight on a 727 aircraft over a section of the fuselage, Figs. 3 and 4. These experiments were conducted in the late 60's and early 70's as a result of the development of rubber wedges at the Boeing Company to control the response and the radiation of a structure excited by a turbulent boundary layer. The wind tunnel experiments indicated that the convected pressure field of the turbulent boundary layer causes the panel structure to vibrate. The resultant surface vibration acts as a radiator of sound. The plate motion shows pronounced oblique ridges and valleys, which are interpreted as waves having the same phase velocity as the convected wall pressure fluctuations. The streamwise space-time correlation, $R(x_1,x_2; \tau)$, along the plate centerline of the displacement of a clamped 16.51 x 91.50 x 0.102 cm. aluminum plate is shown in Fig. 5. The figure indicates that the peak correlations are swept forward and backwards. In addition, the surface wave has a phase velocity of $\pm$ Uc, where Uc = .75 Ue or Uc ~ $(x_1-x_2)/\tau$, where Uc and Ue are the convection and free stream velocities, respectively. The symmetry of the correlation function at the positive and negative time delays was interpreted as waves, one

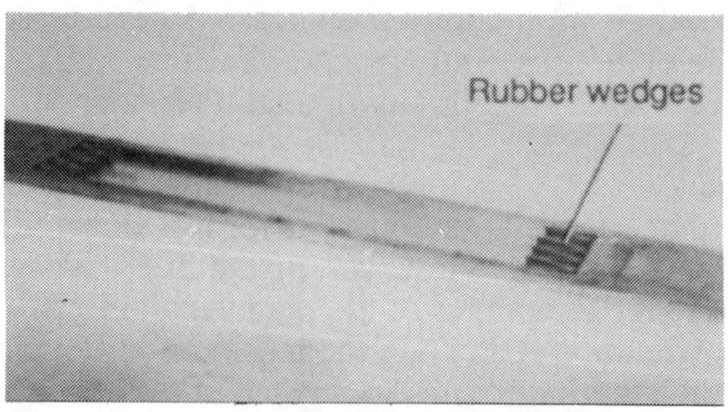

Figure 3 - Plate with rubber wedges

moving or transmitted with the flow, and another against the flow or reflecting from the rigid boundary.

A method of passive control was devised to modify the upstream and downstream boundary conditions to alleviate the reflected waves. The problem was experimentally investigated by damping both the upstream and the downstream boundaries. The damping material consisted of rubber wedges glued across the panel, as shown in Figs. 3 and 4. The space-time correlations of the panel vibration with damped boundary, rubber wedges, in a wind tunnel are shown in Fig. 6. The correlation function of the plate displacement with damped edges is quite different from that of the undamped plate. Notice that for the damped plate the waves moving with + Uc (downstream) remain undamped while the waves moving with - Uc (upstream) are damped.

The sound power levels measured in the anechoic chamber, radiated by the plate with and without damping, are shown in Fig. 7. There is a significant reduction in acoustic power above 250 cps using rubber wedges, indicating that the higher modes have been successfully damped. Rubber wedges add approximately 15 percent to the weight of the bare plate; however, significant reduction in the weight can be achieved with new materials.

A flight experiment was carried out using a 727 airplane. The interior of the fuselage is shown in Fig. 4. The purpose was to verify the wind tunnel measurements on the reduction of the interior noise level using rubber wedges. Flight data also provide a means of identifying dominant noise transmission paths between the exciting boundary layer flow on the outside and the sound pressure inside the fuselage. The noise reduction provided by the rubber wedges was obtained by measuring the noise level difference inside the fuselage before and after the installation of the wedges (Fig. 8). A distinct change in sound pressure level between bare plate and plate with wedges is observed.

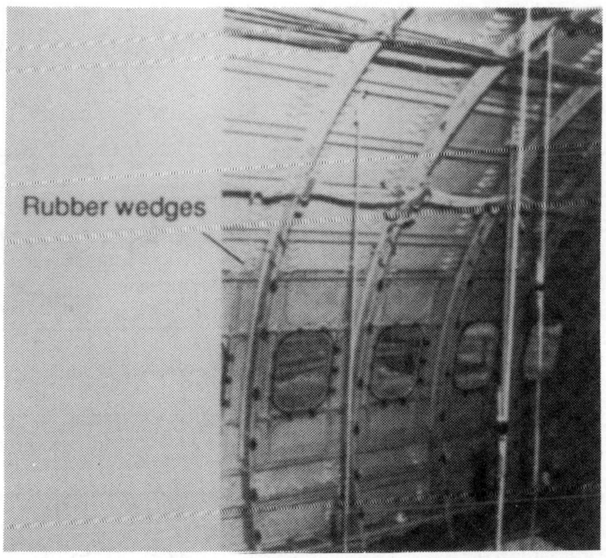

Figure 4 - Rubber wedges on 727 fuselage skin

3B. CONTROL OF THE BOUNDARY LAYER TRANSITION

It is known that in the boundary layer [7 to 15] an acoustic source upstream of the leading edge can induce early transition. In the presence of a flexible wall, the boundary layer becomes even more sensitive to upstream acoustic disturbances. Measurements show the power spectral density of the laminar boundary layer, Fig. 1, to change to turbulent, Fig. 9, as a result of early transition triggered by the upstream acoustic source. The corresponding responses of the plate are shown in Figs. 2 and 10, respectively. The first spike corresponds to the frequency of the square wave input (286 Hz), while the other spikes in the spectrum represent its Fourier harmonies. The Reynolds number and the shape factor at the measurement location change to R=1890, H=1.42, indicating a turbulent state.

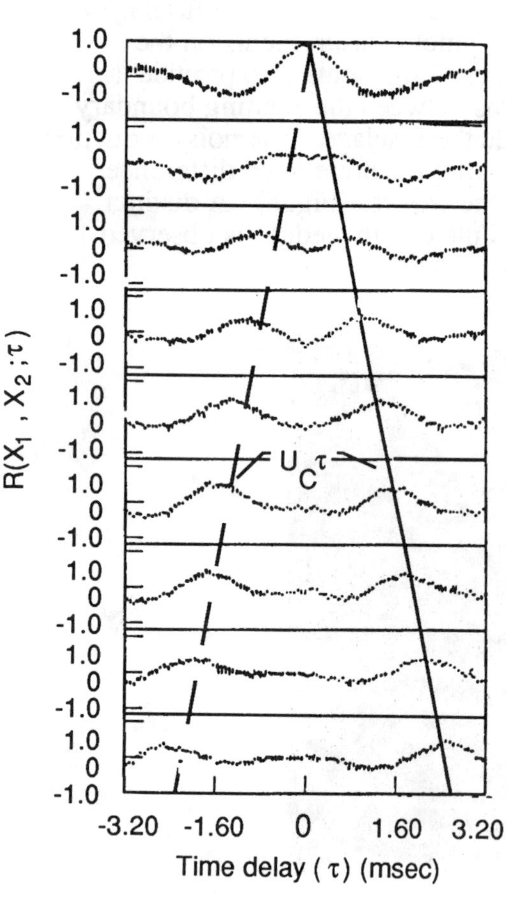

Figure 5 - Space-time correlation of the plate displacement

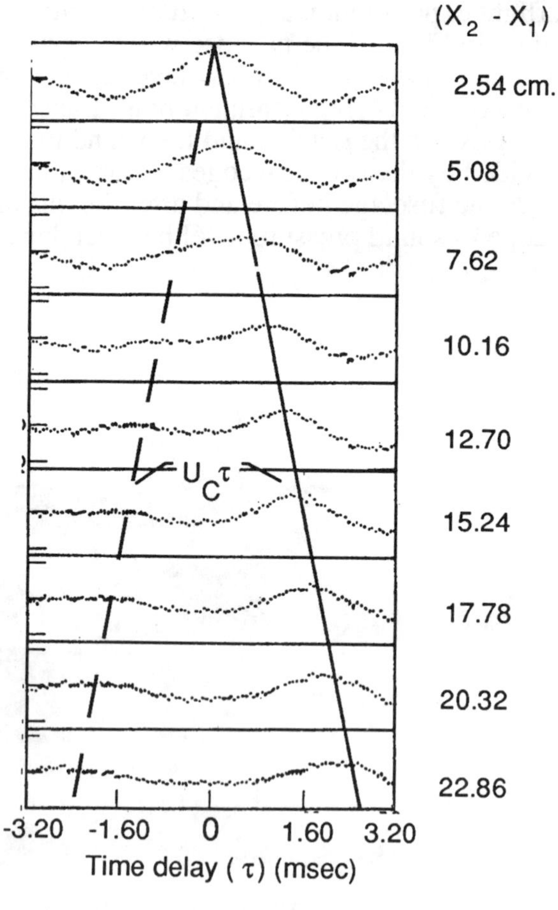

Figure 6 - Space-time correlation of the plate displacement with rubber wedges

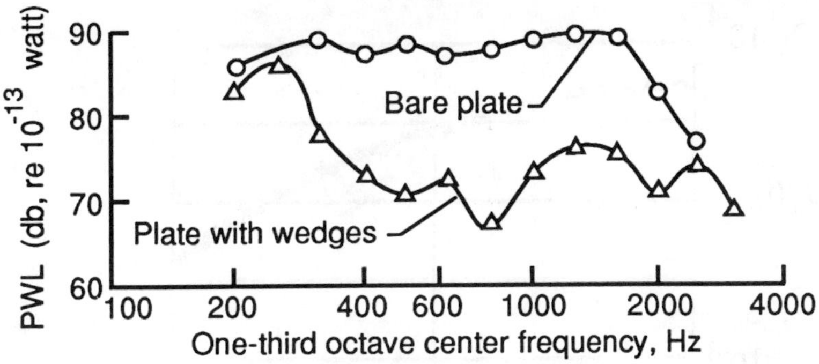

Figure 7 - Sound power radiated from the plate with and without rubber wedges

By localized surface heating placed near a curved leading edge of a rigid or flexible surface downstream, transition delay was observed using either passive or active means [13, 16, 17, 18]. Fig. 11 shows the power spectrum of the velocity fluctuations as a result of upstream surface heating. The spectrum differs from the turbulent spectrum shown in Fig. 9, and resembles the laminar spectrum shown in Fig. 1, except for the finite number of spikes corresponding to the frequencies of the acoustic source upstream. Fig. 12 shows the corresponding spectrum of the plate acceleration which resembles the spectrum without surface heating, as shown in Fig. 10. This indicates that localized surface heating delays transition, but has little effect on the response of the plate induced by an acoustic source. The presence of the upstream acoustic wave is apparent by the distinct peaks in the velocity and acceleration spectral densities. The fact that sound from upstream is present in the

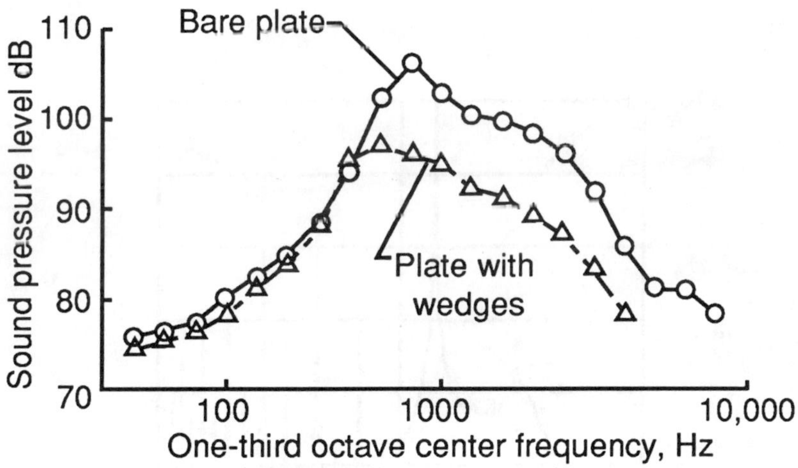

Figure 8 - Sound pressure level between bare plate and plate with rubber wedges

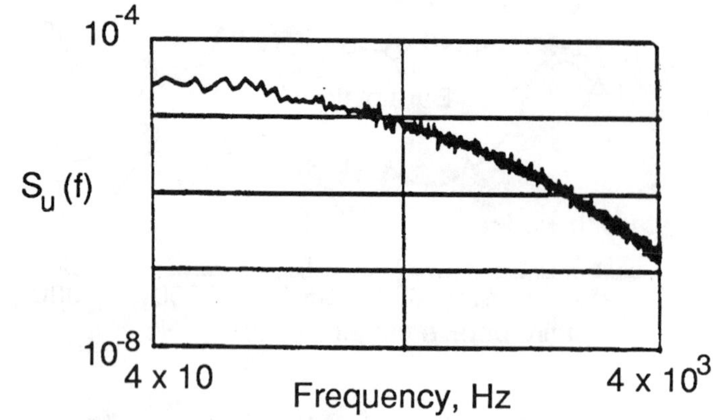

Figure 9 - Power spectral density of the velocity perturbation in the turbulent boundary layer above the flexible plate triggered by upstream sound

boundary layer after the transition delay indicates that the thermal gradient along the direction of flow (in the sublayer) is the stabilizing mechanism. The changes in the mean flow parameters, C_f, R_θ, H, δ, θ, the power required by the heater, P=IE, and the power reduction in the boundary layer from the turbulent state to the laminar state, $D_f U_e$, are shown in Table, where I is the current; E, the voltage; D_f, the skin frictional drag; and U_e, the free stream velocity. The Table shows that the input power to the heater is higher than the power saved by the reduction in skin frictional drag. The power required by an acoustic field to change the transition location is significantly less than the power lost in skin frictional drag. Improvement can be made in the design of the heating system to reduce the power needed.

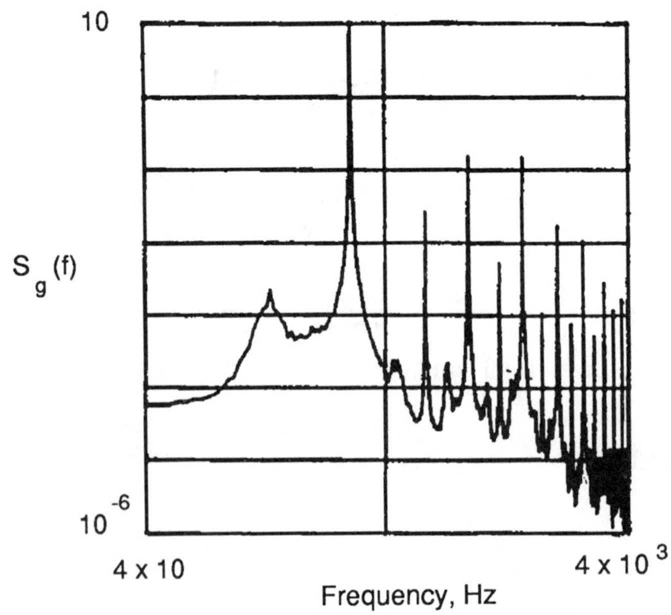

Figure 10 - Power spectral density of the plate acceleration in the turbulent boundary layer triggered by upstream sound

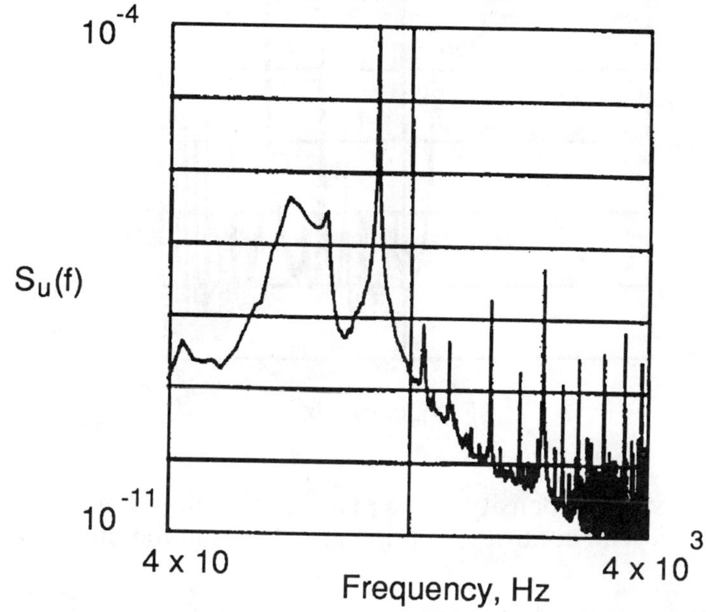

Figure 11 - Power spectral density of the velocity perturbation above the flexible plate as a result of surface heating in the presence of sound

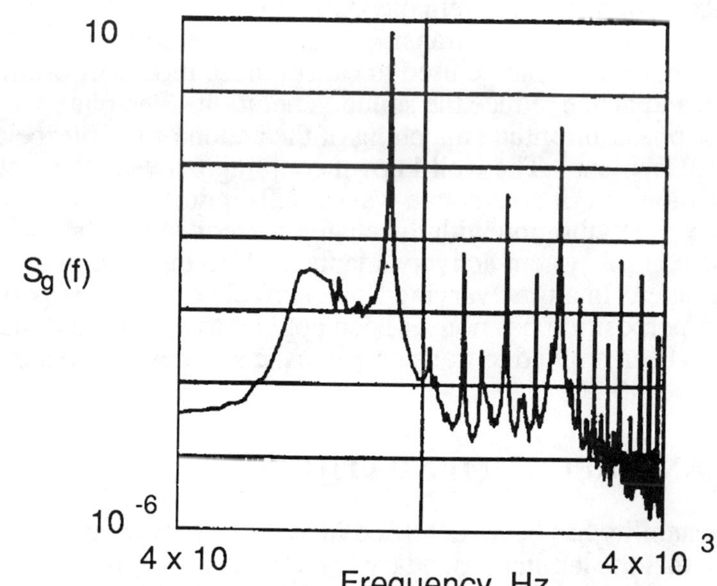

Figure 12 - Power spectral density of the plate acceleration as a result of upstream heating in the presence of sound

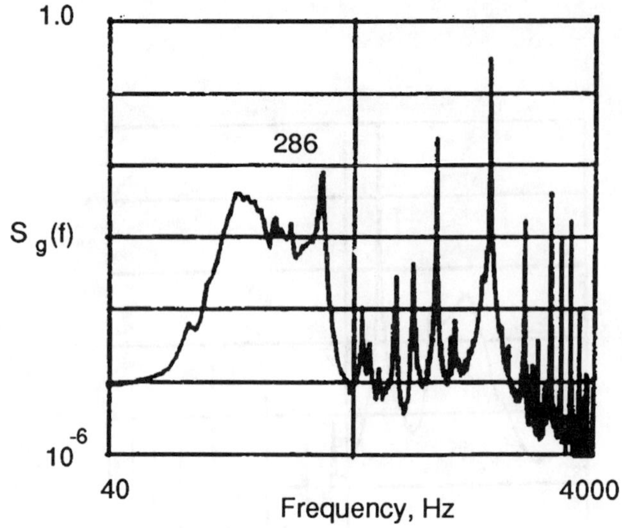

Figure 13 - Power spectral density of the plate acceleration as a result of upstream heating, with control excitation at 286 Hz, in the presence of sound

4. ACTIVE CONTROL OF THE PLATE VIBRATION

In this section, control of the plate oscillation, and hence the sound generated on the unwetted side of the plate, is discussed. A piezoelectric actuator mounted beneath the plate is used to control the vibration of the plate [8]. Fig. 13 shows the acceleration spectrum with the control forcing the plate out of phase with the excitation field at 286 Hz. Comparing with Fig. 12, note that the power spectrum at 286 Hz is reduced by 28dB. The power spectrum of the boundary layer remains the same as that in Fig. 11. Thus, use of localized passive surface heating to delay transition has been demonstrated while an active control at selective frequencies can be used to cancel those modes of oscillation with relatively large amplitude, and hence reduce the sound generation. The plate vibration control relies on the out-of-phase amplitude matching of the action of the piezoelectric actuator with the plate displacement. The well-known coupling between the plate motion and the boundary layer in the form of convective waves [1,2] induces a shift of the resonant frequencies of the plate vibration with the changes in convection velocity of the boundary layer. Thus the control system actively adapts itself to the amplitude and time response of the boundary layer. In a time varying flow, actively controlling perturbations utilizing the principle of superposition has not yet been applied to structural dynamics; however, the technique has been applied to transient acoustic sources and to a transitional boundary layer with success [19,20].

5. CONCLUSIONS AND RECOMMENDATIONS

An important source of instability has been identified in a laminar boundary layer over a flexible plate. It is shown that the laminar boundary layer is very receptive to plate vibration. In addition, experiments are presented to show that local passive surface heating can be used to delay boundary layer transition over a flexible plate. The broad band oscillation of a flexible plate can be controlled passively, without an external energy source, by changing the upstream and downstream boundary conditions of the plate. In addition, active control can be used to cancel the oscillation at specific frequencies using an external energy source. The combination of passive and active controls of the plate oscillation can substantially reduce the sound radiation on the unwetted side of the plate.

A major aspect of the study of controlled dynamic systems involving flow, structure, and sound radiation is the problem of controllability of a coupled unstable system via the boundary conditions. In the experiments conducted, it is shown that there are circumstances in which a coupled nonlinear system can be uncoupled and stabilized. An example is the unstable boundary layer transition which can be stabilized by a gradual spacial surface cooling, even in the presence of an acoustic field which is the original source of instability. For global control of a system involving flow/structure/sound radiation, the combination of passive and active controls is perhaps one of the most effective ways to maintain laminar flow over a flexible surface and small displacement amplitude in the presence of a noisy environment (such as the outer surface of an airplane in flight).

REFERENCES

1. Maestrello, L., 1969. "Use of Turbulent Model to Calculate the Vibration and Radiation Responses of a Panel with Practical Suggestion of Resolving Sound Level." Journal of Sound and Vibration, pp. 405-448.

2. Maestrello, L., September 1968. "Design Criterion of Panel Structure Excited by Turbulent Boundary Layer." Journal of Aircraft, 5(4).

3. Baht, W.V. and Wilby, J.F., 1971. "Interior Noise Radiation by an Airplane Fuselage Subjected to Turbulent Boundary Layer Excitation and Evaluation of Noise Reduction Treatments." Journal of Sound and Vibration, 18():449-464.

4. Fuller, C.R., Metcalf, V. L., Silcox, R.J., and Brown, D.E., June 1989. "Experiments on Structural Control of Sound Transmitted Through an Elastic Plate," Proceedings of Amer. Control Conf., Pittsburgh, PA, June 1989, pp. 2079-2084.

5. Gibbs, P. G. and Fuller C. R., 1990. "Experiments on Active Control of Vibrational Power Flow Using Piezoceramic Actuators and Sensors," AIAA 1990.

6. Redman-White, W., Nelson, P.A., and Curtis, A.R.D., 1987. "Experiments on Active Control of Flexural Waves Power." Journal of Sound and Vibration, 112(1):177-181.

7. Maestrello, L. and Grosveld, F.W. "Transition Control of Instability Waves Over a Flexible Surface in the Presence of an Acoustic Field," AIAA paper 90-4008, 13th AIAA Aeroacoustics Conference,Tallahassee, Florida, October 22-24, 1990.

8. Maestrello, L. and Grosveld, F.W. "Control of Boundary Layer Transition Location and Plate Vibration in the Presence of an External Acoustic Field." Paper to be presented at the 4th International Conference on Recent Advances in Structural Dynamics, Southampton, England, July 1991.

9. Hall, P. and Horseman, N.J. "The Inviscid Secondary Instability of Fully Nonlinear Longitudinal Vortex Structures in Growing Boundary Layers." ICASE Rep. No. 90-71.

10. Liepmann, H.W. "Investigation of Laminar Boundary Layer Stability and Transition on Curved Boundaries." NACA Wartime Report W@ 107.

11. Goldstein, M.E. 1983. "The Evolution of Tollmien-Schlichting Waves Near a Leading Edge." Journal of Fluid Mechanics, 127:59-81.

12. Lowson, M. "Acoustic Forcing of Three Dimensional Shear Layer," AIAA 89-1063.

13. Maestrello, L., 1986. "Active Transition Fixing and Control of the Boundary Layer in Air." AIAA Journal, 24:1577-1581.

14. Gapanov, S.A., May-June 1983. "Excitation by Sound of Tollmien-Schlichting Waves in a Supersonic Boundary Layer." Novosibirsk, translated from Izvestiya Akademii Nauk USSR, Mekhanika Zhidkosti i Gaza, No.3, pp. 59- 66.

15. Maestrello, L. and Nagabushana,K.A. "Relaminarization of Turbulent Flow on a Flat Plate by Localized Surface Heating," AIAA 89-0985. 12.

16. Liepmann, H.W. and Nosenchuck, D.M., 1982. "Active control of Laminar Turbulent Transition." Journal of Fluid Mechanics, 118:201-204.

17. Struminski, V.V., Dovgal, A.V., Lebedev, V.B., Levchenko, B.YA., Timofeev, V.A., and Formichev, V.M., 1987. "Theoretical and Experimental Investigation of the Stability of the Boundary Layer to Unsteady Heating on the Surface." Reprint No. 3-87, Inst. of Theor. and Appl. Mech., Academy of Science, USSR. (Russian Edition).

18. McCroskey, W.J., 1962. "Experimental Investigation of Boundary Layer Transition on a Flat Plate with a Point Heat Source at the Leading Edge." Princeton University Report No. 623.

19. Maestrello, L. and El-Hady, Nabil M. "Radiation and Control of the Acoustic Field Due to a Wave Packet Evolving Along a Concave-Convex Surface." Excerpted for publication in AIAA Journal, AIAA 89-1110.

20. Maestrello, L., 1986. "Active Transition Fixing and Control of the Boundary Layer in Air." AIAA Journal, 24(10):1577-1581.

Advanced Actuators and Sensors

DESIGN OF SHAPE MEMORY ALLOY COILS AND THEIR APPLICATIONS IN VIBRATION CONTROL

C. Liang,[1] C. A. Rogers[1]

ABSTRACT

Shape memory alloys (SMAs) have several unique characteristics, including their Young's modulus-temperature relations, shape memory effects, and damping characteristics. The Young's modulus of the high-temperature austenite of SMAs is about three to four times as large as that of low-temperature martensite. Therefore, a spring made of shape memory alloys can change its spring constant by a factor of four. Since a shape memory alloy spring can vary its spring constant, provide recovery stress (shape memory effect), or be designed with a high damping capacity, it can be very useful in adaptive vibration control. Some vibration control concepts utilizing the unique characteristics of SMAs will be presented in this paper. It is expected that using shape memory alloy springs as a control element will have a significant impact on active/adaptive vibration control.

Shape memory alloy springs have been used as actuators in many applications although its use in the vibration control area is very recent. Since shape memory alloys differ from conventional alloy materials in many ways, the traditional design approach for springs is not complelely suitable for designing SMA springs. Some design approaches based upon linear theory have been proposed for shape memory alloy springs. A more accurate design method of SMA springs based on a new non-linear thermomechanical constitutive relation of SMA is also presented in this paper.

INTRODUCTION

Shape memory alloys have been commercially available since the 1970s. However their applications have been very limited. Most applications of shape memory alloys are force and displacement actuators. Until most recently, Rogers (1988) suggested that shape memory alloy fibers could be embedded into conventional composites such as graphite/epoxy to control the structural response including static deformation, vibration, buckling, and structural acoustic radiation/transmission. Nagaya et al. (1987) suggested that shape memory alloys could be used to control the critical speed of rotating shafts. A great deal of research and development related to SMAs is currently in progress (Dye, 1990). This paper will begin with an introduction to some of the basic

[1]Center for Intelligent Material Systems and Structures, Mechanical Engineering Department, Virginia Polytechnic Institute and State University, Blacksburg, VA 24061, U.S.A.

characteristics of shape memory alloys, then briefly present a design approach for shape memory alloy springs, and finally discuss some of the possible applications of shape memory alloy springs in vibration control.

INTRODUCTION TO SHAPE MEMORY ALLOYS

The first observation of the shape memory effect (SME) was made in 1932 with gold-cadmium. The phase transformation associated with the shape memory effect was later discovered in 1938 with brass (Perkins et al., 1975). It was not until 1962 that Buehler and Wiley (1965) at the Naval Ordinance Laboratory (NOL) discovered a series of Nickel-Titanium alloys that demonstrated this shape memory effect. The shape memory alloy discovered by Buehler et al. was later named Nitinol, and has been made commercially available ever since.

A shape memory alloy is able to "memorize" its original configuration after it has been deformed by heating the alloy above characteristic transition temperatures. This unique effect of returning to its original geometry after a large inelastic deformation (more than 1 percent) is known as the shape memory effect (SME).

If steel in a high temperature austenitic phase is quenched it will generally harden. After polishing and etching, observation with a microscope will show an extremely fine structure, first named martensite after the German metallurgist Adolf Martens. Later this martensitic phase transformation involving shear deformation was found to be diffusionless due to the cooperative atomic movement of the crystal lattice. The phase transformation is accompanied by a shape change (or surface relief) (Clark and Wayman, 1970). Different materials have a different amount of surface relief depending upon the crystal orientation of the parental phase during the martensitic phase transformation. Shape memory alloys possess a martensitic transformation similar to steel, but the characteristic transition temperature is much lower than steel. The lower phase transformation temperature and large amount of surface relief during the martensitic transformation of SMAs play an important role in the deformation mechanism operating in the shape memory effect.

The martensitic transformation is the basic characteristic of shape memory alloys involved in all the unique characteristics of SMAs. The martensitic transformation may be simply illustrated by the change of martensite volume fraction with respect to temperature as shown in Figure 1. The four important transition temperatures are martensite finish (M_f), martensite start (M_s), austenite start (A_s), and austenite finish (A_f).

The mechanical behavior of shape memory alloys is unique. Ordinary materials like steel do not involve phase transformations in the working temperature range that affects its mechanical behavior significantly. However, because the associated phase transformation of shape memory alloys is so unique, the mechanical characteristics of SMAs are likewise significantly different from other engineering materials. Figure 2 shows the stress-strain curves of a nitinol alloy ($Ti_{55}Ni$) at various temperatures (Cross et al., 1970).

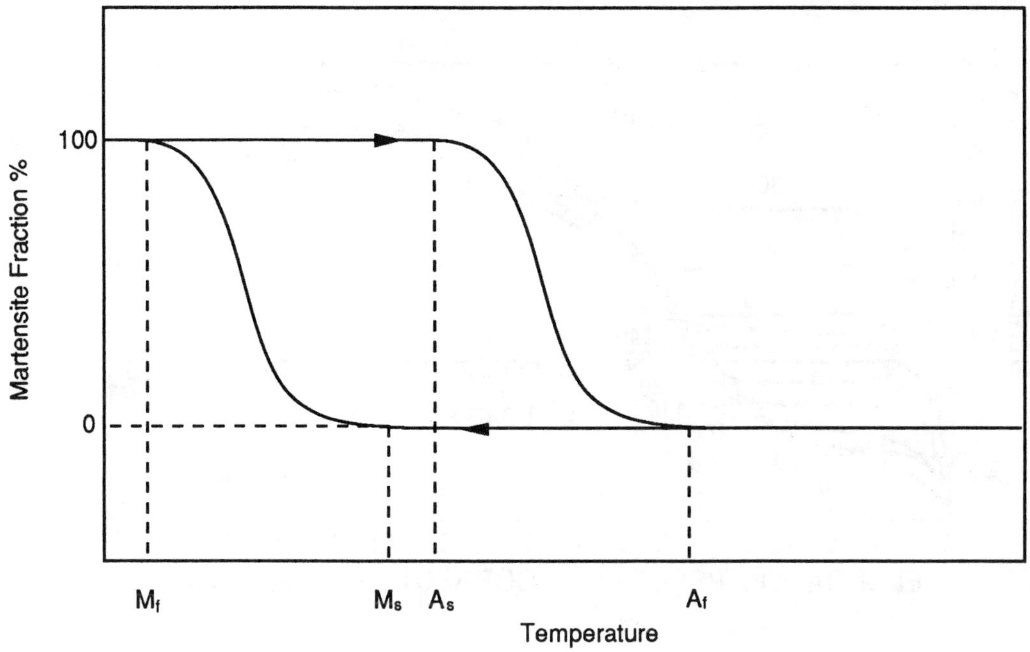

Figure 1: A Schematic Diagram of Martensitic Fraction vs. Temperature

When the testing temperature is much higher than the austenitic finish temperature (A_f), shape memory alloys present a very unique nonlinear behavior, i.e., pseudoelasticity (or super-elasticity in some papers). The pseudoelasticity may be viewed as a combination of elastic and martensitic phase transformation related nonlinear loading and unloading behavior. Figure 3 (Funakubo, 1986) shows the complete loading and unloading process of a nitinol alloy at various temperatures. The main features of the pseudoelastic effect are summarized in Figure 4. The lines OA, $B.C.$, and DO shown in Figure 4 show the elastic loading and unloading process. The AB section of the loading process corresponds to a martensitic phase transformation (austenite to martensite); the CD section of the unloading process corresponds to the reverse phase transformation (martensite to austenite). The stress value of A, B, C, and D may be considered as material constants for a pseudoelastic shape memory alloy. The strain value of B or C can be as high as seven percent. The energy loss due to the static hysteresis is significant compared with ordinary alloys. The damping characteristics of pseudoelastic shape memory alloys differs from that of the plastic deformation of materials for which plastic stress-strain hysteresis will inevitably induce internal damage resulting in limited fatigue life-span. Pseudoelastic shape memory alloy materials, however, have relative high fatigue strength due to the fact that the nonlinear hysteresis is metallurgically reversible and it does less "damage" to the material than plastic deformation.

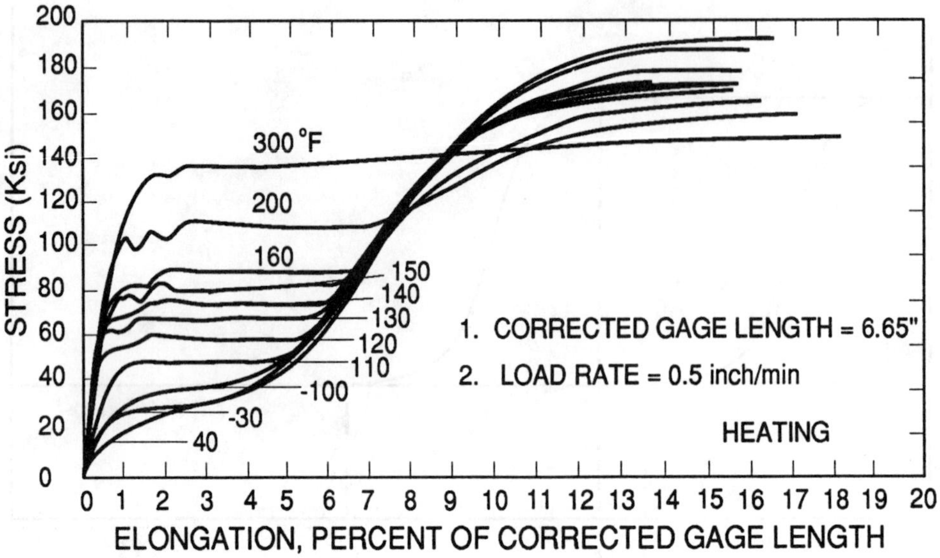

Figure 2: Stress-Strain Curves of a Nitinol ($Ti_{55}Ni$) at Various Temperatures (Cross et al., 1970)

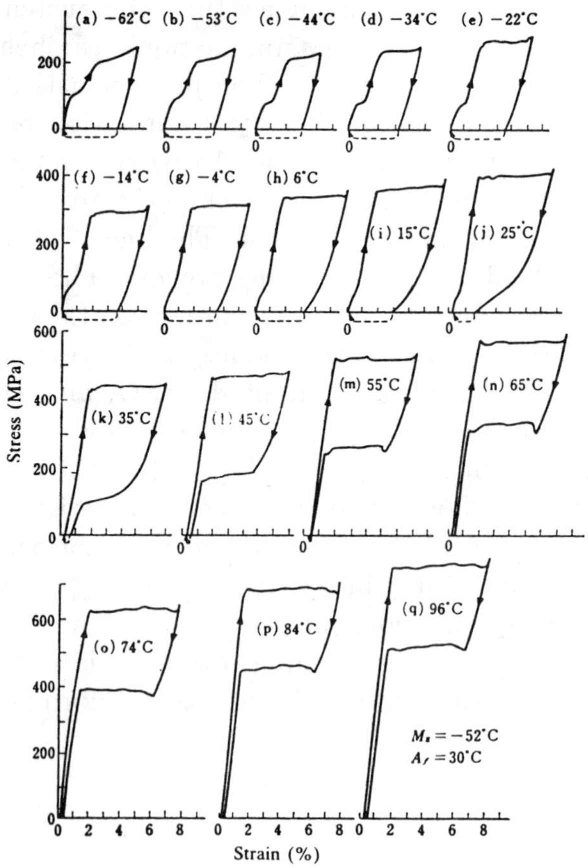

Figure 3: Stress-Strain Curve of a Nitinol Alloy (Funakubo, 1986)

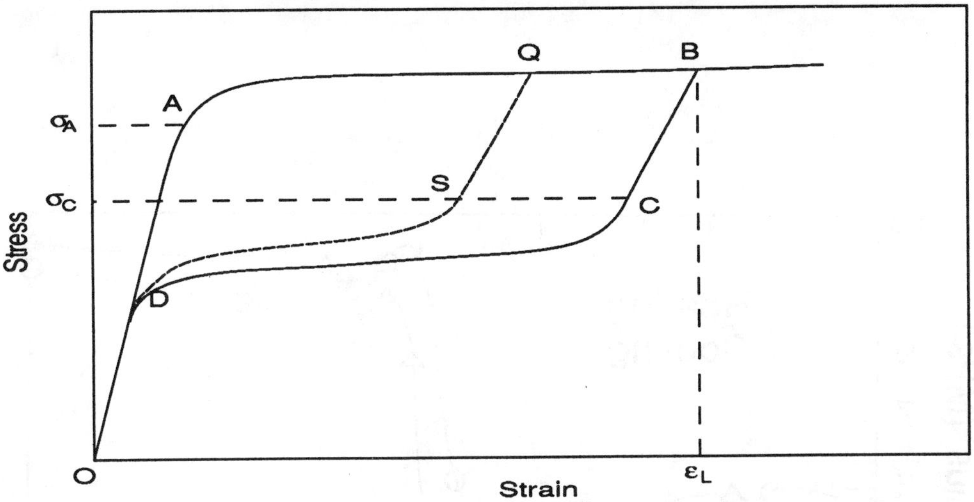

Figure 4: Schematic Diagram of the Pseudoelastic Effects of Shape Memory Alloys

The change of Young's modulus of shape memory alloys is very different from conventional metallic materials. For most metals, the Young's modulus decreases as the temperature increases. However, the Young's modulus of shape memory alloys increases within the phase transformation temperature range for the heating process. The Young's modulus of nitinol increases by 3 to 4 times from temperatures below M_f to temperatures above A_f. This is shown in Figure 5 (Cross et al., 1970). The shear modulus of SMA has a similar temperature behavior.

The shape memory effect does not simply mean that the previous shape will be restored freely upon heating the SMA. Restraining a deformed SMA wire while heating the wire above the transition temperatures will generate a large recovery stress as shown in Figure 6. This unique behavior is one of the fundamental characteristics exploited in many applications of shape memory alloys as actuators.

Besides the damping from the pseudoelastic stress-strain hysteresis, the damping of SMA due to internal friction is also an important characteristic. Figure 7 shows the internal friction δ (logarithmic decrement) as a function of temperature. The low-temperature martensitic phase has a high damping capacity (the loss factor can reach up to 10 percent) while the high-temperature austenitic phase has a lower damping capacity (loss factor about 1 percent). The large damping capacity of martensitic alloys is believed to be caused by the reversible movement of the martensite and/or twin plate boundaries. Similar to other alloy materials, interfacial mobility, transformation hysteresis, stress and strain amplitide and excitation frequency affect the damping characteristics of SMA. The influence of strain amplitude on the damping capacity of SMA martensite is also shown in Figure 7. Frequency does not change the $\delta - T$ profile although it may affect relative magnitude if the martensitic phase transformation time interval is coherent with the frequency.

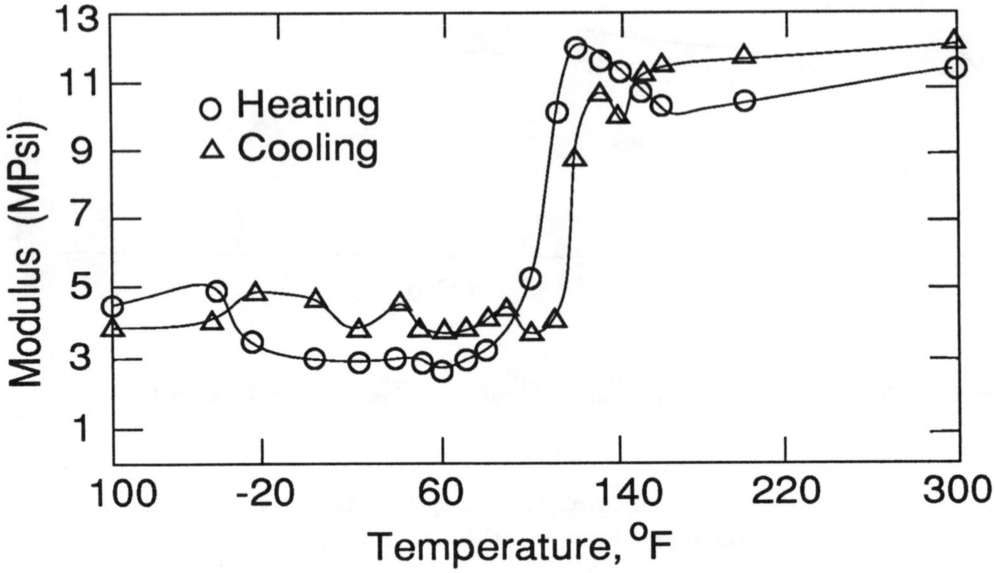

Figure 5: Young's Modulus of a Nitinol vs. Temperature (Cross et al., 1970)

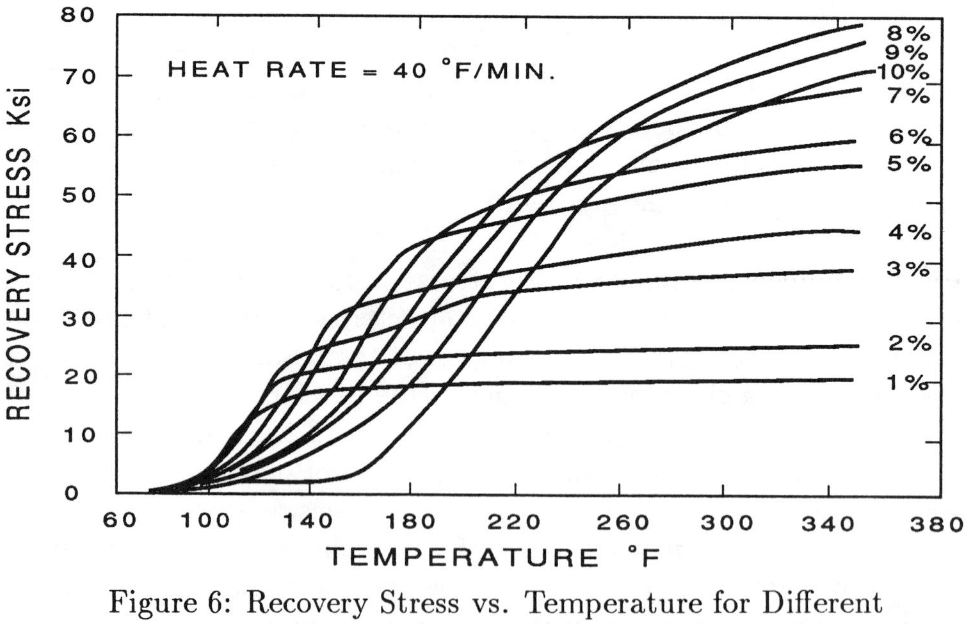

Figure 6: Recovery Stress vs. Temperature for Different Initial Strains (Cross et al., 1970)

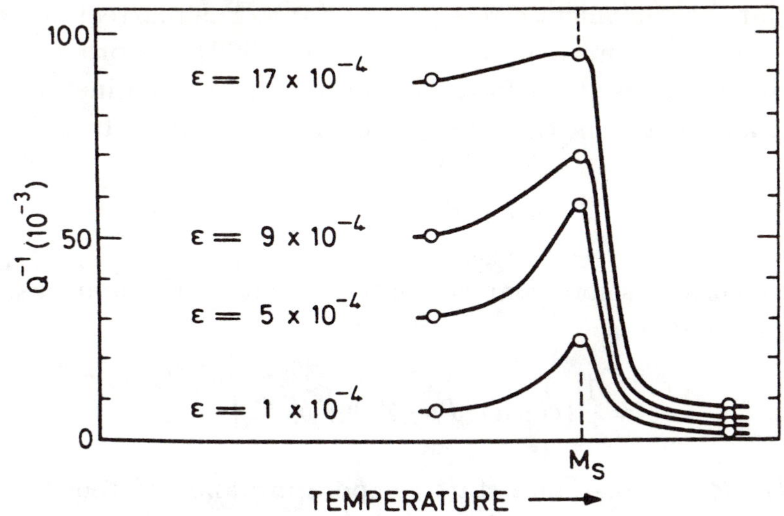

Figure 7: Loss Factor of a Cu-Zn-Al Shape Memory Alloy vs. Temperature for Various Strain Amplitude, (Dejonghe et al., 1977)

INTRODUCTION TO THE CONSTITUTIVE MODEL OF SMAS

The nonlinear behavior of shape memory alloys is a result of their internal phase transformation. Several constitutive models have been proposed to describe the unique characteristics of SMAs, i.e., the stress-strain-temperature relations and shape memory effects (detailed review can be found in Liang (1990)). Among these models, the thermomechanical constitutive model, developed by Tanaka and Nagaki (1982) and later modified by Liang and Rogers (1990a), can be used to predict and describe the mechanical behaviors of SMAs quantitatively. A thermomechanical multi-dimensional model based on the one-dimensional model has also been developed (Liang and Rogers, 1991). A general design method for shape memory alloy actuators has been developed based on this constitutive model (Liang and Rogers, 1990b).

Based on the multidimensional constitutive relation (Liang and Rogers, 1991), the one-dimensional shear stress and strain relation of shape memory alloys can be expressed as

$$\tau - \tau_0 = G(\gamma - \gamma_0) + \frac{\Omega}{\sqrt{3}}(\xi - \xi_0) + \frac{\Theta}{\sqrt{3}}(T - T_0) \tag{1}$$

where G is the elastic shear modulus, Ω the phase transformation tensor which can be calculated from $-D\ \epsilon_L$ where D the elastic modulus and ϵ_L the maximum recoverable strain of SMAs; Θ is the thermoelastic tensor measured from experiments, and T the temperature, and ξ the internal variable which describes the degree of martensitic transformation which is defined as a martensitic fraction (total martensite/total SMA). The martensitic fraction is a function of stress and temperature, and can be generally expressed as follows for multi-dimensional problems.

$$\xi = \Xi(\sigma_{eq}, T) \tag{2}$$

where σ_{eq} stands for the equivalent stress ($\sigma_{eq} = \sqrt{3}\tau$). The martensitic fraction, ξ, may be derived based on transformation kinetics. Liang (1990) has proposed two empirical relations to describe the phase transformation processes. The cosine model is used here. In a martensitic transformation, the martensite fraction is given by

$$\xi = \frac{1}{2}\{\cos[a_M(T - M_f) + b_M\bar{\sigma_{eq}}] + 1\} \tag{3}$$

and for a reverse transformation (martensite to austenite), the martensitic fraction according to the cosine model is

$$\xi = \frac{1}{2}\{\cos[a_A(T - A_s) + b_A\bar{\sigma_{eq}}] + 1\} \tag{4}$$

where A_s and M_f are the austenite start temperature and martensite finish temperature, a_M and a_A can be derived from the transition temperatures, and b_M and b_A may be calculated by assuming a linear relation of transition temperatures and stress. Detailed discussion can be found in Liang and Rogers (1990a) and Liang (1990).

INTRODUCTION TO THE DESIGN OF SMA SPRINGS

The design of conventional springs is standard and can be found in most machine design text books (Shigley, 1977). A general review is briefly given here. The maximum shear stress of a linear elastic spring is

$$\tau_{max} = K\frac{2FR}{\pi r^3} \tag{5}$$

where K is called the Wahl correction factor; F is the external force, R the mean spring radius, and r the radius of the spring wire.

The deflection of a linear elastic helical spring can be derived by considering the deformation of an element of length, dx, cut from a wire of radius r as shown in Figure 8. Consider a line ab on the surface of the wire element which is parallel to the spring axis. After deformation it will rotate through the angle γ and occupy the new position ac. The angle γ, form the Hook's law for torsion, is given by

$$\gamma = \frac{\tau}{G} = \frac{2FR}{\pi r^3 G} \tag{6}$$

where the value of τ can be calculated from Equation (6) and the Wahl correction factor is assumed to be unity (this modification does not induce much error according to Shigley (1977).

The angle $d\alpha$, through which one section of the element rotate with respect the other, is

$$d\alpha = \frac{\gamma(dx)}{r} \tag{7}$$

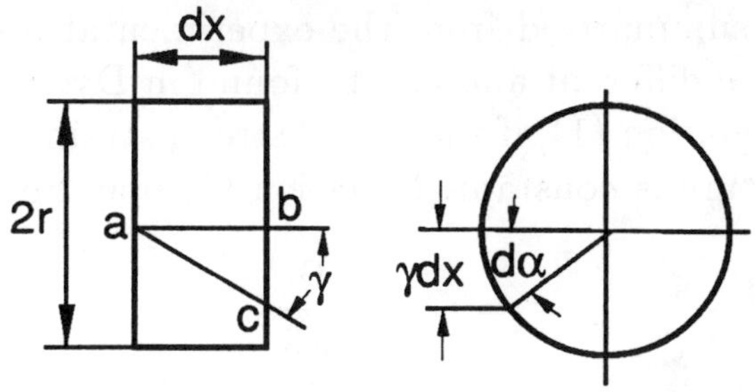

Figure 8: Cross-Section Element of a Helical Spring

The angular deflection of one end of the spring wire with respect to the other is

$$\alpha = \int_0^{2\pi RN} \frac{\gamma}{r} dx = \frac{4FR^2N}{r^4G} \tag{8}$$

where N is the total number of active coils. The total deflection is expressed as

$$y = \alpha r = \frac{4FR^3N}{r^4G} \tag{9}$$

The spring constant, from Equation (10), is

$$k = \frac{r^4G}{4R^3N} \tag{10}$$

The elastic modulus of shape memory alloys varies with temperature. The modulus and temperature generally has the following relations according to Liang (1990).

$$D - D_A + (D_M - D_A)\xi \tag{11}$$

where D, D_A, and D_M are the Young's modulus in any state, austenitic phase, and martensitic phase respectively. This linear relation of $D - \xi$ can nicely fit most $D - T$ relations if the relation of ξ vs. T is substituted into Equation (12). The ratio of D_A to D_M for most commercially available shape memory alloys is about three to four. The shear modulus is given by

$$G = \frac{D}{2(1+\mu)} \tag{12}$$

where μ is the Possion's ratio and for most SMAs is about 0.3 in the elastic range. When the temperature is lower than M_s the shear modulus is the martensite shear modulus, G_M. When the temperature is higher than A_f the shear modulus is the austenite shear modulus, G_A. The ratio of G_A to G_M should be about three to four for most shape memory alloys.

The elastic stress limit of shape memory alloys is a function of phase transformation, i.e., a function of martensitic fraction, and can generally written as

$$\begin{aligned} \sigma_e &= C(T - M_s) & T &\geq M_s \\ \sigma_e &= C'(M_s - T) & M_f &\leq T \leq M_s \\ \sigma_e &= constant & T &\leq M_f \end{aligned} \tag{13}$$

The above equations are summarized from the experimental results of Dye (1990). C' and C in Equation (13) are different and can be found in Dye (1990). From the general constitutive relation of Equation (1), if the initial stress, strain, and martensitic fraction are zero and the temperature is constant, Equation (1) may be rewritten as

$$\tau = G\gamma + \frac{\Omega}{\sqrt{3}}\xi \tag{14}$$

The shear stress of an SMA coil from Equation (6) can still be expressed as

$$\tau = \frac{2FR}{\pi r^3} \tag{15}$$

γ can be solved from Equation (15) as

$$\gamma = \frac{\tau}{G} - \frac{\Omega}{\sqrt{3}G}\xi \tag{16}$$

The deflection of an SMA spring from Eqs. (9) and (10) is

$$y = R\int_0^{2\pi RN} \frac{\gamma}{r}dx = \frac{2\pi R^2 N}{r}(\frac{\tau}{G} - \frac{\Omega}{\sqrt{3}G}\xi) \tag{17}$$

Substituting Equation (16) into Equation (17) yields

$$y = \frac{4KR^3N}{Gr^4}F - \frac{2\Omega N\pi R^2}{\sqrt{3}Gr}\xi \tag{18}$$

The martensite fraction is zero in the elastic range. A stress induced phase transformation starts when the equivalent stress, σ_{eq}, reaches σ_e (the shear stress, τ, equals to $\frac{\sigma_e}{\sqrt{3}}$). The force which corresponds to this elastic limit, F_e, is

$$F_e = \frac{\pi r^3 \sigma_e}{2\sqrt{3}KR} \tag{19}$$

The displacement corresponding to the elastic stress limit, y_e, is

$$y_e = \frac{2\pi N R^2 \sigma_e}{\sqrt{3}KdG} \tag{20}$$

The force-displacement relation of an SMA spring can be normalized to have the following form.

$$\frac{y}{y_e} = K(\frac{F}{F_e} - \frac{\Omega}{\sigma_e}\xi) \tag{21}$$

The Wahl coefficient, K, can be made unity and still achieve accurate displacement results according to Shigley (1977). Therefore the force-displacement relation of an SMA spring can be simplified as

$$\frac{y}{y_e} = \frac{F}{F_e} - \frac{\Omega}{\sigma_e}\xi \tag{22}$$

and the shear stress can be expressed as

$$\tau = \frac{\sigma_e}{\sqrt{3}} \frac{F}{F_e} \tag{23}$$

Thus, the martensite fraction, ξ, can be rearranged to have the following form:

$$\xi = \Xi(T, \frac{\sigma_e}{\sqrt{3}} \frac{F}{F_e}) \tag{24}$$

The force-displacement expression of Equation (23) is very convenient in the design and application of shape memory alloy springs as actuators and other control elements. The phase transformation tensor is independent of temperature. Other parameters, such as y_e, F_e, and σ_e, are temperature-dependent. All the design variable of SMA springs, i.e., SMA radius (r), spring radius (R), Wahl coefficients (K), and the number of active coils are only in the normalization factors, y_e and F_e.

The shape memory effect of SMA springs may also be derived similarly to the derivation of their force-displacement relations. A general form of the force-displacement-temperature relation of SMA springs may be derived as

$$\frac{F}{F_e} = \frac{y}{y_e} + \frac{\Omega}{\sigma_e}(\xi - \xi_0) + \frac{\Theta}{\sigma_e}(T - T_0) \tag{25}$$

The recovery force-temperature relation, recover force-recovery displacement and recovery force-displacement-temperature relations can then be predicted from Equation (26) with different initial conditions, various constraint conditions, and different ξ relations. The pseudoelastic behavior can also be predicted and described. What has been presented here is only a brief introduction. A detailed discussion about the design of SMA springs will be published in another paper.

APPLICATIONS OF SMA SPRINGS IN VIBRATION CONTROL

Active vibration control can be realized by either tuning material properties or applying controllable stimulus. Shape memory alloys have a tunable elastic modulus and recovery force. Since tuning the properties or recovery stress of SMA requires heating or cooling, SMA may not be suitable for high frequency vibration control, but it may be used for low-frequency vibration control and quasi-static response control. The high damping capacity of the martensitic phase of shape memory alloys can be used in passive vibration control. The hysteric damping of pseudoelastic SMAs can be exploited in passive/adaptive vibration control. To demonstrate the possibility and feasibility of using SMA springs in vibration control, we shall discuss four cases. Here, SMA springs will be used, but other configurations such as bars or wires may also be used.

CASE 1: VIBRATION CONTROL UTILIZING THE HIGH-DAMPING CHARACTERISTICS OF SMA

The loss factor of martensite of SMA usually ranges from 0.001 to 0.1 depending on amplitude as shown in Figure 7. Ordinary alloys such as steel and aluminum have loss factors around 0.005. Damping material like acrylic rubber, natural rubber, and urethane, have loss factors ranged from 0.1 to 1.0 depending on the temperature and frequency. However, the Young's modulus of rubber-like damping material is about 10 Kpsi while the Young's modulus can reach as high as 12 Mpsi. The strength of SMA is also much higher than rubber materials. The high damping capacity of SMA can be very useful. For example, shape memory alloy hybrid composites (Rogers, 1987) contains 10 to 40 percent SMA fibers as reinforcement which will add a significant amount of damping to composite structures.

Shown in Figure 9 is the loss factor of a Cu-based SMA vs. stress level. An SMA specimen must be under a state of stress in order to achieve high internal friction. Therefore, SMA components must be preloaded to a certain level if they are to be used as dampers.

Conventional viscous damping model may be used to describe the internal friction characteristics of SMAs. Consider the following one-degree of freedom system as shown in Figure 10. A conventional spring is used to preload the SMA spring which is modeled as a conventional spring with a parallel damper. The damping may also be assumed to have a linear relation with strain rate. The equation of motion for the system shown in Figure 10 is

$$m\frac{\partial^2 x}{\partial t^2} + (K_C + K_S)x + \frac{K_S\eta}{\omega}\frac{\partial x}{\partial t} = F_0 \cos\omega t \tag{26}$$

where $\frac{K_S\eta}{\omega}\frac{\partial x}{\partial t}$ is the resistance of the viscous damper which models the internal friction of SMA (Jones, 1987), η the loss factor (same as Q^{-1}), $F_0 \cos\omega t$ the external forcing function K_C the spring constant of the conventional spring, and K_S the spring constant of the SMA spring.

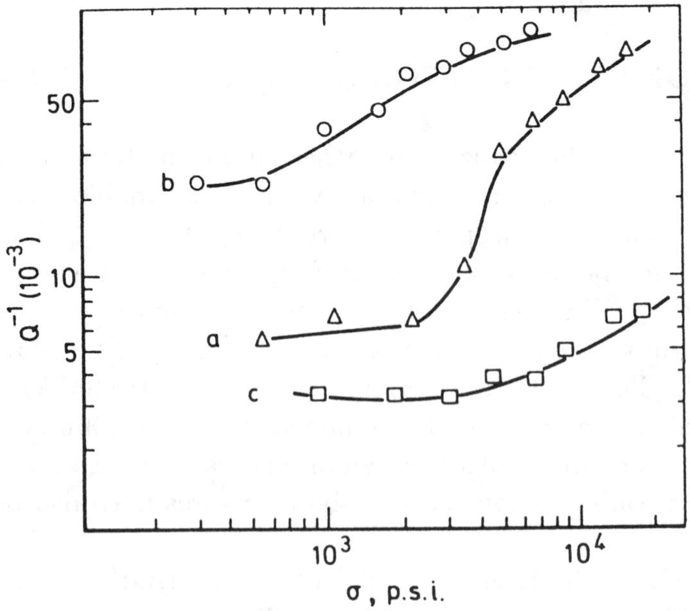

Figure 9: Amplitude Dependence of Damping Capacity in Martensite Region (a), Transformation Region (b), and Austenite Region (c) of a Cu-based SMA, (Dejonghe et al., 1977)

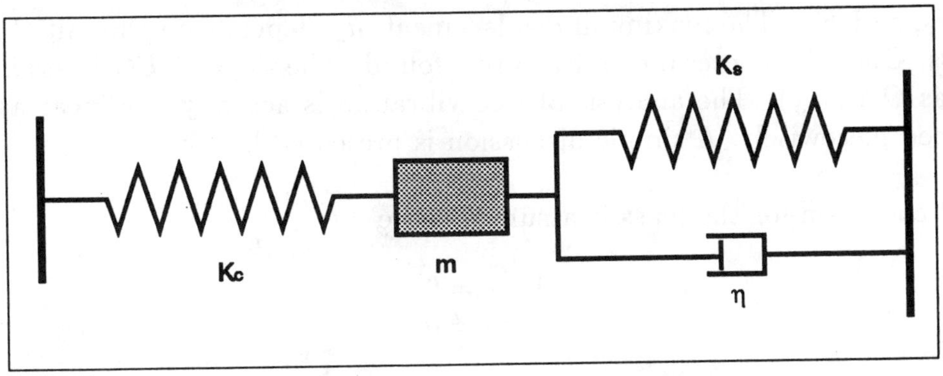

Figure 10: Modeling of the Damping Characteristics of an SMA Spring

In vibration analysis, damping ratio, ξ, is often used instead of loss factor, η. The damping ratio has the following relation with loss factor.

$$\eta = 2\xi \tag{27}$$

The vibration amplitude of the forced vibration is then obtained as

$$A = \frac{F_0}{2(K_c + K_s)\xi\sqrt{1-\xi^2}} \tag{28}$$

This analysis treats the internal friction of SMA as viscous damping. Once the damping ratio of SMA or SMA hybrid composites is determined, the vibration analysis follows the standard approach introduced in any vibration text book.

CASE 2: VIBRATION CONTROL UTILIZING THE HYSTERETIC DAMPING OF SMA

The austensite finish temperature (A_f) of pseudoelastic shape memory alloys is far below room temperature. This results in a complex unloading process as shown in Figure 3. The large hysteresis is due to the irreversible thermodynamic aspects of phase transformation. If the stress is below σ_A, the material is elastic and there is little internal damping. Once the stress is higher than σ_A, nonelastic transformation deformation is generated. The maximum amount of recoverable strain (elastic strain and strain resulting from martensitic de-twinning deformation) is ϵ_L. Unloading before ϵ_L will first follow the elastic path QS. The reverse transformation (martensite to austenite) at higher ambient temperature ($\gg A_F$) starts at about the same stress level, σ_C. Unloading path (CD) corresponds to the reverse transformation process. Energy is dissipated in the form of heat which requires a certain amount of time for the damped energy to dissipate completely. THerefore, the loading frequency is crucial for this type of damping mechanism.

Now, consider the free vibration of a one-degree of freedom system consisting of a pseudoelastic SMA spring and a mass. To simplify the analysis, the pseudoelastic SMA spring is assumed to have a force-displacement relation as shown in Figure 11. F_1 and F_2 are material constants; the spring constant K_S is known, so d_1 and d_4 can be solved from F_1, F_2, and K_S. The maximum displacement, d_2, depends on the initial condition of the mass. Once d_2 is determined, d_3 can be found. The slope of BC is assumed to be the same as OA (K_S). The analysis of free vibration is actually nonlinear and should be performed piecewisely. A simple discussion is presented below.

The initial condition for the mass is assumed to be

$$\begin{cases} x_0 = 0 \\ \dot{x}_0 \neq 0 \end{cases} \tag{29}$$

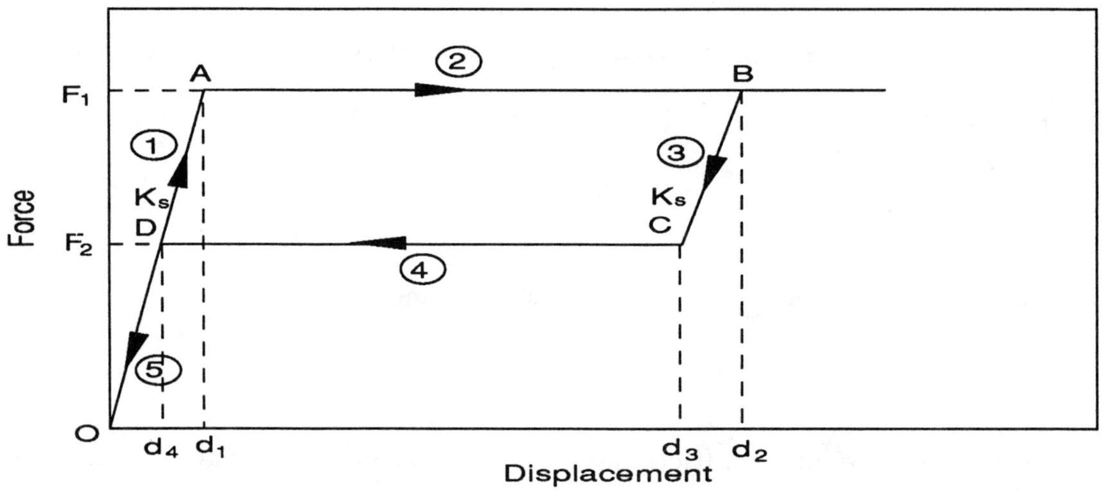

Figure 11: Simplified Pseudoelastic Force-Displacement Curve

If the initial kinetic energy of the mass is greater than the maximum linear potential energy of the SMA spring, $\frac{1}{2}K_S d_1^2$ which is the area of OAd_1, the hysteric damping of the material will be effective. In this case, the initial velocity of the mass, $\dot{x}_0$, must satisfy the following inequality.

$$\dot{x}_0 > \sqrt{\frac{K_S}{m}} d_1 \tag{30}$$

The equation of motion in the OA section is

$$\ddot{x} + px = 0 \tag{31}$$

where $p = \sqrt{K_S/m}$. The general solution to Equation (31) is

$$x = A_1 \sin(pt + \phi_1) \tag{32}$$

The unknown vibration amplitude, A_1, and phase angle, ϕ_1, can be determined from the initial conditions. The time (t_1) when the mass reaches A can then be determined from Equation (32). The velocity of the mass, $\dot{x}_1$, at A may also be determined.

From A to B (B is unknown thus far), the governing equation is

$$m\ddot{x} = -F_1 \tag{33}$$

Its solution is

$$x = -\frac{F_1}{2m}t^2 + a_2 t + b_2 \tag{34}$$

The unknown coefficients, a_2 and b_2, can be solved from the initial conditions, which are

$$\begin{cases} x|_{t=t_1} = d_1 \\ \dot{x}|_{t=t_1} = \dot{x}_1 \end{cases} \tag{35}$$

The maximum displacement, d_2, can be determined from energy balance relation, which is given by

$$\frac{1}{2}m\dot{x}_0^2 = \frac{1}{2}K_S d_1^2 + F_1(d_2 - d_1) \tag{36}$$

Similarly, the time (t_2) when the mass reaches B can be solved from Equation (34). The velocity of the mass at B is zero. At this point the mass begins to move in the opposite direction. The new initial conditions for section BC are

$$\begin{cases} x|_{t=t_2} = d_2 \\ \dot{x}|_{t=t_2} = 0 \end{cases} \tag{37}$$

The equation of motion is the same as Equation (32). The displacement response now is

$$x = A_3 \sin(pt + \phi_3) \tag{38}$$

The unknown coefficient, A_3 and ϕ_3, can be determined from the initial conditions. Position C can be determined once d_2 is known. Following the same procedure, the entire displacement response can be determined. $O \to A$, $B \to C$, and $D \to O$ responses are described by sine functions; $A \to B$ and $C \to D$ responses are described by parabolic functions. After the first complete cycle, the dissipated energy, ΔE^1, is

$$\Delta E^1 = (F_1 - F_2)(d_2^1 - d_1) \tag{39}$$

where the superscript '1' represents the first cycle. Assuming that the pseudoelastic SMA spring has the same compressive force-displacement relation, the next cycle will have a similar energy reduction except d_2^2 will be different. The new d_2^2 can be solved based on the residual kinetic energy (after deduction of ΔE^1). It is simply a mathematical manipulation to solve the number of cycles that is needed to dissipate the excessive amount of energy, ΔE, given by

$$\Delta E = \frac{1}{2}m\dot{x}_0^2 - \frac{1}{2}K_S d_1^2 \tag{40}$$

If there is no other damping, the pseudoelastic SMA spring and mass system will finally stablized to an undamped free vibration. The vibration amplitude will be d_1 and the angular velocity, ω, should be $2\pi\sqrt{K_S/m}$. Schematically, the entire response of this system is shown in Figure 12. Obviously, pseudoelastic SMA material can be used as dynamic dampers. Graesser (1990) suggested that SMA material be made as add-on damping devices. Pseudoelastic materials differ from conventional damping materials, such as rubber, which has higher damping ratios when it is subjected to shear forces. Pseudoelastic SMA has similar damping characteristics whether it is subjected to tension, compression, or shear.

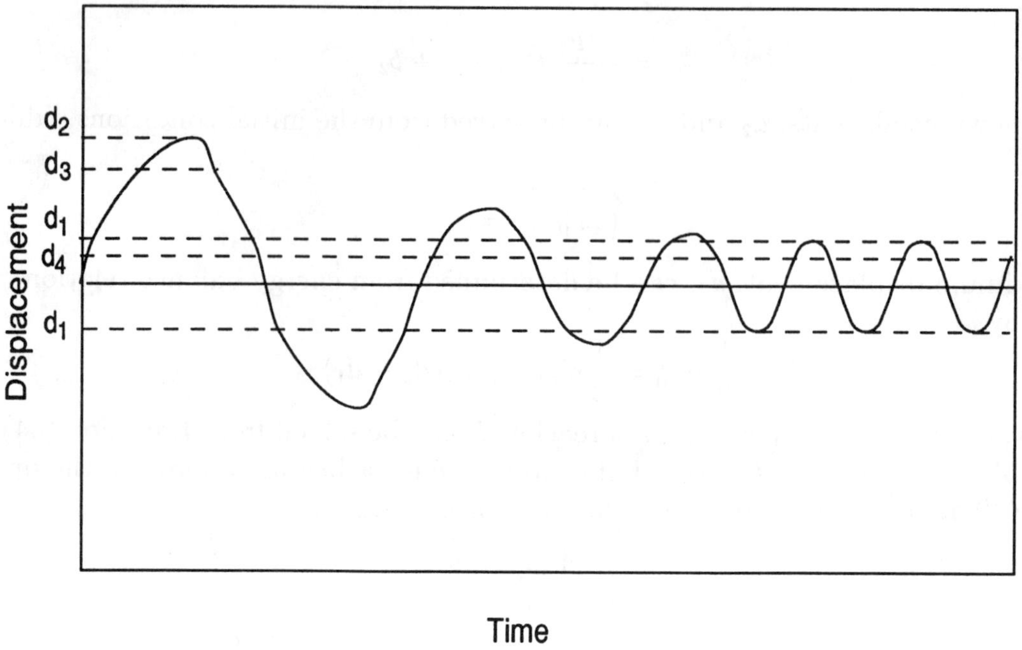

Figure 12: Schematic Response of the Pseudoelastic SMA Spring and Mass System

One application of these unique characteristics is to use SMAs as a seismic suppression device for civil engineering structures. In this application, SMAs may be used as part of the basement support. The load that the SMA support carries can be designed to be closed to its elastic limit (point A as shown in Figure 4). The energy of a large amplitude vibration of the building caused by a major earthquake may very well be dissipated.

Another application of pseudoelastic SMA springs is what we call vibration filters. Shown in Figure 13 is a simply-supported beam which has a pseudoelastic SMA spring as a support. The SMA spring can be preloaded to somewhere in the OA section (Figure 4). Now assume that the SMA spring has been loaded to A. When the simply-supported beam is impacted, many vibration modes will be excited. The number of modes excited is infinite but the energy associated with high-order modes is small. The vibration is basically dominated by low-order modes. The modes that can be excited depend on the type of excitation and the location of the excitation. For example, if a pin force impacts in the middle of the beam, the odd modes will be excited and no even mode can theoretically be excited. The vibration of the beam, in this case, will be dominated by the first, third, and fifth mode, etc. The kinetic energy associated with each mode is inversely proportional to n^2 where n is the order of a vibration mode. Rogers and Liang (1990) indicated in their early study that the total vibration energy of a system can be discretized, each vibration mode has its inherent kinetic or potential energy that is independent of other modes. This principle can be very well demonstrated from the vibration control of the simply-supported beam with a pseudoelastic SMA spring attached at one third of the beam as shown in Figure 13. When a pin force impacts in the middle of the beam, most of the odd modes excited will be damped except the third mode which has its nodal lines at one third of the beam (assuming that the energy associated with other higher order modes is extremely small compared with the third mode). The position and preload level can be adjusted, so that desired modes may be preserved and unwanted modes damped. We should say that ordinary damping material may also be used to perform this type of control, however, SMA material has its obvious advantage as an add-on passive/adaptive control device.

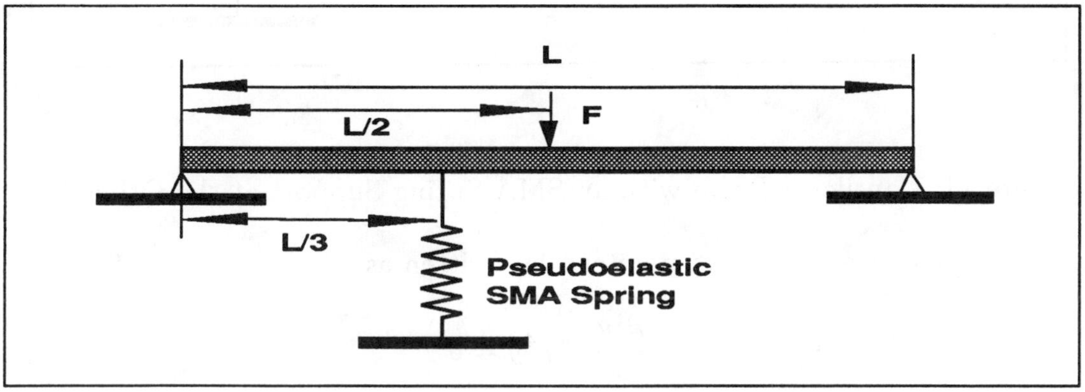

Figure 13: Vibration Filter Using a Pseudoelastic SMA Spring

CASE 3: VIBRATION CONTROL UTILIZING THE VARIABLE STIFFNESS OF SMA SPRINGS

The Young's modulus of shape memory alloys changes as a function of temperature. The modulus of high-temperature austenite is 3 to 4 times higher than that of low-temperature martensite. The shear modulus of SMA has the same relation. The spring constant of an SMA spring, as a linear function of its shear modulus, changes by three times over a small range of temperature. The transition temperature range for the Young's modulus of SMA is about ten to thirty degrees. So far, most of the applications of SMAs as variable stiffness elements, such as shape memory alloy hybrid composites (Rogers, Liang, and Jia, 1990), have utilized only the high-temperature modulus or low-temperature modulus; it is difficult to achieve a intermediate Young's modulus by simply controlling temperatures. The reason is that the Young's modulus of SMA is a function of its martensitic fraction which is affected by both the temperature and stress. In the following discussion of the applications of SMA springs as variable stiffness control elements, it is considered that a SMA spring has only two spring constants, K_S^A (high-temperature austenitic spring constant) and K_S^M (low-temperature martensitic spring constant).

Consider the natural frequencies of a simple structure, a cantilever beam with an SMA spring on the other end of the beam as shown in Figure 14. It is a simple problem and its solutions are given virtually in any vibration text-book. However, this problem is now different in that people have never thought that the elastic support could have a variable stiffness.

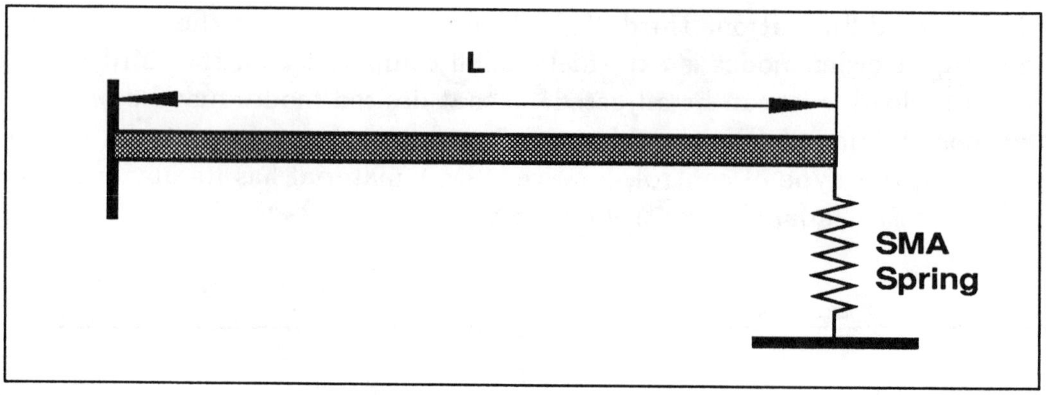

Figure 14: Cantilever Beam with an SMA Spring Support at the Other End

The governing equation for the system can be written as

$$\frac{\partial^4 y}{\partial x^4} - k^4 y = 0 \tag{41}$$

where $k = \omega^2 \rho / EI$. It's boundary conditions are

$$\begin{cases} y|_{x=0} = 0 \\ \frac{\partial y}{\partial x}|_{x=0} = 0 \end{cases} \tag{42}$$

and

$$\begin{cases} \frac{\partial^2 y}{\partial x^2}|_{x=L} = 0 \\ EI\frac{\partial^3 y}{\partial x^3}|_{x=L} = K_S y \end{cases} \tag{43}$$

where E is the Young's modulus of the beam, I the moment of inertia of the cantilever beam, ρ the density of the beam, and K_S the spring constant. The natural frequency of the system can be solved from the following equation.

$$\bar{K}(\sin kL \ \cosh kL - \cos kL \ \sinh kL) + (kL)^3(1 + \cosh kL \ \cos kL) = 0 \tag{44}$$

where $\bar{K} = K_S L^3/EI$ is the ratio of the spring stiffness to the bending stiffness of the beam. The relative natural frequency, $k_n L$, can be solved from Equation (44) and the first five of them are listed in Table 1.

When $\bar{K}$ is zero, the beam system is simply a cantilever beam. If $\bar{K}$ is infinite, the beam system is a clamped-simply supported system. The calculated k_n clearly indicates this point. Another point worth mentioning is that low-order natural frequencies are sensitive to a relatively small change of the stiffness of the support when $\bar{K}$ is small. The high-order natural frequencies, on the contrary, seem not to respond to the change of the stiffness of support as much when $\bar{K}$ is small. The above conclusion will be reversed at large $\bar{K}$ values. For example, when $\bar{K}$ increases from zero to 5, the first natural frequency increases by about 26% and the second natural frequency by only 1% while higher-order natural frequency hardly changes. When $\bar{K}$ increases from 1000 to 10^{10}, there are not many changes for the first natural frequency while changes are obvious for the fourth and fifth mode.

Table 1: The First Five Relative Natural Frequencies of the Beam System

$\bar{K}$	k_nL				
	$n=1$	$n=2$	$n=3$	$n=4$	$n=5$
0	1.88	4.69	7.85	11.00	14.14
0.1	1.89	4.70	7.86	11.00	14.14
0.5	1.95	4.70	7.86	11.00	14.14
1.0	2.01	4.70	7.86	11.00	14.14
5.0	2.37	4.74	7.87	11.00	14.14
20.0	2.97	4.90	7.90	11.01	14.14
50.0	3.40	5.20	7.96	11.03	14.16
100.0	3.64	5.62	8.04	11.07	14.17
1000.0	3.90	6.88	9.55	11.95	14.58
10^{10}	3.93	7.07	10.21	13.38	16.49

CASE 4: VIBRATION CONTROL USING SHAPE MEMORY EFFECTS

Consider the SMA spring which is elongated to point A as shown in Figure 15. The SMA spring is then unloaded and connected to an ordinary spring with a spring constant K_C. The spring and mass system is somewhat similar to the system shown in Figure 10 except that damping of the SMA spring is not considered. When the SMA spring is loaded again, it has a spring constant of K_S. The SMA spring performs linearly if the force on the spring is less than F_A as shown in Figure 15. The angular frequency of the system is

$$\omega = 2\pi\sqrt{\frac{K_S + K_C}{m}} \tag{45}$$

Once the SMA spring is activated upon being heated above the austensite start temperature (A_s), the recovery force generated by the SMA spring will pull the mass to a new equalibrium position B. The dotted line is referred to as the recovery force-displacement curve at a constant temperature. If the system is excited, the mass will oscilate around the new equilibrium position. The system may be modelled as a linear system if its vibration amplitude is small. The angular frequency of the activated system is solved as

$$\omega = 2\pi\sqrt{\frac{K_{SR} + K_C}{m}} \tag{46}$$

Equation (46) is essentially the same as Equation (45) except that K_{SR}, the tangential stiffness at point B is used instead of the normal spring constant, K_S. It is necessary to point out that K_{SR} is at least a hundred times smaller than K_S.

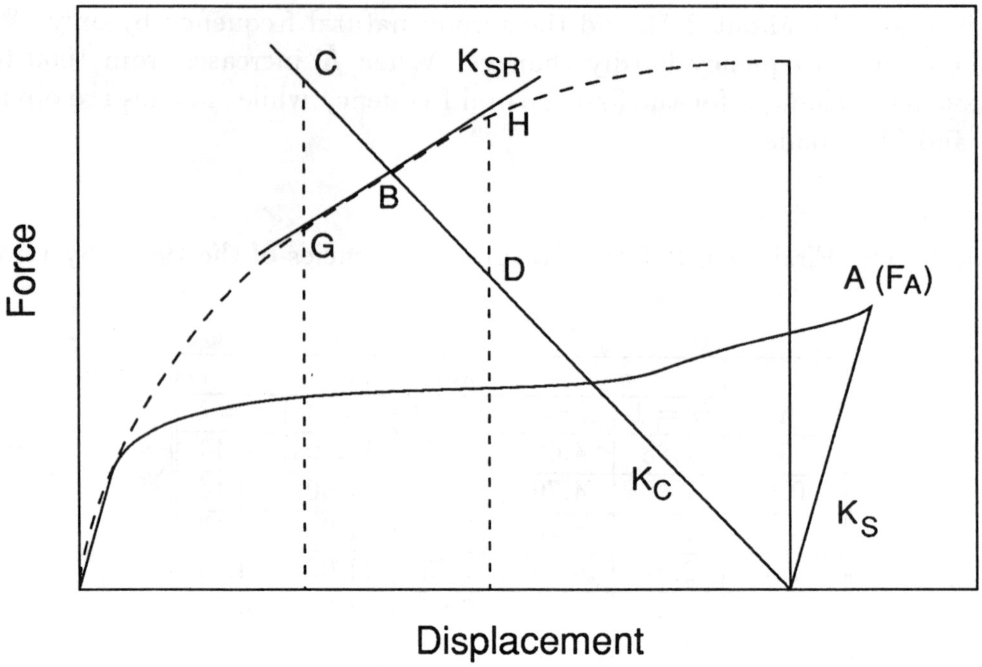

Figure 15: Recovery Stress-Strain Relation and Vibration Control

If the vibration has a relatively large amplitude, nonlinear vibration analysis must be used. The mass will vibrate between C and D. The position of C and D can always be determined in such as way that the area of CBG (potential energy stored in the ordinary spring) is always equal to the area of BHD (potential energy stored in the SMA spring) and equal to the total initial energy (potential energy + kinetic energy).

In the above analysis, activation of an initially elongated SMA spring changes its effective spring constant. The recovery force generated only changes the static equilibrium position of the system. However, the recovery force of an SMA spring can also be tuned as a control force. Rogers (1990) showed that nitinol actuators (wires) attached to the root of a cantilever beam can reduce the free vibration of the beam significantly. The logarithmic decrement was increased from 0.04 for the uncontrolled beam to 0.2. The recovery force may also be used to increase the effective stiffness of a structure utilizing the concept of active strain energy tuning first developed for of shape memory alloy hybrid composites (Rogers et al., 1989).

CONCLUSION

This paper has introduced most of the important characteristics of shape memory alloys that can be used in active and passive vibration control. A theoretical design method of SMA springs is briefly presented. The design method is believed to be convenient and accurate. Some possible applications of SMA springs in vibration control also have been discussed. SMA springs may be used to carry load and dissipate vibration energy. Pseudoelastic SMA springs can be designed as a vibration filter or add-on passive/adaptive control device. The ability of an SMA spring to vary its spring constant is another viable control parameter.

ACKNOWLEDGEMENTS

The author gratefully acknowledges the support for this work by ONR-YIP Grant N00014-88-K0566 and ONR Grant N00014-88-K0721.

REFERENCES

Buehler, W. J. and Wiley, C. R., 1965, Nickle-Based Alloys, US Patent 3,174,851.

Clark, H. M. and Wayman, C. M., 1970, "Surface Relief Effects in Solid State Phase Transformation," Phase Transformation, edited by American Society for Metakls, pp. 59-114, Chapman & Hall Ltd., London.

Cross, W. B., Kariotis, A. H., and Stimler, F. J., 1970, Nitinol Characterization Study, Goodyear Aerospace Corporation Report, No. Ger 14188 (NASA CR-1433), Akron, Ohio.

Dejonghe, W., Delaey, L., De Batist, R., and Van Humbeeck, J., 1977, "Temperature- and Amplitude-Dependence of Internal Friction in Cu-Zn-Al Alloys," *Metal Science,* pp. 523-520, November.

Dye, T., 1990, An Experimental Verification of a One-Dimensional Constitutive Model of Shape Memory Alloys, Master Thesis, Department of Mechanical Engineering, Virginia Polytechnic Institute and State University, Blacksburg, Virginia, August.

Funakubo, H., 1986, Shape Memory Alloys, Gordon and Breach Science Publishers, New York.

Graesser, E. J., 1990, Multi-Dimensional Modeling of Hysteretic Materials Including Shape Memory Alloys: Theory and Experiment, PhD Dissertation, Department of Mechanical and Aerospace Engineering, State University of New York at Buffalo, January.

Jones, D. I. G., 1988, "Application of Damping Treatments," Shock & Vibration Handbook, edited by Harris, C. M., Mcgraw-Hill Book Company.

Liang, C., 1990, Constitutive Modeling of Shape Memory Alloys, PhD Dissertation, Department of Mechanical Engineering, Virginia Polytechnic Institute and State University, Blacksburg, Virginia.

Liang, C. and Rogers, C. A., 1990a, "A One-Dimensional Thermomechanical Constitutive Relation of Shape Memory Materials," *Journal of Intelligent Material Systems and*

Structures, Vol. 1, No. 2, pp. 207-234.

Liang, C. and Rogers, C. A., 1990b, "Design of Shape Memory Alloy Actuators,", Intelligent Structures, Proceedings of the International Workshop on Intelligent Structures, Taipei, Taiwan, 23-26 July, edited by Chong, K. P., Liu, S. C., and Li, J. C., Elsevier, London, pp. 416-438.

Liang, C. and Rogers, C. A., 1991, "The Multi-Dimensional Constitutive Relations of Shape Memory Alloys," Proceedings of the 32nd Structures, Structural Dynamics, and Material Conference, Baltimore, MD, 8-10 April, Paper No. AIAA 91-1165.

Nagaya, K., Takeda, S., Tsukui, Y., and Knmaido, Y., 1987, "Active Control Method for Passing Through Critical Speeds of Rotating Shafts by Changing Stiffness of the Supports with Use of Memory Metals," *Journal of Sound and Vibration,* Vol. 113(2), pp. 307-315.

Perkins, J. (Edited), 1975, Shape Memory Effects in Alloys, Plenum.

Rogers, C. A., 1988, " Novel Design Concepts Utilizing Shape Memory Alloy Reinforced Composites," *Proceedings of the American Society for Composites 3rd Technical Conference on Composite Materials,* Technomic Publishing, Co., Seattle, September.

Roger, C. A., Liang, C., and Jia, J., 1989, "Behavior of Shape Memory Alloy Reinforced Composite Plates - Part I: Proceedings of the 30th Structures, Structural Dynamics, and Materials Conference, Mobile, AL. 3-5, April, Vol. 4, pp. 2011-2017.

Rogers, C. A., 1990, "Active Vibration and Structural Acoustic Control of Shape Memory Alloy Hybrid Composites: Experimental Results," *Journal of Acoustical Society of America,* pp. 2803-2811, December.

Rogers, C. A. and Liang, C., 1990, " Utilization of Intelligent Material Systems in Active Vibration Control," 120th Meeting of the Acoustical Society of American, San Diego, CA., 26-30 November.

Shigley, J. E., 1977, Mechanical Engineering Design, McGraw-Hill, Inc..

Tanaka, K. and Nagaki, S., 1982, " A Thermomechanical Description of Material with Internal Variables in the Process of Phase Transformation," *Ingenieur-Archiv,* vol. 51, pp. 287-299.

ADAPTIVE CHARACTERISTICS OF THE MAGNETOSTRICTIVE ALLOY, TERFENOL-D, FOR ACTIVE VIBRATION CONTROL

Mel J. Goodfriend,[1] K. M. Shoop[1]

ABSTRACT

The recently developed "Giant" magnetostrictive material, Terfenol-D, has been applied to development of a new class of adaptive actuators for vibration control. Terfenol-D offers the highest strain ($\Delta L/L > 10^{-3}$) and energy density ($1.4\text{-}2.5 \times 10^{4}$ J/m^3) of any commercially available material.

This paper reviews the material properties of Terfenol-D, with emphasis on characteristics which adapt to external variables. Performance data of some basic actuators and a discussion of essential design elements is presented, along with engineering data of value to designers of "smart" actuators.

INTRODUCTION

Active actuators have been demonstrated to provide impressive isolation capabilities in the control of vibrations and acoustic emissions. Recently developed alloys with huge linear magnetostriction ($\Delta L/L > 10^{-3}$) are being used to construct actuators which overcome limitations of previous active control devices [1],[2],[3],[4].

The ternary crystalline alloy $Tb_xDy_{1-x}Fe_2$ where $x \approx 0.3$, known as Terfenol-D exhibits the largest room temperature strain of any commercially available material [5]. The name comes from the symbols for the elements in the alloy and for the Naval Ordinance Laboratory (NOL), now Naval Surface Warfare Center (NSWC), where the material was first investigated.

In this paper the material properties of Terfenol-D are reviewed, along with drive level dependencies of characteristics important for maximum energy conversion such as saturation magnetostriction, magnetomechanical coupling factors, elastic compliances and permeabilities. Finally, performance data are presented for basic actuator designs.

[1]ETREMA Products, Inc., a subsidiary of Edge Technologies, Inc., 306 South 16th Street, Ames, IA 50010, U.S.A.

EXPERIMENT

Commercial Terfenol-D drive elements prepared by two patented processes licensed to Edge Technologies, Inc. were used in these studies. Both processes are well documented in the literature [6]. The free stand zone melt (FSZM) technique produces twinned single crystal rods of 3 to 7mm diameter [6]. Grain oriented rods of 8 to 50mm diameter are produced by the modified Bridgman (MB) method [7]. Transducer drive elements from both processes have the magnetostrictive growth axis perpendicular to the rod axis [6].

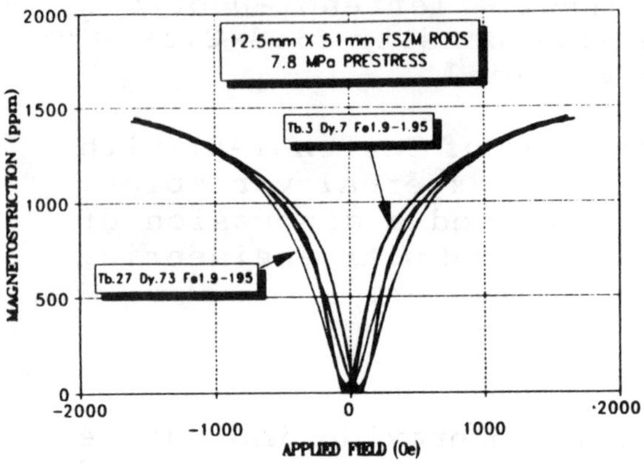

Figure 1. Free Stand Zone Melt (FSZM) Terfenol-D response curves for two stoichiometries.

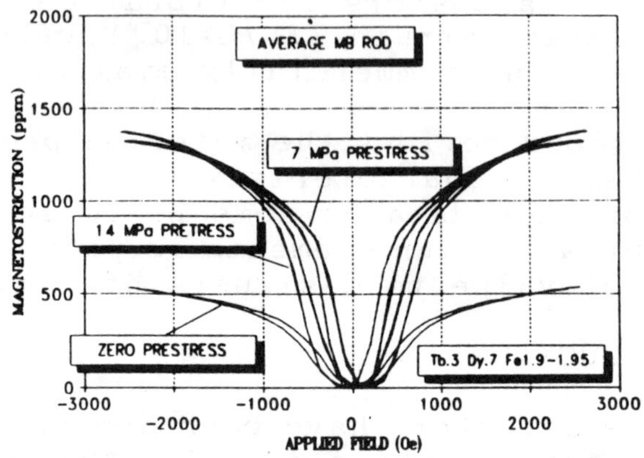

Figure 2. Performance of average modified Bridgman (MB) Terfenol-D drive elements, showing the effect of changing mechanical prestress.

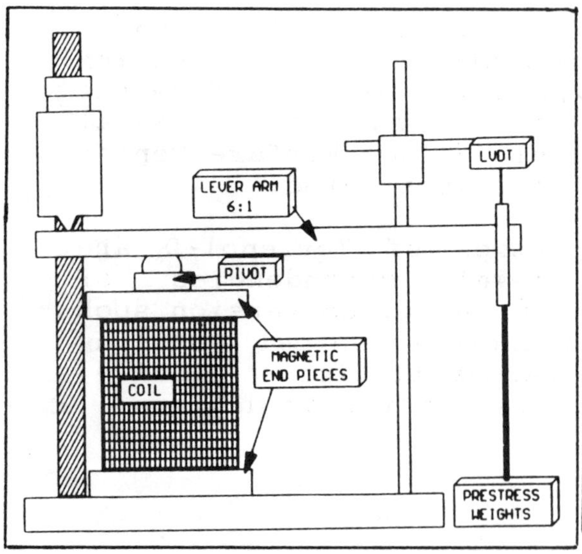

Figure 3. Test apparatus used for Terfenol-D drive elements and actuators.

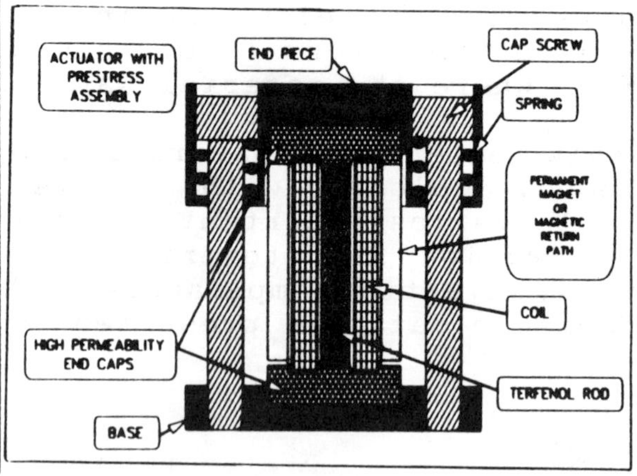

Figure 4. Actuator cross section drawing.

Typical performance curves for FSZM material are shown in figure 1. Plots for 2 similar rods differing in stoichiometry are shown. Figure 2 shows average performance of MB process rods at varying mechanical preloads.

The apparatus used to measure Terfenol-D performance is shown in figure 3. This same apparatus was used to measure actuator displacement by replacing the solenoid coil, magnetic end pieces and Terfenol-D drive element with an actuator or actuator motor section. Various techniques for measuring Terfenol-D performance have been described [8], [9],[10]. The apparatus chosen for these studies was designed to simulate configuration of an actual Terfenol-D driver.

The cross section drawing of actuators commercially available from Edge Technologies, Inc. shown in figure 4 illustrates essential elements of basic Terfenol-D transducers. Maryland Engineering Specialties, Inc., Severna Park, Maryland designed the actuators used in these studies and provided the frequency response curve shown in figure 10. Motor sections consisting of Terfenol-D rod, coil, high permeability end caps and permanent magnets or magnetic return path were removed from the actuator and used in our measurement apparatus to produce the data in figures 7,8 and 9.

RESULTS AND DISCUSSION

In the following plots strain (λ) or displacement due to magnetostriction is expressed as change in length over length ($\Delta L/L$) and is therefore dimensionless. The magnitude of the strain is given as parts per million (ppm). The width of the curve represents hysteresis. Reversing the field produces identical response. Terfenol-D magnetostriction is essentially constant volume. Therefore $\lambda = \Delta L/L \approx 2\,\Delta D/D$, where D is diameter.

Figure 1 shows FSZM Terfenol-D response curves for two stoichiometries. Alteration of Terfenol-D stoichiometry often allows optimization of the strain vs field curve for various applications. Performance data for various stoichiometries are still under study, however extensive documentation can be found in the literature [5], [6]. $Tb_{.27}Dy_{.73}Fe_{1.95}$ stoichiometry produces a more linear response with lower hysteresis. The high slope at low fields of the λ vs H (magnetic field in Oersteds (Oe)) curve produced by $Tb_{.3}Dy_{.7}Fe_{1.95}$ alloy is usually preferred where oscillatory motion and highest energy conversion are desired, such as in active antivibration or sonar applications.

Typical performance of an MB drive element is shown in figure 2. Performance is similar to FSZM material. FSZM Terfenol-D exhibits both a higher strain vs field slope at low to moderate fields and higher saturation strain.

Figure 2 also shows that application of mechanical prestress is important to Terfenol-D performance. Increased prestress results in higher saturation strains and more linear response, although more field may be required to reach a given displacement.

Workers at laboratories sponsored by the United States Navy have recently published data from extensive studies of magnetomechanical parameters which govern conversion of magnetic energy to mechanical energy [9], [10]. These data were collected under the high stress, large magnetic field excursion conditions needed for maximum energy conversion. Table I summarizes results from these studies, presents the drive level dependencies of the relative permeability, piezomagnetic d_{33} coefficient, constant field compliance and coupling factor. Note the dependency of these factors on external conditions.

The data in Table I are for large diameter centerless ground 38.1mm x 114.5mm long modified Bridgman grown rods. Hysteresis loops of extensional strain vs applied magnetic field at the constant stress values of Table I for these same MB rods are shown in figures 5 and 6. Studies performed at NSWC used their test apparatus [10].

Table I. Piezomagnetic Coefficients for Terfenol-D

Bias Condition	Compressive Pre-Stress (ksi)	(MPa)	Magnetic Bias Field (Oe)	Relative Permeability μ^T_{33}/μ_0	Piezomagnetic Coefficient d_{33} (nm/A)	Compliance (Constant Field) s^H_{33} ($10^{-12}Pa^{-1}$)	Coupling Factor $k_{33}=d_{33}(\mu^T_{33}s^H_{33})^{-1/2}$
1	1.01	6.9	150				
2	2.22	15.3	400	8.4	15.5	47.5	0.69
3	3.43	23.6	700	6.1	13.0	39.0	0.75
4	4.64	32.0	1000	4.9	10.3	36.0	0.69
5	5.86	40.4	1300	4.3	9.1	34.5	0.67
6	7.07	48.7	1600	4.0	8.6	34.0	0.66
7	8.28	57.1	1900	3.7	8.1	32.5	0.66
8	9.49	65.4	2200	3.5	7.6	30.5	0.66

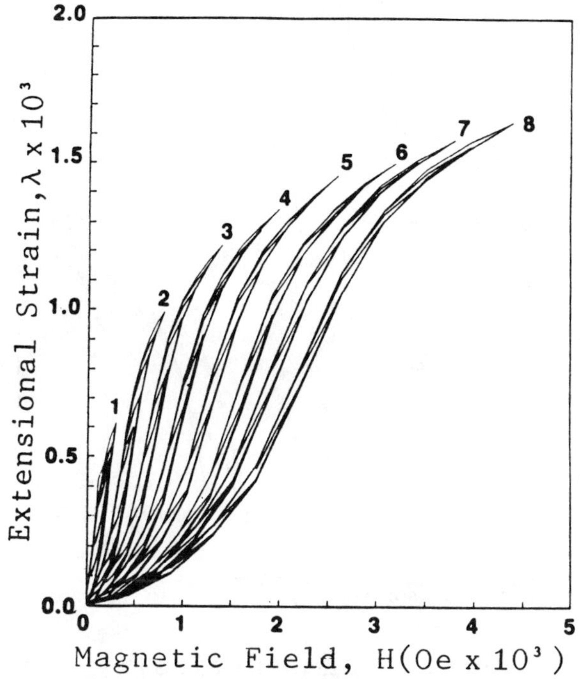

Figure 5. Hysteresis loops of extensional strains vs applied magnetic field for constant stress values of Table I. MB rod 114.5 in. by 38.1 mm diameter.

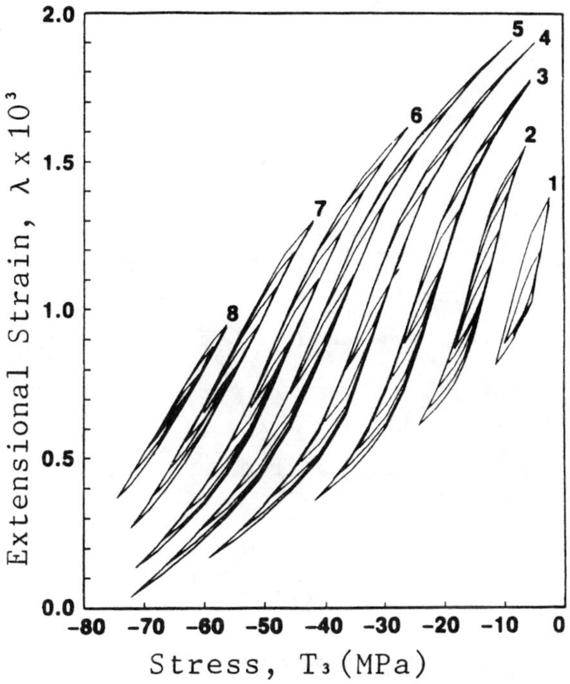

Figure 6. Hysteresis loops of extensional strain vs stress for constant magnetic fields of Table I. MB rod 114.5 mm long by 38.1 mm diameter.

For linear operation under an ac field Terfenol-D driven actuators require magnetic bias [11]. Thus, most magnetostrictive actuators for dynamic or oscillatory outputs employ magnetic bias. Bias may be supplied by a dc field, but use of permanent magnets is more common and economical in the long run. A properly biased and loaded transducer can be made to operate in the linear portion of the strain vs field curve (figure 1) where the slope of the curve is greatest. Net displacement and efficiency will be maximized.

The effect of magnetic biasing is illustrated in figures 7 and 8 which show performance curves for identical actuators, except one is magnetically biased and the other nonbiased. Specifications for these actuators are shown in Table II. The magnetically biased actuator has a permanent magnet which provides a bias field of about 300 Oersteds. Displacement at a given input is higher for the biased actuator. Note that if the biased actuator is "overdriven" the displacement curve will pass through the minimum position.

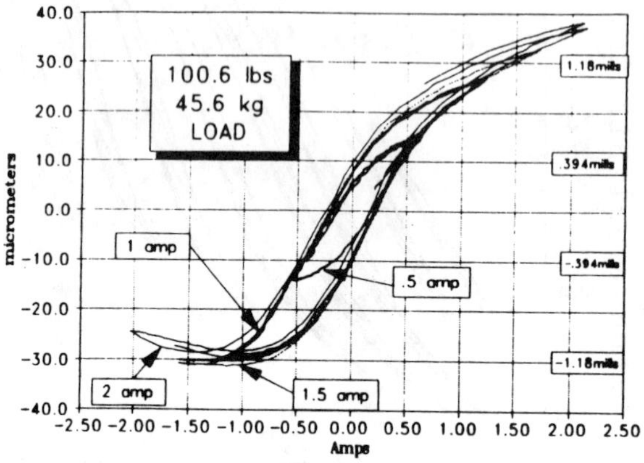

Figure 7. Magnetically biased actuator performance. Displacement at various current input levels. Terfenol-D driver is 51.3 mm long by 6.4 mm diameter.

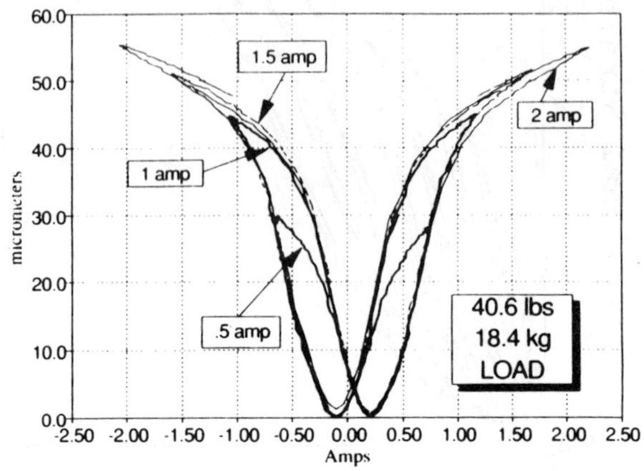

Figure 8. Nonmagnetically biased actuator performance displacement at various current input levels. Terfenol-D driver is 51.3 mm long by 6.4 mm diameter.

TABLE II

ACTUATOR SPECIFICATIONS

Motor Section	
Terfenol-D Length	51.3 mm (2.02 inch)
Terfenol-D Diameter	6.4 mm (0.25 inch)
Terfenol-D Type	Free Stand Zone Melt (FSZM)
Coil Type	Free Standing (Without Bobbin)
Coil Wire Diameter	0.47 mm (AWG #26)
Coil Rating	300 Oersteds/amp (1181 amp-turns)
Motor Section O.D.	25.4 mm (1.0 inch)
Motor Section Length	63.5 mm (2.5 inch)

Overall Dimensions		
Diameter	60 mm	(2.36 inches)
Length	75 mm	(2.95 inches)
Prestress	6.9 MPa	(1000 psi)

If only positive displacement is needed in a Terfenol-D actuator, no magnetic bias is required. When an ac current is supplied to a nonmagnetically biased actuator, expansion of the Terfenol-D will take place on both halves of the input cycle; thus, actuator frequency will be twice the frequency of the input current.

One of the most useful features of Terfenol-D driven actuators is the high energy density. Thus, a compact actuator can produce high forces. Clamped force of a Terfenol-D rod is given by $F=EA\Delta L/L$ where E is Youngs modulus ($3.5x10^{10} N/m^2$) and A is cross section area of the Terfenol-D driver in m^2. More extensive relationships of Terfenol-D actuator force in dynamic operation have been published [11].

The load on the magnetically biased actuator used to prepare the plots in figure 8 was varied and the load stroke relationships plotted in figure 9. These data conform to the theoretical performance expectations developed previously in this paper and show that response of the actuator can be altered by changing the load. Since magnetic bias is supplied by a permanent magnet, changing the shape of the λ vs H curve will shift the rest position of the actuator and will affect net positive vs negative displacement response of an ac signal.

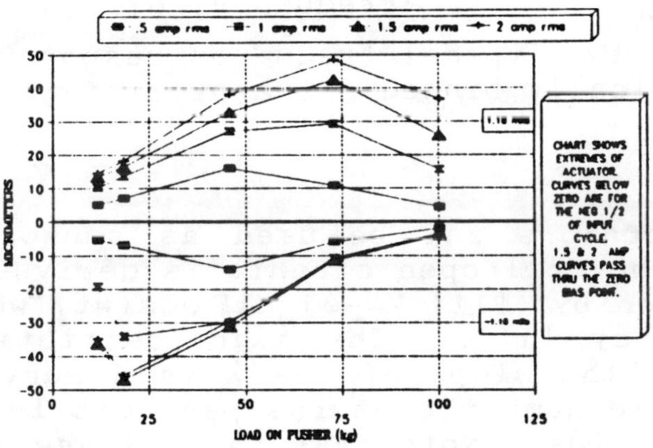

Figure 9. Magnetically biased actuator stroke versus load at various input levels.

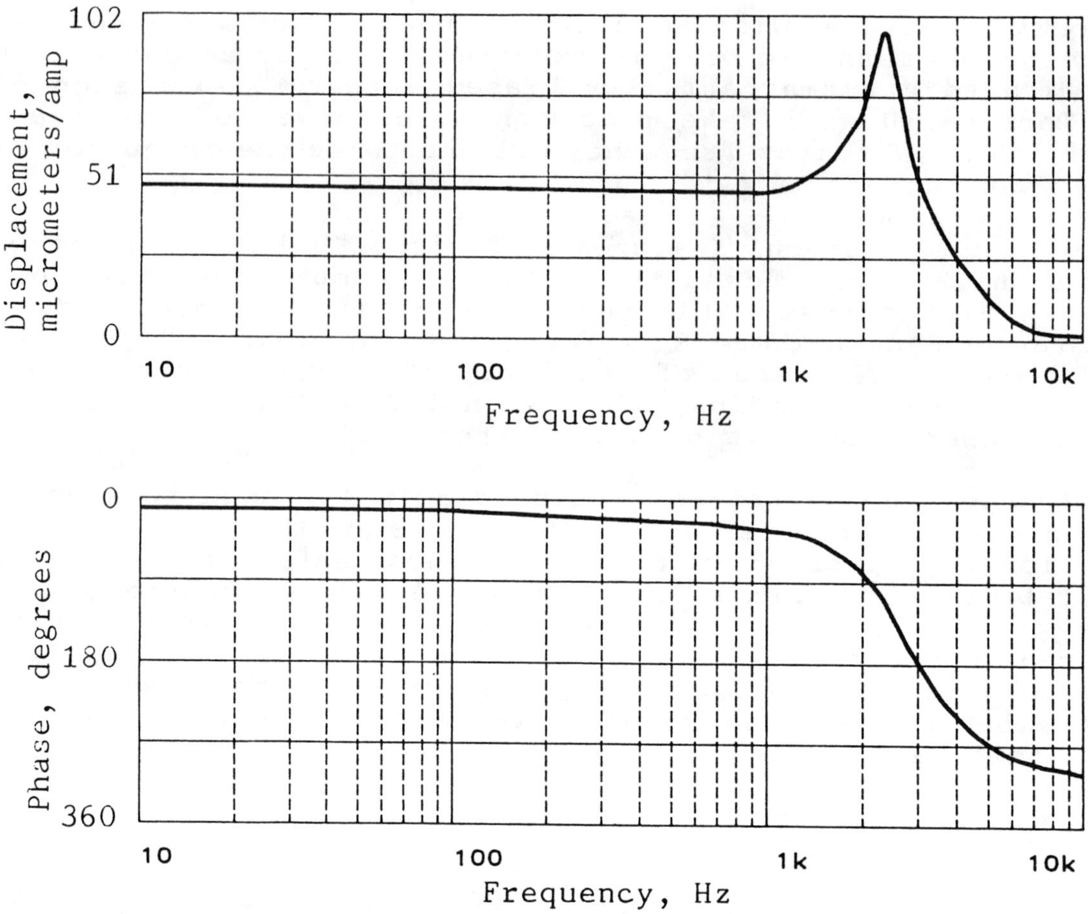

Figure 10. Typical frequency response of magnetically biased actuator.

Terfenol-D actuators may be used as sensors. The output voltage with the coil open circuit is derived from Faraday's law and is given by: [11] $V=-\omega d_{33}nlF_0\cos(wt)$ where $\omega=2\pi f$ and f is the frequency [11]. The magnetostrictive d_{33} constant corresponds to the slope of the λ vs H curve in the linear region, n is the number of turns per unit length l and F_0 is the force amplitude. Note that the voltage is frequency and force dependent.

Terfenol-D may also be used to convert a mechanical force or stress to a current [11]. For a low impedance electrical load, the current generated for a mechanical stress T may be written $I=-gT/n$ with $g=d_{33}/u^T$. Here u^T is the free permeability.

Terfenol-D actuators can be designed for constant response over a broad frequency range. Figure 10 shows typical frequency response of the motor section from the magnetically based actuator at an input of 1.0 amp rms and a constant preload of about 20 Kg [12]. Stiffness of the prestress system was kept very low so that prestress changes very little during actuator operation. In many applications the actuator will be operated in the linear region below the resonance frequency, resulting in relatively constant displacement with changing frequency.

CONCLUSIONS

The extremely high energy density, wide bandwidth and relatively high displacement available from Terfenol-D allows design of active actuators which can provide vibration isolation capabilities in excess of those provided by previously available actuators. Considerable data is now available to engineers for design of efficient devices.

The first generation of prototype actuators utilizing Terfenol-D have successfully demonstrated impressive control capabilities while allowing development of necessary magnetomechanical drive parameters.

The magnetomechanical data now available will allow evolution of more refined magnetostrictive actuators. The data needs to be applied to fully develop a control theory that will exploit the Terfenol-D characteristics to their maximum.

References

1. Reed, R.S., 1988. "Shock Isolation Using an Active magnetostrictive Element," Proceedings of the 59th Shock and Vibration Symposium, Sandia Report Sand 88-2473C, Vol. IV, SAND 88-2473C. UC-13.

2. Reed, R.S., 1988. "Active Vibration Isolation Using a Magneto-strictive Actuator," Modeling and Simulation, Vol 19.

3. Hiller, M.W., M.D. Bryant, J. Umegaki, 1989. "Attenuation and Transformation of Vibration Through Active Control of Magneto-strictive Terfenol," Journal of Sound and Vibration, Vol. 134(3) 507-519.

4. Parvinmehr, A., R. D. Greenough, A.J. Wilkinson and A. Jenner, 1989. "Discrete Variable Structure Control for an Active Anti-Vibration System," Proc. IFAC Symposium on Adaptive systems in Control and Signal Processing, Inst. Meas. Cont., Vol II, pp 547-552.

5. Clark, A.E., 1989. "Giant Magnetostriction From Cryogenic Temperatures to 200^0C," Proceedings of the Tenth International Workshop on Rare Earth Magnets and their Applications" Sponsored by the Society of Non-Traditional Technology, Vol. 1, pp 345. Tokyo, Japan.

6. Verhoeven, J.D., J.E. Ostenson, E.D. Gibson and O.D. McMasters, 1989. "The Effect of Composition and Magnetic Heat Treatment on the Magnetostriction of $Tb_xDy_{1-x}Fe_y$ Twinned Single Crystals," J. Appl. Phys., Vol 66(2), pp 772-779.

7. Verhoeven, J.D., E.D. Gibson, O.D. McMasters and J.E. Ostenson, 1990. "Directional Solidification and Heat Treatment of Terfenol-D Magnetostrictive Materials," Metallurgical Transactions, Vol. 21A, pp 2249-2255.

8. Ostenson, J.E., D.K. Finnemore and D.C. Jiles, 1989. "Pressure Cell for Magnetostrictive Measurements," Rev. Sci Instrum., 60(2) pp. 278-280.

9. Clark, A.E., J.P. Teter, M. Wun-Fogle, M. Moffett and J. Lindberg, 1989. "Magnetomechanical Coupling in Bridgman Grown $Tb_{0.3}Dy_{0.7}Fe_{1.9}$ at High Drive Levels," Proc. 34th Annual Conf. on Magnetism and Magnetic Materials, Boston, MA.

10. Moffett, M.B., A.E. Clark, M. Wun-Fogle, J.F. Lindberg, J.P. Teter and E.L. McLaughlin, 1990. "Characterization of Terfenol-D for Magnetostrictive Transducers," Proc. 117 Annual Meeting Acoustical Society of America, Syracuse, NY.

11. Butler, John L., 1988. Applications Manual for the Design of ETREMA Terfenol-D Magnetostrictive Transducers, available from Edge Technologies, Inc., Ames, IA.

12. Miller, Carl G., Unpublished data. Maryland Engineering Specialties, Inc. Richmond, VA.

PREVENTING TARGET CANCELLATION IN ADAPTIVE BROADBAND MICROPHONE ARRAYS

S. Nordholm,[1] I. Claesson[1]

ABSTRACT

THE CAR ENVIRONMENT IS KNOWN TO BE DIFFICULT FOR HANDSFREE OPERATION OF MOBILE TELEPHONES OR SPEECH DETECTION FOR SPEECH CONTROLLED FACILITIES IN THE CAR. A METHOD TO ENHANCE THE QUALITY OF THE SPEECH SIGNAL EMPLOYING A SPATIALLY FILTERING MICROPHONE ARRAY IN COMBINATION WITH NOISE CANCELLING TECHNIQUES IS PROPOSED AND MEASUREMENT RESULTS ARE PRESENTED.

1 Introduction

A method to increase the signal-to-noise ratio when using mobile telephones in hands-free operation has previously been presented [1]. The method employs conventional noise cancelling techniques in combination with an adaptive microphone array. Since the noise cancelling techniques are well established we have concentrated on the adaptive array, which has been implemented in real time on a multi-processor DSP computer, DSP900.

An adaptive microphone array, ARR900, has been implemented on DSP900. The implementation of ARR900 gave some practical problems, such as nearfield and distributed sources, imperfections in microphones, amplifiers and filters. These problems must be taken into consideration when choosing filter structure, adaptive algorithm and how to eliminate superresolution if target cancellation is to be avoided. Typical cures are to impose linear constraints, such as derivative constraints or hourglass techniques [2, 3]. A method [4] to control the resolution is employing spatial filtering in the blocking matrix of the Genal Sidelobe Canceller, GSC, see figure 1. The problem with spread sources in the nearfield is solved by focusing to an area where the speaker mainly is situated combined with a speech controlled adaptive algorithm, i.e. adaptation takes only place when the speaker is silent. This technique also handles the problem with reflections which give rise to speaker correlated jammers arriving from different directions.

[1]University College of Karlskrona/Ronneby, Department of Signal Processing, S-372 25 Ronneby, Sweden.

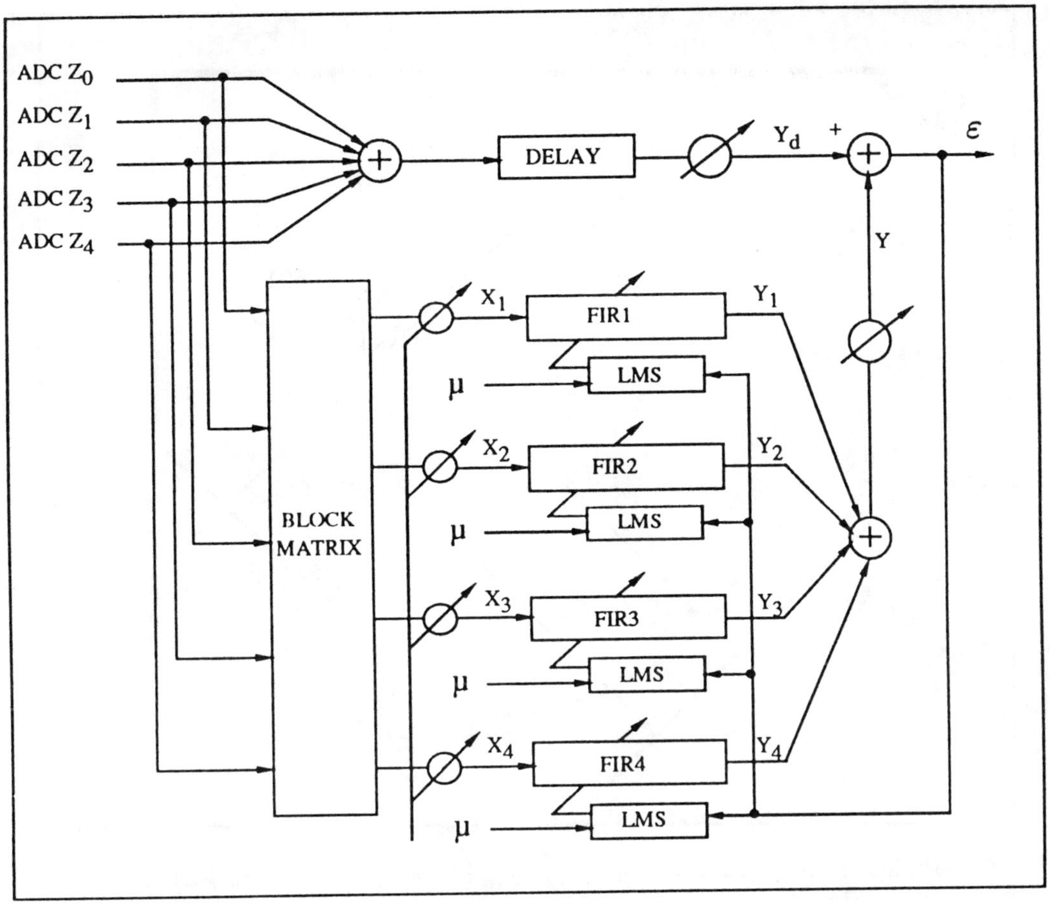

Figure 1: The General Sidelobe Canceller structure

2 Implementation of ARR900

We chose to use a GSC structure [5], see figure 1, in combination with the LMS-algorithm and to use the "hourglass technique" [3], nulling an area of weights around the center, to avoid superresolution. This combination leads to almost perfect parallelism. The updating formula for an arbitrary coefficient is

$$h_{ij}(k+1) = h_{ij}(k) + 2\mu\varepsilon(k)x_{ij}(k) \tag{1}$$

All coupling between coefficients are in $\varepsilon(k)$. The simple weight updating formula hereby facilitates parallel implementations. A 5 microphone array implementation is naturally divided between 5 DSP:s with this type of filter structure and algorithm.

2.1 Measurements

Preliminary measurements have been conducted from an anechoic chamber and in this paper we present measurement results together with a method to control superresolution in both broadband and narrowband adaptive arrays. All measurements have been done at 8 kHz sampling frequency and a single 10 cm loudspeaker

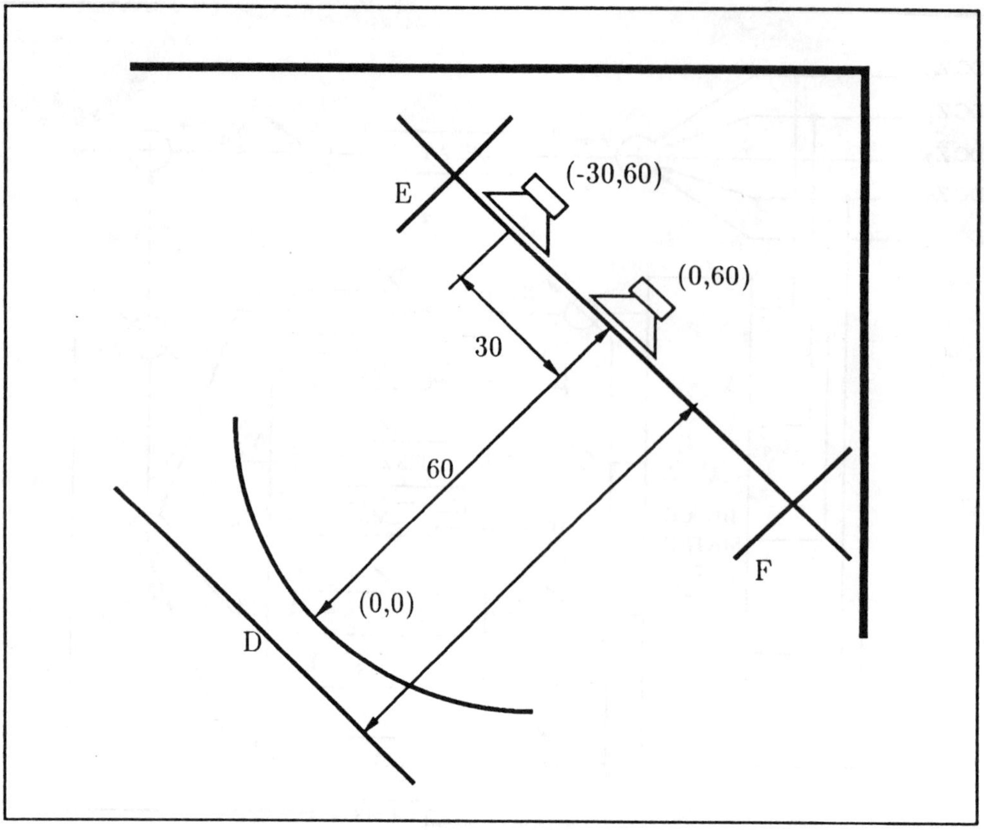

Figure 2: Placement of array, loadspeakers and reflection surfaces.

broadband source covering 300 – 1100 Hz. The microphones were placed on a circle with radius 60 cm with 5 cm distance along the circle between successive elements. The measurements show a possible 20 – 30 dB suppression of the jammer, even with reflecting walls inserted around the microphone array to primitively simulate the situation in a car. The Figs. 3-4 show the achieved suppression over the frequency interval of interest. An interesting result is that the achieveable suppression is strongly limited by the hourglass technique. Similar measurements have been performed in a car, but a serious limitation is that also when the speaker is silent, sound arrives from the speaker direction, thus enabling cancellation also from this direction. Corresponding results of interference cancellation were obtained (20 dB), but the target signal was cancelled with almost 15 dB, leaving an only 5 dB gain. This problem can, however, be eliminated by means of pre-spatial filtering of the signals arriving to the adaptive filters. The spatial filter also eliminates the need for the hourglass technique.

3 Spatial Filtering

A method [4] to control superresolution in adaptive arrays can however be used to avoid this target cancellation. This method employs spatial filtering in the blocking matrix of Generalized Sidelobe Cancellers and uses conventional equiripple

Parks-McClellan filter design to select the spatial filters. The method is based on forming the blocking matrix using a spatial filter and the adaptive array can hence be interpreted as a cascade of two filters, one spatial which is fixed in the spatial domain and one two-dimensional filter-and-sum beamformer which is adaptive both in the spatial and temporal domain. The spatial filtering are only linear constraints imposed on the array in a structured manner. The cancelling part (lower signal track) of the GSC beamformer can also be interpreted as a cascade of two filters in the two-dimensional mixed spatial-temporal domain. First, a spatial filtering is performed by the blocking matrix in front of the array of adaptive filters. Then, a two-dimensional adaptive filter is applied. In the simplest case all input signals have passed a spatial filter consisting of two samples i.e.

$$f(k) = \partial(k) - \partial(k-1) \tag{2}$$

The mapping from continuous temporal frequency to discrete spatial frequency follows from the spatial sampling

$$\Omega = \frac{d\sin\theta}{c} \cdot 2\pi f = T \cdot \omega \tag{3}$$

Hence the filter $F(f,\theta)$ can also be viewed as a discrete filter $F(\Omega)$. The spatial filter obviously cancels <u>all</u> temporal frequencies coming from the look-direction, $\theta = 0$.

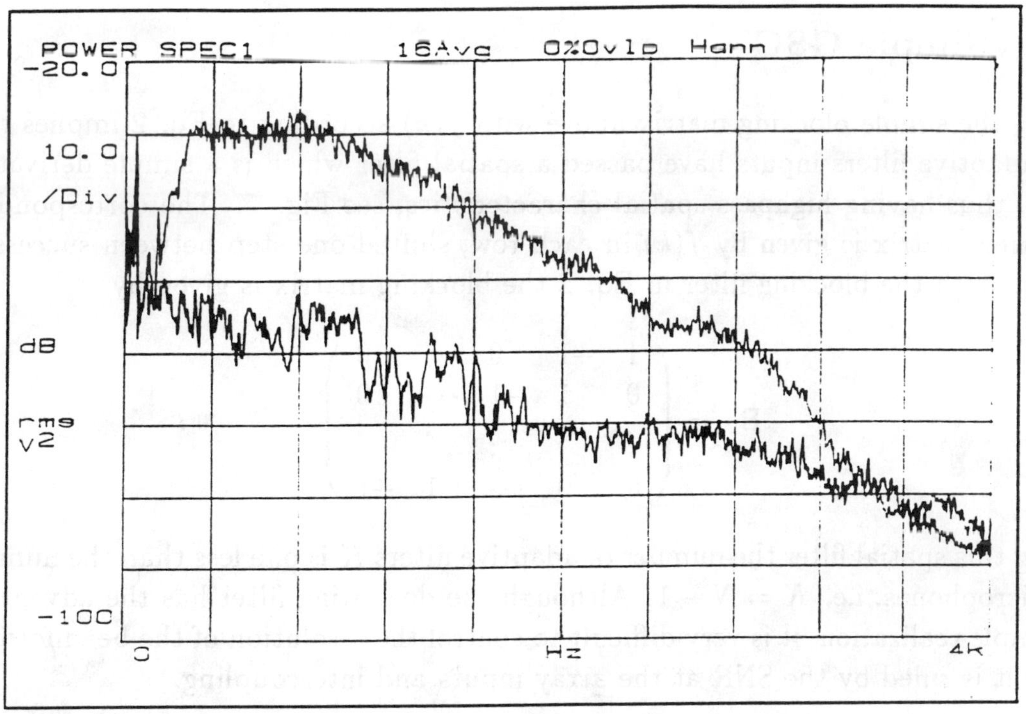

Figure 3: Output power spectrum before and after adaptation. The input signal is WGN filtered through a BP-filter(300-1100 Hz) from direction (20,60), the filter length is 45 and the hourglass is 1 sample.

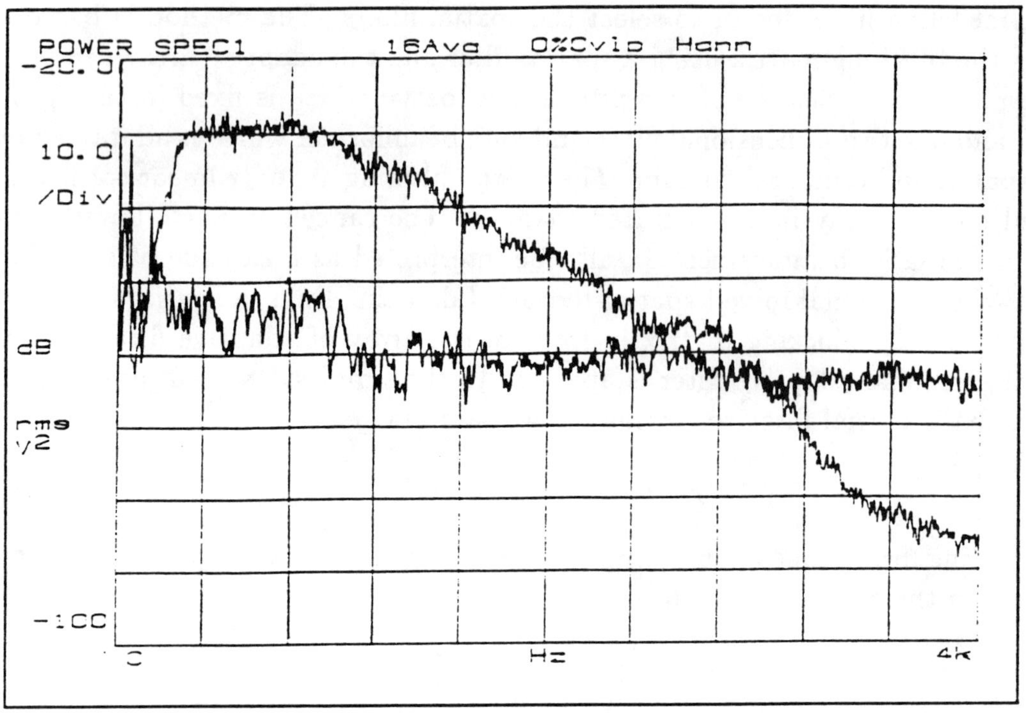

Figure 4: Output power spectrum before and after adaptation. The input signal is WGN filtered through a BP-filter(300-1100 Hz) from direction (20,60), the filter length is 47 and the hourglass is 3 samples.

3.1 Simple GSC

Using the simple blocking matrix above with $f(k)$ according to Eq. 2 implies that the adaptive filters inputs have passed a spatial filter which is a simple derivating filter, thus having highpass spatial characteristics, see Fig. 5. The corresponding blocking matrix is given by $f(k)$ in each row, shifted one step between successive rows. With the blocking filter in Eq. 2 the blocking matrix is given by

$$\mathbf{B} = \begin{pmatrix} 1 & -1 & 0 & \cdots & 0 \\ 0 & 1 & -1 & \cdots & 0 \\ \vdots & \vdots & \vdots & \vdots & \vdots \\ 0 & 0 & \cdots & 1 & -1 \end{pmatrix} \tag{4}$$

Using this spatial filter the number of adaptive filters K is one less than the number of microphones, i.e. $K = N - 1$. Although the derivating filter has the advantage of simple realization, it is very difficult to control the resolution of the beamformer, since it is ruled by the SNR at the array inputs and intercoupling.

3.2 General Method

We instead propose the following procedure to obtain a desired resolution.

1. Determine over which temporal frequency band $[f_l \; f_u]$ the array is aimed to operate.

2. Select an angular interval $[\theta_l \; \theta_u]$ over which the array is not allowed to cancel the target signal. Normally $\theta_l = 0$ since the look-direction is chosen to broadside.

3. Using the relationship in Eq. 3 a spatial filter is designed suppressing the region $[\Omega_l \; \Omega_u]$ sufficiently, where

$$\begin{aligned} \Omega_l &= \frac{d \sin \theta_l}{c} \cdot 2\pi f_l \\ \Omega_u &= \frac{d \sin \theta_u}{c} \cdot 2\pi f_u \end{aligned} \tag{5}$$

The suppression is sufficient if the signal-to-noise ratio at the adaptive filter inputs is negative (approximately -10 dB). If this suppression is chosen sufficiently large no cancellation is obtained within the selected frequency and angular direction intervals. Within these intervals the output will on the whole be the output from the desired beamformer y_d, in this case the summing beamformer output. The number of adaptive filters is reduced by M-1, where M is the number of samples in $f(k)$, i.e. $K = N - (M - 1)$.

3.3 Design examples

The spatial filter is designed with a desired angular passband $[\theta_l, \theta_u]$ for signals within a selected bandwidth $[f_l, f_u]$. This gives the upper and lower cut-off frequencies for the spatial filter according to Eq 5. The necessary filter stopband suppression is determined from the signal-to-noise ratio at the adaptive filter inputs and the number of adaptive filters. With these parameters set and the filter length M chosen, the filter coefficients can be found using Parks-McClellan optimum hilbert FIR filter design [6]. The filter coefficients are then put in the rows of the blocking matrix **B** which becomes a $N \times (N - M + 1)$ matrix.

In Figures 6-7, the array output spectrum $R_{\varepsilon\varepsilon}(f, \theta)$ is plotted versus angle and frequency. The signal-to-noise ratio at the adaptive filter inputs is 35 dB. The spatial filter $F(\Omega)$, see Fig. 5, is a 12 coefficient FIR hilbert type high pass filter suppressing low frequencies 50 dB up to the cut frequency $\Omega_u = 0.025 \cdot 2\pi$, which corresponds to 3° at the highest frequency 3200Hz. This implies that <u>all</u> frequencies up to 3200 Hz cannot be cancelled by the adaptive filters if the source lies within 3° from the look direction. The target source $m(t)$ was moved between 0° and 14° in steps of 0.2 degrees and the optimum filters $\mathbf{H}(f, \theta)$ and the corresponding output spectrum $R_{\varepsilon\varepsilon}(f, \theta)$ were calcultated for each position.

3.4 Implementation Aspects

An simple and straightforward way to obtain a real-time adaptive beamformer with prescribed resolution as above is using the leaky LMS algorithm to update the adaptive weights [7], thus simply controlling the signal-to-noise ratio at the adaptive filter inputs. The leaky LMS algorithm for the adaptive weights update is described by

$$h_{ij}(k) = \gamma h_{ij}(k-1) + 2\mu\varepsilon(k)x_{ij}(k) \tag{6}$$

where γ is a leakage factor. Using a leaky algorithm is equivalent to inject additive white gaussain noise at the adaptive filter inputs when updating the weights, although there is no noise present disturbing the output [7]. The strength of the equivalent white noise is given by

$$\sigma_\eta^2 = \frac{1-\gamma}{2\mu} \tag{7}$$

which, appropriately chosen, dominates over the microphone noises. An implementation using the spatial filtering technique and Leaky LMS is presently under manufacturing and if possible new measurement results will be presented at the conference. In order to employ the spatial filtering methods to a reasonable cost the number of microphones have been increased to about 15, while the number of adaptive filters in the broadband sidelobe canceller have been kept low (5 TMS 320C25 DSP:s).

References

[1] S. Nordholm I. Claesson B. Bengtsson P. Eriksson "Implementation of a Griffiths-Jim Adaptive Microphone Array on a TMS320C25 Multi-DSP" Proc. Acoustics 90, Southampton, England

[2] M. H. Er A. Cantoni "Derivative Constraints for Broad-Band Element Space Antenna Array Processors" IEEE Trans. on ASSP, vol. 31, no. 6, Dec. 1983

[3] B. Widrow J. M. McCool "A Comparasion of Adaptive Algorithms Based on the Methods of Steepest Descent and Random Search" IEEE Transactions on Antennas and Propagation, vol. 24, Sept. 1976

[4] I. Claesson S. Nordholm "A Spatial Filtering Approach to Robust Adaptive Beamforming" Submitted for publication in IEEE Transactions on Antennas and Propagation.

[5] L. J. Griffths C. W. Jim "An Alternative Approach to Linearly Constrained Adaptive Beamforming" IEEE Transactions on Antennas and Propagation, vol. 30, no. 1, Jan. 1982

[6] T. W. Parks C. S. Burrus "Digital Filter Design" Wiley & Sons, 1987

[7] B. Widrow S. D. Stearns "Adaptive Signal Processing" Prentice Hall, 1985

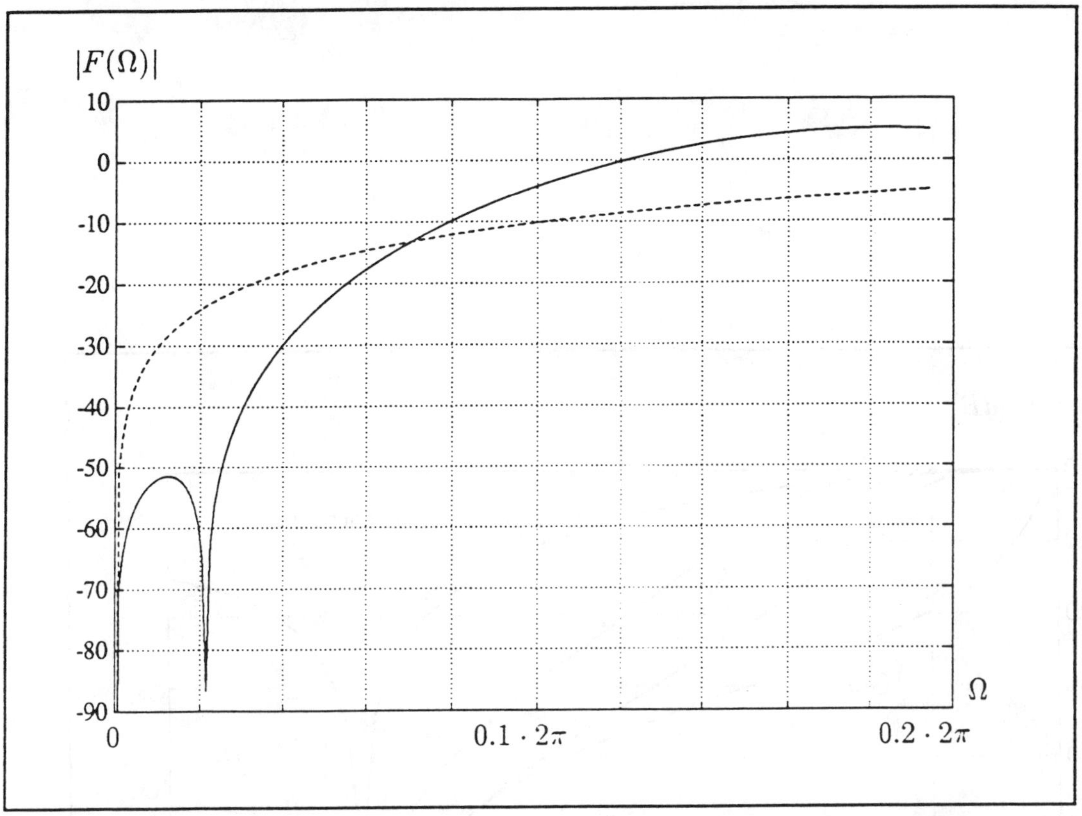

Figure 5: $F(\Omega)$, spatial high pass filter M=12 with passband frequency $0.025 \cdot 2\pi$ corresponding to 3° at frequency 3200Hz with microphone interspace 5cm. Derivating filter [0.5 -0.5] included for comparison (dotted line).

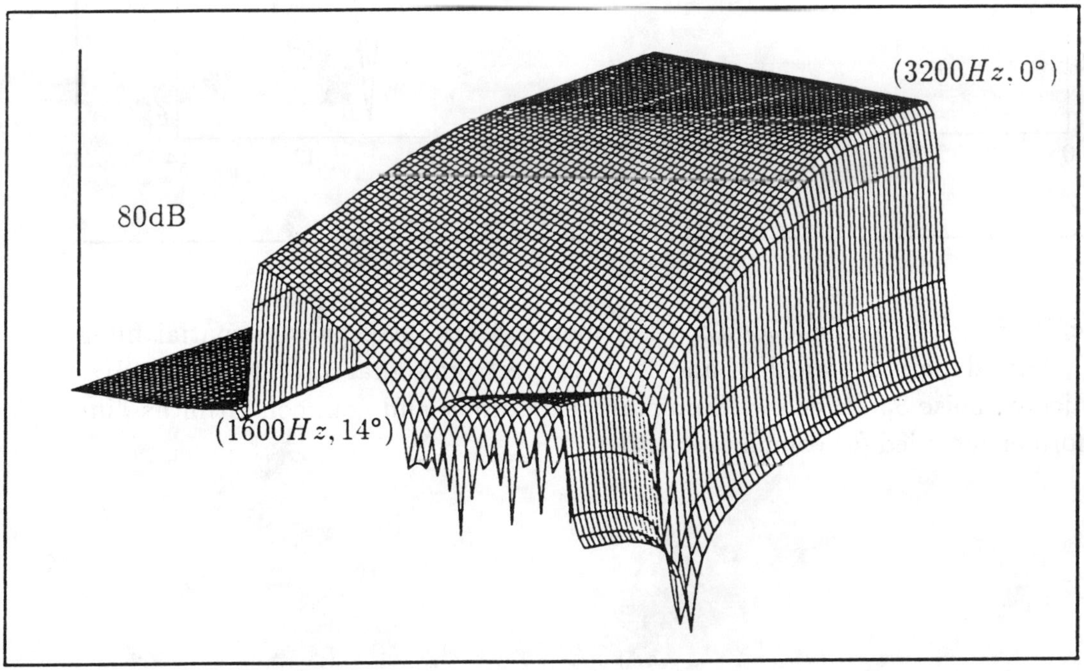

Figure 6: $R_{\varepsilon\varepsilon}(f, \theta)$, 12 elements, 5 cm interspacing, spatial filter M=12, desired mainlobe width 3°. Flat spectrum signal $m(t)$, 1600-3200 Hz, and injected noise 35 dB below.

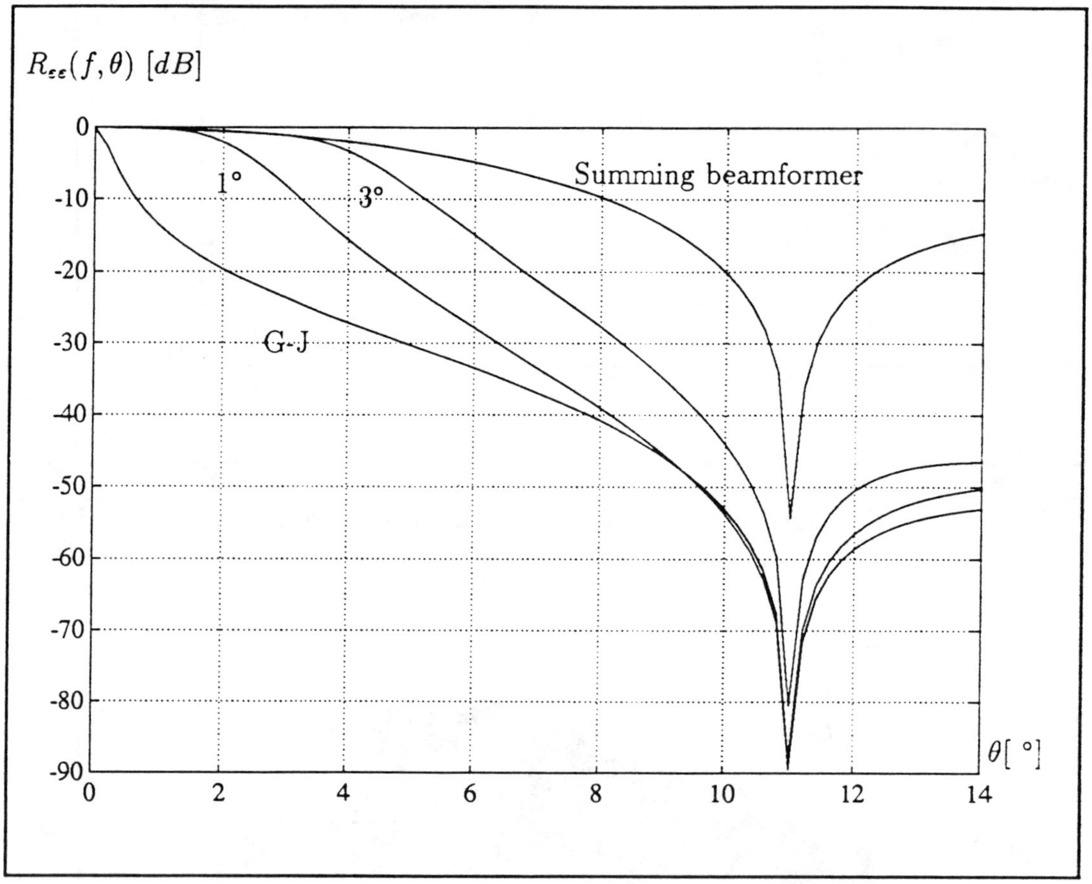

Figure 7: $R_{\varepsilon\varepsilon}(f,\theta)$ for f=3000 Hz, 12 elements, 5 cm interspacing, spatial filter M=12, desired mainlobe width 1° and 3°. Flat spectrum signal $m(t)$, 1600-3200 Hz, and injected noise 35 dB below. Summing beamformer and common Griffiths-Jim beamformer included for comparison.

PIEZOPOLYMER SPATIAL FILTERS FOR ACTIVE STRUCTURAL CONTROL

Simon A. Collins,[1] David W. Miller,[1] Andreas H. von Flotow[1]

ABSTRACT

Distributed sensors and their application to active structural control are studied, highlighting the spatial filtering properties of such sensors. Typical noncausal sensor dynamics, such as gain rolloff without phase lag, are possible in real time because of the spatially distributed nature of the sensor. The effects of the presence of an actuator outside and within the spatial domain of the sensor are discussed. The latter pertains to typical control scenarios. The change in these dynamics as the sensor and actuator approach each other is discussed. Piezoelectric polymer is an excellent candidate material for these sensors, but it may be necessary to transform distributed inertial sensors into equivalent distributed curvature sensors augemented by point inertial measurements to facilitate implementation. A closed-loop disturbance rejection example shows that significant improvements in performance are possible when using a distributed sensor.

INTRODUCTION

Due to the stringent disturbance rejection requirements placed on optical systems, the schemes proposed for active optics usually include a large number of sensors and actuators distributed across the optical surface [1]; a similar control architecture is currently being envisioned for intelligent or smart structures [2]. Coupled with this, there is a growing appreciation of the fact that, at a local level, these structures are distributed-parameter systems, suggesting that spatially distributed transducers may be better suited to build up the set of sensors and actuators than point transducers [3]. In this context, a distributed sensor is one whose output is a function of the structure's response at several locations; these locations may be a series of points or they may be continuously distributed.

Some advantages of using a distributed sensor rather than a point sensor for the closed-loop control of structural vibrations are considered in this paper. The work presented herein was motivated by the results of an active control experiment involving a graphite-reinforced epoxy beam, point sensors and embedded piezoelectric actuators [4]. While this was quite a sophisticated structure, its behavior is believed to be typical of many structures to which active control might be applied. At low frequencies the structure could be described quite accurately by a simple mathematical model, namely a Bernoulli-

[1]Space Engineering Research Center, Department of Aeronautics and Astronautics, Massachusetts Institute of Technology, Cambridge, MA 02139, U.S.A.

Euler beam model. However, at higher frequencies the response was significantly affected by other phenomena, such as transverse modes, amplifier and accelerometer dynamics, and sensor/actuator dislocations. These, and other unexplained factors, destabilized the control systems used on this structure. The experimental solution was to place a low-pass filter in the feedback loop, which resulted in stable closed-loop systems capable of significant improvements in some of the structure's damping ratios. But the addition of the filter markedly reduced the bandwidth of the system.

The spatial filtering property of a distributed sensor offers a way to avoid this trade-off between gain margin and bandwidth. Specifically, it is possible to design a distributed sensor which provides the required gain rolloff and yet does not introduce phase lag. The design and operation of this sensor will be discussed below, and it will be shown how such a sensor allows one to increase the gain margin of the closed-loop system, thereby making greater performance possible, without sacrificing bandwidth. The implementation of a distributed sensor using the piezoelectric polymer polyvinylidene fluoride (PVDF) is also described.

Early work on distributed transducers focused on actuators, with the parallel investigation of piezopolymers [5, 6] and piezoceramics [7, 8]. Both of these piezoelectric materials have been used in many closed-loop control systems, and one is now beginning to see piezopolymer distributed sensors as well [9, 10]. Being a compliant polymer rather than a brittle ceramic, PVDF offers the researcher the ability to shape the distributed transducer. The effect of the transducer's shape on the dynamics of the transducer has been illustrated by a number of workers [11–14]; the spatial distribution will be used in the subsequent sections to design a sensor with apparently noncausal dynamics.

SPATIAL FILTERS

Researchers examining turbulent boundary layers had two distinct phenomena contributing to their pressure measurements. The two pressure components were characterized by their wavenumbers, so transducer systems which incorporated a wavenumber sensitivity were developed [15,16]. A wavenumber, the inverse of a wavelength, can be thought of as a spatial frequency; mathematically, it is the spatial Fourier variable k. An example of a wavenumber or spatial filter is a rectangular membrane: the Fourier transform of this rectangular distribution in the spatial domain is a sinc function, $\sin(k)/k$, in the wavenumber domain (Fig. 1). A sensor acting as a spatial filter, such as a piezoelectric membrane, will respond well to pressure fluctuations at low wavenumbers (long wavelengths) but will attenuate the components at higher wavenumbers (shorter wavelengths) [16]. So gain rolloff can certainly be provided by spatial filters.

The dispersion relation [17] provides the relationship between the wavenumbers and the temporal frequencies which are supported in a material. With the dispersion relation and the magnitude and phase of the sensor's wavenumber response, the sensor dynamics can be converted to frequency domain control design tools such as a Bode plot or a Nyquist diagram.

Wave approaches [18] provide both an easy way to derive dispersion relations and a conve-

$$f(x) = \begin{cases} A & \text{if } |x| < L \\ 0 & \text{if } |x| > L \end{cases}$$

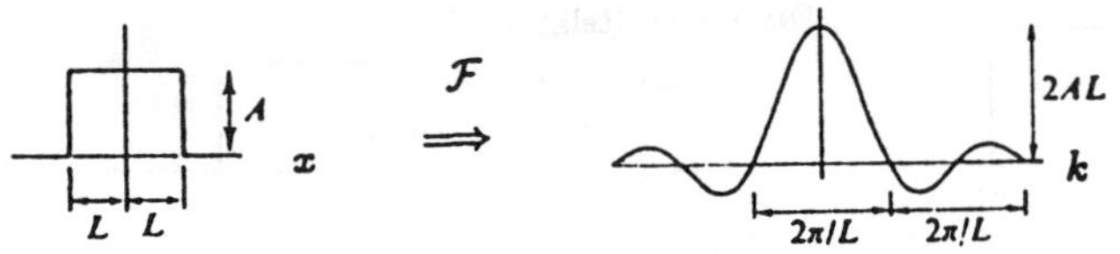

$$2A\frac{\sin Lk}{k}$$

Figure 1: Spatial transform of a rectangular distribution.

nient method to analyze structures which can be viewed as a network of one-dimensional waveguides. For example, substituting the wave solution

$$w(x,t) = Qe^{ikx+iwt} \tag{1}$$

into the governing partial differential equation for an undamped Bernoulli-Euler beam

$$EI\frac{\partial^4 w(x,t)}{\partial x^4} + \rho A\frac{\partial^2 w(x,t)}{\partial t^2} = 0 \tag{2}$$

yields the dispersion relation

$$k^4 = \frac{\rho A}{EI}\omega^2 = \frac{\omega^2}{{c_0}^4} \tag{3}$$

Thus, a Bernoulli-Euler beam can support four wave modes:

$$w(x,t) = (q_{lp}e^{ikx} + q_{le}e^{kx} + q_{rp}e^{-ikx} + q_{re}e^{-kx})e^{iwt} \tag{4}$$

where $w(x,t)$ is the transverse deformation of the beam and q is a wave mode amplitude. Borrowing the terminology of acoustics [19], the subscripts identify the wave modes as either propagating or evanescent. Spatially, evanescent wave modes decay exponentially as they extend away from a junction; they contribute significantly to the structure's near-field response, while the far-field response is dominated by the propagating wave modes. The other subscript indicates rightward or leftward propagating wave modes and rightward or leftward emanating evanescent modes.

A wave perspective also provides insight which helps explain the increased freedom the sensor designer has when choosing the phase response of a distributed sensor. The magnitude and phase characteristics of a temporal real-time filter are tied through the Bode gain-phase theorem [20]; once one is specified, the other is fully determined. Causality is the issue here: if based on a point sensor, a temporal filter only has access to past and present information.

The situation is fundamentally different in the case of a distributed sensor. Rather than requiring past, present and future information, a computation could equivalently require information at the location of the point sensor and to the left and right of it. The response at any given structural cross-section is due to the weighted sum of the wave mode amplitudes at that location. The spatial Fourier components which caused the response in the past are presently travelling away from the cross-section, while those which will determine the future response are presently travelling towards the cross-section. A spatially distributed sensor is affected by *all* of these components, allowing the determination

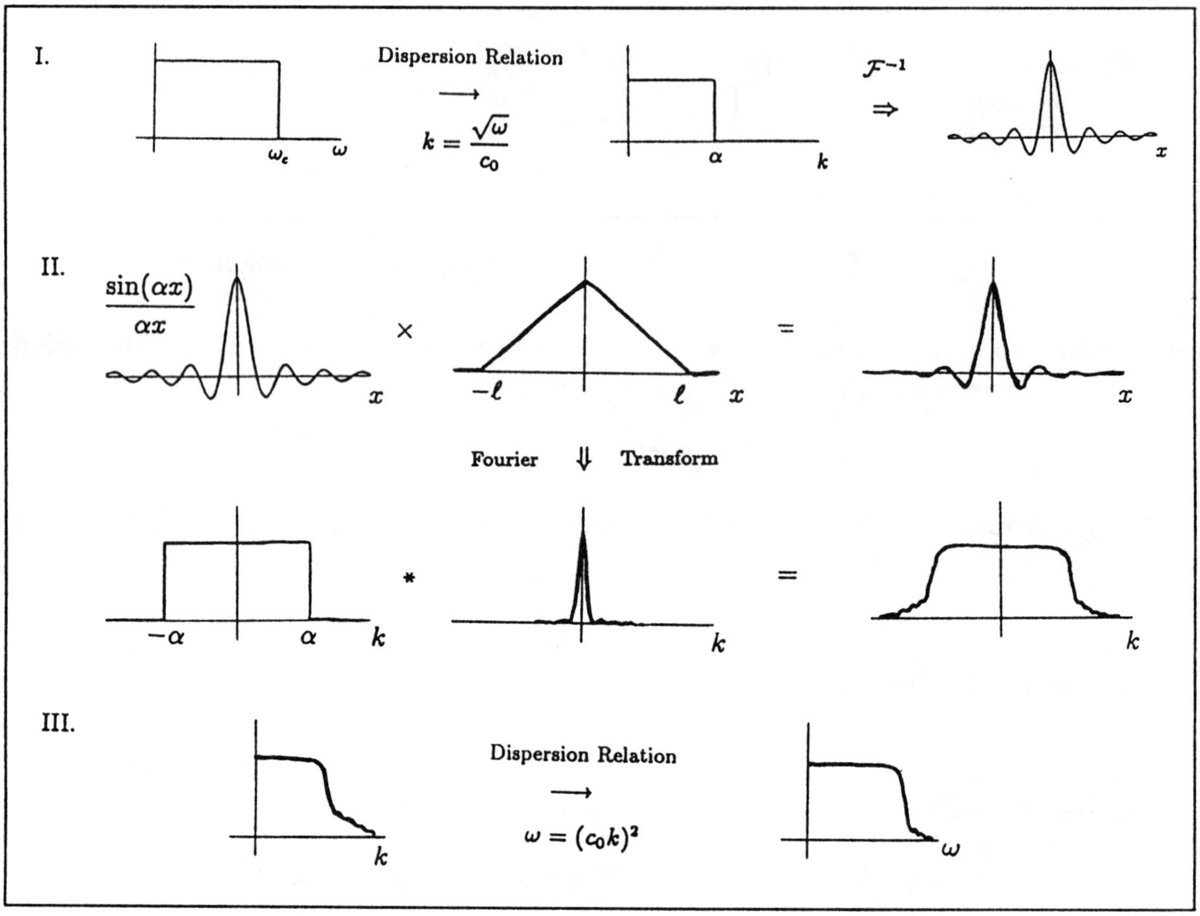

Figure 2: Ideal filter approximation using a distributed sensor.

of past, present and future cross-sectional response. With respect to the cross-section of interest, the spatial filter appears to be non-causal: it could respond *before* a stimulus has reached that cross-section. As Bode's gain-phase theorem only applies to causal systems, it is indeed possible with spatial filters to design gain and phase characteristics independently, *e.g.* gain rolloff without phase lag. Noncausal dynamics can lead to some remarkable results: Ref. [21] demonstrates that resonances in a beam can be eliminated using one actuator, its dual sensor and a noncausal compensator.

SINC FUNCTION SENSOR

Can spatial filters be used to implement an ideal filter? Such a sensor would respond with unity gain to all signals below a cut-off frequency ω_c and would be completely insensitive to signals above ω_c. Figure 2 shows how one can approximate an ideal filter with a distributed sensor.

In Step I, the dispersion relation is used to convert the desired frequency response to a wavenumber response. Assuming the sensor is located in a far-field region of the structure (away from boundaries and actuators), a spatial pattern can be obtained from this response by means of the inverse spatial Fourier transform.

Notice that the spatial pattern, which is a sinc function, is infinite in extent. Unless the structure is also infinite, the pattern must be truncated. Many "windows" have been developed for signal processing applications—a Bartlett, or triangular, window works well here. This truncation (Step II) can be represented as the product of an infinite spatial pattern with a finite window in the spatial domain, which corresponds to the convolution of the ideal wavenumber filter with the transform of the window in the wavenumber domain. The result is the wavenumber response of the (finite) distributed sensor, which can of course be converted to a frequency response via the dispersion relation (Step III).

The frequency response of a sinc function truncated by a rectangular window has been experimentally verified [22], and close agreement with theory was found. The sensor dynamics were a good approximation of an ideal filter, with the approximation improving as a longer sensor is used. However, a series of 180° phase reversals occured due to the rectangular window; these reversals are eliminated by the Bartlett window.

NEAR-FIELD RESPONSE

The results discussed above were based on the assumption that the sensor was located in the far-field of the structure. If, instead, the sensor is close to a boundary or an actuator (as will certainly be the case with a collocated sensor/actuator pair), the near-field frequency response of the distributed sensor must be determined. Both propagating and evanescent wave modes play a role here.

To this end, a wave model of the structure used by de Luis *et al.* can be developed. The model consists of a free-free Bernoulli-Euler beam 1.22 m in length. Curvature rate damping is included, which modifies the dispersion relation. The governing partial differential equation becomes

$$EI\frac{\partial^4 w(x,t)}{\partial x^4} - R\frac{\partial^3 w(x,t)}{\partial t\,\partial^2 x} + \rho A\frac{\partial^2 w(x,t)}{\partial t^2} = 0 \tag{5}$$

and the dispersion relation is

$$k^2 = -i\frac{R}{2EI}\omega \pm \frac{(4EI\rho A - R^2)^{1/2}}{2EI}\omega \tag{6}$$

If $R^2 < 4EI\rho A$ (the case for most structures), the wavenumber can be written

$$k = k_R - ik_I \tag{7}$$

where

$$\begin{aligned} k_R &= \frac{\sqrt{\omega}}{c_0}\left(1 - \frac{R^2}{8EI\rho A}\right)^{1/2} \\ k_I &= \frac{\sqrt{\omega}}{c_0}\left(\frac{R^2}{8EI\rho A}\right)^{1/2} \end{aligned}$$

The transverse deformation can then be expressed in terms of wave modes:

$$w(x,t) = \left(q_{lp}e^{(ik_R+k_I)x} + q_{le}e^{(k_R+ik_I)x} + q_{rp}e^{-(ik_R+k_I)x} + q_{le}e^{-(k_R+ik_I)x}\right)e^{i\omega t} \tag{8}$$

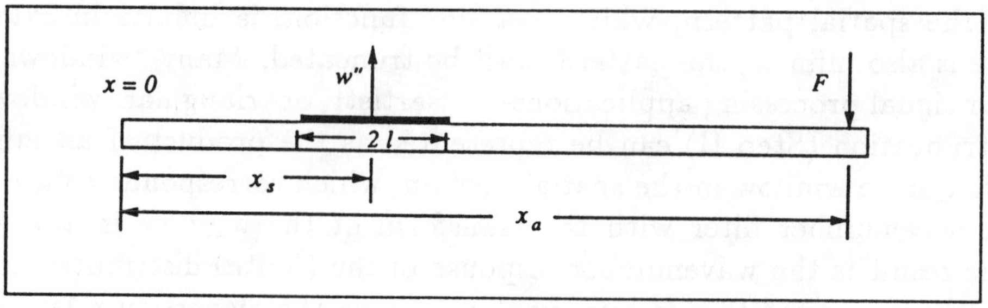

Figure 3: Free-free beam model with actuator and sensors.

The wave modes are no longer purely propagating or evanescent, although they will more strongly represent one type than the other since the damping is light ($k_R \gg k_I$). The values used to model the experimental structure were $EI = 16.59$ N $\cdot$ m^2, $\rho A = 0.7736$ kg/m and $R = 0.07$ kg $\cdot$ m/sec.

A point force actuator was included at x_a and a point curvature sensor was located at x_s (Fig. 3). A distributed curvature sensor of length 2ℓ was centered on x_s. The output of the point sensor was determined using the wave approach described in Ref. [18], while the distributed sensor's output was taken as

$$\overline{w''}(t) = \int_{x_s-\ell}^{x_s+\ell} g(x - x_s)\, w''(x,t)\, dx \tag{9}$$

where $g(\xi)$ defines the sensor's spatial distribution, *e.g.* a windowed sinc function. Numerical integration is used, but the curvature within the integral is determined analytically. For the following examples, a sinc function ($f_c = 2\pi\omega_c = 500$ Hz) and a Bartlett window are used. The total length of the sensor is $2\ell = 0.6$ m and $x_s = 0.4$ m. The forcing function is taken to be white noise of unit amplitude; two actuator locations are studied.

Figure 4 shows the transfer functions from the applied force at $x_a = 1.2$ m to the outputs of the point sensor and the distributed sensor. As described above, at low frequencies the response of the point sensor and the distributed sensor are identical, whereas the gain of the distributed sensor is significantly lower at high frequencies; the two sensors exhibit almost identical phase throughout. The sensor dynamics can be separated from the dynamics of the beam by taking the ratio of the output of the two sensors (Fig. 5). The magnitude plot compares well with Fig. 2. Two differences are seen in the phase plot. The slight phase oscillations (less than three degrees) which occur above f_c are due to the damping in the model which introduces complex wave modes. The spikes in the phase at 71 and 230 Hz correspond to the first two zeros in the force-to-curvature transfer functions; a very slight mismatch of these transfer functions results in these spikes when the ratio is computed.

As the actuator is moved toward the point sensor, a rather different behavior emerges. The evanescent waves in the sensor's domain will now be more significant. For example, applying the force at $x_a = 0.42$ m (near the center of the distributed sensor) results in the frequency responses shown in Fig. 6. The first point to notice is that not only is the gain of the spatial filter attenuated at higher frequencies, the resonant structure of the response is not as pronounced and the zeros are not as deep. Secondly, the noncollocation of the actuator and the point sensor causes accumulated phase lag, whereas the distributed sensor prevents this. Finally, the zeros of the two sensors are no longer aligned. If the sensor dynamics are extracted by taking the ratio of the two outputs, a "noisy" response results (Fig. 7) because of the way the zeros have shifted. In this case, it would appear to be more instructive to examine the transfer functions from force to sensor output rather than these computed sensor dynamics.

The reason for the modified resonant structure in the distributed sensor's high frequency response is not well understood. It is *not* simply due to the greater contribution of the evanescent modes. This can be verified by examining the wave mode components which make up the sensor's response. With $x_a = 1.2$ m, the propagating wave modes completely overwhelm the evanescent components at most frequencies (Figs. 8 and 9). Notice that all components exhibit a resonant structure, both in the case of the distributed sensor and the point sensor.

As would be expected, the leftward emanating evanescent mode makes a significant contribution when $x_a = 0.42$ m (Figs. 10 and 11)—the point sensor is just to the left of the actuator. This wave mode shows little of the beam's resonance, whereas the rightward emanating evanescent wave mode does. This may be explained as follows. The rightward emanating one sensed by the point sensor originated at the left end of the beam by means of scattering or wave mode conversion. So the path to the sensor is much longer than the path taken by the leftward emananting evanescent wave mode, which explains the much lower amplitudes. Along this path the beam, acting as a mechanical spatial filter, has been able to impose a resonant structure on the evanescent mode. The conversion of

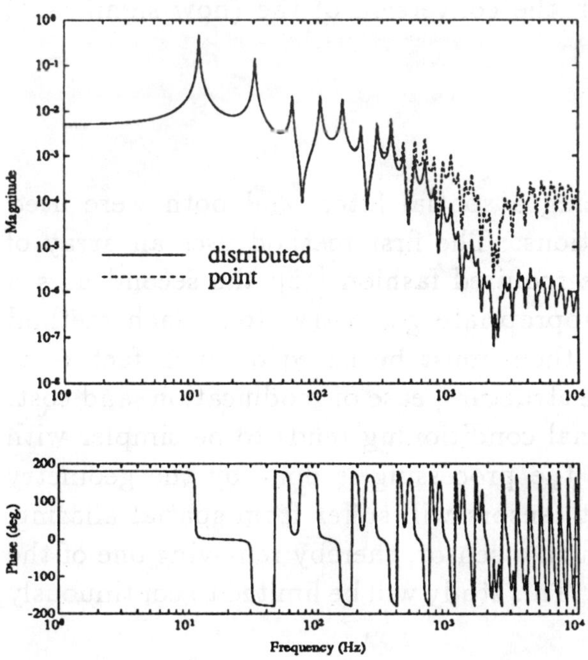

Figure 4: Transfer functions from applied force to sensor output, $x_a = 1.2$ m.

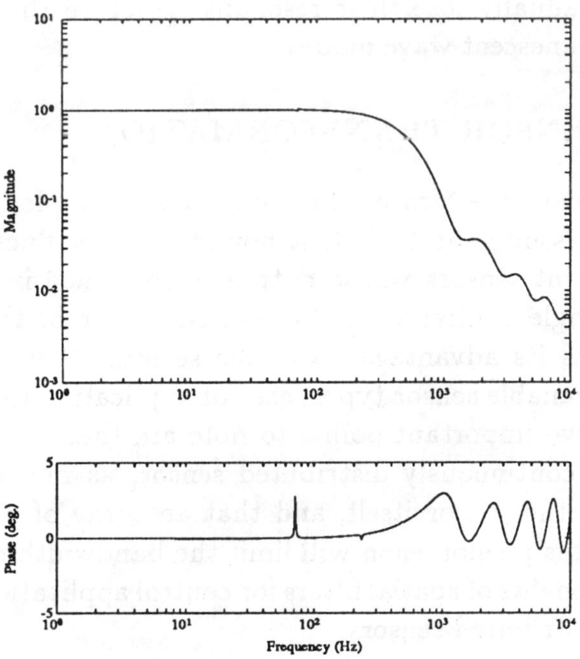

Figure 5: Transfer functions from point sensor to distributed sensor, $x_a = 1.2$ m.

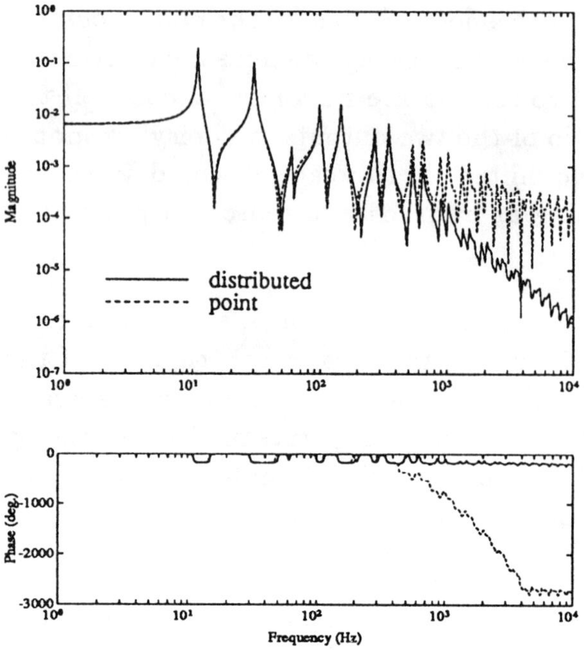

Figure 6: Transfer functions from applied force to sensor output, $x_a = 0.42$ m.

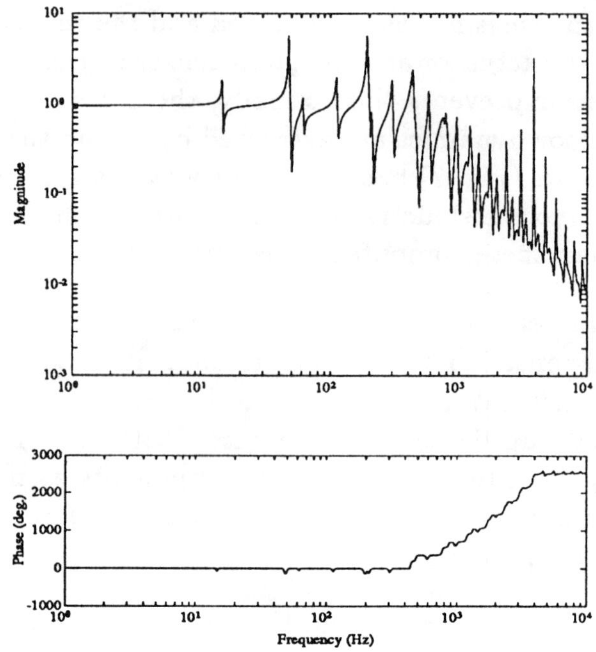

Figure 7: Transfer functions from point sensor to distributed sensor, $x_a = 0.42$ m.

propagating modes to evanescent modes at the free end also affects this. But in the case of the distributed sensor, this evanescent mode is not much weaker, and it displays little in the way of resonance. A possible explanation is that the distributed sensor picks up the rightward emanating evanescent wave mode as it is generated by the actuator without the need for reflections or wave mode conversion. Then the propagating wave modes gradually lose their resonant structure through the conversion of the (now significant) evanescent wave modes.

SENSOR TRANSFORMATION

There are two distinct methods of implementing a spatial filter, and both were used to study the turbulent boundary-layer fluctuations. The first method uses an array of point sensors whose output is combined in the required fashion [15]; the second uses a single continuously distributed sensor of the appropriate geometry [16]. Each method has its advantages, and the selection between them must be based on such factors as available sensor types, ease of application to the structure, ease of modification, and cost. Two important points to note are that the signal conditioning tends to be simpler with a continuously distributed sensor, as much of the processing is done by the geometry of the sensor itself, and that an array of point sensors will suffer from spatial aliasing. This phenomenon will limit the bandwidth of such a sensor, thereby removing one of the benefits of spatial filters for control applications. This study will be limited to continuously distributed sensors.

As suggested in the Introduction, the piezopolymer PVDF is an excellent material from which to fabricate distributed transducers. While there are some concerns about the suitability of the material to serve as an actuator (due to the difference between the modulus of PVDF and that of most structural materials), there is little doubt that PVDF makes an excellent sensor. It is lightweight and flexible, thus having little impact on the structure's dynamics; it has a high piezoelectric stress constant (the ratio of induced electric field to applied stress under open circuit conditions, g_{31}, is about 220 mV $\cdot$ m/N, more than 20 times the value for piezoceramics [23]); and it is commercially available in a form from which complicated sensors can be easily made.

This material is strain-sensitive, as it relies on the piezoelectric effect to produce a charge which is proportional to the stress applied to the piezopolymer. This stress arises from the structure's surface strain (when a good bond is achieved). So strain (or strain-rate) gauges can be constructed from PVDF. A spatial weighting function can be incorporated in the sensor by varying the thickness of the piezo film or by varying the width of the piezoelectrically active region of the sensor. The width can be easily adjusted by cutting the film or by etching the surface electrodes in the desired pattern. The material is

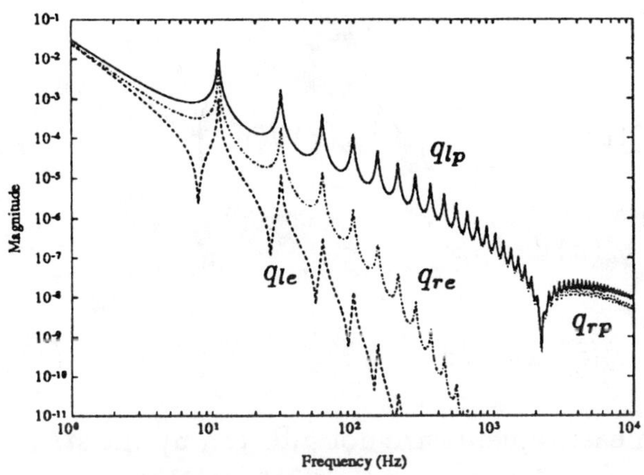

Figure 8: Wave mode components in response of point sensor, $x_a = 1.2$ m.

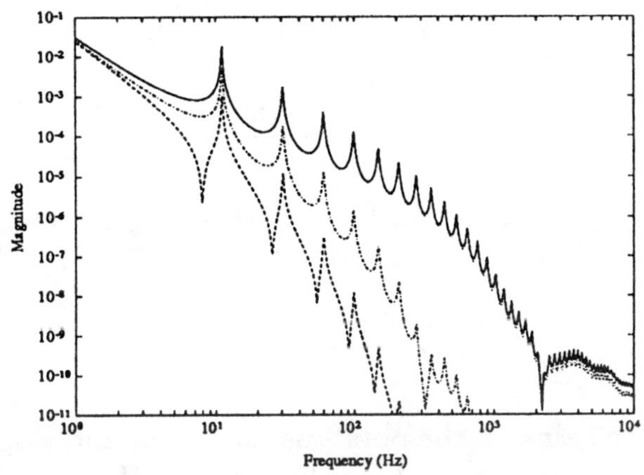

Figure 9: Wave mode components in response of distributed sensor, $x_a = 1.2$ m.

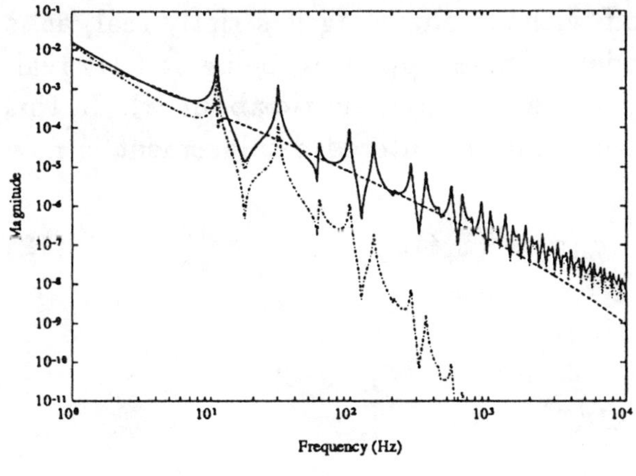

Figure 10: Wave mode components in response of point sensor, $x_a = 0.42$ m.

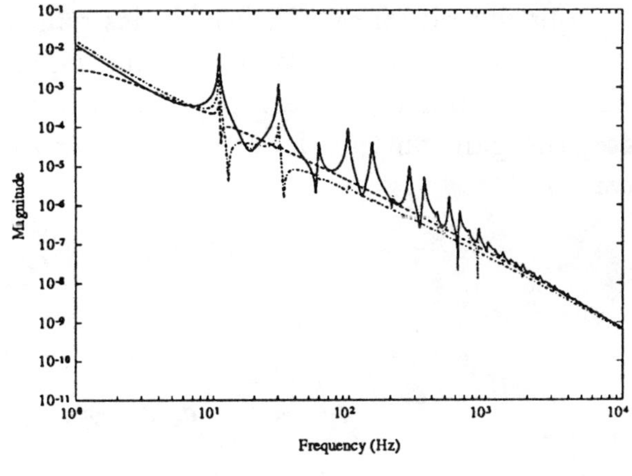

Figure 11: Wave mode components in response of distributed sensor, $x_a = 0.42$ m.

polarized, so the weighting function need not be strictly positive; negative regions can be included by reversing the polarity of the film, by appropriate wiring of the sensor, or by mounting some segments "upside down." This technique was used successfully for the tests reported in Ref. [22].

Keeping the potential control application in mind, what if the desired feedback variable is not strain or strain rate? What if, say, velocity feedback is called for? A transformation of variables is possible from distributed velocity to distributed strain plus point velocity and rotation rate measurements, as shown in Ref. [24]. For simplicity, assume the desired spatial filter is to be sensitive to transverse displacement:

$$\overline{w}(t) = \int_{-\ell}^{+\ell} g_d(x) w(x,t)\, dx \tag{10}$$

where a change of variables was used such that $x_s = 0$. The spatial weighting function $g_d(x)$ is the displacement kernel. Integration by parts can be used to transform this into a curvature kernel, and the resulting boundary terms, which capture the rigid-body motion, can be shifted to x_s. Then

$$\begin{aligned} \overline{w}(t) &= w(0,t) \int_{-\ell}^{+\ell} g_d(x)\, dx \\ &+ \frac{\partial w(0,t)}{\partial x} \left(\int_0^{+\ell} \int_x^{+\ell} g_d(\xi)\, d\xi dx - \int_0^{-\ell} \int_x^{-\ell} g_d(\xi)\, d\xi dx \right) \\ &+ \int_0^{+\ell} \frac{\partial^2 w(x,t)}{\partial x^2} \int_x^{+\ell} \int_\xi^{+\ell} g_d(\eta)\, d\eta d\xi dx \\ &- \int_0^{-\ell} \frac{\partial^2 w(x,t)}{\partial x^2} \int_x^{-\ell} \int_\xi^{-\ell} g_d(\eta)\, d\eta d\xi dx \end{aligned} \tag{11}$$

The gains on the point displacement and slope measurements are not affected by the structure's deformations and need only be computed once. The curvature integrations involve new kernels which similarly need only be computed once; these kernels will determine the shape of the PVDF sensor.

A frequency response which includes gain rolloff without phase lag is strictly real, as is the wavenumber response (in an undamped model). These responses, being strictly real, will have symmetric trnasforms, so the kernel $g_d(x)$ will be symmetric about x_s. In this case, the gain on the slope measurement vanishes and the filtered displacement is now given by

$$\bar{w}(t) = K_p w(0,t) + \int_{-\ell}^{+\ell} g_c(x)\, w''(x,t)\, dx \tag{12}$$

where

$$\begin{aligned} K_p &= 2\int_0^{\ell} g_d(x)\, dx \\ g_c(x) &= \int_{|x|}^{\ell} \int_\xi^{\ell} g_d(\eta)\, d\eta d\xi \end{aligned}$$

The displacement kernel has been transformed into a curvature kernel. Using the truncated sinc function described above as an example displacement kernel, the gain on the point displacement measurement is 0.1109, and the curvature kernel is shown in Fig. 12.

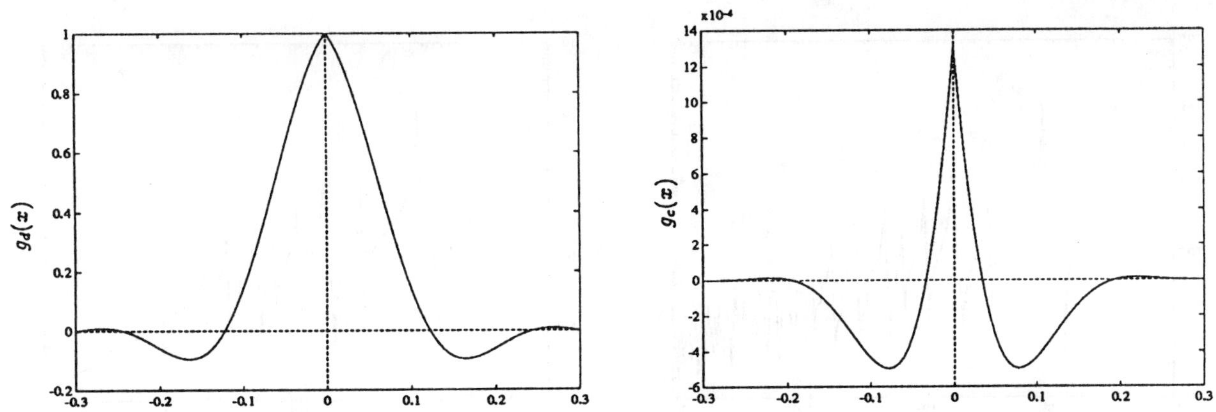

Figure 12: Displacement kernel $g_d(x)$ and corresponding curvature kernel $g_c(x)$.

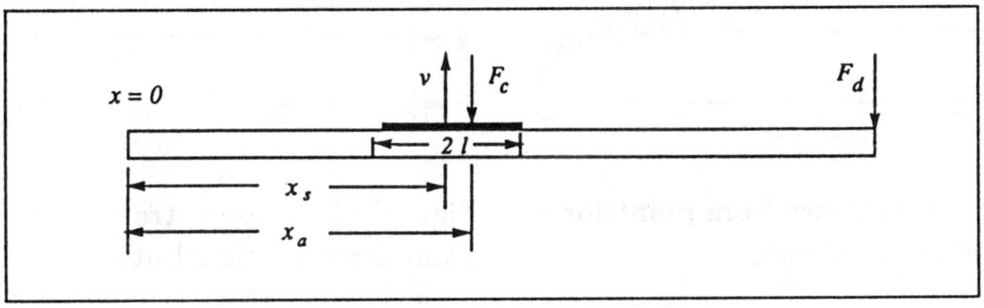

Figure 13: Sensor and actuator placement.

This form of a spatial filter is a composite sensor: it is made up of a distributed sensor and a point sensor. Equation (12) represents a viable way to implement a displacement-sensitive spatial filter since both point displacement sensors and distributed curvature sensors are available. If rate feedback is desired, the displacement sensor becomes a transverse velocity sensor and the distributed sensor must be curvature-rate sensitive, which can be easily accomplished with piezoelectrics by measuring the developed current rather than the voltage. Both K_p and $g_c(x)$ remain the same.

CLOSED-LOOP EXAMPLE

The spatial filter can now be used as a sensor for the feedback loop of a closed-loop system. The plant is the free-free beam model described above. Transverse velocity feedback to a point force actuator was used in an attempt to improve the structure's disturbance rejection behavior. The disturbance was also a point force, spectrally white and of unit amplitude; it was applied to the structure at the right end (Fig. 13). The performance achievable with a distributed sensor is compared to that possible with a point sensor.

The actuator was placed so as to ensure good controllability of the first five modes. A location in the center of the beam (x_a = 0.56 m) was selected to avoid the near-field effects of the boundaries. To mimic the high-frequency instabilities observed in the experiment [4], a deliberate offset was introduced between the sensor and the actuator x_s = 0.52 m. The geometry of the problem is shown in Fig. 13.

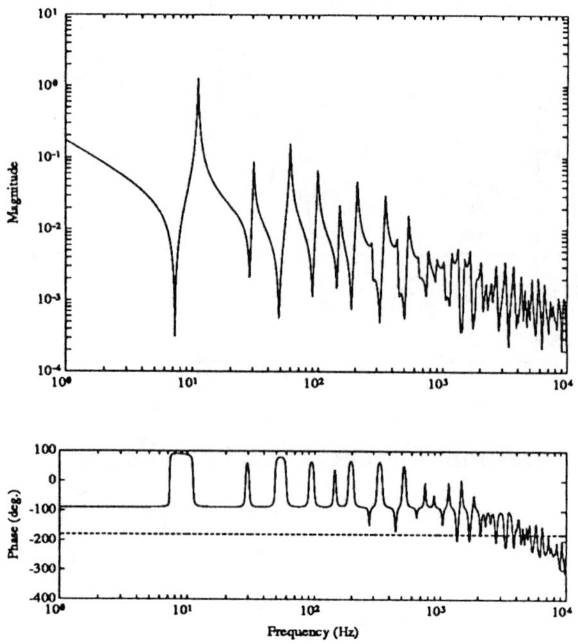

Figure 14: Loop transfer from point force to point transverse velocity.

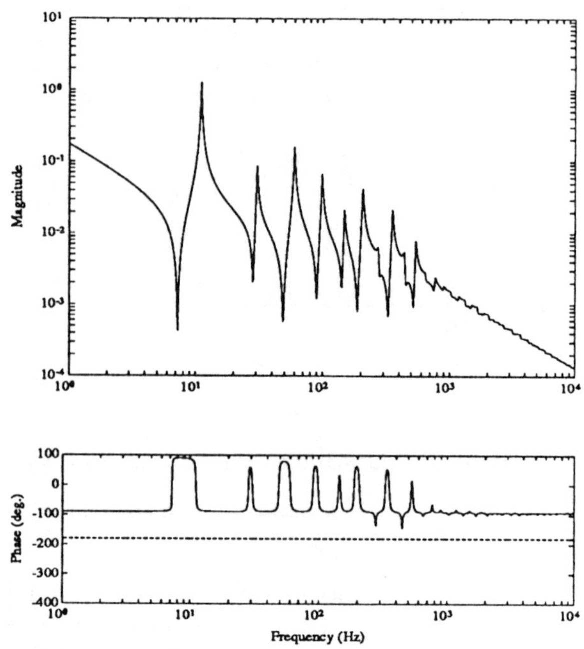

Figure 15: Loop transfer function from point force to distributed velocity.

LOOP TRANSFER FUNCTIONS

The loop transfer function from the control actuator (force) to the point sensor (transverse velocity) is shown in Fig. 14. The phase margin for this system is roughly 90°, and the lowest gain margin is about 250, which occurs at 1700 Hz. At first glance, this suggests the distributed sensor should be designed to roll off "sufficiently" before this frequency; classical rules-of-thumb would place a filter's corner frequency a decade below the instabilty, or around 170 Hz. But reviewing Fig. 6, it can be seen that the phase characteristics are modified by a spatial filter in such a way so as to avoid a $-180°$ crossing altogether. So f_c can be raised to 500 Hz (or more) and the closed-loop system will still be stable; a substantial improvement in bandwidth over any scheme using a temporal filter.

The composite sensor developed in the previous section acts as a spatial filter, is sensitive to transverse displacement and can be implemented using PVDF. The modification needed to sense transverse velocity was described above. The response of the composite sensor will be equivalent to the response of a distributed velocity sensor; all subsequent measurements of the "distributed velocity" are made with the composite sensor. Centering this sensor on $x_s = 0.52$ m, the loop transfer function from the control force F_c to the distributed velocity $\partial w/\partial t = v$ is shown in Fig. 15. From a control point of view, this loop transfer function is obviously a considerable improvement over the one obtained for the point sensor. The phase margin is still roughly 90°, but the gain margin is infinite as the phase of the loop transfer function never crossed $-180°$.

PERFORMANCE METRIC

The performance of the closed-loop system was evaluated by looking at the transfer function from the disturbance force F_d to the transverse point velocity v at x_s. As the *actual* velocity of the beam at this point was to be reduced, the point sensor had to be used when measuring performance. Used in open-loop, the composite sensor would reduce the *observed* velocity without actually changing the motion of the beam. Assuming F_d was white noise, a root-mean-square velocity v_{rms} for the point x_s was also calculated.

The control law was simply velocity feedback with a gain K. The velocity could be the point velocity or the filtered velocity. When the loop was closed with the point sensor and the gain K was increased, the root-mean-square velocity decreased markedly. As the gain approached 300, a noticeable increase in the amplitude of several high frequency (> 1 kHz) modes was observed, which adversely affected v_{rms}. While this gain was somewhat higher than the gain margin predicted by the loop transfer function, the difference could be due to the spacing of the frequency points used to compute the frequency response; the frequency of the instability may not be well-captured. Of course, an actual structure has no such problem and performance would be more severely compromised. The open-loop value of v_{rms} was 14.1 mm/s, while the lowest value achieved with the point sensor before the instability occured was 4.34 mm/s or 31% of the open-loop value. This value was reached when the gain was 125; the magnitude of the velocity (both open-loop and with $K = 125$) is shown in Fig. 16. Note that the amplitude of some of the high frequency modes has increased, especially near the pedicted instability at 1700 Hz..

No such gain limit exists when the composite sensor was used—the value of v_{rms} decreases continuously as the gain is increased until an asymptote of 2.74 mm/s is reached. This value is less than 20% of the open-loop value and roughly 63% of the value obtained with the point sensor. While this does not seem like a very large improvement over the point sensor, it was calculated over a bandwidth that includes many modes hardly affected by either form of feedback; these modes came to dominate the performance. The significant

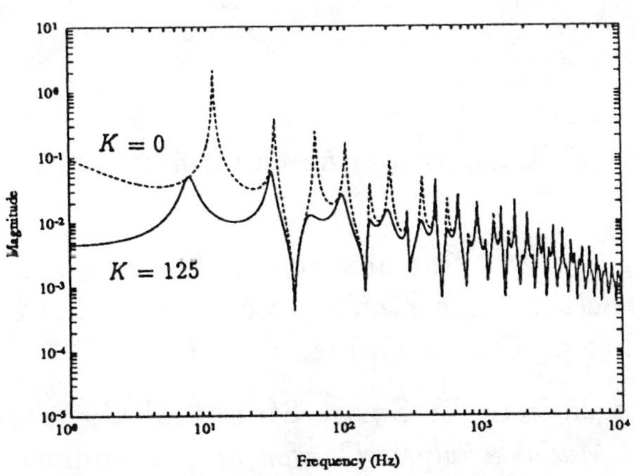

Figure 16: Closed-loop transfer from disturbance to point transverse velocity, using point sensor.

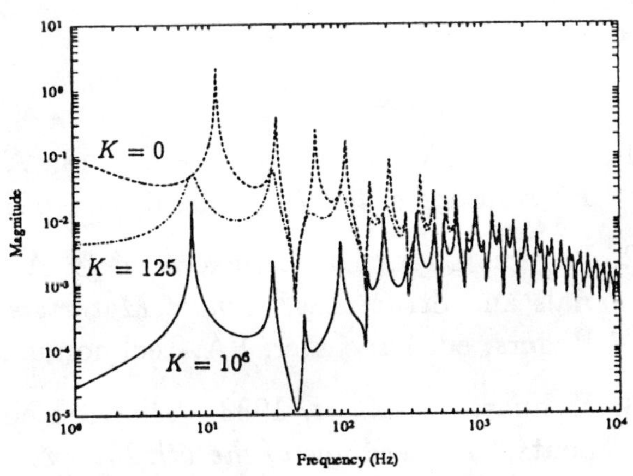

Figure 17: Closed-loop transfer from disturbance to point transverse velocity, using composite sensor.

reduction of $v(x_c)$ below 600 Hz can be seen in Fig. 17 which shows the amplitude of the open- and closed-loop transverse velocity. The closed-loop results represent the the performance for $K = 125$ and the high-gain asymptote. This limit exists because of the fact that the filtered velocity was used for feedback, while the point velocity was used to evaluate the performance. The other factor that limits the achievable performance is the presence of fairly deep zeros in the loop transfer functions, which appear as sharp peaks in the closed-loop transfer function from F_d to $v(x_c)$.

CONCLUSIONS

Distributed sensors provide dynamics which cannot be obtained using point sensors and real-time temporal filters. These additional dynamics can be used to improve the closed-loop performance of a control system. Specifically, this work has shown how a spatial filter can be used to increase disturbance rejection in the control of structural vibrations.

It is also important to realize that these conclusions have practical relevance. Distributed curvature sensors with arbitrary spatial distributions can be manufactured using the piezo-electric polymer PVDF, and a technique has been presented to transform feedback kernels into the necessary forms. While a particular example was studied above, several other applications which could benefit from the use of distributed sensors were mentioned in the course of this example. All of these highlight the new avenues opened by thinking about the spatially distributed nature of the structural control problem, the complete control system (the structure, the sensors and the actuators) and the interaction of the components.

ACKNOWLEDGEMENTS

This work was supported by the Office of Naval Research under Contract No. N00014-88-K-0720 and by the M.I.T. Space Engineering Research Center under NASA Grant NAGW-1335.

REFERENCES

1. Hardy, J. W. 1978. "Active Optics: A New Technology for the Control of Light," *Proceedings of the IEEE*, 66(6):651–697.

2. Rogers, C. A., D. K. Baker and C. A. Jaeger. 1989. "Introduction to Smart Materials and Structures," *Smart Materials, Structures, and Mathematical Issues*, C. A. Rogers, ed. Lancaster, PA: Technomic Publishing Company, Inc., pp. 17–28.

3. Hubbard, J. E. Jr. 1988. "Distributed Transducers for Smart Structural Components," *Proceedings of the 6th International Modal Analysis Conference*, Kissimmee, FL, Vol. 2, pp. 856–862.

4. de Luis, J., E. F. Crawley and S. R. Hall. 1989. "Design and Implementation of Optimal Controllers for Intelligent Structures Using Infinite Order Structural Models," Space Systems Laboratory Report No. 3-89, Massachusetts Institute of Technology, Cambridge, MA.

5. Sato, T., H. Ishida and O. Ikeda. 1980. "Adaptive PVDF Piezoelectric Deformable Mirror System," *Applied Optics*, 19(9):1430–1434.

6. Bailey, T. and J. E. Hubbard, Jr. 1985. "Distributed Piezoelectric-Polymer Active Vibration Control of a Cantilever Beam," *Journal of Guidance, Control, and Dynamics*, 8(5):605–611.

7. Forward, R. L. 1981. "Electronic Damping of Orthogonal Bending Modes in a Cylindrical Mast—Experiment," *Journal of Spacecraft and Rockets*, 18(1):11–17.

8. Crawley, E. F. and J. de Luis. 1987."Use of Piezoelectric Actuators as Elements of Intelligent Structures," *AIAA Journal*, 25(10):1373–1385.

9. Miller, S. E. and J. Hubbard. 1987. "Observability of a Bernoulli-Euler Beam Using PVF_2 as a Distributed Sensor," *Dynamics and Control of Large Space Structures*, L. Meirovitch, ed. Blacksburg, VA, pp. 375–390.

10. Tzou, H. S. 1989. "Integrated Distributed Sensing and Active Vibration Suppression of Flexible Manipulators Using Distributed Piezoelectrics," *Journal of Robotic Systems*, 6(6):745–767.

11. Sato, T, H. Ishikawa and O. Ikeda. 1982. "Multilayered Deformable Mirror Using PVDF Films," *Applied Optics*, 21(20):3664–3668.

12. Burke, S. E. and J. E. Hubbard, Jr. 1987. "Active Vibration Control of a Simply Supported Beam Using a Spatially Distributed Actuator," *IEEE Control Systems Magazine*, 7(4):25–30.

13. Lee, C.-K. and F. C. Moon. 1990. "Modal Sensors/Actuators," *Journal of Applied Mechanics*, 57(2):434–441.

14. Collins, S. A., R. J. Notestine, C. E. Padilla, M. Ramey, E. Schmitz and A. H. von Flotow. 1990. "Design, Manufacture, and Application to Space Robotics of Distributed Piezoelectric Film Sensors," *Proceedings of the AIAA/ASME/ASCE/AHS Structures, Structural Dynamics, and Materials Conference*, Long Beach, CA, pp. 1899-1906. To appear in *Journal of Guidance, Control, and Dynamics.*

15. Maidanik, G. and D. W. Jorgensen. 1967. "Boundary Wave-Vector Filters for the Study of the Pressure Field in a Turbulent Boundary Layer," *Journal of the Acoustic Society of America*, 42(2):494–501.

16. Martin, N. C. and P. Leehey. 1977. "Low Wavenumber Wall Pressure Measurements Using a Rectangular Membrane as a Spatial Filter," *Journal of Sound and Vibration*, 52(1):95–120.

17. Strang, G. 1986. *Introduction to Applied Mathematics*. Wellesley, MA: Wellesley-Cambridge Press, pp. 554–558.

18. von Flotow, A. H. 1986. "Disturbance Propagation in Structural Networks," *Journal of Sound and Vibration*, 106(3):433–450.

19. Dowling, A. P. and J. E. Ffowcs Williams. 1983. *Sound and Sources of Sound*. Chichester, England: Ellis Horwood Limited, pp. 90–95.

20. Bode, H. W. 1945. *Network Analysis and Feedback Amplifier Design*, Princeton, NJ: D. Van Nostrand, Inc.

21. Miller, D. W., S. R. Hall and A. H. von Flotow. 1990. "Optimal Control of Power Flow at Structural Junctions," *Journal of Sound and Vibration*, 140(3):475–497.

22. Miller, D. W., S. A. Collins and S. P. Peltzman. 1990. "Development of Spatially Convolving Sensors for Structural Control Applications," *Proceedings of the AIAA/ASME/ASCE/AHS Structures, Structural Dynamics, and Materials Conference*, Long Beach, CA, pp. 2283–2297.

23. *KYNAR Piezo Film Technical Manual*, 1987. Pennwalt Corp., Valley Forge, PA, 65 pp.

24. Miller, D. W. and M. C. van Schoor. 1990. "Formulation of Full State Feedback for Infinite Order Structural Systems," presented at the First Joint US/Japan Conference on Adaptive Structures, Maui, HI.

Noise Control

THE CONTINUING EVOLUTION OF ACTIVE NOISE CONTROL WITH SPECIAL EMPHASIS ON DUCTBORNE NOISE

L. J. Eriksson[1]

I. BACKGROUND

Active noise control is a technology that has seen continued evolution for over six decades. Following early work on a wide variety of applications, attention focused in the 1970s on the silencing of ductborne noise. As discussed recently in a related paper [1], two basic approaches were used. The feedforward technique, shown in Fig. 1, was the oldest and utilized a microphone upstream of the cancelling loudspeaker [2]. Although potentially capable of producing broadband attenuation of random noise, this technique suffered from instability due to acoustic feedback.

An alternative approach was to place the input microphone at or downstream from the loudspeaker in a feedback configuration, as shown in Fig. 2 [3]. This configuration could provide only limited attenuation over a restricted frequency range for periodic or band-limited signals and also suffered from instability due to the possibility of positive feedback.

Because of these problems, research emphasized the use of transducer arrays, as shown in Fig. 3, to create unidirectional transducers to reduce instabilities due to acoustic feedback from the cancelling loudspeaker source to the input microphone [4-6]. Other work at this

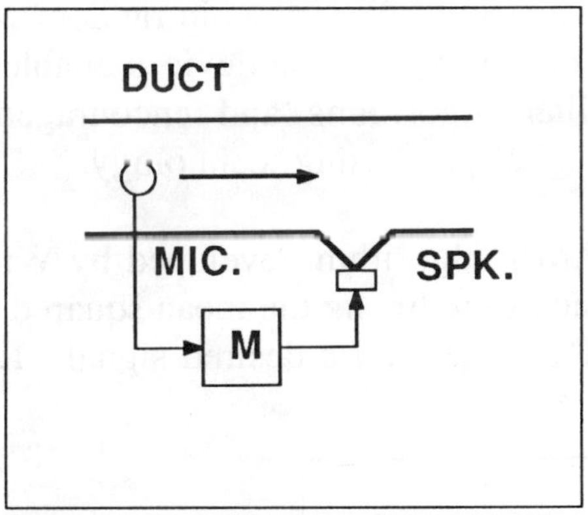

Fig. 1 - Feedforward form of active noise control as presented by Lueg [2].

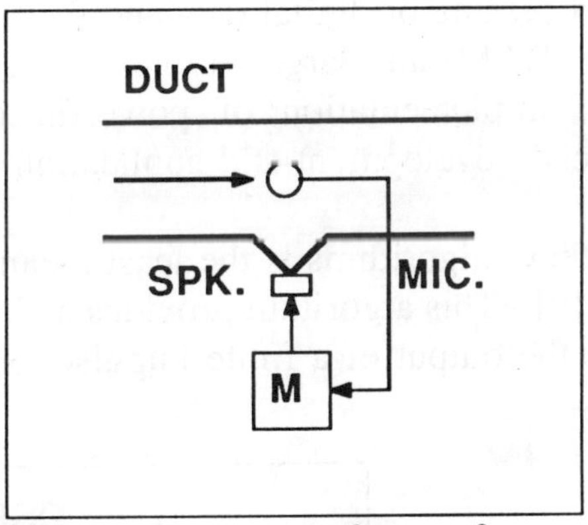

Fig. 2 - Feedback form of active noise control as described by Olson [3].

[1]Nelson Industries, Inc., Highway 51 West, P.O. Box 600, Stoughton, WI 53589-0600, U.S.A.

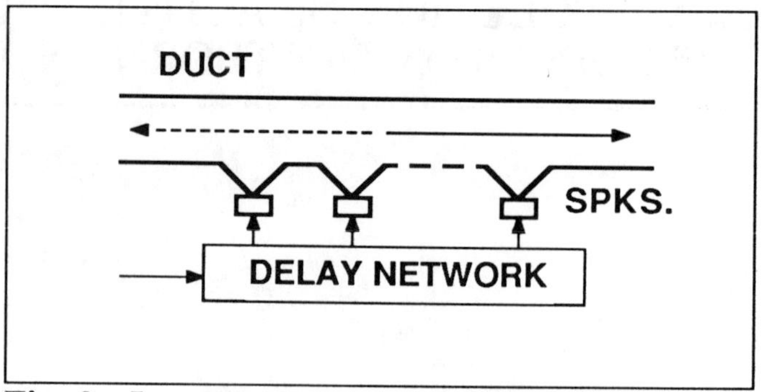

Fig. 3 - Loudspeaker arrays have been utilized to minimize acoustic feedback and create a unidirectional source [4-6].

time eliminated the input microphone and associated acoustic feedback through the use of a synchronizing signal from the primary source as the input signal to the controller [7].

This paper will briefly review the evolution of digital signal processing over the past decade and focus on its impact on active noise control technology. Future trends will be discussed, and a possible scenario for the ultimate application of this technology will be described.

II. SIGNAL PROCESSING

In the 1980s, the research focus was dramatically affected by the development of powerful digital signal processor hardware and the parallel development of adaptive filter theory. The specialized microprocessors that were introduced in this era were designed for very high speed processing of digital data and featured a hardware multiplier, on-chip random access memory (RAM) and large word sizes, as shown in Fig. 4 [8]. These devices enabled the low-cost implementation of powerful adaptive filter algorithms and encouraged the widespread development and application of digital signal processing technology.

One of these algorithms is the least mean squares (LMS) algorithm developed by Widrow in 1960 [9]. This algorithm provides a simple approach to reducing the mean squared error between the output of a finite impulse response (FIR) model and a desired signal. Its

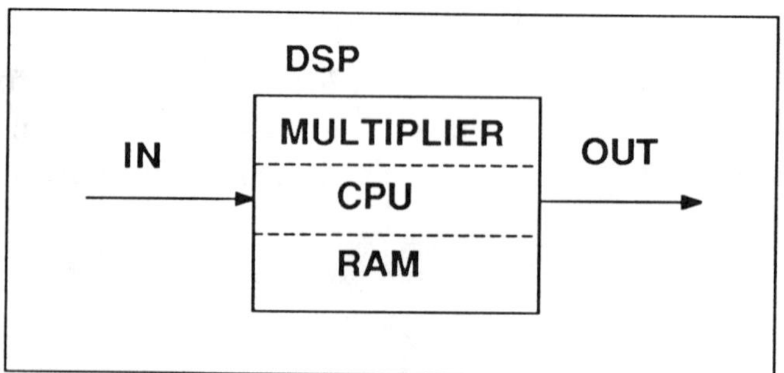

Fig. 4 - Digital signal processors feature multipliers and random access memory (RAM) in addition to large word sizes and fast central processor units (CPU) [8].

application to active noise control was described by Burgess in 1981 [10]. Due to the transfer function, H, following the adaptive model (and including the loudspeaker), the input, X, to the weight update calculation must be filtered by this same transfer function, as shown in Fig. 5. This form of the LMS algorithm has come to be known as the filtered-X algorithm. The completed system has the ability to automatically adapt its response to maintain a minimum error as the physical system or source changes.

In addition to the ability to adapt to system changes, the new digital technology enables complex configurations of acoustical transducers to be replaced by simple monopole systems. Eriksson presented a new adaptive approach based on an infinite impulse response (IIR) filter using a recursive least mean squares (RLMS) adaptive algorithm that reduced the undesired noise while simultaneously compensating for acoustic feedback [11]. Since the generalized input vector, U, to the weight update calculation must be filtered by the transfer function, H, following the adaptive model, as shown in Fig. 6, this form of the RLMS algorithm has been designated the filtered-U algorithm. The on-line determination of the transfer function following the adaptive model, primarily due to the loudspeaker and error path, has been the subject of a recent paper [12].

A more general framework within which to view the filtered-X and filtered-U algorithms is shown in Fig. 7 [13]. In adaptive algorithms based on LMS-related approaches, the weight update results in a small change in the weight vector based on an estimate of the gradient of the mean squared error obtained using the instantaneous value of the error signal. The typical path of the resulting weight vector is shown in Fig. 8 superimposed on contours of constant mean squared error. The path (a) to the minimum error is noisy and can lead to instabilities for large step sizes.

More complex weight update procedures, such as the recursive least squares (RLS) algorithm, are available that use a gain selected such that the direction of the weight update is directly towards the minimum in the mean squared error surface (b) and of the correct

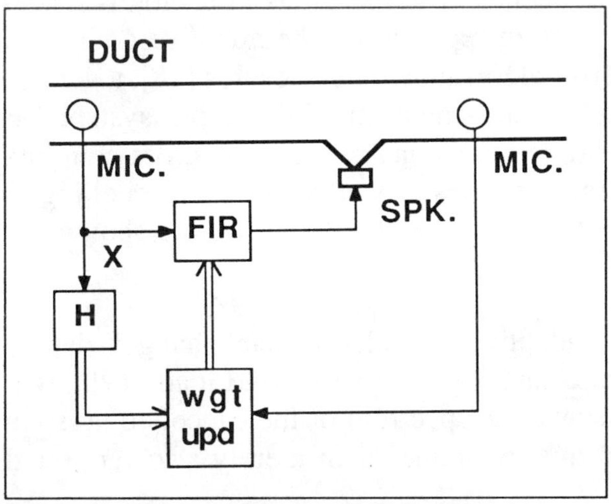

Fig. 5 - Filtered-X algorithm features an FIR filter structure and pre-filtering of the input vector used in the weight update ("wgt upd") calculation [10].

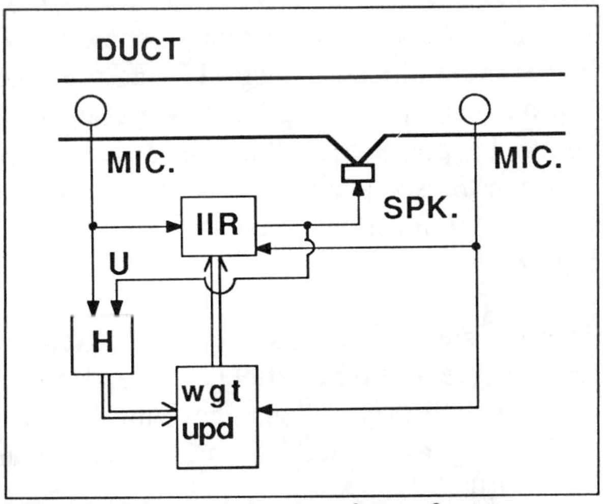

Fig. 6 - Filtered-U algorithm features an IIR filter structure and pre-filtering of the input vector used in the weight update ("wgt upd") calculation [11-12].

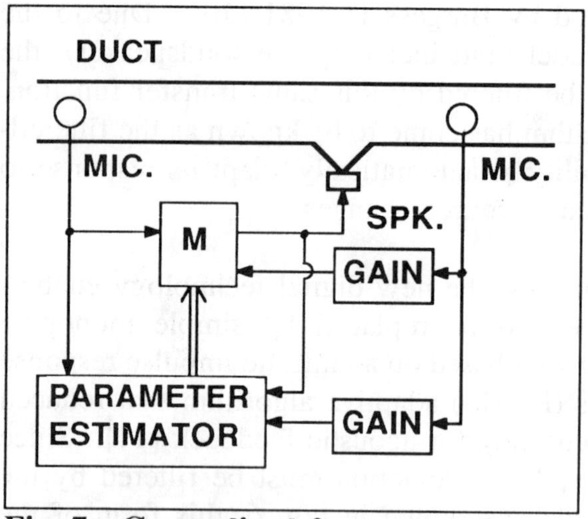

Fig. 7 - Generalized framework for active noise control signal processing [13].

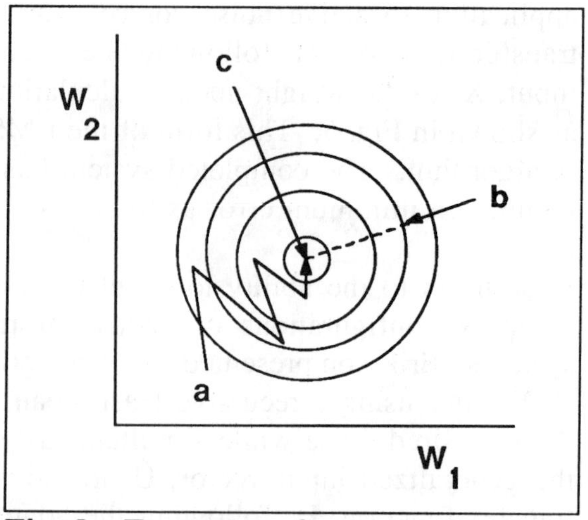

Fig. 8 - Examples of convergence path for two coefficient weight vector to mininum error for a) stochastic gradient ("LMS-like") and b,c) deterministic adaptive algorithms ("RLS-like") [13-16].

magnitude (c) to reach the minimum in one iteration of the algorithm [13-16]. However, the extensive amount of computation that is required reduces the attractiveness of these algorithms for actual applications. It is possible that, as computing costs are reduced and power increased, schemes based on algorithms using the recursive least squares technique and related methods may become practical. These algorithms result in much faster convergence when measured in iterations, but much more computation per iteration.

The future may also bring the widespread use of multiple-input, multiple-output (MIMO) adaptive algorithms. This will allow the application of active noise control to more complex sound fields such as higher order modes in ducts and sound fields in enclosed spaces such as vehicle cabs. Although some work has been reported on the adaptive control of higher order modes in ducts [17] and multiple-error algorithms for enclosures [18], a general approach for fully-adaptive active control of a multiple-input, multiple-output system has yet to be presented. Such an approach would require consideration of cross-coupling between the sound sources and the input/error microphones as well as on-line modelling of the transfer functions necessary for proper adaptation such as the loudspeakers shown in Fig. 9.

These systems must also retain the ability to adaptively model system changes due to temperature and other effects as well as source changes due to speed and load [19]. It is probable that they will also have the ability to shape the spectrum of the cancelled noise in such a way as to produce an improved "sound quality" rather than merely a lower sound level [20].

The active control of room acoustics is a complementary use of active noise control that is another possible application. Techniques have already been discussed for controlling the reverberation time and the sound absorption of a room using active techniques [21]. As will be discussed below, this may lead to the combined use of active duct silencers, room sound cancellation systems, and active sound absorbers to fully control the acoustics of a room.

III. SUPPORTING TECHNOLOGY

In addition to the continued development of signal processing algorithms, improvements in various supporting technologies will allow increasingly widespread application of active noise control technology in ducts and other areas. These include the development of sound sources that are able to withstand harsh environments, including high temperatures, corrosive chemicals, high levels of contaminants, and high moisture content. Sound sources that are compact in size, light in weight, capable of high power output, and very durable are also required for some applications.

Future systems will also include other improvements designed to reduce cost and broaden the range of potential application of active noise control technology. For example, improved sound sensors with high sensitivity that are compact and easy to use as well as improved electronic hardware that is more powerful, smaller and less costly seem to be likely developments.

Summarizing the technical status of today's active noise control systems for ducts, good progress has been made in achieving acceptable attenuation over a broad frequency range for plane wave propagation, overall stability and compensation for feedback, and tracking

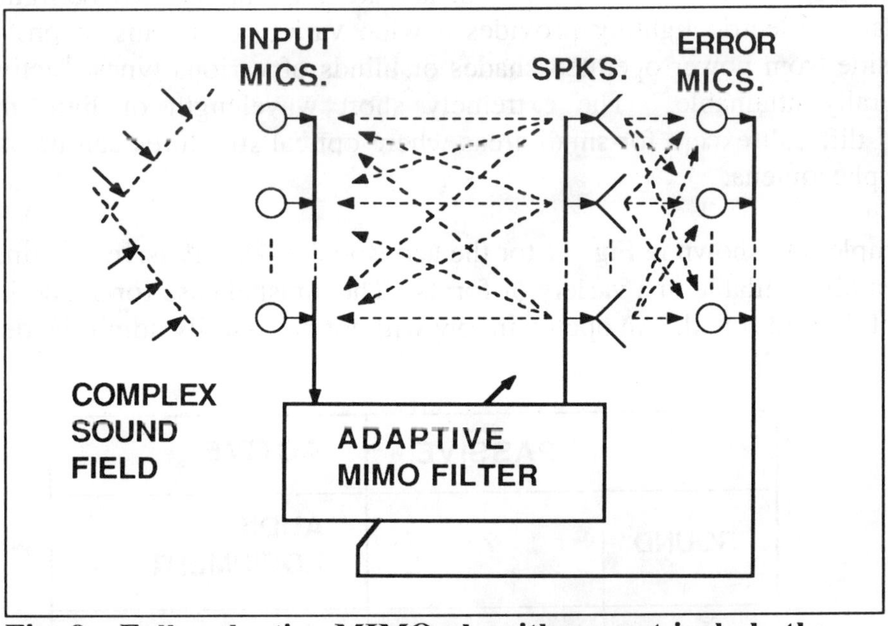

Fig. 9 - Fully adaptive MIMO algorithm must include the effects of transfer functions following the adaptive filter and of cross-coupling between the loudspeakers and the microphones on an adaptive basis.

	PASSIVE	ACTIVE
LIGHT	WINDOWS (DAYTIME)	ARTIFICIAL LIGHTING
DARK	WALLS SHADES	?

Fig. 10 - Passive and active means can be used to control light and dark.

capabilities for plant and source changes. In the future, improvements will be seen in the ability to shape the frequency spectrum after cancellation, to control complex sound fields, to function reliably in harsh environments at high sound levels, and to produce systems at lower cost [22].

IV. "THE SOUND SWITCH"

This continued process of evolution may lead to the use of active noise control in fully engineered environments in the 21st century. The active control of sound will allow the sound environment to be designed and acoustically sculpted much as we have done in this century through the use of artificial lighting. Passive noise control will be used to complement active noise control much as current lighting uses a combination of passive lighting and artificial or "active" lighting, depending on the specific needs of the application.

For example, as shown in Fig. 10, passive means are available to provide light (during the day!) and dark. Electric lighting provides a wide variety of means to provide "active" lighting. Aside from power-operated shades or blinds of various types, "active darkness" is not generally attainable. The extremely short wavelength of light makes "light cancellation" difficult except for small areas where optical structures can be used to create interference phenomena.

Parallel examples are shown in Fig. 11 for the acoustic domain. Passive quieting means are, of course, readily available in a variety of forms. The unusual case for sound is the passive production of desired sound. An open window will not necessarily admit the desired sound

	PASSIVE	ACTIVE
SOUND	?	AUDIO EQUIPMENT
QUIET	WALLS SILENCERS	ACTIVE NOISE CONTROL

Fig. 11 - Passive and active means are evolving to control sound and quiet.

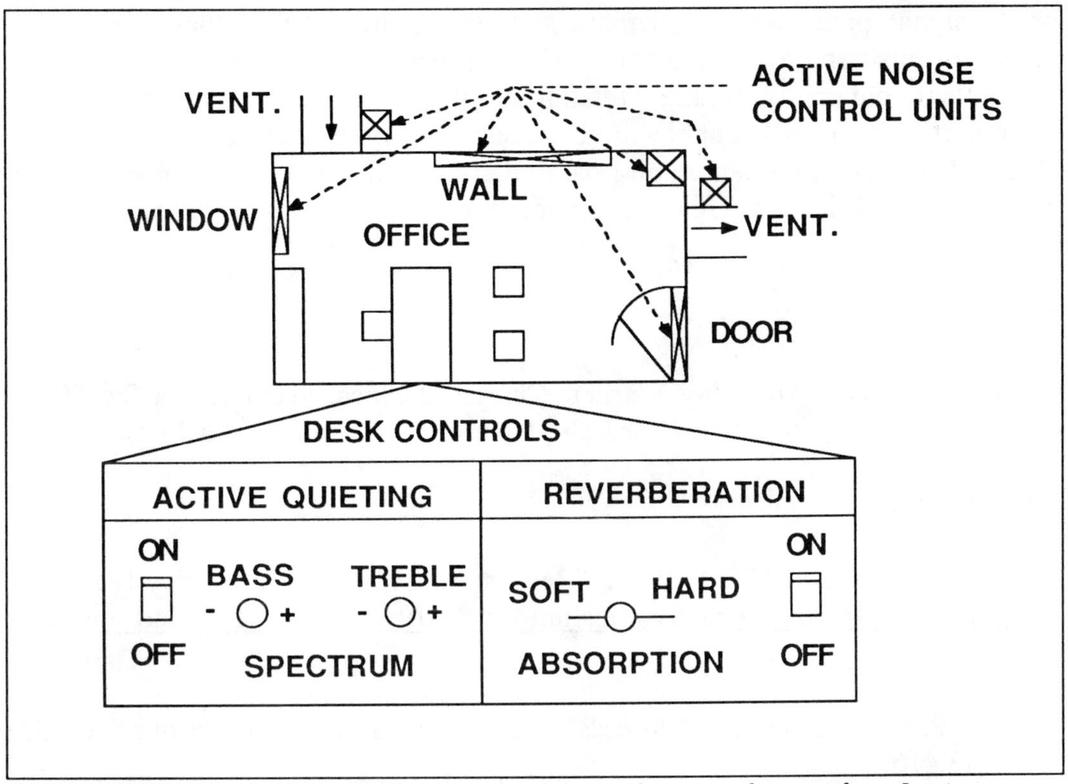

Fig. 12 - A vision of the office of the future integrating active duct silencers, active room sound cancellation, active sound control at doors and windows, and active room reverberation control.

(even in the daylight!). However, active means are now available for the production of sound or quiet. The familiar use of audio equipment to actively create desired sounds is being extended by active noise control to create desired quiet. The relatively long wavelengths of sound make these active processes much more practical than for light.

Some examples of the combined use of active and passive techniques to control sound and quiet are shown in Fig. 12 for an office environment. As with lighting, it seems likely that a wide variety of approaches will be used rather than one single solution. In addition to standard passive techniques, such as walls for sound barriers and carpet for sound absorption, active techniques may be used to control and shape the spectrum of the background noise in the room. Active noise control systems may also be used to control the noise emitted from the ventilation ducts, passing through the doors and windows, and reflecting off the walls. The occupant might have access to a personal panel to control the operation of the various systems, including spectral shape and reverberation time, as shown in Fig. 12. The "sound switch" may come to parallel the light switch.

V. SUMMARY

The 1980s have seen a transition in active noise control from systems based on complex arrangements of transducers to systems based on sophisticated adaptive digital signal processing. This transition has occurred due to the widespread availability of low-cost, powerful microprocessors designed for digital signal processing and to the development of appropriate signal processing algorithms. It is anticipated that the 1990s will see a continuation of this trend to more complex signal processing algorithms, including multiple input, multiple output configurations. In addition, the development of improved transducers and lower cost systems will enable more widespread application of active noise control technology. Ultimately, the active and passive control of sound may come to parallel the use of artificial lighting to create engineered environments with desirable acoustical characteristics.

ACKNOWLEDGMENTS

The assistance of Mark Allie, Jay Warner, Charles Brokish, and Patricia Steaffens of the Nelson Industries, Inc. Corporate Research Department is gratefully acknowledged.

REFERENCES

1. Eriksson, L. J., April 9-11, 1991. "Recursive Algorithms for Active Noise Control," Proceedings International Symposium on Active Control of Sound and Vibration, Tokyo, Japan.

2. Lueg, P., June 6, 1936. "Process of Silencing Sound Oscillations," U. S. Patent #2,043,416.

3. Olson, H. F., 1956. ""Electronic Control of Noise, Vibration, and Reverberation," J. Acoust. Soc. Am., 28: 966-972.

4. Jessel, M.J.M, and G. A. Mangiante, 1972. "Active Sound Absorbers in Air Ducts," J. Sound Vib., 23: 383-390.

5. Swinbanks, M. A., 1973. "The Active Control of Sound Propagation in Long Ducts," J. Sound Vib., 27: 411-436.

6. Eghtesadi, Kh., and H. G. Leventhall, 1981. "Active Attenuation of Noise: The Chelsea Dipole," J. Sound Vib., 75: 127-134.

7. Chaplin, G. B. B., 1983. "Waveform Synthesis - The Essex Solution to Repetitive Noise and Vibration," Proc. Inter-Noise 83: 399-402.

8. Lin, F-S., G. A. Frantz, and R. Simar, Jr., 1987. "The TMS320 Family of Digital Signal Processors," Proc. IEEE, 75(9): 1143-1159. (Special issue on Hardware and Software for Digital Signal Processing.)

9. Widrow, B., and S. D. Stearns, 1985. Adaptive Signal Processing, Englewood Cliffs, NJ: Prentice-Hall.

10. Burgess, J. C., 1981. "Active adaptive sound control in a duct: A computer simulation," J. Acoust. Soc. Am., 70: 715-726

11. Eriksson, L. J., M. C. Allie, and R. A. Greiner, 1987. "The Selection and Application of an IIR Adaptive Filter for Use in Active Sound Attenuation," IEEE Trans. on Acoustic, Speech, and Signal Processing, ASSP-35: 433-437.

12. Eriksson, L. J., 1991. "The Development of the Filtered-U Algorithm for Active Noise Control," to be published in J. Acoust. Soc. Am. 89.

13. Eriksson, L. J., August 1985. "Active Sound Attenuation Using Adaptive Digital Signal Processing Techniques," Ph.D. Thesis, University of Wisconsin-Madison, Wisconsin U.S.A.

14. Gritton, C. W. K., and D. W. Lin, 1984. "Echo Cancellation Algorithms," IEEE ASSP Magazine, 1(2): 30-38.

15. Alexander, S. T., 1986. "Fast Adaptive Algorithms: A Geometrical Approach," IEEE ASSP Magazine, 3(4): 18-28.

16. Haykin, S., 1986. Adaptive Filter Theory, Englewood Cliffs, NJ: Prentice-Hall.

17. Eriksson, L. J., M. C. Allie, R. H. Hoops, and J. V. Warner, 1989. "Higher Order Mode Cancellation in Ducts Using Active Noise Control," Proc. Inter-Noise 89: 495-500.

18. Elliott, S. J., I. M. Stothers, and P. A. Nelson, 1987. "A Multiple Error LMS algorithm and Its Application to the Active Control of Sound and Vibration," IEEE Trans. ASSP, ASSP-35: 1423-1434.

19. Eriksson, L. J., M. C. Allie, C. D. Bremigan, and J. A. Gilbert, 1989. "Active Noise Control on Systems with Time-Varying Sources and Parameters," Sound and Vibration, 29(7): 16-21.

20. Eriksson, L. J., and M. C. Allie, 1990. "Annoyance Reduction Using Active Noise Control with Adaptive Signal Processing," Proc. Noise-Con 90: 163-166.

21. Guicking, Dieter, Käthe Karcher, and Mathias Rollwage, 1985. "Coherent Active Methods for Applications in Room Acoustics," J. Acoust. Soc. Am., 78: 1426-1434.

22. Eriksson, L. J., 1990. "Computer-Aided Silencing: An Emerging Technology," Sound and Vibration, 24(7): 42-45.

DEVELOPMENT OF THE SIMULATION MODEL OF THE MULTIPLE INTERACTING SENSORS AND ACTUATORS (MISACT) FOR AN ACTIVE CONTROL SYSTEM

Kh. Eghtesadi,[1] M. McLoughlin,[1] E. W. Ziegler, Jr.[1]

ABSTRACT

A simulation model was developed to assist in projecting the performance of the full matrix interaction active cancellation algorithm with the MISACT.

The simulation model consists of a model of the environment in which the active cancellation controller is operating and of the equations and logic of the algorithm. The environmental model consists of a noise model and a model of the interactions between actuator signals and the resulting signals at each sensor. The simulation model is a discrete time step model with the time interval chosen by the user. The user also controls the number of actuators, the number of sensors, the fundamental cancellation frequency, the number of harmonics at which cancellation is attempted, the value of the cancellation convergence parameter and the number of samples of the run.

The simulation model has proven to be a valuable tool in analyzing the performance of the algorithm. Simulation results has confirmed stability criteria and expected behavior in real-world operations. The model also was useful in debugging the real-time code.

1. INTRODUCTION

A simulation model was developed to assist in projecting the performance of the full matrix interaction active cancellation algorithm with multiple interacting sensors and actuators (MISACT). The MISACT algorithm enables active cancellation to be applied to multi-point problems by addressing limitations in earlier active cancellation algorithms. These earlier algorithms provided cancellation at a single point, such as in an air duct or a single vibration mount. Multiple points could be handled if there was little interaction between the points, such as in active noise cancelling headphones. Active noise cancellation algorithms of this class include those by Chaplin[1], Warnaka[2], Eriksson[3] and Ziegler[4].

The approaches used by Warnaka and Eriksson incorporated forms of the LMS algorithm described by Widrow[5], Morgan[6] and Burgess[7] and the "filtered x LMS" algorithm of Widrow and Stearns[8]. Algorithms based on this approach can be slow to adapt to changes and subject to variations in multipath and other effects that vary the correlation between the reference sensor and the feedback error sensor. Also, these algorithms tend to use long FIR filters requiring substantial computational resources and there is limited ability to control the cancellation selectivity of these algorithms.

[1]Noise Cancellation Technologies, Inc., 1015 West Nursery Road, Linthicum, MD 21090, U.S.A.

Elliott[9] extended the "filtered x LMS" algorithm to multiple inputs and outputs but limited to cases in which the cancellation output signal does not feed back to the reference signal. This extended algorithm combines multi-tap adaptive FIR filters to generate the cancellation signals with either a delayed reference signal like Warnaka or a generalized filtering of the reference signal like Morgan and Burgess. While this algorithm is capable of cancelling random noise it apparently was intended for use with periodic noise from rotating equipment in which a synchronizing signal is available as described by Chaplin[10].

Ziegler extended the approach utilizing sinusoidal references described by Glover[11] to handle delays in the environment and to eliminate filtering of the reference signal. This can lead to small, low cost, fast adapting active cancellation systems since only two weights are required per frequency to be cancelled and the processing overhead of filtering the reference signal is eliminated. Additionally, precise control over frequencies to be cancelled, or not to be cancelled, is available giving the algorithm the ability to distinguish between desirable and undesirable signals.

In the approach used by Glover and Ziegler, the noise cancellation signal is produced by adapting the weights of the sine and cosine of each frequency to be cancelled. A resulting signal of any amplitude and phase at each frequency can be produced by appropriately selecting these weights. The individual components are summed to produce the composite cancellation signal. Often, many of the frequency components are harmonically related to the fundamental frequency of some piece of operating equipment. A synchronizing signal from the equipment provides the active cancellation system with the fundamental frequency.

The cancellation signal at sample k is computed as

$$y_k = -\sum_n [W0_{k,n} \sin (w_n k T) + W1_{k,n} \cos (w_n k T)] \tag{1}$$

where y_k is the value of the cancellation signal at sample k, $W0_{k,n}$ is the weight of the sine for frequency n at sample k, $W1_{k,n}$ is the weight of the cosine for frequency n at sample k, $\omega_n = 2 \pi f_n$, is the radial value of frequency n and T is the sampling interval.

An error signal is used as the adaptive feedback signal to update the cancellation weights as

$$W0_{k+1,n} = W0_{k,n} + \alpha e_k \sin (\omega_n k T - \theta d_n) \tag{2}$$

$$W1_{k+1,n} = W1_{k,n} + \alpha e_k \cos (\omega_n k T - \theta d_n) \tag{3}$$

where e_k is the value of the error signal at sample k, θd_n is the phase delay of the plant environment at frequency n, α determines the rate of adaptation to changes and the others are as defined above.

Based on the work by Ziegler, NCT has had considerable success in developing selective cancellation algorithms that adapt rapidly to changes and that reduce the computational resources required, and thus the size and cost. These algorithms eliminate the independent reference sensor, and its problems, using instead a timing signal providing the rate at which the noise is being produced. Multiple timing signals can be used when there are multiple noise sources.

Based on this experience, NCT extended the above to a selective active cancellation algorithm that explicitly accounts for the full matrix of interactions between multiple sensors and actuators. Because of the complexity of the algorithm the simulation model described in this paper was developed.

There were three primary objectives of the simulation model:

1. To assess the sensitivity of the performance of the algorithm to algorithm parameters and the operation environment.
2. To compare with simulation results from the real time code to assist in debugging.
3. To compare with actual results to assist in explaining "real-world" observations.

The simulation model proved useful for each of the three objectives. Sensitivity analyses were conducted by varying environmental and algorithm parameters. Comparing detailed computations from the simulation model with the real-time code running on a software simulator simplified the job of debugging the code. The simulation model served as a bench mark giving the expected performance against which actual performance was compared with various combinations of environmental values and algorithm values.

There are two versions of the simulation model. One is written in C for Microsoft's Windows 3 and is used to analyze the impact of changes in the algorithm parameters and the interaction of the algorithm and the environment. The other uses Microsoft's QuickC and is used to analyze changes in the algorithm implementation. This paper focuses on the Windows version.

2. DESCRIPTION OF THE SIMULATION MODEL

The simulation model is a discrete time step model with the time interval chosen by the user. The model consists of a noise model, an environmental model, the algorithm model and the user interface. Typical of software written for Windows, the user interface makes up most of the code. The entire model consists of approximately 90 pages of C source code. Of these 90 pages, the algorithm model is one page; the noise model and environment models together are seven pages and the user interface is eighty-two pages. Figure 1 shows a block diagram of the simulation model. Both inputs and outputs of the environment model consist of multiple signals. It is not necessary for the number of inputs and outputs to be equal.

A relatively simple model is employed for the noise signal. A single sinusoidal is generated with a specified frequency. The amplitude and phase of the noise signal can be varied for each sensor so that the noise source to sensor characteristics can be incorporated into the model.

This simplification is based on our experience with more complex noise models and with real world systems. The behavior of the active cancellation system is sensitive to the number of frequencies to be cancelled but the behavior typically can be observed with only a single frequency actually present in the noise. The inclusion of multiple frequencies complicates the signal but generally adds little to what is learned from the simulation model. Algorithms of this type behave in a predictable fashion to independent frequency components and the behavior at a single frequency is indicative of the behavior at other frequencies with the same external conditions.

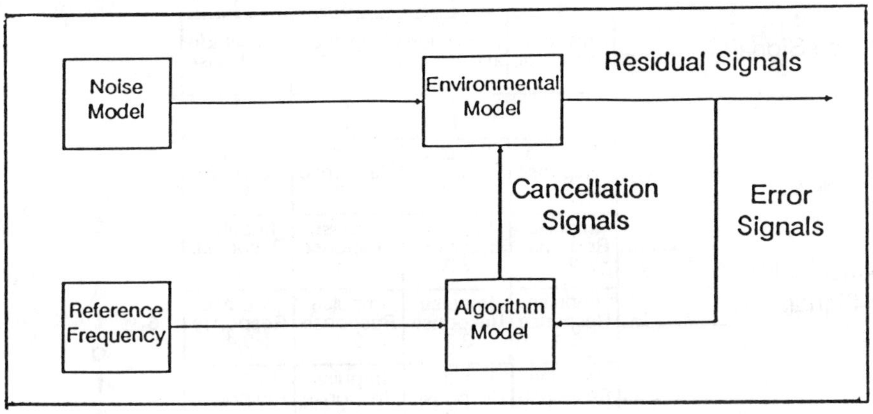

FIGURE 1 - Simulation Model Block Diagram

Other of our simulation models have included pseudo-random noise mixed in with the sinusoidal components. The behavior of algorithms of this type in the presence of non-correlated noise is well known and little additional is learned by the incorporation of such noise in the model. Non-correlated noise sets a floor to the amount of reduction of sinusoidal components.

The noise can be simulated within the model or read from a file. In the simplest case the noise is the same at each sensor with an amplitude set via the user interface and a phase of zero. The frequency also is set via the user interface. Alternatively, the amplitude and phase of the noise can be read from a file containing the real and imaginary components of the noise at each sensor. Again, the frequency is set via the user interface. In addition, the noise signal values at each sample can be read from a file. This method allows the effects of more complex signals or of signals recorded from real-world data to be analyzed.

The relevant aspects of the environment needed in the simulation model are the delays (phase shifts) and amplitude variations, including any resonances and any natural cancellation at various frequencies, between the actuators and sensors. This information is entered into the model via a matrix of cross impulse responses. The impulse responses can be entered as preset data within the model or read from a disk file. The latter method has been used to incorporate real-world data into the simulation model. A block diagram of the environmental model is shown in Figure 2.

Three options are included for selecting the impulse responses. The first option produces impulse responses with a simple delay and amplitude for actuators and sensors with the same number and zeros everywhere else. i.e., there are no interactions. The second option produces impulse responses as follows: 1.) a delay as given via the user and a fixed amplitude of 0.50 between actuator one and senor one and between actuator two and sensor two, 2.) an "echo" with 2 1/2 times the above delay and amplitude of -0.45, 3.) a delay twice the first delay with an amplitude of 0.25 between actuator one and sensor two and between actuator two and sensor one and zero everywhere else.

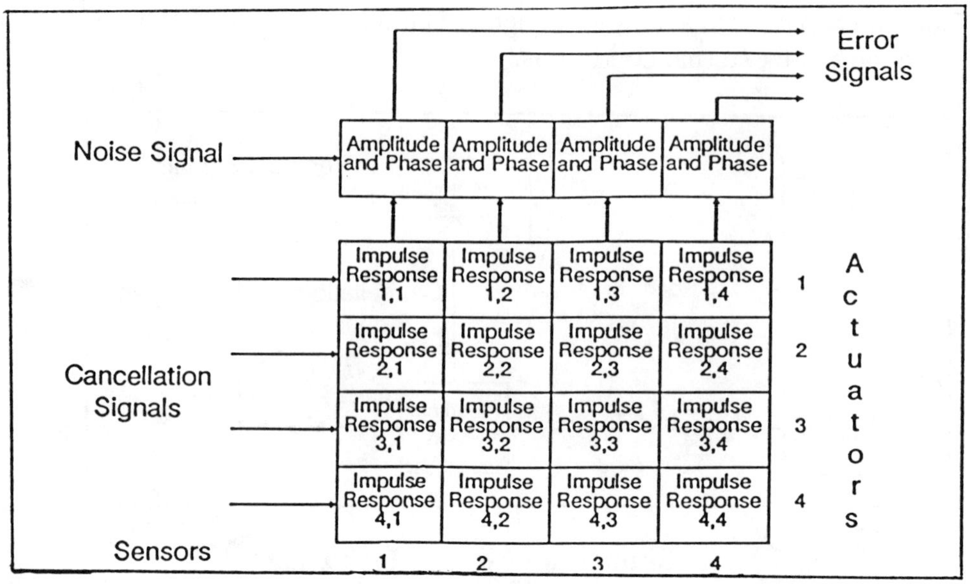

FIGURE 2 - Environmental Model

Finally, an arbitrary set of impulse response data can be read from a file.

Weights are applied to the transfer functions to determine how sensor values will be used in the MISACT algorithm. Changing these weights changes the criteria attempting to be minimized by the algorithm. This allows different objective functions (also called cost or error functions) to be evaluated. The weights are a matrix of real values with the number of rows and columns equal to the number of sensors. This matrix is multiplied by the transfer function matrix to produce the transfer function values used during cancellation.

Two sets of weights are built into the model. The first consists of a matrix with all 1's on the diagonal and zeros elsewhere. This produces the "standard" sum of squares objective function. The second consists of a matrix with all 1's everywhere. This produces an objective function minimizing the square of the sums of the residuals. Additionally, the weight matrix can be read from a file to produce a more complex objective function.

The parameters are displayed in a window for viewing and alteration. This window is shown in Figure 3 and the parameters are described in appendix A. The simulation run is initiated from this window.

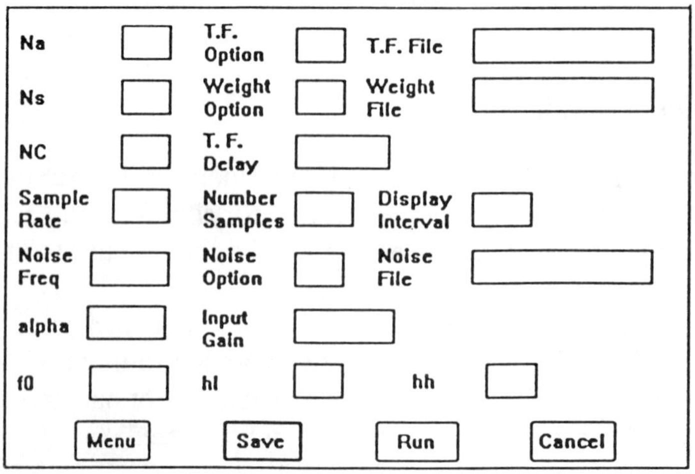

FIGURE 3 - Simulation Parameters Form

A box is displayed during the simulation run that shows the length of the run, the current sample number and the most recent amount of reduction of the sensor signals. The reductions are displayed in dB over the previous display interval as set in the Display Interval parameter. Two values are shown; J, the conventional sum of squares and J2, the square of the sums of the residual signals.

Simulation runs can be cancelled at any time by clicking the Cancel button with the mouse. The simulation model can be run "in the background" to allow other work to be done during long runs. Three beeps are produced when the run is finished.

3.1 SIMULATION OUTPUT

Detailed simulation results are written to files for later display and analysis. This data includes the following at each sample time: 1. the noise signal at each sensor, 2. the residual signal at each sensor, 3. the sum of squares of the sensor signals and 4. the cancellation signal at each actuator.

In addition, the final cancellation for each actuator are written at the end of the run.

3.2 VIEWING SIMULATION RESULTS

The results of previous simulation runs can be viewed. Graphs of the following can be plotted: Transfer functions, Noise signals, Residual signals, Actuator signals and Objective functions.

Three graphs are drawn for the transfer function selected; the impulse response, the phase and the magnitude of the transfer function. The windows in which these graphs are drawn can be manipulated to move them, change the size or close them individually.

The noise, residual and actuator signals are displayed in similar formats. One window is displayed for each sensor or actuator showing the signal. Each window can be manipulated individually.

The objective functions are displayed in two windows, one for the sum-of-squares objective function, J, and one for the square-of-sums objective function, J2.

3.3 SAMPLE RESULTS

Sample results of four simulations are shown in Figures 5 thru 8. The parameters used in these simulation runs are as follows:

Number of actuators and sensors	2,
Number of harmonics cancelled	1 or 32 as noted,
Noise amplitude	1.0 at sensor 1 and 0.4 at sensor 2,
Noise phase	0° at sensor 1 and 22.5° at sensor 2,
Noise frequency (f_o)	16 Hz,
Alpha	1/128 and 1/32 as noted,
Fundamental cancellation freq.	8 Hz or 16 Hz as noted,
Sample rate	256 Hz and
Length of run	1,024 samples.

The environmental model used for these runs included; a four sample delay with an amplitude of 0.50 between each actuator and the sensor of the same number, an "echo" with a ten sample delay and an amplitude of -0.45 between each actuator and the sensor of the same number and an eight sample delay with an attenuation of 0.25 between each actuator and the other sensor.

These impulse responses result in transfer functions with nodes and anti-nodes illustrative of "real-world" situations. For example, the transfer function from an actuator to the sensor of the same number, shown in Figure 4, contains nodes at approximately 43 Hz and 85 Hz and anti-nodes at 22 Hz and 65 Hz.

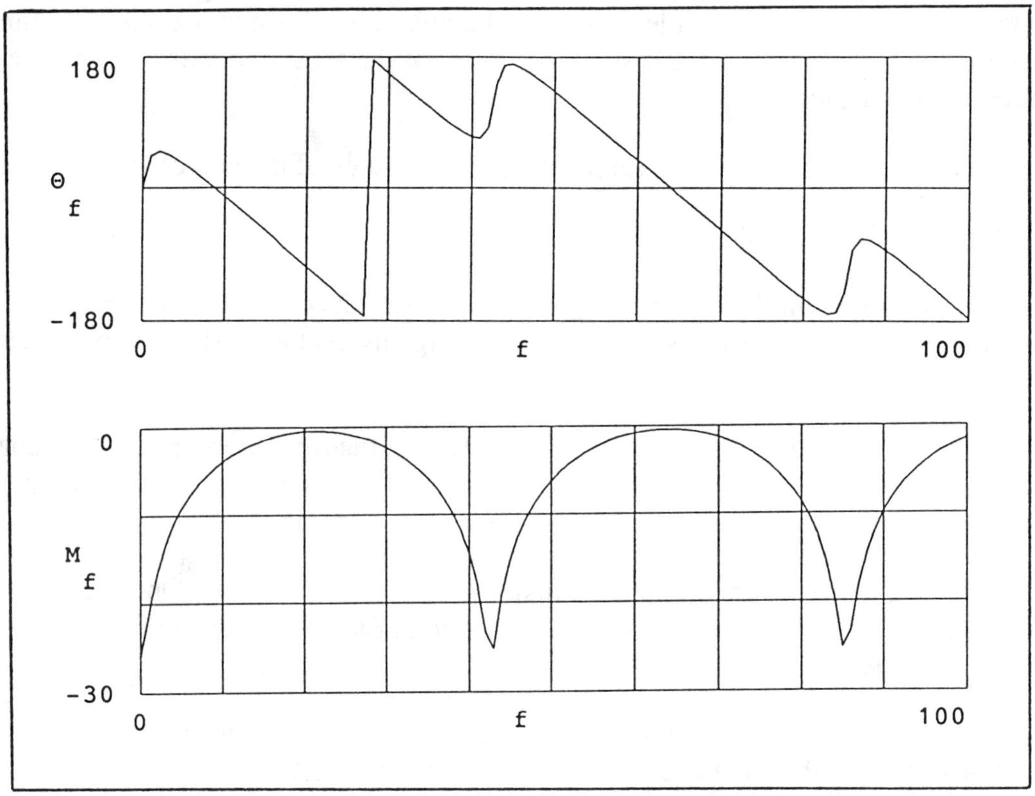

FIGURE 4 - Simulation Transfer Function

This series of simulation results illustrates the interaction between the number of harmonics attempting to be cancelled and the value of the adaptation convergence parameter on the performance of the system. Each Figure contains three graphs showing the residual signal value at the two sensors and the sum of the squares of the two residuals plotted at each sample.

Figure 5 shows the results for the base case cancelling one frequency, 16 Hz, at a sample rate of 256 Hz and Alpha set to 1/128 (about 0.008). In the real-time code Alpha is implemented as binary right shifts and, thus, the use of the fraction in the simulation model. Other values of Alpha can be tested in the model.

In Figure 5 the exponential decay of the residuals can be seen clearly. In Figure 6 the convergence parameter, Alpha, has been increased to 1/32 (about 0.03) and the resulting reduction in adaptation time is evident.

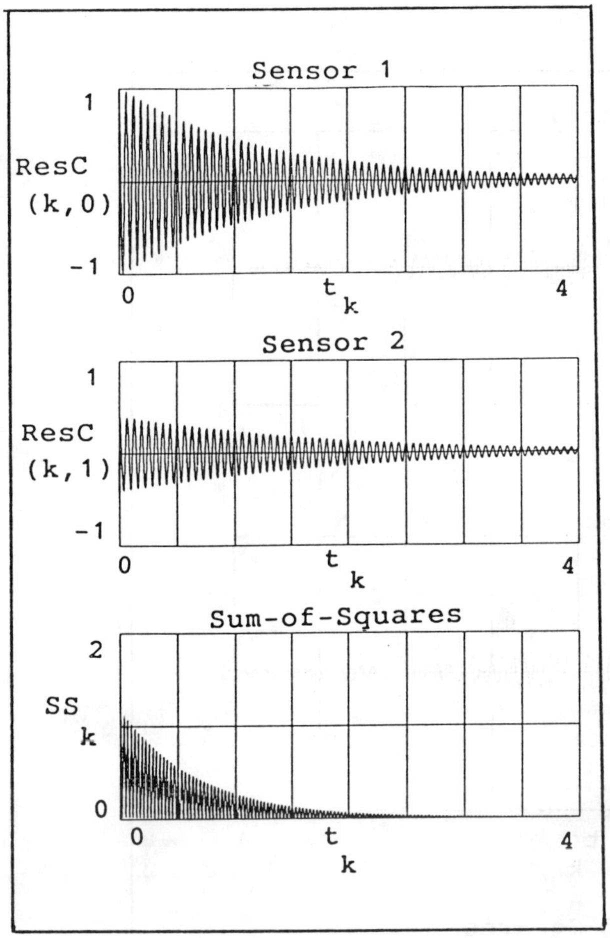

FIGURE 5

Simulation Results for Cancelling

1 Harmonic with 2 Actuators

and 2 Sensors

$f_o = f = 16$ Hz, $S_r = 256$ Hz, $\alpha = \frac{1}{128}$

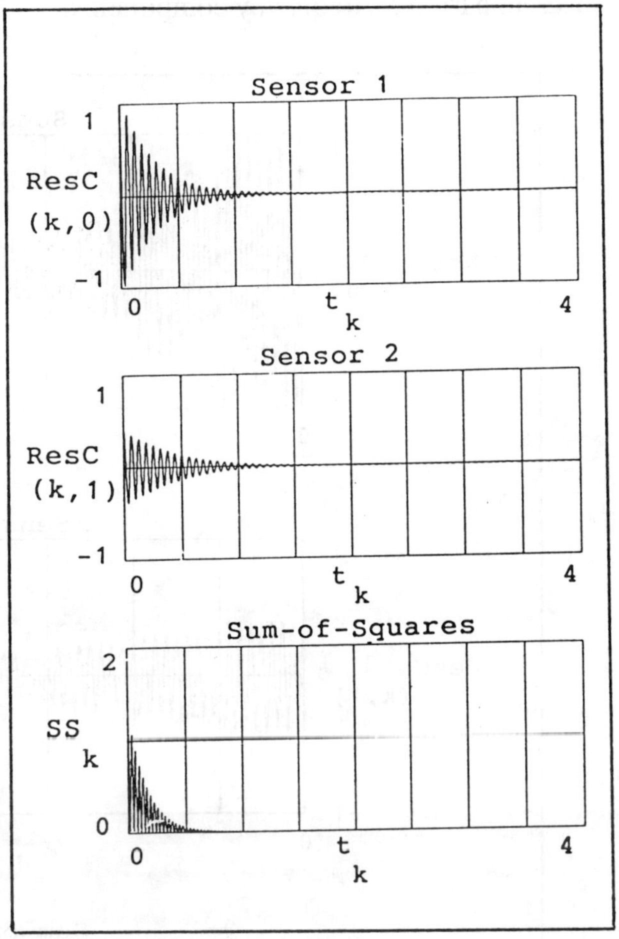

FIGURE 6

Simulation Results for Cancelling

1 Harmonic with 2 Actuators and

2 Sensors

$f_o = f = 16$ Hz, $S_r = 256$ Hz, $\alpha = \frac{1}{32}$

Figure 7 shows the results when the number of harmonics which are attempting to be cancelled is increased to thirty-two. The "apparent block" behavior of the algorithm can be seen in these graphs. The algorithm appears to operate as if the coefficients are being updated every block of $1/f_0$ seconds in spite of actually being updated, in an identical manner, at each sample. This behavior has been confirmed on the real-time system and has been seen in other algorithms using LMS adaptation of the amplitude and phase of multiple discrete harmonic frequency components.

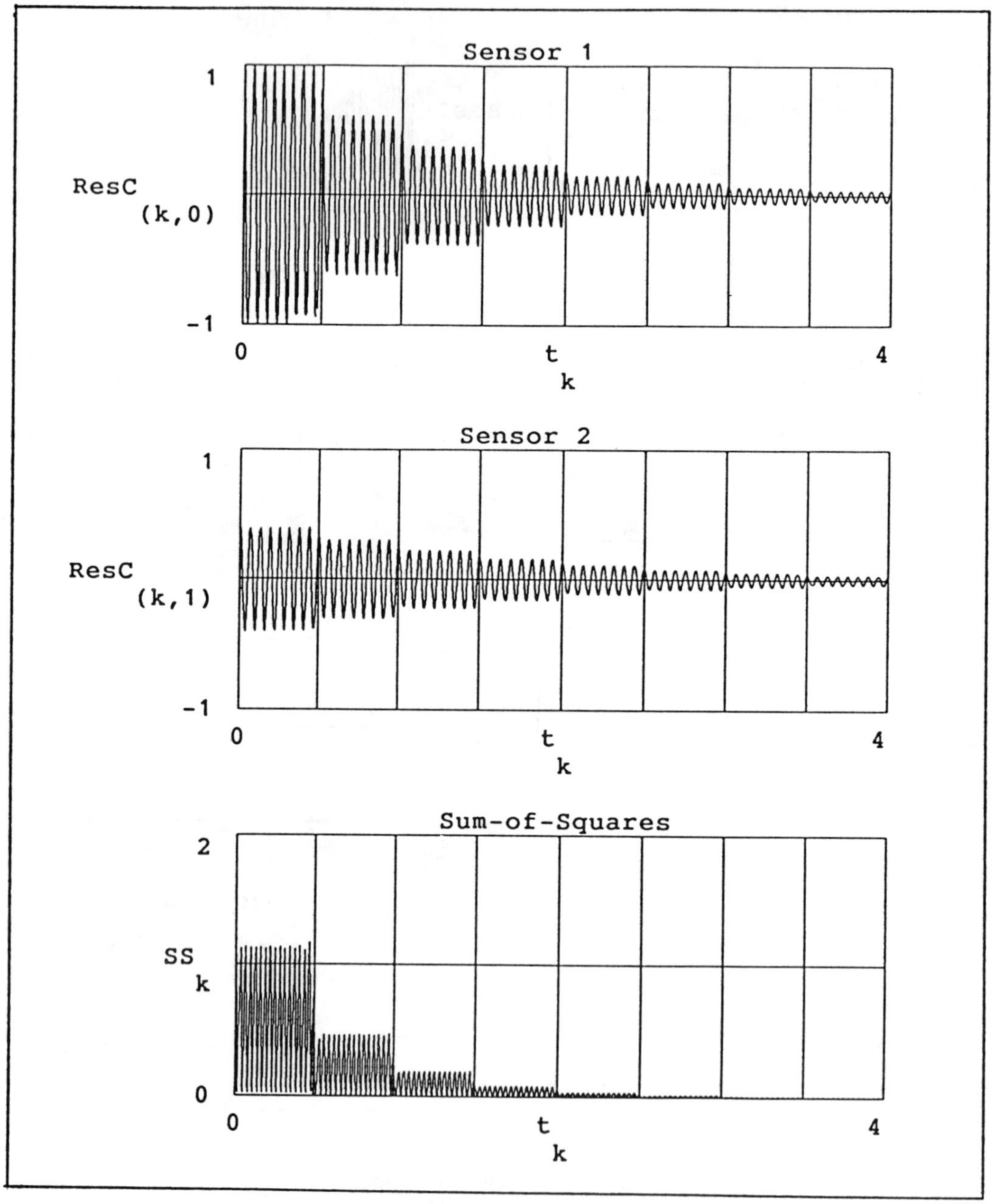

FIGURE 7

Simulation Results for Cancelling 32 Harmonics with 2 Actuators and 2 Sensors

$$f_0 = 2\text{ Hz},\ f = 16\text{ Hz},\ S_r = 256\text{ Hz},\ \alpha = \frac{1}{128}$$

Finally, Figure 8 shows a simulation run in which the algorithm could become unstable because of an inappropriate choice of alpha. In the real-time code a proprietary technique is employed during cancellation to maintain stability.

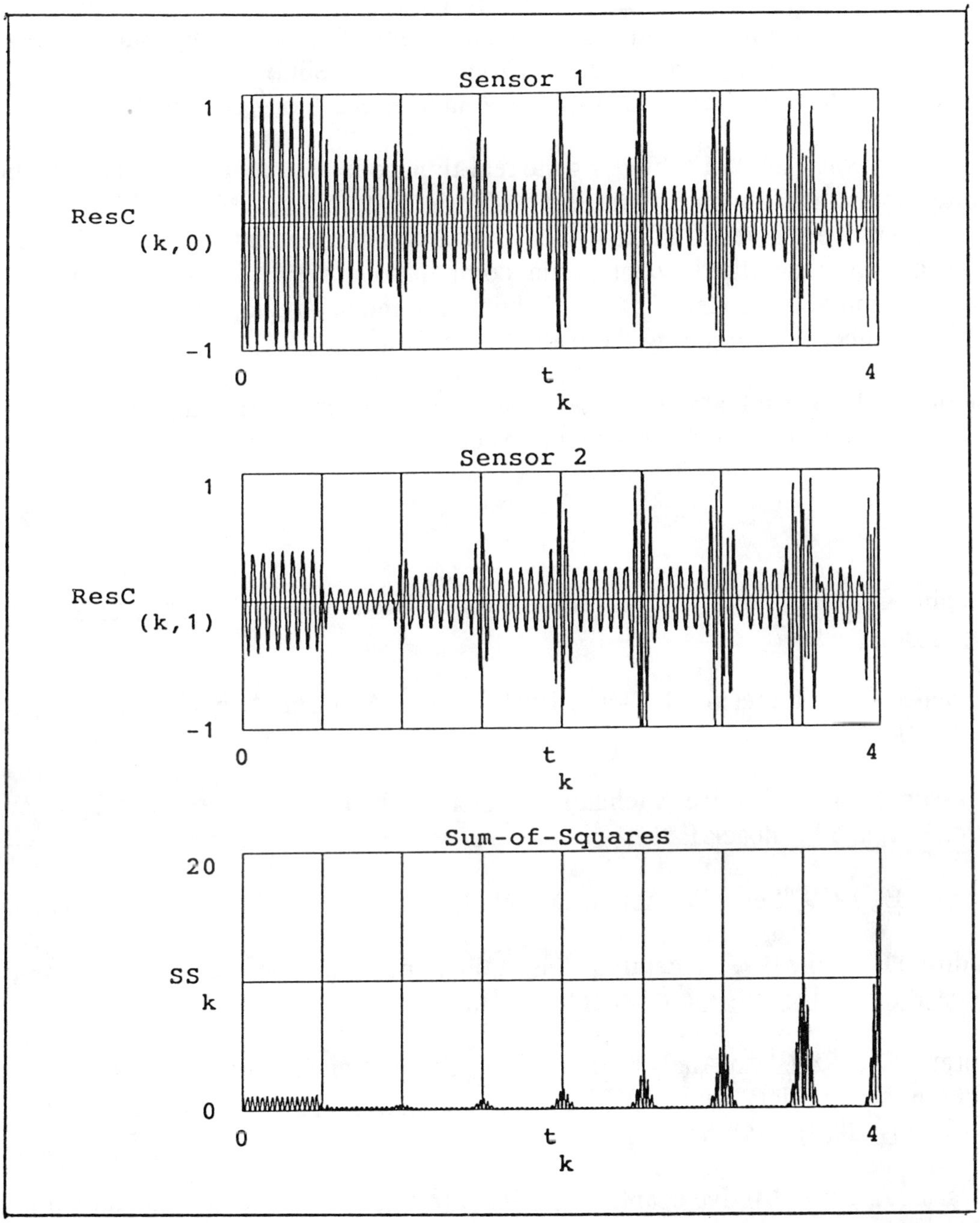

FIGURE 8

Simulation Results of an Unstable System for Cancelling 32 Harmonics with 2 Actuators and 2 Sensors

$f_o = 2$ Hz, $f = 16$ Hz, $S_r = 256$ Hz, $\alpha = \frac{1}{32}$

4. CONCLUSIONS

The simulation model has proven to be a valuable tool in analyzing the performance of the algorithm. Simulation results have confirmed stability criteria and expected behavior in real-world operations. An example of the latter is the apparent block processing effect illustrated in Figure 7. This effect first was observed in the real-time system. The simulation model confirmed that this is normal behavior of the algorithm and not a result of problems in the real-time system nor the environment. This eliminated potentially substantial effort that could have been spent trying to find a "non-problem.".

The model also was useful in debugging the real-time code. The real-time code was run on a software simulator of the object processor, a Texas Instruments TMS320C25. Identical simulated inputs were given to the real-time code and to the simulation model. Errors in the real-time code were identified by comparing computational results between the real-time code and the simulation model. This enabled much of the debugging to take place on a desk top computer rather than requiring the real-time hardware.

The use of Windows resulted in substantial investment in learning programming for Windows but in a user interface expected to be easy for others to learn.

References

1. Chaplin, G. B. B., 1983. "Anti-noise - the Essex Breakthrough", Chartered Mechanical Engineer, pp.41-47.

2. Warnaka, G., L. Poole and J. Tichy, 1984. "Active Acoustic Attenuator", U. S. patent 4,473,906

3. Eriksson, L., 1987 "Active Attenuation System with On-line Modeling of Speaker, Error Path and Feedback Path".

4. Ziegler, E., 1989 "Selective Active Cancellation System for Repetitive Phenomena".

5. Widrow, B. et al., 1975 "Adaptive Noise Canceling: Principles and Applications", Proceedings of the IEEE, 63(12):1692-1716.

6. Morgan, D., 1980. "An Analysis of Multiple Correlation Cancellation Loops with a Filter in the Auxiliary Path", IEEE Transaction on Acoustics, Speech, and Signal Processing, 28(4):454-467.

7. Burgess, J., 1981. "Active adaptive sound control in a duct: A computer simulation", Journal of the Acoustical Society of America, 70(3):715-726.

8. Widrow, B. and S. Stearns, 1985. "Adaptive Signal Processing", Prentice-Hall.

9. Elliott, S. J., I. Stothers and P. Nelson, 1987. "A Multiple Error LMS Algorithm and Its Application to the Active Control of Sound and Vibration", IEEE Transactions on Acoustics, Speech, and Signal Processing, 35(10):1423-1434.

10. Chaplin, G. B. B., 1983. "Method of Reducing the Adaption Time in the Cancellation of Repetitive Vibration", U. S. patent 4,417,098.

11. Glover, J., 1975. "Adaptive Noise Cancelling of Sinusoidal Interferences", Stanford University, Stanford, California, (Ph.D. dissertation).

Appendix A
Simulation Parameters

Na	Number of actuators (max 4),
Ns	Number of sensors (max 4),
NC	Number of points in calibration impulse filter (max 100),
Sample Rate	In Hz,
Number of Samples	Length of simulation run,
Display Interval	Number of samples between the display of simulation results while running,
Noise Freq.	In Hz,
alpha	The LMS adaptation parameter,
f_o	The fundamental cancellation frequency, in Hz,
hl	The lowest multiple of f_o to be cancelled (1 to 32),
hh	The highest multiple of f_o to be cancelled (1 to 32),
Input Gain	Amplitude of internally simulated noise or multiplier for noise from a file,
T. F. Option	0 = Internal, see below, 1 = Internal, see below, 2 = Read impulse response from a file,
T. F. File	File name, with type, when T. F. Option is 2,
Weight Option	0 = sum of square residuals, 1 = square of sum of residuals, 2 = read weights from a file,
Weight File	File name, with type, when Weight Option is 2,
T. F. Delay	The time, in seconds, used for T. F. Option 1, see below,
Noise Option	0 = Internally generated using Noise Freq. and Input Gain, 1 = Read amplitude and phase from file, 2 = Read each noise point from file,
Noise File	File name, with type, when Noise Option is 1 or 2.

ACTIVE CONTROL OF SOUND IN WAVEGUIDES

James D. Stell,[1] **Robert J. Bernhard**[1]

ABSTRACT

This paper presents an investigation of active noise control strategies for cancelling sound in a rigid-walled infinite waveguide. The transfer function of a multiple mode controller that considers acoustic feedback is derived. This controller is then simplified to one that does not consider feedback and further simplified to a controller for a single mode. The effects of feedback on the structure of the controller is investigated and related to simpler plane wave controllers. Some potential problems involved in controlling sound in waveguides are identified using the single mode controllers.

INTRODUCTION

In previous investigations the performance of active control of sound in waveguides has been investigated. These previous investigations have helped develop guidelines for the position of performance sensors and control sources. In this paper, the previous investigations are extended to address issues associated with implementing active control using feedforward controllers commonly used for control of sound in ducts. The controllers used for waveguides will be compared to those used for one dimensional duct applications.

WAVEGUIDE THEORY

The Frequency Response Function (FRF) between a volume velocity input and a pressure output in a finite length, rectangular, rigid-walled, waveguide can be derived using several different methods. One method of deriving the FRF is similar to the development of the pressure equations for a waveguide shown by Doak [1] and restated by Stell and Bernhard [2]. An example rigid-walled, rectangular, waveguide is shown in Figure 1. The waveguide terminations are assumed to be reflective with complex reflection coefficients R_{1mn} and R_{2mn} for the $mn^{\underline{th}}$ mode for the left and right ends respectively. Figure 2 shows the the xz plane of the waveguide with a volume source input location at point 1. The outgoing and reflected waves from this input are represented by the arrows in Figure 2. The complex magnitudes of the four traveling waves are the constants A_{mn}, B_{mn}, C_{mn}, D_{mn}.

If the input at point 1 is assumed to be a simple compact source of complex source strength q_s, the pressure at any point, (x,y,z), within the waveguide can be written as; for $z<z_1$

$$P(x,y,z)\, e^{j\omega t} = \sum_{m,n=0}^{\infty} P_{mn}(x,y) \left[A_{mn} e^{j(\omega t - k_{z_{mn}} z)} + B_{mn} e^{j(\omega t + k_{z_{mn}} z)} \right] \tag{1}$$

and for $z>z_1$

$$P(x,y,z)\, e^{j\omega t} = \sum_{m,n=0}^{\infty} P_{mn}(x,y) \left[C_{mn} e^{j(\omega t - k_{z_{mn}} z)} + D_{mn} e^{j(\omega t + k_{z_{mn}} z)} \right] \tag{2}$$

[1]Ray W. Herrick Laboratories, Mechanical Engineering, Purdue University, West Lafayette, IN 47907, U.S.A.

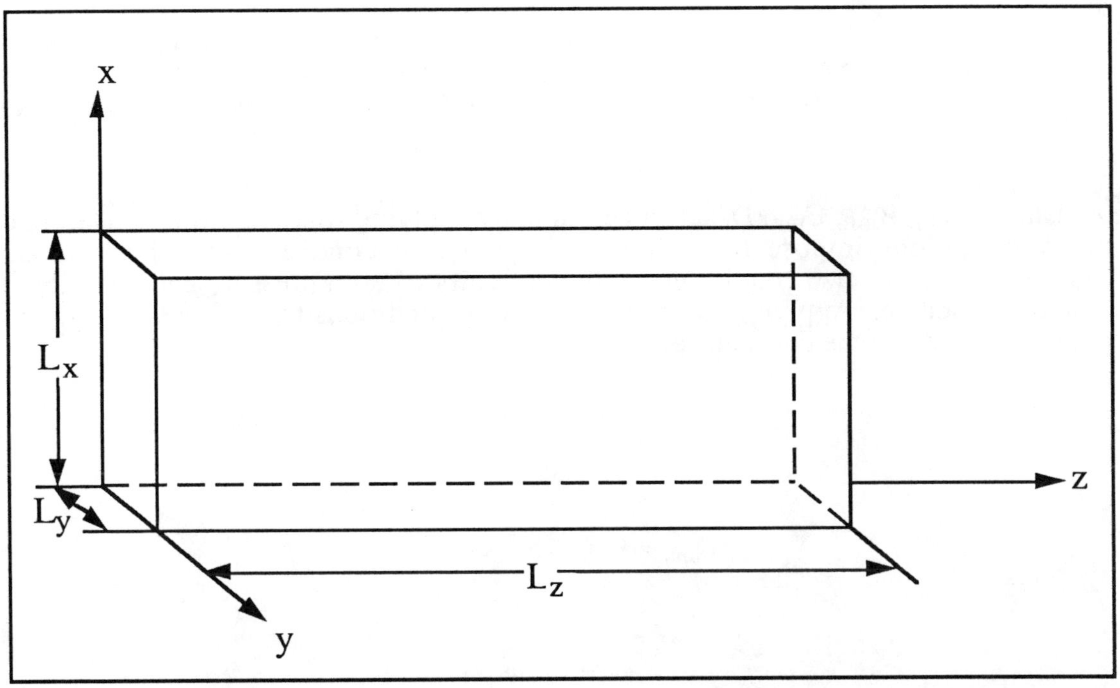

Figure 1. Diagram of waveguide under investigation.

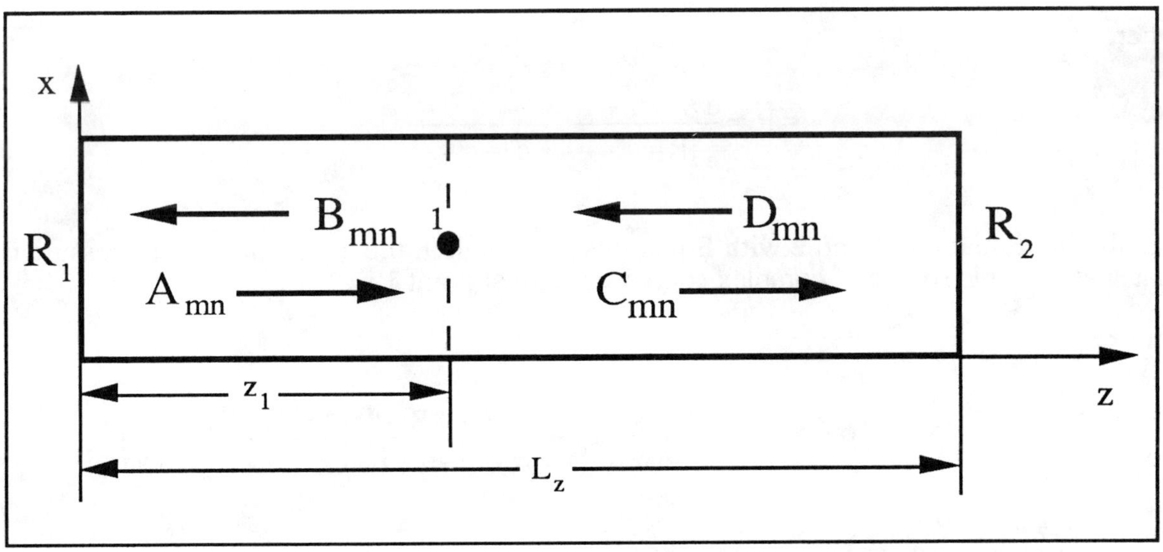

Figure 2. Sideview of the waveguide under investigation.

where for a rigid-walled, rectangular waveguide

$$P_{mn}(x,y) = \sqrt{2-\delta_{0m}}\sqrt{2-\delta_{0n}} \cos\frac{m\pi x}{L_x} \cos\frac{n\pi y}{L_y} \tag{3}$$

and

$$k_{z_{mn}}^2 = k^2 - \left[\left(\frac{m\pi}{L_x}\right)^2 + \left(\frac{n\pi}{L_y}\right)^2\right] \tag{4}$$

The constants A_{mn}, B_{mn}, C_{mn}, D_{mn} can be found by applying four conditions. The first two boundary conditions impose the reflective termination condition at either end of the waveguide. The other two conditions apply continuity of both pressure and velocity at the location of the source. Applying these four boundary conditions following the derivation of Stell and Bernhard [2], the constants are

$$A_{mn} = R_1 B_{mn} \tag{5}$$

$$D_{mn} = R_2 C_{mn} e^{-jk_{z_{mn}} 2L_z} \tag{6}$$

$$B_{mn} = C_{mn}\tilde{R} \tag{7}$$

$$C_{mn} = \left(\frac{k\rho_0 c}{k_{z_{mn}} A}\right)\left[\frac{P_{mn}(x_1,y_1)}{\left(1-R_1\tilde{R}\right)e^{-jk_{z_{mn}} z_1} + \left(\tilde{R}-R_2 e^{-jk_{z_{mn}} 2L_z}\right)e^{jk_{z_{mn}} z_1}}\right] q_s \tag{8}$$

where

$$\tilde{R} = \frac{e^{-jk_{z_{mn}} z_1} + R_2 e^{jk_{z_{mn}}(z_1-2L_z)}}{R_1 e^{-jk_{z_{mn}} z_1} + e^{jk_{z_{mn}} z_1}} \tag{9}$$

Combining Equations 1 and 2 with Equations 5 through 9, the equation for the pressure due to a single simple source of complex source strength at point 1 is;

for $z<z_1$

$$P(x,y,z) = \sum_{m,n=0}^{\infty} P_{mn}(x_1,y_1)\, P_{mn}(x,y) \frac{k\rho_0 c}{Ak_{z_{mn}}}\left(\frac{\tilde{R}\ q_s\left[R_1 e^{-jk_{z_{mn}} z} + e^{jk_{z_{mn}} z}\right]}{\left(1-R_1\tilde{R}\right)e^{-jk_{z_{mn}} z_1} + \left(\tilde{R} - R_2 e^{-jk_{z_{mn}} 2L_z}\right)e^{jk_{z_{mn}} z_1}}\right) \tag{10}$$

and for $z>z_1$

$$P(x,y,z) = \sum_{m,n=0}^{\infty} P_{mn}(x_1,y_1)\, P_{mn}(x,y) \frac{k\rho_0 c}{Ak_{z_{mn}}}\left(\frac{q_s\left[e^{-jk_{z_{mn}} z} + R_2 e^{jk_{z_{mn}}(z-2L_z)}\right]}{\left(1-R_1\tilde{R}\right)e^{-jk_{z_{mn}} z_1} + \left(\tilde{R} - R_2 e^{-jk_{z_{mn}} 2L_z}\right)e^{jk_{z_{mn}} z_1}}\right) \tag{11}$$

The frequency response function (FRF) is the ratio of the pressure at some point, designated point 2, and the complex source strength. For $z<z_1$, the FRF is

$$H(k) = \sum_{m,n=0}^{\infty} P_{mn}(x_1,y_1)\, P_{mn}(x_2,y_2) \frac{k\rho_0 c}{A k_{z_{mn}}} \left(\frac{\tilde{R}\left[R_1 e^{-jk_{z_{mn}} z_2} + e^{jk_{z_{mn}} z_2}\right]}{\left(1-R_1\tilde{R}\right)e^{-jk_{z_{mn}} z_1} + \left(\tilde{R} - R_2 e^{-jk_{z_{mn}} 2L_z}\right)e^{jk_{z_{mn}} z_1}} \right) \tag{12}$$

and for $z>z_1$, the FRF is

$$H(k) = \sum_{m,n=0}^{\infty} P_{mn}(x_1,y_1)\, P_{mn}(x_2,y_2) \frac{k\rho_0 c}{A k_{z_{mn}}} \left(\frac{\left[e^{-jk_{z_{mn}} z_2} + R_2 e^{jk_{z_{mn}} (z_2-2L_z)}\right]}{\left(1-R_1\tilde{R}\right)e^{-jk_{z_{mn}} z_1} + \left(\tilde{R} - R_2 e^{-jk_{z_{mn}} 2L_z}\right)e^{jk_{z_{mn}} z_1}} \right) \tag{13}$$

Previous investigators have found several design guidelines for active noise control systems operating in simple waveguides [2,3,4]. Problems can occur in the waveguide due to evanescent (non-propagating) energy created by the sound sources and detected by the sensors. These studies have shown that problems associated with evanescent modes can be avoided by locating the secondary source as close as possible to the source of the disturbance. The effects of evanescent modes can also be reduced by moving the error sensor as far away from the secondary source as possible. The effects of evanescent modes can also be avoided by placing the sensors and actuators on the nodelines of the dominating evanescent modes. This paper will consider some of the other implementation issues associated with active noise control systems.

CONTROLLER FORMULATION

One type of controller often used for the control of noise in waveguide applications is referred to as a "feedforward controller" [5,6]. Figure 3 shows such a system with a detector sensor at point 1 located at (x_1,y_1,z_1), a secondary actuator at point 2 located at (x_2,y_2,z_2), and an error sensor at point 3 located at (x_3,y_3,z_3). The undesired noise is represented by N. The active noise control system will detect the undesired noise N with the detector sensor, then create a control signal for the secondary actuator such that the pressure sensed at the error sensor is minimized.

Figure 4 shows a block diagram of the system shown in Figure 3. The transfer functions

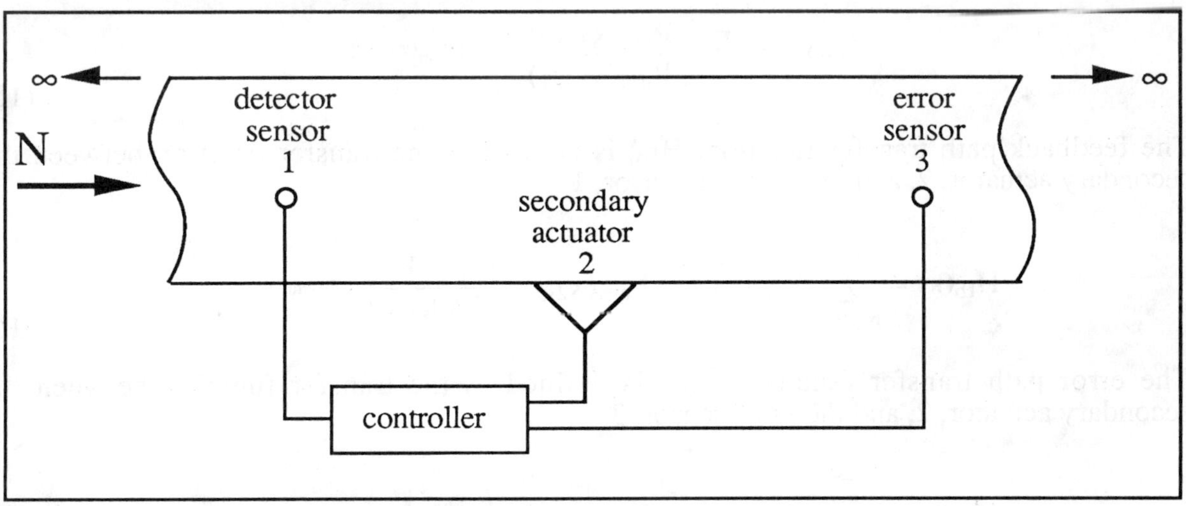

Figure 3. Active noise control system in an infinite waveguide

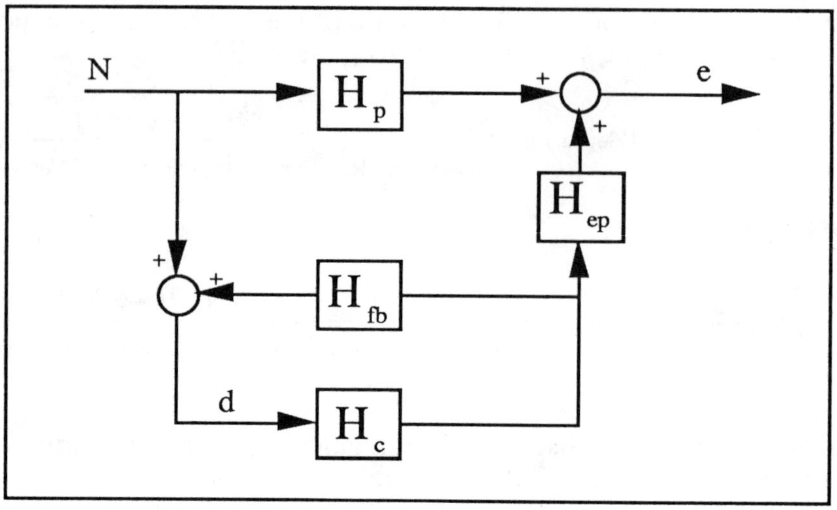

Figure 4. Block diagram of Figure 3.

shown in Figure 4 can be defined using the derivations of the previous section. To simplify the discussion somewhat, an infinite waveguide will be considered here. For an infinite waveguide, the complex reflection coefficients R_1 and R_2 are zero. From Equation 13 the transfer function between a volume velocity at (x_a,y_a,z_a), and a pressure at (x_b,y_b,z_b), where $z_b>z_a$, is

$$H_{ab}(k) = \sum_{m,n=0}^{\infty} P_{mn}(x_a,y_a)\, P_{mn}(x_b,y_b) \left(\frac{\rho_0 c\, k}{2A\, k_{z_{mn}}}\right) e^{-jk_{z_{mn}}(z_b-z_a)} \tag{14}$$

The plant transfer function, H_p, is defined as the transfer function between the detector signal, 1, and the error signal, 3. If it is assumed that the undesired noise N can be created with simple sources, then the plant transfer function is a ratio of transfer functions similar to Equation 14. Thus, H_p is

$$H_p(k) = \sum_{m,n=0}^{\infty} \frac{P_{mn}(x_3,y_3)}{P_{mn}(x_1,y_1)}\, e^{-jk_{z_{mn}}(z_3-z_1)} \tag{15}$$

The feedback path transfer function, H_{fb}, is defined as the transfer function between the secondary actuator, 2, and the detector sensor, 1

$$H_{fb}(k) = \sum_{m,n=0}^{\infty} P_{mn}(x_1,y_1)\, P_{mn}(x_2,y_2) \left(\frac{\rho_0 c\, k}{2A\, k_{z_{mn}}}\right) e^{-jk_{z_{mn}}(z_2-z_1)} \tag{16}$$

The error path transfer function, H_{ep}, is defined as the transfer function between the secondary actuator, 2, and the error sensor, 3

$$H_{ep}(k) = \sum_{m,n=0}^{\infty} P_{mn}(x_2,y_2)\, P_{mn}(x_3,y_3) \left(\frac{\rho_0 c\, k}{2A\, k_{z_{mn}}}\right) e^{-jk_{z_{mn}}(z_3-z_2)} \tag{17}$$

The controller transfer function that forces the error to zero can be derived using techniques similar to those used for plane wave controllers, except that this controller considers high order modes as well as the plane wave. Using the transfer functions represented by Equations 15 through 17 the controller is [7]

$$H_c = \frac{H_p}{H_p H_{fb} - H_{ep}} \tag{18}$$

Figure 5 shows a plot of the magnitude and phase of the controller transfer function which includes three modes, the plane wave, the 1,0 and 2,0 cross modes. Figure 5 shows the typical complexity of a controller with several modes in the controller transfer function.

Active controllers have been developed using unidirectional sources and sensors that allow the acoustic feedback in the system to be neglected [8]. While it is more difficult to develop such transducers for waveguides, some insights can be gained by considering these cases. The controller for a system without acoustic feedback is

$$\widetilde{H}_c = \frac{-H_p}{H_{ep}} \tag{19}$$

Even by neglecting acoustic feedback, the controller is not in a form that could be implemented as the simple sum of modal controllers. This form does however lead to an interesting form for the controller with feedback considered. The controller which considers acoustic feedback, given by Equation 18, can be written in terms of the Equation 19 as

$$H_c = \frac{\dfrac{-H_p}{H_{ep}}}{1 - \dfrac{H_p H_{fb}}{H_{ep}}} = \frac{\widetilde{H}_c}{1 - \dfrac{H_p H_{fb}}{H_{ep}}} \tag{20}$$

This controller has a form similar to that used in commercially available active noise control systems [9]. Such active noise control systems are implemented with a recursive filter with a block diagram shown in Figure 6 and represented by an equation of the form

$$H_{controller} = \frac{B}{1 - A} \tag{21}$$

where A and B are transversal filters which are adaptively identified. Equations 20 and 21 indicate that

$$B = \widetilde{H}_c \tag{22}$$

and

$$A = \frac{H_p H_{fb}}{H_{ep}} \tag{23}$$

Filter B is the controller that would be used for a waveguide without feedback.

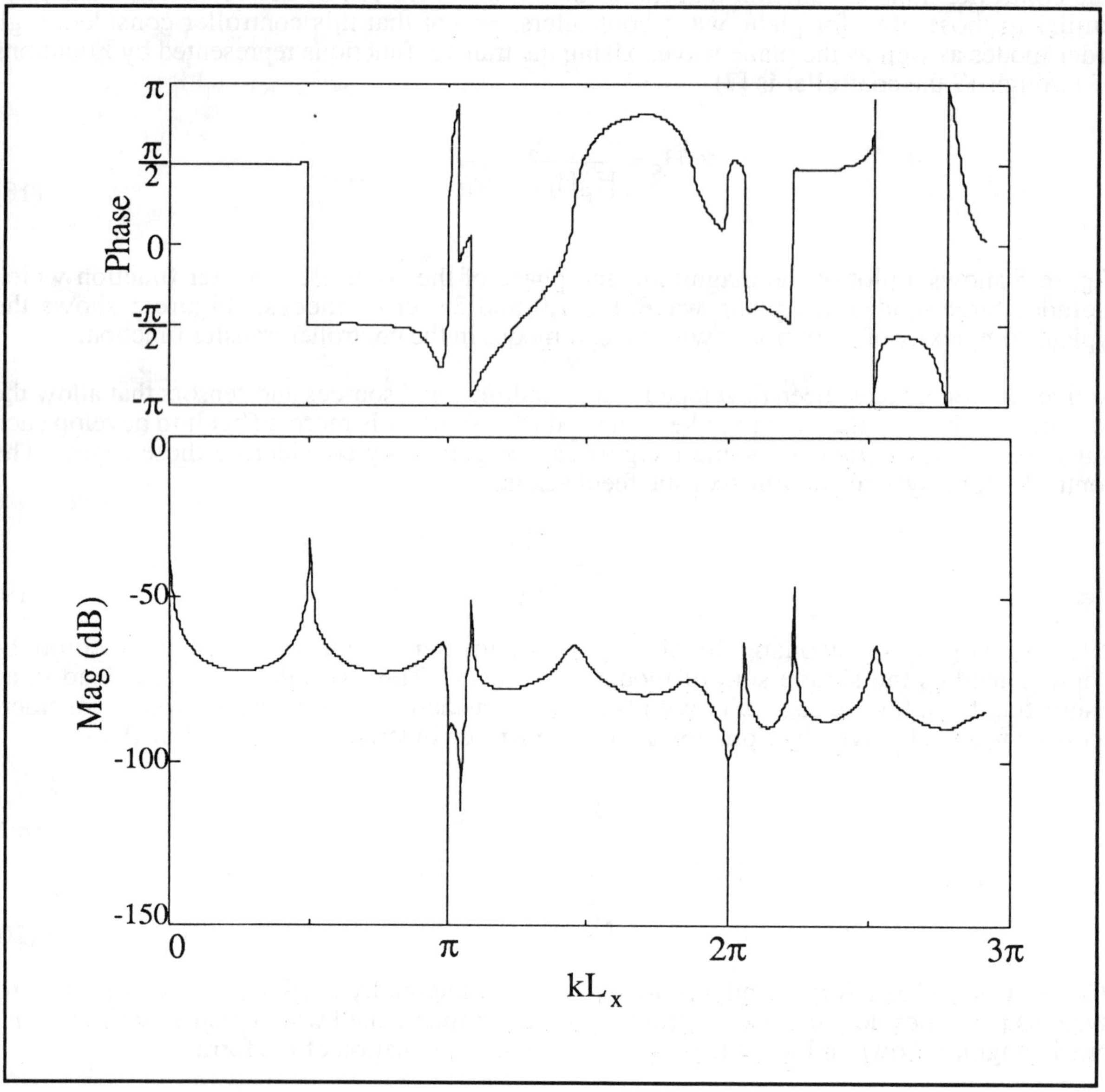

Figure 5. Magnitude and Phase of a controller transfer function which includes multiple modes.

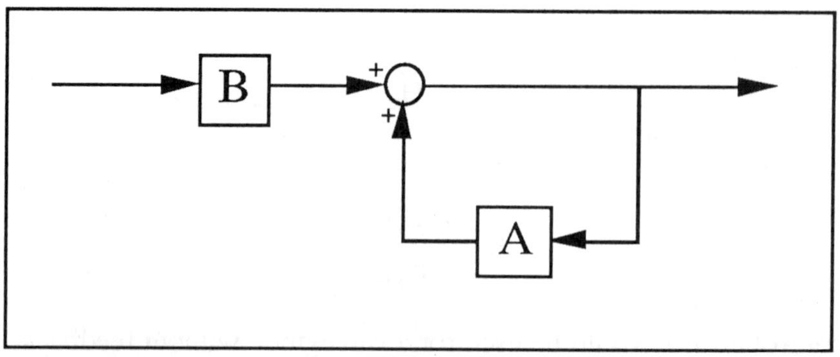

Figure 6. Block diagram of a recursive controller.

SINGLE MODE CONTROLLER

If the detector and error sensors are such that they only detect the energy of a particular mode, then all terms except the terms associated with that mode can be neglected in the controller transfer function. For this case, the controller transfer function can be written in terms of a single modal contribution of each of the transfer functions in Equation 18.

Several single mode controllers will be discussed to give some insight into the sophistication of the controllers that will be necessary to control sound in waveguides. The controller transfer function for the mn$^{\underline{th}}$ mode is

$$H_{c_{mn}} = \frac{1}{P_{mn}(x_1,y_1)\, P_{mn}(x_2,y_2) \left(\dfrac{\rho_0 c\, k}{2A\, k_{z_{mn}}}\right)\left(e^{-jk_{z_{mn}}(z_2-z_1)} - e^{jk_{z_{mn}}(z_2-z_1)}\right)} \tag{24}$$

If the feedback path is negligible the controller transfer function for a single mode reduces to

$$\widetilde{H}_{c_{mn}} = \frac{-1}{P_{mn}(x_1,y_1)\, P_{mn}(x_2,y_2) \left(\dfrac{\rho_0 c\, k}{2A\, k_{z_{mn}}}\right)\left(e^{jk_{z_{mn}}(z_2-z_1)}\right)} \tag{25}$$

Equation 24 can be written in terms of the controller that neglects feedback as

$$H_{c_{mn}} = \frac{-\widetilde{H}_{c_{mn}}}{\left(1 - e^{-2jk_{z_{mn}}(z_2-z_1)}\right)} \tag{26}$$

Equations 25 and 26 give some insight into the mechanisms of active noise control in the two simplified waveguides systems being considered. Figures 7 and 8 are plots of the magnitude and phase verses frequency of controller transfer functions considering a system with and without acoustic feedback for a controller that considers the plane wave and 1,0 cross-mode respectively. The waveguide being considered has cross sectional dimensions of 1.0 and 0.1 in the x and y directions respectively.

The plane wave controller transfer function for the case without feedback, shown in Figure 7, is characterized by a constant magnitude and a phase which varies linearly with frequency. This FRF is essentially a delay. The control signal is proportional to the detected signal delayed in time.

The controller for the plane wave case with feedback does not show the characteristic behavior of a system with delay. However, if plane wave parameters are used in Equation 26, the numerator is the same as the system with delay found for the case without feedback. The second term of the denominator is a delay twice as long as the delay in the numerator. Thus, while delay is not apparent in Figure 7 for the case with feedback, when the controller is examined in recursive form, the delay of the feedforward and feedback paths are apparent.

Acoustic feedback creates large peaks in magnitude of the controller transfer function. Frequencies at which the exponential term in the denominator of Equation 26 goes to zero are frequencies at which the magnitude of the controller gets very large. The frequencies of these peaks are the frequencies at which the left traveling pressure wave being fed back from the secondary actuator is exactly out of phase with the right traveling undesired pressure wave. At these frequencies the detected pressure is very small. Therefore a large control gain is required in order to obtain control. A similar effect is seen in Figure 8, when acoustic

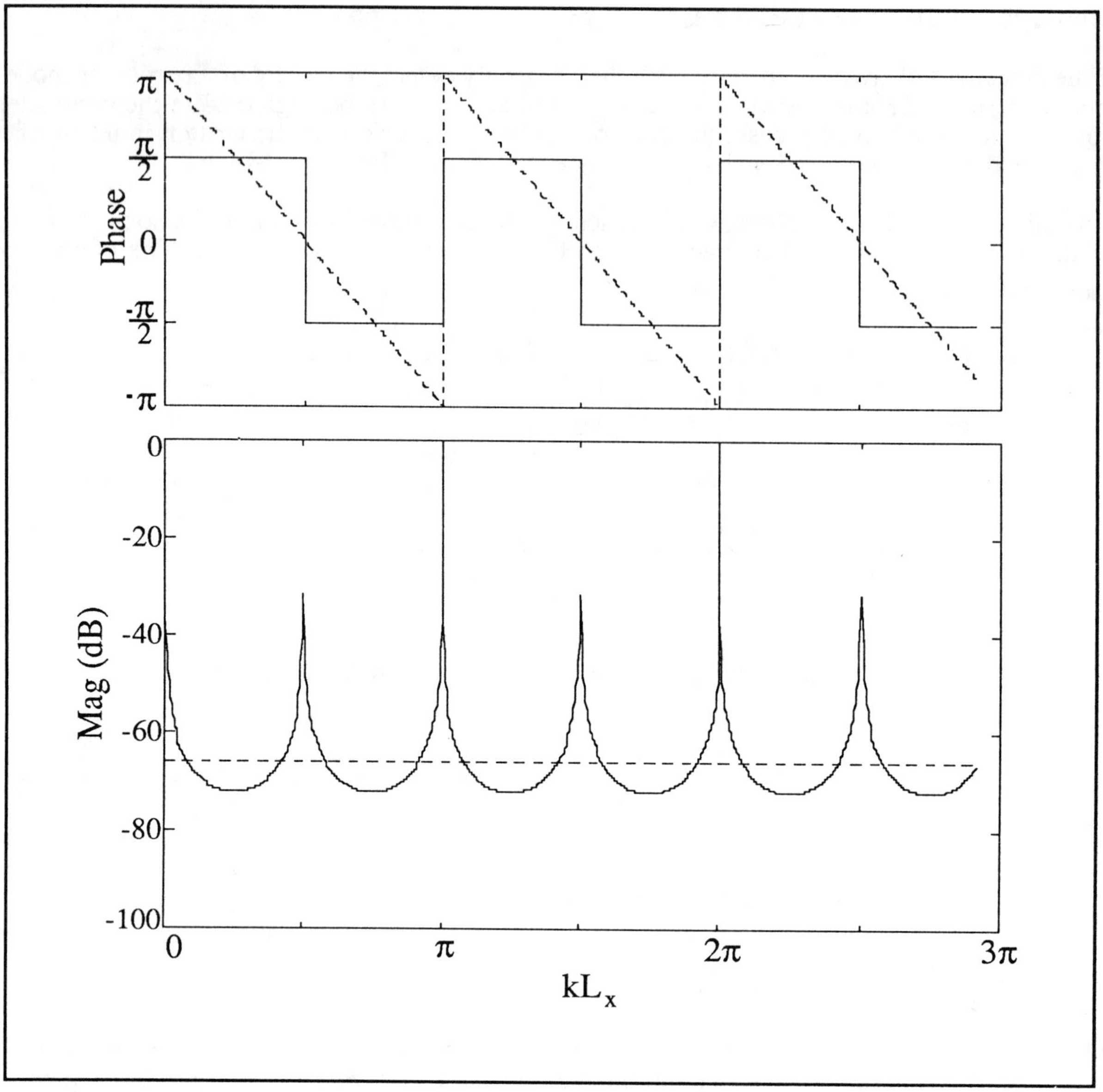

Figure 7. Magnitude and phase of a plane wave controller with (—) and without (...) acoustic feedback.

feedback is considered. These frequencies are frequencies at which there are potential stability problems for both the plane wave and high order mode controllers. The frequencies at which these peaks occur are characterized by

$$f = \sqrt{\left(\frac{\alpha\pi}{z_2 - z_1}\right)^2 + \left(\frac{m\pi}{L_x}\right)^2 + \left(\frac{n\pi}{L_y}\right)^2} \tag{27}$$

where $\alpha = 1,2,3,\ldots$ and $m = n = 0$ for the plane wave controller.

The controller transfer function shown in Figure 8 considers a single mode which is not the plane wave mode. For this controller there are two frequency regions to consider. The first is the region below the cut-on frequency of the mode being considered. The second is the region above the cut-on frequency of that mode. In the frequency region below the cut-on of the mode being controlled, the two controllers behave almost identically. Both Equation 24 and Equation 25 have a $k/k_{z_{mn}}$ term in the denominator. This term dominates the shape of the magnitude curve in this region. The effects of the evanescent modes become less significant as frequency becomes lower. There is a point, however, when the inability of the secondary actuator to generate evanescent energy begins to become important. This effect cause the magnitude in Figure 8 to increase at very low frequencies.

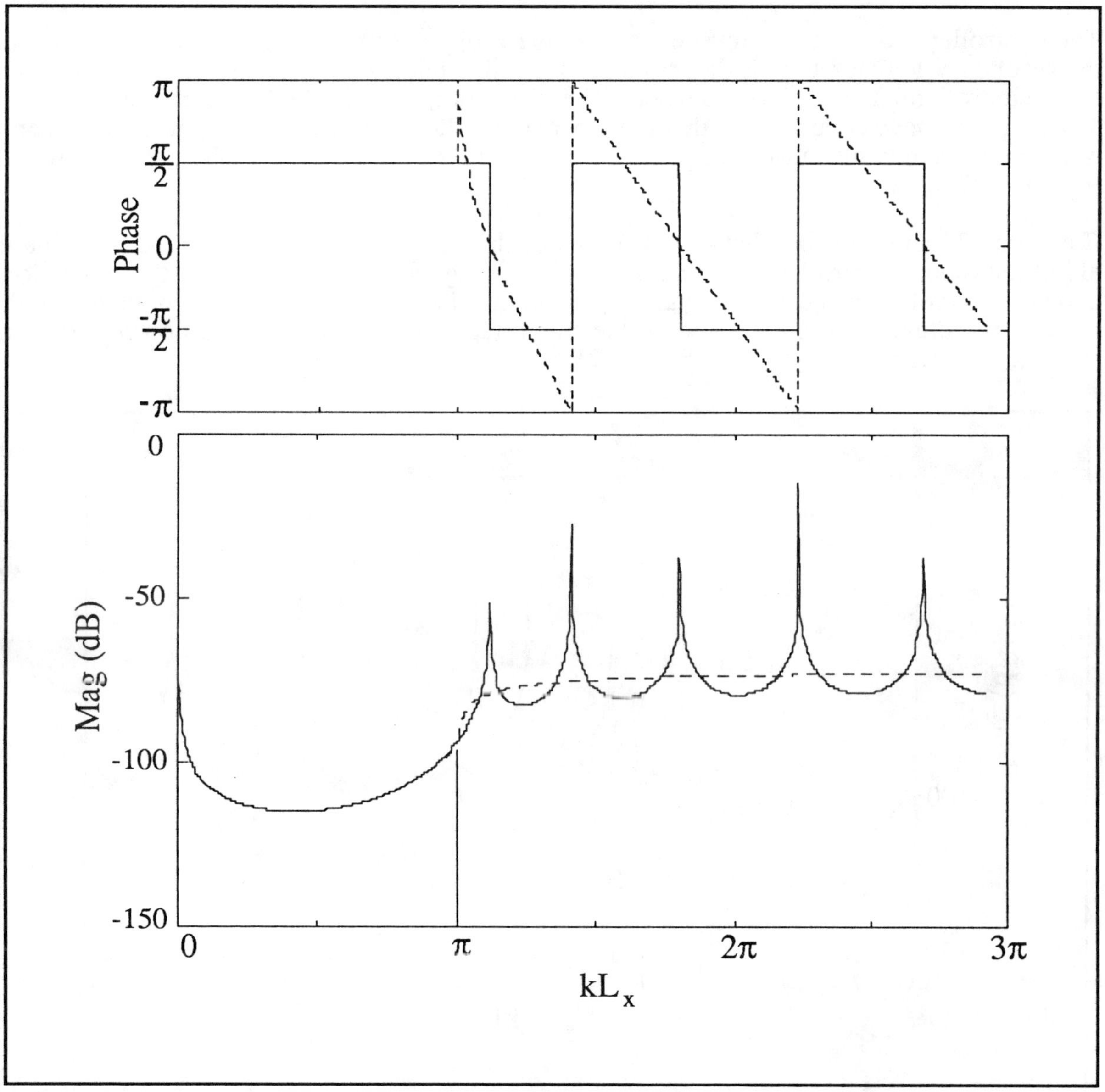

Figure 8. Magnitude and Phase of a high order mode controllers with (—) and without (····) acoustic feedback.

Above the cut-on frequency the two plots differ considerably. For the controller without feedback, there is a notch at the cut-on frequency. The magnitude of this controller at frequencies above the cut-on frequency approaches a constant. The phase of the controller without feedback is a constant at frequencies below the cut-on. Above the cut-on frequency the controller has phase characteristics similar to a plane wave controller, with some nonlinear effects just above the cut-on frequency. The nonlinear effects result in a frequency dependent delay. At frequencies above but near the cut-on frequency, the delay is very long. At frequencies well above the cut-on frequency the the delay is the same as the delay in the plane wave controller that neglects acoustic feedback. The difference between the higher mode controller and the plane wave controller is that the higher mode controller takes into account evanescent effects that occur below the cut-on frequency of the mode. This controller also takes into account the frequency dependent delays in the controller.

The controller shown in Figure 8 that accounts for the effects of acoustic feedback between the secondary actuator and the detector sensor for the modal controller is similar to the plane wave controller that considers feedback. The difference between the higher mode controllers and the plane wave controllers is that the modal controllers are controlling a mode that is only propagating above the cut-on frequency. The spacing of the peaks is unequal as shown in Equation 27.

Equations 24 and 25 show that the locations of the detector sensor and secondary actuator effect the modal controller transfer function. Investigation of the effects of these locations upon the controller can be done by varying the relative location of the detector in relation to the secondary actuator. Figure 9 shows the magnitude and phase of three controllers with acoustic feedback.

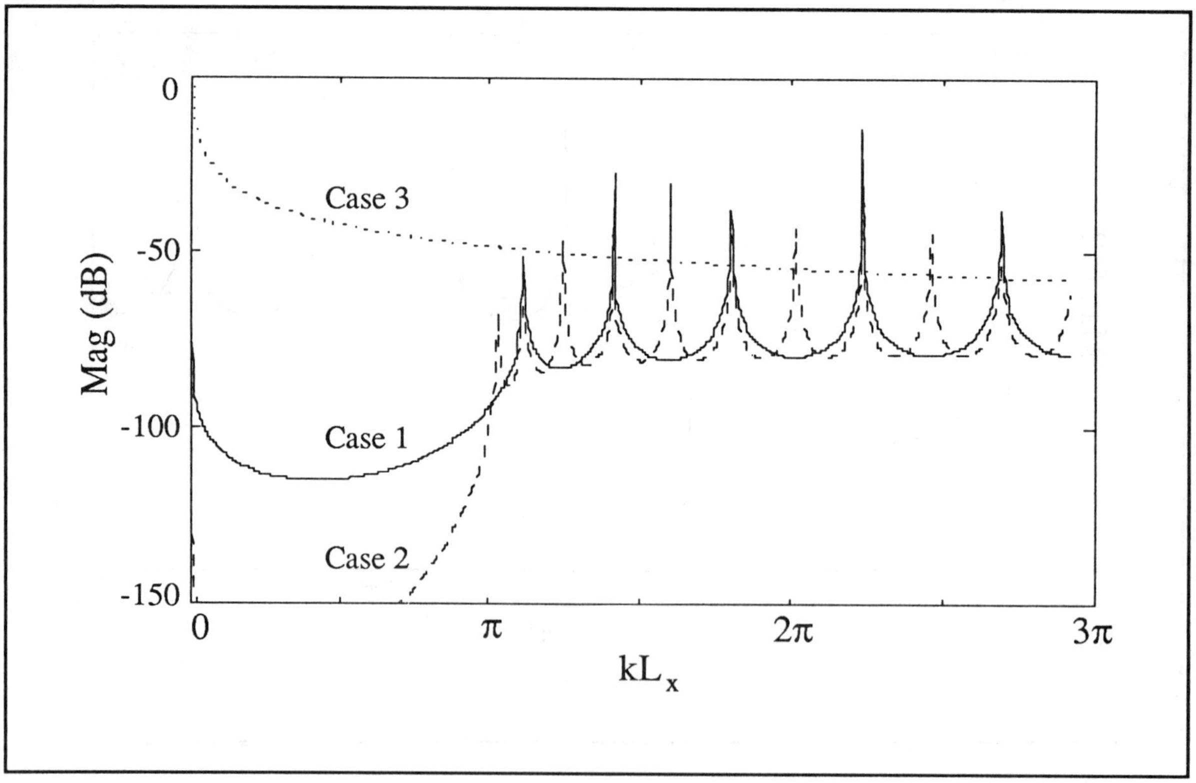

Figure 9. Magnitude of three controllers with different relative locations of detector and secondary actuators.

The three cases shown in Figure 9 differ only in distance between the detector sensor and the secondary actuator. In the nominal case, case 1, this distance was arbitrarily set to 2 units. In case 2, the distance is 4 units, whereas in case 3 the distance is reduced to .01 units. Depending on the type of disturbances expected, tonal or random, different conclusions can be drawn from Figure 9.

For systems in which tonal disturbances dominate, reducing the number of peaks in the magnitude of the controller takes precedence. Since these peaks indicate frequencies at which the controller can have stability problems, fewer peaks are desired. Reducing the number of peaks, however, tends to increase the controller gain at frequencies below the cut-on frequency. For tonal disturbances, this has minimal adverse effect. The benefits of reducing the number of peaks allows the controller to be more robust without sacrificing stability. For example, if a waveguide system were dominated by a single tone disturbance a robust controller can be designed by bringing the detector close to the secondary actuator. The number of peaks in the controller is reduced and the controller can easily track variations in the tonal disturbance over a wide frequency band.

If the disturbance is characterized by a random signal, there is a concern not only about the number of peaks present, but also about the effect that the controller has at frequencies below the cut-on frequency. Another concern for the random controller is computation time. In a controller designed for a random disturbance the detector and secondary actuator must be located sufficiently far apart to allow for the computation time needed in the controller. This is not a concern in the tonal controller since the sound can be predicted. When random disturbances are dominate, it is desirable to both reduce the number of peaks in the magnitude of the controller at frequencies above the cut-on, as well as reduce the overall magnitude of the controller at frequencies below the cut-on frequency while keeping a sufficient distance between the detector sensor and secondary actuator to allow time for the computation needed in the controller. Reducing the number of peaks in the controller reduces the number of unstable regions that the controller sees, whereas reducing the overall magnitude below the cut-on frequency reduces the controller effort. The location of the detector sensor and secondary actuator would have to be selected based on the particular application being considered.

CONCLUSIONS

Several of the characteristics of feedforward active controllers for waveguides have been identified here. Even if it is possible to eliminate the feedback of sound from the control source to the detector or to set up the controller such that it is a "modal" controller, the controllers will be relatively complicated compared to plane wave controllers. The most significant complicating factor is the variation of the group speed with frequency near the cut-on frequency. The result is an infinite impulse response which will limit control when finite digital filters are used to implement these controllers. The other complicating factors are the modal summations. If the detector, error and control transducers cannot be set up to discriminate between modes, the controllers will be much more complicated than the modal controllers emphasized here.

ACKNOWLEDGEMENTS

This investigation was supported by Nelson Industries, Stoughton, Wisconsin.

REFERENCES

1. Doak, P.E., 1973. "Excitation, "Transmission and Radiation of Sound from Source Distribution in Hard-Walled Ducts of Finite Length." Journal of Sound and Vibration, 31(1):1-72.

2. Stell, J.D. and Bernhard, R.J., 1990. "Active Control of High Order Acoustical Modes In a Semi-Infinite Waveguide." Proceedings of ASME 1990 Winter Annual Meeting, NCA-8:131-142.

3. Tichy, J., Warnaka, G.E., and Poole, L.A.,1983. "A Study of Active Control of Noise in Ducts." ASME Journal of Vibration, Acoustics, Stress, and Reliability in Design, 106:399-404.

4. Urusovskii, I.A., 1977. "Active Noise Suppression in a Waveguide." Soviet Physics,-Acoustics, 23(2):170-174.

5. Eriksson, L.J., Allie, M.C., Hoops, R.H., and Warner, J.V., 1989. "Higher Order Mode Cancellation in Ducts Using Active Noise Control." Proceedings of InterNoise 89, pp. 495-500.

6. Silcox, R.J., and Elliott, S.J., May, 1990. "Active Control Of Multi-Dimensional Random Sounds in Ducts." NASA Technical Memorandum #102653.

7. Roure, A., 1985. "Self-Adaptive Broadband Active Sound Control System." Journal of Sound and Vibration, 101(3):429-441.

8. La Fontaine, R.F., and Shepherd, I.C., 1983. "An Experimental Study of a Broadband Active Attenuator for Cancellation of Random Noise in Ducts." Journal of Sound and Vibration, 91(3):351-362.

9. Eriksson, L.J., Allie, M.C., and Greiner, R.A., 1987. "The Selection and Application of an IIR Adaptive Filter for Use in Active Sound Attenuation." IEEE Transactions on Acoustics, Speech, and Signal Processing, ASSP-35(4):433-437.

A UNIFIED CONTROL STRATEGY FOR THE ACTIVE REDUCTION OF SOUND

N. J. Doelman[1]

ABSTRACT

The Generalized Minimum Variance (GMV) control strategy is proposed as an effective strategy for active sound and vibration control systems. The GMV strategy is shown to unify well-known adaptive filtering approaches based on LMS-type algorithms and purely feedback strategies as used in other types of control systems. The short-comings of the LMS adaptive filtering systems for active sound control are clearly uncovered. Pure adaptive filtering arises from the GMV strategy merely when the extraneous noise at the error sensor is Gaussian noise, the controller complexity is restricted and no fast tracking of system changes is required. The necessary additional identification of the secondary acoustic path is an extra problem of the adaptive filtering approach. Simulation experiments with real and synthetic data verify the wide applicability of the GMV control strategy and the limited efficacy of (LMS) adaptive filters for use in active sound reduction systems.

I INTRODUCTION

During the last ten years considerable attention has been paid to the design of control algorithms for active sound and vibration reduction. A lot of this research has been concentrated on the active reduction of low frequency sound in a duct system, as in that case merely a plane wave field has to be controlled. So many of the existing control algorithms are designed for SISO systems. It is felt, that the basic principle of SISO control algorithms can easily be transferred to the MIMO case. Consequently, this paper will discuss the SISO case only.

The main part of the developed control strategies are based on a feedforward control approach. A certain detection sensor measures the incoming primary wave field before it reaches the position of the secondary (anti-sound) source. This detection sensor can be acoustical, optical or otherwise, which depends on the nature of generation of the primary sound. The principle is that the detection sensor provides the controller with information on the disturbance signal (primary sound) at the error sensor. The signal from the error sensor is used to adapt the controller characteristics if necessary. Essentially, this is an adaptive filtering approach.

It is not always possible to find a detection signal, that gives full information on the primary signal. This may be caused by e.g. a significant contribution of extraneous noise at the error sensor. Moreover, it is not necessarily possible to describe the transfer function from detection sensor to error sensor as a deterministic pole-zero model. In these cases an additional feedback control approach can be very effective. Now, the signal from the error

[1]TNO Institute of Applied Physics, P.O. Box 155, 2600 AD Delft, The Netherlands; Laboratory of Acoustics and Seismics, Faculty of Applied Physics, Delft University of Technology.

sensor is fed back into the controller, which drives the secondary source. Olson's sound absorber [1] is a simple, analogue realization of this principle. Also, in vibration control a feedback approach is quite accustomed. In this paper a unified control strategy is proposed, which incorporates both feedforward and feedback control.

II SYSTEM MODELLING

All control strategies discussed in this paper apply to the discrete-time case. Designing a digital control system starts with the choice for the system model. A very popular way of system modelling employs the so-called backward shift operator q^{-1}; [2]. This operator has the following property:

$$q^{-1}x(t) = x(t-1); \qquad q^{-n}x(t) = x(t-n) \tag{1}$$

The variable t is an integer number and stands for the sampling interval index. This means that the sequence of samples of $x(t)$ becomes $[x(t)x(t-1)x(t-2)......]$. An example of a transfer function in q^{-1} is :

$$y(t) = q^{-1}\frac{1+0.8q^{-1}}{1-0.6q^{-1}}x(t)$$

which stands for $y(t) = x(t-1) + 0.8x(t-2) + 0.6y(t-1)$. Now consider the general single-channel control system for the active reduction of sound; Figure 1. The system is capable of reducing plane waves generated by the primary source at the beginning of the duct system. The system modelling is mainly concerned with the description of the signal from the error sensor, $y(t)$. This signal consists of the primary disturbance and the signal form the secondary source. In general, the transfer function form the control signal $u(t)$ to the error signal $y(t)$ can be modelled as a pole-zero operator in q^{-1}. It holds that

$$\begin{aligned} y(t) &= \frac{B(q^{-1})}{A(q^{-1})}u(t-\tau_1) + z(t) \\ A(q^{-1}) &= 1 + a_1q^{-1} + a_2q^{-2} + \cdots\cdots + a_{n_A}q^{-n_A} \\ B(q^{-1}) &= b_0 + b_1q^{-1} + b_2q^{-2} + \cdots\cdots + b_{n_B}q^{-n_B} \end{aligned} \tag{2}$$

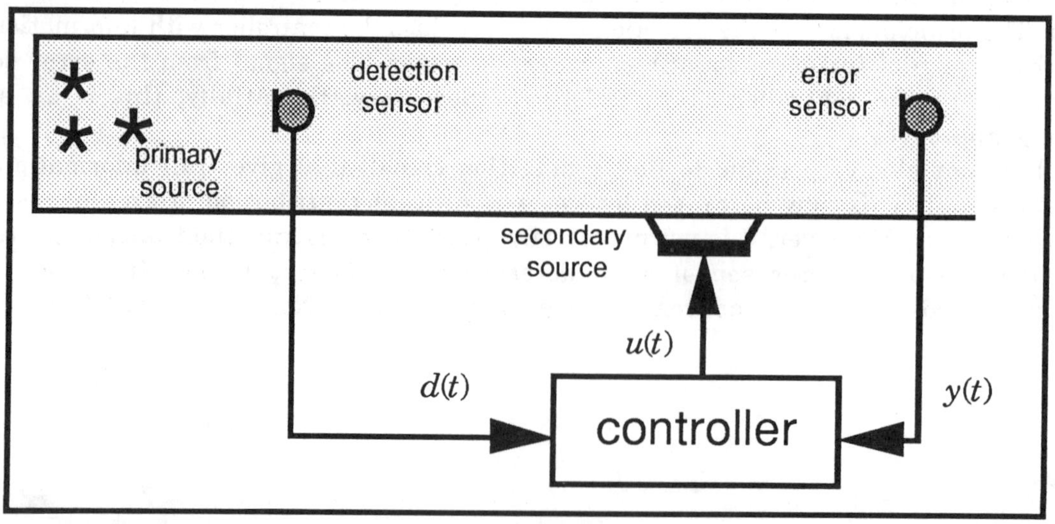

Figure 1; Schematic diagram of a single-channel system for active sound control.

The time delay, which is principally caused by the propagation time from the secondary source to the error sensor, is denoted by τ_1 sampling intervals. Note that the time delay is not necessarily an even multiple of the sampling interval. This acoustic process is also called the secondary acoustic process or secondary path. The signal $z(t)$ denotes the disturbance signal at the error sensor, which is to be cancelled by the secondary sound field. Part of this disturbance signal is correlated with the primary sound $x(t)$, which is measured by the detection sensor. This contribution can be modelled as

$$z_1(t) = \frac{G(q^{-1})}{E(q^{-1})} x(t - \tau_2) \tag{3}$$
$$G(q^{-1}) = g_0 + g_1 q^{-1} + g_2 q^{-2} + \cdots\cdots + g_{n_G} q^{-n_G}$$
$$E(q^{-1}) = 1 + e_1 q^{-1} + e_2 q^{-2} + \cdots\cdots + e_{n_E} q^{-n_E}$$

The remaining part of the signal at the error sensor is not correlated with the primary sound at the detection sensor. It is modelled as a filtered sequence of Gaussian noise

$$z_2(t) = \frac{C(q^{-1})}{D(q^{-1})A(q^{-1})} \xi(t) \tag{4}$$
$$C(q^{-1}) = 1 + c_1 q^{-1} + c_2 q^{-2} + \cdots\cdots + c_{n_C} q^{-n_C}$$
$$D(q^{-1}) = 1 + d_1 q^{-1} + d_2 q^{-2} + \cdots\cdots + d_{n_D} q^{-n_D}$$

in which $\xi(t)$ is randomly distributed noise with mean μ and variance σ^2. The model in (5) is very general, as it is capable of representing both stochastic and deterministic disturbance signals; [2]. In most publications on active sound control this term is left out. However, in many acoustic situations it appears that a stochastic contribution to the primary sound cannot be neglected. In e.g. a vehicle cabin or a ventilation duct there is a substantial part of extraneous noise. Superposition of both disturbance signals and using (2), (3) and (4) yields for the over-all system model:

$$y(t) = \frac{B(q^{-1})}{A(q^{-1})} u(t - \tau_1) + \frac{G(q^{-1})}{E(q^{-1})} x(t - \tau_2) + \frac{C(q^{-1})}{D(q^{-1})A(q^{-1})} \xi(t) \tag{5}$$

This model is used throughout the paper. It allows for the description of both periodic and random primary sound. The signal $x(t)$ in (5) is in fact unknown to the control system as the detection signal $d(t)$ consists of a primary sound contribution **and** a secondary sound contribution. The detection sensor also measures the secondary sound field coming from the secondary source. So it holds for $d(t)$:

$$d(t) = x(t) + \frac{F(q^{-1})}{A(q^{-1})} u(t - \tau_3) \tag{6}$$
$$F(q^{-1}) = f_0 + f_1 q^{-1} + f_2 q^{-2} + \cdots\cdots + f_{n_F} q^{-n_F}$$

It should be emphasized that the transfer function modelling the paths $u(t) \to y(t)$, $u(t) \to d(t)$ and $\xi(t) \to y(t)$ have common poles. This assumption is based on the fact that the poles of transfer functions are determined by the acoustic modes in the enclosure; [3]. All transfer functions apply to the same acoustic enclosure. For the sake of simplicity the numerator polynomial of the transfer function is assumed to contain the dynamics of the electronic and the electro-acoustic path. The transfer operator for the Gaussian noise contains an extra denominator polynomial in order to maintain generality. The poles of the transfer function from $x(t)$ to $y(t)$ are usually not equal to the system poles. This is due to the fact, that actually this transfer function is the fraction of two other transfer functions, see e.g. [4]. The circumstance that there is no source involved in this transfer function eliminates the polynomial $A(q^{-1})$.

III GENERALIZED MINIMUM VARIANCE CONTROL

The generalized minimum variance (GMV) control strategy is extensively described by Clarke in [5] and [6]. It appears to be very suitable for a control system for the active reduction of sound. It will be demonstrated that it is a useful extension of the accustomed approach of adaptive filtering. The basic principle of the control strategy is as follows:

a priori		define a cost function
each sampling instant	1	predict the cost function
	2	choose the control signal $u(t)$ such, that it minimizes the variance of the cost function

In this paper it will be demonstrated, that this type of control corresponds to the well-known method of adaptive filtering. Consider the following auxiliary function:

$$\Phi(t+\tau_1) = P(q^{-1})y(t+\tau_1) + Q(q^{-1})u(t) \tag{7}$$

in which P and Q are polynomials in q^{-1}. The control object of the GMV strategy is to minimize the variance of the auxiliary function $\Phi(t+\tau_1)$. The mere instrument to minimize the variance of the auxiliary function at time $t+\tau_1$ is the control signal at time instant t, $u(t)$. In order to minimize its variance, the auxiliary function at time instant $t+\tau_1$ $\Phi(t+\tau_1)$ has to be estimated. Assuming $u(t-1)$ and previous samples to be known at time t, merely the following part of Φ is not completely known.

$$\Phi_y(t+\tau_1) = P(q^{-1})y(t+\tau_1) \tag{8}$$

The optimal prediction in a least-squares sense of this function is implicitly given by:

$$C\Phi^\circ{}_y(t+\tau_1) = Ny(t) + MBDu(t) + MDA\frac{G}{E}x(t+\tau_1-\tau_2) \tag{9}$$

The polynomials M and N obey the following relation

$$\frac{P(q^{-1})C(q^{-1})}{D(q^{-1})A(q^{-1})} = M(q^{-1}) + q^{-\tau_1}\frac{N(q^{-1})}{D(q^{-1})A(q^{-1})} \tag{10}$$

The order of polynomial M is τ_1-1: $M(q^{-1}) = m_0 + m_1q^{-1} + \cdots\cdots + m_{\tau_1-1}q^{-\tau_1+1}$. The meaning of the M and N polynomials becomes clear when examining the contribution of the signal $\xi(t)$ to the auxiliary function

$$\Phi_\xi(t+\tau_1) = \frac{P(q^{-1})C(q^{-1})}{D(q^{-1})A(q^{-1})}\xi(t+\tau_1) = \underbrace{M(q^{-1})\xi(t+\tau_1)}_{\text{future}} + \underbrace{\frac{N(q^{-1})}{D(q^{-1})A(q^{-1})}\xi(t)}_{\text{past}} \tag{11}$$

At time instant t the contribution of the stochastic noise to $y(t+\tau_1)$ consists of future and past values of $\xi(t)$. The future samples cannot be included in the estimation of $\Phi_y(t+\tau_1)$. The past values, however, are intrinsically present in the values of $y(k)$ for $k \le t$. Application of (10) to the function $\Phi_y(t+\tau_1)$ yields

$$C\Phi_y(t+t_1) = PCy(t+t_1) = MDAy(t+\tau_1) + Ny(t) \tag{12}$$

which equals, using (5)

$$C\Phi_y(t+\tau_1) = Ny(t) + MBDu(t) + \frac{MGDA}{E}x(t+\tau_1-\tau_2) + MC\xi(t+\tau_1) \tag{13}$$

The future part of $\xi(t)$ cannot be estimated. Consequently, the optimal prediction of $Py(t+\tau_1)$ is given by

$$C\Phi^\circ{}_y(t+\tau_1) = Ny(t) + MBDu(t) + \frac{MGDA}{E}x(t+\tau_1-\tau_2) \tag{14}$$

which is identical to (9). The estimation of the complete auxiliary function is now given as

$$C\Phi^\circ(t+\tau_1) = Ny(t) + [MBD + CQ]u(t) + \frac{MGDA}{E}x(t+\tau_1-\tau_2) \tag{15}$$

The difference between the actual and the optimal estimation of the auxiliary function Φ as defined in (15), is nothing but a moving average of the Gaussian noise disturbance. The estimation error made is minimal in a least squares sense.

$$Py(t+\tau_1) - Py^\circ(t+\tau_1) = M\xi(t+\tau_1) = \varepsilon(t+\tau_1) \tag{16}$$

The corresponding variance of the auxiliary function reads

$$J_1 = \mathrm{E}\left\{\left[P(q^{-1})y(t+\tau_1) + Q(q^{-1})u(t)\right]^2\right\} \tag{17}$$

As $\xi(t+k)$ is orthogonal to other terms on the Right Hand Side of (17), this equals

$$J_1 = \mathrm{E}\left\{\left[P(q^{-1})y^\circ(t+\tau_1) + Q(q^{-1})u(t)\right]^2\right\} + \mathrm{E}\left\{\left[\varepsilon(t+\tau_1)\right]^2\right\} \tag{18}$$

Setting the term in the first bracket to zero, in fact minimizes the cost function J_1. The control law thus becomes

$$P(q^{-1})y^\circ(t+\tau_1) + Q(q^{-1})u(t) = 0 \tag{19}$$

Using (15), this can be written as

$$H(q^{-1})u(t) + K(q^{-1})y(t) + L(q^{-1})x(t+\tau_1-\tau_2) = 0$$

$$\text{with} \quad H = BDEM + CEQ \qquad K = EN \qquad L = GDAM \tag{20}$$

under the assumption that C and E are stable polynomials. Clearly, this type of control employs both feedforward and feedback. Globally, the feedforward action is to reduce the measurable disturbance $x(t)$, whereas the feedback loop cancels the unknown disturbance

$\xi(t)$. The GMV-control law (20) can be shown to minimize cost function J_2 as well:

$$J_2 = \mathrm{E}\left\{\left[P(q^{-1})y(t+\tau_1)\right]^2 + \left[\frac{n_0}{q_0}Q(q^{-1})u(t)\right]^2 \middle| t\right\} \tag{21}$$

This type of cost function has also been examined in [7], in context with the adaptive filtering strategy. When the controller possesses the optimal coefficients, as given in (20), the over-all transfer function reads

$$y(t) = \frac{GAQ}{E[BP+AQ]}x(t-\tau_2) + \frac{BDM+CQ}{D[BP+AQ]}\xi(t) \tag{22}$$

The overall system poles are determined by the original system polynomials A and B and the weighting polynomials P and Q. This influence of P and Q can be employed to acquire a desired system behaviour by choosing suitable pole locations. The strategy of pole assignment within the GMV strategy will be demonstrated in section VIII.

IV STOCHASTIC AND PERIODIC PRIMARY SOUND

As well-known, the type of primary noise determines the structure of the active sound reduction system. The system as depicted in Figure 1 appeals to the case in which the primary sound has a random nature, a broadband spectrum. However, when the primary sound is essentially periodic, a simpler system structure is possible. The fundamental frequency of the periodic primary sound can generally be acquired by actually monitoring the action of the rotating noisy apparatus. An tachometer with e.g. an optical or a magnetic sensor will do in that case. Consequently, for periodic primary sound the acoustic detection sensor is not necessary and the acoustic feedback from secondary source to this sensor is eliminated. In Figure 2 a block diagram of this system structure is shown.

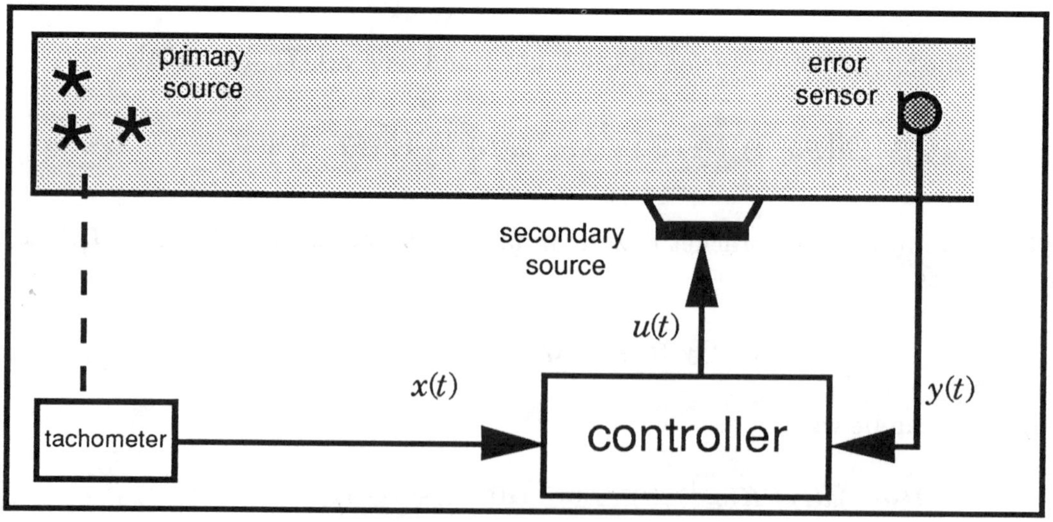

Figure 2; Schematic diagram of an active control system for periodic sound.

The optimal estimation of the auxiliary function is as follows:

Periodic Primary Sound

$$F(q^{-1})=0;\ E(q^{-1})=A(q^{-1})$$
$$C\Phi^{\circ}(t+\tau_1)=[BDM+CQ]u(t)+Ny(t)+GDM\,x(t+\tau_1-\tau_2) \tag{23}$$

For primary sound of a random nature the required signal $x(t)$ is not directly available. Using eq.(6) yields for the auxiliary function:

Random Primary Sound

$$x(t)=d(t)-\frac{F(q^{-1})}{A(q^{-1})}u(t-\tau_3)$$
$$C\Phi^{\circ}(t+\tau_1)=\left[BDM+CQ-q^{\tau_1-\tau_2-\tau_3}\frac{GFDM}{E}\right]u(t)+Ny(t)+\frac{GADM}{E}d(t+\tau_1-\tau_2)$$

It is evident, that for random primary sound the controller structure is more complicated than for periodic primary sound. In Figure 3 a block diagram for this GMV controller is given.

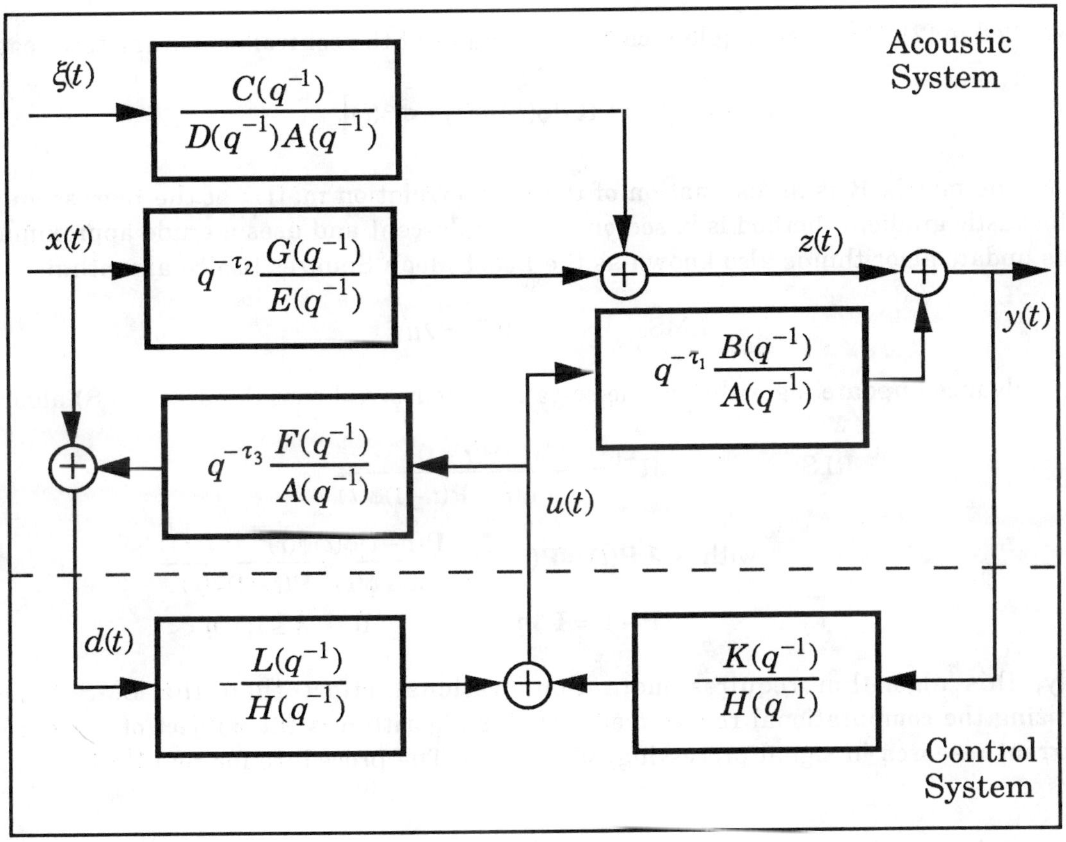

Figure 3; Block diagram of the acoustic system and the GMV controller.

V PARAMETER ESTIMATION

There are several ways to find the optimal control parameters sequentially. The method of sequential regression is a very suitable. Consider the estimation of the parameters of the optimal prediction of the auxiliary function $\Phi(t)$

$$\hat{C}\hat{\Phi}^{\circ}(t) = \hat{K}y(t-\tau_1) + \hat{H}u(t-\tau_1) + \hat{L}d(t-\tau_2) \tag{25}$$

in which $\hat{C}$ is an estimate of C, CA or CEA, depending on the type of GMV-control. The controller parameters of K, H and L are to estimated. Define a parameter vector $\boldsymbol{\theta}$ of the estimated controller parameters:

$$\boldsymbol{\theta}(t) = \left[\hat{k}_0(t) \cdots \hat{k}_{n_K}(t) \quad \hat{h}_0(t) \cdots \hat{h}_{n_H}(t) \quad \hat{l}_0(t) \cdots \hat{l}_{n_L}(t)\right]^T \tag{26}$$

and a regression vector $\boldsymbol{\phi}$ of past signal samples:

$$\boldsymbol{\phi}(t) = \left[y(t-\tau_1) \cdots y(t-\tau_1-n_K) \quad u(t-\tau_1) \cdots u(t-\tau_1-n_H) \quad d(t-\tau_2) \cdots d(t-\tau_2-n_L)\right]^T \tag{27}$$

The estimate of the auxiliary function $\Phi^{\circ}(t)$ can now be written as

$$\hat{\Phi}^{\circ}(t) = \boldsymbol{\theta}(t)^T \boldsymbol{\phi}(t) \tag{28}$$

As described in [2], the most simple recursive estimator of the controller parameters reads

$$\boldsymbol{\theta}(t+1) = \boldsymbol{\theta}(t) + \mathbf{R}^{-1}\boldsymbol{\phi}(t)\left[\Phi(t) - \hat{\Phi}^{\circ}(t)\right] \tag{29}$$

in which the matrix $\mathbf{R}$ is an estimation of the autocorrelation matrix of the regression vector. The stochastic gradient method is based on steepest descent and uses a crude approximation of $\mathbf{R}$. This update algorithm is also known as the Least Mean Squares (LMS) algorithm.

$$\text{LMS:} \qquad \mathbf{R}^{-1} = 2\mu\mathbf{I} \tag{30}$$

A more advanced update algorithm is the so-called Recursive Least Squares (RLS) algorithm.

$$\begin{aligned}
\text{RLS:} \qquad & \mathbf{R}^{-1} = \frac{\mathbf{P}(t-1)}{\lambda + \boldsymbol{\phi}(t)^T\mathbf{P}(t-1)\boldsymbol{\phi}(t)} \\
\text{with} \qquad & \lambda\,\mathbf{P}(t) = \mathbf{P}(t-1) - \frac{\mathbf{P}(t-1)\boldsymbol{\phi}(t)\boldsymbol{\phi}(t)^T\mathbf{P}(t-1)}{\lambda + \boldsymbol{\phi}(t)^T\mathbf{P}(t-1)\boldsymbol{\phi}(t)} \\
& \mathbf{P}(-1) = \mathbf{I}/\eta; \qquad 0 << \lambda \le 1; \quad \eta << 1
\end{aligned} \tag{31}$$

Clearly, this algorithm requires more computational effort than the LMS algorithm. Minimizing the computational requirements of this algorithm is the subject of a great deal of the current research in signal processing; see e.g. [8]. The procedure for real-time control is as follows

1. Calculate the auxiliary function $\Phi(t)$
   $\Phi(t) = P(q^{-1})y(t) + Q(q^{-1})u(t-\tau_1)$
2. Estimate the parameters of $\Phi^{\circ}(t)$
   $C\Phi^{\circ}(t) = K(q^{-1})y(t-\tau_1) + H(q^{-1})u(t-\tau_1) + L(q^{-1})d(t-\tau_2)$
3. Apply control according to (20)
   $u(t) = -\dfrac{\hat{K}(q^{-1})}{\hat{H}(q^{-1})}y(t) - \dfrac{\hat{L}(q^{-1})}{\hat{H}(q^{-1})}d(t+\tau_1-\tau_2)$

In the section VII it will be demonstrated that well-known active sound control algorithms fit to the procedure given above.

VI PURE FEEDFORWARD CONTROL

As exhibited in section 3 the general GMV control strategy employs both feedforward and feedback control. In this section it will be shown, that under certain conditions the feedback control loop is not required. Furthermore, it will be derived that the accustomed adaptive filtering techniques are special cases of an feedforward GMV control strategy. Pure feedforward control arises form the GMV strategy, if a feedback loop does not contribute to a minimization of the cost function. Mathematically, a feedback loop is not required when the polynomial $N(q^{-1})$ in (10) equals zero. Suppose there exists a polynomial $X(q^{-1})$ of finite degree, which equals

$$X(q^{-1}) = \frac{P(q^{-1})C(q^{-1})}{D(q^{-1})A(q^{-1})} \tag{32}$$

For N to equal zero, X has to satisfy

$$X(q^{-1}) = x_0 + x_1 q^{-\tau_1} + \cdots\cdots + x_{\tau_1 - 1} q^{-\tau_1 + 1} = M(q^{-1}) \qquad \text{order}\left[X(q^{-1})\right] \le \tau_1 - 1 \tag{33}$$

This means that DA is factor of PC. The constraints signify that the control signal $u(t)$ cannot deal with the contribution of the stochastic disturbance to the auxiliary function, Φ, at time instant $t+\tau_1$. As can be interpreted from (12), this contribution is completely determined by future values of $\xi(t)$. There is no correlation between $y(t)$ and $\Phi_\xi(t+\tau_1)$. The most likely case is that the extraneous noise contribution is a moving average of white noise samples; D=1 and A is a factor of C. For random primary sound the estimate of the auxiliary function then yields

$$\frac{C}{A}\Phi^\circ(t+\tau_1) = \left[\frac{B}{A}M + \frac{C}{A}Q - q^{\tau_1-\tau_2-\tau_3}\frac{GFM}{E}\right]u(t) + \frac{GM}{E}d(t+\tau_1-\tau_2) \tag{34}$$

Minimizing the variance of the auxiliary function yields a pure feedforward control law in this case. For a system dealing with periodic primary sound, the estimate can be found by substituting: F=0 and E=A. The pure feedforward control strategy arises from the GMV-strategy under the assumption that the stochastic disturbance can not be dealt with. As will be shown below, the adaptive filtering strategy is a special case of the GMV feedforward control strategy. The adaptive filtering strategy is based on the pure minimum variance control strategy: P=1, Q=0. Furthermore, it is required that the stochastic disturbance is white noise; X=1 and $C = A$. Then, the estimation of the auxiliary function becomes

$$\Phi^\circ(t+\tau_1) = \left[\frac{B}{A} - q^{\tau_1-\tau_2-\tau_3}\frac{GF}{E}\right]u(t) + \frac{G}{E}d(t+\tau_1-\tau_2) \tag{35}$$

The estimate of the auxiliary function is the basis of the derivation of adaptive filtering strategies, which are described in [7] and [9].

VII IDENTIFICATION OF ACTIVE SOUND CONTROL STRATEGIES

In most publications on the active control of sound, the primary or disturbance signal is assumed to be perfectly correlated with the detection signal, whether from a microphone or from a tachometer. This also applies to the descriptions of the filtered x LMS algorithm for periodic primary sound, [7], and to the Recursive LMS algorithm for random primary sound, [10]. In the previous section on pure feedforward control strategies were derived especially for the case of a white noise extraneous disturbance. These strategies will be shown to be compatible with the LMS strategies mentioned below.

The RLMS-based control strategy for random primary sound

In [9] Eriksson suggests a solution for the controller polynomials H and L like

$$H = 1 - q^{\tau_1-\tau_2-\tau_3}\frac{GF}{BE}; \qquad L = q^{\tau_1-\tau_2}\frac{GA}{BE} \tag{36}$$

This type of solution for H and L can be readily seen in (35) when it is rewritten:

$$\Phi^\circ(t+\tau_1) = \frac{B}{A}[Hu(t) + Ld(t+\tau_1-\tau_2)] \tag{37}$$

Estimation of the controller parameters can be carried out using a sequential regression algorithm as discussed in the section V. Denote the vectors $\boldsymbol{\theta}$ and $\boldsymbol{\phi}$ as

$$\boldsymbol{\phi} = \left[u^F(t)u^F(t-1)\cdots u^F(t-n_H) \quad d^F(t+\tau_1-\tau_2)d^F(t+\tau_1-\tau_2-1)\cdots d^F(t+\tau_1-\tau_2-n_H)\right]^T$$

$$\boldsymbol{\theta} = \left[\hat{h}_0(t)\hat{h}_1(t)\cdots\hat{h}_{n_H}(t) \quad \hat{l}_0(t)\hat{l}_1(t)\cdots\hat{l}_{n_L}(t)\right]^T \tag{38}$$

with $$u^F(t) = q^{-\tau_1}\frac{B}{A}u(t); \qquad d^F(t) = q^{-\tau_1}\frac{B}{A}d(t) \tag{39}$$

In [10] it is suggested to use a modified version of the RLMS algorithm to estimate the controller parameters. The RLMS algorithm was originally proposed by Feintuch [11] for the purpose of estimating both the non-recursive and recursive coefficients of an IIR filter using an LMS-type algorithm. The modification Eriksson describes in [10] comprises the use of the so-called filtered signal vector, $[u^F \; d^F]$, and the ignorance of the dynamics of the controller parameters. The update equation is given by

$$\boldsymbol{\theta}(t+1) = \boldsymbol{\theta}(t) + 2\mu\boldsymbol{\phi}(t)y(t) \tag{40}$$

Neglecting the changes in the values of the controller parameters seriously restricts the admissible values for the step-size parameter μ. Choosing μ too large will readily result in a diverging algorithm. The required small step-size parameter implies a severe restriction on the rate of convergence. Until now, no simple rules for the setting of the step-size parameter are available. An additional requirement for the "filtered u RLMS" algorithm is the estimation of the secondary acoustic process $q^{-\tau_1}B/A$ in order to generate the filtered u and d vector. In (36) the polynomials H and L are to represent a fraction of polynomials. This requires a sufficient degree for H and L. A restricted length for the polynomials H and L, however, is possible when rewriting (35) as

$$AE\Phi^\circ(t+\tau_1) = \left[BE - q^{\tau_1-\tau_2-\tau_3}FG\right]u(t) + GA\,d(t+\tau_1-\tau_2) = Hu(t) + Ld(t+\tau_1-\tau_2) \tag{41}$$

The LMS-based control strategy for periodic primary sound

For this type of system there is no acoustic feedback, as the detection sensor is not acoustic; F=0, E=A. Hence, the estimate of the auxiliary function according to these characteristics yields

$$\Phi^\circ(t+\tau_1) = \frac{B}{A}u(t) + \frac{G}{A}(t+\tau_1-\tau_2) \tag{42}$$

The resulting type of control filter is that of an Infinite Impulse Response (IIR). A possible simplification, however, is the FIR structure. Since the order of persistent excitation of a periodic signal is limited, the controller structure need not necessarily be recursive. If the periodic signal $x(t)$ consists of M sinusoidal components for instance, the controller needs merely $2M$ coefficients to generate every possible periodic output signal. So the complexity restriction results in the control law:

$$u(t) = W(q^{-1})x(t) \\ \text{with} \quad W(q^{-1}) = w_0 + w_1 q^{-1} + \cdots\cdots + w_{2M-1} q^{-2M+1} \tag{43}$$

The advantage of this structure is the guaranteed stability of the filter. An IIR filter generates an unstable output signal, when poles of its transfer function lie outside the unit circle. The optimal W or (42) obeys the following relation:

$$\left[BW + q^{\tau_1 - \tau_2} G\right] x(t) = 0 \tag{44}$$

The relation of the estimate of the auxiliary function to the control law (43) reads

$$\Phi^\circ(t+\tau_1) = \frac{B}{A}\left[u(t) - W(q^{-1})x(t)\right] \tag{45}$$

Employing a sequential regression algorithm to estimate the controller parameters yields

$$\boldsymbol{\theta}(t) = \left[\hat{w}_0(t)\, \hat{w}_1(t) \cdots\cdots \hat{w}_{n_W}(t)\right]^T \qquad \boldsymbol{\phi}(t) = \left[x^F(t)\, x^F(t-1) \cdots\cdots x^F(t-n_W)\right]^T \tag{46}$$

$$\text{with} \quad x^F(t) = q^{-\tau_1} \frac{B}{A} x(t) \tag{47}$$

The so-called "filtered x LMS algorithm", see e.g. [12], uses the estimate of 2μ for $\mathbf{R}^{-1}$ and also ignores the dynamics of the controller coefficients. The varying of the coefficients of the adaptive filter is assumed to be slow compared to the timescale of the response of secondary acoustic process; [7]. This yields for the filtered x LMS update equation:

$$\boldsymbol{\theta}(t) = \boldsymbol{\theta}(t) - 2\mu\boldsymbol{\phi}(t)y(t) \tag{48}$$

which is identical to (40). The algorithm is applied in many active noise reduction systems, mainly because of its simplicity. Again, the assumption of a slowly varying filter seriously limits the convergence speed of the algorithm. Furthermore, the algorithm requires knowledge of the secondary acoustic process in order to calculate the filtered x signal.

The main drawback of the algorithm is its limited convergence speed. Another LMS-type algorithm, which can achieve a more favourable rate of convergence is described in [13]. It uses a prediction error instead of the true system output to update the controller coefficients. However, stochastic gradient algorithms still suffer in the case of a large eigenvalue spread in the autocorrelation matrix of input samples. The RLS algorithm can deal with this problem, as its rate of convergence is theoretically independent of this eigenvalue spread.

The self-tuning control procedure for periodic and random primary sound, based on adaptive filtering can now be given as:

1 Calculate / measure the auxiliary function; $\Phi(t) = y(t)$
2 Estimate the parameters of the model of the secondary path.
3 Estimate the controller parameters; using (37) or (45).
4 Apply control.

Methods to accomplish step **2** are described in [10] and [14].

VIII SIMULATIONS

In order to compare and to test the control strategies discussed and suggested in this paper a few simulations have been carried out. The simulation both deal with synthetic and measured random and periodic primary sound, in which the common active sound control strategies are applied and the original GMV control strategy. It should be noted that in each simulation the required identification of the secondary path for the adaptive filtering strategies was carried out a priori. Consequently, for each control strategy merely the controller parameters were to be estimated. The characterization of the several control strategies is as follows:

-Adaptive filtering 1	-control law based on (37)	-RLMS estimator (40).
-Adaptive filtering 2	-control law based on (41)	-RLS estimator (31).
-Periodic Control	-control law based on (45)	-LMS estimator (48).
-Feedback control	-control law based on(23,24) with $L = 0$!!	-RLS estimator (31).
-Feedforward/feedback	-control law based on (23,24)	-RLS estimator (31).

Simulation 1

Both the model and the signals are generated synthetically in this simulation. It deals with a primary source radiating periodic and random sound into a duct system. The system transfer functions are modelled using a simple form of the one way wave model, as described in [15]. All time delays are chosen to be even multiples of the sampling instant. The transfer functions of transducers are neglected. The power spectrum of the primary signal is shown in Figure 4. The adaptive filtering strategies are not capable of reducing the complete primary signal, whereas the GMV strategy accomplishes a very good reduction.

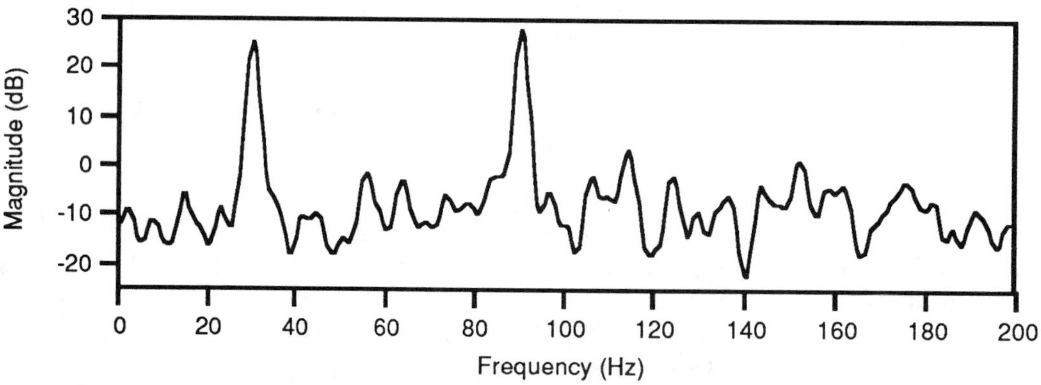

Figure 4; Power spectrum of the primary signal in simulation 1.

The results of the simulation 1 can be found in Figure 5. The system parameters are

$A(q^{-1}) = 1 + 0.36q^{-8}$	$B(q^{-1}) = 0.6 + 0.54q^{-6}$	$\tau_1 = 1$
$C(q^{-1}) = 1 - 1.5q^{-1} + 0.7q^{-2}$	$D(q^{-1}) = 1$	
$E(q^{-1}) = 1 - 0.24q^{-6}$	$G(q^{-1}) = 0.6$	$\tau_2 = 3$
$F(q^{-1}) = 0.6 + 0.66q^{-2} - 0.36q^{-4}$	$\tau_3 = 2$	$P(q^{-1}) = 1 \quad Q(q^{-1}) = 0$
$x(t) = \cos(3\pi t / 20) + 2\cos(9\pi t / 20) + 3\omega(t) - 1.5$		
$\omega(t)$: uniform noise interval [0,1];	$\xi(t)$: Gaussian noise var = 0.02 mean = 0	

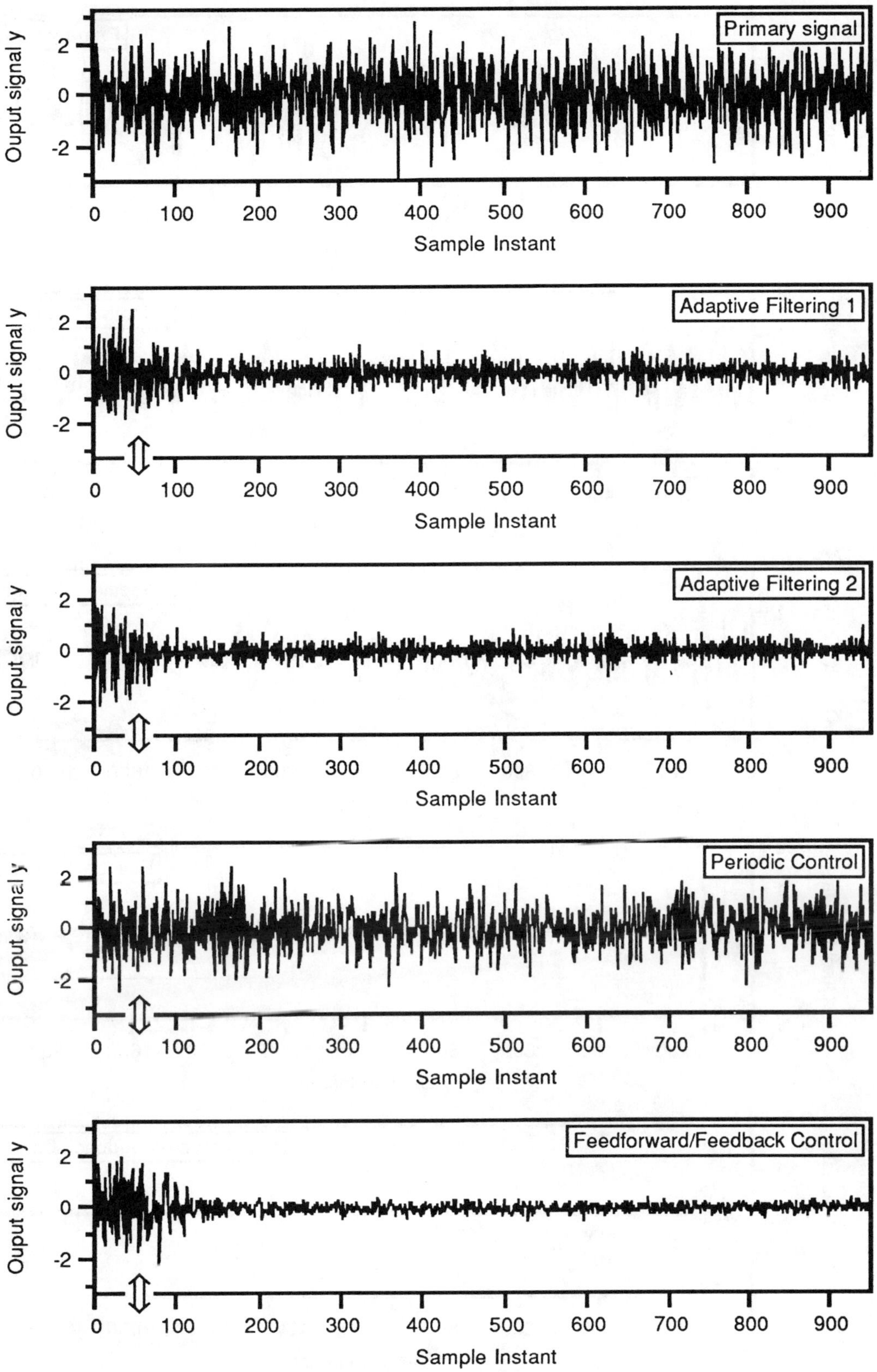

Figure 5; Results of simulation 1. The arrow indicates the start of control action.

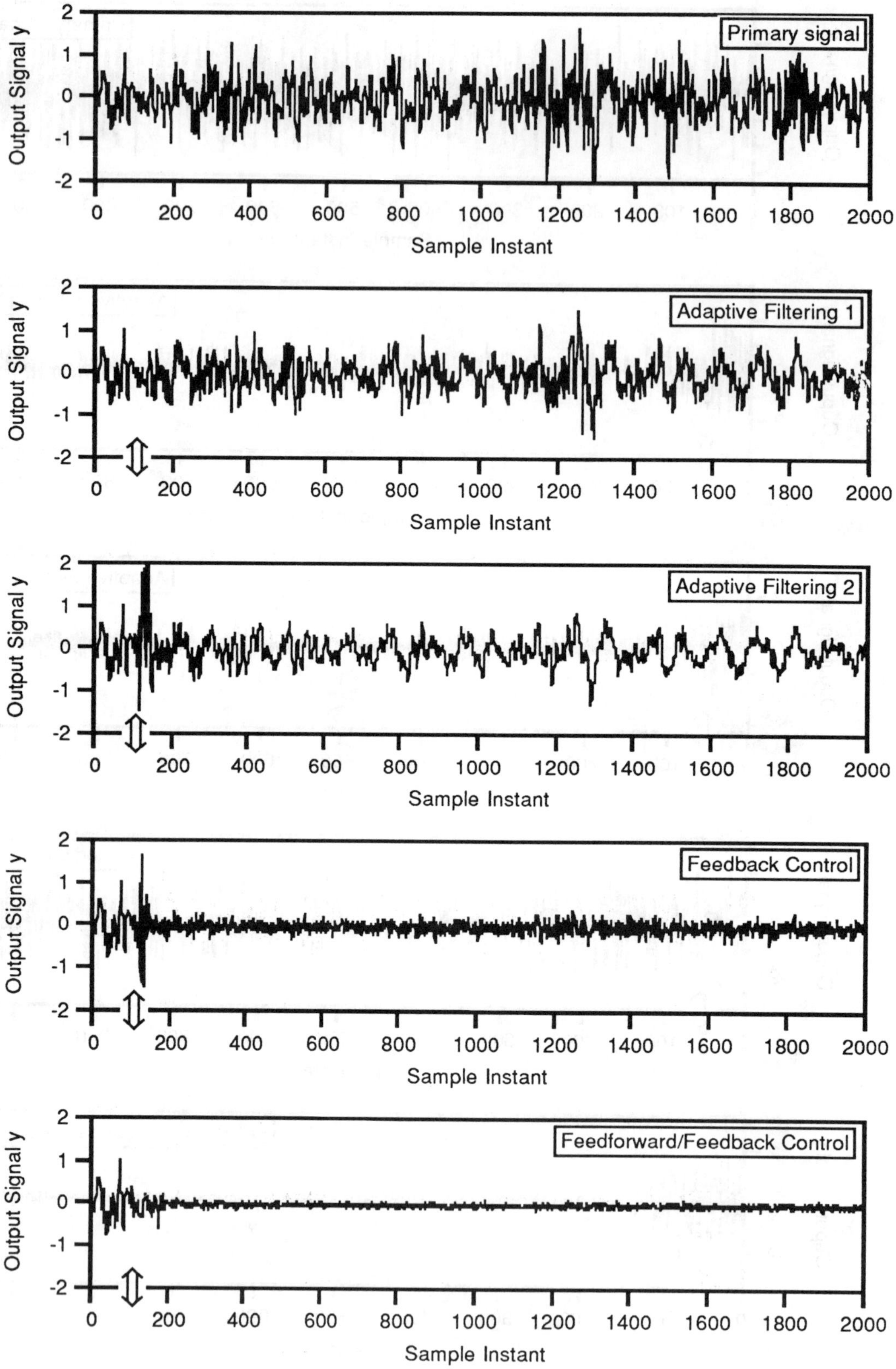

Figure 6; Results of simulation 2. The arrow indicates the start of control action.

Simulation 2

For this simulation experiment the primary signals at the detection and error sensor were measured in ventilation duct. The system transfer functions were modelled as

$A(q^{-1})=1$	$B(q^{-1})=1$	$\tau_1=2$
$F(q^{-1})=0.8$	$\tau_3=3$	
$P(q^{-1})=1$	$Q(q^{-1})=0$	

The simulation results can be found in Figure 6. Again, only the combined feedforward /feedback achieves the optimal attenuation of the primary sound.

Simulation 3

Again the primary signal at the error sensor is measured. It was generated by a pulse combustion steam generator in the so-called "tail pipe". As it was not possible to acquire a tachometer signal, the periodic primary sound could not be handled by a feedforward control strategy.

$A(q^{-1})=1+0.15q^{-1}-0.122q^{-2}-0.877q^{-3}$	$B(q^{-1})=1-0.4q^{-2}$	$\tau_1=2$
$P(q^{-1})=1$	$Q(q^{-1})=0$	

This simulation shows the possible power of a pure feedback controller; see Figure 7.

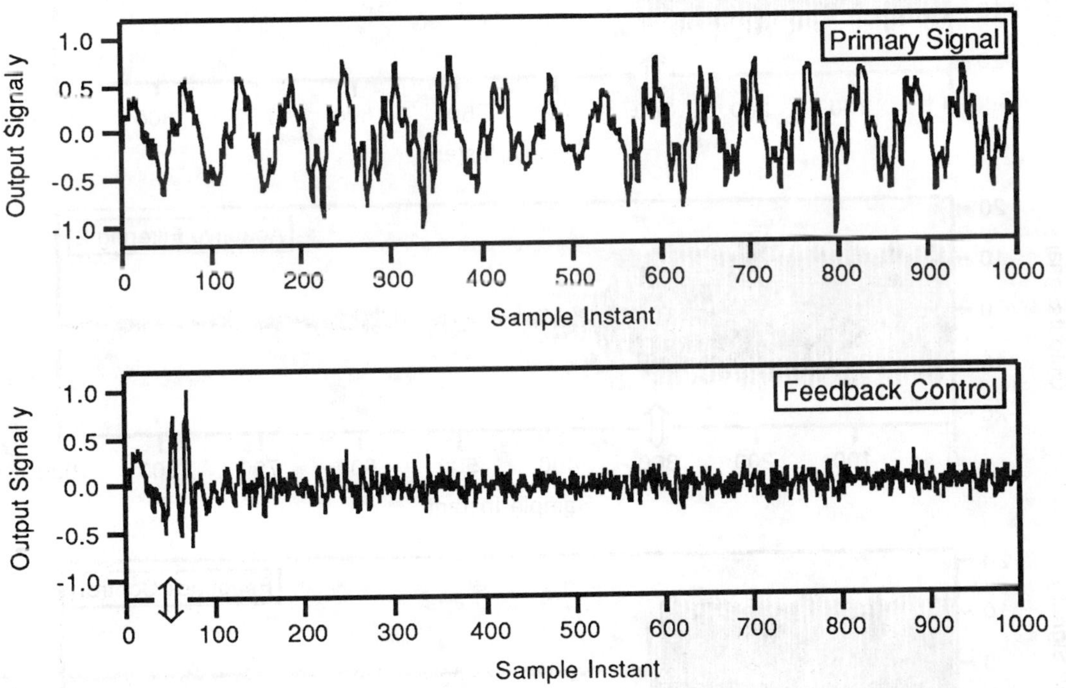

Figure 7; Results of simulation 3. The arrow indicates the start of control action.

Simulation 4

This simulation experiment deals with a periodic primary signal. The simulated enclosure is quite reverberant. The model and signals are given by

$A(q^{-1}) = 1 - 0.3q^{-1} + 0.95q^{-2}$	$B(q^{-1}) = 1 + 1.6q^{-1}$	$\tau_1 = 2$
$C(q^{-1}) = 1$	$D(q^{-1}) = 1$	
$G(q^{-1}) = 4$	$\tau_2 = 0$	
$P(q^{-1}) = 0.5 - 0.7q^{-1} + 0.25q^{-2} - 0.25q^{-3}$	$Q(q^{-1}) = 0.5 + 0.5q^{-2}$	
$x(t) = \cos(\pi t / 2)$ $\quad\xi(t)$:	Gaussian noise var = 0.1	mean = 0

The secondary acoustic process is non-minimum phase, which makes it impossible to invert for this process. Therefore, the polynomial Q is used in order to limit the control signal by putting a weight on its energy. Polynomials P and Q are chosen such that the GMV controller remains stable and converges very fast. This is in contrast to the adaptive filtering strategies which suffer from a slow convergence, as the over-all system poles are determined by the characteristics of A. Figure 8 exhibits the results of this experiment.

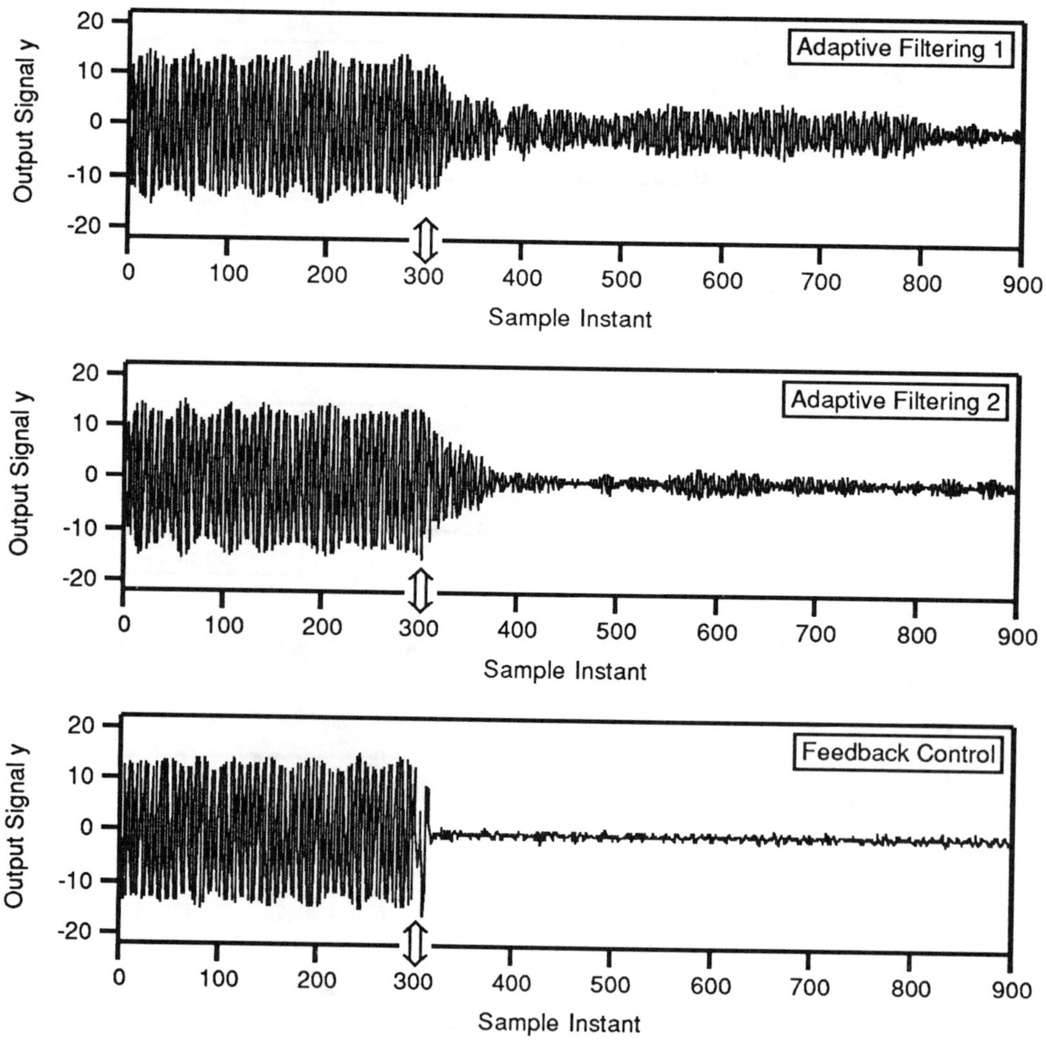

Figure 8; Results of simulation 4. The arrow indicates the start of control action.

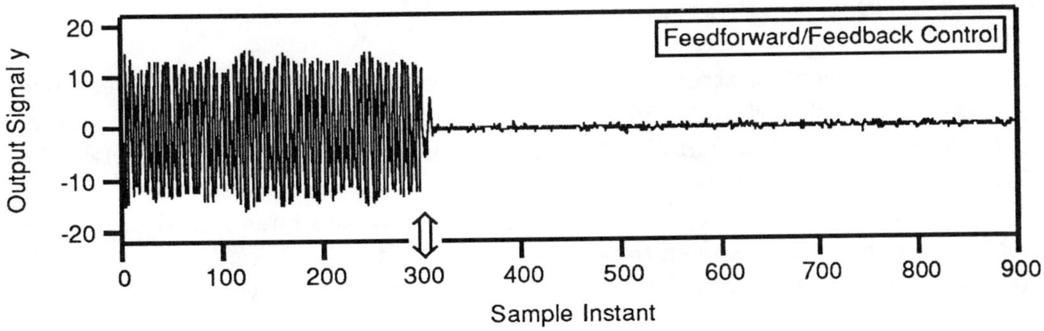

Figure 8 (continued); Results of simulation 4. The arrow indicates the start of control action.

More simulation experiments on systems for periodic primary sound can be found in [14] and [16]. The over-all simulation results are presented in Table 1.

Table 1; The mean residual square output signal of the simulation experiments for several control strategies.

Control strategy \ Simulation	1	2	3	4
Adaptive Filtering 1	0.101	0.1305	–	6.999
Adaptive Filtering 2	0.086	0.0840	–	1.144
Periodic Control	0.526	–	–	–
Feedback Control	0.0123	–	0.01	0.236
Feedforward/ Feedback Control	0.022	0.0015	–	0.133
No Control	1.106	0.2327	0.126	87.18

IX CONCLUSIONS

In this paper a general control strategy for the active reduction of sound has been presented. Although, it was developed during the 70's, [5], for general control purposes, it has scarcely been applied to the active control of sound. The GMV control strategy is completely self-tuning, which is favourable for active sound control systems.

Well-known strategies for active sound reduction like the ones based on the filtered x LMS or RLMS algorithm can be shown to be special cases of the GMV control strategy. For these adaptive filtering strategies to arise from the GMV strategy, the following is required:

1 the extraneous noise at the output is purely white noise.
2 the controller parameters are estimated using a modified LMS algorithm.
3 an a priori or real-time estimation of the secondary acoustic process is carried out.

These three requirements clearly represent the advantages and disadvantages of the popular adaptive filtering strategy for active sound control. The mere advantage is the simplicity of implementation both in hardware and software form. The limitations of the algorithms have been demonstrated in theoretical derivations and in simulation experiments.

1 When the extraneous noise at the error sensor is not Gaussian noise but coloured noise, the control strategy should employ a feedback loop in order to deal with the coloured noise. The performance of the adaptive filtering strategy seriously degrades in this case.
2 The modifications of the LMS algorithms to handle the secondary acoustic process are not ideal, [13]. This results in a decelerated rate of convergence and a limited ability to track changes in system or primary signal characteristics.
3 The required identification of the secondary path is not shared by the main GMV control strategy. All other GMV strategies presented in this paper are **direct**, which means that no extra identification procedure is necessary.

The short-comings of the LMS-type strategies have been elucidated in simulation experiments with both synthetic and real data.

It is shown that the GMV control strategy presented here is is valuable extension to the existing control strategies for active reduction of sound. It is capable of reducing both the deterministic and the stochastic parts of the primary signal at the error sensor by combining feedforward and feedback control. The popular adaptive filtering approach, as described in [7], [10] and [13] is simple and should be applied to uncomplicated sound attenuation problems only. The more general GMV strategies can handle complicated problems quite effectively.

X ACKNOWLEDGEMENTS

These investigations in the program of the Fundamental Research on Matter (FOM) have been supported by the Netherlands Technology Foundation (STW).

XI REFERENCES

[1] Olson, H.F. and E.G. May, 1953. "Electronic sound absorbers". Journal of the Acoustical Society of America. 25: 1130-1136.

[2] Goodwin, G.C. and K.S. Sin, 1984. Adaptive Filtering, Prediction and Control. Prentice Hall, New Jersey.

[3] Koizumi, N. and R.H. Lyon, 1989. "On the model order for the identification of acoustical systems". 118th Meeting of the Acoustical Society of America.

[4] Roure, A., 1985, "Self-adaptive.broadband active sound control system. "Journal of Sound and Vibration, 101: 429-441.

[5] Clarke, D.W. and D.A Gawthrop, 1975. "Self-tuning controller". Proceedings IEE part D, 122: 929-934.

[6] Clarke, D.W., 1981. "Introduction to self-tuning control". in Self-tuning and adaptive control: theory and applications. IEE Control Engineering Series 15.

[7] Elliott, S.J., I.M. Stothers and P.A. Nelson, 1987. "A multiple error LMS algorithm and its application to the active control of sound and vibration." IEEE Transactions on ASSP, 35: 1423-1434.

[8] Haykin, S.S., 1986. "Adaptive Filter Theory", Prentice Hall, New Jersey.

[9] Eriksson, L.J., M.C. Allie, C.D. Bremigan and J.A. Gilbert, 1989. "Weight vector analysis of an RLMS adaptive filter with on-line auxiliary path modelling". Proceedings of the IEEE Transactions on ASSP, paper A2.2.

[10] Eriksson, L.J. and M.C. Allie. 1989. "The use of random noise for on-line transducer modeling in an adaptive attenuation system". Journal of the Acoustical Society of America, 85: 797-802.

[11] Feintuch, P.L., 1976. "An adaptive recursive LMS filter". Proceedings IEEE, 64: 1622-1624.

[12] Widrow, B. and S.D. Stearns, 1985. "Adaptive Signal Processing", Prentice Hall, New Jersey

[13] Doelman, N.J., 1991. "A novel adaptive filter LMS algorithm for use in the active control of sound". submitted to IEEE Transactions on ASSP.

[14] Doelman, N.J., 1990. "Control strategies for active sound reduction". Proceedings Internoise, Göteborg, pp. 1273-1276.

[15] Doelman, N.J., M.M. Boone and P.H.J. Koers, 1988. "A one way wave model for the analysis of active reduction of broadband noise". Proceedings Internoise, pp. 1021-1024

[16] Doelman, N.J. and E.J.J. Doppenberg, 1991. "The application of self-tuning control strategies to the active reduction of sound. Proceedings of IEE Conference Control 91.

THE EFFECTS OF MODELLING ERRORS ON THE PERFORMANCE AND STABILITY OF ACTIVE NOISE CONTROL SYSTEMS

C. C. Boucher,[1] S. J. Elliott,[1] P. A. Nelson[1]

ABSTRACT

The effects of errors within the system model of both single and multichannel filtered-x algorithms for the active control of sound and vibration is presented. In the single channel case we consider the effects of phase errors in the system model on the optimum convergence coefficient and convergence time and how these effects combine with those of delays in the error path.

A statistical approach is used to investigate the effects of errors within the system model of the multichannel algorithm, this includes results from a 32 microphone, 16 loudspeaker active noise control system and computer simulations. The analytic work demonstrates how effort weighting in the cost function may be used to stabilise otherwise unstable systems.

INTRODUCTION

The filtered-x LMS algorithm has been applied to many problems in active noise and vibration control using both its single and multichannel forms. In general it has been found to be fast and very robust, ideally suited to such engineering problems. The algorithms require a model of the transfer response from each secondary source to each error sensor. This paper investigates the sensitivity of these algorithms to errors within this model on both the performance and stability of such systems.

For the single channel case the effects of phase errors are investigated. Provided they are within well defined bounds (approximately $\pm 90°$) and the convergence coefficient is small enough, then the system will be stable. The variation of phase error with optimum convergence coefficient (and hence minimum convergence time) is discussed.

The multichannel work looks at the effect of adding Gaussian random noise to the elements of the system model. Here stability is no longer assured and the probability of instability with increasing error is investigated. A common practice is to provide some form of effort weighting in the cost function in order to avoid excessive power demands. This paper explains how this effort weighting can also be use to stabilise an otherwise unstable system.

[1]Institute of Sound and Vibration Research, University of Southampton, Southampton SO9 5NH, England.

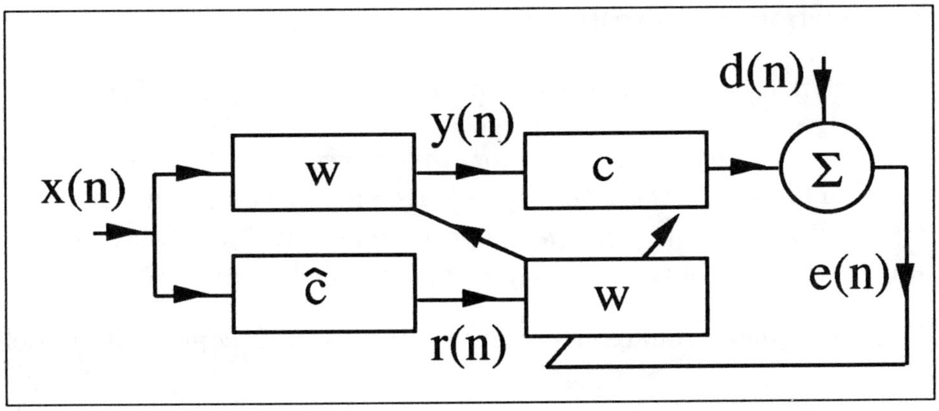

Figure 1: Block diagram of a single channel feedforward control system.

SINGLE CHANNEL FEEDFORWARD CONTROL

THEORY

A block diagram for a simple single channel feedforward control system which uses the filtered-x LMS algorithm [1, 2] is shown in figure 1. The physical system is driven by passing the reference signal $x(n)$ through the Ith order FIR control filter w to give the output signal, $y(n)$, which is used to drive the actuator. The system response in figure 1 is represented by a Jth order FIR filter, denoted c, and will, in practice consist of the combined responses from the actuator, sensor, anti-aliasing filters and data converters along with the transfer response from the actuator to the sensor. The system output, $e(n)$, is defined as the linear sum of the contributions from the primary source, $d(n)$, and the secondary source. The conventional LMS algorithm cannot be used to update w because of the presence of the system response, c, and so the 'filtered-x' LMS algorithm [2] which first passes the reference signal through a model of the system, denoted $\hat{c}$, before adapting the 'slave' filter w is used. The coefficients of this filter are then copied into the controller as indicated to implement the control system.

In vector form, the equations for this system can be written,

$$e(n) = d(n) + \boldsymbol{r}^T(n)\boldsymbol{w} \tag{1}$$

where

$$\boldsymbol{r}^T(n) = [r(n), r(n-1), \ldots, r(n-I+1)] \tag{2}$$

$$\boldsymbol{w}^T = [w_0, w_1, \ldots, w_{I-1}] \tag{3}$$

and the filtered reference signal is given by

$$r(n) = \sum_{j=0}^{J-1} c_j x(n-j). \tag{4}$$

The mean square error can be written as

$$E[e^2(n)] = E[d^2(n)] + 2\boldsymbol{w}^T E[\boldsymbol{r}(n)d(n)] + \boldsymbol{w}^T E[\boldsymbol{r}(n)\boldsymbol{r}^T(n)]\boldsymbol{w} \tag{5}$$

where $E[]$ denotes the expectation operator, this shows that the mean square error is a quadratic function of each of the filter coefficients which has a unique global minimum if the autocorrelation matrix $E[\boldsymbol{r}(n)\boldsymbol{r}^T(n)]$ is positive definite (i.e., there are no more than twice as many coefficients in w as there are spectral components in $x(n)$).

The derivative of the mean square error with respect to each control filter coefficient can be expressed as

$$\frac{\partial E[e^2(n)]}{\partial \boldsymbol{w}} = 2E[\boldsymbol{r}(n)d(n)] + E[\boldsymbol{r}(n)\boldsymbol{r}^T(n)]\boldsymbol{w} \tag{6}$$

$$= 2E[\boldsymbol{r}(n)e(n)]. \tag{7}$$

The filtered-x LMS algorithm is formed by updating the control filter coefficients by the instantaneous version of this derivative [1, 2]:

$$\boldsymbol{w}(n+1) = \boldsymbol{w}(n) - \alpha \boldsymbol{r}(n)e(n), \tag{8}$$

where α is the convergence coefficient.

This paper is concerned with the effects of errors with the system model use to generate the filtered reference signal. If we represent this model by a kth order FIR filter, $\hat{c}_k$, then the reference signal is now given by

$$\hat{r}(n) = \sum_{k=0}^{K-1} \hat{c}_k x(n-k). \tag{9}$$

and so the practical filtered-x LMS algorithm is given by

$$\boldsymbol{w}(n+1) = \boldsymbol{w}(n) - \alpha \hat{\boldsymbol{r}}(n)e(n) \tag{10}$$

where

$$\hat{\boldsymbol{r}}^T(n) = [\hat{r}(n), \hat{r}(n-1), \ldots, \hat{r}(n-K+1)]. \tag{11}$$

THE EFFECTS OF ERRORS

Elliott *et al* [3] have shown how the complete, dynamic response of this system could be represented by an equivalent transfer function given by

$$H(z) = \frac{1 - 2z\cos(\omega_0) + z^2}{1 - 2z\cos(\omega_0) + z^2 + \beta C(z)(z\cos(\omega_0 - \Phi) - \cos\Phi)} \tag{12}$$

in which $\hat{C} = Ae^{j\Phi}$ is the estimate of the true transfer function C at the frequency ω_0 and

$$\beta = \frac{I\alpha A}{2}. \tag{13}$$

For a pure tone reference signal, setting $C(z) = 1$ and in the limit of very slow convergence, $\beta \to 0$; where the dynamic properties of the error path become unimportant, equation 12 becomes

$$H(z) = \frac{1 - 2z\cos(\omega_0) + z^2}{z^2 - (2\cos(\omega_0) - \beta\cos(\omega_0 - \Phi))z + (1 - \beta\cos\Phi)}. \tag{14}$$

By examining whether poles of this second order recursive system lie within the unit circle the systems stability may be investigated. For small β, $H(z)$ will have conjugate poles at a distance of $(1-\beta\cos\Phi)^{1/2}$ from the origin. Since all the terms in β are assumed positive, the distance of the poles from the unit circle can only be greater than 1 if $\cos\Phi$ is negative, so the stability condition [1] must be:

$$\cos\Phi > 0. \tag{15}$$

Therefore, for stability,

$$-90° < \Phi < 90°. \tag{16}$$

This result, that the phase response of the system and that of the system model need only be within 90° of one another for the system to converge, suggests that for slow convergence, the only condition for stability is that on average the update quantity $(\boldsymbol{r}(n)e(n))$ is of the correct sign. Such a formulation also suggests that for finite convergence speeds the maximum convergence coefficient is decreased by a factor of $\cos\Phi$ and the convergence time is increased by a factor of $1/\cos\Phi$ [1]. However, there is a rather subtle interaction between these limits, imposed by the errors within the model, and those imposed by the presence of delays. To investigate this interaction results were obtained from a simple computer simulation.

Simulations of a pure tone filtered-x LMS algorithm (with four samples per cycle) were run in which the optimum convergence coefficient and minimum convergence time were determined empirically for phase errors in the range ±80° and for delays of 1, 2 and 3 cycles (4, 8 and 12 samples). The optimum convergence coefficients are plotted in figure 2 and were found to be approximately one-third of the values for which the system became unstable.

Figure 3 shows the corresponding minimum convergence time. From these results it can be seen that for phase errors above 70° the stability of the system is very dependent on the delay within the system. For phase errors less that ≈ 45° the effects of both the delays and phase errors are comparatively small.

MULTI-CHANNEL FEEDFORWARD CONTROL

THEORY

A block diagram for a typical multichannel active control system [4] is shown in figure 4. It consists of a number of secondary sources and primary source from which a reference signal is derived, which is unaffected by the action of the secondary sources. The secondary sources are driven by adaptive filters fed from this reference signal. Distributed

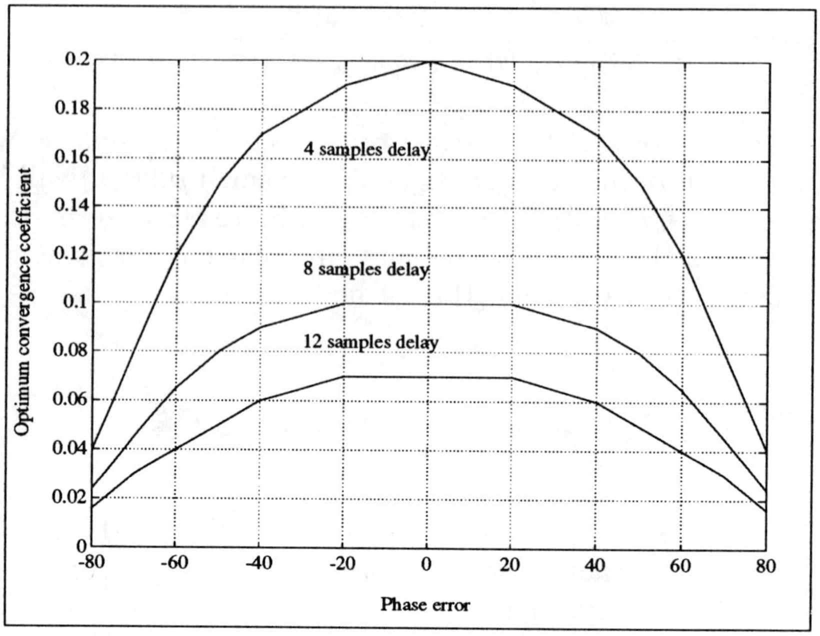

Figure 2: The variation in optimum convergence coefficient with phase error and update delay for a single channel LMS adaptive filter.

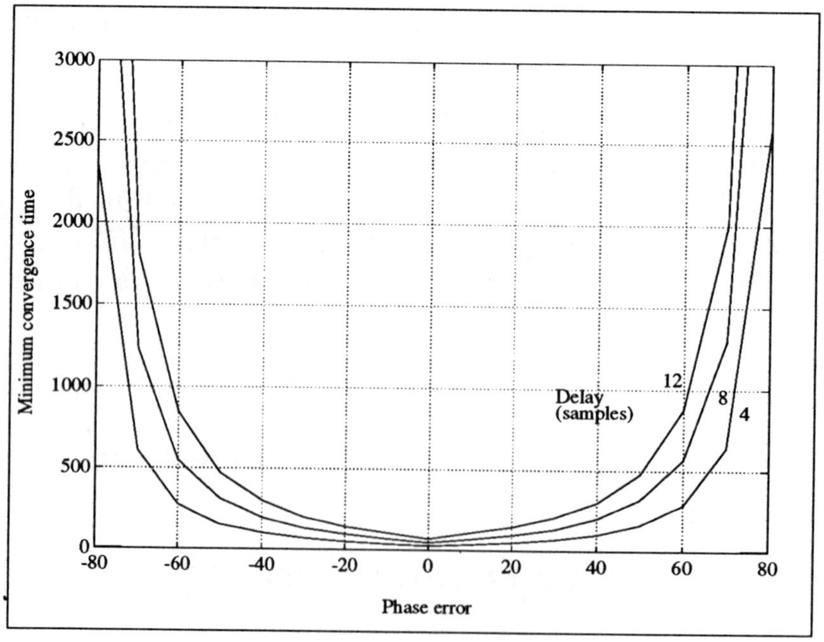

Figure 3: The variation in minimum convergence time for a single channel LMS adaptive filter with phase error and update delay.

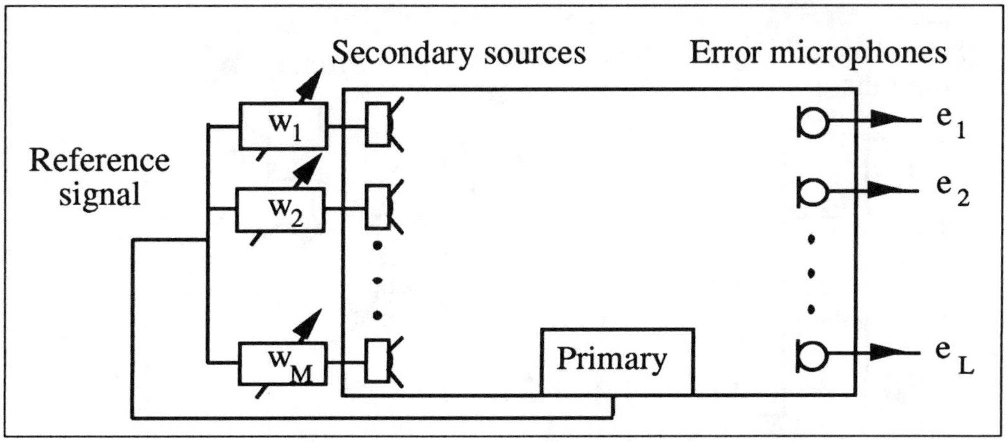

Figure 4: Physical block diagram of a multichannel control system for controlling the sound from a periodic primary source.

throughout the enclosure are a number of error sensors which are usually placed as near to the ears as possible.

We assume that there is one reference signal at a single frequency, ω_0. Under these conditions the controller only has to adjust the amplitude and phase of this signal before it is fed to each of the secondary sources. The secondary source inputs and error sensor outputs can now be characterized, in the steady-state, by vectors of complex quantities. In using a frequency domain formulation for the adaptive algorithm we are ignoring the effects of any dynamics in the physical system under control, such as delays due to, for example the acoustic path from each secondary source to each error sensor. Later we will see that the convergence of control systems with large numbers of channels is generally dominated by effects other than such dynamics.

Figure 4 may now be recast into a frequency domain block diagram, figure 5. The secondary sources are fed by complex elements of the vector $\boldsymbol{q}$ while the error sensors give an output represented by the complex elements of the vector $\boldsymbol{p}$. The matrix $\boldsymbol{Z}$ models the acoustic response of the enclosure and, in a practical system, the response of the anti-aliasing and reconstruction filters. Finally $\boldsymbol{p}_p$ represents the error sensor response due to the primary field alone. These quantities are related as [3]

$$\boldsymbol{p}(\omega_0) = \boldsymbol{p}_p(\omega_0) + \boldsymbol{Z}(\omega_0)\boldsymbol{q}(\omega_0). \tag{17}$$

In choosing a cost function we include a term proportional to the mean square effort of the system as well as the total mean square error,

$$J = \boldsymbol{p}^H\boldsymbol{p} + \beta\boldsymbol{q}^H\boldsymbol{q} \tag{18}$$

where β is a real scalar parameter which normally determines the balance between a control system reducing the total mean square pressure ($\boldsymbol{p}^H\boldsymbol{p}$), and not using an excessive

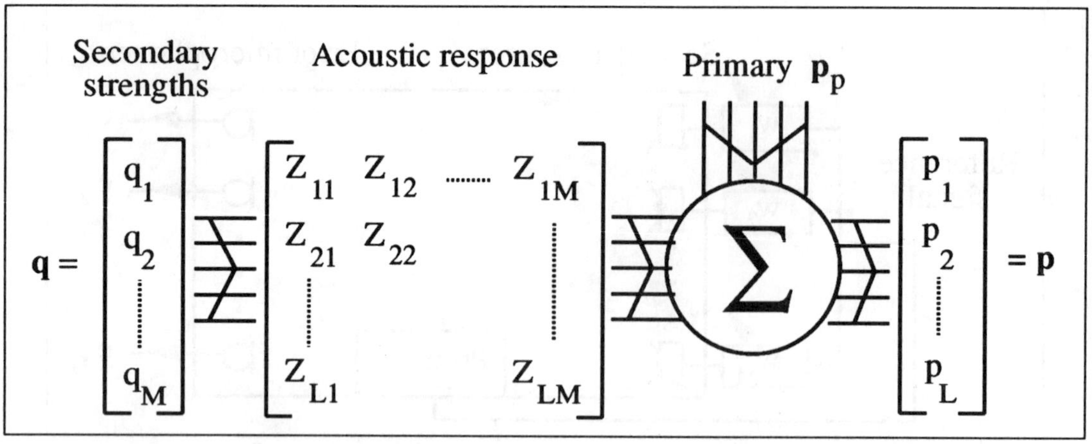

Figure 5: The multichannel control system in the steady-state form.

control effort ($\boldsymbol{q}^H\boldsymbol{q}$) to do so [5] and the superscript H denotes the Hermitian transpose. Therefore

$$J = \boldsymbol{p}_p^H \boldsymbol{p}_p + \boldsymbol{p}_p^H \boldsymbol{Z}\boldsymbol{q} + \boldsymbol{q}^H \boldsymbol{Z}^H \boldsymbol{p}_p + \boldsymbol{q}^H(\boldsymbol{Z}^H\boldsymbol{Z} + \beta \boldsymbol{I})\boldsymbol{q} \tag{19}$$

The explicit dependence on ω_0 has been dropped for convenience. Provided there are at least as many sensors as secondary sources the matrix $(\boldsymbol{Z}^H\boldsymbol{Z} + \beta\boldsymbol{I})$ will be positive definite and the cost function (J) will have a unique global minimum for some value of $\boldsymbol{q}$ the solution for which is given by

$$\boldsymbol{q}_{\text{opt}} = -[\boldsymbol{Z}^H\boldsymbol{Z} + \beta\boldsymbol{I}]^{-1}\boldsymbol{Z}^H\boldsymbol{p}_p. \tag{20}$$

Substituting this into equation 19 gives the minimum error criterion,

$$J_{\min} = \boldsymbol{p}_p^H[\boldsymbol{I} - \boldsymbol{Z}[\boldsymbol{Z}^H\boldsymbol{Z} + \beta\boldsymbol{I}]^{-1}\boldsymbol{Z}^H]\boldsymbol{p}_p. \tag{21}$$

Since the cost function J is a quadratic function of each of the available variables (the real and imaginary parts of $\boldsymbol{q}$), gradient descent methods, if properly implemented and stable, should converge to the global minimum of J i.e., the optimum least squares solution, $J_{\min}$. If, for a system with M secondary sources, the complex gradient vector is defined as

$$\nabla J = \left[\frac{\partial J}{\partial \Re q_1}\frac{\partial J}{\partial \Re q_2}\cdots\frac{\partial J}{\partial \Re q_M}\right]^T + j\left[\frac{\partial J}{\partial \Im q_1}\frac{\partial J}{\partial \Im q_2}\cdots\frac{\partial J}{\partial \Im q_M}\right]^T \tag{22}$$

then using the standard rules of vector differentiation [6] we have, in this case, that

$$\nabla J = 2\boldsymbol{Z}^H\boldsymbol{p}_p + 2(\boldsymbol{Z}^H\boldsymbol{Z}\boldsymbol{q} + \beta\boldsymbol{I})\boldsymbol{q} \tag{23}$$

and so from equation 17 we can express the gradient vector in the form

$$\nabla J = 2(\boldsymbol{Z}^H\boldsymbol{p} + \beta\boldsymbol{q}). \tag{24}$$

Using the standard gradient descent technique of adjusting the secondary source vector at the kth iteration to be equal to the previous $(k-1)$ value minus an amount proportional to the gradient we obtain,

$$\boldsymbol{q}(k) = \boldsymbol{q}(k-1) - \mu \nabla J(k-1) \tag{25}$$

where μ is a convergence coefficient, finally, from equation 24, we may now write the frequency domain version of the multichannel filtered-x algorithm [5]:

$$\boldsymbol{q}(k) = (1 - \alpha\beta)\boldsymbol{q}(k-1) - \alpha \boldsymbol{Z}^H \boldsymbol{p}(k-1) \tag{26}$$

where $\alpha = 2\mu$.

THE EFFECTS OF ERRORS

In any practical controller the model of the true acoustic response $\boldsymbol{Z}$ will clearly not be an exact one and so we must reformulate the theory above to contain the model of the acoustic response, which we shall denote $\hat{\boldsymbol{Z}}$. In this case equation 26 becomes

$$\boldsymbol{q}(k) = (1 - \alpha\beta)\boldsymbol{q}(k-1) - \alpha \hat{\boldsymbol{Z}}^H \boldsymbol{p}(k-1). \tag{27}$$

Note, however, that the error signals are still related to the secondary sources via the physical response matrix $\boldsymbol{Z}$:

$$\boldsymbol{p}(k) = \boldsymbol{p}_p + \boldsymbol{Z}\boldsymbol{q}(k). \tag{28}$$

If we assume that the final solution for this system is now given by the equation

$$\boldsymbol{q}_\infty = -[\hat{\boldsymbol{Z}}^H \boldsymbol{Z} + \beta \boldsymbol{I}]^{-1} \hat{\boldsymbol{Z}}^H \boldsymbol{p}_p \tag{29}$$

then we can write equation 27 in the form

$$(\boldsymbol{q}(k) - \boldsymbol{q}_\infty) = [\boldsymbol{I} - \alpha(\hat{\boldsymbol{Z}}^H \boldsymbol{Z} + \beta \boldsymbol{I})](\boldsymbol{q}(k-1) - \boldsymbol{q}_\infty) \tag{30}$$

therefore

$$(\boldsymbol{q}(k) - \boldsymbol{q}_\infty) = [\boldsymbol{I} - \alpha(\hat{\boldsymbol{Z}}^H \boldsymbol{Z} + \beta \boldsymbol{I})]^k (\boldsymbol{q}(0) - \boldsymbol{q}_\infty). \tag{31}$$

The convergence of this iterative solution is assured provided now that

$$|1 - \alpha(\lambda_m + \beta)| < 1 \quad \text{for all } m \tag{32}$$

i.e.,

$$\alpha < \frac{2(\sigma_m + \beta)}{|\lambda_m + \beta|^2} \quad \text{for all } m \tag{33}$$

where λ_m is the mth eigenvalue of $\hat{\boldsymbol{Z}}^H \boldsymbol{Z}$ and σ_m is the real part of λ_m [1]. When this stability condition is met, $\boldsymbol{q}_\infty$ will be the final steady-state solution for the system. Note however that, with $\beta = 0$, equation 33 requires that the real parts of all the eigenvalues of $(\hat{\boldsymbol{Z}}^H \boldsymbol{Z})$ be positive, which in general will not be the case. However, by setting β

such that $(\sigma_m + \beta)$ is greater than zero for all m, a previously unstable system can be stabilised.

The control system described in [4] for the active control of propeller-induced cabin noise which used 16 secondary sources to minimise the sum of the squared pressures at 32 microphones was set up in a large wooden enclosure measuring $2.2m \times 2.2m \times 6m$. The microphones were set out in a regular 4 by 8 grid at standing head height whilst the loudspeakers were evenly distributed at both floor and ceiling height. The primary field was supplied by another, larger loudspeaker set away from the others at a height of 0.5 meters.

The system was first set up to adaptively 'learn' the acoustic transfer response from each loudspeaker to each microphone, this was done under near 'ideal' conditions with no external noise sources present. The primary field (an 88 Hz pure tone) was then turned on and the system allowed to converge. The measured reduction in the sum of squares of the 32 microphone signals of 33.1 dB was within 0.2 dB of the optimum level predicted by equation 21. This indicates that the initial estimates of the acoustic plant response are reasonably accurate and that the performance is not significantly degraded by excess mean square error caused by the iterative nature of the algorithm. The distribution of the real and imaginary components of $\boldsymbol{Z}$ were approximately due to the large number of acoustic modes which were excited [7].

Simulations were conducted to investigate the effects of errors within the model of the true system ($\hat{\boldsymbol{Z}}$) in which this model was formed by adding a matrix of Gaussian random numbers to the true response, $\boldsymbol{Z}$. By comparing the variance of the two matrices the amount of error could be measured e.g., when the two variances were equal 100% error had been added.

The variation in the probability of a given system being stable with increasing percentage error was calculated using a large number of such simulations and is shown in figure 6. With no effort weighting the system is very sensitive to errors with less than 15% of systems stable with only 10% error. However, as predicted in the theory above, modest amounts of effort weighting can be used to stabilise otherwise unstable system so that, with $\beta = 10^{-2}$ and 50% error, 80% of the systems were stable. From the results obtained in these simulations it is possible to choose a value of β for which, say, 90% of systems will be stable for any given error.

Effort weighting will not only improve the stability of the system but also limit the final steady-state solution attained. The distributions of steady-state reduction in the sum of the squares of the pressures for all the *stable* systems from a sample of 3000 simulations with 20% and 100% average error added to the system model are shown in figures 7 and 8 respectively. In each case β was chosen so as to make 90% of the systems stable.

With 20% average error the spread of reductions is relatively small with a well defined peak of 29 dB, this compares well to the optimum reduction with no error of 33.3 dB. However, with 100% average error the most likely reduction is decreased to 19 dB and the distribution of results is much broader. This change in most likely steady-state level with increasing percentage error is shown in figure 9.

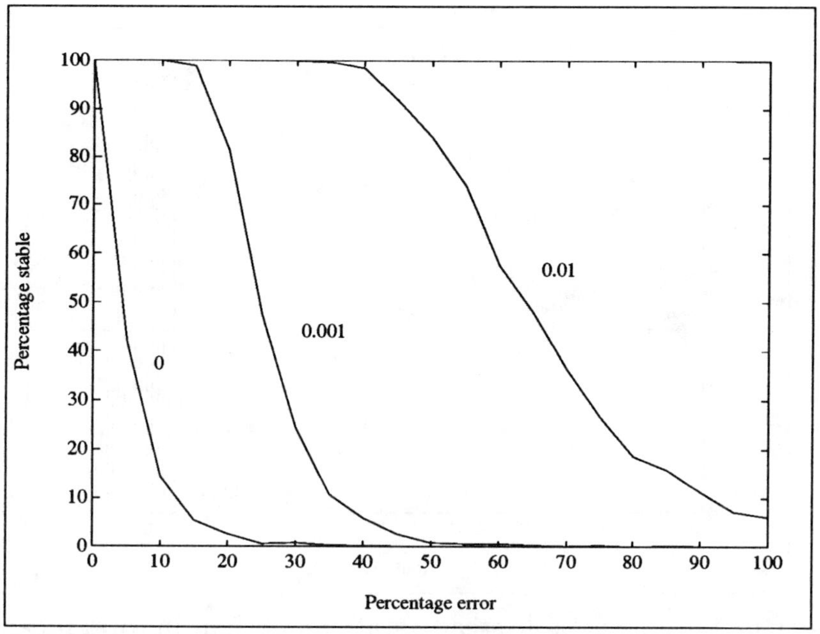

Figure 6: The variation in the percentage of stable systems with the amount of error added to the system model for various values of effort weighting (β).

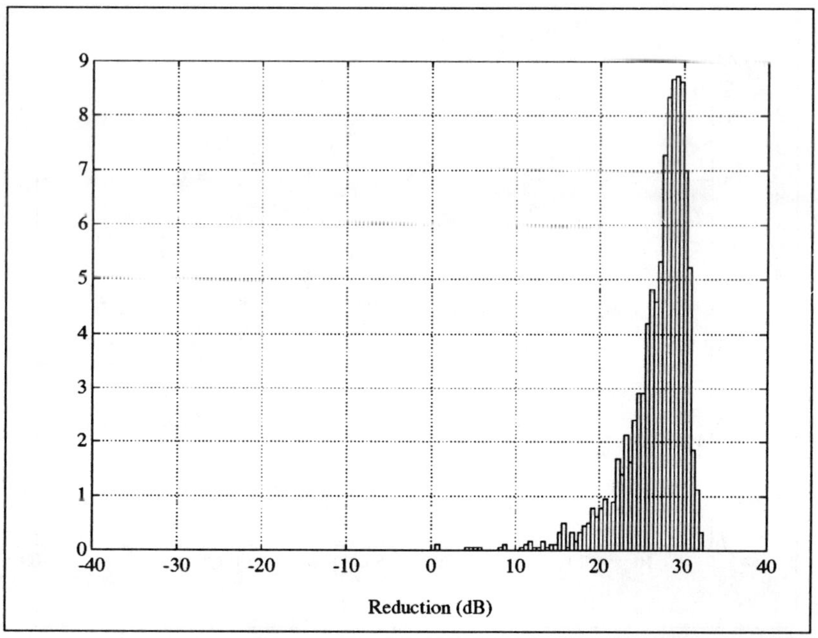

Figure 7: The distribution of stable steady state reductions in total mean square pressure for systems with 20% average plant error and sufficient effort weighting to make 90% of the adaptive control systems stable.

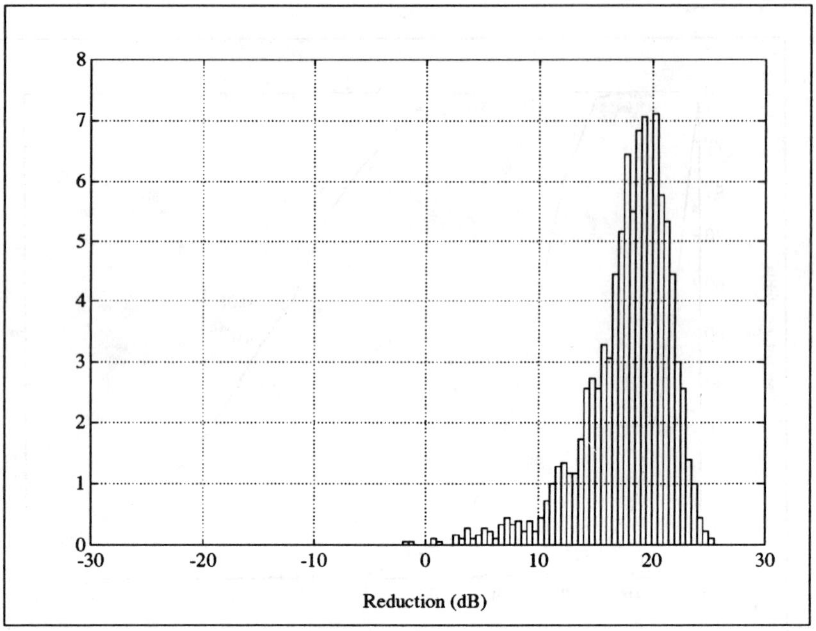

Figure 8: The distribution of stable steady state reductions in total mean square pressure for systems with 100% average plant error and sufficient effort weighting to make 90% of the adaptive control systems stable.

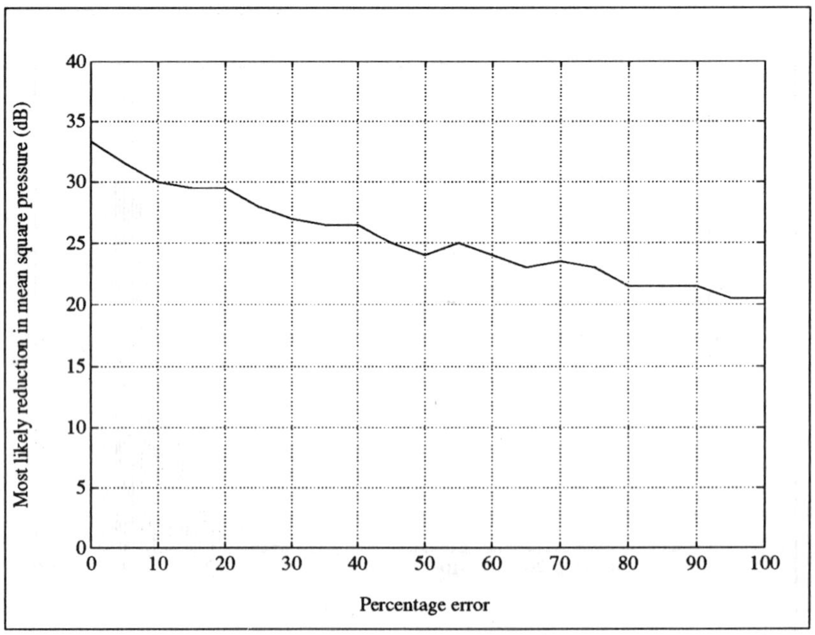

Figure 9: The variation in most probable steady state solution with percentage error within the plant model when sufficient effort weighting is included to make 90% of the systems stable.

CONCLUSIONS

The single channel filtered-x LMS algorithm will converge, in the limit of slow adaption, provided that the phase error in system model is less than 90°. In combination with a delay in the error path it has been shown that, for phase errors in the system model of less than approximately 45°, the algorithm is relatively unaffected.

Theory has been presented which shows how modest amounts of effort weighting can be used to stabilise the multichannel filtered-x LMS algorithm. An analytic bound for this value of effort weighting has been derived in terms of the eigenvalue of the $\hat{\boldsymbol{Z}}^H \boldsymbol{Z}$ matrix with the largest negative real part. Although in practice the true response, $\boldsymbol{Z}$, will not be known, experimental results using data obtained from a 32 microphone, 16 loudspeaker control system have shown that the required value of effort weighting is of the order of 10^{-3} to 10^{-2} depending on the amount of error present. The effect of this effort weighting on the most likely steady-state reduction in mean square pressure has also be presented, and the overall result is a suprisingly modest reduction in attenuation even with very large levels of error.

ACKNOWLEDGEMENTS

This work is partly the result of a programme of work on active control funded by the UK Department of Trade and Industry which supports one of the authors (C. C. Boucher).

References

[1] D. R. Morgan. An analysis of multiple correlation cancellation loops with a filter in the auxiliary path. *IEEE Transactions on Acoustics, Speech, and Signal Processing*, ASSP-28(4):454–467, 1980.

[2] B. Widrow and S. D. Stearns. *Adaptive Signal Processing.* Prentice-Hall, 1985.

[3] S. J. Elliott, I. M. Stothers, and P. A. Nelson. A multiple error LMS algorithm and its application to the active control of sound and vibration. *IEEE Transactions on Acoustics, Speech, and Signal Processing*, ASSP-35(10):1423–1434, 1987.

[4] S. J. Elliott, P. A. Nelson, I. M. Stothers, and C. C. Boucher. In-flight experiments on the active control of propeller induced cabin noise. *Journal of Sound and Vibration*, 140(2):219–238, 1990.

[5] S. J. Elliott, C. C. Boucher, and P. A. Nelson. The behaviour of a multiple channel active control system. Paper Submitted to IEEE Transactions on Acoustics, Speech, and Signal Processing, 1991.

[6] S. Haykin. *Adaptive Filter Theory.* Prentice-Hall, 1986.

[7] M. R. Schroeder. Statistical parameters of the frequency response curves of large rooms. *Journal of the Audio Engineering Society*, 35(5):299–305, 1987.

THE USE OF THE SILENTSEAT® IN AIRCRAFT CABINS

G. P. Eatwell[1]

1. Introduction

The increasing cost of aviation fuel combined with the congestion of major airports has led to a demand for propeller driven aircraft which are larger and have a longer range. This has brought them into competition with jet aircraft. One of the areas of competition is comfort. Propeller aircraft are currently at a disadvantage in this area because of high cabin noise levels.

There has been an increasing amount of interest in the use of active noise control over the last few years. This is because the low frequency propeller noise is difficult to control passively without adding significant weight to the aircraft and thereby reducing its efficiency. Most of the effort has been in the area of 'global control'. This uses loudspeakers built into the trim of the aircraft to produce the cancelling noise which is adjusted in response to microphones on the seatbacks. Whilst this approach can be very effective [1],[2], it is not ideally suited to retrofitting and needs to be specifically designed for each type of aircraft.

An alternative approach, which has received much less attention, is the use of the 'SilentSeat'. This is a modular approach which uses loudspeakers and microphones built into the seatbacks. The main advantages of this approach are that it can be more easily retrofitted and it requires less aircraft-specific design. It is also likely to be cheaper since the controller can be used in other applications and so benefits from volume production. The main disadvantage is that noise reductions are likely to be achieved at seated head height only.

Measurements and calculations for a single SilentSeat have shown that the zone of quiet around the microphones is fairly limited especially at higher frequencies. As a consequence 'global' control has received more attention.

However, when systems are fitted to all the seats in a cabin the

[1]NCT (U.K.) Ltd., St. John's Innovation Centre, Cowley Road, Cambridge, England.

performance is likely to be different because of the influence of adjacent systems on each other.

This paper presents some simple predictions of the performance of a set of SilentSeats and compares them to predictions and measurements for a single seat.

2. The SilentSeat®

The SilentSeat, which has been patented by NCT, is basically a zonal control system. Some of the very earliest publications on active noise control described the use of feedback control with a single microphone to sense the residual noise and a single loudspeaker to generate the cancelling sound [3]. The effect of such a system is to produce a zone of silence around the microphone. The performance of these early systems was limited by the control systems which had a tendency to become unstable.

Modern versions of zonal controllers include adaptive digital feedback control and synchronous controllers. These controllers have overcome the problem of instability and their performance is mainly determined by the acoustics of the system. For example, performance depends upon the positions of the loudspeakers and microphones with respect to each other and with respect to the unwanted noise.

The SilentSeat, with two microphones and two loudspeakers mounted in the seatback, comprises two interacting zonal controllers. Its potential for use in aircraft depends upon the size of the zone of reduction. Some preliminary measurements have been made for a single seat in a propeller driven aircraft [2]. These appear to show that the zone of reduction is too small for the system to be a major benefit.

3. A Simple Model

We begin by considering a single system. The component of the sound field, with wavenumber k, generated by the loudspeakers

$$p(k,\underline{y}) = A_1(k)T(k,\underline{y};\underline{x}_1) + A_2(k)T(k,\underline{y};\underline{x}_2) \tag{1}$$

where A_1 and A_2 are the (complex) amplitudes of the loudspeaker drive signals, $\underline{x}_1$ and $\underline{x}_2$ are the loudspeaker positions and $T(k,\underline{y},\underline{x}_i)$ is the transfer function to the position $\underline{y}$. The optimal drive signals are chosen to give silence at the microphone positions $\underline{y}_1$ and $\underline{y}_2$ so that

$$T_{1,1}A_1 + T_{1,2}A_2 = -p_0(\underline{y}_1) \tag{2}$$

and

$$T_{2,1}A_1 + T_{2,2}A_2 = -p_0(\underline{y}_2) \ , \tag{3}$$

where $T_{i,j} = T(k,\underline{y}_i;\underline{x}_j)$ and p_0 is the unwanted sound.
These equations can easily be solved for A_1 and A_2 to give

$$A_1 = [\ -p_0(\underline{y}_1)T_{2,2} + p_0(\underline{y}_2)T_{1,2}]/\Delta \qquad (4)$$

$$A_2 = [\ -p_0(\underline{y}_2)T_{1,1} + p_0(\underline{y}_1)T_{2,1}]/\Delta \qquad (5)$$

where

$$\Delta = T_{1,1}T_{2,2} - T_{1,2}T_{2,1}. \qquad (6)$$

The residual sound is $p_0 + p$ which depends upon the spacial distribution of the original sound field, the transfer functions from the loudspeaker and the relative positions of the microphones and loudspeakers. These can all be determined from measurements or from computation for a particular application. Examples will be given later for the simplest case where the unwanted sound is a plane wave and the effects of reflections from the trim are neglected.

The performance of a number of interacting seats can be evaluated in the same way by solving the appropriate matrix equation for the loudspeaker drive signals.

In aircraft the seats are arranged on a regular grid. It is expected that the performance of the system will improve as the number of seats is increased. However, since the biggest effect on sound reduction close to one seat will be from loudspeakers on adjacent seats, there will be a limit to the amount of reduction that can be acheived. In the remainder of this section we will look at this limit by considering a periodic array of SilentSeats.

The loudspeakers are assumed to be at positions

$$\underline{x}_{ij} = (iX,\ jY,\ 0) \qquad (7)$$

and the microphones at positions

$$\underline{y}_{kl} = (kX + X_0,\ lY + Y_0,\ 0). \qquad (8)$$

The geometry is shown in figure 1.

The pressure produced by the loudspeakers is

$$p(\underline{y}) = \sum_{ij} A_{ij}T(\underline{y};\ \underline{x}_{ij}). \qquad (9)$$

To get silence at the microphones we need to solve the set of equations

$$p_0(\underline{y}_{kl}) + p(\underline{y}_{kl}) = 0. \qquad (10)$$

The transfer functions are assumed to be independent of translation so that

$$T(\ \underline{y} + (nX,mY,0);\ \underline{x}_{ij}) = T(\ \underline{y};\ \underline{x}_{(i-n)(j-m)}). \qquad (11)$$

This translational symmetry means that the equations can be solved using a modified discrete Fourier transform. The transform and its inverse are defined by

$$G(x,\alpha) = \sum_n \exp(i\alpha nX)F(nX+x) \qquad (12)$$

$$F(x) = \frac{1}{2\pi}\int_{-\pi}^{\pi} G(x,\alpha)d(\alpha X). \qquad (13)$$

This transform has been used in [4] for example. The two dimensional transform can be defined similarly.

The transform of the sound produced by the loudspeakers is

$$P(\underline{y};\alpha,\beta) = \sum_{kl} A_{kl} \sum_{nm} T(\underline{y}+(nX,mY,0);\underline{x}_{kl})\exp(i\alpha nX + i\beta mY)$$

$$= \sum_{kl} A_{kl}\exp(i\alpha kX + i\beta lY)\, S(\underline{y};\alpha,\beta), \tag{14}$$

where

$$S(\underline{y};-\alpha,-\beta) = \sum_{nm} T(\underline{y},\underline{x}_{nm})\exp(-i\alpha nX - i\beta mY). \tag{15}$$

For cancellation at the microphone positions we require

$$P_0(\underline{y}_{pq};\alpha,\beta) + P(\underline{y}_{pq};\alpha,\beta) = 0 \tag{16}$$

This is actually a single equation since the transform has the property

$$G(\underline{y}_{pq};\alpha,\beta) = G(\underline{y}_{00};\alpha,\beta)\exp(-i\alpha pX - i\beta qY). \tag{17}$$

This gives

$$\sum_{rs} A_{rs}\exp(i\alpha rX + i\beta sY) = - P_0(\underline{y}_{00};\alpha,\beta) \,/\, S(\underline{y}_{00};-\alpha,-\beta)$$

or

$$A_{rs} = -\frac{1}{4\pi^2}\int_{-\pi}^{\pi}\int_{-\pi}^{\pi} P_0(\underline{y}_{00};\alpha,\beta) \,/\, S(\underline{y}_{00};-\alpha,-\beta)\, \exp(-i\alpha rX - i\beta sY)\, d(\alpha X)d(\beta Y). \tag{18}$$

This is the most general result. For the special case when the unwanted sound is the plane wave with wave vector $\underline{k}$

$$p_0(\underline{y}) = \exp(i\underline{k}.\underline{y}) \tag{19}$$

the expression (3.18) can be simplified since

$$P_0(\underline{y};,\alpha,\beta) = \exp(i\underline{k}.\underline{y})\sum_{nm} \exp[i\underline{k}.(nX,mY,0) + i\alpha nX + i\beta mY]$$

$$= 4\pi^2\exp(i\underline{k}.\underline{y})[\ \sum_n\delta(k_xX+\alpha X-2n\pi).\sum_m\delta(k_yY+\beta Y-2m\pi)\ ], \tag{20}$$

where δ is the unit delta function. Inserting this into (3.17) gives

$$A_{rs} = -\exp(i\underline{k}.\underline{y}_{rs})/S(\underline{y}_{00};k_x,k_y). \tag{21}$$

The residual sound is

$$e(\underline{y}) = \exp(i\underline{k}.\underline{y}) - \sum_{rs} T(\underline{y};\ \underline{x}_{rs})\exp(i\underline{k}.\underline{y}_{rs})/S(\underline{y}_{00};k_x,k_y)$$

$$= \exp(i\underline{k}.\underline{y})\left[\ 1 - \exp(i\underline{k}.(\underline{y}_{00}-\underline{y}))S(\underline{y};k_x,k_y)/S(\underline{y}_{00};k_x,k_y)\ \right]. \tag{22}$$

4. Results

The performance of a single SilentSeat is compared to that of an array of seats for the case when the unwanted sound is a plane wave, given by (3.19), and the loudspeakers are simple point sources. The transfer functions are then given by

$$T(\underline{y},\underline{x}) = \exp(ikR)/4\pi R, \quad R = |\underline{y} - \underline{x}|. \tag{23}$$

The worst performance is likely to be when the plane wave is travelling in the direction from the microphone to the loudspeaker. In this case the phase of the original sound is increasing with distance while that of the cancelling sound is decreasing, thus producing the largest phase errors. In addition, this is likely to be the predominant direction of wave propagation in an aircraft since the strongest excitation of the fuselage is in the plane of the propellers at the front of the cabin. All of the results presented will be for this case.

The graphs show the reduction in the sound as a function of the distance x from the loudspeaker in the direction $\underline{y}=(x,0,0)$. The geometry is shown in figure 2. The results are for a fundamental frequency of 88Hz and its is first two overtones. This corresponds to the blade passing frequency for a BAe748 aircraft in cruise and was chosen to allow comparison with the experimental results for a single seat [2].

Figure 3 shows the residual sound for a single seat (in dB relative to the original level) as a function of the distance from the seatback (or loudspeaker). The loudspeaker separation, Y, is 20cms. The reference microphone position is $\underline{y}_{00}$= (7cms,0,0) and the plots show that perfect cancellation is achieved at these positions. Between microphone and the loudspeaker the residual increases very rapidly, but the model breaks down very close to the loudspeaker because of the idealised form of the source. Away from the microphone position the performance also decreases rapidly, dropping to below 6dB at about 17cms from the seatback. The performance drops off even more rapidly at the higher frequencies.

Also shown in figure 3 are the corresponding experimental results obtained in [2]. The agreement is good considering the simplicity of the model.

The corresponding results for an array of seats is shown in figure 4. The seat spacing is 75cms. At the fundamental blade passing frequency of 88Hz the reduction is better then 10dB across the whole region. The reduction at the larger distances is strongly influenced by the loudspeaker on the seat in front. This performance is as good as that obtained by a global control system.

The performance at the higher frequencies is not as good, but reductions of at least 5dB are maintained across the seated region.

At 264Hz the seat spacing is a significant fraction of the wavelength and a slightly more complicated pattern of reduction is beginning to emerge.

5. Conclusions

The results from the simple model described here indicate that performance of an array of SilentSeats is not well predicted by the performance of a single seat in isolation. This is largely due to the interaction between adjacent seats. The amount of interaction depends on the acoustics of the cabin and the positioning of the transducers, so the results for an actual array of seats may not be well predicted by the results presented here especially if the cabin is lightly damped. However, the indications are that performance of an array of SilentSeats will be significantly better than that of a single SilentSeat.

There are of course other factors which will affect the performance, one of these is the controller configuration. The results here assume that a multichannel control system is used, taking into account all of the interactions between the seats. In practice it would be simpler if the individual seats or small groups of seats could be considered independently. This will affect not only the optimal performance but also the stability of any adaptive algorithms for attaining that performance. This aspect of the problem is currently under investigation.

6. References

1. Dorling, C.M., Eatwell, G.P., Hutchins, S.M., Ross, C.F., Sutcliffe, S.G.C. 1989. "A Demonstration of Active Noise Reduction in an Aicraft Cabin."Journal of Sound and Vibration, **128**(2), pp. 358-360.

2. Elliot, S.J. et al. 1990. "In-flight Experiments on the Active Control of Propeller Induced Cabin Noise." Journal of Sound and Vibration, **139**(4).

3. Lueg, P. 1936. "Process of Silencing Oscillations." US patent US 2,043,416.

4. Eatwell, G.P. and Willis, J.R. 1982. "The Excitation of a fluid loaded Plate Stiffened by a semi-infinite Array of Beams." I.M.A. Journal of Applied Mathematics, **29**, pp. 247-270.

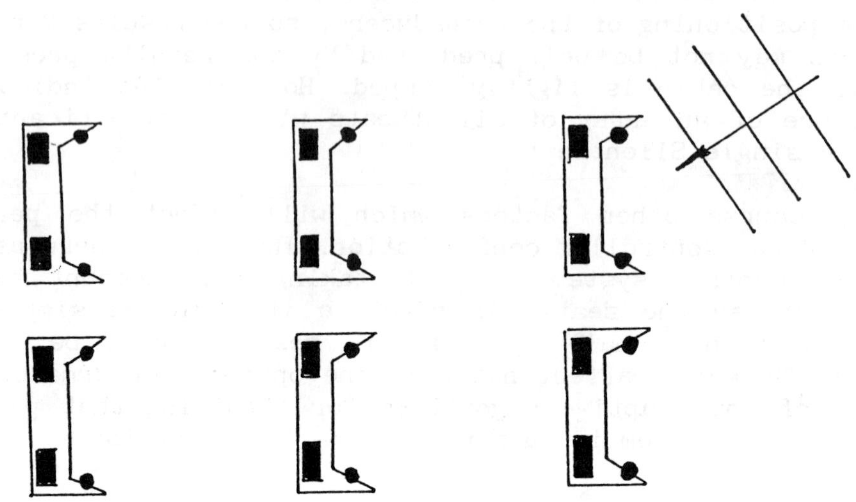

Figure 1. Array of SilentSeats.

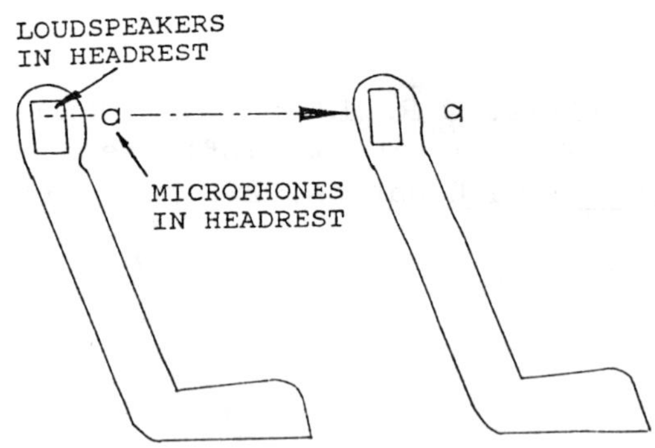

Figure 2. Geometry for results.

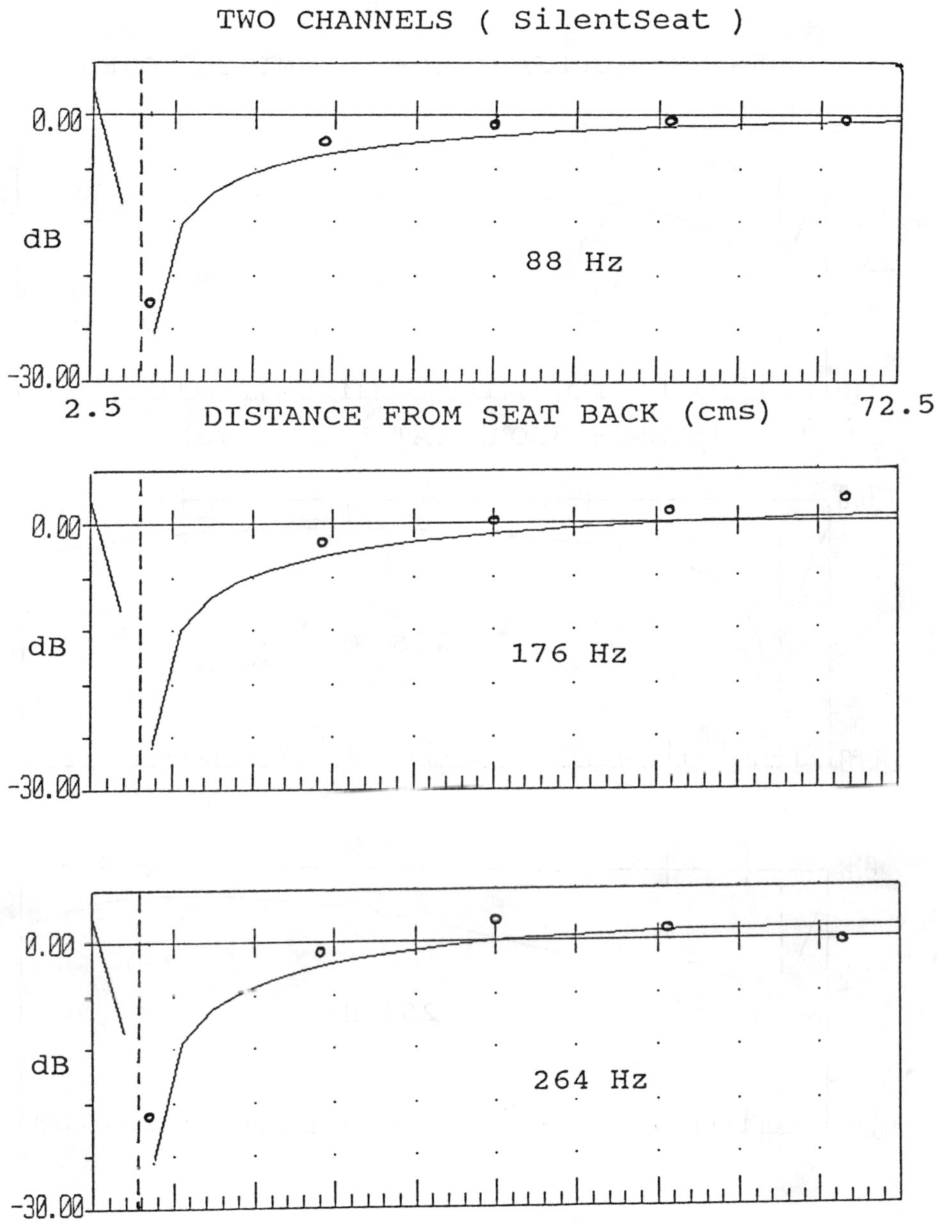

Figure 3. Residual Sound for a single SilentSeat.

(o = experimental results)

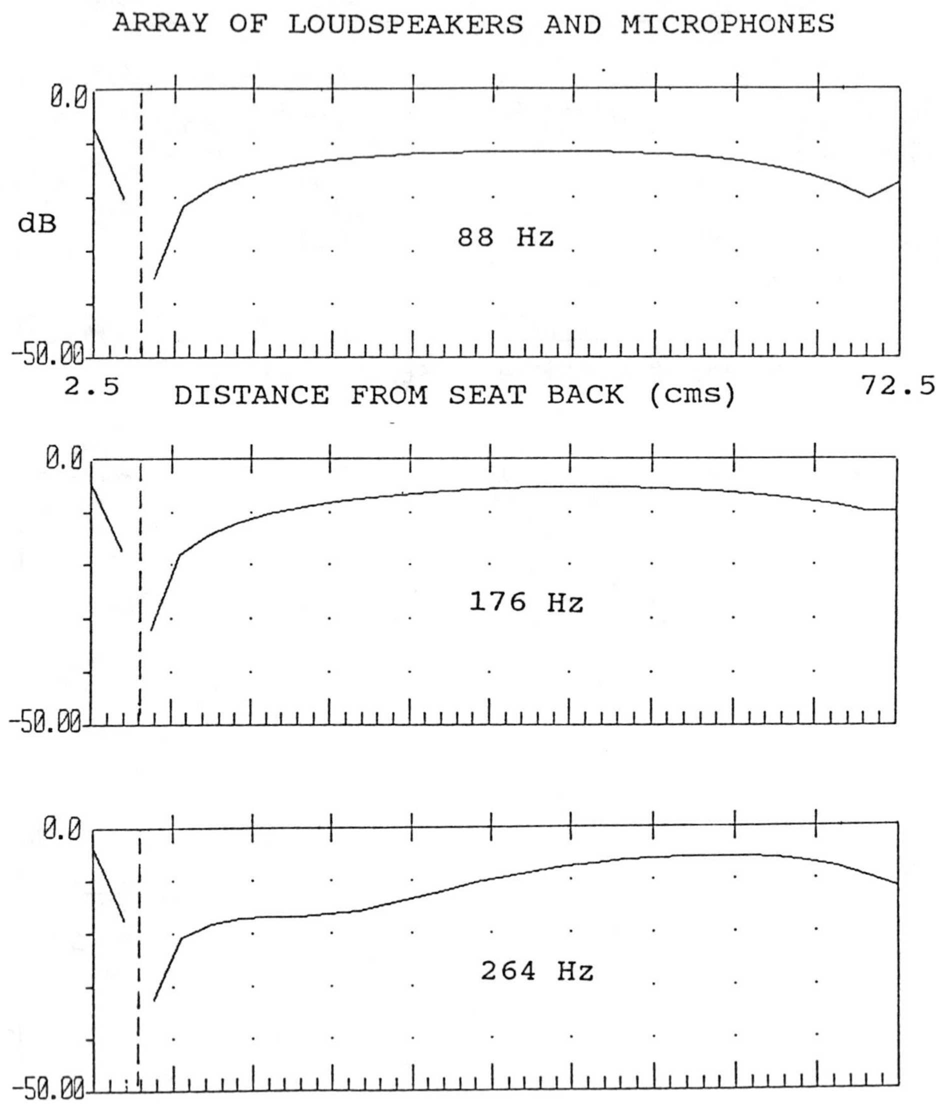

Figure 4. Residual Sound for an array of SilentSeats.

Control Algorithms I

DISTURBANCE REJECTION IN DISTRIBUTED SYSTEMS USING RULE-BASED ADAPTIVE CONTROLLERS

L. G. Lebow,[1] G. L. Blankenship,[1] W. H. Bennett, Jr.[1]

ABSTRACT: In this paper we describe an algorithm for rejection of distributed disturbances using an array of interacting sensors and actuators. The model framework is a multiple input, multiple output (MIMO) control problem with unknown plant models. Each sensor/actuator pair is a loop in the total system or plant. The algorithm uses a separate controller on each loop for local signal rejection. Each controller is supported by a rule–based, real-time expert system which provides a local adaptive control capability to permit adaptive response to locally varying disturbances. Interactions of the individual expert systems are coordinated through a dynamic, "blackboard" database.

MODEL PROBLEM

Consider the scenario sketched in Figure 1 – a material surface is illuminated by a disturbance signal $u(t)$, which may be noise or a pulse stream. The surface response at any point x is $r(t,x)$. The control objective is to generate a distributed response signal $z(t,\cdot)$, based on measurements of $r(t,\cdot)$ which will "cancel" the effects of the disturbance $u(t)$, which is not itself measured directly. If the signal rejection algorithm works perfectly, then $r(t)$ will be reduced to zero, and will remain zero for all inputs.

As suggested in the figure, suppose the response is generated by a collection of active elements distributed across the surface. Suppose each element is capable of reacting to the disturbance signal, generating an appropriate response signal based on local and global information. Suppose each element is capable of communicating with its neighbors. In Figure 2 we show the basic concept at a single element. The disturbance signal $u(t)$ interacts with the local response features of the surface. We suppose that we have only a limited knowledge of those features. For example, we may have information on the basic interaction mechanism, but we may not know the detailed response defined by the aspect angle Θ, the surface–source distance, the relative velocity ν, and other geometric features Ω. Suppose that we are able to detect the local response $r(t)$, perhaps by measuring surface charge distribution or vibration.

The active element's response is computed by the blocks labelled "Controller" and "Adaptive/Learning Algorithm." The controller generates the signal $z(t)$ which interacts with the disturbance signal to produce the net signal $u(t) - z(t)$ acting on the surface. We assume that

[1]Techno-Sciences, Inc., 7833 Walker Drive, Suite 620, Greenbelt, MD 20770, U.S.A.

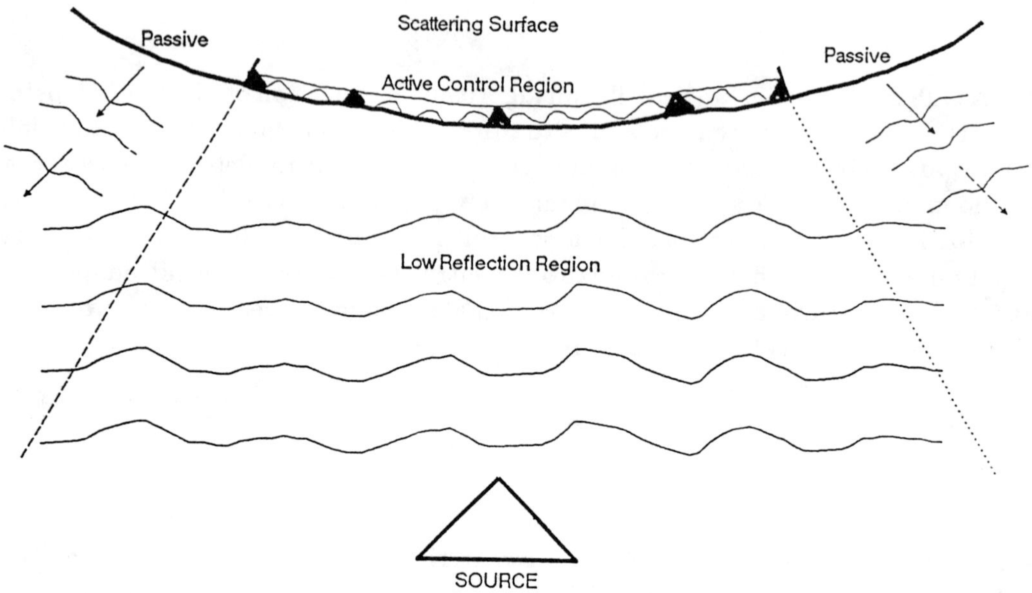

Figure 1: Directional absorption of disturbance fields.

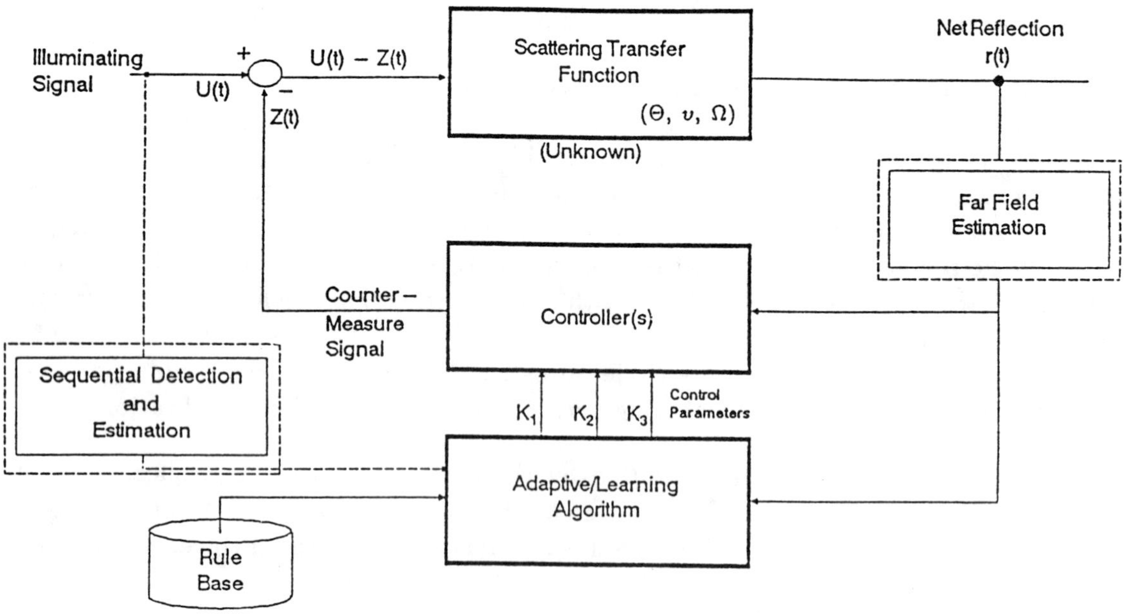

Figure 2: Rule–based disturbance rejection.

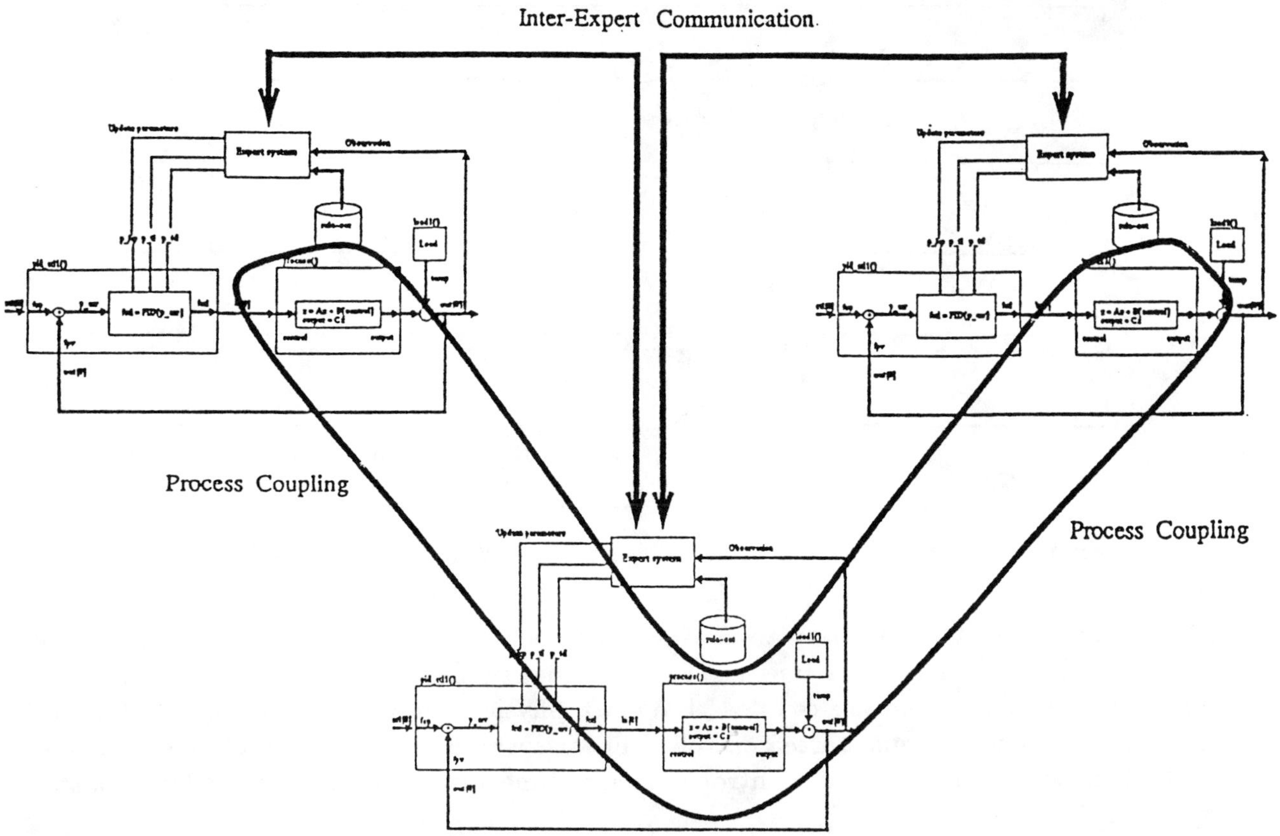

Figure 3: Multi–loop system with coupling and a distributed adaptive algorithm.

the controller has a finite set of parameters $\{k_i, i = 1, \ldots, m\}$ which can be adjusted to adapt to changing conditions. The role of the adaptive/learning function is to make the adjustments based on measurements of the local response $r(t)$ and the processed version of the disturbance signal $u(t)$.

In designing the algorithm presented here, we assume that the surface–disturbance interaction physics are not modelled explicitly, so the transfer function in Figure 2 is unknown, that the disturbance signal $u(t)$ is not measured directly, and that only the local response $r(t)$ can be measured.

Classically, one would use adaptive control techniques to provide the "learning" algorithm; in effect, to determine values for a parametric model of the transfer function and input signal. These algorithms can be computationally demanding, they are slow to converge, and they are strongly dependent on the structure assumed for the parametric model. Rule–based, adaptive algorithms offer certain advantages if the number of "rules" is small. They can be implemented very simply, and the associated computational burden (in real time) is modest. In many instances rule–based algorithms can be implemented using parallel architectures, gaining additional performance. Our studies thus far show that the rule–based algorithms and control architecture shown in Figures 2,3 are capable of recognizing a wide range of signals quickly.

We are interested in designing each of the controllers to respond to reject disturbances measured at the local coordinates. This is comparatively straightforward if the physical coupling effects are small. However, since the disturbances we are attempting to reject are fields dis-

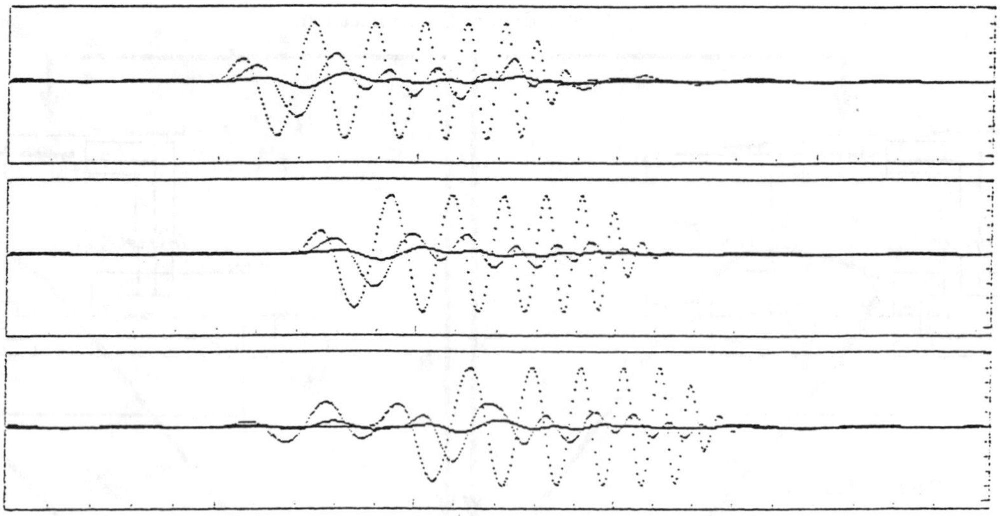

Figure 4: Three element coupled system responding to a chirp pulse signal.

tributed in space, it is necessary to consider the coupling effects, and to "coordinate" the actions of the various active elements. For example, suppose the active elements in a disturbance rejection system are coupled through a physical interaction process. In Figure 3 we suggest this by showing three controlled linear systems coupled through the dynamics. Each of the local elements under control is a linear time–invariant system modeled by a state equation

$$\dot{x}_i = A_i x_i + B_i u_i, \qquad y_i = C_i x_i, \qquad i = 1, 2, \ldots, N \tag{1}$$

where N is the number of elements, and the coefficient matrices (A_i, B_i, C_i) may be different for each elementary system. The coupling is through "cross–terms" among the elements. For example, the (sub)state x_i might interact with (sub)states x_j, x_k, etc. through "off–diagonal" coefficients A_{ij}. Each of the active elements is a complete closed loop control system, with feedback controller $fctl = PID[p_{err}]$ as shown in the figure.

In Figure 4 we show a simulation of a hypothetical three element system responding to reject a pulse signal. The responses of the three elements are shown in the three graphs. The incident signal is the large pulse "chirp" signal, which has an increasing frequency across the pulse width. The incident signal corresponds to the signal $u(t)$ in Figure 2. The next largest signal in each graph is the disturbance rejection signal generated by the controller ($z(t)$ in Figure 2). The middle trace in each graph is the total system response ($r(t)$ in Figure 3). *(The adaptive algorithm is not active in this example.)*

Note the coupling between the responses of elements 1 and 3. Element 3 starts to respond to the control signal generated by element 1, before the disturbance arrives at element 3. The response at element 1 continues after the pulse terminates, due to the activity at element 3. Element 2 is uncoupled from the other elements in this example.

In Figure 5 we show the response of element 1 in detail after the adaptive algorithm has been activated. Note the close alignment of the incident pulse $u(t)$ and the control signal $z(t)$, so the difference $u(t) - z(t)$ is small. This difference leads to the small variations in the middle curve in the graph, which is the effective system response $r(t)$ (recall Figure 2).[1]

The algorithm which achieves these results is a rule-based computation for adapting the three control parameters in the feedback element shown in Figure 2. The simulation system has about 20 rules for adjustment of parameters based on a finite number of "features" extracted from the measured signal ($r(t)$ in Figure 2). Most of the rules can be tested independently of the other rules, and may be executed in parallel (or at least in small groups). Each of the rules can be executed by simple arithmetic operations involving a small number of multiplications and additions; hence, each rule can be executed very quickly.

RULE–BASED ADAPTIVE CONTROL

In the current configuration each control signal $z_k[n]$ is provided by a PID (proportional, integral, derivative) algorithm. In a conventional PID controller, the output is compared to a setpoint and the error signal $e[k]$ is computed and fed back to a compensator that combines it with its integral and derivative to produce a control signal $z[k]$

$$z[k] = k_p \left[e[k] + \frac{\delta t}{t_i} \sum_{i=1}^{n} e[i] + \frac{t_d}{\delta t} (e[k] - e[k-1]) \right] \tag{2}$$

The key parameters in this control law are the coefficients k_p, t_i, t_d. These three coefficients can be manipulated to produce various response curves from a given plant. The whole con-

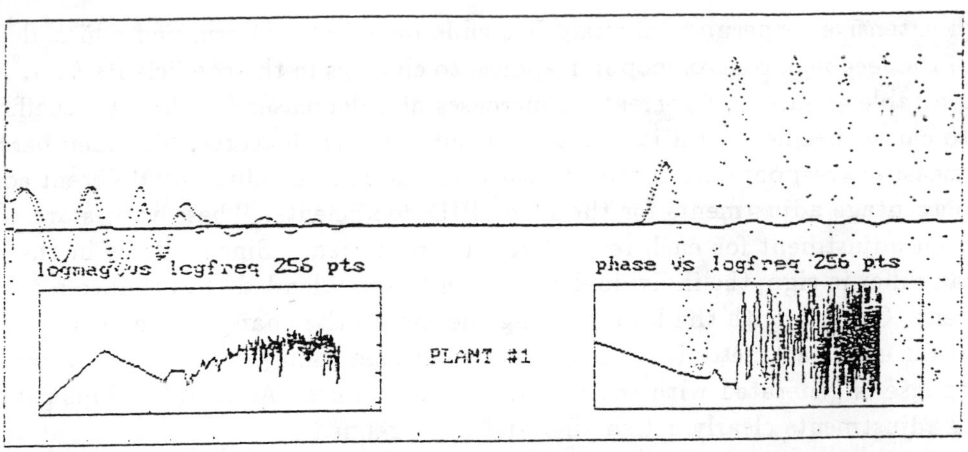

Figure 5: Sensor 1 under active control responding to a chirp pulse signal.

[1]The small graphs shown in the boxes in Figure 5 are the FFT's of the signal $r(t)$. They are not used in the current version of the adaptive algorithm.

cept for adaptive PID control revolves around interactively changing these three values. In equation (2), δt is the sample rate for the digital simulation.

The tuning method used is based on extracting a small number of features from the measured signal, and using a simple *pattern recognition* algorithm to compare the measured features with prespecified desired values. If the measured and desired feature values do not match well, then rules are used to modify the controller parameters, individually in each active loop. When the comparison is *"acceptable,"* the control parameters are fixed.

The determination of whether a given response is "acceptable" is based on the use of compatibility or membership functions derived from fuzzy set theory. The determination of how close a given feature, such as the measured response overshoot, compares with a specification for optimal is found by applying both values in an equation that defines a membership function. A typical membership function is a bell curve where the optimal value is at the peak. Values above and below optimal will fall on this curve somewhere to one side or the other of the peak. In this way a "membership" or "quality" rating is assigned to each measurement.

The algorithm requires specification of the desired output response in terms of six criteria: settle time, percent overshoot, rise time, ratio of a second peak overshoot to the first, average DC offset, and the number of peaks. In addition to the optimal values for each criterion, on must specify an allowable bandwidth, a weighting factor, and whether values on one side or the other of optimal are acceptable (membership of unity). A weighting factor can be set to zero if a criterion is unimportant.

A "score" is defined for a given response as the sum of all the criteria's membership values. Hence, achieving the maximum score of six means that all the criteria fall within the specification. The numerical membership ratings are used to adjust those control parameters which can improve the performance. This algorithm, called the Membership Based Tuning Algorithm (MBTA) is the heart of the adaptive process.

Through extensive, experimental analysis trends have been determined which define the behavior of the feedback control loop in response to changes in the coefficients k_p, t_i and t_d. The result is a table of values of percentage increases and decreases for the PID coefficients k_p, t_i and t_d to cause specified variations in the output response features. The membership ratings from a measured response are combined with this table to produce six different contributions to the percentage adjustments for the three PID coefficients. These values are combined to produce an adjustment for each respective PID coefficient. Since entries in the adjustment tables can differ in sign, the individual adjustments are "blended" to produce a "compromise" adjustment. Criteria with the lowest ratings dominate the changes. The final composite adjustment for each parameter is applied as a percentage change from the previous value. The next response is generated with the new parameter values. As memberships get higher, the resulting adjustments clearly get smaller and more refined.

The adjustments in the MBTA algorithm, based membership ratings, are usually small. In some cases a major adjustment in the parameters must be made to properly initialize the algorithm. Also, instabilities may arise. For these cases groups additional rules are used. The rules are of the form

If <condition> then <action>

where the *condition* is a conclusion about some aspect of the current situation (the output signal is in a certain range, is becoming unstable, etc.) and the *action* is a macroscopic response (stabilize system, stop adaptive process, etc.) or a microscopic response (adjust parameter values). The MBTA algorithm is activated by the firing of one of two rules. The current rules fall into four groups:

Large changes: unstable response, process poorly tuned and not changing, or process getting unstable, etc.

Standard tuning: apply standard MBTA - plant type A or B

Stop tuning: a constant low score and tuning is destabilizing, or plant is well tuned – enter monitor mode.

Resume tuning: plant was tuned but decrease in score > 10% - tuning resumed

Each group of rules can be executed in parallel, so long as there is no conflict in the rules which are ultimately fired. During our simulations, the rules to apply standard MBTA tuning fire the vast majority of the time.

SIGNAL MODELS AND SIMULATION

The models for the plant and the disturbance input are defined by discrete state space realizations based on a common sample rate. Thus, the data (to start the simulation) consists of a system matrix (A), an input matrix (B), an output matrix (C) and a feed forward matrix (D). A plant model response and the disturbance models are simulated by code that numerically implements a state space realization as given just below.

$$\begin{aligned} x[k+1] &= A\,x[k] + B\,(\,d[k] - z[k]\,) \\ y[k] &= C\,x[k] + D\,d[k] \end{aligned} \tag{3}$$

Each of these models has additional parameters to model delays on the inputs and outputs of a given plant.

SIMULATION SYSTEM AND RESULTS

A simulation system has been developed in C which runs on an IBM compatible personal computer. The simulation requires input files to specify the plant and disturbance models, and the criteria for tuning. The simulation produces graphical displays of all signals of interest at each of the active elements. It also produces a tuning history for each active element and a "blackboard" or composite tuning history for all active elements.

SIMULATION SCREENS

Single Element Screen

Refering to Figure 5, the main portion of the screen is an oscilloscope like display. Several traces can displayed in various combinations. One trace represents the plant's output *r[n]* and this can be displayed alone. Other traces available for display are the controller signal *u[n]* and the disturbance input *d[n]*. Tuning status information is displayed as it becomes available. The current settings for the PID coefficients k_p, t_i and t_d are given along with the last adaptive tuning rule to "fire"' and the score from the last pass. Two smaller windows are also included on this display that show the frequency response (magnitude and phase) of the plant output normalized in each case by the plant input. This information is not currently used in tuning but may be in future implementations. This screen also includes a command menu to manipulate the display and the simulation in various ways.

Multiple Plant Screen

Refering to Figure 4, the main portion of the screen is a group of oscilloscope like displays, one for each element in the overall plant model. These windows each are analogous in operation and functionality to the one discussed above for the single plant interface. The traces can be displayed in various combinations.

Tuning and Blackboard Status Screen

Refering to Figure 6, the main portion of this display provides an overview the results of auto tuning. The information found in this display consists of the current controller settings, the number of passes executed, the most recent and best score, and the "trend." The trend is a number that indicates the number of passes that a score has been increasing or deceasing. We also compute the norm of the error signal $(u - z)$ displayed as T_ERR and the norem of the response (r) displayed as O_ERR. (Note the substantial changes in these values after several pulses have been processed.) The active elements operate asynchronously, and data entries are posted when an element completes a tuning pass.

ALGORITHM PERFORMANCE

The algorithm performs well on response transfer functions that have dynamics that are predominantly second order. This is sufficient to describe both EM and acoustic scattering from surfaces. We have simulated a variety of disturbance signals and interaction geometries. In Figures 7-12 we show a particularly challenging case corresponding to two independent disturbance signals incident on a system with three active elements. Each disturbance signal consists of a chain of three frequency/amplitude agile chirp pulse signals. The pattern of three pulses in each disturbance is repeated periodically. The disturbance signals are delayed so that one disturbance arrives first at active element 1, then at 2, then at 3. The arrival delays are reversed for the second disturbance.

In Figure 7 we show the signals at each of the elements prior to the adaptation process. Two distinct disturbance signals are present. In Figure 8 we show the total disturbance incident at each element which is the sum of the two. Clearly, each element sees a very different signal. Each of the three pulse patterns seen by the elements is different.

In Figure 9,10 we show the response of the system after the adaptation process has been running. In Figure 9 the system has seen just one pulse (the associated blackboard entry is in the top half of Figure 6). In Figure 10 the system has seen four pulses. Thus, it has seen the complete three chain sequence once. Note the improvement in the alignment between the disturbance and response signals.

In Figure 11 we show the response of the system after the adaptation process has seen six pulses – the complete pattern twice (see also the bottom half of Figure 6). In Figure 12 the system has seen eight pulses. Note the alignment between disturbance and the response signal at each element is excellent. Note also that the six and the eight (combined) pulses seen by the system are quite different, since the individual disturbance signals are frequency and amplitude agile. In this simulation the three elements achieved perfect scores of six after eight pulses at elements 1 and 3, and nine at element 2. The scores coninued to change after that due to the signal agility effects; however, excellent responses were obtained for all subsequent pulses.

Acknowledgements:

This work was supported in part by the Office of Naval Research Contract N00014-89-0300. Jack Hanson and Jim Smith are the Program Managers. G.L. Blankenship is also with the Electrical Engineering Department and Systems Research Center, University of Maryland, College Park, MD 20742.

```
            This entry resulted from an AUTO_UPDATE related to PLANT #3

     Mode  Trd          No. PID Parameters                  Score   T_Err   O_Err
     ----  ---          --  --------------------------      -----   -----   -----
#1   ACT   +00   Last: 01   kp= 3.50 ti= 0.80 td= 0.40      0.99    8.1648  0.2289
                 Best: 00   kp= 0.00 ti= 0.00 td= 0.00      0.00    0.0000  0.0000
#2   ACT   +00   Last: 01   kp= 3.50 ti= 0.80 td= 0.40      1.00    9.0219  0.2308
                 Best: 01   kp= 3.50 ti= 0.80 td= 0.40      1.00    9.0219  0.2308
#3   ACT   +00   Last: 01   kp= 3.50 ti= 0.80 td= 0.40      1.99    6.7385  0.1207
                 Best: 00   kp= 0.00 ti= 0.00 td= 0.00      0.00    0.0000  0.0000

                                        ***

            This entry resulted from an AUTO_UPDATE related to PLANT #3

     Mode  Trd          No. PID Parameters                  Score   T_Err   O_Err
     ----  ---          --  --------------------------      -----   -----   -----
#1   ACT   +01   Last: 05   kp= 9.42 ti= 1.07 td= 1.89      5.00    0.2606  0.0005
                 Best: 05   kp= 9.42 ti= 1.07 td= 1.89      5.00    0.2606  0.0005
#2   ACT   -02   Last: 06   kp=10.14 ti= 1.19 td= 1.96      3.38    0.4052  0.0015
                 Best: 04   kp= 7.21 ti= 1.06 td= 1.26      3.77    2.2758  0.0102
#3   ACT   +01   Last: 06   kp= 8.94 ti= 0.99 td= 2.00      4.35    0.4561  0.0010
                 Best: 06   kp= 8.94 ti= 0.99 td= 2.00      4.35    0.4561  0.0010
```

Figure 6: Blackboard display after one pulse and after six pulses

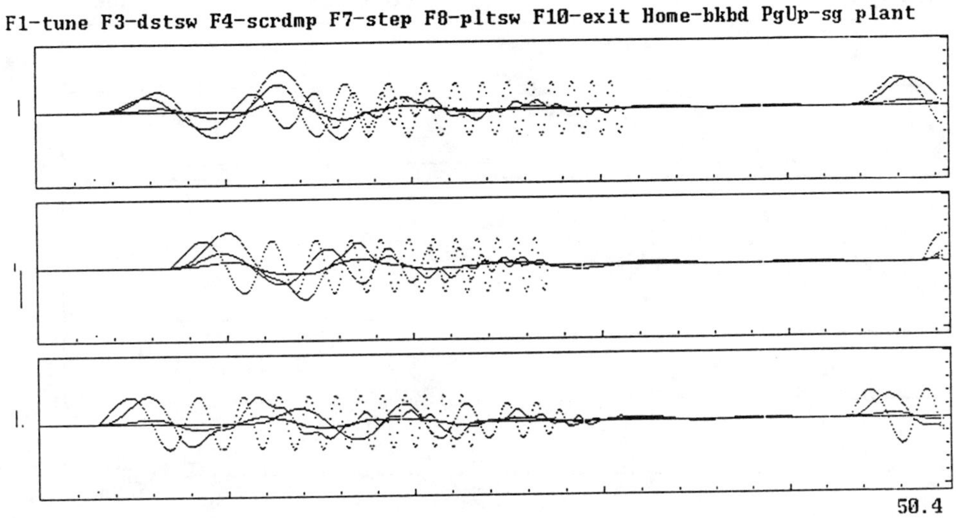

Figure 7: System response to two distinct disturbances prior to adaptation

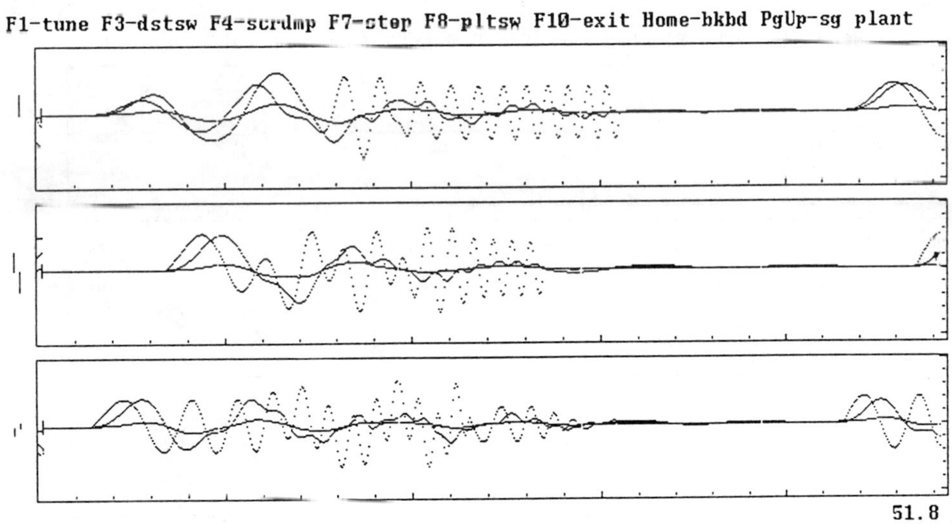

Figure 8: System response to combined disturbances prior to adaptation

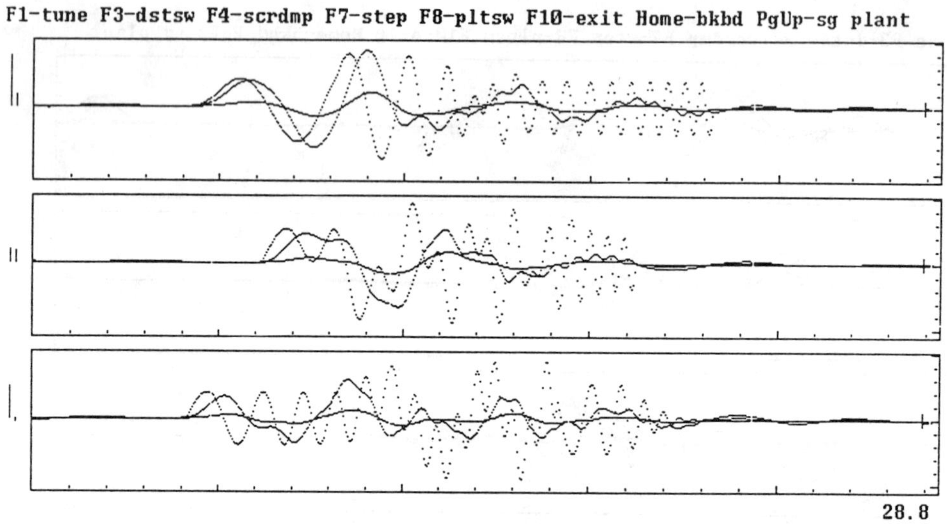

Figure 9: System response to combined disturbances after adaptation – one pulse processed.

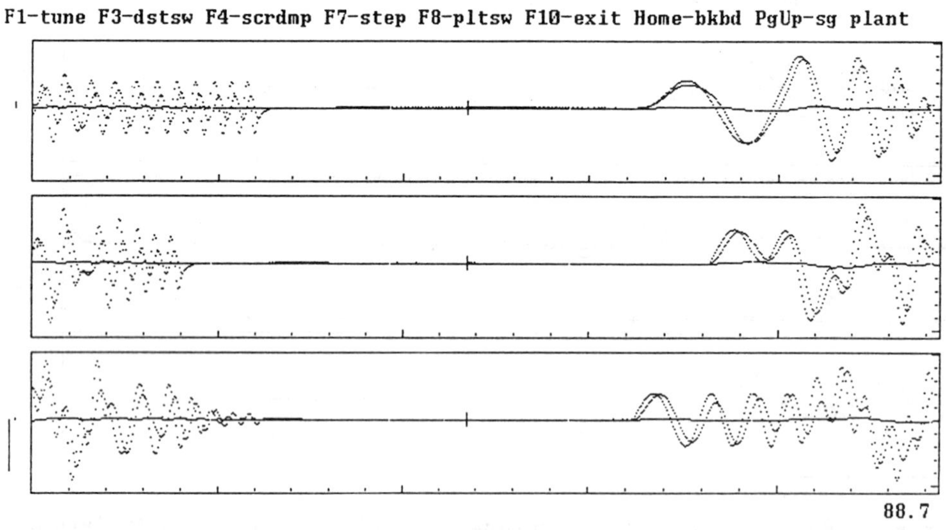

Figure 10: System response to combined disturbances after adaptation – four pulses processed.

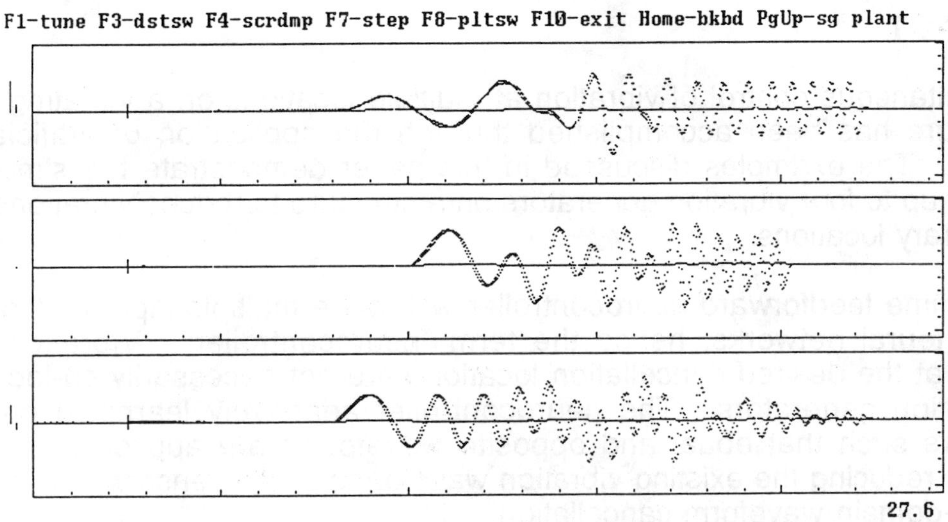

Figure 11: System response to combined disturbances after adaptation – six pulses processed.

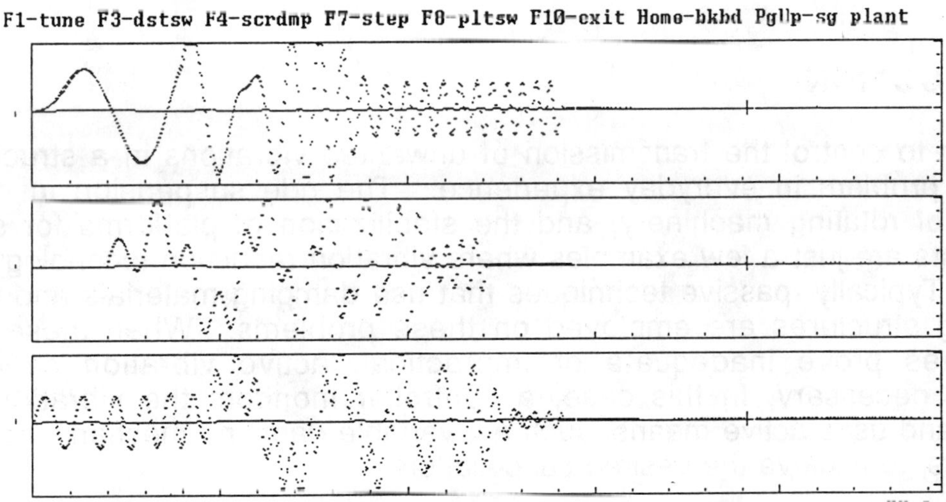

Figure 12: System response to combined disturbances after adaptation – eight pulses processed.

NEUROCONTROLLERS APPLIED TO REAL-TIME VIBRATION CANCELLATION AT MULTIPLE LOCATIONS

Daniel J. Bozich,[1] H. Bruce MacKay[1]

ABSTRACT

The simultaneous control of vibration at multiple locations on a vibrating machine or structure has been accomplished through the application of artificial neural networks. The examples discussed in this paper demonstrate the simultaneous control of up to four vibration generators on a structure to reduce vibrations at up to four arbitrary locations.

The real-time feedforward neurocontroller will utilize multiple-input-multiple-output (MIMO) neural networks, hence the term "neurocontroller". Vibration sensors mounted at the desired cancellation locations are not necessarily co-located with the vibration generators. The neurocontroller adaptively learns to control the generators such that equal and opposite waveforms are applied at the sensor locations, reducing the existing vibration waveforms at the sensors. This results in true time-domain waveform cancellation.

Reductions of up to 20 dB $\pm$ 10 dB have been accomplished within a 200 Hz Bandwidth. Reduction of both discrete-frequency and narrowband-random vibrations have been observed.

INTRODUCTION

The need to control the transmission of unwanted vibrations in a structure is a common problem in everyday experience. The ride suspension in cars, the isolation of rotating machinery, and the stabilization of platforms for sensitive instruments are just a few examples where vibration reduction technology can be applied. Typically, passive techniques that use damping materials and vibration absorbing structures are employed on these problems. When these passive approaches prove inadequate or impractical, active vibration cancellation becomes necessary. In this case, a controller monitors the vibrations to be reduced and uses active means, such as, variable damping mechanisms or force generators, to achieve the desired cancellation.

Design of an analog or digital controller for active vibration cancellation typically requires modeling and analyzing the vibrational modes of the relevant structures.

[1]Science Applications International Corporation, 4161 Campus Point Court, San Diego, CA 92121, U.S.A.

This can be a formidable task when complex structures with high modal densities are involved, or when changes in configuration or loading cannot be predicted. If multiple sensors and actuators must be simultaneously coordinated for adequate performance, the design can exceed the capabilities of conventional control analysis techniques.

One approach to this problem is an adaptive controller that can learn, in situ, the sensor-to-actuator transfer functions necessary for vibration control. Neural networks offer just such learning capabilities, and this paper reports on the actual implementation of a neural network controller for real-time vibration reduction on a complex structure [1],[2]. The neurocontroller techniques developed for these applications were neural network adaptations of classical adaptive control techniques reported in the literature [3].

EXPERIMENTAL APPARATUS

The experimental apparatus , shown in two views in Figure 1, consists of a steel frame table top, approximately 3 feet by 4 feet, of bolted unistrut beam construction, that is isolated from a 2-1/2 foot high base of similar unistrut construction with 4 rubber mounts. This table will accommodate up to five vibration generators, as indicated in Figure 1, one of which serves as the vibration source, and four serve as multiple vibration cancelers. All vibration is in the vertical axis as these particular vibration generators, also known as shakers, are equipped only with vertical suspensions.

This apparatus was especially constructed for research and development of a technique to reduce vibration caused by rotating machinery and other mechanical vibration sources. Figure 1 also contains simple diagram overlays indicating the shakers and the excitation and control instrumentation including response measurement accelerometers, signal generator, cabling, and adaptive neurocontroller system.

The neurocontroller system consists of an IBM-compatible PC 386/AT or 486/AT computer system with the following additions:

- SAIC DELTA II Fast Floating Point Processor (FFPP)
- SAIC DAI 1016 DT Connect Interface Module
- Data Translation Data Acquisition Module, Model DT2841-F16SE, 16 Channel Analog-to-Digital Conversion @ 12 bits/ sample @ 100kHz burst conversion rate
- ICS Data Conversion Module, Model A0B6, 6 channel Digital-to-Analog Conversion @ 100kHz burst conversion rate
- SAIC Real-Time Artificial Neurocontroller System (RTANCS) software, including SAIC DELTA Extended Assembler and OS, and Microsoft Quick C

A signal generator was used to provide the single frequency sinusoidal signals, the sinusoidal sweep signals, and the narrowband random signals for excitation of the source shaker, and, hence, excitation of the table top.

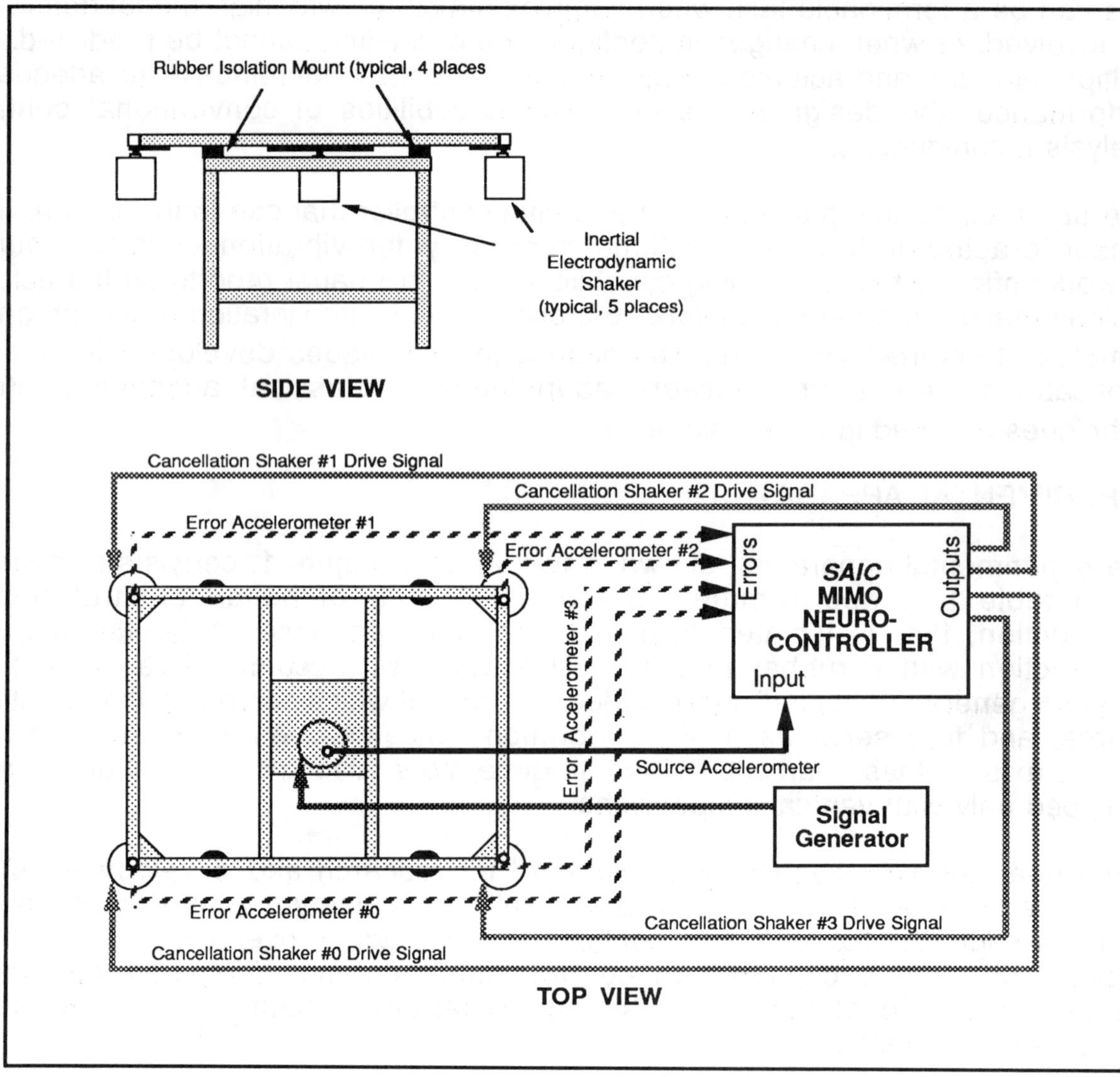

Figure 1. Experimental Apparatus with Measurement and Control Instrumentation.

Five accelerometers are used to measure the response of the source shaker and the response at each of the four locations where vibration reduction is desired. As depicted in Figure 1, the accelerometers at the four reduction points will provide the error signals to the neurocontroller, whereas the source accelerometer will provide the input signal to the neurocontroller. Operation of the experimental apparatus will be explained in more detail in the following discussions. All data used for this paper were obtained from experiments performed using this frame table.

NEURAL NETWORK CONTROLLERS

Two neural network controller architectures were utilized in this work. The first to be described is a single neural network with a feedforward arhitecture and back-propagation learning algorithm [1],[2]. The only modification made to the standard back propagation process is to use the signal from vibration sensor directly as the error signal, instead of the normal practice of subtracting the network output from a target value. In this arrangement, the learning algorithm adjusts the weights of the network, and thus the signal to the shaker, to minimize the vibrations at the sensor location.

Although effective for cancellation, the stability of this single network technique is limited to a narrow frequency band due to the phase shift of the error signal. This motivated a second approach in which multiple networks were configured to correct for the shift and stablize the performance over a wider frequency band.

SINGLE NETWORK CONTROLLER

The neural network for the single controller has an input layer of 40 units and an output layer of 4 units, i.e., one output for each shaker. There are two feedforword paths between the input and output layers. One path has two hidden fully interconnected layers of 10 units each, and the other path fully connects the input to the output layer directly. There are a total of 724 weights in the network, including bias weights. These weighted connections are updated using the standard back propagation algorithm of gradient descent. A sigmoid activation function is used which is adjusted to range between +0.5 and -0.5. Learning rates for the multilayer path and the direct connect path were independently adjustable and typically set to 0.1 and 0.01 respectively.

Figure 2 displays the rough schematic for the single controller. The 40-unit input layer to the neural network is actually broken up into two 20-unit shift registers which provide a time history of the inputs, IN0 and IN1. These inputs are derived from the digitized signal of the source accelerometer which is filtered through a digital high pass filter. The high pass filter has a 2 Hz cutoff to eliminate any DC level. The IN0 receives the gain adjusted signal directly, while IN1 is the gain adjusted first difference of the source signal. That is, the previously sampled source accelerometer signal is subtracted from the current sample to form IN1. This second input has been found to eliminate certain modes where the network stops learning in spite of a large error signal.

Also shown in Figure 2 are four channels of network output and error input pairs. Each output/error-input pair is connected to one of the four shakers and its associated accelerometer. The network output goes through a gain stage to the D/A converter. The D/A analog output is low-pass filtered through a 6-pole analog filter with a cutoff of 250hz, before being sent to the shaker. The accelerometer signal for that shaker is considered the error signal for that shaker's channel. It is digitized by an A/D converter, high-pass filtered for DC elimination, and gain adjusted before being summed with the negative of the DC compensation signal. This DC compensation is derived by filtering the network output with a digital low-pass filter set at 2 Hz to extract the DC component. The DC compensation forms a negative feedback loop to eliminate DC from the network output. Without this, the DC level at the network's output tends to drift and saturate, since there is no DC feedback through the shaker-to-accelerometer path.

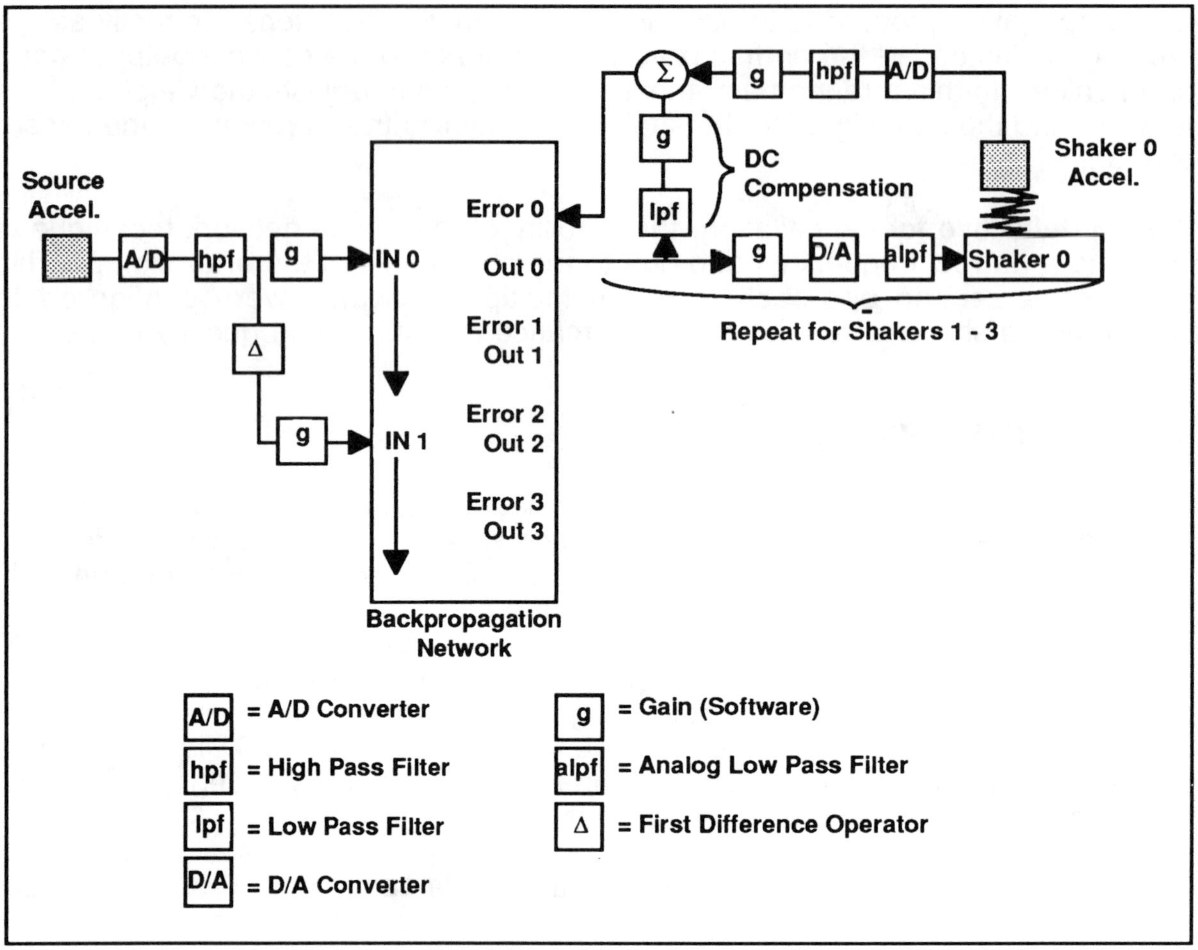

Figure 2. Single Network Neurocontroller for up to Four Shakers.

The gain stages referred to above and shown as boxed g's in Figure 2 represent floating-point multiplication by a stored set of adjustable constants. These gain constants are empirically determined for optimum performance, although in some cases, such as on the network output, the gain is a simple 1 or 0 value for output on/off control. As a rule of thumb, the gain on the input IN1 is typically 3 times that of IN0, with the IN0 gain determined by the strength of the source accelerometer signal. The sign of the gains for the error signals are individually set for stability. With the source shaker running, each cancellation shaker is operated alone, under control of the network, and the sign of its error gain determined for negative feedback operation. This is necessary, since the time delays and structural phase shifts in the signal path between the network output and its error input can be significant for the frequencies of interest. Error gain magnitude is adjusted for desired cancellation performance versus stability. The DC compensation gain is set to a small fractional value, typically 0.1, and seldom adjusted.

DETAILS

All the A/D input channels are sampled at a burst rate of 100Khz every 1.95 ms, which sets the sample period for the system. That is, the A/D channels are sampled, the network feedforward operation is performed, the outputs are sent to the D/A converters, and, if in the learning mode, the error signals are calculated, back propagated, and weights updated all within the 1.95 ms time frame before the next A/D sample burst. There is the capability of running in a batch mode, i.e., accumulating the weight changes over several sample periods before updating the weights, but that mode was not used for the data presented here.

TRAINING

For training from scratch, the network weights are initialized to random values between 0.3 and -0.3. With the source shaker operating, the network operation is enabled in the 'learning' mode in which the weights are updated. Cancellation is achieved within a few seconds to a few minutes, depending on the application, although refinement continues for several minutes afterwards.

RESULTS

A 30 Hz sine was used to excite the source shaker. The neurocontroller trained itself to reduce the vibration at the four locations co-located with the cancellation shakers with the results listed in Table I. Note that there was a reducttion greater than 30 dB at each location. Also, note that the source shaker level itself was reduced by about 8 dB due to the coupled reduction operations of the corner shakers and the stiffness of the frame structure.

We added noise to the 30 Hz sine source excitation signal that was about 10 dB below the sine level. Figure 3 shows the reduction during learning and immediately after vibration control has been established. This result is similar at the other three locations. The spectra before and after reduction are shown in the top and bottom of Figure 4, respectively. In Figure 4, note the additional reduction of the second and third harmonics of the 30 Hz excitation.

MULTIPLE NETWORK CONTROLLER

The filtered-x controller architecture discussed here is derived from the version presented in Reference [3]. The main difference is that backpropagation networks

Table I. Single Frequency Vibration Reductions at Four Table-Top Locations at 30 Hz.

Point Location	No Control (dB) Source	No Control (dB) Point	Control (dB) Source	Control (dB) Point	Reduction Δ (dB)
0	-24.4	-19.7	-32.4	-53.8	-34.1
1	-24.5	-20.4	-32.3	-52.1	-31.7
2	-24.6	-19.1	-32.3	-51.4	-31.7
3	-24.8	-17.1	-32.3	-53.1	-36.0

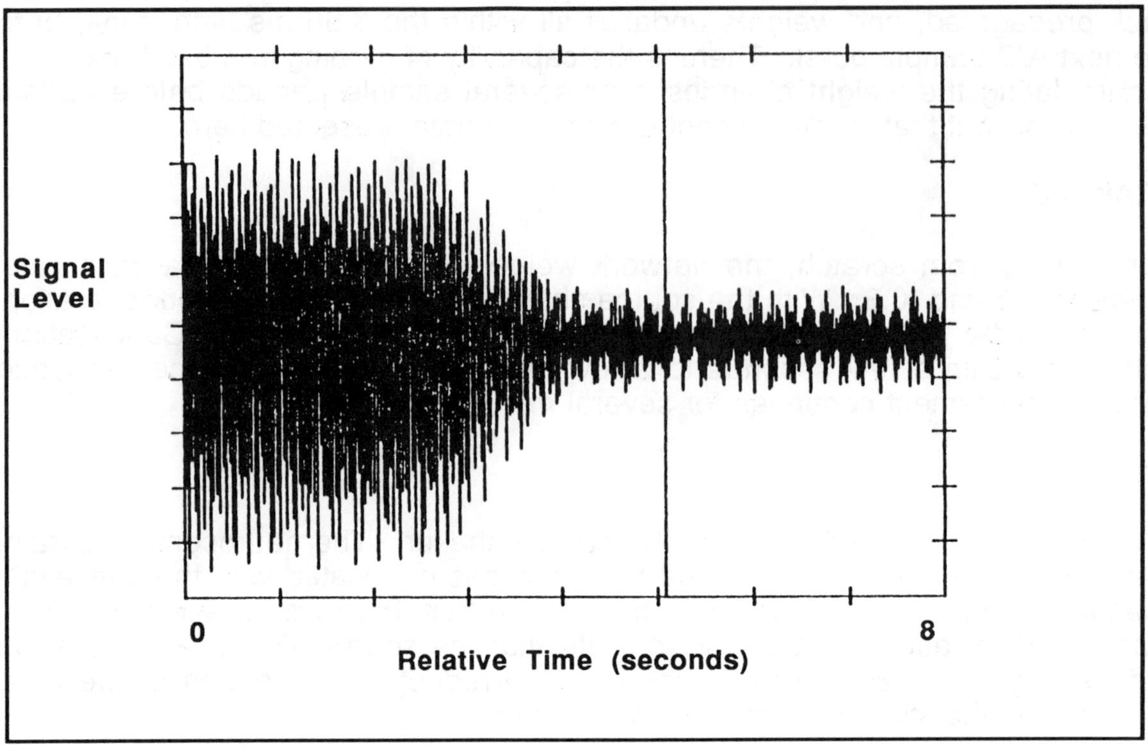

Figure 3. Time History Before and After Reduction.

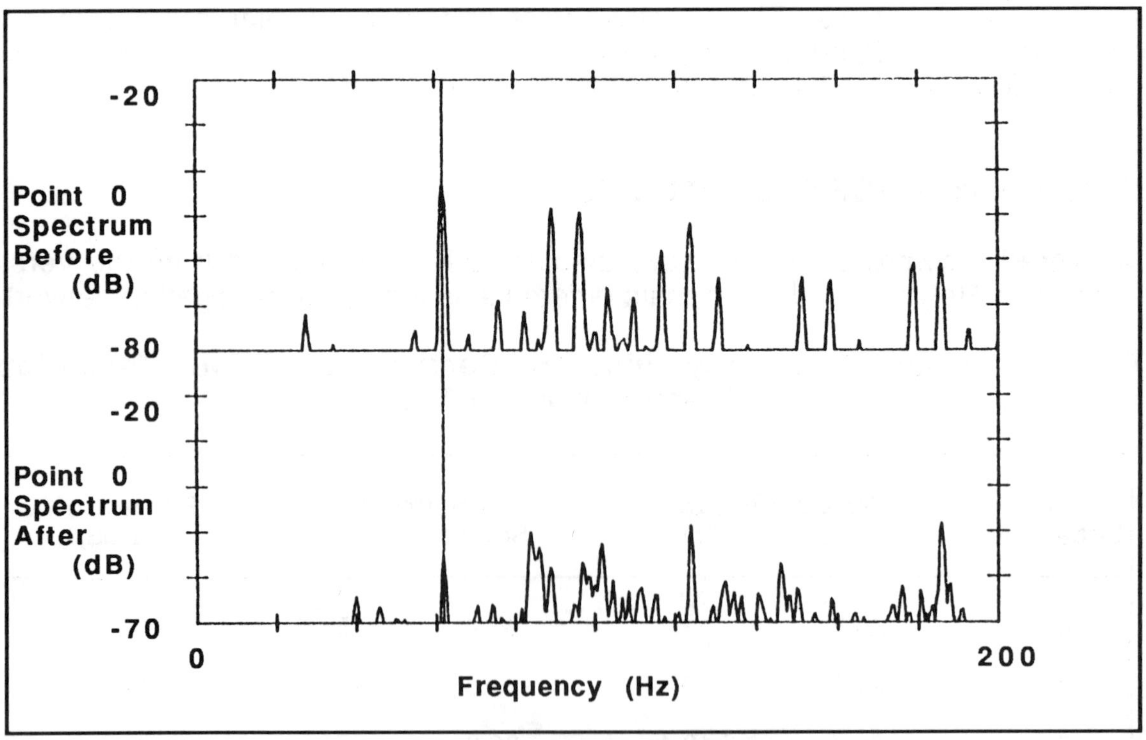

Figure 4. Spectra of 30 Hz Sine with -10 dB Noise Before (upper) & After (lower) Reduction at Location 0.

are substituted for the LMS networks in the reference. As shown in Figure 3, the filtered-x controller uses multiple networks [2]. In this embodiment, these neural networks are all identical to the network used in the single network controller described above, except only one output is enabled from each net and only one shaker is being controlled.

The filtered-x concept is an attempt to avoid the instability due to the phase delay of the error signal returned to the controller. That is, for a given controller output, the resultant error signal is delayed by the transfer function of the shaker and table responses. If the transfer function from the source accelerometer input to the shaker accelerometer response is considered to be the product of the controller transfer function and the shaker/table transfer function, then for a moderately linear system, it would be possible to switch the order of the controller and shaker/table operations and get the same overall transfer function. The difference is that now

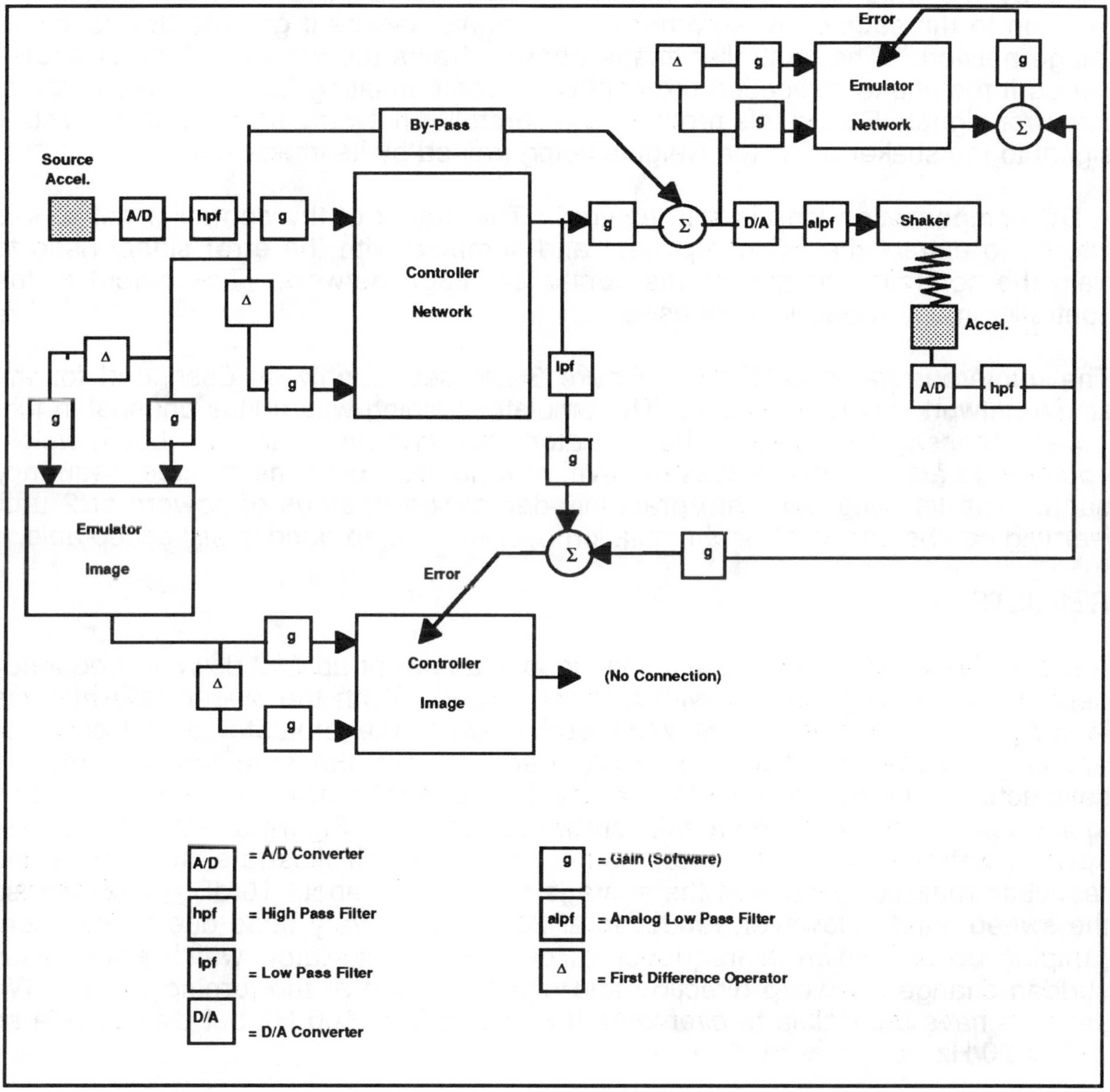

Figure 5. Multiple Network Neurocontroller/Neuroemulator

the controller network is the last operation, meaning the error signal is no longer phase-delayed with respect to the controller output.

To achieve this switch in the order of controller and shaker operations, the shaker/table transfer function is first learned by an emulator network shown in Figure 3. During this emulator training mode, the controller network is disabled and the controller bypass enabled to feed the source accelerometer signal directly to the emulator and shaker inputs. The emulator input is the same as the shaker input, and its output is compared to the shaker accelerometer response. The emulator is thus trained until it matches the shaker/table response over the range of frequencies desired.

Switching from the emulator training mode to the control mode, the controller bypass is disabled and the controller output enabled. The emulator weights are used by the emulator image network which now applies the shaker transfer function to the source accelerometer input signal, before it goes to the controller image network. The controller image network trains the weights of the controller net as if the shaker preceded the controller, compensating for the phase delay of the error signal. During this process, the controller network is providing the actual signal to the shaker using the weights being trained by its image network.

A DC compensation loop is still required. The output of the controller network is filtered to extract the DC component and summed with the error signal used to train the controller weights in the controller image network. The output of the controller image network is not used.

The numerous gains detailed in Figure 3 are set roughly as described for the single network controller above. The emulator training was a little unusual in that it was necessary to decrease the emulator's learning rates, and use batch weight updating to achieve the necessary level of emulation over the desired frequency band. The learning rates are gradually decreased in steps of powers of 2 until learning can be turned off and emulation over the sweep band is still acceptable.

RESULTS

A sinusoidal source shaker excitation signal was swept up and down in frequency between 20 Hz and 100 Hz with a linear sweep. Both the sweep rate and the learning rate of the emulator were varied while the emulator was trained to emulate the shaker within the sweep band. Once the emulator was trained satisfactorily, the neurocontroller was put into controller training mode. Training took a few minutes with the results shown in Figure 6. Figure 6 shows the cross-spectra before control and after control overlayed to illustrate the wide band reduction realized. Note that the average reduction is about 10 dB $\pm$ 5 dB across the sweep band. However, reduction at 20 Hz is not very good due to the linear ramping up and down in frequency of our sweep oscillator, which produces a sudden change in sweep direction and high distortion at the turning points. We seem to have been able to overcome this anomaly at 100 Hz but did not fare so well at 20 Hz.

Now, applying a narrowband noise signal as the source shaker excitation signal, with an effective bandwidth of about 50 Hz, i. e., from about 20 Hz to about 70 Hz,

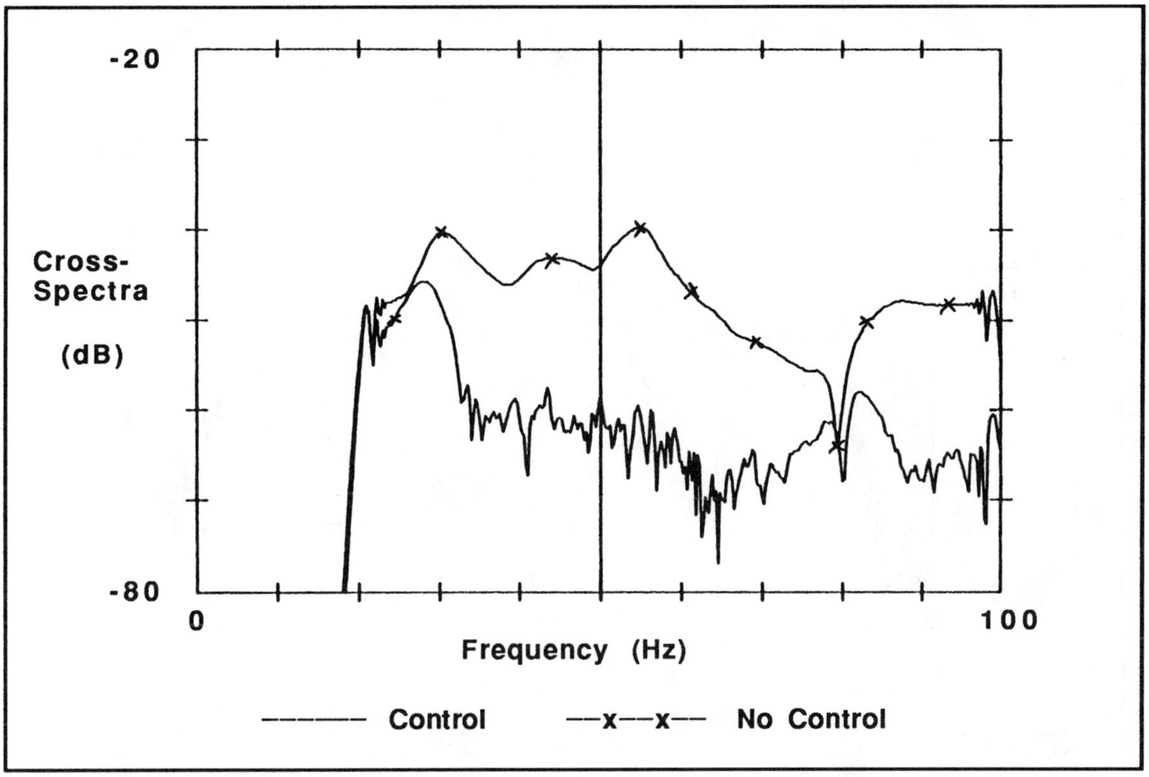

Figure 6. Comparison of Cross-Spectra Obtained from Swept Frequency Excitation Before and After Control Applied

we trained the shaker emulator network and proceeded to train the controller network. The trained controller results are shown in the comparison of the before and after cross-spectra in Figure 7. We realized reductions from about 1 dB to about 15 dB with an average reduction of about 5 dB ± 2 dB across the effective band.

CONCLUSIONS

The single network controller worked very well for single frequency vibration cancellation and a few narrowband applications. However, when swept frequency or narrowband random noise excitations were encountered, the network could not handle the dynamics of the structure over several structural vibration modes, i. e., phase shift and delay variations, and remain very stable. This lack of stable broadband success, led us to consider the inclusion of an emulator network.

After several development generations, we were able to successfully implement the multiple network architecture reported herein. As can be seen from the examples, the emulator network essentially learns to model the dynamics of the system and stabilize the neurocontroller, allowing it to concentrate on reducing the vibration presented as an error to the controller image network.

The small reductions realized with random excitation, as seen in Figure 7, are a result of the limited cancellation shaker energy available being spread over a

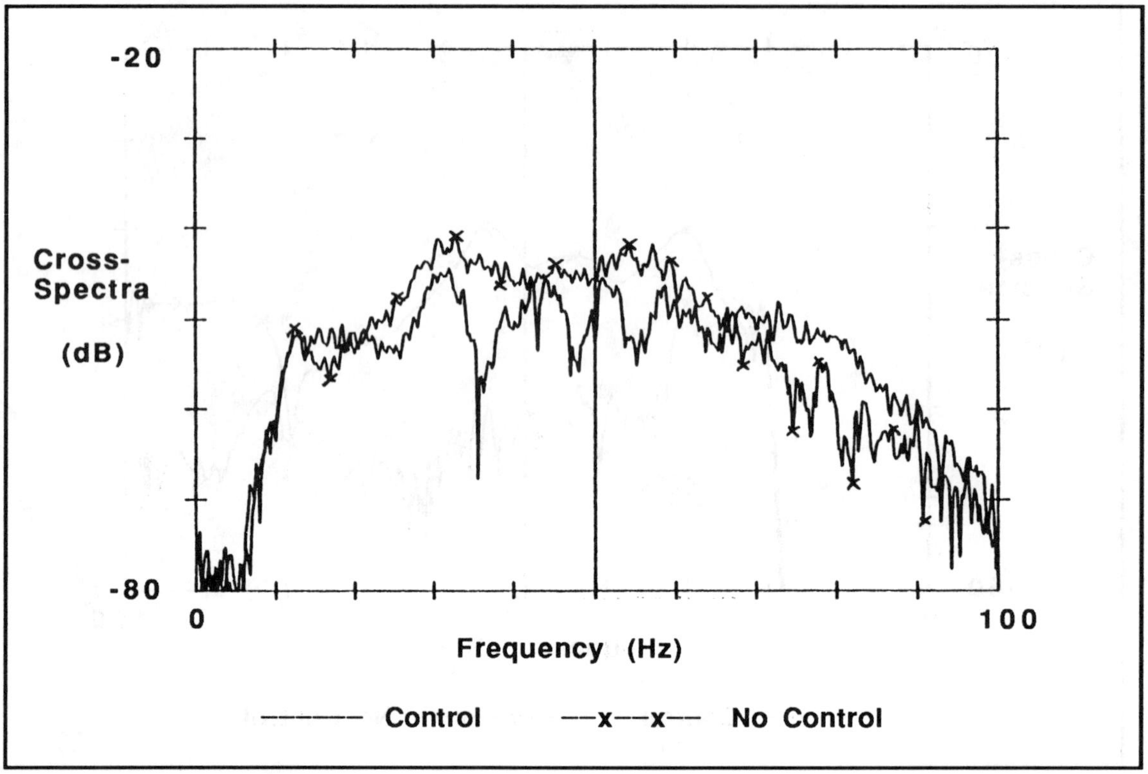

Figure 7. Comparison of Cross-Spectra Obtained with Narrow Band Noise Excitation Before and After Control Applied.

wide band of frequencies. A stronger cancellation shaker may be required to reduce vibrations on structures with a high dynamic magnification factor (high-Q).

Neurocontrollers have been shown to offer an exciting new technology to apply to adaptive control in general and to vibration control in particular. Their ability to learn a control algorithm without the benefit of analysis or modeling can be of great benefit for difficult and complex vibration control applications and especially for nonlinear vibration control applications.

We plan to explore other neural network architectures with the goal of further improving the level of vibration reduction, widening the bandwidth, and increasing the processing bandwidth such that the techniques reported can be applied to a more general class of problems including acoustic noise reduction applications. We also plan to extend the multiple network neurocontroller to multiple cancellation locations.

ACKNOWLEDGMENTS

The authors explorations into the reduction of vibration on line, in real time, and at multiple locations was supported as a company-funded Internal Research and Development (IR&D) project The use of neurocontrollers to reduce vibration was an outgrowth of earlier IR&D projects (1) to linearize nonlinear hydraulic shakers using neurocontrollers and (2) to reduce vibration at a single point using a single hydraulic shaker.

REFERENCES

[1] Bozich, D. J., and MacKay, H. B., *Adaptive Vibration Reduction using Artificial Neural Networks,* SME, Proceedings of AUTOFACT '89, Conference and Exposition on the Automatic, Integrated Factory, October 30 - November 2, 1989.

[2] Bozich, D. J., et al, U. S. Patent Pending.

[3] Widrow, B., and Stearns, S. D., Adaptive Signal Processing, Prentice-Hall, Inc., Englewood Cliffs, N. J. 07632, ISBN 0-13-004029 01, 1985.

AN APPLICATION OF GENETIC ALGORITHMS TO ACTIVE VIBRATION CONTROL

Alan R. D. Curtis[1]

ABSTRACT

A genetic algorithm is developed for the calculation of the time optimal control force for flexible systems. The algorithm is a robust random search method which uses a simple and compact coding of the control signal as a binary string of maximum positive and negative forces. A random population of admissible strings is allowed to evolve through the three processes of selection, reproduction and mutation to produce strings with increasing fitness and optimality. The introduction of stop characters allows the time optimal control problem to be solved in one procedure. A simulation is presented to illustrate the behaviour of the algorithm.

1. INTRODUCTION

A problem in the active control of vibration is the accurate positioning of a flexible system such as a lightweight robot arm. When moving such a system from one location to another the controller has two tasks. The first is to accelerate and decelerate the total mass of the system to achieve the required displacement (the rigid-body motion). The second is to reduce the effect of any residual vibrations induced by the motion of the system.

The many control strategies available for achieving these aims can be divided into two main categories, feedback and feedforward schemes. Feedback controllers recognise that residual vibrations are inevitable and aim to minimise their effect during and after the rigid body motion of the system. Feedforward controllers attempt not to excite the vibration in the first place. In feedback schemes, for example Modal Control [1], the position of the system is continuously monitored and the control is some function of the instantaneous states of the system. The control is continuously updated during motion and its aim is to ensure a fast response of the rigid-body mode and a rapid decay of residual oscillations. In a feedforward schemes, for example Input Shaping [2], a control is calculated for the full duration of the required motion before its commencement and is

[1]Department of Control Engineering, The University of Sheffield, Sheffield, U.K.

implemented without monitoring the system trajectory. The control input signal is shaped so that it achieves the rigid body motion without exciting the vibrational modes.

Both of feedforward and feedback controllers have their advantages and disadvantages. Feedforward controllers have a fast response and, although the calculation of the input shaping may be difficult, once calculated the signal can be easily constructed. However, feedforward controllers are not robust to errors in the model used to calculate the input signal. This lack of robustness can lead to a degradation in performance both of the accuracy of the final position and in the excitation of residual vibration. Feedback controllers are robust to modelling errors but are generally slightly slower than feedforward controllers and are more complicated to implement. A combined feedforward/feedback controller, for example a Sliding Mode controller [3,4], has the advantages of both schemes with the feedforward section ensuring rapid response and the feedback section ensuring accurate positioning and the damping of residual vibration.

2. TIME OPTIMAL CONTROL

One type of feedforward controller is based on the open loop optimal control. There are an infinite number of control signals which will move a given system from one location to another with no residual vibration. An optimal control is one of these set of controls which also minimizes some cost function such as control energy or the total time for the motion. The time optimal control is the control which achieves the motion in the shortest possible time and so has great practical significance. If the control force is unconstrained then there is no solution to the time optimal control problem; for any control which achieves the motion in a given time then an increase the size of the control force will move the system faster. An infinite force, if available, will move the system in no time at all. In practice the size of the control force will be limited and with a magnitude constraint, the time optimal control problem has a solution which is bang-bang, that is the control force at any instant is set at either its maximum positive or maximum negative limits. For vibrating systems, the time optimal control can switch many times between its limits.

There are many problems with the implementation of time optimal controls, one of which is the calculation of the control law for any system apart from the most simple. For a system with more than one degree of freedom the analytical calculation of the control is formidable and a numerical approximation is used.
Commonly used algorithms are based on gradient techniques [5,6] and take the form of a two step procedure. The first step is to fix the duration of the control and to solve the Minimum Miss Problem.

MINIMUM MISS: For a control of fixed duration T, find a control which drives the system closest to the target state. The distance, in the state space of the system, of the final state from the target state is the miss.

The second step is the Minimum Time Problem.

MINIMUM TIME: For an increasing sequence of control durations, find the first T for which the solution to the Minimum Miss Problem is a control which drives the system to target state (the miss is zero, the control hits the target).

At some stage in a numerical calculation, the dynamics of a continuous system is implicitly or explicitly replaced by a discrete approximation. It is convenient when assessing the resulting effects to explicitly approximate the system and to represent it by a discrete state space model of the form

$$\mathbf{x}_{k+1}=\mathbf{A}*\mathbf{x}_k+\mathbf{b}*u_k$$

where $\mathbf{x}_k$ is the vector of system states and u is a (scalar) input. The control is constrained such that

$$|u_k|<=1.$$

For a control signal of length T the constraint $|u_k|<=1$ defines the interior and boundary of a hypercube in control space. An admissible control is represented by the vector **u** of u_k's which represents a point in the hypercube. For each control **u** there is a corresponding final state $\mathbf{x}_F$ in the state space of the system. The set of all points within the hypercube maps to a set of reachable final states in the state space.

For a discrete system the Minimum Miss Problem now becomes

DISCRETE MINIMUM MISS: Find an admissible control **u** for which the corresponding final state $\mathbf{x}_F$ is closest to the target state.

For discrete systems the optimal control is not necessarily bang-bang [7] as some elements of the vector **u** need not be at their extreme values. If the target state is not in the set of reachable states then the Discrete Minimum Miss Problem has a unique solution and the optimal control will be on the boundary of the hypercube. Most elements of **u** will be at their extreme values. However, if the target state is in the reachable set then the solution may not be unique as more than one admissible control may hit the target. None of the elements of **u** need be at their extreme values.

For discrete systems the Minimum Time Problem becomes

DISCRETE MINIMUM TIME: Find the first T for which the reachable set includes the target state.

If the reachable set does not include the target state most elements of **u** will be at their extreme values, the control will be 'almost bang-bang'. In this case a simpler problem to finding the exact minimum miss control is to find the Bang-Bang Minimum Miss,

BANG-BANG MINIMUM MISS: Find the bang-bang control **u**, with all of its elements at their extremes, which has the minimum miss.

The admissible controls are now the set of vertices of the hypercube in control space. If the reachable set does not include the target state, the solution to this simpler problem will be close to the solution to the exact Minimum Miss Problem which will be a point on one of the faces adjoining the vertex. If the reachable set does include the target state, the Bang-Bang Minimum Miss Problem will not necessarily be close to the exact solution and will only produce a zero miss if a vertex of the hypercube maps exactly onto the target state. In nearly all cases the resulting minimum miss will be greater than zero.

A technique which might be suitable for finding the solution to the Bang-Bang Minimum Miss Problem is a genetic algorithm.

3. GENETIC ALGORITHMS.

Genetic algorithms [8] are optimizing search algorithms based on the principle of evolution by random selection and the survival of the fittest. Many of the ideas and terminology of the technique are taken from the study of genetics [9]. Genetic algorithms are used to find the maximum of a given cost function with respect to a set of variable parameters. The variable parameters are coded as a fixed length string of characters or *alleles* which is called a *chromosome*. Each chromosome has an associated *fitness* which is the value of the cost function for that set of parameters. The fitness must be a positive function which reaches a maximum for the optimal chromosome. A chromosome and its fitness define an *individual*. The chromosome will contain sub-strings or *genes* (possibly ill-defined) which as units, contribute in different ways to the overall fitness of the individual. The genetic algorithm proceeds by taking a *population* of different individuals and creating a new population, or *generation*, by combining features of the chromosomes of individuals from the old population with highest fitness. The aim is both to produce individuals with increasing fitness and to increase the average fitness of the population at large. Genes which contribute strongly to the fitness will survive in the new generation, those which do not will be lost.

The simple genetic algorithm used here has three basic operations,

Selection
Reproduction
Mutation.

Selection is the process of natural selection. Individuals are chosen randomly from the population with a frequency proportional to their fitness. Individuals with higher fitness are more likely to be selected.
Reproduction is the creation of new strings by combining two individuals selected from the population. In the simple scheme used here, two new chromosomes are created by cutting each old string at a random location and replacing the tail of one string with the tail of the other.
Mutation is required to avoid the irrecoverable loss of useful genetic information by over zealous selection. It is possible by selection and crossover to lose all copies of a particular allele at a given location, an allele which might be required for the final optimal chromosome. Mutation is the process of randomly changing any allele in the population to ensure that no gene is ever completely lost.

4. A GENETIC ALGORITHM FOR THE BANG-BANG MINIMUM MISS PROBLEM.

For the Bang-Bang Discrete Minimum Miss Problem a very simple and compact coding is possible using binary alleles. A one at a given location decodes as a maximum positive control and a zero as a maximum negative control. A chromosome is decoded by starting at one end, the head, and converting the strings of ones and zeros into a bang-bang control signal. This control is applied to the system and the final state and the miss are calculated. The miss will decrease as the control improves. The genetic algorithm requires an increasing function and so the fitness of the string is assigned as 1/miss. There will be a selection pressure towards sub-strings or genes which drive the system towards the target, leading hopefully to the creation of the optimal minimum miss chromosome.

A simulation of a simple single degree of freedom vibrating system is used to illustrate the behaviour of the genetic algorithm. The system is a unit mass on a unit spring controlled by a force whose magnitude is constrained to be less than 1 and is modelled by the equation,

$$d^2x/dt^2 + x = u$$

With an initial state x=-3, dx/dt=0, and a target state of the origin the exact time optimal solution can be found analytically. The control changes sign twice, the first from positive to negative at a time 0.251, the second from negative to positive at a time 3.392 and has a total duration of 4.838.

Gen	Miss	Best Chromosome
1	0.840	-+-----+----+++-++++
2	0.825	+-+----+----+++-++++
4	0.572	--+--------+--+++-++
6	0.531	-+--+-------+++-++++
8	0.464	+-+----+----+-++++++
9	0.181	+-+---------+++-++++
13	0.159	+-----------+++-++++
15	0.063	+-+---------+-++++++
19	0.050	++----------+-++++++
20	0.031	+-+----------+++++++
23	0.030	+------------++-++++

Table 1. The evolution of the best chromosome in a population of 50 individuals with chromosomes a fixed length of 20 binary alleles.

The system is replaced by a discrete approximation using a time step of 0.25 which should give an time optimal control with between 19 and 20 alleles. The discrete version of the system is

$$x_1(k+1) = 0.969x_1(k) + 0.247x_2(k) + 0.031u(k)$$
$$x_2(k+1) = -0.247x_1(k) + 0.969x_2(k) + 0.247u(k)$$

where x_1 is the displacement x and x_2 is the velocity dx/dt.

The genetic algorithm was started by chosing a random selection of 50 individuals. Each chromosome has a fixed length of 20 binary alleles which corresponds to the length of true time optimal control rounded up to the nearest integer. In several runs with different initial populations, the best individual in the population converged to the optimal individual in 15 to 25 generations. A typical evolution is shown in Table 1. More details of the parameters of the algorithm are given in the appendix.

The miss decreases with each generation as expected and converges to the optimal control in 23 generations. The best Bang-Bang control has four switches. The first two at 0.25 and 3.25 correspond to the two switches of the true time optimal control at 0.251 and 3.392. The extra double switch at 3.75 and 4.00 is a small 'shuffle' to get nearer the origin. The Bang-Bang control is restricted to the vertices of the constraint hypercube in control space and is selecting the vertex closest to the origin. The exact solution lies within the hypercube. However, the minimum miss is well within the minimum step length of the discrete system represented by the norm of the vector **b** (0.4950).

Length	Miss	Best Chromosome
1	8.876	-
3	7.841	+--
5	5.859	+----
7	3.788	+------
7	3.788	+------
9	2.137	+--------
11	1.303	-----------
13	1.023	-------------
15	0.819	-------------++
17	0.304	+------------++++
19	0.062	+-------------+++++
20	0.030	+------------++-++++
21	0.001	+------------++++-+++
23	0.001	+-------+-+---+++++++++
25	0.001	-+----------+-++++++-++-+

Table 2. The optimal chromosome for increasing lengths.

5. A GENETIC ALGORITHM FOR THE MINIMUM TIME PROBLEM.

A common method of computing a solution to the Minimum Time Problem is to solve the Minimum Miss Problem for a sequence of durations and to take the first duration for which the minimum miss is zero. For example, Table 2 shows the decrease in minimum miss for increasing lengths of chromosomes for the system presented above.

No minimum miss is equal to zero. The first length to get closer to the target than one step length is 17 with a miss of 0.304. A length of 20 gives a miss of 0.030 and a length of 21 a miss of 0.001. However, even a chromosome of length 25 is unable to get closer than 0.001 because, as before, the control is restricted to the vertices of the hypercube whereas the exact solution lies within the hypercube.

Using this two stage method the Minimum Miss Problem must be solved separately for each duration which can be very computationally expensive for longer durations. However, with a small modification, the genetic algorithm can produce a solution to the Minimum Time Problem in one process. The key feature is the introduction of an additional allele which acts as a signal to the decoding algorithm to stop; it starts decoding the string at the head end as before but stops when a stop allele is encountered. The active length of any chromosome is then variable and is determined by the first occurrence of a stop allele. Initially a chromosomes will have many stop bits and so will have short effective lengths.

If chromosomes with longer effective lengths are more likely to have greater fitness then they are more likely to be selected. There will be a selection pressure towards longer effective lengths and as generations pass, the effective lengths of individuals will grow. There will also be a counter selection pressure acting against longer effective strings which will arise as follows. The crossover mechanism selects a crossover site at random. If each site is equally probable then genes at the tail end of the chromosome will be swapped more often than genes towards the head end. Thus shorter strings are more likely to survive intact than longer but equally fit strings.

In the Minimum Time Problem all strings above a certain length will get close to the target and so will have similar fitnesses. The chromosome with the shortest effective length which hits the target is the time optimal control and is the chromosome most likely to survive the counter pressure. In practice the effective lengths of strings will grow until the optimal length is reached when there will be a zero or negative selection pressure acting against further growth. Longer strings, if now generated, will be less likely to survive than the short string with the same fitness.

The simulation was rerun with a fixed length chromosome of 30 but with the additional stop allele. Table 3 shows the evolution the best individual for a typical population.

The best chromosome grows over 70 generations to one with an effective length of 21. The algorithm proceeds to the maximum number of generations at 100 finding only a small improvement in miss but the effective length does not grow past 21 as longer but equally fit chromosomes are destroyed by reproduction and cannot get a foothold in the population. There is at least one very similar chromosome of length 21 with a smaller miss which was discovered in the simulation of Table 2. This chromosome may yet be found by this simulation but after 100 generations the diversity of the population will

Gen	Length	Miss	Best Chromosome
1	4	6.893	+---....--+-.+.+.-.+.-.+.-++++
10	6	4.802	+-----..++..++...-.+..++.+++++
20	10	3.081	+------+--.+.++-.+.-+.--.-+-++
30	10	1.601	+---------.+.++-.+.-+..-.-+-.+
40	15	1.169	+----------+-++..+.-+.--.---.+
50	14	1.151	+------------+.+.-.-+.-...+-+-
60	18	0.490	+------------+++-+.++.....--+-
70	21	0.017	+------------+++-++++..+..--.+
80	21	0.017	+------------+++-++++..+..--.+
90	21	0.014	+------------+++++-++.-.---..-
100	21	0.014	+------------+++++-++.-.---..-

Table 3. The evolution of the best chromosome for a population with stop alleles.

be small, there will be many individuals that are very similar, and we must rely on a chance mutation to give the small increase in performance which is expected.

In early generations, most chromosomes contain a lot of genetic information in the tail end which is not decoded (or expressed to used the genetic term). There is a strong precedent for this in nature where in higher organisms the majority of the genetic material is made up of 'spacer' or 'junk' DNA which has no obvious function. True genes are delineated by stop and start sequences of DNA. I do not wish to speculate on the purpose of this extra genetic material in nature as its existence is one of the great unsolved problems of genetics. However, it would seem to be useful in finding time optimal controls.

6. CONCLUSIONS

A simple genetic algorithm has been developed which can solve the time optimal control problem for flexible systems. In solving the Minimum Miss Problem the algorithm converges rapidly to the optimal control. The Minimum Time Problem can be solved by increasing the control duration until a optimal control is found which hits the target. However, the simple modification of the addition of a stop allele allows the time optimal control to be found in a single process. Unlike other algorithms [10] which use a single process, the genetic algorithm is robust, starts from a random population with random control durations and does not require a good initial guess to converge.

As the genetic algorithm is a random search, it is not guaranteed to converge to the true solution. However, its robust and global power could be used in conjunction with other more sensitive algorithms, such as gradient descent, to provide accurate solutions. The genetic algorithm would provide an approximate control which would then be fine tuned by the gradient descent algorithm.

A feature of genetic algorithms not exploited in this work is their natural advantages for parallel implementation. At a basic level each individual could be allocated its own processor. The algorithms developed here were simple and were not optimized for speed. There is some scope to tune the parameters and structure of the genetic algorithm (particularly if parallel processing is available) to increase its efficiency and this should be done before undertaking a fair trial with against other more established methods.

7. REFERENCES

[1] Meirovitch L., Baruh H., "Control of self-adjoint distributed parameter systems.", J. of Guidance, Control and Dynamics **5** pp60-66, 1982.
[2] Singer N.C., Seering W.P., "Preshaping command inputs to reduce system vibration.", Trans. ASME J. of Dynamic Systems, Measurement and Control **112** pp76-81, 1990.
[3] Iyer A., Singh S.N., "Sliding mode control of flexible spacecraft under disturbance torque.", Int. J. of System Science **21**(9) pp1755-1711, 1990
[4] Newman W.S., "Robust near time optimal control.", IEEE Trans. **AC35**(7) pp841-843, 1990.
[5] Ho Yu-Chi, Brentani P.B., "On computing optimal control with inequality constraints.", J. SIAM Control Ser. A. **1**(3) pp 319-348, 1963.
[6] Eaton J.H., "An on line solution to sampled data time optimal control.", J. Electron. Control **15**(4), pp.333-341, 1963.
[7] Desoer C.A., Wing J., "A minimal time discrete system.", IRE Trans. **AC6** pp111-125, 1961.
[8] Goldberg D.E., "Genetic algorithms in search, optimization, and machine learning.", Addison-Wesley ISBN 0 201 15767 5, 1989.
[9] Rosenfield I., Ziff E., Van Loon B., "DNA for beginners.", Writers and Readers ISBN 0 86316 022 0, 1983.
[10] Plant J.B. and Athens M., "An iterative procedure for the computation of time-optimal controls.", Proc. of the Third IFAC Congress, paper 13D,London, 1965

APPENDIX

Population Size	=	50
Probability of Crossover	=	0.700
Probability of Mutation	=	0.020
Maximum Number of Generations	=	100

Exact System Parameters.

A[1,1]	=	0.9689124217
A[1,2]	=	0.2474039593
A[2,1]	=	-0.2474039593
A[2,2]	=	0.9689124217
b[1]	=	0.0310875783
b[2]	=	0.2474039593

EXPLOITATION OF CHAOS FOR ACTIVE CONTROL: AN EXPERIMENT

M. L. Spano,[1] W. L. Ditto,[1] S. N. Rauseo[1]

ABSTRACT

We have implemented a control scheme to exploit chaotic behavior in a physical system. The method of control requires only small time dependent perturbations of a single system parameter and does not require model equations for the dynamics. We have demonstrated the power of the method by controlling an experimental chaotic system around unstable periodic orbits and exploiting the underlying chaos to switch the motion between these orbits at will. The robustness of the method is tested by adding white noise to our experimental system during control to obtain information on the effect of noise on control failures.

INTRODUCTION

Developments in nonlinear physics have indicated that chaotic behavior is quite common in real world physical systems [1-2]. While most of chaos research has concentrated on the identification and categorization of typical types of chaotic behavior, a recent control scheme has been proposed that demonstrates the potential power of exploiting chaos in physical systems [3-4]. This control scheme, which is straightforward in application and concept, presents a mechanism by which systems can exploit chaos for adaptation.

A prototypical system exhibiting chaotic behavior is provided by a new class [5-15] of adaptive materials (transversely annealed amorphous metallic glass). They exhibit large changes of Young's modulus (a factor of 10 or more) with small changes (~1 Oe) in the applied magnetic field. This effect is used to drive gravitationally a vertical ribbon of the material, clamped at the bottom and free at the upper end. The ribbon is driven by varying the Young's modulus of the material periodically, thereby causing the ribbon to buckle and unbuckle. Broad band *chaotic* oscillations are observed in the vibrating ribbon.

The recently developed theory for *exploiting* chaos for active control is implemented in real time to control the chaotically oscillating ribbon. The method requires only small time dependent perturbations of a *single* system

[1]Naval Surface Warfare Center, 10901 New Hampshire Ave., Silver Spring, MD 20903-5000, U.S.A.

parameter (in our case a magnetic field) and no *a priori* model of the system. Only a brief history of the motion of the system is initially required. Additionally we demonstrate the potential of the method for smart control of oscillating systems by stabilizing the chaotically oscillating ribbon about unstable periodic motions of period one, two, and four, exploiting the chaos to switch between these orbits. We have attained control of these chaotic oscillations with mean time between failures greater than 4 days. The method is found to be very robust experimentally and is far from being a control that requires experimentally unattainable precision. We believe this method can be widely implemented in a variety of systems including adaptive structures and materials.

THE MATERIAL

A new class of material, transversely-annealed amorphous magnetic $Fe_{81}B_{13.5}Si_{3.5}C_2$ (Metglas+ 2605 SC), has been developed that exhibits large reversible changes of Young's modulus, E(H), with small magnetic field changes (Fig. 1). In previous experiments [16-18] a vertically aligned ribbon of this material, clamped at the bottom and free at the top, is subjected to a vertical magnetic field. When the applied magnetic field is varied, Young's modulus is altered and this causes the ribbon to buckle and unbuckle under the influence of gravity. We have exploited this buckling to drive the ribbon with an alternating magnetic field. Initial quantitative observations of the behavior of a driven ribbon have presented us with a rich variety of behavior that includes periodic motion, period doubling cascades, quasiperiodic motion, transient chaotic motion, intermittent chaotic motion, and sustained chaotic motion.

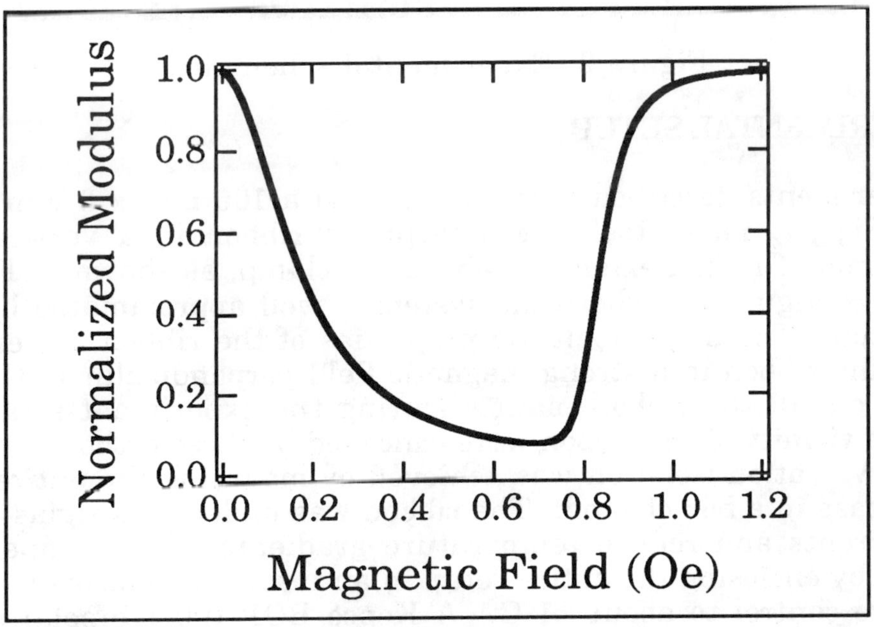

Figure 1. Measured Young's modulus (normalized) versus magnetic field for the ribbon used in the experiment.

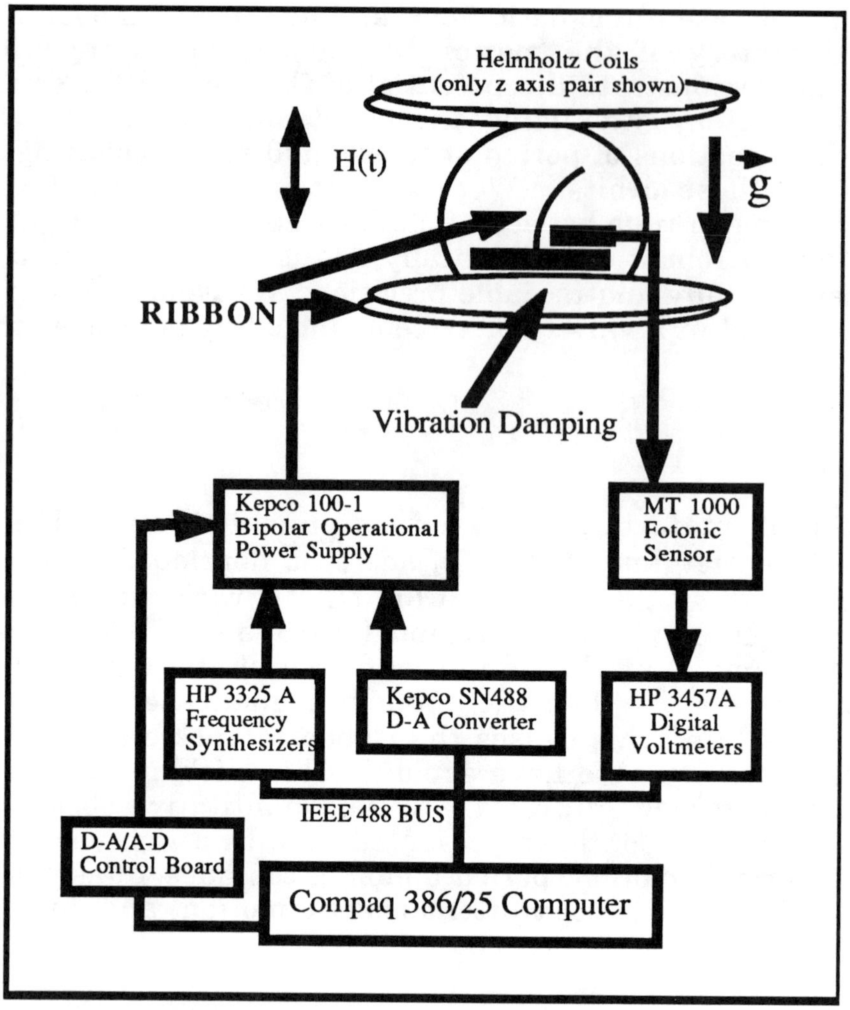

Figure 2. Experimental schematic.

THE EXPERIMENTAL SETUP

In the experiments described here we clamped a 100 mm x 3 mm x .025 mm $Fe_{81}B_{13.5}Si_{3.5}C_2$ magnetoelastic amorphous ribbon in a vertical position, leaving 65 mm of ribbon extending above the clamp, as shown in Fig. 2. This produces a configuration where the system is well approximated by a double-well potential. The magnetoelastic properties of the ribbon are enhanced by annealing the ribbon in a strong magnetic field perpendicular to the length of the ribbon but in the ribbon plane. During the experiment stray magnetic fields in the vicinity of the ribbon were cancelled by three orthogonal Helmholtz coil pairs. Vibration isolation was achieved by mounting the entire system on a massive base in a box of sand. The ribbon was under a plexiglas box to keep out air currents and reduce temperature gradients. The temperature was kept stable by enclosing the entire setup in an insulating enclosure with active temperature control to about ±1 C. A Kepco BOP 100-1 bipolar operational power supply driven by a Kepco SN-488 D-A converter and a Hewlett Packard HP 3325A frequency synthesizer applied a vertical magnetic field of the form $H = H_{dc} + H_{ac} \cos(2\pi ft)$. Thus there are three independent parameters that control the dynamics of this system. For the noise experiments a programmable D-A board was used to provide a source of white noise. The frequency is typically set between 0.1 and 2.0 Hz (stable to better than 10^{-5} Hz). The displacement of the ribbon was measured at a point 6 mm above its clamped base by an MTI 1000 Fotonic Sensor and was output as a voltage that is monotonically related to the displacement of the ribbon from the vertical.

A BRIEF TUTORIAL ON CHAOS

Chaotic motion is not random motion. Random motion has, by definition, no intrinsic structure or order. It is non-deterministic. On the other hand chaotic motion is highly structured and deterministic. It may be thought of as a combination of a possibly infinite number of unstable periodic motions, some of which are visited more often than others. The key to controlling chaos is to cull out only the periodic motion that you desire.

Consider a series of measurements of the position and momentum of a dynamical system. In general the position (as well as the corresponding momentum) is a multiple component vector. The space comprised of both the position coordinates and the momentum coordinates is the *phase space* of the dynamical system. If the measurements are spaced closely enough, they will trace out a trajectory in phase space. For deterministic systems (which include chaotic systems), the position in the phase space uniquely determines the state of the system at a given time. (Implicit is the assumption of no orbit self intersections as trajectories in phase space may not intersect each other due to uniqueness.) Now cut the phase space with an appropriate plane. If properly chosen, the plane intersects the curve at a set of points and represents a Poincaré section. Consequently we have converted a continuously evolving system into a map with discrete time steps [1].

For a periodic system the Poincaré section will exhibit a finite number of discrete points whose number describes the periodicity of the trajectory. For a chaotic system the Poincaré section generally shows a fractal geometric structure called a strange attractor [1,2].

THE METHOD OF CONTROL

In a recent letter [3], Ott, Grebogi and Yorke (OGY) demonstrated that one can convert the motion of a chaotic dynamical system to periodic motion by controlling the system about one of the many unstable periodic orbits embedded in the chaotic attractor, through only small time-dependent perturbations in an accessible system parameter. They demonstrated their method numerically by controlling the Hénon map [19].

The method is based on the observation that unstable periodic orbits are dense (in the sense that periodic orbits pass through any neighborhood, however small, of any point on the attractor) in a typical chaotic attractor [3-4]. Their method assumes only the following four points:

(1) The dynamics of the system can be represented as arising from an n-dimensional nonlinear map (*e.g.*, by a surface of section or time one return map), the iterates given by $\xi_{n+1} = \mathbf{f}(\xi_n, p)$ where p is some accessible system parameter.

(2) There is a specific periodic orbit of the map which lies in the attractor and around which one wishes to stabilize the dynamics.

(3) There is a maximum perturbation $\delta p*$ in the parameter p, by which it is acceptable to vary p from the nominal value p_0.

(4) The position of the periodic orbit is a function of p, but the local dynamics about it do not vary much with the allowed small changes in p.

Note that while the dynamics is assumed to arise from a map, one needs no model for the global dynamics. These assumptions would seem to allow for the control of any chaotic system for which a faithful Poincaré section can be constructed. The construction of a map from and the location of periodic orbits in experimental data is a straightforward process [16-18, 20].

To control chaotic dynamics one only needs to learn the *local* dynamics around the desired periodic orbit by observing iterates of the map near the desired orbit and fitting them to a local linear approximation of the map **f**. From this, one can find the stable and unstable eigenvalues as well as the local stable and unstable manifolds (given by the eigenvectors) [20-22]. Next, by changing p slightly and observing how the desired orbit changes position, one can estimate the partial derivatives of the orbit location with respect to p.

To control the chaos, one attempts to confine the iterates of the map to a small neighborhood of the desired orbit. When an iterate falls near the desired orbit, we change p from its nominal value p_0 by δp, thereby changing the location of the orbit and its stable manifold, such that the *next* iterate will be forced back toward the stable manifold of the *original* orbit for $p=p_0$. (Fig. 3 illustrates this method for the case of a saddle fixed point located at $\xi_F(p_0)$).

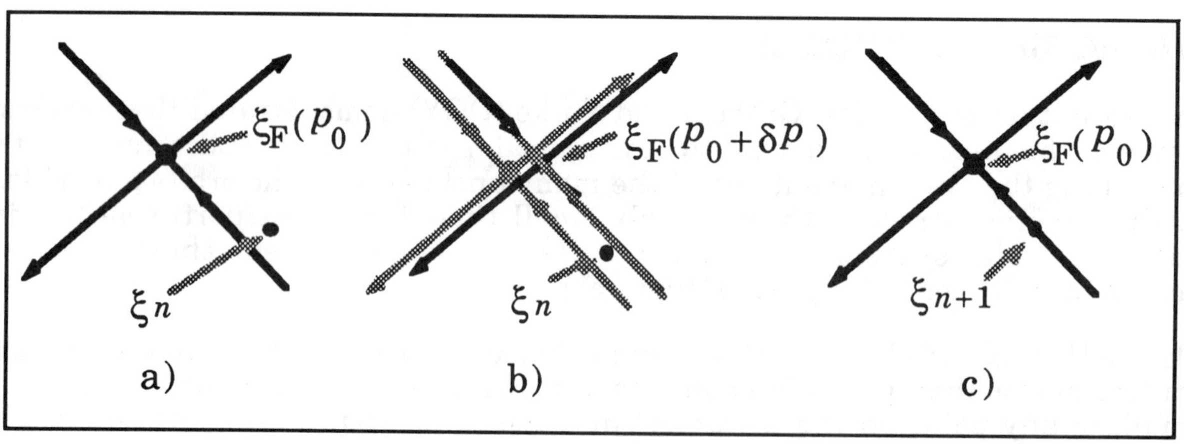

Figure 3. Outline of OGY control algorithm for a saddle fixed point: (a) the nth iterate ξ_n falls near the fixed point $\xi_F(p_0)$. (b) Turn on the perturbation of p to move the fixed point. (c) The next iterate is forced onto the stable manifold of $\xi_F(p_0)$. Turn off the perturbation.

EXPERIMENTAL REALIZATION OF CONTROL

In practice it is not always possible to measure all of the variables that comprise the phase space of a dynamical system. If you are limited to one measurement as a function of time, it is possible to reconstruct the phase space of your system using a *delay coordinate embedding* [23]. In this method, the sensor's output voltages $V_i = V(i\Delta)$, where Δ is the sampling time, are arranged into delay coordinate vectors whose d components at time $i\Delta$ are $(V_i, V_{i+n}, V_{i+2n}, \ldots, V_{i+(d-1)n})$. The tip of this vector traces out a phase portrait in a phase space of dimension d.

Before attempting control we determined the position of the target fixed point in the Poincaré section (i.e., the position of the unstable periodic orbit in phase space). Also we found the stable and unstable eigenvectors, as well as the change in the fixed point position with respect to a change in the system parameter we chose for control - the dc magnetic field, H_{dc}. All of these could easily have been determined "on-line" in a more sophisticated experiment.

With these results in hand we attempted to control the system via the OGY algorithm. The Poincaré section is easily obtained in an analog fashion by triggering the voltmeter directly from the frequency synthesizer. Consequently, our phase space consists of vectors of the form (V_i, V_{i+1}) which define a two dimensional map. The sampling time Δ is the period of the ac magnetic drive field. We wait for the system to visit a region of phase space that is within some radius, ε, of the target fixed point. Then it is only necessary to form the delay coordinate vector, calculate the change in the position of the fixed point (along the unstable direction) that is desired, and finally find the adjustment to the dc magnetic field that is required to achieve that change. The results are shown in the next section.

EXPERIMENTAL RESULTS

The chaotic attractor is shown in gray in Fig. 4. (H_{ac} = 2.05 Oe, H_{dc} = 0.112 Oe (=p_0), and f = 0.85 Hz.) When the control is activated, we wait for the system to move into the region near the period 1 fixed point. (Note that, since a phase plot has the current position versus the previous position, the fixed point is easily found by observing the intersection of the attractor with the diagonal.) Then the control is achieved (by varying H_{dc} by a maximum of 0.008 Oe), as shown by the dark region in Fig. 4(a). Period 2 fixed points are located by plotting the current position versus the next-to-last position and so forth for higher period fixed points. The period 2 control is shown in Fig. 4(b). While we could have controlled about both period 2 fixed points, we chose to control around only one of them. Note that the control is still quite good, despite the fact that we are only using the control algorithm every other drive cycle.

We have also controlled about the period 4 fixed points, in this case exercising the control procedure every 4th drive cycle. The control remained quite good, although occasional failures did occur. Evidently the higher period orbits require control about several or all of their constituent fixed points.

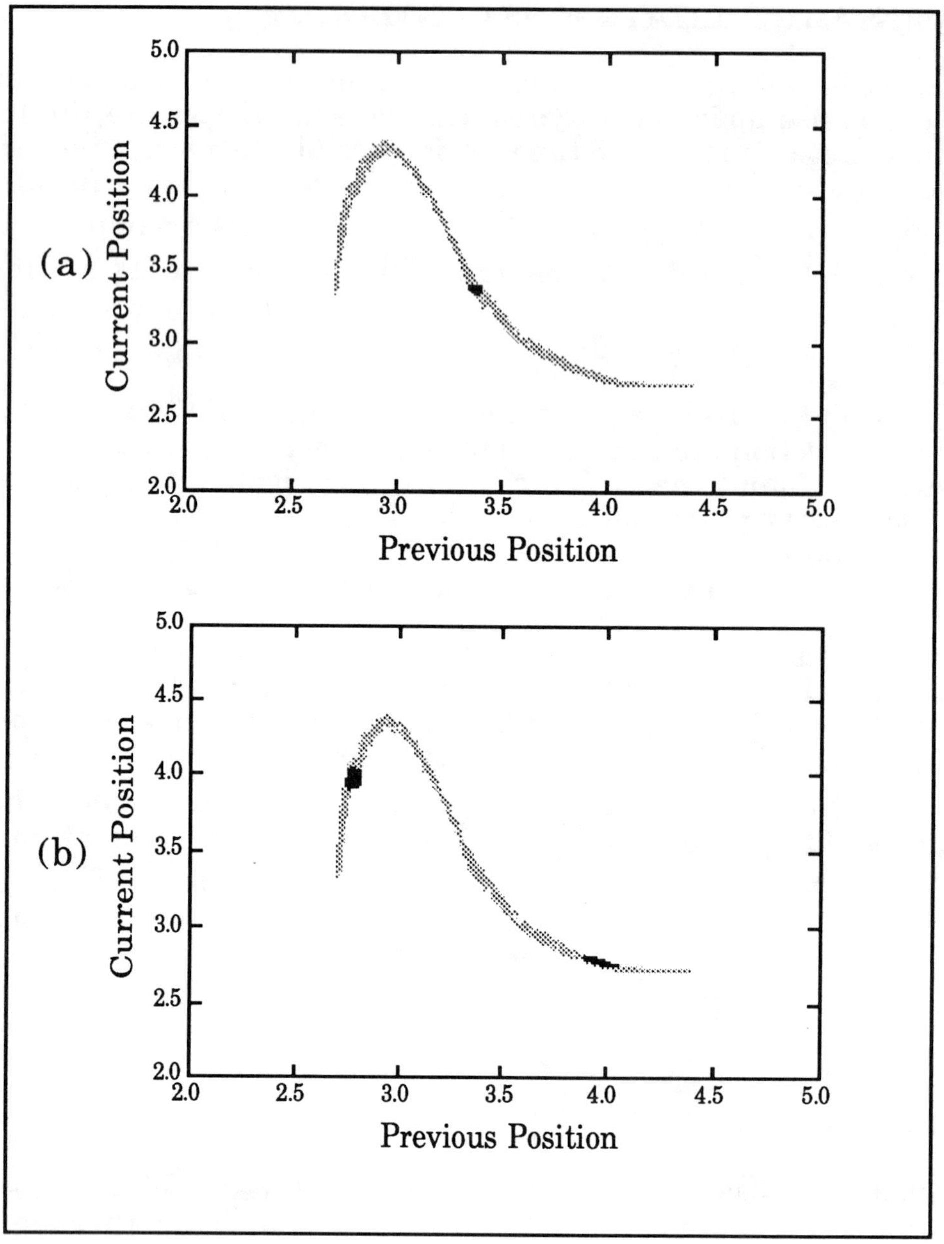

Fig. 4. The current ribbon position V_i is plotted versus the previous position V_{i-1} yielding a phase portrait of the attractor. (a) Period one control (black). (b) Period two control (black). In both cases the grey represents the uncontrolled chaotic behavior.

SWITCHING CONTROL

As a demonstration of the versatility of the method, Fig. 5 shows time series data while the system was switched between no control, control about the period one fixed point, period two orbit, and the period four orbit.

NOISE STUDIES AND ROBUSTNESS OF THE METHOD

In order to test the robustness of the OGY method, we added noise to the driving field in a controlled fashion [22]. Fig. 6 shows how the controlled signal increases in width as the noise is increased. Note the escapes from control at the higher noise values. If we take a cross-section of the control region in phase space, we find that the distribution of the controlled signal about the fixed point is quite gaussian, as shown in Fig. 7 for several values of the noise. We have tested our system from no noise to enough noise to cause the control to break down more than half the time. For all of these cases the data are well described by a gaussian. The peak widths of the fits are shown in Fig. 8 and the reciprocals of the peak heights are plotted in Fig. 9. The widths are well described by a straight line, as shown on the graph. The reciprocals of the peak heights are also fit by a straight line, although the scatter is somewhat greater than for the widths.

Knowing that the distribution of the data around the fixed point is gaussian allows us to calculate the probability that the control will fail at any given time step. The probability of failure turns out to be the complementary error function:

$$\text{Probability of failure at a given iteration} = \text{erfc}\,(1.665\,\varepsilon\,/\,\text{gaussian width})$$

where ε is the size of the control region as mentioned previously. In our experiment we have chosen an ε of 0.18 in units of position. (cf. Fig. 4) For our

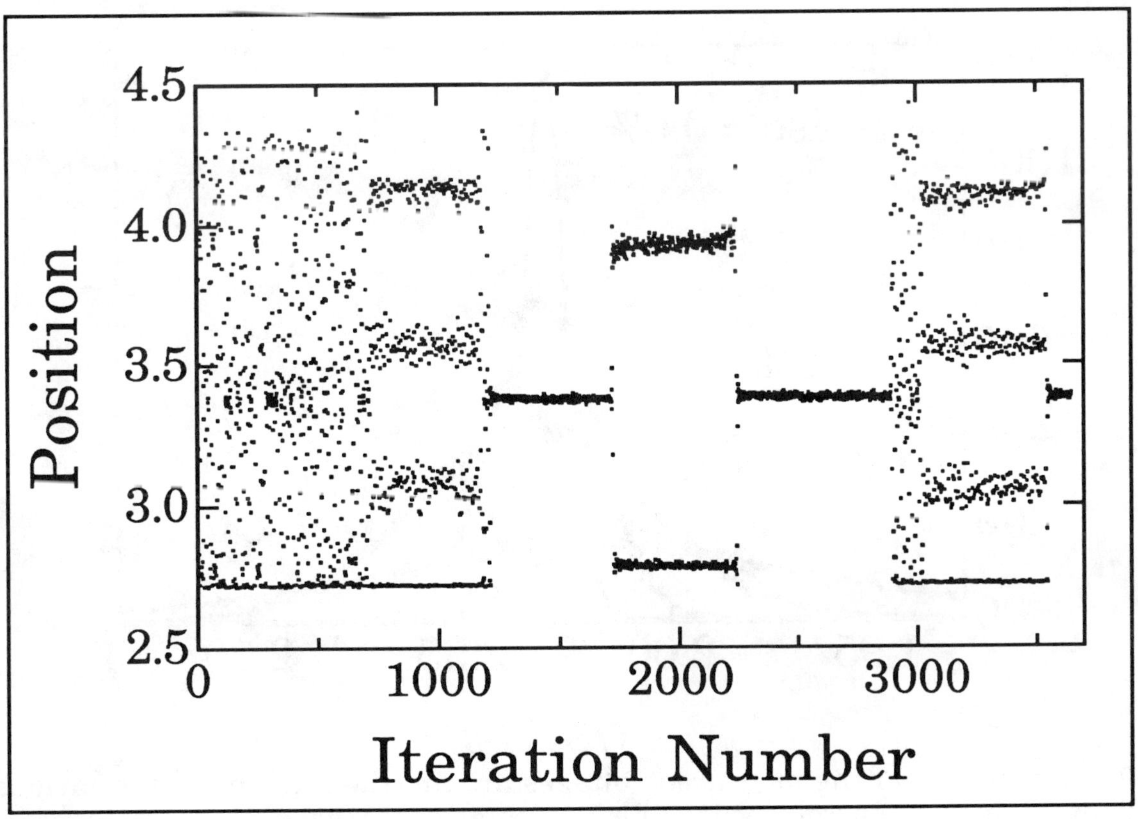

Fig. 5. Switching between different periodic orbits as described in the text.

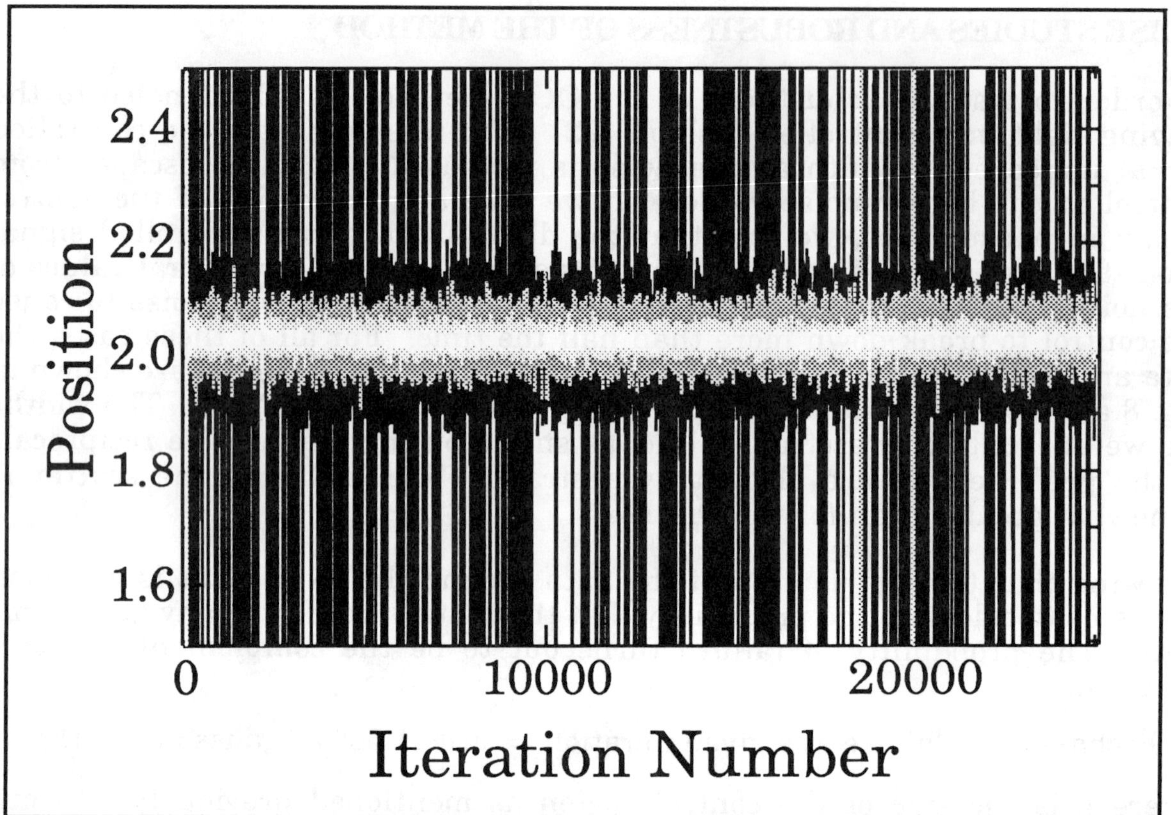

Fig. 6. Efficacy of the control for various noise levels. Noise amplitude = 1.6% of the ac drive (black), 0.8% (dark grey), 0.4% (light grey) and 0.0% (white).

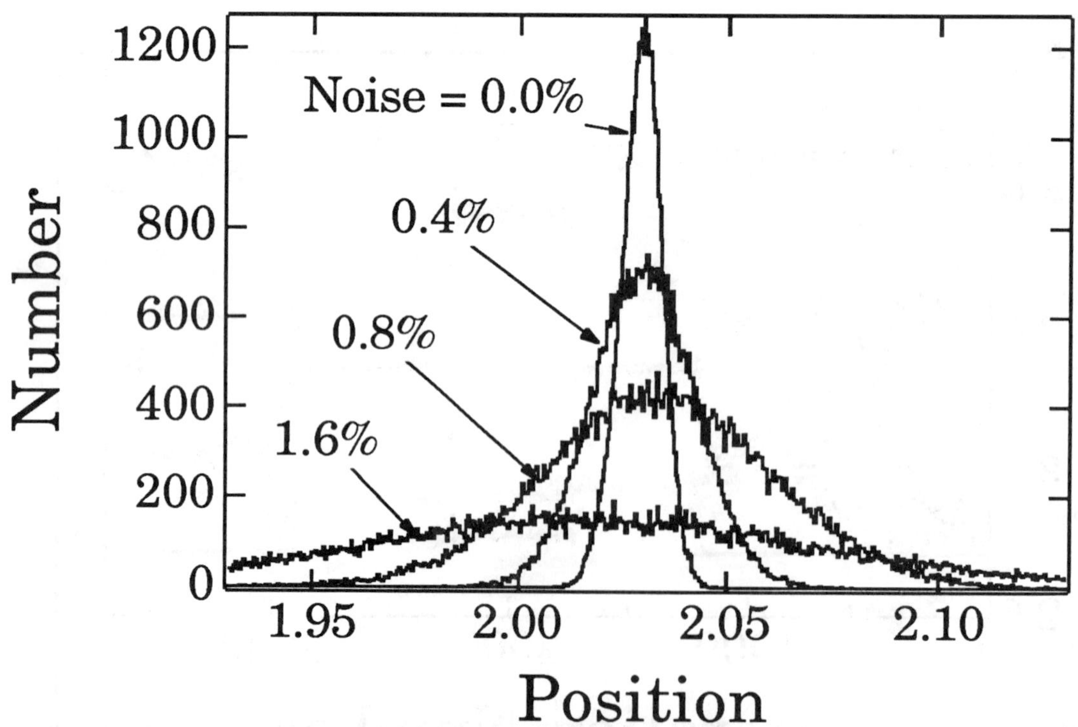

Fig. 7. Distribution of the measured voltages around the fixed point for various noise levels during control. Noise amplitudes are expressed as percentages of the ac drive amplitude.

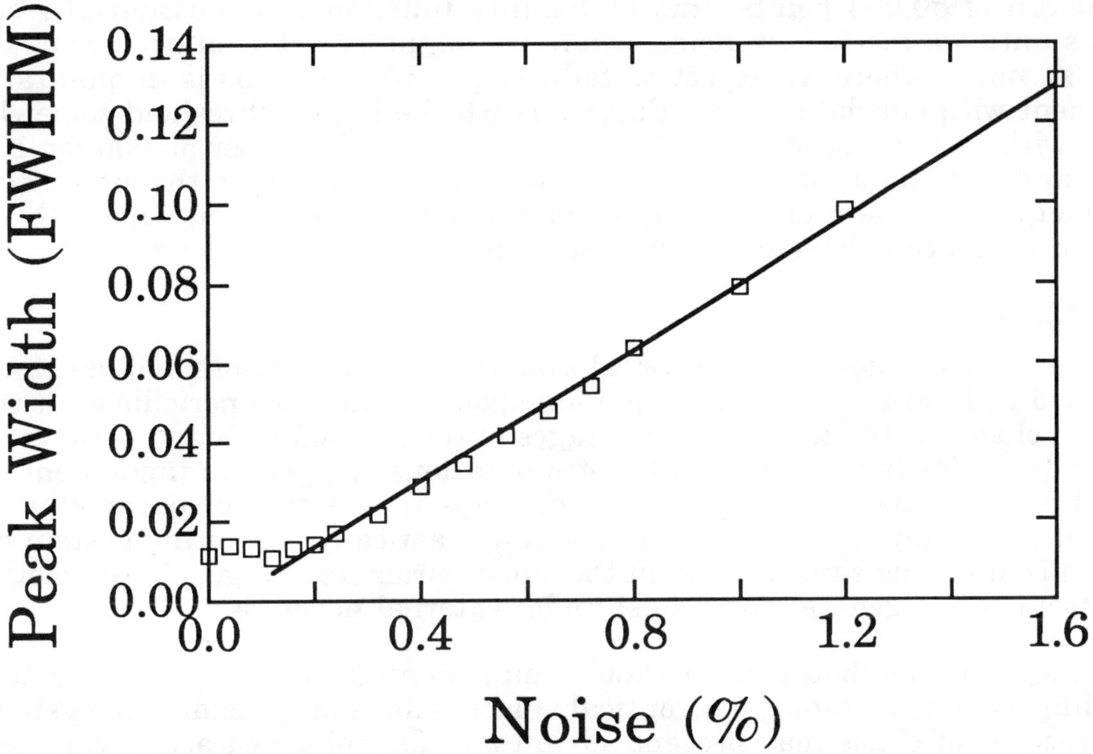

Fig. 8. Gaussian peak full widths at half maximum for various noise levels. Noise amplitudes are expressed as percentages of the ac drive amplitude. The line is a fit of the data described by FWHM = -0.00286 + 0.0828 * noise.

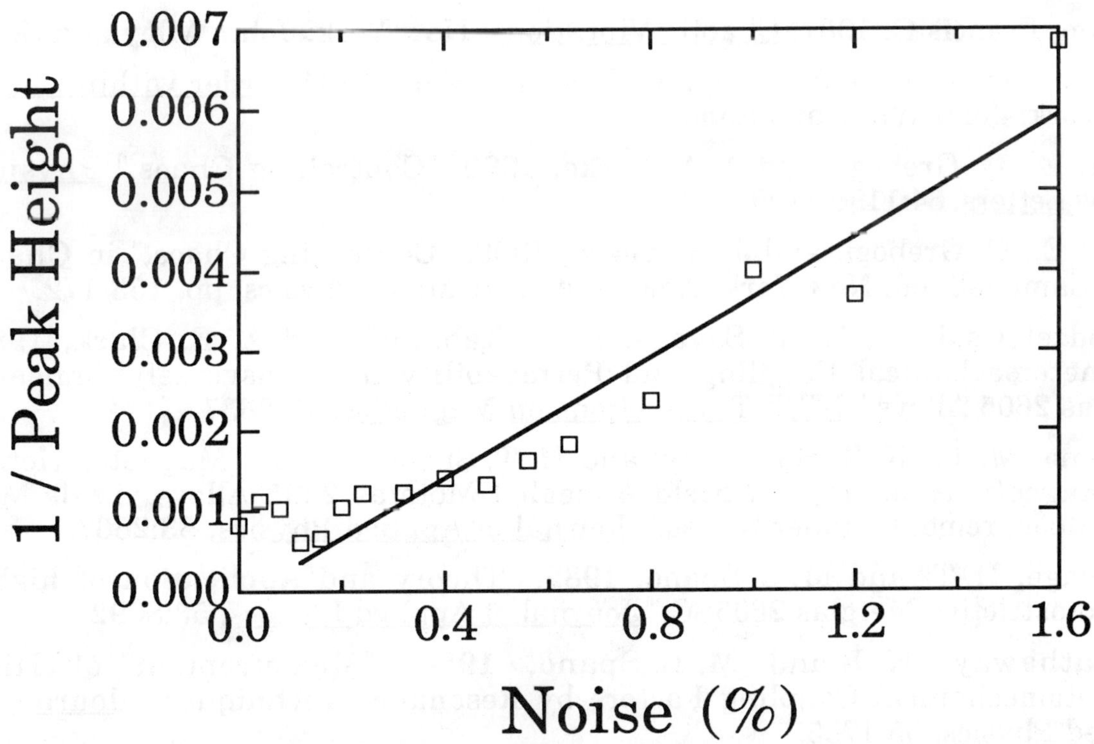

Fig.9 Reciprocal of the gaussian peak heights for various noise levels. Noise amplitudes are expressed as percentages of the ac drive amplitude. The line is a fit of the data described by 1/height = -0.000101 +0.00378 * noise.

typical run of 50,000 points, this probability function yields essentially zero failures until we reach 1.2% noise, where we might expect 1 failure per 50,000 and 1.6% noise, where we expect 58 failures per 50,000. This is in qualitative agreement with our data, except that we reach the failure threshold somewhat earlier with 10 failures per 50,000 for 1% noise, 84 failures per 50,000 for 1.2% noise, and thousands of failures for 1.6% noise. However the qualitative agreement with the sudden turn-on of the failures is consistent with the behavior of the complementary error function.

CONCLUSION

We have demonstrated that a physical system exhibiting chaotic motion can be controlled by localizing the chaotic motion about an unstable periodic orbit that lies on a chaotic attractor. The advantages of the method include: 1) no model for the dynamics is required; 2) the computations required to implement the control are minimal; 3) the required changes in the control parameter are quite small; 4) different periodic orbits in a chaotically oscillating system can be stabilized for the same system in the same parameter range; 5) the method of control is quite general and robust under external noise.

We believe this method can be widely implemented in a variety of systems including chemical, biological, optical, electronic and mechanical systems. The presence of chaos may present advantages for enhanced active control of nonlinear systems.

REFERENCES

+ Metglas is a registered trademark of the Allied Corporation.

1. Moon, Francis C. 1987. Chaotic Vibrations, New York: John Wiley and Sons.

2. Berge, Pierre, Yves Pomeau, and Christian Vidal,1984. Order within Chaos, New York: John Wiley and Sons.

3. Ott, E., C. Grebogi, and J. A. Yorke, 1990. "Controlling Chaos." Physical Review Letters, 64:1196-1199.

4. Ott, E., C. Grebogi, and J. A. Yorke, 1990. "Controlling Chaos," in Chaos, D. K. Campbell, ed. New York: American Institute of Physics, pp. 153-172.

5. Modzelewski, C., H. T. Savage, L. T. Kabacoff, and A. E. Clark, 1981. "Magnetomechanical Coupling and Permeability in Transversely Annealed Metglas 2605 Alloys." IEEE Transactions on Magnetics, 17:2837.

6. Spano, M. L., K. B. Hathaway, and H. T. Savage, 1982. "Magnetostriction and Magnetic Anisotropy of Field Annealed Metglas 2605 Alloys via dc M-H Loop Measurements under Stress." Journal of Applied Physics, 53:2667.

7. Savage, H. T. and M. L. Spano, 1982. "Theory and Application of highly Magnetostrictive Metglas 2605SC." Journal of Applied Physics, 53:8092.

8. Hathaway, K. B.and M. L. Spano, 1984. "Measurement of High Magnetomechanical Coupling Factors by Resonance Techniques." Journal of Applied Physics, 55:1765.

9. Wun-Fogle, M., A. E. Clark, and K. B. Hathaway, 1986. "Permeability in "Frozen" High Magnetostrictive Coupling Amorphous Ribbons." Journal of Magnetism and Magnetic Materials, 54-57:893.

10. Kabacoff, L. T., H. T. Savage, and M. Wun-Fogle, 1986. "Thermal, Magnetic and Magnetomechanical Properties of Amorphous Magnetron Sputtered $Fe_{78}B_{13}Si_9$." IEEE Transactions on Magnetics, 22:427.

11. Wun-Fogle, M., H. T. Savage, and A. E. Clark, 1987. "Sensitive, Wide Frequency Range Magnetostrictive Strain Gage." Sensors and Actuators, 12:323.

12. Wun-Fogle, M., H. T. Savage, L. T. Kabacoff, K. B. Hathaway, J. M. Merchant, and B. Beihoff, 1988. "Magnetoelastic Effects in Amorphous Wires and Amorphous Ribbons with Nonmagnetic Thin-Film Coatings." Journal of Applied Physics, 64:5405.

13. Wun-Fogle, M., H. T. Savage, L. T. Kabacoff, M. L. Spano, J. R. Cullen, G. A. Jones, and D. G. Lord, 1989. "Effect of Applied Stress on the Magnetization of Amorphous Magnetoelastic Wires." IEEE Transactions on Magnetics, 25:3617.

14. Clark, A. E., and M. Wun-Fogle, 1989. "A New Method of Magnetostrictivity and Magnetostriction Measurement." IEEE Transactions on Magnetics, 25:3611.

15. Wun-Fogle, M., H. T. Savage, and M. L. Spano, 1989. "Enhancement of Magnetostrictive Effects for Sensor Applications." Journal of Materials Engineering, 11:103.

16. Ditto, W. L., S. Rauseo, R. Cawley, C. Grebogi, G.-H. Hsu, E. Kostelich, E. Ott, H. T. Savage, R. Segnan, M. L. Spano, and J. A. Yorke, 1989. "Experimental Observation of Crisis-Induced Intermittency and Its Critical Exponent." Physical Review Letters, 63:923.

17. Savage, H. T., W. L. Ditto, P. A. Braza, M. L. Spano, S. N. Rauseo, and W. C. Spring III, 1990. "Crisis-Induced Intermittency in a Parametrically Driven, Gravitationally Buckled, Magnetoelastic Amorphous Ribbon Experiment." Journal of Applied Physics, 67:5619.

18. Ditto, W. L., M. L. Spano, H. T. Savage, S. N. Rauseo, J. Heagy, and E. Ott, 1990. "Experimental Observation of a Strange Nonchaotic Attractor." Physical Review Letters, 65:533.

19. Henon, M., 1982. "On the Numerical Computation of Poincaré Map." Physica, 5D:412.

20. Sommerer, J. C., W. L. Ditto, C. Grebogi, E. Ott, and M. L. Spano, 1991. "Experimental Confirmation of the Theory for Critical Exponents of Crises." to appear in Physics Letters.

21. Ditto, W. L., S. N. Rauseo, and M. L. Spano, 1990. "Experimental Control of Chaos." Physical Review Letters, 65:3211.

22. Sommerer, J. C., W. L. Ditto, C. Grebogi, E. Ott, and M. L. Spano, 1990."Experimental Confirmation of the Scaling Theory for Noise-Induced Crises." submitted to Physical Review Letters.

23. Packard, N. H., J. P. Crutchfield, J. D. Farmer, and R. S. Shaw, 1980. "Geometry from a Time Series." Physical Review Letters, 45:712.

ADAPTIVE EQUALIZATION OF THE ACOUSTIC RESPONSE IN THE NASA LANGLEY SONIC BOOM CHAMBER

Donald E. Brown,[1] Brenda M. Sullivan[1]

ABSTRACT

Two methods are presented for equalizing simulated sonic booms inside a test chamber constructed at the NASA Langley Research Center. Human test subjects can be evaluated regarding their annoyance to various sonic boom signatures which are simulated by modified N-waves. However, the test chamber distorts these N-waves before they are heard by the test subjects. The solution to this distortion problem is similar to that for classical equalization problems. This involves estimating the transfer function of the test chamber and producing its inverse function. The N-wave is then pre-filtered with the inverse function before being played into the chamber, resulting in a sonic boom that is shaped more like the original N-wave, thus almost nullifying the distortion effects of the chamber. Accurately estimating the transfer function of the test chamber involves measuring the acoustic response of the chamber at a number of discrete frequencies depending on the desired frequency resolution. Typically this type of problem is solved in the frequency domain but a time domain technique is used as well. This paper investigates and compares the properties of the frequency domain inverse Wiener filtering technique and the adaptive time domain Widrow-Hoff least mean-square (LMS) estimation technique as methods for equalizing the response of the sonic boom test chamber. It is shown that while both techniques are successful at reducing distortions in the N-waves, the LMS technique requires significant computer time for adaption while the inverse Wiener filter requires knowledge of the noise variance.

INTRODUCTION

Public annoyance to sonic booms has restricted high-speed civil aircraft to flying at supersonic speeds only over water. Supersonic overland operation is important for the economic viability of future high speed air transportation systems. Different aircraft configurations have been found to produce different sonic boom signatures [1]. To evaluate the effect of these different signatures on public acceptance, a test chamber has been constructed at the NASA Langley Research Center in which human test subjects will rate various types of sonic booms [2]. Sonic boom signatures can be simulated using modified N-waves [3]. However due to the frequency response of the test chamber, N-waves played into the chamber are distorted as illustrated in figure 1.

[1]Lockheed Engineering and Sciences Company, NASA Langley Research Center, Mail Stop 463, Hampton, VA 23665-5225, U.S.A.

It is possible to minimize the distortion effects of the test chamber by pre-filtering the desired N-wave with an equalization filter. Previous work by Niedzwiecki and Ribner [3] utilized frequency domain techniques to equalize sonic boom signatures. Elliott and Nelson [4] equalized the acoustic response of a room using adaptive time domain filters. Although time domain optimization algorithms are gaining in popularity and are simple to implement, the basic principles of these algorithms are not widely understood. Therefore this paper will present the results of using both frequency domain and time domain techniques to equalize sonic booms. The first method uses a time domain technique called the Widrow-Hoff least mean-square (LMS) adaptive algorithm [5]. The second technique uses the frequency domain inverse Wiener filter [6]. The merits of each filter's equalization capabilities will be discussed.

TIME DOMAIN TECHNIQUES

The digital analysis techniques used in this paper assume that a time sequence $x(n)$ is known at regularly spaced intervals Δt *sec* apart, where Δt is the inverse of the sample rate. The integer n is used in this paper instead of time t.

In figure 2 the output of the sonic boom test chamber $y(n)$ is modeled as

$$y(n) = \sum_{m=0}^{M-1} p(m)x(n-m) + e(n). \tag{1}$$

The input $x(n)$ is defined for $n = 0, 1, \cdots, \infty$. The function $p(m)$ is a finite impulse response estimate of the actual impulse response of the test chamber. The constant M is the filter length of $p(m)$ which is defined for $m = 0, \cdots, M-1$. Background noise in the test chamber and electronic noise are denoted by $e(n)$. The noise is assumed to be minimal and white, *i.e.* having a zero mean, constant variance and being uncorrelated over time.

The variable m in equation (1) is a variable used to index the values of $p(n)$ for the convolution definition. The output $y(n)$ can be characterized as a weighted moving average of the input $x(n-m)$, whose weights are specified by the coefficients $p(m)$.

Equation (1) can be interpreted in either the discrete domain, where the input $x(n)$ is infinitely long and non-periodic, and all calculations take place in the time domain [7], or in the 'classical' periodically discrete domain, where input records are fixed length and required to repeat periodically [8]. The first interpretation of equation (1) is used when all operations are in the time domain. The second interpretation will be assumed when calculations are performed in the discrete frequency domain.

LMS plant estimate

In the time domain, an estimate of the test chamber impulse response is made using the LMS adaptive algorithm [5]. As illustrated in figure 3 this algorithm requires knowledge of the plant input and output sequences $x(n)$ and $y(n)$ respectively. The LMS algorithm minimizes the mean-square error $|y(n)-y_p(n)|^2$ by iteratively updating

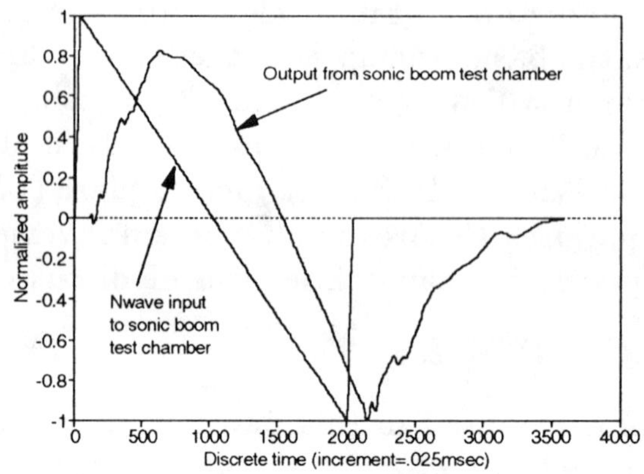

Figure 1. A comparison of an undistorted N-wave and the resultant distorted output from the sonic boom chamber.

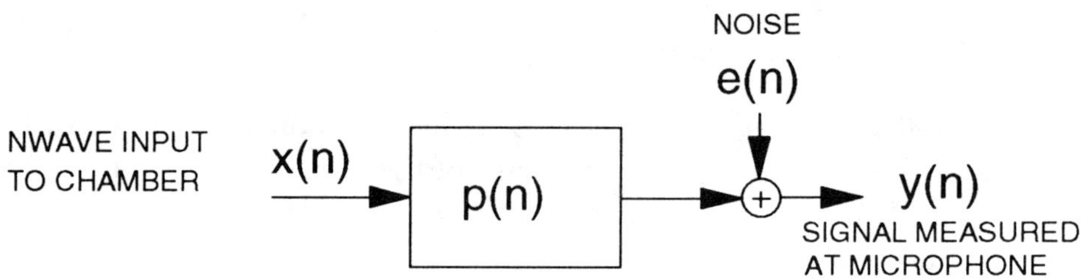

Figure 2. A model of the sonic boom test chamber.

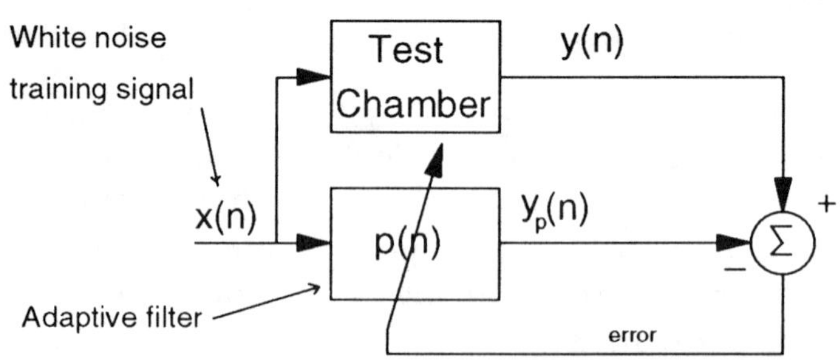

Figure 3. Model used to make an estimate $p(n)$ of the actual impulse response of the sonic boom test chamber.

the coefficients $p(m)$, resulting in an estimate of the impulse response of the sonic boom test chamber.

To generate the coefficients $p(m)$, for $m = 0, \cdots, M-1$, the LMS algorithm uses the error signal $y(n) - y_p(n)$ at time n to iteratively adapt the coefficients $p(m)$. The iteration is performed by adding the incremental value $\Delta p(m)$ to the previous value of $p(m)$ where

$$\Delta p(m) = 2\mu[y(n) - y_p(n)]x(n-m) \quad for\ m = 0, \cdots, M-1. \tag{2}$$

Between iterations an updated value of $y_p(n)$ is calculated using equation (1). Then a new value of $[y(n) - y_p(n)]$ is calculated before the iteration is repeated again.

The constant μ is a convergence parameter used to control the rate of convergence. A value too large will cause the algorithm to become unstable. Therefore a small initial value of μ is chosen and slowly increased until the error decreases below a predetermined level. This level is called the minimum mean-square error [5] and is a function of the convergence parameter, the extraneous noise and the number of coefficients adapted.

The total number of coefficients is equal to the filter length M. The exact value to use for M is a function of the frequency characteristics of the acoustic signal, the frequency response of the test chamber, the amount of noise present, and the amount of error permissible.

LMS equalization filter

Figure 4 illustrates the model used to generate a digital filter $r(n)$ that will equalize the frequency response of the sonic boom test chamber. The LMS algorithm of the preceeding section is easily adapted to estimate the coefficients $r(m)$ by merely substituting the error signal $[x_d(n) - x_r(n)]$ for $[y(n) - y_p(n)]$. Iteratively updating the coefficients $r(m)$ generates a filter that will equalize the frequency response of the test chamber. The delay modeled by $x_d(n)$ should be longer than the delay through the test chamber to allow a causal solution.

Once the equalization filter $r(n)$ has been determined, it is used to pre-filter N-waves [4] as illustrated by figure 5. The LMS equalization process used in this paper consists of first identifying the equalization operator $r(n)$ and convolving N-waves with $r(n)$ before they are played into the test chamber.

FREQUENCY DOMAIN TECHNIQUES

The discrete Fourier transform (DFT) converts a record of discrete time data $x(n)$ into the periodically discrete frequency spectrum $X(f)$. The DFT is also used to convert a digital filter $r(n)$ into the frequency response $R(f)$. The DFT $R(f)$ is defined as [8]

$$R(f) = \frac{1}{N} \sum_{n=0}^{N-1} r(n) e^{\frac{-i2\pi fn}{N}}. \tag{3}$$

For a filter $R(f)$ of length N, the frequency resolution is $\frac{1}{N\Delta t}$. A sample rate of $40kHz$ is used for all the examples presented in this paper. This results in a frequency resolution of $10Hz$ for a filter length of $N = 4096$.

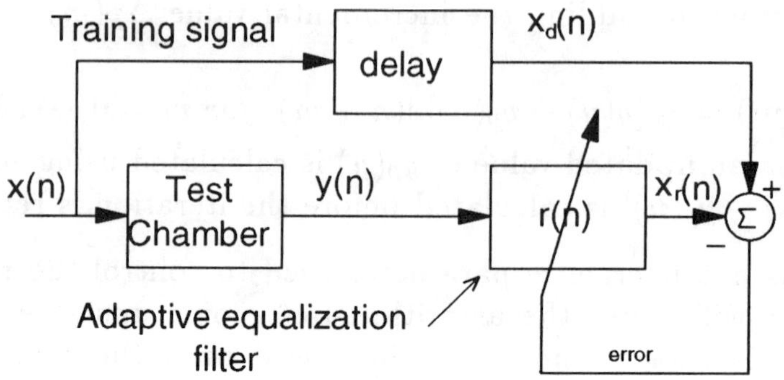

Figure 4. Model used to estimate an equalization filter $r(n)$ that that will nullify the distortion effects of the sonic boom test chamber.

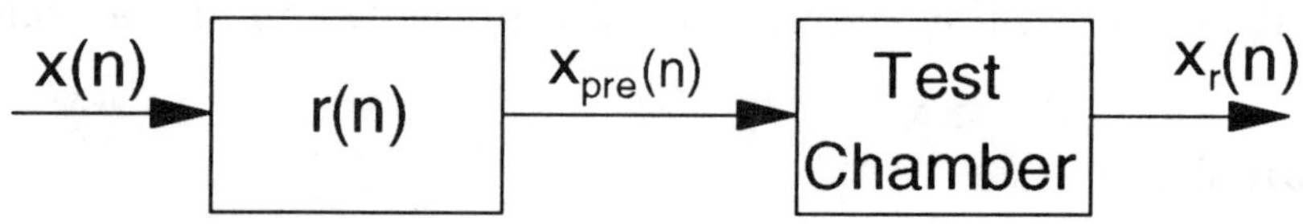

Figure 5. The equalization filter $r(n)$ as used to pre-filter N-waves $x(n)$ before being played into the sonic boom test chamber.

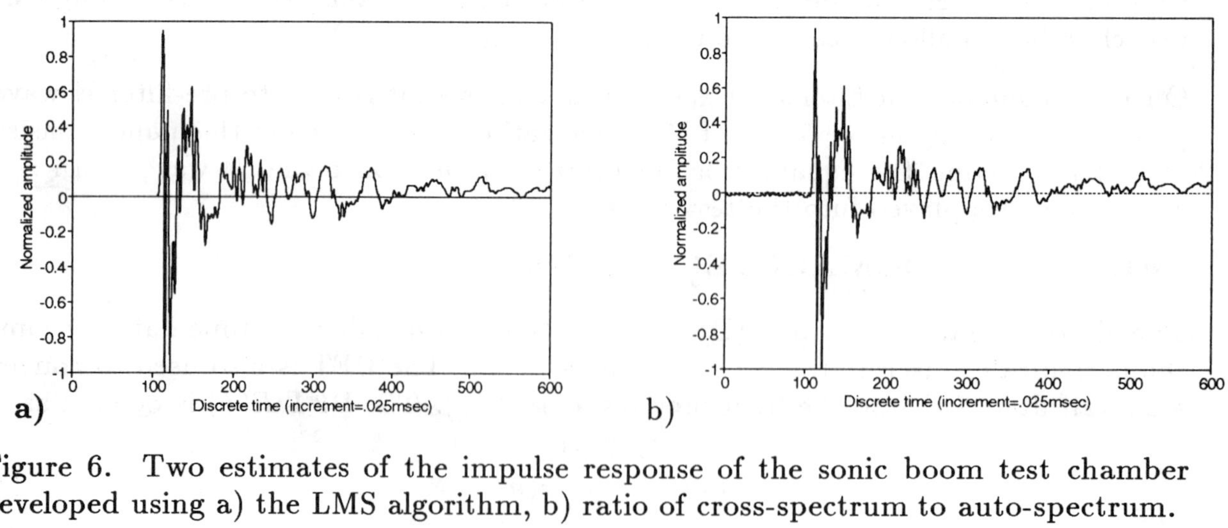

Figure 6. Two estimates of the impulse response of the sonic boom test chamber developed using a) the LMS algorithm, b) ratio of cross-spectrum to auto-spectrum.

The DFT of equation (1) yields the following fundamental principle of linear system theory

$$Y(f) = P(f)X(f) \tag{4}$$

which states that the spectrum of the output is equal to the spectrum of the input times the frequency response of the plant.

It can be demonstrated that $R(f)$ is periodic with a period N and that it displays $R(f) = R^*(N - f)$ symmetry [8], resulting in $\frac{N}{2}$ unique frequency values. The maximum frequency $f = \frac{N}{2}$ is called the Nyquist frequency and is equal to one half the sample rate. Because of periodicity, multiplication in the frequency domain using equation (4) circularly convolves data near the ends of the record with data from the other end. To avoid circular convolution most discrete frequency techniques require the data to be padded with zeroes before performing the DFT.

Plant frequency response

An estimate of the frequency response $P(f)$ is made in the frequency domain using stochastic analysis techniques. Using the model in figure 2 where all plant noise is assumed to be located at the output, the frequency response $P(f)$ is determined from the ratio of the cross spectrum $G_{xy}(f)$ and the auto-spectrum $G_{xx}(f)$ [9]

$$P(f) = \frac{G_{xy}(f)}{G_{xx}(f)}. \tag{5}$$

Because these terms are expected values, ensemble averaging is required to yield the expected frequency response.

Before performing the DFT to calculate the spectral terms of equation (5), the data is zero padded by 50%. This zero padding will allow the use of the inverse DFT to make an unbiased estimate of $p(n)$ from $P(f)$ [10].

Inverse DFT

The inverse discrete Fourier transform is used to convert a frequency function $R(f)$ into the time domain function $r(n)$. The inverse DFT of $R(f)$ is defined as [8]

$$r(n) = \sum_{f=0}^{N-1} R(f) e^{\frac{i2\pi fn}{N}}. \tag{6}$$

When estimating impulse responses using frequency domain techniques, a correction for the 'bow-tie' effect of the frequency response $R(f)$ must be performed following the inverse DFT operation to make $r(n)$ an unbiased estimate of the true impulse response [10].

Wiener equalization filter

In the absence of noise (figure 4), the ideal equalization filter $R(f)$ would be the inverse of $P(f)$ [11]. However, the presence of noise and the possibilities of singularities in the inverse of $P(f)$ make using an actual inverse impractical. Instead, mean-square estimation techniques are used to generate $R(f)$.

Using Parseval's equation it can be shown that minimizing $|x_d(n) - x_r(n)|^2$ in figure 4 can be equivalently solved in either the frequency domain or time domain [8], *i.e.*

$$\frac{1}{N}\sum_{n=0}^{N-1}|x_d(n) - x_r(n)|^2 = \sum_{f=0}^{N-1}|X_d(f) - X_r(f)|^2. \tag{7}$$

By minimizing the error $|x_d(n) - x_r(n)|^2$ in the frequency domain and assuming a delay of zero, the classical inverse Wiener filter $R(f)$ can be derived [6].

$$R(f) = \frac{P(f)^*}{|P(f)|^2 + \frac{\sigma^2}{N|X(f)|^2}} \tag{8}$$

The terms $P(f)$, $X(f)$ and $R(f)$ are expected values requiring ensemble averaging to yield the expected equalization filter. The noise variance, denoted as σ^2, must be either known or estimated from the data $x(n)$ and $y(n)$.

The Wiener equalization algorithm used in this paper consists of pre-filtering N-waves in the frequency domain using equation (4) and then inverse Fourier transforming the results to the time domain before they are played into the test chamber. This is in contrast to the LMS equalization technique where all computations take place in the time domain.

TEST SETUP

The sonic boom test chamber, with approximate dimensions of $4' \times 4' \times 5'$, is constructed of concrete building blocks and lined on the inside with acoustical foam to minimize reverberations [2]. The acoustic signal enters the chamber through 4 low-frequency and 4 mid-frequency speakers mounted on the door of the chamber. A chair and light are placed inside for use by the test subject. A 1 *inch* condenser microphone inside the test chamber measures the sound heard by the subject.

Typical durations for N-waves representing sonic booms range from 100 *msec* to 300 *msec* [3]. The leading edge of the N-wave is described by the *rise* time and a typical value is 3 *msec*.

All digital computations are performed on a mini computer. A custom designed A/D and D/A board allows the synthesis of digital wave forms and the measurement of signals in the test chamber.

Discrete white noise is used as a training signal to generate the output $y(n)$ in figures 3 and 4. Both the input and output sequences $x(n)$ and $y(n)$ are stored in a data file. They are retrieved at later time to calculate the LMS and Wiener equalization filters. During a sonic boom test session, N-waves are pre-filtered by the computer and played directly into the chamber.

RESULTS AND DISCUSSIONS

All of the figures presented in this paper show data that has been normalized. The relative levels of plots within the same figure (*e.g.* 7.*a* and 7.*b*) can be compared, while the levels in different figures cannot.

Identification

Figure 6.a displays the LMS time domain estimate $p(n)$ of the test chamber impulse response using figure 3. The number of coefficients adapted was 4096. The input signal $x(n)$ used during the adaption process was a discrete white noise training signal. Figure 6.b illustrates the inverse DFT of the frequency domain estimate $P(f)$. Note that figure 6.b compares closely to figure 6.a. This demonstrates, at least in respect to plant identification, that both the time domain and the frequency domain estimates are equivalent.

Figures 7.a and 7.b illustrate the frequency responses of the two previous estimates. Although, the amplitudes differ, the responses are similar.

Equalization filters

Figure 8.b displays the inverse Wiener filter $R(f)$ for $\sigma = .00001$ and $N = 4096$. The values for $P(f)$ were taken from figure 7.b. An estimate of $X(f)$ was taken from the discrete white noise training signal, resulting in a broadband equalization filter. The spectrum of an N-wave could have been used, but the resulting equalization filter would have been significant for only that N-wave.

Figure 8.a shows the frequency response of the LMS equalization filter $r(n)$. A comparison with the inverse Wiener filter of figure 8.b shows a significant difference in the frequency response of the two filters. Part of the difference can be attributed to an error in the estimation of σ. While the Wiener filter is a closed form solution, estimating σ can quite easily become an iterative process in itself. Another reason for the difference might be a procedural error in the calculations. Work continues on finding the explanation of the differences in the two estimates of $R(f)$.

One feature of the LMS technique is that it removes the effects of any noise $e(n)$, but at the cost of many iterations. As explained before, the Wiener filter requires an accurate estimate of σ^2.

The average number of iterations required to generate the LMS estimate of $r(n)$ was on the order of half a million. Considering the fact that 4096 coefficients are updated for each iteration, the computational burden is quite extensive. The number of iterations required by the LMS algorithm to generate $p(n)$ was considerably less than that required to generate $r(n)$. Reasonably accurate estimates of $p(n)$ where obtained after only $10,000$ iterations.

Both these equalization filters are broadband in nature and are able to equalize any signal consisting entirely of frequencies below the Nyquist frequency ($20kHz$).

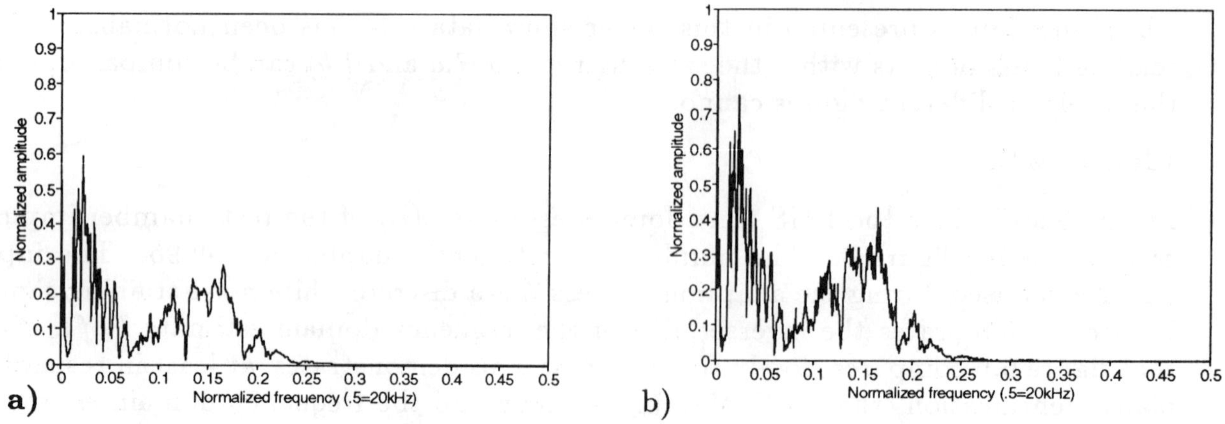

Figure 7. Two estimates of the the frequency response of the sonic boom test chamber developed using a) LMS algorithm, b) ratio of cross-spectum to auto-spectrum.

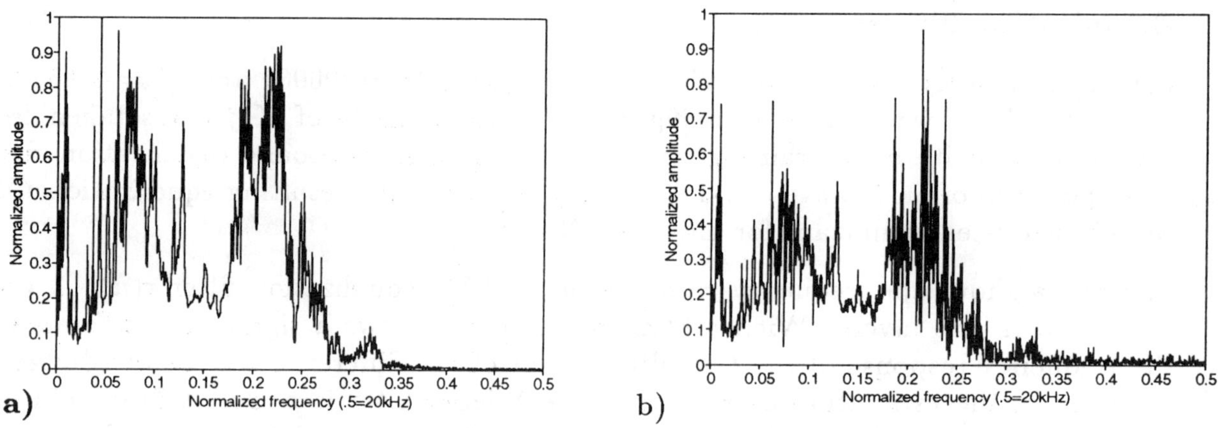

Figure 8. Amplitude of the frequency response of two equalization filters designed using a) LMS algorithm, b) inverse Wiener filter.

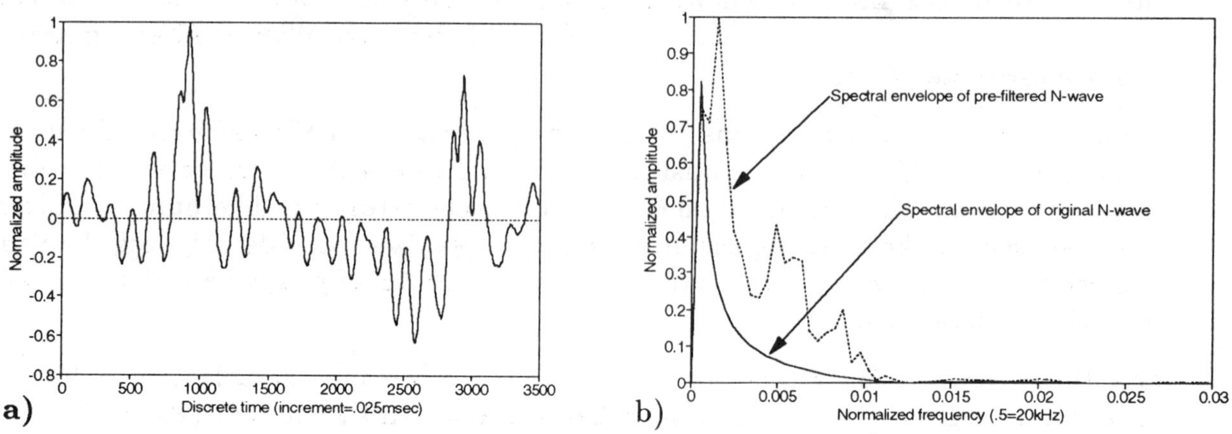

Figure 9. a) An N-wave pre-filtered with the inverse Wiener filter. b) The spectral envelope of the original N-wave and the pre-filtered N-wave.

Pre-equalization

Figure 9.a shows the original N-wave from figure 1 (simulating a 31 *msec* sonic boom with a 1 *msec* rise time) pre-filtered with the inverse Wiener filter $R(f)$ before it will be played into the test chamber. The pre-filtering was done in the frequency domain using equation (4) and inverse Fourier transformed to the time domain.

Figure 9.b shows two spectral envelopes. Because each is normalized, only the general shape of each is significant. The first is the envelope of an N-wave and the second is the envelope of an N-wave pre-filtered by the inverse Wiener filter. The equalization filter minimizes the distortion of the N-wave by increasing the frequency spectrum of the original N-wave .

Figure 10.a shows the equalized N-wave measured after playing the LMS pre-filtered N-wave into the test chamber. All data processing including adaption and filtering took place in the time domain. The rise time has been largely restored to its original value of 1 *msec* and the overall shape of the equalized N-wave more closely resembles the undistorted N-wave in figure 1.

Figure 10.b shows the equalized N-wave measured after playing the frequency domain pre-filtered N-wave into the test chamber. The pre-filtering occurred in the frequency domain using the inverse Wiener filter and the result was inverse discrete Fourier transformed to the time domain before being played into the test chamber.

Both figures 10.a and 10.b show a drop in the amplitude of the positive portion of the equalized N-wave. A possible explanation might be a limitation in the speaker performance. Work continues on identifying the cause of this discrepancy.

Figure 11 displays the frequency spectra of three different N-waves. The solid line represents the spectrum of the original N-wave. The dashed line is the spectrum of an equalized N-wave pre-filtered using the LMS equalization filter. The dotted line is the spectrum of an equalized N-wave pre-filtered using the inverse Wiener filter. Neither one of the equalized spectra exactly matches the spectrum of the original N-wave, yet each is reasonably successful in restoring the original envelope of the desired N-wave as illustrated in figures 10.a and 10.b.

Figure 12 displays the equalized N-wave obtained after pre-filtering a 250 *msec* long N-wave (10,000 time increments) with a shorter equalization filter of 4096 coefficients. Note that the normalized amplitude is less than expected beyond the time period of approximately 4096 increments. This time period equals the length of the LMS equalization filter convolved with the original N-wave. This effect might be avoided by using an equalization filter with more coefficients, but the adaption time would become prohibitive.

CONCLUSIONS

The LMS equalization filter has been demonstrated to be effective at equalizing N-waves in the NASA Langley sonic boom test chamber. Though not in complete agreement with, the LMS equalization filter compares well with the classical inverse Wiener filtering technique. Truncation errors in the time-frequency transformation procedures

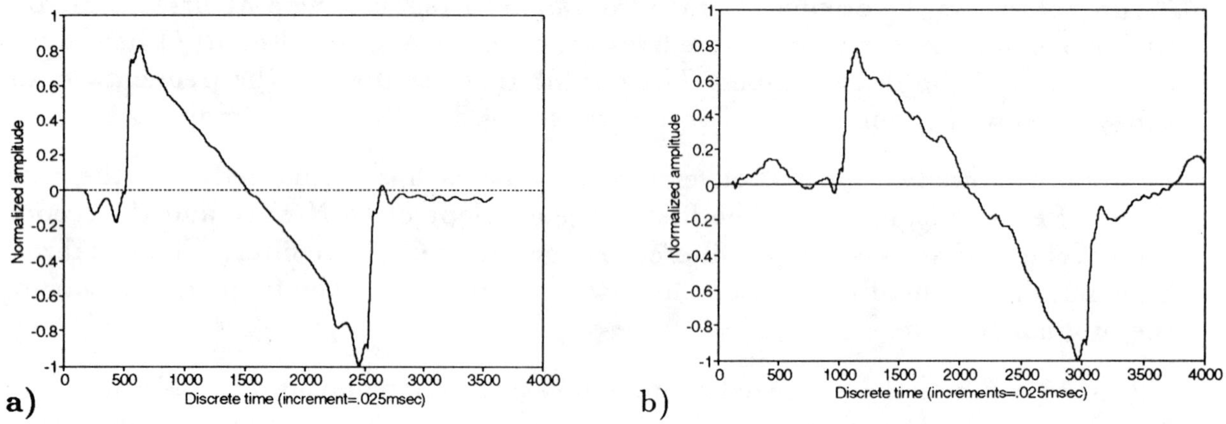

Figure 10. Amplitude of the two equalized N-waves pre-filtered using a) LMS equalization filter, b) inverse Wiener filter.

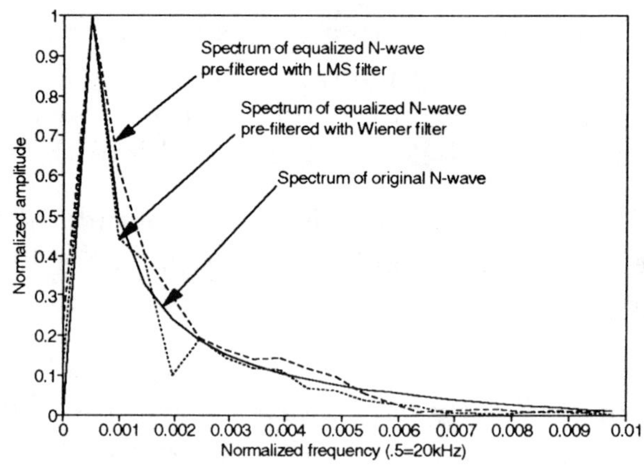

Figure 11. Amplitude of the frequency spectra of the original N-wave and two equalized N-waves.

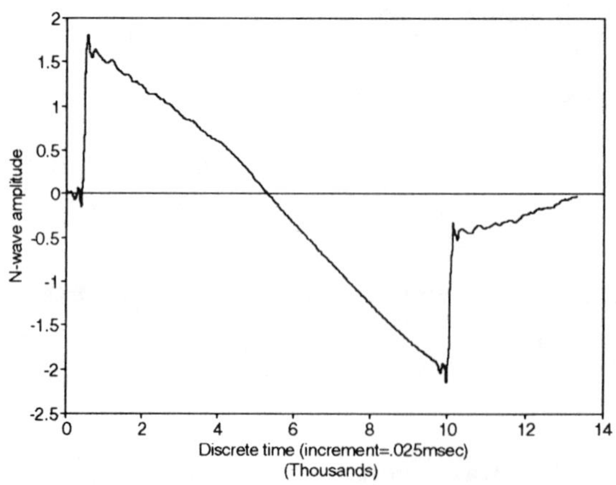

Figure 12. Example of an equalized N-wave of $250msec$ duration using an LMS equalization filter of 4096 coefficients.

or an error in the estimation of σ^2 are possible explanations for differences in the frequency response of the LMS equalization filter and the inverse Wiener filter. The LMS algorithm averages out the effects of small amounts of noise, but a greater number of iterations is required to develop an optimal equalization filter. The Wiener filter requires an accurate estimate of the noise variance. This requirement leads to inaccuracies when working under real-noise conditions, where estimating the noise variance is difficult. Further investigation is needed to explain the differences in the resulting equalized N-waves.

Neither the LMS equalization filter nor the inverse Wiener filter were able to completely neutralize the distortion effects of the test chamber. Both filters fell short in restoring the amplitude level of the positive portion of the N-wave. This might be due to limitations in the speaker performance and will be the subject of further study.

Acknowledgements

This work was performed under NASA Contract NAS1-19000, Richard J. Silcox, technical monitor.

References

1. Darden, C. M. 1975. "Minimization of Sonic Boom Parameters in Real and Isothermal Atmospheres", NASA TN D-7842.

2. Leatherwood, J. D, K. P. Shepherd and B. M. Sullivan, 1991. "Sonic Boom Simulation and Testing", NASA TM 16912.

3. Niedzwiecki, A. and H. S. Ribner, 1978. "Subjective loudness of 'minimized' sonic boom waveforms", *J. Acoust. Soc Am.*, 64(6):1622-1626.

4. Elliott, S. J. and P. A. Nelson, 1989. "Multiple-Point Equalization in a Room Using Adaptive Digital Filters", *J. Audio Eng. Soc.*, 37(11):899-907.

5. Widrow, B. and D. S. Sterns, 1985. *Adaptive Signal Processing*, Englewood Cliffs, NJ: Prentice Hall Inc.

6. Slepian, D. 1967. "Linear Least-squares Filtering of Distorted Images", *J. Opt. Soc. Am.*, 57:918-922.

7. Hamming, R. W. 1989. *Digital Filters*, Englewood Cliffs, NJ: Prentice Hall, Inc.

8. Gonzalez, R. C. and P. Wintz, 1977. *Digital Image Processing* Reading, Massachusetts: Addison-Wesley.

9. Hardin, J. C. 1986. *Introduction to Time Series Analysis*, Hampton, VA: NASA Reference Publication 1145.

10. Bendat, J. S. and A. G.Piersol, 1971. *Random Data: Analysis and Measurement Procedures*, NY, NY: John Wiley & Sons, Inc.

11. Andrews, H. C. and B. R.Hunt, 1977. *Digital Image Restoration*, Englewood Cliffs, NJ: Prentice Hall, Inc.

THE IMPLEMENTATION OF A LATTICE FILTER STRUCTURE AND SUITABLE ALGORITHM FOR AN ADAPTIVE ACTIVE NOISE CONTROL SYSTEM

Neil C. Mackenzie,[1] **Colin H. Hansen**[1]

ABSTRACT

Currently it is common for the electronic controller of an active noise or active vibration control (ANVC) system to utilize a transversal filter and the LMS algorithm. However, it has been shown [1] that the effects of the time varying time delay and electro-acoustic transfer functions, inherent in such a system, can result in system instability. Conventional means of overcoming this problem consist of using on-line time delay and transfer function measurements which are then taken into account in the control algorithm. However this increases system complexity and in some systems reduces the potential noise reduction which can be obtained, due to the need to inject periodic random noise (or chirps) into the system.

It is proposed in this paper to present a preliminary analysis of the properties of lattice filters, as applied to the field of adaptive active noise control, with the view to applying the results to the design of a multi-channel adaptive active controller which is phase insensitive and thus will not require an on-line system identification network.

INTRODUCTION:

A simple active noise or active vibration control (ANVC) system may be divide into an "acoustic plant" and an "error plant" [5,6]. The acoustic plant consists of the transfer function between a reference signal, which is well correlated to the primary disturbance, and the primary disturbance at the input of the error sensor. The error plant consists of the transfer function between the output of the controller and the secondary disturbance at the input of the error sensor.

In practice, the reference signal may be derived from a microphone, an accelerometer or in some cases a tachometer measuring the rotating speed of the machine generating the noise. The error sensor may be an accelerometer, a microphone or a number of these arranged to measure acoustic power flow or acoustic potential energy (in an enclosure, for example).

[1]Department of Mechanical Engineering, University of Adelaide, GPO Box 498, Adelaide, South Australia 5001, Australia.

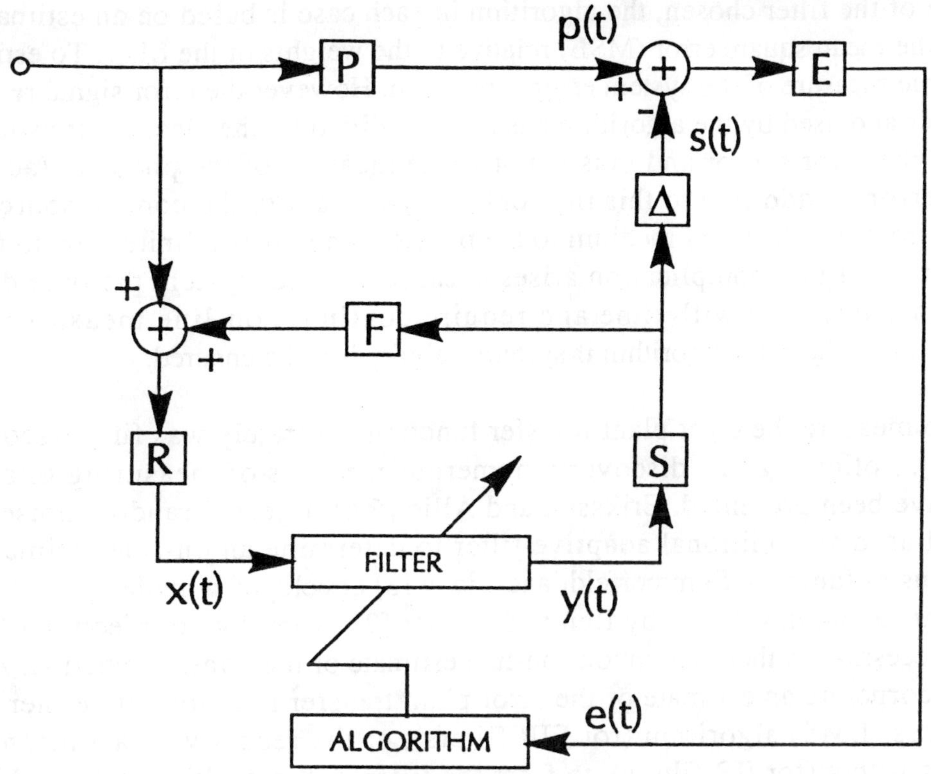

Figure 1. General block diagram of a simple ANVC system involving a single control source and error sensor. The transfer functions represent E = error sensor, Δ = time delay, S = control source, F = feedback plant, P = acoustic plant and R = reference sensor. The signals represent x(t) = reference, y(t) = control, e(t) = error, p(t) = primary disturbance and s(t) = secondary disturbance.

The general block diagram of an ANVC system shown in figure 1 subdivides the error plant into an acoustic time delay (or the time for the secondary disturbance emanating from the control source to reach the error sensor) and a control source transfer function. Also shown is a feedback transfer function due to feedback (input contamination) from the control source to the reference signal.

The filter shown in the figure has in the past been implemented in the form of a tapped delay line (transversal filter). Where there has been no feedback an FIR (finite impulse response) filter (standard transversal filter or tapped-delay-line, with no zeros) has been implemented successfully. However, where input contamination is present, it has been necessary to use an IIR (infinite impulse response) filter (which is recursive and hence non-linear, with both poles and zeros) as the system with input contamination is non-linear. The IIR filter consists of two transversal filters, one feeding the signal forwards and the other feeding it backward.

Irrespective of the filter chosen, the algorithm in each case is based on an estimate of the gradient of the mean square error (MSE) relative to the weights of the filter. To estimate the gradient a true measure of the system error is required. However the error signal received by the controller and used by the algorithm has been modified by the electro-acoustic transfer function of the error sensor and thus is not a true measure of the physical (acoustic or vibration) error. In addition to this the control signal sent to the control source is also transferred from the electrical medium to the physical and takes a finite time to reach the error sensor. A further complication arises because in a real system the time delay and transfer functions vary with time and require continual on-line measurement and incorporation into the filter algorithm if system stability is to be ensured.

The need to measure the error plant transfer function accurately was first discovered by Burgess [4]. Following this discovery, numerous methods of measuring this transfer function have been presented. Eriksson and Allie [5,6], injected random noise into the system and used an additional adaptive filter to determine an on-line estimate of the required transfer function. Sommerfeldt and Tichy [3] developed a method, which appears to be similar to that developed by Ren and Kumar [2], whereby no injection of random noise was necessary in their continuous on-line estimate of the transfer function. All of the methods incorporate an estimate of the error plant transfer function into either the least mean squares (LMS) algorithm (for FIR filters), or the recursive least mean squares (RLMS) algorithm (for IIR filters), to form the filtered-X algorithm, all of which were developed by Widrow and Stearns [8].

The influence of transducer transfer functions and acoustic time delays on the filtered-X algorithm has been investigated separately by Snyder and Hansen [1], and Ren and Kumar [2]. Snyder and Hansen found that errors in estimation of both the phase and magnitude of the transfer function reduced the allowed bounds of the convergence coefficient for stability, and in some cases could lead to instability regardless of this coefficient. Similarly if the weights of the adaptive filter are continuously updated at every sample the allowed bounds of the convergence coefficient for stability were found to be dependent upon the time delay. Ren and Kumar used a different analysis and found that a very crude approximation of the transfer function would ensure the convergence of the algorithm. However they did not consider the change in the allowed bounds on the convergence coefficient caused by errors in the estimation of the transfer function.

The primary aim of this paper is to introduce an alternative algorithm which eliminates the need for on-line measurement of electro-acoustic transfer functions and acoustic time delays, and the many practical difficulties and performance limitations associated with this measurement. To achieve this aim, the introduction of a new type of filter, referred to as a tapped-lattice, is necessary. As will be shown, this filter comprises a lattice and transversal filter. The lattice filter is responsible for the transformation of the reference signal into as many uncorrelated signals as there are weights in the transversal filter. The use of uncorrelated signals in the lattice as opposed to delayed reference signals means that the weights of the transversal filter may be adjusted independently. This forms the basis of the alternative algorithm.

The tapped-lattice filter has been implemented in the field of electronics for channel equalization and noise cancelling [13,14,15,17]. The introduction of a transfer function after the adaptive filter, a necessity in ANVC systems, has not previously been considered for any types of lattice filter structures.

This paper begins with a brief discussion of the LMS algorithm (a stochastic algorithm) before continuing with an analysis of the basic lattice filter structure and its efficient use of an exact algorithm. Following this a brief discussion of the tapped-lattice structure leads to the development of an alternative algorithm. Finally, the simulation of a simple acoustic system involving both time delays and transfer functions is shown.

For simplicity, a single control source and single error sensor is considered. It is assumed that input contamination is not present and that both the reference and error signals are stationary.

REVIEW OF EXISTING FILTER STRUCTURES AND ALGORITHMS:

Adaptive filters consist of a filter structure and an adaptive algorithm. The filter structure most commonly used, due to its simplicity, is the tapped delay line (or transversal filter) which represents a finite impulse response (FIR) transfer function and is shown in figure 2.

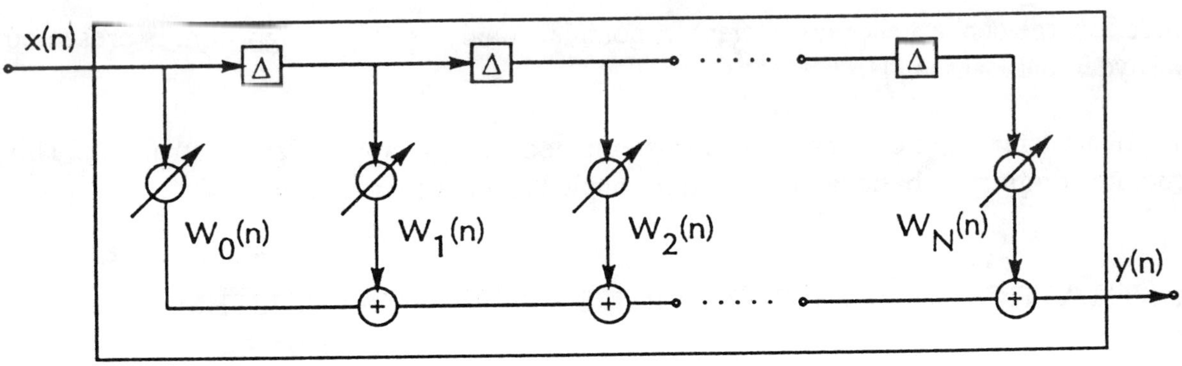

Figure 2. Tapped-Delay-Line (Transversal Filter).

An adaptive algorithm attempts to adjust the filter weights to minimize a predefined cost function, based on a measure of the error signal. The error signal may be calculated as:

$$e_N(n) = d(n) - X_N^T(n)W_N(n-1) \quad (1)$$

where n is the current sample number, N is the number of weights and:

$e_N(n)$ = error signal (a priori),
$d(n)$ = desired signal (primary disturbance),
$X_N^T(n)W_N(n)$ = estimated desired signal (secondary disturbance),

where

$$X_N(n) = [\ x(n),\ x(n-1),\ ..,\ x(n+1-N)\ ]^T \quad (2)$$

= Vector of reference signal samples,

$$W_N(n) = [\ w_0(n),\ w_1(n),\ ..,\ w_{N-1}(n)\ ]^T \quad (3)$$

= Filter weights.

Previous active noise control systems have used a stochastic cost function known as the mean square error (MSE) defined by:

$$J_N(n) = E[e_N^2(n)] \quad (4)$$

where E[] denotes the expectation operator. The commonly used LMS algorithm is derived from an estimate of the gradient of this cost function yielding the well known weight update equation:

$$W_N(n+1) = W_N(n) + 2\mu X_N^T(n)e_N(n+1) \quad (5)$$

where μ is the convergence coefficient. A complete analysis of this algorithm is presented by Snyder and Hansen [1].

An FIR transfer function may also be implemented using a tapped- lattice structure. This structure comprises a basic lattice structure coupled to a transversal filter.

BASIC LATTICE STRUCTURE PROPERTIES AND EXACT ALGORITHM

The complex derivation of this structure will not be undertaken here; however, a complete derivation of the structure may be found in [7,9,10]. It was originally derived with linear prediction in mind [7,8,9,10], hence the signals comprising the structure are known as forward and backward prediction errors. The coefficients within the structure are known as PARCOR coefficients, since their values are related to the partial correlation of the forward and backward prediction errors. It can be shown [11] that each stage within a lattice structure is decoupled from the previous stages and the coefficients in each stage may thus be adapted independently. The following recursion relationships hold for both the forward and backward prediction errors [7,8,9,10]:

$$e_{fm}(n+1) = e_{f(m-1)}(n+1) - K_{bm}(n)e_{b(m-1)}(n) \quad (6)$$
$$e_{bm}(n+1) = e_{b(m-1)}(n) - K_{fm}(n)e_{f(m-1)}(n+1) \quad (7)$$

where n is the sample number, m is the filter stage number, K_{fm} and K_{bm} are the PARCOR coefficients and e_{fm} and e_{bm} are the forward and backward prediction errors respectively.

The structure of a stage of the lattice filter can be represented as shown in figure 3.

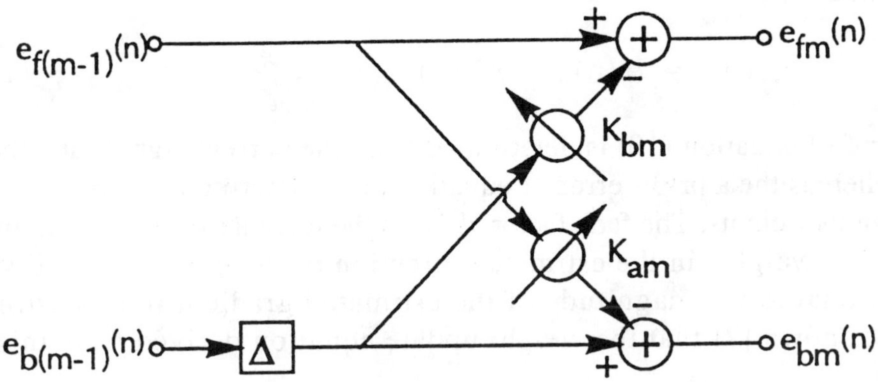

Figure 3. Basic structure of a stage within a lattice filter.

The ratio of K_{fm} to K_{bm} is a measure of the stationarity of the reference signal [11]. Although it can be shown that the backward prediction errors are uncorrelated, the same cannot be said of the forward prediction errors [11]. Specifically the expected value of the product of any two backward prediction errors is given by:

$$E[e_{bi}(n)e_{bj}(n)] = \begin{cases} J_{bi} & \text{for } i=j \\ 0 & \text{for } i \neq j \end{cases} \quad (8)$$

where J_{bi} represents the power of the backward prediction error signals. The overall lattice structure for a filter of (N+1) stages (there is a pseudo stage 0) can be represented as shown in figure 4.

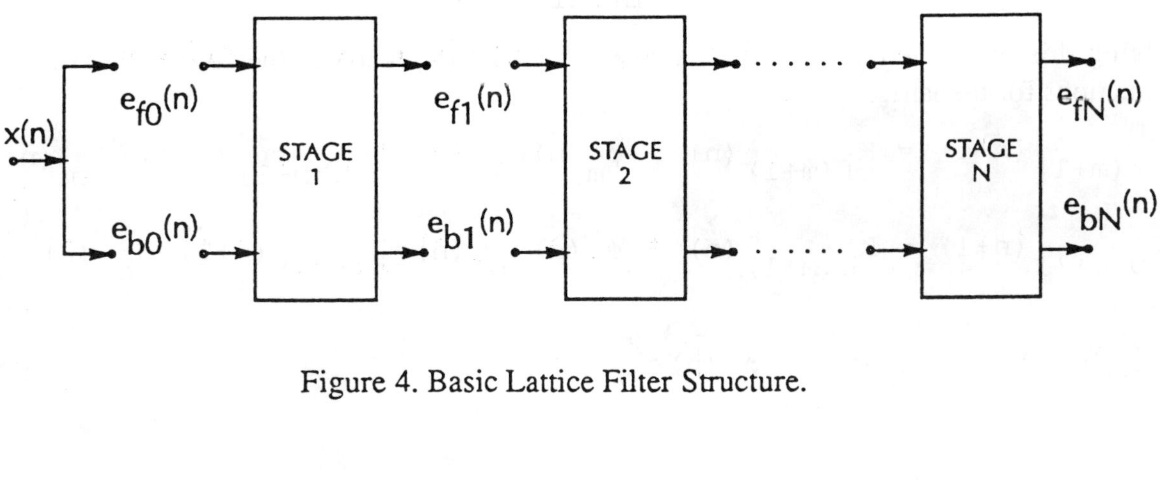

Figure 4. Basic Lattice Filter Structure.

The decoupling of each stage in the lattice structure makes it ideally suited for use with the exact algorithm. The exact algorithm uses a cost function defined by the exact data as:

$$J_N(n) = \sum_{p=1}^{n} \gamma^{n-p} \epsilon_N^2(p) \tag{9}$$

where n is the total number of samples taken by the system at the current time, γ represents a "forgetting factor" or "fade factor", N represents the order of the filter and ϵ_N is the a posteriori error defined by:

$$\epsilon_N(n) = d(n) - X_N^T(n) W_N(n) \tag{10}$$

The a posteriori error of equation (10) is determined from the current signals and the just adjusted weights whereas the a priori error of equation (1) is determined from the current signals and the previous weights. The fade factor defines the memory of the algorithm. The algorithm updates the weights in the estimated direction of the optimum MSE with a magnitude proportional to the magnitude of the estimated gradient in this estimated direction. It can be shown [7] that the weight update equation derived from this cost function is given by:

$$W_N(n+1) = W_N(n) + R_N^{-1}(n) X_N(n+1) e_N(n+1) \tag{11}$$

where

$$R_N(n) = \sum_{p=1}^{n} \gamma^{n-p} X_N(p) X_N^T(p) \tag{12}$$

Since it is not practically possible to efficiently determine the inverse of the sample autocorrelation matrix, $R_N(n)$, a recursive equation can be formulated [7,9,10] as:

$$R_N^{-1}(n+1) = \frac{\phi_N(n+1)}{\gamma} R_N^{-1}(n) \tag{13}$$

where

$$\phi_N(n+1) = \epsilon_N(n+1)/e_N(n+1) \tag{14}$$

$$= \text{Error Ratio}$$

Using this basic theory, a detailed analysis [7,9,10,16] leads to the following recursive equations for the lattice:

$$K_{f(m+1)}(n+1) = K_{f(m+1)}(n) + \phi_m(n) e_{fm}(n+1) e_{b(m+1)}(n+1)/J_{fm}(n+1) \tag{15}$$

$$K_{b(m+1)}(n+1) = K_{b(m+1)}(n) + \phi_m(n) e_{bm}(n) e_{f(m+1)}(n+1)/J_{bm}(n) \tag{16}$$

$$J_{fm}(n+1) = \gamma J_{fm}(n) + \phi_m(n) e_{fm}^2(n+1) \tag{17}$$

$$J_{bm}(n+1) = \gamma J_{bm}(n) + \phi_m(n+1) e^2_{bm}(n+1) \tag{18}$$

$$\phi_{m+1}(n) = \phi_m(n) - \phi^2_m(n) e^2_{bm}(n) / J_{bm}(n) \tag{19}$$

An alternative analysis can be used to determine the recursive update equations based on the LMS algorithm, from which it can be shown that the differences between the stochastic and exact algorithms are:

(a) The step change in the PARCOR coefficients (see equations (14) and (15)) represented by the division of the product of the forward and backward prediction errors by $J_{bm}(n)$ and $J_{fm}(n+1)$ respectively, results in estimates of the forward and backward prediction signal powers, which for independent weights is an estimate of the eigenvalues of the forward and backward prediction error autocorrelation matrices; thus the division maintains the same adaptive time constant for each stage of the lattice [12].

(b) It can be shown [7,9] that the ratio $\phi_N(n)$ can be represented by:

$$\phi_N(n) = 1 - \theta_N(n) \tag{20}$$

where

$$\theta_N(n) = X^T_N(n) R^{-1}_N(n) X_N(n) \tag{21}$$

It has been shown [9] that $\theta_N(n)$ is a measure of the likelihood that the N most recent input data samples come from a Gaussian process with a sample autocorrelation $R_N(n)$ determined from all available past observations. A value of $\phi_N(n)$ close to unity indicates that the recent data samples are likely observations from the Gaussian process described above; however, a value of $\phi_N(n)$ close to zero indicates that the observations are unexpected.

When $\phi_N(n)$ is close to zero, it indicates that sudden changes in the observations have occurred and the magnitude of the step change in the coefficient update algorithm increases.

Since division occurs in equations (15), (16), and (19), it is necessary to initialize the algorithm so that the estimated value of the energies of the forward and backward prediction errors are non-zero. Any initialization signal may be used but due to lack of space details will not be discussed here.

TAPPED-LATTICE STRUCTURE

The lattice structure shown in the previous section can be used for linear prediction only; however one advantage is that it creates a set of uncorrelated signals which may thus be tapped (as in tapped-delay-lines) forming a transversal structure as shown in figure 5.

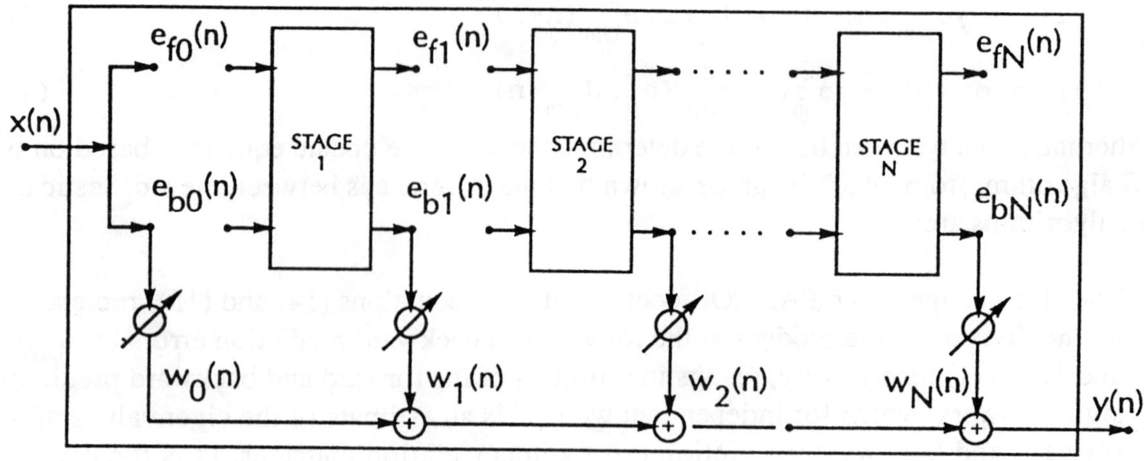

Figure 5. General FIR implementation using a Lattice Filter in conjunction with a Transversal Filter.

From figure 5 it can be seen that a general error represented by equation (1) can be written as:

$$e_N(n) = d(n) - E_{bN}^T(n) W_N(n-1) \quad (22)$$

where

$$E_{bN}(n) = [\, e_{b0}(n),\ e_{b1}(n),\ ..,\ e_{b(N-1)}(n) \,]^T \quad (23)$$

= Vector of backward prediction errors,

$$W_N(n) = [\, w_0(n),\ w_1(n),\ ..,\ w_{N-1}(n) \,]^T \quad (24)$$

= Transversal filter weights.

These weights can be adapted using the exact algorithm; however, it will firstly be shown that they are independent and following this a new algorithm will be developed.

Substituting equation (22) into the cost function given by equation (4) yields:

$$J_N(n+1) = e[d^2(n+1)] + W_N^T(n) E[E_{bN}(n+1) E_{bN}^T(n+1)] W_N(n)$$
$$- 2W_N^T(n) E[d(n+1) E_{bN}(n+1)] \quad (25)$$

However, since the backward prediction errors are uncorrelated this equation can be reduced to:

$$J(n+1) = \phi_d + W_N^T(n) tr[J_{bi}(n+1)] W_N(n+1) - 2W_N^T(n) C_N^{d,e_b}(n+1) \quad (26)$$

where

$$\phi_d = E[d^2(n+1)], \tag{27}$$

$$C_N^{d,e_b}(n+1) = E[d(n+1)E_{bN}(n+1)] \tag{28}$$

and $tr[J_{bi}(n+1)]$ represents an N-dimensional diagonal matrix of the predicted powers of the backward prediction errors.

Differentiating equation (26) with respect to the weights W_N yields the optimum weight values as:

$$W_{Nopt} = tr[J_{bi}^{-1}(n+1)]C_N^{d,e_b}(n+1) \tag{29}$$

Substituting (29) into (26) gives the optimal cost function as:

$$J_{Nopt} = \phi_d - W_{Nopt}^T C_N^{d,e_b}(n+1) \tag{30}$$

Substituting this back into (26) gives the form of the cost function as:

$$J_N(n+1) = J_{Nopt} + [W_N(n) - W_{Nopt}]^T tr[J_{bi}(n+1)][W_N(n) - W_{Nopt}] \tag{31}$$

Equation (31) highlights the independence of the weights which leads to the development of an alternative algorithm.

DEVELOPMENT OF AN ALTERNATIVE ADAPTIVE ALGORITHM

The use of independent tap weights enables the derivation of an efficient curve-fitting algorithm. If this algorithm were applied to a filter structure whose weights were not independent, it would be inefficient (in terms of convergence time and achievable minimum MSE), as is shown in figures 6a and 6b below.

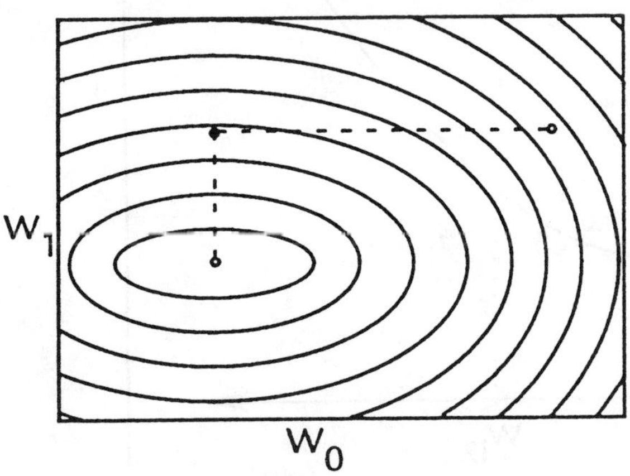

Figure 6a. Independent Weight Path to optimum.

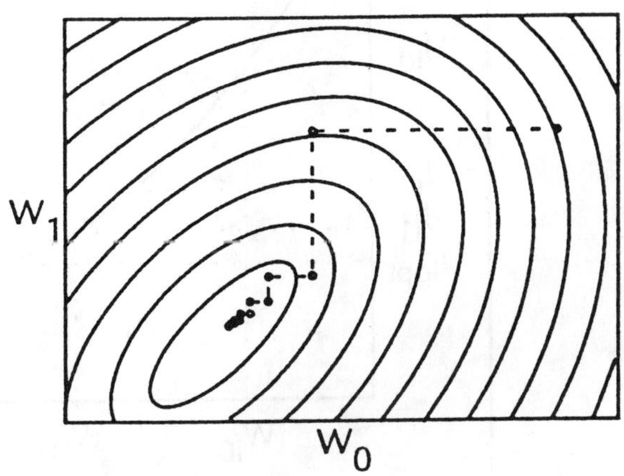

Figure 6b. Non-Independent weight path to optimum.

In viewing figure 6a, it can be seen that the independence of the weights enables the optimum MSE to be obtained using only two curve-fits. However, in figure 6b the non-independence of the weights indicates that a large number of curve-fits must be performed before a reasonable minimum MSE may be achieved.

From the figures it can be seen that the curve-fitting algorithm can be developed to ensure efficient adaption as follows:

(a) Delay estimating the MSE by a number of samples equivalent to a time greater than the acoustic time delay.

(b) Perform a sufficient average of the MSE dependent upon the accuracy (ie. minimum MSE) required.

(c) Adjust the particular weight and perform (a) to (c) twice more.

(d) Determine the optimum value for the particular weight by curve-fitting the known points to a parabola.

(e) Change to a new weight and repeat (a) to (e).

The lattice structure will use the exact algorithm discussed previously, and it is only the transversal filter which uses the curve-fitting algorithm. The algorithm is not as fast as other algorithms since it requires a number of samples delay followed by a sufficient number of averaged samples; however, it is more stable since it does not require the estimation of the transfer functions and time delay. The minimum mean square error will be dependent upon the number of averages taken.

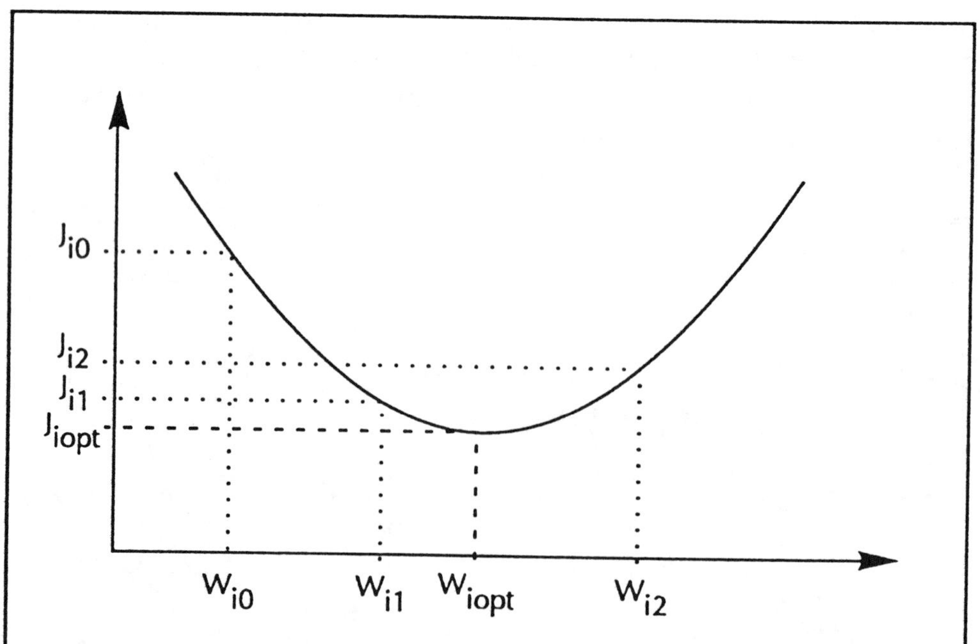

Figure 7. MSE curve for an independent weight.

It can be shown, by curve-fitting the points shown in figure 7 to a parabola, that the optimal value of the ith independent weight is given by:

$$W_{iopt} = 0.5 \frac{J_{i0}(W_{i1}^2 - W_{i2}^2) + J_{i1}(W_{i2}^2 - W_{i0}^2) + J_{i2}(W_{i0}^2 - W_{i1}^2)}{J_{i0}(W_{i1} - W_{i2}) + J_{i1}(W_{i2} - W_{i0}) + J_{i2}(W_{i0} - W_{i1})} \quad (32)$$

where J_{i0}, J_{i1} and J_{i2} represent averaged MSE values corresponding to the independent weight values given by W_{i0}, W_{i1}, and W_{i2}. From equation (32) a formal definition of the algorithm may be made incorporating an independent weight step proportional to the just averaged MSE. Due to lack of space this will not be presented.

SIMULATION

A simple set of simulations will show the current limitations of the curve-fitting algorithm. The simulations will consider reference signals consisting of a sinusoidal signal, white noise or a combination of both. The transfer functions will be made up of two coefficients and hence the tapped-lattice will have an order of two. The simulations show a MSE contour map onto which is projected the adaptive path of the weights as they approach their optimal values for minimum MSE.

VARIABLES

The acoustic plant transfer functions used were 5.767 and -4.576. The number of samples delay used was 5. The error plant transfer function coefficients used were 3.078 and -3.078. The delay before averaging was 10 and the number of averages taken was 2000. There was no white noise in the error signal.

Case 1 : Sinusoidal Reference Signal

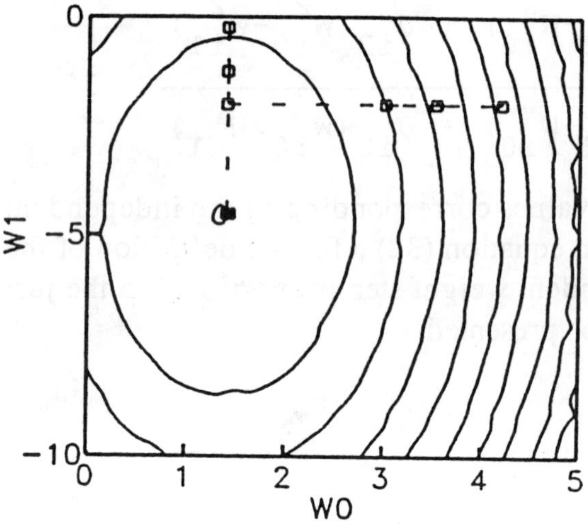

Figure 8(a).

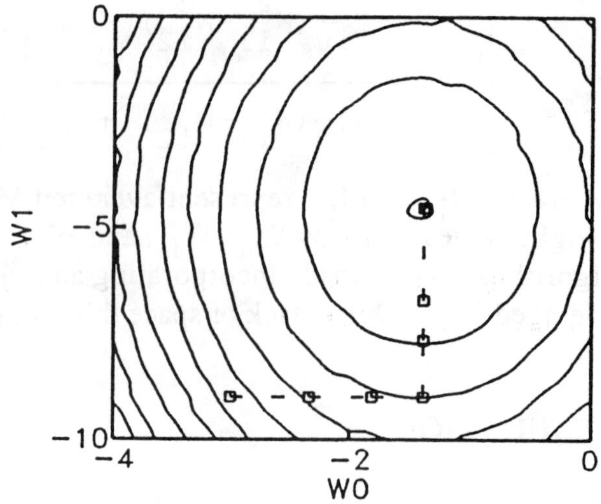

Figure 8b.

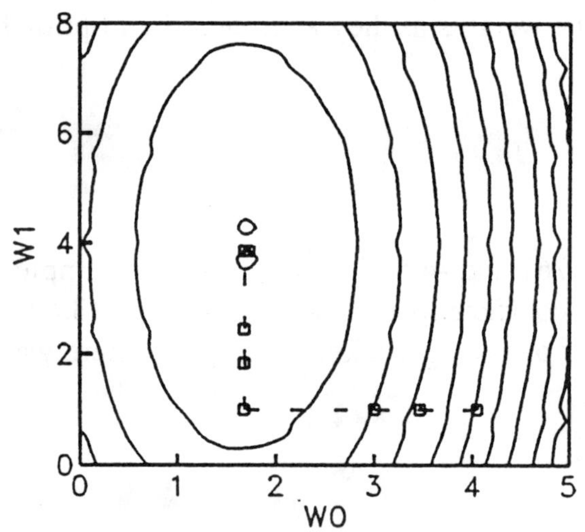

Figure 8(c).

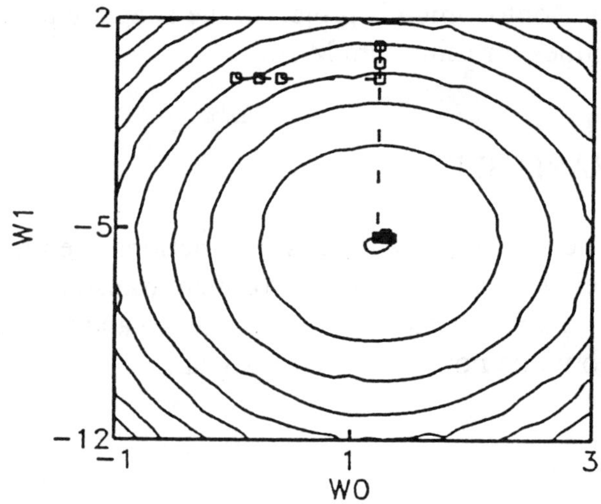

Figure 8(d).

Figure 8. Path of weights to the optimum MSE, projected onto the ideal MSE contour map. Paths found using (a) no transfer function and no delay, (b) delay and no transfer function, (c) transfer function and no delay and (d) delay and transfer function.

——— Ideal MSE contours. - - -□ Weight path.

From figure 8 it can be seen that irrespective of whether the transfer function and time delay are included, the weights are independent. The minimum achievable MSE is dependent upon the step size and the number of averages.

Figure 9. Path of weights to the optimum MSE, projected onto the ideal MSE contour map. Paths found using (a) no transfer function and no delay, (b) delay and no transfer function, (c) transfer function and no delay and (d) delay and transfer function.

——— Ideal MSE contours. - - -□ Weight path.

Since the delay of 5 samples is greater than the number of weights within the tapped-lattice, the minimum MSE in figures 9b to 9d is not zero as in the previous simulation. In these figures, after having reached the optimum MSE, the algorithm adapts the weights by a step proportional to the magnitude of the MSE for the current weight setting, hence it appears as a large offset from the optimum. It appears that when a transfer function is applied with a white noise reference signal, the weights of the tapped-lattice lose their independency.

Case 2 : White Noise Reference Signal

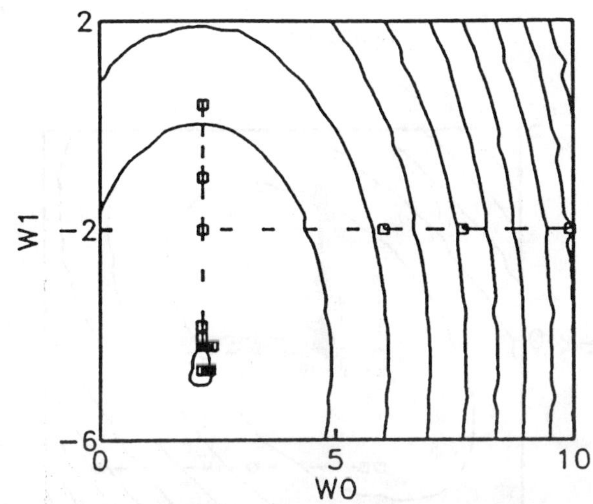

Figure 9a.

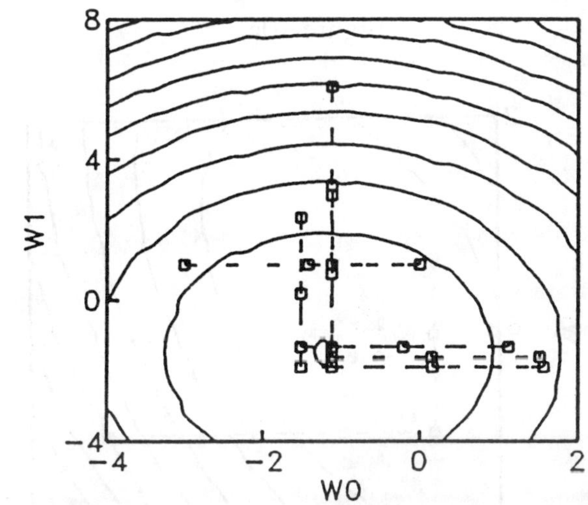

Figure 9b.

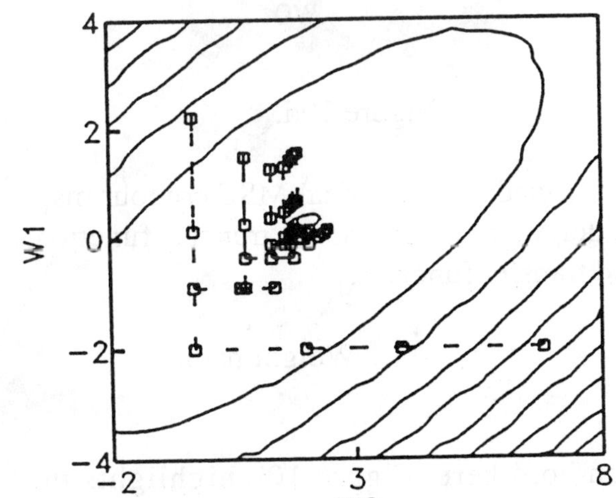

Figure 9c.

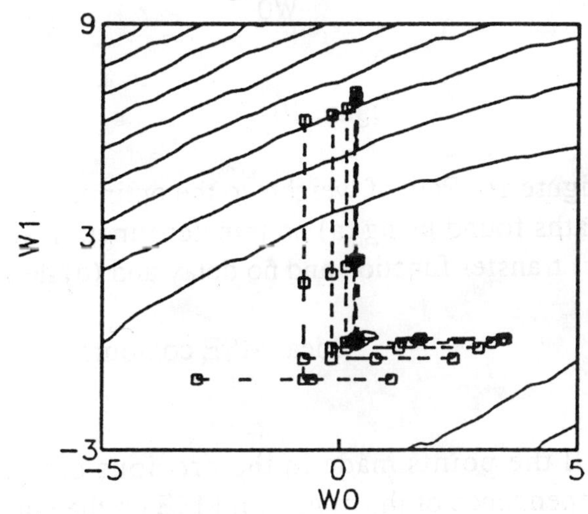

Figure 9d.

Case 3 : Combination of Sinusoidal and White Noise Signals

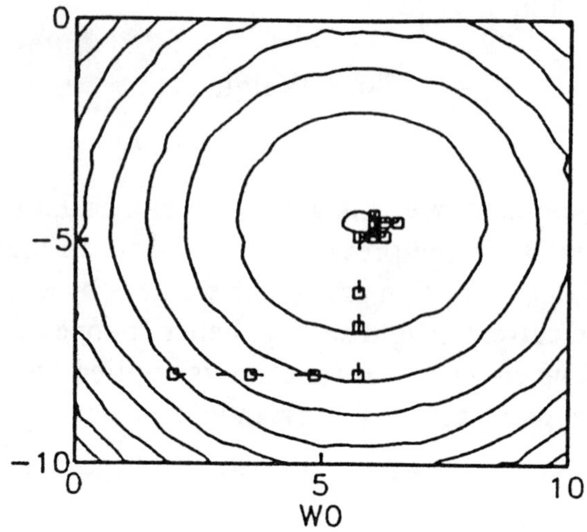

Figure 10a.

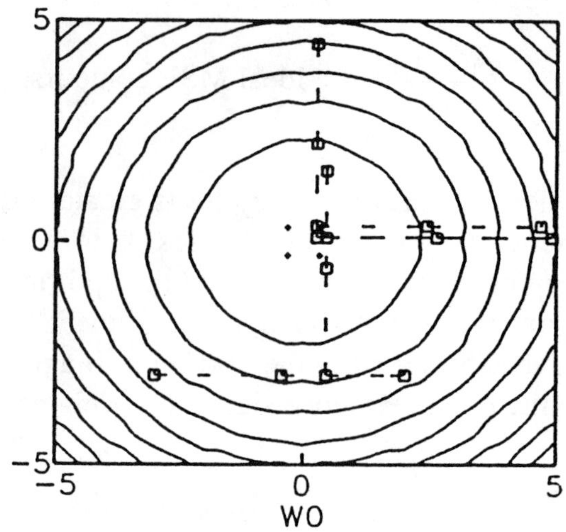

Figure 10b.

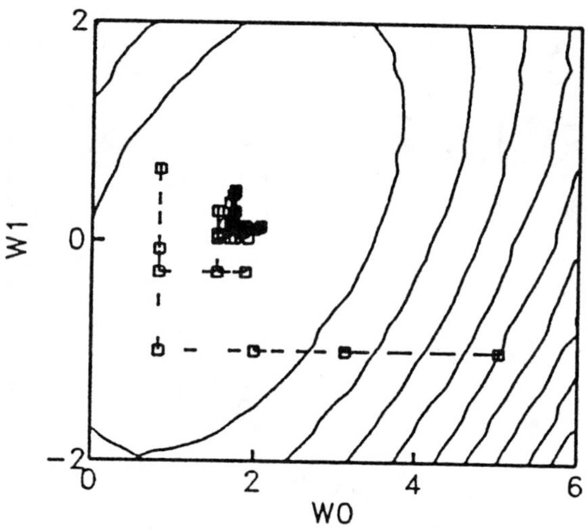

Figure 10c.

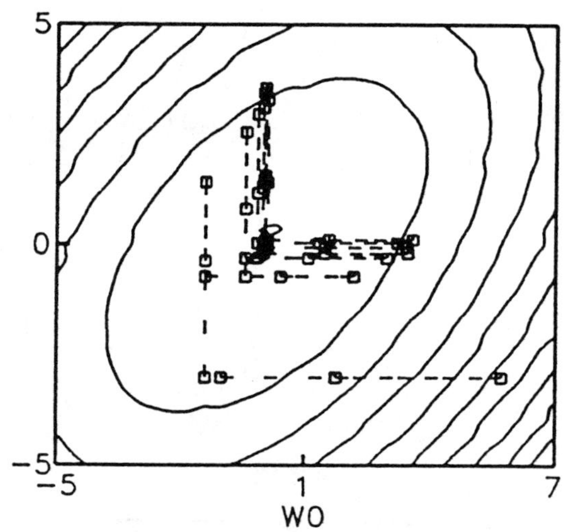

Figure 10d.

Figure 10. Path of weights to the optimum MSE, projected onto the ideal MSE contour map. Paths found using (a) no transfer function and no delay, (b) delay and no transfer function, (c) transfer function and no delay and (d) delay and transfer function.

——— Ideal MSE contours. - - -□ Weight path.

All the points made in the previous disscussion hold here. Figure 10b highlights the dependence of the minimum MSE on the number of averages and the step size.

CONCLUSIONS

An alternative adaptive algorithm has been introduced for use with a tapped-lattice filter which eliminates the need to measure the acoustic time delay and electro/acoustic transfer functions in an active noise or vibration control system. It eliminates the instabilities associated with the use of other algorithms and the required estimation of the acoustic time delay and electro/acoustic transfer functions, which are often estimated inaccurately.

Early simulation studies appear encouraging, and suggest the following disadvantages in the curve fitting algorithm:

(a) The requirement of a reasonably high signal to noise ratio in the reference signal.

(b) The long convergence time required.

These deficiencies are currently under investigation and it is hoped that further modification to this algorithm will yield a more acceptable algorithm for use in active noise and vibration control.

ACKNOWLEDGEMENTS

The authors would like to acknowledge financial support from the Sir Ross and Sir Keith Smith Fund, the Australian Research Council and the Australian Electrical Industry Research board.

REFERENCES

1. Snyder, S.D. and Hansen, C.H., 1990. "The Influence of Transducer Transfer Functions and Acoustic Time Delays on the Implementation of the LMS Algorithm in Active Noise Control Systems," Journal of Sound and Vibration, 140(3):409-424.

2. Ren, W. and Kumar, P.R., 1989. "Adaptive Active Noise Control: Structures, Algorithms and Convergence Analysis," Proceedings of Inter-Noise, pp. 435-440.

3. Sommerfeldt, S.D. and Tichy, J., 1990. "Adaptive Control of a Two-Stage Vibration Isolation Mount,"Journal of the Acoustical Society of America, 88(2):938-944.

4. Burgess, J.C., 1981. "Active Adaptive Sound Control in a Duct: A Computer Simulation," Journal of the Acoustical Society of America, 70(3):715-726.

5. Erikkson, L.J. and Allie, M.C., 1989. "Use of Random Noise for On-Line Transducer Modelling in an Adaptive Active Attenuation System," Journal of the Acoustical Society of America, 85(2):797-802.

6. Eriksson, L.J., Allie, M.C. and Greiner, R.A., 1987. "The Selection and Application of an IIR Adaptive Filter for use in Active Sound Attenuation," IEEE Transactions on Acoustics Speech and Signal Processing, ASSP-25(4):433-437.

7. Bellanger, M.G. 1987. Adaptive Digital Filters and Signal Analysis, Marcel Dekker, Inc..

8. Widrow, B. and Stearns, S.D., 1985. Adaptive Signal Processing, Prentice-Hall, Inc..

9. Cowan, C.F.N. and Grant, P.M., 1985. Adaptive Filters, Prentice-Hall, Inc..

10. Alexander, S.T., 1986. Adaptive Signal Processing: Theory and Applications, Springer-Verlag, Inc..

11. Makhoul, J. 1978. "A Class of All-Zero Lattice Digital Filters: Properties and Applications," IEEE Transactions on Acoustics Speech and Signal Processing, ASSP-26(4):304-314.

12. Griffiths, L.J., 1977. "A Continuously Adaptive Filter Implemented as a Lattice Structure," Proceedings of the IEEE conference on Acoustics Speech and Signal Processing, pp. 829-867.

13. Griffiths, L.J. 1978. "An Adaptive Lattice Structure for Noise Cancelling Applications," Proceedings of the IEEE conference on Acoustics Speech and Signal Processing, pp. 87-90.

14. Satorius, E.H. and Alexander, S.T., 1979. "Channel Equalization using Adaptive Lattice Algorithms," IEEE Transactions on Communications, COM-27(6):899-905.

15. Satorius, E.H. and Pack, J.D., 1981. "Application of Least Squares Lattice Algorithms to Adaptive Equalization," IEEE Transactions on Communications, COM-29(2):136-142.

16. Satorius, E.H. and Shensa, M.J., 1980. "Recursive Lattice Filters - A brief Overview," Proceedings of the IEEE conference on Decision and Control, pp. 955-959.

17. Satorius, E.H., Smith, J.D. and Reeves, P.M., 1979. "Adaptive Noise Cancelling of a Sinusoidal Interference using a Lattice Structure," Proceedings of the IEEE conference on Acoustics Speech and Signal Processing, pp. 929-932.

Active Mounts

VIBRATION CONTROL OF SUBMERGED ROTORS USING MAGNETIC BEARINGS

H. Ming Chen,[1] C.-P. Roger Ku[1]

ABSTRACT

The fluid annulus between a rotor and casing generates large, synchronous, hydrodynamic forces that cause casing vibration. A scheme to control this vibration uses accelerometers on the casing as sensors and magnetic bearings as actuators. The scheme applies an influence-coefficient method for determining proper gains and phases of signals injected to the bearing power amplifiers. It is an open-loop control with periodic vibration monitoring and control update.

INTRODUCTION

Over the last two decades, magnetic bearings have evolved from a laboratory curiosity to working mechanisms that support turbomachinery rotors. Compared to conventional bearings, such as oil-film and rolling-element types, magnetic bearings have no metal contact, need no lubrication, and consume less power. Furthermore, stiffness and damping in magnetic bearings can be manipulated in order to cope with instability and imbalance sensitivity, which are common problems in rotating equipment.

Recently, there has been a growing interest in applying magnetic bearings as actuators, in addition to their basic function as bearings, to control machine casing vibrations. This paper presents an analytical study of such an application involving motor rotors submerged in fluids.

Some hermetic centrifugal pumps contain submerged motors [1, 2]. The annular fluid has a large inertia effect due to the relatively large clearance between the rotor and stator. This effect is manifested as an attached mass to the rotor, or equivalently, a negative spring that grows proportionally to the square of the rotating speed. Synchronous excitations due to imbalance and/or manufactured rotor eccentricity can generate large hydrodynamic forces through the fluid annulus to shake the stationary parts, i.e., the motor stator and casing.

Chen and Lewis [3] have discussed the control of hydrodynamic forces using a relatively simple model. A detailed study has since been conducted and a computer program has been coded for analyzing this type of controlled pump system.

[1]Mechanical Technology Incorporated, 968 Albany-Shaker Road, Latham, NY 12110, U.S.A.

FORMULATION OF SYSTEM DYNAMICS

Figure 1 outlines the essential components in submerged rotordynamics. The following paragraphs discuss the mathematical modeling of these components and their interrelationships.

ROTOR

The rotor is treated as a flexible beam because some of its natural bending modes may be in the control bandwidth of the magnetic bearings. Mathematically, the rotor is divided into a series of concentric, uniform cylinders with different lengths and diameters. The connections of these cylinders are called stations, where concentrated masses and the moments of inertia of the impellers, the thrust disk, the coupling, etc., are attached. The centerlines of a magnetic bearing and its displacement sensor coincide with two different stations. Each station is given four degrees of freedom (DOFs) of motion - two lateral and two angular - designated as $(X, \phi_x, Y, \phi_y)'$. Using the finite-element method [4], each cylinder is represented by an 8×8 mass matrix and an 8×8 stiffness matrix. The angular momentum of an element due to shaft rotation resists angular motions. The resisting moments are proportional to the angular vibration velocities. Thus, an 8×8 element damping matrix is formed to account for the gyroscopic effect, which is a function of rotor speed and polar moment of inertia. By stacking element matrices of the same type with overlaps, a rotordynamic model is represented by Equation (1).

$$[M_r]\ \{\ddot{U}_r\} + [C_r]\ \{\dot{U}_r\} + [K_r]\ \{U_r\} = \{F_r\} \tag{1}$$

where

$\{U_r\} = (X_1, \phi_{x1}, Y_1, \phi_{y1}, \ldots)'$

$[M_r], [C_r], [K_r]$ = rotor mass, damping, and stiffness matrices, respectively

$\{F_r\}$ = rotor forcing vector including imbalance forces.

The axial flexibility of the rotor is not modeled because the associated natural modes are high in frequency and are above the rotordynamic interest. Treating the rotor axially as a rigid

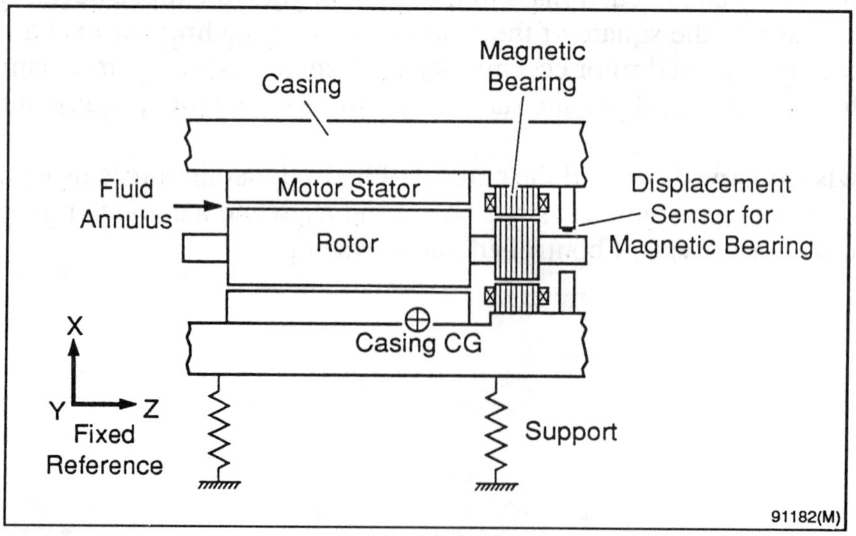

Figure 1. Essential Components in Submerged Rotordynamics

body constrained by the stiffness and damping of a thrust bearing, the associated one DOF will be coupled to the casing DOFs.

CASING

Machine casings are generally much stiffer than their rotors. In addition, the predominant natural frequencies of the casings are usually well above the range of rotordynamic interest. Casing mass affects system rotordynamics, especially when the casing is light and is not rigidly tied to a large foundation. To account for the mass effect, it is adequate to model the casing as a rigid body with six DOFs (x, y, z, θ_x, θ_y θ_z). The casing is assumed to be mounted on flexible supports, e.g., elastomer mounts, tied to the ground. Piping attached to the casing contributes to both the casing mass and its support stiffness.

The six DOFs are expressed at the casing's center of gravity (CG). Each flexible support is represented by stiffness and damping coefficients in three directions. The locations of the supports relative to the CG determine angular stiffness and damping. The equations of motion for the casing excluding the rotor are shown below:

$$[M_c]\ \{\ddot{U}_c\} + [C_c]\ \{\dot{U}_c\} + [K_c]\ \{U_c\} = \{F_c\} \qquad (2)$$

where

$\{U_c\} = (x, y, z, \theta_x, \theta_y, \theta_z)'$
$[M_c]$ = Diag. $(m_c, m_c, m_c, I_{xx}, I_{yy}, I_{zz})$
m_c = casing mass
I_{xx}, I_{yy}, I_{zz} = mass moments of inertia about X, Y, and Z axes at CG, respectively
$[K_c]$, $[C_c]$ = 6×6 casing stiffness and damping matrices, respectively
$\{F_c\}$ = casing forcing vector containing forces and moments at CG.

ROTOR/CASING COUPLING THROUGH BEARINGS AND FLUID ANNULUS

Forcing vectors $\{F_r\}$ and $\{F_c\}$ contain internal force components of magnetic bearings and the fluid annulus. Since the fluid annulus behaves like a cylindrical journal bearing, it will be referred to as one of the bearing connections. The internal forces in terms of stiffness and damping coefficients are functions of relative motion between the rotor and casing. By coupling Equations (1) and (2) into a dynamic system, the required coordinate transformations are described below.

The lateral motions at a bearing center are related to five casing DOFs by Equation (3). Note that the rotor displacement planes are aligned with the control axes of radial magnetic bearings.

$$\begin{Bmatrix} X_b \\ Y_b \end{Bmatrix} = [T] \begin{Bmatrix} x \\ y \\ \theta_x \\ \theta_y \\ \theta_z \end{Bmatrix} \qquad (3)$$

where

$[T] = [T_2]\ [T_1] = 2 \times 5$ transformation matrix

$$[T_1] = \begin{bmatrix} 1 & 0 & 0 & \Delta Z & -\Delta Y \\ 0 & 1 & -\Delta Z & 0 & \Delta X \end{bmatrix}$$

$$[T_2] = \begin{bmatrix} \cos\alpha & -\sin\alpha \\ \sin\alpha & \cos\alpha \end{bmatrix}$$

α = orientation angle of magnetic bearing control axis, 0° or 45°

$\Delta X, \Delta Y, \Delta Z$ = relative distances from casing CG to bearing center, as defined in Figure 2.

The axial motion of a thrust bearing is related to three casing DOFs using Equation (4).

$$Z_b = (1, \Delta Y, -\Delta X) \begin{Bmatrix} z \\ \theta_x \\ \theta_y \end{Bmatrix} \tag{4}$$

MAGNETIC BEARINGS

As shown in Figure 3, radial magnetic bearings have laminated rotors and stators similar in structure to motors. The stator poles are divided into four identical sectors, each having two or more poles. The magnetizing coils of each sector are driven by one or more current-source power amplifiers. An opposite pair of sectors forms a lateral control axis. During control when current in a sector is increased, the current in the opposite sector is reduced at the same time and by the same amount. All poles apply attractive forces on the rotor proportional to the square of the current-gap ratio. In order to achieve a push-pull action and to use a linear

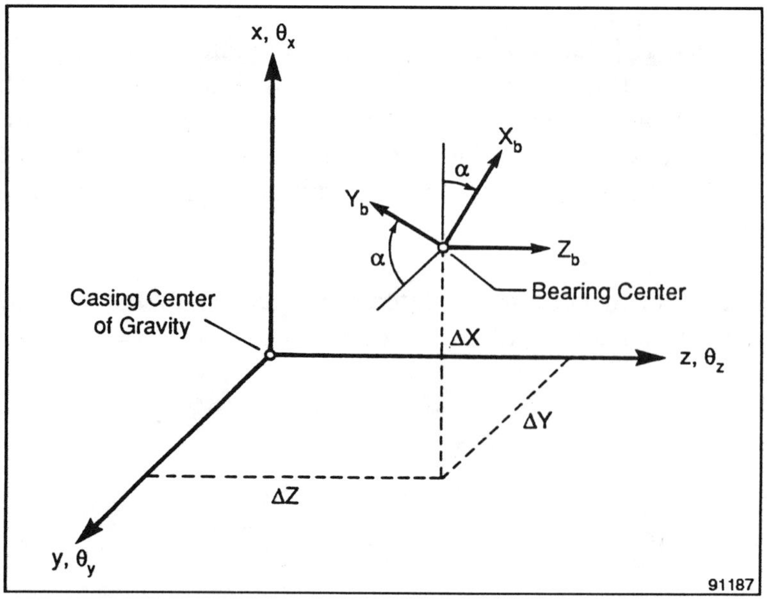

Figure 2. Definitions of Bearing and Casing Coordinates

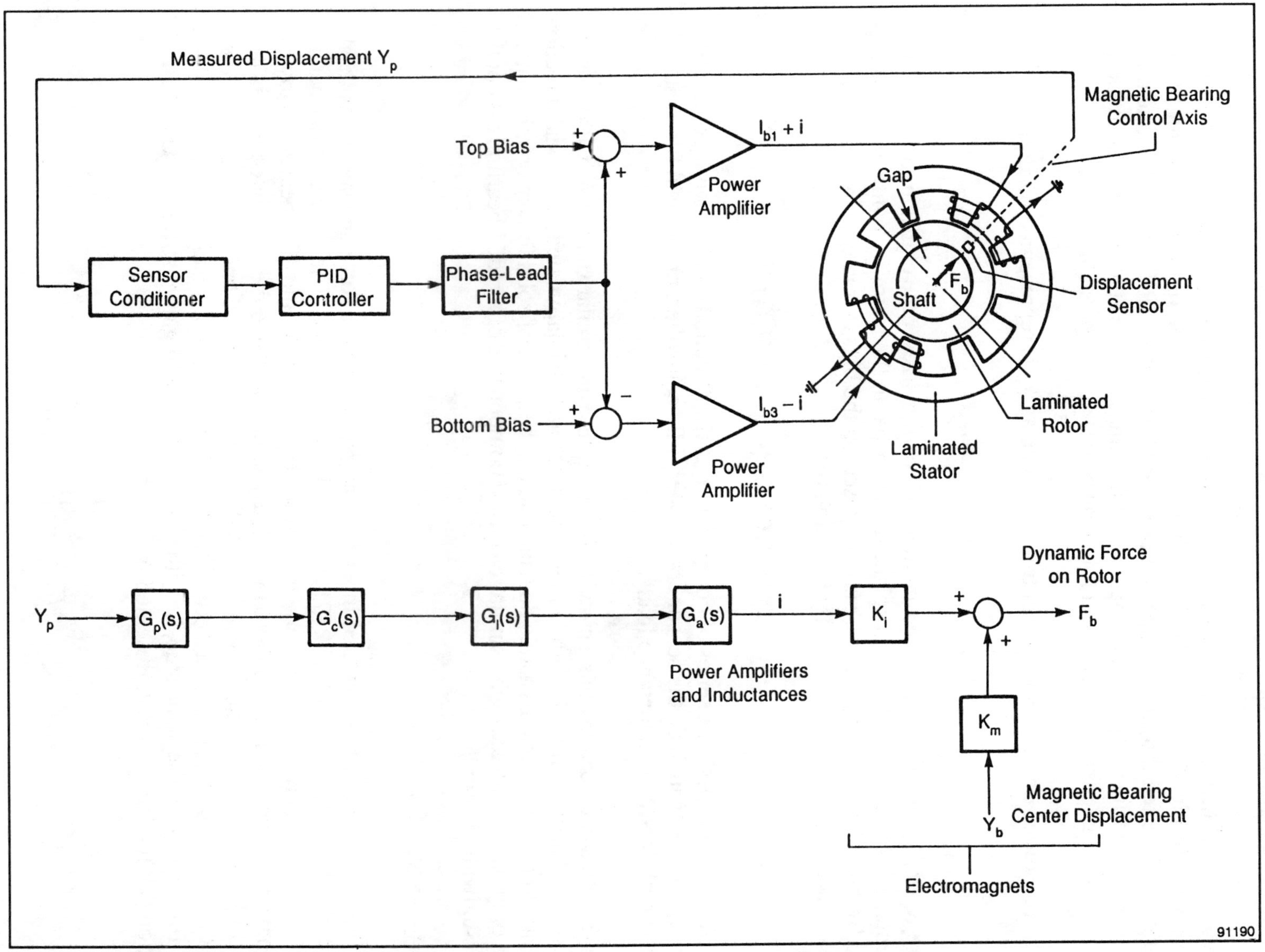

Figure 3. Typical Magnetic Bearing and Its Control Block Diagram

feedback controller, bias (dc) currents are induced in the coils. The control currents are smaller than the bias currents and are superimposed on them. The gap between the rotor and the stator poles is usually designed to be 15 to 30 mil. Control axes are normally oriented in a 45° angle to share the rotor weight.

Each axis has either an inductance or capacitance type displacement sensor. As shown in Figure 3, the displacement signals, which indicate the deviation of the rotor center from the stator center, are fed through a proportional-integral-derivative (PID) controller and a phase-lead filter. The filter output with proper gains and phases drives the power amplifiers, which regulate the magnetizing currents and force the rotor back to the stator center.

The force exerted on the rotor in a control axis, such as Y, is represented by Equation (5).

$$F_b = K_i\ i + K_m\ Y_b \tag{5}$$

where K_i (lb/A) and K_m (lb/in.) are functions of bearing sizes, pole configuration, and bias currents. Magnetic bearing rotor displacement relative to the stator in Y direction is designated by Y_b. The control current, i, superimposed on the bias is related to a measured rotor displacement, Y_p, near the bearing by Equation (6).

$$i = G(s)\ Y_p = G_p(s)\ G_c(s)\ G_l(s)\ G_a(s)\ Y_p \tag{6}$$

Transfer function G(s) is a series of component transfer functions multiplied together [5]. These components include the sensor conditioner, PID controller, phase-lead filter, and power amplifiers with inductance loads.

The thrust magnetic bearing has a circular disk rotor and an annular stator pole with coils at each side of the disk. It is difficult to make the stator and disk laminated in an axial direction. The thrust bearing has less dynamic demand and can be made of solid ferromagnetic material. It has only one control axis with similar electrical components of a radial bearing, except the total bandwidth is smaller.

FLUID ANNULUS

The dynamic effect of fluid trapped in an annular gap between a rotor and stator has been investigated by Fritz [6] and Black [7] among other researchers. The clearance-diameter ratio is usually an order of magnitude larger than the ratio of a fluid-film bearing. However, the "bearing" effect cannot be ignored in rotordynamics because the associated area is large. Particularly significant is the flow inertia. When a rotor sways to one side and forces the fluid to flow circumferentially, a Bernoulli effect is created. The resultant pressure pulls the rotor further toward the same side, i.e., it is equivalent to a negative stiffness. Using Black's notation [7], the stiffness is represented by Equation (7).

$$K_f = \frac{-m_a\,\omega^2}{4} + j\bar{K}\,\omega \tag{7}$$

Thus, the hydrodynamic force vector is

$$F_f = K_f \Delta \tag{8}$$

where

$m_a = \rho\pi L r^3/C$
$\overline{K} = 6\pi\mu_e L(r/C)^3$
$\mu_e = 0.0053\ \mu R_r^{0.75}$ = effective turbulent viscosity
μ = fluid viscosity
ρ = fluid density
L = submerged rotor length
r = submerged rotor radius
C = radial clearance of fluid annulus
$j = \sqrt{-1}$
Δ = relative displacement vector between rotor and casing at the annulus.

Taking into consideration the rotor bending mode shapes, the length of a fluid annulus should be divided into several axial sections. At the end sections, where axial flows are significant, a short-bearing correction factor can be applied to Equation (7).

SYNCHRONOUS EXCITATIONS

Combining Equations (1) through (8) and rearranging the bearing force terms to the left-hand side of the equations, the system dynamics are represented by Equation (9).

$$[M]\ \{\ddot{U}\} + [C]\ \{\dot{U}\} + [K]\ \{U\} = \{F\} \tag{9}$$

where

[M], [C], [K] = system mass, damping, and stiffness matrices, respectively
{U} = system state vector = $(X_1, \phi_{x1}, Y_1, \phi_{y1}, \ldots x, y, z, \theta_x, \theta_y, \theta_z, Z_r)'$
Z_r = axial DOF of rotor

Included in the system forcing vector, {F}, are three types of steady-state synchronous forces:

- Imbalance
- Fluid annulus
- Casing control.

The rotating imbalance force vector, F_i, can be represented by Equation (10).

$$F_i = u\omega^2\ e^{j(\omega t + \Theta)} \tag{10}$$

where

u = imbalance (mass × eccentricity)
ω = rotating speed (rad/sec)
Θ = imbalance location angle relative to a rotor reference mark.

Imbalance makes the rotor whirl eccentrically in the fluid annulus and generates hydrodynamic force. The hydrodynamic force is included in the left-hand side of Equation (9).

Fluid annulus force occurs if a manufactured rotor offset, δ, exists, as shown in Figure 4. The corresponding hydrodynamic force vector on the rotor is

$$F_f = K_f \delta e^{j(\omega t + \Theta)} \tag{11}$$

Casing control force is created by injecting a sinusoidal voltage signal at the input of the bearing power amplifiers. For example, if the signal had an amplitude of E_o and a phase angle of Θ, the corresponding control current would be:

$$i = G_a(j\omega)\; E_o\; cos\,(\omega t + \theta) \tag{12}$$

where

$G_a(j\omega)$ = transfer function of power amplifiers and coil inductance.

The casing control force on the rotor from a magnetic bearing axis is

$$F_m = K_i\, i \tag{13}$$

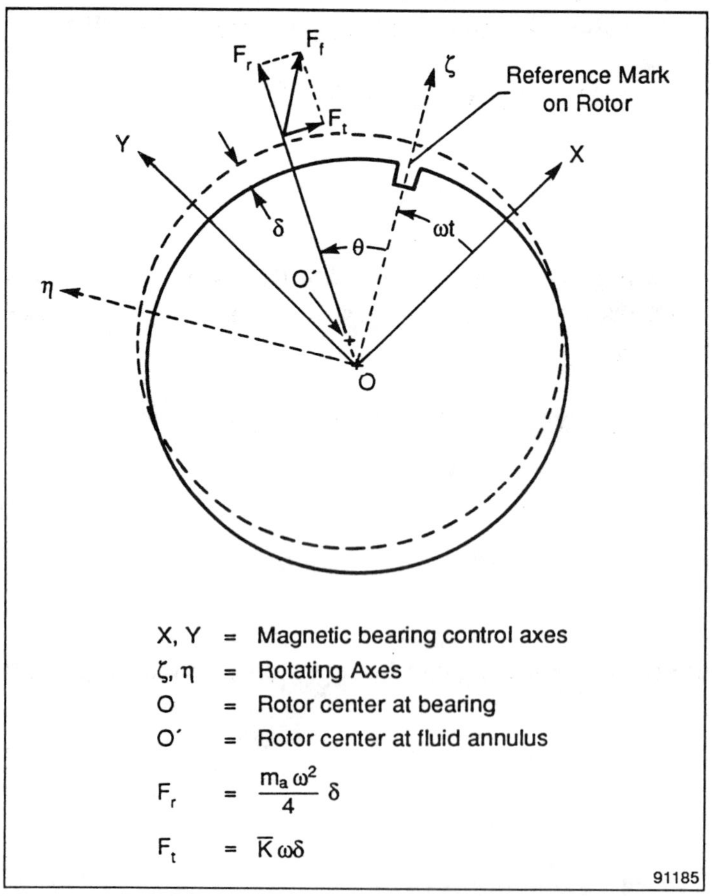

Figure 4. Rotor Offset at Fluid Annulus and Associated Hydrodynamic Forces

The associated forces and moments on the casing CG due to the last two types of excitation can be derived from Equations (11) through (13) using transformation matrix [T].

CASING VIBRATION CONTROL SCHEME

Figure 5 shows a casing vibration control scheme using four accelerometers to define uniquely lateral rigid-body casing vibrations. The scheme also includes a rotor speed sensor that generates once-per-revolution pulses. There are two essential components in the control loop from the accelerometers to the power amplifier: the influence-coefficient generator (ICG) and the sinusoidal signal generator (SSG). The ICG is a mirocomputer equipped with hardware of digital signal processors and software of discrete fast Fourier transform (DFFT) among others. At the beginning of a control process when the rotor is levitated and running, the ICG injects a trial signal to one bearing axis at a time, and records the amplitude and phase changes of the casing vibrations measured by the accelerometers. After all the bearing axes are tried, the recorded data are used to calculate an influence-coefficient matrix, defined by Equation (14).

$$\{E\} = [C_i]\ \{A\} \tag{14}$$

where

$\{E\} = [E_1, E_2, \ldots E_n]' =$ vector of signals injected to power amplifiers
$[C_i] = n \times n$ influence-coefficient matrix
$\{A\} = [A_1, A_2, \ldots A_n]' =$ vector of casing accelerations
$n =$ number of control axes or accelerometers.

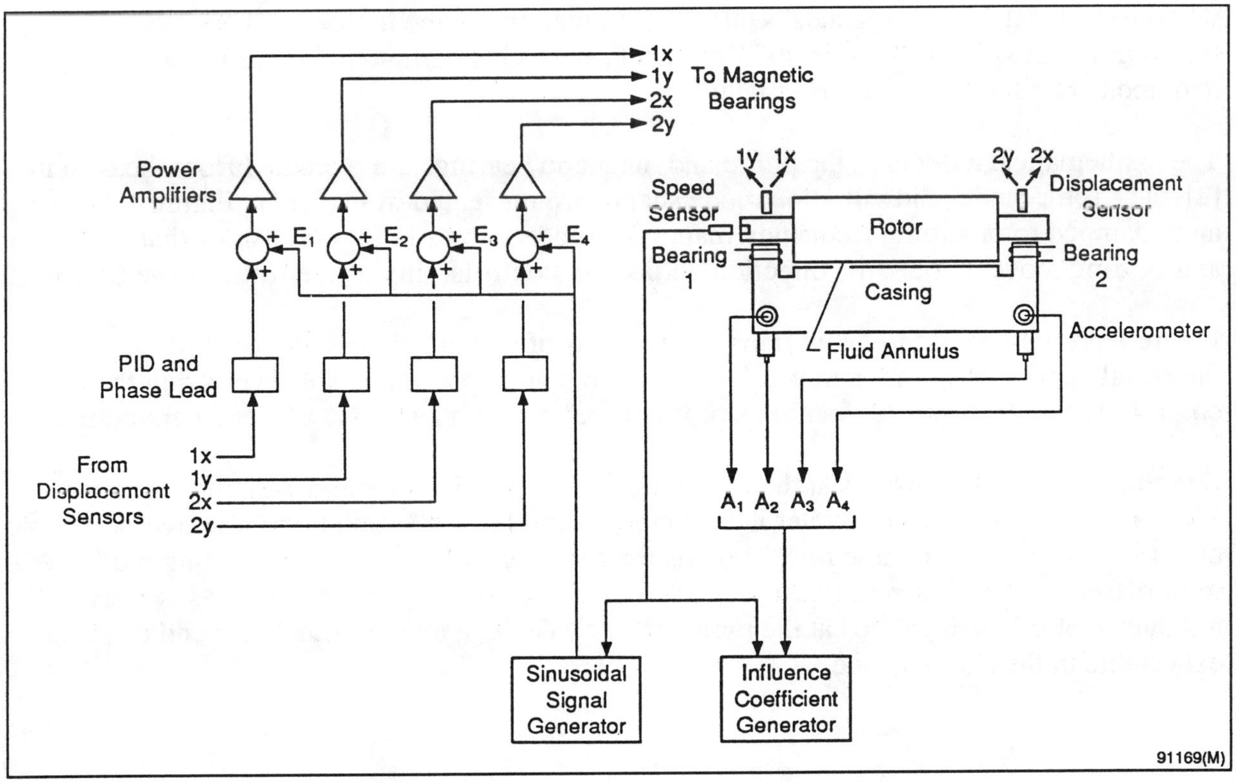

Figure 5. Casing Vibration Control Scheme

Using matrix $[C_i]$ and the uncontrolled casing accelerations, the appropriate amplitudes and phases of the injected signals are calculated and sent to the SSG. The SSG then produces sinusoidal signals accordingly at a frequency synchronous to the pulses from the speed sensor.

The casing vibration control scheme is intended for compensating internal forces generated by the rotor due to mass imbalance or the hydrodynamic effect. The scheme's speed-tracking mechanism provides a narrow control bandwidth so that the magnetic bearings will not be overloaded by external disturbances.

The internal forces are not expected to change fast or frequently. When the casing control is in process, the ICG will continue to monitor the casing acceleration levels. If changes exceed a certain predetermined threshold value, the injected signals will be updated by incremental amounts determined by Equation (14). Therefore, strictly speaking, the scheme is an open-loop control, which does not create instability problems for the magnetic bearings.

In terms of adaptive vibration cancellation [8, 9], the vibration control scheme is a frequency domain control of a sinusoidal excitation. In the example presented herein, the number of sensors (accelerometers) is equal to the number of actuators (bearing control axes). Ideally, the casing control can be made into a time domain process by adding four 2-point adaptive LMS filters. This should be achievable if the closed-loop system stability is ensured.

APPLICATION OF CASING VIBRATION CONTROL TO AN INDUSTRIAL PUMP

A 10-hp centrifugal pump retrofitted with magnetic bearings was presented in Reference [2]. The pump operates at 3600 rpm and delivers 150 gpm at 135 ft of head. Process fluid is allowed to leak through the bearings and the motor gap, thereby achieving motor cooling without the need for a mechanical seal. The laminations of the motor and bearings are protected from corrosion using thin stainless steel cans. This pump will be used as an example to demonstrate the casing control method.

The mathematical models of the pump and magnetic bearings are recreated from the data in [2]. The rotor model and vibration mode shapes are presented in Figure 6. Figure 7 contains an undamped rotor natural frequency map. The mode shapes in Figure 6 show that the sensor and bearing move in the same direction, indicating controlability of the rotor by the bearings.

Figure 8 presents the component transfer functions of a magnetic bearing control loop. All the radial bearing loops are similar. The thrust magnetic bearing is not involved in the casing control. Figure 9 shows the locations of four accelerometers and other casing parameters.

The fluid annulus dimensions at the motor are 3.75 in. in diameter and 7 in. in length, with 1/16 in. of radial clearance. Water is the assumed fluid in the simulation. The hydrodynamic effect is distributed into three rotor stations for considering the first rotor bending mode. A rotor offset of 1.0 mil at zero phase angle is assumed arbitrarily at the motor, as is a residual imbalance of 0.1 oz-in. at 90° at the motor center. The amounts of imbalance and offset are expectable in the real machine.

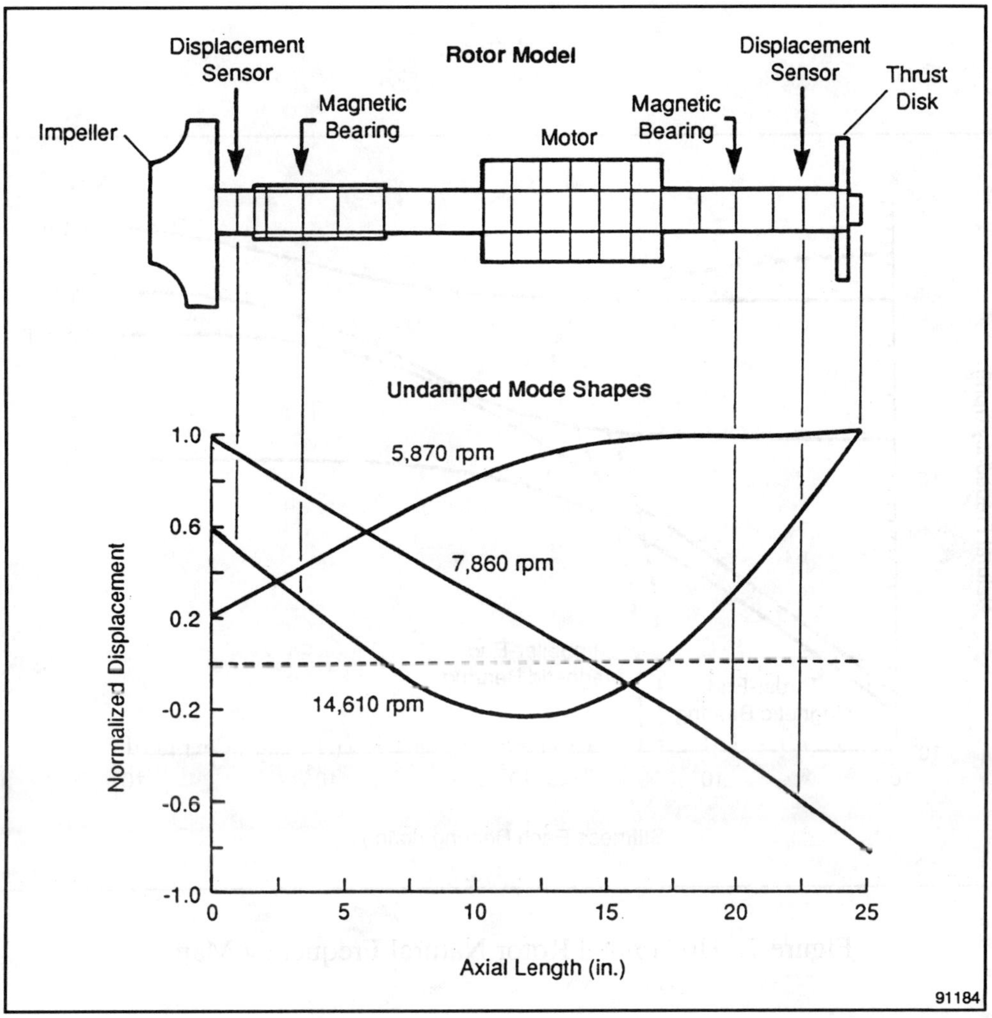

Figure 6. Rotor Model and Vibration Mode Shapes

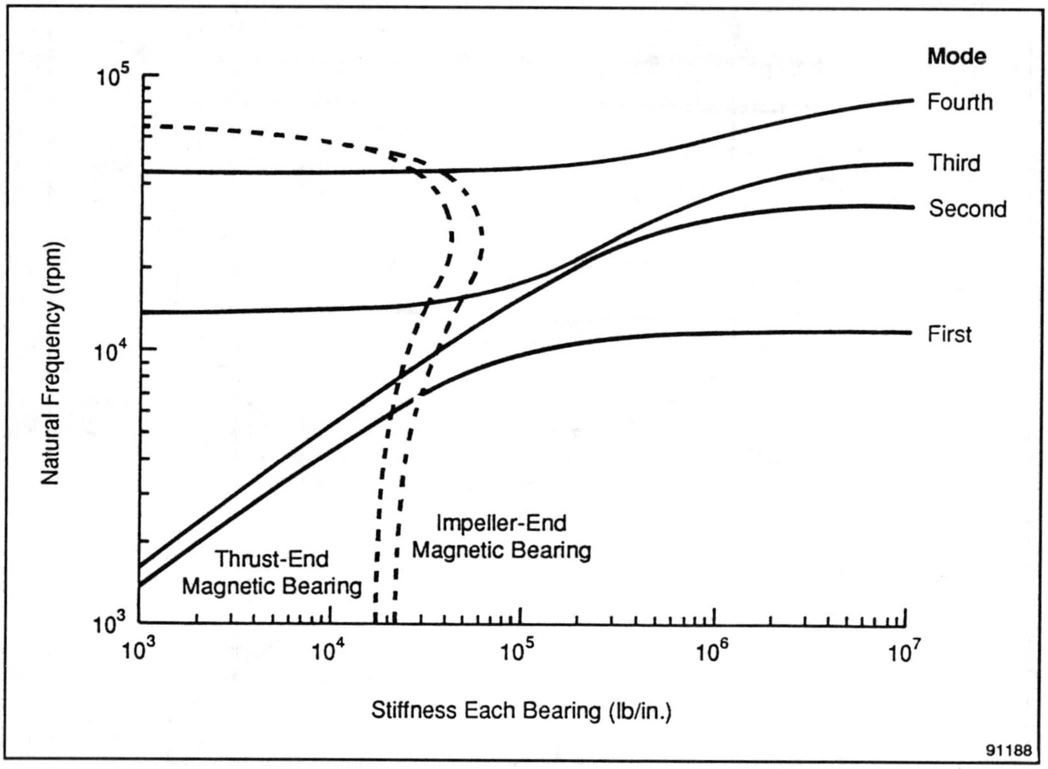

Figure 7. Undamped Rotor Natural Frequency Map

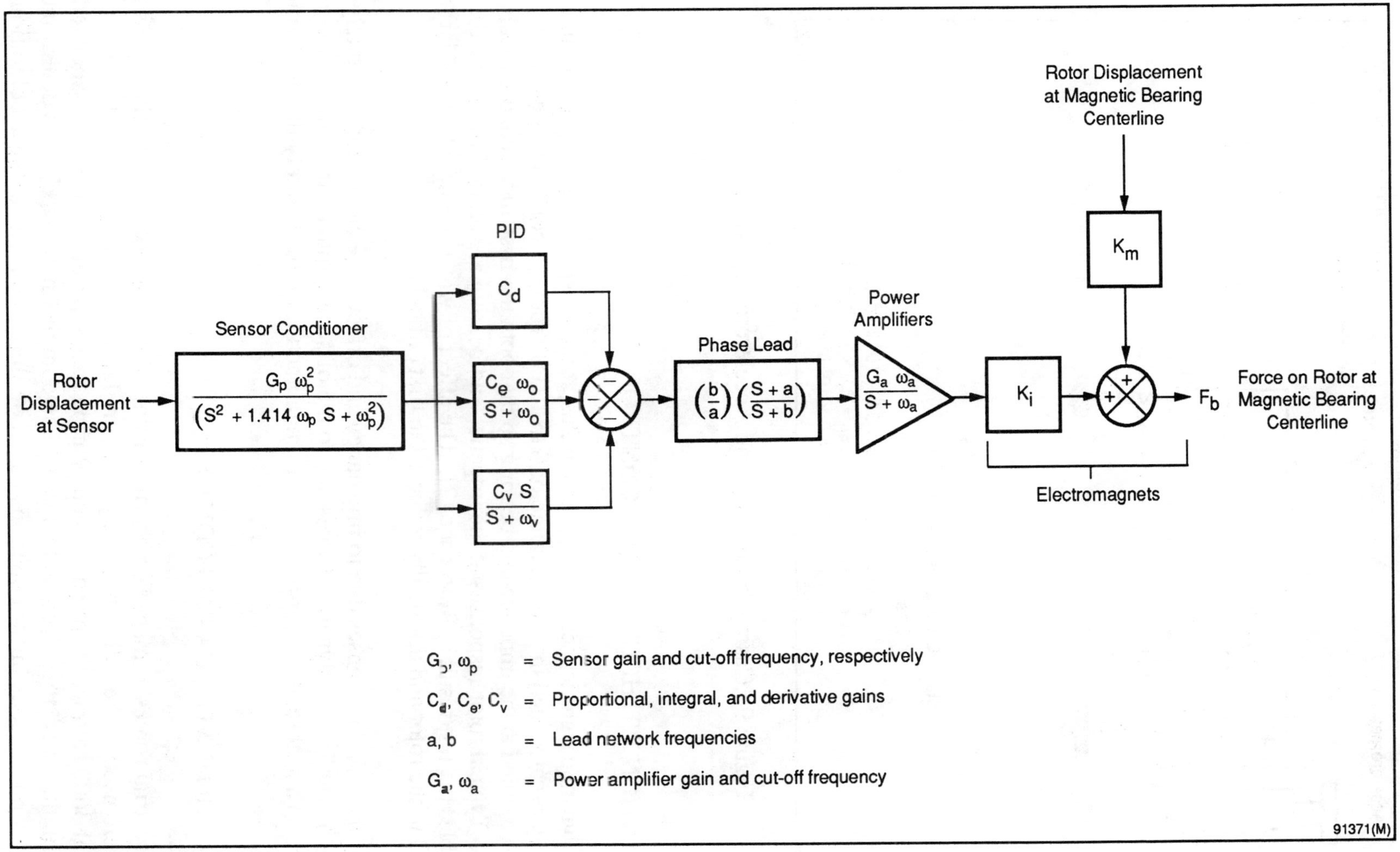

Figure 8. Component Transfer Functions of Magnetic Bearing Control Loop

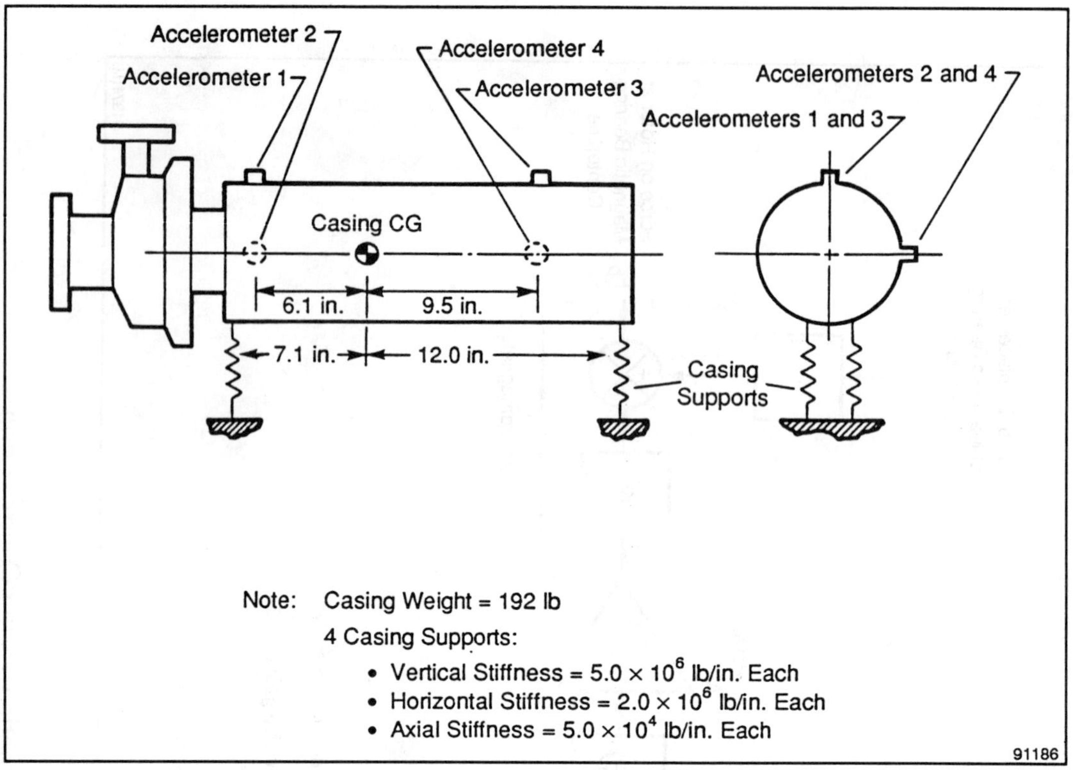

Figure 9. Casing Parameters and Accelerometer Locations

PUMP VIBRATION WITHOUT CASING CONTROL

To show the dynamic significance of the fluid annulus, forced responses of the system in air and in water were calculated in a speed range of 0 to 10,000 rpm. A constant force amplitude of 11.4 lb was applied at the impeller-end bearing so that both the first and second criticals were excited. Typical rotor responses at the impeller-end bearing are plotted in Figure 10, which resembles the test result in Reference [2]. The effect of water is equivalent to a large mass attached to the rotor and drastically reduces the first critical speed.

At 3600 rpm, the "wet" responses due to imbalance and offset were calculated separately. As compared in Table I, the dynamic forces due to the offset are much larger than those of the imbalance. Therefore, the rotor offset due to manufactural error is likely the predominant excitation.

PUMP VIBRATION WITH CASING CONTROL

For simplicity, trial forces to magnetic bearings rather than trial voltage signals to power amplifiers were used in the following analytical simulation. A trial force of 10 lb with zero phase angle at 3600 rpm was applied to each bearing axis separately, and the associated responses with the "wet" rotor were calculated. The complex matrix $[C_i]$ was determined by solving 16 simultaneous linear equations based on the calculated casing accelerations and the given trial forces. The bearing forces required to calm the casing were equal to the matrix multiplied by the accelerations due to the imbalance or the offset. Applying the

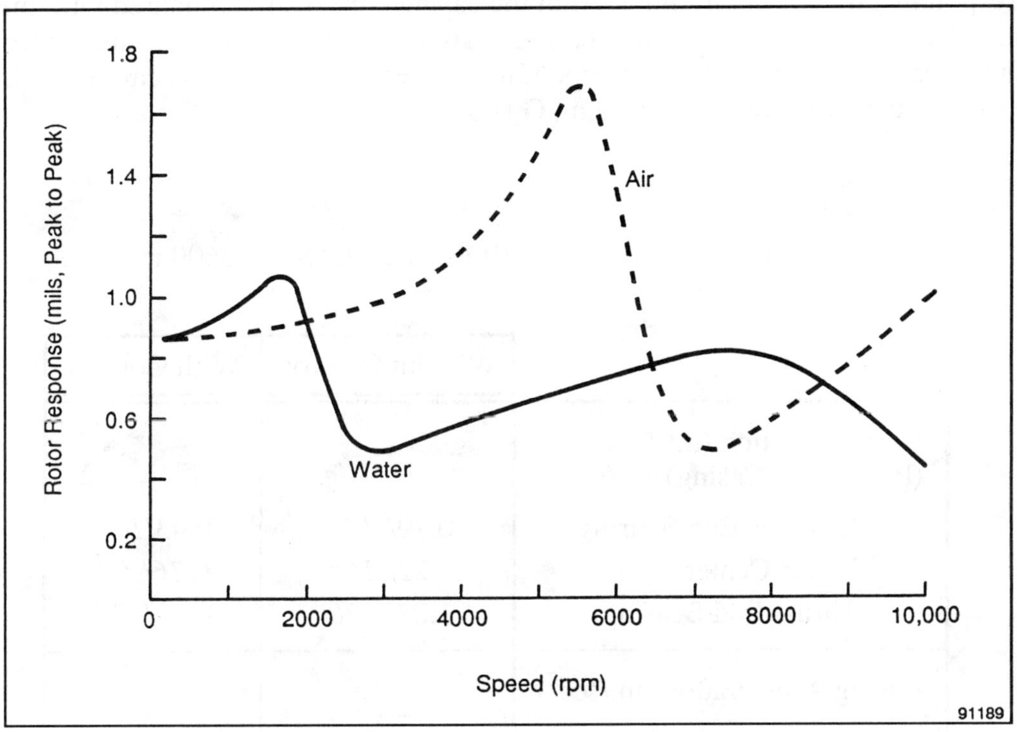

Figure 10. Forced Responses at Impeller-End Bearing Showing Fluid Annulus Effect

TABLE I – INTERNAL FORCES AT 3600 rpm

	Force (lb) Due to Imbalance (0.1 oz-in.)	Force (lb) Due to Rotor Offset (1 mil)
Impeller-End Bearing	0.57	13.6
Fluid Annulus	2.90	17.3
Thrust-End Bearing	0.76	18.1

bearing forces as additional excitation to the system, the controlled responses are presented in Table II for a combined excitation of the imbalance and rotor offset. The table compares results with and without the casing control. When the casing vibration is eliminated by the control, some internal forces are increased, but the rotor vibrations at the bearings and fluid annulus are reduced.

The corresponding dynamic currents used in the casing control are compared with the allowable dynamic currents of the magnetic bearing designs in Table II. The casing control is well within the bearing capacities. The corresponding power amplifier inputs can be back-calculated from the dynamic currents using $G_a(j\omega)$.

TABLE II – CASING CONTROL RESULTS AT 3600 rpm

	Without Control	With Control
Rotor Vibration (mil/°) (Relative to Casing)		
Impeller-End Bearing	0.79/*-171*	0.43/*144*
Motor Center	1.22/*-168*	0.76/*-56*
Thrust-End Bearing	1.07/*-170*	0.23/*179*
Casing Acceleration (in./sec^2)		
Accelerometer 1	0.12	0
Accelerometer 2	1.38	0
Accelerometer 3	0.08	0
Accelerometer 4	0.90	0
Internal Force (lb)		
Impeller-End Bearing	13.3	28.7
Fluid Annulus	18.2	88.4
Thrust-End Bearing	17.8	15.1
Dynamic Current (A)		
Impeller-End Bearing	0.23	0.28 (1.6)*
Thrust-End Bearing	0.68	1.46 (4.0)

*Numbers in parentheses indicate dynamic current limits.

CONCLUSIONS

A submerged rotor generates large hydrodynamic forces due to rotor imbalance and/or rotor offset in a fluid annulus. These internal forces cause casing vibrations which in some applications are undesirable. Using magnetic bearings to support the rotor provides the advantage of applying additional control to the casing. An open-loop, influence-coefficient casing control scheme has been presented herein. A dynamic simulation of the scheme to an industrial pump shows the feasibility of the scheme and reveals the following:

1. Compared to imbalance, rotor offset contributes more to casing vibration.
2. Controlling casing vibration using magnetic bearings does not necessarily lead to larger rotor vibration.

The casing control can be extended to a real-time, closed-loop scheme by using four 2-point LMS filters. The inverse of the influence-coefficient matrix will become the coupled transfer functions of the "secondary" loops, and the speed sensor signal will be the "reference" to the filters.

REFERENCES

1. Allaire, P. E., et al., 1989. "Design, Construction, and Test of Magnetic Bearings in an Industrial Canned Motor Pump." Proceedings of the Sixth International Pump Users Symposium, Texas A&M University, pp. 65-73.

2. Blair, B. J., et al., August 1990. "A Canned Pump with Magnetic Bearings for Industrial Use - Laboratory Testing." Proceedings of 25th IECEC, Reno, Nevada, pp. 475-81.

3. Chen, H. M. and P. Lewis, April 1990. "A Magnetic Bearing Control Scheme for Canceling Submerged Rotor Vibratory Forces." Presented at the Third International Symposium on Transport Phenomena and Dynamics of Rotating Machinery, Honolulu, Hawaii.

4. Zorzi, E. S. and H. D. Nelson, January 1977. "Finite-Element Simulation of Rotor-Bearing Systems with Internal Damping." ASME J. Engineering for Power, pp. 71-76.

5. Chen, H. M., et al., July 1990. "Stability Analysis for Rotors Supported by Active Magnetic Bearings." Proceedings of the Second International Symposium on Magnetic Bearings, Tokyo, Japan, pp. 325-28.

6. Fritz, R. J., 1970. "The Effect of an Annular Fluid on the Vibration of a Long Rotor." ASME Series D, Vol. 92, pp. 923-29.

7. Black, H. F., 1979. "Effects of Fluid-Filled Clearance Spaces on Centrifugal Pump and Submerged Motor Vibrations." Proceedings of 8th Turbomachinery Symposium.

8. Fuller, C. R., et al., 1989. "Experiments on Structural Control of Sound Transmitted Through An Elastic Plate." Presented at AIAA American Control Conference, Pittsburgh, Pennsylvania.

9. Elliott, S. J., et al., October 1987. "A Multiple-Error LMS Algorithm and Its Application to the Active Control of Sound and Vibration." IEEE Trans, Vol. ASSP-35, No. 10, pp. 1423-34.

AN INTEGRATED DIGITAL CONTROLLER FOR MAGNETIC BEARINGS

John K. D'Addio,[1] Ronald D. Williams,[1] F. Joseph Keith,[1] Steven J. Fedigan[1]

ABSTRACT

Magnetic bearings used to support rotors permit many advantages over conventional fluid-film bearings. These advantages can only be fully realized through the use of digital controllers and the complex algorithms they support. This paper briefly discusses the history of magnetic bearing digital control research at the University of Virginia. Related research concerning switching power amplifiers is also discussed. An advanced digital signal processor based controller design tailored to the magnetic bearing application is presented. The controller's hardware architecture, functionality, and real-time operating system are discussed.

INTRODUCTION

Over the past decade, one of the most significant advances in vibration control for rotating machinery has been the use of magnetic bearings to support rotors. Magnetic bearings employ a circular ring of independent electromagnets to apply forces to a rotating shaft thereby supporting the shaft with no contact or friction. There are several advantages offered by magnetic bearings over conventional fluid-film bearings. The most obvious advantage is the elimination of the lubrication systems necessary for fluid-film bearings. Magnetic bearings can also be immersed in many process fluids thereby eliminating the need for the seals used in many conventional bearing applications. Perhaps the most significant advantage of magnetic bearings is their ability to provide active control over the bearing's dynamic characteristics and thus permit the modification of a rotor's dynamic properties while in operation. This capability has become more important because many modern rotor designs use more flexible shafts thus increasing the number of modes encountered during normal operation.

BACKGROUND

A typical magnetic bearing support system used for laboratory experimentation is shown in Figure 1. This configuration uses two radial bearings and a single thrust bearing to apply force to the shaft along two sets of x and y axes and a single z axis. Magnetic bearings are inherently open-loop unstable and thus require some type of electronic controller to regulate the bearing forces and stabilize the system. The controller relates rotor position to bearing coil current: position measurement is accomplished by shaft displacement sensors located along each

This work was supported in part by the Virginia Center for Innovative Technology.

[1]Department of Electrical Engineering, University of Virginia, Charlottesville, VA 22903-2442, U.S.A.

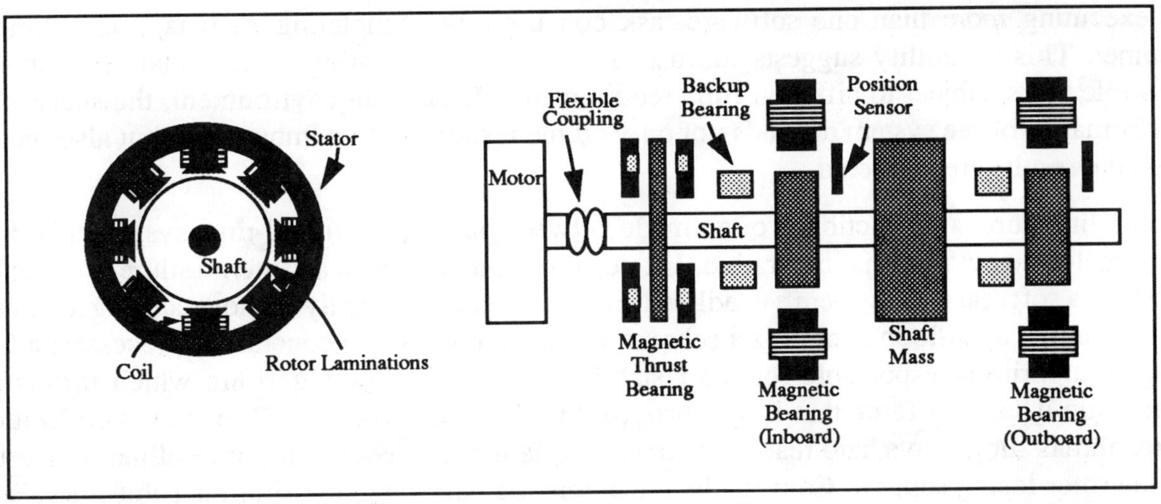

Figure 1: Generic rotor/bearing system with one thrust and two radial bearings

control axis, and the controller output current request is passed to a transconductance power amplifier which develops the requested current in the bearing coils.

BEARING CONTROLLERS

Analog controllers are effective for simply stabilizing a rotor using independent axis proportional-derivative (PD) and proportional-integral-derivative (PID) type algorithms to control the stiffness and damping of each bearing axis [1]. If simple stabilization is sufficient for a particular system, an analog controller can impart many of the advantages of magnetic bearings and may be a good design choice. However, the dynamic advantages of magnetic bearings can only be fully realized via more complex algorithms which can be difficult if not impossible to implement using analog electronics. Such algorithms can, however, be realized with relative ease by implementing the controller with a digital computer.

Adaptive, open-loop and multi-axis state space control schemes are examples of techniques which require the flexibility and computational capability of a digital computer for successful implementation. For example, one type of adaptive control automatically adjusts feedback coefficients as the rotor speed is varied. This modifies the dynamic characteristics of the rotor system and adjusts the rotor's critical operating speeds to prevent the system from operating near (or passing through) critical speeds during rotor run-up or steady state operation. In addition, certain abrupt changes in operating conditions, such as a sudden imbalance caused by blade loss on a turbine wheel, can necessitate a complete change in controller algorithm. This change is effected with a digital controller by merely ceasing execution of the "normal" control algorithm and invoking an "emergency" algorithm. MIMO (multiple-input, multiple-output) state-space algorithms which incorporate models of the rotor can also provide significant advantages over simpler control schemes. Such algorithms estimate rotor behavior at locations along the rotor not directly measured by sensors or directly acted upon by a bearing [2]. This information can then be used to exercise control over a section of the rotor not easily accessed by a sensor or a magnetic actuator.

CONTROLLER MULTITASKING

In addition to the often extensive computational requirements imposed by these advanced control algorithms, such techniques are best implemented on a digital control system capable of executing more than one software task concurrently, completing each task in a timely manner. This capability suggests the use of a real-time operating system that can support multiple tasks subject to different time requirements. In such an environment, the successful performance of the system depends not only on the results of the computation, but also on the time the results are delivered.

In the literature, a distinction is often made between hard and soft real-time systems [3]. In a hard real-time system it is imperative that deadlines are met, else a system failure will occur, while in a soft real-time system a deadline can be missed occasionally without causing a failure. For example, open-loop control is a scheme which embodies two concurrent processes: a fast control algorithm responsible for rotor stabilization and a slow algorithm which injects an automatic balancing force that is synchronized to the rotor speed [4]. The rotor stabilization algorithm is subject to a hard real-time constraint; failure to meet its output deadline will result in the rotor losing support. Conversely, the automatic balancing algorithm is subject to a soft real-time constraint, since its output deadline can be missed occasionally with little consequence. As another example, diagnostic information for both the rotor and the controller is often desired [5]. Such information can be extracted easily from a controller capable of executing the diagnostics algorithms concurrently with the main control algorithm and capable of distinguishing between high and low priority tasks.

Developing software to support multiple tasks running concurrently with time constraints of varying criticality demands the use of a real-time operating system. A real-time operating system not only handles task scheduling with attention to time constraints, but also manages other resources like the system's memory. When a high priority task such as the primary control loop needs to deliver data to the actuator drivers while a low level task is running, the high level task preempts the low level one [3].

ACTUATORS

Active magnetic bearings generate a force on the supported structure (the rotor) through magnetic actuators. The force generated by these magnetic actuators is generally modeled as functionally related to the electric current in the electromagnet coils. In class A applications, opposing electromagnets are prebiased either with a bias current or with permanent magnets and the force generated is essentially proportional to a perturbation current applied to both opposing magnets [1][6][7][8]. In class B operation, only one of an opposing pair of magnets is activated at a time, and the force is related to the square of the current [7]. The underlying assumption in control schemes developed for either of these two modes of operation, and for other control schemes as well, is that the current in the electromagnet coils of the bearing can be controlled explicitly. Thus, some means of directly controlling the coil currents must be provided.

ACTUATOR DRIVE

Explicit control of actuator current is accomplished through a power amplifier that drives a current through a load. This current is proportional, over the amplifier bandwidth, to a control signal that is typically a voltage. Since the impedance of a magnetic bearing coil is generally a strong function of the operating conditions, and the output stage of a power amplifier typically applies a voltage to the attached load rather than a current, high gain current feedback is required to achieve the required true transconductance behavior. In addition, amplifiers used in magnetic bearing applications must possess sufficient voltage overhead to provide adequate current slewing, and thus bearing force slewing, capability in the typically highly inductive bearing coils [9]. As a result, the required supply voltage will typically be well in excess of the voltage needed merely to provide the coil bias current.

PREVIOUS WORK

Extensive magnetic bearing research has been in progress at the University of Virginia since 1984 [10]. This research has included interest areas such as analog controllers and algorithms [1], digital controllers and algorithms [11][12], power amplifiers [13], industrial applications [14][15][16], permanent magnet biased magnetic bearings [17][18], and magnetic bearing/rotor system dynamics [19].

The first magnetic bearing digital controller developed at the University of Virginia was based on an IBM-type personal computer and was used to support a small flexible three-mass rotor [8]. This controller consisted of the personal computer, a multiplexed analog-to-digital (A/D) subsystem for digitizing position sensor inputs, multiplexed digital-to-analog (D/A) converters and a multichannel linear transconductance power amplifier. A block diagram depicting the major components of the control loop is shown in Figure 2. The limited computational capability of the 8086 CPU restricted the controller sampling rate to approximately 3 kHz, but in spite of this limitation, the controller successfully supported a small flexible rotor at speeds up to 14,000 RPM.

While the first controller design demonstrated the viability of digital controls for magnetic bearings, its obvious performance limitations restricted its use as a working design tool. To address these limitations, a second design effort was undertaken to create a controller architecture that was customized for the magnetic bearing application [20]. Although the basic control loop remained the same as that in the first controller, the resulting two-level hierarchical structure of the second controller employed several identical digital control modules (DCM) for executing primary control algorithms and a supervisory computer for handling communication and diagnostic tasks. Each DCM included a TMS320C25 fixed-point digital signal processing (DSP) chip and A/D-D/A capability, while the supervisory computer was a commercially available single board computer based on the Motorola MC68020 microprocessor. This second controller was designed to provide a reliable and powerful digital controller that would support multiaxis control, using the specialized digital signal processors

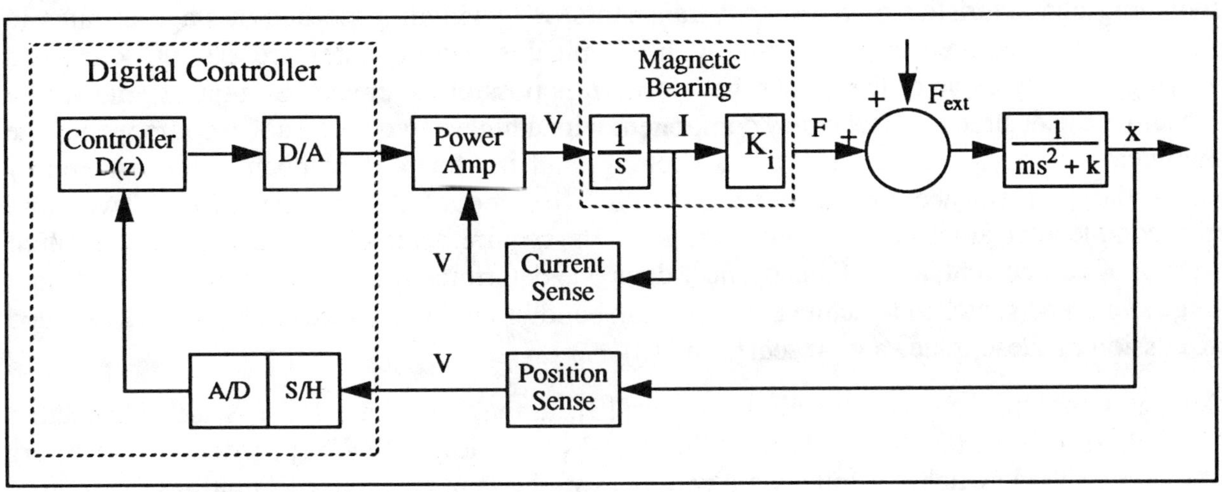

Figure 2: Control loop diagram for first and second controllers

to supply the needed computational capacity at a relatively low cost and size. A prototype of this second design was built using wire-wrap construction techniques, and it successfully demonstrated that powerful digital control could be attained with inexpensive microcomputer technology.

Although this second controller was a successful design, it had a significant drawback. Since the original prototype was constructed by hand wiring all connections on a wire-wrap board, the modules were not easily replicated. Thus, constructing additional copies of the entire system as designed would have required a great deal of tedious, error-prone assembly time. Additionally, the distributed nature of this second controller complicated the implementation of MIMO algorithms, which would have to be parsed into parallel tasks to be run on each processor. So although MIMO control would be possible, successful implementation would require a user to become intimately familiar with the hardware architecture. Furthermore, advances in the available digital signal processing computer chips offered the opportunity for significant improvements in the design.

AMPLIFIERS

Early amplifier design at the University of Virginia involved the design and construction of a linear transconductance amplifier. This amplifier continues to operate several small experimental test rigs successfully, but its intrinsic inefficiency prohibits operation with larger, higher current bearings. Linear amplifiers achieve a variable output voltage by controlling the resistance of output transistors which are in series with the load between a fixed power supply and ground. At any output voltage intermediate between the power supply voltage and ground, the output transistors must dissipate power resistively. For this reason, and because linear amplifiers must be tuned to the specific load to ensure stability of the current feedback loop, it was concluded that switching amplifiers would provide a better approach.

Switching power amplifiers avoid this power dissipation by only operating the output transistors in a saturated "on" (low resistance) state or an "off" (very high resistance) state, thus restricting the output to two fixed voltage levels. Power dissipation occurs in these amplifiers primarily while the transistors are switching from one state to the other, with some power also being dissipated in the on state due to the non-zero on-resistance of the transistors. Because most of the power dissipation in these amplifiers occurs during the state transitions, the efficiency relies on keeping the switching rate below some threshold which depends upon the switching characteristics of the output transistors. By switching the output stage at rates in excess of the required amplifier bandwidth (typical switching rates are 10-100 kHz) and varying the duty cycle of the output waveform, it is possible to create an output signal which combines the desired low frequency component with a higher frequency noise component. The nonlinear compensator regulates this switching, and the form of compensation can greatly effect the performance of the overall circuit. The central design issue for a switching transconductance amplifier then becomes: given the requirements of load insensitivity, explicit control of coil current, and efficiency, how does one determine the manner in which the output stage should be switched to achieve robust, high bandwidth, low distortion operation? Detailed discussion of these issues are presented in [13].

Several switching schemes are discussed in [13]; among them are Pulse-Width-Modulated (PWM), Sample/Hold, Hysteresis and Minimum-Pulse-Width (MPW) approaches. Amplifiers utilizing a PWM switching strategy are widely available, but frequency limitations inherent in the PWM design restrict its usefulness in closed-loop transconductance applications,

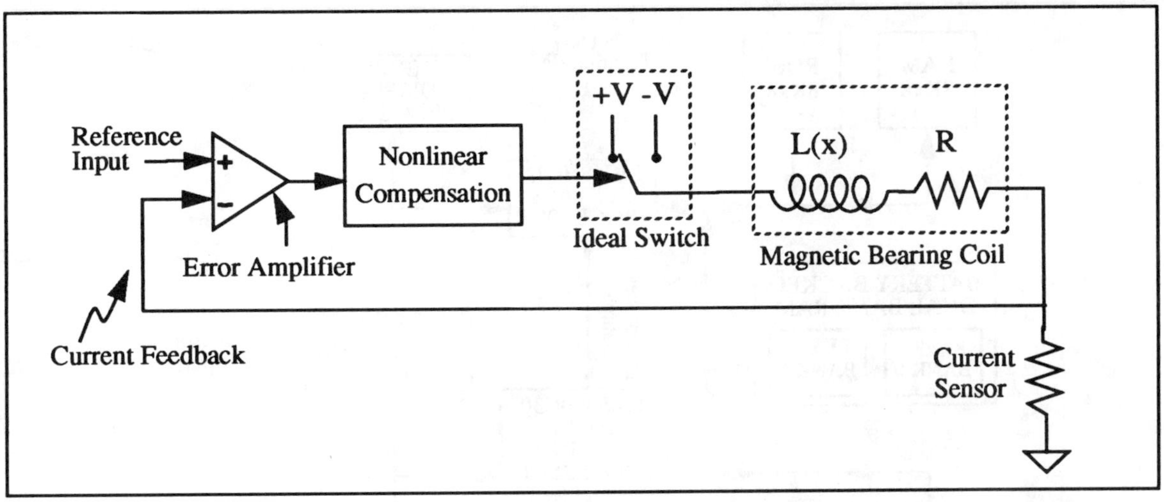

Figure 3: Generic closed-loop transconductance amplifier

particularly when rapid response to large current errors is desired. Bang-bang algorithms such as Sample/Hold and Hysteresis provide explicit transconductance behavior with far simpler circuitry than PWM, but the Sample/Hold scheme tends to distort low amplitude output waveforms, and the noise output level and nominal switching rate of a hysteresis amplifier depend on the load characteristics. A scheme referred to as Minimum Pulse Width (MPW) control shares most of the beneficial aspects of sample/hold and hysteretic control while addressing their shortcomings; MPW control is discussed extensively in [13]. All of these methods are implemented in a configuration similar to that shown in Figure 3, where the algorithm being considered (sample/hold, MPW, etc.) is chosen as the type of nonlinear compensation used to develop the switching decisions.

INTEGRATED DIGITAL CONTROLLER

The digital controller currently under development has been designed to provide a powerful platform for executing real-time digital control algorithms while also being easy to use, expandable, and flexible. The design is tailored to the typical magnetic bearing application shown in Figure 1. This configuration requires five axes of control including two radial bearings with eight poles each and a double acting thrust bearing. The architecture, however, is easily expandable to accommodate more axes of control and an increased number of bearings and poles. Although there are commercially available computers that could be used as controllers, their designs are not optimized for the magnetic bearing application, and they would be significantly more expensive and contain unnecessary hardware.

Attaining the highest possible throughput is a primary design goal when developing a digital controller architecture. Thus a controller must sample and digitize any number of input signals, perform all required control calculations, and provide an updated output in a limited period of time. The maximum rate at which these tasks can be accomplished is determined by the computational capability of the processor executing the control algorithms. Thus the sampling rate and the bandwidth of the controller are directly related to the computational capability of the processor.

As can be seen in Figure 4, the system hardware architecture is based on a single digital signal processor, the Texas Instruments TMS320C30. Since this processor is optimized for the execution of finite difference equations, which are required for many control applications, it is

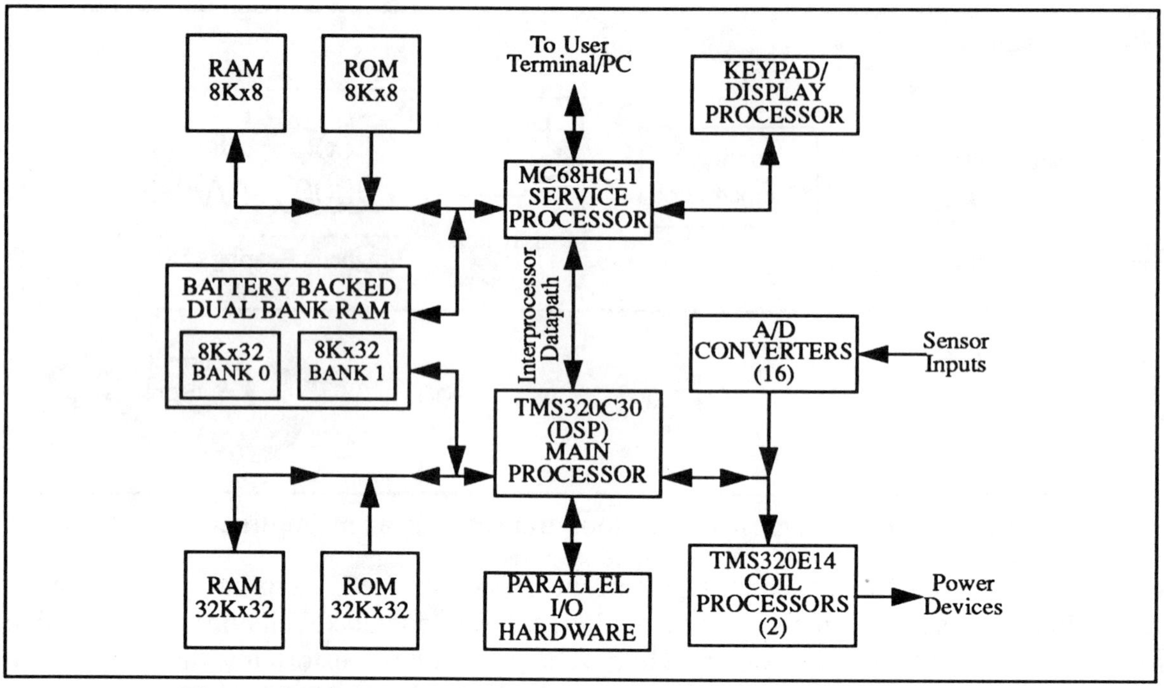

Figure 4: Integrated controller hardware architecture diagram

an ideal choice for the main control processor. At the time of design inception, the TMS320C30 was the most advanced floating point DSP available.

One requirement for this digital controller is the ability to digitize sixteen analog input signals: five position sensor signals, ten current sensor signals, and one perturbation signal. Several analog to digital (A/D) converter configurations were investigated to determine the optimum for both price and performance. Throughput requirements dictated an upper bound on conversion time of approximately 10μs for conversion of all sixteen channels. Initially, a conventional A/D configuration utilizing one fast A/D multiplexed across all sixteen channels was considered. However, the resulting sample skew would complicate state-space algorithms. An alternative configuration uses sixteen sample and holds multiplexed into one A/D; all signals are sampled simultaneously and then digitized successively by the A/D converter. Although this scheme eliminates the sample skew in the previous configuration, the cost of sixteen sample and hold devices, the high-speed A/D, and the associated multiplexing circuitry made the configuration impractical. A more practical solution, employs, in parallel, sixteen slower A/D converters with integral sample and holds. This configuration is cost effective and actually has higher throughput than the sequential digitizing configuration. The A/D chosen for this system is the Analog Devices AD7870, which is capable of 12-bit resolution and a maximum conversion time of 11μs. Thus, this A/D permits controller throughputs as high as 90 kHz/channel, subject to algorithm processing time. Rapid sampling rates are necessary to permit the control of higher modes of high speed rotors.

The controller must easily interface with the outside world and be readily configured by the user. A single processor design would require these interface functions to be carried out by the TMS320C30, a responsibility which could interfere with the processing of the control algorithms. To avoid this interference, a second processor, the Motorola 68HC11 microcontroller, is used in the system as a service processor to carry out functions not directly related to control algorithm execution. The 68HC11, like all microcontrollers, is essentially a microprocessor with on-chip memory, input/output facilities, timers, and other peripherals. It provides a minimum component solution to many of the design's requirements. Although it is quite capable, when compared to other conventional processors the 68HC11 is not particularly fast or computationally efficient and therefore would not be appropriate for executing magnetic bearing control algorithms. For example, an integer multiply operation on the 68HC11 would take 5μs as compared to 60ns for the TMS320C30 to accomplish a parallel floating point multiply and accumulate, a commonly used instruction when executing finite difference equations.

Two independent interfaces to the 68HC11 are available: an asynchronous serial port which permits communication to an external computer via an RS-232 port, and a synchronous serial port which interfaces the controller to a keypad/display processor. The external computer interface permits major modifications to the controller's operation by performing such functions as parameter adjustment and algorithm selection/modification. Additionally, the RS-232 connection permits a user to download an entirely new algorithm or set of data coefficients into the controller's semi-permanent memory (discussed below). While the serial connection to a terminal is intended for making major changes to the controller setup and operation, the keypad/display combination is intended to support minor controller changes while the controller is in operation. The keypad/display processor is used to display an operational menu on a liquid crystal display and also interpret commands entered on a sixteen key keypad, both of which are located on the controller front panel. Thus, a user can perform parameter adjustment and algorithm selection from the controller front panel without attaching the controller to a terminal. The display is also useful for notifying the user of controller and plant status conditions while in operation.

Under certain rare conditions, it is possible for a microprocessor to lose track of the program it is executing. For a magnetic bearing digital controller, such an event would lead to erroneous outputs which would create an unstable system. This is entirely unacceptable because it would result in loss of rotor support and its potentially destructive consequences. To guard against this possibility, the architecture includes a "watchdog" function through which the main control processor (TMS320C30) confirms its proper operation by periodically signaling the service processor (68HC11). This is accomplished by using the 68HC11 to setup a software timer for some reasonable amount of time, such as 50μs, after which the 68HC11 resets the TMS320C30. As part of the main control loop, the TMS320C30 must send a timer-reset message to the 68HC11 within the 50μs period. In the event of a main processor program runaway, the reset message would fail to be sent, and the service processor would reset the main processor. This reset would restart the main control loop and return the controller to proper operation. Due to the mass of the rotor, the system can tolerate erroneous outputs for this limited time period before the rotor falls from support. Fault detection and recovery within this time allows the system to achieve some degree of fault tolerance.

In addition to the inter-processor data path, the architecture contains a dual-bank random access memory (RAM) to facilitate information transfer between processors. Two distinct memory banks comprise the dual-bank RAM, and are accessible in both the 68HC11's memory space and the TMS320C30's memory space. At any given time, each processor has exclusive read and write access to one of the memory banks, thus providing a fast and efficient way to transfer large amounts of data or program code between processors without interrupting either processor's operations. For example, new control algorithms or coefficients can be transferred from a user's personal computer, through the 68HC11, into the memory space of the

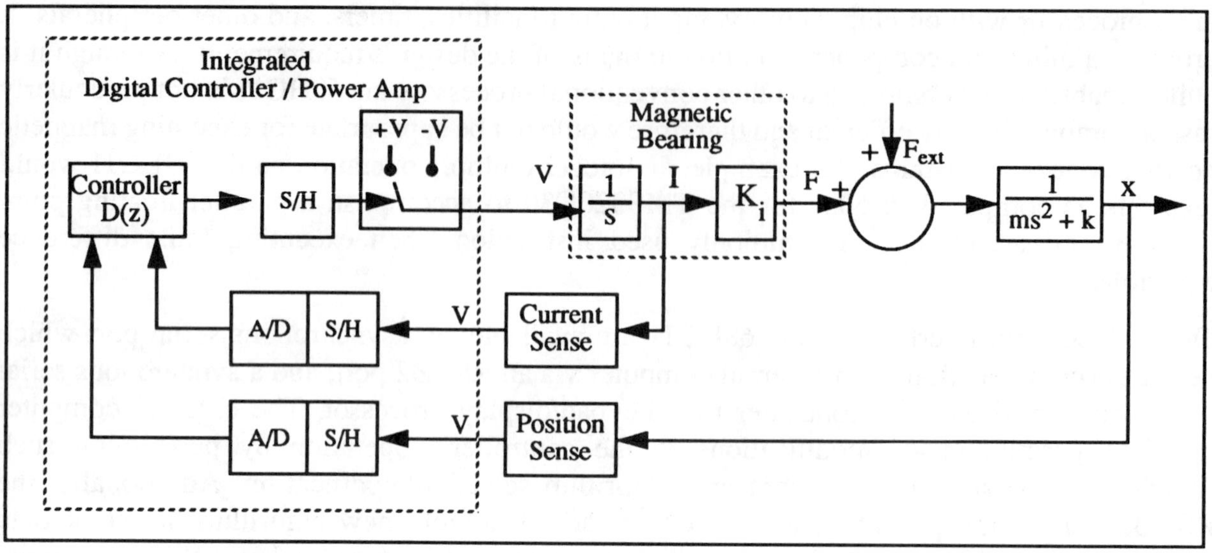

Figure 5: Control loop diagram for integrated digital controller

TMS320C30 without interrupting control operations. This is useful for carrying out adaptive control schemes in which an entirely new algorithm is to be executed or existing algorithm parameters are to be modified. Additionally, the memory space can be used to store data acquired by the TMS320C30, which can then be uploaded to a user's personal computer by the 68HC11. This allows the main processor to carry out on-line data acquisition with minimal throughput reduction due to data transfer operations. In order to preserve data and program integrity when the system power is removed, the dual-bank RAM also includes battery back-up, freeing the user from having to reload programs or data that is stored in the RAM between controller uses.

In addition to the main and service processors, the architecture also contains two additional processors which are used to execute the minor loop coil current control. These processors, Texas Instruments TMS320E14's, are combination DSP-microcontrollers that have fast computational ability similar to other DSP's and also contain a number of peripherals like most microcontrollers. However, they are neither as computationally powerful as the TMS320C30 nor as functionally capable as the 68HC11. The coil processors generate digital output drive signals in accordance with a software programmable switching algorithm; these signals are used to directly drive the power modules which regulate the coil currents. The power drive modules switch the output transistors as dictated by the binary inputs so that the current in a given magnetic bearing coil will slew up or down until the next state transition occurs. Because the signal path between the digital control computer and the power output stage is now a differential digital signal, this arrangement offers much greater noise immunity than the standard configuration in which the digital controller output is converted to an analog voltage which is then used as a reference for a switching transconductance power amplifier. This integrates the controller and the power amplifier into essentially one unit which is depicted the control loop block diagram of Figure 5.

Performing the amplifier compensation in software permits the implementation of a variety of switching schemes not readily accomplished through analog hardware, and it also offers the flexibility to use different compensation methods for different circumstances through simple software changes. For example, the sample/hold bang-bang algorithm mentioned previously is implemented in software by merely sampling the output currents periodically, comparing the measured value to the requested value, and setting the corresponding output bits to the power modules to the desired logical values. This bang-bang algorithm could also be readily modified to include a pseudo-linear operating region; with this scheme the output is regular bang-bang for large current errors but is open-loop PWM for small errors. Such an arrangement retains the rapid slew rate characteristics of the sample/hold scheme while exhibiting the small signal accuracy of PWM. A sliding-mode algorithm of this sort has the potential to reduce the magnitude of the current switching noise generated in the magnetic bearing coils, since the output devices will potentially be switched at a higher rate than the sampling frequency of the main control loop. Alternately, the sample/hold algorithm could be augmented with a small amount of dither, a common technique used to randomize the otherwise highly correlated noise generated by the synchronous sample/hold algorithm.

In addition to controlling the output devices directly, the system also permits computer-controlled adjustments of the supply voltage seen by the bearing coils. Such control could be exercised to provide a high voltage when rapid coil current slewing is needed and to reduce this voltage when the coil current error is small. This supply voltage reduction directly decreases the amplitude of the high frequency current noise in the magnetic bearing coils.

Implementing the described architecture in physical hardware was facilitated by CAD tools for schematic capture and printed circuit layout. This enabled the design to be reviewed at the schematic level and then transferred to a circuit board while minimizing the possibility of introducing errors into the design. More importantly, printed circuit implementation enables the entire controller to be replicated quite easily. To support the design goals of flexibility and expandability, the controller is constructed in a modular fashion on three different circuit boards interconnected by a high-speed bus backplane. The first circuit board is comprised of the main and service processors, all system memory, and the processor interface circuitry. The second contains the A/D converters and their analog and digital buffering circuitry. The third board consists of the coil processors and their differential drive circuits. If it is later decided that the controller must support a rotor system with more bearings, coils, or control axes, all that is necessary is to replicate the coil processor board and the A/D board and install them in the backplane for increased I/O capability.

To support multi-tasking and scheduling processes subject to a variety of time constraints, the system employs a real-time operating system on both the main and service processors. A preemptive scheduling algorithm is used because both high priority tasks, such as the primary control loop, and low level tasks, such as system diagnostics, are executed concurrently. Thus, low priority tasks can be suspended while higher priority tasks are processed to meet more critical timing deadlines. For example, the open-loop control scheme previously discussed is easily supported by the main processor operating system. Additionally, the service processor operating system assigns message decoding from the main processor as its highest priority so that watchdog reset messages from the main processor are handled promptly. This prevents unnecessary resets of the main processor.

Other features of the operating system include a rapid context switch, a capability for inter-task and inter-processor communication, and small memory requirements. A rapid context switch minimizes the timing overhead needed by the operating system. Inter-task communication is needed to synchronize transitions between tasks or to pass messages between them. Small memory requirements permit a maximum number of tasks to be running concurrently. All of the features which are incorporated into the operating system are directed at enhancing the controller's flexibility.

In addition to process task management, the operating system also insulates the user from many system hardware details by providing predefined routines for such functions as data input and output, interprocessor communication, and A/D management. This frees the user from concerns about actual controller hardware, allowing a focus on control algorithms and their effect on rotor dynamics. Furthermore, the availability of an optimizing C compiler for both the main and service processors allows both the operating system and the control algorithms to be implemented in a high level language. This facilitates easier algorithm development, debugging, and testing than systems restricted to programming in assembly language.

CONCLUSION

This paper presented a brief synopsis of digital control research for magnetic bearing applications at the University of Virginia and introduced the design of a new digital controller currently under development. Previous controllers established the viability of digital controls for magnetic bearing applications and demonstrated the usefulness of digital signal processing technology for magnetic bearing controllers. Recent advances in switching power amplifier design and DSP capabilities have driven the design of the new digital controller which incorporates the more advanced technology. This enabled the typically separate controller and power amp to be integrated into one unit while also maintaining a high throughput rate. The entire controller design has been optimized for magnetic bearing applications, which is most evident in the centralized architecture of the controller and the many features it contains including the highly parallel A/D conversion capabilities and the dual bank memory. The use of a multi-tasking real-time operating system in conjunction with the specialized hardware provides a capable test environment for magnetic bearing research and control algorithm development.

REFERENCES

[1] Humphris, R. R., R. D. Kelm, D. W. Lewis and P. E. Allaire, "Effect of Control Algorithms on Magnetic Bearing Journal Properties," *Journal of Engineering for Gas Turbines and Power*, Vol. 108, October 1986.

[2] Stanway, R. and C. R. Burrows, "Active Vibration Control of a Flexible Rotor on Flexibly-Mounted Journal Bearings," *ASME Journal of Dynamic Systems, Measurement and Control*, Vol. 103, December, 1981, pp. 383-388.

[3] Burns, A. and A. Wellings, *Real Time Systems and Their Programming Languages*, New York, Addison-Wesley, 1990.

[4] Humphris, R. R., C. Knospe and S. Sundaram, "Flexible Rotor Balancing Using Magnetic Bearings," Presented at the Conference on Recent Advances in Active Control of Sound and Vibration, VPI&SU, Blacksburg, Virginia, April 15-17, 1991.

[5] Humphris, R. R., P. E. Allaire, D. W. Lewis and L. E. Barrett, "Diagnostic and Control Features with Magnetic Bearings," *Proceedings of the 24th Intersociety Energy Conversion Engineering Conference (IECEC)*, Vol. 3, No. 899441, Crystal City, VA, August 7-11, 1989, p. 1491-1498.

[6] Hebbale, K., "A Theoretical Model for the Study of Nonlinear Dynamics of Magnetic Bearings," Cornell University Thesis, January 1985.

[7] Bleuler, H., "Decentralized Control of Magnetic Rotor Bearing Systems," Ph.D. Dissertation, Swiss Federal Institute of Technology, ETH report number 7573, Zurich, Switzerland, 1984.

[8] Keith, F. J., "Digital Control System Design for Active Magnetic Bearings," M. S. Thesis, University of Virginia, May, 1988.

[9] Maslen, E., P. Hermann, M. Scott and R. R. Humphris, "Practical Limits to the Performance of Magnetic Bearings: Peak Force, Slew Rate, and Displacement Sensitivity," *ASME Journal of Tribology*, Vol. 111, April 1989, pp. 331-336.

[10] Allaire, P. E., R. R. Humphris, R. D. Kelm and D. W. Lewis, "Magnetic Bearings for Vibration Reduction and Failure Prevention," *Proceedings of the 40th Meeting of the Mechanical Failures Prevention Group*, National Bureau of Standards, Gaithersburg, Maryland, April 16-18, 1985.

[11] Williams, R. D., F. J. Keith and P. E. Allaire, "Digital Control of Magnetic Bearings," *IEEE Transactions on Industrial Electronics*, Vol. 37, No. 1, February, 1990, pp. 19-27.

[12] Keith, F. J., R. D. Williams, P. E. Allaire and R. M. Schafer, "Digital Control of Magnetic Bearings Supporting a Multimass Flexible Rotor," *Tribology Transactions*, Vol. 33 (1990), No. 3, pp. 307-314.

[13] Keith, F. J., E. H. Maslen, R. R. Humphris and R. D. Williams, "Switching Amplifier Design for Magnetic Bearings," *Proceedings of the 2nd International Symposium on Magnetic Bearings*, July 12-14, 1990, Tokyo, Japan, pp. 211-218.

[14] Allaire, P. E., J. Imlach, J. P. McDonald, R. R. Humphris, D. W. Lewis, B. B. Banerjee, B. J. Blair, J. Clayton, R. D. Flack, "Design, Construction and Test of Magnetic Bearings in an Industrial Canned Motor Pump," *Proceedings of the Sixth International Pump Users Symposium*, Texas A&M University, Houston, Texas, May, 1989.

[15] Blair, B. J., R. R. Humphris, P. E. Allaire, D. W. Lewis and L. E. Barrett, "A Canned Pump with Magnetic Bearings for Industrial Use - Laboratory Testing," Presented at the 25th Intersociety Energy Conversion Engineering Conference (IECEC), Reno, Nevada, August 13-17, 1990.

[16] Maslen, E. H., P. E. Allaire and M. A. Scott, "Magnetic Bearing Design for a High Speed Rotor," *Magnetic Bearings*, Proceedings of the First International Symposium on Magnetic Bearings, ETH Zurich, Switzerland, June 6-8, 1988.

[17] Sortore, C. K., P. E. Allaire, E. H. Maslen and R. R. Humphris, "Design of Permanent Magnet Biased Magnetic Bearings for a Flexible Rotor," Proceedings of the 44th Mechanical Failures Prevention Group Meeting, Virginia Beach, Virginia, April 2-5, 1990.

[18] Sortore, C. K., P. E. Allaire, E. H. Maslen, R. R. Humphris and P. A. Studer, "Permanent Magnet Biased Magnetic Bearings - Design, Construction and Testing," *Proceedings of the 2nd International Symposium on Magnetic Bearings*, July 12-14, 1990, Tokyo, Japan, pp. 175-182.

[19] Maslen, E. H. and J. R. Bielk, "Implementing Magnetic Bearings in Discrete Flexible Structure Models," Submitted to *ASME Transactions on Dynamic Systems and Controls*, June, 1989.

[20] Yates, S. W. and R. D. Williams "A Fault-Tolerant Multiprocessor Controller for Magnetic Bearings," *IEEE Micro*, Vol. 8, No. 4, August, 1988, pp. 6-17.

FLEXIBLE ROTOR BALANCING USING MAGNETIC BEARINGS

Carl R. Knospe,[1] Robert R. Humphris,[1] Subramanian Sundaram[1]

ABSTRACT

Balancing rotating machinery is typically done by adding correction weights to counter residual unbalance. A condition of balance can be obtained for each critical speed using modal balancing. However, rotor balancing in the field is usually time consuming. Also, overspeeding or system variation will in general lead to poor performance. This paper describes an experiment with rotor balancing achieved through magnetic bearings via synchronous rotating magnetic forces. These rotating magnetic forces, in the same manner as correction weights, effectively balance the shaft to reduce the synchronous response. Stable magnetic suspension is achieved through feedback control. The balancing control signal is injected open loop into the magnetic bearing actuators. Thus, open loop control enables two plane balancing without a system shutdown. The number of balancing planes is limited to the number of magnetic bearings. Both theoretical and practical issues of the implementation of this method are presented. Results demonstrate over 50 dB of orbit attenuation.

INTRODUCTION

The synchronous response of rotating machinery is often the worst vibration problem encountered during operation. This problem is alleviated in conventional machinery through mechanical balancing – the process of adding correction weights to the shaft to eliminate the residual unbalance. Rotor balancing in the field, unfortunately, is usually time consuming and costly. The down–time incurred can also be very expensive. Additionally, if the unbalance condition is a function of operation, as in centrifuges, mechanical balancing will not be successful.

Many researchers have investigated the use of magnetic bearings to reduce rotor synchronous response. Two methods of synchronous vibration reduction have been advocated: feedback control and open loop control. These two methods are often confused, in that both have been referred to as "automatic balancing".

Feedback control reduces the synchronous response by altering the eigenstructure of the rotor system. The simplest and commonest form is the placement of a notch filter at the running speed in the feedback path used to achieve stable rotor suspension. This results in reduced effective stiffness and damping at the operating speed. The rotor thus tends to spin about its inertial axis and the harmonic vibration transmitted through the bearings to the foundation is greatly reduced [1]. This method has many disadvantages. It is difficult to employ at more than one operating speed or over a range of speeds. It cannot achieve a high level of attenuation in the neighborhood of the system criticals and it may result in poor transient and asynchronous response [2].

[1]University of Virginia, Center for Magnetic Bearings, Dept. of Mechanical and Aerospace Engineering, Thornton Hall, Charlottesville, VA 22901, U.S.A.

Many researchers have investigated other feedback control methods of suppressing synchronous vibration [3,4,5,6]. Many of these employ recent control theory (disturbance accommodation, frequency weighted performance functions) to reject the unbalance response at a particular operating frequency. These methods also pose design and implementation problems. The magnetic bearing controller must be designed simultaneously to achieve both stable suspension (including good transient response) and synchronous disturbance rejection. If a state space approach such as the Linear Quadratic Regulator (LQR) with frequency weighted performance functions is to be employed to achieve synchronous disturbance rejection, then the same state space (i.e. LQR) form must be used in achieving the transient response specifications. Thus, the methodology of solution of the transient problem is determined by the synchronous problem. This may be undesirable, depending on the application. These methods also require the addition of pseudo–states to the plant model and make controller synthesis procedures more computationally difficult. Robustness may also be lost via the introduction of the synchronous response requirement into the controller design procedure. Finally, these methods only allow unbalance response attenuation at one operating speed.

Open loop control permits the rejection of rotor unbalance response through the introduction of synchronous forces through the magnetic bearings. Since the synchronous response is highly correlated, it may be altered without feedback. Thus, the system stability and transient response are not affected by the open loop control. This distinction has been noted by several researchers, including Burrows, Sahinkaya, and Clement, who investigated adaptive open loop control via a single magnetic actuator on an oil–film bearing rotor rig [7,8]. Vibration reductions of approximately 20 dB were obtained at the first rotor critical speed. The actuator employed was not used to support the rotor or to implement feedback control.

Higuchi et al. employed an adaptive open loop control on a magnetic bearing spindle [9]. Reductions of approximately 75% (12 dB) were obtained in the rotor vibration and in the housing vibration.

Kanemitsu, Ohsawa, and Watanabe [10] used a personal computer interfaced to a digital signal processor based controller to adaptively minimize rotor synchronous response via open loop control. The proper synchronous control was computed through a two–plane influence coefficient balancing method. At the first bending critical speed, approximately 14 dB of synchronous vibration attenuation was achieved.

In this paper, we demonstrate that much higher levels of attenuation can be achieved with open loop control than have been previously reported. Therefore, this technology may be applied to many rotating machinery problems, including those where very low vibration levels are required.

OPEN LOOP CONTROL

A rotor supported in magnetic bearings is described by the dynamical equations

$$[M]\{\ddot{x}\} + [C]\{\dot{x}\} + [K]\{x\} = \{f_u\} + [B]\,[\{f_f\} + \{f_o\}] \tag{1}$$

where $\{x\}$ is the displacement vector, $[M]$, $[C]$, and $[K]$ are the $n\times n$ mass, rotor damping, and rotor stiffness matrices respectively, $\{f_u\}$, $\{f_f\}$, and $\{f_o\}$ are the $n\times 1$ unbalance force vector, the $m\times 1$ feedback control force vector, and the $m\times 1$ open loop control force vector and $[B]$ is the $n\times m$ bearing selection matrix. Using the Laplace transform, Eqn.(1) becomes

$$\Big[[M]s^2 + [C]s + [K] + [B]\,[G(s)]\Big]\{x(s)\} = \{f_u(s)\} + [B]\,\{f_o(s)\} \tag{2}$$

where [G(s)] is transfer function matrix of the feedback controller. Since the unbalance force vector and open loop control force vector are harmonic at the operating speed ω, Eqn. (2) can be evaluated for the forced response

$$\{X\} = \Big[-[M]\omega^2 + j[C]\omega + [K] + [B][G(j\omega)]\Big]^{-1}\Big[\{F_u\} + [B]\{F_o\}\Big] \tag{3}$$

or

$$\{X\} = \{X_u\} + [T]\,\{F_o\} \tag{4}$$

where $\{X_u\}$ is the n$\times$1 uncontrolled force response vector and [T] is the n$\times$m influence coefficient matrix relaying the m synchronous open loop control force magnitudes to the n synchronous displacement magnitudes. If the vibration at m locations are chosen to be suppressed with the open loop control, then these m rows of Eqn. (4) can be chosen to yield

$$\{X\}_m = \{X_u\}_m + [T]_m\{F_o\} \tag{5}$$

If the open loop control force vector is chosen to be

$$\{F_o\} = -[T]_m^{-1}\,\{x_u\}_m \tag{6}$$

then the vibration amplitudes at the m locations are zero. This is analogous to conventional m plane balancing for m points [11].

The open loop vibration control scheme is represented in Figure 1. Synchronous open loop control currents are superimposed on the stabilizing feedback control currents to drive the magnetic bearing coils. Then, these electromagnets exert the desired synchronous magnetic forces on the rotor and thus produce the required rotating force field. For the two radial magnetic bearings pictured, four synchronous control currents must be used.

EXPERIMENTAL RIG

The rotor, suspended in the magnetic bearings, has a 12.7 mm (0.5 inch) diameter and 508 mm (20 inches) bearing span. Laminated iron disks 58.4 mm (2.3 inches) in diameter and 25.4 mm (1.0 inch) long are placed over the shaft at the bearing locations. A 4 Kg (1.8 lb) disk is located at the center of the bearing span. The magnetic bearings consist of four horseshoe electromagnets distributed radially around the laminated iron disk with a nominal radial clearance of 1.0 mm (0.039 inches). This geometry is pictured in Figure 2. Eddy current position sensors were located vertically and horizontally, 32 mm (1.25 inches) outboard from the center of each bearing and 51 mm (2 inches) inboard from the center of the midspan disk. The rotor was driven by a 1/10 hp motor, through a flexible coupling. The test rig is illustrated in Figure 3.

The control system used local proportional–derivative (PD) analog control circuits. The horizontal and vertical sensor signals at each bearing are passed through proportional and rate amplifiers in parallel and the resulting signals are summed. The gain of each of the amplifiers was adjusted with a ten–turn potentiometer. After summing, the signal passes through a lead network to the position and current or power amplifier. Connected to this final power amplifier is a perturbation input, which was used to introduce the open loop control signals.

To produce the desired rotating magnetic force field via this open loop control technique, a reference rotating signal, or key phasor, was used to synchronize two

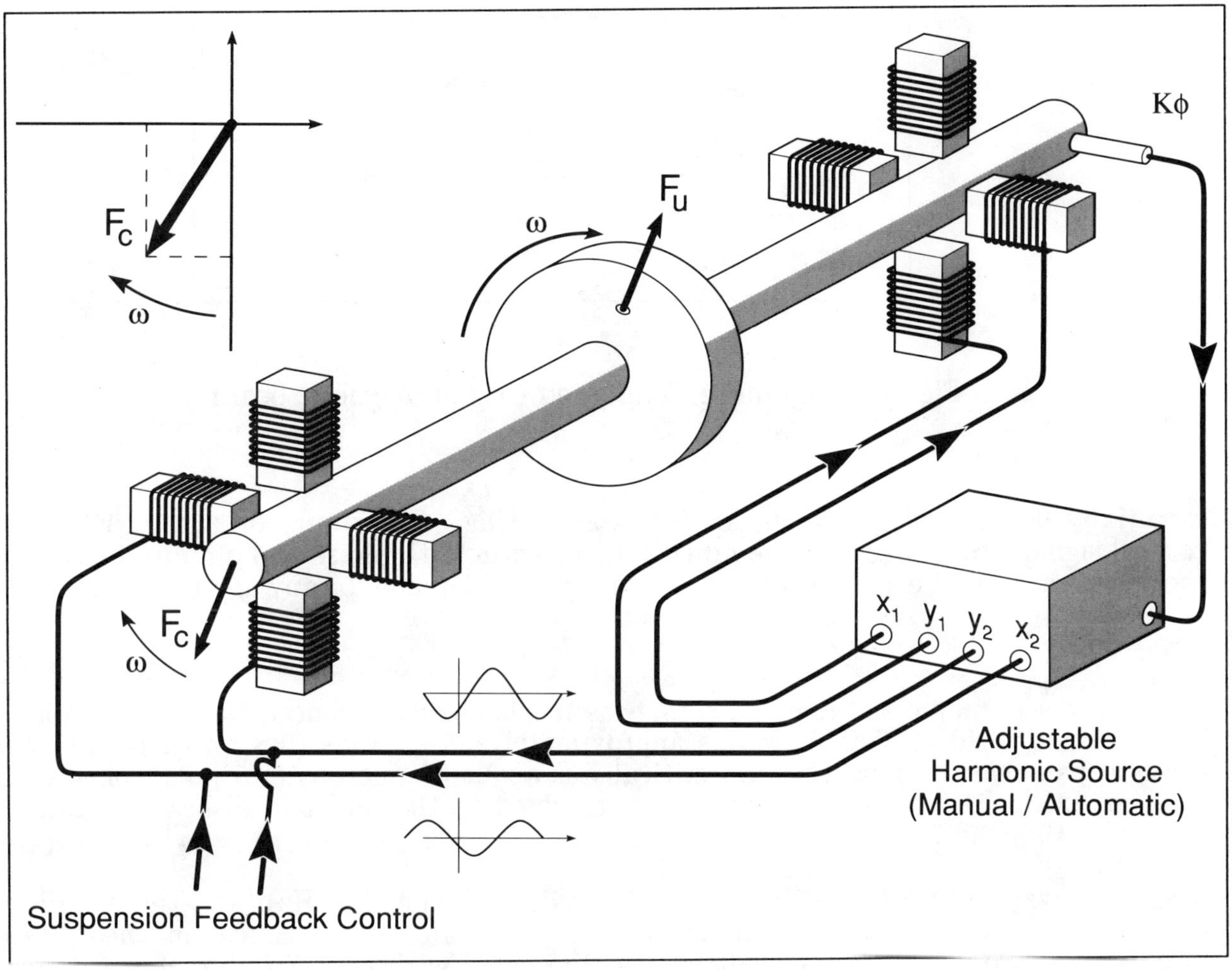

Fig. 1: Block diagram of open loop control of rotor synchronous response

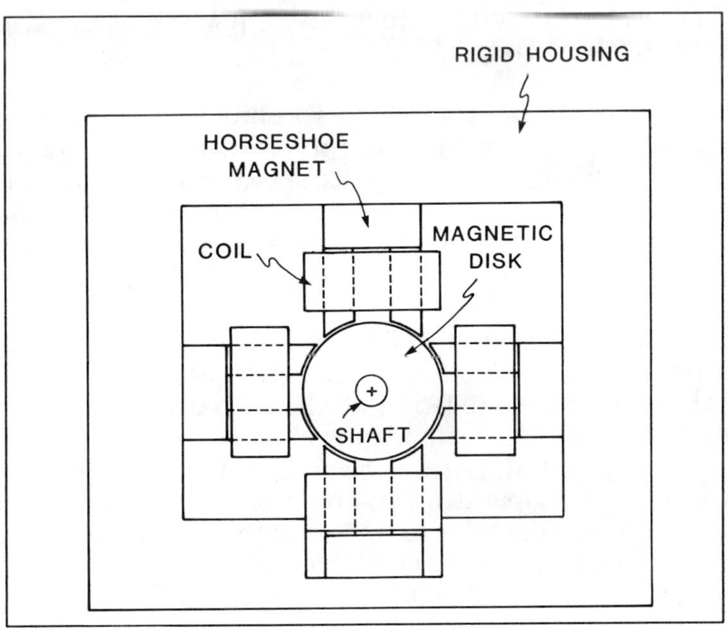

Fig. 2: Magnetic bearing geometry

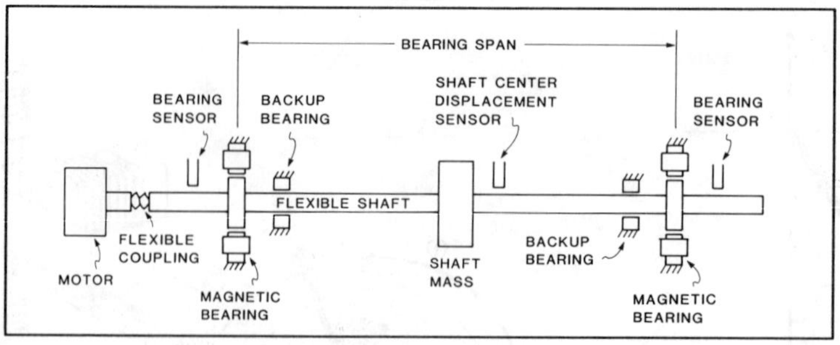

Fig. 3: Diagram of flexible rotor rig in magnetic bearings

essentially 90° out–of–phase signals for each bearing. These out–of–phase, open loop control signals, could each be individually adjusted in both phase and amplitude, so that there were a total of eight open loop adjustments to be made for any given speed.

RESULTS

Shaft bow was minimized and the rotor was mechanically balanced for the first mode. The midspan residual shaft bow was approximately 28 microns (0.8 mils), pk–pk, and the midspan shaft orbit at the first critical was about 38 microns (1.5 mils), pk–pk. A 1.22 gm (2.7 x 10^{-3} ounce) weight was then added to the midspan disk at a radius of 30 mm (1.19 inches).

The rotor was run up to the first critical speed, 2155 rpm (36 Hz), and the open loop control currents were adjusted to reduce the midspan orbit. Since, in theory, the vibration at two points on the shaft can be eliminated with two bearings, this one point vibration reduction may be accomplished with many combinations of the control settings. In this test, rather than choosing another point on the shaft, the controls were adjusted to also minimize (elimination is not possible) the orbits at the two bearings. By manual adjustment, this process took just a few minutes. Using a digital computer, this could be done much more quickly.

After the open loop control was adjusted to eliminate the first critical speed midspan vibration, the rotor was run up to 4000 rpm with the open loop control "on" and then again with it "off". The amplitude versus speed information was recorded. The open loop control signal was synchronous with the operating speed. The magnitude and phase (in relation to the key phasor) of this control signal were constant over the entire speed range.

Figure 4 presents the run up results for the inboard, midspan, and outboard locations. Note the sharp reduction in the midspan vibration at the first critical speed, 2155 RPM. The first critical synchronous vibration at the inboard and outboard locations were also reduced by over 60%. The small amplitude change noted at 2500 RPM is due to a structure resonance. This figure illustrates that open loop control attenuates the unbalance response in a wide range of operating speeds near the first critical speed for which it was tuned. If the open loop control was re–adjusted for each frequency, the synchronous vibration could be attenuated over the entire range of operating speeds. Furthermore, if a digital controller was used to identify, on–line, the $[T]_m$ matrix and $\{x_u\}_m$ vector at an operating speed, the proper open loop control could be rapidly calculated and employed. Since an open loop control, which greatly reduces unbalance response at one speed, also greatly reduces the response at nearby operating speeds, an open loop control computed in this manner, even during a run–up at an approximate rate of 50 RPM/sec, should perform well. Note that reduction is obtained over an approximately 300 RPM range, even though the plant description ($[T]_m$, $\{x_u\}_m$) changes most quickly at the critical speeds.

The frequency spectrum of the midspan sensor signal, when operating at the first critical speed, 2155 rpm, is shown in Figure 5. Without open loop control, the midspan orbit is about 254 microns (0 dB). With the open loop control turned on, the midspan vibration was reduced by 52 dB, a factor of 400. Incidentally, the twice synchronous component was also reduced by about 20 db, a factor of 10. This indicates some nonlinearities in the plant. An adaptive open loop controller could reduce the 2X component as well as the 1X, however, the nonlinearities present would require that cross–harmonic terms be present in the resulting influence coefficient matrix $[T]_m$.

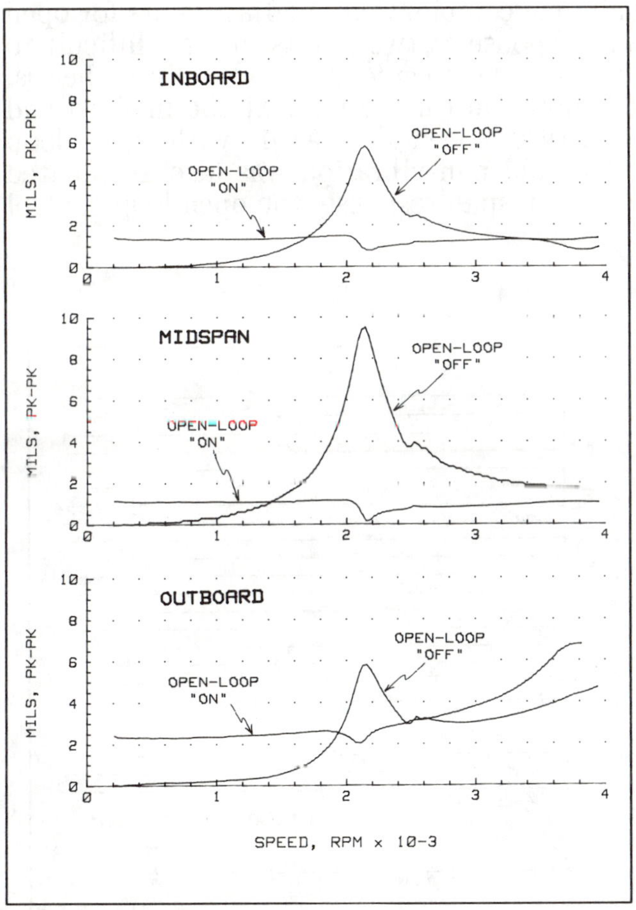

Fig. 4: Amplitude response of unbalanced rotor at horizontal midspan, without and with open loop control set for minimum midspan vibration amplitude

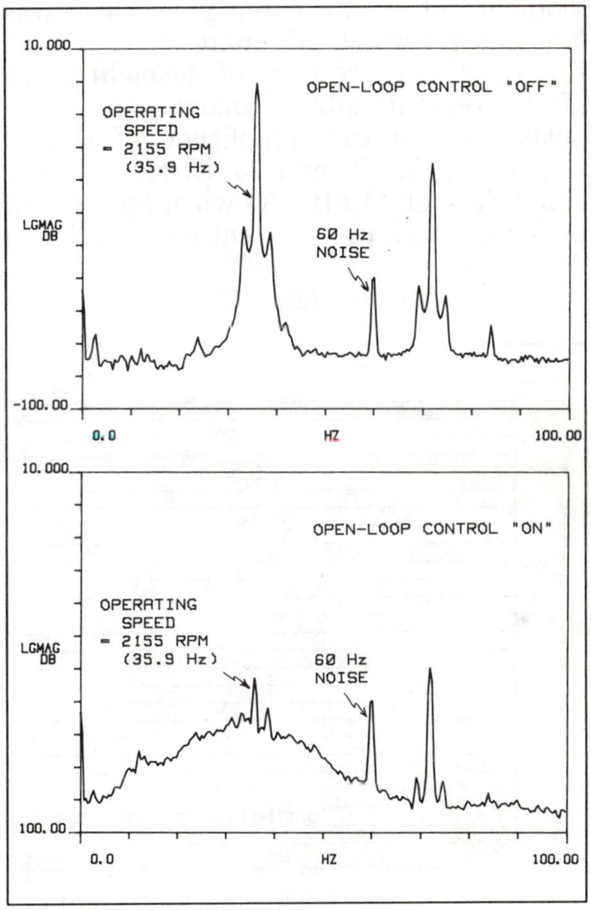

Fig. 5 Frequency spectrum of 50 averages of horizontal midspan response with and without open loop control for constant operating speed

Figure 6 shows a cascade or waterfall plot for the midspan vibration, both with and without open loop control. Note the general suppression of the 1X vibration and the complete absence of any response at the critical speed with open loop control. At some speeds, a small component of second harmonic of the running speed is observed and this indicates a system nonlinearity.

The orbits with and without open loop control at inboard, midspan, and outboard at the first critical speed are presented in Figure 7. As shown, the midspan does not have an orbit when the open loop control is used. The inboard vibration is also greatly attenuated. The outboard vibration is of the same magnitude with and without open loop control. (This picture was not taken at the same time as the other data presented; the settings were different and therefore results differ slightly from those noted in Figure 4.)

Figure 8 illustrates the transient in first critical speed midspan vibration when the open loop control is turned on. In less than 20 rotor revolutions, the unbalance response is almost completely attenuated from its original large amplitude of approximately 254 microns (10 mils), pk–pk.

The open loop control was also adjusted to minimize the unbalance response at the two bearings at the first critical speed. Figure 9 presents the run–up with and without open loop control for this case. Since there was only one combination of adjustments for open loop control which eliminates the synchronous response at two points, it was difficult to find these exact settings of the eight adjustments of the open loop control. Nevertheless, after several minutes of manual "tuning", it is noted in this figure that the inboard and outboard vibration amplitude of the first critical speed decreased, with open loop control, by 29 dB and 23 dB respectively. The midspan vibration was also attenuated about 80% or 14 dB. As when set for minimum midspan response, the open loop control performed well over the entire operating speed range.

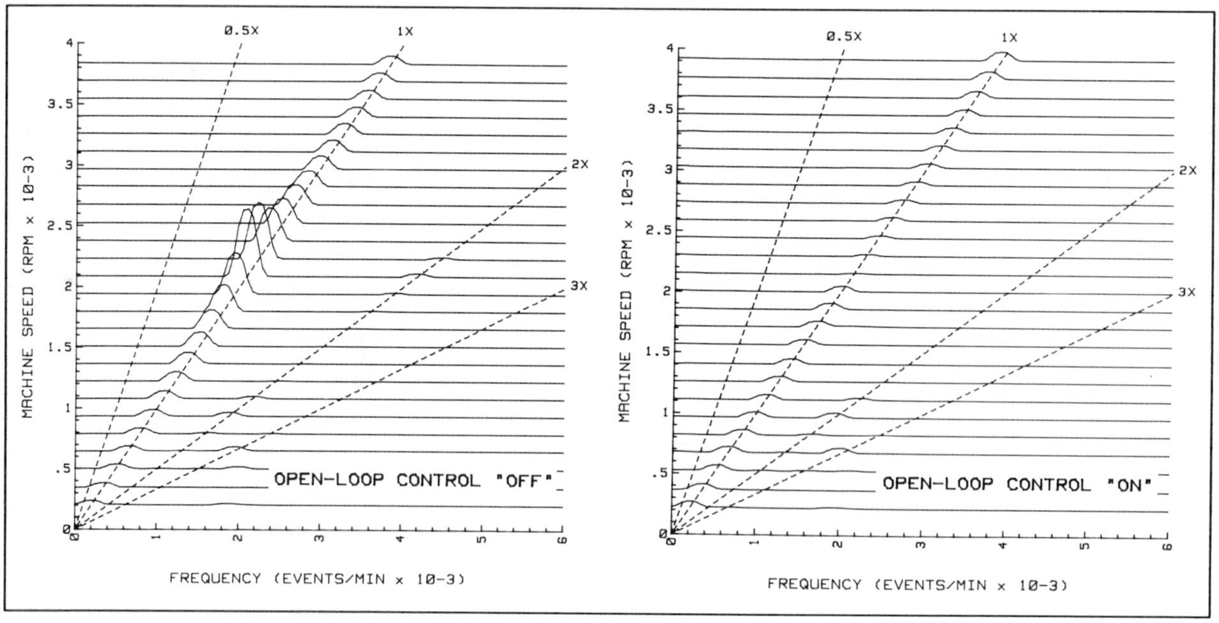

Fig. 6: Cascade plot of horizontal midspan response with and without open loop control

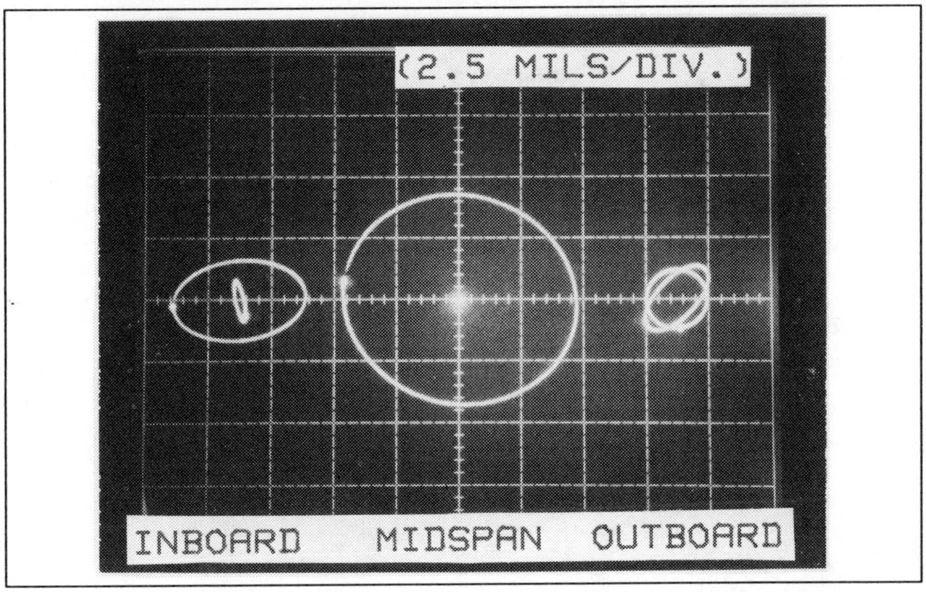

Fig. 7: Photograph of shaft orbits at the first critical speed, 2155 RPM, with and without open loop control

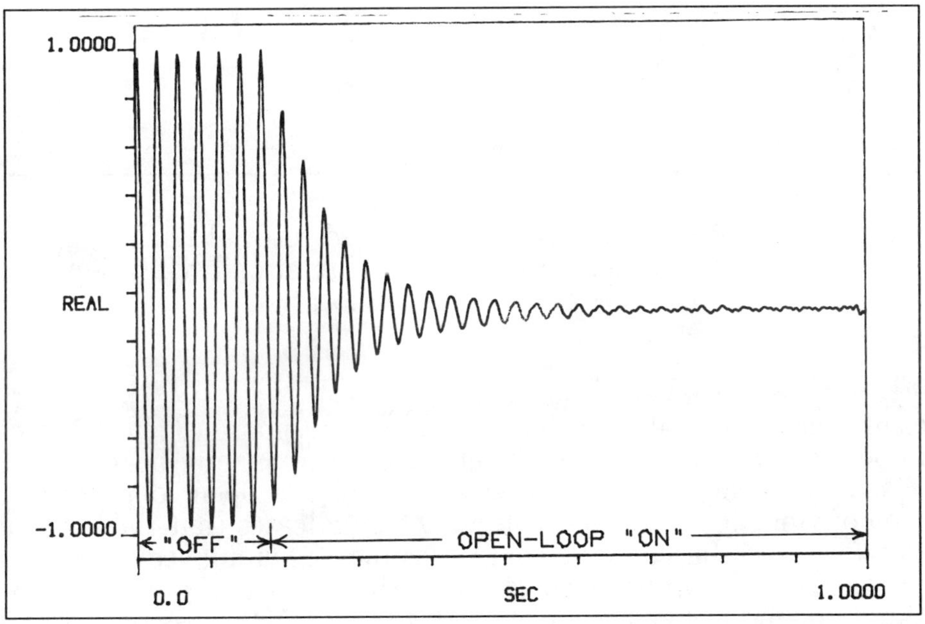

Fig. 8: Time trace of horizontal midspan response at first critical speed, showing effect of open loop control

CONCLUSIONS

The open loop control method was shown to reduce synchronous vibration at the first critical speed by up to 52 dB. The control was also shown to greatly reduce unbalance response over a wide range of operating speeds around the speed at which the open loop control was set. These reductions in vibration were obtained without altering the system's stability because the normal closed loop operation, which controls the system stability and dynamics, has not been changed in any manner by the open loop control.

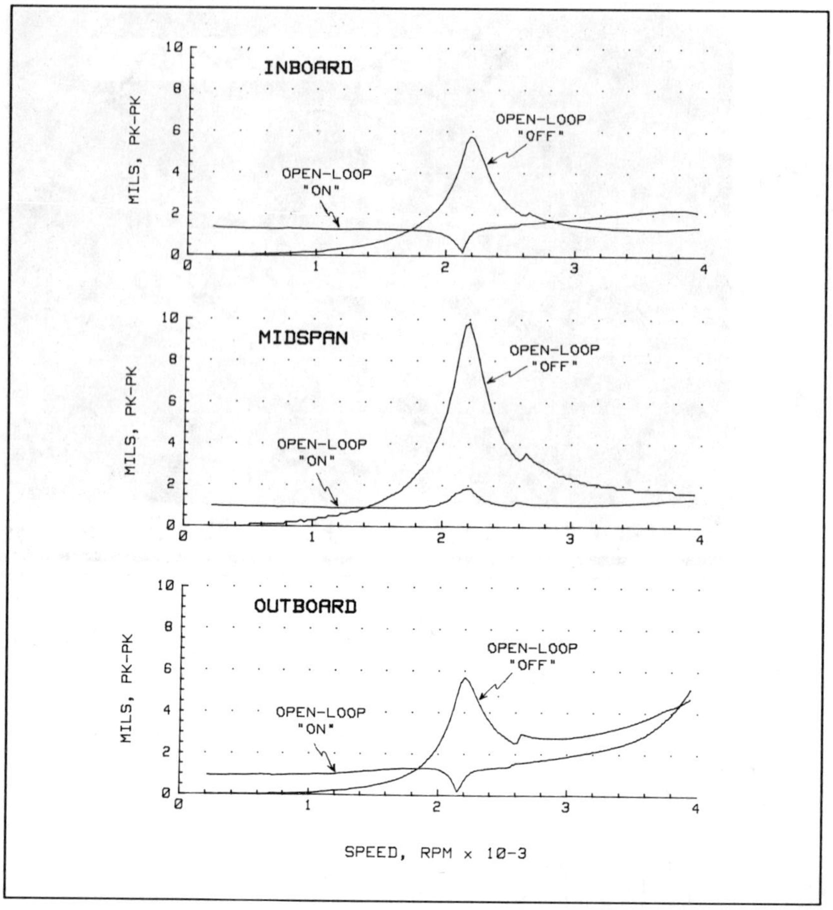

Fig. 9: Amplitude response of unbalanced rotor at horizontal midpsan, without and with open loop control set of minimum vibration amplitudes at each bearing location

These results indicate that very high unbalance response attenuation can be obtained through open loop control on a rotor supported in magnetic bearings. There is a wide range of speeds, around a fixed speed setting of the open loop control, in which the attenuation of vibration is quite effective. This indicates that the rotor need not be at a fixed speed for successful adaptive control, as may be employed with a digital computer. Adaptive control only requires that a near optimal control be found before the operating speed changes by more than the width of this range of effective operating speeds.

ACKNOWLEDGEMENTS

This research was funded in part by the Commonwealth of Virginia's Center for Innovative Technology.

REFERENCES

[1] Haberman, H. and Brunet, M., "The Active Magnetic Bearing Enables Optimum Damping of Flexible Rotors," ASME Paper 84–GT–117.

[2] Knospe, C. and Allaire, P., "Magnetic Bearing Control of Rotor Synchronous Response: Notch Filter Stability and Performance" submitted to the ASME Journal of Vibration and Acoustics, December 1990.

[3] Reinig, K. and Desrochers, A., "Disturbance Accommodating Controllers for Rotating Mechanical Shaft," ASME Journal of Dynamic Systems, Measurement and Control, Vol. 108, p. 24–31, March 1986.

[4] Higuchi, T., Mizuno, T., and Tsukamoto, M., "Digital Control System for Magnetic Bearings with Automatic Balancing," 2nd International Symposium on Magnetic Bearings, Tokyo, Japan, July 12–14, 1990, pp. 27–32.

[5] Matsumura, F., Fujita, M., and Okawa, K., "Modeling and Control of Magnetic Bearing Systems Achieving a Rotation around the Axis of Inertia," 2nd International Symposium on Magnetic Bearings, Tokyo, Japan, July 12–14, 1990, pp. 273–280.

[6] Mizuno, T. and Higuchi, T., "Design of Magnetic Bearing Controllers based on Disturbance Estimation," 2nd International Symposium on Magnetic Bearings, Tokyo, Japan, July 12–14, 1990, pp. 281–288.

[7] Burrows, C. and Sahinkaya, M., "Vibration Control of Multi–Mode Rotor–Bearing Systems," Proceedings of the Royal Society–London, Vol. 386, 1983, pp. 77–94.

[8] Burrows, C., Sahinkaya, M., and Clements, S., "Active Vibration Control of Flexible Rotors: An Experimental and Theoretical Study," Proceedings of the Royal Society – London, Vol. 422, p. 123–146.

[9] Higuchi, T., Otsuka, M., Mizuno, T., and Ide, T., "Application of Periodic Learning Control with Inverse Transfer Function Compensation in Totally Active Magnetic Bearings," 2nd International Symposium on Magnetic Bearings, Tokyo, Japan, July 12–14, 1990, pp. 257–264.

[10] Kanemitsu, Y., Ohsawa, M., and Watanabe, K., "Real Time Balancing of a Flexible Rotor Supported by Magnetic Bearings," 2nd International Symposium on Magnetic Bearings, Tokyo, Japan, July 12–14, 1990, pp. 265–272.

[11] Tessarzik, J. M., Badgley, R. H., and Anderson, W. J., "Flexible Rotor Balancing by the Exact Point–Speed Influence Coefficient Method," Journal of Engineering for Industry, ASME Transactions, Series B, Vol. 94, No. 1, February 1972, p. 148.

FORCE REQUIREMENTS OF AN ACTIVE ISOLATION MOUNT

Michael Burke,[1] **Mohamed Khalid Abdelhamid**[2]

ABSTRACT

The optimum control force for a single mount active isolation system is derived using frequency response methods. Two approaches are taken in the analysis. The first approach uses the compliances of the unconnected equipment, mount, and foundation. In this form, the control force can be discussed in terms of an equivalent source representation of the vibrating equipment. The second approach uses the compliances of the assembled system. The second form is useful for estimating the required force using quantities that can be measured on an existing system.

A lumped parameter system is used to illustrate the characteristics of the control force. The effect of equipment resonance and damping on the control force magnitude is demonstrated. This example is also used to illustrate the response of the system with and without the control force acting. The ability of the control force to stop the motion of all of the below mount points is demonstrated and explained by considering power flow.

The maximum force required is compared to the optimum force. The maximum force is observed by performing a simulation in which a filtered-x LMS algorithm is used as the implementation of the controller. The time history of the control force generated by the algorithm is plotted as a path along the error surface. A path is also plotted for a case in which the magnitude of the control force is constrained.

INTRODUCTION

Machines produce unsteady forces that result in unwanted vibration and noise. Typically machinery sources are periodic, exhibiting a harmonic spectrum with many lines. Vibration isolation is required over a wide range of frequencies. Passive isolation mounts are ineffective at low frequencies and at system resonances. Active isolation systems have been discussed by several authors [1][2][3], and offer hope of improved performance.

[1]Noise Cancellation Technologies, Inc., 1015 West Nursery Road, Suite 101, Linthicum, MD 21090, U.S.A.
[2]University of Maryland, College Park Campus, Room 2157 Engineering Building, College Park, MD 20742, U.S.A.

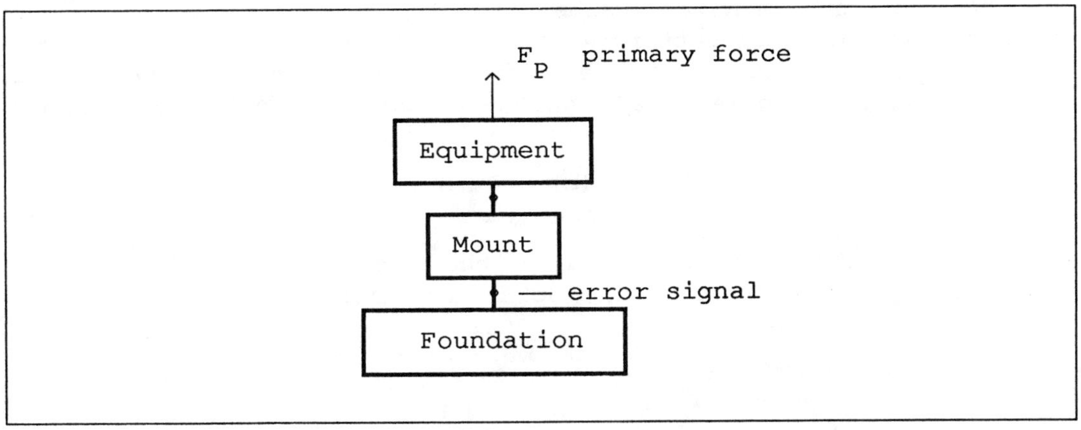

Figure 1 Schematic of Single Mount System

The current investigation focuses on active control of periodic disturbances because these are most relavent to machinery isolation problems. A single sinewave is used as the disturbance in this paper. The extension to multiple harmonics is discussed by others [4][5].

The system shown in Figure 1 is the subject of the current study. A primary force is applied to the equipment somewhere. Forces caused by rotating and reciprocating unbalance are examples of what is meant by primary forces. An error signal (residual) defines the performance of the mount. Both displacement and force error signals are discussed. The mount in Figure 1 is an active mount, which means that it can apply a secondary force (control force) to the system.

The main objective of the paper is to show how the dynamics of the mechanical system, particularly flexibility in the equipment, effect the magnitude of the optimum secondary force. A simulation is performed to verify that an adaptive controller converges to the analytical optimum solution. Several other active isolation issues are touched on, such as the differences in the plant due to the choice of error sensors, and the system response under control.

The analysis that follows is typical of analysis performed in connection with adaptive feedforward control systems in that it is a study of superposition in a physical system. Analysis of this kind is useful, but it does not specifically address the stability of the controlled system. The location of the closed loop poles must be examined to investigate stability.

OPTIMUM CONTROL FORCE

The optimum control force is that which minimizes the chosen cost function. For a single mount system the cost function, $J(\omega)$, has a simple form. The error at frequency ω is $e(\omega)$. The superscript, *, indicates the complex conjugate. A frequency domain formulation is used because it is most convenient for the analysis of periodic disturbances.

$$J(\omega) = |e|^2 = e(\omega)^* e(\omega) \tag{1}$$

In this case, the minimum value of $J(\omega)$ is zero, and the optimum condition is defined by setting $e(\omega)$ equal to zero. The form of $e(\omega)$ is obtained by superposition of the residual signal caused by the primary source, $e_p(\omega)$, and the residual introduced by the secondary source.

$$e(\omega) = e_p(\omega) + T(\omega)\, F_s(\omega) \tag{2}$$

The transfer function, $T(\omega)$, defines the residual signal produced by a unit secondary force, $F_s(\omega)$. The optimum secondary force is;

$$F_{opt}(\omega) = -T(\omega)^{-1} e_p(\omega) \tag{3}$$

The cost function can be plotted as a surface over a plane defined by the real and imaginary parts of the secondary force. An example of such an error surface is shown in Figure 2. For a SISO system the error surface is a quadratic bowl. The optimum force locates the bottom of the bowl. At each frequency the transfer function $T(\omega)$ is different, so there is a different optimum force and a different error surface shape. The shape of the bowl is of interest because it effects the convergence of adaptive algorithms. Because the bowl moves around in the control force plane as a function of frequency, it is more convenient to define a normalized error surface on which results can be presented. Dividing J by $e^*_p e_p$ and defining

$$R = F_s / F_{opt} \tag{4}$$

a normalized cost function N can be written as;

$$N = J/(e^*_p e_p) = \left| 1 + \frac{TF_s}{e_p} \right|^2 \tag{5}$$

$$N(\omega) = \left| 1 - R(\omega) \right|^2 \tag{6}$$

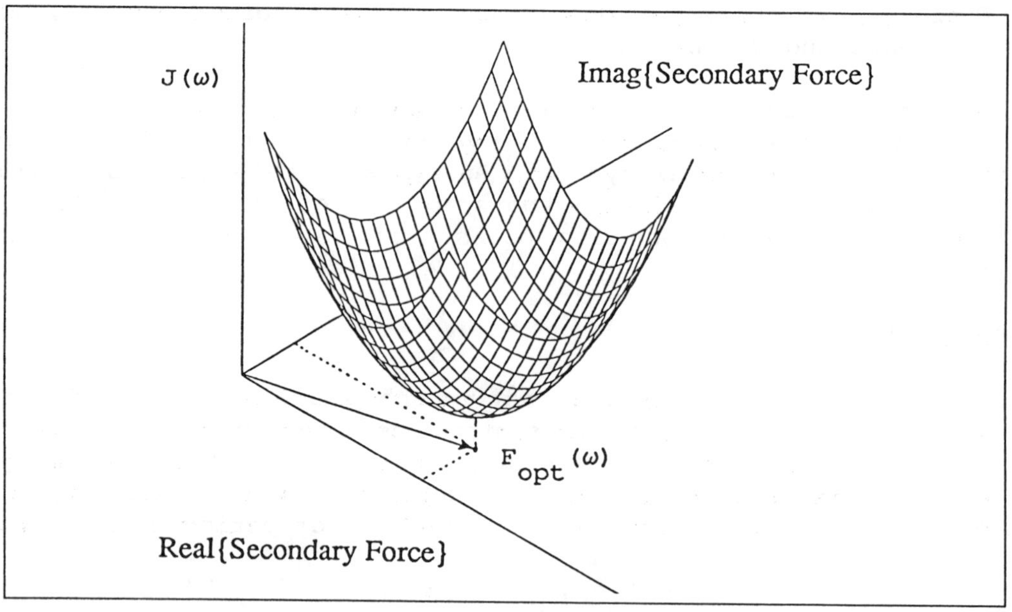

Figure 2 Example Error Surface

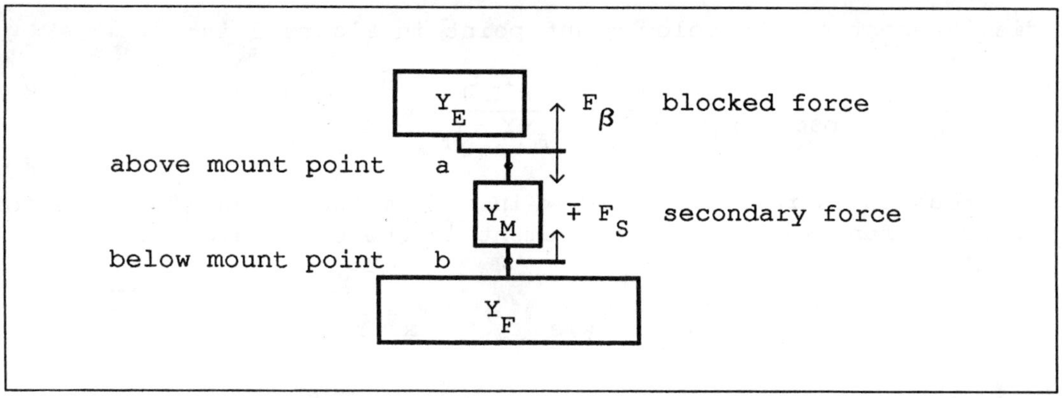

Figure 3 Equivalent Source Representation

The normalized error surface is a circular bowl centered at R = 1 and of unit height at R = 0. This is not the surface that the adaptation is taking place on, but it is a convenient way to present results at several different frequencies.

SINGLE MOUNT FREQUENCY RESPONSE MODELS

The details of the optimum force and the transfer function, $T(\omega)$, are investigated from a sub-system point of view and a system point of view. The above and below mount response is also examined.

SUB-SYSTEM MODEL

The sub-system model describes the above and below mount response in terms of the compliances of the unconnected equipment, mount, and foundation. The compliance at frequency ω is the complex ratio of displacement to force. The compliance of the equipment and the foundation are arbitrary, but the mount is restricted to being a complex stiffness to simplify the analysis.

The equipment is vibrating due to the action of the primary force. The exact nature of the primary force, even the points of application, are generally unknown. It is convenient to represent the vibrating equipment by an equivalent source defined by the equipment compliance and blocked force as shown in Figure 3. Also shown is the active element, a secondary force applied in parallel to the passive mount. The secondary force is considered to be synonymous with the controller output. This ignores the dynamics of the actuator system, but it makes it easier to focus on the influence of the dynamics of the mechanical system.

Several subscripts are used in the analysis. The subscript, p, indicates that the displacement is due to the action of the primary force only (or its blocked force representation). The subscript, sp, indicates that both the secondary and primary forces are acting. The above and below mount points are identified by the subscripts a, b. In analyzing mechanical sources the terms blocked force and free displacement are often used. The subscript, β, is used to indicate blocked and the subscript, ϕ, is used to indicate free.

The displacement of the below mount point in Figure 3 is;

$$X_{bsp} = \frac{F_\beta Y_E Y_F}{Y_E+Y_M+Y_F} - \frac{F_s Y_E Y_F}{Y_E+Y_M+Y_F} + \frac{F_s Y_F (Y_E+Y_M)}{Y_E+Y_M+Y_F} \tag{7}$$

This expression is re-written below in a form that identifies the form of the transfer function when displacement is the error signal.

$$X_{bsp} = X_{bp} + T_x F_s \quad \Rightarrow \quad T_x = \frac{Y_F \; Y_M}{Y_E+Y_M+Y_F} \tag{8}$$

In a similar manner, the transfer function using below mount force as the error signal can be obtained;

$$F_{bsp} = F_{bp} + T_f F_s \quad \Rightarrow \quad T_f = \frac{Y_M}{Y_E+Y_M+Y_F} \tag{9}$$

Both equations 8 and 9 can be manipulated to yield expressions for the optimum force. Both results are identical, so only one approach is examined in detail. Setting the displacement of equation 8 equal to zero,

$$X_{bsp} = 0 = Y_F(F_s Y_M + F_\beta Y_E) \tag{10}$$

Either the foundation compliance is equal to zero, anti-resonance, or

$$F_s Y_M + F_\beta Y_E = 0 \tag{11}$$

The optimum secondary force is,

$$F_{opt} = -F_\beta \frac{Y_E}{Y_M} \tag{12}$$

Equation 11 can be re-written by defining the two groups of terms to be the free displacement of the mount and the free displacement of the equipment.

$$X_{M\phi} + X_{E\phi} = 0 \tag{13}$$

It is significant to note that the properties of the foundation do not enter into this result. Perfect control occurs when the free displacement of the mount cancels the free displacement of the equipment. The equipment/mount system can be represented by an equivalent source with finite compliance and zero free displacement. Such a source is incapable of doing work on any load, as shown in the following expression for the power delivered to the foundation.

$$P = \frac{1}{2} \operatorname{Re}\{ F_b V_b^* \} = \frac{1}{2} \operatorname{Re}\{ i\omega Y_F \} \frac{|X_{E\phi} + X_{M\phi}|^2}{|Y_E+Y_M+Y_F|^2} \tag{14}$$

The relationship between the above mount displacement with and without the secondary force acting can also be obtained.

$$X_{asp} = \left[1 + \frac{Y_E}{Y_M + Y_F} \right] X_{ap} \tag{15}$$

Inspection of this result and the expression for the secondary force required (equation 12) suggests that difficulties are caused by an equipment with a large compliance. A large secondary force is required and the above mount displacement will increase when the equipment compliance is large.

An active mount effectiveness, a measure of the reduction in power transmission due to the secondary force, can be defined using the below mount displacements with and without the secondary force acting.

$$E = \frac{X_{bp}}{X_{bsp}} = \frac{1}{1 + \dfrac{F_s Y_M}{F_\beta Y_E}} = \frac{1}{1 - R} \quad \text{where} \quad R = \frac{F_{secondary}}{F_{optimum}} \tag{16}$$

The sub-system analysis is useful for understanding the behavior of a mounted system, but it requires information that is not likely to be available. In the following section the optimum force result is reformulated in terms of system parameters.

SYSTEM MODEL

The system model describes the above and below mount response in terms of the compliances of the connected equipment, mount, and foundation. It is possible to define a matrix of compliances that relate the displacements and forces at any set of points in a system.

$$(X) = [Y](F) \tag{17}$$

Refering to Figure 3, the displacements of points a,b can be related to the forces at a,b as follows;

$$X_{asp} = Y_{aa}(F_\beta - F_s) + Y_{ab}F_s \tag{18}$$

$$X_{bsp} = Y_{ba}(F_\beta - F_s) + Y_{bb}F_s \tag{19}$$

By setting $X_{bsp} = 0$, the second equation can be solved for the optimum secondary force as a function of the blocked force of the machine.

$$F_{opt} = \frac{Y_{ba}}{Y_{ba} - Y_{bb}} F_\beta \tag{20}$$

The secondary force required can also be expressed in terms of the displacements of the passively operating equipment, X_{bp}.

$$F_{opt} = \frac{X_{bp}}{Y_{ba} - Y_{bb}} \tag{21}$$

This expression has the advantage that all of the terms are directly measurable on a representative piece of equipment. No modeling is required.

LUMPED PARAMETER EXAMPLE

SYSTEM MODEL

The analytical results of the preceeding section are explored by means of a numerical example. The four degree of freedom lumped mass system shown in Figure 4 is used in the calculations. This is the simplest system with an equipment and a foundation that exhibit a resonance (pole) and an anti-resonance (zero).

The parameters of the system are chosen to be reasonable for a mounted equipment system. The mount stiffness, k_2, is one tenth the stiffness of the foundation, $k_3k_4/(k_3+k_4)$. The fixed base natural frequency of the equipment mass on the mount stiffness, $\sqrt{k_2/(m_1+m_2)}$, is arbitrarily set equal to unity. The parameters of the system are listed in Table I. The three damping columns, cb, ce, cm, correspond to the baseline damping configuration, double equipment damping, and double mount damping.

The frequencies of the poles and zeros of the equipment and foundation are presented in Table II along with the system natural frequencies. All of these frequencies are for the undamped system. Small amounts of damping, as shown in Table I, are used in the calculations.

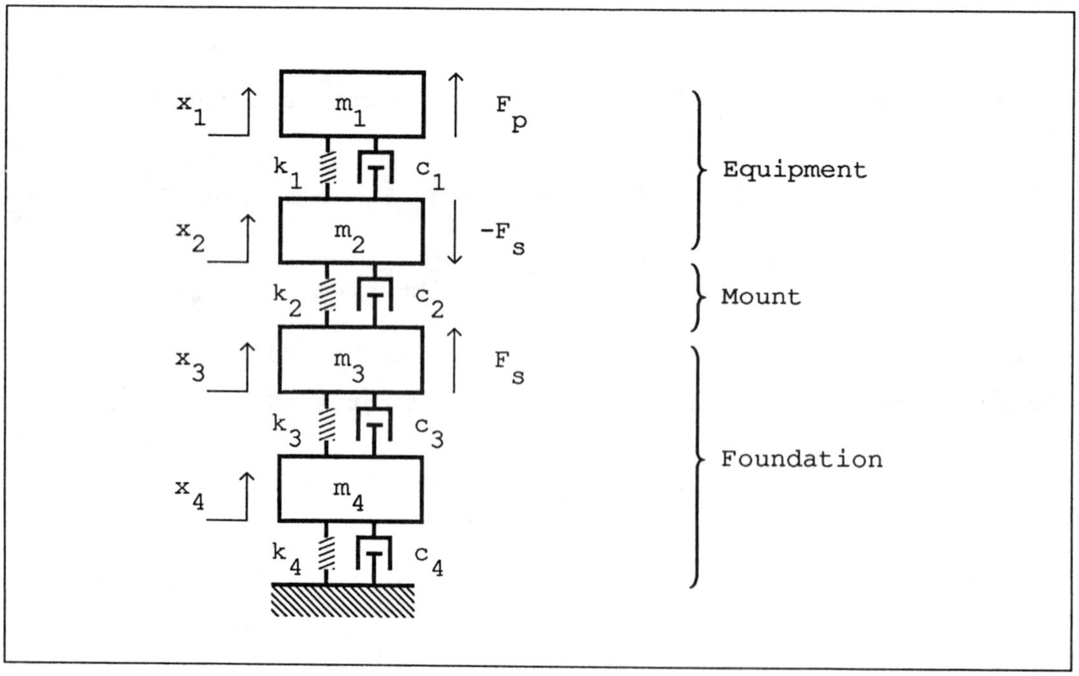

Figure 4 Four Degree of Freedom Model

TABLE I - PARAMETERS FOR THE FOUR DEGREE OF FREEDOM MODEL

i	m_i	k_i	cb_i	ce_i	cm_i
1	0.50	6.25	0.0625	0.1250	0.0625
2	0.50	1.00	0.0500	0.0500	0.1000
3	1.00	20.0	0.2500	0.2500	0.2500
4	1.00	20.0	0.2500	0.2500	0.2500

FREQUENCY DOMAIN RESULTS

The secondary force required to null the below mount displacement is calculated using the formulas from the sub-system section. The magnitude of the secondary force, normalized by the primary force, is plotted as a function of the disturbance frequency in Figure 5. The scale of the y-axis is chosen to show the details of the behavior near the equipment resonance. The information that is clipped off of the graph is easily described. The secondary force continues to increase below $\omega=1$ because of the second order pole of the equipment system at $\omega=0$. The secondary force continues to decrease monotonically above $\omega=6$ due to the inertia of the equipment.

What is more interesting is the peak at $\omega=5$, the equipment pole. If the equipment were a rigid mass this peak would not exist. The secondary force magnitude would decrease with $1/\omega^2$. Three curves are displayed coresponding to the three damping configurations. The solid curve is the baseline case, cb. The dashed curve shows the result of doubling the equipment damping, ce, which lowers the peak force. The dotted curve shows the result of doubling the mount damping, cm, which increases the peak force. The magnitude of the secondary force is important because it determines actuator requirements. Another important property is the location of the optimum force in the complex plane.

As shown in Figure 2, the real and imaginary part of the secondary force locate the bottom of the bowl. At each frequency the bottom of the bowl is at a different location in the secondary force plane. The locus of optimum secondary forces for the example problem is plotted in Figure 6

TABLE II - FREQUENCIES OF INTEREST FOR THE FOUR DEGREE OF FREEDOM EXAMPLE

Equipment	Zero		3.54		
	Pole	0.00		5.00	
Foundation	Zero				6.32
	Pole		2.76		7.24
System	Pole	0.93	2.90	5.11	7.26

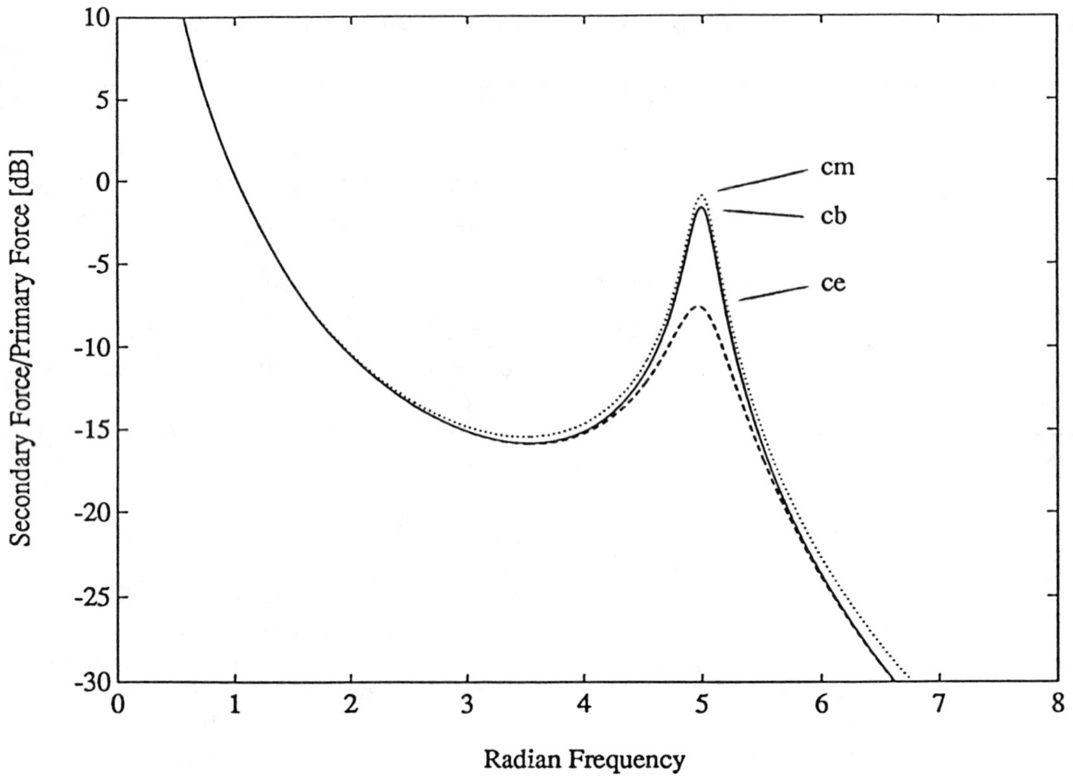

Figure 5 Secondary Force Magnitude

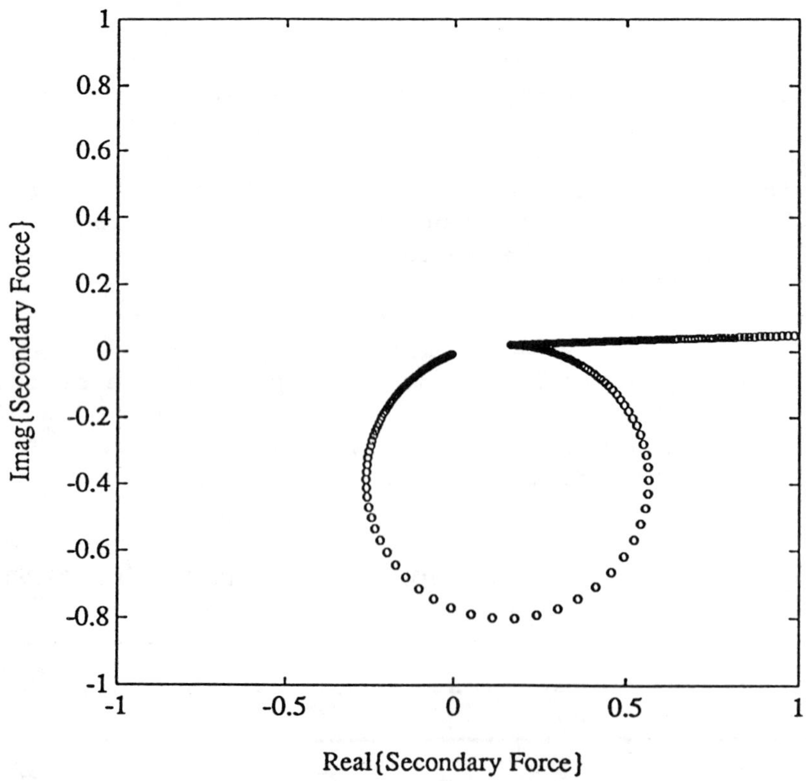

Figure 6 Locus of Optimum Secondary Forces

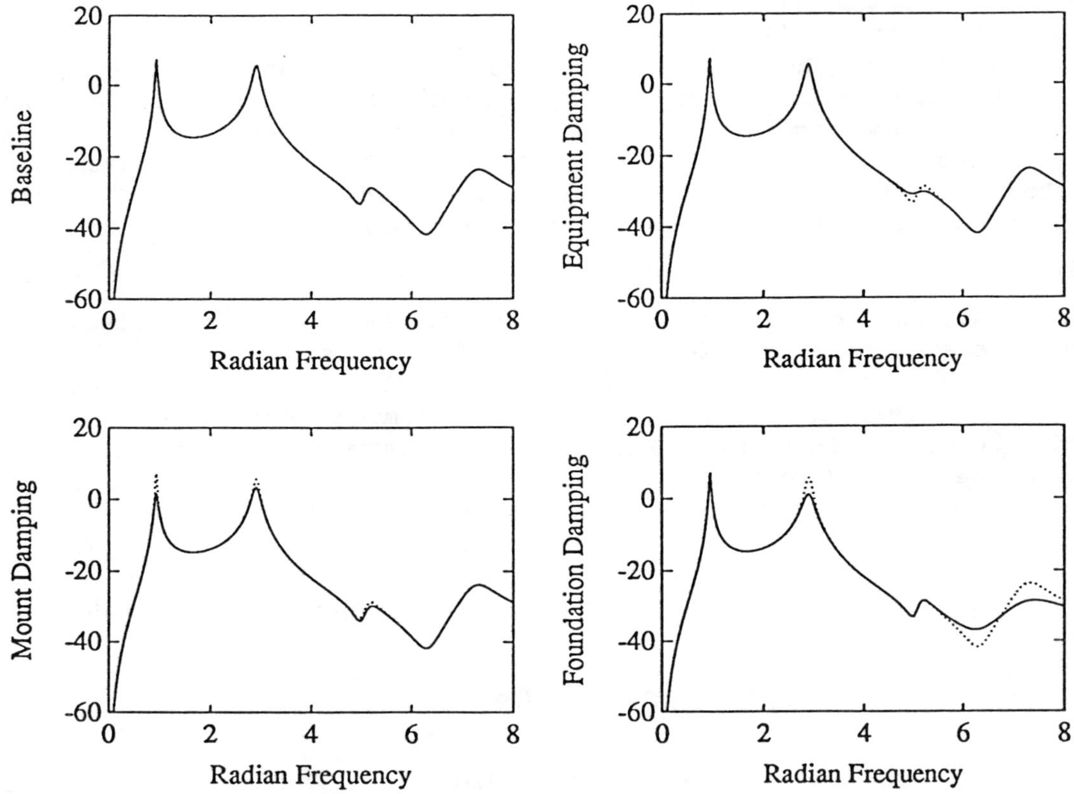

Figure 7 Displacement Transfer Functions

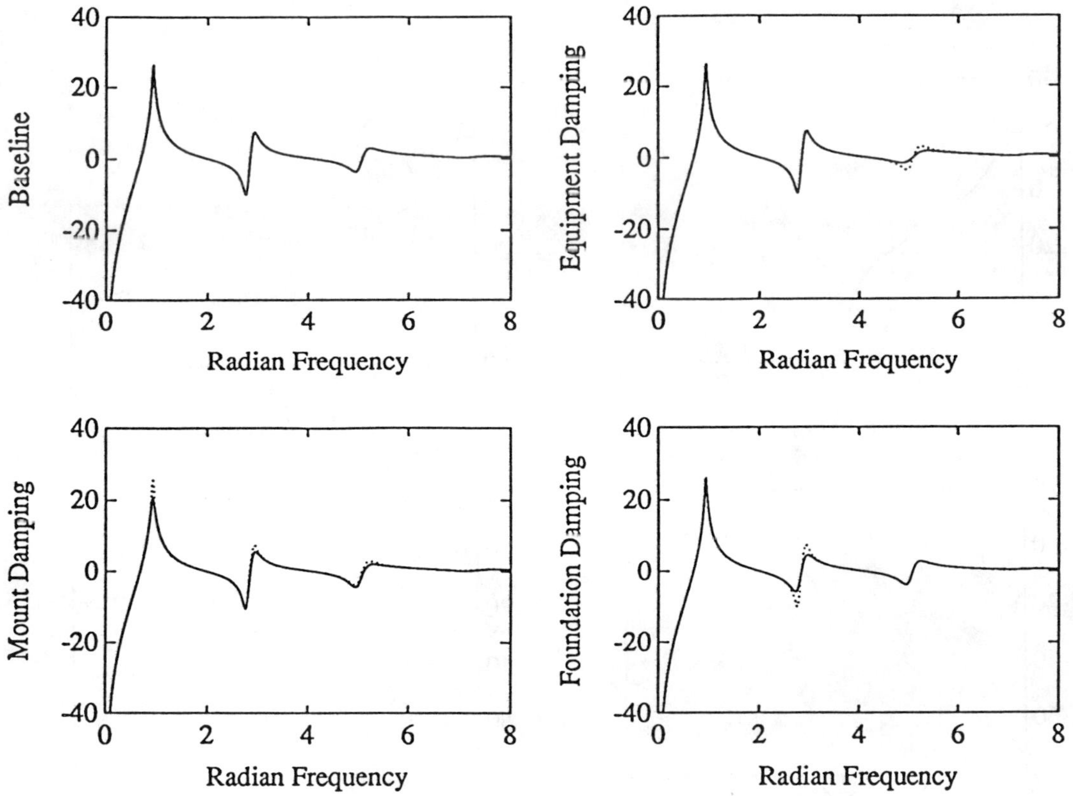

Figure 8 Force Transfer Functions

for $1 \le \omega \le 8$ with the baseline damping. The upper right corner is $\omega=1$. As frequency increases, the bowl moves toward the origin until the equipment resonance becomes dominant and causes it to circle clockwise. Above the resonance, the bowl again moves toward the origin. The locus of the optimum forces is important because it indicates the difficulty associated with tracking a disturbance that varies in frequency. Another point that can be illustrated is the nature of the transfer functions T_x (eqn 8) and T_f (eqn 9). These transfer functions are usually the only knowledge that the controller has about the system being controlled. The displacement and force transfer functions are plotted in Figures 7 and 8 for several damping configurations. The force transfer function is noticeably flatter, that is, it has less dynamic range in the control bandwidth ($\omega \ge 1$). This fact alone may be important when trying to implement the controller in hardware. The most important aspect of the transfer function is that it determines the instability boundary of the controller.

The response of the system at the disturbance frequency is calculated by solving the equation;

$$[z](x)=(f)_p+(f)_s \tag{22}$$

$$[z]=[k]+i\omega[c]-\omega^2[m] \tag{23}$$

at each frequency. The primary force vector is $(f)_p = [1\ 0\ 0\ 0]^T$ and secondary force vector is $(f)_s=[0\ -f_s(\omega)\ f_s(\omega)\ 0]^T$.

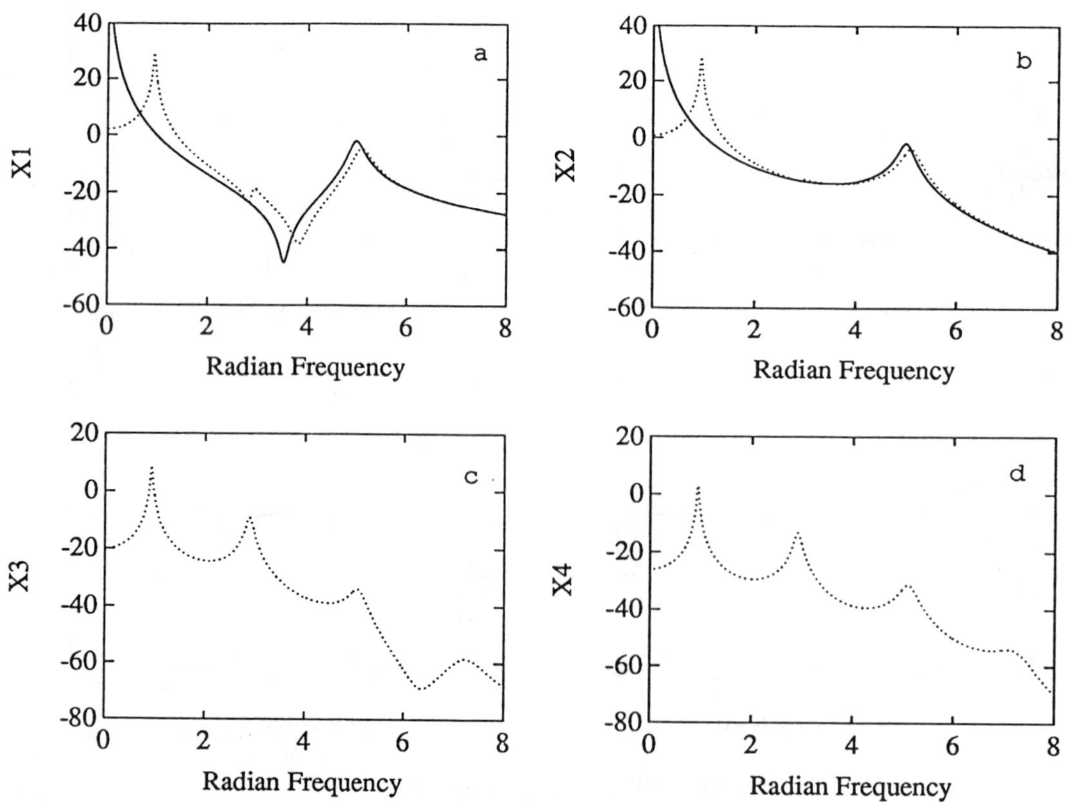

Figure 9 Secondary Force in Parallel

The displacements are shown in Figures 9a,b,c,d. The dotted line is the displacement with only the primary force acting. The solid line is the displacement with both primary and secondary forces acting. Figures 9a,b show that the displacement of the equipment system could increase or decrease due to the action of the secondary force. Figures 9c,d show that the secondary force of equations 12 and 21, which is designed to produce a null at mass m_3, also produces a null at mass m_4. This is because both the displacement of the below mount point and the force transmitted to the below mount point are nulled by the same secondary force. The action of the secondary force is to null the power transmitted to the foundation.

ADAPTIVE CONTROL SIMULATION

STATE SPACE MODEL

The reponse of the four degree of freedom model is calculated using state space methods. First the continuous time state equations are written as;

$$(\dot{X}) = [A](X) + [B](U) \tag{24}$$

$$(Y) = [C](X) + [D](U) \tag{25}$$

$$[A] = \begin{bmatrix} [0] & [I] \\ -[m]^{-1}[k] & -[m]^{-1}[c] \end{bmatrix} \qquad [B] = \begin{bmatrix} [0] \\ [m]^{-1} \end{bmatrix} \tag{26}$$

$$(X) = \begin{pmatrix} (x) \\ (\dot{x}) \end{pmatrix} \qquad (U) = (f) \tag{27}$$

The output vector, (Y), is chosen to include the displacements of the four masses and the force transmitted to mass m_3.

$$[C] = \begin{bmatrix} [I] & [0] \\ 0\ k_2\ -k_2\ 0 & 0\ c_2\ -c_2\ 0 \end{bmatrix} \qquad [D] = \begin{bmatrix} [0] \\ 0\ 0\ 1\ 0 \end{bmatrix} \tag{28}$$

The continuous time equations are discretized using a zero order hold sampling. The result is the following relationship between time samples.

$$(X)_{k+1} = [\Phi](X)_k + [\Gamma](U)_k \tag{29}$$

$$(Y)_k = [C](X)_k + [D](U)_k \tag{30}$$

The primary force is defined to be a unit amplitude sinusoid. The simulation is started with an initial state that coresponds to the steady state condition for this disturbance.

$$(X)_o = \mathrm{Im}\left\{ \left[e^{i\omega\Delta t}[I] - [\Phi] \right]^{-1} [\Gamma][1\ 0\ 0\ 0]^T \right\} \tag{31}$$

The residual signal is chosen from the output vector to be either the displacement of mass m_3 or the force transmitted to it. Whichever residual is chosen, it is minimized using a filtered-x LMS algorithm that is specialized to periodic disturbances.

ADAPTIVE CONTROLLER

Adaptive controllers based on LMS algorithms are discussed in several references [4][5][6] and are only outlined here. The secondary force at time sample k is generated by passing a reference signal through an FIR filter.

$$F_{s,k} = W_k^T X_k \tag{32}$$

For a periodic disturbance, the reference signal usually consists of cosine and sine elements at the disturbance frequency.

$$X_k = [\cos(\omega t_k) \quad \sin(\omega t_k)]^T \tag{33}$$

The FIR filter consists of two weights, the values of which are time varying.

$$W_k = [\ wo_k \ w1_k]^T \tag{34}$$

The output of the controller is a linear combination of cosine and sine elements. By varying the filter weights a secondary force of any magnitude and phase can be generated. The weights are adapted so as to minimize the cost function and find the optimum solution. The weight update equation is written below.

$$W_{k+1} = W_k - 2\mu e_k X_{f,k} \tag{35}$$

The new weight vector is obtained from the old weight vector by moving a small amount, controlled by the parameter μ, in the direction opposite of the instantaneous value of the gradient of the error surface. The term, $X_{f,k}$, in this equation is the filtered reference signal. When the disturbance is periodic the filtered reference is usually defined using a steady state approximation to the response of the system being controlled.

$$X_{f,k} = T(\omega) X_k \tag{36}$$

Simulations using this sort of control scheme were performed at several frequencies to investigate the system response and the secondary force magnitude, which is defined by the filter weights.

RESULTS

Simulation results are presented for two disturbance frequencies, ω=2.7 a foundation resonance, and ω=5 an equipment resonance. Two kinds of information, displacement and secondary force, are displayed for each case. Displacements of the four masses are plotted versus time sample number to show the system response while the controller converges on the solution. Because of the density of the points on the graphs, only the envelope of the time series can be clearly identified. The time history of the secondary force is plotted as a path along the normalized error surface defined by equation 6. The secondary force plots are essentially a view looking down the axis of the bowl in Figure 2.

Figure 10 shows the time response at a foundation resonance, ω=2.7. In this case the equipment response, X1 and X2, is essentially unchanged. The foundation response, X3 and X4, decays exponentially. The secondary force path for this case is plotted in Figure 11. The fine structure of the path cannot be seen because of the size of the symbols used. The circular contours mark intervals of 12 dB.

Figure 12 shows the time response at an equipment resonance, ω=5. The equipment has a significant transient response. The steady state solution in the controlled state exhibits larger equipment displacements than the passive response. The foundation responses still decay exponentially. The secondary force path is plotted in Figure 13. In this case the instantaneous value of the secondary force overshoots the optimum value. More control force is required than may have been allowed for. Either there is enough headroom in the controller and the actuators or clipping will result.

An alternative to increasing the capacity of the controller/actuator system is to provide a constraint on the controller output. An output limited simulation (at ω=5) is shown in Figures 14 and 15. The time response is actually superior to the unlimited case. The secondary force path shows the output limit clearly.

CONCLUSIONS

The optimum secondary force can be estimated theoretically, if the equivalent source representation of the equipment is known, or experimentally, if a representative physical system is available. The force is defined by the properties of the equipment and the mount only. The magnitude of the secondary force is large at frequencies at which the compliance looking into the equipment is large. Increasing damping in the equipment structure decreases the secondary force magnitude. Increasing damping in the mount increases the required force.

At a given disturbance frequency the response of the equipment structure can increase or decrease due to the application of the secondary force. The response of the foundation structure decreases. The optimum secondary force produces a null in the power transmitted to the foundation. This guarantees global control in the foundation structure.

The maximum force required may exceed the optimum force. The control system must either be sized to accomodate these overshoots, or output limiting must be designed into the controller. For best performance, the output limit should exceed the optimum force. Otherwise, performance will be degraded in the manner presented in equation 16. All of these options require that an estimate of the optimum force be made.

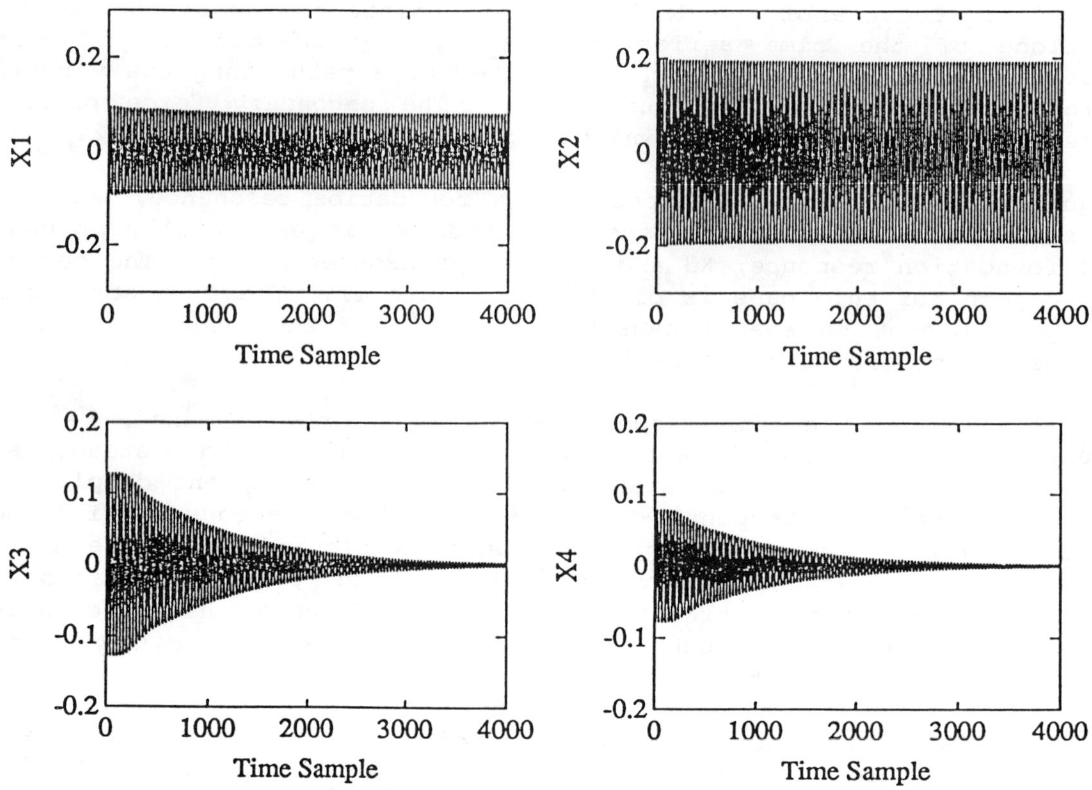

Figure 10 Displacement Time Series (w=2.7)

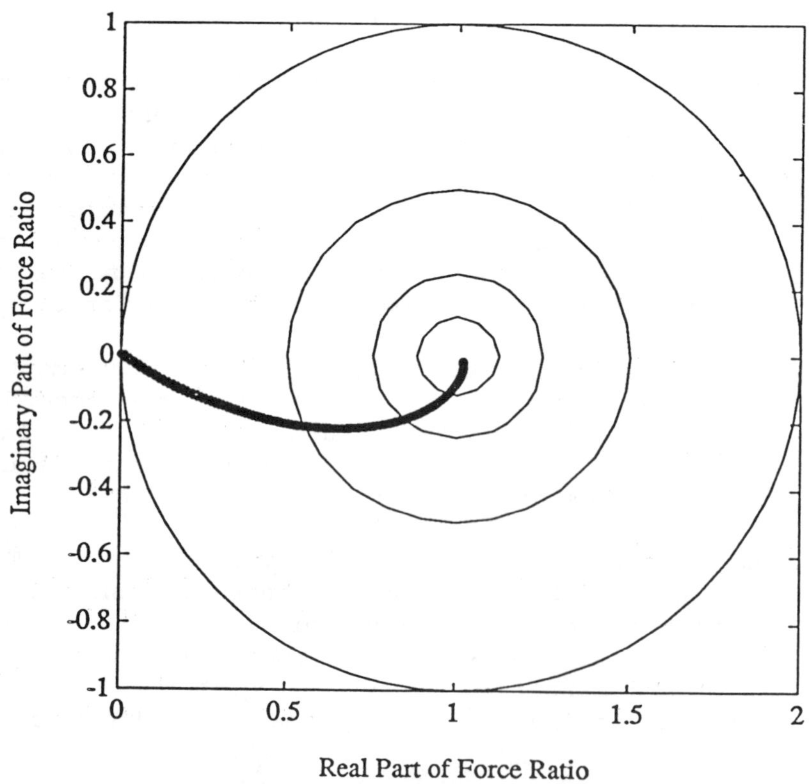

Figure 11 Secondary Force Path (w=2.7)

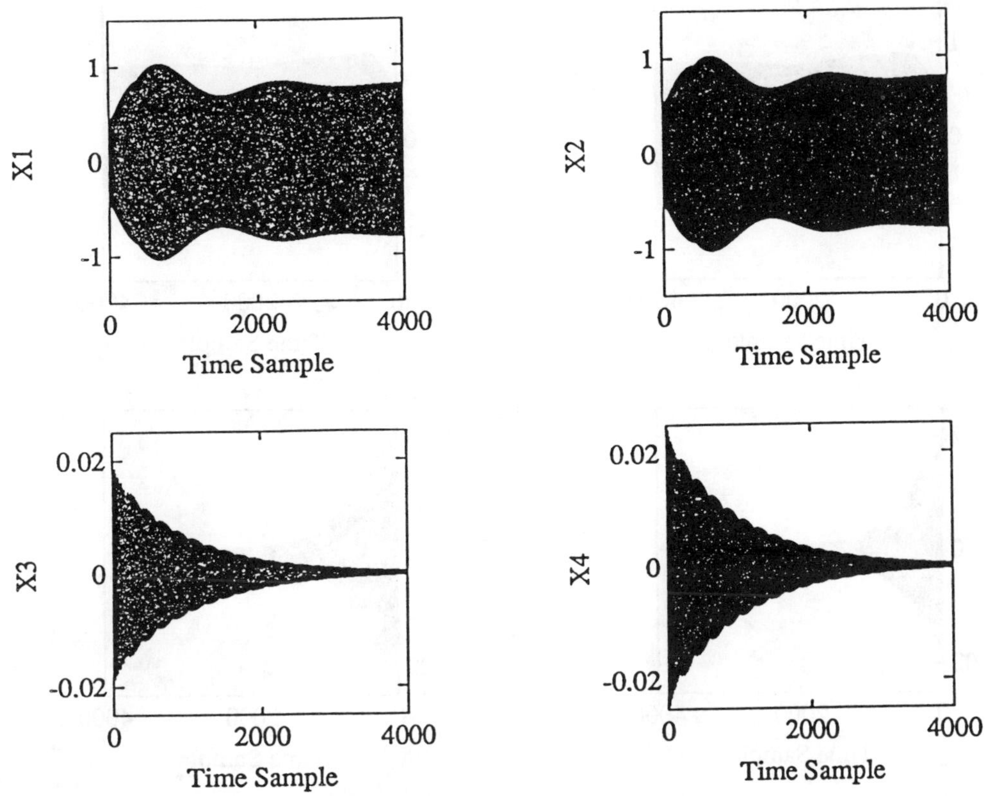

Figure 12 Displacement Time Series (w=5)

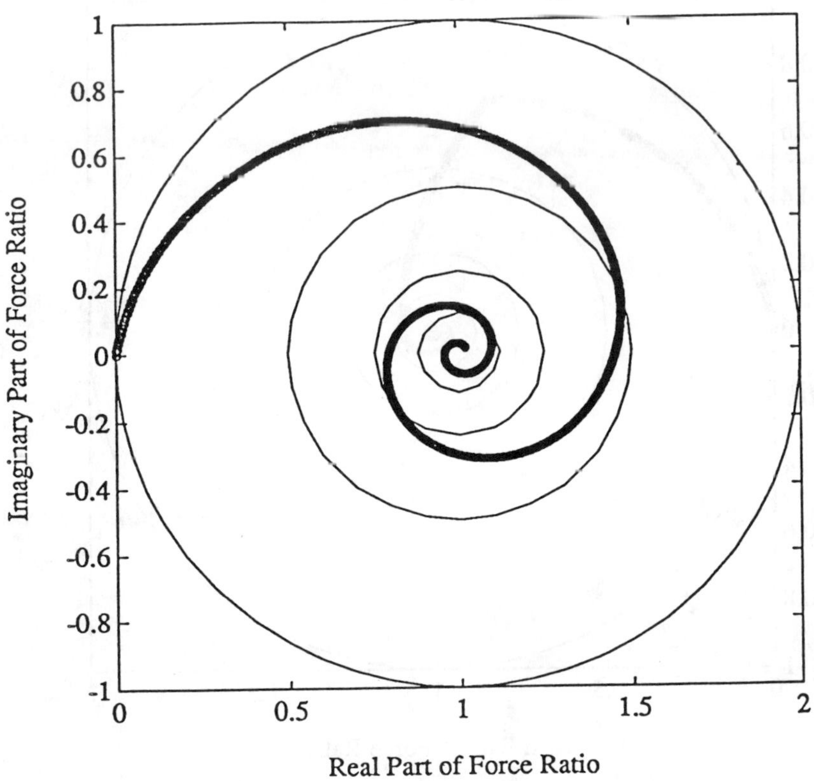

Figure 13 Secondary Force Path (w=5)

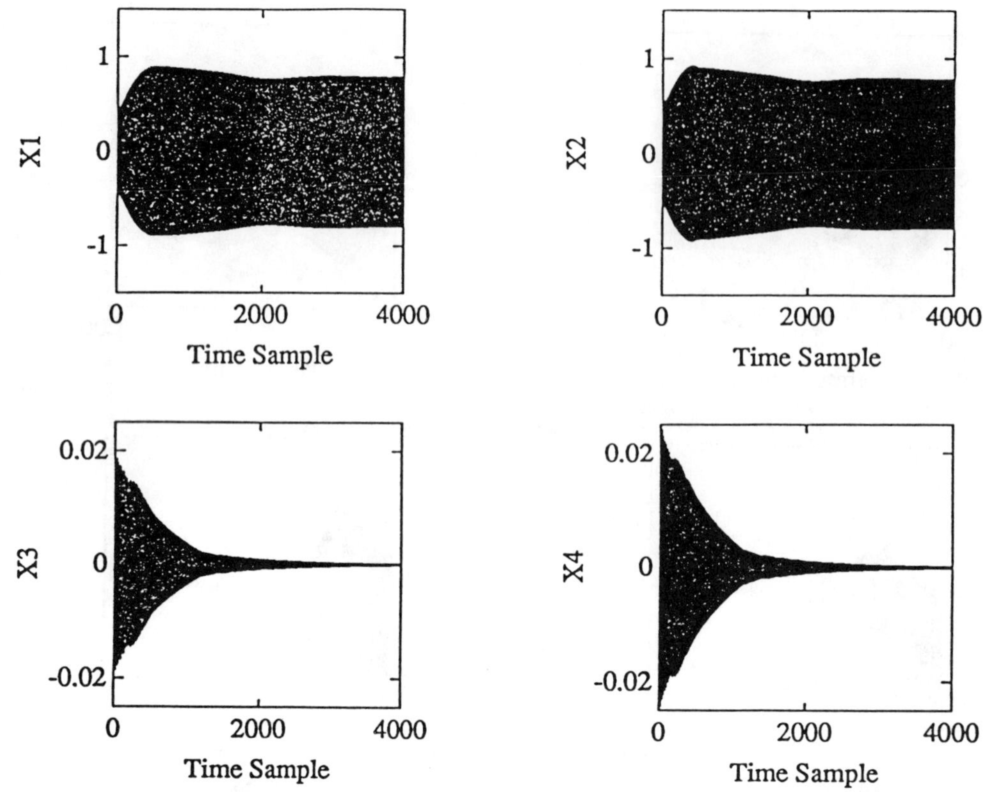

Figure 14 Displacement Time Series, Output Limited (w=5)

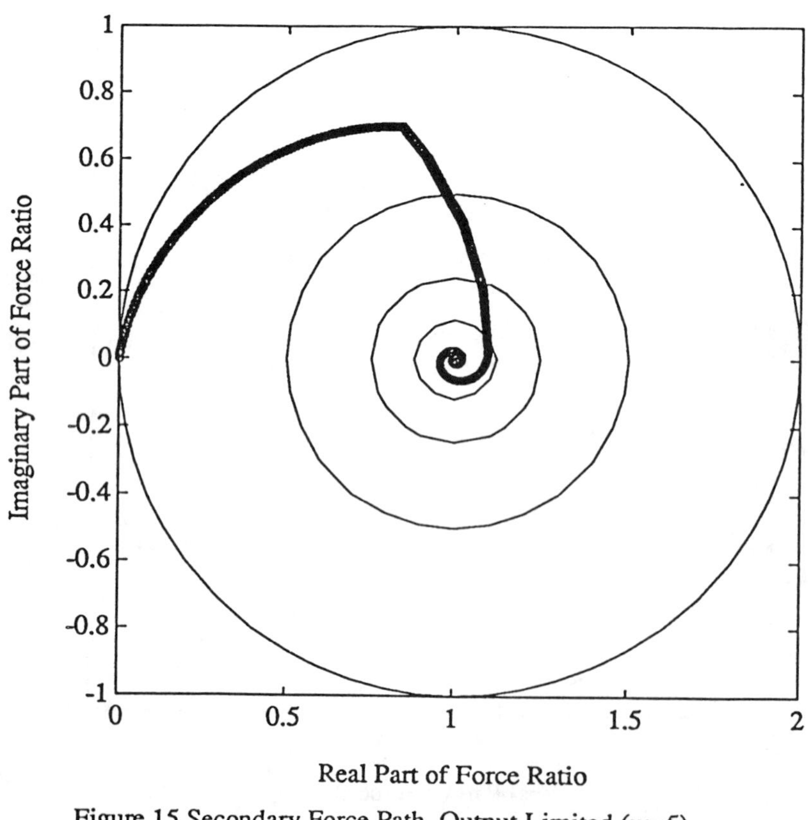

Figure 15 Secondary Force Path, Output Limited (w=5)

References

1. P.A. Nelson, M.D. Jenkins, S.J. Elliot, "Active Isolation of Periodic Vibrations", NOISE-CON 87, pp. 425-430.

2. M.D. Jenkins, P.A. Nelson, S.J. Elliot, "Active Isolation of Periodic Machinery Vibration from Resonant Substructures", ASME Winter Meeting , 1990, pp. 95-99.

3. K.B. Scribner, L.A. Sievers, A.H. von Flotow, "Active Narrowband Vibration Isolation of Machinery Noise from Resonant Substructures", ASME Winter Meeting, 1990, pp. 101-111.

4. B. Widrow, S.D. Stearns, Adaptive Signal Processing, Prentice-Hall, 1985.

5. J.R. Glover, "Adaptive Noise Cancelling of Sinusoidal Interferences", PhD thesis, Stanford, 1975.

6. S.J. Elliot, I.M. Stothers, P.A. Nelson,"A Multiple Error LMS Algorithm and its application to the Active Control of Sound and Vibration",IEEE Vol ASSP-35, No. 10,OCT 1987.

CHARACTERISTICS OF AN ACTIVE VIBRATION ISOLATION SYSTEM USING ABSOLUTE VELOCITY FEEDBACK AND FORCE ACTUATION

Dale W. Schubert[1]

ABSTRACT

Isolation of floor and ground vibration relative to a vibration sensitive body is generally achieved by supporting the body on air or gas filled springs combined with pendulum like structures. The combined action of such isolation mechanisms reduce the vibration level of the supported payload at high frequencies in all six degrees of freedom. However, vibration isolation systems of this type display a region in the low frequency domain wherein the vibration level of the supported payload can be considerably larger than the vibration source level. This frequency region of vibration amplification is caused by the isolation system resonance. The resonance of the vibration isolation system can generate an overall RMS vibration level of the supported payload which is more than the overall RMS vibration level of the vibration source. A solution which lowers the overall RMS vibration level of the supported payload requires elimination of the isolation system resonance region without degrading the vibration isolation at higher frequencies. An active vibration isolation system has been developed and improved over the past four years, which eliminates the degrading characteristics of conventional passive vibration isolation systems, while at the same time greatly improving the overall vibration level of the supported payload from floor vibration sources as well as from vibratory forces applied directly to the supported payload.

ACTIVE ISOLATION SYSTEM CONFIGURATION

The active isolation system performance presented is for a system using absolute velocity feedback of the supported payload, amplified, and applied to a force motor acting on the supported payload. The control force applied to the supported payload is thus proportional to the absolute velocity of the supported payload, thereby acting as a mechanical viscous damper coupled between the supported payload and fixed motionless space. As a result of this analogy the active isolation system can be called an electronic damping system. The mechanical analog of this electronic

[1]Barry Controls, 40 Guest Street, Brighton, MA 02135-9105, U.S.A.

system can be called a sky-hook damper since one end of the damper is attached to fixed motionless space; thus the sky-hook.

The electronic damping system presented in this text is constructed using a conventional design pneumatic air spring and mechanical height control valve used to support the static weight of the supported payload. The air spring design provides for a 4.0 Hz resonance frequency in the vertical axis when operating as a passive system. A set of pendulum rods support the payload weight and are designed to provide passive vibration isolation in radial directions with a resonance frequency of 2.5 Hz. The payload is supported by four of these isolators.

The active components of the electronic damping isolation system consists of four isolators of the type described above with each isolator having three velocity sensors, each attached to the payload side of the isolator, a controller to electronically amplify and alter the velocity signals from all four isolators, and three force motors for each isolator connected between the ground side of the isolator and the payload side of the isolator. The velocity sensors consist of one sensing vertical vibration and two sensing radial vibration; for the fore-aft and the lateral directions. The three force motors per isolator are aligned in the same manner as the velocity sensors for the vertical, fore-aft, and lateral directions.

The velocity sensors consist of a device called a geophone, which is made up of a coil of wire supported by very soft springs within a magnetic field. The magnet side of the geophone is attached to the supported payload. One end of the support springs is attached to the supported payload with the other end attached to the coil of wire. The magnet moves with the vibration of the supported payload while the inertia of the coil tends to hold it fixed in space. As the magnet moves relative to the stationary coil, a voltage is induced in the coil; the magnitude of the induced voltage is proportional to the velocity of the magnet relative to the coil. The geophone is selected as a velocity sensor for two reasons; first, it can sense very small vibrations; and second, the electronic noise output from the coil is very small. The voltage output from the geophone coil is equal to the absolute velocity of the supported payload relative to fixed space when the coil remains stationary. In reality, the effect of the coil support springs fixes the coil motion equal to the supported payload motion at zero frequency. With equal coil and payload motions there is no output from the geophone because the coil is not moving relative to the magnetic field. As the vibration frequency of the supported payload increases, the inertia of the coil starts to deflect the support springs thus generating relative movement between the magnet and coil resulting in a voltage output. The velocity sensor mechanical dynamics thus provide for a voltage output from the coil which acts as a second order high pass filter.

The velocity signals are processed in the electronic damping isolation system controller to form a signal which is applied to power amplifiers to operate the twelve system force motors. The force motors consist of a wire coil held within a magnetic field, such that a force is applied to the coil as current from the power amplifier flows through the wire. The force motor construction is that of a high quality loud speaker with some modifications allowing attachment of the magnet portion to the ground side of the isolators, and the coil to the supported payload side of the isolators. The coil is modified to allow heat generated from current flowing through its wire to flow into a large heat sink which makes up part of the housing of the supported payload side of the isolators. This modification allows for the generation of more than the rated force from the force motors. The force motor coils are protected from overheating by a simple circuit in the force motor power amplifiers which lowers the current applied to the coils as a function of the current magnitude time history. The selection of the loud speaker type design for the force motors is based on zero friction between the coil and magnet, and the very large dynamic range of force which can be generated.

The selection of the geophone as the velocity sensor and the loud speaker design as the force motor is used in conjunction with a power amplifier design having very low noise output is used to obtain wide dynamic range. The combination of the geophone, force motor and power amplifier allow for a system operating dynamic range of over 120 dB. This large dynamic range is required for the electronic damping active isolation system to isolate floor vibrations of a fraction of a micro meter, while at the same time lowering supported payload motions produced by payload mechanical moving components.

The operation of the electronic damping active isolation system is performed by the operation of six servo control channels making up the controller. The six servo control channels process the velocity signals and operate the force motors in a manner that provides for all six degrees of active isolation of the supported payload. Three controller servo channels operate from vertical velocity sensors and power vertical force motors. The operation of these three channels provide for the active isolation of vertical, pitch, and roll motions of the supported payload. Two of the vertical control channels operate on a single isolator, while the third channel uses an averaged velocity signal from two isolator sensors, located on a common side of the supported payload, and outputs a common force motor signal to the isolator force motors. Two controller channels operate from the fore-aft velocity sensors and force motors. The operation of these two channels provide for active isolation of fore-aft and yaw motions of the payload. Both of these channels use averaged velocity signals from two isolators sensors, located on common sides of the supported payload, and provide a common force signal to the two isolator force motors. One controller channel operates from all four of the lateral velocity sensors and force motors. The operation of this controller channel provides for lateral active isolation of the supported payload. This channel operates from the average signal of the four lateral velocity sensors and provides a common force motor signal to the four lateral force motors.

GEOPHONE TRANSFER FUNCTION

The geophone velocity sensors weight the true velocity of the supported payload by their transfer function, which is formed from the action of the coil mass and the coil attachment springs. The resulting transfer function weighting the true velocity signal is a second order high pass filter. The coil mass and support spring system of the geophone produce a natural frequency of 4.5 Hz. The damping of the transfer function can be adjusted by placing a resistor between the output of the geophone coil wires. The damping action of this resistor acts as follows. The damping resistor causes current flow through the geophone coil as voltage is induced from vibration. The current flow through the coil produces a force proportional to the current flow which acts to resist movement of the coil relative to the magnet. The current flow through the coil is proportional to the relative velocity between the coil and magnet, thus the resisting force is also proportional to this velocity. A resisting force which is proportional to velocity is a viscous damping force. The geophones used in the electronic damping active isolation system use coil shunting resistors which set the fraction of critical damping of the geophone transfer function to unity. Thus the true absolute velocity of the supported payload is weighted by a second order high pass filter having a natural frequency of 4.5 Hz with critical damping.

ACTIVE SYSTEM STABILITY

The transfer function of the geophone is in cascade with the transfer function of the supported payload and air springs. The resulting open loop transfer function thus exhibits gain and phase shifts which act as two second order filters in cascade; one is critically damped with a 4.5 Hz natural frequency and results from the action of the geophone velocity sensor; the second has much less damping with a 4.0 Hz natural frequency and results from the action of the supported payload and the isolator air springs.

The effect of the transfer function of the geophone produces lagging phase, starting at +180 degrees at zero frequency, and reaching 0 degrees slightly above its natural frequency of 4.5 Hz. The effect of the passive frequency response of the isolation system produces an additional 180 degree phase lag as frequency sweeps up through the region of the passive natural frequency.

The gain from the velocity sensor increases at a rate of 6 dB per octave with increasing frequency. While the phase angle from the combined action of the geophone and air filled isolators starts at +180 degrees at zero frequency, lags by 180 degrees from zero frequency to the geophone natural frequency and lags by an additional 180 degrees in the region of the passive isolation system natural frequency. The increasing gain from the velocity sensors are filtered by the low pass filter action of the passive isolation system, which produces a large resonance amplification at its natural frequency, followed by a decreasing gain of -6 dB per octave as frequency increases. The natural action of the geophone sensor and passive isolation system alone result in a very unstable active isolation system, even with very low gain. The active isolation system is stabilized by lowering the geophone phase lag and gain characteristics in frequency by at least two orders of magnitude. In this manner, the geophone lagging phase occurs in at frequency region where the total gain is very low.

CONTROLLER FREQUENCY RESPONSE

Filters within the controller act on the geophone velocity signals to achieve the required signal processing to obtain a highly stable active isolation system for all six channels. Figure 1 presents the frequency response of a typical active isolation channel controller. The controller transfer function is obtained from two second order lag-lead type filters, each having two zeros, and two poles. The gain of the controller transfer function amplifies the geophone signal at low frequencies to

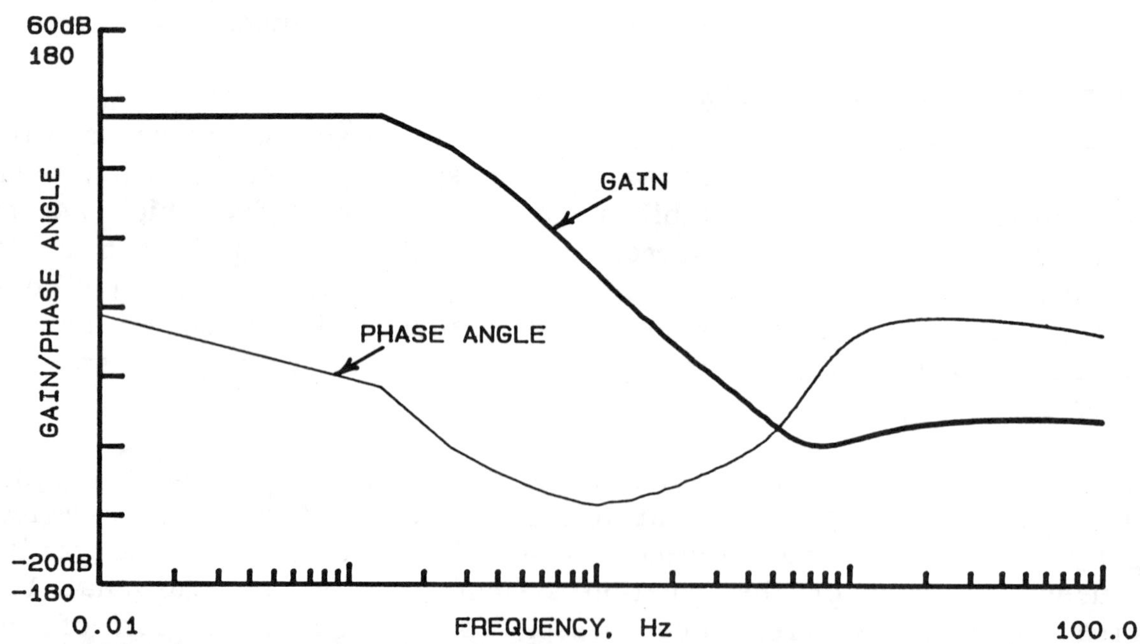

FIGURE 1: Frequency response of isolation system controller relating the absolute velocity sensor output signal to the force motor power amplifier command signal.

obtain a flat geophone velocity signal from about 0.05 Hz and up. While the phase shifting response of the controller transfer function lowers the phase angle of the geophone signal from 0.05 Hz up to the natural frequency of the geophone. The net effect of the controller transfer function is to electronically lower the natural frequency of the geophone to 0.05 Hz. The synthesis of a very low natural frequency for the geophone transfer function results in gain and phase shifts for the geophone filter response to take place in a low frequency region. In this low frequency region the open loop transfer function gain of the active isolation control loops are very low.

In addition to the low frequency signal processing of the controller, additional signal processing is performed at very high frequencies, also to maintain the active isolation system channel stability. The geophone has a small resonance around 300 Hz and other resonances occur within the payload in this frequency region and above, which if not attenuated would cause high frequency oscillations of the isolation system channel. The attenuation of these resonances are performed using simple minimum phase lag-lead compensators to lower the gain by 10 to 20 dB as required. In cases where more than 10 dB attenuation is required, two or more lag-lead compensators are used, each having 10 dB or less attenuation. The frequency response curve presented on Figure 1 represents actual data measured for one of the three vertical active isolation control channels of the electronic damping system. All six active isolation control channels present almost identical frequency response characteristics, except at very high frequencies where the lag-lead compensation may differ from channel to channel.

THE OPEN LOOP TRANSFER FUNCTION

The open loop transfer function is used to measure the degree of stability of the six channels of the electronic damping vibration isolation system. The degree of stability determines the location of the closed loop resonance frequency and the degree of amplification at the resonance frequency. In this electronic damping active vibration isolation system, the open loop transfer function characteristics are adjusted such that the closed loop resonance frequency is between 0.5 and 1.0 Hz with less than 1.5 dB of vibration amplification at the resonance.

At high frequencies, the payloads typically carried by this system often contain resonances in the 10 to 1000 Hz frequency range. The electronic damping effect can lower the amplitude of such resonances within the supported payload provided that the open loop gain at the resonance is more that 0 dB and the phase angle at the resonance peak is more than 60 degrees. For frequencies between 10 and 100 Hz, such resonances can always be damped with their peak resonance lowered by as much as 20 dB when the loop is closed. Above 100 Hz, such resonances usually can be attenuated in the closed loop by as much as 10 dB. However, in

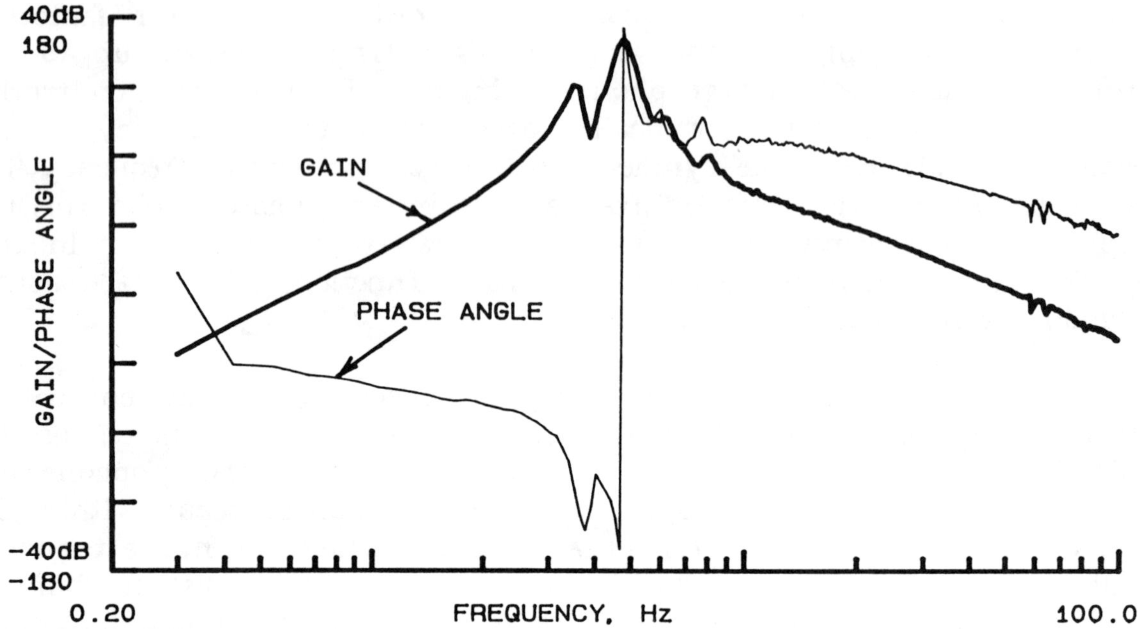

FIGURE 2: Open Loop Transfer Function of one isolation system channel relating the signal from the controller to the signal applied to the force motor power amplifier.

cases where many resonances occur in cascade, it may not be possible to attenuate all of them. Under such conditions, the high frequency lag-lead compensation is fixed in a manner that does not amplify any of the payload resonances when the control loop is closed.

Figure 2 presents an open loop transfer function measured for one of the vertical active isolation channels. The resonance frequency is seen as the peak in the gain curve along with the rapid change of phase that results from the lightly damped passive system transfer function. In the frequency region of the passive system resonance frequency, secondary peaks in gain are caused by mechanical coupling between the electronic damping system degrees of freedom, mainly between the vertical and radial directions.

The gain response for frequencies less than the resonance frequency increases at a uniform rate of 6 dB per octave, which is the ideal response. At frequencies above the resonance frequency the gain decreases at a uniform rate of -6 dB per octave. This is also the ideal response. At frequencies less than the resonance frequency, the open loop transfer function phase angle response starts near zero degrees of phase shift. This non-lagging phase shift is caused by a very low frequency lead filter used to keep DC voltages coming from the output of the geophone preamplifier from entering the channel signal processor. As frequency increases, the phase angle is lowered by the controller signal processing used to decrease the effective geophone natural frequency from 4.5 Hz to less than 0.1 Hz. At the resonance frequency of the passive isolation system, the phase lags rapidly by -180 degrees. As frequency increases beyond the resonance of the passive system, the phase angle falls off only slightly up to 100 Hz. This decrease in the phase angle at high frequencies is caused by two high frequency lag-lead gain lowering filters required for high frequency stability. These two stabilizing filters are located well above 100 Hz, and act on one high frequency geophone resonance and two very high frequency payload resonances.

THE CLOSED LOOP TRANSFER FUNCTION

The closed loop transfer function for the electronic damping active vibration isolation system is measured by inserting a velocity command signal into one of the controller channels, then measuring the transfer function of the payload velocity to the velocity command signal. Figure 3 presents the closed loop transfer function for one of the three vertical channels. This transfer function was measured with all other controller channels running in the closed loop mode. The closed loop gain has a small resonance near 1.0 Hz with a peak amplification of about 1 dB. The active bandwidth where active isolation results is from 0.5 up to about 100 Hz. The phase angle for an ideal system would be -180 degrees below the passive resonance frequency, and +180 degrees above this frequency. The actual closed loop transfer function indicates nearly

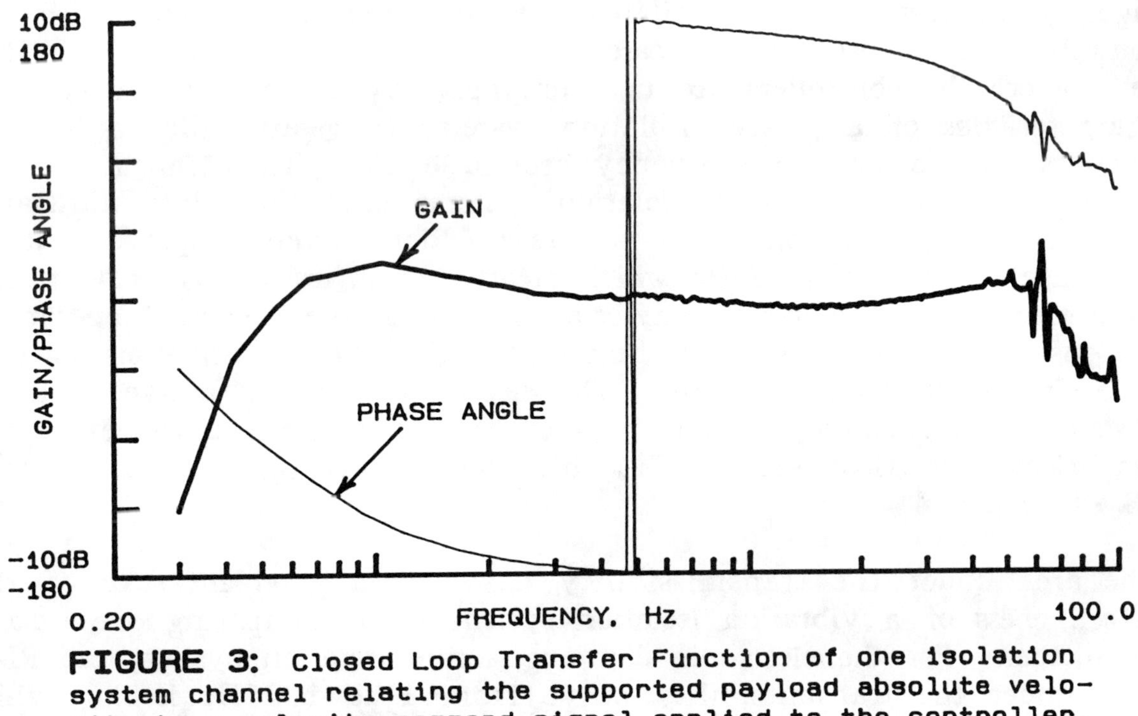

FIGURE 3: Closed Loop Transfer Function of one isolation system channel relating the supported payload absolute velocity to a velocity command signal applied to the controller.

ideal operation between the frequencies of 1 Hz to 70 Hz. The small resonances indicated by this figure near 100 Hz are not in the payload or geophone, but are in the isolator base structure and do not show up in measurements of the vibration isolation system characteristics.

The overall flatness of the gain curve indicates another possible use for this active vibration isolation system. With all controllers operating in the closed loop mode, the vibration level on the supported payload is very small. By inserting a velocity command signal into one or more of the system controllers, a vibration signal is generated. Using this mode of operation the electronic damping active isolation system can be used as a vibration generating device which will produce very small amplitude vibrations with a high degree of accuracy. This effect is useful in the calibration of vibration sensors used to measure low amplitude vibration.

PASSIVE AND ACTIVE TRANSMISSIBILITY

The effectiveness of a vibration isolation system is measured by the term transmissibility. In the case of the electronic damping active vibration isolation system the vibration excitation comes from naturally occurring floor vibration. The system transmissibility in this case is measured by using the velocity frequency spectrum occurring on the supported payload divided by the velocity frequency spectrum occurring on the floor. This measurement was made using a two channel FFT analyzer, to obtain the two frequency spectrums, and velocity sensors designed to measure very small vibrations over a frequency range from 0.25 to 500 Hz.

Figure 4 presents the transmissibility measurements for the case with the controllers turned off, and the case with the controllers running. In the case where the controllers are not running, the isolation system has the characteristics of a passive isolation system using air filled vibration isolators with a natural frequency around 4 Hz. This transmissibility curve is typical of passive isolation systems used to isolate naturally occurring floor vibration. The passive isolation system displays a large resonance around 3.5 to 5 Hz which greatly amplifies the floor vibration transmitted to the supported payload. In the active mode of operation, the passive resonance region is eliminated with vibration isolation starting at a frequency less that 1 Hz. The electronic damping active isolation system has a resonance region between 0.25 and 0.5 Hz. However, the amplification of floor vibration in this region is very small, less than 1 dB.

The area under the transmissibility curve is a good measure of the effectiveness of a vibration isolation system when comparing one system to another. For the electronic damping active isolation system, the RMS area under the transmissibility curve from 1 to 100 Hz is 0.6, while with the controller turned off the area under the transmissibility curve is 6.6. The ratio of the electronic damping active isolation system

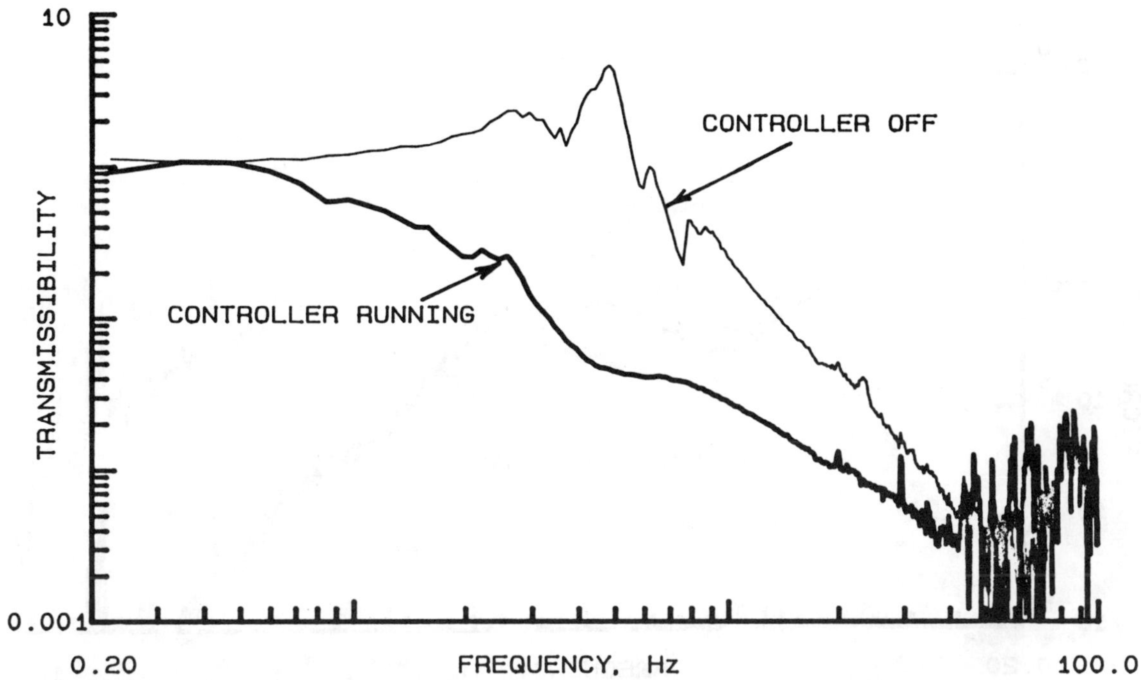

FIGURE 4: Transmissibility as a function of frequency measured using naturally occuring floor vibration for the active isolation system controller running and turned off.

transmissibility area to the passive vibration isolation system transmissibility area is 0.1 indicating that for a uniform floor vibration spectrum the active vibration isolation system will have 10 times less vibration than the passive system. The majority of the RMS transmissibility magnitude for the controller off state comes from the region of the isolation system resonance. This shows clearly that it is very important in any vibration isolation system to maintain the resonance amplification as close to unity as possible.

Figure 5 presents the floor and supported payload displacement spectrum from which the transmissibility curves of Figure 4 were obtained. The total RMS displacement for the floor from 1 to 100 Hz is 0.3 micro meters, while the RMS displacement for the supported payload with the controllers running is 0.011 micro meters. This data indicates an overall reduction of the supported payload RMS displacement of 0.036 which represents 96 percent vibration isolation.

Figure 6 presents the analog time history for the supported payload velocity with the controllers running and off. This figure gives a better picture of the effectiveness of the electronic damping active isolation system over conventional air operated vibration isolation systems. Of note is the large amplitude of the signal with the controllers off. This large amplitude signal results from the combination of the resonance frequency

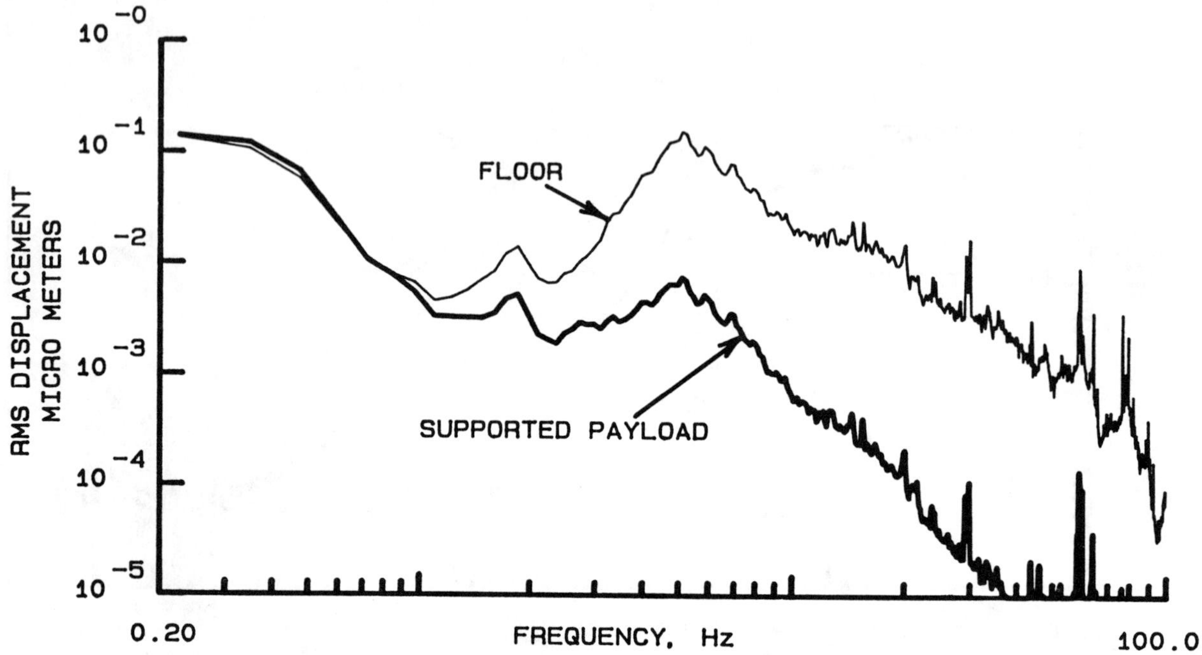

FIGURE 5: Supported Payload and Floor RMS displacement spectrums measured in micro meters [Microns] with the active vibration isolation system controller running.

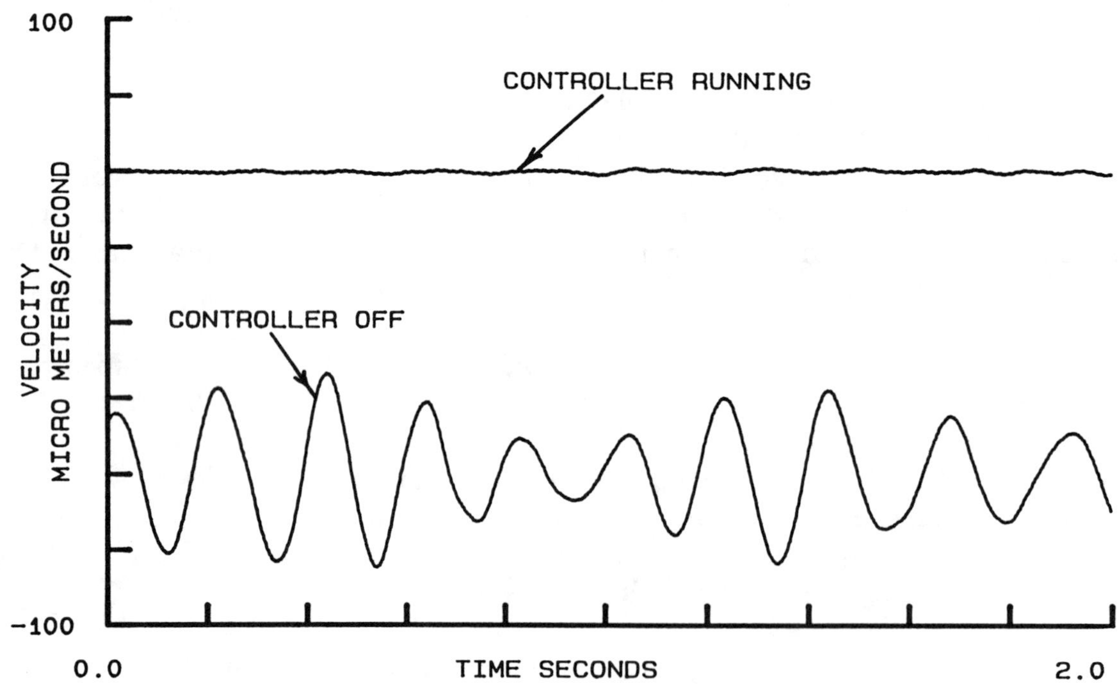

FIGURE 6: Supported payload velocity time response to naturally occuring floor vibration with the active vibration isolation system controller running and off.

of the passive system nearly equaling the frequency of the peak spectral amplitude of the floor; which is a common occurrence.

MOBILITY TRANSFER FUNCTION

The mobility transfer function measures the supported payload displacement response to forces applied to it. Most vibration isolation systems are subject to forces acting directly on the isolated payload. Generally these forces result from the movement of mechanical components which make up part of the supported payload. A typical example of this is the semiconductor lithography machine wherein a stepper must move the product from point to point as successive exposures are produced. The acceleration of the stepper mass causes a reaction force to be applied to the supported payload resulting in movement of the payload. In this type of application, it is important that the total payload motion be small and that the motion which does result damp out fast. Such action allows the payload machinery to proceed to the next process step without delay. Even in applications where the payload has no moving mechanical components, such as a gravity antenna, or electron microscope, forces from air currents and air pressure pulses from HVAC equipment can apply forces that cause motion of the supported payload.

Figure 7 presents the magnitude of the mobility transfer function for the electronic damping active isolation system.

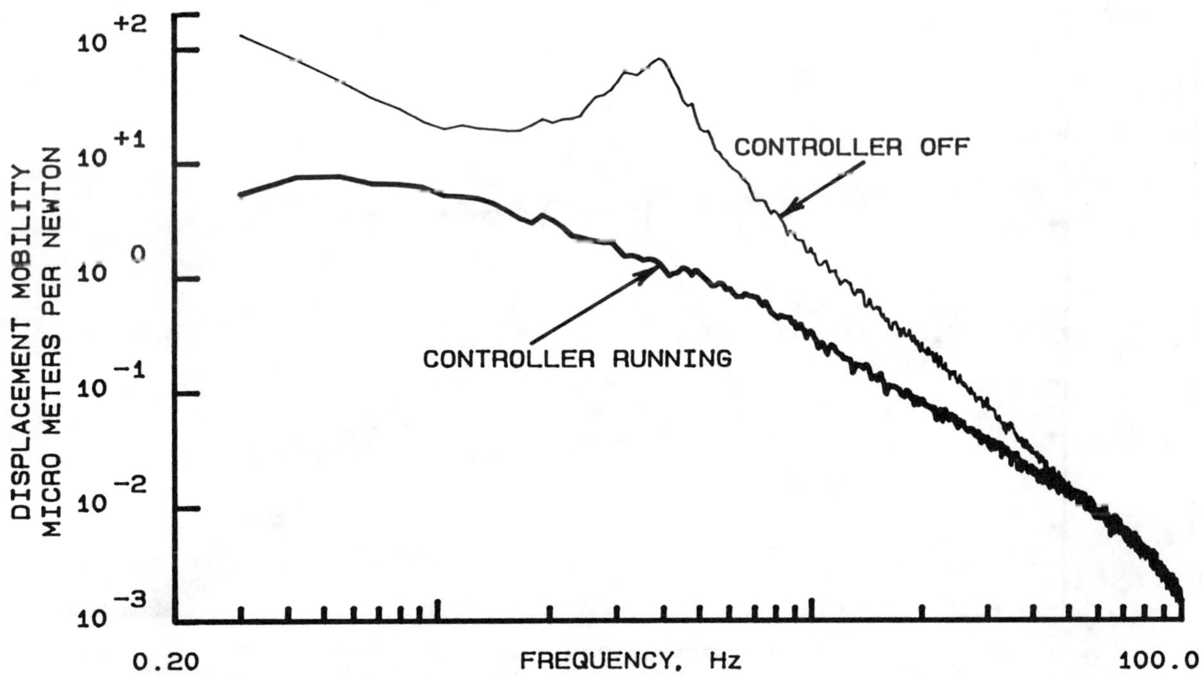

FIGURE 7: Displacement Mobility in Micro Meters [Microns] per Newton measured for the conditions of the active isolation system controller running and off.

The data presented on this figure was obtained by attaching four force actuators to the supported payload, one at each corner, and oriented to generate a vertical force. These force actuators were of the same design as used for the force motors of the electronic damping active isolation system. The current signal applied to the four actuators was adjusted to obtain a total vertical acting force of one newton RMS. Frequency response data was obtained for the case with the controller running, and the controller turned off.

The effectiveness of the electronic damping isolation system in lowering mechanical mobility is due to the sky-hook damping effect of the mechanical analog for the active isolation system. The data presented in this text uses a velocity feedback gain which gives an equivalent fraction of critical damping for the sky-hook damper of about 10. With a damping value this high the electronic damping isolation system clamps the supported payload to fixed space quite firmly.

The mobility curve on figure 7 with the controller off represents a total RMS displacement for the 1 newton force of 170 micro meters, and for the same force spectrum with the controller running the total RMS displacement is 8.4 micro meters. The total RMS values indicate that the displacement mobility of the electronic damping active isolation system is

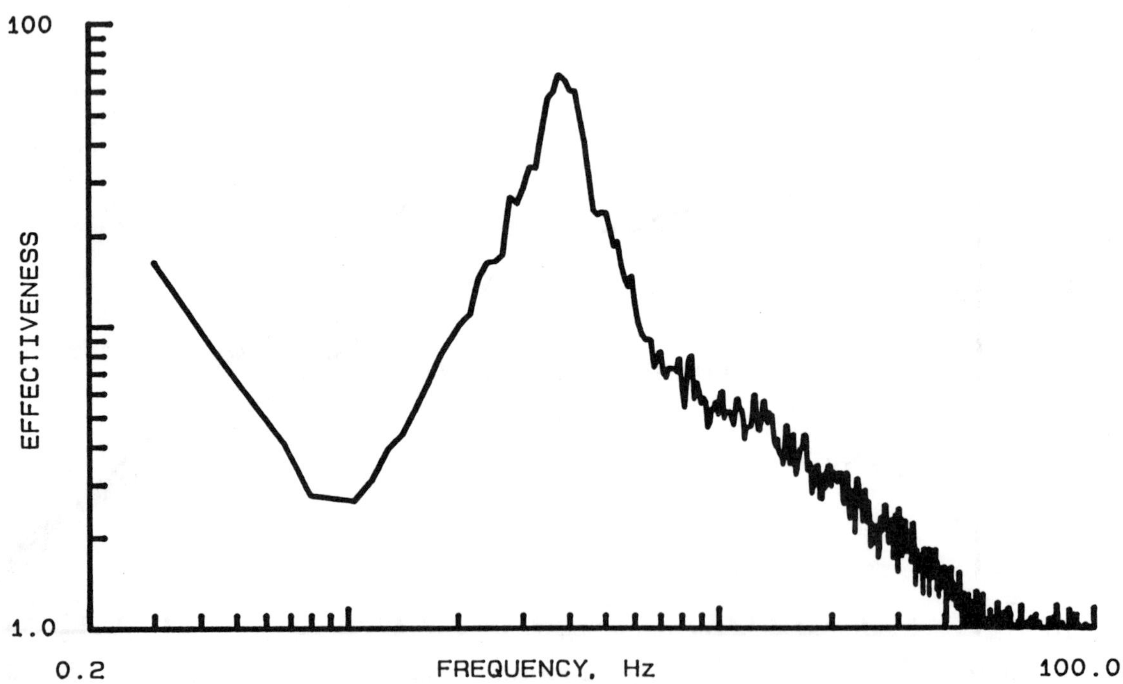

FIGURE 8: Ratio of the mobility curve with the controller off to the controller running is defined as Effectiveness. Values above 1.0 represents a reduction of payload movement.

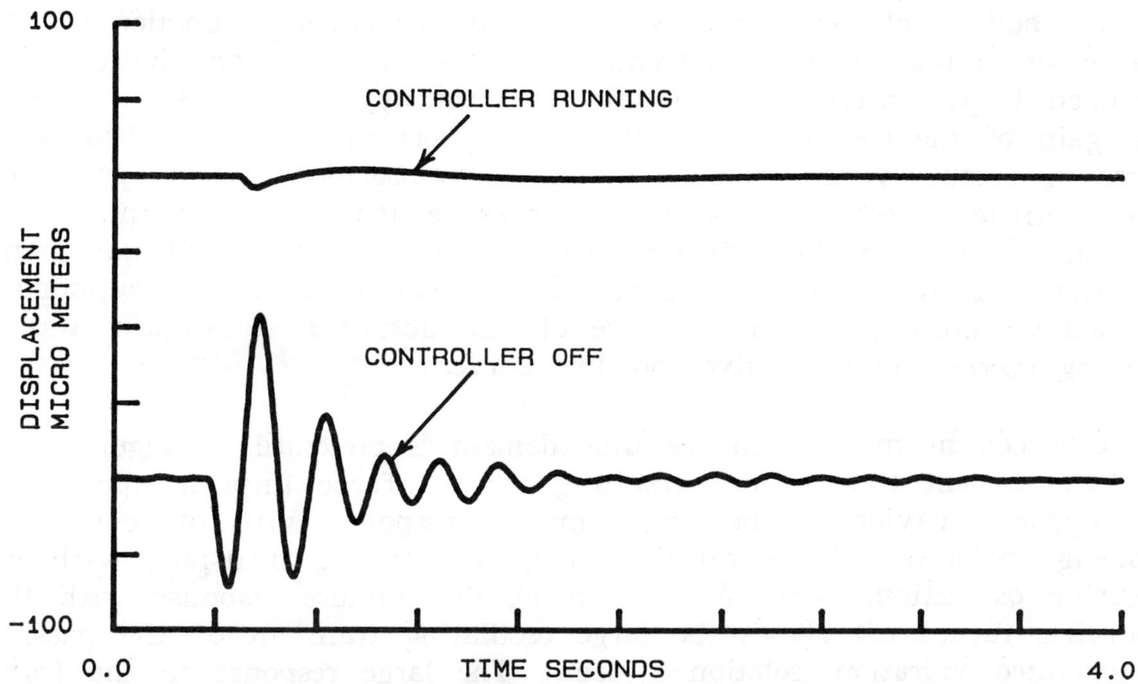

FIGURE 9: Payload displacement time history response to a force impulse applied to the supported payload with the controller running and the controller off.

20 times less than for the passive air operated isolation system alone, as represented by the controller off state.

Dividing the controller off mobility curve by the controller running mobility curve shown on figure 7 presents a curve indicating the frequency regions where the electronic damping active isolation system reduces the displacement mobility to the greatest degree. Figure 8 presents the results of such an operation.

The curve presented on Figure 8 indicates that the active isolation system lowers the displacement mobility at all frequencies within its bandwidth, which is up to about 75 Hz. The curve starts at a ratio value of about 20 then reaches a minimum around 0.8 Hz. This dip in the ratio curve results from the resonance characteristics of the electronic damping system, as was presented on the active transmissibility curve shown on figure 4. The ratio curve then climes to a large maximum value around 3 Hz. This maximum results from the resonance characteristics of the passive air operated isolation system for the controller off state. The maximum is as high as it is because the effective fraction of critical damping of the passive air operated system is nearly 100 times less than the effective fraction of critical damping for the fully active isolation system. The peak magnitude of this curve almost exactly matches the ratio of the active to passive fractions of critical damping.

As the frequency increases above 3 Hz, the gain of the electronic damping active isolation system is lowered by the transfer function of the passive air operated isolation system. This was indicated on figure 2 by the open loop transfer function gain vs frequency curve. As the open loop gain of the electronic damping system becomes less and less with increasing frequency, the mobility of the active isolation system approaches and eventually equals that of the passive isolation system. For frequencies above about 100 Hz, the passive and active displacement mobility curves are equal except where resonances in the supported payload structure cause an increase of the active system gain and a resulting decrease in the active mobility curve.

The displacement mobility in the time domain is presented on Figure 9 as the displacement time history resulting from a force impulse applied to the supported payload. The force impulse response with the controller running indicates only a minimal response to the transient with no lingering oscillation. On the other hand, the impulse response with the controller turned off shows the large oscillating transient of the passive air operated vibration isolation system. The large response to the force impulse with the controller off is consistent with the peak magnitude of the effectiveness curve presented on figure 8 at the passive system natural frequency.

The force impulse transients presented on figure 9 shows the rapid response of the electronic damping active isolation system in bring the supported payload to rest after a transient event. Such action is critical when the supported payload machinery must wait for vibration to decay after each operation prior to the next operation starting.

CONCLUSIONS

The use of absolute velocity feedback and force actuation to produce an active isolation system having electronic damping has been shown to be an effective means of obtaining a nearly motionless supported payload. Configuring feedback sensors and force motors for six degrees of freedom of active isolation provides a dynamic environment that is nearly free from support structure motion while at the same time greatly reducing the dynamic response from moving payloads components and adding active damping to resonant payload components. When compared to conventional air operated vibration isolation systems the electronic damping active system will provide more than an order of magnitude reduction of supported payload motion resulting from support structure vibration, payload induced vibration, and mechanical resonances within the isolated structure.

References

1. Schubert, Dale W. 1988 Shock and Vibration Handbook, McGraw-Hill, Inc. Chapter 33.

2. Schubert, Dale W. 1989 Active Vibration Isolation System, United States Patent No. 4796873. Assigned To Barry Wright Corp.

MULTI-DEGREE OF FREEDOM VIBRATION DETUNERS: THE POTENTIAL OFFERED BY ACTIVE TECHNIQUES

E. Rider,[1] W. R. Hodson[2]

ABSTRACT

The use of auxiliary detuners (passive dynamic absorbers) to control the vibration of a main system in one degree of freedom is widespread. The additional benefits of a single detuner capable of controlling vibration in more than one degree of freedom are described in this paper. The attendant complexity of the design process is outlined via an analysis of a particular geometry.

Using a two degree of freedom detuner as an example it is demonstrated that its effectiveness is increased if use were made of active techniques to control the dynamic properties of the detuner. The possibilities on offer and potential constraints and problems are described.

INTRODUCTION

Auxiliary detuners (also referred to as passive dynamic absorbers) are commonly used as a means of vibration control. The theory and practice of such detuners is widely published - see, for example, [1], [2].

Detuners are, however, generally used to control vibration at fixed frequency in one degree of freedom only. Figure 1 shows a schematic of the usual arrangement.

[1]Vickers Shipbuilding and Engineering Ltd., Barrow, England.
[2]Topexpress Ltd., Cambridge, England.

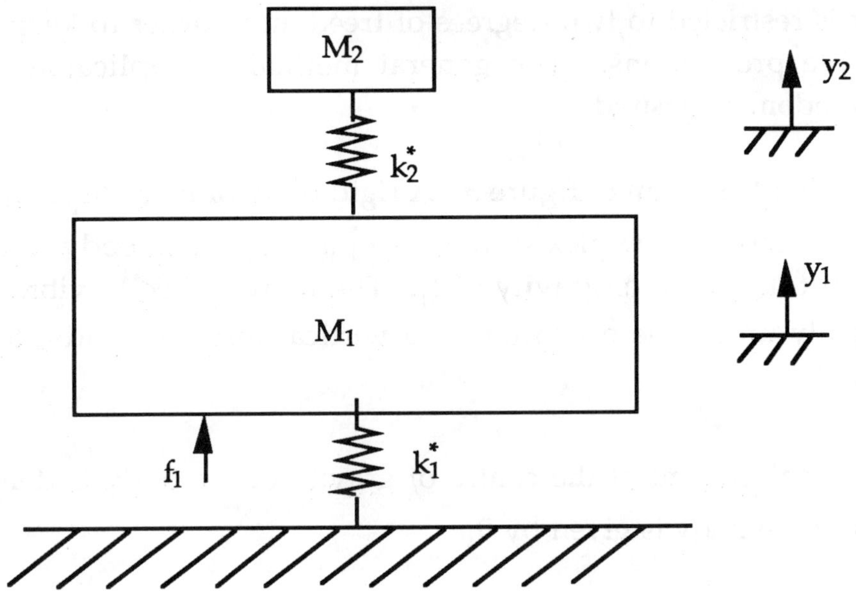

Figure 1: Single degree of freedom detuner

M_1 is is the primary mass, excited by a sinusoidually varying vertical force, f_1, on a mount of complex stiffness k_1^*. M_2 is the detuner mass attached to M_1 via a complex stiffness k_2^*. M_2 is chosen to resonate on k_2^* at a frequency close the resonance of M_1 on k_1^*. Figure 2 demonstrates the reduction in vibration of M_1 that a well-designed detuner can attain across a range of excitation frequencies.

The vibration control performance of the detuner is useful but limited in scope. If the excitation was at changing or multiple frequencies (or both) or in more than one degree of freedom then the detuner mass is simply a weight penalty offering little or no benefit.

The aim of this paper is to demonstrate how these limitations may be overcome with particular reference to active techniques.

A MULTI-DEGREE OF FREEDOM DETUNER

This section describes a multi-degree of freedom detuner for a simple geometry. The following section goes on to consider improving performance via active techniques.

The detuner is restricted to two degrees of freedom in order to keep the algebra to manageable proportions. The general method is applicable to as many degrees of freedom as desired.

Consider the simple system of figure 3. A rigid body of mass M_1 and length 2l is supported at its ends by complex stiffnesses k_1^* and k_2^*. The body has a moment of inertia about its centre of gravity of I_1. The body is free to vibrate vertically and rotationally under the excitation of a vertical force, f_1, and a torque, t_1, as marked.

The vertical displacement of the centre of gravity of the body is denoted by y_1. The rotation of the body is given by θ_1.

The body has two degrees of vibrational freedom.

Furthermore, the two degrees of freedom are cross-coupled (a linear displacement in the y-direction will produce a force and a torque and similarly for a rotation).

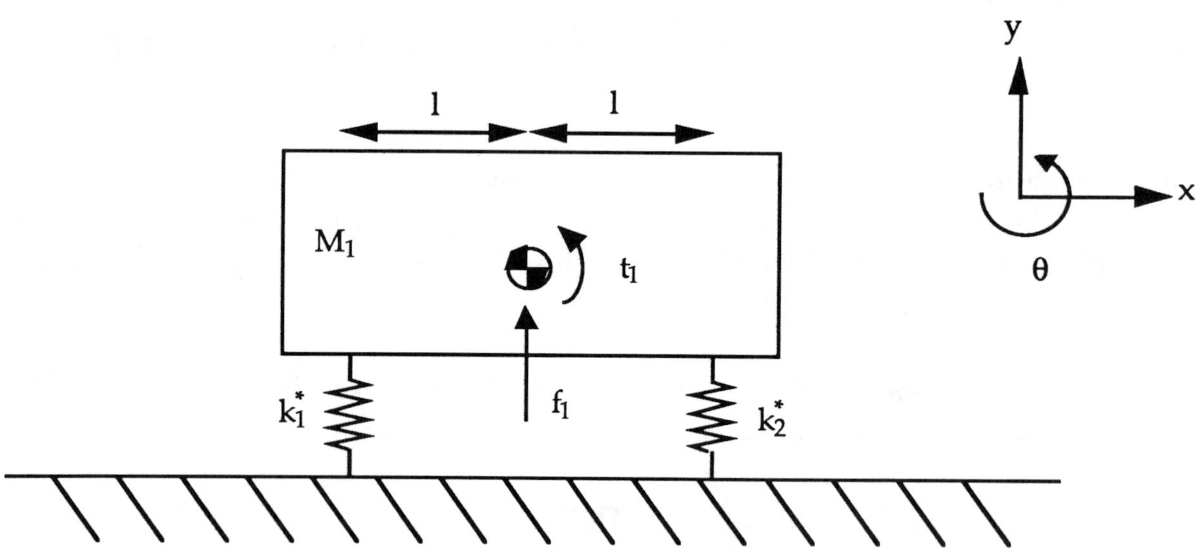

Figure 3: Simple two degree of freedom system

The following simple analysis describes the motion of the body:

Vertically:

$$M_1\ddot{y}_1 = f_1 - k_1^*\left(y_1 - l\theta_1\right) - k_2^*\left(y_1 + l\theta_1\right) \tag{1}$$

Which, for harmonic excitation at frequency w becomes:

$$\left(k_1^* + k_2^* - w^2M_1\right) y_1 = f_1 + k_1^* l\theta_1 - k_2^* l\theta_1 \tag{2}$$

Rotationally:

$$I_1\ddot{\theta}_1 = t_1 + k_1^*\, l\left(y_1 - l\theta_1\right) - k_2^*\, l\left(y_1 + l\theta_1\right) \tag{3}$$

For harmonic excitation:

$$\left(k_1^*\, l^2 + k_2^*\, l^2 - w^2 I_1\right)\theta_1 = t_1 + k_1^*\, l\, y_1 - k_2^*\, l\, y_1 \tag{4}$$

Combining (2) and (4) in matrix form:

$$\begin{pmatrix} k_1^* + k_2^* - w^2M_1 & l\left(k_2^* - k_1^*\right) \\ l\left(k_2^* - k_1^*\right) & k_1^*\, l^2 + k_2^*\, l^2 - w^2 I_1 \end{pmatrix} \begin{pmatrix} y_1 \\ \theta_1 \end{pmatrix} = \begin{pmatrix} f_1 \\ t_1 \end{pmatrix} \tag{5}$$

Write (5) as:

$$\left(\underline{K}_1 + \underline{K}_2 - w^2\underline{M}_1\right)\underline{Y}_1 = \underline{F}_1 \tag{6}$$

Where:

$$\underline{K}_1 = \begin{pmatrix} k_1^* & -lk_1^* \\ -lk_1^* & l^2k_1^* \end{pmatrix}$$

$$\underline{K}_2 = \begin{pmatrix} k_2^* & lk_2^* \\ lk_2^* & l^2k_2^* \end{pmatrix}$$

$$\underline{M}_1 = \begin{pmatrix} M_1 & 0 \\ 0 & I_1 \end{pmatrix}$$

$$\underline{Y}_1 = \begin{pmatrix} y_1 \\ \theta_1 \end{pmatrix}, \quad \underline{F}_1 = \begin{pmatrix} f_1 \\ t_1 \end{pmatrix}$$

Hence the motion of the body is given by:

$$\underline{Y}_1 = \left(\underline{K}_1 + \underline{K}_2 - w^2 \underline{M}_1\right)^{-1} \underline{F}_1 \tag{7}$$

The elements of $\left(\underline{K}_1 + \underline{K}_2 - w^2 \underline{M}_1\right)$ which lie off the leading diagonal describe the cross-coupling between degrees of freedom.

Figures (4) and (5) show the variation in the leading diagonal elements of $\left(\underline{K}_1 + \underline{K}_2 - w^2 \underline{M}_1\right)^{-1}$ with frequency for the artificial parameter set:

$M_1 = 1$ kg
$I_1 = 0.2$ kg m^2
$k_1^* = 10 + 0.1j$ Nm^{-1}
$k_2^* = 5 + 0.1j$ Nm^{-1}
$l = 1$ m

Both figures show two clear resonant peaks. The peak at lower frequency is the translational resonance. The peak at higher frequency is the rotational resonance.

Now introduce a detuner mounted above the centre of gravity of the main body:

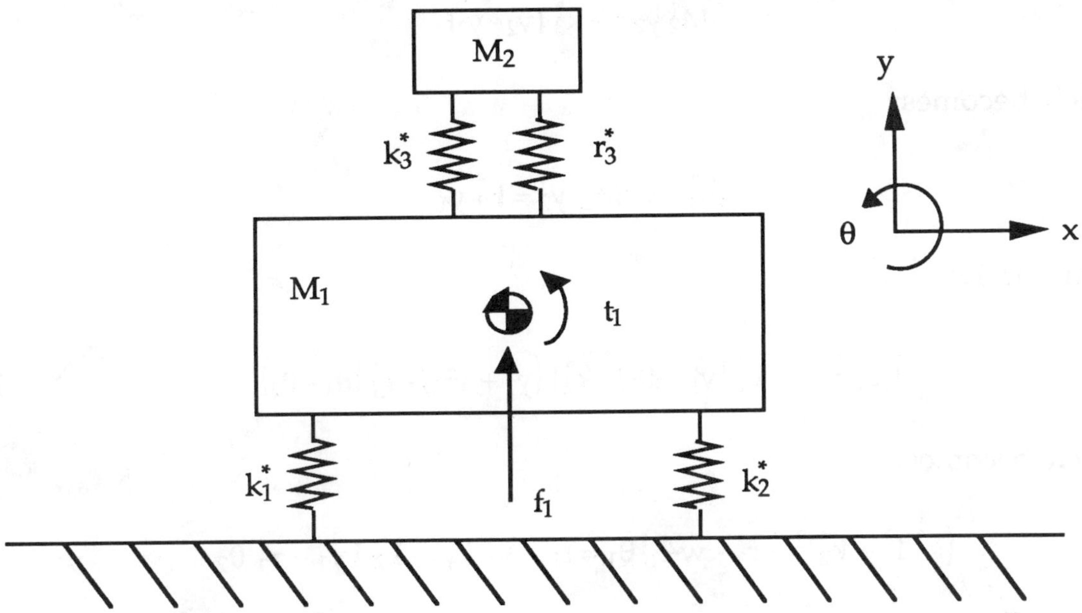

Figure 6: Detuner applied

The detuner has mass M_2 and moment of inertia I_2 and is mounted via a linear complex stiffness k_3^* and a rotational complex stiffness r_3^*. Its physical dimensions are assumed small compared to the main body. The detuner has co-ordinates y_2, θ_2, as for the main body.

The detuner thus has two degrees of freedom. It should be possible to control the two degrees of freedom of the main body.

Repeating the analysis with the detuner in place:

Vertically:

$$M_1\ddot{y}_1 = f_1 - k_1^*\left(y_1 - l\theta_1\right) - k_2^*\left(y_1 + l\theta_1\right) - k_3^*\left(y_1 - y_2\right) \tag{8}$$

Which becomes:

$$\left(k_1^* + k_2^* + k_3^* - w^2M_1\right)y_1 = f_1 + k_1^*\, l\theta_1 - k_2^*\, l\theta_1 + k_3^* y_2 \tag{9}$$

$$M_2 \ddot{y}_2 = - k_3^* (y_2 - y_1) \quad (10)$$

Which becomes:

$$\left(k_3^* - w^2 M_2\right) y_2 = k_3^* y_1 \quad (11)$$

Rotationally:

$$I_1 \ddot{\theta}_1 = t_1 + k_1^* \left(y_1 - l\theta_1\right) - k_2^* l \left(y_1 + l\theta_1\right) - r_3^* \left(\theta_1 - \theta_2\right) \quad (12)$$

Which becomes:

$$\left(k_1^* l^2 + k_2^* l^2 + r_3^* - w^2 I_1\right) \theta_1 = t_1 + k_1^* l\, y_1 - k_2^* l\, y_1 + r_3^* \theta_2 \quad (13)$$

$$I_2 \ddot{\theta}_2 = - r_3^* \left(\theta_2 - \theta_1\right) \quad (14)$$

Which becomes:

$$\left(r_3^* - w^2 I_2\right) \theta_2 = r_3^* \theta_1 \quad (15)$$

Combining (9) and (13) into matrix form:

$$\begin{pmatrix} k_1^* + k_2^* + k_3^* - w^2 M_1 & l\left(k_2^* - k_1^*\right) \\ l\left(k_2^* - k_1^*\right) & k_1^* l^2 + k_2^* l^2 + r_3^* - w^2 I_1 \end{pmatrix} \begin{pmatrix} y_1 \\ \theta_1 \end{pmatrix}$$

$$= \begin{pmatrix} f_1 \\ t_1 \end{pmatrix} + \begin{pmatrix} k_3^* & 0 \\ 0 & r_3^* \end{pmatrix} \begin{pmatrix} y_2 \\ \theta_2 \end{pmatrix} \quad (16)$$

Write (16) as:

$$\left(\underline{K}_1 + \underline{K}_2 + \underline{K}_3 - w^2 \underline{M}_1\right) \underline{Y}_1 = \underline{F}_1 + \underline{K}_3 \underline{Y}_2 \quad (17)$$

Combining (11) and (15) into matrix form:

ACTIVE TECHNIQUES

A two degree of freedom detuner has been shown to be potentially useful. It represents a significant step forward over a single degree of freedom detuner. It is, however, still limited in scope. This section considers the potential for extending that scope by the use of active techniques.

Figure (9) is equivalent to figure (7) except that k_3^* and r_3^* now have no imaginary part (both the springs are undamped).

At the detuner's linear resonance frequency of just over 0.5 Hz the response level is over 30 dB lower than for the damped detuner used for figure (7). There are, however, also two large response peaks closely adjacent to the trough.

The reason for the large trough in response is clear from equation (21). If $(\underline{K}_3 - w^2\underline{M}_2) = \underline{0}$ then the response as plotted disappears to zero. (It fails to do this in figure (9) due to the resolution of the plot). $w^2\underline{M}_2$ is a wholly real matrix. Hence $\underline{K}_3$ must be wholly real also for $(\underline{K}_3 - w^2\underline{M}_2) = \underline{0}$. If the elements of $\underline{K}_3$ have imaginary parts (ie if the springs are damped) then $\underline{K}_3 - w^2\underline{M}_2$ will not vanish to zero at the detuner's resonance. Equally, the twin peaks on either side of the trough in response at ~0.5 Hz in figure (9) will be damped to lower levels.

Figure (7) was produced by striking a compromise between the magnitude of response reductions and the bandwidth over which they are attained by choosing a suitable spring damping value. The selection of passive detuner parameters is a compromise across the frequency range.

Better performance could be attained by actively controlling the dynamic properties of the detuner. Suppose, for instance, that it is possible to specify k_3^* as a function of frequency as desired.

If $$k_3^*(w) = w^2M_2 \qquad (22)$$

$$\begin{pmatrix} k_3^* - w^2M_2 & 0 \\ 0 & r_3^* - w^2I_2 \end{pmatrix}\begin{pmatrix} y_2 \\ \theta_2 \end{pmatrix} = \begin{pmatrix} k_3^* & 0 \\ 0 & r_3^* \end{pmatrix}\begin{pmatrix} y_1 \\ \theta_1 \end{pmatrix} \tag{18}$$

Write (18) as:

$$(\underline{K_3} - w^2\underline{M_2})\,\underline{Y_2} = \underline{K_3}\,\underline{Y_1} \tag{19}$$

Hence:

$$\underline{Y_2} = (\underline{K_3} - w^2\underline{M_2})^{-1}\ \underline{K_3}\,\underline{Y_1} \tag{20}$$

Substituting into (17) and rearranging:

$$\underline{Y_1} = \left[(\underline{K_1} + \underline{K_2} + \underline{K_3} - w^2\underline{M_1}) - \underline{K_3}(\underline{K_3} - w^2\underline{M_2})^{-1}\ \underline{K_3}\right]^{-1}\ \underline{F_1} \tag{21}$$

Equation (21) describes the motion of the main body with the detuner in place.

Figures (7) and (8), correspond to figures (4) and (5) and demonstrate the vibration control performance of the detuner in two degrees of freedom - linear and rotational - for the following parameters:

$M_2 = 0.1$ kg
$I_2 = 0.02$ kg m^2
$k_3^* = 1.05 + 0.5j$ Nm^{-1}
$r_3^* = 1.15 + 0.66j$ Nm / rad

Figures (7) and (8) both include the original responses of figures (4) and (5) with sharp resonant peaks. The detuned responses show reductions in these peaks of up to 20dB.

The detuner used to generate these results has no cross-coupling between degrees of freedom. A real multi-degree of freedom detuner is likely to have such cross-coupling. This would be modelled by non-zero elements off the leading diagonal in $\underline{K}_3$.

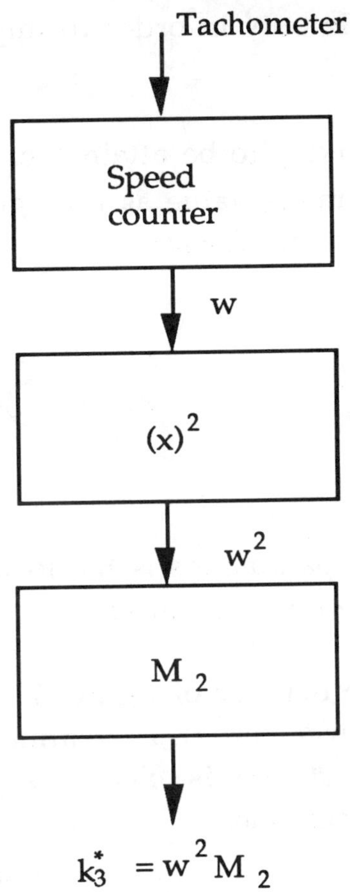

Figure 13: Outline control scheme

then the linear resonance of the main mass should disappear entirely since k_3^* is wholly real and thus undamped. Additionally, the twin peaks of the passive undamped detuner in figure (9) should disappear also since the detuner is tuned to operate effectively at all frequencies. Figure (13) outlines how k_3^* would be derived from a tachometer signal.

It should be noted at this point that many passive materials would provide a complex stiffness that varies with frequency. It is, however unlikely that a passive material could be found with a particular desired dynamic characteristic.

Figure (10) demonstrates the effect of specifying $k_3^*(w)$ as above. The response levels are tiny (and are in fact due to numerical calculation noise). Note that figure (10) is plotted from 0-1 Hz only in order to improve resolution at and around the linear resonance.

In practice, zero damping is unlikely to be attainable. Figure (11) introduces damping into k_3^* and uses the same r_3^* value as for figure (7), to which figure (11) corresponds. The parameter set used was:

$M_2 = 0.1$ kg
$I_2 = 0.02$ kg m^2
$k_3^*(w) = w^2 M_2 + 0.05\,j$ Nm^{-1}
$r_3^* = 1.15 + 0.66\,j$ Nm/rad

The response produced by these parameters is the lower curve in figure (11). The upper curve is the response with no detuner.

Figure (12) compares the passive detuner of figure (7) (upper curve) with the active detuner of figure (11) (lower curve) as treatments for the same main mass mounting arrangement. There is clearly significant performance potential available from active techniques.

(In the example, active techniques are used in one of two degrees of freedom only).

An active detuner is a compromise between passive and fully active mounting arrangements. It is an attractive first step towards active mounts because of its relative simplicity. It is an ideal retro-fit to a known problem: no major redesign of existing mounting arrangements is required. It also permits better vibration control without the use of high detuner masses.

The most likely problems that will be encountered in the construction of a practical active detuner are actuator related. Any particular problem will dictate force, displacement and bandwidth requirements upon an actuator. Different possible actuators have different strengths and weaknesses. A hydraulic actuator will produce large forces but over only a limited bandwidth. Electromagnetic shakers can achieve wider bandwidth but with much smaller forces. Piezoelectric materials can achieve large forces over a wide bandwidth but cannot produce large displacements (about 100 microstrain is typical). Additionally, piezoelectric materials often require high control voltages which can lead to practical problems. Some of the newer, "rare earth", magnetostrictive materials can also produce large forces over a wide bandwidth. Displacements of up to 1000 microstrain are attainable for reasonably small control voltages. Actuator choice (or design) depends entirely upon the particular application.

CONCLUSION

A simple two degree of freedom system has been used to demonstrate the performance of two degree of freedom vibration detuner. The analysis used is applicable to three and more degrees of freedom. The vibration control performance attained by the two degree of freedom detuner is shown to offer benefits over simple detuners. The potential benefits of using active techniques for additional performance are demonstrated via the same example system. The potential for active techniques is shown to be substantial.

ACKNOWLEDGEMENTS

The authors would like to acknowledge the help of Chris Dorling and Martin Myers of Topexpress during the work reported here. The work was funded by VSEL plc.

References

[1] Snowdon J., Vibration and Shock in Damped Mechanical Systems, Wiley 1968.

[2] Seto K., Hara F., Yamashita S., Vibration Control in Piping System by Dual Dynamic Absorber, JSME International Journal 1989.

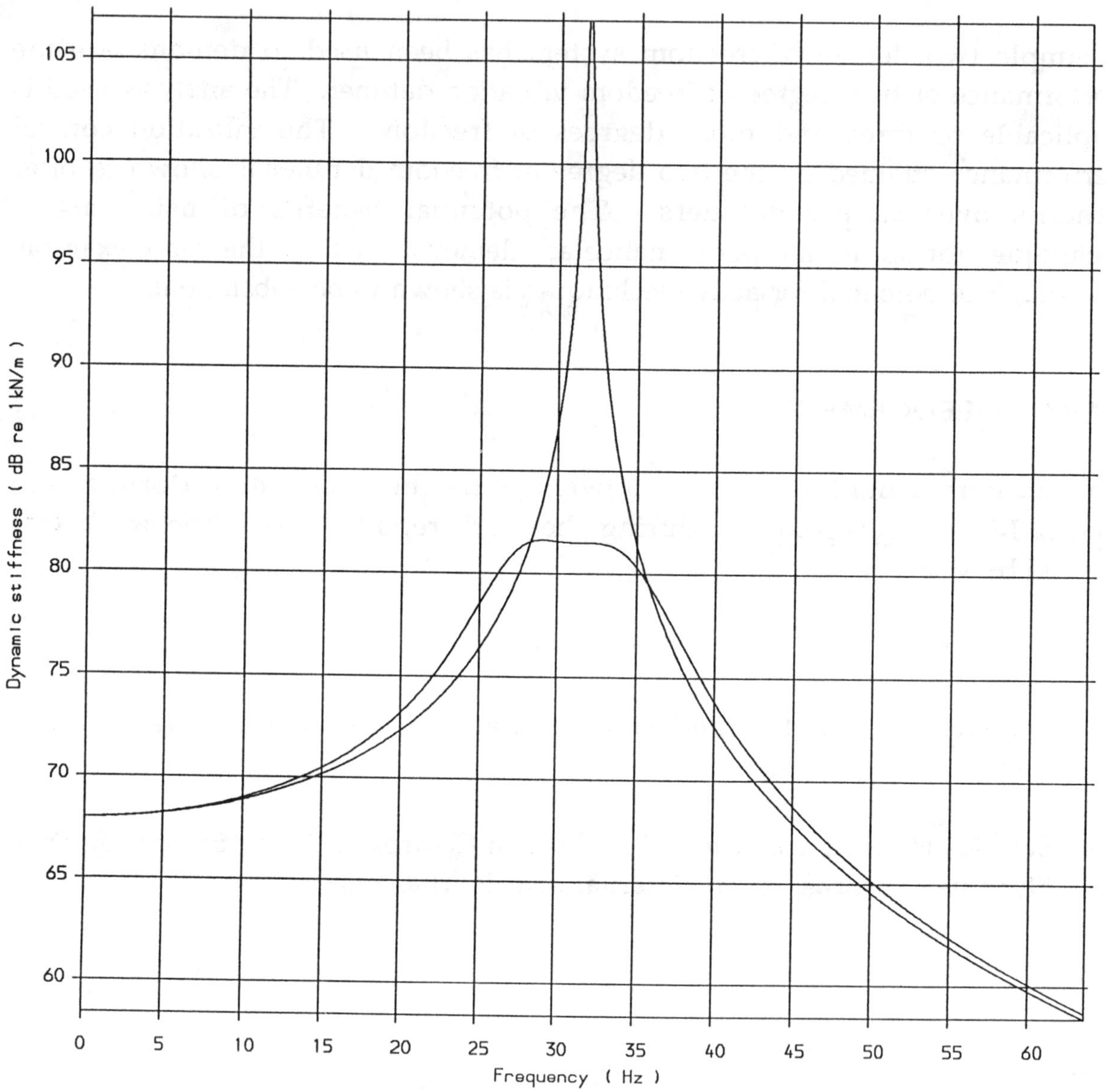

Figure 2 : Simple detuner behaviour

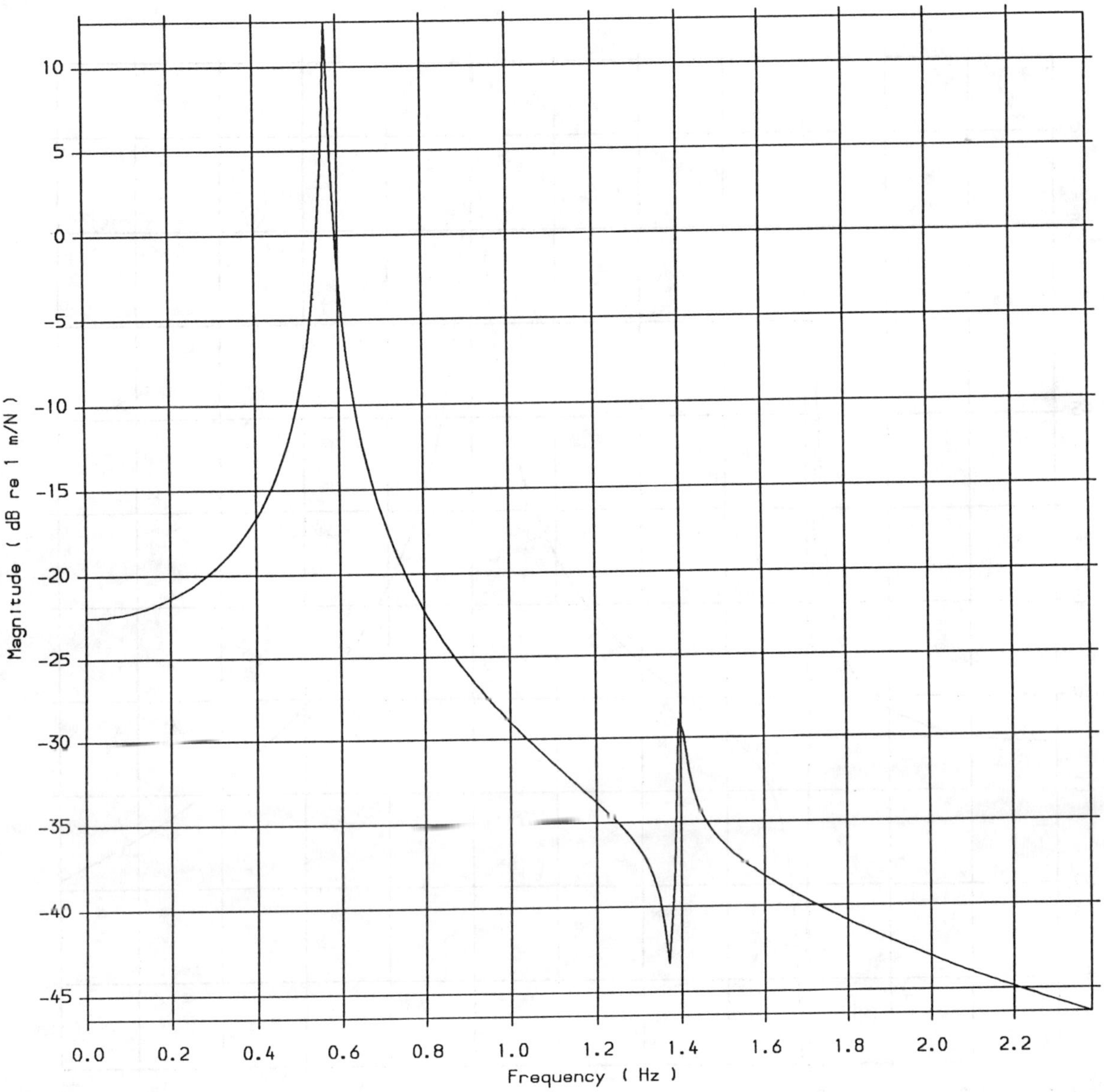

Figure 4 : no detuner

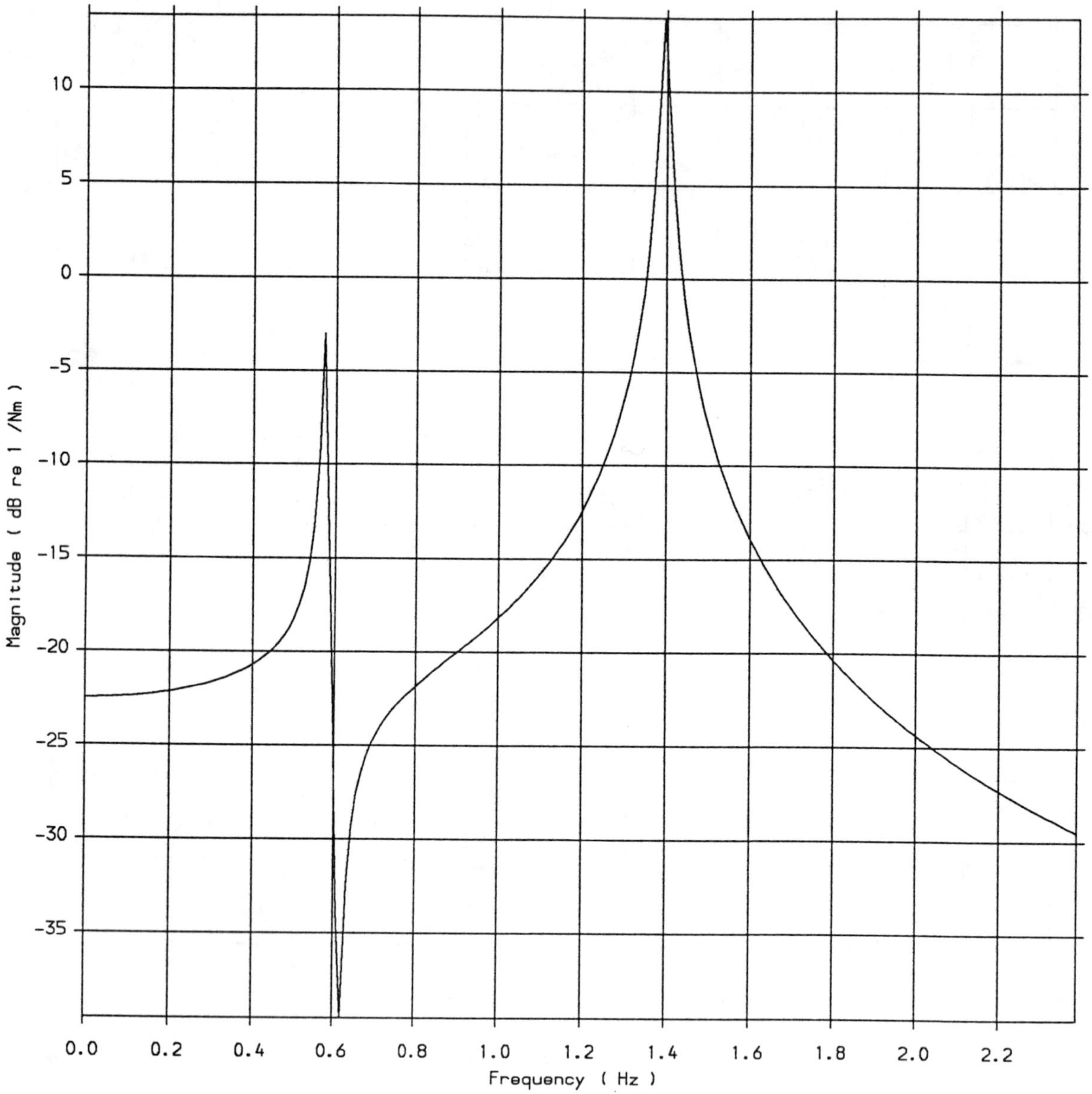

Figure 5 : no detuner

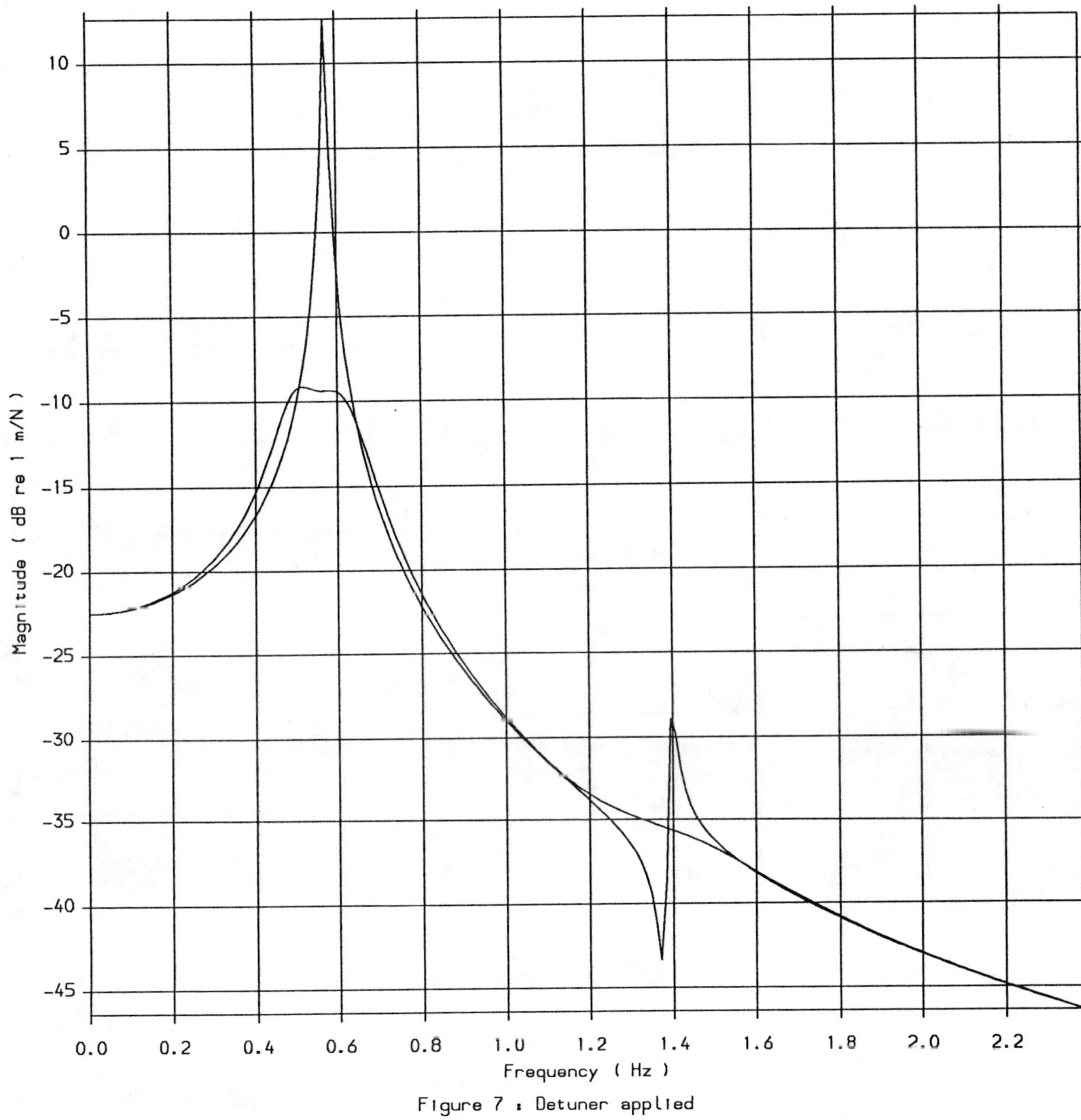

Figure 7 : Detuner applied

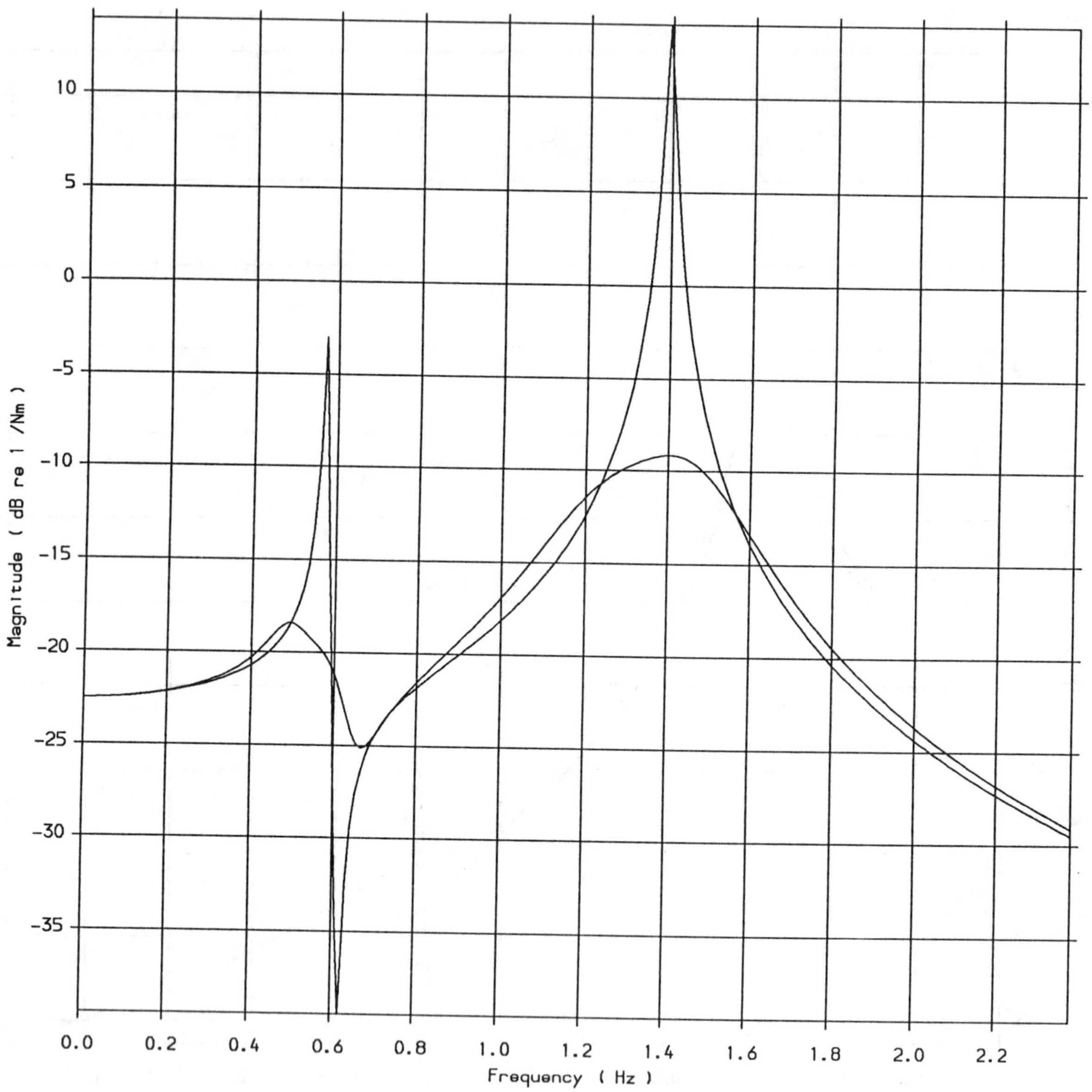

Figure 8 : Detuner applied

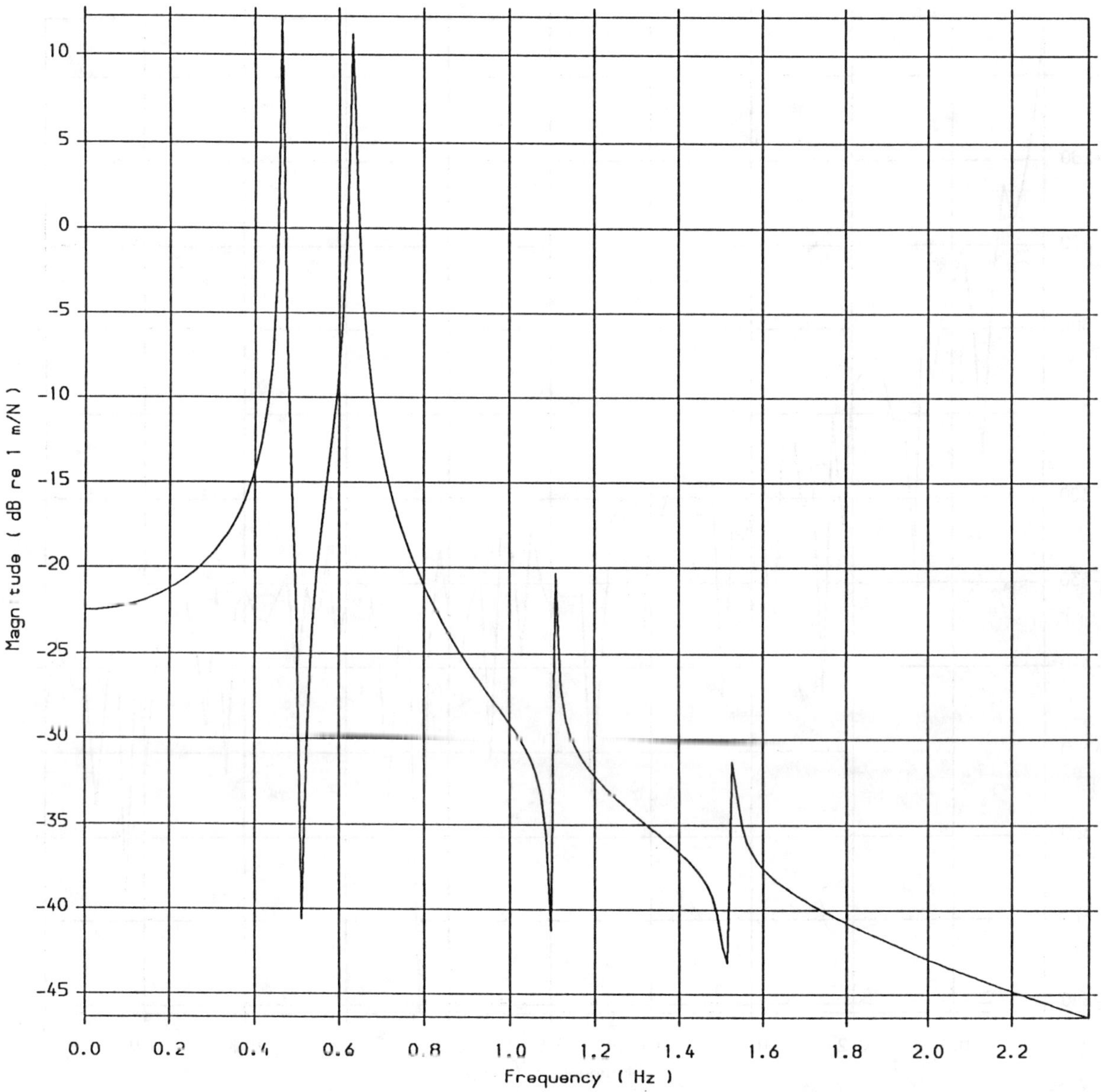

Figure 9 : Passive undamped detuner

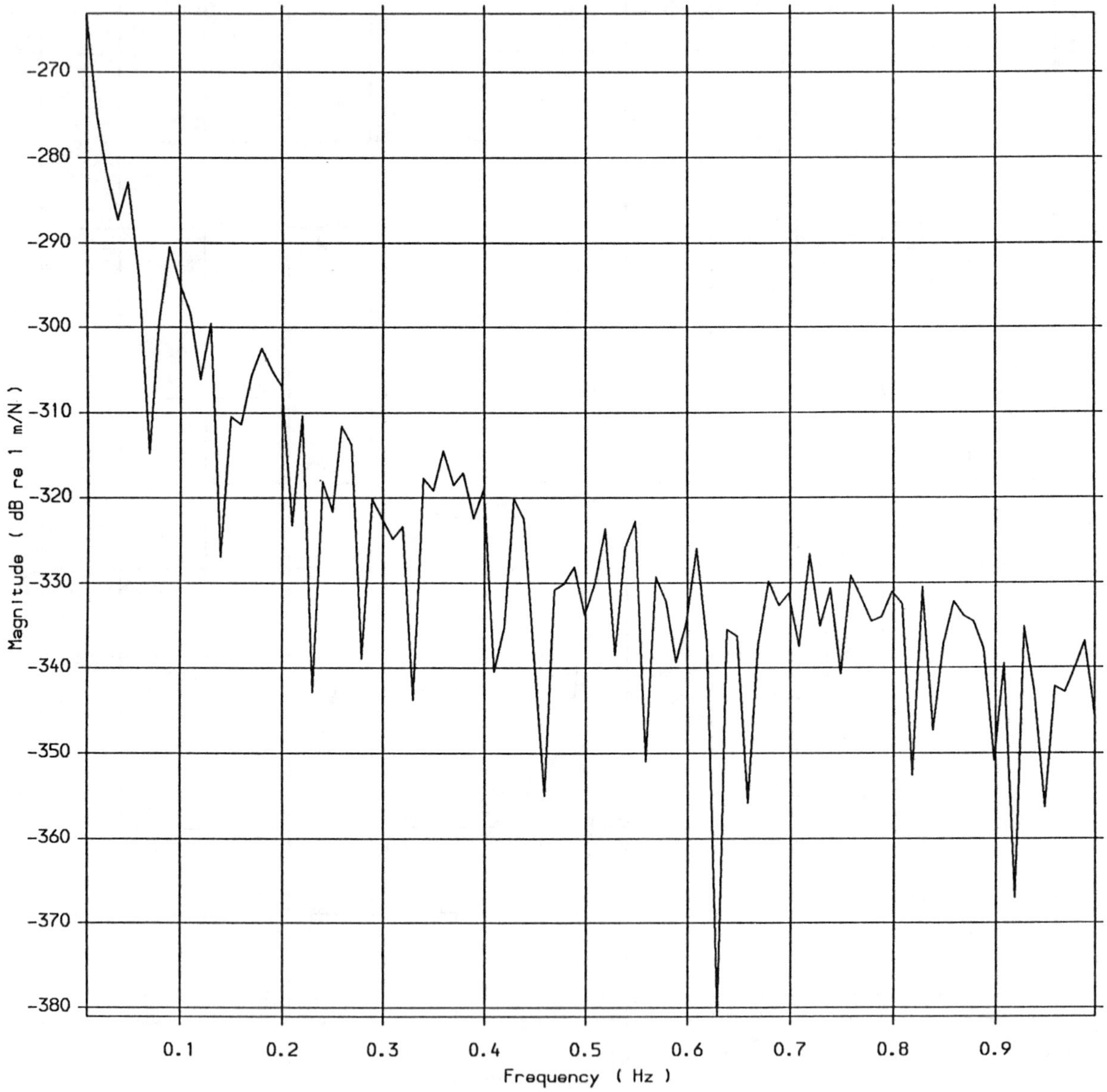

Figure 10 : Active undamped detuner

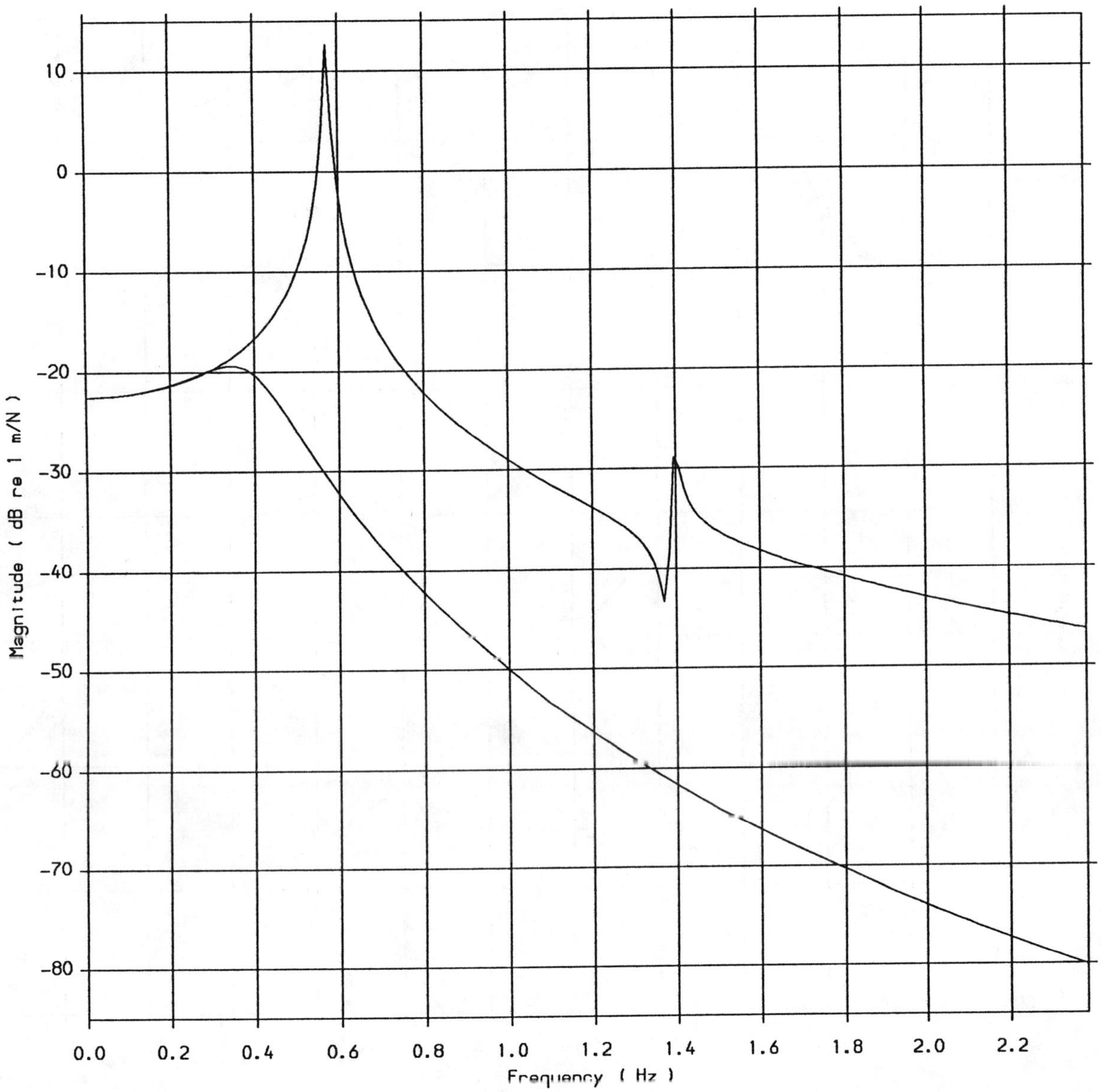

Figure 11 : Active damped detuner

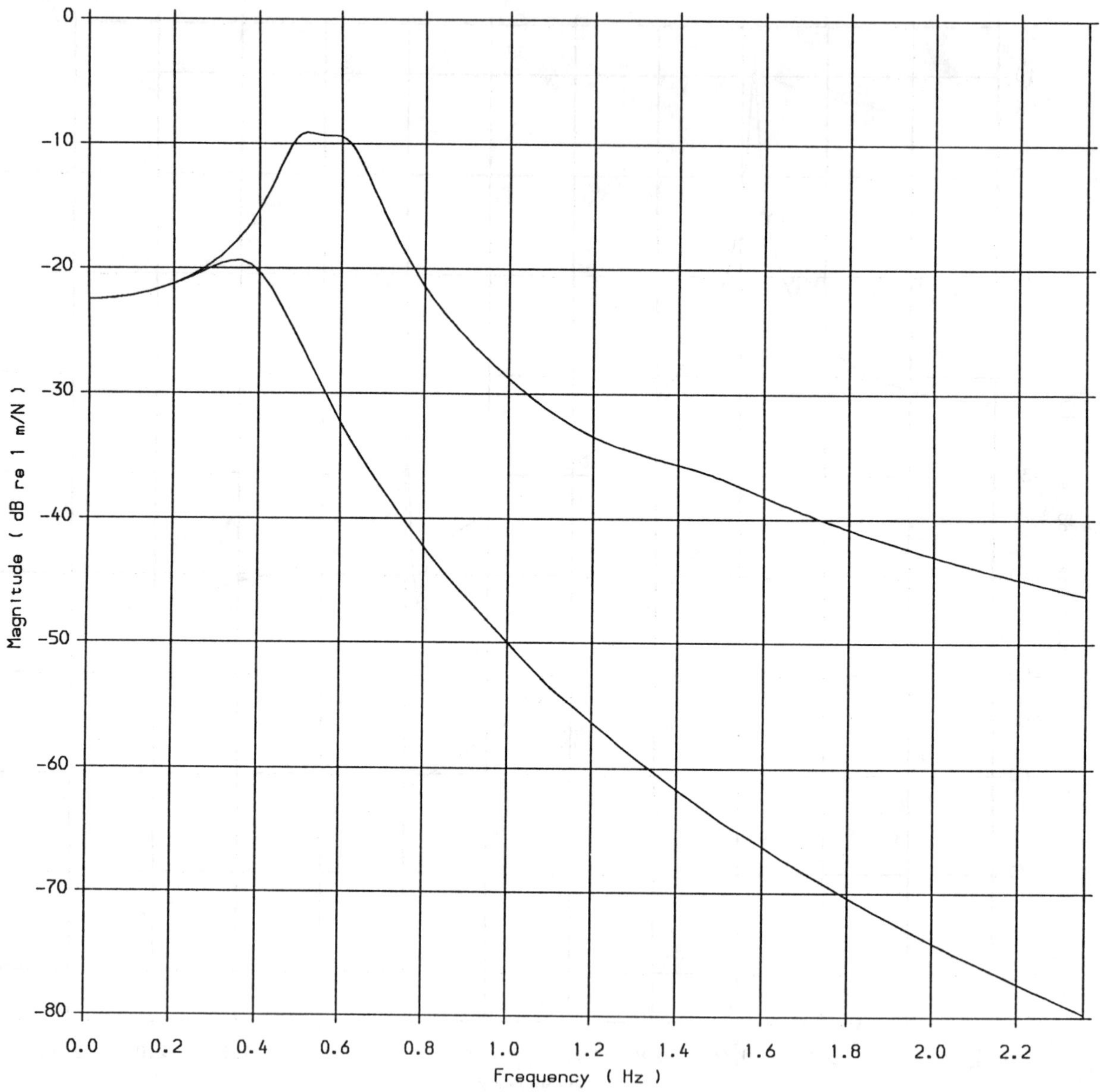

Figure 12 : Active vs Passive Detuner

Structurally Radiated Sound

EFFECT OF GEOMETRIC AND STRUCTURAL/ACOUSTIC VARIABLES ON THE ACTIVE CONTROL OF SOUND RADIATION FROM A VIBRATING SURFACE

Colin H. Hansen,[1] Scott D. Snyder[1]

ABSTRACT

Results are presented which facilitate the understanding of the physical mechanisms involved when active control is applied to a harmonically vibrating surface radiating sound. The model chosen for the purposes of the illustration is a simply supported rectangular panel radiating into free space. The analytical model describing this arrangement has been developed and verified previously for the use of either acoustic or vibration control sources. Here the analytical model is used to study the effect which various geometric and structural/acoustic parameters (such as panel size, modal density, damping, response type - resonant or forced, error sensor type, control source type, and performance of the electronic controller) have on the maximum achievable sound power reduction. This information is essential for the optimization of both the control source and error sensor configuration in the design of active systems to control structurally radiated noise.

INTRODUCTION

The design of active adaptive control systems for the reduction of noise radiated by vibrating structures can be divided into two parts; the design of the control source and error sensor configuration (physical control system design) and the design of the electronic control system. Poor design of either of these parts will result in poor overall noise reduction performance, even if the other part is designed optimally. Of particular importance is that it is possible to optimize the physical arrangement of the control sources and error sensors independently of the electronic control system.

The number and location of the control sources will set an upper bound to the maximum reduction in radiated sound power which can be achieved. The number, type, and location of the error sensors will determine how close to this optimum that it is possible to approach if a perfect electronic control system is available. Thus the performance of the electronic controller can be evaluated by comparing the actual reduction achieved to the maximum that it is possible to achieve with the particular control source and error sensor arrangement.

Previous work by the authors [1-3] resulted in the development of the required analytical models and numerical search procedures to optimize the physical arrangement of a specified

[1]Dept. of Mechanical Engineering, University of Adelaide, GPO Box 498, Adelaide, South Australia, 5001, Australia.

number of control sources and error sensors to achieve the optimum reduction in sound power radiated by a vibrating structure. Work reported elsewhere [4,5] has involved consideration of the design optimization of the electronic control system and the optimization of the number of error sensors for a given number of control sources.

Here, use is made of the previous work to examine several aspects of the physical control system design for a simply supported harmonically vibrating panel, mounted in a large baffle and radiating into free space. In particular, the effect of various parameters such as control source location, error sensor location, panel size, modal density, structural damping, excitation frequency, and panel response type (resonant or forced) are evaluated in terms of their effect on the maximum achievable noise reduction and the physical mechanisms [6] which are responsible. Also the effect of reduced performance of the electronic control system on the maximum overall achievable reduction in sound power is considered.

The work presented here is restricted to consideration of a single point vibration control source. The use of a single control source allows the effect of various parameters on the optimum achievable noise reduction to be evaluated without the added complication of multiple sources. However, the techniques used here are equally applicable to multiple control sources, to acoustic control sources and to combinations of acoustic and vibration control sources. Consideration of a single acoustic control source is the subject of another publication [7].

1. ANALYTICAL MODELS

The general analytical models to be used to obtain the numerical results described in the following sections have been developed previously [2,3] so only the final relevant equations will be summarized here.

The velocity response, $v(\vec{x}_i,\omega)$ of the panel at a vector location $\vec{x}_i$, to a point harmonic input force $F(x_F,\omega)$ of frequency ω and location $\vec{x}_F$ on the panel is given by:

$$v(\vec{x}_i,\omega) = F(\vec{x}_F,\omega) \sum_{r=1}^{\infty} \frac{\psi_{mn}(\vec{x}_i)\,\psi_{mn}(\vec{x}_F)}{M_{mn}(\omega_{mn}^2 + j\eta_{mn}\omega_{mn}^2 - \omega^2)} \tag{1}$$

where $\psi_{mn}(\vec{x}_i)$ is the mode shape function for the mnth mode evaluated at vector location $\vec{x}_i$ $(= x_i, y_i)$, ω_{mn} is the resonance frequency of the mnth mode and η_{mn} is the structural loss factor of the mnth mode.

For a simply supported panel the modal mass, M_{mn} for the mnth mode is given by:

$$M_{mn} = \iint_A \rho_s h\,\psi_{mn}^2\,dA = \rho_s hA/4 \tag{2}$$

where A is the panel area, ρ_s is the panel density and the mode shape function ψ_{mn} is defined by:

$$\psi_{mn} = \sin\left(\frac{m\pi x}{L_x}\right) \sin\left(\frac{n\pi y}{L_y}\right) \tag{3}$$

where m,n is the mode order, ((m-1) is the number of vertical nodal lines and (n-1) is the number of horizontal nodal lines on the panel), L_x, L_y are the panel dimensions in the x and y directions respectively (see Figure 1) and the origin of the coordinate system is at the lower left corner of the panel.

The sound pressure radiated by the mn panel mode to a point (r,θ,ϕ) in the far field (see Figure 1) is given by:

$$\vec{p}_{mn}(r) = \frac{jv_{mn}k\rho_o c}{2\pi r}\exp(-jkr)\frac{L_xL_y}{mn\pi^2}\left[\frac{(-1)^m\exp(-j\alpha)-1}{\left(\frac{\alpha}{m\pi}\right)^2-1}\right]\cdot\left[\frac{(-1)^n\exp(-j\beta)-1}{\left(\frac{\beta}{n\pi}\right)^2-1}\right] \tag{4}$$

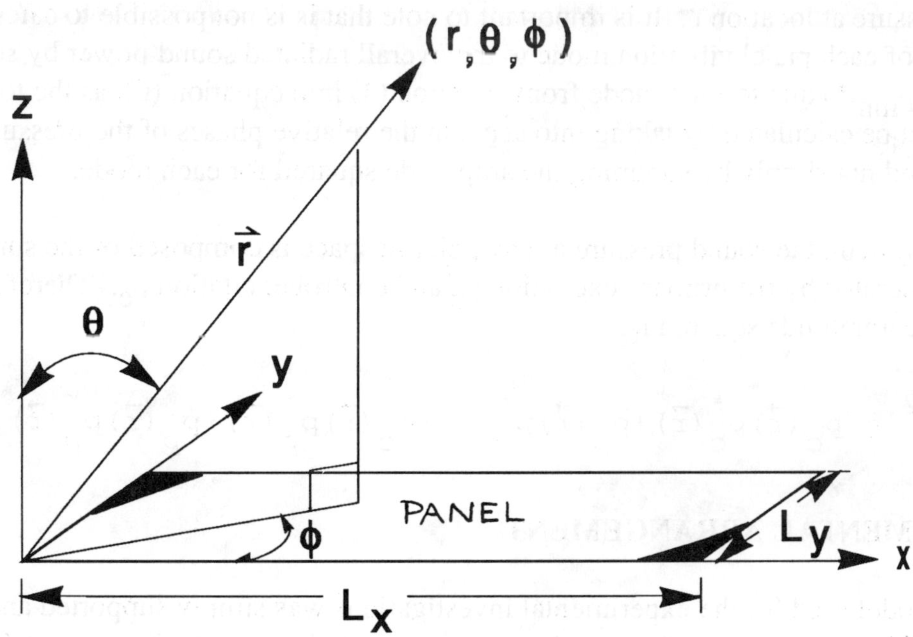

Figure 1. Panel arrangement.

where $\alpha = kL_x \sin\theta \cos\phi$, and
$\beta = kL_y \sin\theta \cos\phi$,

k is the acoustic wave number and ρ_o is the density of the fluid medium. Postive time dependence of the form, $e^{j\omega t}$ is assumed, and $\vec{r}$ is the vector location of the point (r,θ,ϕ) in space.

Note that p_{mn} will be an r.m.s. quantity if $F(\vec{x}_F,\omega)$ of equation (7) is an r.m.s. quantity.

The modal resonance frequencies ω_{mn} are given by:

$$\omega_{mn} = \left[\frac{Eh^2}{12\rho_s(1-\nu^2)}\right]^{1/2}\left[\left(\frac{m\pi}{L_x}\right)^2 + \left(\frac{n\pi}{L_y}\right)^2\right] \tag{5}$$

The pressure at any location in the far field may be found by using equation (4) and summing the contributions from each mode. Note that the pressure is represented as a complex number and must be treated a such in the summing process.

The total sound power radiated by the vibrating panel is given by:

$$W_{tot} = \int_0^{2\pi}\int_0^{\pi/2} \frac{|p(\vec{r})|^2}{2\rho_o c_o} |\vec{r}|^2 \sin\theta \, d\theta \, d\phi \tag{6}$$

where $\vec{r}$ is the vector location in space of the point (r,θ,ϕ) and $|p(\vec{r})|$ is the modulus of the complex pressure at location $\vec{r}$. It is important to note that is is not possible to calculate the contribution of each panel vibration mode to the overall radiated sound power by substituting the pressure $p_{mn}(\vec{r})$ due to each mode from equation (4) into equation (6), as the total pressure must be calculated by taking into account the relative phases of the pressures due to each mode and not simply by summing the amplitude squared for each mode.

For a linear system, the sound pressure at any point in space is composed of the sum of the pressures generated by the primary excitation p_p and control excitation p_c. Therefore, the total pressure amplitude squared is:

$$|p(\vec{r})|^2 = p_c(\vec{r})p_c^*(\vec{r}) + p_p(\vec{r})p_c^*(\vec{r}) + p_c(\vec{r})p_p^*(\vec{r}) + p_p(\vec{r})p_p^*(\vec{r}) \tag{7}$$

2. EXPERIMENTAL ARRANGEMENT

The panel model used for the experimental investigations was simply supported and mounted in a large baffle of dimensions (4.8m x 2.4m) in an anechoic room of dimensions (4.79m x 3.90m x 3.94m).

The panel, of dimensions 0.380 x 0.300 x 0.0018 m thick was mounted in a heavy steel frame using shim steel strips to achieve the simply supported boundary condition [8]. For the baseline case, the panel was excited by a primary point force at (0.063, 0.150)m (with respect to the left lower corner of the panel) which corresponded to an outer antinode of the (3,1) mode. The force was applied using a 0.006m diameter rod connected to a small electrodynamic shaker.

The modal velocity amplitudes on the panel were determined using data from twelve 0.5 gram and eight two gram accelerometers mounted at random locations over the panel, and the modal decomposition method outlined in an earlier paper [9] was used to determine the relative amplitudes and phases of the first twenty modes. However, only the results for the first seven or eight modes were actually used here, as the accuracy of the higher order modal amplitudes and phases is limited. To ensure that the modal decomposition results were accurate, the accelerometer responses were calibrated (in amplitude and phase) relative to a reference accelerometer at each test frequency by attaching each accelerometer along with the reference accelerometer to a 0.02 x 0.02 x 0.02m block of steel mounted on a Bruel and Kjaer mini-shaker. The differences in amplitude and phase response were included in the analysis of the data. It was found that significant errors resulted if the accelerometer calibrations determined at the standard 50 Hz frequency of a typical calibration instrument were used for all test frequencies.

All numerical calculations derived from the theoretical analysis refer to the same panel as used for the experimental work. Only a small number of test cases were examined experimentally, as it was much more efficient to use the theoretical model for many of the detailed investigations. Note that the same limitations on the number of modes which could be resolved experimentally did not apply to the numerical calculations, as these used a much greater number of data points.

3. DEMONSTRATION OF PHYSICAL CONTROL MECHANISMS

When a vibration control source is used to reduce the noise radiated by a structure, the physical control mechanism will be one of two well defined mechanisms or a combination of both [9]. One mechanism is characterized by a reduction in the amplitude of the vibration modes contributing significantly to the radiated sound pressure field. The second mechanism involves relative rearrangement of the modal phases in the time domain or the spatial domain or both. For the case of the simply supported panel, the phase rearrangement occurs in the time domain, as the spatial phases are fixed by the boundary constraints. This latter mechanism can result in a reduction in the radiated sound power by causing a reduction in the structure's overall velocity levels, and/or by reducing the sound radiation efficiency of the structure.

The first of the two mechanisms described above is most likely to occur when the system is radiating at a frequency near a structural resonance; the second is most likely to occur when the panel response is forced (that is, the excitation frequency is not close to any of the panel resonance frequencies). Each of these mechanisms is affected differently by various structural/acoustic parameters, as will be shown in the following sections.

To demonstrate these two mechanisms, both theoretical and experimental results will be presented for the baseline panel described in section 2. The panel was excited by a unit primary force at a frequency of 320 Hz which lies between the (2,2) and (3,1) modes which had resonances at 317 and 323 Hz respectively (note that the (2,2) mode was not excited significantly). The mass loading effect of the shakers and accelerometers was not taken into account in the analysis and resulted in slightly different resonance frequencies for the experimental panel. However, the purpose of the work described here is to demonstrate physical mechanisms and the effect of various parameters on these, and it is not essential that the theoretically predicted resonance frequencies exactly match the measured resonance frequencies.

The theoretical model was used to predict the maximum possible sound power attenuation that could be achieved (assuming an ideal electronic controller) as a function of the location of a single point control force on the panel (Figure 2). Note that the maximum possible sound power attenuation is based on the premise that the error sensor arrangement was capable of measuring the total radiated sound power, as opposed to the sound pressure at one or more specific locations. In practice, this is achieved to a close approximation by using sound pressure measurements at one or more carefully selected locations which will be a function of the control source location.

From Figure 2 it can be seen that there are three distinct areas for location of a single point control force which correspond to a local maximum in the attenuation of the radiated sound power, one area in the middle (corresponding to 22.4 dB attenuation) and two towards the vertical edges of the panel (near the antinodes of the (3,1) mode, locations (0.063, 0.150)m and (0.317, 0.150)m). The maximum attenuation corresponding to the right hand edge (away

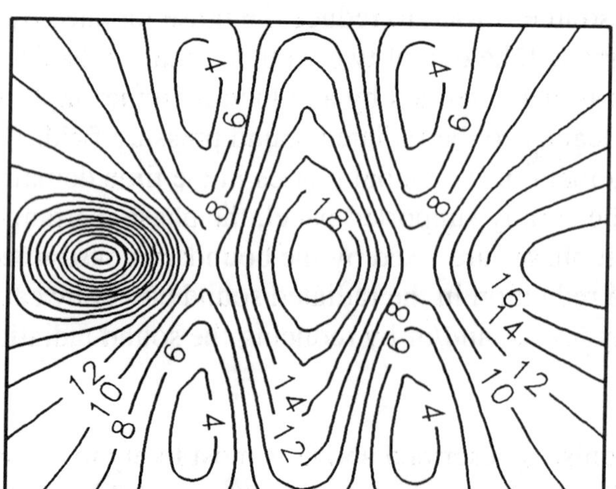

Figure 2. Maximum calculated achievable levels of sound power attenuation (dB) as a function of the vibration source location on the panel, 0.38 x 0.3 x 0.0018 m, with light damping ($\eta = 0.04$) and harmonic excitation at 320 Hz.

from the primary excitation source) is 18.6 dB. The lack of a single location corresponding to a local attenuation maximum has implications for the choice of gradient search routine used for determining optimum control source placement. If a gradient based method is used, the starting point will determine which location corresponding to a local attenuation maximum is selected, even though there may be only one clear optimum location. Clearly, the best location is for the control source to be mounted at the same location as the driving force. This, however, is a trivial case, and one would expect to obtain the abnormally high sound attenuations shown in the figure because the control source is simply counteracting the primary driving force, and the panel dynamics assume a secondary role.

The effect of controlling the panel radiation with a single point force at the location (0.317, 0.150)m, upon the relative amplitudes and phases (referenced to the left edge of the panel) of the first five non zero panel modes is shown in Figure 3(a), for the case of sound pressure minimization at an azimuthal error microphone location of 90°. Here it can be seen that the amplitudes of the nearly-resonant (3,1) mode (resonant at 323 Hz) and the (1,1) mode (resonant at 84 Hz) are substantially reduced. These two modes, based on their velocity levels and radiation efficiencies at this frequency, have the greatest potential for sound power generation. Their reduction will cause a significant reduction in the total radiated sound power, which is the primary control mechanism producing the 18.5 dB sound power attenuation achieved here. Note that this is a little less than the theoretical optimum of 18.6 dB, either because the error microphone is not exactly in its optimal location or because the minimization of the sound pressure at a single (optimum) error microphone location can only only approximate sound power level minimization, or it may be a combination of both reasons.

The modal data measured experimentally for the same case as just discussed are shown in Figure 3(b). As can be seen by comparison of Figures 3(a) and 3(b), the trends shown by theory and experiment are similar. Both the (1,1) and (3,1) modal amplitudes are reduced because the location of the control force is symmetrical with the primary exciting force in terms of both of the (3,1) and (1,1) mode shapes. Thus if the control force is equal and opposite to the exciting force, one would expect the amplitudes of both modes to be reduced. If the frequency were such that a non symmetrical mode (such as the (2,1) mode) were contributing significantly to the radiated sound field, then it would not be possible to reduce the amplitudes of all three of the modes simultaneously with the same control force and two control forces would probably be necessary.

The effect which using a vibration source at the center of the panel, location (0.190, 0.150)m, and an error microphone at an azimuthal location of 90°, to control sound power radiation has upon the first five non-zero amplitude panel modes is shown in Figure 3(c). Here the nearly resonant (3,1) mode, has a reduction in amplitude of approximately 10 dB. This would appear to be offset, however, by an increase in the amplitude of the (1,1) and (3,1) modes of approximately 10 dB. Just in viewing these results, it would seem unlikely that the 21.9 dB reduction in radiated sound power would be achieved only by a reduction in the modal vibration amplitudes of the radiating modes. However, it can be seen in Figure 3(c) that the relative amplitudes and phases of the modes (and thus the overall radiation efficiency of the panel) have changed significantly and this is responsible for the extent of the sound

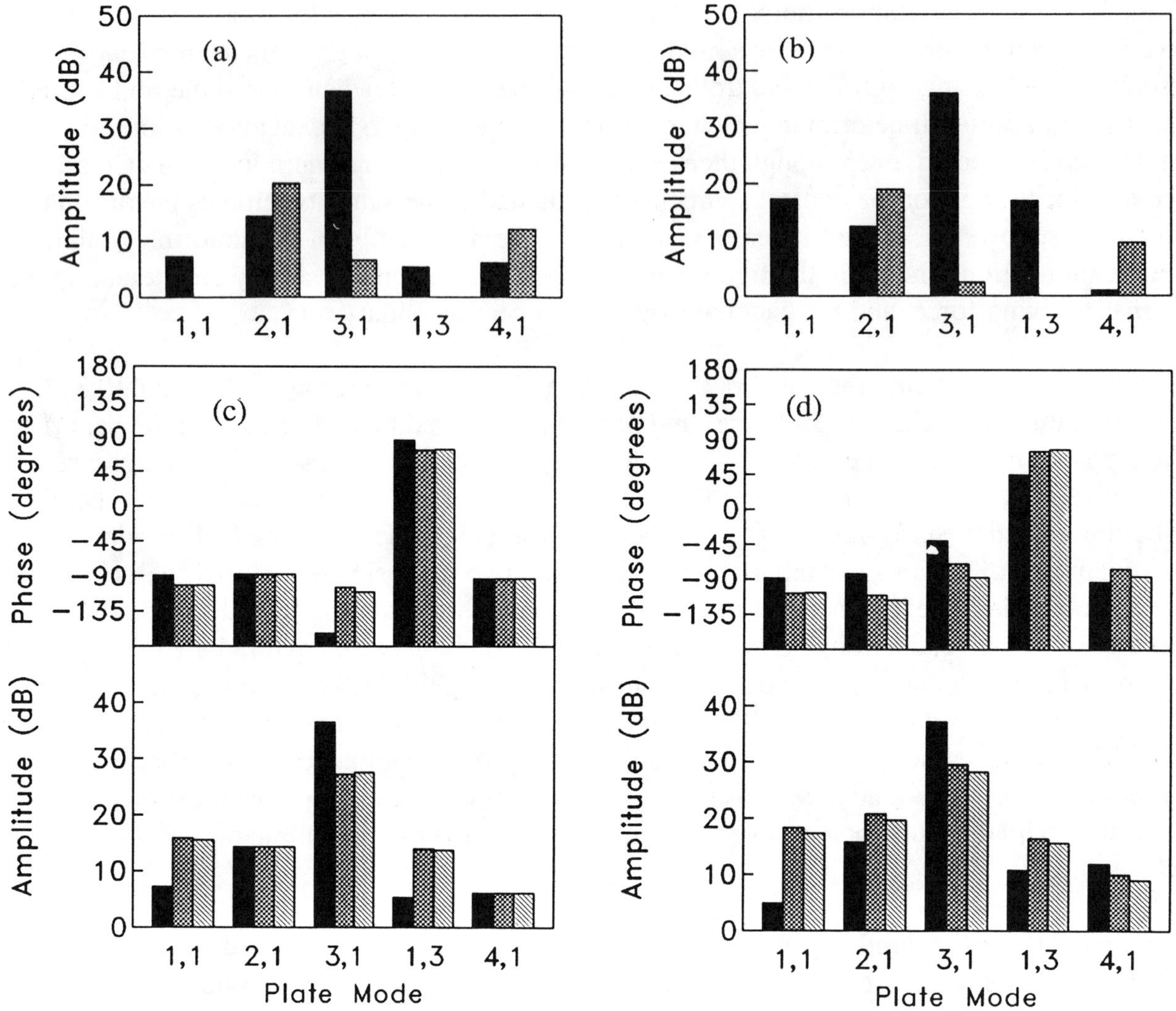

Figure 3. Comparison of calculated modal velocity levels on the lightly damped ($\eta = 0.06$) panel, 0.38 x 0.3 x 0.0018 m, harmonically excited at 320 Hz, before and after control.
(a) Control at (0.317, 0.150)m, calculated data.
(b) Control at (0.317, 0.150)m, measured data.
(c) Control at (0.190, 0.150)m, calculated data.
(d) Control at (0.190, 0.150)m, measured data.
■ = primary only
▒ = primary + control.

power reduction achieved. The experimentally determined modal data, plotted in Figure 3(d) supports this conclusion. Note that this is a little less than the theoretical optimum of 22.4 dB, either because the error microphone is not exactly in its optimal location or because the minimization of the sound pressure at a single (optimum) error microphone location can only only approximate sound power level minimization, or it may be a combination of both reasons.

With the control force in the panel centre, it will be symmetrical with the primary excitation force for the (1,1) mode but antisymmetrical for the (3,1) mode. Thus it will not be possible to control the amplitudes of both modes simultaneously. However, it is possible to change the relative phases of the two modes because the (3,1) modal transfer function will be similar in magnitude for the primary force and control force, but the (1,1) modal transfer function will be greater for the control force than for the primary force (due to the differences in value of the mode shape functions at the shaker attachment points). This also explains why the amplitude of the (1,1) mode increases after control is achieved and why the relative phase between the (3,1) and (1,1) modes (referenced to the center of the panel) is 180° after control. Note that the phase shown in the figures is referenced to the left edge of the panel. Thus the physical control mechanism demonstrated here is modal phase rearrangement.

For the case of excitation at the resonance frequency of a single mode which dominates the radiated sound field, one would expect control to be achieved by the modal amplitude suppression mechanism only, as only one mode is involved. This was indeed shown to be true, but an example is not shown here due to lack of space.

As mentioned previously the rearrangement of the relative amplitudes and phases of the modes can have two possible effects; the total panel velocity levels can decrease, and/or the radiation efficiency of the panel can be reduced. A comparison of the mean square panel velocity distribution, before and after control at (0.190, 0.150)m, shown in Figures 4(a) and 4(b), depicts an overall reduction in amplitude of approximately 4 dB. This is because the reduction in amplitude of the (3,1) mode is greater than the increase in amplitude of the (1,1) mode. This, however, is not enough to account for the 21.9 dB of radiated sound power attenuation. However, examination of the relative phase of the velocity distribution over the panel (not illustrated here) reveals that the vibration pattern has become more complex with a corresponding smaller radiation efficiency which is primarily responsible for the reduction in radiated sound power.

(a) (b)

Figure 4. Calculated mean square velocity levels (dB) on the (0.38 x 0.3 x 0.0018) panel surface for harmonic excitation at 320 Hz and light damping ($\eta = 0.06$). (a) Primary only, (b) Primary + control.

From Figure 2, it is clear that the choice of control source location will determine the maximum sound power attenuation possible under ideal conditions. The placement of the error sensor (providing a feedback measure of the residual sound field) will determine how close to this value the achievable level of attenuation is. The influence of the angular location of a single microphone error sensor located at a radius of 1.8 meters (the sound pressure is minimized at this location), upon the sound power attenuation possible when the control source is located at (0.317, 0.150)m is shown in Figure 5(a), plotted with both the primary radiated sound pressure field and the residual controlled sound pressure field under "ideal" control (maximum achievable control of total radiated sound power for the specified control source arrangement, which is 18.6 dB in this case). Note that this radius will be used for all radiation plots throughout this paper unless otherwise stated. As can be seen, the location of the error microphone influences the maximum achievable level of sound power reduction. The optimum angular location is at the point of greatest difference between the primary and the optimally controlled residual sound pressure field. In this case, the residual sound field is dominated by the (2,1) panel mode, so the optimum error microphone angular location is nearly normal to the panel center (azimuthal location of 90°). Minimization of the sound pressure at this location results in close to the maximum possible sound power attenuation of 18.6 dB (for this control source location) being achieved.

A similar plot for the control source located at (0.190, 0.150)m is given in Figure 5(b). In

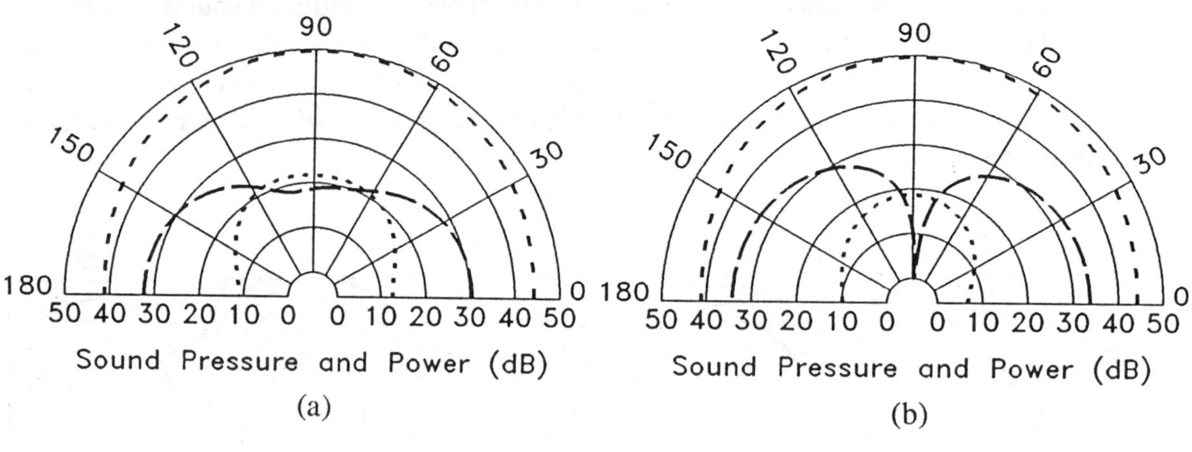

Figure 5. Calculated sound pressure levels (dB) before control and after "ideal" control as a function of azimuthal angle in a horizontal plane normal to the mid point of the panel surface; and the calculated maximum achievable reductions (dB) in radiated sound power as a function of angular location of a single error microphone in the same plane. The panel size is 0.38 x 0.3 x 0.0018, lightly damped ($\eta = 0.04$), and it is harmonically excited at 320 Hz. (a) Vibration control source located at (0.317, 0.150)m, (b) Vibration control source located at (0.190, 0.150)m.

- - - - = primary radiated,

—— = calculated residual radiated after control,

······ = sound power reduction as a function of azimuthal location of a single error microphone.

this case the optimum angular location for the microphone is 105°, corresponding to a sound power attenuation close to the maximum possible of 22.4 dB.

Calculated results for the radiated sound field at 320 Hz, both before and after control, are shown in Figure 6(a) for the control source at (0.317, 0.150)m and the error microphone (at which the sound pressure is minimized) at 90°. Measured data for the same case are shown in Figure 6(b). The trends shown by the theory are matched by the experimental data. Note that for this case, substantial global sound pressure level control is achieved with a resulting

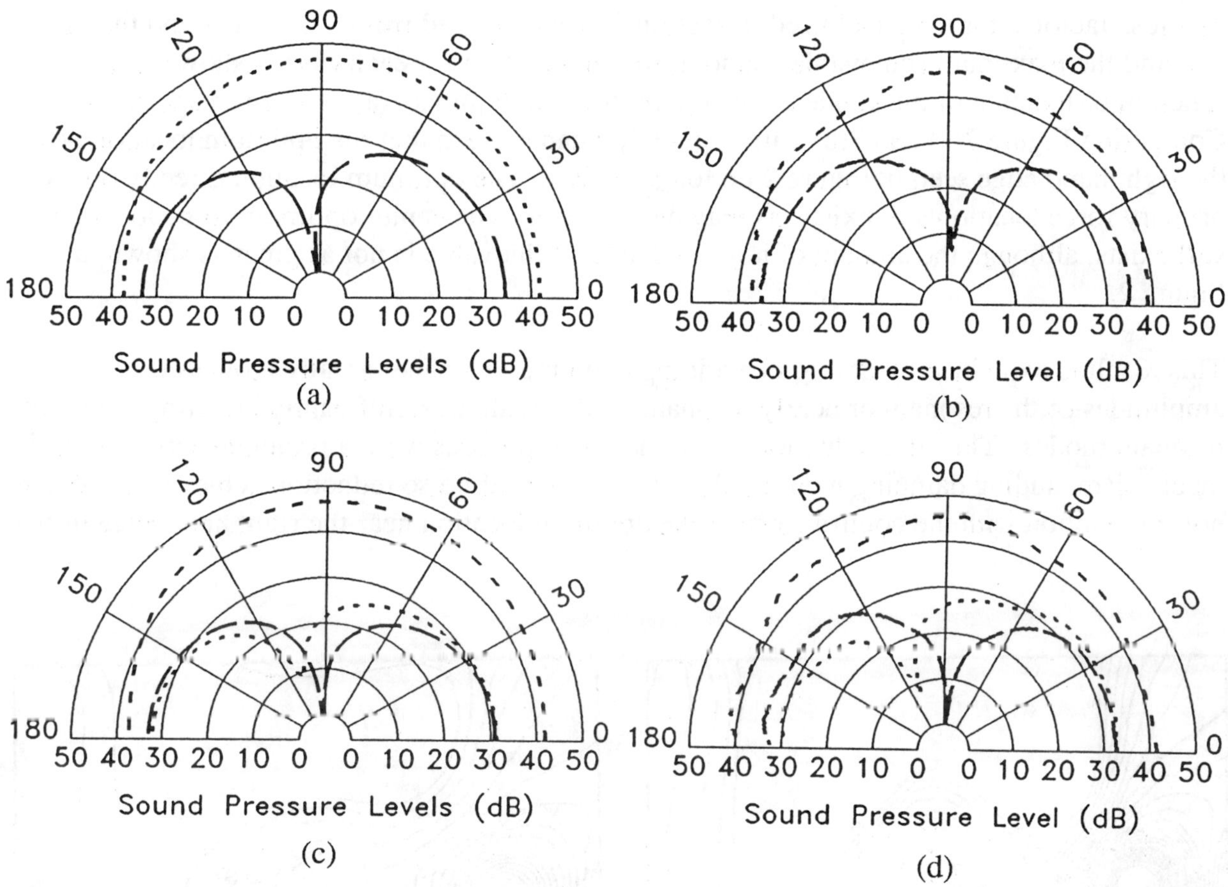

Figure 6. Calculated sound pressure levels (dB) before control and after maximum control with an error microphone at an azimuthal location of 90° as a function of azimuthal angle in a horizontal plane normal to the mid point of the panel surface. The panel size is 0.38 x 0.3 x 0.0018, lightly damped ($\eta = 0.06$), and it is harmonically excited at 320 Hz.

(a) Vibration control source located at (0.317, 0.150)m, calculated data.
(b) Vibration control source located at (0.317, 0.150)m, measured data.
(c) Vibration control source located at (0.190, 0.150)m, calculated data.
(d) Vibration control source located at (0.190, 0.150)m, measured data.

- - - - = primary radiated,
—— = calculated residual radiated after control.
· · · · · = reduced control at error microphone

15.6 dB reduction in total radiated sound power. This is slightly less than the attenuation of 18.5 dB shown in Figure 5(a), as the loss factor in this case is 0.06 rather than 0.04.

Similar data to those shown in Figures 6(a) and 6(b) are shown in Figures 6(c) and 6(d), except that in the latter two figures, the control source is located at (0.190, 0.150)m, the center of the panel. Similar sound pressure and sound power attenuation levels (18.8 dB as opposed to 15.6 dB) are obtained; however, as explained earlier, the physical mechanism responsible has changed from modal phase rearrangement to modal amplitude reduction. The third curve in Figures 6(c) and 6(d) will be discussed later.

4. EFFECT OF STRUCTURAL DAMPING

The loss factor η for the panel used in section 3 was increased from 0.04 to 0.2 and then to 0.4, and the maximum achievable sound power attenuation for each case is shown as a function of location of the single control point force in Figures 7(a) and 7(b) respectively. Comparing Figure 7(a) and 7(b) with Figure 2, it can be seen that the optimum location near the right hand edge seen in Figure 2 no longer exists. The optimum location directly on the primary force location still exists, as may be expected. The center optimum force location still exists, although the amount of noise reduction achievable is not as great as shown in Figure 2.

This result is explained by noting that adding damping to a panel greatly reduces the amplitudes of the resonant or nearly resonant modes without significantly affecting the non resonant modes. Thus if a reduction in the modal amplitudes were a principle active control mechanism, adding damping would reduce the amount of noise reduction achievable. As the noise reduction with the control force at the optimum location near the right hand edge of the

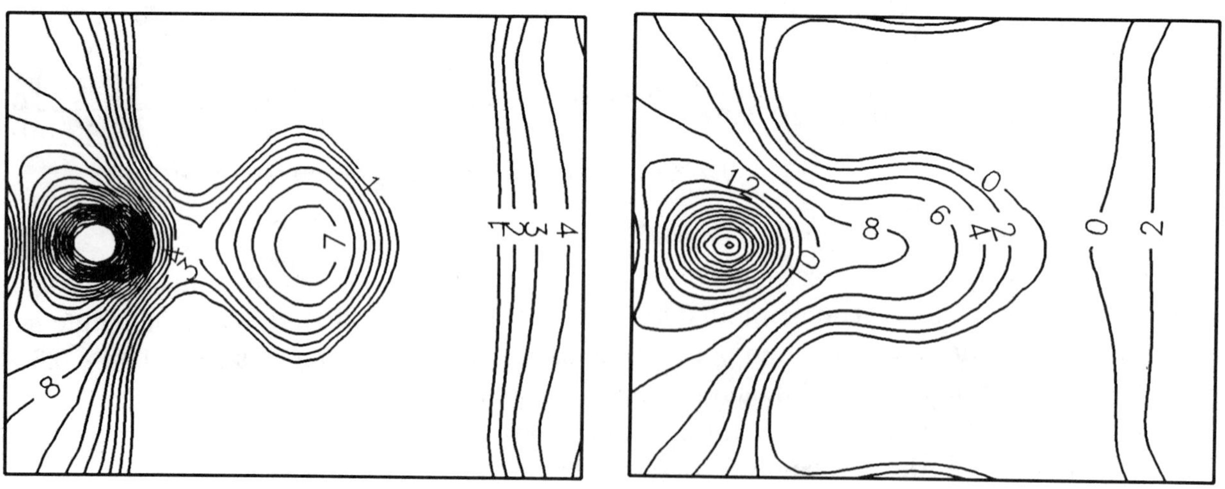

(a) (b)

Figure 7. Calculated maximum achievable levels of sound power attenuation (dB) as a function of the vibration source location on the panel, 0.38 x 0.3 x 0.0018 m, with heavy damping and harmonic excitation at 320 Hz. (a) $\eta = 0.2$ (b) $\eta = 0.4$.

panel relied upon the modal amplitude reduction mechanism, then one might expect this optimum location to essentially disappear with the addition of a large amount of damping and this is indeed what has happened.

5. INFLUENCE OF MODAL DENSITY

The previously considered panels have a relatively low modal density. To determine whether or not a single vibration source could produce significant levels of sound power attenuation with an increased modal density, the thickness of the original (lightly damped) panel was reduced from 0.0018 m to 0.0005 m, with the edge dimensions kept constant. Results showing maximum possible sound power attenuation levels as a function of control source location on the 0.0005 m thick panel are given in Figure 8. From this figure, it can be seen that there are areas on the panel where the application of a single vibration control source can produce significant levels of sound power attenuation, even with the increase in modal density. There is one location in the center of the panel and a further one on each of the top and bottom edges.

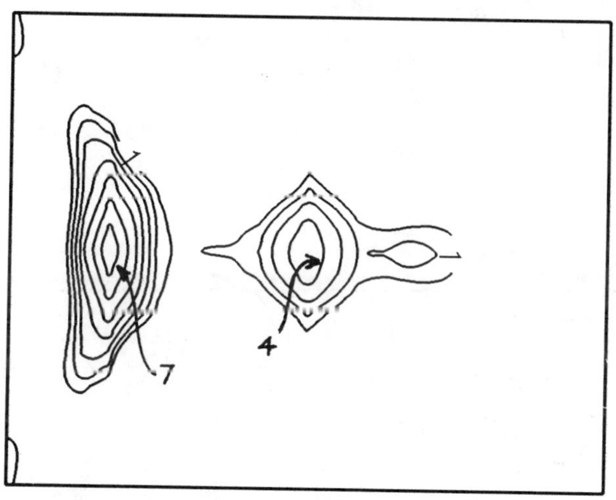

Figure 8. Calculated maximum achievable levels of sound power attenuation (dB) as a function of the vibration source location on the panel, of dimensions 0.38 x 0.3 x 0.0005 m, with light damping ($\eta = 0.04$) and harmonic excitation at 320 Hz.

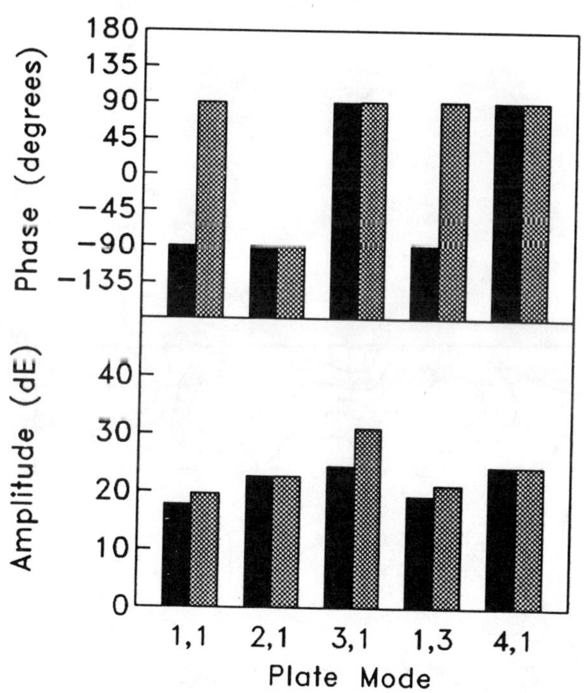

Figure 9. Comparison of calculated modal velocity levels on the lightly damped ($\eta = 0.04$) panel, 0.38 x 0.3 x 0.0005 m, harmonically excited at 320 Hz, before and after control,

■ = primary only

▒ = primary + control at (0.190, 0.150)m.

The calculated modal velocity levels corresponding to a control force in the center of the panel are shown in Figure 9. There is an increase in the amplitude, and a change in phase, for many of these modes (note especially the efficient radiating (1,1), (3,1), and (1,3) modes). The associated overall panel velocity distribution before and after control is shown in Figure 10(a) and 10(b). Comparing the before and after panel velocity levels shown in Figures 10(a) and 10(b), it can be seen that there has been a general slight increase in velocity level under the action of control. Thus the control mechanism is a reduction in radiation efficiency resulting from the increase in complexity of the velocity phase distribution (not shown) over the panel.

6. EFFECT OF PANEL SIZE

As the modal phase rearrangement mechanism achieves sound power reduction by changing how the structure interacts with the (nearfield) acoustic space, it would seem reasonable to suggest that the size of the structure relative to the acoustic wavelength at the frequency of interest would have an influence upon the effectiveness of this mechanism. To examine this, the base case panel was extended proportionally in size (to keep the same mode order). The panel thickness was increased, and the material density decreased, to maintain the same modal density and modal overlap. The relationship between the the panel area, thickness, and material density required to do this is:

$$\frac{\rho_{s1}}{\rho_{s2}} = \left(\frac{h_2}{h_1}\right)^4 = \left(\frac{A_2}{A_1}\right)^{4/3} \tag{8}$$

(a) (b)

Figure 10. Calculated mean square velocity levels (dB) on the (0.38 x 0.3 x 0.001m) panel surface for harmonic excitation at 320 Hz and light damping ($\eta = 0.04$). (a) Primary excitation only, (b) Primary + control excitation at (0.190, 0.150)m.

The final dimensions considered here are 1.258m x 0.944m x 0.004m, resulting in a panel size of approximately one wavelength across at the frequency of interest (320 Hz), in contrast to almost one third for the original panel. The panel density in this case was 320 kg/m^3, compared to the base panel density of 7800 kg/m^3. It should be noted that these alterations have the effect of reducing the panel critical frequency, thus increasing all of the modal radiation efficiencies. However, the frequency at which both the base case panel and this extended panel is studied is well below the critical frequency, and the increase in modal radiation efficiencies is not important. The primary point force was relocated (proportionally) to (0.210, 0.472)m, the center of the panel. The maximum levels of sound power attenuation plotted as a function of control source position are given in Figure 11 for this panel.

In comparing Figure 11 to the base case Figure 2, it is apparent that the pattern of optimal attenuation contours is similar, but that the levels of attenuation are significantly reduced. For the single control source in the centre of the panel, the maximum achievable reduction in sound power level has been reduced from 22.4 dB to 16.9 dB. The effect upon the modal amplitudes and phases of minimizing the radiated sound power by applying a control force at the panel center is shown in Figure 12. It is clear that the amplitudes of the (3,1) mode has

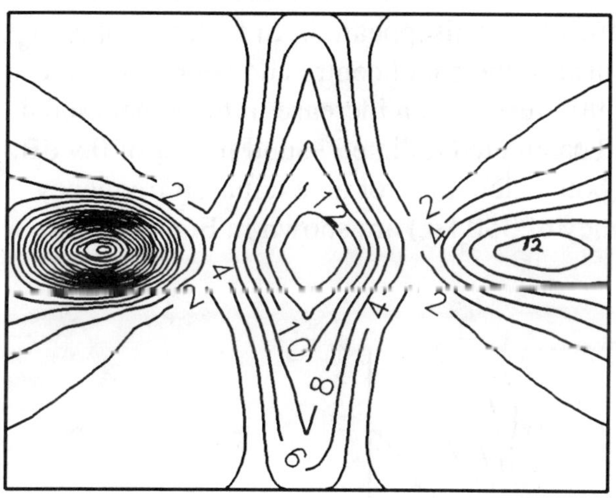

Figure 11. Calculated maximum achievable levels of sound power attenuation (dB) as a function of the vibration source location on the panel, 1.075 x 0.8485 x 0.004 m, with light damping ($\eta = 0.04$) and harmonic excitation at 320 Hz.

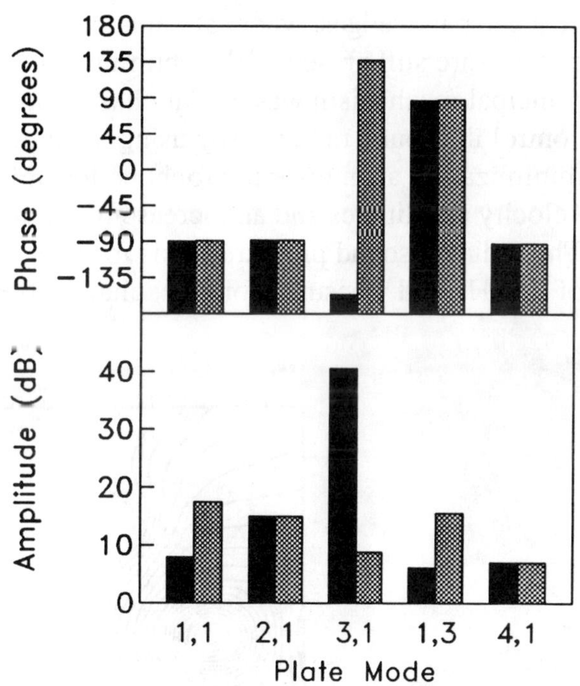

Figure 12. Comparison of calculated modal velocity levels on the lightly damped ($\eta = 0.04$) panel, 1.075 x 0.8585 x 0.004 m, harmonically excited at 320 Hz, before and after control at (0.190, 0.150)m.

■ = primary only

▒ = primary + control.

been reduced by an amount much greater than the increase in the (1,1) modal amplitude and the relative phases of the (1,1) and (3,1) modes have only changed slightly. This indicates that the principal mechanism of control has now changed from modal phase rearrangement to modal amplitude control, due to the increase in the size of the structure.

7. EFFECT OF FREQUENCY ON THE MODAL REARRANGEMENT MECHANISM

It was shown in the base case examination that modal rearrangement was the principal mechanism by which sound power attenuation was achieved when a vibration control source was placed at the panel center. This was because the two principal radiating modes, the (3,1) and (1,1), were out of phase at this point. At 320 Hz (the frequency examined in the base case) the (3,1) mode had not yet passed through resonance. When this does happen, the (3,1) and (1,1) modes will no longer be out of phase, but rather in phase. This should have a significant effect upon the sound power attenuation achieved by applying vibration control at the panel center.

The effect of increasing the frequency to 365 Hz (above the (3,1) resonance of approximately 323 Hz) can be deduced by viewing Figure 13 which shows the maximum sound power attenuation as a function of control source location. The optimum control source locations near the panel edges, where attenuation was achieved mainly through modal amplitude control, are still present. The optimum location in the panel center, however, where the principal mechanism was modal phase rearrangement, has disappeared. In fact, attempts to control the sound radiation by using a control source at the panel centre and sound pressure minimization at an error microphone location of 90° result in an increase in the panel modal velocity amplitudes and an increase in the overall calculated radiated sound power of 0.9 dB. The radiated sound pressure fields for this case (both calculated for an excitation frequency of 365 Hz and measured for an excitation frequency of 350 Hz) are shown in Figures 14(a)

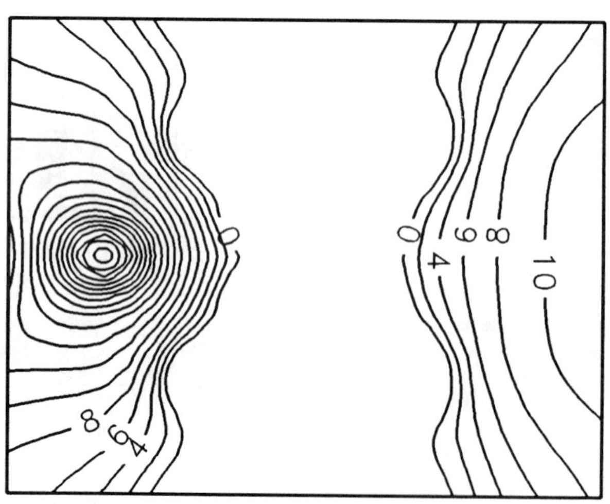

Figure 13. Calculated maximum achievable levels of sound power attenuation (dB) as a function of the vibration source location on the panel, 0.38 x 0.3 x 0.0018 m, with light damping (η = 0.04) and harmonic excitation at 365 Hz.

and 14(b), which illustrate the general increase in sound power level for the controlled case.

8. EFFECT OF REDUCED SOUND PRESSURE ATTENUATION AT THE ERROR MICROPHONE

Thus far it has been assumed that it is possible to completely minimize the sound pressure at the error microphone position. In theory, this means perfect cancellation. In practice, however, such levels of reduction are not achievable by the electronic control systems. For periodic sound attenuation, reductions at a point of the order of 25 dB are more reasonable. The effect which this will have upon the total radiated sound power attenuation needs to be considered.

Figure 6(c) depicts the calculated primary radiated sound field for the base case 0.0018 m panel with the control source located at the panel centre (0.190, 0.150)m, and two calculated residual sound fields produced by minimizing the sound pressure at an azimuth angle of 90° at a radius of 1.8 meters. One residual sound field is produced by perfect cancellation at the error microphone location, the other by only a 25 dB sound pressure level reduction at this point. The two residual sound field patterns are similar in shape, with the 25 dB case slightly greater in the front lobe, and slightly less on the sides. In fact, the 25 dB case has produced a sound power attenuation level of 18.4 dB, while the perfect cancellation case has achieved 18.8 dB of sound power control. Thus, the effect of reduced sound pressure level control at

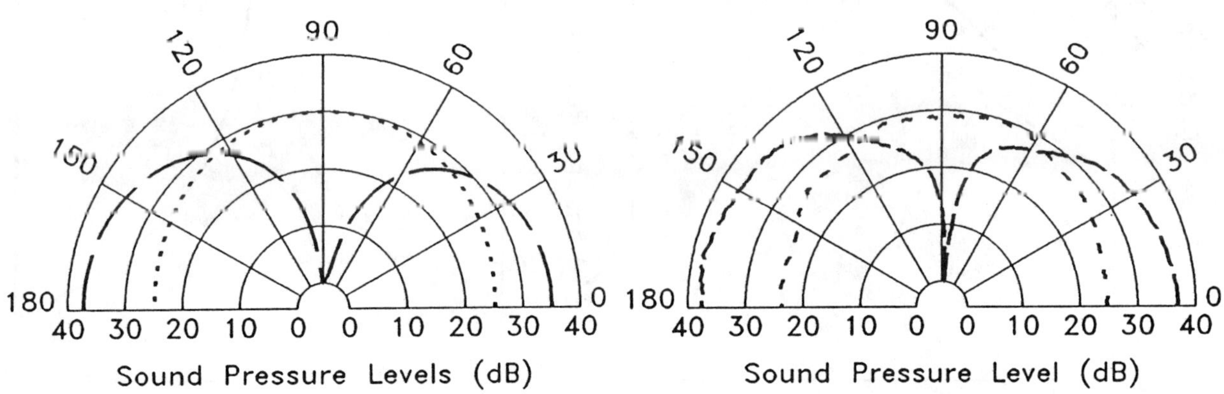

(a)

(b)

Figure 14. Sound pressure levels (dB) before control and after maximum control with an error microphone at an azimuthal location of 90°, as a function of azimuthal angle in a horizontal plane normal to the mid point of the panel surface. The panel size is 0.38m x 0.3m x 0.0018m, lightly damped ($\eta = 0.06$), harmonically excited at 365 Hz, and the single vibration control source is located at (0.190, 0.150)m. (a) Calculated data, (b) Measured data.

- - - - = primary radiated,

— — = residual radiated after control.

the error microphone location is minimal, as the dramatic pressure reductions are due only to local cancellation. Note also that the minimum in the controlled sound pressure field for the 25 dB reduction case no longer corresponds to the error microphone location of 90°.

Figure 6(d) shows the measured data corresponding to the calculated data in Figure 6(c). It can be seen that the measured data supports the general trends of the calculated data.

9. TOLERANCE TO PHASE ERROR

When the noise source is periodic in nature, it may be advantageous to implement the electronic control system in the frequency domain. Unless the sampling rate of the controller is synchronized with the (acoustic) frequency of interest, there will be phase errors associated with the bandwidth of the bins of the Fourier transform. If a frequency domain scheme is to be viable, the effect of phase errors on the sound power reduction must be known.

The effect which holding the optimum amplitude, but deviating the phase, of the control source vibration signal has upon the maximum levels of sound power attenuation is shown in Figure 15, and upon the levels of sound pressure reduction at the error microphone in Figure

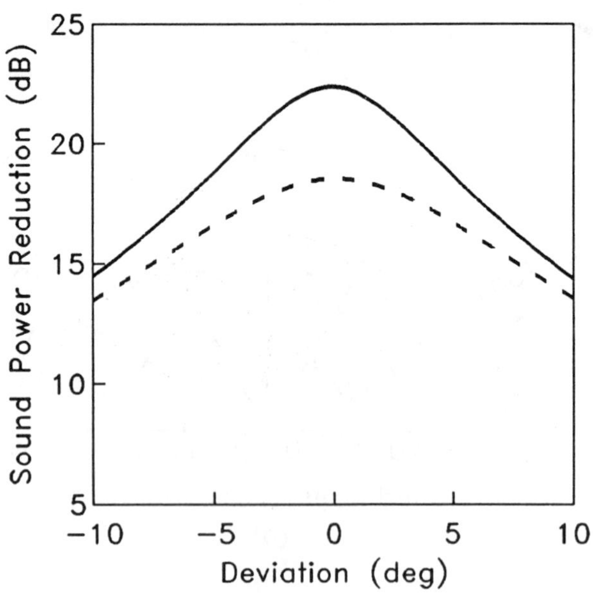

Figure 15. Maximum achievable sound power reduction (dB) as a function of phase error in the signal used to drive the control source for a lightly damped panel (0.38 x 0.3 x 0.0018m) harmonically excited at 320 Hz.
= control at (0.190, 0.150)m,
= control at (0.317, 0.150)m.

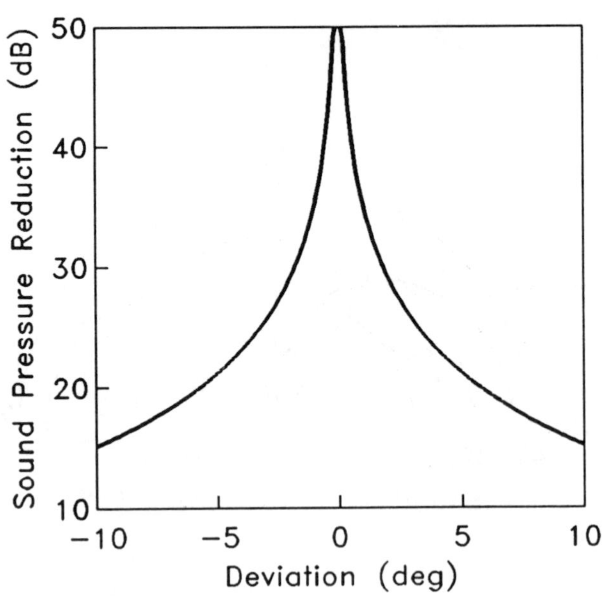

Figure 16. Maximum achievable sound pressure reduction (dB) at the error microphone as a function of phase error in the signal used to drive the control source for a lightly damped panel (0.38 x 0.3 x 0.0018) harmonically excited at 320 Hz, with control at either (0.190, 0.150)m or at (0.317, 0.150)m.

16. In comparing these, it is obvious that the pressure reduction at the error microphone is much more sensitive to phase errors than the overall reduction in radiated sound power. This further indicates that the dramatic levels of sound pressure reduction possible at the error microphone are due to local cancellation. It would also appear from Figure 15 that the modal rearrangement mechanism, which is dominant when the control source is placed at (0.190, 0.150)m, is more sensitive to phase errors than the modal control mechanism, which is dominant when the control source is placed at (0.317, 0.150)m.

10. CONCLUSIONS

The physical mechanisms associated with the use of an actively controlled vibration source to reduce the sound power radiated by a harmonically vibrating panel have been demonstrated both theoretically and experimentally. Control is provided in two different ways; by reduction of the vibration amplitudes of the modes which control the sound radiation and by rearrangement of the relative phases and amplitudes of the contributing modes. The second mechanism can also be associated with an overall increase in the panel velocity levels. With this mechanism noise reduction is achieved by an effective change in the radiation efficiency of the panel.

It has also been observed that the optimum location of the control force is affected by various structural and geometric variables such as the panel damping, the size of the panel relative to the acoustic wavelength of interest, the frequency of excitation, the panel modal density, the location of the control source and the location of the error sensor. These parameters also affect the maximum noise reduction achievable with an ideal electronic controller, as well as the physical mechanisms by which the noise reduction is achieved.

The results indicate that for the control of total radiated sound power, the phase accuracy requirement for the signal generated by the electronic controller is not nearly as critical as it is for minimizing the sound pressure at a particular location. This has important consequences for the detailed design of the digital electronic controller, in particular the number of bits necessary to obtain good results.

ACKNOWLEDGEMENTS

The authors would like to acknowledge financial support from the Sir Ross and Sir Keith Smith Fund and The Australian Research Council.

REFERENCES

1. Snyder, S.D. and Hansen, C.H., "Using multiple regression to optimise active noise control system design". Accepted for publication in the Journal of Sound and Vibration (1990).

2. Snyder, S.D. and Hansen, C.H., "A general approach to the design of active systems to control sound radiated by vibrating structures, Part 1: Analytical models". Submitted for publication in the J. Acoust. Soc. of Am. (1990).

3. Pan, J., Snyder, S.D., Hansen, C.H. and Fuller C.R., "Active control of farfield sound radiated by a rectangular panel: A general analysis". Submitted for publication in the J. Acoust. Soc. of Am. (1990),

4. Snyder, S.D. and Hansen, C.H., "Design requirement for active noise control systems implementing the multiple input, multiple output LMS algorithm". Submitted for publication in the J. Sound Vib., (1990).

5. Snyder, S.D., Clark, R.L. and Hansen, C.H., "Convergence characteristics of the multiple input, multiple output LMS algorithm", submitted for publication in the J. Intell. Mat. Sys. Struct. (1990).

6. Hansen, C.H., Snyder, S.D. and Fuller, C.R., "Reduction of noise radiated by a vibrating rectangular panel by use of active sound sources and active vibration sources: a comparison.", Proceedings of Noise and Vibration '89, Singapore, August 1989.

7. Snyder, S.D. and Hansen, C.H., "A general approach to the design of active systems to control sound radiated by vibrating structures, Part 2: Radiation into free space". Submitted for publication in the J. Acoust. Soc. of Am. (1990).

8. Ochs, J. and Snowden, J., "Transmissibility across simply supported thin plates, 1. Rectangular and square plates with and without damping layers". J. Acoust. Soc. of Am. 58, 832-840 (1975).

9. Snyder, S.D. and Hansen, C.H., "Mechanisms of active noise control using vibration sources". J. Sound Vib. Accepted for publication (1990).

ACTIVE STRUCTURAL ACOUSTIC CONTROL WITH ADAPTIVE STRUCTURES INCLUDING WAVENUMBER CONSIDERATIONS

Robert L. Clark,[1] C. R. Fuller[1]

ABSTRACT

The structural acoustic control of a baffled simply supported plate was modeled analytically to determine the feasibility of using the model to design shaped sensors. Results from the analytical study were compared to those obtained in experiments previously conducted which implemented either microphones or polyvinylidene fluoride (PVDF) strips as error sensors. Since the experimental data was subjected to a 30 dB dynamic range (due to A/D of controller as well as noise floor of instrumentation), the theoretical control approach was modified such that the analytical model would be limited to the same dynamic range. To evaluate the success of the structural acoustic control, radiated sound pressure was measured or computed in an arc perpendicular to the plane of the plate. Analytical predictions for sound radiation before and after control correlated well with the data obtained in the experimental investigation. In an effort to explain the physical mechanism of control, the k-transform of the structural response was computed. Results from this analysis indicate that structural acoustic control is achieved when the supersonic wavenumber components are reduced. Based on this conclusion, the possibility of implementing a cost function in the k-domain was investigated, and results suggest that sensors shaped for observing only the supersonic wavenumber components of the 2-D k-transform would yield a minimization of the acoustic response of the structure.

INTRODUCTION

Interest in development of adaptive structures implementing piezoelectric actuators and polyvinylidene fluoride (PVDF) sensors in structural acoustic control approaches has prompted many theoretical and experimental investigations [1-5]. Much has been documented concerning the modeling and application of piezoelectric actuators for control of structural vibration as well as sound radiation from structures [6-8]. While many studies have investigated the possibilities of using PVDF as sensors and actuators for control of structural vibration [9-14], little has been done to implement dedicated sensors for structural acoustic approaches. In a recent study by Clark and Fuller [15-16], PVDF strip sensors were mounted to the surface of a simply supported plate in an effort to observe the more efficient acoustic radiators of the structure. By observing the more efficient acoustic radiators, (or averaging out the high subsonic structural wavenumber components) control effort is dedicated to reducing the structural response contributing to the radiated sound, allowing spillover into the less efficient radiating modes.

[1]Department of Mechanical Engineering, Virginia Polytechnic Institute and State University, Blacksburg, VA 24061, U.S.A.

By designing shaped sensors for structures with the acoustic field in mind, much of the post processing of the sensor output can be eliminated from the design of the controller. This process increases the complexity of the chosen sensor design; however, a reduction in the complexity of the control system is achieved. After determining optimal sensor shapes and actuator locations for a given structural acoustic control application, an LMS algorithm can be implemented to achieve control. By this technique, a gradient decent algorithm is implemented to "seek" the optimal control voltages for each piezoelectric actuator given the cost function formulated in terms of the response of the shaped sensors. By placing sensors on the surface of the structure, microphones located in the radiated field can be eliminated from the control approach, resulting in an adaptive structure.

In designing the shaped sensors, the physical mechanisms of sound radiation must be investigated and understood. The most graphical method for describing sound radiation from structures involves obtaining the k-transform of the structural response. This method of analysis will be reviewed in this paper as it pertains to the understanding of structural acoustic control. In addition, the radiation efficiency of structural modes will be discussed, providing greater insight into the importance of considering the response in the acoustic field as opposed to the response of the structure when choosing sensors or designing controllers.

The following discussion and results will be based on analytical models of a simply supported plate residing in an infinite baffle. Previous studies characterizing the nature of structural response when utilizing piezoelectric actuators justify the use of the analytical models for studying the physical mechanisms of control [14]. The filtered-x version of the adaptive LMS algorithm was modeled such that experiments previously conducted could be simulated. Since the converged value of the control voltages from this approach are identical to those obtained from optimal control theory, the optimal control approach was chosen as the most efficient method of obtaining results for the purpose of this study. Analytical studies often result in unrealizable results since infinite attenuation of an error signal is not physically possible. To alleviate this problem, the analytical model incorporating the optimal control approach will be modified such that levels of attenuation are comparable with those observed in the lab due to the noise floor of the instrumentation.

THEORY

In developing the analytical model for the purpose of this study, one must derive the response of the plate to a point force representing the disturbance, and a piezoelectric actuator representing the control input. In addition, analytical models with a computed response proportional to the response of the chosen error sensors (i.e. microphones and PVDF strip sensors) must be derived. After developing the sensor and actuator equations, the control approach must be modeled mathematically. For this particular study, the optimal control approach as well as a multi-channel filtered-x version of the adaptive LMS algorithm were chosen to obtain the optimal control voltages at the piezoelectric actuators. To convey this information, subsections are devoted to actuator models, sensor models and control approaches. Due to constraints on the length of this paper, only the necessary equations to construct the computational model are presented; however, references are included for a more detailed analysis.

ACTUATOR EQUATIONS

As stated earlier, the structure chosen for this pariticular study is a simply supported plate with the following response from modal analysis [17].

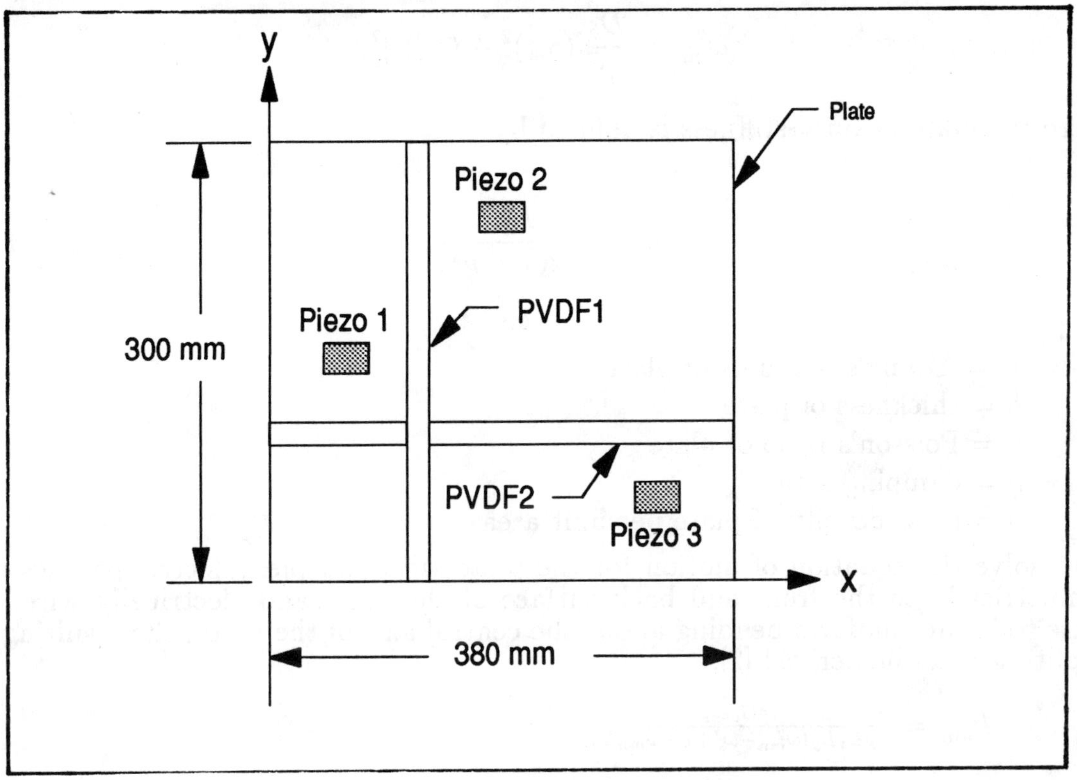

Figure 1: Schematic of Plate

$$w(x,y,t) = \sum_{m=1}^{\infty}\sum_{n=1}^{\infty} W_{mn}\sin(\gamma_m x)\sin(\gamma_n y)\exp(j\omega t), \tag{1}$$

where:

$\gamma_m = \frac{m\pi x}{L_x}$
$\gamma_n = \frac{n\pi y}{L_y}$
W_{mn} = modal amplitudes
L_x = length of plate in x direction
L_y = length of plate in y direction
ω = driving frequency
t = time.

A schematic of the plate is presented in Figure 1 with representative coordinates for all actuators and sensors located on the plate. Solving the equation of motion for the plate subjected to harmonic excitation of a point force located at x-y coordinates (c, d) yields the following expression for the modal amplitudes.

$$F_{mn} = \frac{4F}{\rho L_x L_y}\frac{\sin(\gamma_m c)\sin(\gamma_n d)}{(\omega_{mn}^2 - \omega^2 + j2\eta\omega\omega_{mn})}. \tag{2}$$

The resonant frequencies of the simply supported plate expressed in the previous equation are given by:

$$\omega_{mn}^2 = \frac{D_e}{\rho}[(\gamma_m)^2 + (\gamma_n)^2]^2, \tag{3}$$

where the plate flexural stiffness is defined by

$$D_e = \frac{Eh^3}{12(1-\nu^2)}, \tag{4}$$

and

E = Young's modulus of plate
h = thickness of plate
ν = Poisson's ratio of plate
η = damping ratio
ρ = mass density of plate per unit area.

If we solve the equation of motion for the plate with two piezoelectric patches bonded symmetrically to the front and back surface of the plate and electrically wired out of phase to induce uniform bending about the central axis of the plate, the resulting modal amplitudes can be derived [2].

$$P_{mn}^p = \frac{4C_o\epsilon_{pe}}{\rho L_x L_y(\omega_{mn}^2 - \omega^2 + j2\eta\omega_{mn}\omega)} \times \left[-\frac{\gamma_m^2+\gamma_n^2}{\gamma_m\gamma_n}\left(\cos(\gamma_m x_1^p) - \cos(\gamma_m x_2^p)\right)\left(\cos(\gamma_n y_1^p) - \cos(\gamma_n y_2^p)\right)\right]. \tag{5}$$

Constants used in the above equation are defined as follows.

$$\epsilon_{pe} = \frac{d_{31}V^p}{t}. \tag{6}$$

$$C_o = -\frac{Eh}{6}\frac{(1+\nu)}{(1-\nu)}\frac{P}{[1+\nu-(1+\nu_{pe})P]}. \tag{7}$$

$$P = -\frac{E_{pe}}{E}\frac{(1-\nu^2)}{(1-\nu_{pe}^2)}K. \tag{8}$$

$$K = \frac{6th(h+t)}{(h^3+8t^3)+6ht^2}, \tag{9}$$

where:

(x_1^p, y_1^p) = coordinates of lower left corner of pth actuator
(x_2^p, y_2^p) = coordinates of upper right corner of pth actuator
V^p = complex voltage of pth actuator
t = thickness of piezo patch
E_{pe} = Young's modulus of piezo patch
ν_{pe} = Poisson's ratio of piezo patch.

To obtain the combined response of excitation by the point force disturbance and piezoelectric actuators, superposition is applied. Based on equation (1), the modal amplitude for the summed response can be written as below.

$$W_{mn} = F_{mn} + \sum_{p=1}^{na} P^p_{mn}, \tag{10}$$

where:

na = number of actuators.

SENSOR MODELS

In experiments previously peformed [15,16], two types of sensors were implemented in the control approach, PVDF strip sensors and microphones. Lee and Moon presented an equation representing the electrical response of PVDF as a result of strain induced in the structure [14]. This expression was formulated in terms of charge, and the resulting voltage is proportional to the impedance of the chosen instrument for measuring the voltage. As long as this analytical signal is proportional to the measured electrical signal, terms for predicting a given electrical output can be ignored for the purpose of developing an analytical model of the control system. With this said, the expression for the response of a rectangular strip of PVDF film distributed over the surface of the simply supported plate depicted in Figure 1 can be written as follows.

$$\begin{aligned} q_u(t) = \; & C \textstyle\sum_{m=1}^{\infty} \sum_{n=1}^{\infty} W_{mn} [e^o_{31} \frac{mL_y}{nL_x} + e^o_{32} \frac{nL_x}{mL_y}] (\cos(\gamma_m x^u_4) - \cos(\gamma_m x^u_3)) \\ & \times (\cos(\gamma_n y^u_4) - \cos(\gamma_n y^u_3)) \exp(j\omega t), \end{aligned} \tag{11}$$

where:

(x^u_3, y^u_3) = coordinates of lower left corner of uth PVDF strip
(x^u_4, y^u_4) = coordinates of upper right corner of uth PVDF strip
e^o_{31} = charge per unit area in x-direction
e^o_{32} = charge per unit area in y-direction
C = constant of proportionality.

Observe that while the previous equation represents the response of a single rectangular strip of PVDF, sensors of arbitrary shape can be modeled with this equation by simply dividing the irregular shape into subdivisions of rectangular elements and summing the response from each element, with the correct phasing.

If a microphone is chosen as the error sensor in the far-field, an expression for the pressure can be used in developing the cost function to model the sensor [18].

$$\begin{aligned} p(r, \theta, \phi) = \; & \frac{k\rho_o c L_x L_y}{2\pi^3 r} \exp(-jkr) \textstyle\sum_{m=1}^{\infty} \sum_{n=1}^{\infty} \frac{W_{mn}}{mn} \left[\frac{(-1)^m \exp(-j\alpha) - 1}{(\alpha/m\pi)^2 - 1}\right] \\ & \times \left[\frac{(-1)^n \exp(-j\beta) - 1}{(\beta/n\pi)^2 - 1}\right], \end{aligned} \tag{12}$$

where:

$$\alpha = kL_x \sin(\theta) \cos(\phi), \tag{13}$$

$$\beta = kL_y \sin(\theta) \sin(\phi), \tag{14}$$

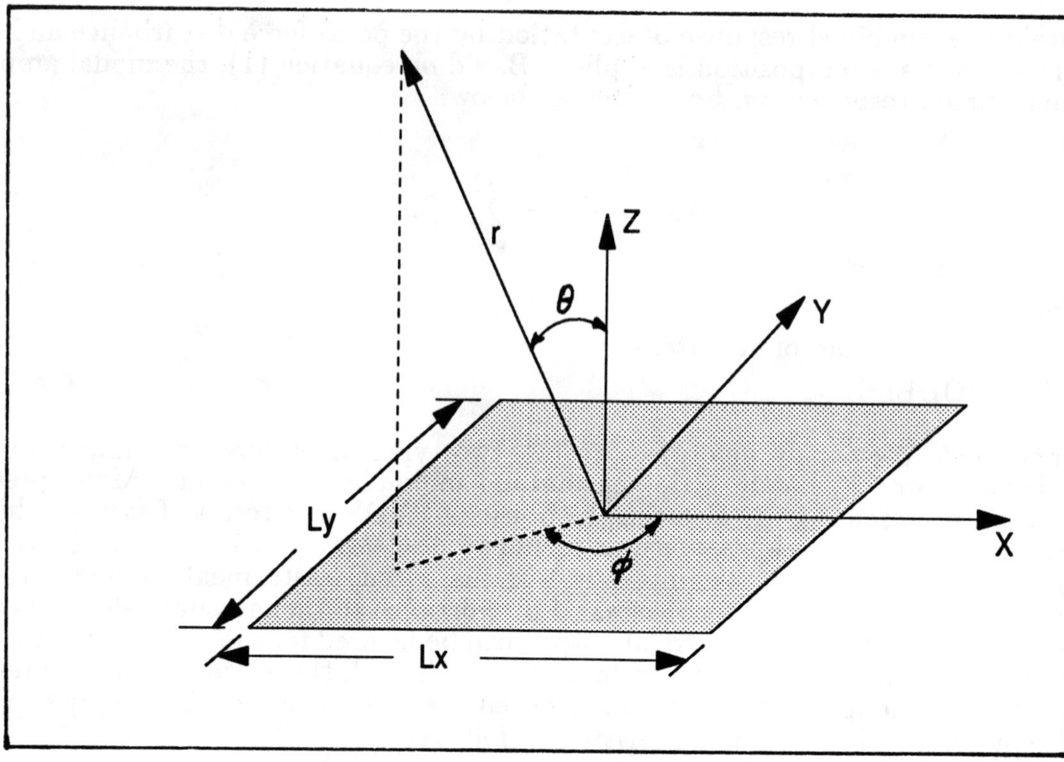

Figure 2: Coordinate System for Computing Pressure

and:

r = radius from plate
k = wavenumber in medium
c = speed of sound in medium
ρ_o = density of medium.

The coordinate system for the far-field expression for pressure is presented in Figure 2.

CONTROL APPROACHES

In experiments previously conducted, the multi-channel filtered-x version of the adaptive LMS algorithm was implemented in the control approach. The experiments can be simulated given the previous actuator and sensor equations in conjunction with the adaptive LMS algorithm. This algorithm was previously described by Elliot et.al. [19], and will be briefly outlined in this section.

The output of an error sensor can be modelled at the n'th time step as:

$$e_l(n) = d_l(n) + \sum_{m=1}^{M} \sum_{j=0}^{N-1} P_{lmj} \sum_{i=0}^{N-1} w_{mi}(n-j)x(n-i-j), \tag{15}$$

where:

n = time step number
m = actuator number
l = error sensor number
i = filter coefficient number
$d_l(n)$ = desired response of lth error sensor
$x(n)$ = input reference source
w_{mi} = coefficients of the adaptive fixed impulse response (FIR) for the mth actuator and the ith coefficient
P_{lmj} = j^{th} coefficient of the transfer function between the output of the mth adaptive filter and the lth error sensor
M = number of control actuators
N = number of filter coefficients.

In the LMS algorithm, the mean square error signal is defined by:

$$J = E\left[\sum_{l=1}^{L} e_l^2(n)\right], \tag{16}$$

where E is the expectation operator. Since this error function is quadratic, only one minimum solution exists. The outputs of the fixed filters, P_{lmj}, at each time step n, were used by the LMS algorithm to minimize the mean square error signal by modifying the coefficients of the adaptive filter as follows:

$$w_{mi}(n+1) = w_{mi}(n) - \mu \sum_{l=1}^{L} e_l(n) r_{lm}(n-i), \tag{17}$$

and

$$r_{lm}(n-i) = \sum_{j=0}^{N-1} P_{lmj} x(n-i-j). \tag{18}$$

Upon achieving control, the adaptive fixed filter coefficients converge to a steady state value. In the above algorithm, P_{lmj} is define by the transfer functions between the control actuators and the error sensors chosen. For modeling purposes, the input voltage to the actuator should be set equal to 1, and the response from each error sensor computed, defining the necessary transfer function. The initial error is defined by the response of the sensor due to the disturbance alone. As the adaptive coefficients begin to converge, the error signal is driven to a minimum. Upon converging, the adaptive coefficients w_{mi} contain the necessary information for computing the optimal control voltage to each actuator.

To convert the time domain coefficients to the frequency domain, simple trigonometry can be used. For a sinusoidal input, two coefficients are required to model the signal. In this case, the transformation can be respresented by the following equations.

$$Re(V_m) = w_{m1} + w_{m2}\cos(\omega \Delta t), \tag{19}$$

and,

$$Im(V_m) = w_{m2} \sin(\omega \Delta t), \tag{20}$$

where:

Δt = time between discrete samples
Im() = imaginary part of the complex voltage
Re() = real part of the complex voltage.

In addition to modeling the adaptive LMS algorithm, an optimal control approach was taken to compute the control voltages for each piezoelectric actuator. In this approach, the cost function (function to be minimized) is formulated as the square of the chosen error sensors.

$$C = \sum_{b=1}^{ne} q_b q_b^*, \tag{21}$$

where q_b^* is the complex conjugate of q_b.

$$q_b = (\sum_{s=1}^{na} T_{bs} \hat{V}_s) + T_{bD} F, \tag{22}$$

where:

T_{bs} = transfer function between the sth actuator and the bth error sensor
$\hat{V}_s$ = complex voltage of sth actuator
T_{bD} = transfer function between the disturbance and the bth actuator
F = input force.

Taking the partial derivative of the cost function with respect to the real and imaginary part of each control voltage results in the following system of linear equations which must be solved.

$$\sum_{s=1}^{na} \left\{ \sum_{b=1}^{ne} (\sum_{l=1}^{na} T_{bl} \hat{V}_l + T_{bD} F) T_{bs}^* = 0 \right\}. \tag{23}$$

The previous system of equations can be represented with linear algebra.

$$[A][\hat{V}] = [b]. \tag{24}$$

Now if $ne > na$, then [A] is of dimension $ne \times na$. To find the optimal solution, a least mean squares approach is taken.

$$[\hat{V}] = [[A]^*[A]]^{-1} [A]^*[b]. \tag{25}$$

While the above equation yields the optimal solution, it may not be the most computationally stable method of solving for the complex control voltages. In addition, the number of error sensors must be greater than or equal to the number of actuators or the system of equations is under determined.

Based on results from both control approaches, the LMS algorithm was observed to converge to the same control voltages as predicted from optimal control theory. Therefore, results generated in this paper were computed from the optimal control approach to save computational time. For example, the LMS algorithm took 3.2×10^6 iterations to converge for a 2 channel control case. The simulation does provide insight into the structural and acoustic response during the control process, which cannot be determined from the optimal control approach; however, a graphics video of the results is required to meaningfully represent the data.

WAVENUMBER ANALYSIS

In a previous experimental study, the k-transform of the structural response of a simply supported plate taken before and after control provided significant insight into the physical mechanism of control [16]. Results from the experiments indicate that upon achieving control in the acoustic field, the amplitudes of the wavenumber spectrum decreased in the supersonic (propagating) region. In addition, a theoretical study by Fuller and Burdisso [20] suggested the possiblity of developing a controller to minimize selective wavenumber amplitudes in the k-domain to achieve directional control of a simply supported beam.

As a result of the previous efforts, the k-transform will be used to interpret results in this analytical study. In addition, a method of uniformly minimizing the sound radiation over the hemisphere of the plate will be outlined by choosing the amplitudes of the wavenumber spectrum in the supersonic region as the cost function. Before proceeding further with details of this study, the k-transform will be outlined to clarify understanding of the subject.

For those familiar with Fourier transforms of time dependent signals utilized in signal processing, the k-transform (i.e. wavenumber transform) is analogous to obtaining the frequency spectrum from a time-dependent signal [18]. To obtain the wavenumber transform, a Fourier integral transform of the spatial response is evaluated.

$$F(k_x, k_y) = \int_{-\infty}^{+\infty} \int_{-\infty}^{+\infty} f(x,y) \exp(-jk_x x) \exp(-jk_y y) dy dx \tag{26}$$

Replacing f(x,y) with the spatial response of the plate equations (1), and evaluating the integral over the boundaries of the plate, the wavenumber transform is obtained.

$$W(k_x, k_y) = \sum_{m=1}^{\infty} \sum_{n=1}^{\infty} W_{mn} \frac{(m\pi/L_x)[(-1)^m \exp(-jk_x L_x) - 1]}{[k_x^2 - (m\pi/L_x)^2]} \times \frac{(n\pi/L_y)[(-1)^n \exp(-jk_y L_y) - 1]}{[k_y^2 - (n\pi/L_y)^2]} \tag{27}$$

Plotting the magnitude of this function yields physical insight into the structural acoustic response. An example plot is presented in Figure 3. The acoustic wavenumber, defined as follow:

$$k = \frac{\omega}{c}, \tag{28}$$

is the critical number defining the region between waves which propagate to the far-field and waves that simply create near-field disturbances [21]. The propagating region is termed the supersonic region and is the shaded portion in Figure 3 where the structural wavenumber is less than the acoustic wavenumber [18]. The nonpropagating region is termed the subsonic region and is defined when the structural wavenumber is greater than the acoustic wavenumber.

Using this technique to describe the relationship between the structural and acoustic response of the plate facilitates interpretation of results. For the case of "modal reduction" [23], the magnitude of the wavenumber transform is simply reduced after applying

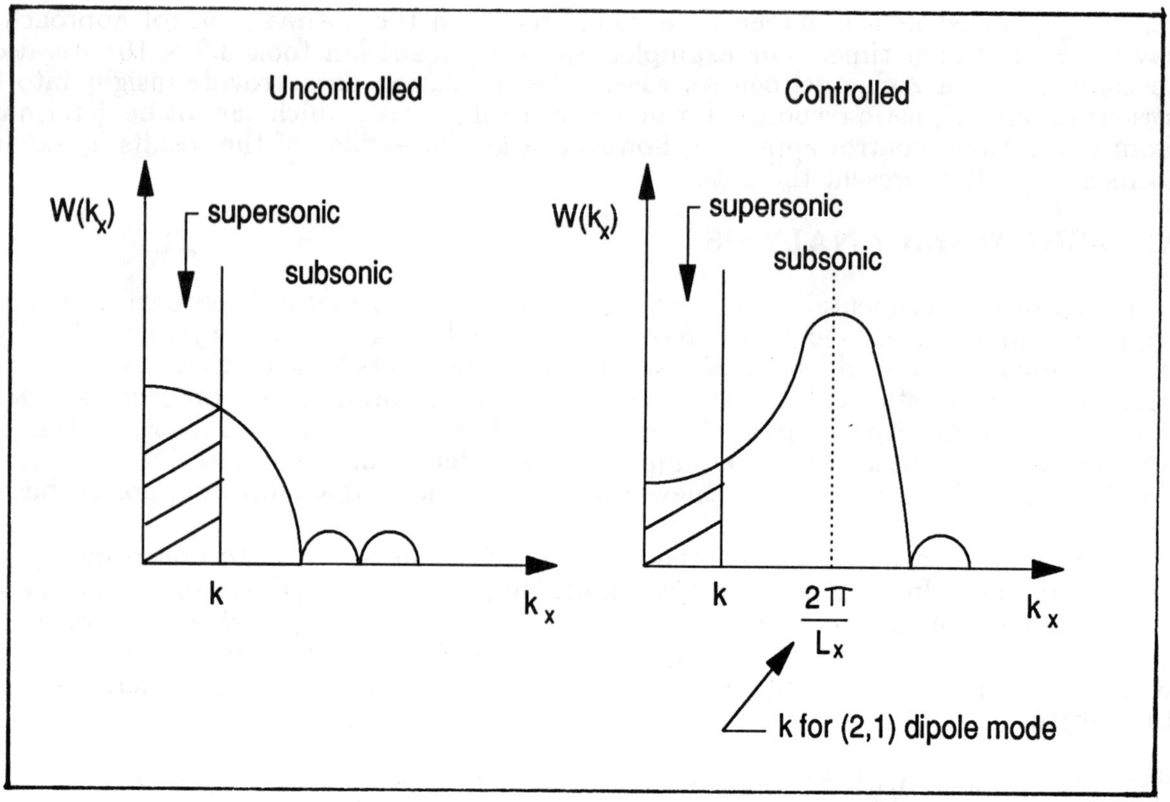

Figure 3: Example of 1-D k-transform

control. However for the case of "modal restructuring" [23], the energy is observed to shift from the supersonic (propagating) region to the subsonic (nonpropagating) region as illustrated in Figure 3. For the case presented, a cross section of the wavenumber transform was illustrated for simplicity (i.e. $k_y = 0$). In reality, a three-dimensional figure is required to illustrate the results since the structural wavenumber is a function of both the x and y-direction as illustrated in equation (27).

In the following section discussing the results, a subsection is devoted to implementing the complex wavenumber spectrum as a cost function. To minimize the acoustic radiation uniformly, the 2-D k-transform must be reduced in the supersonic region. Since an infinite number of spectral lines exist in this region, the area will be divided into discrete sections and the amplitude will be minimized at each chosen coordinate, (k_x, k_y).

PLATE CONSTRAINTS AND CONFIGURATION

For the purpose of this study, the simply supported plate was assumed to be located in an infinite baffle. The plate model includes material properties for steel, and the dimensions will correspond to those used in previous experiments, 380 x 300 x 1.96 mm. All theoretical studies will be conducted with the point force disturbance located at coordinates (240,130) mm, which corresponds to coordinates chosen for shaker location in previous experiments [15,16]. For all test cases conducted, the input force was chosen at 1.28 N as measured in previous experiments with a force transducer. Piezoelectric actuators and PVDF strip sensors were located on the plate as illustrated in Figure 1, corresponding to locations in previous experiments. Material properties of the G1195 piezoelectric material used in this study can be found in the "Piezoelectric Motor/Actuator Kit Manual" supplied by Piezo Systems [22].

Before presenting results, the resonant frequencies of the simply supported plate for the chosen dimensions and material properties are presented. The following table enables the reader to determine modes of interest for the chosen off resonance cases in the following summary of the results.

Table I. Theoretical Resonant Frequencies

Mode	Theoretical f_n (Hz)
(1,1)	88
(2,1)	189
(1,2)	250
(2,2)	351
(3,1)	357
(1,3)	519
(3,2)	520
(4,1)	592
(2,3)	621

RESULTS

Test cases were chosen based on results from experiments published in earlier work which included either microphones or PVDF strip sensors as error sensors [15,16]. In addition, the supersonic region on the k_x axis in the wavenumber domain was chosen as a cost function and the resulting controlled acoustic field was computed. Since attenuation of sound from experimental results was limited to approximately 30 dB due to the noise floor on the instrumentation, a study was performed to determine the effect of a finite noise floor on structural acoustic control.

EFFECT OF NOISE FLOOR

In the optimal control approach, a limit was placed on the attenuation of the error to determine the effect on the residual acoustic field. By subtracting an offset error proportional to the uncontrolled error signal, the amount of attenuation could be limited to any desired value. Note this is equivalent to a model reference control approach, which is also a topic of further work at VPI&SU. A test case was chosen with three piezoelectric actuators and three microphones as error sensors. The optimal control voltages were computed for the unlimited control case and cases where acoustic attenuation was limited to -30 dB and -60 dB. Results from this study are presented in Figure 4. The driving frequency was chosen at 400 Hz, which is an off resonance case for the simply supported plate. As can be observed from the graph, the most significant effect of limiting acoustic attenuation is observed at the error sensors. The residual acoustic field (regions other than microphone locations) remained relatively unaffected by the limited acoustic attenuation. This result implies that if the theoretical attenuation of the residual acoustic field is less than the noise floor of the control system, the global sound attenuation will remain relatively unaffected by the noise floor of the controller.

MICROPHONE ERROR SENSORS

The second case was also chosen based on a test conducted at 400 Hz, which is an off resonance case. Three microphones again were implemented as error sensors and up to

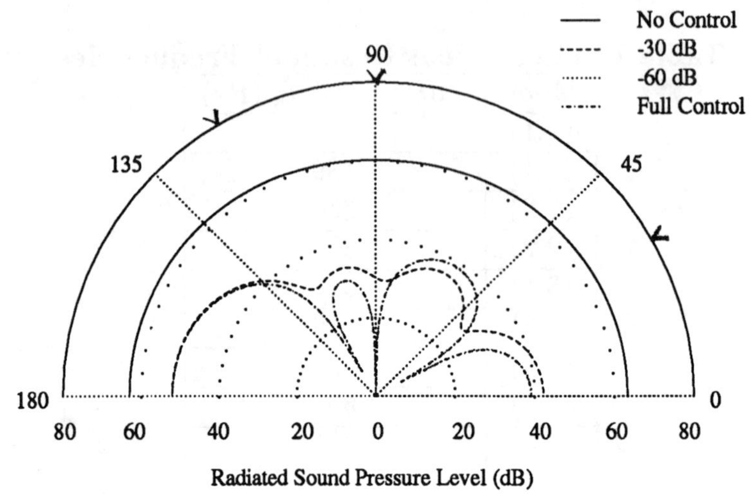

Figure 4: Comparison of Theoretical Attenuation Levels

three control actuators were utilized as control inputs. The microphone locations are indicated on the perimeter of the graph presented in Figure 5. A comparison between the predicted and measured sound field [15] is presented with the same control actuator and error sensor locations. As indicated in the graphs, trends from the theoretical response follow those of the measured data, especially when considering the increase in number of control actuators from one to two. For the theoretical case, the acoustic attenuation at the error sensors was limited to 30 dB such that the theory and experiment would be subjected to the same noise floor. Plans have been made to repeat the experiments with a more accurate instrumentation system in an effort to obtain a better noise floor and residual radiation plots with the controller activitated.

The k-transform of the structural response was computed to study the effect in the supersonic region after control. As depicted in Figure 6, the amplitudes of the wavenumber spectrum were observed to decrease after applying control. This observation was made for both experimental and theoretical cases. In effect, the controller has modified the structural response such that the structural energy is removed from the region which is responsible for sound radition (supersonic region) and shifted it into the subsonic region. As stated previously, the results of Figure 6 correspond to "modal restructuring".

PVDF ERROR SENSORS

As in the previous test case, the theoretical study was conducted such that results from previous experiments could be compared. For this case, the plate was driven at 320

Hz, and up to two control acuators and two PVDF strip sensors were implemented in the control approach. The noise floor of the theoretical control was again limited to 30 dB such that the theoretical model would be subjected to the same constraints as the physical system. As illustrated in Figure 7, the predicted acoustic field is very similar to the measured results. Theory predicts approximately 10 dB of attenuation, and this was observed in the experiments. The physical interpretation of the control can be described after viewing the k-transform of the structural response illustrated in Figure 8. Limited acoustic attenuation was achieved during control, and this result is evident when considering the wavenumber spectrum. Portions of the supersonic region decreased during control and portions increased. In addition, portions of the subsonic regions increase upon achieving control. The method of control can again be attributed to modal restructuring; however, results suggest that a more optimal sensor shape could be chosen such that all structural supersonic wave components are decreased.

K-DOMAIN COST FUNCTION

As indicated in results from both experiment and theory, the physical mechanism of control can be described by viewing the supersonic region of the k-transform. With this in mind, the supersonic region of the k_x axis of the k-transform was chosen as the cost function. A finite number of the spectra were chosen to create the cost function, namely seven. The k_x axis was broken into six discrete intervals from -k to +k, resulting in seven points to minimize the wavenumber spectrum. Note that this will only minimize radiation towards the plane $\phi = 0^o, 180^o$ (i.e. for $k_y = 0$) which is the arc represented in the following radiation plots. Total radiation control would require minimization of a 2-D integral evaluated over the k_y axis from -k to +k as well. In chosing $V(k_x)$ as the error variable one is in effect minimizing $|V(k_x)|^2$ as shown in equation (16). As explained by Fahy [18], the total power radiated from the plate is directly proportional to the integral of $|V(k_x, k_y)|^2$ over the supersonic region in the 2-D transform plane. Hence minimizing $V(k_x)$ over the region -k to +k is equivalent to minimizing the acoustic line power being radiated through the $\phi = 0^o, 180^o$ planes at angles of θ. Results from this test case are presented in Figure 9. The acoustic attenuation achieved was approximately 20 dB along the x-axis of the plate. Results from the k-transform indicate that the amplitudes of the wavenumber spectrum were indeed minimized upon applying control. While the

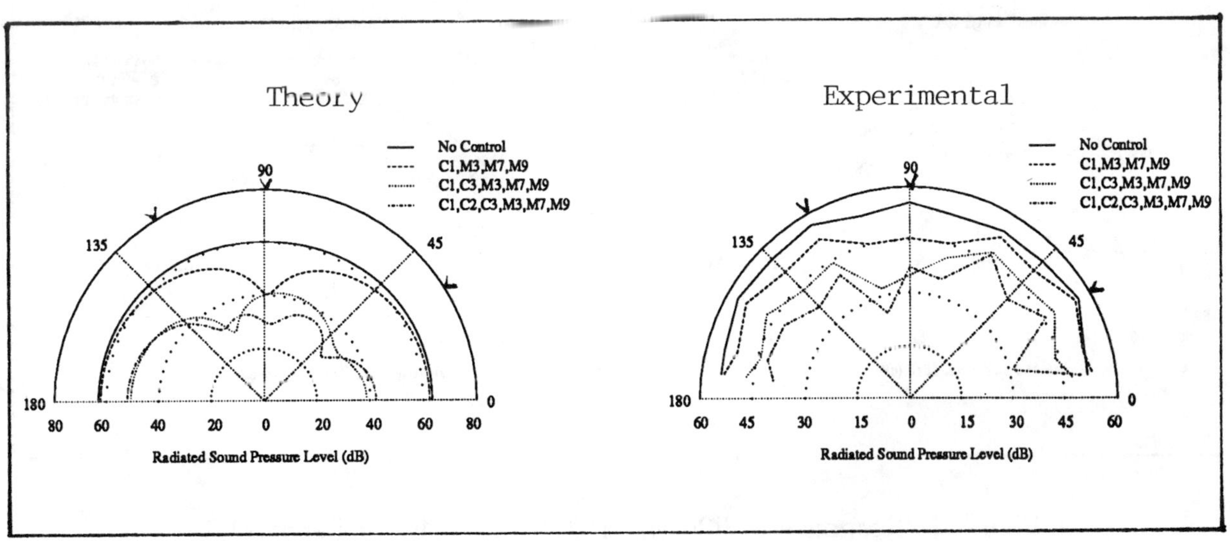

Figure 5: Response at 400 Hz (Microphone Error Sensors)

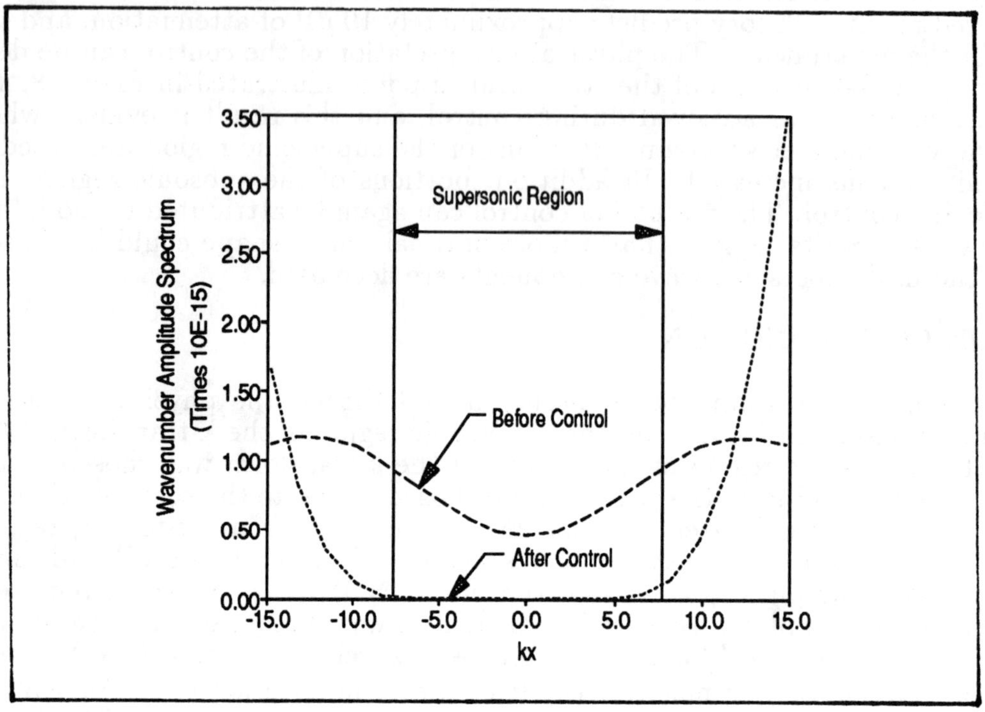

Figure 6: k-transform at 400 Hz ($k_y = 0$)

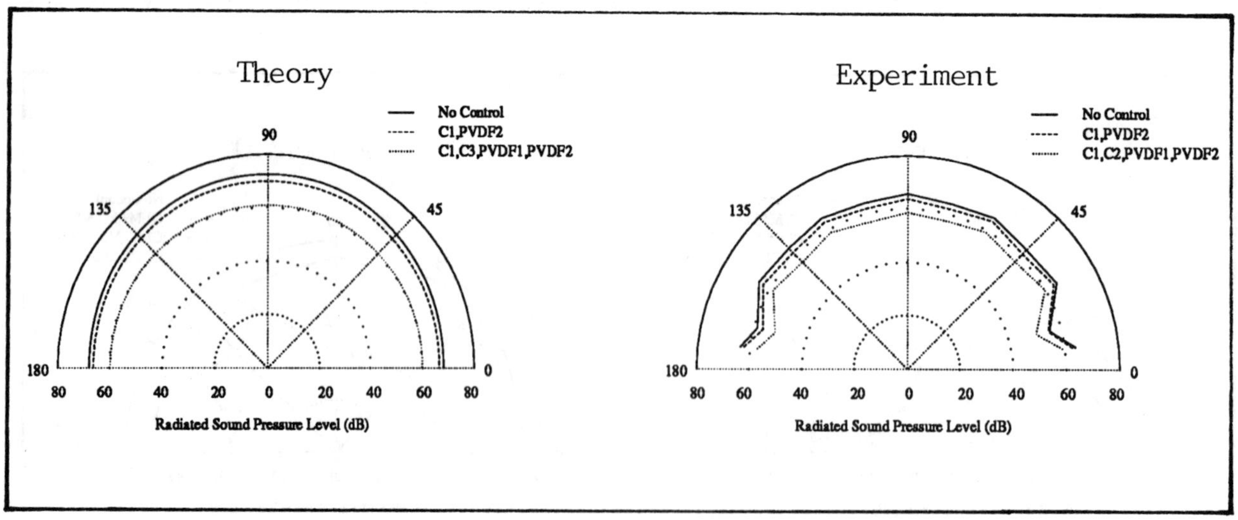

Figure 7: Response at 320 Hz (PVDF Strip Error Sensors)

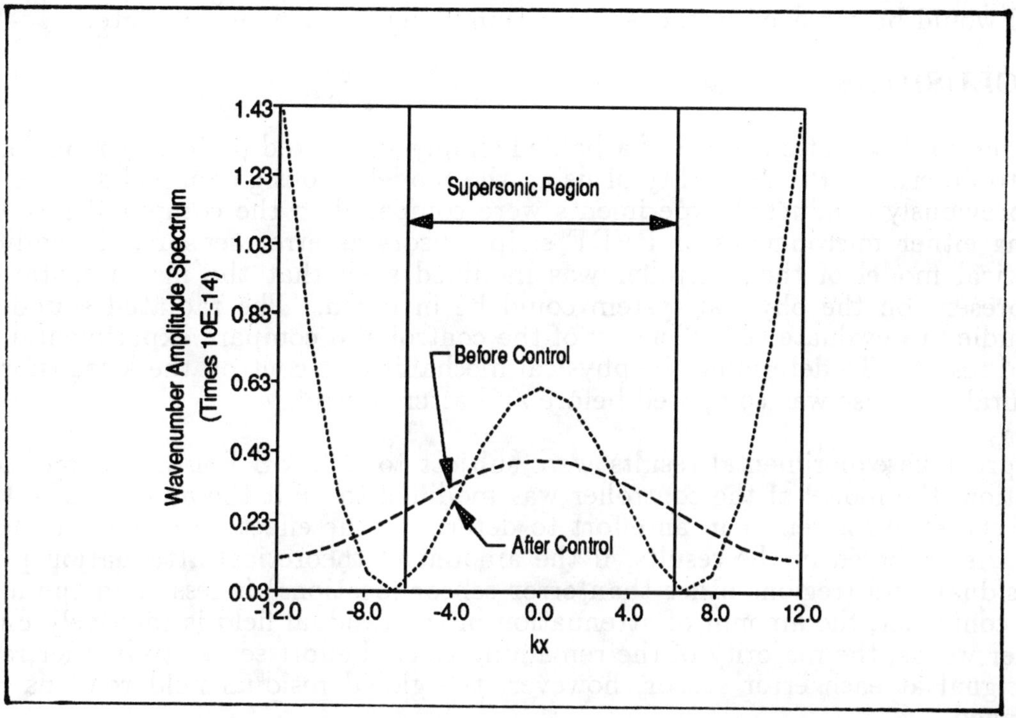

Figure 8: k-transform at 320 Hz ($k_y = 0$)

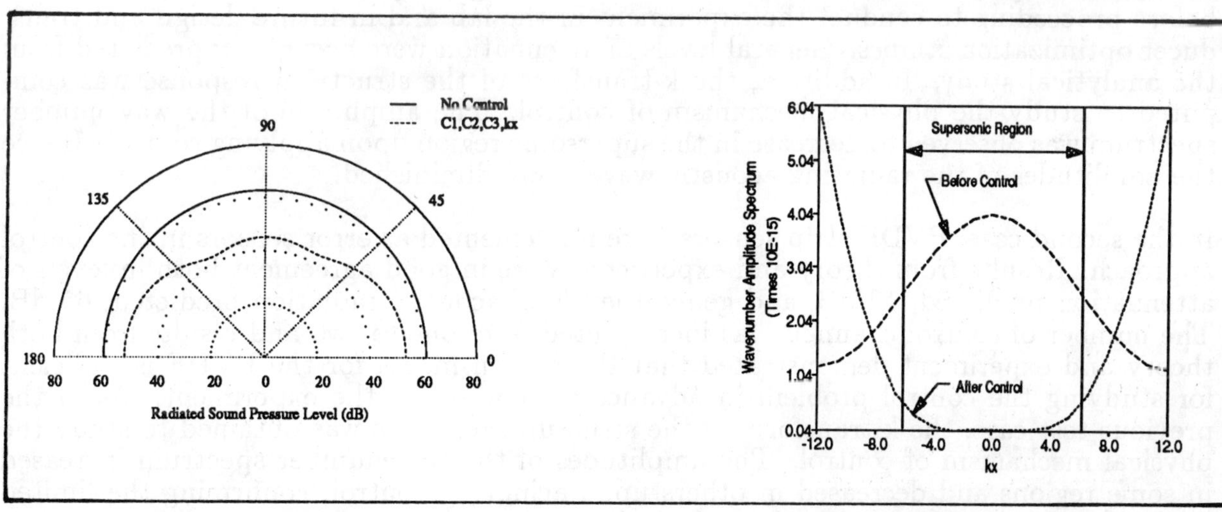

Figure 9: Response at 320 Hz (k-transform Cost Function)

wavenumber transform is actually three dimensional, the power of this method is readily observed from the case presented. In addition to global control, directional control can be achieved by simply limiting control to a specific region of the wavenumber domain. For example, if the cost function were constructed from 0 to -k along the k_x axis, directional control would be achieved in the +x direction along the axis of the plate.

CONCLUSIONS

The structural acoustic control of a baffled simply supported plate was modeled analytically to determine the feasibility of using the model to design shaped sensors. Results from previously conducted experiments were compared to the computed results implementing either microphones or PVDF strip sensors as error sensors. In addition, the theoretical model of the controller was modified such that the instrumentation noise floor present on the physical system could be included. The radiated sound pressure was studied to evaluate the efficiency of the control and compare experimental and theoretical results. To determine the physical mechanism of control, the k-transform of the structural response was compared before and after control.

Since previous experimental results were subject to a 30 dB noise floor on the instrumentation, the model of the controller was modified to limit the amount of attenuation at the chosen error sensor in an effort to determine the effect on overall sound attenuation. As observed in the results, if the amount of theoretical attenuation possible in the residual field (regions other than error sensor locations) is less than the noise floor of the controller, the amount of attenuation in the residual field is relatively unaffected. In other words, the majority of the remaining control effort serves to further reduce the error signal at each error sensor, however, the global residual field remains relatively unchanged.

After determining the effect of the noise floor on the resulting acoustic response, a test case was studied for the off resonance case at 400 Hz. Microphones were implemented as error sensors, and theoretical and experimental results were compared while increasing the number of control actuators. As demonstrated in the results, general trends between theory and experiment compared well, demonstrating the potential of using the model before proceeding to conduct the experiment in the lab and in future design and transducer optimization studies. General levels of attenuation were accurately predicted from the analytical study. In addition, the k-transform of the structural response was computed to study the physical mechanism of control. The amplitude of the wavenumber spectrum was observed to decrease in the supersonic region upon applying control. Hence the amplitudes of the radiating acoustic waves were diminished.

In the second case, PVDF strip sensors were implemented as error sensors in the control approach. Results from theory and experiment were in good agreement both in levels of attenuation predicted, 10 dB, and general levels of acoustic radiation predicted, 65 dB. The number of control channels was incremented from one to two and results from both theory and experiment demonstrated that the chosen model for the PVDF is sufficient for studying the control problem in advance of conducting the experiment. As in the previous test case, the k-transform of the structural response was obtained to study the physical mechanism of control. The amplitudes of the wavenumber spectrum increased in some regions and decreased in others upon achieving control, confirming the limited sound attenuation observed both analytically and experimentally.

Results from these test cases as well as others not presented in this paper suggest that if the supersonic region of the k-transform were chosen as the cost function in the control approach, the radiated sound could be minimized. In an effort to evaluate this hypothesis, a case was conducted whereby the k_x axis of the 2-D transform was minimized between -k and +k. As expected, acoustic attenuation was achieved along the $\phi = 0^o, 180^o$ plane of the radiation field, corresponding to the x-axis. In addition, the structural acoustic control can be made directional by simply minimizing a specific section of the wavenumber spectrum in the supersonic region.

Based on results of this study, future work will be devoted to developing spatially convolving sensors dedicated to observing the supersonic wave components on the surface of the structure in an effort to reduce sound radiation without locating microphones in the radiated field.

ACKNOWLEDGEMENTS

The authors gratefully acknowledge the support of this work by DARPA and the Office of Naval Research under grant ONR-N00014-88-K-0721.

REFERENCES

1. Fuller, C.R., Hansen, C.H., and Snyder, S.D., 1989. "Active Control of Structurally Radiated Noise Using Piezoceramic Actuators," *Inter-Noise 89*, pp. 509-511.

2. Dimitriadis, E.K., Fuller, C.R., and Rogers, C.A., 1989. "Piezoelectric Actuators for Distributed Noise and Vibration Excitation of Thin Plates," *Proceedings of ASME 8th Biennial Conference on Failure Prevention and Reliability*, Montreal, Canada, pp. 223-233.

3. Wang, B.T., Dimitriadis, E.K., and Fuller, C.R., 1990. "Active Control of Structurally Radiated Noise Using Multiple Piezoelectric Actuators," *Proceedings of AIAA SDM Conference*, AIAA Paper 90-1172-CP, Long Beach, California.

4. Fuller, C.R., Gibbs, G.P., and Silcox, R.J., 1990. "Simultaneous Active Control of Flexural and Extensional Waves in Beams," *Journal of Intelligent Material Systems and Structures*, 1 (2), pp. 235-247.

5. Clark, R.L. and Fuller, C.R., 1990. "Experiments on Active Control of Structurally Radiated Sound Using Multiple Piezoceramic Actuators," Accepted for publication in the *Journal of Acoustical Society of America*, (also presented at the 120th ASA meeting, Nov. 1990).

6. Crawley, E.F., and de Luis, J., 1989. "Use of Piezoelectric Actuators as Elements of Intelligent Structures," *AIAA Journal*, 25 (10), pp. 1373-1385.

7. Fanson, J.L., and Chen, J.C., 1986. "Structural Control by the Use of Piezoelectric Active Members," *Proceedings of NASA/DOD Control-Structures Interaction Conference*, NASA CP-2447, Part II.

8. Clark, R.L., Fuller, C.R., and Wicks, A.L., 1990 "Characterization of Multiple Piezoelectric Actuators for Structural Excitation," Submitted to *Journal of Acoustical Society of America*, (also presented at the 119th ASA meeting, April 1990).

9. Bailey, T., and Hubbard, J.E., 1985. "Distributed Piezoelectric-Polymer Active Vibration Control of a Cantilevered Beam," *AIAA Journal of Guidance and Control*, 6 (5), pp. 605-611.

10. Burke, S.E., and Hubbard. J., 1987. "Active Vibration Control of a Simply Supported Beam Using a Spatially Distributed Actuator," *IEEE Contr. Syst. Mag.*, pp. 25-30.

11. Plump, J.M., Hubbard, J.E., and Bailey, T., 1987. "Nonlinear Control of a Distributed System: Simulation and Experimental Results," *J. Dyn. Sys., Mea. and Contr.*, Vol. 109, pp. 133-139.

12. Lee, C.K., Chiang, W.W., and O'Sullivan, T.C., 1989. "Piezoelectric Modal Sensors and Actuators Achieving Critical Damping on a Cantilever Plate," *AIAA Paper*, No. 89-1390.

13. Miller, D.W., Collins, S.A., and Peltzman, S.P., 1990. "Development of Spatially Convolving Sensors for Structural Control Applications," *AIAA Paper*, 90-1127-CP, pp. 2283-2297.

14. Lee, C.K., and Moon, F.C., 1990. "Modal Sensors/Actuators," *ASME JAM*, Vol. 57, pp. 434-441.

15. Clark, R.L., Fuller, C.R., 1990. "Control of Sound Radiation with Adaptive Structures," Presented at *First Joint U.S./Japan Conference on Adaptive Structures*, November 13-15, 1990, Maui, Hawaii.

16. Clark, R.L. and Fuller, C.R., 1990. "Modal Sensing of Efficient Acoustic Radiators with PVDF Distributed Sensors in Active Structural Acoustic Approaches," Submitted to *Journal of Acoustical Society of America*, (also presented at the 120th ASA meeting, Nov. 1990).

17. Meirovitch, Leonard, 1967. Analytical Methods in Vibrations, Macmillan Publishing Co., New York, New York, pp. 179-185.

18. Fahy, Frank, 1987. Sound and Structural Vibration, Academic Press, Orlando, Florida, pp. 61-63.

19. Elliott, S.J., Stothers, I.M., and Nelson, P.A., 1987. "A Multiple Error LMS Algorithm and Its Application to the Active Control of Sound and Vibration," *IEEE Transaction on Acoustic Speech and Signal Processing*, ASSP-35, 1:1423-1434.

20. Fuller, C.R., and Burdisso, R.A., 1990. "A Wave Number Domain Approach to the Active Control of Structure-Borne Sound", Accepted for publication in the *Journal of Sound and Vibration.*

21. Maidanik, G., 1974. "Vibrational and Radiative Classification of Modes of a Baffled Finite Panel," *Journal of Sound and Vibration*, 34(4), pp. 447-455.

22. "Piezoelectric Motor/Actuator Kit Manual," Piezo Systems, 186 Massachusetts Avenue, Cambridge, MA 02139.

23. Fuller, C.R., Hansen, C.H., and Snyder, S.D., 1991. "Active Control of Sound Radiation From a Vibrating Rectangular Panel by Sound Sources and Vibration Inputs: An Experimental Comparison", to be published in the *Journal of Sound and Vibration.*

ACTIVE CONTROL OF RESPONSE OF RIBBED PANELS

J. Dickey,[1] G. Maidanik[1]

ABSTRACT

This paper presents an analytical foundation for the active control of the response of a ribbed panel. The analysis is based on an impulse response function for the panel; hence the response and/or radiation due to any prescribed external drive is easily formulated. Similarly, a second external drive, or array of drives which reduce the effect of the first, may be specified . The two examples presented use a membrane simulating a thin plate which is ribbed with a finite number of parallel ribs. In the first example, the response of the panel due to an external line drive is calculated as a function of the spatial coordinate in the panel normal to the ribs. A frequency, ω_0, at which the control is desired, and a position for the application of a line control source x_c are picked, and the formalism is used to specify a control drive which will reduce the response at some prescribed position, x_0, (parallel to the ribs), on the panel. The control drive is specified in terms of the original external drive, the system impulse response, ω_0, x_c, and x_0. In the second example, the response of the panel due to an external drive associated with an incident pressure wave is calculated as a function of the wavenumber in the plane of the panel and the frequency. A control frequency, ω_0, and a position, x_c, (or array of positions) are again chosen for the application of a control drive and this drive (or drives) for this system is specified in terms of the original external drive, the system impulse response, the frequency ω_0, x_c and the wavenumber k_0, at which the response is to be reduced. The reduction is verified by calculation for both examples.

INTRODUCTION

Recent advances in computer technology and control systems have generally outpaced the advances in analytical modeling of structures; yet it is asserted that the modeling is equally essential to the active control of structural vibration. This paper will demonstrate this assertion and support other generalizations regarding the role of modeling in the intelligent application of active controls to structural systems. The demonstrations are presented as examples which employ a recently developed analytical model of a ribbed panel [1]. The examples are designed to illustrate several specific points; e.g., the utility of the impulse response function (or Green's function) in specifying a control drive, and some often undesirable side effects which may occur in the application of controls to improperly modeled systems.

[1]David Taylor Research Center, Annapolis, MD 21402, U.S.A.

THE PARADIGM

The impulse response function, g, completely describes the dynamics of a structural system. g is algebraic and is a functional of the spatial and temporal coordinates (or their Fourier conjugates) of the dynamic system. Attention will be restricted to systems which are stationary in time and are one-dimensional in the sense that discontinuities which scatter traveling waves occur in only one direction, the x coordinate. Together, these restrictions allow the following abbreviation:

$$g(x \mid x', y \mid y', t \mid t') \rightarrow g(x \mid x', \underset{\sim}{\omega}) \tag{1}$$

where $\underset{\sim}{\omega} = \{k_y, \omega\}$ with ω being the frequency and k_y being the conjugate variable to y, the spatial direction of stationarity. The $\underset{\sim}{\omega}$ dependence will be the same for all variables and will hereafter be dropped. Consideration will further be limited to linear systems which allow one to formulate the response due to an external drive, $p_e(x)$, as

$$v(x) = \int g(x \mid x')\, dx'\, p_e(x') \tag{2}$$

where the drive is termed external in the sense that it cannot change the characteristics of the system, i.e., $g(x \mid x')$. The integration over the drive variable of the product of the impulse response function and the external drive will be written as the operator $\mathscr{h}$; thus Eq. 2 becomes,

$$v(x) = \mathscr{h}(x \mid x')\, p_e(x') \quad . \tag{3a}$$

Equation 3a may also be written in the following Fourier conjugate spaces:

$$v(x) = \tilde{\mathscr{h}}(x \mid k')\, P_e(k') \quad , \tag{3b}$$

$$V(k) = \mathscr{H}(k \mid k')\, P_e(k') \quad , \tag{3c}$$

$$\text{and} \quad V(k) = \tilde{\mathscr{H}}(k \mid x')\, p_e(x') \quad . \tag{3d}$$

The preference of a particular version of Eq. 3 is made on the basis of the space in which the response (and later, the control) is sought, and on the space in which the external drive is more conveniently expressed.

With Eqs. 3, the specification of a control drive, $p_c(x')$ (or $P_c(k')$), which renders a null response at some point x_0 (or k_0) when applied to a system which is otherwise driven by an external noise drive $p_n(x')$ (or $P_n(k')$), is straightforward. It is a control drive which satisfies

$$0 = \mathscr{h}(x_0 \mid x')\, p_n(x') + \mathscr{h}(x_0 \mid x')\, p_c(x') \quad . \tag{4}$$

It is important to note that there are seven other similar equations derivable by Fourier transforming the spatial variables in either, or both, terms. [cf. Eq. 3.] Equation 4 under specifies the control drive; i.e., it does not specify a unique drive. An example of a further specification imposed on the control drive is that it is applied at the spatial point, x_c; i.e.,

$$p_c(x') = p_{0c}\, \delta(x' - x_c) \quad . \tag{5}$$

This choice renders the specification in Eq. 4 unique and does not limit the purpose of the paper. With Eqs. 5 and 3a,

$$0 = \mathscr{h}(x_0 \,|\, x')\, p_n(x') + g(x_0 \,|\, x_c)\, p_{0c} \quad , \tag{6a}$$

which is algebraic; hence Eq. 4 yields

$$p_{0c} = -[g(x_0 \,|\, x_c)]^{-1}\, \mathscr{h}(x_0 \,|\, x')\, p_n(x') \quad . \tag{6b}$$

This is the specification of the drive applied to the point x_c which will cause a null response at the point x_0. For the case of the external noise drive being localized, e.g., in space

$$p_n(x') = p_{0n}\, \delta(x' - x_n) \quad , \tag{7a}$$

the specification of the control amplitude becomes particularly simple; e.g.,

$$p_{0c} = -[g(x_0 \,|\, x_c)]^{-1}\, g(x_0 \,|\, x_n)\, p_{0n} \quad . \tag{7b}$$

Again, Eqs. 6 and 7 are one of a set of 8 possible equations. Another member of this set which will be used in one of the examples in the following section is:

$$p_{0c} = -[\tilde{G}(k_0 \,|\, x_c)]^{-1}\, G(k_0 \,|\, k_n)\, P_{0n} \quad , \tag{7c}$$

and relates the amplitude of a control drive applied at the point x_c, to the amplitude of an external drive that is specified at the wavenumber k_n, such that they will together produce a null in the response of the system at the wavenumber k_0.

TWO EXAMPLES

Two examples are given to illustrate Eq. 4. Both examples use a one-dimensional linear structural system in accordance with the precepts of Eq. 4; this system is chosen to be a ribbed panel. The panel is modeled as a plane infinite isotropic membrane with nine ribs attached where the ribs are mass-like line impedances which are taken to be identical, parallel, and equally spaced. The masses of the ribs are taken to be 0.3 times the mass of the membrane between ribs. The x-coordinate axis is taken to be in the plane of the panel and normal to the ribs; hence the system is homogenous in y (as in Eq. 1) and considered as one-dimensional in x. The loss factor for the panel in the x-direction is $\eta_p = 0.05$.

The response being considered is chosen to be the flexural response with the dispersive characteristics of a thin plate, i.e., $k_p^2 = \omega\,\omega_c/c^2$, k_p being the free-wavenumber in the panel and c the sound speed in the fluid. The ribs are separated by a distance b, such that $k_c b = 16 = (\omega_c/c)b$.

In the figures, frequency is always given as the dimensionless variable (ω/ω_c) where ω_c is the critical frequency of the panel relative to a fluid of characteristic impedance ρc. It should be kept in mind however, that the panel is not fluid-loaded in the sense that rib/rib or rib/panel interactions are allowed to occur through the fluid.

EXAMPLE 1

The configuration of the panel is shown in Fig. 1a. It is assumed that there is an external noise drive applied at $x = x_n = 1.3b$ (with $x = 0$ being at the first rib on the left) as shown. The drive is applied along an infinite line parallel to the ribs. Figure 1b shows the magnitude of the response amplitude of the panel as a function of x at a frequency of $(\omega/\omega_c) = 0.2$. This frequency is slightly above the second bay resonance frequency which occurs at $bk_p = 2\pi$ or $(\omega/\omega_c) \simeq 0.16$. The rib positions are also indicated in the figure and the standing wave pattern between adjacent pairs of ribs attests to this near resonance condition; also, the general decline of the response away from the drive position is attributable to the loss factor.

We now pick a point at which the response is to be cancelled by the addition of a control drive at some, as yet, unspecified point in the system, and ask for the amplitude and phase of that control drive which would be required as a function of where that drive was applied. The point to be controlled is arbitrarily chosen to be $x = x_0 = 2.7b$ and Fig. 1c shows the magnitude of the amplitude of the required control relative to the amplitude of the noise drive, again at $(\omega/\omega_c) = 0.2$. One sees immediately that there are "good" places and "bad" places to apply a control drive. If the required amplitude were the sole consideration, then applying the control drive at the point to be controlled is shown to be the best choice. Applying the control at the noise drive requires the same amplitude as the noise drive. Driving near a rib is not a good choice, etc.

Unfortunately, the required control drive may not be the only criterion; thus it is supposed, in providing an example, that circumstances require that the control drive be placed at $x = x_c = 6.3b$. Figure 1d shows the magnitude of the response, still at $(\omega/\omega_c) = 0.2$, which results when the control drive is added to the system. The first important feature is that the response is indeed zero at the point to be controlled. Another important feature is that near the control drive, the response is larger than anything in Fig. 1b. These features will be true at other frequencies as well, although the details of the responses and the required control amplitude will be different. Note that the phase of the required control drive is equally important in the control process and could also be readily calculated from the impulse response function, but the phase is not shown.

The required control amplitude and the controlled response for the system with the same locations of noise drive, point to be controlled, and control drive are shown with frequency as a parameter in Fig. 2. Here, the logarithms of the parameters shown in Figs. 1c and d are shown at equal increments of (ω/ω_c) from

0 to 0.4, and the progression in frequency is spread in the vertical direction. Several interesting patterns and generalizations emerge. For example, the development of the bay resonances with frequency show that picking a "good" control point is highly frequency dependent. Another interesting feature is the pass and stop band structure as manifested by the changes in the average (i.e., standing wave effects averaged over) slope in both the required control drive and the response. The pass-bands are characterized by relatively small slopes (the slope being attenuation vs. distance) and, therefore the control process will generally require less control drive at frequencies which are in the pass-bands.

If the positions of the noise drive, the point to be controlled, and the control drive are given, one may want to know the required control amplitude as a function of frequency. This is shown in Fig. 3 for two choices of control location. One choice is $x = x_c = 6.3b$ as in Figs. 1d and 2b, and the other is a "better" choice, $x = x_c = 2.3b$. The first choice is "poor" because it is further from the point to be controlled than is the noise source; in this case, a control amplitude on the order of 100 times the noise amplitude is required at some frequencies, i.e., the stop bands. The valleys in this curve indicate the pass bands. The second choice is "better" because it places the control drive closer to the point to be controlled. In this case the high attenuation in the stop bands works to advantage with regard to the control drive amplitude.

EXAMPLE 2

The configuration of the panel, is shown in Fig. 4, and is the same as the first example except that the drive is an incident field with wavenumber k_n, and the point of application of the control drive is chosen to be different. The incident wavenumber is chosen to represent a plane wave incident on the panel at an angle $\theta_n = 0.5$ from the normal. The quantity to be controlled in this example is the radiation, by the ribs, into a particular angle, $\theta_0 = 0.6$. The total reflection, i.e. including the specular component, could be done as easily.

The reflected field is readily calculated from the impulse response function [2] and the magnitude of the amplitude of the non-specular part of this field for a single frequency, $(\omega/\omega_c) = 0.2$, is shown as a function of angle in Fig. 5. The magnitude and phase of a control drive to be applied to the point $x = x_c = 2.3b$ is calculated using Eq. 7c with $k_0 = (\omega/c) \sin \theta_0$. The control drive is applied and the controlled response, also in Fig. 5, does indeed go to zero at the prescribed angle. One quickly notices, however, that with the particular choice of parameters, the response in other directions is typically ten-fold higher. This leads to some conclusions.

SOME CONCLUSIONS

Some of the points illustrated and inferred in this paper may seem intuitive to many readers. If this is the case, the paper may serve to put some of these intuitions on a better foundation. It is asserted. however, that the magnitude of the effects, their sensitivity to the many parameters involved, and their interaction are beyond intuition. Furthermore, the model used in the example is a fairly simple one; more complete models would bring out subtler and less intuitive phenomena. Finally, it is hoped that the paper has demonstrated the utility of an analytical model in the consideration of active controls for structural systems.

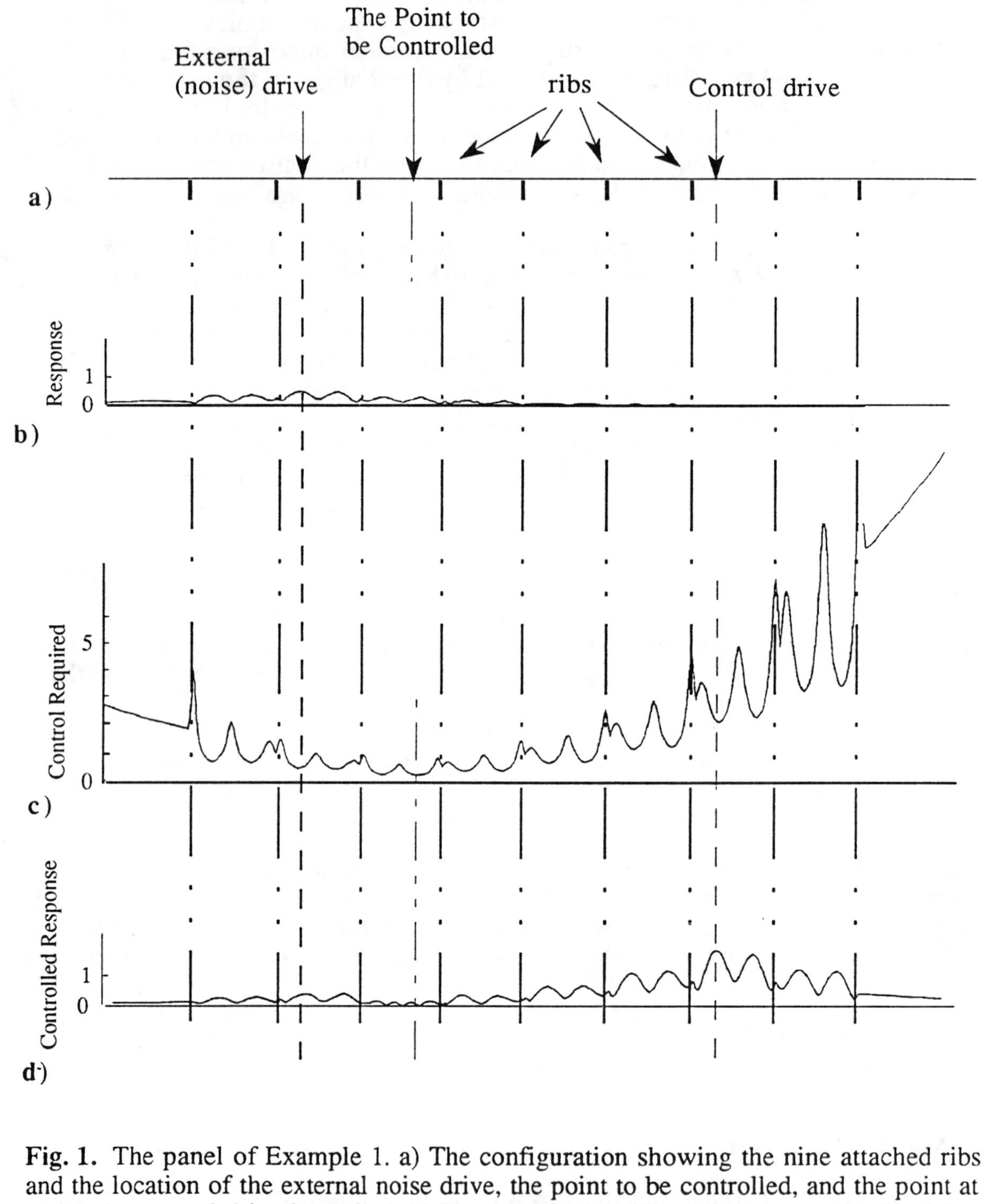

Fig. 1. The panel of Example 1. a) The configuration showing the nine attached ribs and the location of the external noise drive, the point to be controlled, and the point at which the control drive is applied. b) The magnitude of the response amplitude of the panel to the noise drive alone. All response quantities are normalized by the magnitude of the drive. c) The magnitude of the control drive which would be required to cancel the response at the point to be controlled: i.e. when the control point is specified, this function determins the required amplitude. d) The response of the panel after a control point is picked (cf top figure) and the control drive is added to the system.

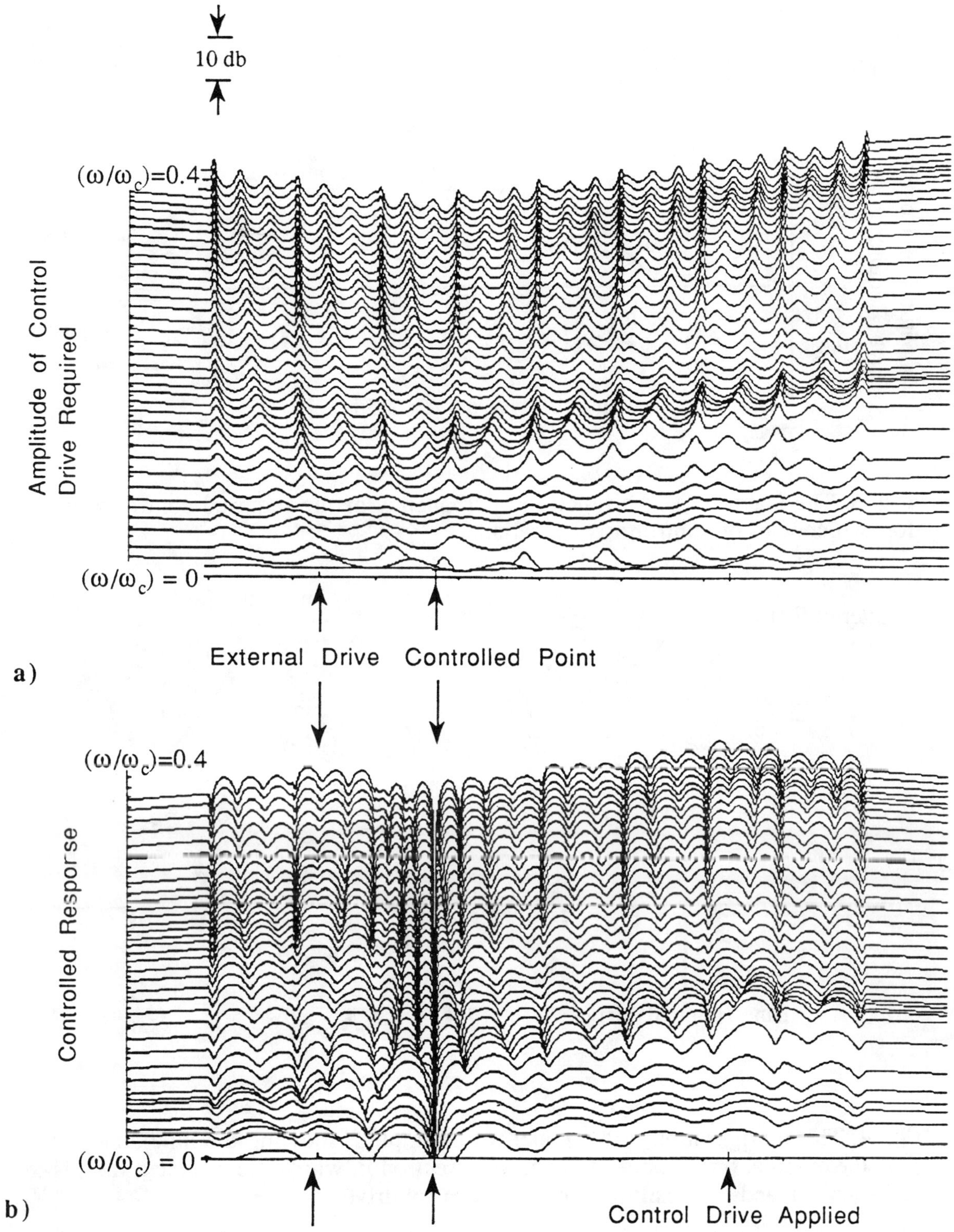

Fig. 2. The log of the magnitudes of the amplitudes of a) the control drive required, and b) the controlled response once a control point is picked. Both functions are shown for equal increments of the frequency variable as shown and each successive curve is displaced in the vertical direction. The calibration of the log scale is shown above a).

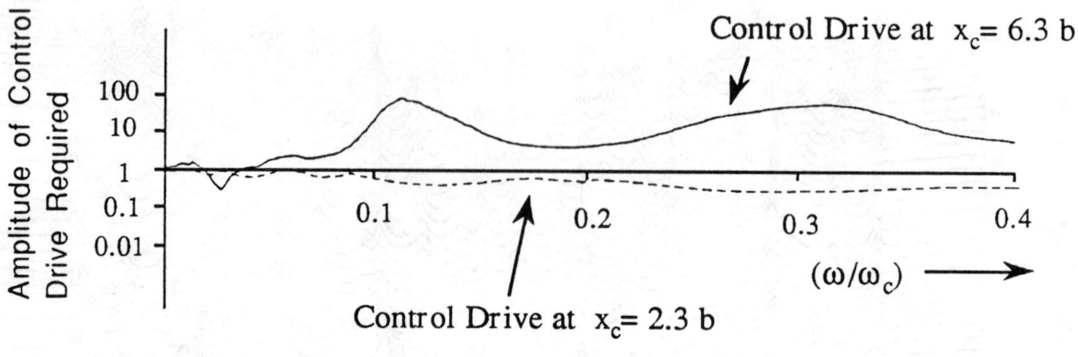

Fig. 3. The magnitude as a function of frequency of the required control drive for two choices of control drive location.

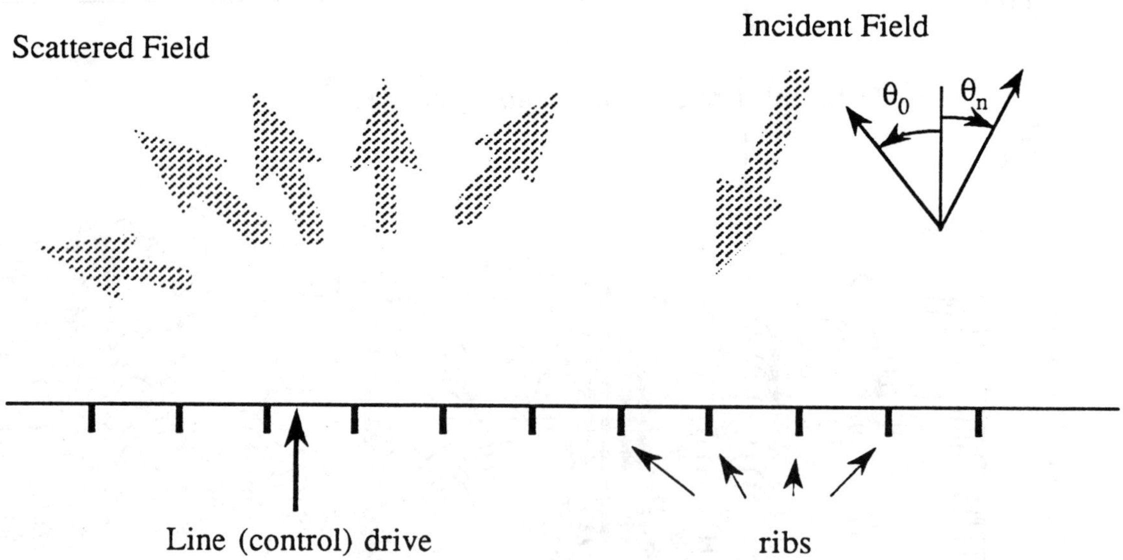

Fig. 4. The configuration of the panel for Example 2 showing the nine attached ribs, the external drive incident field, the direction in which the reflected field is to be cancelled, and the position of the line control drive.

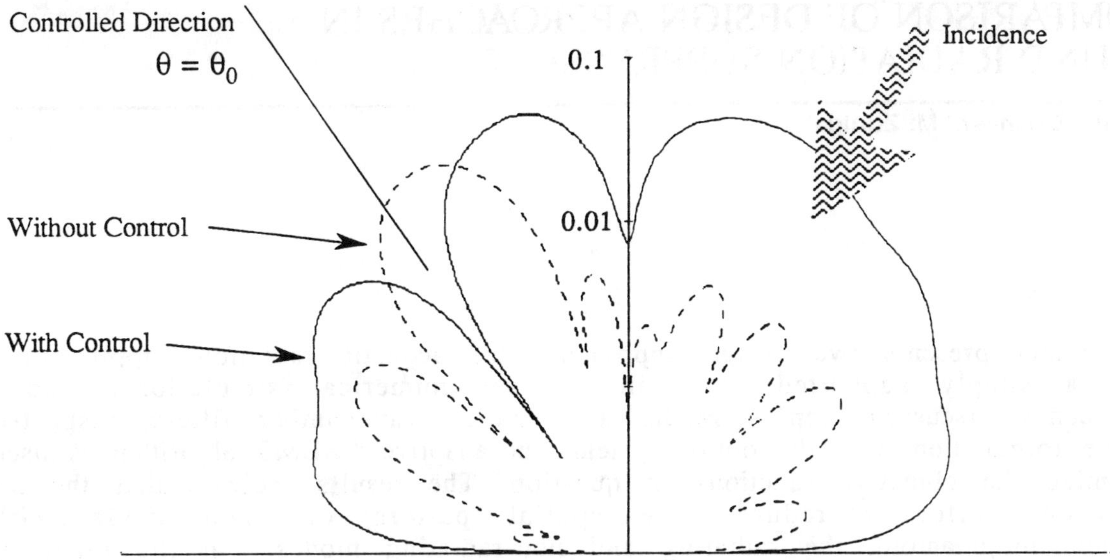

Fig. 5. The non-specular part of the reflection of an incident plane wave from the panel shown in Fig. 4, due to the incident field alone and due to the combined incident field and control drive applied as a line drive shown in Fig. 4.

References

1. Maidanik, G., and J. Dickey, 1988. "Acoustic behavior of ribbed panels." Journal of Sound and Vibration, 123: 293-307.

2. Maidanik, G., and J. Dickey. "Reflection of incident pressure waves by ribbed panels." In preparation.

COMPARISON OF DESIGN APPROACHES IN SOUND RADIATION SUPPRESSION

J. Thi,[1] E. Unver,[1] M. Zuniga[1]

ABSTRACT

This paper presents two design approaches for acoustic radiation suppression and uses a simply supported plate in air for numerical simulations. The two approaches discussed employ modal and spatial wavenumber filters, respectively. In the formulation of both control systems an adaptive FXLMS algorithm is used to minimize the objective functions in question. The results indicate that the modal approach uniformly reduces the spatial pattern of each mode without distinguishing among the radiating and non-radiating modes. On the other hand, wavenumber control alters the shapes of the radiating modes so that radiation cancelation occurs.

1. INTRODUCTION AND SUMMARY

Recently, substantial contributions have been made in the field of acoustic radiation suppression using adaptive techniques [2-5]. The main focus of these investigations is the use of secondary acoustic noise sources to cancel the undesired acoustic signal of a primary source. It is generally recognized that active suppression of acoustic radiation from a vibrating elastic body can be achieved by appropriately controlling the normal velocity distribution on the surface of the body. One approach is to independently control the normal modes of vibration [6] which offers a number of practical advantages, such as minimizing the number of actuators, sensors, and corresponding dimensionality of the controller. This is especially true if the vibration problem is known to involve only a few significant modes. Another approach is to control the vibrations associated only with surface velocity patterns that radiate well. It is known that for acoustic suppression, an active control system need only suppress the supersonic wavenumber components of the structural vibration and not necessarily the complete surface vibration distribution. Consequently, a wavenumber control system is expected to modify the structure vibration such that only the supersonic components are minimized over the frequency band of interest.

The objective of this paper is to compare the two aforementioned active control approaches in the design of a radiation suppression system. The acoustically radiating body studied here is a simply supported flat plate in air. The FXLMS filter algorithm [7] is used to adaptively minimize an objective function which represents, in the first approach, the power associated with three different modes of vibration, and in the second approach, the power of the supersonic velocity

[1]AT&T, Bell Laboratories, 1919 S. Eads Street, Arlington, VA 22202, U.S.A.

components. In the latter case, the wavenumber transformation is used to estimate the vibration associated with the supersonic components. A method for cascading three FXLMS notched filters is used for obtaining a wider cancellation bandwidth. Although a direct cascade of the FXLMS notched filter is not possible due to the cancellation path transfer function which does not allow access to the cancellation node, it is possible to configure a "pseudo cascade" arrangement that ideally performs as a true cascade but only requires one physical entry point [7].

Comparisons of the residual vibration signals for the two approaches demonstrate that wavenumber control indeed modifies the structural vibration such that only the acoustically efficient velocity patterns are reduced, whereas independent modal control minimizes the vibration associated with the selected modes to be controlled. For comparison, the uncontrolled and controlled plate response is presented in the spatial and wavenumber domains together with the associated acoustic farfield. It is found that both control approaches provide a means of acoustic suppression without acoustic farfield measurements.

2. MODAL FILTER APPROACH

For independent spatial modal control it is desirable to form linear combinations of the sensor signals in order to extract measurements of the mode shapes [8]. A modal model for the simply supported plate is derived in the Appendix. If L sensors are employed, then samples of the continuous eigenfunctions $C_\Phi(x)$ in Equation (A16) form a set of L dimensional discrete vectors $\Phi_1, \Phi_2, \ldots, \Phi_N$. Thus, for L error sensors and M modes to be controlled, the modal filter must take L inputs to produce M outputs as shown in Figure 1a.

A detailed formulation of an adaptive modal control system is described in Reference [6]. In this study, 19 sensors, 3 actuators and 10 modes are considered in the dynamics of the simply supported plate presented in the Appendix (see Equations (A13) and (A14)). Out of the 10 modes included in the dynamics of the plate, only 3 of them will be considered for control. The modal filters will be designed to minimize the average normalized cross-response power. Therefore the objective function described in Equation (1a) will be minimized with respect to f_n, subject to constraint of Equation (1b), where $f_1, f_2, f_3, \ldots, f_M$ represent a set of N-dimensional filter vectors and M denotes the number of actuators in the control system:

$$J = \sum_{m-1}^{M} (f_m^T \Phi_n)^2, \; n=1,2,\ldots,N \tag{1a}$$

$$f_n^T \Phi_n = 1 \tag{1b}$$

Using vector differentiation with the method of Lagrange multipliers as described in Reference [6], the solution to Equation (1) is given by

$$F = (C_\Phi C_\Phi^T)^{-1} C_\Phi \Lambda \tag{2}$$

$$F = [F_{m1}^T \, F_{m2}^T, \ldots F_{m10}^T]^T \tag{3}$$

where F_{mi} is the modal filter which allows the extraction of mode i from the L sensor measurement, Λ being a diagonal scaling matrix that is calculated to enforce the constraint of Equation (1b). A more detailed analysis of the modal based approach including weighted sum modal powers and performance degradation for a limited number of sensors can be found in reference [6]. In general, an actuator will not produce a single pure mode, but rather will generate a spectrum of spatial modes, therefore requiring a set of modal de-coupling filters. This problem is not considered here, however, as ideal actuator response is assumed.

3. FORMULATION OF SPATIAL WAVENUMBER FILTER

It is known in the structural acoustic literature that acoustic radiation can be reduced by altering the structural vibration pattern and not necessarily by eliminating it. Therefore, the objective of the wavenumber controller will be to minimize the wavenumber outputs $w_{nc}(t)$ and $w_{ns}(t)$, for discrete values of $k=k_n$, defined by the following cosine/sine wavenumber transform pairs:

$$w_{nc}(t) = \sum_{i=1}^{L} y(x_i,t)\ \cos(x_i k_n) \tag{4a}$$

$$w_{ns}(t) = \sum_{i=1}^{L} y(x_i,t)\ \sin(x_i k_n) \tag{4b}$$

where $y(x_i,t)$ is the plate displacement and, x_i is the location of the error sensor.

The linear combination of the sensor signals in Equation (4) can be considered as spatial filters that will pass the radiating components of the vibration signal but will reject the non-radiating components. The filter can be represented in the matrix form as

$$\begin{bmatrix} w_{1c} \\ w_{Nc} \\ ---- \\ \cdot \\ \cdot \\ ---- \\ w_{1s} \\ w_{Ns} \end{bmatrix} = \begin{bmatrix} \cos x_1 k_1 & \cdots & \cos x_L k_1 \\ \cdots & \cdots & \cdots \\ \cos x_1 k_N & \cdots & \cos x_L k_N \\ \sin x_1 k_1 & \cdots & \sin x_L k_1 \\ \cdots & \cdots & \cdots \\ \sin x_1 k_N & \cdots & \sin x_L k_N \end{bmatrix} \begin{bmatrix} y(x_1,t) \\ \cdot \\ \cdot \\ \cdot \\ \cdot \\ y(x_L,t) \end{bmatrix} \tag{5}$$

$$W = F_w^T\, Y \tag{6}$$

The selection of the maximum and minimum values of k is based of the radiating acoustic wavenumbers which is a function of the speed of sound c of the media and the maximum frequency of interest f_{max} and is given below:

$$|k| \le \frac{2\pi f_{max}}{c}. \tag{7}$$

4. FORMULATION OF THE ADAPTIVE FXLMS ALGORITHM:

In this section, a formulation of the control system using the FXLMS algorithm to minimize an objective function for both the modal filter and wavenumber filter approach is developed. Furthermore a cascade of three FXLMS notch filter is used to obtain a wider cancellation bandwidth. Although a direct cascade of the FXLMS notch filter is not possible due to cancellation path transfer function which does not allow access to the cancellation node, it is possible to configure a "pseudo cascade" arrangement that ideally performs as a true cascade but requires one physical entry point. The analysis presented for the FXLMS algorithm assumes a sinusoidal reference signal and the cancellation path transfer function is given by the dynamics of a simply supported plate. The state equation which describes the dynamics of the plate is derived in the Appendix. The parameters of the plate are also given in the Appendix. The 9 sinusoidal references were selected around the first three modal frequencies with 3 sinusoidal reference frequencies centered around each mode.

A. Modal Control

For a single reference signal consisting of one tone, the open loop transfer function is:

$$H_{mi}(\omega) = \frac{C_i(\omega)\, \mu_i\, (F_{mi}^T\, \hat{C}_i(\omega_i))^H\, F_{mi}^T}{j(\omega - \omega_i)\, T} \tag{8}$$

$$C_i(\omega) = C_\Phi (j\omega I - A)^{-1}\, B_{Fi} \tag{9}$$

where:

F_{mi}^T is the modal filter for extracting mode i from L linear measurements,

B_{Fi} is the B vector given in the Appendix with x_a as the actuator i^{th} location,

μ_i is the convergence gain for the ith reference sinusoidal,

$C_i(\omega)$ is the vector transfer function for actuator i to L sensors.

T is the sampling period and ω_i is the i reference sinusoidal frequency.

$\hat{C}_i(\omega)$ is the estimated vector transfer function from actuator i to L sensors. The derivation of Equation (8) is given in Reference [7]. $\hat{C}_i(\omega_i)$ can usually be experimentally determined by direct measurement. If 3 actuators and 3 reference sinusoidal signals are used, then the open loop transfer function from Equation (8) becomes:

$$H_{mc1}(\omega) = \sum_{i=1}^{3} \frac{C_i(\omega)\, \mu_i\, (F_{mi}^T\, \hat{C}_i(\omega_i))^H\, F_{mi}^T}{j(\omega - \omega_i)\, T} \tag{10}$$

The closed loop response is given by

$$G_{mc1}(\omega) = [I + H_{mc1}(\omega)]^{-1} \tag{11}$$

Figure 1a shows the system block diagram of Equation (11) and Figure 1b shows its equivalent representation which can be analyzed using standard techniques. Since the open loop gain is infinite for $\omega = \omega_i$, the closed loop response will exhibit a deep notch at the reference frequency. Thus, for a sinusoidal reference signal, the FXLMS algorithm forms an adaptive notch filter. For nine reference signals located at the first 3 modal frequencies, ω_i (i=1,2,3) consisting of the cascade of 3 tones, $\omega_i, \omega_{ii}, \omega_{iii}$, where ω_{ii} and ω_{iii} are adjacent to ω_i, the closed loop transfer function in Equation (11) becomes:

$$G_{mc}(\omega) = \left[I + \sum_{i=1}^{3} \frac{C_i(\omega)\, \mu_i\, (F_{mi}^T\, \hat{C}_i(\omega_i))^H\, F_{mi}^T}{j(\omega - \omega_i)\, T} \right]^{-1}$$

$$\cdot\left[I+\sum_{i=1}^{3}\frac{C_i(\omega)\,\mu_i\,(F_{mi}^T\,\hat{C}_i(\omega_{ii}))^H\,F_{mi}^T}{j(\omega-\omega_{ii})\,T}\right]^{-1}$$

$$\cdot\left[I+\sum_{i=1}^{3}\frac{C_i(\omega)\,\mu_i\,(F_{mi}^T\,\hat{C}_i(\omega_{iii}))^H\,F_{mi}^T}{j(\omega-\omega_{iii})\,T}\right]^{-1} \tag{12}$$

$$\cdot\, C_\Phi(j\omega I-A)^{-1}\,B_d$$

where, B_d is the B matrix given in the Appendix with x_a as the disturbance force location. Figure 3 shows the three stage FXLMS adaptive filter given by Equation (12).

B. Wave-Number Control

A similar formulation to the modal control above can be derived for the wave-number control system where Equation (8) is replaced by:

$$H_{wi}(\omega)=\frac{C_i(\omega)\,\mu_i\,I_R\,(F_w^T\,\hat{C}_i(\omega_i))^H\,F_w^T}{j(\omega-\omega_i)\,T} \tag{13}$$

where , F_w^T is the spatial wave number filter given by Equation (6) and, $I_R = [1\ \ 1 \ldots\ 1]_{1x37}$, a 1 x 37 row vector. The dimension 37 corresponds to the number of sine and cosine components of Equation (4) (with L=19) where one of the $\sin(x_i k_0)$ terms for $k_0 = 0$ is equal to zero.

For three reference signals and three actuators, the open loop response similar to Equation (10) for the wave number control approach can be derived as:

$$H_{wc1}(\omega)=\sum_{i=1}^{3}\frac{C_i(\omega)\,\mu_1\,I_R\,(F_w^T\,\hat{C}_i(\omega_i))^H\,F_w^T}{j(\omega-\omega_i)\,T} \tag{14}$$

For nine reference signals centered at three modal frequencies, the closed loop response becomes:

$$G_{wc}(\omega)=\left[I+\sum_{i=1}^{3}\frac{C_i(\omega)\,\mu_i\,I_R\,(F_w^T\,\hat{C}_i(\omega_i))^H\,F_w^T}{j(\omega-\omega_i)\,T}\right]^{-1}$$

$$\cdot\left[I+\sum_{i=1}^{3}\frac{C_i(\omega)\,\mu_i\,I_R\,(F_w^T\,\hat{C}_i(\omega_{ii}))^H\,F_w^T}{j(\omega-\omega_{ii})\,T}\right]^{-1} \tag{15}$$

$$\cdot \left[I + \sum_{i=1}^{3} \frac{C_i(\omega)\,\mu_i\, I_R\, (F_w^T\, \hat{C}_i(\omega_{iii}))^H\, F_w^T}{j(\omega-\omega_{iii})\,T} \right]^{-1}$$

$$\cdot\, C_\Phi(j\omega I-A)^{-1}\, B_d$$

Fig. 2a shows the system block diagram of Equation (15) and Fig. 2b shows its equivalent representation.

5. ACOUSTIC RADIATION:

Once the surface velocity distributions of interest are calculated the corresponding farfield pressure can be determined. Since this paper considers active sound radiation suppression of a plate in air, the mass loading of the surrounding medium on the plate can be neglected. The farfield pressure generated by this plate can be found by using Rayleigh's formula [9]:

$$p(R) = \left(\frac{\rho_0}{2\pi R}\right) \int_l e^{-jk|R-R_0|}\, \ddot{y}(R_0)\, dL(R_0), \qquad R_0 \ll R \qquad (16)$$

where $\ddot{y}(x)$ is the plate acceleration and $|R-R_0|$ is the distance between a field point and a point on the plate. Based on equation (1) the radiated farfield pressure of the plate shown in Figure 4 can be written as follows:

$$p(\theta) = \frac{\rho_0}{2\pi R}\, e^{-jk(R+l\sin\theta/2)} \int_0^L e^{jkx\sin\theta}\, \ddot{y}(x)\, dx \qquad (17)$$

where k is the acoustic wavenumber. Discretizing the integration in Equation (17) and noting that $\ddot{y}(x) = -\omega^2\, y(x)$, the farfield pressure $p(\theta)$ may be expressed as:

$$p(\theta) = \sum_{p=1}^{n_p} E'(\theta,x_p)\; y(x_p) \qquad (18)$$

where n_p is the number of surface sensors, x_p the coordinate of a surface sensor, and ω is the angular frequency and,

$$E'(\theta,x) = \frac{\rho_0}{2\pi R}\, e^{-jk(R+l\sin\theta/2)}(-\omega^2) \sum_{p=1}^{n_p} e^{jkx_p\sin\theta}\;. \qquad (19)$$

Here, $E'(\theta,x)$ is the matrix operator which maps the plate response, y(x) to the farfield pressure.

6. RESULTS

The simply supported plate described in the Appendix is used to illustrate the design approaches discussed in the preceding sections. The system considered consists of 3 modes, 19 sensors, and 3 actuators. Although the dynamic state equation of the plate derived in the Appendix includes 10 modes, only the first three modes are considered for control investigations. The locations of actuators sensors and disturbance signals are given in the Appendix. Equations (12) and

(15) are simulated using a step size gain of $\mu_i = 0.003$ with $T = 0.0005$ for $i = 1,2,3$. The multiple notches are overlapped at each modal frequency with a tone spacing of 0.25Hz. Although one could calculate the cancellation band at each mode and theoretically compute the optimal reference spacing which covers that bandwidth [7], in this study the spacing of the reference tones are not optimized for each mode.

Figure 5 shows the frequency response from the disturbance to sensor 8. Figure 6 shows the frequency response of the first three modes with and without modal control. Figure 7 shows the frequency response from disturbance to modal filter output. Figure 8 shows the frequency responses from disturbance to wavenumber filter output for cosine and sine components, respectively, as defined in Equations (4).

The spatial responses with and without modal control are depicted in Figure 9, which shows overall vibration level reductions across the plate such that the mode shapes are preserved. Figure 10 shows the effect of wavenumber control and demonstrates how the spatial responses are rearranged rather than suppressed. Out of the three modes selected the first and third modes are odd modes and the second mode is an even mode. Figure 10 depicts that the acoustically more efficient odd modes are affected much more significantly by the wavenumber control, than the acoustically inefficient second mode. Since the radiation cancellation process occurs when the in-phase and out-of-phase half waves negate [11] , the half cells of an even mode will largely cancel, whereas those of an odd mode will add. In fact, the effect of wavenumber control on the first mode especially reflects this process, where the spatial pattern of this mode is altered to form two half waves of opposite phases. The second mode is not altered as much since it is acoustically inefficient consisting of two half waves of opposite phases. Finally, for the third mode, where the half waves on the edges of the plate remain uncancelled, the wavenumber control changed the mode shape so that the areas under the waves of opposite phases are equal and that the cancellation could occur.

The wavenumber responses with and without control are presented in Figures 11 and 12 for each control approach. The former illustrates uniform reductions of wavenumber components, for each mode, throughout the whole wavenumber range considered. The latter shows that wavenumber domain control does not provide any significant reductions to mode 2 which does not radiate efficiently.

Finally, Figures 13 and 14 show the farfield pressures with and without modal and wavenumber control, respectively. As expected the performances obtained are similar to those obtained in the supersonic region of the wavenumber domain.

7. CONCLUSIONS

This paper presents two design approaches for acoustic radiation suppression and uses a simply supported plate in air for numerical simulations. The two approaches discussed employ modal and spatial wavenumber filters, respectively. The modal approach is designed to reduce the power of the selected modes of vibration, whereas the wavenumber approach is designed to reduce the supersonic components of plate vibration. In the formulation of both control systems an FXLMS algorithm is used to minimize the objective functions in question exploiting a cascading technique to obtain a wider cancellation bandwidth. Once the residual surface displacement patterns are obtained, they are multiplied by a radiation operator to determine the effects of control on the farfield pressure field.

The results indicate that the modal approach uniformly reduces the spatial pattern of each mode without distinguishing among the radiating and non-radiating modes. On the other hand, wavenumber control alters the shapes of the radiating modes so that radiation cancelation occurs.

APPENDIX . SIMPLY SUPPORTED PLATE MODEL

In this Appendix, the simply supported plate model (Figure 4) used for the analysis in the main text will be defined. It is assumed that the plate is a uniform of length L and width b. The mass per unit length m is assumed constant, with modules of elasticity E. The undamped homogeneous partial differential equation for the plate is given by

$$EI\frac{\partial^4 y}{\partial x^4} + m\frac{\partial^2 y}{\partial t^2} = 0 \qquad \text{(A1)}$$

The eigenfunction mode shapes of this equation are given by [10]

$$\Phi_n(x) = A_n \sin(\frac{n\pi x}{L}) \qquad \text{(A2)}$$

where A_n is a normalized constant chosen such that

$$\int_0^L \Phi_n^2(x)\, dx = 1 \qquad \text{(A3)}$$

and

$$\{A_n\} = \sqrt{\frac{2}{L}} \qquad \text{(A4)}$$

with corresponding natural frequencies

$$\omega_n = (\frac{n\pi}{L})^2\sqrt{\frac{EI}{m}} \qquad \text{(A5)}$$

where I is the second area moment.

When damping and loading are introduced, Equation (A1) is modified to [1]

$$EI\frac{\partial^4 y}{\partial x^4} + c_s I\frac{\partial^5 y}{\partial x^4 dt} + m\frac{\partial^2 y}{\partial t^2} + c\frac{\partial y}{\partial t} = f(x,t) \qquad \text{(A6)}$$

where c_s is the strain velocity damping coefficient (internal), c is the absolute velocity damping coefficient (external), and f(x,t) is the loading force per unit length as a function of location x. and time t. The displacement and the loading force is decomposed into modal components

$$f(x,t) = \sum_{n=1}^{N} F_n(t)\, \Phi_n(x) \qquad \text{(A7)}$$

$$y(x,t) = \sum_{n=1}^{N} y_n(t)\, \Phi_n(x) \qquad \text{(A8)}$$

and the Φ_n are defined in Equation (A2). By substituting Equation (A7) and Equation (A8) into (A6), multiplying by $\Phi_n(x)$, integrating, and using the orthogonality conditions results in the time differential equation [1]

$$\ddot{y}_n(t) + 2\zeta_n\omega_n\dot{y}_n(t) + \omega_n^2\, y_n(t) = F_n(t,x)/m \tag{A9}$$

where

$$\zeta_n = \frac{c_s\omega_n}{2E} + \frac{c}{2m\omega_n} \tag{A10}$$

is the damping coefficient. For the special case of a line force excitation at location x_a, the force is expressed as

$$f(x,t) = f(t)\ \delta(x-x_a) \tag{A11}$$

where each

$$F_n(t,x_a) = f(t)\Phi_n(x_a) \tag{A12}$$

The response y(x,t) can be expressed in the state space representation using Equations (A8), (A9), and (A12) as following,

$$\dot{x} = A\ x(t) + B\ f(t) \tag{A13}$$

$$y(x,t) = C_\Phi'(x)\ x(t) \tag{A14}$$

where

$$A = \begin{bmatrix} A_1 & 0 & 0 & 0 \\ 0 & A_2 & 0 & 0 \\ \cdots & \cdots & \cdots & \cdots \\ 0 & 0 & A_n & 0 \\ 0 & 0 & 0 & A_N \end{bmatrix}, \qquad B = \frac{\sqrt{2}}{m}\begin{bmatrix} -\frac{1}{2\beta_1}\sin(\pi x_a/L) \\ 0 \\ ------------ \\ \cdots\cdots\cdots \\ ------------ \\ -\frac{1}{2\beta_N}\sin(N\pi x_a/L) \\ 0 \end{bmatrix} \tag{A15}$$

$$C_\Phi(x) = [\ \Phi_1(x) \mid \Phi_2(x) \mid \ldots \mid \Phi_N(x)],$$
$$C_\Phi'(x_0) = [\ 0 \mid \Phi_1(x) \mid \ldots \mid 0 \mid \Phi_N(x)], \tag{A16}$$

$$An = \begin{bmatrix} \alpha_n & \beta_n \\ -\beta_n & \alpha_n \end{bmatrix}, \tag{A17}$$

$n = 1,2,\ldots,N$

$$\alpha_n = -\zeta_n\ \omega_n \tag{A18}$$

$$\beta_n = \omega_n\sqrt{1-\zeta_n^2} \tag{A19}$$

A special case for N=10 is used in this paper. The following parameter values were used in the numerical calculations.

Dimensions:

- Length, l = 1m
- Width, w = 0.01m
- Thickness, t = 0.002m
- Second moment of area, $I = wt^3/12 = 2/3\times10^{-11}\ m^4$

Material Properties:

- Modules of Elasticity, $E = 2\times10^{11}\ N/m^2$
- Mass density per unit volume, $\rho = 8\times10^3\ kg/m^3$

Damping Parameters:

$c_s/E = 7.75 \times 10^{-6}$ s

$c/m = 0.202\ s^{-1}$

$m = \rho wt = 0.16$ kg/m

with the above values, the natural frequencies are then calculated from Equation (A5) as

$$\{f_n\} = \{\omega_n/(2\pi)\} = \{4.5344,\ 18.1379,\ 40.8104,...\} \qquad (A20)$$

The above damping parameters were selected so that $\zeta_1 = \zeta_{10} = 0.01$ from Equation (A10) for the first and tenth modal frequencies since only the first ten modes are considered in here.

The plate was discretized using 161 points with:
Actuator locations at points 40, 70 and 110
Sensor locations at points n*8, n=1,19 and,
Disturbance location at point 60.

REFERENCES

[1] R.W. Clough, J. Penzien, *Dynamics of Structures* (McGraw-Hill, New York, 1975).

[2] P. A. Nelson, A. R. D. Curtis, S. J. Elliott, and A. J. Bullmore, "The Active Minimization of Harmonic Enclosed Sound Fields, Part II: A Computer Simulation, "*J. Sound Vibration.* 117, 35-58 (1987).

[3] S. J. Elliott, I. M. Stothers, and P. A. Nelson, "A Multiple Error LMS Algorithm and Its Application to the Active Control of Sound and Vibration," *IEEE Trans. Acoust.*, Speed, Signal Processing ASSP-35, 1423-1434 (1987).

[4] L. J. Erickson, M. C. Allie, "System Considerations for Adaptive Modeling Applied to Active Noise Control," IEEE Int. Symp. on Circuits and Systems.

[5] L. J. Erickson, M. C. Allie, "The Use of Random Noise for On-line Transducer Modeling in an Adaptive Active Attenuation System, "*J. Acoust. Soc. Am.* 80 SII (A), (1986).

[6] D. R. Morgan "An Adaptive Modal-based Active Control System" *J. Acoust. Soc. Am.*, 89, 248-256, (January 1991).

[7] D. R. Morgan, J. Thi "A Multi-tone Pseudo-casade Filtered-X LMS Adaptive Notch Filter," to be presented at 1991 IEEE Int. Symp. on Acoust. Speed and Signal Processing.

[8] L. Meirovitch and J. K. Bennighof, "Modal Control of Traveling Waves in Flexible Structures," *J. Sound Vibration.* III, 131-144 (1986).

[9] M. C. Junger and D. Feit, *Sound Structures and Their Interaction* (MIT, Cambridge, MA, 1982) 2nd ed.

[10] L. Meirovitch and H. Baruh, "On the Problem of Observation Spillover in Self-Adjoint Distributed-Parameter Systems," *J. Optimization Theory and Applications* 39, pp. 269-291 (1983).

[11] F. Fahy, *Sound and Structural Vibration* (Academic Press, 1985).

This work was supported by the Office of Naval Research under Contract N00014-90-C-0258

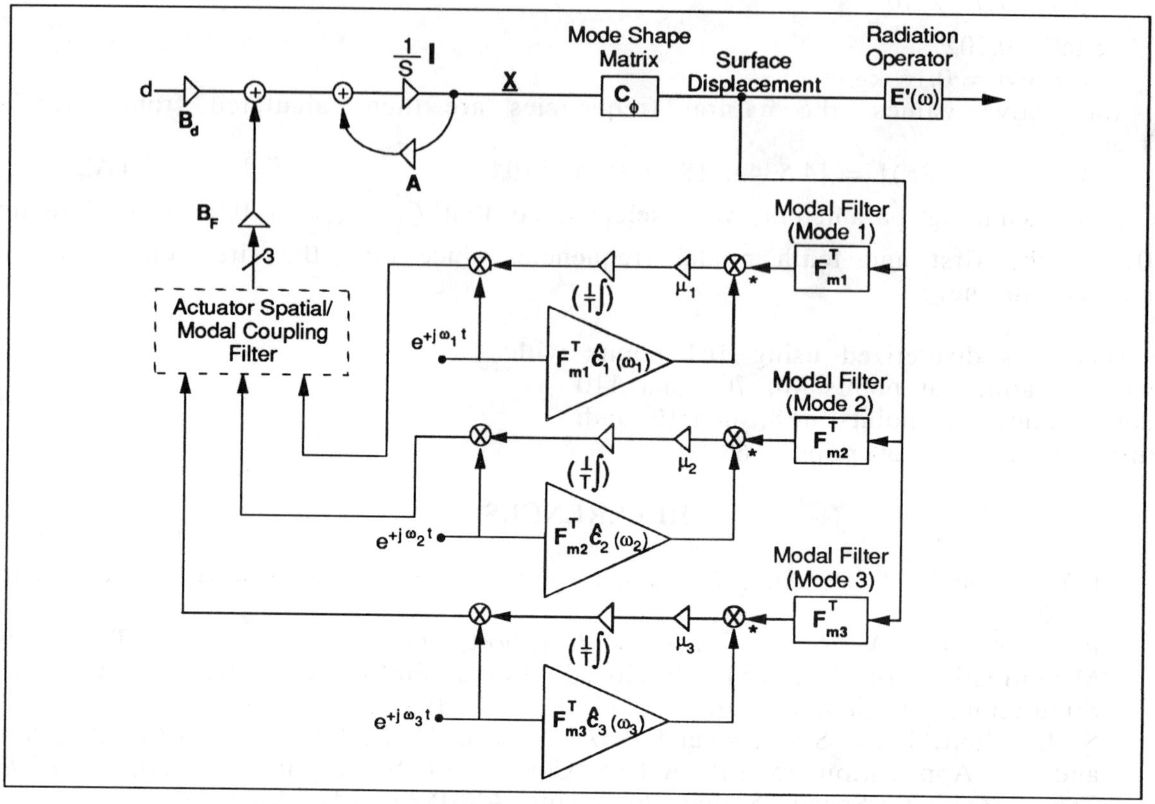

Figure 1a. Three Tone Adaptive FX LMS Algorithm
With Modal Filter

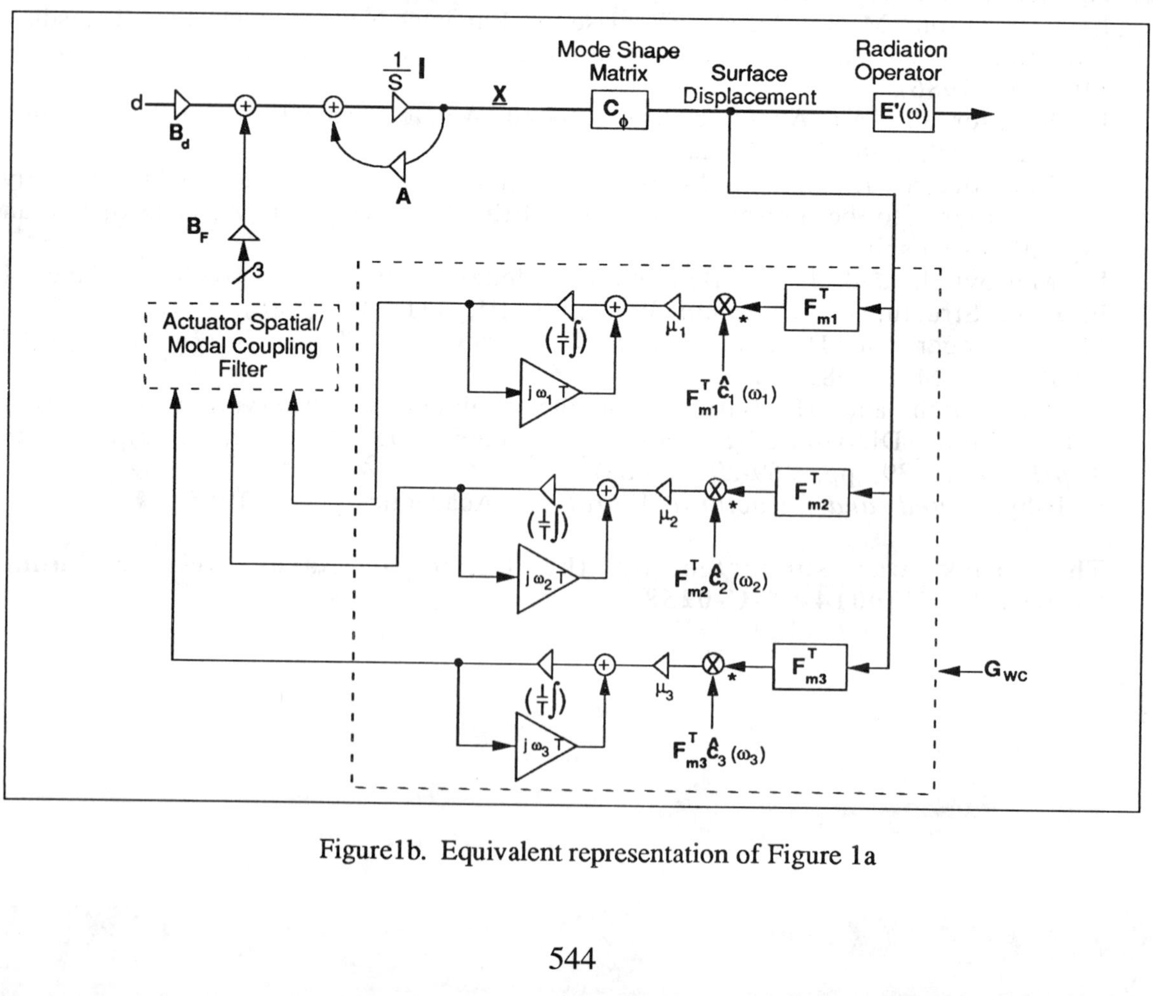

Figure1b. Equivalent representation of Figure 1a

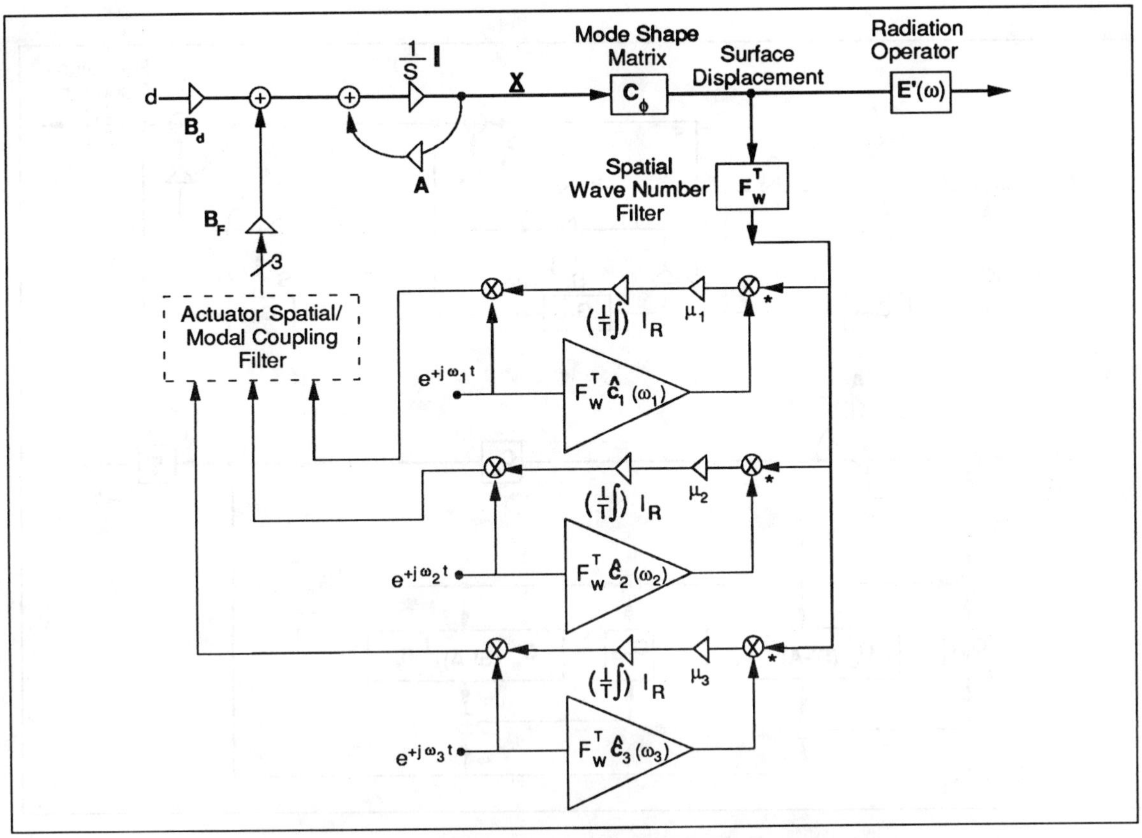

Figure 2a. Three Tone Adaptive FX LMS Algorithm With Spatial Wave Number Filter

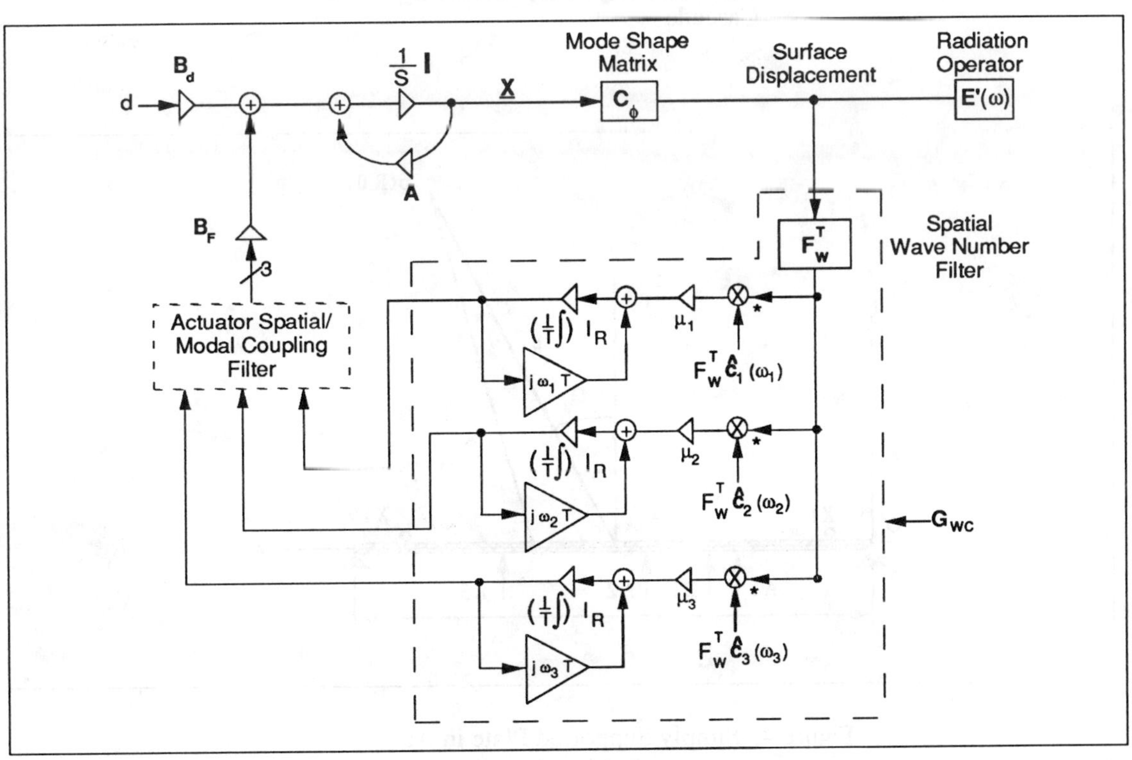

Figure 2b. Equivalent representation of Figure 2a

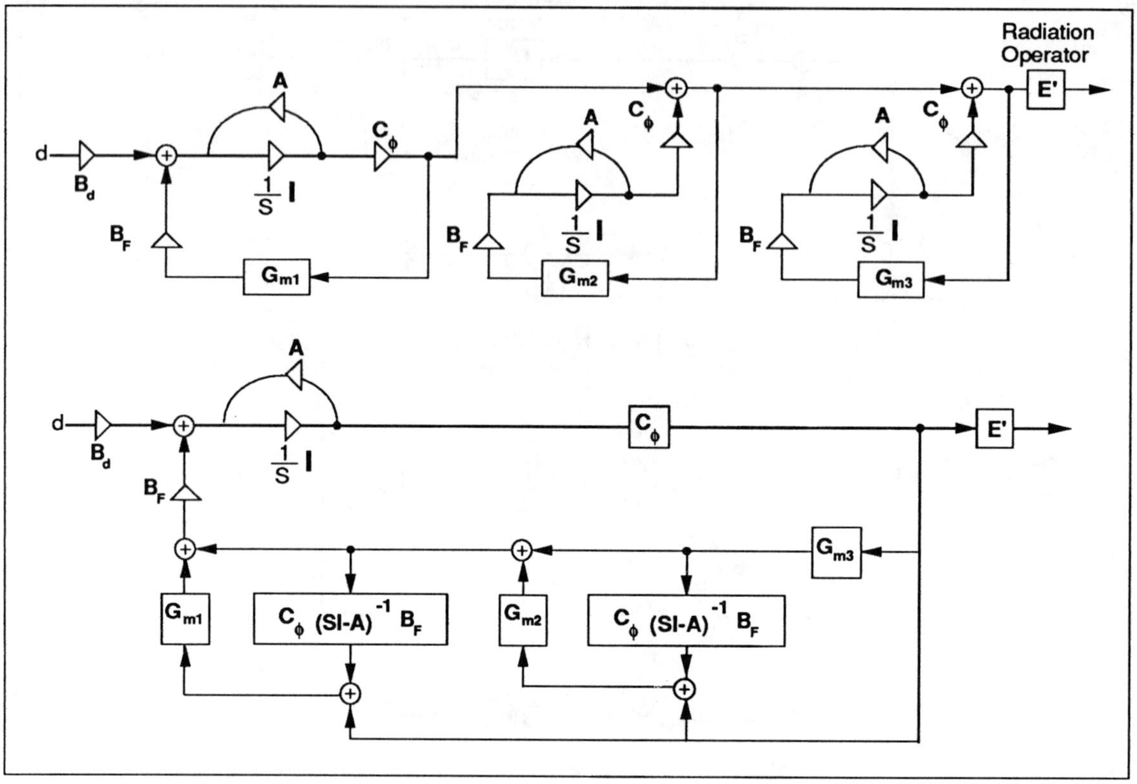

Figure 3. Three-Stage FX LMS Adaptive Filter
(a) True Cascade (b) Alternative Pseudo
Cascade

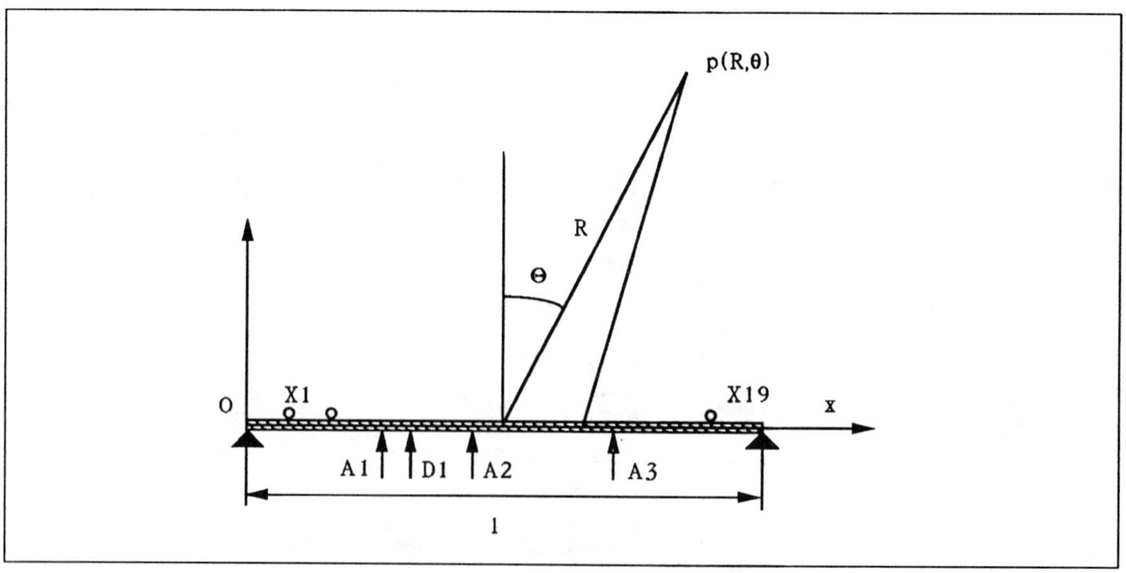

Figure 4. Simply Supported Plate in Air

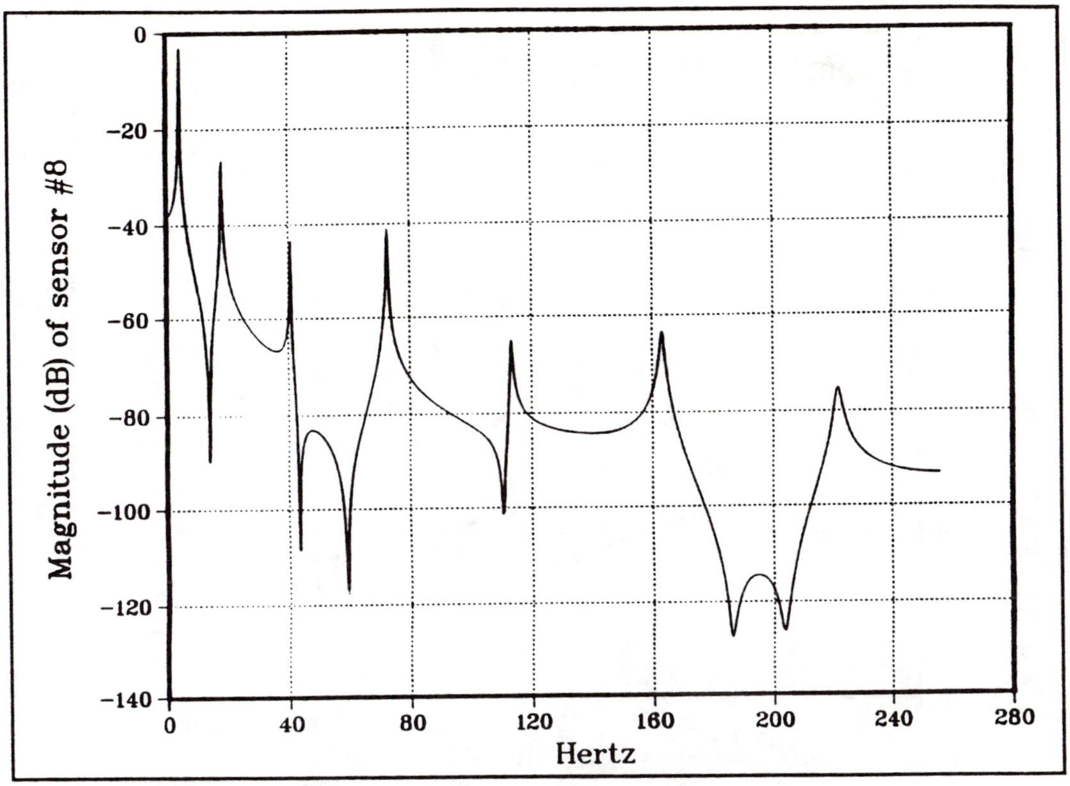

Figure 5. Frequency Response from Disturbance to sensor 8

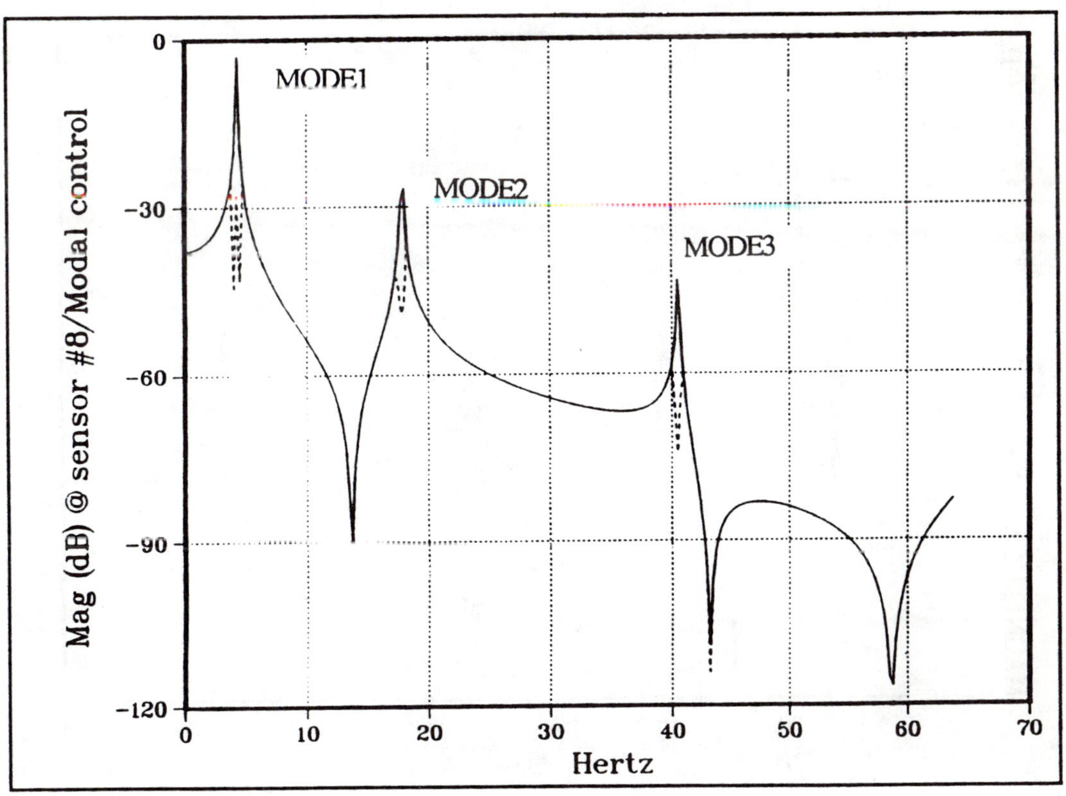

Figure 6. Frequency Response from Disturbance to sensor 8
with and without control.

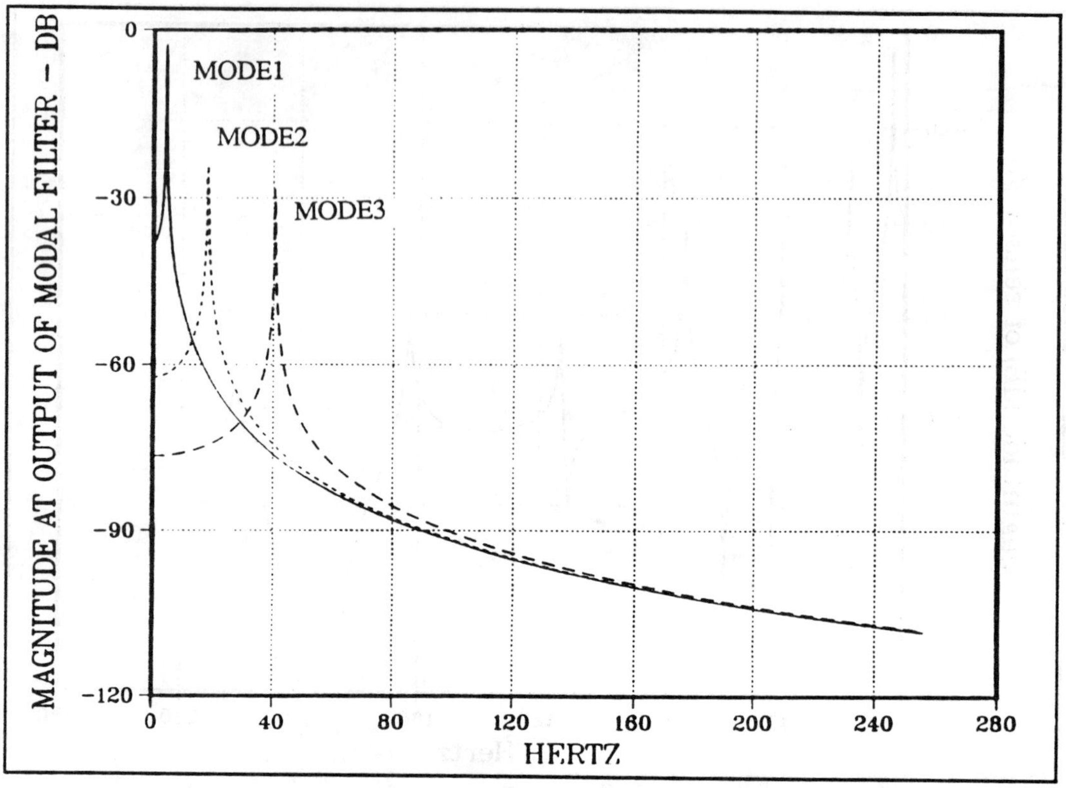

Figure 7. Frequency Response from Disturbance to modal filter outputs

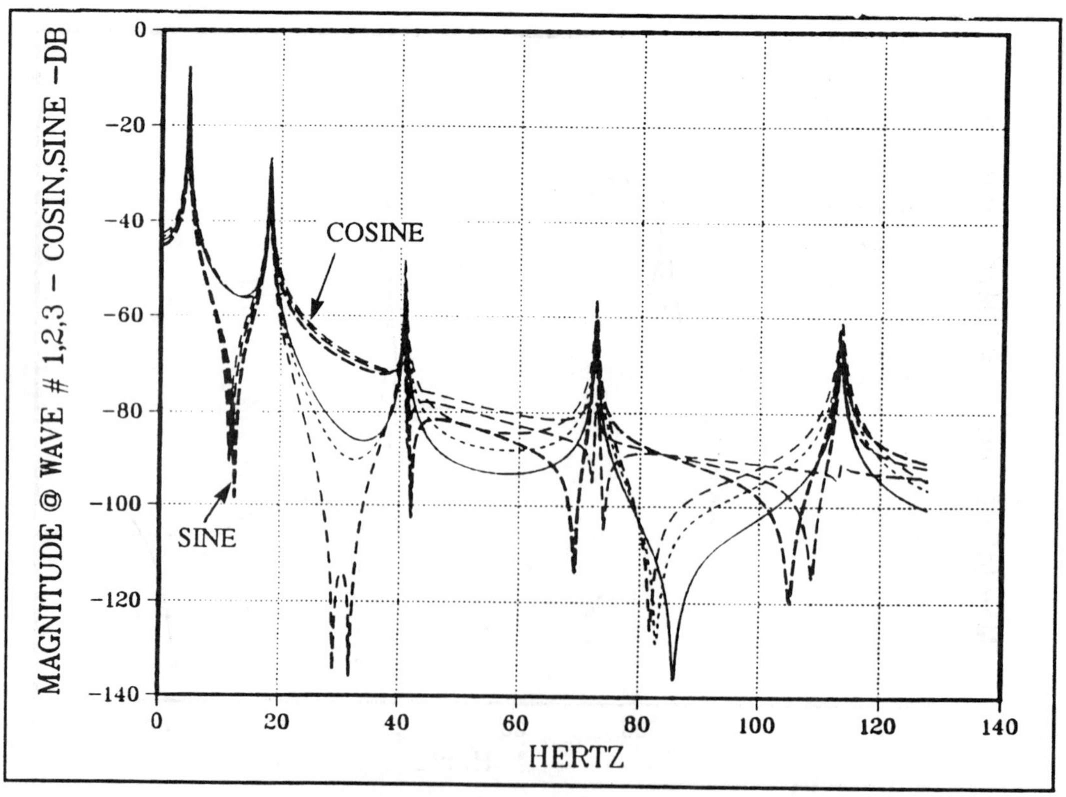

Figure 8. Frequency Responses from Disturbance to wavenumber filter outputs 1, 2, 3 (cosine and sine)

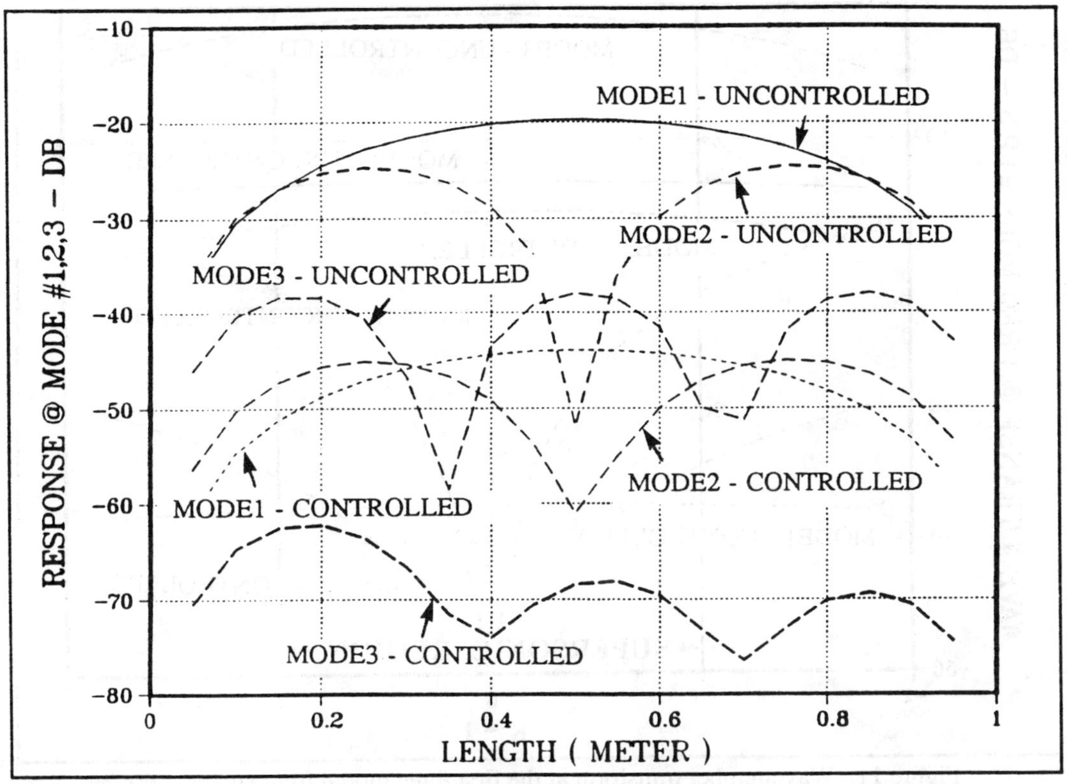

Figure 9. Spatial Response with and without modal control

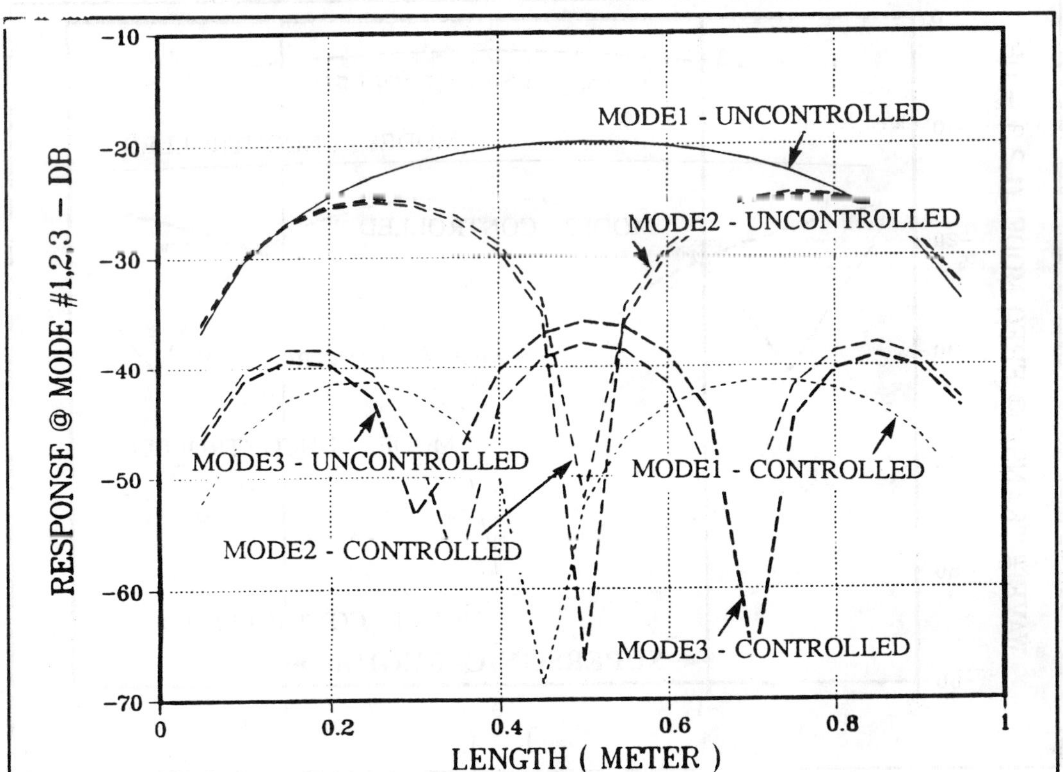

Figure 10. Spatial Response with and without wavenumber control

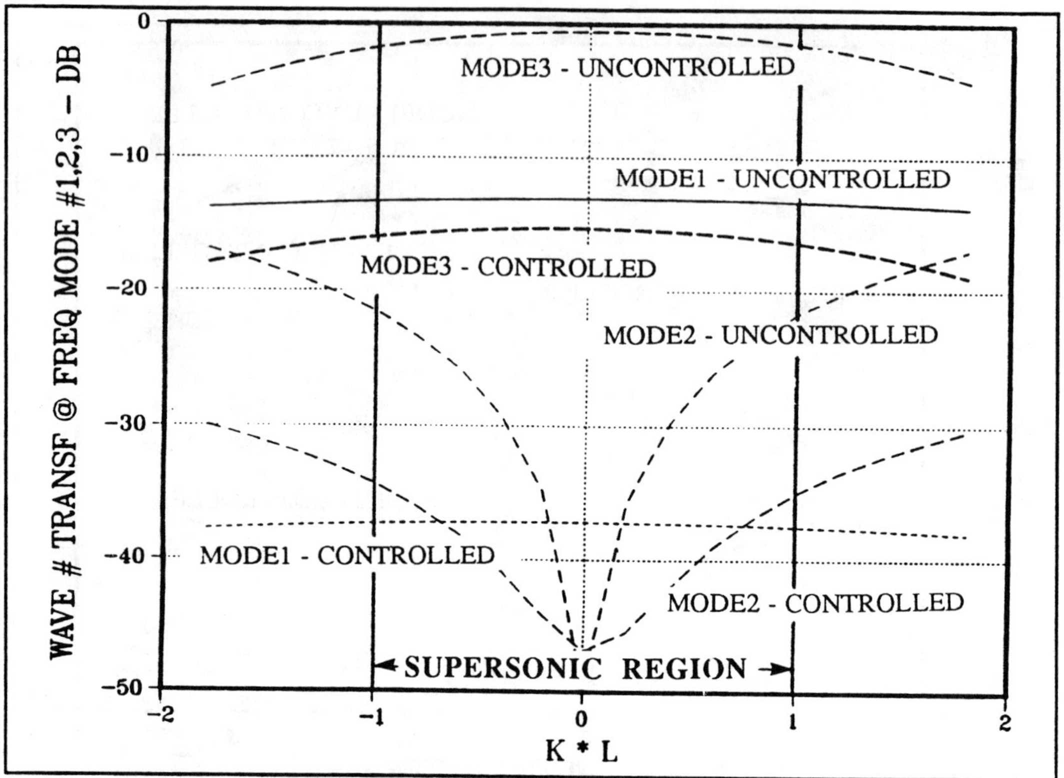

Figure 11. Wavenumber transform at the first three modal frequencies with and without modal control

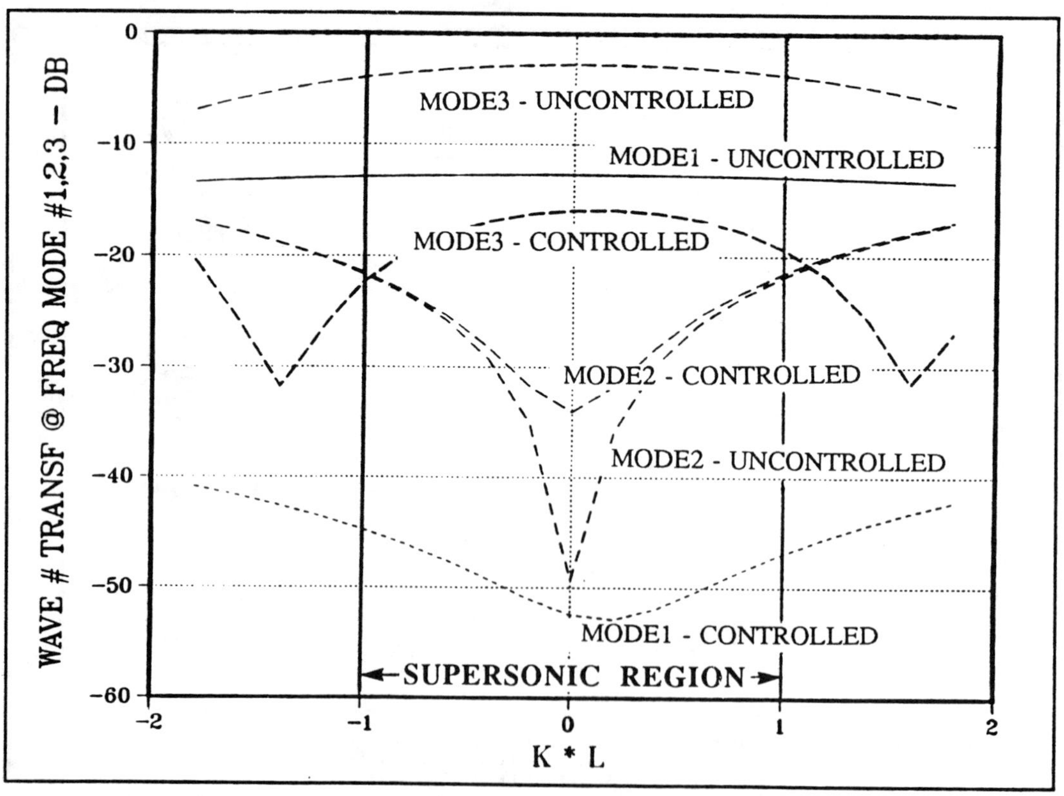

Figure 12. Wavenumber transform at the first three modal frequencies with and without wavenumber control

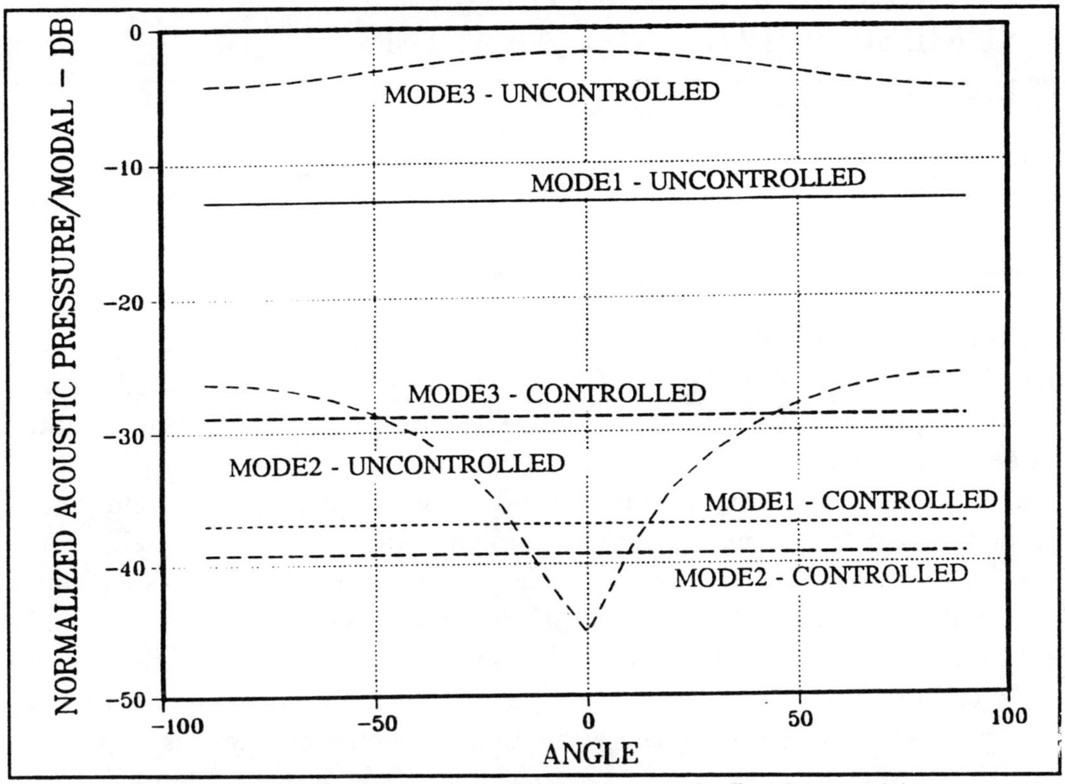

Figure 13. Normalized acoustic farfield pressure at the first three modal frequencies with and without modal control

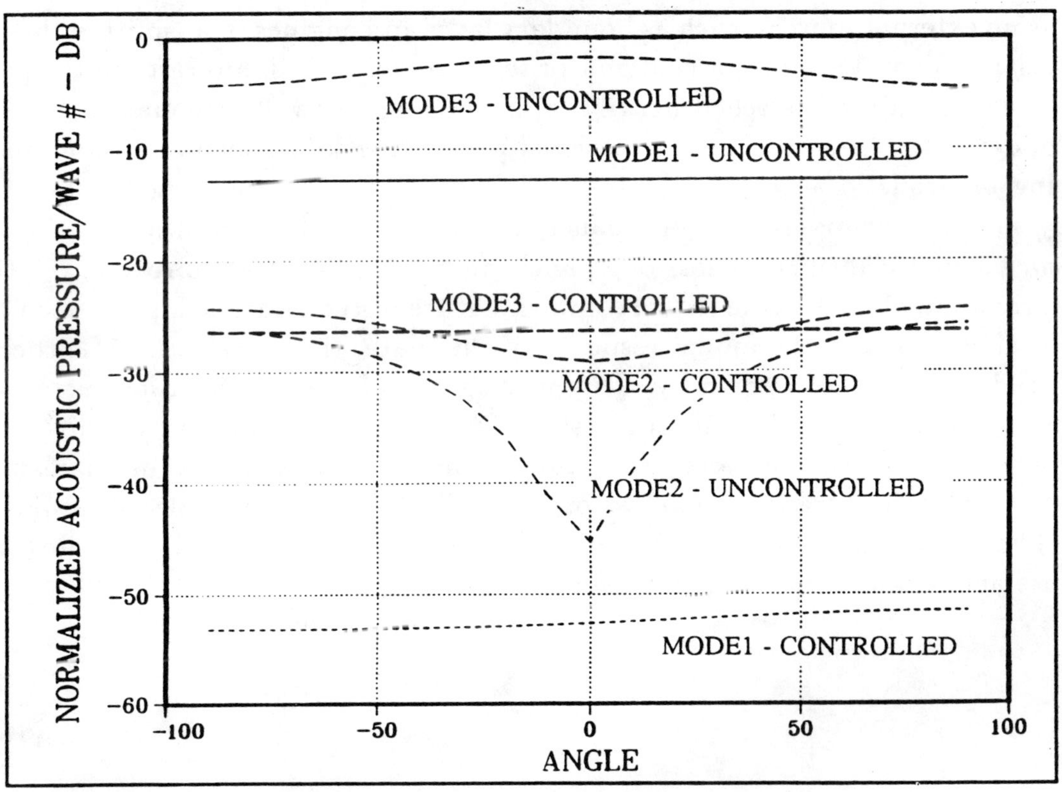

Figure 14. Normalized acoustic farfield pressure at the first three modal frequencies with and without wavenumber control

A STUDY OF STRUCTURAL CONTROL OF PLATE TRANSMISSION LOSS USING WAVENUMBER–FREQUENCY ANALYSIS

Gopal P. Mathur,[1] Boi N. Tran[1]

ABSTRACT

The mechanism of improving the sound transmission loss of a plate by single or multiple vibratory inputs acting on the vibrating structure is studied using the wavenumber-frequency analysis approach. The vibration response of a thin, flat plate due to multiple, point mechanical inputs and a uniform acoustic field is determined using the modal expansion method. The radiated sound field from the plate is then calculated using a nearfield implementation of the Rayleigh integral based on the boundary element method. The wavenumber spectrum is obtained by two-dimensional Fourier transformation of the spatial vibration response and the radiated sound field. The analytical results presented in this paper show that the effects of structural input levels, phase and locations on the sound transmission loss of a plate can be evaluated using the wavenumber-frequency analysis approach.

INTRODUCTION

Noise from external sources, such as boundary layer and engines, transmitted through the fuselage sidewall is a major contributor to cabin noise in all aircraft. In the past, interior noise control has relied heavily on improving sidewall attenuation by passive "add-on" treatments. The conventional passive methods, such as adding mass, damping or acoustic absorption, have been ineffective in improving the low frequency sound transmission loss of a panel. Recent studies have shown that the low frequency sound transmission loss of a panel can be significantly improved by active structural inputs [1,2,3]. Due to the complex nature of the relationship between the structural control and the modal response of the plate, the mechanism of airborne sound radiation and consequently of improving sound transmission loss using active inputs is not yet sufficiently understood [4]. The main objective of this paper is to present a wavenumber analysis approach of understanding the mechanism of improving the sound transmission loss of a panel by single or multiple vibratory inputs. The analytical results on the effects of structural control of airborne vibration response and sound radiation from a finite plate are presented.

[1]Douglas Aircraft Company, 3855 Lakewood Blvd.; MC 36-60, Long Beach, CA 90846, U.S.A.

ANALYSIS

We consider the problem of controlling airborne sound radiation from a distributed system, such as a finite plate, by means of single or multiple, discrete actuators placed at points on the vibrating surface. These actuators are used in an open loop control scheme. The governing differential equation of motion for the transverse vibration displacement, $w(x, y, t)$, of a simply-supported uniform finite thin isotropic plate (dimensions: 1.7526m × 1.143m × 1.27mm) mounted in an infinite baffle may be written as:

$$D\nabla^4 w(x, y; t) + \lambda \dot{w}(x, y; t) + \mu \ddot{w}(x, y; t) = f(x, y; t) \tag{1}$$

where the biharmonic operator is:

$$\nabla^4 = \left\{ \frac{\partial^2}{\partial x^2} + \frac{\partial^2}{\partial y^2} \right\}^2 \tag{2}$$

and D is flexural rigidity of the plate, μ is mass per unit area of the plate, λ is damping coefficient, $f(x, y; t)$ represents forcing function, and $\dot{w}$ and $\ddot{w}$ are, respectively, the first and second derivitives of w with respect to time t.

The airborne excitation considered is in the form of a spatially uniform, normally incident acoustic field, f_a. The control inputs are simulated by a set of point loads, $\{f_{sl}\}_{l=1}^{N_s}$. The forcing functions may be represented by:

$$f(x, y; t) = f_a e^{j\omega t} \tag{3}$$

and

$$f(x, y; t) = \sum_{l=1}^{N_s} f_{sl}\, \delta(x - a_l, y - b_l) e^{j\omega t} \tag{4}$$

where the mechanical inputs f_{sl} are applied at locations defined by (a_l, b_l), the symbol δ denotes the Dirac delta function, N_s is the number of mechanical point forces, and ω is radian frequency.

The steady state solution for the transverse vibration response of the rectangular plate due to combined acoustic and structural inputs can be described by the application of modal analysis method as follows:

$$w(x, y; t) - \sum_m \sum_n (A_{mn} + S_{mn}) \sin\left(\frac{m\pi x}{L_x}\right) \sin\left(\frac{n\pi y}{L_y}\right) e^{j(\omega t - \gamma_{mn})} \tag{5}$$

where L_x and L_y are dimensions of the plate, γ_{mn} is the phase angle associated with the mn^{th} mode due to the modal damping, and A_{mn} and S_{mn} represent airborne and structure-borne influence coefficients for the mn^{th} mode respectively.

The airborne influence coefficient A_{mn} of the mn^{th} vibrational mode is [5]:

$$A_{mn} = \frac{4f_a e^{-j\phi_a}}{\rho h \omega_{mn}^2 mn\pi^2} \frac{(1-\cos m\pi)(1-\cos n\pi)}{\sqrt{(1-\frac{\omega^2}{\omega_{mn}^2})^2 + (2\xi_{mn}\frac{\omega}{\omega_{mn}})^2}} \tag{6}$$

and structure-borne influence coefficient S_{mn} of the mn^{th} vibrational mode is

$$S_{mn} = \frac{4\sum_{l=1}^{N_s} f_{sl} e^{-j\phi_{sl}} \sin(\frac{m\pi a_l}{L_x}) \sin(\frac{n\pi b_l}{L_y})}{\rho h \omega_{mn}^2 L_x L_y \sqrt{(1-\frac{\omega^2}{\omega_{mn}^2})^2 + (2\xi_{mn}\frac{\omega}{\omega_{mn}})^2}} \tag{7}$$

where ρ is density of the plate, h is thickness of the plate, ξ_{mn} is modal damping coefficient, and ϕ_a and ϕ_{sl} are phase angles associated with airborne and structure-borne inputs respectively. The modal natural frequencies ω_{mn} of the plate are given by:

$$\omega_{mn}^2 = \frac{D}{\rho h}\left[\left(\frac{m\pi}{L_x}\right)^2 + \left(\frac{n\pi}{L_y}\right)^2\right]^2 \tag{8}$$

The modal wavenumbers are given by:

$$\begin{aligned} k_{mn}^2 &= k_m^2 + k_n^2 \\ &= \left(\frac{m\pi}{L_x}\right)^2 + \left(\frac{n\pi}{L_y}\right)^2 \end{aligned} \tag{9}$$

where $k_m = \frac{m\pi}{L_x}$ and $k_n = \frac{n\pi}{L_y}$ are the modal wavenumbers in the x and y directions respectively.

The wavenumber, k_{mn} , may also be expressed as follows:

$$k_{mn}^2 = k_b^2 = \sqrt{\frac{\omega_{mn}^2 \rho h}{D}} \tag{10}$$

where k_b is the free bending wavenumber of the plate.

In this study, the plate is modeled as a series of flat rectangular elements. The Rayleigh integral is evaluated numerically in a piecewise manner on an element by element basis, similar to the boundary element method. The well-known Rayleigh integral equation, in its normal form, describes the acoustic pressure field generated by a normal velocity distribution on an infinite surface. For the case of a finite flat plate in an infinite rigid baffle, the integral needs to be evaluated only on the plate surface because the vibration velocity is assumed to be zero on the baffle. The sound pressure radiated by the finite plate measured at a distance $\mathbf{r}$ is then given by

$$\mathbf{p}(\mathbf{r},\omega) = \frac{j\omega\rho_0}{2\pi}\int_S \mathbf{v}_n(\mathbf{r}_S,\omega)\frac{e^{-jkR}}{R}dS \tag{11}$$

where:
$\mathbf{v}_n(\mathbf{r}_S, \omega)$ = normal velocity of dS,
$R = | \mathbf{r} - \mathbf{r}_S |$,
$\mathbf{r}$ = position vector of the observation point,
$\mathbf{r}_S$ = position vector of the elemental surface dS,
S = surface area of the plate,
ρ_0 = density of ambient medium.

The spatial transform of the radiated surface pressure of the plate is

$$P(k_x, k_y; \omega) = \int\int \mathbf{p}(x, y; \omega)\ \mathrm{e}^{-j(k_x x + k_y y)}\ \mathrm{d}x\mathrm{d}y \tag{12}$$

The total sound power radiated from the plate is

$$\pi(\omega) = \int\int_{k_b < k_0} |P(k_x, k_y; \omega)|^2 \sqrt{1 - \left(\frac{k_x^2 + k_y^2}{k_0^2}\right)}\ \mathrm{d}k_x \mathrm{d}k_y \tag{13}$$

where k_0 is the acoustic wavenumber.

The change in total sound power radiated from the plate due to combined excitations versus the sound power radiated due to only acoustic excitation is defined by $\Delta(\omega)$ as follows:

$$\Delta(\omega) = 10\ log_{10}\left(\frac{\pi(\omega)}{\pi_a(\omega)}\right) \tag{14}$$

where $\pi_a(\omega)$ is the power radiated with airborne excitation only.

If $\Delta(\omega) > 0$; then there is an increase in radiated sound power, whereas $\Delta(\omega) \leq 0$ indicates reduction in radiated sound power.

RESULTS AND DISCUSSION

The vibration response and sound radiation from the simply supported plate for several cases of individual and combined excitations were computed in both spatial, (i.e. (x, y)), and wavenumber (k-space , i.e. k_x and k_y) domains using the analytical model. The vibration velocity field was predicted over the actual plate area. The sound pressure field in the near vicinity (1.25cm from plate surface) of the plate was calculated over an area which is twice the plate surface area. The coincidence frequency of the plate is 9450 Hz. The frequency range of interest in the present investigation is much below the coincidence frequency.

Figure 1 shows the k-space spectrum of the radiated acoustic pressure (at 277 Hz) when the plate is subjected to uniform acoustic excitation, impinging normal to the plate. The incident sound pressure level is set at 100 dB. The wavenumber pressure spectrum for the airborne case exhibits a dominant peak at zero wavenumber which occurs due to spatial coupling between the lower order nonresonant modes of the plate and the uniform acoustic field [5]. The k-space spectrum of the acoustic pressure due to a single mechanical input force, shown in Figure 2, indicates that the nearfield sound radiation for the mechanical excitation is mostly from acoustically slow resonant modes. The lower order acoustically fast modes also appear to contribute to the overall sound radiation from the plate.

The effects of the number, location, amplitude and phase of the control forces, on the sound radiated due to uniform acoustic excitation were investigated using several configurations or arrangements of the control actuators. The results from some typical cases are discussed here. The wavenumber plots of the radiated pressure for the combined input of a single mechanical control force of amplitude 0.185 N ($\frac{1}{24}$ lb) acting 180 degrees out of phase with the uniform acoustic field is shown in Figure 3. The application of a single control force at an arbitrary location appears to be quite ineffective in reducing the low wavenumber component of the radiated sound field.

The k-space pressure spectrum due to a combination of four mechanical control forces; each of amplitude 0.185 N ($\frac{1}{24}$ lb) and acting 180 degrees out of phase with the unform acoustic field is shown in Figure 4. The overall reduction in sound power radiated for this case is about 4.6 dB over the airborne excitation acting alone. The wavenumber plot, shown in Figure 4, illustrates this phenomenon in terms of reduced pressure field at or near zero wavenumber ($k_b < k_0$). The effect of changing phase of the four control forces with reference to the acoustic field can be seen in Figures 5 and 6. The control forces appear to be effective only around the phase angle of 180 degrees. For other combinations of phase angles between each control force and acoustic field, the control forces appear to lose their effectiveness in improving the sound transmission loss of the plate.

The combined effect of varying the amplitude and phase of control forces acting at specified locations on the plate is shown in Figure 7. All four control inputs were assigned the same amplitude and phase for these computations. The four control forces with amplitude of 0.185 N ($\frac{1}{24}$ lb) and the phase angle of 180 degrees provided an optimum combination for reducing radiated sound power due to the airborne excitation of 100 dB amplitude. The radiated sound power is found to increase over that of the airborne case for some combinations of phase angles, particularly below 90 degrees. As seen from Figures 7 and 8, an increase in the amplitude of the control forces need not produce maximum reduction of the radiated pressure field.

The effect of increasing number of control forces was also investigated using the wavenumber-frequency analysis. A summary of results showing overall reduction in radiated sound power is illustrated in Figure 9. It may be observed from this figure that an increase in the number of control forces does not necessarily result in an optimum solution, i.e., more reducion in the radiated pressure field.

CONCLUSIONS

The wavenumber-frequency analysis presented in this paper shows that the mechanism of improving sound transmission characteristics of a plate using discrete control forces can be explained in k-space . The spatial distributions of multiple point forces acting on the finite plate were shown to have an important effect on the radiated sound field. The airborne sound radiated by the panel is mostly concentrated at low wavenumbers since the acoustic field can spatially couple with the lower order nonresonant modes. Consequently, the cancelling structural inputs must be applied in such a way as to simulate a spatial excitation field that can couple with the lower order modes and induce out-of-phase nonresonant vibration response in the plate. The application of a single out-of-phase structural input applied at an arbitrary location was found to be quite ineffective in reducing airborne sound radiation from the plate. The results presented in this paper show that the effects of structural input levels, phase and locations on the sound transmission loss of a plate can be evaluated and optimized using the wavenumber analysis approach.

ACKNOWLEDGEMENTS

This research work was conducted under the McDonnell Douglas Independent Research and Development program.

References

[1] Fuller, C. R., "Active Control of Sound Transmission/Radiation From Elastic Plates by Vibrational Inputs: I. Analysis," *Journal of Sound and Vibration*, v136, 1990, pp. 1-15.

[2] Meirovitch, L. and Thangjitham, S., "Active Control of Sound Radiation Pressure,", *Journal of Vibration and Acoustics*, Trans. ASME, v112, 1990, pp. 237-244.

[3] Hansen, C. H., Snyder, S. D. and Jie, Pan, "A Methodology for the Design of Systems to Actively Control Sound from Vibrating Structures," *Proc. International Congress on Recent Developments in Air- and Structure-Borne Sound and Vibration*, Auburn University AL, March 6-8, 1990, pp. 517-524.

[4] Thomas, D. R., Nelson, P. A. and Elliott, S. J., "Experiments on the Active Control of the Transmission of Sound Through A Clamped Rectangular Plate," *Journal of Sound and Vibration*, v139, 1990, pp. 351-355.

[5] Mathur, G. P., Gardner, B. K., Phillips, J. E. and Burge, P. L., "An Experimental Technique of Separating Airborne and Structure-Borne Noise Using Wavenumber-Frequency Spectrum," *Paper presented at the 13th AIAA Aeroacoustics Conference* held at Tallahasse FL, October 22-24, 1990, AIAA Paper 90-3969.

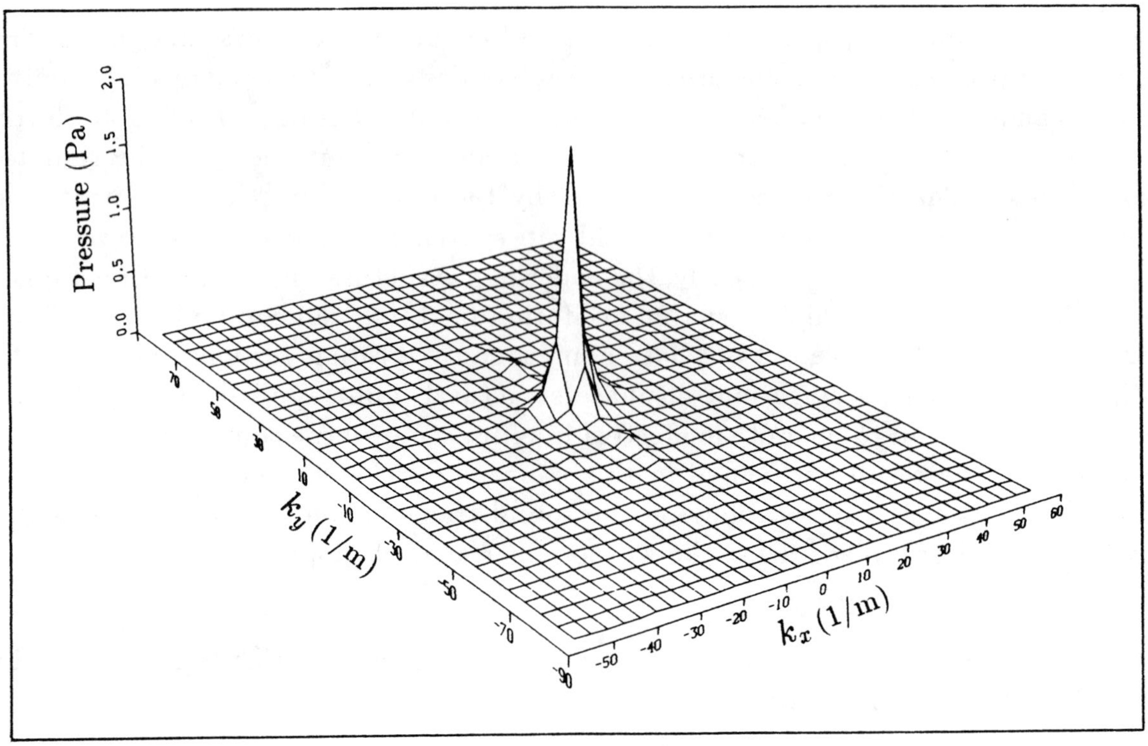

Figure 1. K-Space Pressure Spectrum for Airborne Excitation (100 dB) at 277 Hz.

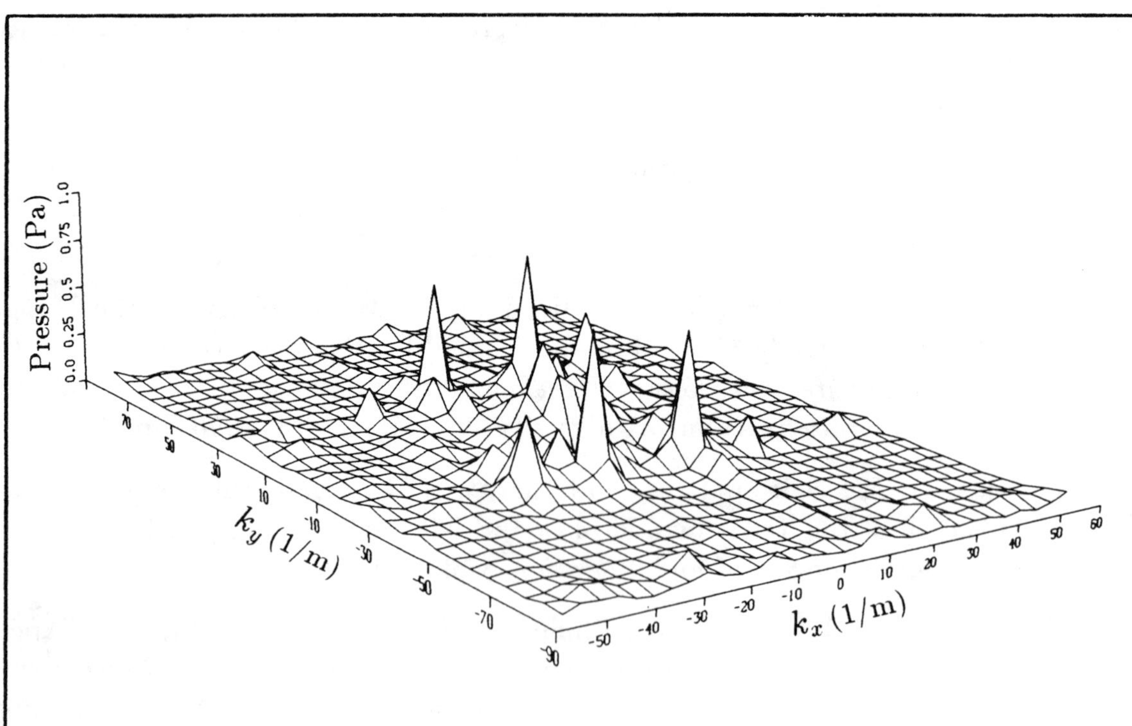

Figure 2. K-Space Pressure Spectrum for Single Mechanical Input (0.185 N) at 277 Hz.

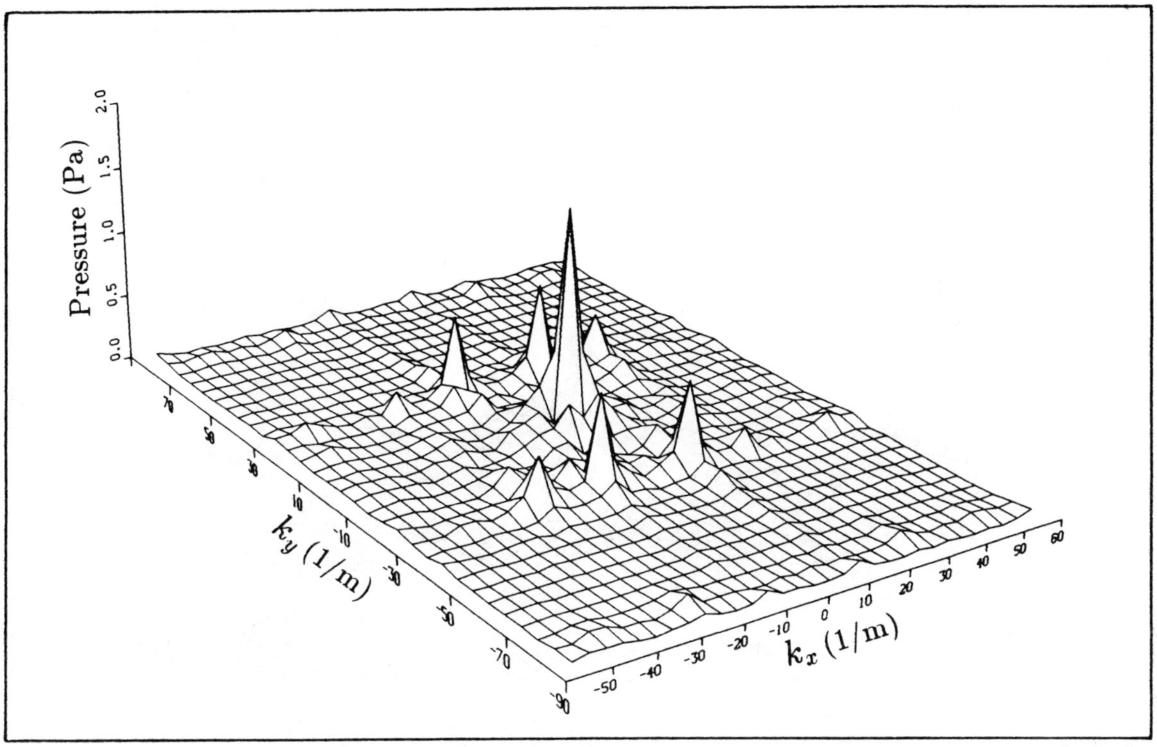

Figure 3. K-Space Pressure Spectrum for Airborne Excitation (100 dB) and Single Control Input (0.185 N) at 277 Hz.

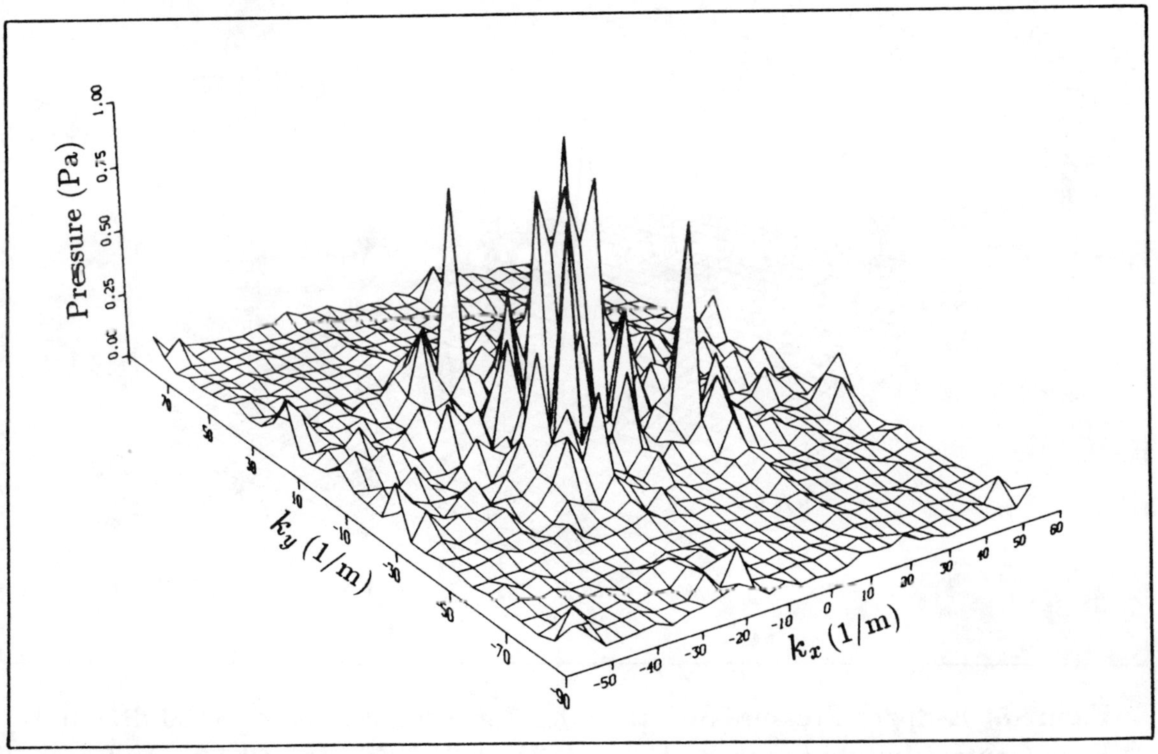

Figure 4. K-Space Pressure Spectrum for Airborne Excitation (100 dB) and Four Control Inputs (Each with 0.185 N and 180 Degree Phase Angle) at 277 Hz.

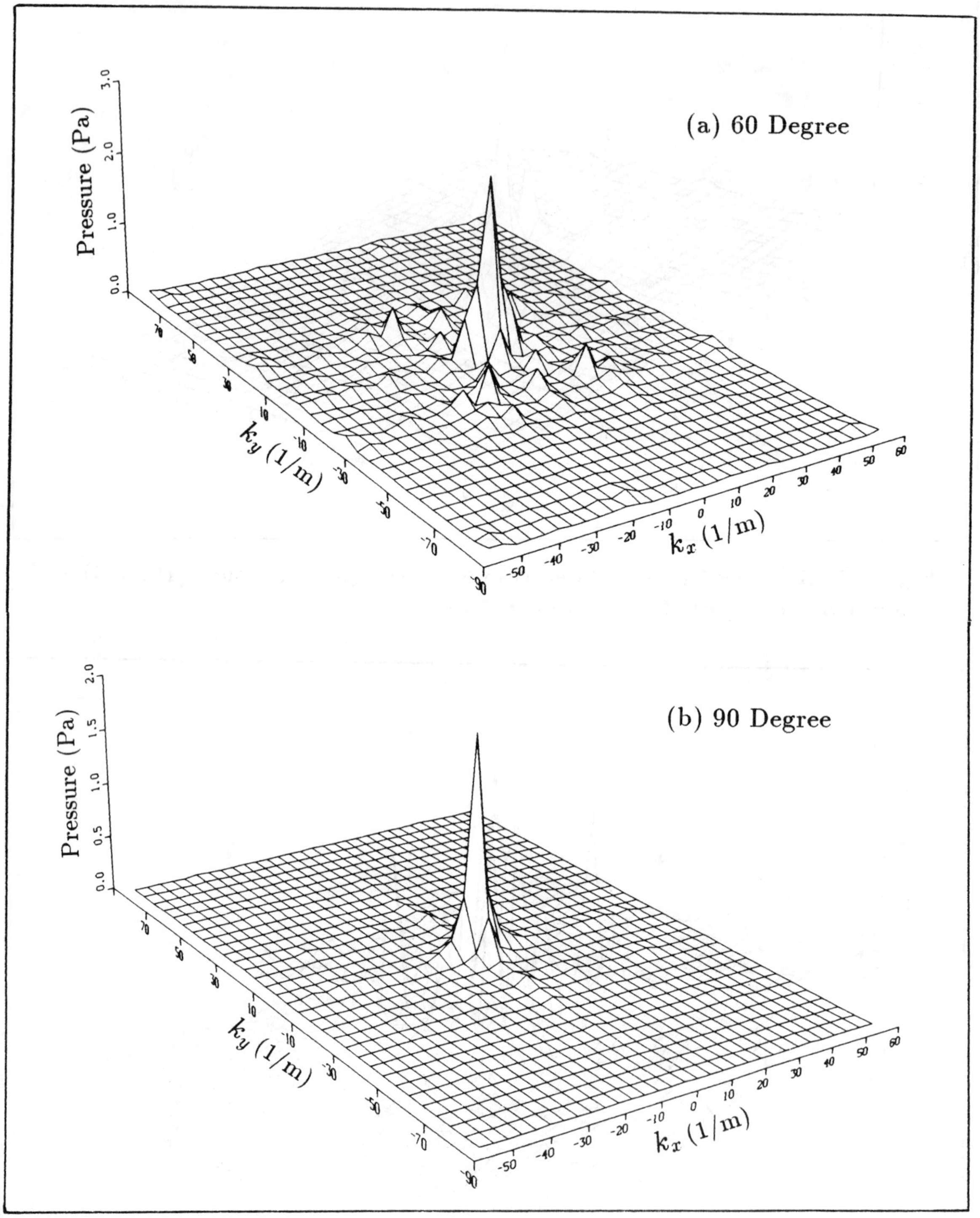

Figure 5. K-Space Pressure Spectrum for Airborne Excitation (100 dB) and Four Control Inputs (Each with 0.185 N) at 277 Hz. (a) 60 Degree Phase Angle, and (b) 90 Degree Phase Angle.

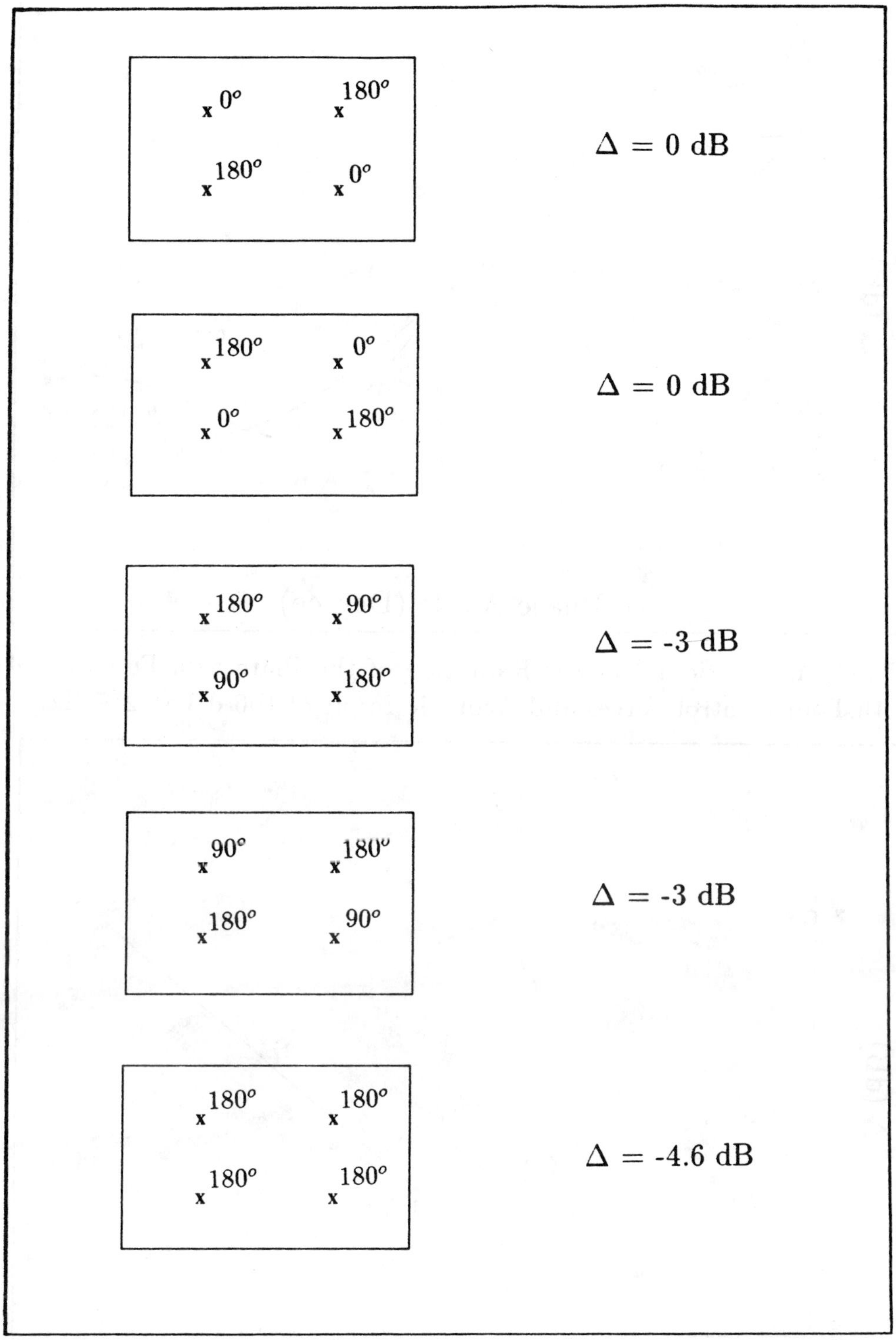

Figure 6. Sound Power Reduction as a Function of Phase Angle (4 Control-Point Forces, each with 0.185 N, and Acoustic Input of 100 dB at 277 Hz).

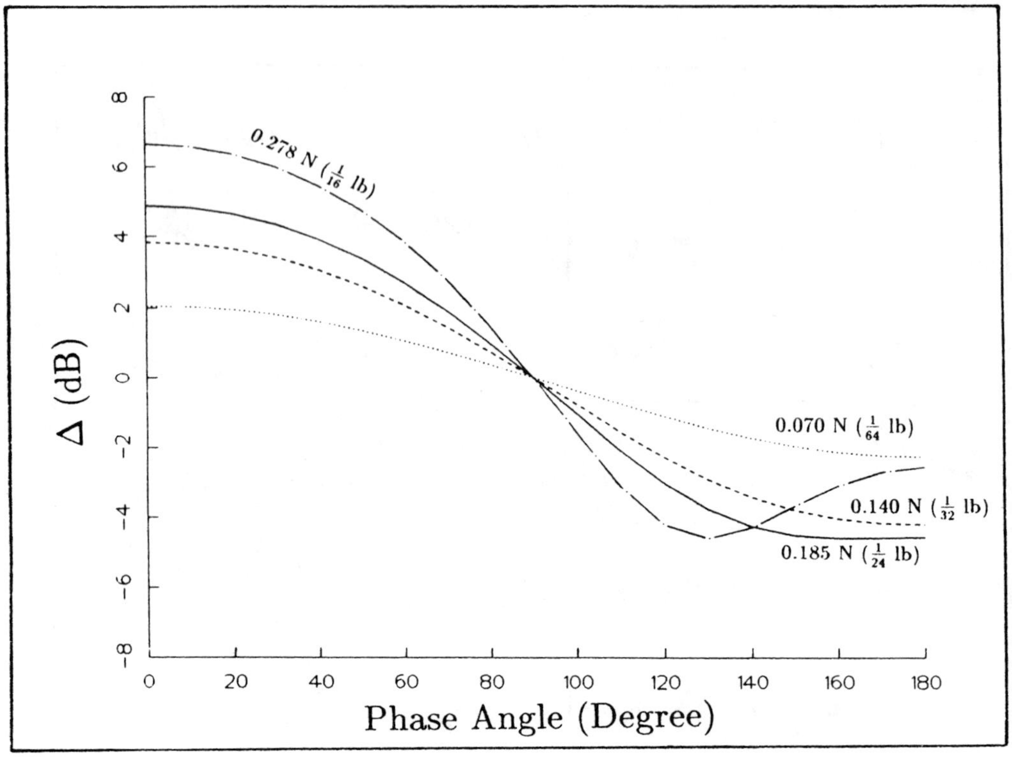

Figure 7. Change in Sound Power Radiation of the Plate as a Function of Phase Angle with Four Control Forces and Acoustic Input of 100 dB at 277 Hz.

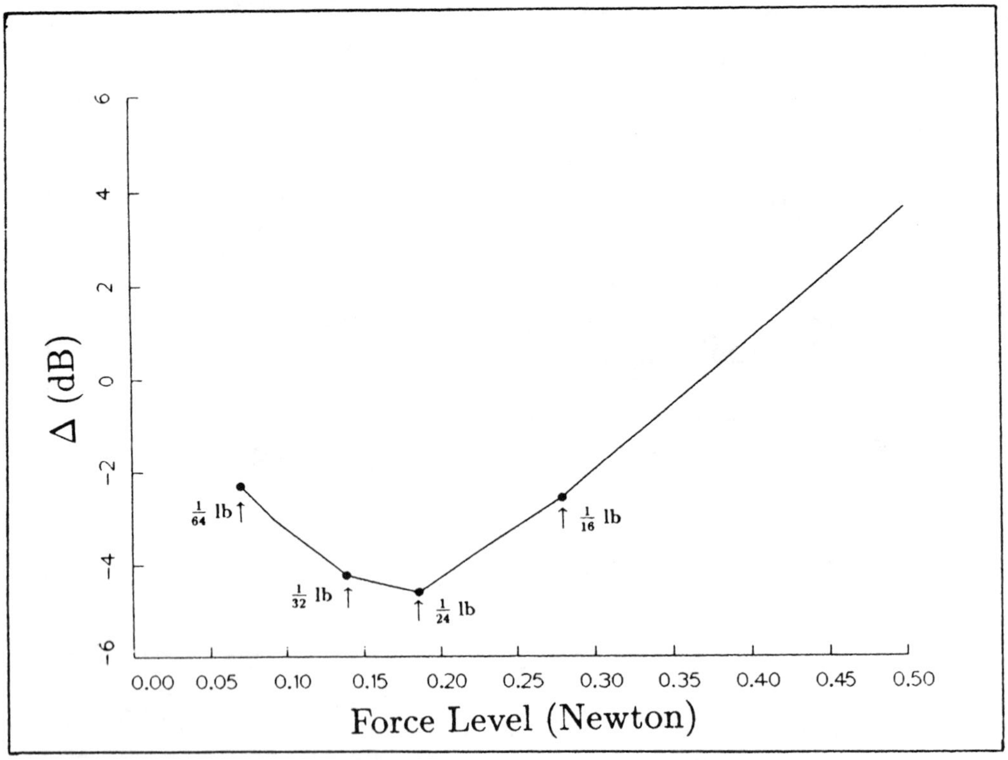

Figure 8. Change in Sound Power Radiation of the Plate as a Function of Control Force Level with Four Control Forces (180 Degree Phase) and Acoustic Input of 100 dB at 277 Hz.

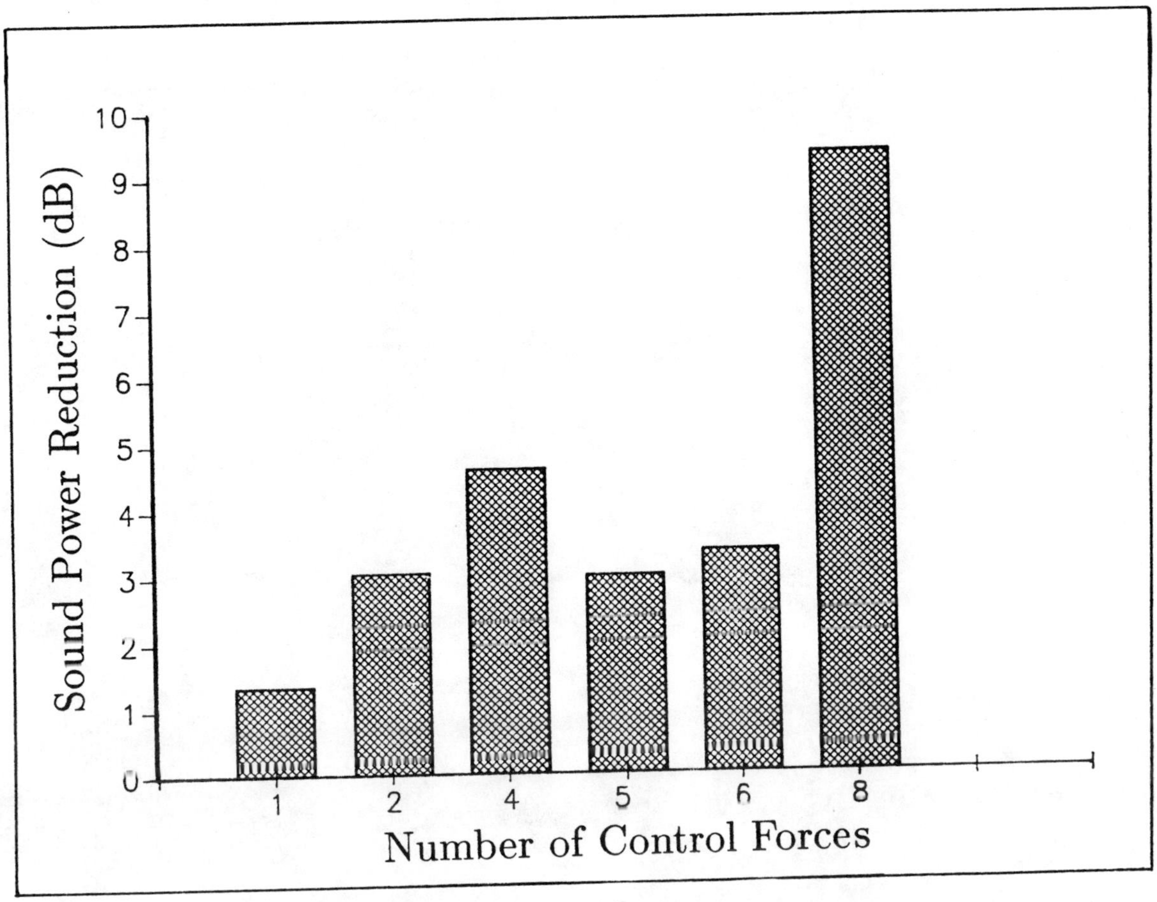

Figure 9. Sound Power Reduction of the Plate as a Function of Number of Control Forces for Acoustic Input of 100 dB at 277 Hz.

Flexible Structures I

AN ACTIVE SUSPENSION SYSTEM FOR SPACE STRUCTURE DYNAMIC TESTING

Dexter Johnson,[1] Daniel J. Inman,[2] Garnett C. Horner[3]

ABSTRACT

With the dynamic testing and analysis of proposed space structures currently in progress, there arise many problems that must be investigated and solved. One such problem that has been observed involves eliminating or minimizing the undesired effects of suspension system dynamic interaction with test structure dynamics. This paper reports on an investigation of this problem and proposes a possible solution. In particular, the use of active suspension control is investigated and is incorporated into a large angle slewing experiment of an eleven foot steel I-beam.

The results of this work show that, for this particular application of active suspension control, the undesired effects of the suspension system on the suspended test structure can be minimized and practically eliminated in a dynamic slewing maneuver.

INTRODUCTION

With extensive ground testing of various proposed large space structures currently underway it has become necessary to examine the many effects that earth bound testing can incur. In particular, the effects of the suspension system of suspended test structures is of interest here. In the dynamic testing of space structures, or as most often the case their scaled models, the structure is usually supported on a platform or suspended by long cables. The suspension dynamics consequently become coupled with the structure's dynamics and pollute the actual dynamics of the structure. This occurs especially when the suspension system is passive and has limited action or is fixed in its position. Most research involving suspension/structure dynamic interaction has only examined passive or fixed suspension systems. Such work has been performed by the authors of [2,6,7,9-11]. The predominant effect that is observed in this investigation is the dynamic coupling between the pendulum mode of a suspended structure and its rigid body mode. Another effect that is commonly investigated is the dynamic interaction of the cable modes and the suspended structure's pendulum mode.

An alternative to long cable suspension systems is an active control suspension system whereby the suspension changes position with respect to a change in position

[1]Ph.D Candidate, Department of Mechanical and Aerospace Engineering, State University of New York at Buffalo (currently employed as an aerospace engineer at NASA Lewis Research Center, 21000 Brookpark Rd., Cleveland, OH 44135, U.S.A.).
[2]Professor and Chairman of the Department of Mechanical and Aerospace Engineering, State University of New York at Buffalo, Buffalo, NY, U.S.A.
[3]Research Engineer at the NASA Langley Research Center, Hampton, VA, U.S.A.

of the test structure. An example of an active suspension system is presented here. A particular active suspension design was implemented on a large stiff beam (11 ft. steel I-beam) in a slewing maneuver. The beam was hinged at one end by a universal joint attached to a fixed support and cable suspended at the other end. The motion of interest is the slewing motion about the vertical axis of the hinge (see Fig. 1.). The suspension system sits on a cart above the cable suspended end of the beam with the cable connected to the cart. The cart moves circumferentially along a circular track with its radius approximately equal to the length of the beam. The cart is designed with four wheels, two pulleys, and a short timing belt. Two of the wheels are located so as to support the weight of the cart and beam and rolls along the top of the rim of the track. Another wheel is positioned to sit against the inside circumference of the track and is spring loaded to assure firm contact of the cart to the track. The last wheel is positioned to roll on the platform and help align the cart belt with the track timing belt. The pulleys are situated, with the small timing belt around them, to fit flush against the strip of timing belt on the track. The cart is driven by a d.c. motor connected to the shaft of one of the pulleys and therefore is enabled to move about the circumference of the track.

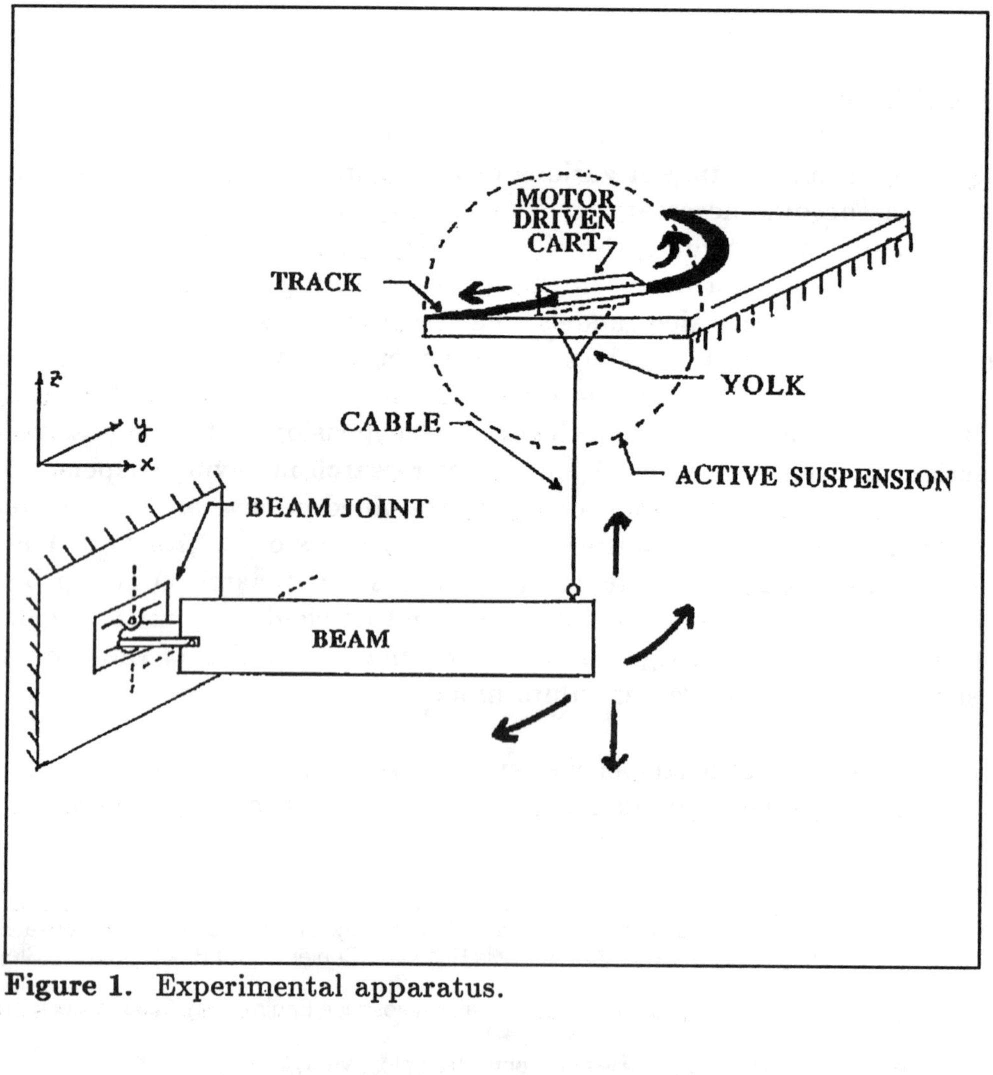

Figure 1. Experimental apparatus.

The cable connection is designed with three cables, two of which are short and are connected in a v-shape, which is called a yolk, along with a long third cable to form a y-shaped cable system. The top part of the y-shaped cable system is connected to the cart and the bottom part is connected to the beam.

BEAM-SUSPENSION MODEL

The beam suspension model is obtained by considering the dynamic moment equilibrium of the previously described system. The model is given in the familiar state equation form as

$$\dot{x}=Ax+Bu \tag{1}$$

where the state vector, **x**, is given by

$$x=\begin{bmatrix}\beta_1 & \dot{\beta}_1 & \beta_2 & \dot{\beta}_2\end{bmatrix}^t \tag{2}$$

and

$$u=\begin{bmatrix}u_1 & u_2\end{bmatrix}^t \tag{3}$$

$$A=\begin{bmatrix} 0 & 1 & 0 & 0 \\ \frac{-K_b}{J_b} & \frac{-C_b}{J_b} & \frac{K_b}{J_b} & 0 \\ 0 & 0 & 0 & 1 \\ \frac{K_b}{J_c} & 0 & \frac{-K_b}{J_c} & \frac{-C_c}{J_c} \end{bmatrix} \tag{4}$$

$$B=\begin{bmatrix} 0 & 0 \\ \frac{1}{J_b} & 0 \\ 0 & 0 \\ 0 & \frac{1}{J_c} \end{bmatrix} \tag{5}$$

where β_1 and $\dot{\beta}_1$ are the angular position and the angular velocity of the beam, respectively. β_2 and $\dot{\beta}_2$ are the angular position and the angular velocity, respectively, of the suspension cart. **u** is the input torque (or moment) vector with components u_1 and u_2. **A** is the state matrix whose elements involve the beam rotational inertia, J_b, the cart rotational inertia, J_c, the effective rotational beam stiffness, K_c, the effective rotational beam damping, C_b, and the effective rotational cart damping, C_c.

The suspension system is intended to simulate slewing dynamics of a microgravity environment for space application. Thus it is desired that the active control suspension-beam system model behavior ideally approach the zero gravity model behavior. The model that describes the behavior of the suspension system in zero gravity is given by

$$\dot{x}_s = A_s x_s + B_s u_s \tag{6}$$

where

$$x_s = [\beta_1 \ \ \dot{\beta}_1]^t \tag{7}$$

$$u_s = [u_1] \tag{8}$$

$$A_s = \begin{bmatrix} 0 & 0 \\ 0 & -\dfrac{C_b}{J_b} \end{bmatrix} \tag{9}$$

$$B_s = \begin{bmatrix} 0 & \dfrac{1}{J_b} & 0 & 0 \end{bmatrix} \tag{10}$$

The simulation of this model with a specified impulse torque is shown in Fig. 2.

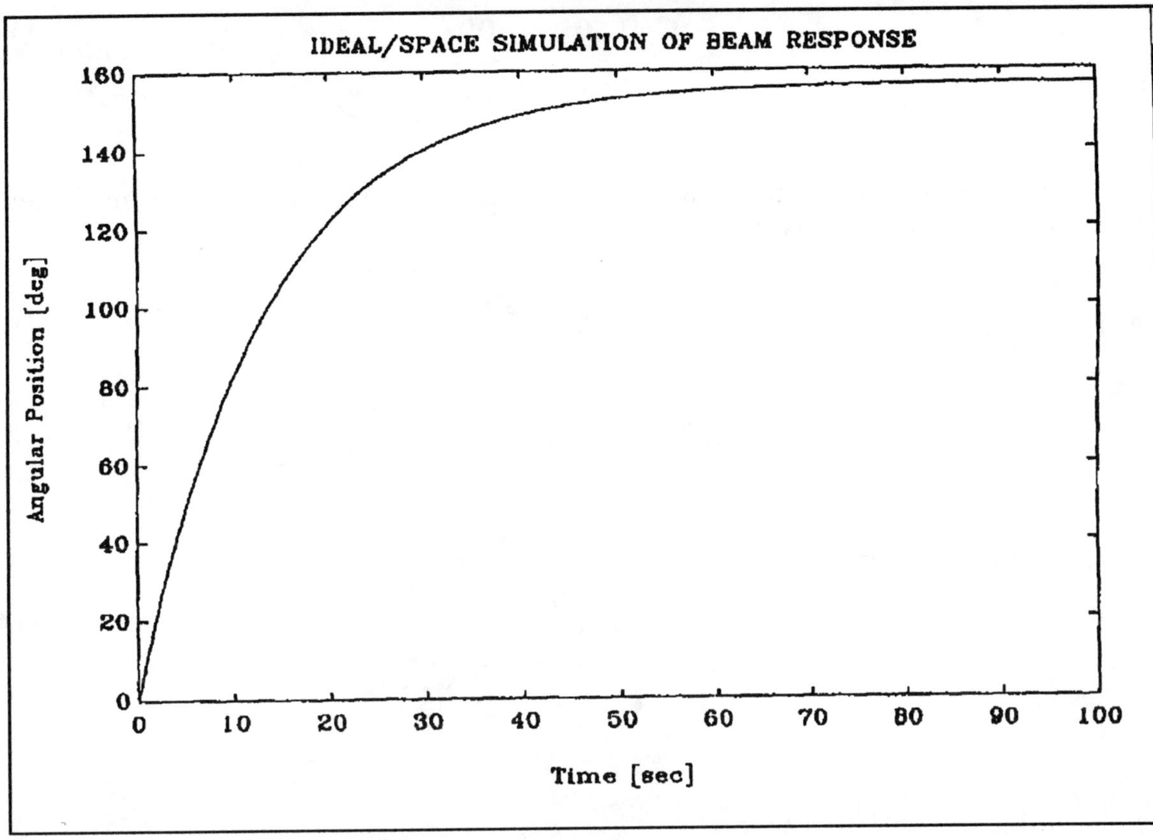

Figure 2. Ideal beam position response.

CONTROL DESIGN USING OUTPUT FEEDBACK

The control law for this control application is in the form of output feedback control with information about the relative angular position of the cart and beam being fed back. More specifically, the control implemented is directly proportional to the difference between the angular position of the cart and beam. This difference is experimentally measured with the use of strain gauge force transducers located judiciously in the yolk of the suspension cables (see Fig. 1.) to detect any net horizontal force in the cables. If a net horizontal force exists then the suspension system is introducing an undesired force on the beam. The form of the feedback equation is obtained from Eqs. 1.-5. and the measurement equation

$$y=Cx \tag{11}$$

where

$$C=[1 \;\; 0 \;\; -1 \;\; 0] \tag{12}$$

and **y** is the output vector and **C** is the output matrix. The control law implemented is relative position feedback, given by

$$u=Ky+N\bar{u} \tag{13}$$

where

$$K=\begin{bmatrix} 0 \\ K_1 \end{bmatrix} \tag{14}$$

The closed loop equations become

$$\dot{x}=\bar{A}x+\bar{B}\bar{u} \tag{15}$$

where

$$\bar{A}=A+BKC \tag{16}$$

$$\bar{B}=BN \tag{17}$$

$$N=[1 \;\; 0]^t \tag{18}$$

$$\bar{u}=[u_1] \tag{19}$$

$$\bar{A}=\begin{bmatrix} 0 & 1 & 0 & 0 \\ -\frac{K_b}{J_b} & -\frac{C_b}{J_b} & \frac{K_b}{J_b} & 0 \\ 0 & 0 & 0 & 1 \\ \frac{(K_b+K_1)}{J_c} & 0 & -\frac{(K_b+K_1)}{J_c} & -\frac{C_c}{J_c} \end{bmatrix} \tag{20}$$

$$B=\begin{bmatrix} 0 & \frac{1}{J_b} & 0 & 0 \end{bmatrix}^t \tag{21}$$

The goal here is to chose the control gain K_1 such that the response of the closed loop system mimics the response expected in a microgravity environment.

CONTROL ANALYSIS

In analyzing the given control we first look at the open loop characteristic equation, F(s), which is found by obtaining the determinant of (sI-**A**), and setting it equal to zero. s is the Laplace Transform variable. The values of s which satisfies

$$F(s)=s^4+(\frac{C_b}{J_b}+\frac{C_c}{J_c})s^3+(\frac{C_b}{J_b}*\frac{C_c}{J_c}+\frac{K_b}{J_c}+\frac{K_b}{J_b})s^2+(\frac{K_b}{J_c}*\frac{C_b}{J_b}+\frac{K_b}{J_b}*\frac{C_c}{J_c})s=0 \tag{22}$$

are the eigenvalues of the system.

For the closed loop characteristic equation, $F_c(s)$, the same method is used as above, using $\bar{\mathbf{A}}$ instead of **A**. Thus,

$$F_c(s)=s^4+(\frac{C_b}{J_b}+\frac{C_c}{J_c})s^3+(\frac{C_b}{J_b}*\frac{C_c}{J_c}+\frac{(K_b+K_1)}{J_c}+\frac{K_b}{J_b})s^2+(\frac{(K_b+K_1)}{J_c}*\frac{C_b}{J_b}+\frac{K_b}{J_b}*\frac{C_c}{J_c})s=0 \tag{23}$$

yields the closed loop eigenvalues.

In order to assure that the control design will be an effective one, several classical performance and stability techniques are used in evaluating the control law. A root locus analysis, which shows the variation of the eigenvalues and eigenvectors for various feedback gain **K**, was performed. The range of gains values chosen for analytical analysis was from 1 to 10000. The range of gains used in the experiment was 1103 to 3677, which yields a range of eigenvalues from -1.18 +/- 6i to -1.0 +/- 4.0i, -.76 to -.35, and zero. These results show the one negative real eigenvalue becomes less negative as **K** increases and approaches a constant value, which means the exponential mode becomes more and more dominating. The complex conjugate pair changes, with an increase in **K**, such that the magnitude of the imaginary part tends to increase and approach infinity as **K** approaches positive infinity, while the real part becomes more negative and approaches a constant value. This shows that the oscillations increase and their duration decreases. The eigenvalues at zero stay at zero, implying an always present rigid body mode. In general, the whole system response tends towards a more exponential response as **K** is increased.

For stability analysis the Routh-Hurwitz criterion was applied (see [4]). This criterion uses the coefficients of the characteristic equation of the closed loop system to determine what constraints on the feedback gain yield eigenvalues of the system which are in the left-half of the complex plane. This would then assure an asymptotically stable system. For this control system the application of the Routh-Hurwitz criterion yields

$$\frac{K_1}{J_c} > -12.037 \tag{24}$$

providing a stability constraint.

EXPERIMENTAL AND ANALYTICAL RESULTS

OPEN LOOP DYNAMIC SIMULATION

The following results were obtained using numerical methods from the software package PC_MATLAB (see [12]). The excitation used in all the simulations was an impulse torque of 300 ft-lb/s. The choice of this value was determined from an estimate of the actual experimental excitation.

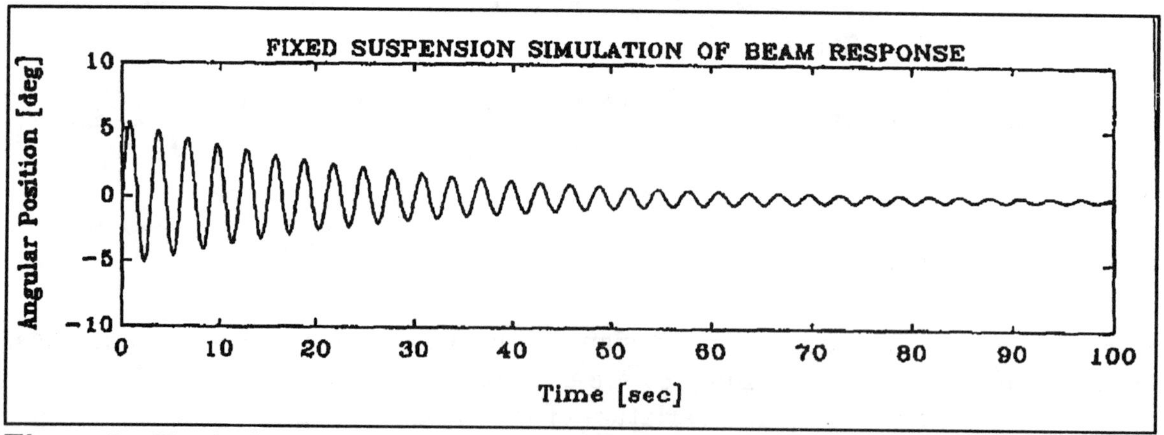

Figure 3. Typical response for a fixed suspension system.

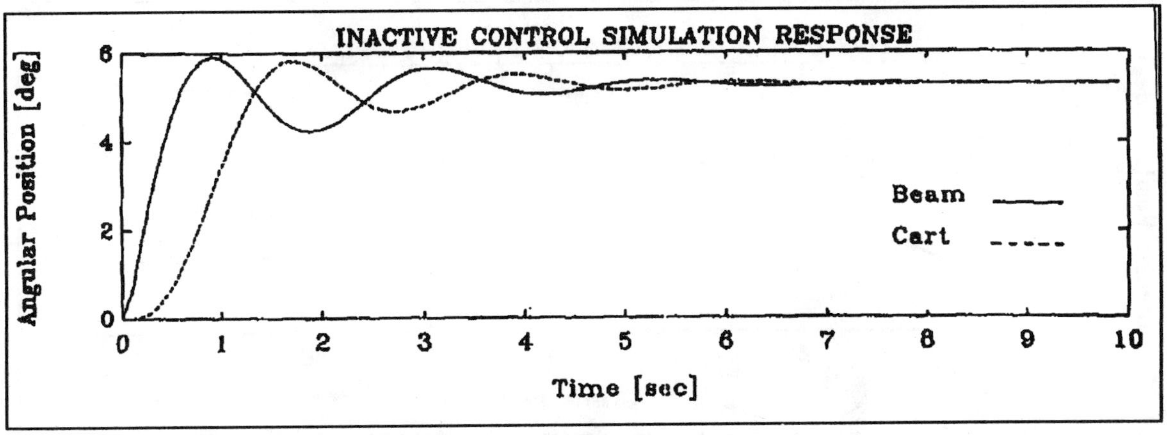

Figure 4. Open loop simulation.

The simulation results for a fixed suspension system is given in Fig. 3. This shows only the beam position over time because the cart is assumed fixed and unable to move. The response is an oscillatory response with a decaying amplitude. The response can be described as the pendulum mode of the beam. The simulation results of the open loop system, or the system without control, is shown in Fig. 4. In the figure the time histories of the beam and cart angular positions are plotted together. This response shows the cart and beam positions crossing each other for several cycles until they finally come to rest at the same position. This indicates substantial interaction between the suspension system dynamics and the structures motion. In Fig. 2. the ideal, zero gravity, space simulation is shown. This illustrates the movement of the beam as if it was unsuspended in space, where the effects of gravity are negligible. The form of the response is exponential in nature, with no oscillations, as it approaches a steady state value. The results of the space simulation are the desired results of the active suspension system. If the response of the system with the closed loop active suspension system operating matches the ideal response of Fig. 2. then the design is successful at simulating a space environment in a ground test.

CLOSED LOOP DYNAMIC SIMULATION

The results of one of several dynamic simulations with various feedback gains is shown in Fig. 5., for a feedback gain of 3677. This result, as well as the others, have the same basic dynamic form as the open loop response. The differences are seen as an improvement in the ability of the cart to track the beam causing the beam and cart responses to have decreased oscillatory amplitudes, a more pure exponential form, and a greater final resting position. These results suggest that as K approaches infinity the position of the cart almost instantaneously approaches the position of the beam. Basically, the result of using proportional relative position feedback control has been to negate the torque on the beam due to the suspension. It minimizes the amount of torque due to the suspension and the time it acts on the system, consequentially reducing its influence on the system. If the gain approaches infinity the amount of torque due to the control is large and is applied so quickly to the cart that the effect of the suspension is almost non-existent.

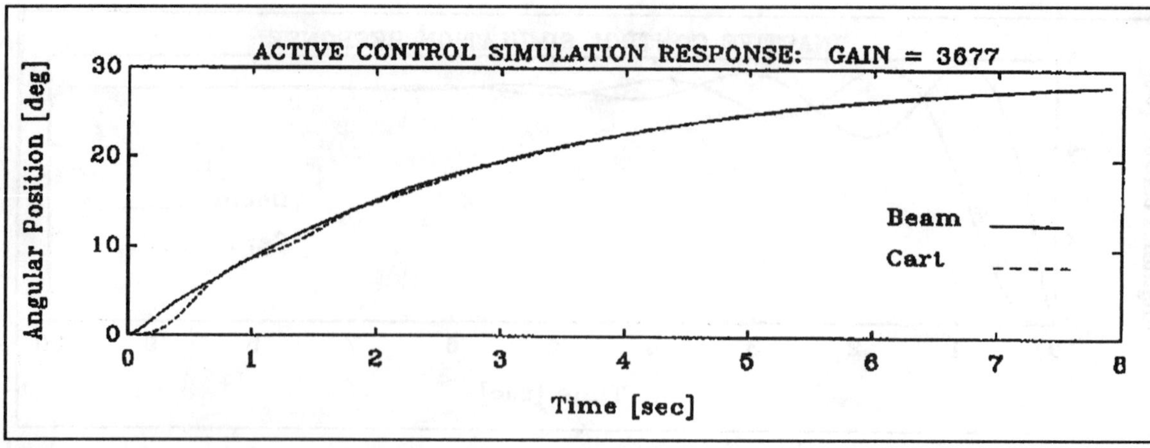

Figure 5. Active control simulation.

EXPERIMENTAL RESULTS

One of the many results obtained experimentally is shown in Fig. 6., and is particularly for a gain of 3677. The results are in very good agreement with those obtained analytically. They show all the basic trends that were predicted by analytical methods. Of course an infinite feedback gain was unattainable. However, if it were possible to achieve a very high gain a substantially better performance could be obtained. From the analysis it is clear that the effects of the suspension system was effectively decreased and that the expected and desired results were realized in this investigation.

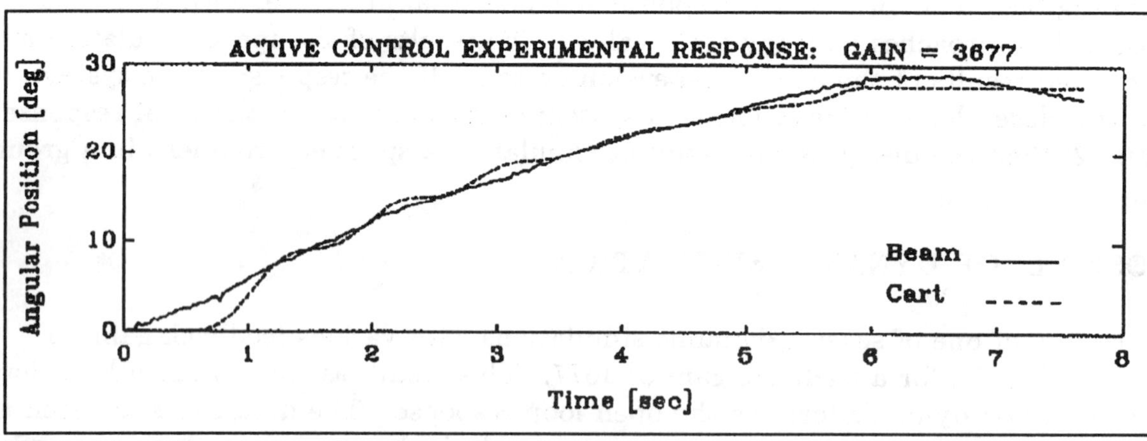

Figure 6. Experimental active control response.

ALTERNATIVE CONTROL METHODS

Other types of control laws were considered for this system but were determined not to be very useful in this particular controls application. One type of control involved was relative velocity feedback in conjunction with the relative position feedback. This was not helpful because it had the effect of adding damping to the system. This is not desired because the final resting position of the structure would never approach the ideal resting position and also this does not produce the effect of removing any influence of any undesired force due to the suspension system. Also an LQR scheme was considered but also introduced damping. One possible alternative that may be used but will not be discussed here is the use of proportional integral control.

CONCLUSION

This application of active suspension control illustrates that with the use of active suspension, as opposed to fixed suspension, the undesired effects of structure/suspension dynamic interaction could be minimized and possibly eliminated. Analysis of the experimental results showed that using proportional relative position feedback, the undesirable effects of the suspension system was effectively decreased. The reduction of the pendulum effect was not as extensive as desired but the results still proved very useful. It showed clearly how performance was improved and displayed quite convincingly that active suspension systems can be very valuable in large structure dynamic testing, especially when concentrating on their rigid body modes. As stated earlier, if a higher feedback gain was attainable the results would be improved but this is not possible due to hardware limitations.

ACKNOWLEDGEMENTS

This work was supported by the National Aeronautics and Space Administration through their Graduate Student Researchers Program, grant no. NGT-70030, with the research performed primarily at The NASA Langley Research Center under the direction of Dr. Garnett C. Horner in collaboration with the Mechanical Systems Laboratory at the State University of New York at Buffalo.

REFERENCES

1. Shih, C., Chen, J. C., and Garba, J., "Vibration of Large Space Beam Under Gravity Effect", AIAA Journal, vol. 24, no. 7, July 1986. p 1212-1216.

2. Belvin, K. W., and Edighoffer, H. H., "Dynamic Analysis and Experiment Methods for a Generic Space Station Model", Journal of Spacecraft and Rockets, vol. 24, May-June 1987, p 270-276.

3. Juang, J., Pappa, R. S., "An Eigensystem Realization Algorithm for Modal Parameter Identification and Model Reduction", AIAA Journal of Guidance Control, and Dynamics, vol. 8, no. 5., Sept-Oct 1985, p 260.

4. Palm, W. J., III, Control Systems Engineering, John Wiley and Sons, 1986.

5. Inman, D. J., Vibration with Control Measurement, and Stability, Prentice Hall, 1989.

6. Letchworth, R., McGowan, P. E., and Gronet, M. E., "COFS III - Multibody Dynamics and Control Technology", NASA/DOD Control/Structures Interaction Technology, 1986, p 757-765.

7. Gronet, M. J., Pinson E. D., Voqui, H. L., Crawley, E. F., and Everman, M. R., "Preliminary Design and Analysis and Costing of a Dynamic Scale Model of the NASA Space Station", NASA Contractor Report 4068, 1987.

8. Garcia, E., System Identification in the Microcomputer Environment, Master Thesis, State University of New York at Buffalo, July, 1987.

9. Hanks, B. R., "Dynamic Verification of Very Large Structures", 2nd International Symposium on Aeroelasticity and Structural Dynamics, Aachen, West Germany, April 1-3, 1985, Bonn, Dutche Gesellschaft, Fuer Luft-Und Raumfahrt, 1985, p 648-655.

10. Ashley, H. "Some Considerations on Earthbound Dynamic Testing of Large Space Structures", AIAA/ASME/ASCE/AHS 27th Structures, Structural Dynamics, and Materials Conference - Part 2, A Collection of Technical Papers, San Antonio, Texas, May 19-21, 1986, p362-375.

11. Woodard, S. E., and Housner, J. M., "The Nonlinear Behavior of a Passive Zero-Spring-Rate Suspension System", Structures, Structural Dynamics, and Materials Conference, 29th Williamsburg, Va., Apr. 18-20, 1988, Technical Papers, Part 2 (a88-321766666612-39), Washington, D. C., AIAA, 1988. p 884-889.

12. The MathWorks, Inc., PC-MATLAB, Copyright 1985-89.

13. Johnson, D., An Application of Active Suspension in Large Angle Slewing Experimentation, Master Thesis, State University of New York at Buffalo, July, 1989.

DISTRIBUTED ACTIVE VIBRATION CONTROLLERS

M. J. Newman[1]

ABSTRACT

This paper contains the theoretical analysis of a PVDF-steel-PVDF actively controlled beam where the electrode pattern has been etched to form three distributed controllers. Experimental studies have been carried out to investigate the ability of the controllers to damp the vibration of a simply supported beam. For the simply supported beam all the controllers have been shown to be able to extract energy from the beam. The passive damping added by gluing the PVDF to the steel beam increases the modal loss factors by upto a factor of 10. The active control system can inaddition increase the loss factor by a factor of 7.

THEORY

Bailey & Hubbard [1] have used a rectangular polyvinyl-idenefluoride (PVDF) piezoelectric plastic actuator to control the vibration of a cantilever beam. The actuator was glued to the side of the beam. The signal used to generate the control was from an accelerometer placed at the tip of the beam. The control voltage applied across the PVDF actuator was the sign of the tip velocity times a constant. With the accelerometer mounted on the same side of the beam as the actuactor, the constant was negative.

Figure 1, shows the beam configuration to be considered here, with PVDF layers applied to both sides of the beam, one as an actuator and the other as a sensor. The analysis will use the Bernoulli-Euler equation for modelling the motion of the beam. The geometry used when describing the beam motion is also shown in figure 1.

[1]Acoustics Research Center, ELAB-RUNIT, SINTEF, N-7034 Trondheim, Norway.

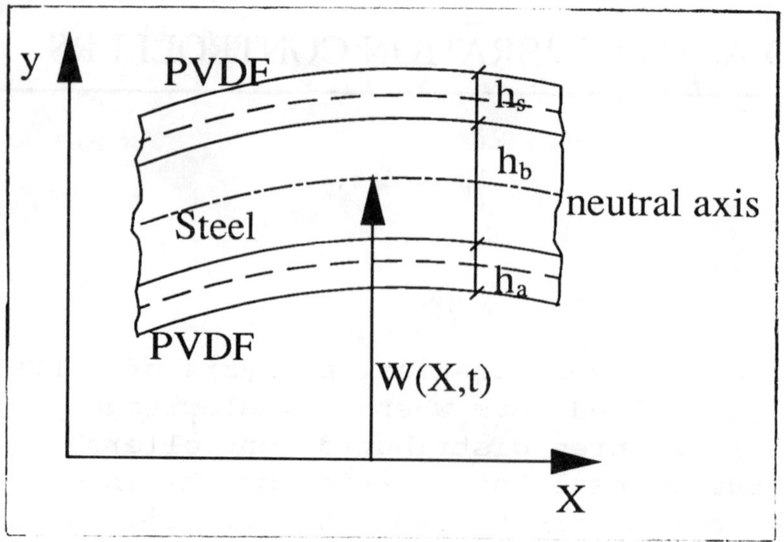

Figure 1. Detail of Beam Cross Section

DISTRIBUTED ACTUATOR

When a voltage is applied to a piece of PVDF that is unrestrained, the strain produced is:

$$\epsilon_a = v_a \ (d_{31}/h_a) \tag{1}$$

With the actuator attached to the beam on the opposite side to the sensor the strain in equation (1) has two effects. the first is to produce the longitudinal strain ϵ_l in the x direction. Equating forces in the x direction gives the strain to be:

$$\epsilon_l = \frac{E_a h_a}{E_a h_a + E_b h_b + E_s h_s} \epsilon_a \tag{2}$$

The second effect is the production of a bending moment. The net force in each layer multiplied by the distance from the layer mid plane to the neutral axis gives the bending moment. The thickness and elastic moduli of the sensor are the same as the actuator, the bending moment can then be simplified to:

$$M = E_a h_a b(\epsilon_a - 2\epsilon_l) \left[\frac{h_a + h_b}{2}\right] \tag{3}$$

and substituting (1) and (2) into (3) gives:

$$M = E_a b v_a d_{31} \left[1 - 2\left(\frac{E_a h_a}{2E_a h_a + E_b h_b}\right)\right] \left[\frac{h_a + h_b}{2}\right] \tag{4}$$

Hence the moment is a constant times the voltage applied across the actuator, the constant being dependent on the geometry and material properties of the composite beam. The governing partial differential equation describing the motion for a thin beam is:

$$\frac{\partial^2}{\partial x^2}\left[EI\frac{\partial^2 w}{\partial x^2}\right] + \rho A\frac{\partial^2 w}{\partial t^2} = 0 \tag{5}$$

The control force added by the piezoelectric actuator acts as an extra spring force and is incorporated into (5) as:

$$\frac{\partial^2}{\partial x^2}\left[EI\frac{\partial^2 w}{\partial x^2} - K_a v_a\right] + \rho A\frac{\partial^2 w}{\partial t^2} = 0 \tag{6}$$

Non-dimensionalizing (6) using:

$$X = x/L \qquad W = w/L$$

$$t = t.(EI/\rho A L^4)^{1/2} \qquad V = v_a K_a L/EI$$

$$EI = E_a I_a + E_b I_b + E_s I_s$$

gives the governing non-dimensional partial differential equation:

$$\frac{\partial^4 W}{\partial X^4} + \frac{\partial^2 W}{\partial t^2} = \frac{\partial^2 V}{\partial X^2} \tag{7}$$

The total energy in the beam motion is:

$$F(t) = \frac{1}{2}\int_0^1 \left[\left(\frac{\partial^2 W}{\partial X^2}\right)^2 + \left(\frac{\partial W}{\partial t}\right)^2\right]dX \tag{8}$$

The first term is the curvature squared, which is proportional to the potential energy and the second term is the velocity squared, which is proportional to the kinetic energy. By differentiating equation (8) with respect to time, the rate of energy extraction is:

$$\frac{\partial F}{\partial t} = \int_0^1 \frac{\partial^2 W}{\partial X^2}\cdot\frac{\partial^3 W}{\partial X^2 \partial t} + \frac{\partial W}{\partial t}\cdot\frac{\partial^2 W}{\partial t^2}\, dX \tag{9}$$

Substituting using the governing partial differential equation

(7) gives:

$$\frac{\partial F}{\partial t} = \int_0^1 \frac{\partial^2 W}{\partial X^2}\cdot\frac{\partial^3 W}{\partial X^2 \partial t} + \frac{\partial W}{\partial t}\cdot\frac{\partial^2 V}{\partial X^2} - \frac{\partial W}{\partial t}\cdot\frac{\partial^4 W}{\partial X^4}\, dX \tag{10}$$

Integrating the last term twice by parts gives

$$\frac{\partial F}{\partial t} = \int_0^1 \frac{\partial W}{\partial t}\cdot\frac{\partial^2 V}{\partial X^2}\, dX - \left.\frac{\partial W}{\partial t}\cdot\frac{\partial^3 W}{\partial X^3}\right|_0^1 + \left.\frac{\partial^2 W}{\partial X \partial t}\cdot\frac{\partial^2 W}{\partial X^2}\right|_0^1 \tag{11}$$

Equation (11) is just one of many forms that the derivative of the energy functional can be presented in, by carrying out integrations by parts other boundary terms can be formed. In its present form the terms of (11) can be examined as descriptive parameters of the beam and controller. The integral is of the velocity of the beam multiplied by the second spatial derivative of the controller. The second term is the velocity multiplied by the "shear force" (/EI) at the ends of the beam. The final term is the angular velocity multiplied by the "bending moment" (/EI) at the ends of the beam. For the standard boundary conditions ie beams that are simply supported, cantilever, free-free, clamped-clamped etc all the boundary terms are zero.

The displacement of the beam can be rewritten using the normal mode model to separate the variables:

$$W(X,t) = \sum_{r=1}^{\infty} A_r \psi_r(t)\cdot\phi_r(X) \tag{12}$$

Separating the variables of the controller gives:

$$V(X,t) = v_{a_{max}} \Lambda_a(X)\cdot\rho(t) \tag{13}$$

By setting the last two terms of equation (11) to zero, as for the standard beam boundary conditions, substituting for velocity and the second spatial derivative of the controller, it becomes:

$$\frac{\partial F}{\partial t} = \sum_{r=1}^{\infty} A_r v_{a_{max}} \rho(t) \frac{\partial \psi_r}{\partial t} \int_0^1 \phi_r(X) \frac{\partial^2 \Lambda_a}{\partial X^2}\, dX \tag{14}$$

The integral term of equation (14) is the controllability/ actuator quality factor for each mode. This is governed by the mode shape $\phi(X)$ and the shape of the actuator Λ_a.

DISTRIBUTED SENSOR

When a stress is applied to a piece of PVDF a charge density is produced:

$$q = -d_{31}\sigma \tag{15}$$

Using the simple beam bending relationships and the geometry of the beam:

$$\frac{M}{I} = \frac{\sigma}{y} \qquad M = EI\frac{\partial^2 W}{\partial X^2} \qquad y = \frac{h_s+h_b}{2}$$

the charge produced by the PVDF sensor is:

$$Q = -d_{31}E\left(\frac{h_s+h_b}{2}\right)\int_0^1 \frac{\partial^2 W}{\partial X^2}\Lambda_s(X)\,dX \tag{16}$$

Using the standard equations for capacitance:

$$v = Q/C \qquad C = \epsilon_0\epsilon_r 2A_s/h_s$$

the voltage produced by the sensor is:

$$v_s = -\frac{d_{31}E\left(\frac{h_s+h_b}{2}\right)h_s}{\epsilon_0\epsilon_r 2A_s}\int_0^1 \frac{\partial^2 W}{\partial X^2}\Lambda_s(X)\,dX \tag{17}$$

Hence for a specific sensor and beam combination:

$$v_s = -K_s\int_0^1 \frac{\partial^2 W}{\partial X^2}\Lambda_s(X)\,dX \tag{18}$$

Substituting a modal representation for the motion of the beam, the same one as in equation (12), gives:

$$v_s = -\sum_{r=1}^{\infty} K_s\psi(t)A_r\int_0^1 \frac{\partial^2\phi_r}{\partial X^2}\Lambda_s(X)\;dX \tag{19}$$

The integral term is the observability/sensor quality factor for each mode.

ENERGY EXTRACTION

The main purpose of an active vibration control system is to extract energy from a structural system or change its response to a vibration input. Equation (14) is the rate at which the distributed controller extracts energy from a beam. Using velocity feedback, the signal from the sensor becomes:

$$\rho(t) = -\sum_{r=1}^{\infty} K_s A_r \frac{\partial \psi}{\partial t} \int_0^1 \frac{\partial^2 \phi_r}{\partial X^2} \Lambda_s(X)\, dX \tag{20}$$

Substituting for $\rho(t)$ in equation (14) gives:

$$\frac{\partial F}{\partial t} = \sum_{r=1}^{\infty} -K_s v_{a_{max}} A_r^2 \left(\frac{\partial \psi_r}{\partial t}\right)^2 \int_0^1 \frac{\partial^2 \phi_r}{\partial X^2} \Lambda_s(X)\,dX \int_0^1 \phi_r(X) \frac{\partial^2 \Lambda_a}{\partial X^2}\, dX \tag{21}$$

Hence for each mode the rate of energy extraction is a constant times the modal velocity squared times the sensor-times the actuator quality factors.

SENSOR AND ACTUATOR QUALITY FACTORS

The actuator and sensor electrode patterns used for the experimental active beam were the same and are shown in figure 2. The electrode pattern was connected together to form: a full length triangle, a half length triangle and a quarter length trapisoid. The half length triangle and quarter length trapisoid were formed from two areas so that they would only act in the longitudinal direction of the beam.

The integral term of equation (14) is used to assess the actuator quality factor, L_r:

$$L_r = \int_0^1 \phi_r(X) \frac{\partial^2 \Lambda_a}{\partial X^2}\, dX \tag{22}$$

For the full length triangular actuator the shape function is:

$$\Lambda_{a_{1/1}} = (1-X)[H(X)-H(1-X)] \tag{23}$$

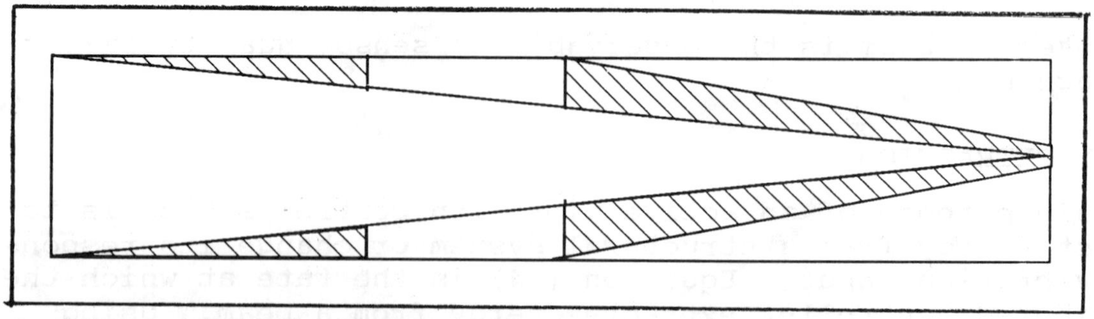

Figure 2. PVDF Electrode Pattern

$$\text{and} \quad \frac{\partial^2 \Lambda_a}{\partial X^2} = -\delta(X) + \delta'(X) + \delta(X-1) \tag{24}$$

Hence:

$$L_r = \phi_r(1) - \phi_r(0) - \frac{\partial \phi_r}{\partial X}_{X=0} \tag{25}$$

The mode shape for the simply supported beam is $\phi_r = \sin(r\pi X)$, as there is no displacement at the ends, $\phi_r(0\&1)=0$, $L_{r_{1/1}} = -r\pi$.

The sensor quality factor is given by the integral term of equation (20):

$$O_r = \int_0^1 \frac{\partial^2 \phi_r}{\partial X^2} \Lambda_s(X)\ dX \tag{26}$$

The full length triangular sensor shape function is as for the actuator (26) but the step functions can be omitted when carrying out the integration:

$$O_{r_{1/1}} = \int_0^1 \frac{\partial^2 \phi_r}{\partial X^2} (1-X)\ dX$$

$$O_{r_{1/1}} = \phi_r(1) - \phi_r(0) - \frac{\partial \phi_r}{\partial X}_{X=0} \tag{27}$$

Which is the same as the actuator quality factor, hence the full length triangular sensor observes all the modes of the simply supported beam.

The energy extraction for the combined controller would be proportional to $(r\pi)^2$ for the simply supported beam.

The quality factors for the other shapes of actuator and sensor can be calculated in a similar manner. These calculations show that, both the half length and quarter length controllers can control the first ten modes of simply supported beam.

EXPERIMENTAL CONFIGURATION AND RESULTS

CONSTRUCTION OF ACTIVE BEAM AND END CLAMPS/HINGES

The PVDF piezoelectric plastic material was supplied in 474x50mm sheets of 52μm thick film with nickel-copper metallization on both sides. It was supplied by Penwalt Piezo Film Ltd. The electrode pattern shown in figure 2 was etched on one of the metallization layers for both actuator and sensor. The PVDF layers were then glued to a 1mm thick steel "beam" 540x52mm using Fasson 310 double sided tape, in addition self-adhesive copper foil was then used to connect

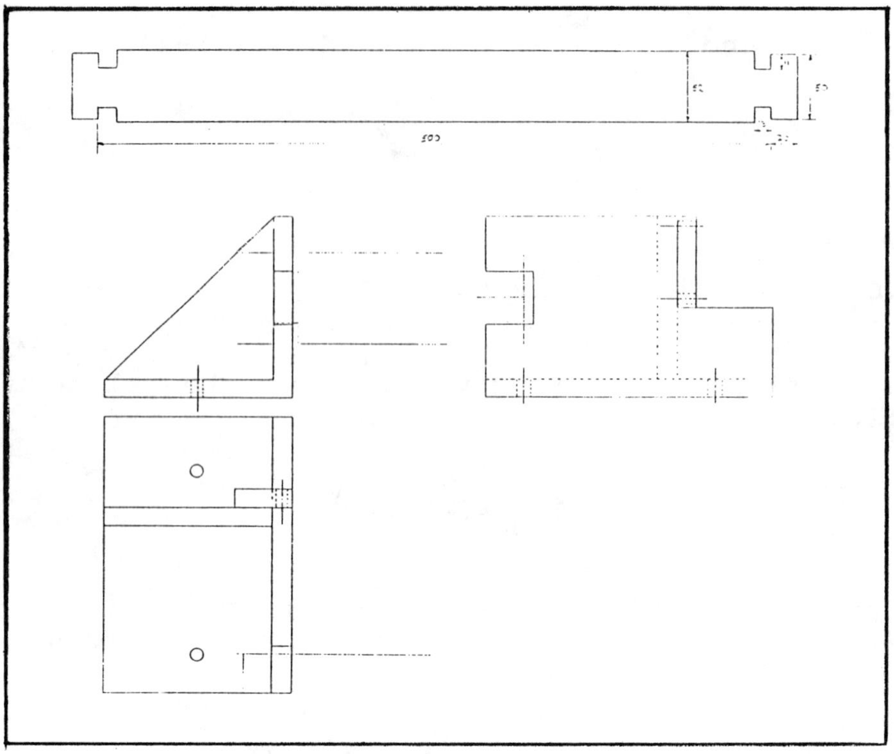

Figure 3. Drawings of Support Brackets and Beam

the metallization to flexible 50Ω coaxial wires. The actuators were connected to the amplifiers with "load" resistors of 22kΩ. The sensors were loaded with 5k, 10K & 27kΩ resistors to act as differentiators before the pre-amplifiers.

The brackets used for holding the active beam were made in such a way that they could approximate either hinges or clamped boundary conditions. Figure 3 shows drawings of the brackets and beam. The brackets were bolted to a machine bed plate with 100mm spaced 'T' slots. The machne bed was supported on partially inflated fenders to isolate it from building vibration.

INSTRUMENTATION

Initial tests were carried out on a steel only beam. For these tests an ultrasonic displacement transducer was used to sense the motion of the beam. For all the experiments a variable reluctance transducer supplied with 1/1 or 1/3 octave noise was used to drive the beam. The simply supported beam configuration was driven just outside the hinge support.

Figure 4 shows a block diagram of the instrumentation used for the experiments. The instrumentation consists of three sections in addition to the beam under test, these are: (i) the driver, (ii) the analysis and (iii) the control feedback loops. The driver section consisted of noise generator, power amplifier and variable reluctance transducer driving the beam. The signal from the noise generator was also fed to the FFT analyser to act as the reference signal. The feedback loops consisted of a load resistor connection box to act in combination with the PVDF sensor capacitance to differentiate the signal, pre-amplifier, filter and power amplifier connected to load resistors and the actuator. A FFT analyser was used for the analysis, the reference signal from the noise generator and the signal from the long triangle sensor were analysed.

TEST METHODS

The loss factor of the beam was used as a measure of the ability of the active system to increase the damping of the beam. The beam was excited by 1/1 or 1/3 octave noise in the bands containing each of its first ten resonances. Two methods were used to determine the loss factor in the first ten

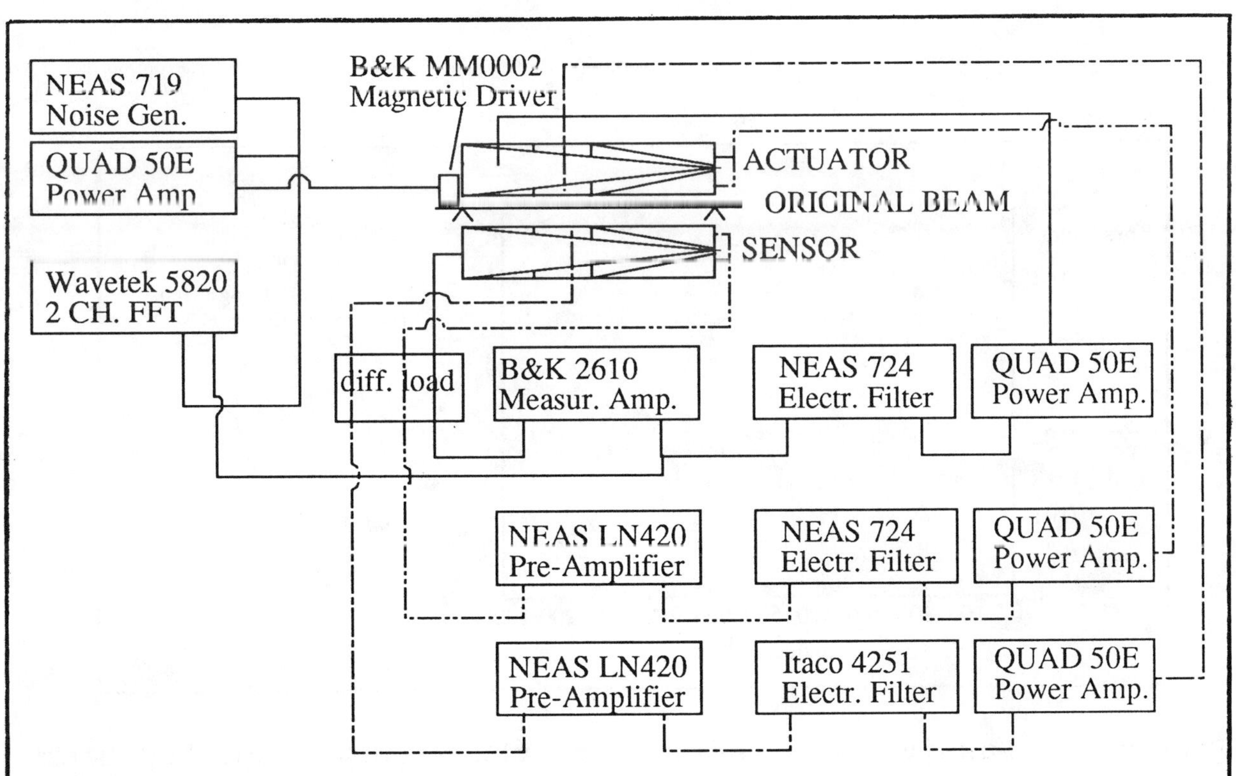

Figure 4. Equipment Configuration

vibrational modes of the beam. The bandwidth of the -3dB points for the resonance gives the loss factor to be:

$$\eta_r = \frac{f_d - f_c}{f_r} \tag{28}$$

The magnitude of the transfer function between the input and the long triangle signals were used to give the values of f's for the loss factor calculation.

The rate of change of phase was the second method used to determine the loss factor, the equation used was:

$$\eta_r = \frac{2}{f_r\left(\frac{d\Phi}{df}\right)_r} \tag{29}$$

The two methods are listed by Bert [2].

RESULTS

Figures 5-7 show the loss factors for the simply supported beam. All the graphs are for the first ten modes of the beam

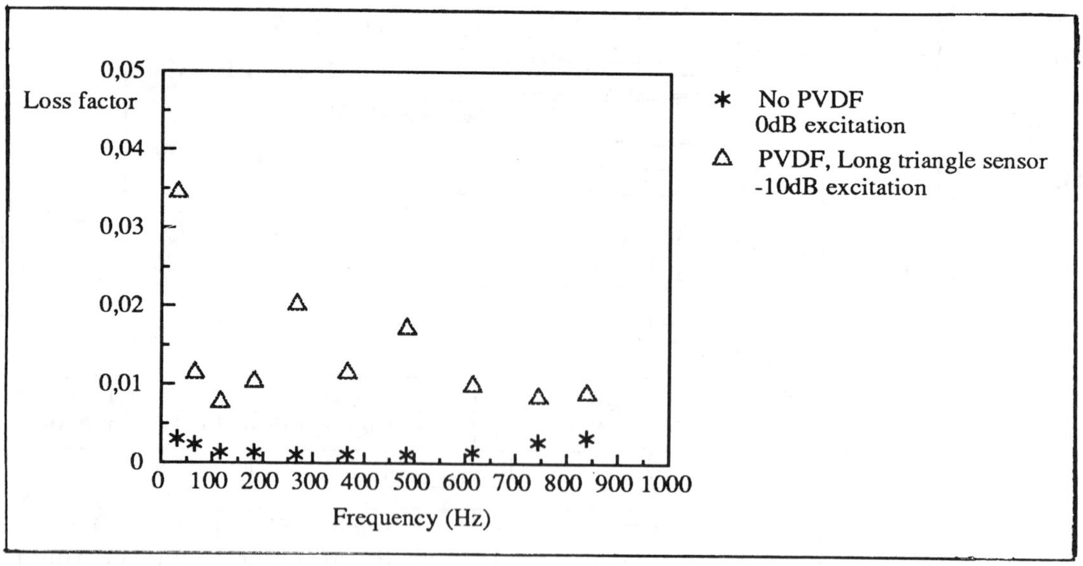

Figure 5. Passive Loss Factors for the Simply Supported Beam With and Without the PVDF Layers & Connection Leads

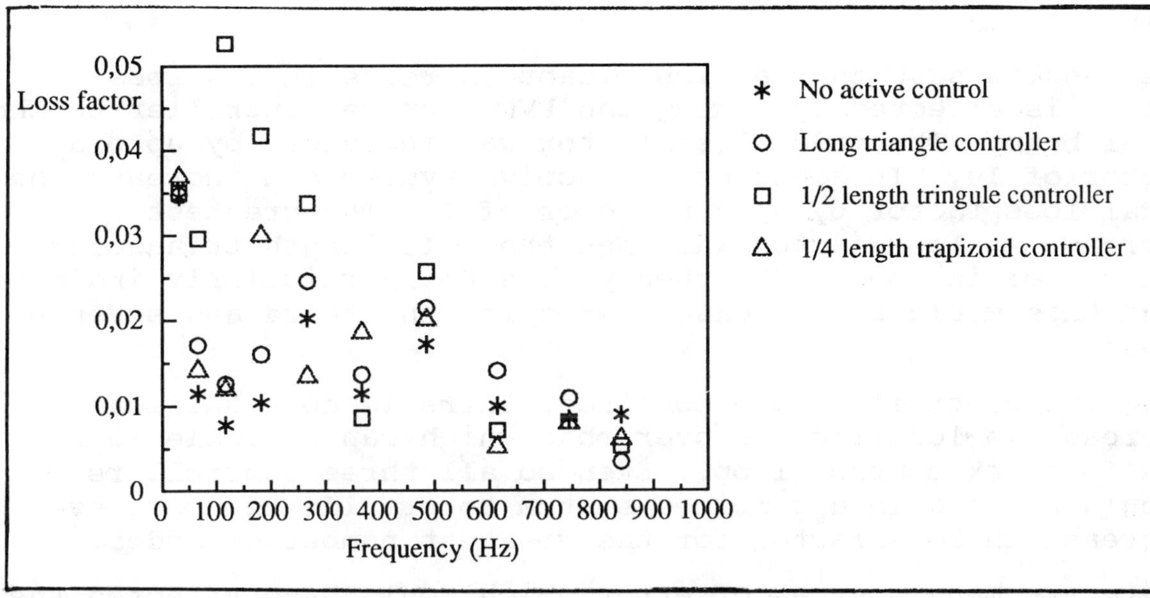

Figure 6. Passive and Active Beam Modal Loss Factors for the Simply Supported Beam Using Single Feedback Control Loops

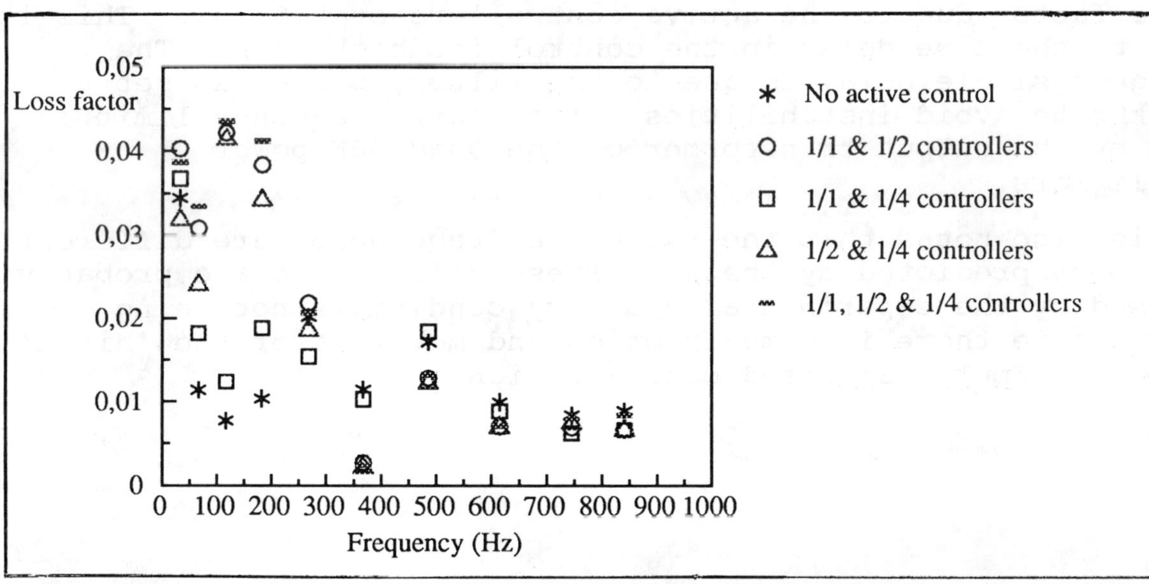

Figure 7. Passive and Active Modal Loss Factors for the Simply Supported Beam Using Multiple Feedback Loops

DISCU SSION

The analysis indicates that the simple three element distributed controller is capable of observing and controlling the first ten modes of the simply supported beam. The experimental results have verified the analysis over a smaller number of modes.

The results show that a significant increase in the loss factor is effected by gluing the PVDF active controller to the steel beam. The modal loss factor was increased by upto a factor of 10. In addition the active system can increase the modal loss factor by upto a factor of 7. The greatest increase in loss factor was when the half length triangular controller is used. The theory does not particularly indicate that this would be the case except for the third and seventh modes.

When the controllers are combined, there is no significant increase in loss factors over that which can be achieved by one feedback control loop. Even so all three controllers should be used in a combined system to provide the most even increase in loss factor for the greatest number of modes.

There are many sources of error which could have affected the experimental results so that they can not fully be predicted by the theory. In section 4.4 the loss factor scatter diagrams indicate that in general there is no increase in the loss factor due to the active controllers over 500Hz. This is due to the time delay in the control feedback loop. The largest single delay is due to the filter, which was set to 1.2kHz to avoid instabilities. The lower frequency limit is set by the output transformer of the Quad 50E power amplifiers.

It is also noted that the resonance frequencies are different to those predicted by theory. These differences are probably caused by the experimental boundary conditions not being perfect ie there is some friction and movement of the 'hinge' for the simply supported configuration.

NOMENCLATURE

A = modal amplitude constant

A_s = area of sensor

$\mathcal{A}$ = cross sectional area of beam

$*_a$ = refers to actuator

b = width of beam

$*_b$ = refers to original beam

C = capacitance

c = lower -3dB point

d = upper -3dB point

d_{31} = static electric piezoelectric constant, in x direction with voltage applied in y direction

E = modulus of elasticity

F(t) = functional, proportional to beam energy

f = frequency

h_* = layer thickness

I = second moment of area

K = constant

L = beam length

L_r = controllability/actuator quality factor

M = moment

O_r = observability/sensor quality factor

Q = charge

q = charge density

r = mode number

$*_s$ = refers to sensor

t = time

v = voltage

V = controller function

w = displacement of beam

β_r = mode frequency parameter

ϵ = strain

ϵ_r = relative permittivity

ϵ_o = permittivity of free space

η_r = specimen modal loss factor

$\Lambda(X)$ = spatial function (describing element shape)

ρ = density (equations 7 & 8)

$\rho(t)$ = controller time function

σ = stress

$\phi(X)$ = modal spatial function

Φ = phase

$\psi(t)$ = modal time function

REFERENCES

1. T. Bailey & J.E. Hubbard Jr. "Distributed Piezoelectric-Polymer Active Vibration Control of a Cantilever Beam" Journal of Guidance, Control and Dynamics Vol. 8 No. 5 Sept-Oct 1985 pp. 605-611.

2. Bert, C.W. "Material damping: an introductory review of mathematical models, measures and experimental techniques" Journal of Sound and Vibration Vol. 29 (1973), pp. 129-153.

3. White, R.G. & Walker, J.G. "Noise and Vibration" chapter 25 by Mead, D.J.

THE NEED FOR PASSIVE DAMPING IN FEEDBACK CONTROLLED FLEXIBLE STRUCTURES*

A. H. von Flotow,[1] D. W. Vos[1]

ABSTRACT

It is well established that an undamped infinite dimensional *mathematical model* of a flexible structure cannot be stabilized with finite bandwidth control; the required gain stabilization beyond the bandwidth is negated by the infinite structural gain at each resonance. Thus, even a mathematical model with no modeling uncertainty will show that passive damping is critical to enabling active control. What is less well known are the benefits of passive damping for the robust control of *real structures*. There has been a tendency, in the research literature, to define the research problem to consist of developing control approaches for the broadband control (many modes in the control bandwidth) of poorly modeled, lightly damped, modally dense structures. There is ample reason to believe that such control is practically unachievable and that the attribute "lightly damped" is one of the most easily and readily remedied characteristics of such a problem structure.

This paper reports upon the enabling effect of passive damping in the control of uncertain flexible structures, particularly with non–collocated actuators and sensors. Quantative results are all single–input single–output and the benefits of passive damping are then understandable in terms of classical ideas of gain and phase stabilization. The paper derives approximate expressions for the minimum acceptable level of passive damping in terms of modeling uncertainty and desired bandwidth. These relationships can then be interpreted as specifying either a minimum level of passive damping or a minimum level of modeling fidelity. If the requirement is not met, robust control with the bandwidth including uncertain flexible dynamics is not possible with linear time invariant (LTI) compensation.

INTRODUCTION

In this paper, "bandwidth" implies "that frequency range over which the loop transfer function magnitude ratio varies by no more than 3dB." This definition is sensible for reference command following and for disturbance rejection when the respective input command and output referenced disturbance signals are expected to be broadband. It is possible that neither condition is met in an application of feedback control to a lightly damped flexible structure. A common goal for flexible structures is disturbance rejection rather than command following. Typical output–referenced disturbances are narrow band, either because they were generated by a time periodic process (typically an operating machine) or because they have been filtered through the resonant structure. With reference to rejection of such disturbances, another definition of bandwidth may be appropriate.

This paper addresses the problem statement: "The control bandwidth must include many poorly modeled, lightly damped, closely spaced modes." This problem statement is figuratively depicted in Figure 1a. The fundamental point made in the paper is that no linear time invariant compensation exists which will robustly (in the sense of stability

---

* This paper is an edited version of a previous draft presented at the 61st Shock and Vibration Symposium, Pasedena, CA, Oct 1990 and in present form will be presented at the VPI+SU Conference on Recent Advances in Active Control of Sound and Vibration, April 1991 and the Air Force Damping '91 Conference, San Diego, Feb 1991.

---

[1]Massachusetts Institute of Technology, Cambridge, MA, U.S.A.

robustness as opposed to performance robustness) achieve what is suggested by Figure 1a, unless a significant amount of passive damping is present. The required level of passive damping is sketched in Figure 1b.

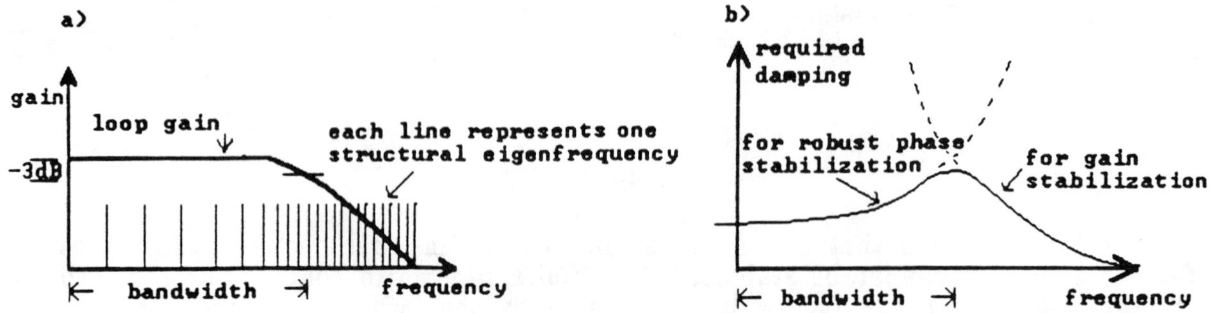

Figure 1. (a) Figurative depiction of problem statement for bandwidth to include many poorly modeled, lightly damped, closely spaced modes. (b) Required level of passive damping to meet problem specification.

The obvious approach to eliminating undesirable dynamics from the controlled plant is to generate control signals which cancel these motions [Rossi]. This "plant inversion" philosophy is what all control design techniques, both linear and nonlinear, SISO and MIMO [Williams], strive to realize when the aim is to achieve bandwidth greater than the frequency of the dominant modes. In benign systems, the effect of reasonable levels of model uncertainty is not such as to yield instability and is thus tolerable. For lightly damped oscillatory systems which typify the broadband structure control problem, however, the undesirable result of high gain/performance plant inversion, is instability of the closed loop system when sufficient model mismatch ocurs. A question which arises is how much damping is required to maintain closed loop stability if the structural dynamics are known only with a given level of uncertainty?

Some specialized literature dealing with design of controllers for uncertain lightly damped structural dynamics exists [ACC, Bryson, Bontsema, Rosenthal, Rossi, Wedell, Wie, et al], but the authors assume a plant given and unable to be modified, leading to a disregard of prescribing increased passive damping as part of the control solution. "Modern" (LQG, H_2, H^∞, L_1) control design techniques revolve around parameterizing of the uncertainty into the design model [Bontsema, Bryson, Wie, Wedell], often leading to rather convoluted design and robustness evaluation procedures. Notable for also ignoring passive damping is a large body of literature on combined optimal design of the structural dynamics and the active control system [JPL, Venkayya, Rao, Belvin]. Much of this literature implicitly defines structural dynamic design as the appropriate selection of stiffness and mass parameters. Passive damping levels are assumed to be very low and beyond specification by the designer. This leads to elegant mathematics and numerics, but also leads to results of questionable engineering relevance.

Novel nonlinear techniques, not LTI, such as that due to [Balas] or [Kopf], where use is made of phase–locked loops to track modal frequency and adjust compensation accordingly, also exist. These techniques may offer much promise for the broadband control of poorly modeled lightly damped structural dynamics, but are beyond the scope of this paper.

This paper makes the strong quantitative statement that sufficiently lightly damped and poorly modeled structural dynamics can never be within the bandwidth of an LTI control system. This statement is supported with some SISO case studies, which serve to quantify the level of passive damping required to permit robust phase stabilization of uncertain structural poles. Uncertainty in eigenfrequencies and in mode shapes is considered. Mode shape uncertainty is recognized as more serious since it is the mode shapes which govern the location of plant zeros. In the common situation of lightly damped poles and zeros such mode shape uncertainty makes the location of the zeros highly uncertain, leading to the strong possibility of transfer function phase uncertainty of 180^0 or greater for dislocated sensor/actuator pairs.

This paper implicitly assumes that the poles of the plant transfer function are spaced much more widely than their bandwidth; $\omega_i - \omega_{i+1} >> \pi\xi\omega_i$. In this case of low modal overlap, each pole can be considered practically in isolation, interacting at most with local plant zeros. The uncertainty in structural dynamic transfer function gain and phase as a function of modal overlap has been studied from a statistical perspective by [Lyon], and is inversely related to modal overlap.

GAIN STABILIZATION BEYOND THE BANDWIDTH

The result we report here is well known and has been reported elsewhere in many different forms [Gran, Spanos, Hughes, von Flotow]. The basic idea is that a poorly modeled or unmodeled flexible mode beyond the bandwidth must be gain stabilized. Since the gain of a flexible structure (from applied force or torque to displacement, rotation etc.) is maximized near each resonance at a value inversely proportional to damping ratio, the conclusion emerges that an undamped flexible mode can never be gain stabilized. How much damping ratio is required to ensure gain stabilization of modes beyond the control bandwidth depends then upon the gain roll–off of the loop, upon the spectral separation between the modal natural frequency and the control bandwidth and upon the modal participation or residue. The relationship is rather obvious and has been reported elsewhere [Gran, Spanos, Hughes, von Flotow]. It is this requirement that leads to one curve sketched in Figure 1b.

PHASE STABILIZATION WITHIN THE BANDWIDTH

If, for the purposes of disturbance rejection or command following, the control bandwidth must be broad and include one or several flexible modes, then each of these flexible modes must be robustly phase stabilized. In the absence of modeling error, this presents no particular difficulty; notch compensation or some other equivalent inversion of the structural dynamics is then an acceptable approach. Unfortunately, modeling errors are always present. In single–input, single–output LTI systems, the effect of these modeling errors upon control design can be summarized in terms of uncertainty in the location of the structural dynamic transfer function poles and zeros. Broadband control is then enabled by a sufficient level of passive damping, since approximate plant inversion is then feasible.

ROBUST POLE–ZERO CANCELLATION

If the closed loop bandwidth (bandwidth as defined in Figure 1a) is to extend past the eigenfrequency of a number of flexible modes, the only solution is pole–zero cancellation. In the absence of modeling errors, all flexible modes may be canceled and the closed loop bandwidth extended arbitrarily. When uncertainty regarding the eigenfrequencies exists, however, extra care needs to be taken in this strategy. A root locus departure angle argument serves well to explain the effect of uncertainty in eigenfrequency on the stability of the closed loop system where pole–zero cancellation is employed in compensation. Here, a single oscillatory mode system is considered and the damping is considered to be very small. Figure 2 assumes that the loop phase due to all other dynamics is -90^0 at the nominal value of the eigenfrequency in question.

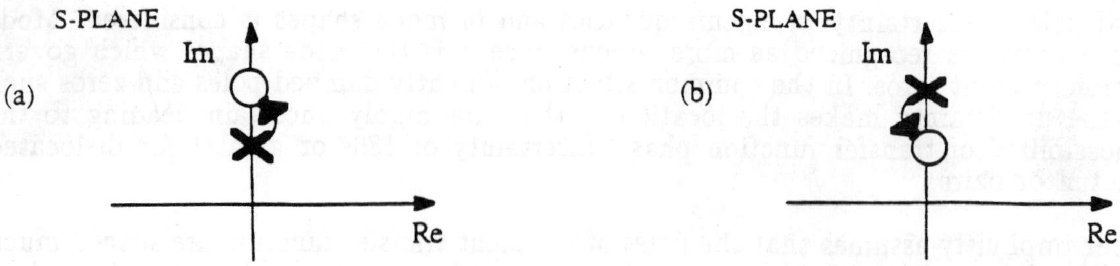

Figure 2. Departure angles due to pole–zero cancellation of single oscillatory mode.

Clearly, Figure 2b does not yield instability for small gains, as the departure angle is away from the imaginary axis. The phase lead added by the zero at frequency lower than the modal frequency ensures this. In Figure 2a, however, the departure angle is toward the imaginary axis, such that for sufficient uncertainty and sufficiently lightly damped poles, the closed loop system will be unstable.

CRITICAL DAMPING FOR ROBUST POLE–ZERO CANCELLATION

Considering the simple case of a single oscillatory mode, it is desirable to implement pole–zero cancellation as compensation in order to achieve a desired closed loop bandwidth. This, however, leads to possible instability if insufficient passive damping exists for a given level of uncertainty in modal frequency. This section serves to quantify ,to first order in relative uncertainty of the eigenfrequency, how much passive damping is necessary for stability (when employing this form of compensation) over the range of eigenfrequency uncertainty.

Phase Gradient with respect to Frequency

First, it is necessary to evaluate the "phase gradient" with respect to frequency, ω. Consider the single oscillatory mode represented by the transfer function

$$G(s) = \frac{1}{(s^2 + 2\xi\omega_n s + \omega_n^2)}$$

setting $s = j\omega$ and evaluating the phase at frequency ω

$$\phi(\omega) = -\tan^{-1}\left(\frac{2\xi\omega_n\omega}{\omega_n^2 - \omega^2}\right)$$

The gradient of this phase evaluated at $\omega = \omega_n$ can be shown to be

$$\frac{d\phi}{d\omega} = \frac{-1}{\xi\omega_n} \tag{1}$$

representing the phase gradient with respect to frequency at the frequency of maximum amplification of the amplitude ratio (vis–a–vis the Bode magnitude plot) as illustrated in Figure 3. Now, if the eigenfrequency is unknown to an uncertainty $\delta\omega = \omega_n - \omega_{actual}$, then the phase at the nominal eigenfrequency, ω_n, will be uncertain by as much as

$$\delta\phi = \frac{-\delta\omega}{\xi\omega_n} \tag{1a}$$

to first order in $\delta\omega/\omega_n$.

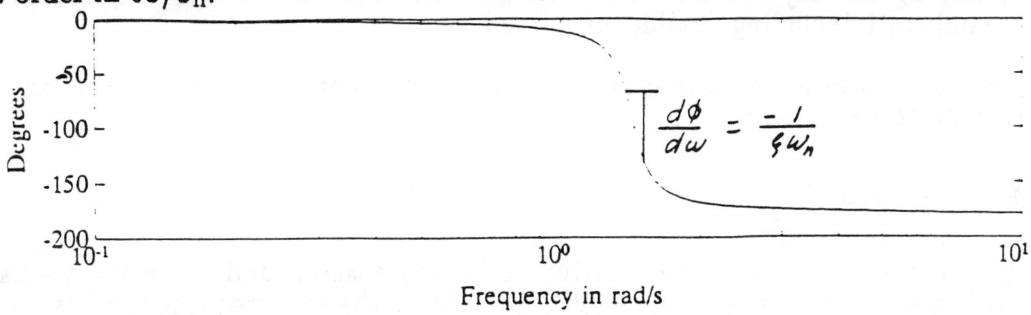

Figure 3. Bode plot illustrating the phase gradient wrt ω.

This allows estimates of local phase contributions due to oscillatory modes. Clearly, the phase gradient due to non–minimum phase oscillatory zeros of frequency ω_n and damping ξ (evaluated at ω_n), is also exactly (1), while that due to minimum phase zeros is the absolute value of (1).

Phase Excursions in Approximate Pole–Zero Cancellation

Using the relation (1a) it is now a simple matter to determine the relationship between the phase excursion due to inexact pole–zero cancellation. For the first uncertain mode in the system to be controlled, the Bode phase plot appears as in Figure 4, where the phase contribution due to the mode as well as the canceling zeros is depicted. Since $\delta\omega$ is assumed positive, the actual pole occurs at a frequency lower than that of the canceling zero and local phase lag is introduced.

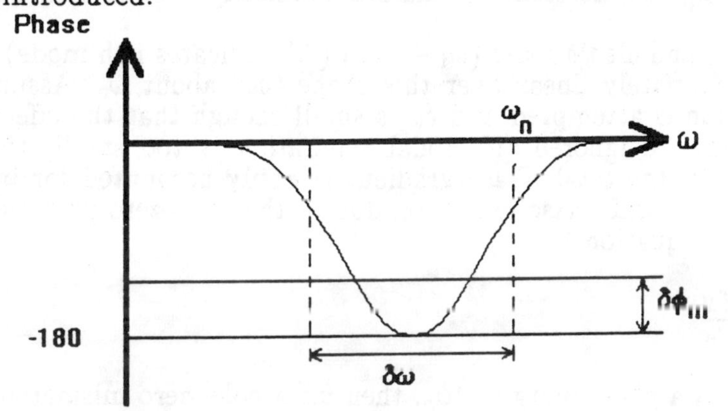

Figure 4. Bode phase plot of uncertain mode approximately canceled by compensator zeros in broadband structural control problem.

If $\delta\phi_m$ represents the "phase margin" at the canceled mode: i.e. the amount of unmodeled phase lag tolerable at this frequency, and if the linear approximation to the phase gradient (1) is assumed, a slightly pessimistic estimate of allowable $\delta\omega$ for a known level of passive damping, is given by this relation

$$\delta\omega \quad \leq (\delta\phi_m)\xi\omega_n$$

and if $\delta\phi_m = 1\text{rad}$

$$\delta\omega \quad \leq \xi\omega_n$$

This defines a first estimate of permissible modal eigenfrequency uncertainty in terms of passive damping for any system with flexibility between the sensors and actuators and required bandwidth including flexible eigenfrequencies.

A more precise expression for the maximum phase excursion due to pole–zero cancellation can be verified to be [Garcia]

$$\delta\phi = 2\tan^{-1}\left(\frac{p_i - z_i}{2\xi_i p_i}\right)$$

This expression is valid for large relative pole–zero spacing and correctly yields phase excursions limited to the range $\delta\phi \in (-180^0, 180^0)$. Since the zero frequency (z_i) is nominally chosen to cancel the pole, $z_i = \omega_n$ and the actual pole location is $p_i = \omega_{actual}$, then for small angles

$$\delta\phi = -\frac{z_i - p_i}{\xi_i p_i}$$

$$\simeq \frac{-\delta\omega}{\xi_i \omega_n}$$

which is the expression (1a).

HOW MUCH ξ IS SUFFICIENT IN THE BANDWIDTH?

For a system compensated by pole–zero cancellation, an estimate (to first order) of the passive damping ($\xi_{required}$) necessary to maintain stability over $\delta\omega$ may be determined.

Assuming $\delta\omega << \omega_n$ and also $\delta\omega << (\omega_i - \omega_{i-1})$ ("i" indicates i'th mode) ,ensures that the phase plot is approximately linear over this range ($\delta\omega$) about ω_n. Assuming further that pole–zero cancellation is attempted and $\delta\omega$ is small enough that the effect of neighbouring poles and zeros may be ignored (if modal separation is too small, the effect of other contributing modes to the total phase gradient is easily accounted for by including these extra terms), then the local phase excursion due to the pole–zero pair under consideration is approximately (by equation 1a)

$$\delta\phi = \frac{\delta\omega}{\xi\omega_n}$$

If the design provides a phase margin, $\delta\phi_m$, then for a pole–zero mismatch of $\delta\omega$ a damping ratio of

$$\xi \geq \frac{1}{\delta\phi_m}\frac{\delta\omega}{\omega_n} \qquad (2)$$

will suffice.

An alternative viewpoint is available by considering the root locus in the vicinity of a nearly canceling pole–zero pair:

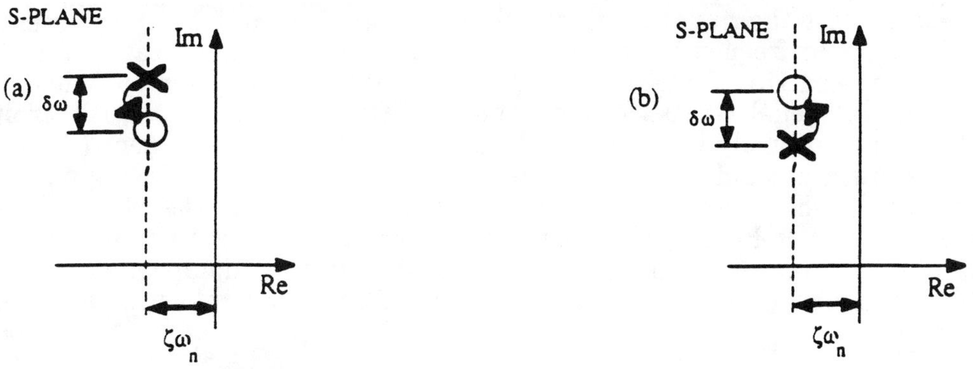

Figure 5. Root locus description of minimum damping ratio required to ensure stability of near pole–zero cancellation when assumed pole location is in error by $\delta\omega$.

PLANT ZERO UNCERTAINTY AND APPROXIMATE POLE–ZERO CANCELLATION

The preceding sections have implied that the plant contributes only poles which must be compensated with compensator zeros. Often the plant will contribute zeros, although their location, depending upon mode shapes, is typically more poorly known than the pole locations. Figure 6, from Rosenthal's PhD [Rosenthal], shows the possibility of a pole–zero "flip" when the inertia of one of the disks in this four degree of freedom system is changed by a factor of two. Near 20Hz, the plant phase is uncertain to ± 180^0. In such a situation phase stabilization is impossible.

Although the precise sequence of poles and zeros in a transfer function through a flexible structure is very uncertain, their approximate location can be known with more certainty [Lyon]. The plant phase can thus be viewed as an average phase related to a pole/zero density (with respect to frequency) and local phase excursions from this average phase. These excursions must be either small or well modeled if the plant is to be phase stabilized. If a close pole/zero pair is identified, but the sequence is uncertain (as in the case of [Rosenthal]) then a sufficient level of passive damping is required.

In this section the arguments of the preceding section are directly applied to such pole–zero cancellation, suggesting that approximate plant pole–zero cancellation is effectively perfect cancellation if sufficient levels of passive damping are present.

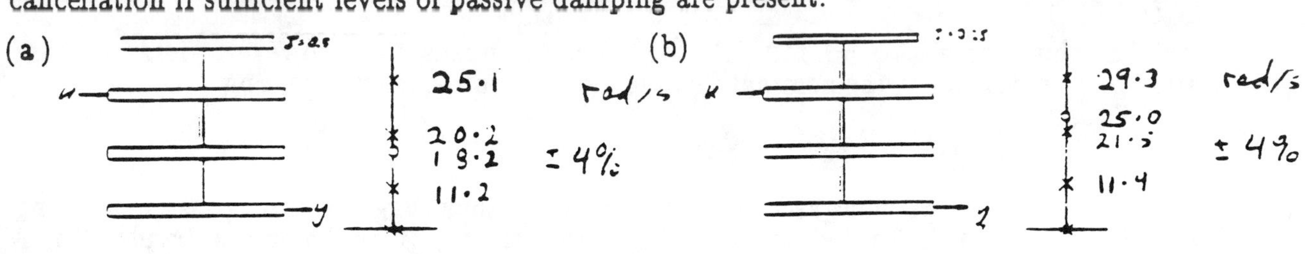

Figure 6. 4–Disk example by Rosenthal.

Poles and zeros in the loop transfer function are indistinguishable with respect to their origin; plant or compensator. We can thus apply the results of the preceding sections directly. If the plant pole–zero pair are known to be within $\delta\omega$ of one another, then a passive damping level of

$$\xi \approx \frac{\delta\omega}{\omega_n}\frac{1}{\delta\phi_m}$$

will suffice to ensure that their contribution will not destabilize the loop. For Rosenthal's system, the pole and zero frequencies were such that a passive damping ratio of 10 to 23 percent (for cases (a) and (b) of Figure 6, respectively) would have been sufficient. Such damping levels would then have permitted robust control of this plant with arbitrary bandwidth. For the intentionally low damping levels of Rosenthal's experiment ($\xi \simeq 0.004$), the closed loop bandwidth could never be made to approach this uncertain region near 20 Hz. Note that even for this laboratory system, the exact location of the plant poles and zeros was never known. For Rosenthal, this measurement uncertainty amounted to 1 to 4 percent. This implies that even for carefully identified lightly damped structures, a minimum passive damping level of 1 to 4 percent is needed to permit robust plant inversion in feedback control.

An Example. Bong Wie benchmark problem with pole–zero cancellation

The Bong Wie benchmark problem [ACC] has transfer function

$$G(s) = \frac{K(\xi\omega_n + \omega_n^2/2)}{s^2(s^2 + 2\xi\omega_n s + \omega_n^2)}$$

with uncertainty in modal eigenfrequency such that ω_{actual} may lie in the range $1.0 < \omega_{actual} < 2.0$ (ω in rad/s) and nominally $\omega_n = \sqrt{2}$ rad/s.

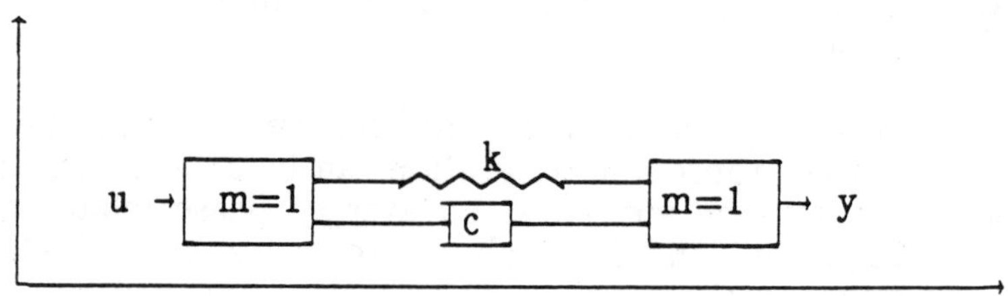

Figure 7. Bong Wie benchmark problem with dashpot providing passive damping.

For this eigenfrequency uncertainty, the estimate (1a) of necessary damping is evaluated as

$$\xi \geq (\sqrt{2} - 1)/\sqrt{2} \qquad = 0.29$$

Note that this assumes $\delta\phi_m = 1$ rad. The rigid body compensation, however, yields a phase margin of $\delta\phi \simeq 1.3$ rad at the nominal flexible mode eigenfrequency, yielding

$$\xi \geq 0.22.$$

The guidelines of the previous section, linear in $\delta\omega/\omega_n$, thus suggest that a passive damping level of 22 percent is needed for this huge uncertainty, if the nominal bandwidth is extended to beyond ω_n while maintaining stability.

It is necessary to verify if this estimate of the required level of passive damping is adequate. To this end, Figure 8 displays a root locus versus gain of the closed loop dynamics where broadband compensation using pole–zero cancellation is used. In the spirit of classical control design, the rigid body modes are first compensated by means of a low frequency zero. The flexible mode is then canceled by zeros at the nominal location of the flexible poles, with suitable high frequency poles yielding the desired roll–off, for a compensator transfer function

$$K(s) = \frac{(6.66s + 1)(s^2 + 2\xi_z\omega_n s + \omega_n^2)16}{(s^2 + 5.656s + 16)\omega_n^2}\,.$$

where the damping ratio of the compensator zero is

$$\xi_z = 0.19$$

The loop gain is chosen to yield closed loop bandwidth of 4 rad/s, greater than the uncertain mode frequency. The system is evaluated for stability when the modal frequency is $\omega_{actual} = 1$ rad/s (worst case, but including 22% passive damping ratio) and found to be stable (figures (8a and 8b)). Clearly the estimate yields a good first guess; the root locus of figure 8b doesn't quite cross into the right–half plane. Unfortunately, the lightly damped closed loop dynamics remaining due to imperfect pole–zero cancellation in Figure 8b limit the bandwidth to less than ω_n (bandwidth defined as in the introduction).

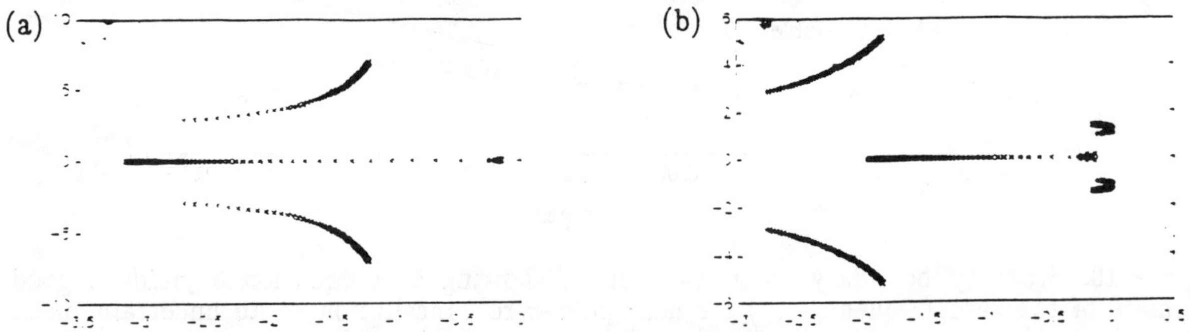

Figure 8. Root loci for Bong Wie problem. (a) shows the nominal case and (b) the nominal compensator implemented for the plant with eigenfrequency at the lower end of the uncertainty range.

Since it may not be practicable to implement the required amount of passive damping to achieve stability robustness for perfect cancellation of the nominal plant flexible modes, it is possible to exploit whatever passive damping is guaranteed by placing the compenastor zero somewhere inside the uncertainty bound for the achievable damping and ensuring stability over thc cntire uncertainty range as well as improving the nominal performance over conservative compensation with the zero at the lower end of the uncertainty range.

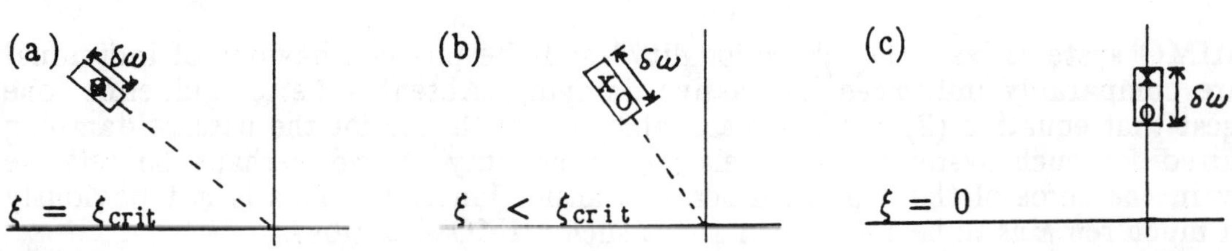

Figure 9. S–plane plots showing intermediate pole–zero cancellation achieving stability robustness and improved nominal performance over undamped uncertain design.

Figure 9 shows the transition in designs from exact pole–zero cancellation of the nominal flexible modes to approximate cancellation at the lower end of the eigenfrequency uncertainty range. For each case the uncertainty is the value $\delta\omega$ and the zero frequency for cancellation may be estimated using equation 1a and the guaranteed level of passive damping ξ. Notice that in the case of this example, the rigid body compensation yields "phase margin" of approximately 75^0 ($\simeq$ 1.3 rad) in the frequency region near the uncertain poles. The frequency of the compensator zero should then be

$$\omega_z \leq \omega_{lower} + \delta\phi\xi \tag{3}$$

where ω_{lower} is the lower bound on the eigenfrequency range. A sequence of such designs for the Bong Wie benchmark problem yield Figure 10, showing the limiting value for the compensator zero as a function of plant damping ratio. This suggests that use of equation 3 is warranted for estimating the required zero frequency when a fixed level of passive damping is available and stability robustness with near pole–zero cancellation is desired.

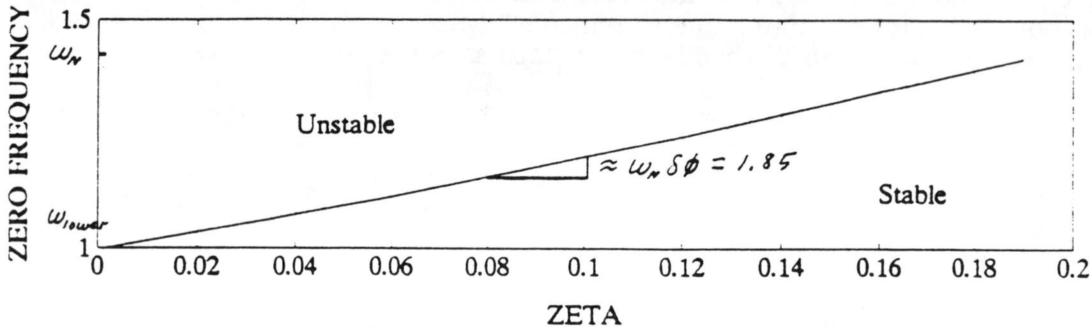

Figure 10. Stability boundary on ω_z vs ξ curve showing that equation 3 yields a good estimate of the zero frequency (ω_z) for near pole–zero cancellation of an uncertain mode where the minimum guaranteed level of passive damping is ξ.

INFERENCES FOR MIMO CONTROL

The preceding sections have presented simple arguments supporting the claim that broadband control of uncertain structural dynamics is impossible with LTI compensation unless a sufficient level of passive damping is present. Although it is not possible to extrapolate these arguments to MIMO systems with confidence, there appears to be little reason to believe that such systems are more benign. Each scalar transfer function of a MIMO system has pole–zero patterns like those described in the preceding sections. Worse, the mapping from actuators to sensors is a strong function of frequency, changing drastically in the near vicinity of each zero of any of the individual scalar transfer functions. The rate of change with frequency of these directions varies inversely with passive damping ratio.

Thus, in MIMO systems both the "directionality" and the phase behaviour of individual channels are comparably influenced by passive damping. Absent a better guideline, one might suggest that equation (2) yields a reasonable rule of thumb for the passive damping levels required for such systems. Here, relative uncertainty should perhaps be relative uncertainty in the zeros of the individual scalar transfer functions. This is not obviously correct and much remains to be understood about such MIMO situations.

SUMMARY

This paper has argued, simply but quantitatively, that a critical level of passive damping can be specified which will permit robust non–collocated LTI control of structural dynamics with the control bandwidth including many flexible eigenfrequencies. We have suggested that this level is as high as 1 to 4 percent, even for carefully identified laboratory structures and much higher for uncertain structures. Figure 1 perhaps summarizes the thesis of this paper most succintly.

REFERENCES

ACC, "A Benchmark Problem for Robust Control Design," special session at American Control Conference, May 1990, San Diego, CA.

Balas,M.J., Quan, R. D. and Das, B., "Low–Order Control of Large Aerospace Structures Using Residual Mode Filters," CU–CSSC–88–05, College of Engineering, University of Colorado, April 1988.

Belvin, W. K. and Park, K. C., Structural Tailoring and Feedback Control Synthesis: An Interdisciplinary Approach," J. Guidance and Control, Vol. 13, No. 3, May–June 1990, pp. 424–429.

Bontsema, J., "Control of Flexible Structures: Robustness, Model Reduction and Spillover," ESTEC Workshop Proceedings, paper no. esa WPP–09, 15–16 June 1989, Noordwijk, The Netherlands.

Bryson, A.E., Jr., Hermelin, S. and Sun, J., "LQG Controller Design for Robustness," presented at ACC, Seattle, Washington, June 1986.

Garcia, J., "Broadband Positioning of Small Optical Components on a Resonant Substructure," Proceedings, ASME WAM, Dallas, TX, Nov 1990.

Gran, R., "Finite Dimensional Controllers for Hyperbolic Systems," Proceedings of 3rd VPI and SV/AIAA Symposium on Dynamics and Control of Large Flexible Spacecraft, June 15–17, 1981.

Hughes, P.C., "Stability of Proportional–PLus–Derivative–Plus–Integral Control of Flexible Spacecraft," J Guidance and Control, Vol. 2, No. 6, Nov–Dec 1979, pp. 499–503.

JPL Internal Report, Milman, M., Salama, M., Scheid, R., Bruno, R., Gibson, J.S., "Integrated Control–Structure Design: A Multiobjective Approach," JPL D–6767, Jan 1990.

Kopf, E.H., Brown, T.K., Marsh, E.L., "Flexible Stator Control on the Galileo Spacecraft," AAS Conf., Paper no. 79–161, Provincetown, MA, June 1979.

Lyon, R. H., "Statistics of Phase and Magnitude of Structural Transfer Functions," contribution to: Random Vibrations – Status and Recent Developments, Eds. I. Elishakoff and R.H. Lyon, Elsevier, Amsterdam, 1986, pp 201–209.

Rao. S.S., Venkayya, V. B. and Khot, N. S., "Game Theory Approach for the Integrated Design of Structures and Controls," AIAA Journal, Vol. 26, No. 4, April 1988, pp. 463–469.

Rosenthal, D. E., "Experiments in Control of Flexible Structures with Uncertain Parameters," PhD thesis, Stanford University, Dept of Aeronautics and Astronautics, March 8, 1984.

Rossi, J. A., "A Robust Compensator Design by Frequency–Shaped Estimation," J. Guidance and Control, Vol. 8, 1985, pp. 541–544.

Spanos, J. T., "Control–Structure Interaction in Precision Pointing Servo Loops," J. Guidance and Control, Vol. 12, No. 2, March–April 1989, pp. 256–263.

Venkayya, V. B. and Tischler, V. A., "Frequency Control and Its Effects on the Dynamic Response of Flexible Structures," AIAA Journal, Vol. 23, No. 11. Nov. 1985, pp. 1768–1774.

von Flotow, A.H., "Control–Motivated Dynamics Tailoring of Truss Work Structures," Proc. of AIAA Guidance and Control Conference, Williamsburg, V.A., Aug, 1986.

Wedell, E., Chuang, C.–H. and Wie, B., "Computational Analysis of a Stability Robustness Margin for Structured Real–Parameter Perturbations," AIAA paper no. 89–3505, AIAA GN&C Conference, August 1989.

Wie, B., Byun, K.–W., "New Generalized Structural Filtering Concept for Active Vibration Control Synthesis," J. Guidance and Control, Vol. 12, No. 2, March–April 1987, pp. 147–154.

Williams, T. and Juang, J.–N., "Pole/Zero Cancellations in Flexible Space Structures," J. Guidance and Control, Vol. 13, No. 4, July–August 1990, pp. 684–690.

STRUCTURAL CONTROL USING ACTIVE BROADBAND IMPEDANCE MATCHING

Douglas G. MacMartin,[1] David W. Miller,[1] Steven R. Hall[1]

ABSTRACT

The analytical and experimental research into impedance matching that has been conducted at the M.I.T. Space Engineering Research Center is presented here in a coherent framework. A travelling wave approach and a dereverberated mobility approach are discussed, and shown to be strongly connected. Control design using these models is based on either an $\mathcal{H}_2$ or an $\mathcal{H}_\infty$ maximization of the power dissipated. The applicability of a local model for control design, either using the wave, or the dereverberated mobility approach is demonstrated experimentally, and the similarities and differences between the $\mathcal{H}_2$ and $\mathcal{H}_\infty$ optimal solutions are discussed. Both solutions obtained greater damping levels than that achievable by rate feedback.

INTRODUCTION

Many envisioned precision structures require high levels of vibrational energy dissipation in order to meet mission requirements associated with orientation, alignment and jitter. Active techniques provide one approach to achieving this dissipation. The formulation of control algorithms for active dissipation of vibrational energy generally requires an accurate model of the structural system. Global models of complex structures, perhaps acquired through finite element methods, are prone to inaccuracy while global measurement based models are time consuming to generate and often unreliable due to suspension coupling during testing. However, since energy dissipation is a spatially local objective, rather than a global objective such as the alignment of two remote points, then provided that the sensors and actuators are collocated and dual, local structural models are sufficient for designing control laws to maximize energy dissipation.

There are several ways in which a local model of the structure can be obtained. One approach is to model the control location in terms of waves, see for example [1–3]. An alternative approach is to use the dereverberated driving point mobility [4,5] at the actuator location. This model can be obtained from a wave model, or directly from experimental data. A similar average impedance model has also been used by Chen *et al.* [6]. The optimal control law for a local objective can be derived in terms of the local model, so that the closed loop system will be stable for any uncertainties in the rest of the structure.

[1]Space Engineering Research Center, Department of Aeronautics and Astronautics, Massachusetts Institute of Technology, Cambridge, MA 02139, U.S.A.

Maximizing the power dissipated by the control action corresponds to matching the structure's drive point impedance as closely as possible, given the available actuators [6–8]. An exact match at every frequency is not always possible [9], as the required compensator is in general non-causal and cannot be implemented. If the frequency weighted power flow is minimized using a Wiener-Hopf [2] or LQG approach, then the resulting compensator will be causal. The main disadvantage of this $\mathcal{H}_2$ formulation is that because the local model does not include the reverberances present in the real structure, the optimal compensator may generate power at one frequency in order to better dissipate power at another, which could lead to instability when implemented on the real structure. If the worst case power flow, rather than the average power flow, is minimized, then stability is guaranteed [4]. This can be formulated as a standard $\mathcal{H}_\infty$ optimal control problem. Both the $\mathcal{H}_2$ and $\mathcal{H}_\infty$-optimal control laws developed with local models have been implemented, and the achievable performance verified on a pinned-free beam in the laboratory [3, 5].

MODELLING

Using the travelling wave model in [2], the structure is modelled as a network of one-dimensional waveguides such as beams, rods, etc. The control input is applied at a junction, where one or more of these waveguides, or members, intersect as shown in Figure 1. The cross-sectional variables (displacements and forces) in each member can be related to the amplitudes of independently travelling wave modes via a frequency dependent transformation

$$\begin{bmatrix} q \\ f \end{bmatrix} = \begin{bmatrix} Y_{qi} & Y_{qo} \\ Y_{fi} & Y_{fo} \end{bmatrix} \begin{bmatrix} w_i \\ w_o \end{bmatrix} \tag{1}$$

where q and f are the cross-sectional variables which can be measured and commanded, respectively, and w_i and w_o are the incoming and outgoing wave mode amplitudes.

The junction can then be described by a matrix S of scattering coefficients which relate outgoing waves to incoming waves, and a generation matrix Ψ relating the outgoing waves to the control input u. With s being the Laplace variable,

$$w_o(s) = S(s)w_i(s) + \Psi(s)u(s) \tag{2}$$

Hence the measurable cross-sectional variables at a junction are given by

$$q = (Y_{qo}\Psi)u + (Y_{qi} + Y_{qo}S)w_i \tag{3}$$

Equation (3) allows a control law which is a feedforward of incoming wave modes to be rewritten as a feedback of the cross-sectional variables, and vice versa.

For arbitrarily complex junctions, such as that in Figure 2, deriving a wave model can be impractical. An alternative to a wave model for obtaining a local model of the structure near the actuator is to use the dereverberated drive point mobility [4]. The structural response due to the actuator can be divided into two parts: a *direct field*, due to the local dynamics; and a *reverberant field*, which is created by reflections from other parts of the structure. The direct field corresponds to that part of the outgoing waves w_o which is due to the control u, while the reverberant field consists of the incoming waves w_i, and that part of w_o which is due to w_i. The term "dereverberated" implies that the "reverberant"

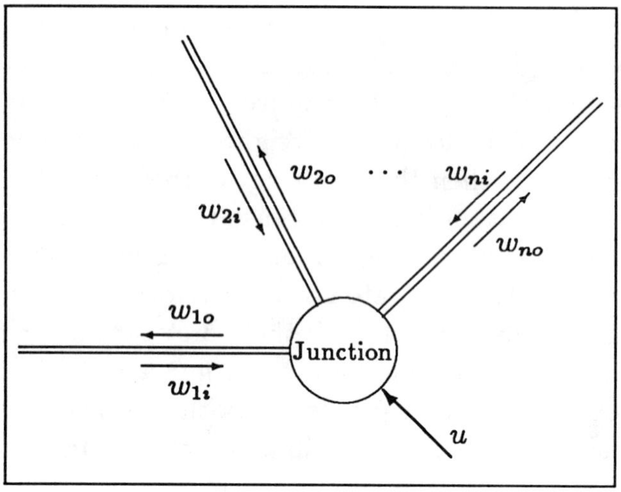

Figure 1: Generic wave junction

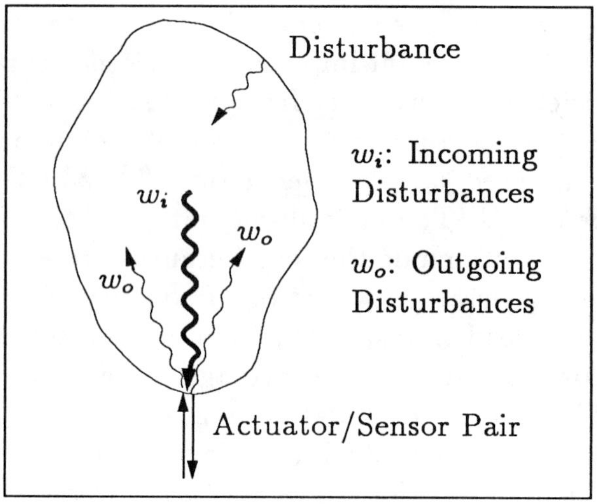

Figure 2: Arbitrary structure

part of the response has been removed before computing the mobility. The dereverberated mobility may be approximated by taking a logarithmic average of the reverberant drive point mobility [10], or as will be shown, from a wave model.

If $G(s)$ is the dereverberated mobility, then the response is of the form

$$y(s) = G(s)u(s) + d(s) \tag{4}$$

where d is a disturbance input. This equation has the same form as Equation (3) derived for waves, except that y is now a velocity, and includes only those cross-sectional variables which are dual to the control forces in u. So $y = sTq$ where the matrix T is present to select the appropriate elements of q corresponding to each element of y. A dereverberated model is obtained by assuming that there are no reflections in the model, hence that the incoming waves w_i do not depend on the control u. Then comparing Equations (3) and (4),

$$G = sT(Y_{qo}\Psi) \quad \text{and} \quad d = sT(Y_{qi} + Y_{qo}S)w_i \tag{5}$$

The wave model provides more physical insight, and a more complete description of the junction than the dereverberated mobility. The junction wave model can also be combined with a wave description of transmission along members to generate a model of the entire structure [11]. The dereverberated mobility approach is motivated by a wave model, but does not require the wave modes of the structure to be identified, as the model can be obtained directly from experimental data by averaging. As a result, it may be easier to apply to complicated structures. Some of the analytical results are also simpler using this model, since all of the relevant information about the junction is contained in a single transfer function matrix. For this reason, the dereverberated mobility framework will be used to derive optimal compensators. One could equally well use the wave framework; after converting back to the cross-sectional variables from wave mode variables, the compensators would be the same.

CONTROL

A possible control objective that requires only a local model is to maximize the impinging vibrational power that is dissipated by the control action. Since y and u in Equation (4) are collocated and dual, the average power flow into the structure at each frequency is of the form [11]:

$$\mathbb{P}(\omega) = \frac{1}{2}\left(y^H(\omega)u(\omega) + u^H(\omega)y(\omega)\right) \tag{6}$$

Optimal compensators of the form $u(s) = K(s)y(s)$ are desired.

A standard result from electric circuits (e.g. [12]) is that the optimal impedance to maximize power dissipation is the conjugate of the load impedance. If this condition is achieved, then one has a "matched termination", and there is no reflected signal. The same result is true in structures. (In fact, if evanescent waves exist, the maximum dissipated power solution is better than the matched termination, since the possibility exists to draw energy to the junction more quickly than it would naturally arrive [2].) The load impedance is the inverse of the mobility of the structure, and the transfer function of the compensator (from velocity to force) is itself an impedance function. Hence the optimal compensator is

$$K(s) = (G^{\sim}(s))^{-1} \quad \text{where} \quad G^{\sim}(s) = G(-s)^T \tag{7}$$

This compensator can also be obtained by optimizing Equation (6) at each frequency.

The matched termination condition is equivalent to continuing the structure at the junction with an identical structure, so that there is no effective discontinuity. If at the end of a beam, the control mimics the forces and moments that would be present if the beam continued indefinitely, then there will be no reflection, and all of the incoming power will be absorbed. This however requires that all of the cross-sectional forces at the junction can be independently controlled. If this is not true, then all of the incoming power cannot be absorbed with a causal compensator. Indeed, unless the dereverberated mobility is a constant, the compensator in Equation (7) is noncausal.

$\mathcal{H}_2$-OPTIMAL SOLUTION

One approach to guaranteeing a causal compensator is to minimize the integral over frequency of the power flow in Equation (6) using a Wiener-Hopf or LQG approach. This is a standard $\mathcal{H}_2$ problem. A penalty on the control effort could also be included.

First, introduce the notation for the spectral factors of a function Φ as

$$\Phi = \Phi_R\Phi_L \quad \text{and} \quad \Phi = \Phi_+ + \Phi_- \tag{8}$$

for the right half plane analytic (RHPA) and left half plane analytic factors of Φ, and the positive and negative time parts of Φ respectively. With disturbance feedforward $u = Hd$, the frequency weighted power flow being minimized is given by Equations (4) and (6) as

$$J = \int_{-\infty}^{\infty} \operatorname{tr}\left\{\Phi_{dd}\left[H^{\sim}(I+GH) + (I+GH)^{\sim}H\right]\right\} d\omega \tag{9}$$

where Φ_{dd} is the spectrum of the disturbance d. The feedforward compensator H must be stable, hence to be causal it must be RHPA. Then for a causal optimum, the first order variation in J with respect to H must be zero for all RHPA variations δH. Solving for the RHPA compensator that satisfies this condition yields the optimal disturbance feedforward compensator as

$$H = -(G + G^{\sim})_R^{-1} \left[(G + G^{\sim})_L^{-1} (\Phi_{dd})_R\right]_+ (\Phi_{dd})_R^{-1} \tag{10}$$

from which the feedback law $u = -Ky$ can be determined via $K = -H(I + GH)^{-1}$. Note that K is causal if and only if H is causal.

As was noted in Reference [2], the $\mathcal{H}_2$ approach suffers from the fact that it does not guarantee a stabilizing compensator. This is because the local model does not include knowledge of the fact that departing energy will eventually return, that is, that the incoming disturbance is correlated with the outgoing waves, and therefore with the control. As a result, the approach will allow power to be generated at some frequencies in order to allow greater dissipation at other frequencies.

$\mathcal{H}_\infty$-OPTIMAL SOLUTION

To guarantee dissipation at all frequencies, the worst case power dissipation will be minimized, which can be cast as an $\mathcal{H}_\infty$ minimization problem.

With the optimal noncausal compensator from Equation (7), the closed loop power flow into the structure is

$$\mathbb{P} = -d^{\sim}(G + G^{\sim})^{-1} d \tag{11}$$

If $d = G_0 w$ where $G_0 G_0^{\sim} = G + G^{\sim}$, then the optimal noncausal compensator will dissipate power $w^{\sim}w$. Since this is the maximum power that can be dissipated, adding it to the power flow into the structure at each frequency yields a positive cost, given by

$$\begin{aligned} \text{Cost}(\omega) &= w^{\sim}w + u^{\sim}y + y^{\sim}u \\ &= |G_0^{\sim}u + w|^2 = |z|^2 \end{aligned} \tag{12}$$

where $z = G_1 G_0^{\sim} u + G_1 w$, and $G_1(s)$ is an inner function ($G_1^{\sim} G_1 = I$) defined so that $G_1 G_0^{\sim}$ is stable. Combining this with the system equation (4) and the definition of w, then

$$\begin{Bmatrix} z \\ y \end{Bmatrix} = \begin{bmatrix} G_1 I & G_1 G_0^{\sim} \\ G_0 & G \end{bmatrix} \begin{Bmatrix} w \\ u \end{Bmatrix} \tag{13}$$

The compensator from y to u that minimizes the $\mathcal{H}_\infty$ norm of the transfer function T_{zw} from w to z will minimize the maximum power flow into the structure. Because of the definition of the cost in Equation (12), if power is being dissipated at any frequency ω, then $T_{zw}(\omega) < 1$. Hence $\|T_{zw}\|_\infty < 1$ implies that power is dissipated at all frequencies, which implies that the closed loop system will be stable.

Closed loop stability can also be shown using a positivity argument. From Haddad and Bernstein [14], a stable transfer function $H(s)$ has bounded magnitude $\|H(s)\|_\infty < 1$ if and only if the function $(I - H)^{-1}(I + H)$ is positive real. Substitution of T_{zw} as the lower linear fractional transformation of the system in Equation (13) with $-K$ yields that $\|T_{zw}\|_\infty < 1$ if and only if the corresponding compensator K is positive real. Since the structural transfer function between collocated and dual actuators and sensors is also positive real, the closed loop system will be stable.

Note that the $\mathcal{H}_2$ problem solved earlier by a Wiener-Hopf approach can also be represented by Equation (13); the $\mathcal{H}_2$ optimal compensator is that which minimizes $\|T_{zw}\|_2$, which is a standard LQG problem.

WEIGHTING

Since a perfect impedance match at all frequencies is not always possible due to the causality constraint, weighting matrices can be introduced into the design process to specify which frequencies and directions are most important for power dissipation.

In the $\mathcal{H}_2$ problem, the power spectral density of the disturbance can be used to introduce frequency weighting into the problem; more importance is attached to a certain frequency range by increasing the incoming disturbance power in that range. Either the incoming wave mode spectrum $\Phi_{w_i w_i}$ or that of the disturbance Φ_{dd} can be specified, since the two are related from Equation (5).

In the $\mathcal{H}_\infty$ problem, including a weighting on w will destroy the guarantee of stability for $\|T_{zw}\|_\infty < 1$. Rather than weighting the sum of the disturbance input power and the power input by the control, as in Equation (12), the cost is defined to be the sum of the disturbance power and some frequency weighted control input power, as

$$\text{Cost}(\omega) = w^{\sim}w + W_1^{\sim}(u^{\sim}y + y^{\sim}u)W_1 \tag{14}$$

For $\|W_1\|_\infty \leq 1$, this can again be factored as $\text{Cost}(\omega) = |z|^2$ and therefore written as a standard $\mathcal{H}_\infty$ problem [4]. The closed loop power dissipation as a function of frequency will be inversely proportional to $|W_1(\omega)|^2$. More importance is given to a frequency range where W_1 is small, since more control effort is required to reduce the cost at that frequency, resulting in more power being dissipated.

$\mathcal{H}_\infty$-optimal solutions guarantee dissipation at all frequencies, however, if modal frequencies of the system are reasonably well known, then it is not necessary to require the compensator to dissipate power at all frequencies, but only at those frequencies where the system could amplify a disturbance. Using the $\mathcal{H}_2$ approach, a much better impedance match at several modal frequencies can be obtained by allowing an amplification of the response in between these modes. This will be demonstrated by example in the next section.

In many structures, low vibration levels may be required only in certain parts of the structure. Thus more effort should be placed on minimizing the power flow into this region than out of it. Selecting a weighting function that places more emphasis on incoming wave modes from one direction than another results in a controller that acts like an energy diode, allowing power to flow more easily in one direction than in the other. A similar "energy valve" has been discussed by Hagedorn and Schmidt [1].

EXAMPLE

As an example of the optimal impedance matching approaches, consider the design of compensators from rotational velocity to a collocated torque actuator at the lefthand, pinned end of a uniform undamped Bernoulli-Euler beam. The partial differential equation describing the beam vibration is

$$\rho A \frac{\partial^2 v}{\partial t^2} + EI \frac{\partial^4 v}{\partial x^4} = 0 \tag{15}$$

where EI is the flexural rigidity, and ρA is the mass per unit length. Then with the wave number $k = \sqrt[4]{\rho A/EI}\sqrt{\omega}$, Equation (1) becomes

$$\left\{ \begin{array}{c} v' \\ EIv''' \\ \hline v \\ -EIv'' \end{array} \right\} = \left[\begin{array}{cc|cc} ik & k & -ik & -k \\ -iEIk^3 & EIk^3 & iEIk^3 & -EIk^3 \\ \hline 1 & 1 & 1 & 1 \\ EIk^2 & -EIk^2 & EIk^2 & -EIk^2 \end{array} \right] \left\{ \begin{array}{c} w_{lp} \\ w_{le} \\ \hline w_{rp} \\ w_{re} \end{array} \right\} \tag{16}$$

The four wave modes supported by the beam are leftward (incoming) and rightward (outgoing) propagating waves (w_{lp} and w_{rp}), and leftward and rightward emanating evanescent waves (w_{le} and w_{re}). For a pinned end with moment actuation, the scattering and wave generation matrices are given by

$$S = \begin{bmatrix} -1 & 0 \\ 0 & -1 \end{bmatrix} \qquad \Psi = \frac{1}{2EIk^2} \begin{bmatrix} 1 \\ -1 \end{bmatrix} \tag{17}$$

The dereverberated mobility may be calculated from the wave information from Equation (5) as

$$G = sTY_{qo}\Psi = \frac{\sqrt{s}}{\sqrt{2}(\rho A)^{1/4}(EI)^{3/4}} \tag{18}$$

The matrix T only selects the first element of $Y_{qo}\Psi$, since only the rotational velocity is required for determining the power dissipated with the torque actuator.

With no causality constraint, the compensator that best matches the input impedance and thus dissipates the maximum possible power at all frequencies is, from Equation (7),

$$K(s) = \frac{\sqrt{2}(\rho A)^{1/4}(EI)^{3/4}}{\sqrt{-s}} \tag{19}$$

This is the ideal compensator for the structure, but cannot be implemented, as $1/\sqrt{-s}$ is noncausal. The $\mathcal{H}_\infty$-optimal compensator with unity weighting has the same magnitude, but a 90° phase shift at all frequencies:

$$K(s) = \frac{\sqrt{2}(\rho A)^{1/4}(EI)^{3/4}}{\sqrt{s}} \tag{20}$$

An $\mathcal{H}_2$ optimal solution which emphasizes frequencies around 40 rad/sec can also be obtained. The power spectral density of the incoming wave modes $\Phi_{w_i w_i}$ is chosen to be

$$\Phi_{w_i w_i} = \frac{a^2}{(s - \omega_n)^2} \begin{bmatrix} 1 & 0 \\ 0 & 1 \end{bmatrix} \frac{a^2}{(s + \omega_n)^2} \tag{21}$$

with $\omega_n = 40$. The resulting optimal compensator is

$$K(s) = (\rho A)^{1/4}(EI)^{3/4}\left(\frac{8s^{3/4}}{(\omega_n + 5s)\sqrt[4]{\omega_n}} - \frac{\sqrt{2}}{\sqrt{s}}\right) \tag{22}$$

This compensator is not positive real, and hence may destabilize the structure by adding power at certain frequencies. In order to guarantee stability, the compensator is approximated by a positive real form:

$$K(s) = \frac{8(\rho A)^{1/4}(EI)^{3/4}s}{5\sqrt[4]{\omega_n}\sqrt[4]{s}(s+92)} \tag{23}$$

An $\mathcal{H}_\infty$ compensator was also designed to emphasize a narrow frequency band near 35 rad/sec. The weighting function used is

$$W_1 = \frac{(s + a\omega_n)(s + \omega_n/a)}{(s + b\omega_n)(s + \omega_n/b)} \tag{24}$$

with $\omega_n = 35$, $a = \sqrt{2}$, $b = 2\sqrt{2}$. The resulting optimal compensator is found to be well approximated by

$$K(s) = 2.62\frac{\sqrt{2}(\rho A)^{1/4}(EI)^{3/4}}{\sqrt{s}}\left(\frac{s^2 + 38.5s + 466}{s^2 + 100s + 3210}\right) \tag{25}$$

The $\mathcal{H}_2$, weighted $\mathcal{H}_\infty$, unweighted $\mathcal{H}_\infty$, and unconstrained optimal compensators are plotted in Figure 3, with ρA and EI corresponding to the beam in the experiment. Note that at the frequency weighted most heavily, the $\mathcal{H}_\infty$ solution achieves the magnitude of the unconstrained optimal compensator, but not quite the phase, while the $\mathcal{H}_2$ optimal solution achieves the phase, but not the magnitude.

The amount of power dissipated at the weighted frequency can be increased in both the $\mathcal{H}_2$ and $\mathcal{H}_\infty$ cases at the expense of broadband performance by changing the weighting functions. For the $\mathcal{H}_2$ case, the poles in the weighting function (21) could be made complex. Then as the damping in these poles approaches zero, the impedance match at the pole frequency becomes perfect.

The weighting in the $\mathcal{H}_2$ case could also be modified to obtain a good response at two distinct frequencies, by selecting, for example,

$$\Phi_{w_iw_i} = \frac{a^2(s^2 - 2\zeta\omega_2 s + \omega_2^2)}{(s^2 - 2\zeta\omega_1 s + \omega_1^2)(s^2 - 2\zeta\omega_3 s + \omega_3^2)}\begin{bmatrix} 1 & 0 \\ 0 & 1 \end{bmatrix}\frac{a^2(s^2 + 2\zeta\omega_2 s + \omega_2^2)}{(s^2 + 2\zeta\omega_1 s + \omega_1^2)(s^2 + 2\zeta\omega_3 s + \omega_3^2)}$$

The resulting compensator with $\zeta = 0.01$, $\omega_1 = 20\pi$, $\omega_3 = 40\pi$, and $\omega_2 = \sqrt{\omega_1\omega_3}$ is shown in Figure 4. The compensator obtains a near perfect impedance match at both ω_1 and ω_3 by allowing the phase to wrap around in between these frequencies. The compensator is not positive real, and hence adds power to the structure between ω_1 and ω_3. Since $\mathcal{H}_\infty$-optimal compensators are constrained to be positive real, a close impedance match at discrete frequencies is only possible if there is a much larger frequency separation between them.

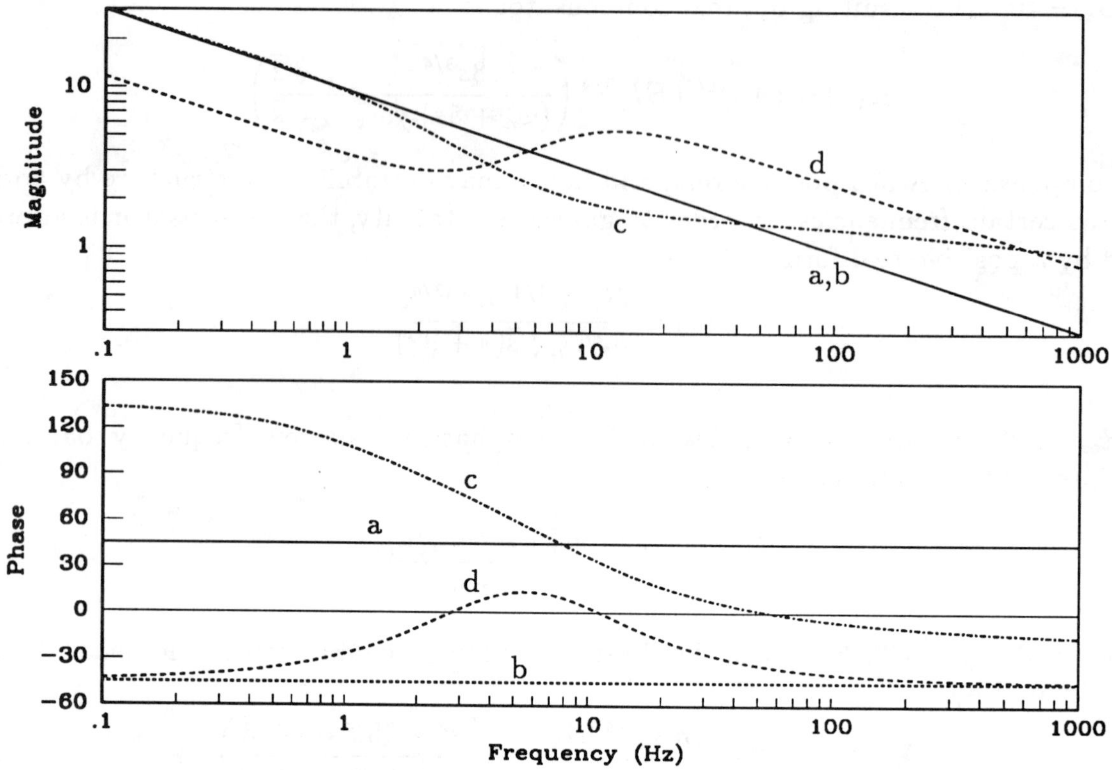

Figure 3: Optimal compensator transfer functions: unconstrained (non-causal) (a), unweighted $\mathcal{H}_\infty$ (b), $\mathcal{H}_2$ (c), and weighted $\mathcal{H}_\infty$ (d).

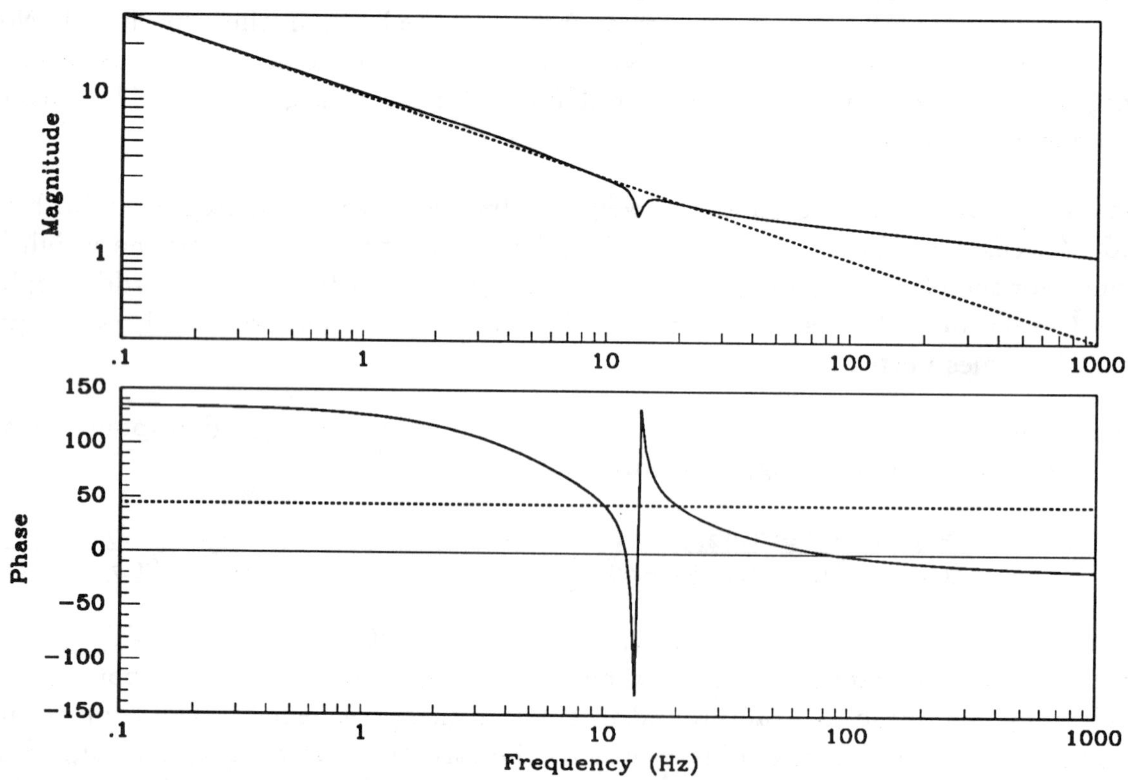

Figure 4: $\mathcal{H}_2$-optimal compensator to match impedance at 10 Hz and 20 Hz (solid) and unconstrained optimum (dashed).

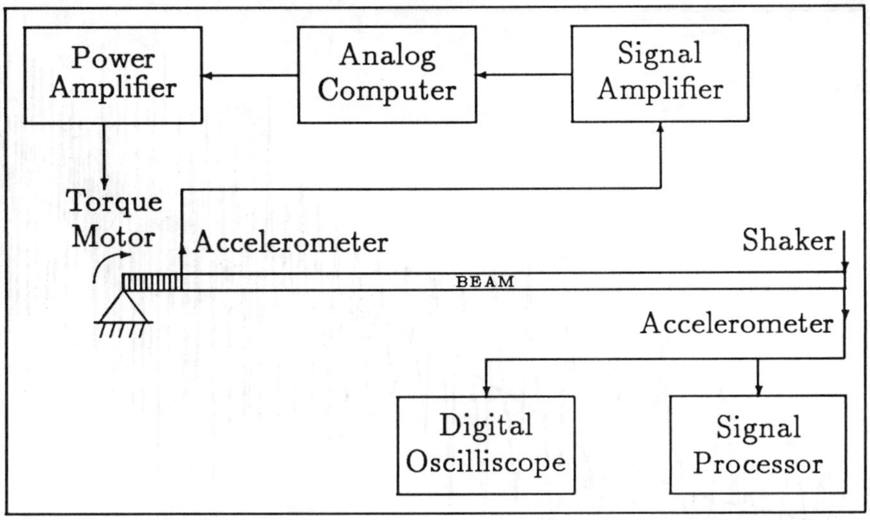

Figure 5: Schematic of Experimental Setup

EXPERIMENT

Laboratory experiments have been performed on a 24 foot pinned-free brass beam to verify the effectiveness of some of the $\mathcal{H}_2$ and $\mathcal{H}_\infty$ optimal compensators developed with local models in the previous section. The setup is shown schematically in Figure 5. The mass per unit length and flexural rigidity of the beam are given by $\rho A = 2.85$ kg/m and $EI = 31.1$ Nm^2 respectively. The open-loop damping of the first 17 modes, up to a frequency of approximately 30 Hz, average about 0.3%.

Control is applied through a torque motor at the pinned end, and sensing is provided by a linear accelerometer mounted a short distance from this end. The member connecting the sensor to the tip is much stiffer than the beam, and can be assumed to be rigid in the frequency range of interest, so that the sensor provides a rotational acceleration measurement collocated with the moment actuator. The compensator between these two was implemented on an analog computer. In addition to the control hardware, a shaker and collocated accelerometer were mounted at the free end of the beam. The closed loop transfer function between these two was used as an indication of the performance achieved. Detailed information on the characteristics of the sensors and actuators can be found in [3].

The dereverberated mobility for this beam can be calculated by averaging the measured transfer function between the control actuator and sensor, as shown in Figure 6. The analytical calculation based on the wave model closely approximates this average, and hence the compensators designed in the previous section could be used in the experiment. The unweighted and weighted $\mathcal{H}_\infty$ designs, and the positive real approximation to the $\mathcal{H}_2$ design were implemented. For comparison with the optimal compensators, a rate feedback compensator was also implemented, with a gain selected to achieve maximum power dissipation at 10 Hz.

Since all of the implemented compensator designs are positive real, they are guaranteed to be stable for any positive real structure. However, the transfer function from the

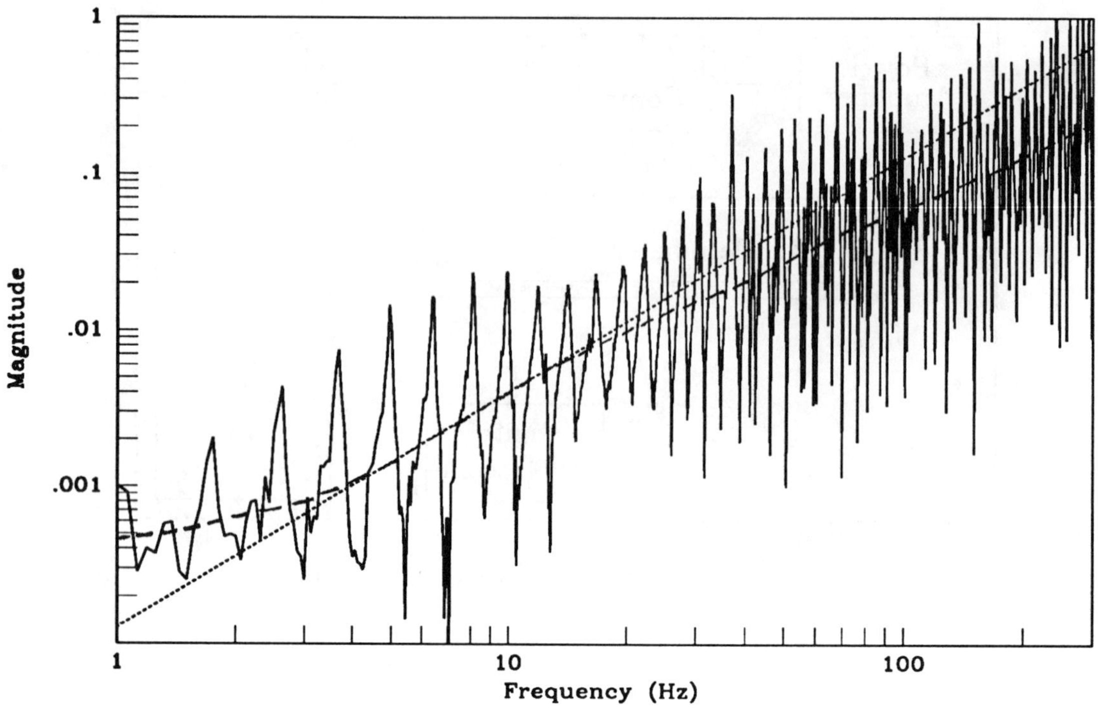

Figure 6: Transfer function at controlled end of beam (solid), dereverberated mobility (dotted) and logarithmic average (dashed).

actuator to the sensor of this beam is not positive real at high frequencies, due to the non-collocatedness and additional dynamics of the sensor and actuator, and any time delays in the system. In order to guarantee stability, the compensator must roll off at high frequency, so low-pass filters were added to all of the designs, with corner frequencies well above the region of interest. The available measurement in the experiment was proportional to angular acceleration, and thus an integration was necessary to obtain angular rate. This integrator was rolled off at DC to prevent saturation and drift problems. The half and quarter integrators ($\frac{1}{\sqrt{s}}$ and $\frac{1}{\sqrt[4]{s}}$) were implemented using the approach of Reference [13]. The approximations are excellent up to about 700 Hz, well above the region of interest.

Open and closed loop transfer functions from force at the free end to collocated displacement with the various compensators are given in Figure 7. Reasonable agreement is obtained between these transfer functions and predictions based on a wave model of the complete beam. The unweighted $\mathcal{H}_\infty$ design achieves broadband damping but sacrifices some of the narrowband damping achieved by velocity feedback. Velocity feedback, however, will be ineffective at sufficiently high and low frequencies, though the narrow frequency range present makes it difficult to judge the broadband behaviour of any of the compensators. The weighted $\mathcal{H}_\infty$ design achieves excellent narrowband damping in the desired frequency-range, while maintaining some damping everywhere. The positive real approximation to the $\mathcal{H}_2$-optimal compensator also achieves good narrowband performance, though the performance drops off dramatically at lower frequencies. In both cases, the narrowband performance is better than that of rate feedback, and is achieved by closely matching the structure's impedance in a narrow frequency band. In fact, the modes near the centre frequency of the weighting can be virtually eliminated if an exact match is obtained.

CONCLUSIONS

For uncertain, modally dense structures, a model that includes only the dynamics near an actuator can still be effectively used for control design. This local model can be derived from a travelling wave approach, or from an equivalent dereverberated mobility approach. The latter approach allows the model to be determined directly from an experimental transfer function.

Maximizing the power dissipation associated with the control corresponds to matching the

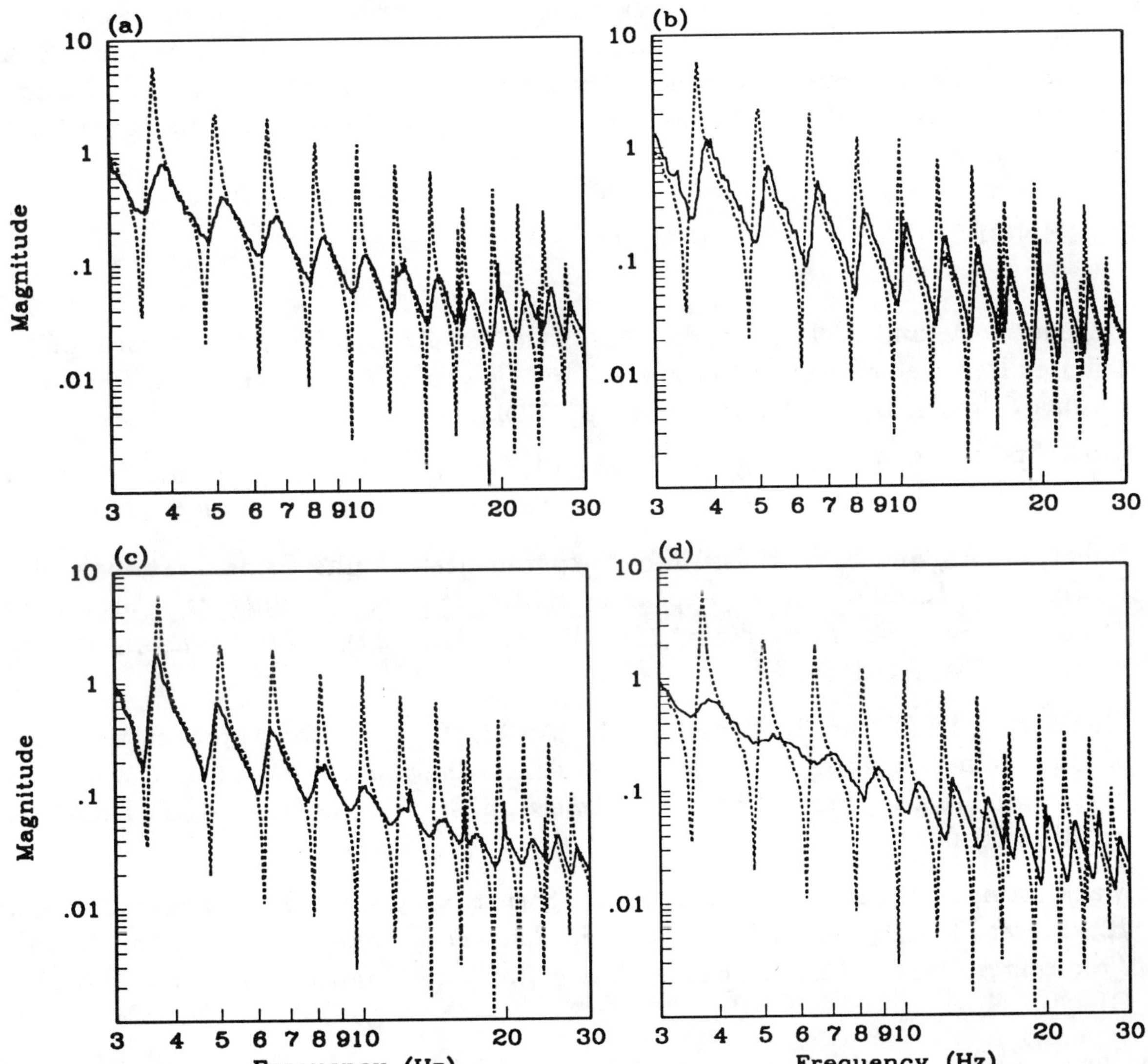

Figure 7: Experimental open (dotted) and closed loop (solid) transfer functions using (a) velocity feedback, (b) unweighted $\mathcal{H}_\infty$ design, (c) $\mathcal{H}_2$ design, and (d) weighted $\mathcal{H}_\infty$ design.

input impedance of the structure. An exact match at all frequencies is not always possible, as it may require a noncausal compensator. Causal compensators can be obtained from an $\mathcal{H}_2$ or $\mathcal{H}_\infty$ optimization of the power flow. Flexibility can be introduced into the design procedure by including weighting functions to emphasize a single narrowband frequency range, multiple narrowband regions, or to introduce a directional preference into the energy dissipation.

Compensators designed using a local model, and both the $\mathcal{H}_2$ and $\mathcal{H}_\infty$ approaches were implemented experimentally and achieved predicted levels of damping greater than that achievable through rate feedback.

ACKNOWLEDGEMENTS

Thanks to Dr. Dennis Bernstein for suggesting that we prove that bounded magnitude power flow results in a positive real compensator. This work was supported by the U.S. Air Force Office of Scientific Research under Grant no. AFOSR-88-0029, by Sandia National Laboratory under contract 69-4391 and by the MIT Space Engineering Research Center under NASA grant NAGW-1335.

REFERENCES

1. Hagedorn, P., and J. T. Schmidt, 1987. "On the Active Vibration Control of Distributed Parameter Systems." *Dynamics and Control of Large Structures, Proceedings of the* 6th VPI&SU/AIAA *Symposium,* pp. 359-373.

2. Miller, D. W., S. R. Hall and A. H. von Flotow, 1990. "Optimal Control of Power Flow at Structural Junctions." *Journal of Sound and Vibration.* 140(3):475-497.

3. Miller, D. W., and S. R. Hall, 1991. "Experimental Results Using Travelling Wave Power Flow Techniques." To appear in *AIAA Journal of Guidance, Control, and Dynamics.* See also, *Proceedings, 1989 Winter Annual Meeting of the ASME,* San Francisco, CA, AD-15:35-42.

4. MacMartin, D. G. and S. R. Hall, 1991. "An $\mathcal{H}_\infty$ Power Flow Approach to Control of Uncertain Structures." To appear in *AIAA Journal of Guidance, Control, and Dynamics.* See also, *Proceedings, 1990 American Control Conference,* San Diego, CA, pp. 3073-3080.

5. MacMartin, D. G., and S. R. Hall, 1991. "Structural Control Experiments Using an $\mathcal{H}_\infty$ Power Flow Approach." To appear in *Journal of Sound and Vibration.* See also *Proceedings, 1990 AIAA Guidance, Navigation and Control Conference,* Portland, OR.

6. Chen, G-S., B. J. Lurie, and B. K. Wada, April 1989. "Experimental Studies of Adaptive Structures for Precision Performance." *Proceedings,* 30th *AIAA/ASME/ASCE/AHS/ASC Structures, Structural Dynamics and Materials Conference,* Mobile, AL, pp. 1462-1472.

7. Guicking, D., J. Melcher, and R. Wimmel, 1989. "Active Impedance Control in Mechanical Structures." *Acustica,* 69:39-52.

8. Chen, G-S., B. J. Lurie, and B. K. Wada, April 1990. "Bridge Feedback for Active Damping Augmentation." *Proceedings, AIAA Dynamics Specialist Conference,* Long Beach, CA, pp. 502-510.

9. Fano, R. M., Jan. and Feb. 1950. "Theoretical Limitations on the Broadband Matching of Arbitrary Impedances." *Journal of the Franklin Institute,* pp. 57-83, 139-154.

10. Hodges, C. H. and J. Woodhouse, 1986. "Theories of Noise and Vibration Transmission in Complex Structures." *Reports on Progress in Physics.* 49:107-170.

11. Miller, D. W. and A. H. von Flotow, 1989. "Power Flow in Structural Networks." *Journal of Sound and Vibration.* 128(1):145-162.

12. Van Valkenburg, M. E., and B. K. Kinariwala, 1982. *Linear Circuits,* Prentice-Hall, pp. 297-298.

13. Carlson, G. E., and C. A. Halijak, June 1964. "Approximation of Fractional Capacitors $(1/s)^{1/n}$ by a Regular Newton Process." *IEEE Transactions on Circuit Theory.* 11:210-213.

14. Haddad, W. M., and D. S. Bernstein, 1990. "Robust Stabilization with Positive Real Uncertainty: Beyond the Small Gain Theorem." *Proceedings, 29th IEEE Conference on Decision and Control,* Honolulu, HI, pp. 2054-2059.

DEMONSTRATION OF A LQG VIBRATION CONTROLLER FOR A SIMPLY-SUPPORTED PLATE*

Stephen P. Rubenstein,[1] William R. Saunders,[1] Graham K. Ellis,[1] Harry H. Robertshaw,[1] William T. Baumann[1]

ABSTRACT

The effectiveness of a low-order vibration controller formulated using Linear Quadratic (LQ) modern control theory is demonstrated. A transputer-based parallel processing control system is implemented to reject steady disturbances of the fundamental mode of a simply-supported plate. A discussion of the control law is provided. A two-mode compensator is used to reject either harmonic or narrowband point force disturbances. Also, the transient suppression provided by the compensator is demonstrated. Experimental and simulated results are presented in a time domain format. The steady-state fundamental mode amplitude is attenuated by more than an order of magnitude at the disturbance frequency. Additionally, the compensator adds approximately twice as much damping to this mode compared to the open-loop.

INTRODUCTION

Active vibration control has been the focus of extensive research in recent years. Two popular control approaches dominate many of the existing studies on active vibration control. First, adaptive feedforward methods have been employed by various researchers to design vibration controllers for a variety of structures. Experimental results for the feedforward controllers are more prevalent than analytical or numerical investigations of the adaptive control methods. Recently, a number of feedforward-based vibration control experiments have been reported by Fuller et.al. [1-3]. Corresponding analytical work has seemingly lagged behind the experimental results, which is not surprising given the nonlinear nature of the feedforward approach. We should also mention that the feedforward methods do not require any modeling of system dynamics so simulations of adaptive vibration controllers are not popular. However, several authors have made progress in developing analytical tools for the adaptive control methods. Elliot and Nelson presented a discussion of important properties for a LMS adaptive control scheme [4]. Burdisso reported on the effective modification to system eigenproperties during feedforward control [5].

*This work was supported under ONR contract N0014-88-K-0721

[1]Center for Intelligent Material Systems & Structures, Mechanical Engineering Department, Virginia Polytechnic Institute and State University, Blacksburg, VA 24061, U.S.A.

The second popular approach to vibration control is based on modern control methods, which are feedback approaches. Posed in state-space form, numerous authors have reported on feedback controllers to provide transient suppression [6-7]. A recent survey paper by Sievers and vonFlotow discussed two important disturbance rejection methods[8].

Recently, there has been some interest in the active control of sound radiation from elastic structures. Fuller initially posed the structural acoustic control problem as a specific vibration control problem using feedforward control [9]. Baumann et.al. reported on a feedback structural acoustic controller designed to optimally minimize acoustic energy radiated from a structure subject to transients and/or steady-state disturbances of nearly arbitrary waveforms [10-11]. The feedback control approach was also posed as an applied vibration control problem.

A review of the literature shows that the distribution of experimental and analytical investigations using modern control approaches is weighted heavily towards analytical results. This unequal division has been caused by the relative ease of designing control laws using linear systems theory. Thus, simulated transient suppression or disturbance rejection results dominate the vibration control literature. However, implementing the controllers on real structures requires significant efforts in the areas of data acquistion hardware and computational performance. Few references are available which discuss demonstration results of modern control disturbance compensation techniques. This paper is an attempt to highlight the performance of feedback controllers on a real structure.

In this paper, we demonstrate the effectiveness of a low-order LQG vibration controller used to reject disturbance inputs to a simply-supported plate. The paper is divided into three main parts. First, a general description of the LQG compensator is provided for the continuous-time representation of the plate. Second, we discuss the implementation aspects of the experiment. Details of the plant, sensors, actuators, data acquisition hardware, computational hardware and control law design are provided. Third, simulated and experimental results illustrating the disturbance rejection and transient suppression provided by the compensator are shown. Results are presented for both harmonic and narrowband steady-state disturbances applied to the plate. Response of the closed-loop system to impulse forces is included. Finally, the conclusions of the experiment are summarized.

LQG COMPENSATOR DESIGN

Consider the general case of a flexible structure which may be modeled by a first-order differential equation. For simplicity, we let the plant be a continuous system subject to a disturbance force and a control force. It is clear that the digital implementation of the compensator discussed below requires a sampled-data representation. However, the governing equations for the continuous-time controlled plant will be presented here for clarity. The state-space description of this system is

$$\dot{q} = Fq + Gu + Lv \tag{1}$$

where $q \in R^{2n}$ is the state of the system in modal coordinates, $u \in R^m$ is the actuator input used to control the structure, and $v \in R^s$ is the disturbance input. The waveform of the disturbance is arbitrary in Eq. 1 but we assume for now that v is zero mean uncorrelated (white) noise.

The chosen measurement equation is given by

$$y = Cq + D_p u + D_d v + \theta \tag{2}$$

where $y \in R^n$ is a vector of measured modal accelerations and $\theta \in R^n$ is zero mean Gaussian measurement noise. Direct feedthrough of the applied control and disturbance forces is present because of the choice of acceleration outputs.

The quadratic cost function to be minimized is

$$J = \int_{-\infty}^{\infty} (q^T Q q + \rho u^T u) dt \tag{3}$$

where Q is a positive semi-definite weighting matrix and ρ is a scalar weighting parameter which specifies the amount of control effort expended. The optimal compensator is state feedback, with the form

$$u = -K_c q \tag{4}$$

where $K_c \in R^{mx2n}$. Notice that this formulation of the controller requires that all states be measured. Since we are using acceleration outputs, no system states are measured and a full state estimator is required.

The optimal estimator in the presence of the specified measurement noise is a Kalman-Bucy filter. The relevant results of the optimal estimator equations for this formulation are

$$\dot{\hat{q}} = F\hat{q} + Gu + K_f(y - \hat{y}) \tag{5}$$

with

$$K_f = PC^T R^{-1} \tag{6}$$

where P is the state error covariance matrix, C is the output matrix, and R is the spectral density matrix of the measurement noise.

Equations 1 through 6 define the standard LQG control formulation for the continuous time representation of the simply-supported plate experiment. Before we discuss the discrete-time implementation of the controller in the actual experiment, we will modify the previous equations to provide rejection of steady-state correlated plant disturbances.

The disturbance input, v, may be represented as a "colored" noise source. For the plate experiment, this assumption provides narrowband disturbance rejection. This technique of disturbance modeling is formulated by driving a "shaping filter" with uncorrelated process noise. In practice, the choice of the shaping filter must correspond to the signal characteristics of the disturbance actually applied to the plant. The general disturbance model in state space form is

$$\dot{w}_d = F_d w_d + G_d \delta \tag{7}$$

with output equation

$$v = C_d w_d. \tag{8}$$

F_d, G_d, and C_d are the disturbance model system matrices and $w_d \in R^l$ are the disturbance states. As stated above, the input to the shaping filter, δ, is zero mean, Gaussian noise. By augmenting the disturbance dynamics (Eq. 7-8) to the plant dynamics (Eq. 1-2), the augmented system equations become

$$\begin{bmatrix} \dot{q} \\ \dot{w}_d \end{bmatrix} = \begin{bmatrix} F & LC_d \\ 0 & F_d \end{bmatrix} \begin{bmatrix} q \\ w_d \end{bmatrix} + \begin{bmatrix} G \\ 0 \end{bmatrix} u + \begin{bmatrix} 0 \\ G_d \end{bmatrix} \delta \tag{9}$$

with the augmented output

$$y = [C \; D_d C_d] \begin{bmatrix} q \\ w_d \end{bmatrix} + D_p u + \theta. \tag{10}$$

Equations 9 and 10 model the simply-supported plate subject to a colored noise disturbance. The disturbance model used in the experiment was

$$\frac{s}{s^2 + 2\zeta\omega_d + \omega_d^2} \tag{11}$$

Through selection of an appropriate damping value ζ, this disturbance model can be used to represent either harmonic or narrowband disturbances.

The LQG controller design for a sampled data system follows the essential equations developed above. After the choice of a sampling interval, the appropriate differential equations are converted to corresponding difference equations. The form of the difference equations for the control experiment will be presented in the next section.

EXPERIMENTATION

One of the stated goals of this paper is to apply the modern controls approach to a real structure, in this case a simply-supported plate. In the following sections, the implementation issues of the experiment are addressed, including the setup and the procedures that were used.

EXPERIMENTAL SETUP

The experimental setup is comprised of three major components: the experimental plant, the sensors and actuators, and the data acquisition and control law hardware. Figure 1 shows a schematic of the setup.

An experimental plant was fabricated at the Mechanical Engineering shop at VPI&SU. A cold-rolled steel plate, 0.6 m long, 0.5 m wide, and 0.003 m thick, was

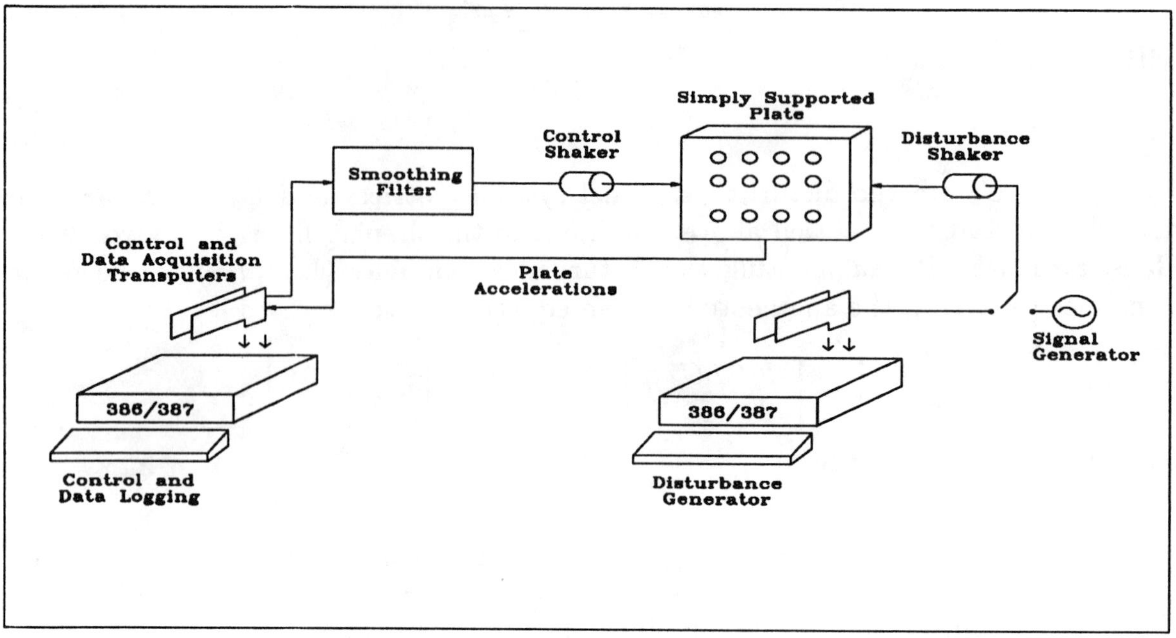

Figure 1: Experimental Setup

sheared. The plate is attached along its edges to thin steel shim stock, and the opposite edge of the shim stock is bolted into a steel frame. The shim stock is joined to the plate through machine screws and adhesive in order to emulate the simply-supported boundary condition. Table 1 shows that the experimentally obtained natural frequencies of the plate for the first four modes of vibration are within 5% of those calculated using the closed form solution for a simply supported plate.

The disturbance input to the plate was applied as a point force. An electromagnetic vibration shaker was driven by a signal generator for harmonic disturbances or a 386/387 style computer configured with a transputer-based digital to analog conversion (D/A) system for narrowband disturbance inputs. The response was measured using twelve piezoelectric accelerometers arranged in a nonsymmetric array. Accelerometer locations were chosen to ensure observability of the first twelve plate vibration modes.

Table 1: EIGENPROPERTIES OF THE SIMPLY-SUPPORTED PLATE

Mode	Analytical	Experimental	
n	ω_n Hz	ω_n Hz	ζ_n % crit.
1	47.9	49.5	0.8
2	106.9	108.9	1.2
3	132.8	130.3	0.8
4	191.8	188.5	0.3

Digital control of the plate was implemented using a 386/387 style computer equipped with a parallel processing transputer-based system to perform both the control law calculation and the analog to digital conversion (A/D) and D/A operations. The transputer data acquisition board (TransDAC) consists of sixteen simultaneously sampled channels of input and 16 simultaneous channels of output. The TransDAC was designed and built at VPI&SU to sample many channels quickly. (A commercially available data acquisition system, used in previous control experiments, was limited to four simultaneous A/D channels and a sample speed of 450 Hz for a two mode compensator.) Emperical results have shown that good modal identification requires sample frequencies approximately ten times faster than the plate motions of interest. The TransDAC design provided this capability for the low-order controller experiment. Additionally, the large number of inputs available allowed spatial oversampling, i.e. more sensors than identified modes. This increased the expected accuracy of the modal acceleration estimates required for control.

The TransDAC design, as well as the performance specifications of the TransDAC, are reported by Ellis [12]. The multitasking ability of the control system architecture, shown in Fig. 2, allowed the control law to be calculated on a single T800 transputer while the TransDAC, with a T222 transputer, performed the data acquisition. The two parallel processes only communicated when the control law needed the input or when it was ready to send the control. Using the TransDAC control system architecture, it was possible to achieve a sample rate of 3300 Hz for the same two-mode controller which had been implemented on the commercially available data acquisition system mentioned previously. (This sample time corresponded to four simultaneously sampled A/D channels).

The staircase nature of the control signal is smoothed in a second-order low pass filter before driving the control shaker. The smoothing filter limits the frequency content of the control signal above the bandwidth of the compensator. The procedure for the vibration suppression experiments and the discrete time control law implemented are presented below.

EXPERIMENTAL PROCEDURE

The vibration suppression experiments were conducted on two types of steady disturbances: harmonic and narrowband. Also, a transient disturbance was rejected to show this inherent property of feedback controllers. As these experiments were performed, the system was numerically simulated to predict its response and to confirm the behavior found through experimentation.

The general procedure for the experiments follows. First, the disturbance signal was generated and sent to the disturbance shaker. The plate response was measured using the twelve accelerometer array and the TransDAC hardware discussed earlier. The measurements were linearly transformed through the modal filter into the modal do-

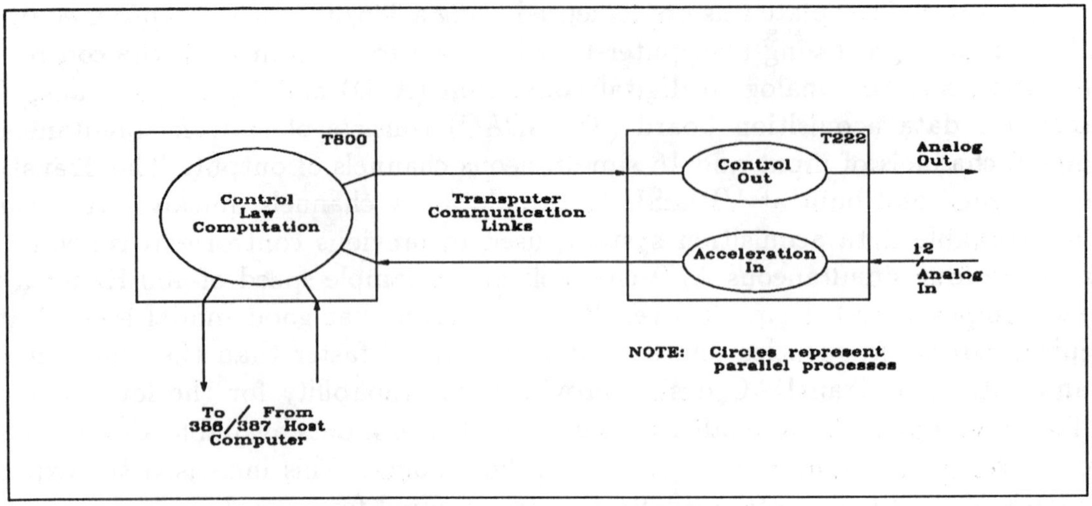

Figure 2: Control System Architecture

main. Then, the two modal accelerations were used in the updated Kalman Filter equations. The loop was closed by the feedback of the predicted Kalman Filter states through the LQR gains and the resulting control was sent through the smoothing filter to the control shaker.

The computational loop shown in Fig. 3 was performed on the T800 transputer. A sample rate of 2100 Hz was achieved for the twelve-channel input, single output control architecture. The equations shown in Fig. 3 are the discrete-time equivalent difference equations for the standard LQG problem formulated earlier.

The simulated and experimental results presented below are for a two- mode compensator which had an eighth-order state-space model. The plant model contained four states, the smoothing filter model and the disturbance model each contained two states. A two-mode model was chosen to ensure that the spatial and temporal sampling requirements discussed earlier could be met by the control system.

SIMULATION AND EXPERIMENTAL RESULTS

This section presents simulated and experimental time histories of the simply-supported plate for various configurations. First, simulation results for a 60 Hz harmonic disturbance compensation are shown. The simulation data were created using MATLAB software installed on a 386 computer. Next, experimental data from different control scenarios are presented. Three different experiments were performed to demonstrate the effectiveness of the LQG compensator. First, results for disturbance compensation of a 60 Hz harmonic force are presented. Second, disturbance compensation of a narrowband force (centered at 60 Hz) is shown. Finally, the inherent transient suppression provided by the LQG compensator is illustrated.

HARMONIC DISTURBANCE COMPENSATION

Simulations of the LQG compensator were performed using the identical controller structure as that implemented in the experiment, i.e. an eighth order controller as discussed previously. For the simulation, the plant model was truncated to four modes. This provided a two-mode controller operating on a four-mode plant with optimal disturbance compensation targeted for one mode only. An additional control actuator would have allowed optimal rejection of both modes.

Simulated time histories of the first two plate modes are shown in Fig. 4. The figures show modal accelerations for open-loop steady state response (dashed line) to a 5 N peak disturbance, followed by the closed-loop response. The controller was turned on at 0.19 secs. Notice that the controller, which was designed to optimally suppress mode 1 response, performed well. The amplitude of mode 2 was driven slightly higher because of the relative phasing of the two modes at the disturbance input location.

Experimental time histories for the first two plate modes are illustrated in Fig. 5. Excellent disturbance rejection was provided by the compensator for the first mode. Notice that the experimental time histories were nearly identical to the simulated time histories. This highlights an important issue for the controller design. Since the bandwidth of the disturbance and the compensator was substantially within the bandwidth of the lower order plate modes, it was possible to design a low order con-

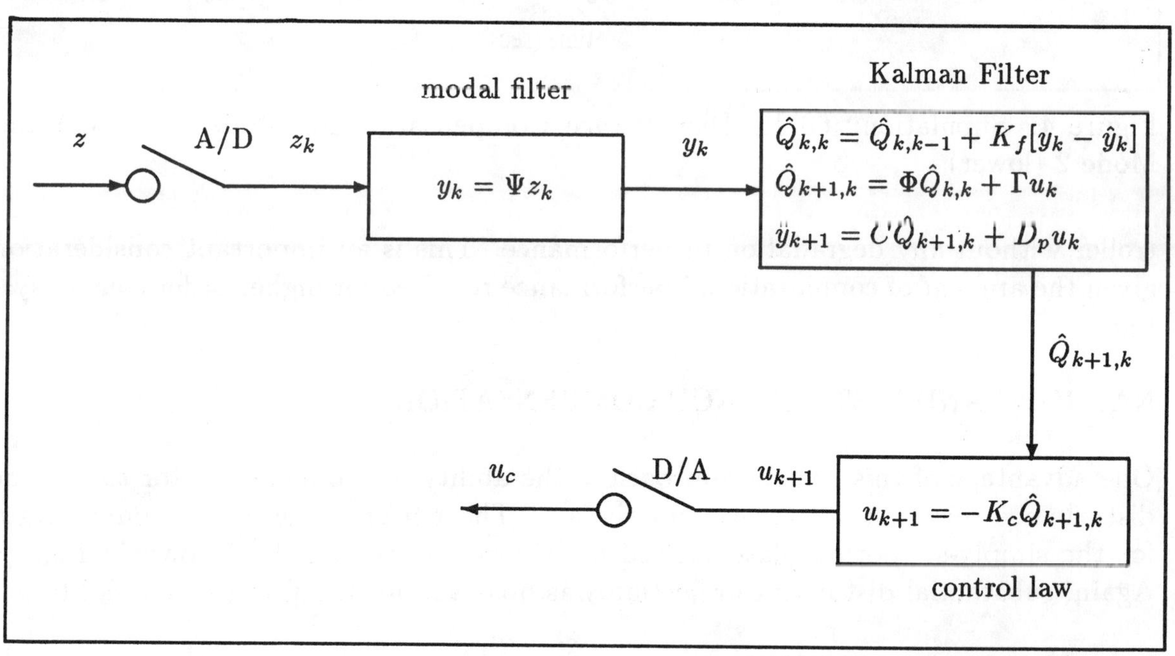

Figure 3: Computational loop

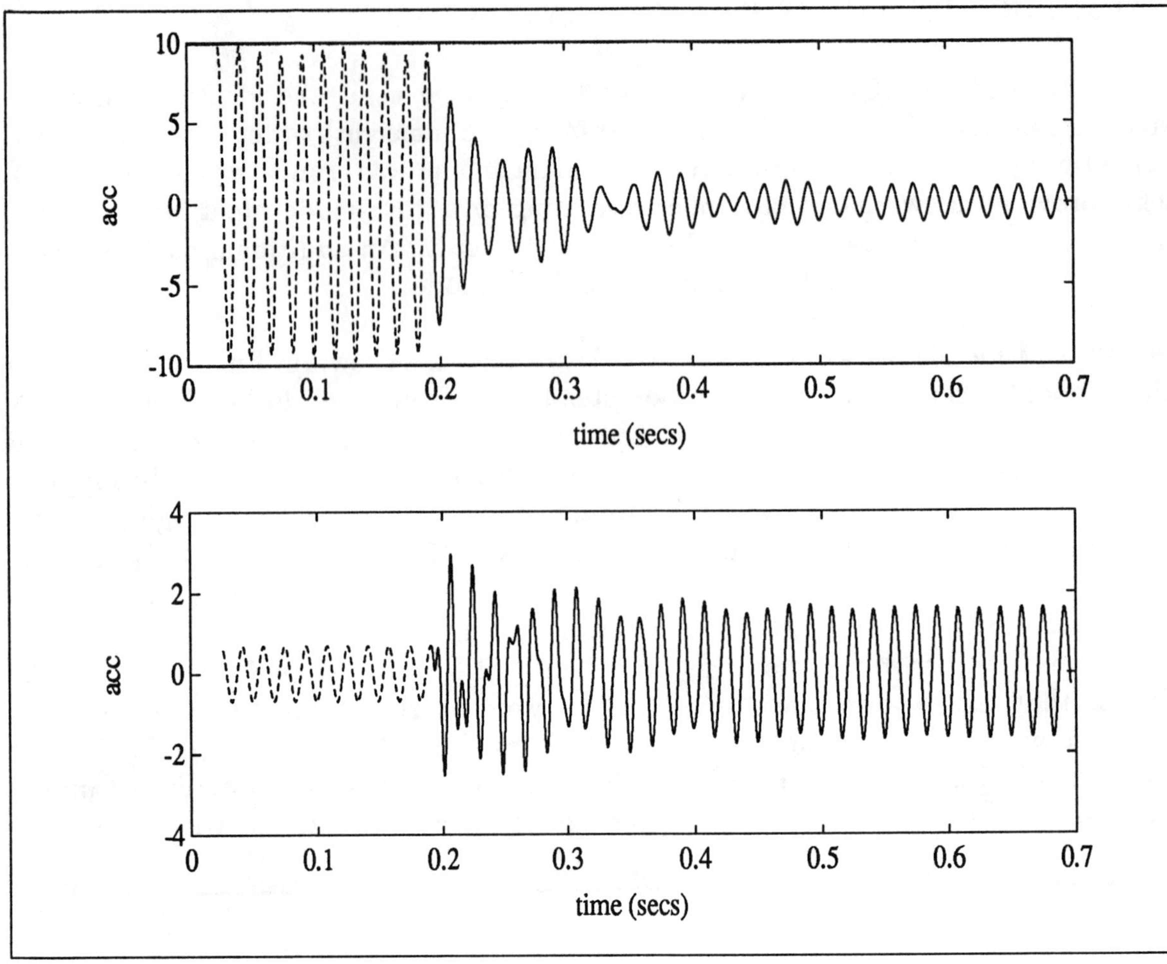

Figure 4: Simulation: 60 Hz Disturbance Compensation for Mode 1 (upper) and Mode 2 (lower)

troller without any degradation in performance. This is an important consideration given the amount of computational performance required for higher order control systems.

NARROWBAND DISTURBANCE COMPENSATION

One advantage of this control approach is the ability of the compensator to provide disturbance rejection of narrowband signals. The performance of the compensator for the simply-supported plate excited by a narrowband signal is shown in Fig. 6. Again, substantial disturbance rejection was observed for the first mode amplitude.

TRANSIENT SUPPRESSION

Transient suppression by the compensator was demonstrated by applying an impulse to the center of the plate during closed-loop control. A modal test hammer was used to deliver an impulsive force. A time history of the fundamental modal acceleration during this procedure is shown in Fig. 7.

Notice that the disturbance was minimized before the transient was applied to the plate. As the impulse was applied to the plate, the mode 1 response damped out quickly, at a time scale corresponding to the closed-loop eigenvalue. The amount of additional damping introduced by the controller is obvious in the comparison plot of Fig. 8. The compensator introduced approximately twice as much damping to the fundamental mode compared to the open-loop decay.

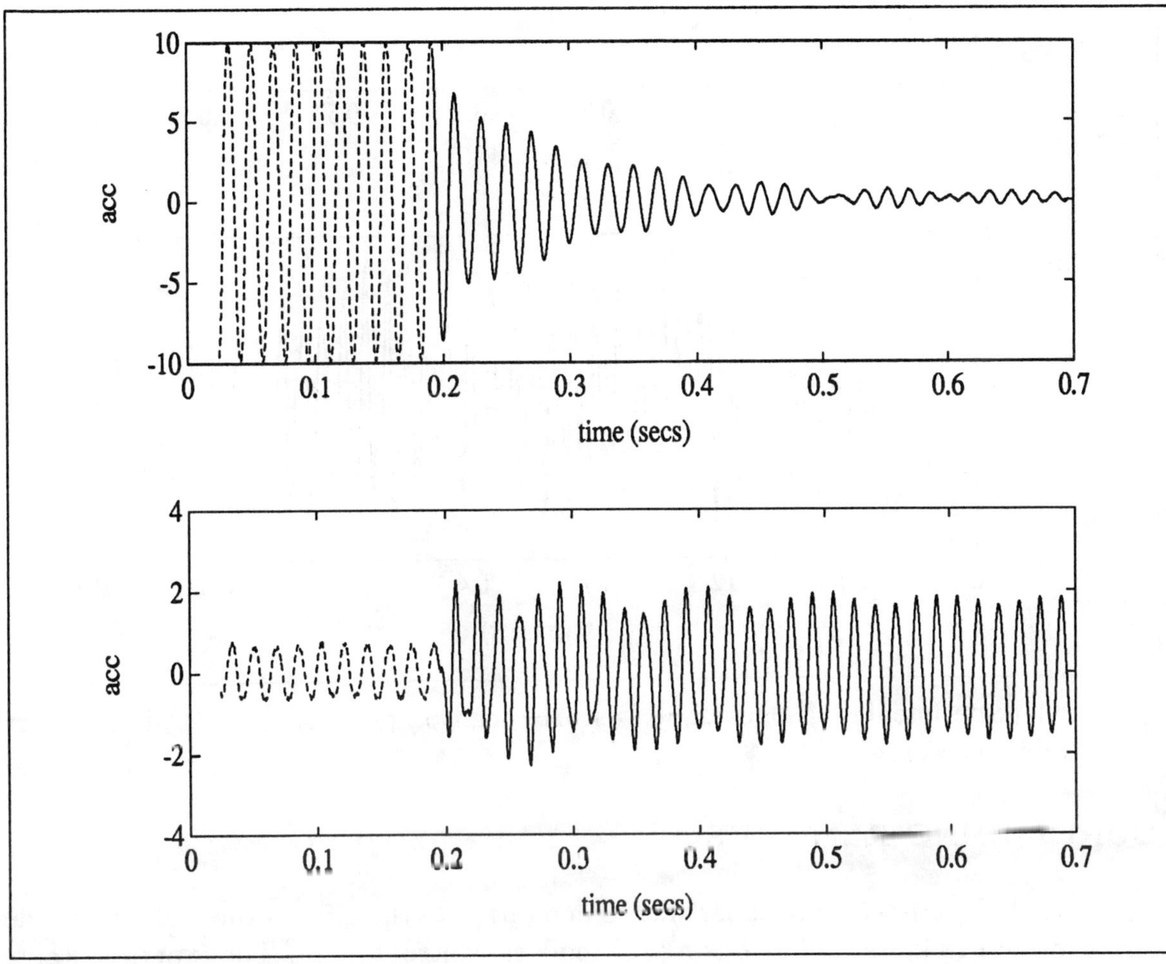

Figure 5: Experimental: 60 Hz Disturbance Compensation for Mode 1 (upper) and Mode 2 (lower)

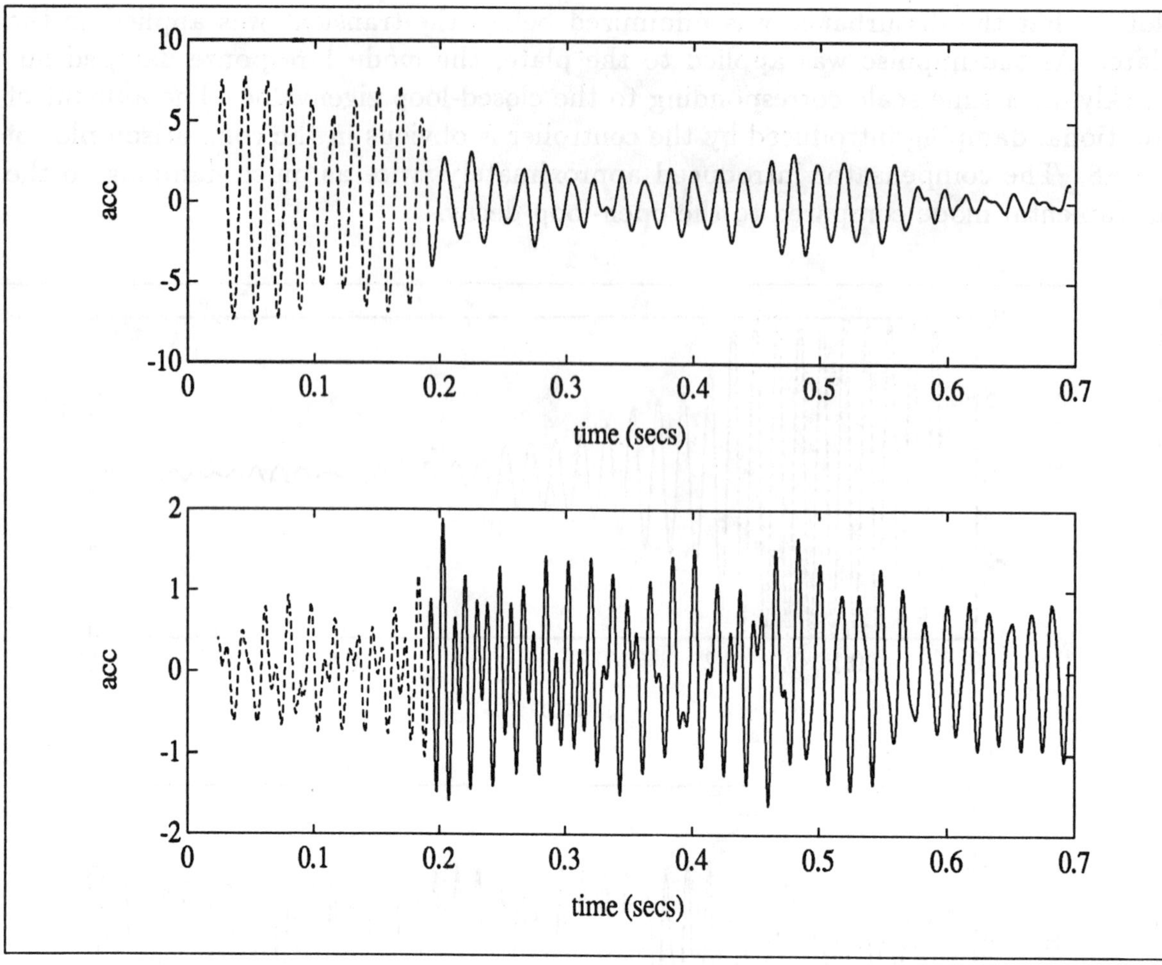

Figure 6: Experimental: Narrowband Disturbance Compensation for Mode 1 (upper) and Mode 2 (lower)

CONCLUDING REMARKS

The stated objective of this paper was to demonstrate the effectiveness of a low-order vibration controller formulated using LQ modern control theory. The experimental results presented in this paper were generated using an identical controller design as the acoustic controller developed by Baumann, with the only difference being the choice of a vibration cost function versus the acoustic cost function [11]. It was shown that the two mode controller rejected both harmonic and narrowband steady disturbances of the simply-supported plate, and provided transient suppression. Further work is under way to implement MIMO controllers with both a vibration cost function and an acoustic cost function. Computational issues of implementing these higher order compensators are also being addressed.

Acknowledgements

The authors gratefully acknowledge the support of this work by DARPA and the Office of Naval Research under grant ONR-N00014-88-K-0721.

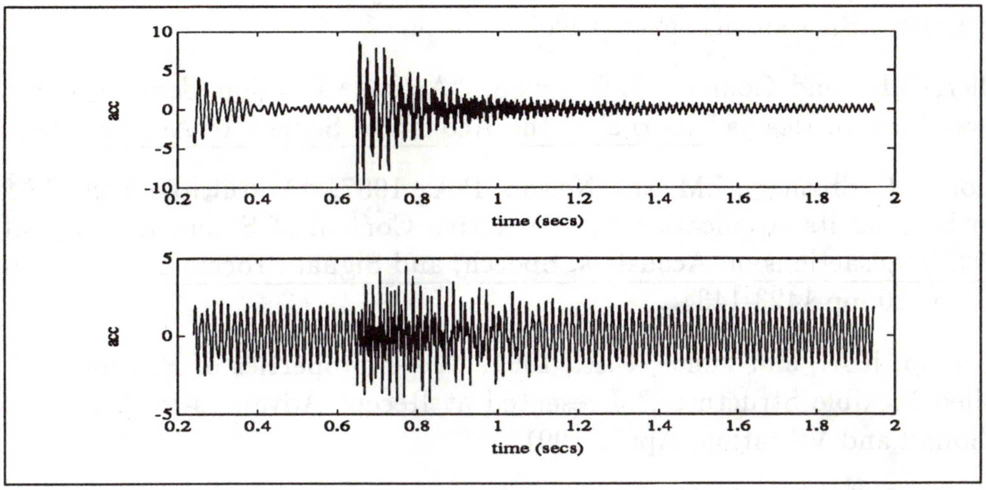

Figure 7: Mode 1 and Mode 2 Closed-Loop Transient Response

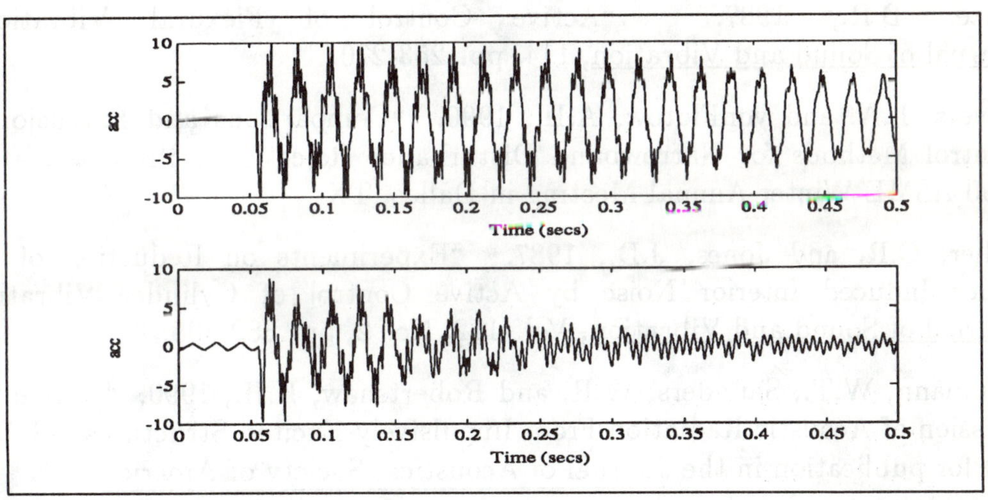

Figure 8: Mode 1 Transient Response, open-loop (upper) and closed-loop(lower)

REFERENCES

1. Fuller, C.R., Gibbs, G.P., and Silcox, R.J., 1990. "Simultaneous Active Control of Flexural and Extensional Waves in Beams," Journal of Intelligent Material Systems and Structures, 1 (2), pp. 235-247

2. Clark, R.L. and Fuller, C.R., 1990. "Experiments on Active Control of Structurally Radiated Sound Using Multiple Piezoceramic Actuators," Accepted for publication in the Journal of Acoustical Society of America, (also presented at the 120th ASA meeting, Nov. 1990).

3. Fuller, C.R., and Gonidou, L.O., 1988. "Active Vibration Control of Flexural Power Flow in Beams," Journal of the Acoustical Society of America, 84, S47.

4. Elliot, S.J., Stothers, I.M. and Nelson, P.A., 1987. "A Multiple Error LMS Algorithm and its Application to the Active Control of Sound and Vibration," IEEE Transactions on Acoustics, Speech, and Signal Processing, Vol. ASSP-35, No. 10, pp.1423-1434.

5. Burdisso, R.A., and Fuller, C.R., 1991. "Eigenproperties of Feedforward Controlled Flexible Structures," Presented at Recent Advances in Active Control of Sound and Vibration, April. 1991.

6. Meirovitch, L., Baruh, H., and Oz, H. 1983. "A Comparison of Control Techniques for Large Flexible Systems,", Journal of Guidance and Control and Dynamics, Vol.6, pp. 302-310.

7. Mace, B.R., 1987. "Active Control of Flexural Vibrations", Journal of Sound and Vibration, 114, pp. 253-270.

8. Sievers, L.A. and vonFlotow, A.H., 1990. "Comparisons and Extensions of Control Methods for Narrowband Disturbance Rejection", Presented at the 1990 ASME Winter Annual Meeting in Dallas, TX.

9. Fuller, C.R. and Jones, J.D., 1987. "Experiments on Reduction of Propeller Induced Interior Noise by Active Control of Cylinder Vibration", Journal of Sound and Vibration, Vol. 112, No. 2, pp. 389-395.

10. Baumann, W.T., Saunders, W.R. and Robertshaw, H.H., 1990. "Active Suppression of Acoustic Radiation From Impulsively Excited Structures," Submitted for publication in the Journal of Acoustical Society of America.

11. Baumann, W.T., Ho F.S. and Robertshaw, H.H., 1990. "Active Acoustical Control of Broadband Structural Disturbances," Presented at the 120th ASA meeting, Nov. 1990.

12. Ellis, G.K. 1990. "Design and Programming of a Medium-Performance Data Acquisition System Based on Transputers," Submitted for publication as Center for Intelligent Material Systems & Structures internal report.

APPLICATION OF THE MODIFIED INDEPENDENT MODAL SPACE CONTROL (M.I.M.S.C.) METHOD FOR ACTIVE CONTROL OF VIBRATIONS OF A CANTILEVERED BEAM: A NUMERICAL STUDY

Ching I. Chen,[1] Marcello R. Napolitano,[2] Victor Mucino[3]

ABSTRACT

This document presents a numerical study related to the application of a control system which uses piezoelectric actuators to reduce the vibration level of a beam structure. The implemented control strategy is the so called M.I.M.S.C. method. Among the interesting features of this technique there is the ability to handle the spill-over effect, that is to eliminate the increase in magnitude of the vibrations associated with the uncontrolled modes since only few modes are selected to be controlled. This is accomplished with a time-sharing technique which will make a reduced number of piezoelectric actuators able to control a larger number of modes. Also, with this approach, a method is used for the determination of the optimal locations of the actuators. A cantilevered steel beam with 5 elements and 2 actuators has been used for a numerical simulation. The method is shown to give a very desirable damping of the structural vibrations. The importance of being able to control the spill-over effect is confirmed in this study especially for high values of the control forces furnished by two actuators. Besides the problem of the saturation of the piezoelectric actuators has been here considered.

INTRODUCTION

To avoid spectacular catastrophic failures of structures, vibration controls by active or passive means are necessary. Active control methods that utilize "intelligent structures" have several advantages over time-proven passive-control techniques. For example, noise reduction inside aircraft and submarines can be achieved by controlling vibrations of the fuselage and hull panels by means of active controllers. Piezoelectric ceramic materials have shown potential for damping dynamic structures. One very efficient control algorithm that has been introduced only recently is the Modified Independent Modal-Space Control (M.I.M.S.C.) method [6,9,10].

[1]Research Associate, Department of Mechanical and Aerospace Engineering, West Virginia University, Morgantown, WV 26506-6101, U.S.A.

[2]Assistant Professor, Department of Mechanical and Aerospace Engineering, West Virginia University, Morgantown, WV 26506-6101, U.S.A.

[3]Associate Professor, Department of Mechanical and Aerospace Engineering, West Virginia University, Morgantown, WV 26506-6101, U.S.A.

THEORETICAL BACKGROUND

Consider a piezoelectric element with dimensions a and b along x and y respectively and thickness t along the z axis. Because of the properties of the piezoelectric materials, if a difference potential V_z applies along the z axis, the strain along x and y is given by [1-3]

$$\varepsilon_x = d_{zx} V_z / t_c \tag{1}$$

$$\varepsilon_y = d_{zy} V_z / t_c \tag{2}$$

where d_{zx} and d_{zy} are constants of piezoelectric charge. For an isotropic piezoelectric material we have

$$d_{zx} = d_{zy} \tag{3}$$

This is known as the "Inverse Piezoelectric Effect" and it allows to use the piezoelectric material as an actuator.

Consider again the piece of piezoelectric material; if there are two equal and opposite forces F acting along the x and y direction, the difference of potential V_z along the z direction is given by

$$V_z = (\varepsilon_x + \varepsilon_y) g_{zx} E_c t_c \tag{4}$$

where E_c = Young modulus of the piezoelectric element

g_{zx} = constant of piezoelectric difference of potential.

Note that for the isotropic piezoelectric material we have

$$g = g_{zx} = g_{zy} \tag{5}$$

The constant g is related to the constant d as

$$d = e_{zz} g \tag{6}$$

$$d = d_{zy} = d_{zx} \tag{7}$$

where e_{zz} = dielectric constant of the piezoelectric material.

This is known as the "Direct Piezoelectric Effect" and it allows to use the piezoelectric materials as sensors.

By looking at the above relations one can understand how these piezoelectric materials are able to transmit strains and deformations to structural elements to which they are connected in someway. There are essentially two ways of integrating the piezoelectric materials to the working elastic structure, that is surface bonded and embedded piezoelectric material. The last approach is more suitable for applications with composite materials. The first approach will generally imply lower production costs of the structural elements and it guarantees a certain flexibility in changing the location of the piezoelectric materials, both as actuators and sensors. Also we have to consider that the piezoelectric materials work essentially in two modes, namely extension and bending. In the bending mode the tensional field inside the piezoelectric depends on whether it is bounded on the surface or it is embedded in it, as it is shown in Figure 1.

Without any loss of generality, a piezoelectric material bonded to the elastic substructure will be considered in this study and it is shown in Figure 2. More particularly the case of piezoelectric elements perfectly bonded to the elastic structure is considered. For small values of t_s (around 0.1 mm, which is practically feasible) this assumption is realistic. It can be shown that under this particular condition the strain ε and the force F transmitted by a piezoelectric element to an elastic structure are given by the following equations

$$\varepsilon = \frac{\alpha}{\Psi + \alpha} \Lambda \tag{8}$$

$$F = - \frac{1}{\Psi + \alpha} \Lambda \tag{9}$$

with associated moment applied to the beam

$$M_o = F \, t_B \tag{10}$$

where $\Psi = E_B \, t_B / E_C \, t_C$ (11)

with α = constant; $\alpha = 6$ for bending and $\alpha = 2$ for extension.

Under the assumption of piezoelectric element perfectly bonded on the structure the entire shear force will be concentrated at the end of the piezoelectric material, with coordinate $x = \mp L/2$. Note that in Eqs. (8) and (9) the ability to strain or apply stresses to the substructure is represented by the term $\Lambda = d_{zx} \, V_z \, / \, t_c$. In the following of this document the application of the piezoelectric material as an actuator is considered.

THE PIEZOELECTRIC MATERIAL-BEAM SYSTEM

A typical structural model of a piezoelectric-beam system is shown in Figure 4a. The beam is assumed to be made of several elements with constant thickness and length and made of the same material. However, these conditions do not have to be respected. The degrees of freedom of the nodes are in terms of vertical linear translations y and angular rotations θ. The piezoelectric actuator is bonded to element 'i' of the beam with a certain bonding layer. External forces or moments can be applied at the connections between the different elements. When an electric voltage is applied across the piezoelectric actuator it will

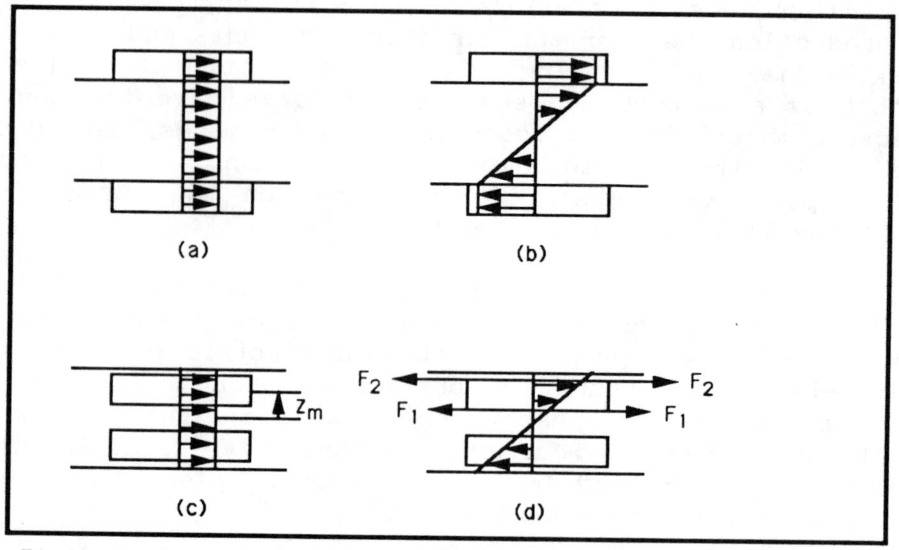

Figure 1. Assumed Piezoelectric-Substructure Strain Distributions: a) Surface-Bonded Extension , $\alpha=2$; b) Surface-Bonded Bending, $\alpha=6$; c) Embedded Extension ; d) Embedded Bending.

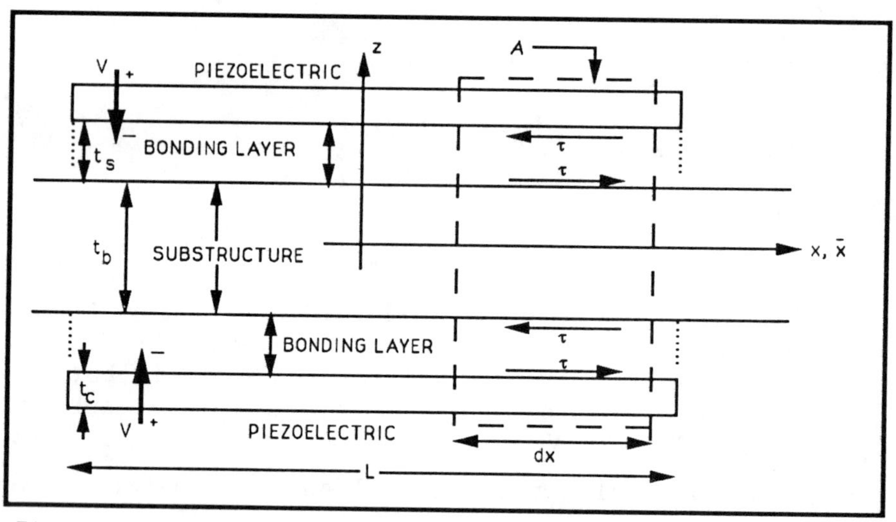

Figure 2. Geometry of Piezoelectrics Bonded to an Elastic Substructure through a Finite Thickness Bonding Layer .

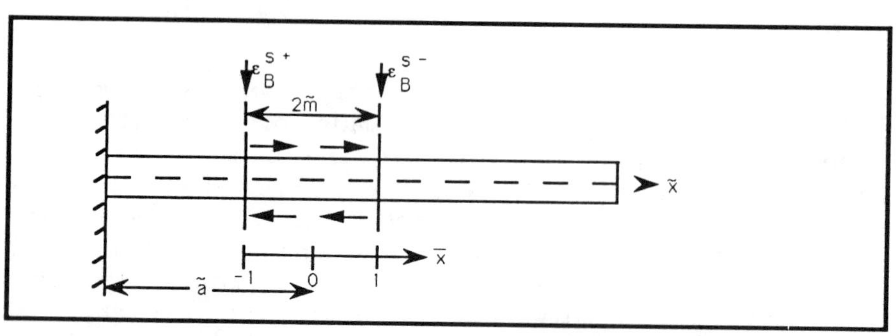

Figure 3. Cantilevered Beam under Surface-Localized Shear Loading.

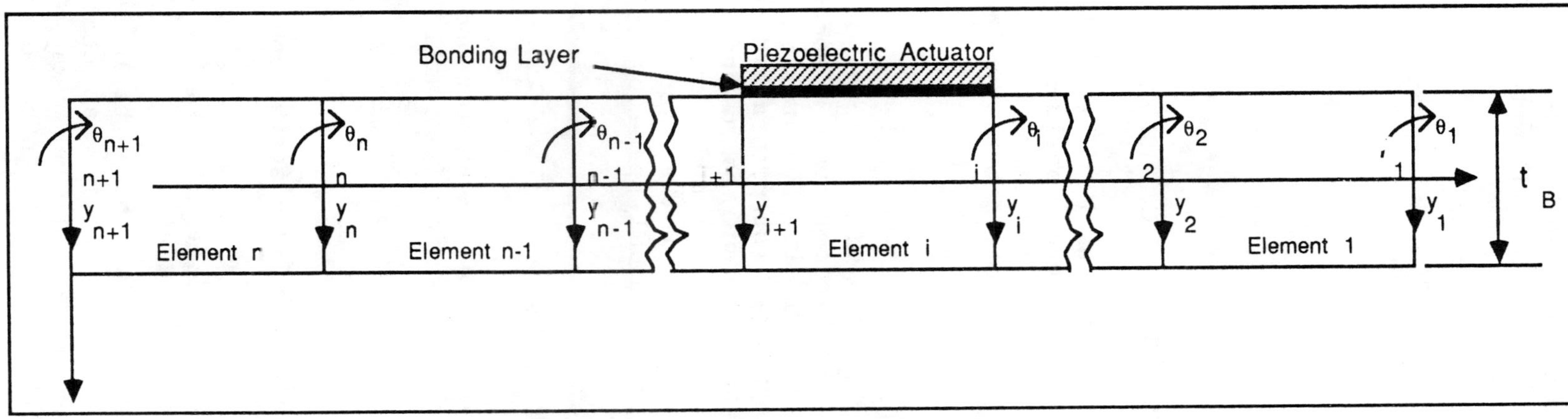

Figure 4a. Piezoelectric Material - Bonding Layer - Beam System

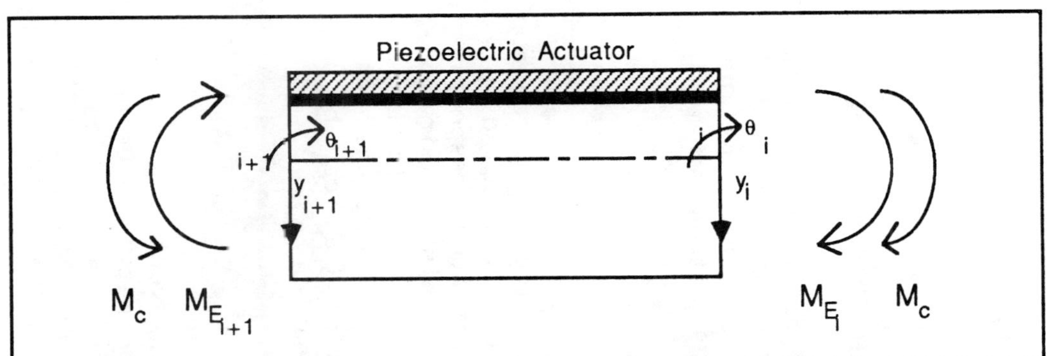

Figure 4b. Representation of Control and External Moments on a Single Element with Piezoelectric Actuator

expand or contract depending on the generated electric field being aligned or opposed with respect to the material polarization. Such expansion or contraction of the piezoelectric material will create a bending moment M_c on the element, as shown on Figure 4b, which is added to the preexisting external moment acting on the element, if any. The ultimate goal is to generate these control bending moments M_c in such a way to counteract the external moments induced by the structural vibrations. The procedures to be used to generate the finite element model of the piezoelectric, bonding, layer and beam system are reported in [6],[9] and [10].

THE MODIFIED INDEPENDENT MODAL SPACE CONTROL (M.I.M.S.C.) METHOD

The control algorithm used in this study is known as Modified Independent Modal Space Control (M.I.M.S.C.) method. If one considers the beam structure whose vibrations need to be controlled and reduced, the finite-element equations of motion of the piezoelectric beam system can be written as :

$$[M]\ \{\ddot{\delta}\} + [K]\ \{\delta\} = \{F\} \tag{12}$$

where {F} is the control force applied by the Piezoelectric elements, $\{\delta\}$ is the vector of the nodal displacements, and [M] and [K] are the inertia and stiffness matrices, respectively. The natural (structural) damping is not included in this dynamic model. The addition of the natural damping will only improve the performance of the controller. A modal analysis will be performed to obtain the natural frequencies and the associated mode shapes. Equation (12) can be rewritten in the modal space by using the modal transformation

$$\{\delta\} = [\Phi]\ \{U\} \tag{13}$$

where {U} is the vector of the modal coordinates of the system and $[\Phi]$ is the modal shape matrix. Using such transformation Eq. (12) will then be reduced to

$$\{\ddot{U}\} + [\lambda]\ \{U\} = \{f\} \tag{14}$$

where $[\lambda]$ is a diagonal matrix of the eigenvalues of the system and {f} is the modal force vector given by

$$\{f\} = [\Phi]^T\ \{F\} \tag{15}$$

Let us assume that there are C number of actuators which can be implemented in the control structure. From the control theory one can control an equivalent number C of vibrational modes and leave a number r of residual uncontrollable modes with r + C = N. Therefore Eq. (15) can be represented as

$$[f] = \begin{bmatrix} f_c \\ -- \\ f_r \end{bmatrix} = \left[\begin{array}{ccc|ccc} \phi_1(\ell_1) & \ldots & \Phi_1(\ell_C) & \Phi_1(\ell_{C+1}) & \ldots & \Phi_1(\ell_N) \\ \ldots & & \ldots & \ldots & & \ldots \\ \Phi_C(\ell_1) & \ldots & \Phi_C(\ell_C) & \Phi_C(\ell_{C+1}) & \ldots & \Phi_C(\ell_N) \\ \hline \Phi_{C+1}(\ell_1) & & \Phi_{C+1}(\ell_C) & \Phi_{C+1}(\ell_{C+1}) & . & \Phi_{C+1}(\ell_N) \\ \ldots & & \ldots & \ldots & & \ldots \\ \Phi_N(\ell_1) & \ldots & \Phi_N(\ell_C) & \Phi_N(\ell_{C+1}) & \ldots & \Phi_N(\ell_N) \end{array}\right] \begin{bmatrix} F_c \\ - \\ F_r \end{bmatrix} \tag{16}$$

where f_c and f_r are the modal forces associated with the controlled and residual modes, respectively; F_c and F_r are the physical forces associated with the controlled and residual nodes, respectively, acting on the system; and $\Phi_i(l_j)$ is the modal shape associated with mode i at location l_j. Equation (16) can also be rewritten as :

$$\begin{bmatrix} f_c \\ f_r \end{bmatrix} = \begin{bmatrix} \Psi_{cc} & \Psi_{cr} \\ \Psi_{rc} & \Psi_{rr} \end{bmatrix} \begin{bmatrix} F_c \\ F_r \end{bmatrix} \tag{17}$$

If only C modes are to be controlled with a C number of control forces F_c, then $F_r = 0$. This implies that

$$f_c = \Psi_{cc} F_c \tag{18}$$

$$f_r = \Psi_{rc} F_c \tag{19}$$

In a previous version of the M.I.M.S.C. method, the so-called I.M.S.C. method [8], it was assumed that there was no spill-over effect. In other words, it was assumed that the control forces F_c did not contribute to the excitation of the residual higher modes. Such assumption is not realistic when the number of uncontrolled residual modes r is much larger than the number of the controlled C modes and where there is not much difference between the frequencies of the controlled and uncontrolled modes. The M.I.M.S.C. keeps into account such effect with a particular procedure later explained. Using an optimal control approach, the C modal control forces can be calculated as :

$$f_i = -(K_1 \omega_i u_i + K_2 \dot{u}_i)/R \qquad i = 1,\ldots,C \tag{20}$$

where R is a weighting control, ω_i is the natural frequency associated with the i-th mode of the C modes and u_i and $\dot{u}_i$ are the modal displacement and velocity with respective feedback gains K_1 and K_2. These gains are calculated by minimizing the performance index J given by

$$J = \int_0^\infty [(\omega_i^2 u_i^2 + \dot{u}_i^2) + R f_i^2] dt \tag{21}$$

It has been shown [6-8] that J reaches a minimum when K_1 and K_2 are given by

$$K_1 = - \omega_i R + \left[(\omega_i R)^2 + \omega_i^2 R \right]^{1/2} \tag{22a}$$

$$K_2 = \left[2 R \omega_i K_1 + \omega_i^2 R \right]^{1/2} \tag{22b}$$

Using the modal displacement and velocity u_i and $\dot{u}_i$, the modal control force f_i can be calculated Eqs. (20), (22a) and (22b). Once this has been done for all the C actuators controlling the C modes, the applied control forces F_c can be calculated as

$$F_c = \Psi_{cc}^{-1} f_c \tag{23}$$

For the dynamic simulation of the process one can integrate Eq. (14) with respect to the time in order to determine the modal displacements and velocities which can be used to calculate f_c for the next time step and so on. Also, from Eq. (13), the actual displacements for the overall system can be calculated. At this point the approach taken to handle the spill-over problem in the M.I.M.S.C. algorithm will be explained. In the dynamic simulation the modes of vibrations of the system can be ranked according to their modal energy level E_i which is given by

$$E_i = \omega_i u_i^2 + \dot{u}_i^2 \tag{24}$$

With the available C piezoelectric actuators, one can control the C modes with the highest modal energy and reduce their modal energy. However, during that time, the uncontrolled modes will be excited due to the spill-over effect and their modal energy will increase. When the modal energy of the uncontrolled modes exceeds that of the controlled modes, the actuators can be switched to control these high energy modes. In other words one can implement a time sharing of the C actuators among the modes, such that with only C actuators one is able to control an higher number of vibrating modes.

The M.I.M.S.C. algorithm also addresses the problem of the selection of optimal locations for the actuators. In order to better understand this point it is very important to recall that the magnitude of the modal control forces f_c depends mainly on the natural frequencies of the controlled modes and the modal displacements and velocities. On the other side, the magnitude of the actual physical control forces F_c depends mainly, for a given controlled mode, on the point of application of these forces or moments, as it can be seen from Eq. (18) and defined by the matrix Ψ_{cc}. Also, it is immediate to understand that an optimal location of the actuators should minimize the actual displacements δ's of the modes of the structure. For the optimal positions we intend the locations of the actuators that minimize a weighted sum J_w of the vibration amplitudes and the generated control forces given by

$$J_w = w_1 A + w_2 B \tag{25}$$

where w_1 and w_2 are weighting factors such that $w_1 + w_2 = 1$. The parameter A is called the displacement index and it is given by

$$A = \sum_{t=0}^{T_{MAX}} \sum_{i=1}^{N} \delta_i^2 \, \Delta t \tag{26}$$

in which T_{MAX} is the maximum time integration, Δt is the integration time step and N is the number of degrees of freedom of the system. The parameter B is called control force index and it is given by

$$B = \sum_{t=0}^{T_{MAX}} \sum_{i=1}^{N} F_i^2 \, \Delta t \tag{27}$$

The w_1 and w_2 are selected such that a larger importance can be given to one of the two indices. Note that the control forces are proportional to the applied voltages as shown in the previous sections.

COMPUTATIONAL ASPECTS OF THE M.I.M.S.C ALGORITHM

The flow-chart of the algorithm is shown in Figure 5. It is assumed that the piezoelectric material to be used as an actuator and its thickness have already been selected. Therefore the parameters d_{zx}, V_{Cmax} and t_c are known, assigned constants. First, the necessary inputs are given. They include:

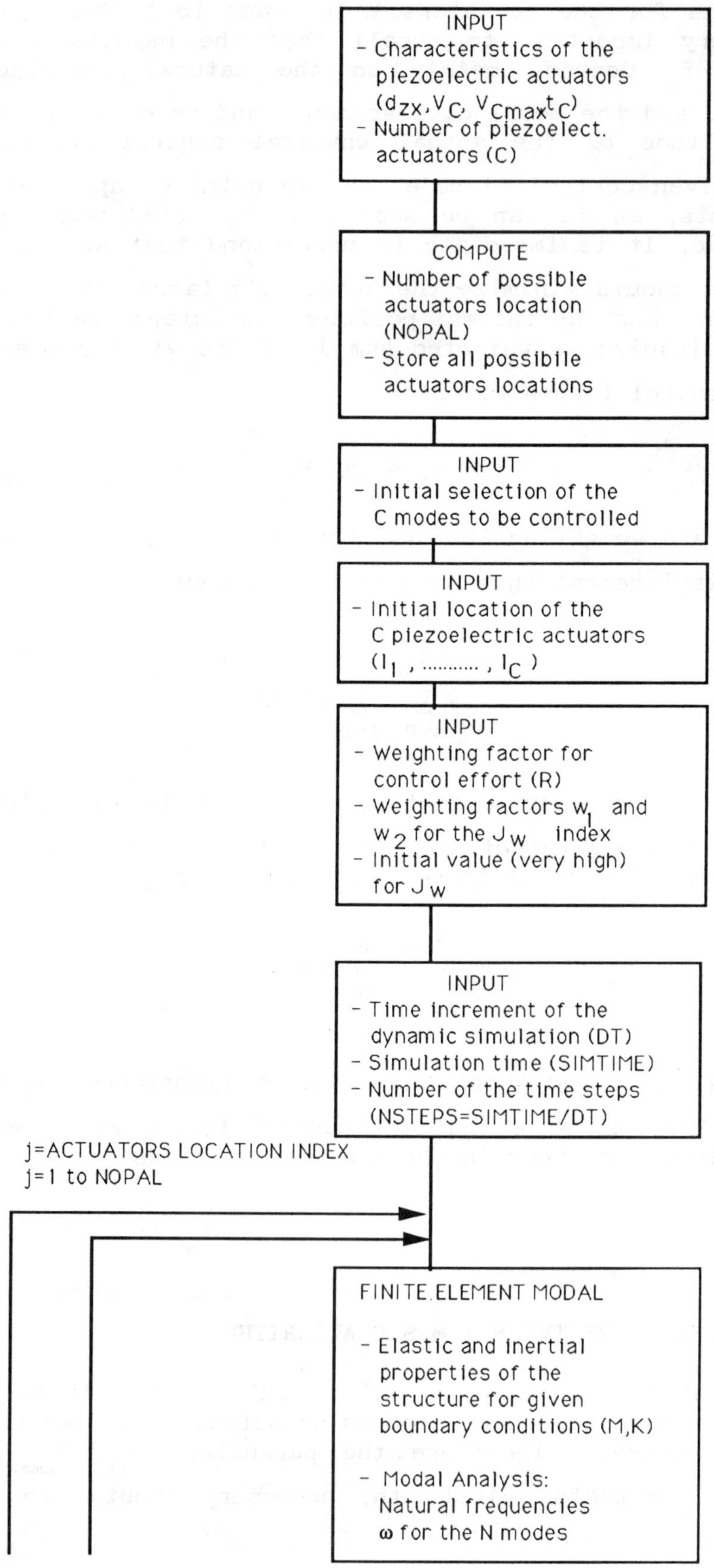
INPUT
- Characteristics of the piezoelectric actuators (d_{zx}, V_C, V_{Cmax} t_C)
- Number of piezoelect. actuators (C)
COMPUTE
- Number of possible actuators location (NOPAL)
- Store all possibile actuators locations
INPUT
- Initial selection of the C modes to be controlled
INPUT
- Initial location of the C piezoelectric actuators (l_1, ..., l_C)
INPUT
- Weighting factor for control effort (R)
- Weighting factors w_1 and w_2 for the J_W index
- Initial value (very high) for J_W
INPUT
- Time increment of the dynamic simulation (DT)
- Simulation time (SIMTIME)
- Number of the time steps (NSTEPS=SIMTIME/DT)
j=ACTUATORS LOCATION INDEX
j=1 to NOPAL
FINITE ELEMENT MODAL
- Elastic and inertial properties of the structure for given boundary conditions (M,K)
- Modal Analysis: Natural frequencies ω for the N modes

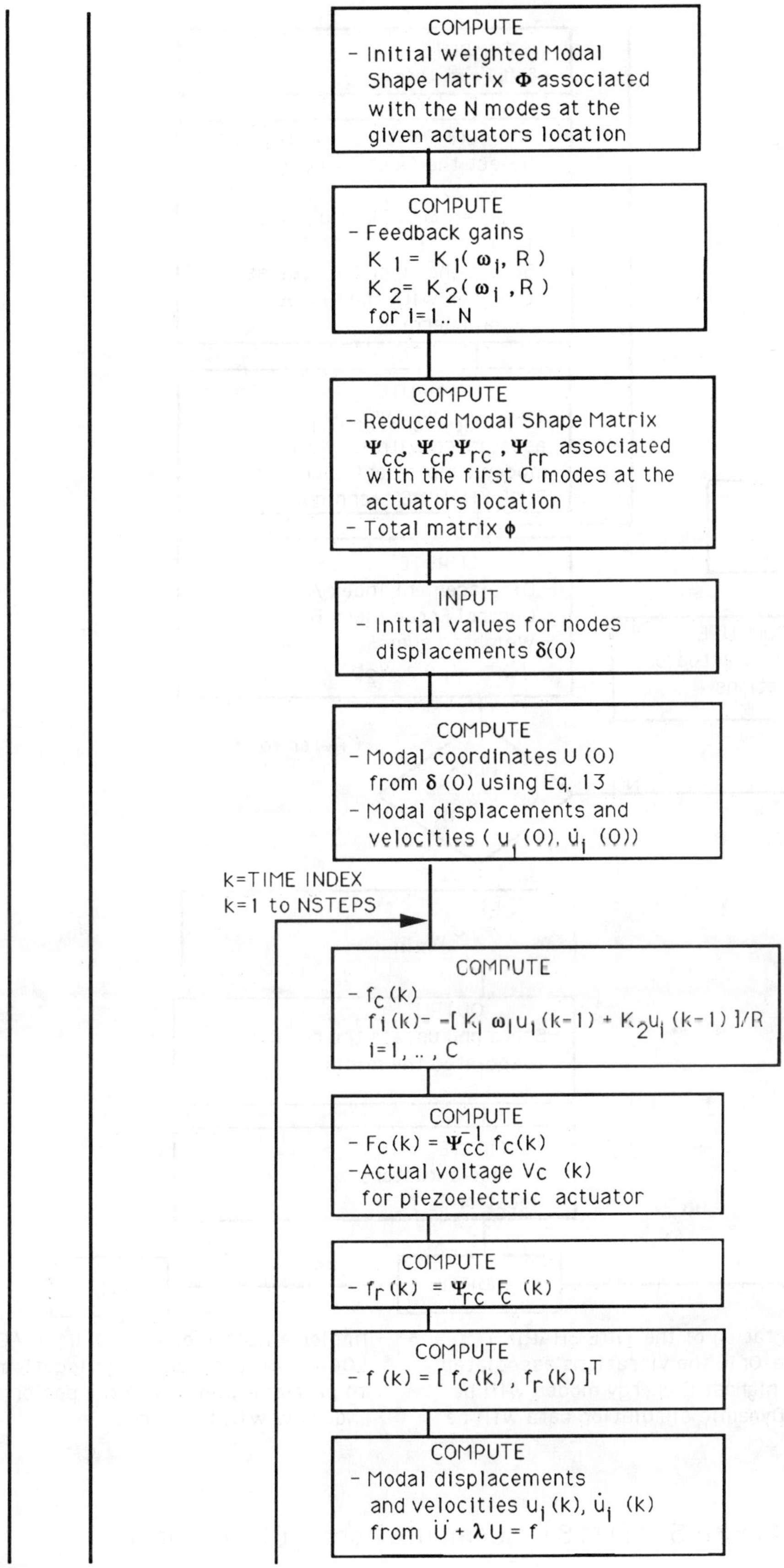
COMPUTE
- Initial weighted Modal Shape Matrix Φ associated with the N modes at the given actuators location
COMPUTE
- Feedback gains
K 1 = K1(ωi, R)
K 2= K2(ωi , R)
for i=1.. N
COMPUTE
- Reduced Modal Shape Matrix Ψcc, Ψcr, Ψrc , Ψrr associated with the first C modes at the actuators location
- Total matrix φ
INPUT
- Initial values for nodes displacements δ(0)
COMPUTE
- Modal coordinates U (0) from δ (0) using Eq. 13
- Modal displacements and velocities (ui (0), u̇i (0))
k=TIME INDEX
k=1 to NSTEPS
COMPUTE
- fc (k)
fi (k)= -[Ki ωi ui (k-1) + K2ui (k-1)]/R
i=1, .. , C
COMPUTE
- Fc(k) = Ψcc-1 fc(k)
-Actual voltage Vc (k) for piezoelectric actuator
COMPUTE
- fr (k) = Ψrc Fc (k)
COMPUTE
- f (k) = [fc (k) , fr (k)]T
COMPUTE
- Modal displacements and velocities ui (k), u̇i (k) from Ü + λ U = f

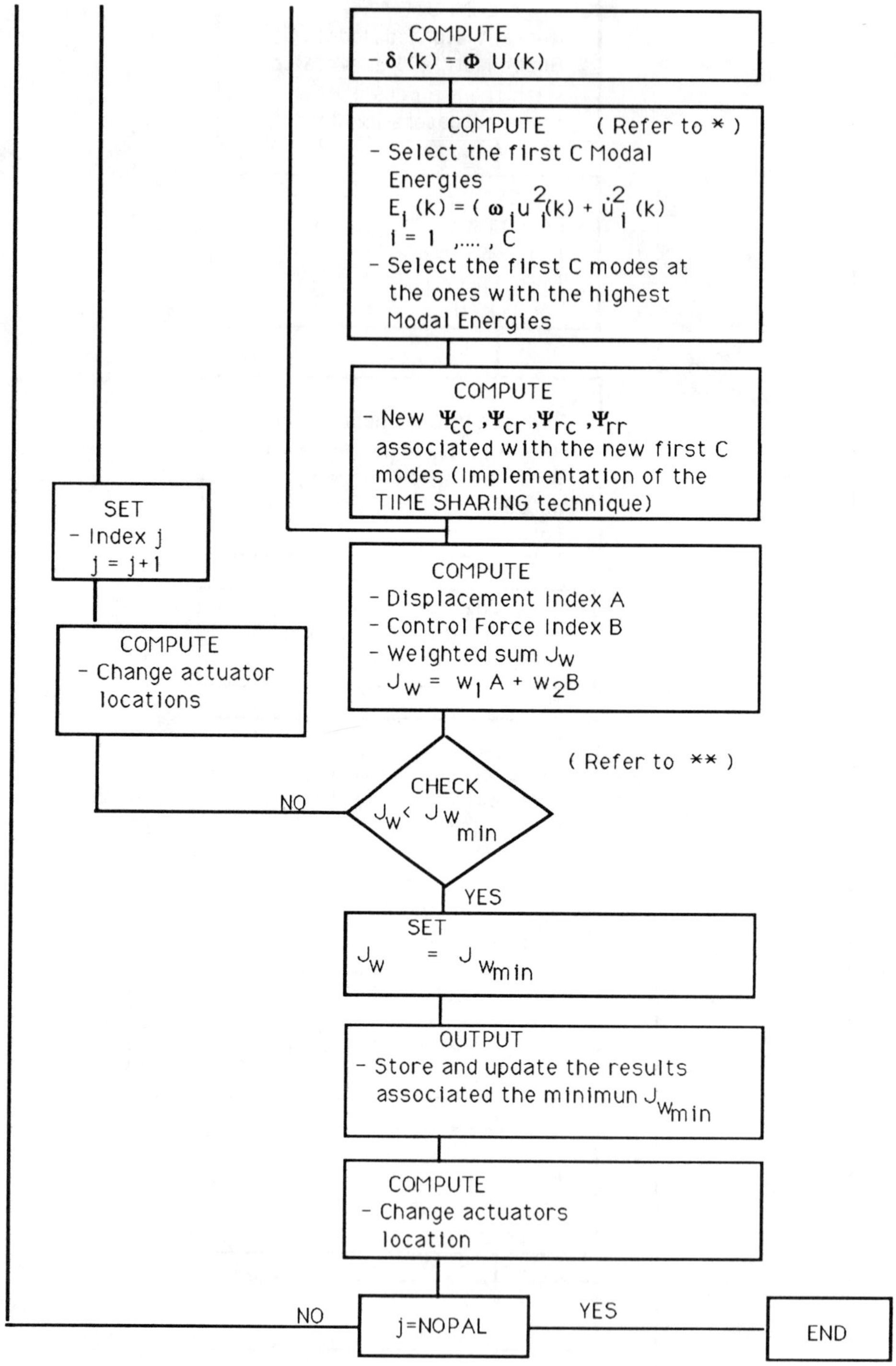

* - Implementation of the TIME SHARING technique. Only the vibrations associated with the highest C energy modes will be damped. Dynamic simulation data will be stored.

** - Implementation of the OPTIMAL ACTUATORS LOCATION technique. Only locations relative to partial minimuns of the performance index Jw will be stored.

Figure 5. M.I.M.S.C. Computational Flow Chart

- characteristics of the piezoelectric actuators (d_{zx}, V_C, V_{Cmax}, t_c);
- number of piezoelectric actuators (C);
- initial locations of the C actuators (l_1, ... ,l_C)
- weighting factor for the control effort (R);
- weighting factors w_1 and w_2 for the optimal actuators location index;
- initial (very high) value of J_w;
- time increment of the dynamic simulation (DT) and simulation time (SIMTIME).

The first external loop is associated with the variations of the actuators locations. one can calculate the N pair of feedback gains with each pair associated to a natural frequency of the structure system. Each value of j is, of course, associated with a particular actuators configuration through a pre-analysis mapping. Remember that the goal is first to find an optimal location of the actuators. A particularly numerical method for the search of the minimum of the parameter J_w can eventually be implemented, as opposed to a random change of actuators locations, for the search of the optimal location for the C actuators. The first operation in this loop is the determination of the reduced modal shape matrices Ψ_{cc}, Ψ_{cr}, Ψ_{rc}, Ψ_{rr}. In order to excite some motion the system is given some initial arbitrary displacements δ(0), with associated modal displacements and velocities (U(0), U(0)) and then leave the structure free to vibrate. This is achieved with the application of an impulsive force.

Next, the internal loop of the dynamic simulation for a certain number of time steps (NSTEPS) is performed. The previously introduced time sharing technique is implemented at each time step when the modal energies are ranked. Then, select the C modes with the C highest modal energies and recalculate the residual modal shape matrices Ψ_{cc}, Ψ_{cr}, Ψ_{rc} and Ψ_{rr}. Note that at this point in the algorithm the output data from the dynamic simulation can be updated and stored in a data file if the J_w is small than the previous actuators position. Since the optimal location results has been saved, it is unnecessary to perform the dynamic simulation associated with the minimum J_w at the end of all possible actuators locations.

If the analysis does not furnish acceptable results, for example in terms of V_{Cmax} or in terms of physical displacements, then the weighting factors R, w_1 and w_2 may be changed, depending on the problem. As conclusive statement of this section, for a given piezoelectric material with given values of d_{zx}, V_{Cmax} and given thickness t_c applied on a given structure, the final results should be:

- Determination of the feedback gains structure.
- Determination of the optimal piezoelectric actuators locations.
- Dynamic simulation associated with actuators at optimal locations.

RESULTS OF THE STUDY

A study showing the application of the algorithm illustrated in Figure 6 to a simple cantilevered steel beam has been performed. The analyzed beam has been modeled with 5 elements and 2 piezoelectric actuators. The effects of the thickness of the bonding layer were already considered in [6]; therefore the thickness of the bonding layer is considered to be constant. The properties of the beam, actuator and bounding layer are given in Table I. The structure has been excited with an external impulsive force at the tip of 3.0 N. The reason for this relatively high value of the impulsive force is that it was desired to have some physically significant displacements. For the assigned load conditions and number of actuators the following points are of particularly interest

1 - Variation of:
- o optimal location of the 2 piezoelectric actuators,
- o time histories of vertical displacement of the tip node of the beam,
- o time histories of the voltages of the two piezoelectric actuators,
- o hystogram of the controlled modes,

as function of the control factor R. The following numerical inputs are considered with R = 1, 10, 50, 100 and $w_1 = w_2 = 0.5$.

2 - Variation of:
- o optimal location of the 2 piezoelectric actuators,
- o vertical displacement of the tip node of the beam,
- o time histories of the voltages of the two piezoelectric actuators,
- o hystogram of the controlled modes,

as function of the displacements and forces weighting factors w_1 and w_2.

RESULTS OF STUDY #1

The numerical results associated with STUDY #1 have been reported in Figures 7-14. A first result is given by the determination of the optimal location of the 2 piezoelectric actuators. It has been found that for all values of the control factor R the optimal location of the actuators is at the 2 elements close to the root of the cantilevered beam (elements 5 and 4), as expected and as shown in Figure 7. The associated values of the index J_w (weighted sum of the displacements and control forces indices) are shown in TABLE II for R=10. Figure 8 shows the displacement of the node at the tip of the structure for different values of the control factor R. As one can see, the case for R=1 gives a very desirable damping of the vibration caused by the external force but, as one may expect, this has to be paid with an 'expensive' control, in terms of the voltages associated with the 2 piezoelectric actuators, as it may be argued by looking at Eqs. (18) and (20). The voltages associated with actuator 1 (located at element 5) are shown in Figure 9 (for R=1,10) and Figure 10 (for R=50,100). No effort has been made, up to this point, to consider the saturation problem of the piezoelectric actuators. It can be noticed that the voltages strongly decrease with the increase of the control factor R.

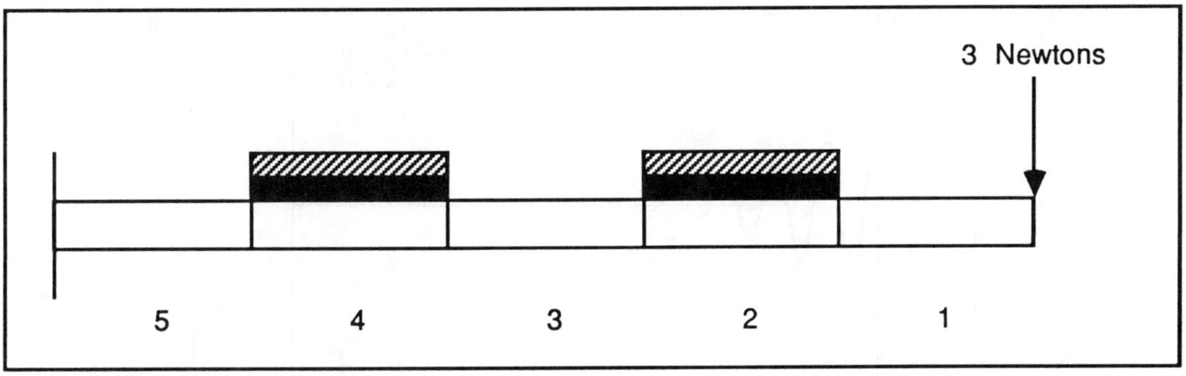

Figure 6. Steel Beam-Piezoelectric-Bonding Layer Model

TABLE I Geometry of Beam-Piezoelectric-Bonding Layer Model

	Type	Thickness	Width	Length	Modulus
Actuator	PZT	2.1 mm	12.5 mm	50 mm	1.39e+5 N/mm/mm
Bonding Layer	Isotac	0.1 mm	12.5 mm	50 mm	1.1 N/mm/mm
Beam	Steel	2.1 mm	12.5 mm	250 mm	2.10e+5 N/mm/mm

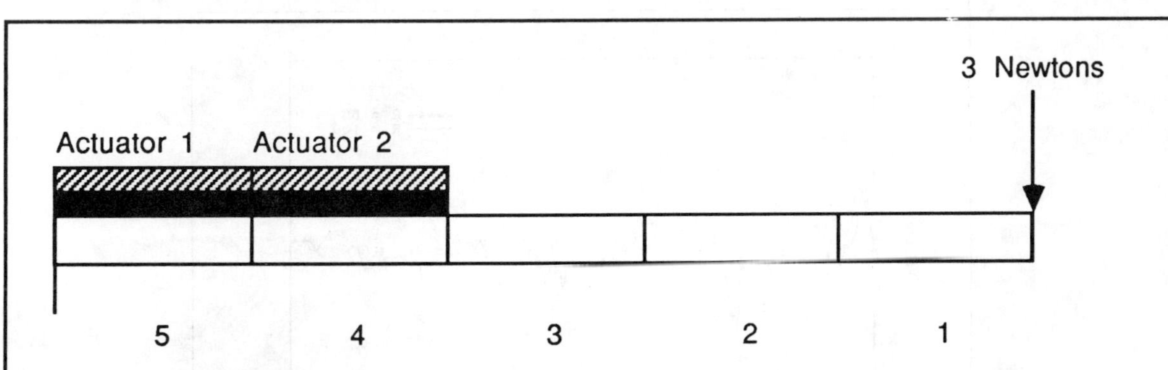

Figure 7. Optimal Location of the Piezoelectric Actuators for R=1,10,50,100 with w1=0.5 and w2 = 0.5

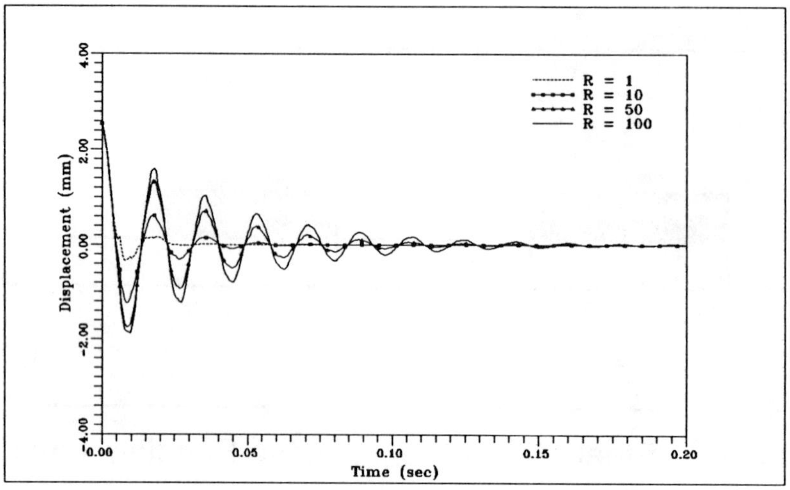

Figure 8. Vertical Displacement of the Tip Node for w1=0.5 and w2=0.5

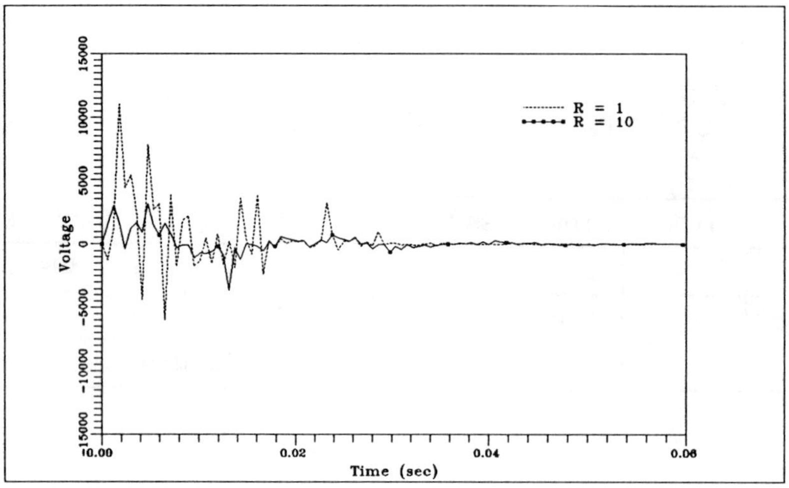

Figure 9. Voltages of Actuator 1 for w1=0.5 and w2=0.5

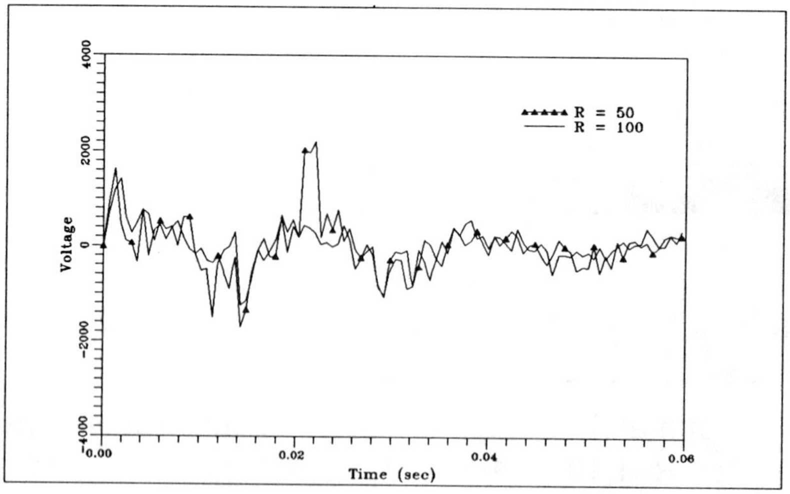

Figure 10. Voltages of Actuator 1 with w1=0.5 and w2=0.5

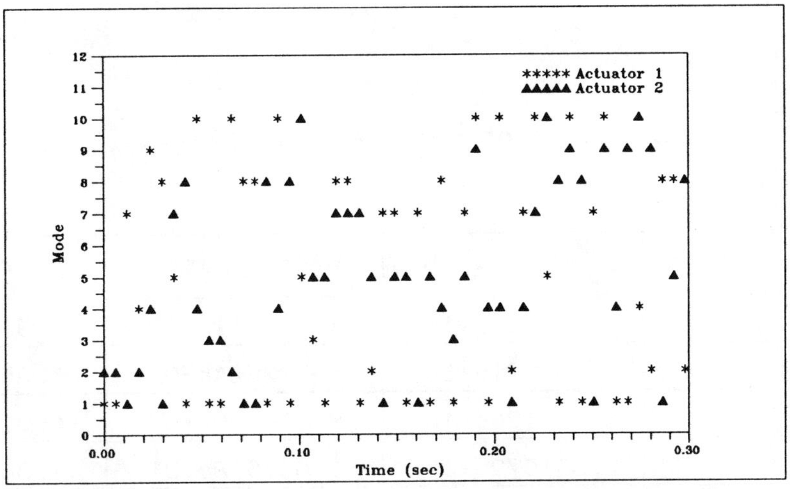

Figure 11. Modes Controlled by the 2 Piezoelectric Actuators for R=1, w1=0.5 and w2=0.5

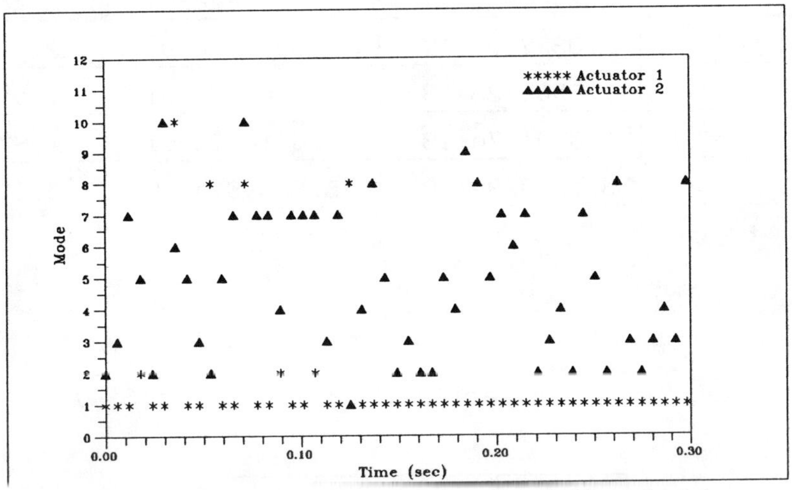

Figure 12. Modes Controlled by the 2 Piezoelectric Actuators for R=10, w1=0.5 and w2=0.5

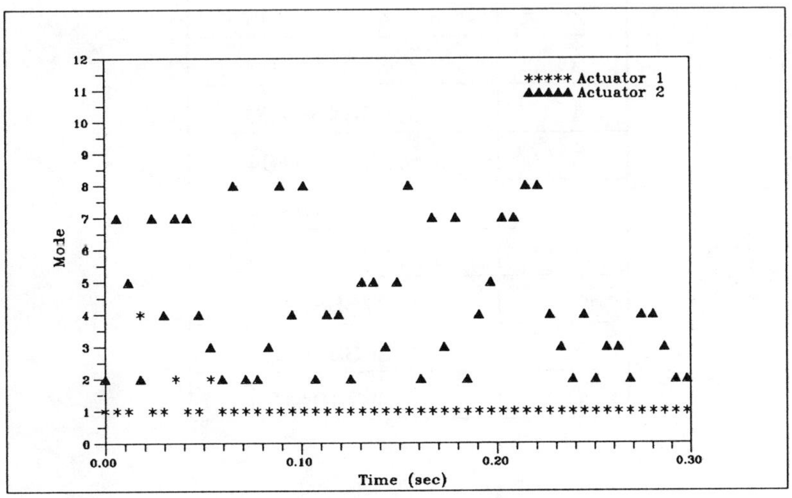

Figure 13. Modes Controlled by the 2 Piezoelectric Actuators for R=50, w1=0.5 and w2=0.5

TABLE II (for R=10, w1=0.5, w2=0.5)

Actuator Location	J_w=w1 A+w2 B	A=disp. index	B=force index
5, 4	2.579616e+15	1.677311e-02	2.579616e+15
5, 3	1.354731e+16	4.722824e-02	1.354731e+16
5, 2	6.915136e+16	8.210030e-02	6.915136e+16
5, 1	2.301679e+16	1.057518e-01	2.301679e+16
4, 3	4.260669e+20	9.287227e-01	4.260669e+20
4, 2	2.284665e+17	2.410384e-01	2.284665e+17
4, 1	2.035889e+17	2.935772e-01	2.035889e+17
3, 2	1.5568723+18	5.298467e-01	1.5568723+18
3, 1	1.870117e+18	9.031493e-01	1.870117e+18
2, 1	3.127029e+20	5.523981e+00	3.127029e+20

TABLE III (for actuators at location 5,4)

Mode	Frequency
1	3.54041e+02
2	1.48935e+03
3	3.84311e+03
4	8.04046e+03
5	1.27863e+04
6	2.07874e+04
7	3.29487e+04
8	4.37558e+04
9	7.30384e+04
10	8.38740e+04

Another less apparent but equally important result can be read in Figures 11-14, which show the modes controlled by the two actuators instant by instant, with the previously introduced time-sharing technique. Before we actually discuss these figures let us recall that, given the structure and the assigned impulsive load, one may expect that mode 1 (1st bending) is the predominant mode with secondary modes possibly mode 2 and 3 (2nd and 3rd bending). The frequencies of the vibrating modes are shown in TABLE III. It can be realized from Figures 11-14 that the spill-over effect is a big factor for the lowest value of R (R=1). In fact, we can see that the strong control forces acting close to the root of the structure continuously excite higher vibrating modes. This explains the apparently chaotic order of Figure 11. Also notice that the spill-over takes place regardless of the relatively big frequency separation between the vibrating modes of the structure. Therefore it is concluded that, even for simple structures like this, the spill-over effect caused by the control forces is a factor and needs to be properly accounted for. This proofs, once again, the validity of this method (M.I.M.S.C.) with respect to the previous version (I.M.S.C.) which simply neglected the spill-over effect. Given that the size of the control forces decrease with higher values of R (10,50,100), the spill-over effect is still present but less dominant for those values of R, as shown in Figures 12-14.

RESULTS OF STUDY #2

In this study we were interested in understanding the effect of the weighting factors w_1 and w_2 on the performance of the M.I.M.S.C. algorithm. Recall that w_1 is associated with the displacements index A while w_2 is associated with the control forces index B. It turned out that the index B is much bigger than the index A, as it can be seen in TABLE II. Therefore it was irrelevant to analyze the performance of the algorithm at intermediate values of w_1 and w_2, (such as w_1=0.25, w_2=0.75 or w_1=0.75, w_2=0.25), given that A and B were not comparable at all (B>>A). Thus only the two extreme cases were considered (with R=10) :

$w_1 = 1$, $w_2 = 0$, $J_w = A$

$w_1 = 0$, $w_2 = 1$, $J_w = B$

The result is that the position of the actuators at element 5 and 4 of the beam is the optimal position in both cases and therefore the associated dynamic simulation will be the same as the one shown in Fig. 8,9,12. for R=10. In other words the optimal location of the actuators is the same (at elements 5 and 4) regardless if we like to minimize the node displacements or the control forces for damping the vibrations induced by the external force. This result, along with the fact that B >> A, questions the need itself for using w_1 and w_2 in the expression for J_w.

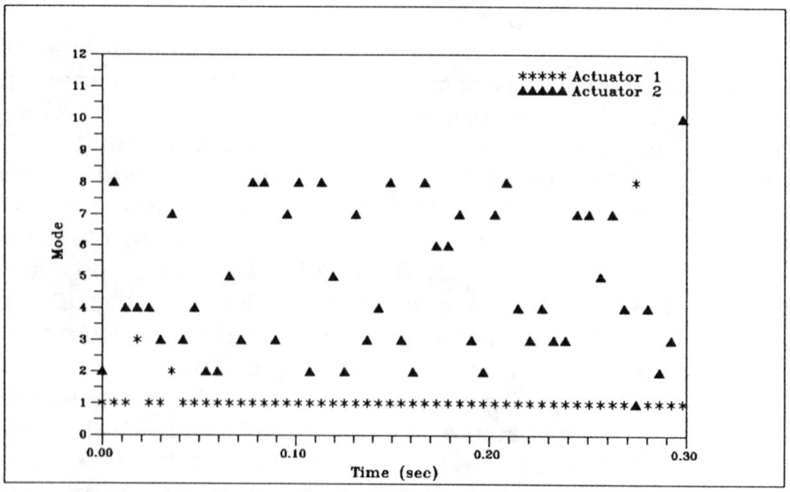

Figure 14. Modes Controled by the 2 Piezoelectric Actuators for R=100, w1=0.5 and w2=0.5

EFFECTS OF THE SATURATION OF THE PIEZOELECTRIC ACTUATORS

An immediate result of STUDY #1 was that extremely high values of the voltages were associated with R=1 while more reasonable lower values of the voltages were calculated for R=10,50,100, as it can be noticed in Figure 9 and 10. At this point, in order to have a more realistic dynamic simulation, the saturation problem of the two piezoelectric actuators needed to be considered. Particularly, a constraint had to be put on the maximum allowed voltage to be applied on the piezoelectric actuator. Consistently with [1] and [6] the value of V_{MAX} = 1000 V is considered for the PZT type piezoelectric actuators. Also, as result of STUDY #2, it was felt that it was no longer necessary to use two separate indices (displacements and control forces indices) A and B for J_w. Therefore a different but simpler expression for J_w was introduced [9] and [10]

$$J_w = \sum_{t=0}^{T_{MAX}} \sum_{i=1}^{N} | \delta_i F_i | \Delta t \qquad (28)$$

The dynamic simulation with a V_{MAX} = 1000 V for the 2 piezoelectric actuators has been performed. While this does not have any effect on the optimal location of the 2 actuators, the vertical displacements of the tip node are shown in Figures 15, 16 and 17 for R=10,50,100 respectively. As one may expect, the actuators saturation has a smaller effect as R increases because the voltages generally decrease with R. The time histories of the voltages associated with actuator 1 are shown in Figures 18,19 and 20, again for R=10,50,100 respectively.

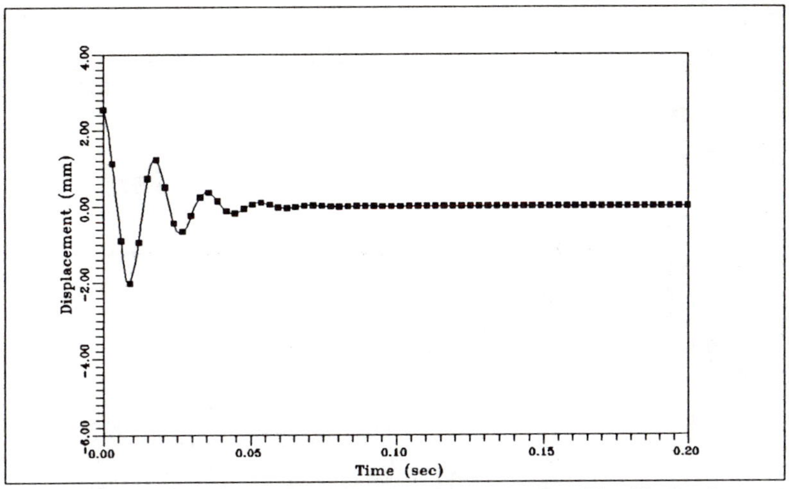

Figure 15. Vertical Displacement of the Tip Node for R=10 with Vmax=1000 V

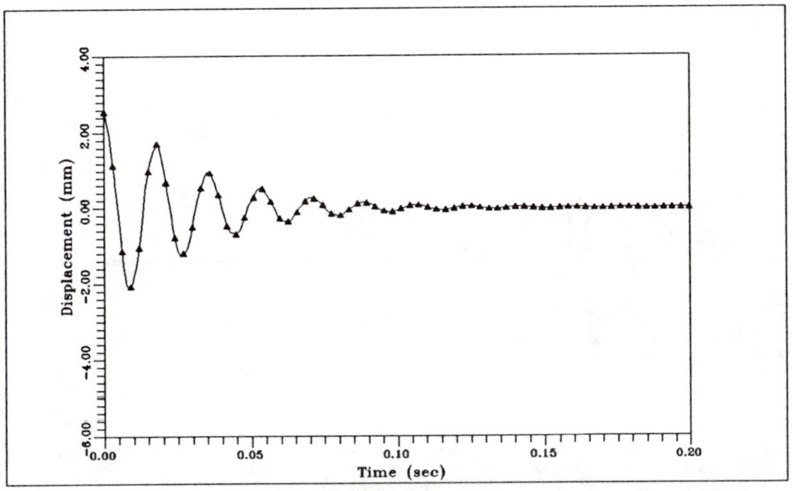

Figure 16. Vertical Displacement of the Tip Node for R=50 with Vmax=1000 V

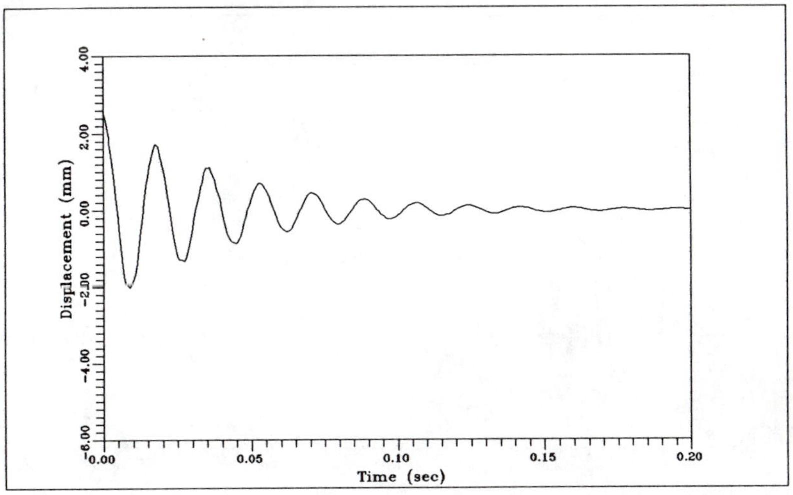

Figure 17. Vertical Displacement of the Tip Node for R=100 with Vmax=1000 V

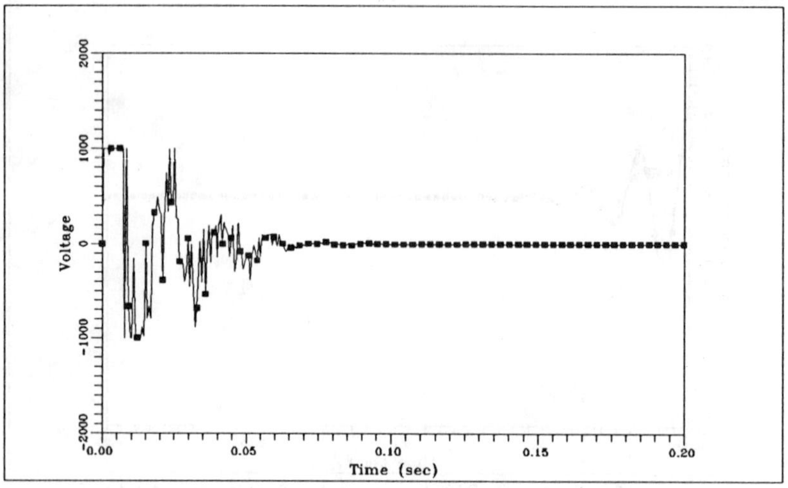

Figure 18. Voltages of Actuator 1 for R=10 with Vmax = 1000 V

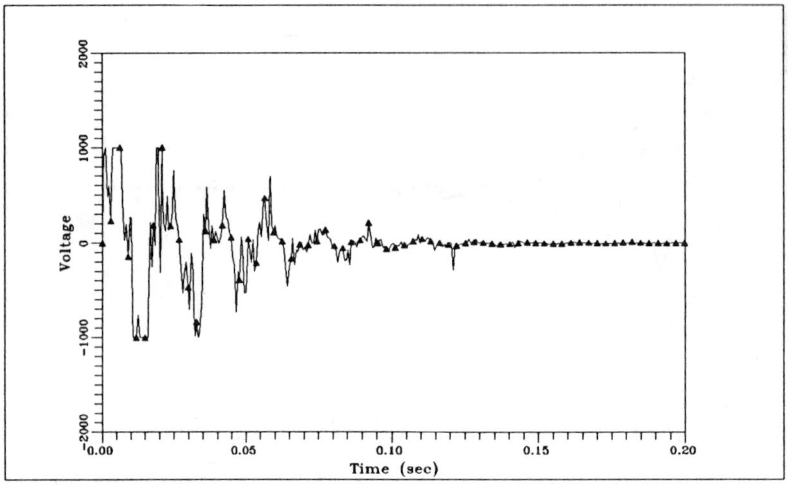

Figure 19 Voltages of Actuator 1 for R=50 with Vmax = 1000 V

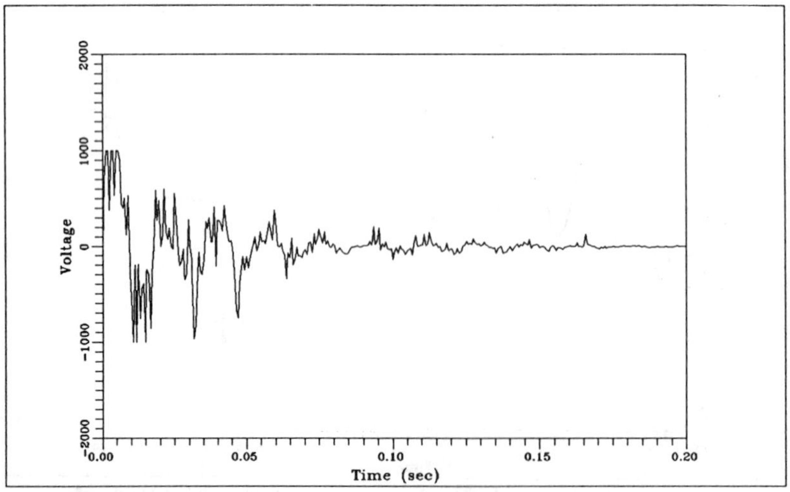

Figure 20. Voltages of Actuator 1 for R=100 with Vmax = 1000 V

CONCLUSIONS AND RECOMMENDATIONS

The overall goal of this study was to implement the M.I.M.S.C. algorithm, to gain inside views of the functioning of the numerically algorithm itself through an application on a simple vibrating structure, before implementing it on a plane-structure (simulating an aircraft fuselage panel) or a rotating beam structure (simulating a rotating helicopter blade). Several conclusions have been drawn from this study. The main one is that when we use the control forces to the full extent (that is when we have low values of the factor R) the spill-over effect is an important factor; this factor is coupled with the actuators saturation problem. It was felt that the M.I.M.S.C. method is an excellent control strategy for active reduction of the vibrations of a structure. The next logic step is the application of the algorithm to more complex vibrating systems of much more engineering interest.

References

1. Crawley, E. F., De Luis J. "Use of Piezoelectric Actuators as Elements of Intelligent Structures," AIAA Journal, Vol.25, No.10, pp.1373 Oct. 1987.

2. Bayley, T.L., Hubbard, J.E. "Distributed Piezoelectric Polymer Active Vibration Control of a Cantilever Beam," Journal of Guidance, Control, and Dynamics, Vol.8, pp. 605, Sept-Oct. 1985.

3. Tsan, H.S., Godre, M., "Active Vibration Isolation by Polymeric Piezoelectric with Variable Feedback Gains," AIAA Journal, Vol. 26, No. 8, pp. 1014, Aug. 1988.

4. Burke, S.E., Hubbard, J.E. Jr., "Distributed Actuator Control Design for Flexible Beams," Automatica, Vol. 24, No.5, pp.619, May 1988.

5. Tzou, H.S., Gadre, M., "Theoretical Analysis of a Multi-Layered Thin Shell Coupled with Piezoelectric Shell Actuators for Distributed Vibration Controls," Journal of Sound and Vibration, Vol. 132, No. 3, pp. 433, 1989.

6. Baz, A., Poh, S., "Performance of an Active Control System with Piezoelectric Actuators," Journal of Sound and Vibration, Vol. 126. No. 2, pp. 327, 1988.

7. Meirovitch, L., Baruh, H., Oz, H., "Comparison of Control Techniques for Large Flexible Systems," Journal of Guidance and Control, Vol. 6, No. 4, pp. 302, July 1983.

8. Oz, H., Meirovitch, L., "Optimal Modal-Space Control of Flexible Gyroscopic Systems", Journal of Guidance and Control, Vol. 3, No.3, pp. 218, May 1980.

9. Baz, A., Poh, S., "Modified Independent Modal Space Control Method for Active Control of Flexible Systems," NASA Technical Report Number N87-23980.

10. Baz, A, "Static Deflection Control of Flexible Beams by Piezo-electric Actuators," NASA Technical Report Number N87-13788.

Interior Noise

ACTIVE ADAPTIVE NOISE CONTROL IN CARS

D. Guicking,[1] M. Bronzel,[1] W. Böhm[1]

ABSTRACT

Noise in the interior of cars is composed of two contributions: one is synchronized with the engine rotation or other (quasi)periodic sources and therefore repetitive in nature, the other part is caused by stochastic inputs such as road surface irregularities or turbulent air flow. Active control of stochastic noise requires adaptive feedforward control, while the cancellation of discrete frequency noise is possible with simpler techniques such as tracking notch filters. The fundamentals of active noise control for both types are outlined in the paper, and some experimental results—partly from a practical realization in a car and partly from model experiments—are presented.

1. INTRODUCTION

Passive noise control, by sound absorption or sound shielding, is greatly restricted in vehicles by mass and space limitations, particularly for low frequencies. Active noise control has generally made considerable progress during the last few years, and some technical applications have been realized for acoustically "simple" situations. Concerning the interior space of cars, three aspects can be considered beneficial for a practical success of active control systems: the (quasi)periodicity of engine-related noise, the comparatively small volume of the cabin which leads to small overlap of the resonant modes in the lower audio frequency range, and the fact that noise reduction is only required in that part of the space where passenger heads are typically located. One major complicating effect is to be seen in the fact that any controller has to cope with rapidly changing transfer functions so that fast signal processing is a crucial prerequisite for any active noise control concept to be applied in a car.

In Section 2 of this paper, the fundamentals of engine-related repetitive noise and possibilities of its active cancellation are outlined, followed by some experimental results. Section 3 summarizes some problems and the present state-of-the-art for the active cancellation of stochastic noise in small interior spaces. Some preliminary results of model experiments are also presented.

[1]Third Institute of Physics, University of Göttingen, Bürgerstr. 42-44, D-3400 Göttingen, Germany.

2. REPETITIVE NOISE

2.1. PROBLEM STATEMENT

The most important primary source of repetitive interior noise in cars is the engine. The passenger comfort can therefore often be improved by active cancellation of those components which are related to the engine speed and its harmonics. The mechanism of the noise generation is outlined in the following.

Fig. 1 is a schematic diagram of the system consisting of piston, connecting rod, and crankshaft for one cylinder of an internal combustion engine, showing also the principal force F_{os} acting in the running engine. To a first order approximation, the periodic motion of the connecting rod and the piston is that of a reciprocating point mass m_0 the inertial force F_{os} of which is transmitted to the engine support via the crankshaft. We have the following relationships, from Fig. 1:

$$s = r + l + s_0 \,, \tag{1}$$

$$s_0 = r \cdot \cos\varphi + l \cdot \cos\alpha \,, \tag{2}$$

$$r \cdot \sin\varphi = l \cdot \sin\alpha \,. \tag{3}$$

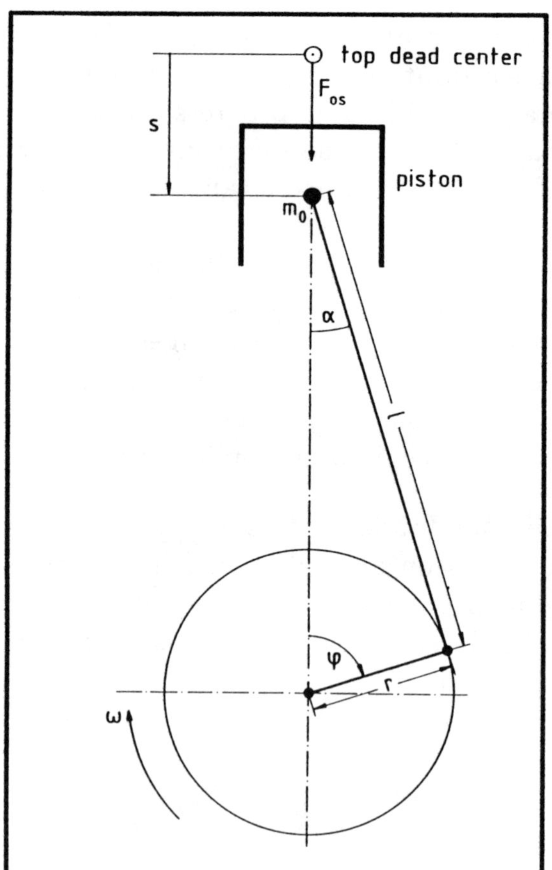

Fig. 1. Schematic diagram of the piston, connecting rod, and crankshaft system of an internal combustion engine. F_{os}: oscillating force, m_0: oscillating mass, s: piston displacement, φ: crank shaft angle, ω: angular frequency.

Introducing the stroke/connecting-rod ratio, $\varepsilon = r/l \ll 1$, eqs. (1) through (3) yield

$$s = (r+l) - r\left(\cos\varphi + \frac{1}{\varepsilon}\sqrt{1-\varepsilon^2\sin^2\varphi}\right), \tag{4}$$

and by series expansion of the root term,

$$s = (r+l) - r\left(A_0 + \cos\varphi + \frac{A_2}{4}\cos 2\varphi + \frac{A_4}{16}\cos 4\varphi + \frac{A_6}{36}\cos 6\varphi + \cdots\right) \tag{5}$$

where

$$A_0 = \frac{1}{\varepsilon} - \frac{1}{4}\varepsilon^3 - \frac{5}{256}\varepsilon^5 - \cdots \approx \frac{1}{\varepsilon}, \tag{6}$$

$$A_2 = \varepsilon + \frac{1}{4}\varepsilon^3 + \frac{15}{128}\varepsilon^5 + \cdots \approx \varepsilon, \tag{7}$$

$$A_4 = -\frac{1}{4}\varepsilon^3 - \frac{3}{16}\varepsilon^5 - \cdots \approx -\frac{1}{4}\varepsilon^3, \tag{8}$$

$$A_6 = \frac{9}{128}\varepsilon^5 + \cdots \approx \frac{9}{128}\varepsilon^5. \tag{9}$$

The inertia force, F_{os}, is equal to the product of the oscillating mass and the second time derivative of the piston displacement:

$$F_{os} = m_0 r\omega^2\left(\cos\varphi + A_2\cos 2\varphi + A_4\cos 4\varphi + \cdots\right). \tag{10}$$

However, due to the assumption $\varepsilon \ll 1$, the spectrum of the oscillating force shows relevant components only for $\cos\varphi$ and $\cos 2\varphi$. These components are sometimes called *first order force* and *second order force*, respectively.

In multicylinder engines, the oscillating forces of the individual cylinders are superimposed and reinforce or diminish each other, depending on the engine type. In a *four-cylinder engine*, the first order forces cancel one another, while the second order forces are enhanced by superposition. This is why the second harmonic of the engine speed is usually a strong component in the noise spectrum within cars of this type. In *six-cylinder engines*, both first and second order forces cancel each other [1]. However, the so-called gas forces remain and cause a (weaker) noise component at the 3rd harmonic of the engine speed. Generally, six-cylinder engines exhibit much lower noise components related to engine orders.

Beside these noise components which are caused directly by engine forces, contributions may occur which are indirectly related to the engine speed (e. g., gear noise), often with fractional orders.

2.2. ALGORITHMS

Engine-order related spectral components of the interior noise field can be canceled in the following manner. The engine speed is picked up by a sensor providing, e. g., a pulse sequence from which a sinusoidal reference signal at the desired frequency is formed by some analog or digital electronics. Magnitude and phase of this reference signal are adjusted in a controller which feeds one or more loudspeakers serving as secondary sources and canceling the corresponding noise component. In many modern cars a signal is available which controls the fuel injection in dependence on the engine speed; it can therefore be used as a sensor signal.

In order to provide an on-line adjustment of the compensation signal even under rapidly changing ride conditions, a fast algorithm is necessary. In [2], adaptive digital FIR (transversal) filters have been applied with coefficient updating according to the Widrow-Hoff LMS algorithm [3] where two coefficients should suffice to adjust magnitude and phase of a sinusoid. We have found out that the robustness and the speed of convergence can be improved by the following modification. Employing a Hilbert transformer, the reference signal is split into two orthogonal components which are separately weighted and then summed up (Fig. 2). The compensation signal is therefore

$$y[k] = w_1[k] \cdot C \sin \omega_0 kT + w_2[k] \cdot C \cos \omega_0 kT \tag{11}$$

where T is the sampling period, k the time index, and C the amplitude of the reference signal. The coefficients w_1 and w_2 are adjusted by the LMS algorithm,

$$w_1[k+1] = w_1[k] + 2\mu\, e[k]\, C \sin \omega_0 kT\,, \tag{12}$$

$$w_2[k+1] = w_2[k] + 2\mu\, e[k]\, C \cos \omega_0 kT \tag{13}$$

where μ is the convergence parameter and $e[k]$ the error signal. Such a system can be considered a closed loop controller with a transfer function $H(z)$ from the primary input, $d[k]$, to the error output, $e[k]$, given by [3]

$$H(z) = \frac{z^2 - 2z\cos(\omega_0 T) + 1}{z^2 - 2(1-\mu C^2)z\cos(\omega_0 T) + 1 - 2\mu C^2}\,. \tag{14}$$

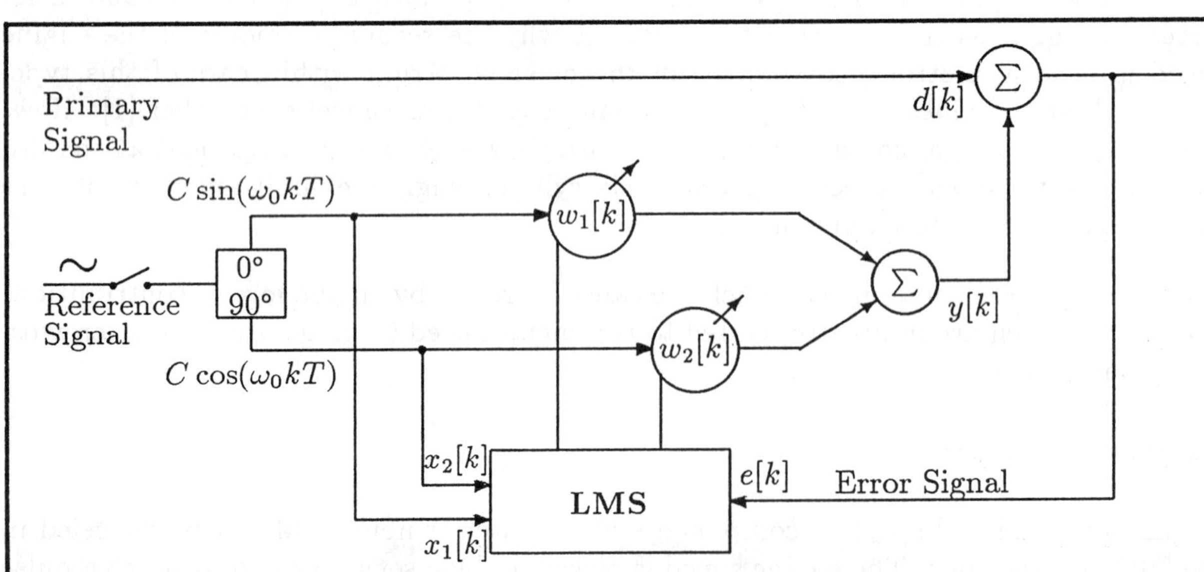

Fig. 2. Block diagram of the modified FIR adaptive filter.

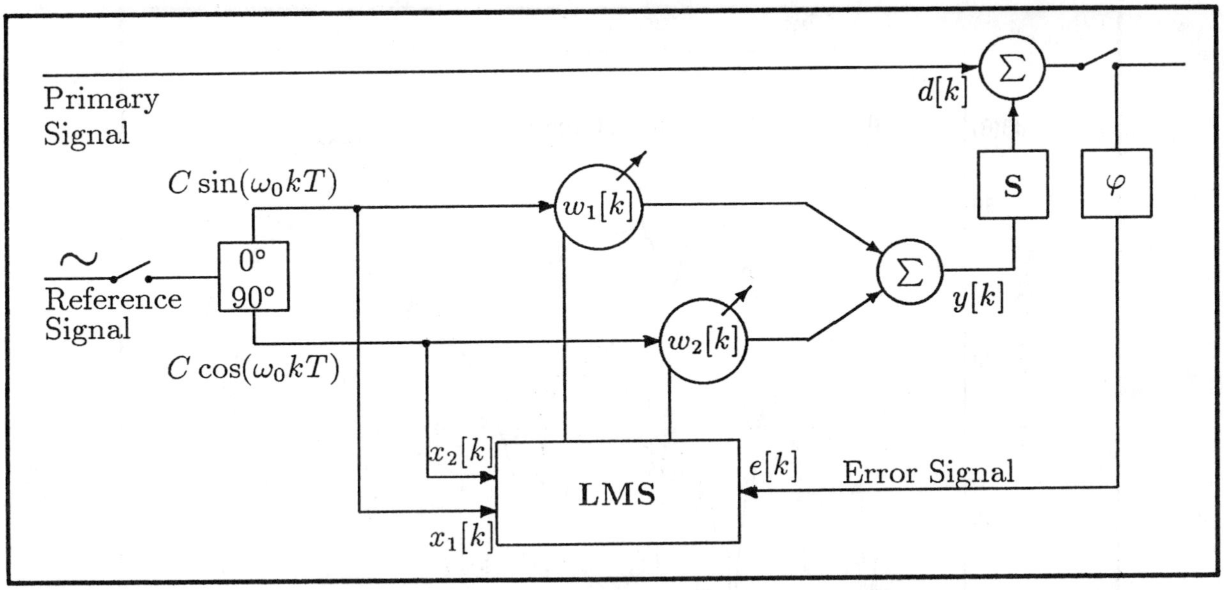

Fig. 3. Block diagram of the modified FIR adaptive filter with secondary acoustic transfer path.

This transfer function represents a notch filter at the frequency ω_0 (which can be adjusted easily by the external reference signal). From eq. (14), the bandwidth can be estimated as

$$\mathrm{BW} \approx \frac{2\mu C^2}{T}, \tag{15}$$

and the stability condition as

$$\mu < \frac{1}{C^2}. \tag{16}$$

Applying such a notch filter to a certain acoustic noise cancellation problem, we have to take into account that in general the cancellation signal is not introduced at the site of the error sensor but with an acoustic transfer path between the secondary speaker and the error sensor. This secondary path, in Fig. 3 represented by a transfer function $S(z)$, can result in a decorrelation of the compensation signal and the error signal so that the algorithm may diverge. One possibility of compensation for this decorrelation is an appropriate phase shift of the error signal (in Fig. 3, the box "φ").

In the case of more than one engine order being present in the interior noise spectrum, an according number of notch filters can be arranged in parallel. The filter output signals can be summed up and fed to a common cancellation loudspeaker.

2.3. EXPERIMENTAL RESULTS

Figs. 4 and 5 show B-rated sound pressure levels measured near the driver's head position in a car with a six-cylinder engine as functions of the engine speed (in rpm). All data have been recorded driving in fourth gear at maximum acceleration and constant engine load, with and without active cancellation.

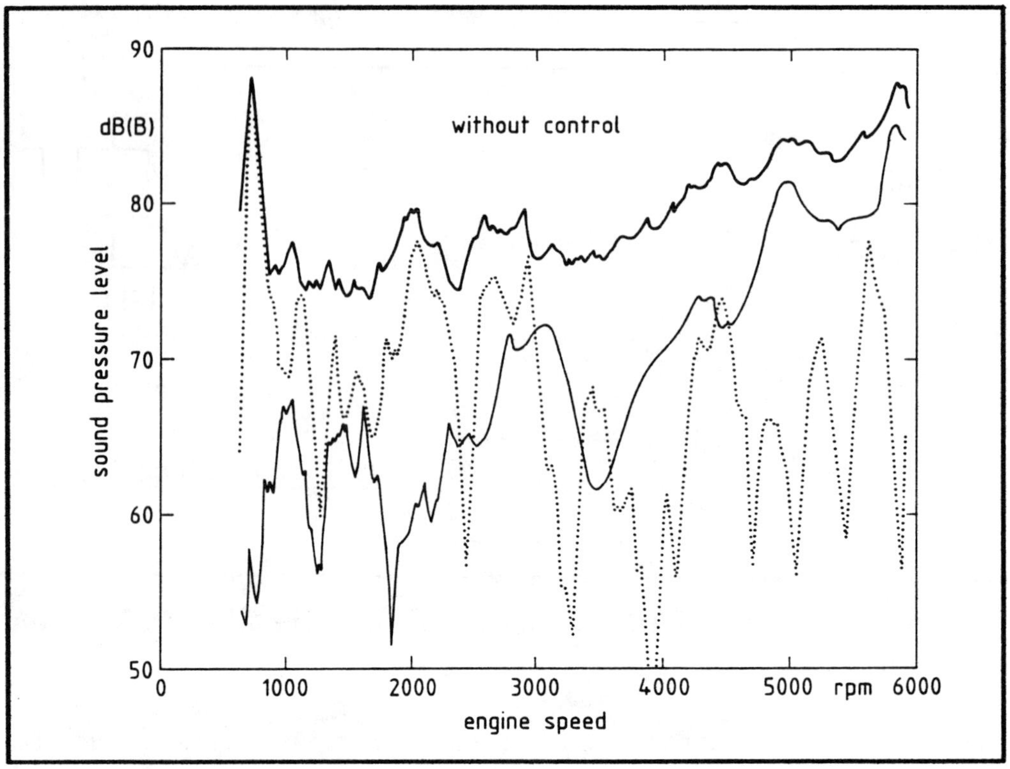

Fig. 4. B-rated sound pressure level in a car without active control as a function of the engine speed. Heavy solid line: total sound pressure level, dotted line: sound pressure level at the 3rd engine order, thin solid line: same, but for 1.5th engine order.

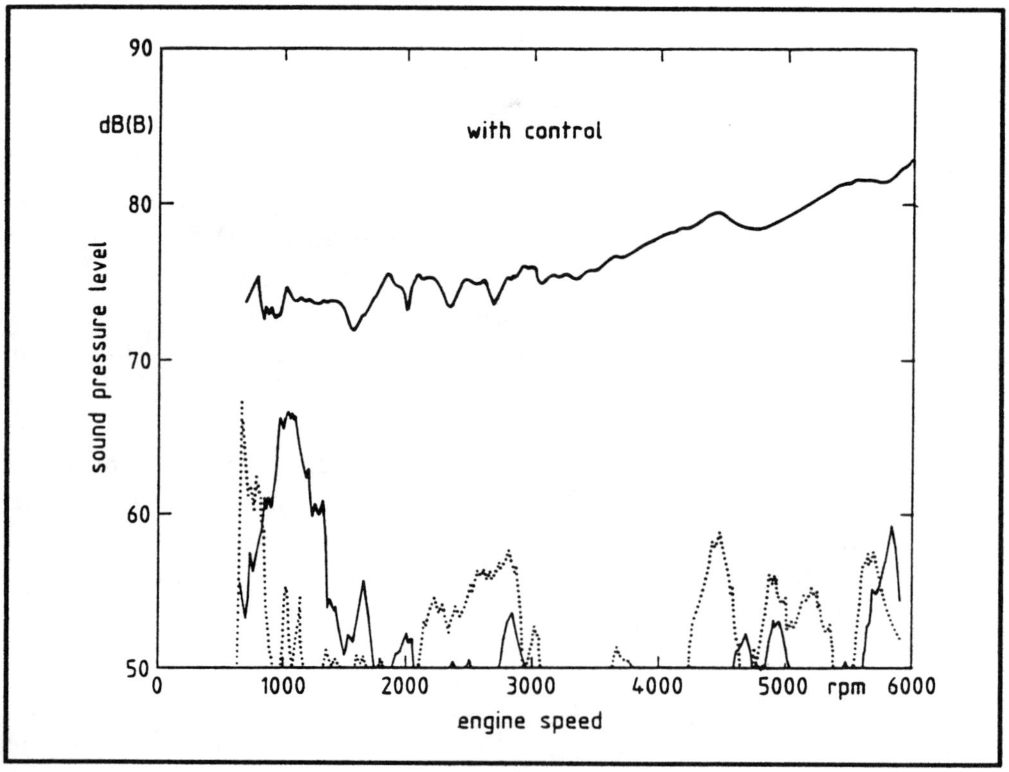

Fig. 5. Same as Fig. 4, but with active control.

The diagrams show the total sound pressure level together with the partial levels at the 3rd and the 1.5th engine order which are the most prominent harmonic noise components in this type of cars. While the 3rd order noise is directly caused by the engine, the 1.5th order noise originates from gear noise.

Fig. 5 shows that both engine orders can simultaneously be compensated for through the whole speed range so that their contribution to the total noise level becomes negligible. However, it can be seen from Fig. 4 that the interior noise does not exhibit a pronounced engine order spectrum through the whole range of speed. By the tracking notch filter controller, a considerable reduction of the total interior noise is only achieved when one of the prominent engine orders coincides with a resonance frequency of the interior space or of the car structure. A cancellation of engine- or gear-related noise components can therefore increase the passenger comfort in such cars only under certain driving conditions. An overall improvement of the noise situation will be obtained only by a cancellation also of the stochastic noise which requires broadband control.

3. BROADBAND NOISE

3.1. GENERAL REMARKS

Broadband active noise control in enclosures can be described in a system identification framework. An active noise canceller has to model certain transfer functions between the reference inputs and the error sensor locations where the noise attenuation should be realized. As the controller has to drive system components on-line, its output will be a time-domain signal and its characteristic can therefore be described conveniently by the impulse response related to the corresponding transfer function. Satisfactory cancellation is only possible if the reference signals are highly coherent with the according error signals. The achievable noise level reduction can be described in terms of the (multiple) coherence function. Another important aspect in three-dimensional active noise control is that of causality. Since both coherence and causality have been treated elsewhere in some detail they will be touched only briefly in this paper.

3.2. ALGORITHMS

Adaptive FIR Filters

Broadband active noise control with acceptable levels of reduction can reasonably be achieved only with digital filters. These filters have to be adaptive to cope with time-varying environmental conditions. Fig. 6 shows a simplified block diagram (without amplifiers, A/D and D/A converters etc.) of the set-up as it has been used in our experiments. $x[k]$ is the primary noise input signal, P the structural/acoustical primary path to the error sensor, and A the adaptive filter. The other symbols will be explained later in the text. We have used so far the well-known robust and stable LMS algorithm [3]. This is a stochastic gradient algorithm and adapts the coefficients of an FIR filter using the initial correlation between the reference signal x[k] and the error signal e[k]. After convergence, these signals are decorrelated [4]. Let

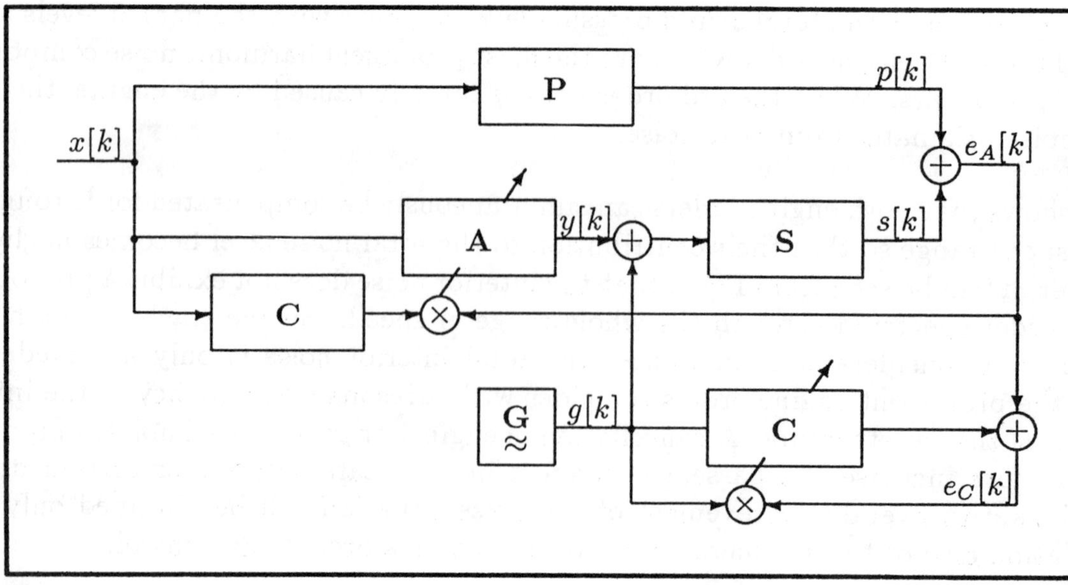

Fig. 6. Block diagram of the experimental set-up. $x[k]$: input signal, P: primary (acoustic) transmission path, $p[k]$: primary signal at the error sensor, A: adaptive cancellation filter, S: secondary transmission path, $s[k]$: cancellation signal, C (left-hand): prefilter, $e[k]$: error signals, G: auxiliary noise source, C (right-hand): adaptive filter, modeling S.

$$y[k] = \sum_{i=0}^{N} a_i \, x[k-i] \tag{17}$$

be the output of the adaptive filter which drives a secondary source to produce the compensating sound field at the error sensor location. To ensure convergence of the adaptive filter algorithm, the (electro-acoustic) transfer function S between the filter output and the error signal has to be accounted for by inserting a model C of S in the update branch of the adaptive filter (left-hand box C in Fig. 6). The algorithm reads now

$$s[k] = \sum_{j} c_j \, y[k-j] \tag{18}$$

which is the "filtered-x" algorithm [4] where the filter weights are updated according to

$$a_i[k+1] = a_i[k] - \mu \, e[k] \sum_{j} c_j \, x[k-i-j]. \tag{19}$$

Using this algorithm in a model experiment, we attained a reduction of the primary noise level up to 20 dB in the frequency range below 500 Hz. We have accomplished this by an adaptive FIR filter with 400 coefficients.

On-Line Adaptation

The path from the primary noise sources (engine, gear, road surface irregularities, fans, exhaust, wind, etc.) to the error sensor(s) and also the secondary path from the filter output to the error sensor(s) can be time dependent. For example, the temperature inside a car and therefore the sound velocity may change, or the coupling between the structural vibrations and the sound field (i. e., the radiation impedance) may vary depending on the actual ride conditions which causes fluctuations of the transfer function. From this reason, also the transfer function of the secondary path has to be adapted on-line to assure stability of the algorithm. At present, the only viable way to do this seems to be by application of an additional low-level white noise test signal which is incoherent with the primary noise and serves to adapt the weights of a filter modeling the secondary path. As a consequence of the incoherence, the additionally radiated signal will be the residual noise after convergence of the system and must therefore be kept at as low a level as possible in comparison with the primary noise. This technique has first been applied by Eriksson [5] to stabilize a one-dimensional noise canceling system operating under time-varying conditions. Our studies [6] have shown that also three-dimensional sound fields in small enclosures can be canceled in this manner if the transfer functions are varying slowly (as compared to the adaptation time). In Fig. 6, G is the auxiliary noise source, and the right-hand filter C is adapted by the LMS algorithm. Its coefficients are copied into the left-hand "prefilter" C.

Leaky LMS Algorithm

Another point of concern is the sound pressure level to be canceled which may also lead to complications enforcing a modification of the algorithm. The unconstrained algorithm tries to minimize the sound field at certain sensor locations. High noise levels—about 90 dB(B) and more if the primary noise activates low frequency resonances inside the passenger compartment (often referred to as the "boom")—may cause nonlinear distortion by overloading the loudspeakers. It is convenient to use the on-board radio loudspeakers in a car as secondary sources for an active noise control system, but their limited range of linear response has to be accounted for. An obvious solution to this problem is the introduction of output power constraints. For example, the weighted squared controller output can be added to the cost function which has to be minimized during adaptation [7]. A similar performance can be obtained by minimizing the filter weights. Defining a reference signal vector $\vec{x}_k$ and the coefficient vector $\vec{a}$ by

$$\vec{x}_k := (x[k], x[k-1], \ldots, x[k-N])^{\mathsf{T}}, \tag{20}$$

$$\vec{a} := (a_0, a_1, \ldots, a_N)^{\mathsf{T}}, \tag{21}$$

eq. (17) can be rewritten in vector notation as

$$y[k] - \vec{x}_k^{\mathsf{T}}\vec{a} = \vec{a}^{\mathsf{T}}\vec{x}_k\,. \tag{22}$$

With this notation, we can define a new cost function by

$$\xi_\lambda = \langle e^2[k]\rangle + \lambda\,\vec{a}^{\mathsf{T}}\vec{a} \tag{23}$$

($\langle\ldots\rangle$ indicates ensemble or time average, λ is a weighting factor) which will be minimized by the algorithm

$$a_i[k+1] = a_i[k] + \mu\,(e[k]\,x[k-i] - \lambda\,a_i[k]) \tag{24}$$

$$= \theta\,a_i[k] + \mu\,e[k]\,x[k-i]. \tag{25}$$

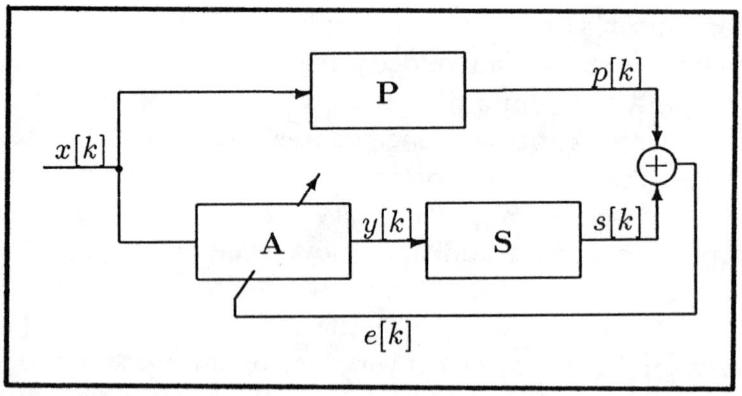

Fig. 7. Block diagram of a general system identification scheme with primary path P and secondary path S.

This algorithm is referred to as the "leaky LMS" algorithm which has been applied before to other problems [4] and tends to fade out the coefficients a_i when e vanishes. We have used this algorithm to limit the output power of loudspeakers in a car while canceling an extremely strong boom and thereby avoided nonlinear distortion. Of course, the level reduction is smaller than without the output power constraint, but the subjective impression is much better without nonlinear distortion.

Causality

The original LMS (and generally most adaptive filters) have been used for system identification purposes. In active noise control applications, the existence of control and error paths demands not only for modifications such as the filtered-x algorithm but leads to a further complication. In a general single-channel active control set-up with secondary path S as in Fig. 7, the adaptive filter A should realize the optimal transfer function

$$A^*(z) = -\frac{P(z)}{S(z)} \tag{26}$$

in the z-domain. So it has to invert the mainly acoustic transfer function $S(z)$ between the filter output $y[k]$ and the error signal $e[k]$. It is, however, impossible to invert the inherent delay due to the anti-aliasing filters and the sound propagation through S if the primary path P does not contain a delay of at least equal length. This is the overall limiting *causality constraint* in broadband feedforward control systems. But moreover, the secondary path S often contains non-minimum phase zeros. It can be shown [4] that even in this case a good approximation can be found to a *delayed* inverse model of this transfer function if the delay is long enough. We have verified this experimentally in a test chamber of dimensions similar to those of a small car [8].

Observability and Controllability

If the transfer function of the secondary path exhibits an approximate zero at the specific location of the error microphone, this corresponds to a zero near the unit circle which is problematic to invert. This is the problem of observability of the secondary sound transmission path. Even if a stable adaptive filter can be realized under these circumstances, it is not possible to get a significant noise cancellation at this frequency. An experimental result from our test chamber measurements is plotted in Fig. 8. The upper solid line shows the primary noise spectrum as monitored by the error microphone. The lower solid line indicates the spectrum of the white noise auxiliary signal, again as measured by the error microphone. It is equal to the frequency response of the secondary path, S, and exhibits a zero at 170 Hz. After convergence of the adaptive noise canceler, this zero has caused a peak in the residual noise spectrum (dotted line) which is elsewhere very close to the (uncontrollable) auxiliary signal spectrum. Obviously, the primary noise is only controllable if the secondary transmission path is observable. This problem can be overcome by using multiple error microphones or multiple secondary sources without common zeros of their transfer functions in the z plane [9].

Multichannel Systems and Coherence

Beside controllability problems, multichannel control will also be necessary for an effective broadband noise cancellation in cars since a) reference inputs have to be taken from all relevant primary sources, b) error microphones have to be placed near all possible passenger head positions, and c) several loudspeakers are required because simultaneous cancellation at different sites is impossible with a single secondary source. Some problems of multichannel broadband control have been addressed in [11]. Let x_r, $r = 1, \ldots, m$ be the reference input signals and y_s a specific error output. In the case of

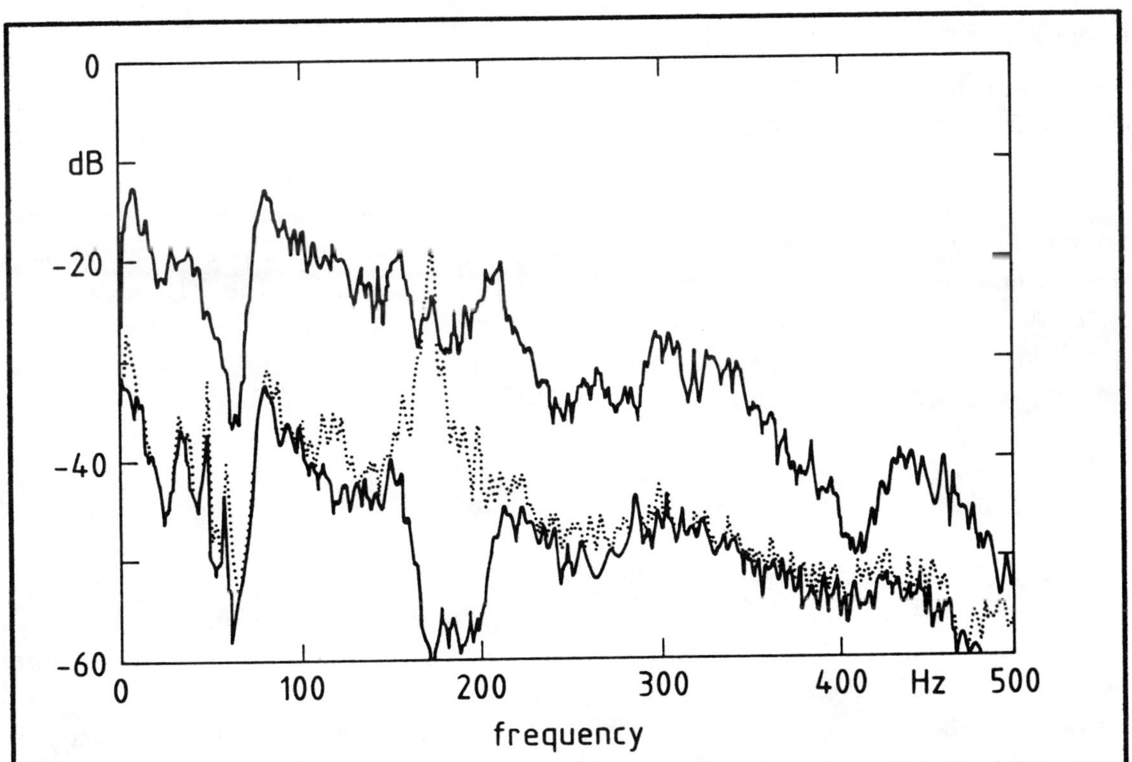

Fig. 8. Measured noise spectra in a three-dimensional test chamber. Upper solid line: primary noise; lower solid line: frequency response of the secondary path; dotted line: residual noise spectrum after convergence of an adaptive noise canceler with on-line error path adaptation.

stochastically independent primary sources, the multiple magnitude squared coherence function between y_s and the totality of the x_r is given by

$$\gamma^2_{y_s:x} = \sum_{r=1}^{m} \gamma^2_{x_r y_s}(\omega) \tag{27}$$

where $\gamma^2_{x_r y_s}(\omega)$ is the ordinary magnitude squared coherence function between an input x_r and the error signal y_s. Herewith, the maximum possible noise level reduction by an adaptive feedforward system is given by [12, 13]

$$\Delta L = 10 \lg \left(1 - \gamma^2_{y_s:x}\right)^{-1} \text{ dB}. \tag{28}$$

If the input signals of an adaptive multichannel controller are not independent of each other (which is likely in a car due to propagation of structure-borne sound through the car body), the analysis must be modified by introducing the concept of partial coherence [10]. Instead of eq. (27), the multiple coherence function reads in this case

$$\gamma^2_{y_s:x} = 1 - \left[(1 - \gamma^2_{x_1 y_s})(1 - \gamma^2_{x_2 y_s \cdot 1}) \cdots (1 - \gamma^2_{x_m y_s \cdot 12 \ldots (m-1)})\right] \tag{29}$$

where the γ^2 on the right-hand side are *partial coherence functions* which can be derived from the measurable signals x_r and y_s. The development and implementation of multi-input–multi-output (MIMO) algorithms requires a thorough understanding of the underlying relationships.

Recursive Algorithms

It is an interesting question if a certain reduction of a primary noise level can be obtained with fewer filter weights using an IIR filter instead of an FIR filter. IIR filters are more complex and tend to get unstable unless handled carefully. For a specific problem, it is therefore appropriate to verify first if better results can be obtained by using an ARMA model approach. From initial experiments with our test chamber we know the optimal impulse response of the controller. We have chosen (in the FIR case) a very high filter order so that residual errors from AR type components of our system should be negligible. Given this impulse response $h[i]$, we have tried to fit an ARMA model to it. Denoting the controller output as

$$y[k] = \sum_{i=0}^{N} a_i\, x[k-i] + \sum_{j=1}^{M} b_j\, y[k-j] \tag{30}$$

where N and M are the transversal and recursive order of our model approach, the approximation error can be defined as the sum of squares of the deviation from the desired impulse response. Thus we find an optimal approximation in the Least Squares (LS) sense. The results are shown in Fig. 9. It can be seen that an MA model (recursive order $M = 0$, realizable by an FIR filter) slowly reduces the error with increasing order whereas an AR model (transversal order $N = 0$) does not fit to our problem. This is mainly due to the inherent time delay in the transfer function which cannot be modeled correctly with AR type filters. Only the combined ARMA model is able to reduce the approximation error significantly. These results show that it is possible to reduce the size of the controller if a stable realisation of an adaptive IIR filter can be found.

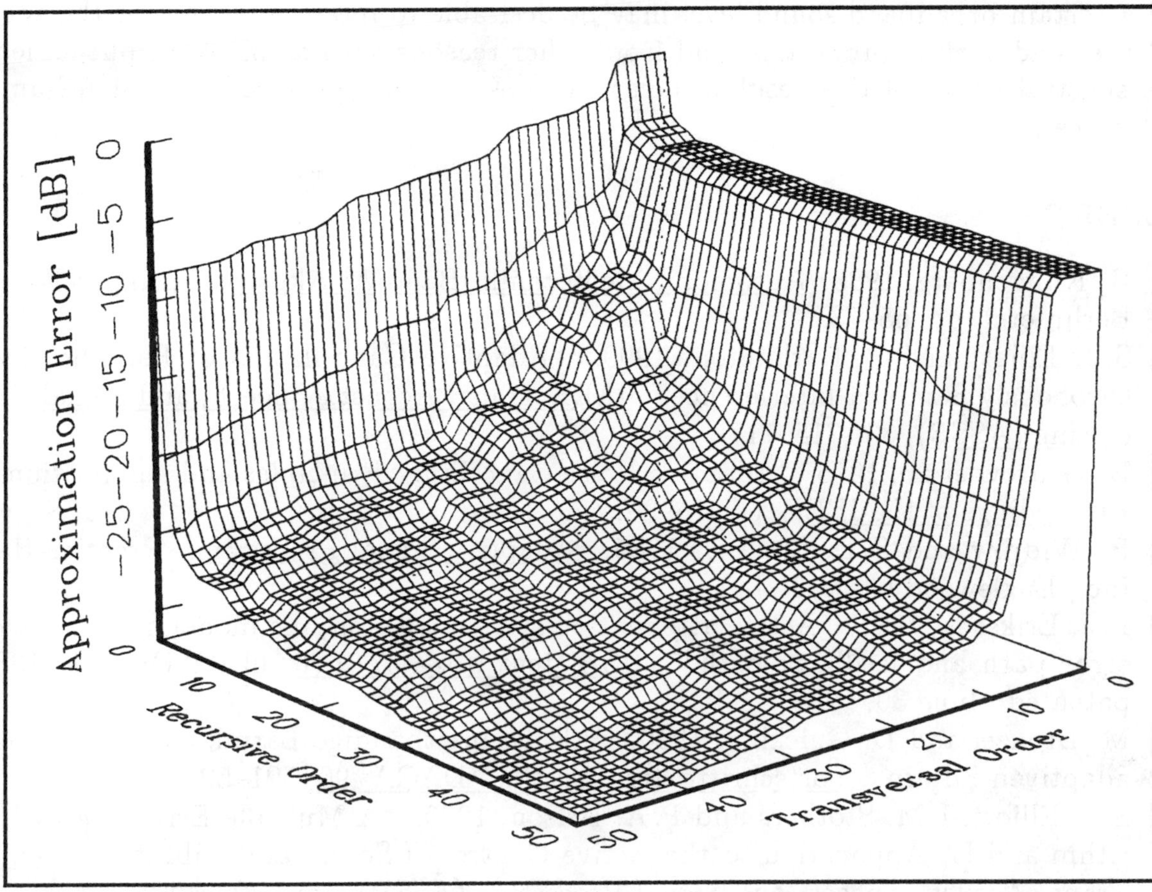

Fig. 9. Residual approximation error as a function of the recursive order (M) and the transversal order (N) of an ARMA filter.

4. OUTLOOK

The active cancellation of repetitive noise inside cars by single frequency tracking controllers has reached a state that makes an installation in series production cars likely within foreseeable future. Broadband noise cancellation by multiple adaptive feedforward controllers requires more research and development work but can be expected to become technically available within the next few years. It can be assumed that a multidimensional coherence analysis in a car including the measurement of a set of transfer functions will help design a viable control concept, concerning not only optimal locations of the reference sensors, control loudspeakers and error sensors, but also the choice of the best-suited multichannel adaptive controller algorithm where the application of recursive filters will possibly reduce the required signal processing power. The on-line adaptation of the secondary acoustic tranfer functions appears to present no major difficulties. Acoustic feedback cancellation will not be necessary if the reference signals are taken from outside the passenger compartment.

In a final application it will be advisable to combine tracking control filters for repetitive, and feedforward control for broadband noise. It will probably also become necessary to include aspects of psychological acoustics concerning the character of the residual spectrum since, e. g., engine-related sound yields information about proper performance, and a certain broadband sound level may be desirable to retain a feeling for the speed and the road surface properties, and from other reasons such as masking phenomena. The spectral shape of this residual noise can possibly be optimized to match human preference.

REFERENCES

[1] H. Klingenberg, 1988. Automobil-Meßtechnik. Band A: Akustik, Springer-Verlag, Berlin etc., pp. 65–73.

[2] S. J. Elliott and P. Darlington, 1990. "Adaptive Cancellation of Periodic, Synchronously Sampled Interference." IEEE Trans. Acoust., Speech, and Signal Processing (ASSP), 33: 715–717.

[3] B. Widrow et al., 1975. "Adaptive Noise Cancelling: Principles and Applications." Proceedings of the IEEE, 63: 1692–1716.

[4] B. Widrow and S. D. Stearns, 1985. Adaptive Signal Processing, Prentice-Hall, Inc., Englewood Cliffs, N.J.

[5] L. J. Eriksson, 1987. "Active attenuation system with on-line modeling of speaker, error path and feedback path." US Patent No. 4,677,676, filed: Feb. 11, 1986, patented: June 30, 1987.

[6] M. Bronzel and D. Guicking, 1990. "Aktive breitbandige Lärmkompensation mit adaptiven Filtern." Fortschritte der Akustik – DAGA '90, 521–524.

[7] S. J. Elliott, I. M. Stothers and P. A. Nelson, 1990. "A Multiple Error LMS Algorithm and Its Application to the Active Control of Sound and Vibration." IEEE Trans. Acoust., Speech, and Signal Processing (ASSP), 35: 1423–1434

[8] M. Bronzel, 1989. "Untersuchungen zur dreidimensionalen lokalen Schallfeldkompensation in kleinen Räumen." Diplomarbeit (unpublished Master's Thesis), Göttingen, F. R. Germany 1989.

[9] M. Miyoshi and Y. Kaneda, 1988. "Inverse Filtering of Room Acoustics." IEEE Trans.Acoust., Speech, and Signal Processing (ASSP), 36: 145-152.

[10] J. S. Bendat and A. G. Piersol, 1980. Engineering Applications of Correlation and Spectral Analysis, J. Wiley & Sons, New York etc. (particularly Chapters 8 and 10).

[11] D. Guicking and M. Bronzel, 1990. "Multichannel Broadband Active Noise Control in Small Enclosures." Internoise 90, Göteborg, Sweden, Aug. 1990. Proceedings: pp. 1255–1259.

[12] C. F. Ross, 1980. "Active Control of Sound." Ph. D. Thesis, Queen's College, Cambridge, UK.

[13] P. A. Nelson, T. J. Sutton and S. J. Elliott, 1990. "Performance Limits for the Active Control of Random Sound Fields from Multiple Primary Sources." Proceedings of the Institute of Acoustics (IOA), 12, Part 1: 677-687.

THE CONTROL OF NOISE INSIDE PASSENGER VEHICLES

Colin F. Ross[1]

ABSTRACT

This paper summarises the developments of active control systems for the control of sound in passenger vehicles, describing their use in aircraft, helicopters, cars and tanks. The major design issue for these systems is the ability to achieve a good spatial match to the uncontrolled sound. The types of algorithm which have been used for the control are discussed and compared. The paper concludes that the performance of large-scale global control systems may be limited by the algorithms available and that smaller, seat-based systems are likely to be more robust.

1. INTRODUCTION

Active control of sound inside passenger vehicles is becoming increasingly researched and shows the potential for commercial exploitation. The control of engine order sound has been demonstrated inside car cabins [3,5]. Here the repetitive component of the car cabin noise, which is due to the engine rotation, is controlled using loudspeakers positioned under the seats and on the rear parcel shelf. The aim of these systems is to reduce the annoyance of the 'booms' created by the excitation of resonances by the engine tones. The booms are particularly annoying as the car accelerates and the sound level can rise and fall by up to 20dB.

The control of propeller noise in aircraft has also been the focus of attention for a number of years and has resulted in demonstrations of the feasibility of such cabin control systems in flight [1,2]. The commercial impetus here has been the increasing demand for short haul aircraft, particularly in Europe, with the expectation that the comfort will be comparable to that experienced in transatlantic jets. Reduction of the tonal components of the propeller noise is becoming increasingly sought. The most effective passive measures for controlling these tones are tuned dampers which are heavy.

[1]Noise Cancellation Technologies Ltd., St. John's Innovation Centre, Cowley Road, Cambridge CB4 4WS, England.

Military vehicles, both tanks and jet aircraft, have very high levels of sound which make command communication difficult or inaccurate. This has led to the requirement for active systems for the passengers. However, since the majority of the passengers already carry headsets or at least helmets the active control systems have been incorporated in the headset [11]. Headset systems are also being sold commercially but their take-up is limited to passengers who already wear a headset and require better noise attenuation. All of the systems reported so far provide broadband noise control.

The noise inside the passenger compartment of the helicopter is well known. Increasingly as the military market is declining the major helicopter manufacturers are aiming to make their product more acceptable to the commercial traveller with a view to expanding the commercial market. This requires a significant investment in measures to attenuate the noise. Studies have shown the effectiveness of active systems for the control of the rotor and gear-box meshing frequencies [4].

The types of active control system that have been used in these passenger compartments can be split into three broad categories. Global control systems, SilentSeatsR, and headset based systems. The global systems require many loudspeakers to create the anti-sound, and many microphones positioned close to the passengers' ears to monitor the residual. These systems need to be integrated into the cabin of the vehicle and are thus expected to be sold with the vehicle. The seat-based systems, on the other hand, is much more easily installed after the vehicle has been built. It would typically require only limited modifications to the cabin fittings. The headset systems are the ultimate in portability however the discomfort and 'Spock' effect make them less attractive.

In the design of any active sound control system there are two major issues to be addressed. The temporal and spatial matching of the sound field. The design of any system is much simplified if the two issues can be tackled separately. This allows the positioning of the anti-sound sources and the residual sensors to be considered separately from the design of the input sensor system. This separation can be achieved when causality is not a constraint on the performance of the system. In that case the temporal design becomes one of finding a coherent signal and the spatial problem becomes one of positioning the anti-sources best to match the original sound field. In the systems which have been demonstrated to date causality has not been an issue and the design process can be split. Furthermore, the temporal limit of coherence has not been active when the input signal has been taken from a shaft tachometer. Consequently, the major issues of debate in the control of cabin noise centre round the ability of the system to provide sufficient spatial matching.

This paper discusses the impact of the spatial matching issues on the control systems dynamic performance.

2. CONTROL ALGORITHMS FOR CABIN NOISE CONTROL

Here we consider the common algorithms that have been used for the control of passenger cabin noise. However, firstly, we will describe the various configurations which have been used for the heart of these control systems, the feedforward element, and we will show that the configurations are all basically the same.

THE FEEDFORWARD PATH

There are four types, which come from the combination of two distinguishing features, the use of a synchronous time-base, as opposed to a uniform time-base, for the digitally sampled processing of the feedforward filter, and the use of amplitude information at the input to the feedforward filter. Almost all the systems that have been used do not have feedback from the actuators creating the anti-sound to the feedforward input. In the few cases where this does exist the optimal controller design strategy is to subtract the signal due to feedback from the input.

Synchronous Sampling

Many of the systems which are used to control the sound from rotating machines use a time-base for the feedforward part of the control system which is synchronised to the period of rotation of the machine. This can have advantages as the number of coefficients required for the feedforward filter is then directly proportional to the number of harmonics being controlled and the filter length need only be one machine cycle long. If the number of coefficients is chosen to be a power of two then transformation to the frequency domain for processing is efficient using the FFT algorithm. However, since the adaption of the coefficients depends upon the transfer function of the physical system of the actuator-to-residual-sensor path the periodic, sampled-data representation of the transfer function will change with sampling frequency. This changing of the transfer function of the sampled-data transfer function can require extra computation when the speeds are changing rapidly and may make the uniformly sampled systems favourable here. However, apart from the minor differences in computational power, it is interesting to note that the uniform time-base systems developed by Nelson & Elliott [2] and the synchronously sampled systems used by Eatwell & Ross [1] give similar performances.

Amplitude Input to the Feedforward Filter

The systems used by Chaplin [6] for the control of machinery noise have the feedforward filter described as a waveform generator. This is because the only input to the feedforward part is timing information from the machine. This timing information is used to stimulate the feedforward filter as if the input were a single, unit-amplitude pulse once per machine cycle. The advantage of using this type of input is that it is not contaminated with noise unrelated to the machine noise.

Alternative systems have been proposed by Nelson & Elliott [7] which sample a harmonic signal, derived from a tachometer by band-pass filtering. This sampled harmonic signal is used as an input to the feedforward filter. However, since its amplitude is not related to the sound source strength it is not providing any extra information. Consequently, there are only minor differences between the processing of these systems and those of Chaplin.

Systems have been described by Salikuddin [8] which are genuinely responsive to the source strength and thus whose feedforward filter characteristic models the system transfer characteristics. Further publication on this work has been difficult to track down and it is likely that systems of this type which can be disturbed by extraneous input noise are not robust enough for development for the control of cabin noise.

The adaption of the coefficients of the feedforward filter is necessary in all of the systems described above since the input is not changing whilst the output is required to match the sound to be controlled.

ADAPTION OF THE FEEDFORWARD FILTER

The processing for this adaption can be achieved in the time, or frequency, domain and the choice depends upon the details of the hardware that is available, the number of harmonics requiring control, and other minor details. The following discussion assumes that the processing is being undertaken in the frequency domain, but the choice is made here purely because the algebra is clearer.

All systems have multiple outputs (N actuators) and use multiple monitors (M sensors) to monitor the residual field. The single input (in the case of a car engine), or pair of inputs in the case of a twin-propeller aircraft, drive the input of one feedforward filter for each of the outputs. Here we describe the adaption of the coefficients of all of the feedforward paths for one frequency (a harmonic of the source). For this frequency the DFT of the coefficients of one filter is represented by w. All of the complex numbers, w, representing all of the feedforward paths form a single N-long vector, W. The N values in this vector of feedforward filter coefficients are changed as the filters are adapted, and a subscript k represents the iteration number. The uncontrolled sound monitored by the residual sensors is similarly written as a M-long vector, Y, for the same harmonic frequency. The residual sound is formed into a M-long vector, ϵ.

Almost all of the systems which have been described in the published literature have algorithms for the adaption of the vector W which follow the general form,

$$W_k = \alpha W_{k-1} - \mu \mathbf{B} \langle \epsilon \rangle_{k-1}, \qquad (1)$$

where α is a leak parameter which controls the amount of influence the past value of the coefficient has on the new value. μ is a small parameter which controls the amount of new information which is used in the adaption of W. **B** is a matrix which transforms the residual error, ϵ, and $\langle \cdot \rangle$ is the expectation operator.

The elements of equation 1 which vary for the different reported algorithms are α, μ, **B**, and the length of time take to measure the expectation $\langle \epsilon \rangle_{k-1}$.

There are two major groups of algorithms, the gradient descent method and the direct method. Each one of these algorithm groups is split into a range of different algorithms which find favour with the various exponents of active noise control.

1. The Gradient Descent

Here B is chosen to be an estimate of the complex conjugate transpose of the transfer function matrix between the actuators and residual sensors, **A**. $\mathbf{B} = \langle \mathbf{A} \rangle^*$. This ensures that the changes to the vector W are along a projection of the negative gradient of the surface of $\epsilon'\epsilon$. α is chosen to provide some form of constraint on the amplitude of the actuator drive signals in addition to the minimisation of the residual sensor signals. The value of α being chosen with reference to the particular system configuration. The value of μ is chosen to maximise the response of the system to changes in the sound to be cancelled within the limits of stability and so that noise on the estimate of the residual does on perturb the weights too much. Its value is scaled on the singular values of the matrix **A**. The Noisy Gradient Descent algorithm uses the current value of ϵ without averaging, whereas the normal one provides some smoothing.

2. The Direct Method

In this algorithm **B** is chosen to be the inverse (or least squares inverse) of the estimate of **A**. $\mathbf{B} = (\langle \mathbf{A} \rangle^* \langle \mathbf{A} \rangle)^{-1} \langle \mathbf{A} \rangle^*$. This ensures that the changes in W are directly towards the optimum value for W. α may be set to be less than 1.0 in order to provide a constraint on the actuator drive amplitudes. μ is chosen with the same objectives as described above, but it is independent of the matrix **A**. The direct method has the potential for adapting the coefficients of the feedforward filters much more quickly to the optimum solution. However, there are limitations on its performance which form the core of the following discussion.

3. A COMPARISON OF THE DIFFERENT ALGORITHMS

The aim of all of the algorithms is to minimise the residual error $\epsilon'\epsilon$. It is thus appropriate to look at the difference equation for the residual, ϵ, which the different algorithms produce. In the general form the residual sensor signals are given by the following equation,

$$\epsilon_k = \mathbf{A}W_k + Y_k. \tag{2}$$

Substituting the update algorithm (from equation 1) gives,

$$\epsilon_k = \alpha\mathbf{A}W_{k-1} - \mu\mathbf{AB}\epsilon_{k-1} + Y_k. \tag{3}$$

The form of equation 2 for the previous iteration is then used to give,

$$\epsilon_k = \alpha(\epsilon_{k-1} - Y_{k-1}) - \mu\mathbf{AB}\epsilon_{k-1} + Y_k. \tag{4}$$

The difference equation is thus:

$$\epsilon_k = \underbrace{(\alpha - \mu\mathbf{AB})\epsilon_k}_{\text{1st term}} + \underbrace{Y_k - \alpha Y_{k-1}}_{\text{2nd term}}. \tag{5}$$

The first term in the difference equation determines the rate of convergence of the algorithm to the optimum solution. The second term shows how variation in the sound to be controlled perturbs the residual.

All of the algorithms require some estimate of **A**, and for the purposes of the following discussion we will assume that the estimate of **A**, <**A**>, is accurate in its estimate of the singular vectors of **A**, but inaccurate in the size of the singular values. Thus

$$\mathbf{A} = \mathbf{UDV}^*, \text{ and}$$

$$\langle\mathbf{A}\rangle = \mathbf{U}\langle\mathbf{D}\rangle\mathbf{V}^*,$$

where **A** and **U** are M by N matrices, and **D** and **V** are N by N matrices. The columns of **U** are orthogonal vectors, **D** is a diagonal matrix and **V** is unitary.

We can form an M by M matrix **X** which is **U** augmented with M-N columns to make it unitary, and a M by M diagonal matrix **S** which is the matrix **D** with additional zeros on the diagonal.

1. The Gradient Descent Method

Substituting these forms into equation 5 for the first algorithm we have

$$\epsilon_k = \mathbf{X}(\alpha\mathbf{I}-\mu\mathbf{S}<\mathbf{S}>^*)\mathbf{X}^*\epsilon_{k-1} + Y_k - \alpha Y_{k-1}. \quad (6)$$

Normally μ is chosen to be close to the reciprocal of the square of the largest singular value of **A**.

If we first consider the case when α is 1 and the sound to be controlled is invariant then equation 6 reduces to

$$\epsilon_k = \mathbf{X}(\mathbf{I}-\mu\mathbf{S}<\mathbf{S}>^*)\mathbf{X}^*\epsilon_{k-1} \quad (7)$$

Writing β for $\mathbf{X}^*\epsilon$,

$$\beta_k = (\mathbf{I}-\mu\mathbf{S}<\mathbf{S}>^*)\beta_{k-1} \quad (8)$$

β represents a vector of spatial coefficients of the residual sensor signals whose convergence is not interdependent. The element of β which corresponds to the largest singular value of **A** will converge in one iteration as the corresponding part of equation 8 in brackets is close to zero. On the other hand the elements of β which correspond to the augmented zeros will not change. This represents the part of the sound to be controlled which is not controllable with the set of actuators. The elements of β which correspond to the smaller singular values of **A** will have a slow convergence. It can be thus seen that with the gradient descent algorithm the convergence of the elements of β vary quite significantly. This shows that part of the cancelling noise will converge rapidly but other parts will converge slowly. The rapidly convergent parts will be those parts of $\mathbf{X}^*Y$ which correspond to the largest singular values of **A** and can be described as the controllable part of the sound field. The slowly convergent parts will be those parts of $\mathbf{X}^*Y$ which correspond to the smallest singular values of **A** and can be described as the poorly controllable parts of the sound field. The overall convergence of the control system will depend upon the relative sizes of the elements of $\mathbf{X}^*Y$. Unless the system of actuators is very well matched to the sound to be controlled, so that the element of $\mathbf{X}^*Y$ corresponding to the largest singular value is dominant, then the convergence can be quite poor. The only way to ensure that the actuators are well matched to the sound is to superpose the sources and anti-sources. This is, generally, not possible in cabin noise control. It has been found [1] that the positioning of the actuators is often limited, by practical constraints, to positions which are far from ideal in matching the sound. The importance of this factor in the control of cabin noise can be quite significant since systems with a large number of actuators and sensors are often described and this inevitably leads to systems with a large spread of singular values.

So far we have discussed the convergence of the algorithm. The performance in attenuating the original sound is also of importance. We can deduce the performance of the algorithm from equation 6 where we see that the reduction in the size of ϵ will be perturbed by the variation in the original sound field ($\delta Y = Y_k - Y_{k-1}$). This variation in Y will be reflected directly in the residual for the components corresponding to the largest singular values. The perturbations will be accumulated for the components corresponding to the augmented zeros, this signifying no control. The perturbations corresponding to the smaller singular values will find the accumulation outstripping the convergence and so will be poorly attenuated. Since we can expect that the typical variation, δY, will be in a similar 'direction' to the uncontrolled sound, Y, (and thus have a similar weighting of spatial components) the performance of the system can be judged from the relative weighting of the spatial components in the original field.

The conclusion is that the gradient descent algorithm is quite likely to be not very useful for the control of sound in cabins unless the sound field and the actuators and be very well matched.

2. The Direct Method

The direct method of adaption is different and thus has different characteristics. In the notation used in the previous discussion the algorithm is

$$\epsilon_k = \mathbf{X}(\alpha \mathbf{I} - \mu \mathbf{S}/\langle \mathbf{R} \rangle^*)\mathbf{X}^* \epsilon_{k-1} + Y_k - \alpha Y_{k-1}, \qquad (9)$$

where **R** is the same as **S** except that the augmented diagonal elements are non zero. The value of μ would normally be chosen to be 1.0 and thus the convergence of all elements of the spatial decomposition of the original sound field would be equivalent: they would all get there in one step. The performance, measured as overall attenuation of the original sound field, would be close to optimal. The only residual would be the unpredictable change in the original field.

This method seems to have great advantages over the gradient descent method, however, performance as just described is only possible if the estimate of the singular values of **A**, $\langle \mathbf{R} \rangle$, are accurate. In practice, it is not possible to achieve this accuracy, and there is severe consequence of using this algorithm with an inaccurate estimate of the singular values of **A**: instability. Whilst it is likely that the larger singular values of **A** will be relatively invariant as they correspond to the gross cabin acoustics, the smaller values will be significantly perturbed by the acoustical variations caused by passenger movement. If the ratio $\mathbf{S}/\langle \mathbf{R} \rangle^*$ exceeds 2 then the update factor $(\mathbf{I} - \mu \mathbf{S}/\langle \mathbf{R} \rangle^*)$ will cause the adaption to diverge for a value of μ=1.0. If the value of μ is reduced to give the algorithm more stability margin then the performance will be correspondingly reduced. The extent of the reduction in performance caused by the inaccurate estimate of **A** is dependent upon the details of the directions of the singular vectors and the original sound, the singular values, and the level of variation. It is, therefore, only possible to summarise the practical experience which suggests that for large systems (ie multiple sensors and multiple actuators) the spread of sizes of singular values can be very large and the sensitivity to inaccurate estimates of **A** can be significant for this algorithm.

The overall conclusion for cabin control systems is that both of the algorithms as described above will have a limited ability to respond to rapid changes in the original sound field and will have poor overall attenuation when the spread of singular values is large and when the original sound direction does not correspond to the well controlled direction. It may be possible to overcome some of these problems by very careful positioning of the anti-sound sources but this is not likely to be possible in a commercial environment. The typical variability of the internal sound filed of two cars following on another on the production line is around 10dB in overall level, and so it can be expected that the details of the spatial distribution will be very different from one car to the next. It would not be practical to have the anti-sound sources tailored to the particular car. There may also be computationally intensive methods of reducing the sensitivity of the systems to variability in **A** but none have been reported to be successful and the computational cost may be prohibitive.

The best way of overcoming these shortfalls seems to be to use a local seat-based system which has two anti-sources and two monitoring sensors. This ensures that the scale of the system is small (there are only two singular values) and the spread of the singular values is small. Whilst the variation in the transfer functions caused by passenger head movement is not insignificant the resultant perturbation of the singular vectors and singular values is substantially smaller than for the larger scale systems. This SilentSeat^R configuration has been developed [9], and has been shown to give effective levels of attenuation in both small cabins and many adjacent seats in larger cabins[2,10].

5. THE SILENTSEAT^R

The typical configuration of the SilentSeat^R has two loudspeakers positioned in the headrest 20cm apart and two microphones 7cm in front of the loudspeakers. Figures one and two of reference [10] in the proceedings show the details of the configurations. The matrix **A** for this configuration is

$$\begin{bmatrix} \frac{e^{ikr_1}}{r_1} & \frac{e^{ikr_2}}{r_2} \\ \frac{e^{ikr_2}}{r_2} & \frac{e^{ikr_1}}{r_1} \end{bmatrix}$$

This matrix has only two singular values and the difference in the values is not large. For a frequency of 88Hz the ratio of larger to smaller is 1.93:1 and for the third harmonic frequency of 264Hz the ratio is 1.64:1. Results for the dynamic performance of the SilentSeat^R in flight will be published shortly.

6. CONCLUSIONS

It should be clear that the most robust system that is available for the control of passenger cabin noise is the seat-based system. The control system adaption can be made significantly more robust to changes in the physical hardware. The configuration of the seat-based system lends itself far more easily to installation in a wide range of passenger transportation systems. We are likely to see this form of cabin control system in commercial service more rapidly than any other system.

7. ACKNOWLEDGEMENTS

SilentSeat^R is a registered trade mark of Noise Cancellation Technologies Inc.

References

1. Eatwell, G. P. *et al.* 1989 'A demonstration of Active Noise Reduction in an Aircraft Cabin.' Journal of Sound and Vibration, **128**(2), pp. 358-60.

2. Elliott, S.J. *et al.* 1990 'In-flight Experiments on the Active Control of Propeller Induced Cabin Noise'. Journal of Sound and Vibration, **1**39(4).

3. Ross, C.F. 1989 'Active Cancellation of Cabin Noise'. 20th International Symposium on Automotive Technology & Automation, Florence, Italy, May 1989.

4. Eatwell, G. P. et al. 'Active Control of Helicopter Cabin Noise'. 14th European Rotorcraft Forum. Milano, Italy, 1988.

5. Ross, C. F. 'Active Methods for Automobile Refinement'. Future Automotive Microelectronics, Torino, Italy Dec. 1988. Associazione Technica Dell'Automobile.

6. Chaplin, G.B.B. UK patent No. 1,577,322.

7. Nelson, P.A. & Elloitt, S.J. International patent application No. PCT/GB87/00706

8. Salikuddin 1985. US Patent No. 4,689,821.

9. Ziegler, E. 1990. 'Silent seat' US patent No. 4,977,600.

10. Eatwell, G.P. 1991 'The use of the SilentSeatR in Aircraft Cabins'. Recent Advances in Active Control of Sound and Vibration. Virginia Tech, USA, April 1991.

11. Wheeler, P.D. 1981 'An active noise reduction system for aircrew helmets'. NATO AGARD CP-311 Soesterberg, March 1981.

USE OF NONLINEAR CONTROLLERS IN THE ACTIVE ATTENUATION OF ROAD NOISE INSIDE CARS

T. J. Sutton,[1] S. J. Elliott,[1] I. Moore[2]

INTRODUCTION

Work reported previously [1] has demonstrated that active control of engine noise inside cars is a practical proposition over a wide range of engine speeds. The engine rotation gives rise to a periodic, deterministic sound inside the car at the engine firing frequency and its harmonics. This unwanted sound can be attenuated by a linear feedforward control system in which a reference signal at the engine firing frequency is taken directly from the ignition circuit. In practical systems the reference signal is passed through an array of adaptive filters to drive up to six loudspeakers inside the car. The coefficients of the filters are adapted to minimise the sum of the squared outputs from up to eight microphones. Reductions of 10-15 dB or more have been reported for systems of this type. Such a system can be designed to minimise the sum of the squared pressures at not only the fundamental firing frequency but also the second harmonic.

Other more recent research [2,3] has indicated that broadband, random noise such as road noise can also be attenuated in vehicle interiors using feedforward control. Simulation studies indicate that reductions of around 5 dB could be achieved by a system in which the reference signals are taken from accelerometers attached to the vehicle structure. The work showed that the available noise reduction depends strongly on the coherence between the reference signals and the interior noise to be attenuated.

This paper presents experimental measurements of a real-time practical implementation of such a broadband feedforward control system in which the low-frequency road noise is significantly reduced in a vehicle. The paper also sets out the limitations imposed by a linear control strategy and considers a possible nonlinear approach to road noise control.

LINEAR CONTROL OF ROAD NOISE

PRINCIPLE

Figure 1 shows the basic block diagram of a linear multiple-channel feedforward system for control of random sound. A set of K reference signals $\mathbf{x}(n)$ are passed through an array of FIR control filters $\mathbf{h}$ generating a further set of M signals $\mathbf{y}(n)$, each driving a loudspeaker inside the car. The reference signals are chosen to represent the sources of noise as closely as possible. They may for example be from accelerometers attached to the vehicle structure. The electroacoustical paths from the speakers through the car interior to the (L) microphones are represented in the diagram as an array of filters $\mathbf{c}$. The primary sound sources (road and tyre noise, wind noise *etc.* in a car) give rise to signals $\mathbf{d}(n)$ at the microphones. The signals $\mathbf{z}(n)$ at the microphones are due to the action of the control system. The control filters $\mathbf{h}$ are adapted continually to minimise the sum of the mean squared pressures $E\{\mathbf{e}^T\mathbf{e}\}$ at the microphones.

[1]Institute of Sound and Vibration Research, University of Southampton, Southampton SO9 5NH, England.
[2]Lotus Engineering, Hethel, Norwich NR14 8EZ, England.

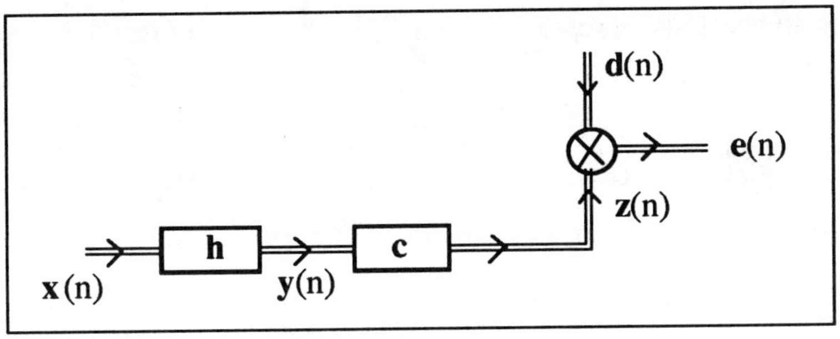

FIGURE 1. ELEMENTS OF MULTICHANNEL FEEDFORWARD CONTROLLER

FILTER ADAPTION

The filtered-x LMS algorithm [4] is used to adapt the control filters. In this algorithm the basic gradient-descent method is modified to take account of the transfer functions of the filters **c** interposed between the control filters **h** and measured errors **e**. Further modification to the algorithm is required to handle the multiple-error case, and this is set out in [5]. To implement the Multiple-Error filtered-x LMS algorithm it is first necessary to obtain an estimate of the vehicle interior acoustic transfer functions, filters **c**. In practice this can be done by applying a white noise to each loudspeaker in turn and measuring the resulting transfer functions to the microphones. Each transfer function is represented by an FIR filter with coefficients c_{lmj} (mth loudspeaker, l-th microphone, jth coefficient) whose response is an estimate of **c**. During on-line noise control, the reference signals **x**(n) are passed through the appropriate individual filters modelling the error path to create a set of filtered reference signals r_{lmk}

$$r_{lmk}(n) = \sum_{j=0}^{J-1} c_{lmj} x_k(n-j) \tag{1}$$

The multiple-error filtered-x LMS algorithm for updating the ith coefficient of the FIR control filter from the k-th reference signal to the m-th loudspeaker is:

$$h_{mki}(n+1) = h_{mki}(n) - \mu \sum_{l=1}^{L} e_l(n) r_{lmk}(n-i) \tag{2}$$

PERFORMANCE LIMITS

Coherence of reference signals

It is clear that the chosen reference signals must be reasonably well-correlated with the interior noise to be cancelled, since the cancellation sound field is derived from the reference signals by a set of linear filters. This requirement can be expressed exactly at a single microphone location in terms of the multiple coherence between the reference signals and the interior noise signal at one of the microphones inside the car. The maximum available sound reduction (in dB) can be expressed at each frequency by [6]:

$$\Delta(\omega) = -10 \log_{10}[1 - \gamma^2_{x:d}(\omega)] \tag{3}$$

where $\gamma^2_{x:d}(\omega)$ is the multiple coherence function [7] between the reference signals **x** and the primary sound signal at one of the microphones. Thus if $\gamma^2_{x:d} = 0.9$ (90% coherence) the maximum available reduction is 10 dB. This maximum reduction is defined by frequency domain considerations, so takes no account of the constraint of causality; the control filter in practice must be both causal and of finite length. For realisable linear time-invariant systems the noise reduction may be less than Δ, but it will not be more.

Time delays

A second and unavoidable limitation of a road noise control system is the time taken to process the signals. The reference signals must not only be linearly coherent with the sound inside the car but must anticipate it so that there is sufficient time for the computation, for the acoustic delay inside the car (about 4ms) and also for delays inherent in the anti-aliasing and reconstruction filters (4-5 ms). Figure 2 shows a prediction of the sound reduction in a small four-seater car. The results have been obtained by driving the car over a coarse road surface at 50 mph. Six measured accelerometer signals and the (uncontrolled) interior noise have been used in a computer simulation which calculates the residual pressure after the application of an appropriately adapted feedforward controller in which the processing time can be varied. The Figure shows the uncontrolled interior noise spectrum along with predictions of the noise level when the control system is active. The simulation shows the effect of a processing delay of 5ms which is introduced in addition to the acoustic delay path through the vehicle interior. It is clear that the attenuation achieved is somewhat degraded as the reference signals are subjected to increasing delay, particularly at higher frequencies. The low frequency reductions are relatively unaffected.

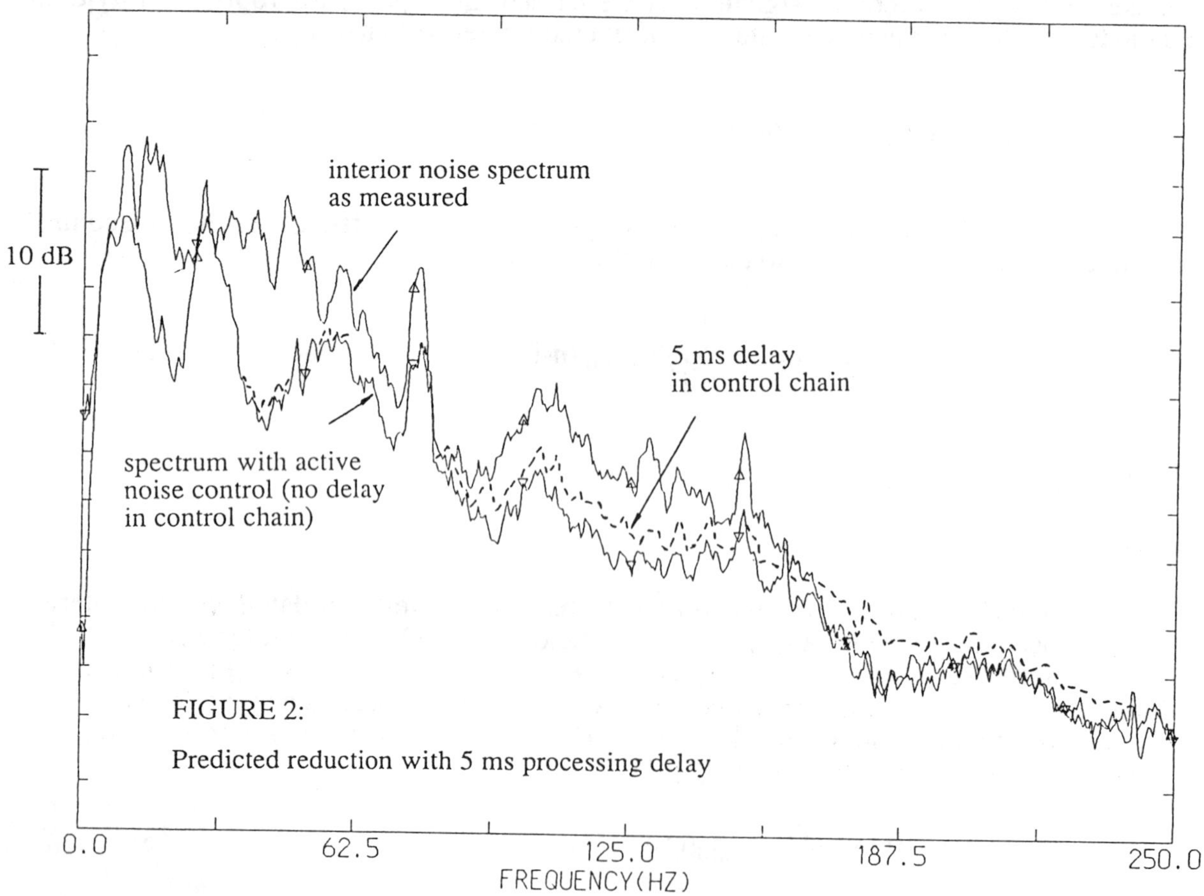

FIGURE 2:

Predicted reduction with 5 ms processing delay

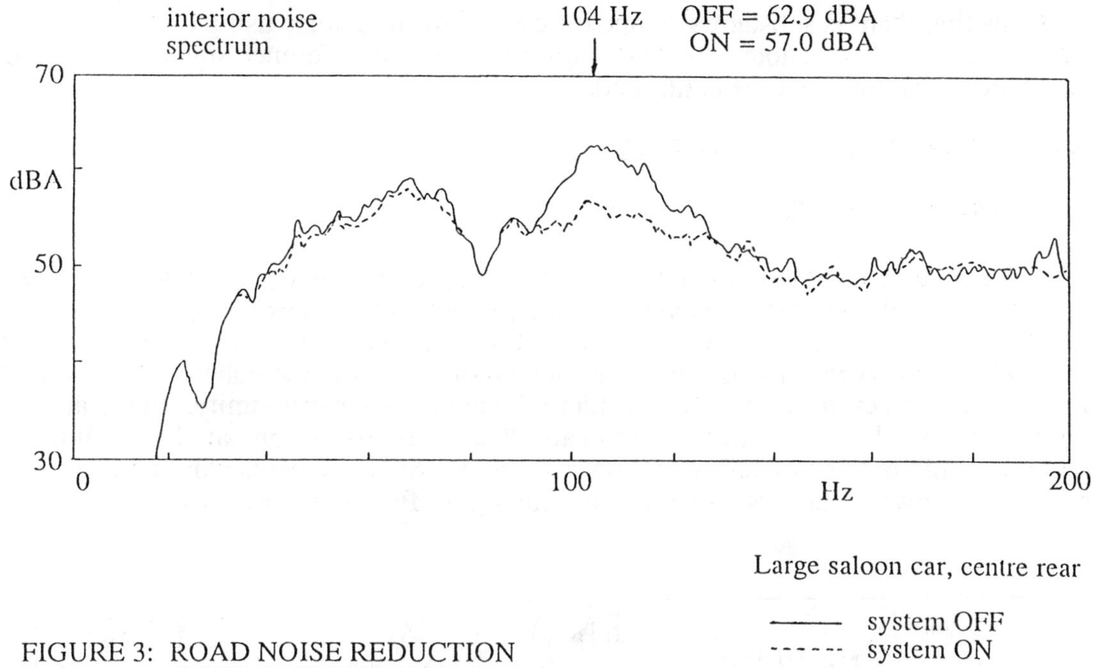

FIGURE 3: ROAD NOISE REDUCTION WITH ACTIVE CONTROL

NOISE CANCELLATION EQUIPMENT

A road noise active control system has been constructed with provision for two reference signals, two loudspeakers and two interior microphones. The required fast digital signal processing is carried out by a single processor, the Texas Instruments TMS320C25. The filters representing the vehicle interior acoustics are identified during an initial setting-up phase after the equipment has been switched on. During this initial phase a white noise signal is applied to each loudspeaker in turn. During on-line noise control the **c** filters are held fixed. The multiple-error filtered-x LMS algorithm is implemented at each time-step for rapid adaption to changing road conditions.

EXPERIMENTAL RESULTS

The road noise control system was installed in a large four-seater saloon car and measurements were made under normal driving conditions. The two reference signals were taken from accelerometers suitably located on the underside of the vehicle. The accelerometer locations were chosen to give the best multiple coherence with respect to the interior noise. Both the two loudspeakers and the two microphones were located at the back of the vehicle, one on each side.

The A-weighted noise spectra inside the car with and without active road noise control are shown in Figure 3. For this test the vehicle was driven for 4 miles at about 70 mph over a range of road surfaces. The solid line shows the interior noise with the system off. The spectrum is dominated by a peak at about 100 Hz which is thought to be due to a structural resonance of the rear suspension which is excited by wheel vibration. There was a reduction of nearly 6 dB at this frequency when the control system was operating.

The results show that there is substantial scope for control of road noise using linear feedforward controllers, but as noted above it is necessary to find reference signals which are coherent with the interior noise to be controlled.

NONLINEAR CONTROL STRATEGY

NONLINEAR VIBRATION TRANSMISSION

Tests on a range of vehicles have shown that the coherence is often poor, less than 30% say, between individual accelerometers attached to a car structure and the noise inside. This is partly because the road-induced vibration has multiple sources: each of the four wheels makes a contribution. However, even if there is only one source of vibration, the coherence between an accelerometer on the structure and an interior microphone is still far from unity. In the absence of other sources of sound and vibration this indicates that the transmission path between the accelerometer and microphone includes nonlinear elements. Such elements may include rubber bushes, hydraulic dampers and hysteresis at loose linkages. Figure 4 shows the

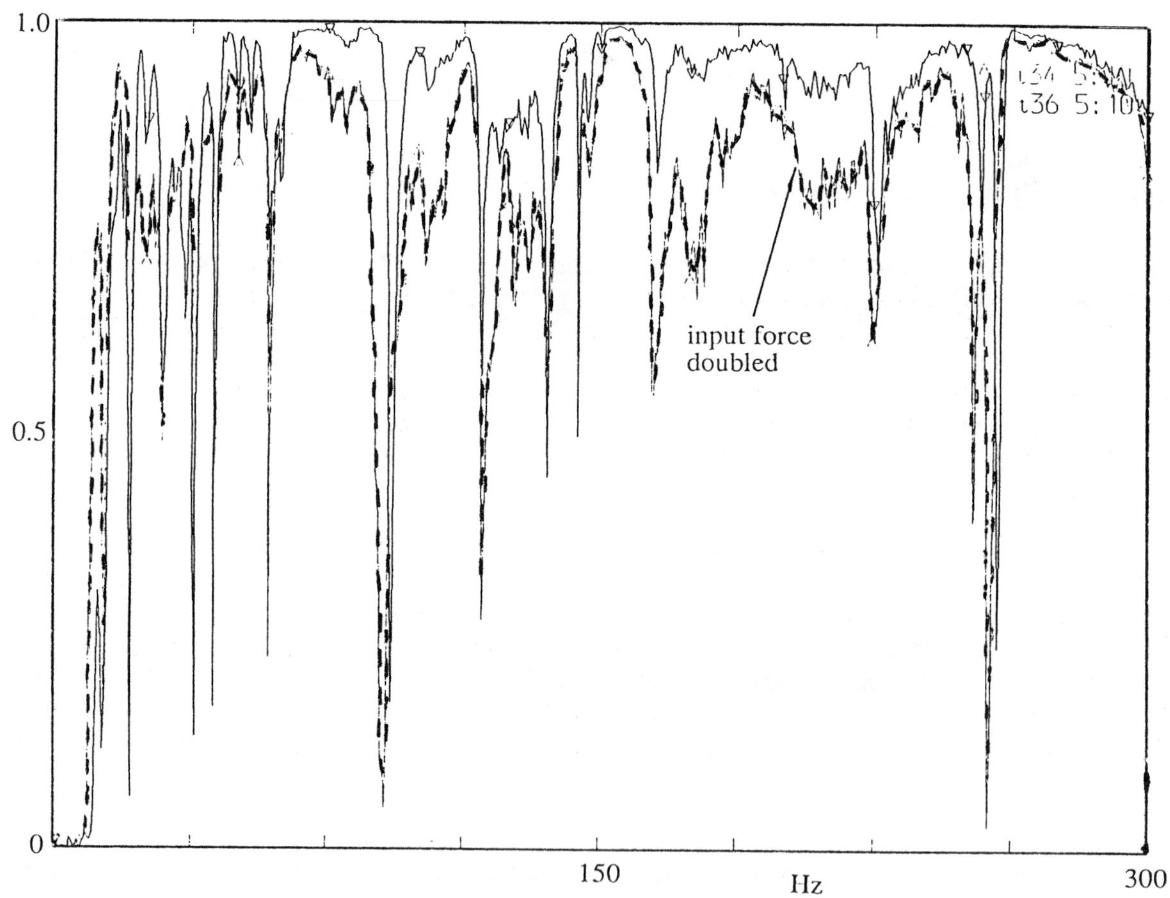

4-seat front-wheel-drive car
Shaker applied to nearside front wheel
Input force spectrum: white noise

FIGURE 4: COHERENCE BETWEEN HUB ACCELERATION (VERTICAL) AND INTERIOR MICROPHONE

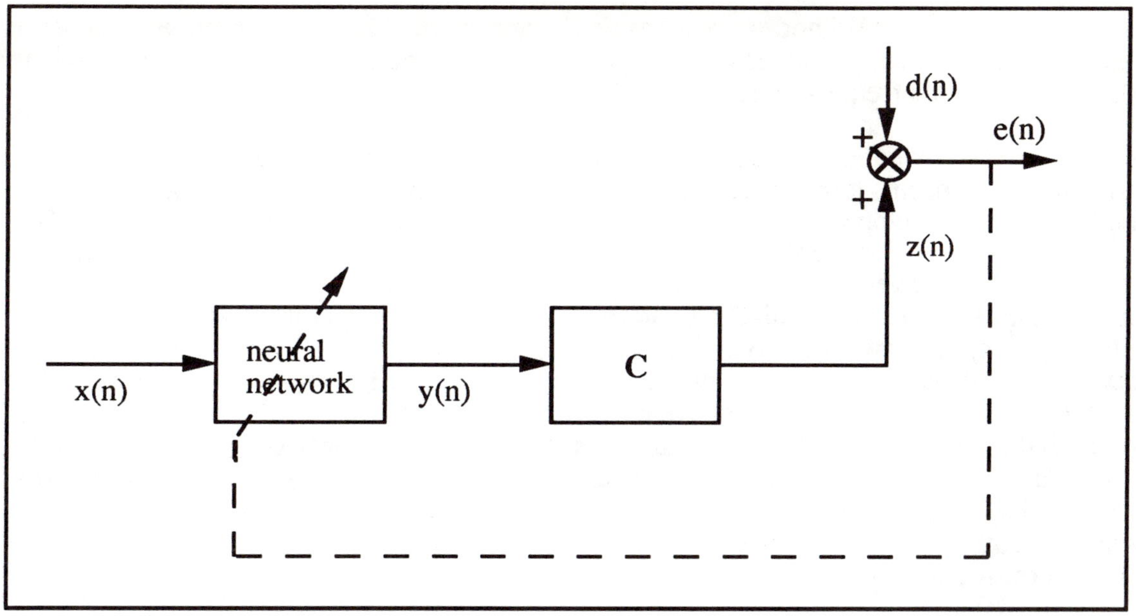

FIGURE 5: FORM OF NONLINEAR CONTROL SYSTEM

coherence between an accelerometer attached to the front suspension of a front-wheel drive car and a microphone inside the car. The vibration source in this case was an electromagnetic shaker placed under one front wheel under laboratory conditions so that the shaker was the only significant source of low-frequency noise inside the car. Also shown in Figure 4 is the coherence between these two signals when the shaker input force was doubled. It is clear that the coherence deteriorates as the amplitude is increased. This is an indication that vibration transmission through the vehicle structure is nonlinear. As explained above, low coherence limits the effectiveness of a linear road noise control system.

NEURAL NETWORK CONTROLLER

Self-organizing neural networks offer a possible means of modelling the nonlinear elements of a vehicle structure, and so effectively improving the coherence between a reference accelerometer and a microphone inside the car. Figure 5 shows how such a system might be implemented.

In this case the linear filter **h** has been replaced by a two-layer neural network of the type outlined by Narendra and Parthasarathy [8] but in this case with symmetric sigmoid (tanh) nonlinearities on the hidden units and a linear output unit. The network inputs are the present and previous values of the reference signal **x**(n) taken from a tapped delay line (Figure 6).

The network can be seen as a generalization of the linear FIR control filter which it replaces in Figure 1. The update algorithm for the complete control system using such a network would be an appropriate modification to the backpropagation algorithm.

As an initial step in the investigation of such a nonlinear controller the ability of the neural network to model simple analytic nonlinearities has been investigated. Figure 7 shows the block diagram of these computer simulations. Initially the system to be modelled was a simple cubic function with no dynamics. The neural network had an input layer consisting of a single node, a hidden layer with 5 nodes and a single linear output. After convergence the level of the spectrum of the error signal (e(n)) was 40 dB below that of the white noise excitation signal (x(n)) over the whole frequency range.

This exercise was repeated with the input/output data from one of the shaker tests on a car reported above. For this test a single time-series (the acceleration of the driving shaker) was applied to a 32-stage tapped delay line (Figure 6). The delay line provided 32 inputs to the neural net. The desired (target) time series was wheel-hub fore-aft acceleration. The network had a single hidden layer comprising 90 hidden units. The training set consisted of a 250 000-point time series. After the training set had been passed three times through the network the weights in the network were fixed and the residual error measured. Figure 8 shows that the residual error after cancellation with the neural net was slightly lower than with a simple 32-point FIR (linear) filter over a narrow frequency range close to 60 Hz. We must emphasise however that the improvement in cancellation is very modest and only occurs over a rather narrow frequency band with this data. In other trials no improvements over a linear controller have been observed. Work in this area is continuing to assess the scope for using neural networks to compensate for nonlinear effects which tend to degrade the reference signals used for active control of systems.

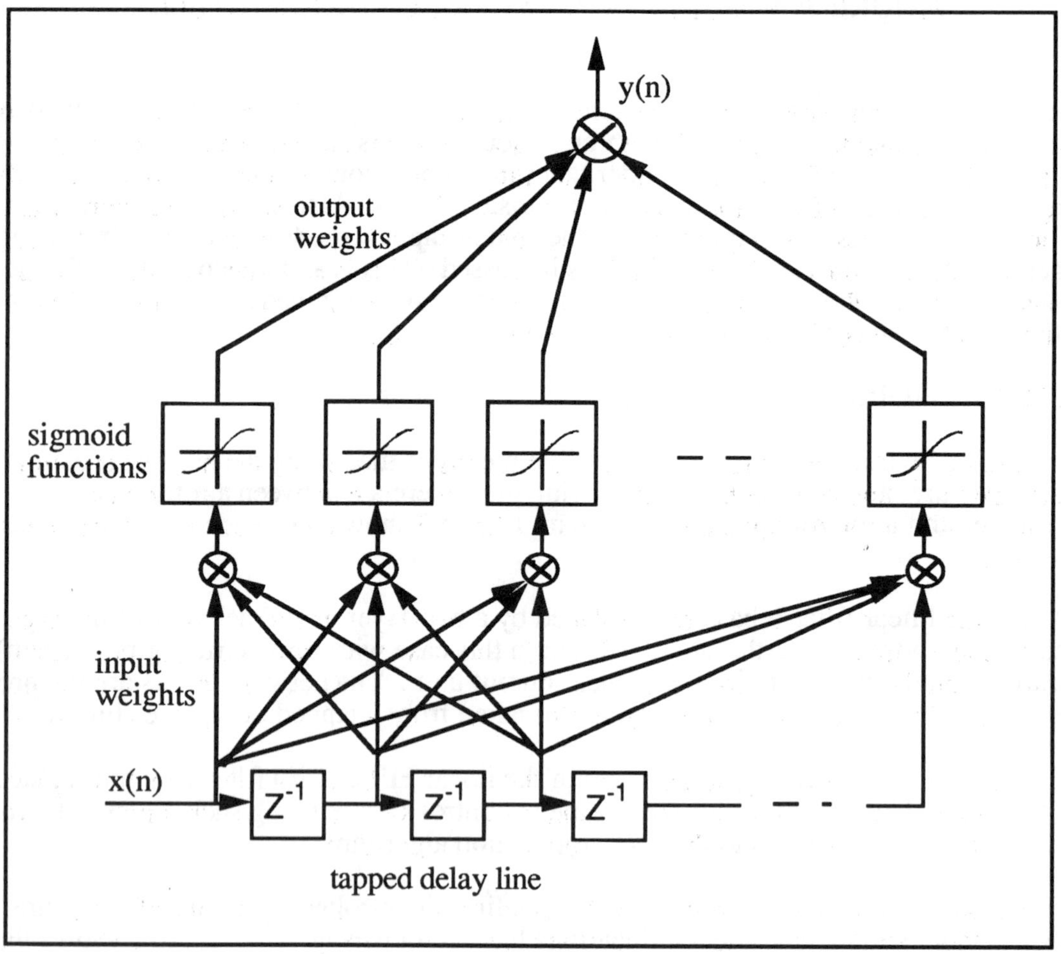

FIGURE 6: TWO-LAYER NEURAL NETWORK

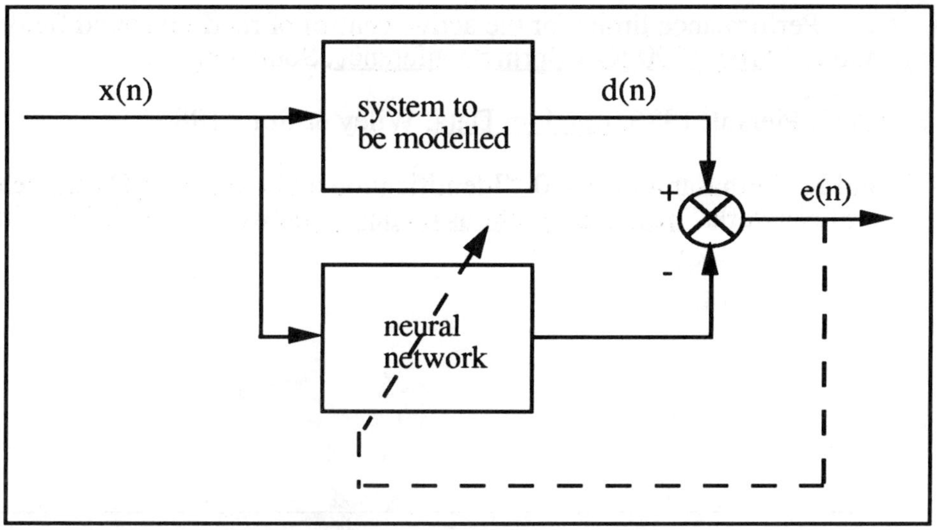

FIGURE 7: SCHEME FOR NEURAL NETWORK INVESTIGATION

CONCLUSIONS

Active reduction of interior road noise with a multichannel, adaptive linear controller has been demonstrated in road tests on a saloon car. A reduction of some 6 dB was measured at around 100 Hz, where the noise was strongest on an A-weighted basis.

The amount of noise reduction which can be achieved depends on the coherence between the reference signals used for control and the interior noise. Nonlinear controllers may offer a means of compensating for nonlinearities in the transmission path, so giving improved coherence and the possibility of improved interior noise reduction.

ACKNOWLEDGEMENT

This work is supported by Lotus Engineering, England, who also supplied the test vehicles, equipment and technical support for the tests.

REFERENCES

1. Elliott, S.J. *et al.* 1988 "The active control of engine noise inside cars" Proc Internoise 88 2, pp987-990

2. Nelson, P.A. *et al.*,1990 "Active control of stationary random sound fields" J Acoust Soc Am 87(3) March 1990 pp 963-975

3. Sutton, T.J. *et al.*, 1990 "The active control of road noise inside vehicles" Proc Internoise 90 2 pp1247-1250

4. Widrow, B and S.D. Stearns, 1985. Adaptive Signal Processing Englewood Cliffs, NJ: Prentice Hall.

5. Elliott, S.J. *et al* ., 1987. "A Multiple Error LMS Algorithm and Its Application to the Active Control of Sound and Vibration." IEEE transactions on acoustics, speech and signal processing Vol ASSP-35 No 10 Oct 1987 pp1423-1434

6. Nelson, P.A. *et al* "Performance limits for the active control of random sound fields from multiple primary sources" Proc. 1990 IOA Spring conference, Southampton

7. Bendat, J.S and A.G. Piersol, 1986. Random Data, Wiley & Sons 1986

8. Narendra, K.S. and K. Parthasarathy, 1990. "Identification and control of Dynamical Systems Using Neural Networks" IEEE transactions on neural networks Vol 1, No 1 March 1990

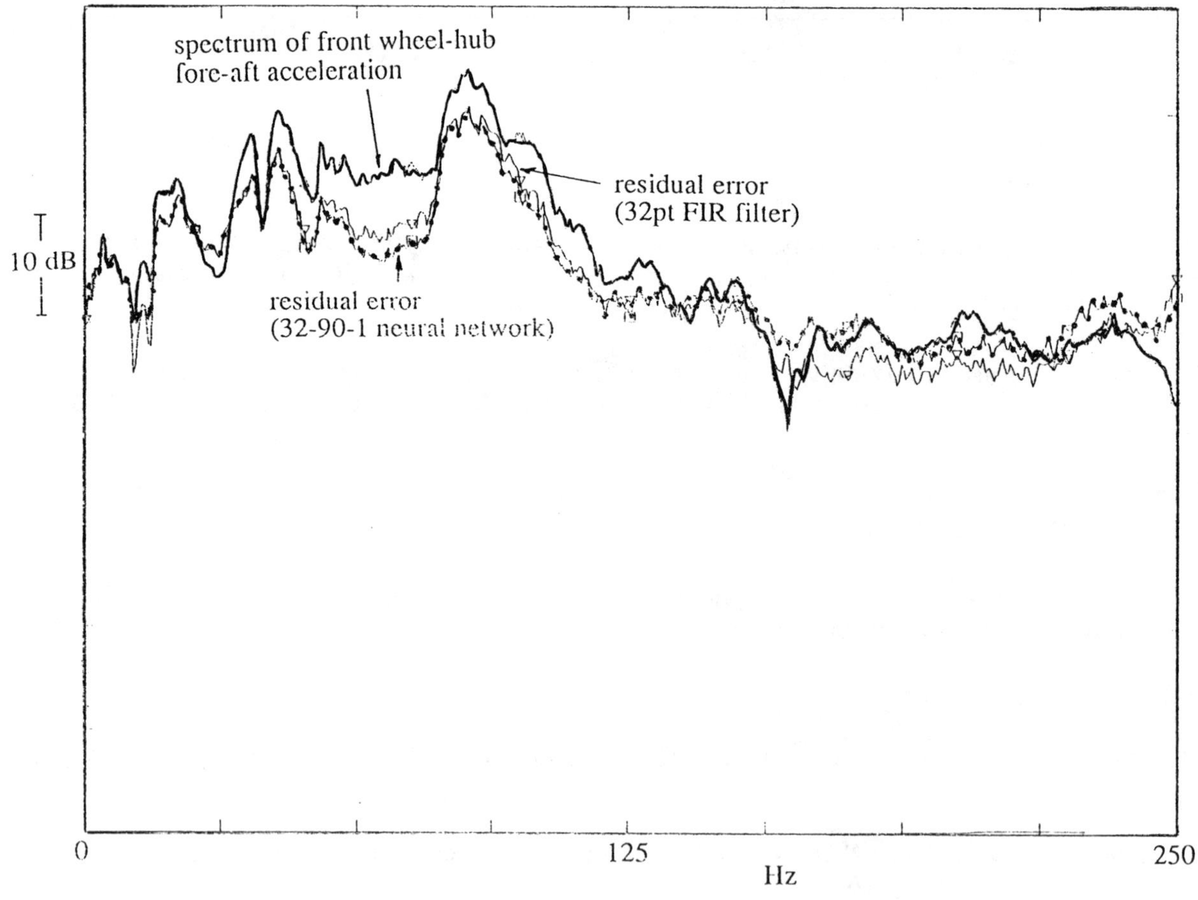

Input to filter and neural network: shaker vertical acceleration

FIGURE 8: Spectrum of wheel-hub fore-aft acceleration (target time-series); residual error after cancellation by (1) linear filter, (2) neural network

ADAPTIVE ACTIVE NOISE CONTROL IN A 3-D REVERBERANT ENCLOSURE

C. Bao,[1] P. Sas,[1] H. Van Brussel[1]

ABSTRACT

A multichannel active noise controller has been developed based on adaptive filtering, which can effectively work in 3-D reverberant sound field. The controller consists of adaptive filters to generate the proper anti noise signals and an on-line estimator to track changes in environment so that successful operation of the controller in changing environment can be ensured. Therefore the controller is fully adaptive and does not require any pre-training or separate modelling. The performance of the controller has been verified on a test set-up. The result was found to be promising.

1. INTRODUCTION

Active noise control (ANC) in 3-D enclosures has since long attracted the attention of researchers in the noise control area, because it has vast prospective applications in many areas such as car and aircraft industry. Comparing with active noise control in 1-D applications such as in ducts, active noise control in 3-D reverberant enclosures introduces significantly greater complexity. For instance more control channels might be required compared with normally one control channel in ducts. Moreover, since the reverberant sound field consists of many superimposed reflections, it is difficult, and often impossible, to obtain an instantaneous measure of the primary noise source. This precludes the simple "delay-and-outphase" method which is often used in ducts. Many have studied the mechanism of ANC in 3-D reverberant enclosures and tried to apply signal processing and control techniques in such environment [1-9]. Some ANC controllers have been tested in aircraft [6-8] and in cars [9]. Few of them were fully adaptive. A priori or separate system modelling were required to some extent. This is only applicable in those applications where the system characteristics do not change or do not change much during operation of the ANC controller. However in many applications the system characteristics do alter with, e.g., temperature, pressure, wind speed and general ride or flight conditions. Therefore a fully adaptive ANC controller which is able to simultaneously perform system identification and control is highly desirable. The development of an ANC controller with such feature is presented in section 2 of this

[1]Mechanical Engineering Department, Katholieke Universiteit Leuven, Celestijnenlaan 300B, B-3001 Leuven, Belgium.

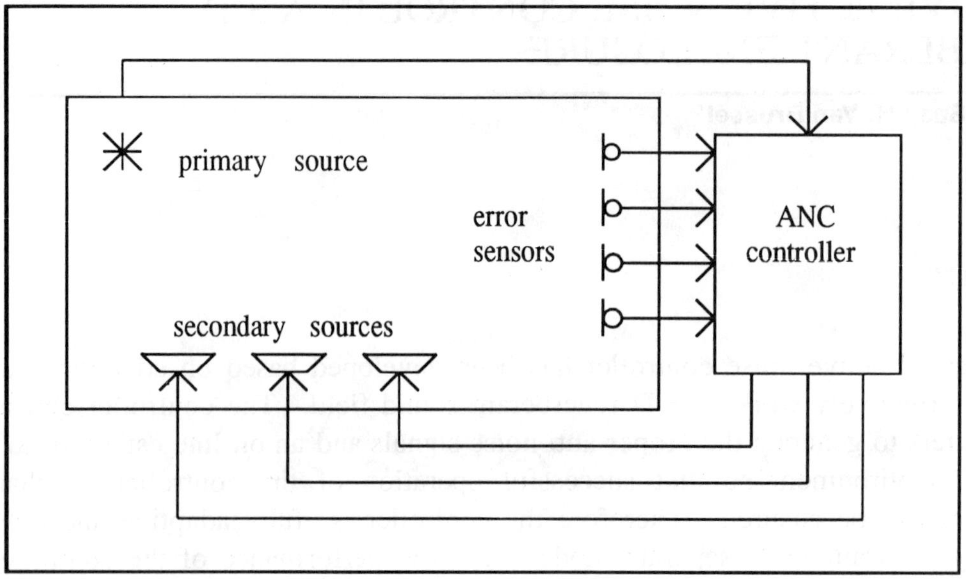

Figure 1. The ANC system in an enclosure.

paper and the experimental results of implementation of the controller are reported in section 3.

2. DEVELOPMENT OF A FULLY ADAPTIVE ANC CONTROLLER

2. 1. THE CONTROL PART

In general the practical set-up of an ANC system in an enclosure can be characterized as in figure 1. A primary noise field is generated by a certain primary source which can be either inside (e.g., machinery noise in a workshop) or outside the enclosure (e.g., propeller noise in an aircraft cabin). The controller which will be developed in this section drives N secondary sources (normally loudspeakers) in order to cancel this primary noise field. As suggested by Nelson *et al* [10], this primary field can be attenuated by minimizing the sum of squares of sound pressures which are measured by error sensors (typically microphones) in M suitable locations within the enclosure. Hereunder we will demonstrate how this minimisation can be realized by filtering techniques.

Define the output of the *m*th error sensor at the sampling instant k as $e_m(k)$, which is equal to the superposition of $d_m(k)$, the primary noise presented at this sensor, and $z_{mn}(k)$, the signal due to each of the secondary sources, i.e.,

$$e_m(k) = d_m(k) + \sum_{n=1}^{N} z_{mn}(k) \qquad (1)$$

Define the input to the *n*th secondary source at the sampling instant k as $y_n(k)$, which is obtained by filtering the reference signal x(k) (which should be somehow correlated with the primary source.) using a finite impulse response (FIR) filter whose *i*th coefficient is w_{ni}, i.e.,

$$y_n(k) = \sum_{i=0}^{Ln-1} w_{ni} x(k-i) = \mathbf{X}_n^T(k)\mathbf{W}_n \quad (2)$$

where $\mathbf{X}_n^T(k) = [x(k)\ x(k-1)\ \dots\ x(k-L_n-1)]$, $\mathbf{W}_n^T = [w_{n0}\ w_{n1} \cdots w_{nLn-1}]$ and L_n is the length of the *n*th FIR filter. Let the transfer function between $y_n(k)$ and $z_{mn}(k)$ be modelled as an infinite impulse response (IIR) filter in general, i.e.,

$$z_{mn}(k) = \sum_{j=1}^{G-1} a_{mnj} z_{mn}(k-j) + \sum_{j=0}^{H-1} b_{mnj} y_n(k-j) \quad (3)$$

In terms of delay operator q, we define

$$A_{mn}(q) = \sum_{j=1}^{G-1} a_{mnj} q^{-j}, \qquad B_{mn}(q) = \sum_{j=0}^{H-1} b_{mnj} q^{-j}$$

so that expression (3) can be rewritten in a more convenient form as follows,

$$z_{mn}(k) = \left(\frac{Bmn(q)}{1 - Amn(q)}\right) y_n(k) \quad (4)$$

Substituting (2) into (4) and defining

$$g_{mn}(k) = \left(\frac{Bmn(q)}{1 - Amn(q)}\right) \quad (5)$$

yields

$$z_{mn}(k) = g_{mn}(q)\mathbf{X}_n^T(k)\mathbf{W}_n = \mathbf{Q}_{mn}^T(k)\mathbf{W}_n \quad (6)$$

where $\mathbf{Q}_{mn}^T(k) = [g_{mn}(q)x(k)\quad g_{mn}(q)x(k-1)\quad \dots\quad g_{mn}(q)x(k-L_n-1)]$. Substituting (6) into (1) yields

$$e_m(k) = d_m(k) + \sum_{n=1}^{N} \mathbf{Q}_{mn}{}^T(k)\mathbf{W}_n$$
$$= d_m(k) + \mathbf{Q}_m{}^T(k)\mathbf{W} \quad (7)$$

where $\mathbf{Q}_m{}^T(k) = [\mathbf{Q}_{m1}{}^T(k)\ \mathbf{Q}_{m2}{}^T(k) \ldots \mathbf{Q}_{mN}{}^T(k)]$ and $\mathbf{W}^T = [\mathbf{W}_1{}^T\ \mathbf{W}_2{}^T \ldots \mathbf{W}_N{}^T]$. If we further define

$$\begin{aligned} \mathbf{e}^T(k) &= [e_1(k)\ e_2(k) \ldots e_M(k)] \\ \mathbf{d}^T(k) &= [d_1(k)\ d_2(k) \ldots d_M(k)] \\ \mathbf{Q}^T(k) &= [\mathbf{Q}_1(k)\ \mathbf{Q}_2(k) \ldots \mathbf{Q}_M(k)] \end{aligned} \quad (8)$$

then the sum of squares of the error signal can be expressed as a function of the controller coefficient set $\mathbf{W}$, i.e.,

$$\sum_{m=1}^{M} e_m(k)^2 = \mathbf{e}^T(k)\mathbf{e}(k) = \mathbf{d}^T(k)\mathbf{d}(k) + \mathbf{W}^T\mathbf{Q}^T(k)\mathbf{Q}(k)\mathbf{W} + 2\mathbf{d}^T(k)\mathbf{Q}(k)\mathbf{W} \quad (9)$$

Let's take the expected value of (9). This gives the total error J as

$$J = E\{\mathbf{e}^T(k)\mathbf{e}(k)\} = J_0 + \mathbf{W}^T\mathbf{R}\mathbf{W} + 2\mathbf{P}^T\mathbf{W} \quad (10)$$

where $J_0 = E\{\mathbf{d}^T(k)\mathbf{d}(k)\}$, $\mathbf{R} = E\{\mathbf{Q}^T(k)\mathbf{Q}(k)\}$ and $\mathbf{P}^T = E\{\mathbf{d}^T(k)\mathbf{Q}(k)\}$. From expression (10) it can be clearly identified that the total error J is a quadratic function of the controller coefficient set $\mathbf{W}$. Thus, if $\mathbf{R}$ is positive definite (which is normally the case in practice.), J will have a unique global minimum. By setting the differential of (10) with respect to $\mathbf{W}$ to zero, the so called Wiener optimal solution can be obtained as $\mathbf{W}^* = -\mathbf{R}^{-1}\mathbf{P}$, which minimizes the total error J to

$$J_{min} = J_0 - \mathbf{P}^T\mathbf{R}^{-1}\mathbf{P} = J_0 + \mathbf{P}^T\mathbf{W}^* \quad (11)$$

Now the question is how to obtain $\mathbf{W}^*$. Because of the quadratic nature of J the optimal solution of $\mathbf{W}$ may be derived adaptively by using a gradient descent method, i.e.,

$$w_{ni}(k+1) = w_{ni}(k) + \mu(-\nabla_{wni(k)}J) \quad (12)$$

where μ is the convergence factor and

$$\nabla_{wni(k)}J = \frac{\partial}{\partial w_{ni}(k)} E\{\mathbf{e}^T(k)\mathbf{e}(k)\}$$

To get ∇J we take the instantaneous value of $[\mathbf{e}^T(k)\mathbf{e}(k)]$ as an estimation of J. Expression (12) thus becomes

$$w_{ni}(k+1) = w_{ni}(k) - 2\mu \sum_{m=1}^{M} e_m(k)\nabla_{wni(k)}e_m(k) \tag{13}$$

To deduce the final form of (13) let's first examine the result of $\nabla_{wni(k)}e_m(k)$.

$$\begin{aligned}\nabla_{wni(k)}e_m(k) &= \nabla_{wni(k)}[d_m(k) + \sum_{n=1}^{N} z_{mn}(k)] \\ &= \nabla_{wni(k)}z_{mn}(k) \\ &= \sum_{j=1}^{G-1} a_{mnj}\, \nabla_{wni(k)}z_{mn}(k-j) + \sum_{j=0}^{H-1} b_{mnj}\, \nabla_{wni(k)}y_n(k-j)\end{aligned} \tag{14}$$

Assuming $w_{ni}(k) \approx w_{ni}(k-1) \approx \ldots \approx w_{ni}(k-p)$ ($p = \max\{G-1, H-1\}$), which corresponds to a slow adaptation, the last equation of (14) can be replaced by

$$\nabla_{wni(k)}z_{mn}(k) \approx \sum_{j=1}^{G-1} a_{mnj}\, \nabla_{wni(k-j)}z_{mn}(k-j) + \sum_{j=0}^{H-1} b_{mnj}\, \nabla_{wni(k-j)}y_n(k-j) \tag{15}$$

Again using delay-operator notation and definition (5) and noticing that $\nabla_{wni(k-j)}y_n(k-j) = x(k-i-j)$, expression (15) can be rewritten as

$$\nabla_{wni(k)}z_{mn}(k) \approx g_{mn}(q)x(k-i) \qquad (16).$$

Also equation (14) reduces to

$$\nabla_{wni(k)}e_m(k) \approx g_{mn}(q)x(k-i) \qquad (17).$$

Substituting (17) into (13) we finally obtain an adaptive form of deriving $\mathbf{W}^*$,

$$w_{ni}(k+1) = w_{ni}(k) - 2\mu \sum_{m=1}^{M} e_m(k)(g_{mn}(q)x(k-i)) \qquad (18).$$

Using the vector and matrix notation defined in (8) this form can also be compactly written as

$$\mathbf{W}(k+1) = \mathbf{W}(k) - 2\mu\mathbf{Q}(k)^T\mathbf{e}(k) \qquad (19).$$

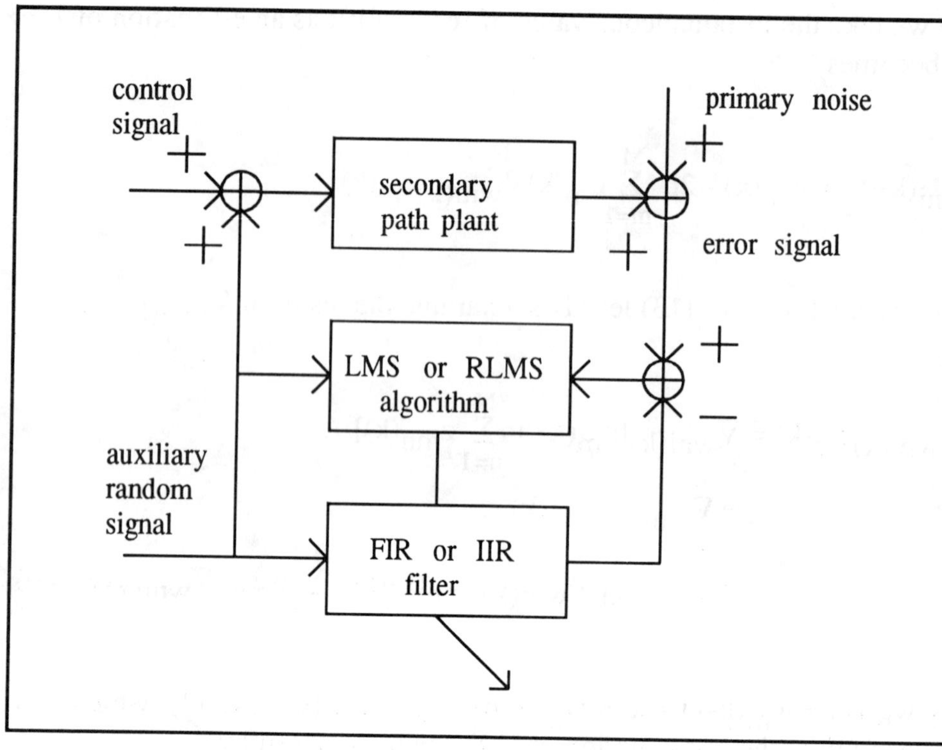

Figure 2. The structure of the identification part in one channel.

By setting $A_{mn}(q)$ in (5) to zero, which corresponds to the transfer functions of the secondary paths being modelled as FIR filters, the adaptive algorithm (19) will exactly correspond to the multiple error LMS algorithm proposed by Elliott *et al* [11]. And for a single-input-single-output system ($M = N = 1$) the algorithm becomes the "filtered-x LMS" algorithm discussed by Widrow and Stearns [12].

2. 2. THE IDENTIFICATION PART

From (18) it can be seen that the adaptive algorithm of the control part needs to utilize the transfer functions of the secondary paths. In some cases these transfer functions are time invariant and can be identified on an off-line basis. In many other cases however they might be time dependent and have to be identified on-line. Many methods can be used for on-line identification of the secondary paths. In this study the RLMS (in case the secondary paths being modelled as IIR filters) or LMS (the secondary paths being modelled as FIR filters) algorithm was chosen for the task because of their simplicity and their relative good performance in the condition of a very low signal to noise ratio. The latter point is essential to the ANC system. Due to the lack of sufficient richness (or persistent excitation) of the control signals in the ANC system, they normally may not be used for the identification. Therefore auxiliary random signals which are independent from the primary noise have to be added to the secondary sources for identification purposes. Furthermore the power of these signals should be kept low in comparison with the power of the primary noise so that the final performance of the ANC system will not be significantly affected by these signals. The structure of one of the channels of the identification part would thus look like in figure 2.

In figure 2 if the primary noise and the control signal are not present, it represents a standard adaptive filter discussed by Widrow and Stearns [12]. For time being let's only consider the LMS algorithm case. The adaptive mechanism of the LMS algorithm can be expressed as follows by following the notation of Widrow and Stearns [12],

$$\begin{aligned}\mathbf{W}(k+1) &= \mathbf{W}(k) + 2\mu e(k)\mathbf{X}(k) \\ &= \mathbf{W}(k) + 2\mu[y(k)-\hat{y}(k)]\mathbf{X}(k)\end{aligned} \tag{20}$$

where $\mathbf{W}(k)$ is the coefficients of the adaptive FIR filter in the identification part, $\mathbf{X}(k)$ is the auxiliary signal, $y(k)$ is the output of the unknown plant and $\hat{y}(k)$ is the output of the adaptive filter. Under such mechanism $\mathbf{W}$ will eventually reach its optimal solution which makes $\hat{y}(k)$ approximate to $y(k)$, i.e., the best approximation of the unknown plant. We will see what happens if the primary noise and the control signal are present. In such case (20) will extend to

$$\begin{aligned}\mathbf{W}(k+1) &= \mathbf{W}(k) + 2\mu_n e_n(k)\mathbf{X}(k) \\ &= \mathbf{W}(k) + 2\mu_n[\, y(k)-\hat{y}(k)+n(k)\,]\mathbf{X}(k) \\ &= \mathbf{W}(k) + 2\mu_n[\, y(k)-\hat{y}(k)\,]\mathbf{X}(k) + 2\mu_n n(k)\mathbf{X}(k)\end{aligned} \tag{21}$$

where $e_n(k)$ is the error signal with the primary noise and the control signal, $n(k)$ is the noise present in $e_n(k)$ due to the primary noise and the control signal (being called as plant noise) and μ_n is the convergence factor with plant noise. Although it could be easily proven that the expected value of $\mathbf{W}$ in (21) converges to its optimal solution $\mathbf{W}^*$ provided that $n(k)$ and $x(k)$ are independent and the expected value of $x(k)$ is equal to zero, this in general does not imply that $\mathbf{W}$ also converges to $\mathbf{W}^*$. It is obvious that $2\mu_n n(k)\mathbf{X}(k)$ term in (21) is a noise term, which will degrade the performance of the adaptive filter. At worst, it will make $\mathbf{W}$ not converge or not converge to $\mathbf{W}^*$. If the convergence factor μ_n is selected sufficiently small, however, $\mathbf{W}$ will eventually converge towards a reasonable approximation of $\mathbf{W}^*$. Now we will deduce the condition that μ_n should meet to ensure above convergence. Suppose at a certain moment k_0 $\mathbf{W}$ somehow reaches $\mathbf{W}^*$ and stays around it, which implies that $y(k)-\hat{y}(k) \approx 0$. Thus expression (21) becomes

$$\begin{aligned}\mathbf{W}^*(k_0+k+1) &\approx \mathbf{W}^*(k_0+k) + 2\mu_n n(k_0+k)\mathbf{X}(k_0+k) \\ &\approx \mathbf{W}^*(k_0+k-1) + 2\mu_n[n(k_0+k)\mathbf{X}(k_0+k)+n(k_0+k-1)\mathbf{X}(k_0+k-1)] \\ &\quad \vdots \\ &\approx \mathbf{W}^* + 2\mu_n \sum_{i=0}^{k} n(k_0+i)\mathbf{X}(k_0+i)\end{aligned} \tag{22}$$

where $\mathbf{W}^*(\,)$ denotes the reasonable approximation of $\mathbf{W}^*$. Moving $\mathbf{W}^*$ of (22) to the left hand side of the equation we have

$$\mathbf{W}^*(k) - \mathbf{W}^* = 2\mu_n \sum_{i=0}^{k} n(i)\mathbf{X}(i) \tag{23}$$

In (23) we omit k_0 since n(k) and x(k) are supposed to be stationary. Define the maximum tolerant estimation error of $\mathbf{W}^*$ as $\Delta w^* = \max\{|\mathbf{W}^*(k)-\mathbf{W}^*|, k=0,1,2,..\}$ and

$$C_{nx} = \max\{|\sum_{i=0}^{k} n(i)\mathbf{X}(i)|, \ k = 0, 1, 2, ...\} \tag{24}$$

Noting that since x(k) is a random signal and independent from n(k), C_{nx} must not be infinite. So, if

$$\mu_n < \Delta w^* / 2C_{nx} \tag{25}$$

then the presence of plant noise will not significantly influence the convergence of **W** towards the reasonable approximation of $\mathbf{W}^*$. However this is not to say that the convergence process of (20) is not significantly affected by plant noise. Since normally μ_n defined by (25) is smaller than μ in (20) (in the case where μ_n is greater than μ, μ_n should be set to μ to ensure the convergence of **W**.), the convergent speed with plant noise will be slower than that without plant noise. When plant noise n(k) is much larger than the auxiliary signal x(k), the convergence speed of **W** in (21) might be tens or even hundreds of times slower than that in (20). Such slow adaptation is obviously not preferable in most applications. There are several ways to increase μ_n, so to the adaptation speed. First from (25) we can see that μ_n could be increased by increasing Δw^*. This is possible since, as suggested by Elliott *et al* [11] and also verified by our studies [13], the control algorithm discussed in section 2. 1 is very robust to the estimation error of $\mathbf{W}^*$, although it should be noted that increasing Δw^* would certainly degrade the performance of the control part of the ANC system. Secondly, noting that as the control part becomes effective the plant noise term n(k) in (24) which is due to the primary noise and the control signal will decrease, so will C_{nx}. Therefore if μ_n is made adjustable such that it is increasing as $e_n(k)$ decreases, the convergence speed of **W** could eventually improve. The third way to improve the adaptation speed is using another adaptive filter or so called adaptive line enhancer (ALE) to attenuate the plant noise n(k) in $e_n(k)$ before feeding $e_n(k)$ to the adaptive identification filter. This would however add an extra computation burden to the system.

Similarly we could state the convergence condition for the RLMS algorithm case with plant noise as follows if Feintuch's RLMS algorithm [14] is used,

$$\mu_b < \Delta b^* / 2C_{nx} \tag{26}$$

$$\mu_a < \Delta a^* / 2C_{nxf} \tag{27}$$

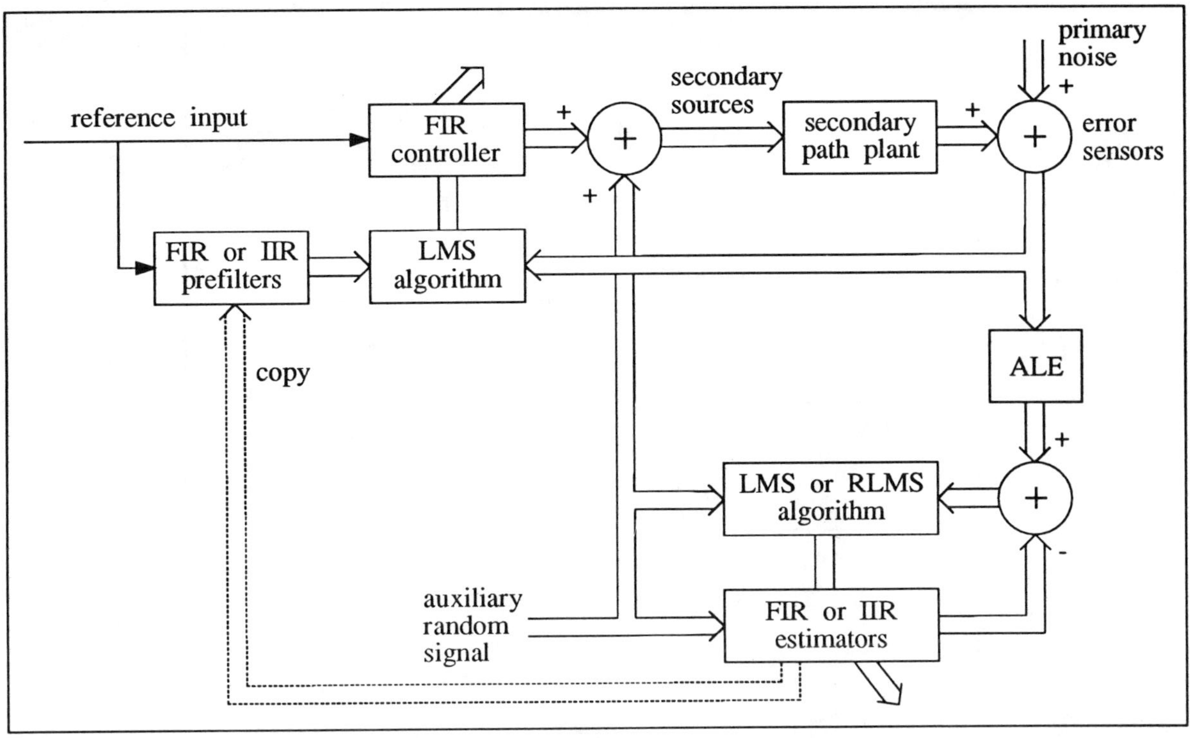

Figure 3. The complete structure of the ANC controller.

where Δb^* and Δa^* are the maximum tolerant estimation error of b^* and a^* respectively and C_{nxf} is defined as

$$C_{nxf} = \max\{\left|\sum_{i=k_0}^{k} n(i)(g(i,q)\mathbf{X}(i))\right|,\ k=k_0, k_0+1, k_0+2, \ldots\} \tag{28}$$

All discussions made for the LMS case are also applicable to the RLMS case. Noting that in (27) we have C_{nxf} instead of C_{nx}. Since in many cases C_{nxf} is greater than C_{nx}, this makes the RLMS algorithm with plant noise more difficult to converge than the LMS algorithm.

Finally, combining the identification part with the control part, the complete structure of the ANC controller looks like in figure 3.

3. EXPERIMENTS ON A TEST SET-UP

The ANC controller developed in section 2 was implemented in a test set-up. The main structure of the set-up is an enclosed cavity made of PVC plastic, which is shaped to simulate a car cabin on a one-third scale. The dimensions of the cavity are 850 x 400 x 400 mm (length x width x height). The lower resonance frequencies of the cavity were found to be 60-120-210-410-440-480-550-600Hz. The acoustic "reverberation time"

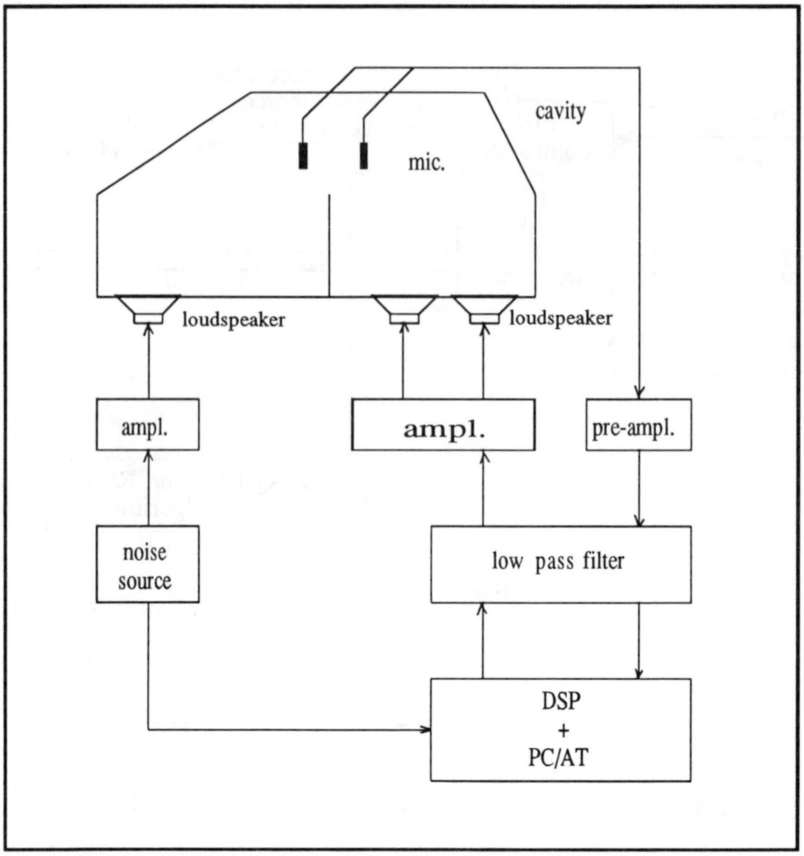

Figure 4. The block diagram of the test set-up.

(i.e., the time taken for the sound to decay by 60dB) at those frequencies are around 0.3 second. The ANC control algorithm was programmed on a digital signal processing (DSP) board, which is equipped with a 100ns instruction cycle TMS320C25 signal processor and has two analog inputs and two analog outputs. This DSP board is inserted in a PC/AT computer, which functions both as a development system of the DSP board and as a monitor of the ANC system. The block diagram of the test set-up is shown in figure 4.

3. 1. EXPERIMENTS ON LOCAL NOISE REDUCTION

The objective of the local noise reduction experiments was to verify the effectiveness of the ANC controller in terms of its convergence behaviour. For this purpose it is more convenient to use a single channel configuration (i.e., one secondary speaker with one error microphone). In the experiment the control bandwidth was chosen from 50 to 650Hz, which demands a sampling frequency of 1600Hz. The length of the adaptive filter in the control part was chosen as 32 (32=1600Hz/50Hz). A typical frequency response of the secondary path looks like in figure 5. An IIR filter of 16th order which gives 32 coefficients or a 64-coefficient FIR filter was used to model the secondary path.

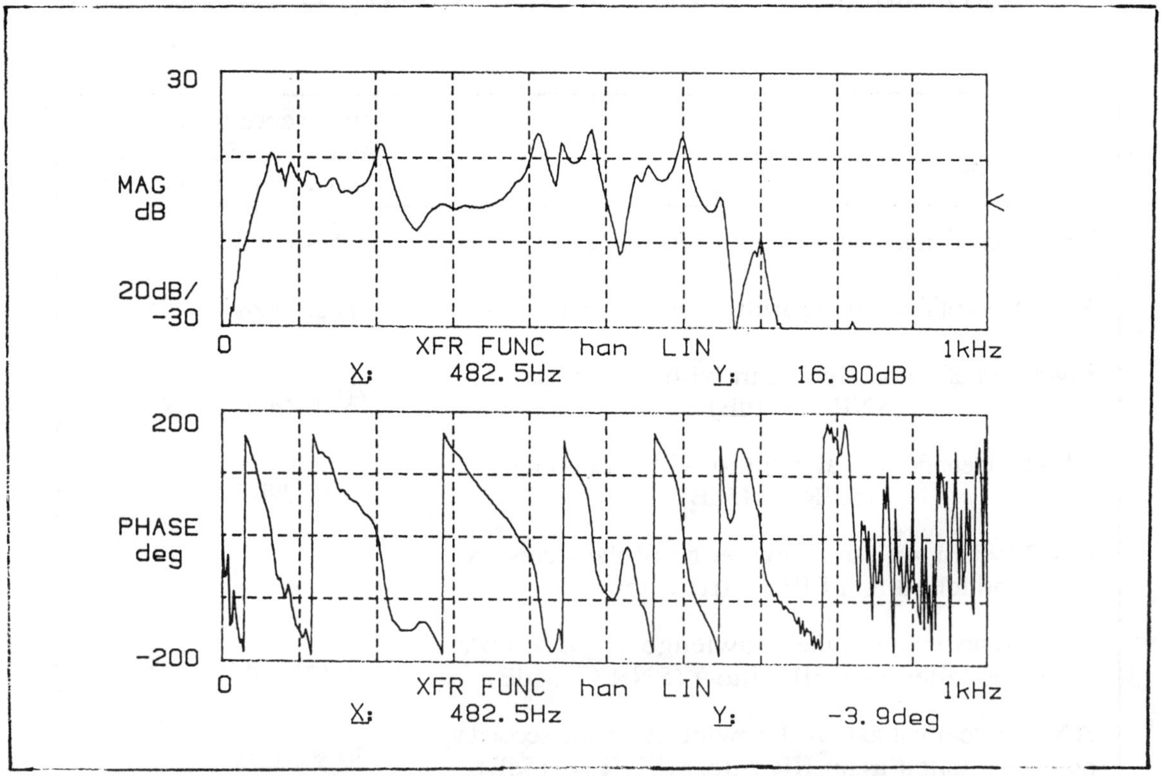

Figure 5. A typical frequency response of the secondary path.

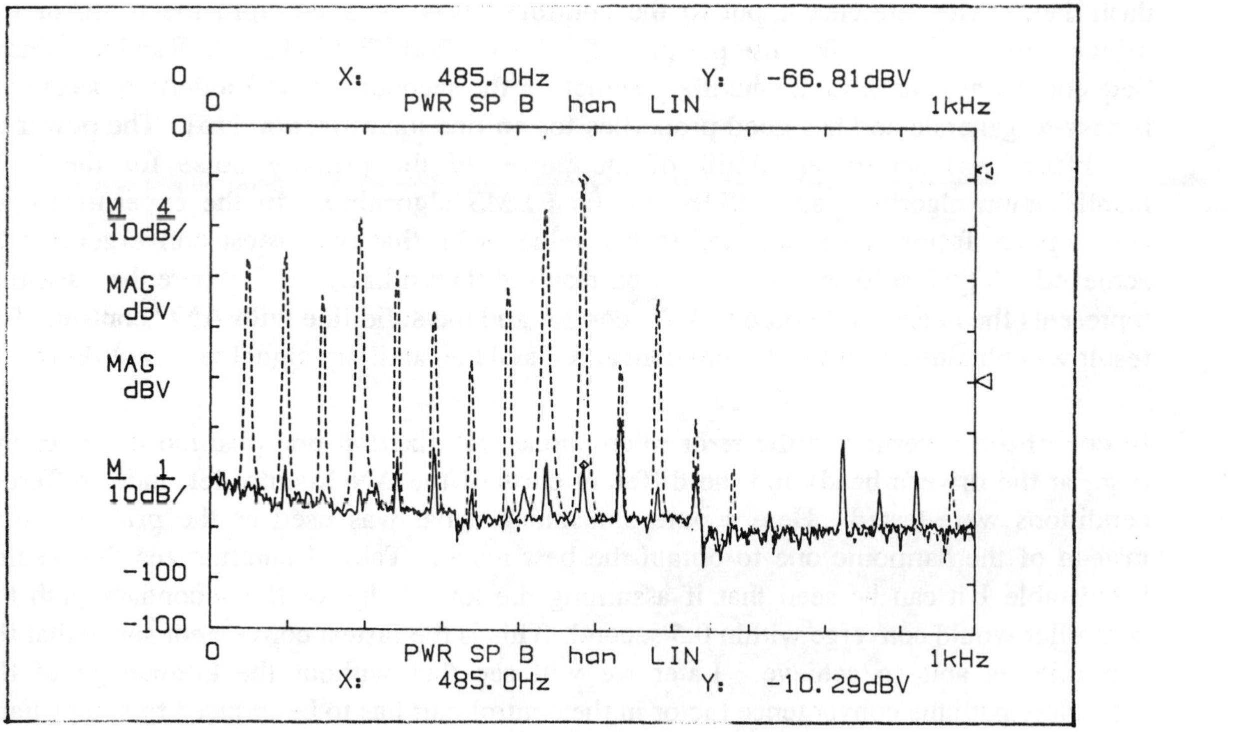

Figure 6. A typical local noise reduction result.

TABLE I - EXPERIMENTAL RESULTS ON THE CONVERGENCE BEHAVIOUR.

conditions	convergence time (reach to 95% of the final values in average)
LMS identification algorithm without plant noise	1.6 seconds
RLMS identification algorithm without plant noise	60 seconds
LMS identification algorithm with plant noise (SNR = -31dB)	60 seconds
RLMS identification algorithm with plant noise (SNR = -15dB)	30 minutes
ANC system with the knowledge of the secondary path (modelled as a FIR or IIR filter)	0.5 second
ANC system without the knowledge of the secondary path (modelled as a FIR filter) (SNR = -31dB)	4 seconds
ANC system without the knowledge of the secondary path (modelled as an IIR filter) (SNR = -15dB)	30 seconds

The primary noise was a periodic one whose power spectrum is shown in figure 6 (the dash-line). The reference input to the controller was obtained from the tacho of the primary noise (i.e., one impulse per period). A so called PRBS (Pseudo Random Binary Sequence) was chosen as the auxiliary signal for the secondary path modelling because it is easy to generate and has good properties for on-line identification [15]. The power of the PRBS was chosen as -31dB of the power of the primary noise for the LMS identification algorithm and -15dB for the RLMS algorithm. In the experiments all convergence factors were adjusted to the values such that the fastest convergence was achieved. A typical local noise reduction result is shown in figure 6, where the dash-line represents the noise level without ANC control and the solid line with ANC control. The result was obtained after the system converged and the auxiliary signal was shut down.

In one of the experiments the error microphone was fixed at one position in the cavity (e.g., at the driver's head) and the different parts of the ANC controller under different conditions were tested. Here a pure sinusoidal signal was used as the primary noise instead of the harmonic one to obtain the best result. Table I summarizes the results. From table I it can be seen that if assuming the knowledge of the secondary path the controller would converge within 0.5 second. This is the fastest convergent speed that the controller is able to achieve. Later we will see that without the knowledge of the secondary path the convergence factor in the control part has to be reduced to somewhat a tenth of the fastest one, due to the need for stabilization. This alone will considerably reduce the response time of the controller to changes in environment, not mention the time required by the identification part. Let's have a look at the behaviour of the identification part. Without plant noise it would converge within 1.6 seconds for the LMS algorithm and 60 seconds for the RLMS algorithm. With plant noise the LMS algorithm would converge within 120 seconds and the RLMS algorithm within 30 minutes. The advantage of the LMS algorithm over the RLMS algorithm in this case is clearly identified. Even using the LMS algorithm, the 120-second convergence speed is obviously too large for most applications. When combining the identification part and the control part together, however, the primary noise could be attenuated within 4 seconds in average. This is because the attenuation of the primary noise by the control part reduces plant noise in the identification part so that the speed of identification can be improved by increasing the convergence factor μ_n as indicated in section 2. 2.

Another interesting experiment carried out was to move the error microphone slowly around within the cavity to emulate changes in the environment when the controller is in operation. The result was rather satisfactory. Without the on-line identification part the controller would become unstable, while with the identification part the controller always stayed stable and the primary noise at the error microphone was attenuated all the time.

3. 2. EXPERIMENTS ON GLOBAL NOISE REDUCTION

It is instructive to estimate the number of (adaptive) filters required in the ANC controller developed in section 2. Assuming m error microphones and n secondary speakers, we have n adaptive filters and m x n pre-filters in the control part and m x n adaptive filters and m adaptive line enhancers for the secondary path modelling. This gives the total of ($2m\times n+m+n$) filters, ($m\times n+m+n$) of which have to be adaptive. For instance, a relatively simple system of two secondary speakers and four error microphones already demands 14 adaptive filters and 8 non-adaptive filters. This certainly requires lots of computation power. Therefore it is always preferable to achieve an acceptable result by using a minimum number of channels. This is the objective of the experiments on global noise reduction.

To find the global effect of the ANC controller, the acoustic pressures with and without ANC were measured in 66 evenly distributed points in the cavity. These measurement points were located in three planes, i.e., the bottom, the middle and the top plane. The distances between the points were approximately a fifth of the wavelength of the highest controllable frequency (600Hz here). Many tests were carried out under two different configurations of the ANC controller, i.e., one secondary speaker with one error microphone and one secondary speaker with two error microphones. The locations of these speakers and microphones were arranged according to the results of a modal analysis of the test cavity, such that better noise reduction could be achieved [16]. A pure sinusoid was used as the primary noise whose frequency was chosen among the resonance frequencies of the cavity (i.e., 60, 210, 410, 440, 480, 550 and 600Hz), since the global effect was expected to be most pronounced around those frequencies. Figure 7 to 9 show the typical global noise reduction at 60, 480 and 600Hz respectively. The top part of these figures show the noise reduction level with the one microphone configuration and the bottom part is the noise reduction level with the two microphone

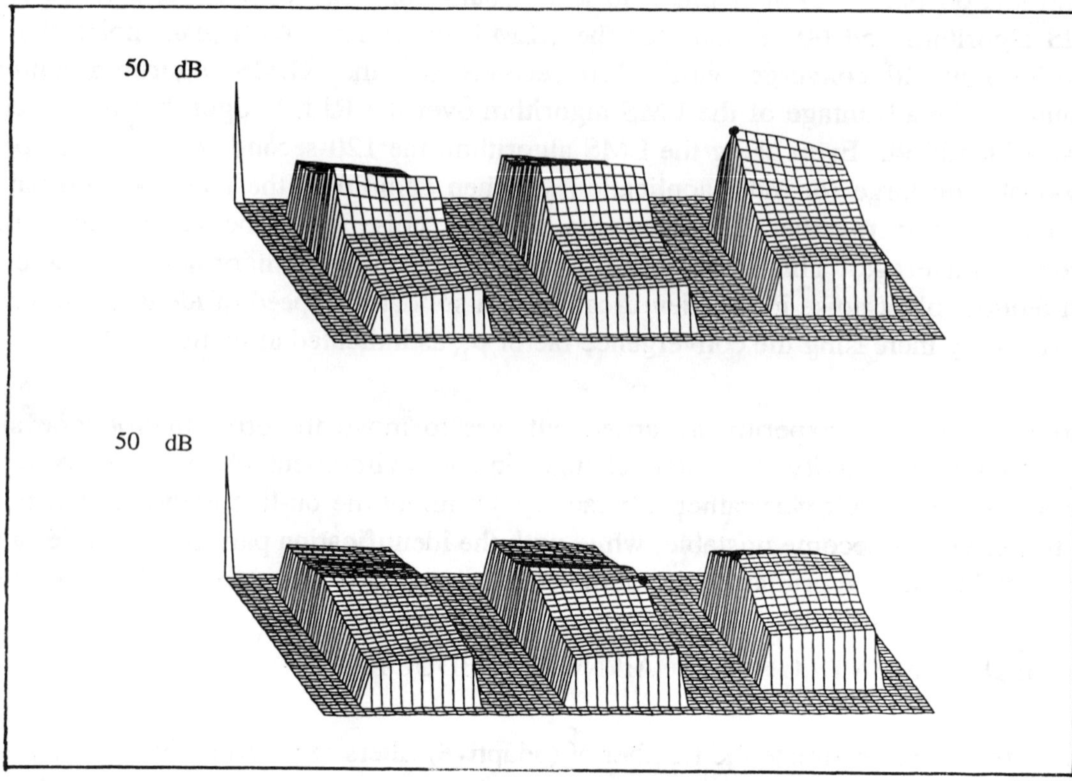

Figure 7. A typical global noise reduction at 60Hz.

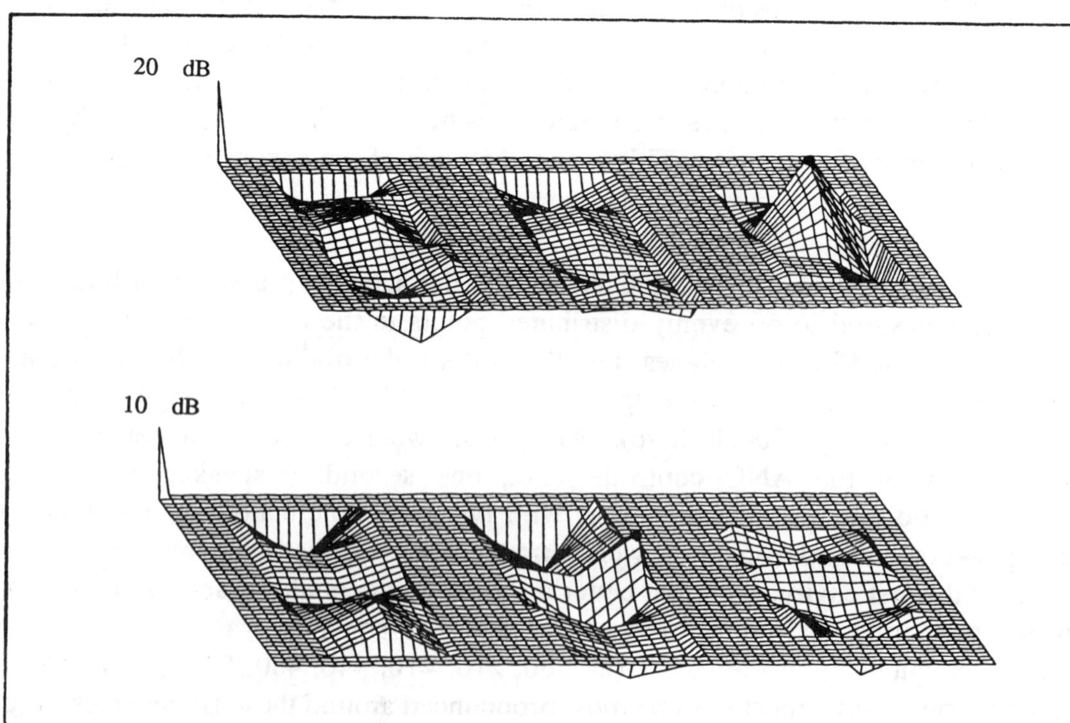

Figure 8. A typical global noise reduction at 480Hz.

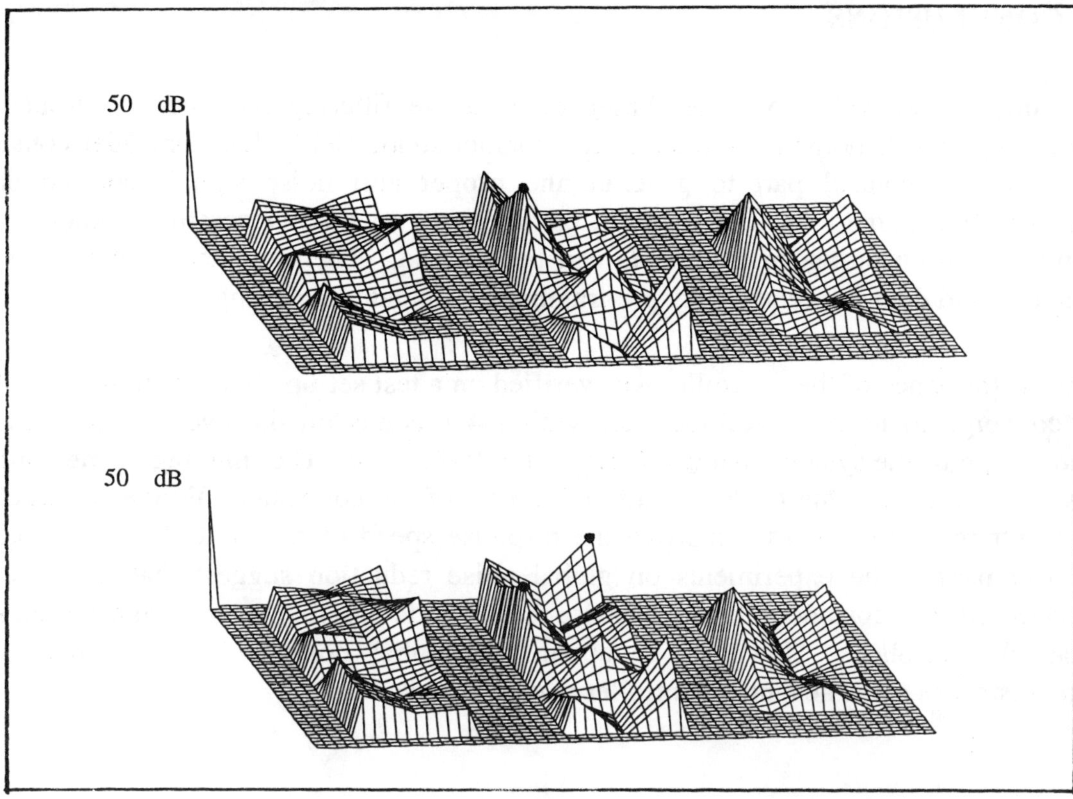

Figure 9. A typical global noise reduction at 600Hz.

configuration, where the dark points indicate the positions of the error microphones. In the figures the bottom plane of the cavity is shown on the left and the top plane on the right.

Following observations can be derived from the experiments. At low frequencies (not necessarily being the resonance frequencies) where the acoustic wavelength is larger than twice the cavity dimension, a considerable global noise reduction has been achieved even using a single channel configuration. Using more speakers and microphones did not yield much improved reductions but made the reduction plane more flat (see figure 7). At higher frequencies, the shape of noise reduction level became rather irregular for the single channel configuration as shown in figure 9. At some points the reduction level could be as high as 50dB while at some other points the reduction was very low, in some cases even noise amplification occurred. However, if the resonance frequency is clearly dominated by only a single acoustic mode, a global reduction was still possible when averaged over the whole cavity. Adding microphones at points showing small noise reduction would increase the reduction level at those points but at same time cause less reduction at the points which originally had higher reduction levels. At those resonance frequencies, dominated by several coupled acoustic modes, a global noise reduction could not be achieved by a single channel system (see figure 8) and a multichannel system had to be used.

4. CONCLUSIONS

A multichannel ANC controller based on adaptive filtering has been developed. Its efficiency was demonstrated in a 3-D reverberant sound field. The controller consists of two parts, a control part to generate the proper anti noise signals and an on-line identification part to track changes in environment so that successful operation of the controller in changing environments can be ensured. Therefore the controller is fully adaptive and does not require any pre-training or separate modelling.

The performance of the controller was verified on a test set-up. The controller was found to converge to its best working state within 4 seconds on the average, without prior knowledge of the system characteristics of the test set-up. The time lag of the controller response is mainly due to the identification part of the controller. Further research will concentrate on this part to improve the response speed of the controller to a changing environment. The experiments on global noise reduction suggest that, under certain circumstances a considerable global noise reduction is possible, even with a simple single channel controller. However, to obtain the global noise reduction under general circumstances a multichannel controller was required.

References

1. Nelson, P. A., A. R. D. Curtis, S. J. Elliott and A. J. Bullmore, 1987. "The active minimisation of harmonic enclosed sound fields, Part I : Theory." J. of Sound and Vibration, 117(1):1-13.

2. Curtis, A. R. D., P. A. Nelson, S. J. Elliott and A. J. Bullmore, 1987. "Active suppression of acoustic resonance." J. Acoust. Soc. Am., 81(3): 624-631.

3. Bullmore, A. J., P. A. Nelson and S. J. Elliott, 1990. "Theoretical studies of the active control of propeller-induced cabin noise." J. of Sound and Vibration, 140(2):191-217.

4. Berge, T., 1983. "Active noise cancellation of low frequency sound inside vehicle cabs." Proc. of Internoise, 83:457-460.

5. Miyoshi, M. and Y. Kaneda, 1988. "Active noise control in a reverberant three-dimensional sound-field." Proc. of Internoise, 88:983-986.

6. Simpson, M. A., T. M. Luong, M. A. Swinbanks, M. A. Russell and H. G. Leventhall, 1989. "Full scale demonstration test of cabin noise reduction using active noise control." Proc. of Internoise, 89:459-462.

7. Warner, J. V. and R. J. Bernhard, 1990. "Digital control of local sound fields in an aircraft passenger compartment." AIAA J., V28(2).

8. Elliott, S. J., P. A. Nelson, I. M. Stothers and C. C. Boucher, 1990. "In-flight experiments on the active control of propeller-induced cabin noise." J. of Sound and Vibration, 140(2):219-238.

9. Elliott, S. J., I. M. Stothers, P. A. Nelson, A. M. McDonald, D. C. Quinn and T. Saunders, 1988. "The active control of engine noise inside cars." Proc. of Internoise, 88(2):459-462.

10. Nelson, P. A., A. R. D. Curtis and S. J. Elliott, 1985. "Quadratic optimization problems in the active control of free and enclosed sound fields." Proc. Inst. Acoust., 7:45-54.

11. Elliott, S. J., I. M. Stothers and P. A. Nelson, 1987. "A multiple error LMS algorithm and its application to the active control of sound and vibration." IEEE Tans. on ASSP, vol. ASSP-35(10):1423-1434.

12. Widrow, B. and S. D. Stearns, 1976. Adaptive signal processing, Prentice-Hall.

13. Bao, C. and P. Sas, 1990. "Active noise cancellation in a reverberant enclosure by adaptive filtering." Proc. of 15th ISMA, 1:255-275.

14. Feintuch, P. L., 1976. "An adaptive recursive LMS filter." Proc. of IEEE, vol. 64:1622-1624.

15. Graupe, D., 1972. Identification of system, Van Nostrand Reinhold New York, char.4.

16. Creemers, G. and L. Cremers, 1989. "Experimenteel en numeriek akoestisch onderoek van een gesloten ruimte." (in Dutch), Master thesis, K.U.Leuven, Belgium.

THE EFFECT OF MODAL COUPLING CHARACTERISTICS ON ONE MECHANISM OF ACTIVE NOISE CONTROL

Scott D. Snyder,[1] Colin H. Hansen[1]

ABSTRACT

When using vibration sources to actively control sound transmission into coupled enclosures, two physical control mechanisms are possible. The first is an increase in the input impedance presented to the primary forcing function of the dominant coupled structural mode(s), which results in a decrease in their amplitude. The second is an alteration in the relative phasing and amplitudes of the coupled structural modes. The ability of this second mechanism to provide global sound control is strongly dependent upon the structural/acoustic coupling characteristics of the system. This dependency is demonstrated analytically by comparing systems of differing coupling characteristics; namely, a plain cylindrical enclosure and a cylindrical enclosure with an integral floor structure.

INTRODUCTION

The majority of previous work concerned with the active control of noise originating from vibrating structures has been directed towards the use of acoustic sources to attenuate the total system acoustic power flow. Recently, however, there has been a body of work directed towards the use of vibration sources to provide this active noise control.

When acoustic sources are used in an active noise control system, the physical mechanism which produces global sound attenuation is known to be a reduction in the radiation impedance of the primary noise source and/or the control source [1,2]. Vibration control sources, however, provide global sound attenuation by a combination of two physical mechanisms; a reduction in the amplitudes of the primary radiating structural modes (modal control), and/or an alteration of the relative amplitudes and phases of these radiating structural modes (modal rearrangement) [3]. This second mechanism has the effect of reducing the radiation impedance seen by the structure, thereby reducing the total radiated sound power.

The relative importance of the two vibration source control mechanisms in providing global sound attenuation is influenced by a number of geometric and structural / acoustic system variables. Studies of the control of free space sound radiation from a baffled, rectangular panel [4,5] have shown that where a single structural mode dominates the radiated sound field, modal control is the principal physical mechanism employed. As the modal density

[1]Dept. of Mechanical Engineering, University of Adelaide, GPO Box 498, Adelaide, South Australia 5001, Australia.

begins to increase, however, the importance of the modal rearrangement mechanism (where the number of significant structural modes exceeds the number of control sources) begins to increase. Studies of the control of sound transmission through a rectangular panel into a rectangular enclosure [6] have shown the division of total coupled structural / acoustic system energy to also influence the control mechanism employed by the use of vibration sources. If the majority of the total system energy is contained in the structure, such as near a structure-dominated system resonance (structure-controlled mode), modal control was found to be the principal mechanism employed. However, if the majority of the energy was contained in the acoustic space, such as near an acoustic space dominated system resonance (acoustic-controlled mode), modal rearrangement was found to be the principal physical mechanism employed.

These two sets of results, for free space sound radiation and for sound transmission into a coupled rectangular enclosure, would appear to differ somewhat from the original experimental work conducted on the use of vibration sources to control sound transmission into a cylindrical enclosure [7]. In this latter work, only the modal control mechanism was found to be at work, regardless of the division of coupled structural / acoustic system energy. No explanation of this apparent discrepancy in results has yet been put forward. However, this discrepancy will be discussed and explained later in this paper.

The purpose of this paper is two-fold. The first aim to analytically examine the mechanisms of control of sound transmission into a cylindrical enclosure using vibration sources. In this way it will be shown that the physical control mechanisms employed in this arrangement are consistent with results obtained elsewhere for similar vibration source active noise control systems. The second aim is to analytically examine the use of vibration sources to control sound transmission into an arrangement more representative of an aircraft fuselage, a cylindrical enclosure with an integral floor structure. This will provide a more accurate assessment of the mechanisms employed when vibration sources are used to control sound transmission into an actual aircraft interior.

ANALYTICAL MODELS

To conduct the analytical study, quadratic optimization theory [8], combined with modal coupling theory [9] to predict the total structural / acoustic system response under both primary and control excitation, was utilized. A general analytical formulation of this problem has been undertaken previously [10], and the ability of this combination of models to predict the effect of the application of active vibration control has been verified experimentally for the case of sound transmission into a rectangular enclosure [11] and, to a limited extent, into a cylindrical enclosure [12]. Modal coupling theory has also been shown to be accurate in predicting the total response of the two weakly coupled systems of interest for the general sound transmission case [13,14].

For the modelled active noise control problems, the object is to minimize the squared value of sound pressure at a discrete point or points in the enclosed acoustic space. For the case of sound transmission into a coupled enclosure, the sound pressure at any discrete point can be estimated by considering the response of m *in vacuo* structural modes and n rigid-walled acoustic modes, written in matrix form as

$$p_i(\vec{x}) = [\Phi]^T [Z_a] [Z_I]^{-1} [\Gamma] \tag{1}$$

where $[\Phi]$ is the (n x 1) vector of acoustic mode shape functions evaluated at the point of interest, $[Z_a]$ is the (n x m) matrix of modal radiation transfer functions, whose (q,j)th term is

$$z_{a_{q,j}} = \frac{i\,\rho\omega S}{\Lambda_q\,(\kappa_q^2 - k^2)}\,B_{q,j} \tag{2}$$

where ρ is the density of the fluid medium (air here), ω is the excitation frequency in rad/s, S is the structural / acoustic contacting area, Λ_q is the volume integral of the qth acoustic mode shape function squared

$$\Lambda_q = \iiint_V \phi_q^2(\vec{y})\,d\vec{y} \tag{3}$$

where $\vec{y} = (r,\theta,z)$ is a coordinate locationin the enclosed acoustic space. κ_q is the eigenvector of the qth acoustic mode (taken to be complex here to introduce acoustic damping), k is the wavenumber of the frequency of interest, and $B_{q,j}$ is the coupling factor between the qth acoustic mode, ϕ_q, and jth structural mode, ψ_j, defined as

$$B_{q,j} = \frac{1}{S}\iint_S \phi_q(\vec{x})\,\psi_j(\vec{x})\,d\vec{x} \tag{4}$$

where $\vec{x} = (\theta,z)$ is a coordinate on the structure at the structural / acoustic interface. $[Z_I]$ is the (m x m) matrix of structural modal input transfer functions, whose terms are

$$\text{diagonal:}\quad z_{j,j} = i\,\rho S^2\omega \sum_{q=1}^{n} \frac{B_{q,j}\,B_{q,j}}{\Lambda_q\,(\kappa_q^2 - k^2)} + \frac{i\,M_j Z_j}{\omega}$$

$$\text{off-diagonal:}\quad z_{\imath,j} = i\,\rho S^2\omega \sum_{q=1}^{n} \frac{B_{q,\imath}\,B_{q,j}}{\Lambda_q\,(\kappa_q^2 - k^2)} \tag{5}$$

where M_j is the modal mass of the jth structural mode

$$M_j = \iint_S \rho_s(\vec{x})\,h(\vec{x})\,\psi^2(\vec{x})\,d\vec{x} \tag{6}$$

ρ_S is the density of the structure material, and h is its thickness, and Z_j is the input impedance of the jth structural mode, given by

$$Z_j = \omega_n^2\,(1 + i\eta) - \omega^2 \tag{7}$$

where η is the structure loss factor. $[\Gamma]$ is the (m x 1) vector of (structure) modal generalized forces, whose jth element is given by

$$\gamma_j = \iint_S \psi_j(\vec{x})\ p_e(\vec{x})\ d\vec{x} \tag{8}$$

where p_e is the external excitation force expressed in units of pressure.

Modelling the j control forces as point forces, the sound pressure squared at a set of r discrete interior points can be found, using equation (1), from

$$[p^2] = [F_c]^H [a]\ [F_c] + [F_c]^H [b] + [b]^H [F_c] + [c] \tag{9}$$

where $[p^2]$ is the (r x 1) vector of sound pressures squared at the points of interest, $[F_c]$ is the (j x 1) vector of complex control forces, and

$$[a] = [\Psi_c]^H \{[Z_I]^{-1}\}^H [Z_a]^* [\Phi_e]^* [\Phi_e]^T [Z_a]^T [Z_I]^{-1} [\Psi_c]$$

$$[b] = [\Psi_c]^H \{[Z_I]^{-1}\}^H [Z_a]^* [\Phi_e]^* [\Phi_e]^T [Z_a]^T [V_p]$$

$$[c] = [V_p]^H [Z_a]^* [\Phi_e]^* [\Phi_e]^T [Z_a]^T [V_p] \tag{10}$$

where $[V_p]$ is the (m x 1) vector of structural modal amplitudes under primary excitation, $[\Phi_e]$ is the (1 x r) column vector acoustic mode shape vectors, $[\Phi]$, evaluated at the error sensing locations, and $[\Psi_c]$ is the (m x j) matrix of structural mode shape functions evaluated at the control force application points. In this form, the optimum control force, which will minimize the squared sound pressures at the points of interest, is [8]

$$[F_c]_{opt} = -\ [a]^{-1}\ [b] \tag{11}$$

The general form of equation (9) needs to be specialized for the cylinder and cylinder with integral floor considered here. For the plain cylinder, the structural mode shape function is

$$\psi_{M,N}(z,\theta) = \sin(\frac{M\pi z}{L})\ (\cos(N\theta) + \sin\ (N\theta)) \tag{12}$$

where M,N are the axial and circumferential modal indices, respectively, and L is the length of the cylinder. The acoustic mode shape function is

$$\phi_{q,n,s}(r,\theta,z) = J_n(\alpha_{ns} r/a)\ \cos(n\theta)\ \cos(\frac{q\pi z}{L}) \tag{13}$$

where q,n are the axial and circumferential modal indices, respectively, J_n is a Bessel function of the first kind of order n, and α_{ns} is the sth zero of the derivative of the Bessel function of order n. Evaluating equation (4), the coupling between the acoustic and structural mode shape functions is found to be [13]

$$B_{MN,qns} = \frac{J_n(\gamma_{ns})\ \epsilon_n \kappa_m (1-(-1)^{M+q})}{2L(\kappa_m^2 - \kappa_q^2)} \qquad n=N,\ \frac{M+q}{2} \neq \text{integer}$$

$$= 0 \qquad \text{otherwise} \tag{14}$$

where

$$\epsilon_n = \begin{cases} 2 & q,n = 0 \\ 1 & q,n > 0 \end{cases}$$

$$\kappa_m = \frac{M\pi}{L}, \quad \kappa_q = \frac{q\pi}{L}$$

For the cylinder with the integral floor structure, the structural mode shape functions can be found using component mode synthesis [15]. The basis function for the total structural mode shape is then a combination of the mode shape basis functions for the plain cylinder and floor (plate) structures. Simply supported boundary conditions are assumed for the longitudinal ends of both the cylinder and floor structure, so that the axial mode shape function of the structure is

$$\psi_M = \sin \frac{M\pi z}{L} \tag{15}$$

The acoustic mode shape functions can be found by using a finite difference implementation of the Helmholtz equation [14]. The acoustic mode shape functions are then represented as eigenvectors of a set of values at a discrete set (grid) of points. Linear interpolation can be used to assess the value of the mode shape at any off-grid point. The axial mode shape function is equal to that of a 1-dimensional enclosure

$$\phi_q = \cos \frac{q\pi z}{L} \tag{16}$$

Coupling between the structural and acoustic modes can be evaluated numerically.

PLAIN CYLINDER TESTS

Consider first the problem of controlling sound transmission into a plain cylindrical enclosure. Referring to figure 1, the cylinder is fabricated of aluminum, of length L=1.212m, radius a=0.254m, and thickness h=1.6mm. Both the structure and acoustic space are assumed to be "moderately" damped, with a structural loss factor of 0.07 and

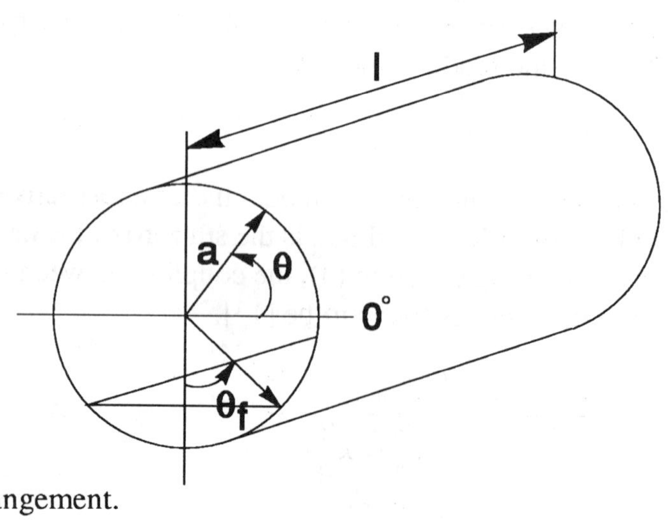

Figure 1. Cylinder arrangement.

acoustic loss factor of 0.01. The primary exciting force is modelled as a point force, located at $(z,\theta) = (0.606, 0°)$. The first 100 structural and 100 acoustic modes were used in the calculations.

The first case to consider is a structure controlled mode. Figure 2 illustrates the interior response at the cylinder midsection to an excitation frequency of 272 Hz, which is near the resonance frequency of the (*in vacuo*) structural (1,2) mode. The influence of the circumferential $\cos(2\theta)$ structural component is reflected in the 4-lobed pattern of the sound

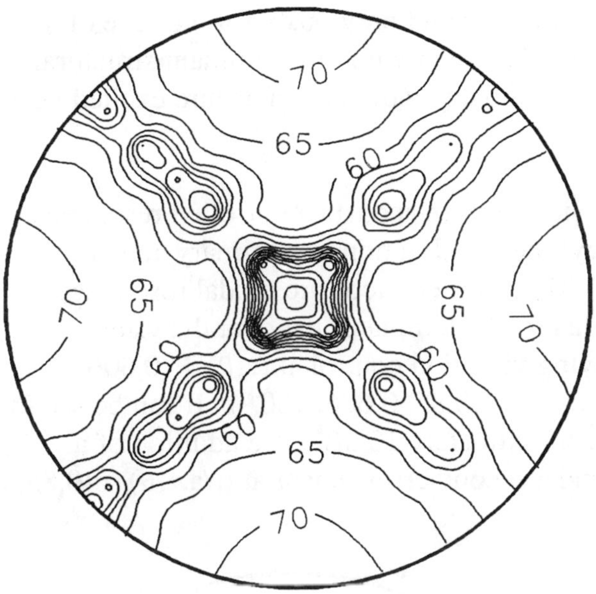

Figure 2. Primary excited sound field, 272 Hz.

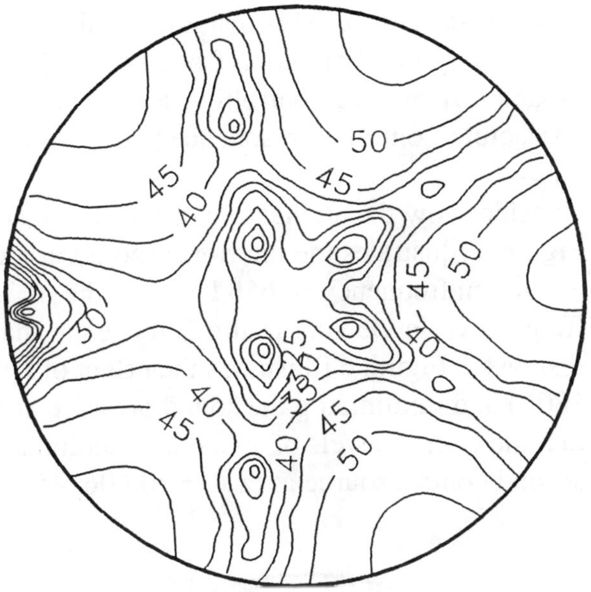

Figure 3. Controlled sound field, 272 Hz, control source at 180°, error sensor at $(r/a, \theta) = (0.85, 180°)$.

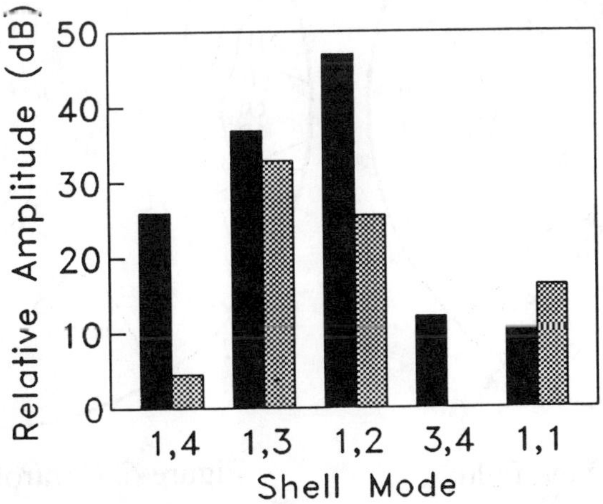

Figure 4. Structural modal amplitudes, 272 Hz. ▬ = primary excitation, ▒ = controlled.

field (note that, as outlined in equation (14), only acoustic modes of equal circumferential indice will be excited by a given structural mode). The residual sound field after applying vibration control diametrically opposite the primary exciting force, at 180°, and minimizing the sound pressure at $(r/a,z,\theta) = (0.85, 0.606, 180°)$ is shown in figure 3. From this it is clear that significant global sound attenuation has been achieved. Figure 4 shows the modal amplitudes of the 5 dominant coupled structural modes before and after the application of active control. In viewing this, it is apparent that the physical mechanism at work is modal control, as the amplitude of the dominant (1,2) structural mode has been significantly reduced. (It should be noted that this reduction in modal amplitude of the principal coupled structural modes is accompanied by an overall reduction in the vibration level of the cylinder structure). This result is consistent with those found previously for the case of controlling sound radiation from a baffled, rectangular panel with one dominant structural mode [4,5], and for controlling sound transmission into a rectangular enclosure excited near a structure-controlled resonance [6].

Consider now the attenuation of sound transmission near an acoustic controlled resonance. Figure 5 illustrates the interior response at the cylinder midsection to a primary source excitation frequency of 656 Hz, near the (0,2,1) (rigid walled) acoustic modal resonance frequency. In viewing figure 5, it is clear that this mode dominates the overall system response. Figure 6 illustrates the effect of applying vibration control at $(z,\theta) = (0.606, 180°)$ and minimizing the sound pressure at $(r/a,z,\theta) = (0.85, 0.606, 180°)$. It can be seen that the overall levels of sound attenuation are approximately 10 dB. The addition of a second control source at $(z,\theta) = (0.606, 45°)$, and a second error sensor at $(r/a,z,\theta) = (0.85,$

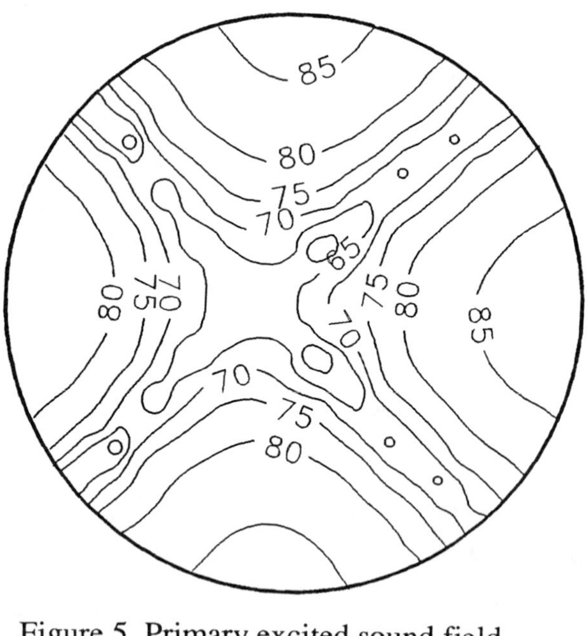

Figure 5. Primary excited sound field, 656 Hz.

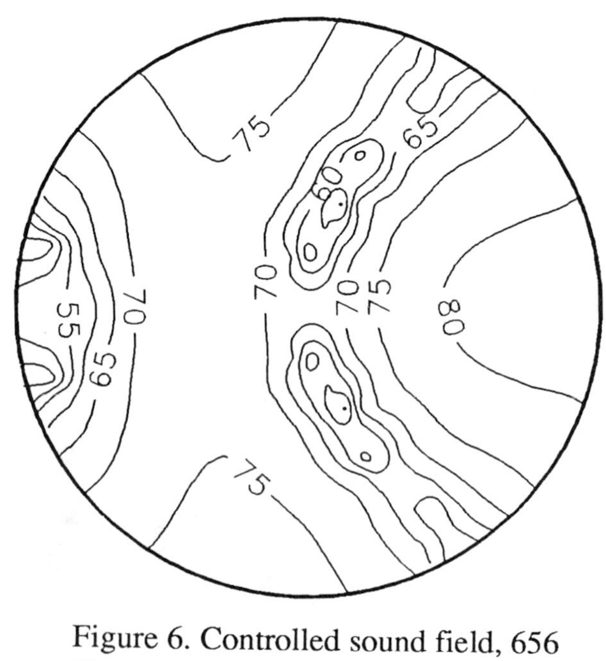

Figure 6. Controlled sound field, 656 Hz, control source at 180°, error sensor at $(r/a,\theta) = (0.85, 180°)$.

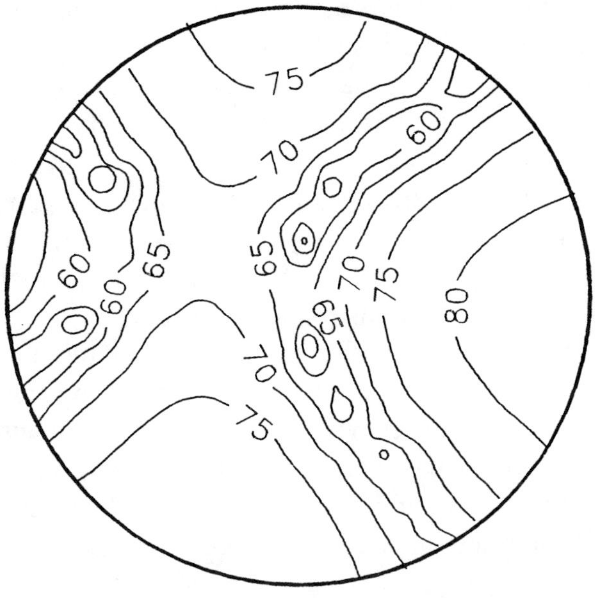

Figure 7. Controlled sound field, 656 Hz, control sources at 45° and 180°, error sensor at $(r/a, \theta)$ = (0.85, 135°) and (0.85, 180°).

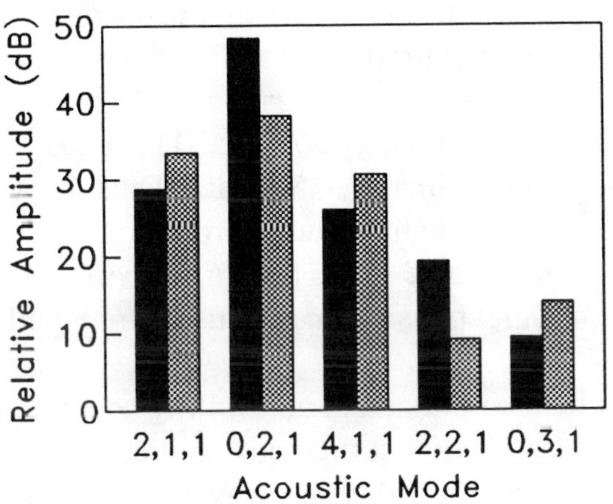

Figure 8. Acoustic modal amplitudes, 656 Hz. ■ = primary excitation, ▒ = controlled.

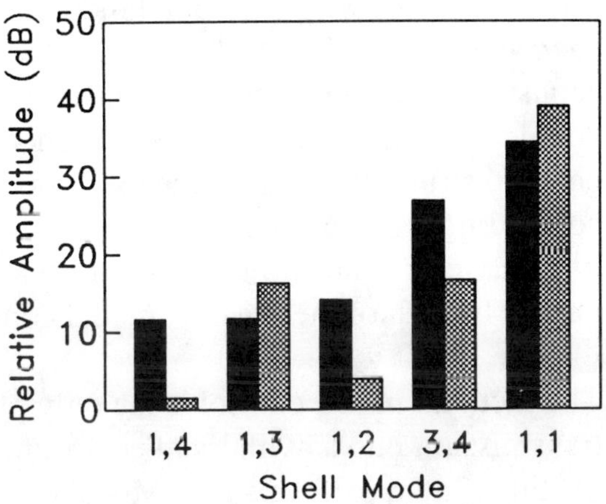

Figure 9. Structural modal amplitudes, 656 Hz. ■ = primary excitation, ▒ = controlled.

0.606, 135°) does not improve upon this result markedly, as shown in figure 7. The amplitudes of the 5 dominant acoustic modes under both primary excitation and single control source / error sensor control (figure 6) are shown in figure 8. It can be seen here that under the action of vibration control, there is a general decrease in the acoustic modal amplitude of the dominant (0,2,1) mode, which is reflected in the decrease in overall sound pressure. This trend is reflected in the dominant coupled structural modes, shown in figure 9 (it should be noted that the (1,2) structural mode is the one principally coupled to the dominant (0,2,1) acoustic mode).

These results indicate that modal control is again the dominant physical mechanism at work. The question that now must be answered is why was modal control the dominant mechanism in this instance, while the similar acoustic-controlled case for a rectangular enclosure involved principally modal rearrangement [6]. The answer lies in the difference in modal coupling characteristics of the two structural / acoustic systems. For the rectangular panel / rectangular enclosure system, even indice panel modes couple with odd indice cavity modes, and vice versa. Hence, a given acoustic mode can couple to a relatively high number of significantly excited structural modes, which offers the potential of modal rearrangement of the relative phases and amplitudes of these coupled structural modes to reduce the effective overall radiation transfer (impedance) of the structure to the dominant acoustic mode. This is similar to the case of controlling free space radiation from a rectangular panel of high modal density with a small number of control sources (such as a single control source) [4,5]. For both of these cases, modal rearrangement is the principal physical mechanism of control. Also, in both cases the overall structure vibration may be increased, despite a reduction in the radiated sound pressure.

The structural / acoustic coupling characteristics of the cylindrical enclosure system are very different from those of the rectangular panel / rectangular panel system. From equation (14), it can be seen that the circumferential indices of the acoustic and structural modes must be equal, and the only structural modes with odd axial modal indices will couple with acoustic modes with even axial modal indices, and vice versa. This criterion means that each acoustic mode "sees" very few significantly excited structural modes. For the (0,2,1) acoustic mode considered previously, resonant at approximately 656 Hz, the primary coupled structural mode is the (1,2) mode, resonant at approximately 272 Hz. The next coupled structural mode is the (3,2) mode, resonant at approximately 1500 Hz. Thus, for controlling a given acoustic mode, it is essentially a single structural mode problem, similar to the case of controlling free field sound radiation from a single mode dominated vibrating panel. In both of these cases, modal control is the principal physical mechanism employed.

The fact that modal control is the only physical mechanism utilized in controlling sound transmission into a cylindrical enclosure with a vibration control source has implications for overall system performance. Modal control, in general, requires the use of as many control sources and error sensors as there are contributing structural modes to the primary disturbance. Therefore, for guaranteed attenuation of sound transmission into a cylindrical enclosure across a wide operating frequency range, a large number of control sources and error sensors may be required. A single control source and error sensor will only be useful in providing sound attenuation to the extent which (effectively) a single structural mode dominates the system response; off resonant performance will be reduced.

This fact is illustrated in figures 10 and 11, which depict the primary sound field at the cylinder mid-length for the off-resonant frequency of 600 Hz, and the residual sound field after applying control at $(z,\theta) = (0.606, 180°)$, and minimizing the sound pressure at $(r/a,z,\theta) = (0.85, 0.606, 180°)$. While high levels of local sound attenuation are achieved, the global result is poor. The acoustic modal decompositions of figure 12, combined with the structural modal decompositions of figure 13, indicate that the cylinder response is dominated by the response of two modal sets. Even with the addition of both a second control source / error sensor set (control source at $(z,\theta) = (0.606, 110^\circ)$, error sensor at $(r/a,z,\theta) = (0.85, 0.606, 110^\circ)$), shown in figure 14, and a further third control source / error sensor set (control source at $(z,\theta) = (0.606, 260^\circ)$, error sensor at $(r/a,z,\theta) = (0.85, 0.606, 260^\circ)$), shown in figure 15, the results are still not as good as with the on-resonance case.

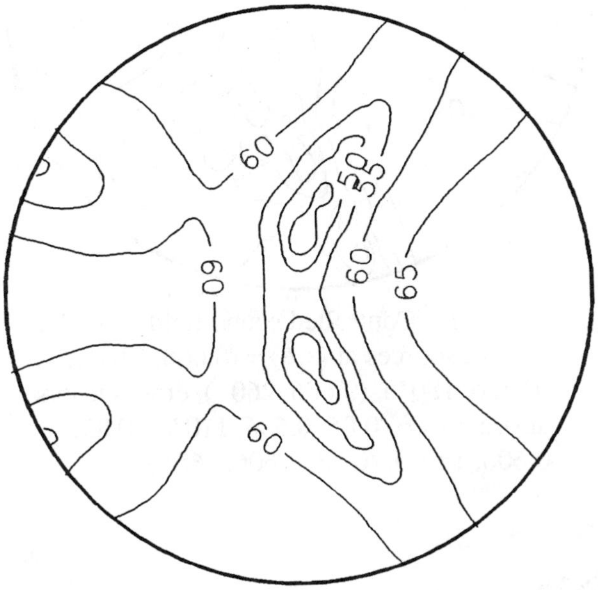

Figure 10. Primary sound field, 600 Hz.

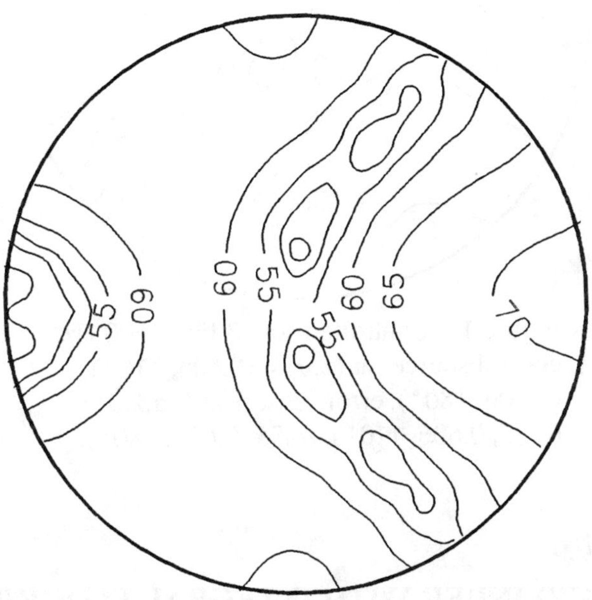

Figure 11. Controlled sound field, 600 Hz, control source at $(z,\theta) = (0.606, 180^\circ)$, error sensor at $(r/a,z,\theta) = (0.85, 0.606, 180^\circ)$.

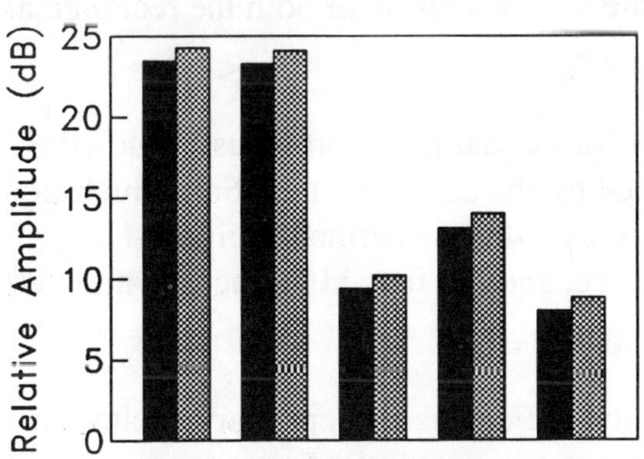

Figure 12. Acoustic modal amplitudes, 600 Hz. ▬ = primary excitation, ▒ = controlled.

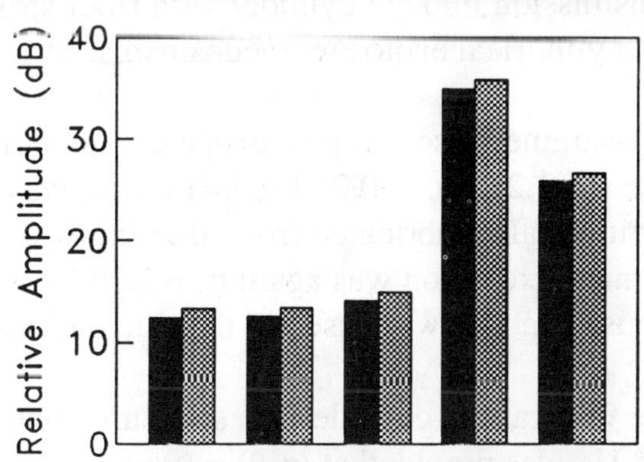

Figure 13. Structural modal amplitudes, 600 Hz. ▬ = primary excitation, ▒ = controlled.

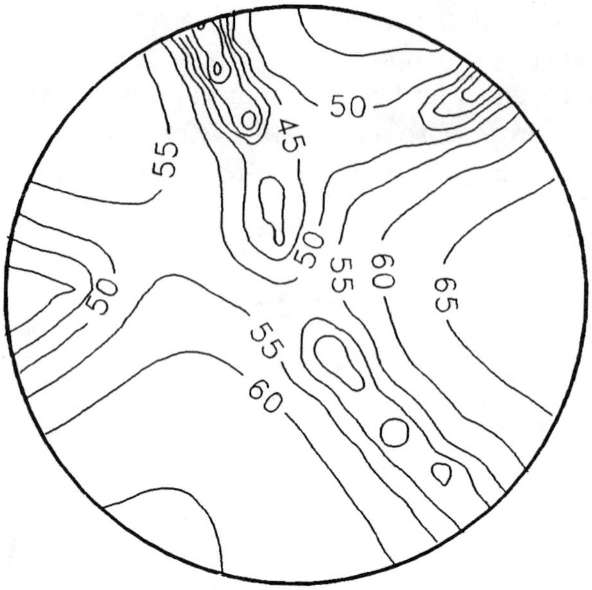

Figure 14. Controlled sound field, 600 Hz, control sources at (z,θ) = (0.606, 110°), (0.606, 180°), error sensors at (r/a,z,θ) = (0.85, 0.606, 110°), (0.85, 0.606, 180°).

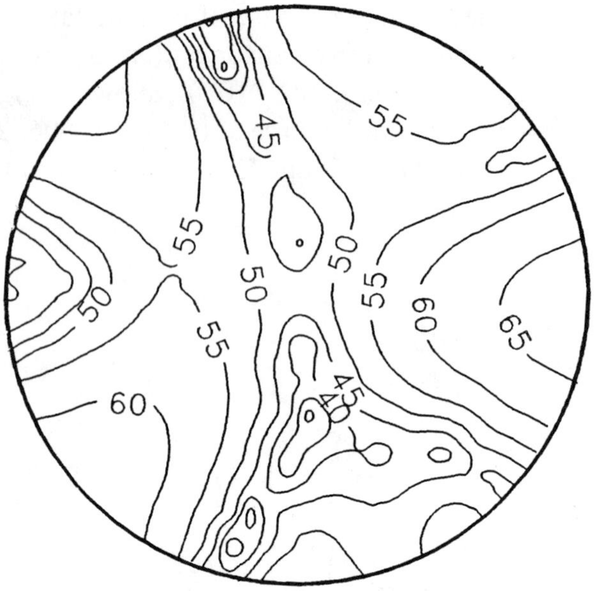

gure 15. Controlled sound field, 600 Hz, control sources at (z,θ) = (0.606, 110°), (0.606, 180°), (0.606, 260°), error sensors at (r/a,z,θ) = (0.85, 0.606, 110°), (0.85, 0.606, 180°), (0.85, 0.606, 260°).

CYLINDER WITH INTEGRAL FLOOR TESTS

A cylindrical enclosure with an integral floor structure is a more accurate model of an aircraft fuselage than either a plain cylindrical enclosure or a rectangular enclosure. In fact, it may be seen as a being combination of the two. As such, it may be expected that the physical control mechanisms employed when vibration sources are used to control sound transmission into the cylinder with floor structure are a combination of both the rectangular and cylindrical enclosure mechanisms.

To examine these mechanisms, a cylinder of dimensions equal to the previously studied case (L=1.212m, a=0.254m, h=1.6mm) was modified by the addition of a 1.6mm thick floor partition, also fabricated from aluminum, at an angle $\theta_f = 45°$ as defined in figure 1. Primary excitation was again provided by a point force, and the first 110 structural and 100 acoustic modes were used in the calculations.

The first case to consider is a structure controlled mode. For this case, primary excitation at 228 Hz was provided at (z,θ) = (0.606, 0°). The frequency is near the first resonance (longitudinal modal indice equal to 1) of the structural mode whose circumferential shape is shown in figure 16. The resulting sound pressure distribution at the cylinder mid-length is shown in figure 17.

The effect of applying vibration control at (z,θ) = (0.606, 180°), and minimizing the sound pressure at (r/a,z,θ) = (0.85, 0.606, 180°), is shown in figure 18. As with the plain cylinder case, global sound attenuation is achieved for the case where the response is dominated by a structure controlled mode. This is accompanied by a reduction in both the modal amplitude of the dominant (near-resonant) structural mode and the overall structural vibration levels. This mechanism is consistent with both the plain cylinder and rectangular enclosure structural / acoustic system mechanisms.

The next case to consider is an excitation frequency near the resonance frequency of an acoustic controlled mode. The frequency chosen for this was 720 Hz, which is near the cut-

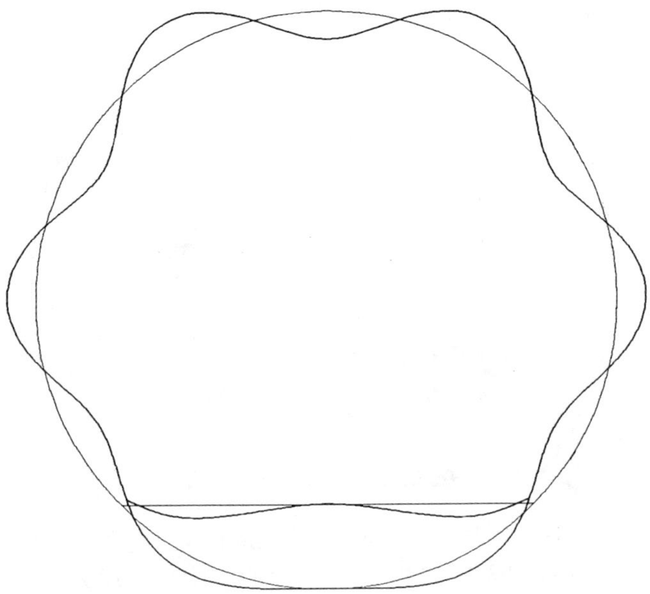

Figure 16. Cylinder with floor structure circumferential mode shape, 228 Hz.

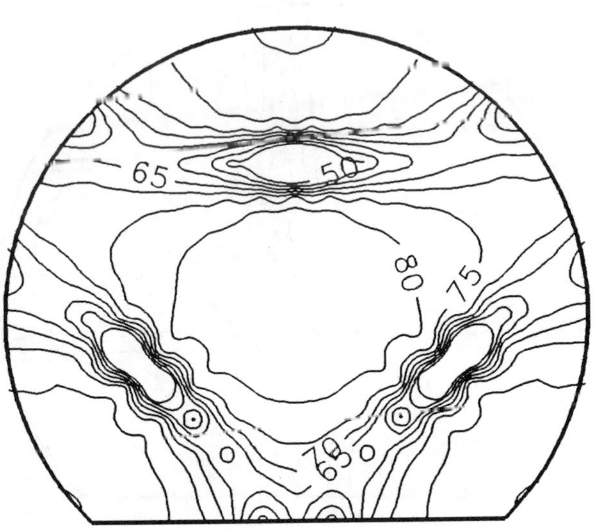

Figure 17. Primary excited sound field, 228 Hz.

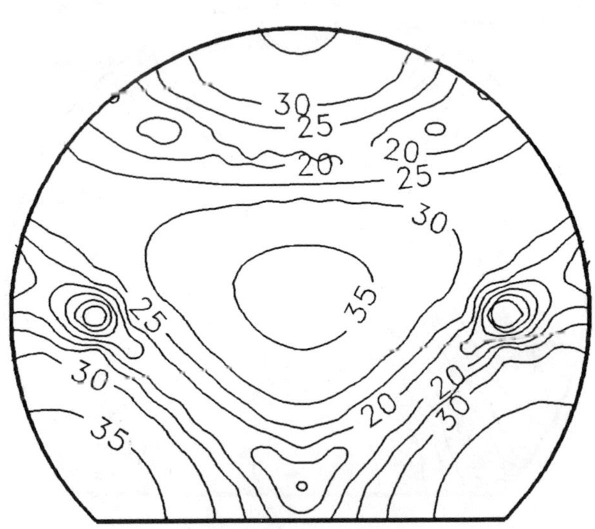

Figure 18. Controlled sound field, 228 Hz, control source at 180°, error sensor at $(r/a, \theta) = (0.85, 180°)$.

on frequency (0 longitudinal modal indice) of the acoustic mode whose circumferential shape is shown in figure 19. For a primary exciting force was applied to a position $(z, \theta) = (0.606, 0°)$, the resulting sound field at the cylinder mid-length is shown in figure 20. The effect of applying a single control force at $(z, \theta) = (0.606, 180°)$, and minimizing the sound pressure at $(r/a, z, \theta) = (0.85, 0.606, 180°)$ is illustrated in figure 21. It is clear here that the greatest levels of sound attenuation are again in the vicinity of the error sensor, accompanied by global sound attenuation in the order of 20 dB.

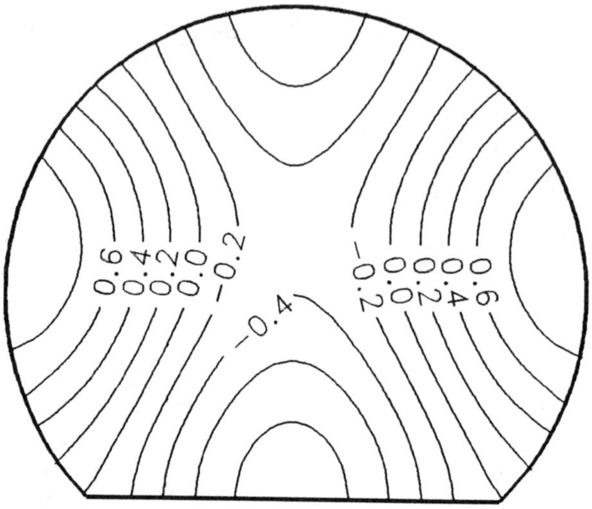

Figure 19. Cylinder with floor circumferential acoustic mode shape, 720 Hz.

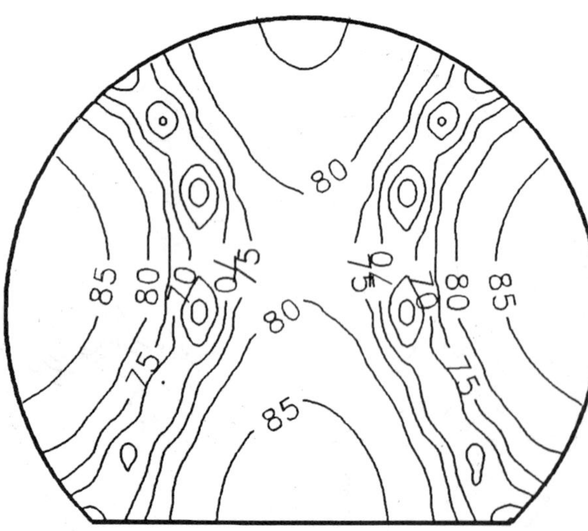

Figure 20. Primary excited sound field, 720 Hz.

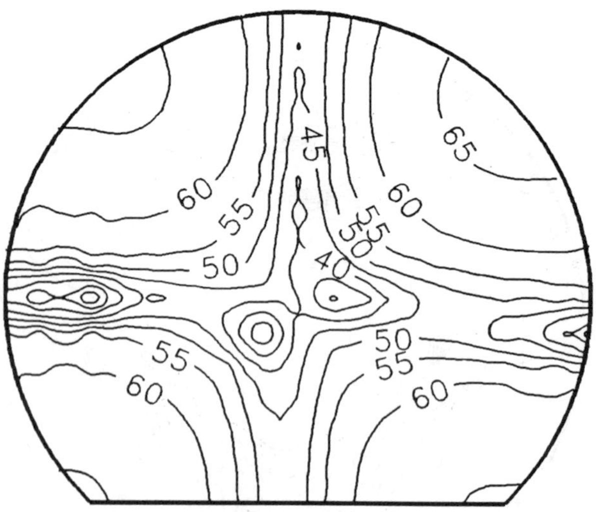

Figure 21. Controlled sound field, 720 Hz, control source at 180°, error sensor at $(r/a, \theta) = (0.85, 180°)$.

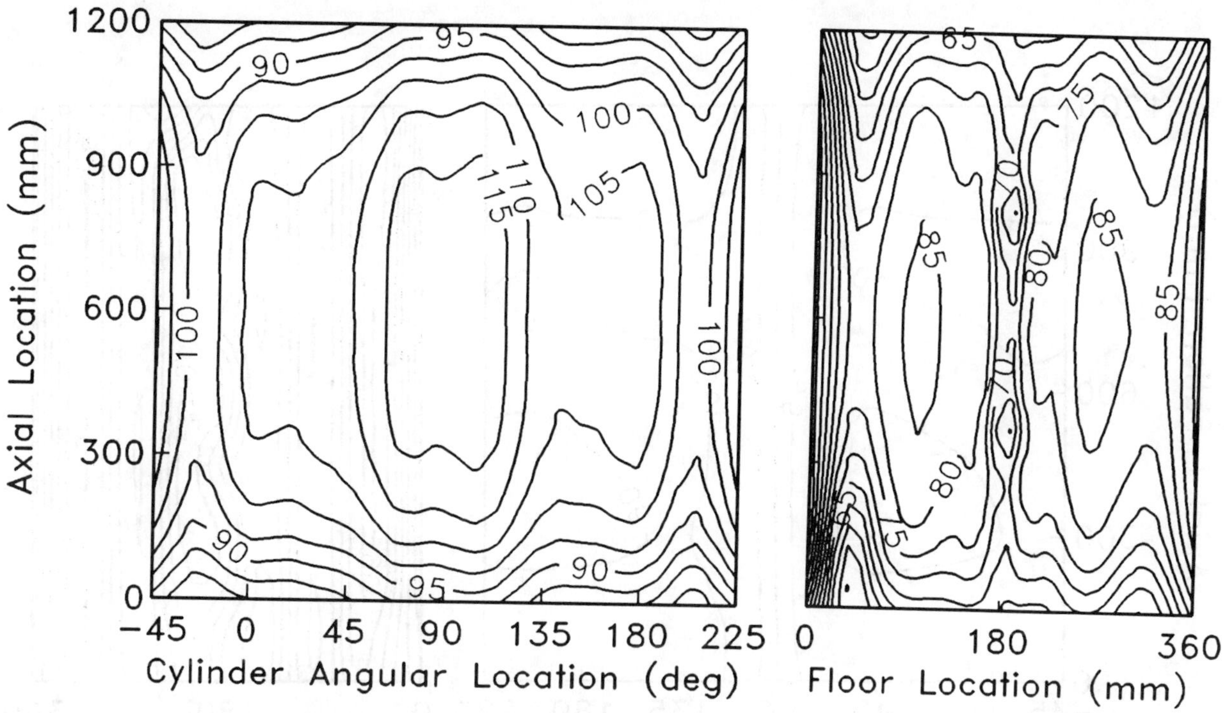

Figure 22. Primary excitation structural velocity levels, 720 Hz.

Despite the significant levels of global sound attenuation achieved by the single control source, single error sensor system, the amplitudes of the primary coupled structural modes were not reduced. Rather, their relative amplitudes and phases were changed. This has the effect of altering the overall vibration pattern of the structure, as seen in figures 22-25. Comparing the initial (primary) vibration levels of figure 22 with the controlled vibration levels of figure 24, it can be seen that the overall vibration levels have, in fact, increased under the action of active vibration control. However, when viewed with the phase plots of figures 23 and 25, it is evident that, under the action of active control, the vibration pattern of the structure has "broken up" into a number of small positive / negative pockets, which has a reduced radiation transfer efficiency compared to the initial vibration pattern. This is the same effect that occurs when the modal rearrangement mechanism is utilized for controlling free space sound radiation [4,5] and sound transmission into a rectangular enclosure through a rectangular panel [6].

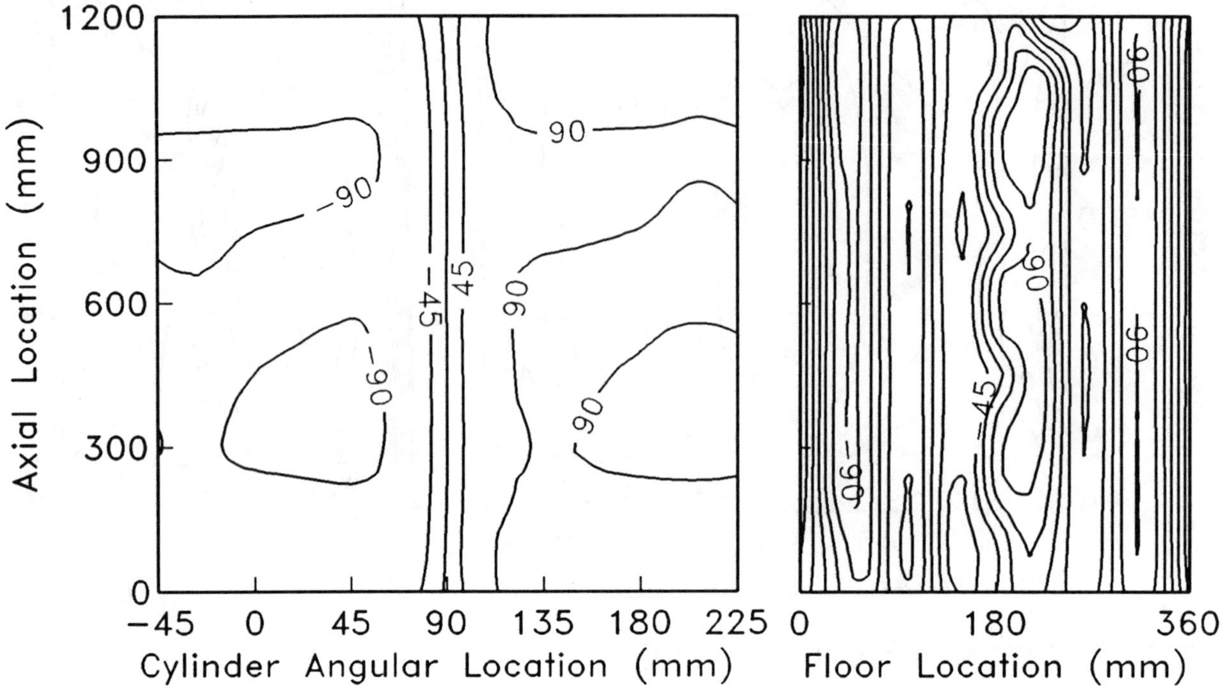

Figure 23. Primary excitation structural phases, 720 Hz.

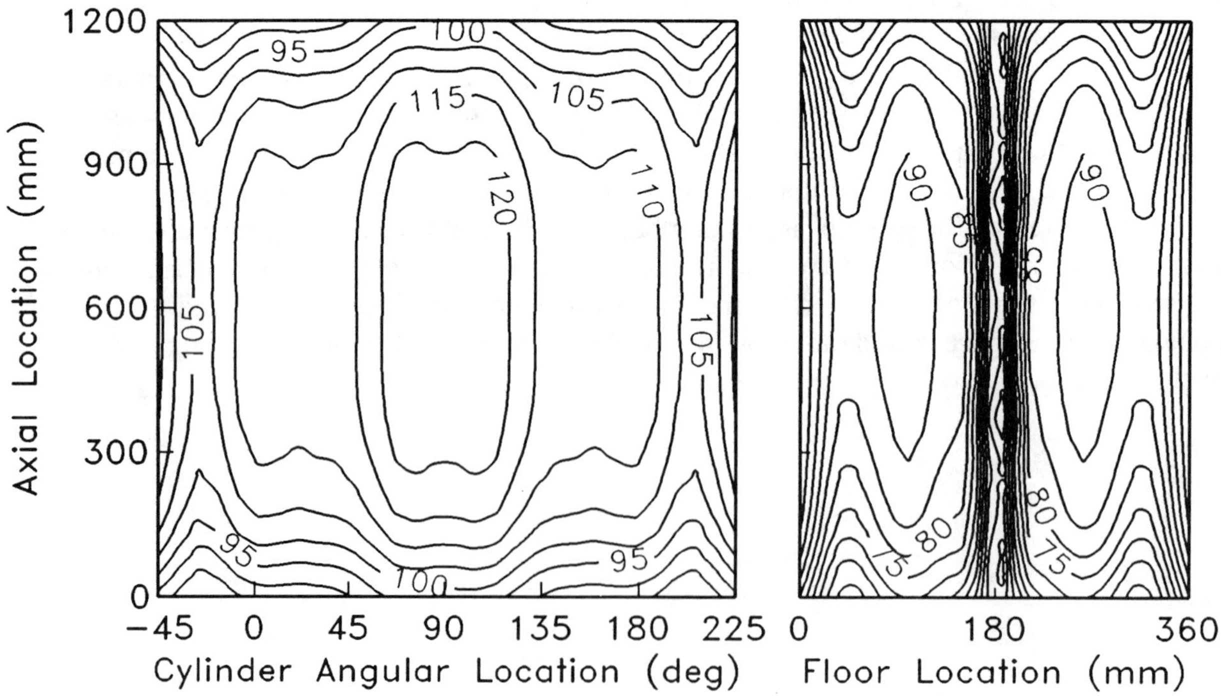

Figure 24. Controlled structural velocity levels, 720 Hz.

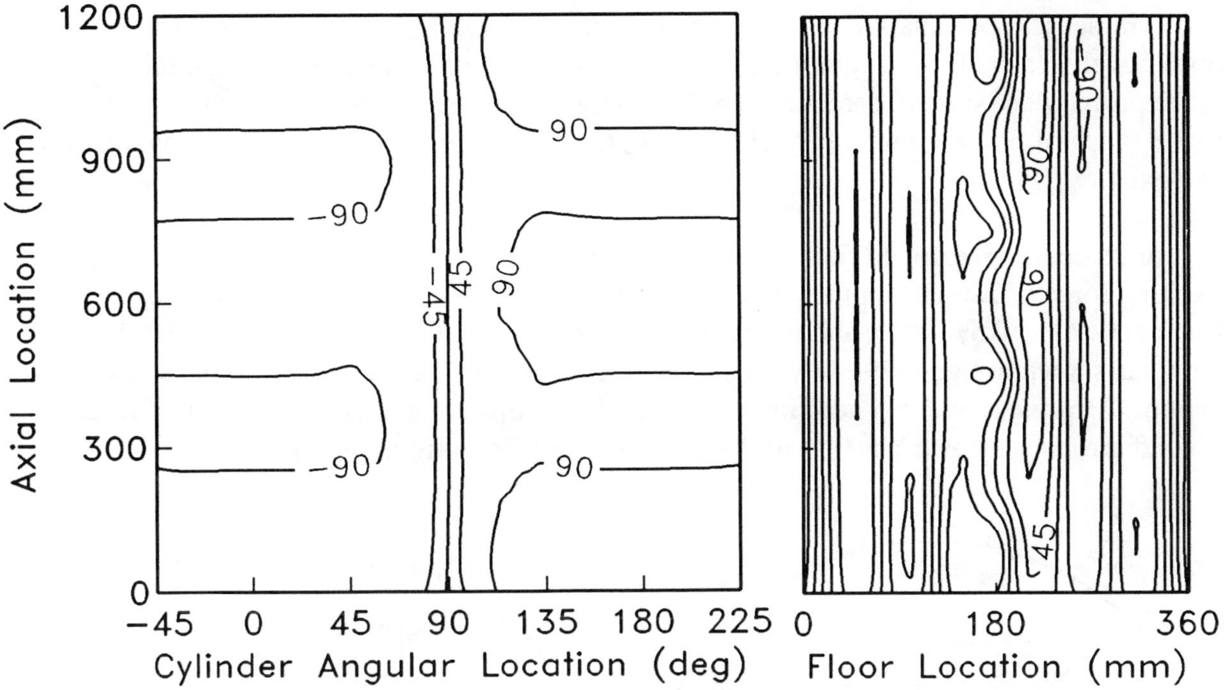

Figure 25. Controlled structural phases, 720 Hz.

The question to be answered now is why was the modal rearrangement mechanism utilized as a global control mechanism for the cylinder with floor structure, and not the plain cylinder? Again, the answer lies in the modal coupling characteristics of the structural / acoustic system. For the cylinder with floor enclosure the majority of the symmetric circumferential acoustic modes couple with the majority of symmetric circumferential structural modes, and the same for the anti-symmetric case (note here again that only odd axial indice structural modes will couple with even axial indice acoustic modes, and vice versa). Therefore, as with the rectangular enclosure case, each acoustic mode for the cylinder with floor case can couple with several significantly excited structural modes, enabling the modal rearrangement mechanism to be effective in achieving global sound control in this instance.

The fact that both the modal control and modal rearrangement mechanisms may be utilized effectively in controlling sound transmission into the cylinder with floor system means that, in some instances, a sound field driven by a very large number of structural modes may be controlled by very few vibration sources. To illustrate this, consider again the 720 Hz sound transmission problem. There are, in fact, 3 acoustic resonances within 3 Hz of this frequency. Besides the mode shape shown in figure 19, the circumferential mode shape shown in figure 26, with an axial modal indice of 3, and the circumferential mode shape shown in figure 27, with an axial modal indice of 1, are resonant at approximately 719 Hz and 722.5 Hz, respectively. Because the mode shape of figure 27 is anti-symmetric, and the two symmetric mode shapes of figures 19 and 26 have even and odd axial modal indices, respectively, each of these acoustic modes will couple to different sets of structural modes. If the primary and control forces are applied midway along the cylinder length, only the mode shape of figure 19 will be excited; away from this point all three near-resonant modes can exist.

Figure 28 depicts the sound field at a length 0.5 meters along the cylinder for a primary exciting force of 720 Hz applied at $(z,\theta) = (0.5, -30°)$. The effect of applying a control force at $(z,\theta) = (0.5, 140°)$, and minimizing the sound pressure at $(r/a,z,\theta) = (0.9, 0.5, 140°)$ is shown in figure 29. In viewing this it can be seen that the global sound attenuation is poor. However, with the addition of a second control source at $(z,\theta) = (0.5, 180°)$, and a second error sensor at $(r/a,z,\theta) = (0.9, 0.5, 180°)$, shown in figure 30, the global

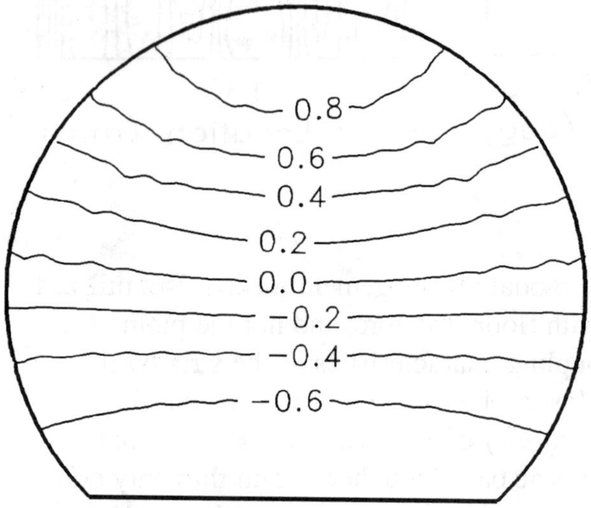

Figure 26. Cylinder with floor circumferential acoustic mode shape, 719 Hz.

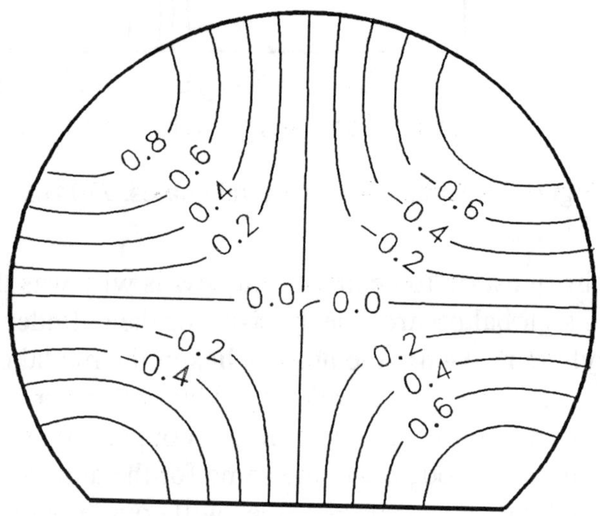

Figure 27. Cylinder with floor circumferential acoustic mode shape, 722.5 Hz.

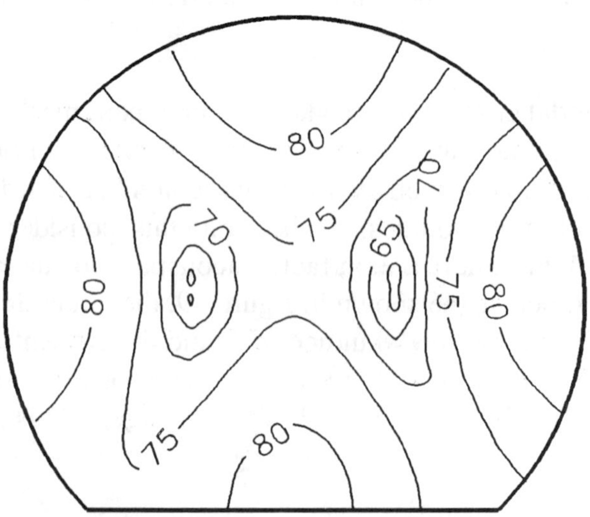

Figure 28. Primary excited sound field at L=0.5m, 720 Hz.

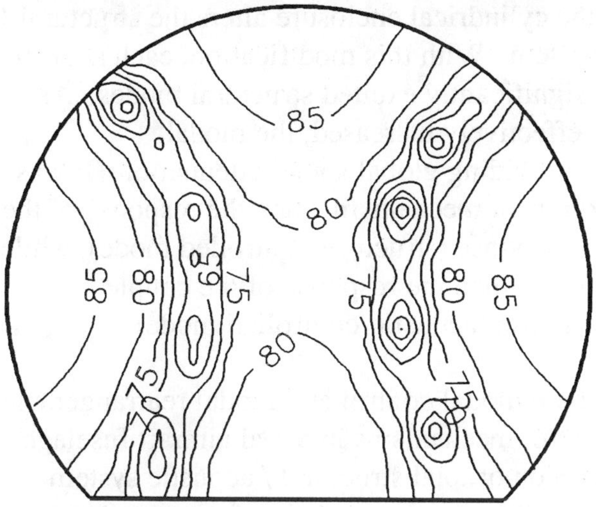

Figure 29. Controlled sound field at L=0.5m, 720 Hz, control source at (z,θ) = (0.5, 140°), error sensor at $(r/a,z,\theta)$ = (0.9, 0.5, 140°).

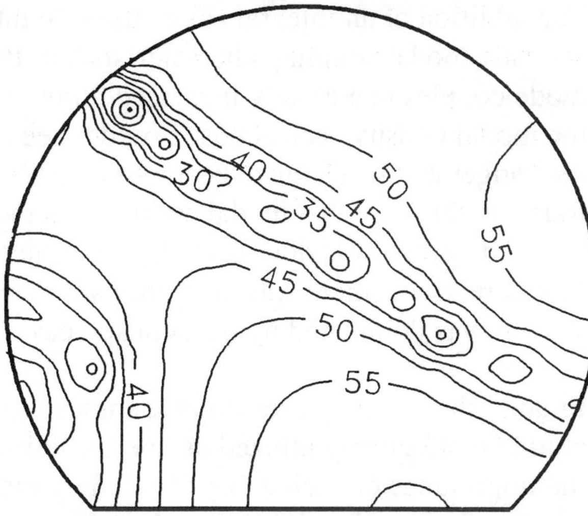

Figure 30. Controlled sound field at L=0.5m, 720 Hz, control sources at (z,θ) = (0.5, 140°), (0.5, 180°), error sensors at $(r/a,z,\theta)$ = (0.9, 0.5, 140°), (0.9, 0.5, 180°).

sound attenuation is greatly improved. Comparing this result to that of the plain cylinder with multiple acoustic modes, shown in figures 11, 14, and 15, suggests that the change in the structural / acoustic modal coupling characteristics of the system due to the addition of the floor partition has resulted in an increase in the effectiveness of vibration control sources.

CONCLUSIONS

When vibration sources are used to actively control structurally radiated noise, two physical control mechanisms are possible; modal control, and modal rearrangement. Modal control operates by reducing the amplitude(s) of the principal radiating structural modes, thereby reducing the amplitude of the resulting sound field. Modal rearrangement operates by altering the relative phases and amplitudes of the dominant structural modes, thereby forcing the structure into a less efficient overall vibration / radiation pattern.

When controlling sound radiation into a weakly coupled structural / acoustic enclosure, the coupling characteristics of the total system have a significant influence upon which control mechanism is utilized, and what level of control is achieved. For a plain cylindrical enclosure, where each acoustic mode is essentially only coupled to one structural mode, modal control is the only effective control mechanism. This is because, as the effective modal density each acoustic mode "sees" is so limited, there are not enough modes to enable modal rearrangement to be effective. Further, as each coupled structural / acoustic mode set is essentially independent, there must be as many control sources as dominant modal sets for global sound attenuation to be assured; a single control source is only effective in providing global sound control to the level which a single coupled modal set dominates the total system response.

The addition of an integral floor structure into the cylindrical enclosure alters the structural / acoustic modal coupling characteristics of the system. With this modification, each acoustic mode couples to a greatly increased number of significantly excited structural modes. As the modal density each acoustic mode "sees" is effectively increased, the modal rearrangement mechanism becomes effective in providing global sound attenuation. It was found in this case that modal control was the dominant mechanism when the response of the coupled system was dominated by the structural response (structure-controlled mode), while modal rearrangement was the dominant mechanism when the response of the coupled system was dominated by the acoustic cavity response (acoustic-controlled mode).

Finally, the results presented here indicate that both modal control and modal rearrangement could be effectively utilized as interior noise control mechanisms in a real aircraft fuselage, the importance of each being dictated by the division of total structural / acoustic system energy. In fact, the global sound attenuation achieved over the operating frequency range would probably be better than may be predicted by a simple, bare cylinder model.

REFERENCES

1. Nelson, P.A., A.R.D. Curtis, S.J. Elliott, and A.J. Bullmore, 1987. "The Minimum Power Output of Free Field Point Sources and the Active Control of Sound". Journal of Sound and Vibration, 116, 397-414.

2. Snyder, S.D. and C.H. Hansen, 1989. "Active Noise Control in Ducts: Some Physical Insights". Journal of the Acoustical Society of America, 86, 184-194.

3. Snyder, S.D. and C.H. Hansen, 1990. "Mechanisms of Active Noise Control Using Vibration." Journal of Sound and Vibration, accepted for publication.

4. Hansen, C.H. and S.D. Snyder, 1991. "Effect of Geometric and Structural / Acoustic Variables on the Active Control of Sound Radiation From A Vibrating Surface". Paper to be presented at the conference on Recent Advances in Active Control of Sound and Vibration, VPI & SU, April 15-17, 1991.

5. Snyder, S.D. and C.H. Hansen, 1990. "A General Approach to the Design of Active Systems to Control Sound Radiation From Vibrating Structures, Part 2: Radiation Into Free Space." Journal of the Acoustical Society of America, submitted for publication.

6. Pan, J., C.H. Hansen, and D.A. Bies, 1990. "Active Control of Noise Transmission Through A Panel Into A Cavity: I. Analytical Study." Journal of the Acoustical Society of America, 87, 2098-2108.

7. Fuller, C.R. and J.D. Jones, 1987. "Experiments on the Reduction of Propeller Induced Interior Noise By Active Control of Cylinder Vibration." Journal of Sound and Vibration, 112, 389-395.

8. Nelson, P.A., A.R.D. Curtis, and S.J. Elliott, 1985. "Quadratic Optimization Problems in the Active Control fo Free and Enclosed Sound Fields." Proceedings of the Institute of Acoustics, 7, 45-53.

9. Pope, L.D., 1971. "On the Transmission of Sound Through Finite Closed Shells: Statistical Energy Analysis, Modal Coupling, and Nonresonant Transmission." Journal of the Acoustic Society of America, 50, 1004-1018.

10. Snyder, S.D. and C.H. Hansen, 1990. "A General Approach to the Design of Active Systems to Control Sound Radiation From Vibrating Structures, Part 1: Analytical Models." Journal of the Acoustical Society of America, submitted for publication.

11. Pan, J. and C.H. Hansen, 1990. "Active Control of Sound Transmission Through A Panel Into A Cavity: II. Experimental Study." Journal of the Acoustical Society of America, accepted for publication.

12. Snyder, S.D., 1990. "A Fundamental Study of Active Noise Control System Design." Ph.D. Dissertation, University of Adelaide.

13. Pope, L.D., D.C. Rennison, and E.G. Wilby, 1980. "Analytical Prediction of the Interior Noise for Cylinder Models of Aircraft Fuselages for Prescribed Exterior Noise Fields, Phase I." NASA CR-159363.

14. Pope, L.D. and E.G. Wilby, 1982. "Analytical Prediction of the Interior Noise for Cylinder Models of Aircraft Fuselages for Prescribed Exterior Noise Fields, Phase II." NASA CR-165869.

15. Peterson, M.R. and D.E. Boyd, 1978. "Free Vibrations of Circular Cylinders With Longitudinal, Interior Partitions." Journal of Sound and Vibration, 60, 45-62.

ACKNOWLEDGEMENTS

The authors would like to acknowledge financial support from the Sir Ross and Sir Keith Smith Fund and The Australian Research Council.

Control Algorithms II

PASSBAND DISTURBANCE IN ACTIVE SOUND CONTROL SYSTEMS

Paul Darlington[1]

ABSTRACT

The time variance inherent in adaptive filters causes modulation of frequency components of the filter's input. This modulation is fundamental to the operation of adaptive cancellers. When adaptive cancellers are applied to control problems the combination of the modulation and the dynamics of the plant introduce additional effects which are not seen in the behaviour of mean square estimators. These effects are described in the context of the synchronous control of sinusoidal disturbance, in which case significant passband ripple is introduced into the equivalent transfer function of the system. These passband phenomena are further demonstrated to be influenced by compensation strategies currently used in the adaptive control of acoustic systems.

INTRODUCTION

When adaptive signal processing techniques are applied to the estimation or cancellation of narrowband signals, or to the control of such disturbances in dynamic systems, a "reference" signal is used as the input to the adaptive filter. This reference is designed to contain all of the frequency components which the filter is to estimate. The controllable parameters of the adaptive filter are automatically adjusted such that the filter's output performs the required function optimally with respect to some user defined cost function. The on-line adjustment of the filter parameters makes the adaptive filter explicitly time variant. This time variance can produce a modulation effect, which allows the filter output to contain frequency components which are not present in its reference input.

If the adaptive filter is considered in its cancelling role, operating on a reference signal comprised of a sum of discrete sinusoids the cancelling operation generates a series of "stopbands" in the transmission of the signal "d" through the summing node to the system output (usually called the "error" signal, "e", as it represents the instantaneous estimation error). These stopbands are centred upon the frequencies of the reference signal's components. The time variance of the adaptive filter allows some modifications of the transmission of information over the summing node at frequencies other than those defined by components of the reference. In the cancelling context, these other frequencies would define the "passband" of the system.

This paper describes the passband disturbances introduced by adaption for both the adaptive estimator and for the signficantly more complicated case of the adaptive controller operating in a limited but important deterministic signal environment.

[1]Department of Applied Acoustics, University of Salford, Salford M5 4WT, U.K.

THE EQUIVALENT TRANSFER FUNCTION

It is convenient to describe the operation of an adaptive canceller, such as that illustrated in Figure 1, when operating upon a sinusoidal reference signal as attempting to prevent signal transmission from "d" to "e" at the frequency of the reference sinusoid, whilst passing other frequencies. Such a description can be formalised to define a transfer function between "e", considered as system output and "d", the system input. Under certain sampling conditions and in the context of reference signals formed of sinusoids, sums of sinusoids and periodic signals the resulting "equivalent transfer function" has been shown to follow a linear time invariant form which completely describes all aspects of the performance of the adaptive system [1, 2].

The synchronous sinusoidal estimator is able to heterodyne the reference frequency up or down to other nearby frequencies (by assuming a time variant weight trajectory) and so perform a limited amount of cancelling at frequencies other than that of the reference. The degree of such passband attenuation is controlled by the amount of mixing which is possible – a wide notch in the magnitude transfer function requires a considerable amount of weight modulation, which, in turn, requires a large update parameter. This explanation of the notch filter form was given in the early paper on sinusoidal cancelling by Glover [3] and is reiterated here simply as an introduction to the more complicated passband effects experienced when simple adaptive filters are applied to control applications.

EQUIVALENT TRANSFER FUNCTIONS OF ADAPTIVE CONTROL SYSTEMS

In order to study the passband effects of an LMS based estimator applied to the control of narrowband disturbance in linear dynamic systems it is necessary to derive a new equivalent transfer function which includes the effects of the system under control and any compensating filters needed to stabilise the adaptive filter. The effects of a dynamic system in the feedback loop of an adaptive filter have been investigated by the author and others [4, 5], although the primary motive in most of the previous work has been to establish stable bounds for the update process.

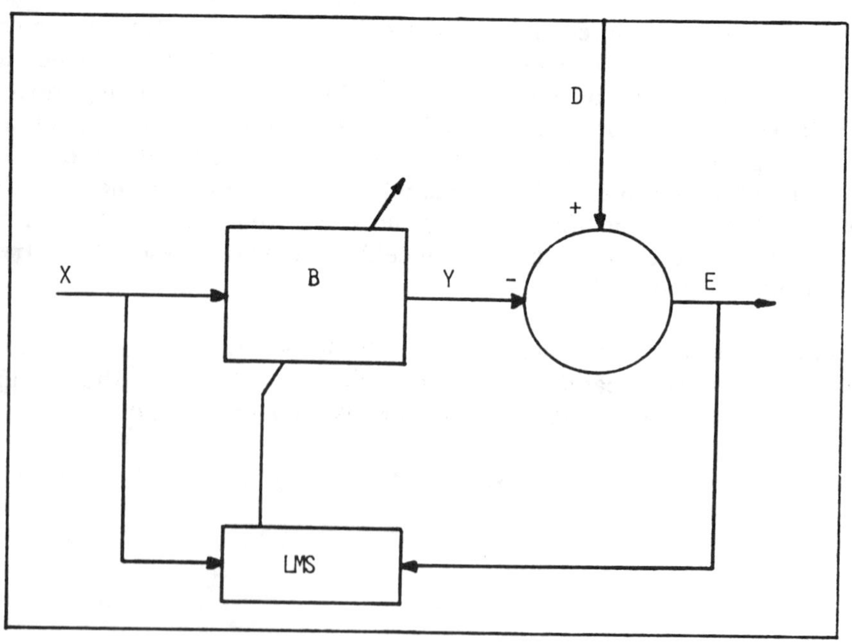

Figure 1 The Adaptive Estimator

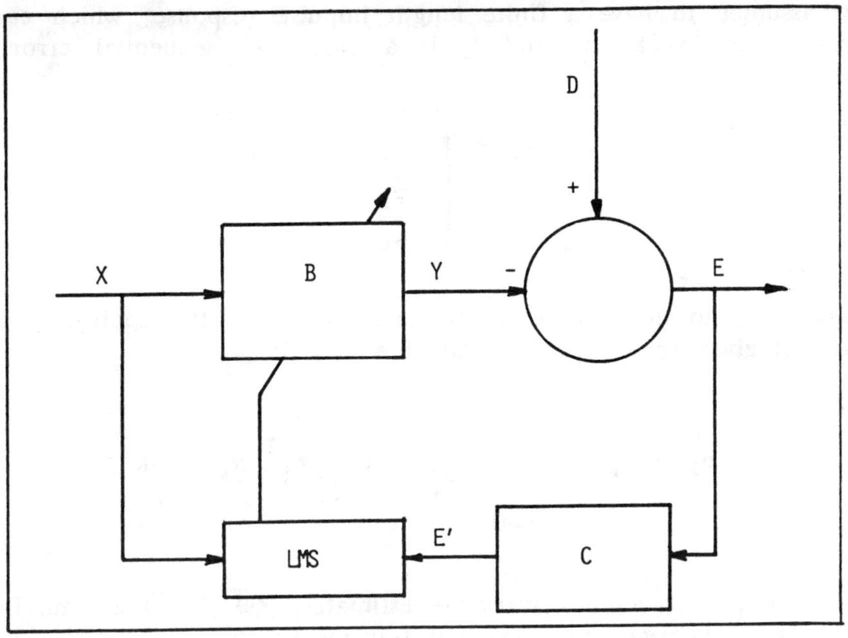

Figure 2 The Adaptive Controller

The equivalent transfer function of an adaptive filter in a control application will be introduced in the context of the system illustrated in Figure 2.

The filter C(z) in Figure 2 includes the dynamics of the system under control, which is assumed, for simplicity, to have a SISO form. The discussions of this paper can be extended to cover the adaptive control of the MIMO systems typical of practical active control of bounded acoustic spaces. Importantly, the filter C(z) also includes the dynamics of any actuators or sensors in the control loop (and their associated amplifiers and signal conditioning electronics) and any compensating filters. Such compensators are introduced to preserve stability [4] and often attempt to force C(z) towards unity, such that the control system of Figure 2 is equivalent to the estimator, Figure 1, whose stable bounds are well documented for the signal environments used in this paper.

The adaptive filter in Figure 2 is updated using the LMS algorithm, but an "observed error" signal (e' in Figure 2) is used to define the update steps, such that the weight update equation is:

$$\underline{B}_{k+1} = \underline{B}_k + \alpha\, e'_k\, \underline{X}_k \qquad (1)$$

where the observed error is related to the instantaneous estimation error by the filter C:

$$E'(z) = C(z).E(z) \qquad (2)$$

The weight trajectory from the initial point $\underline{B} = \underline{0}$ is given by:

$$\underline{B}_k = \alpha \sum_{i=0}^{k=1} \left[\underline{C}_i^{\,T}\, \underline{E}_i \right] \underline{X}_i \qquad k > 1 \qquad (3)$$

in which C is assumed to have a finite length impulse response, which can be written as the L elements of the vector $\underline{C}$ and $\underline{E}$ is a vector of sequential error samples, again having L elements:

$$\underline{E}_k = \begin{bmatrix} e_k \\ e_{k-1} \\ e_{k-(L-1)} \end{bmatrix} \qquad (4)$$

Following similar steps to those used in the development of the equivalent transfer function of an estimator [2] gives an expression relating e and d:

$$e_k = d_k - \alpha \sum_{i=0}^{k=1} \left[\underline{C}_i^T \underline{E}_i \right] \underline{X}_i^T \underline{X}_k \qquad k > 1 \qquad (5)$$

which, assuming the reference autocovariance estimates, $\underline{X}_i^T \underline{X}_k$, are time invariant, can be Z transformed to give the required equivalent transfer function:

$$\frac{E(z)}{D(z)} = \frac{1}{1 + \alpha L R(z) C(z)} \qquad (6)$$

$$\text{where} \qquad R(z) = Z \left\{ \frac{\underline{X}_i^T \underline{X}_k - \underline{X}_k^T \underline{X}_k}{L} \right\}$$

Equation (6) reveals how the introduction of a linear filter in the feedback loop of an LMS based estimator, as is the case when the adaptive filter is applied to control problems, changes the behaviour of the system with respect to the simple estimator of Figure 1.

C(z) does not displace the estimating/cancelling zeros of the transfer function, which are defined by the poles of R(z). The error observation filter does, however, modify the pole structure of the equivalent transfer function; this change introduces a new set of stability constraints and introduces effects to the passband of the system which are significantly more complicated than the notches seen in the synchronous estimation of sinusoids.

EQUIVALENT TRANSFER FUNCTION OF THE "FILTERED-X" ALGORITHM

Although equation (6) can reveal all aspects of the performance of the system depicted in Figure 2, it does not include a compensation strategy which is particularly important in acoustic applications of adaptive filtering - the "Filtered-X" algorithm.

In acoustic control problems the physical distance between the controlling loudspeakers (the "secondary sources") and the microphone used to detect the residual pressure (the "error") in the controlled field introduces a time delay component into C(z) which cannot be removed by casual inverse filtering. The resulting pure delay term between e and e' (in terms of Figure 2) can be compensated by the introduction of an equal delay into the reference input to the update process. A generalization of this time aligning idea resulted in the "filtered-X" compensation strategy [6] in which an additional compensating filter is introduced in the reference path, as shown in Figure 3.

The new compensating filter, F(z), results in a modification of the equivalent transfer function which was described by the author in [4, 7]; the result is repeated here:

$$\frac{E(z)}{D(z)} = \frac{1}{1 + \alpha\ L\ R(z)\ C(z)\ F^{*}(z)} \tag{7}$$

where, if the elements of the steady state response vector $\underline{F}'$ are:

$$\underline{F}' = \begin{bmatrix} f_0 \\ f_1 \\ f_2 \\ \cdot \\ \cdot \\ \cdot \end{bmatrix}$$

then:

$$F^{*}(z) = f_0 + f_1 z^{+1} + f_2 z^{+2} \quad \cdot\ \cdot\ \cdot\ \cdot\ \cdot\ \cdot$$

Note that, if the synchronous sampling conditions necessary to ensure that the reference autocovariance estimates are time invariant are obeyed, $\underline{F}$ can never have more than N elements.

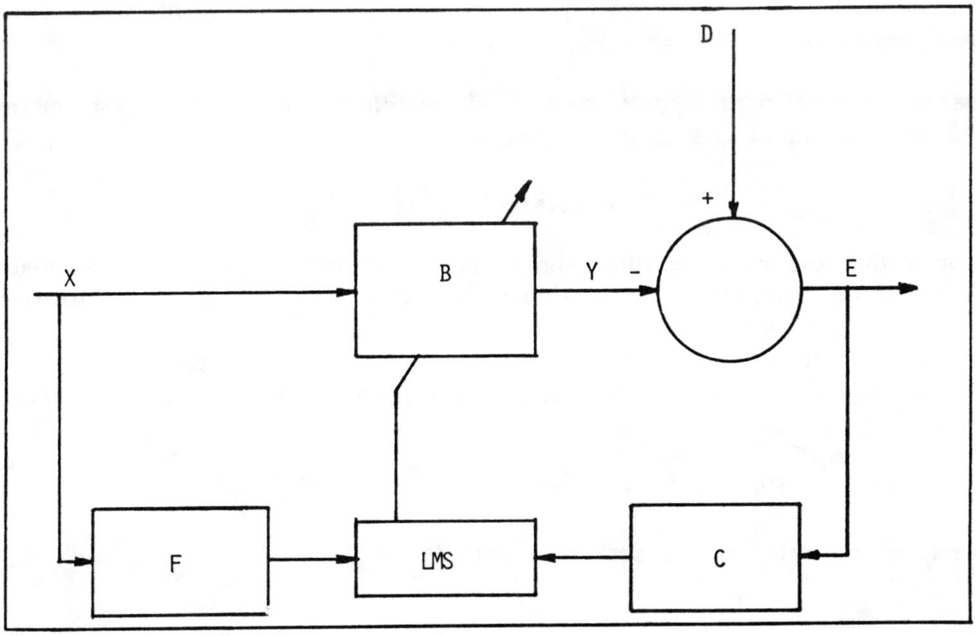

Figure 3 The Adaptive Controller with Filtered-x Compensation

An equivalent transfer function of an LMS based compensated estimator in a SISO control application can be written:

$$\frac{E(z)}{D(z)} = \frac{1}{1 + \alpha\ L\ R(z)\ \Gamma(z)} \tag{8}$$

in which

$$\Gamma(z) = C(z).F^{*}(z)$$

and compensation by both "inverse filtering" and "filtered−x" approaches can be described.

PASSBAND EFFECTS IN A SYNCHRONOUS ESTIMATOR WITH DELAYED COEFFICIENT UPDATE

If the filter C(z) in Figure 3 is a pure delay of an integer number of sample periods, perhaps as a result of exact inverse filtering of the minimum phase component of the response of the system under control:

$$C(z) = z^{-Dl} \tag{9}$$

and an equal compensating delay is introduced into the reference input path:

$$F(z) = z^{-Dl} \tag{10}$$

then the update steps are identical to those which would be taken in the absence of C and F, but are delayed by Dl samples. This coefficient update delay is a key feature of all "filtered−x" compensated controllers.

The introduction of an update delay does not make the system inherently unstable, although it does reduce the stability margin, such that the update parameter alpha may have to be reduced to preserve stability. The update delay does, however, greatly change the passband performance [7].

The influence of coefficient update delay will be illustrated for the case of a sinusoidal reference at one fourth of the sample frequency:

$$x_k = A\sin\ (k\pi/2) \tag{11}$$

which requires that the adaptive filter should have an even number of controllable weights, N, in order that the reference autocovariance estimates, $\underline{X}_i^T\ \underline{X}_k$, are time invariant.

The steady state response of any reference filter F(z) to the excitation defined in equation 11 can be characterised by a first order system, such that F' and F* will have two elements:

$$F'(z) = f_0 + f_1\ z^{-1} \tag{12}$$

If, for example, the delay in the reference path is Dl = 3, then

$$F'(z) = -1z^{-1}$$

whilst the steady state response of the filter $F(z) = z^{-4}$ could be represented by:

$$F'(z) = 1$$

The equivalent transfer function of this synchronous sinusoidal estimator with delayed update is obtained by substitution in equation 7:

$$\frac{E}{D} = \frac{1}{1 + \frac{\alpha LA^2}{2} F^*(z) z^{-Dl} \left[\frac{1}{z^2 + 1} \right]} \qquad (13)$$

Note that the acausal F*(z) in the transfer function above is introduced by the delay in the reference path due to F(z). The magnitude equivalent transfer function of equation 13 is shown plotted in Figure 4, for various values of delay, Dl.

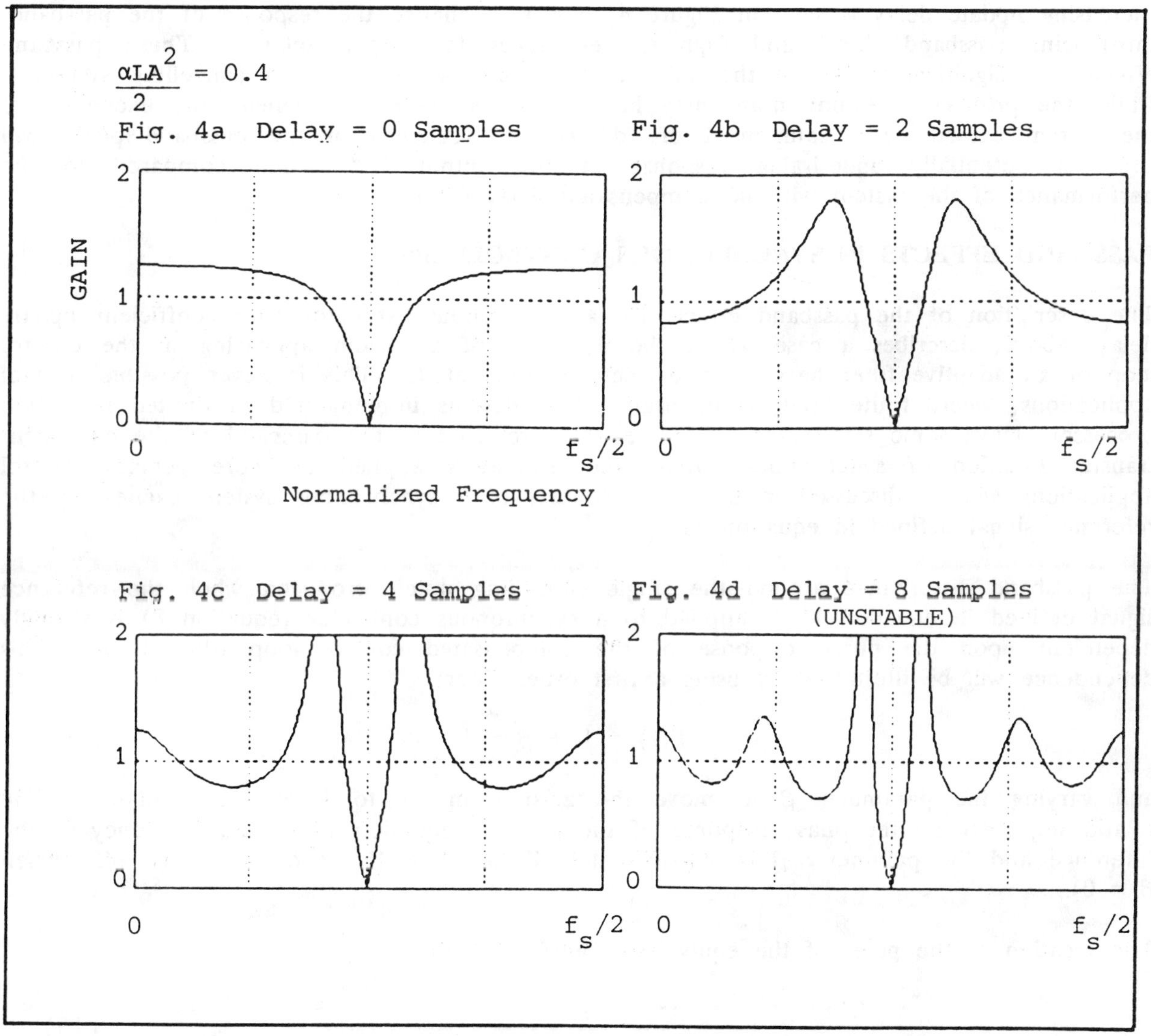

Figure 4 Magnitude Equivalent Transfer Function of the Synchronous Sinusoidal Estimator with Delayed Coefficient Update

Figure 4 shows that the filtered-x compensation technique is able to preserve the "cancelling notch" centred upon the reference frequency. The stability margin of the system is, however, reduced as the compensated delay is increased. The system is actually unstable when the compensated delay is 8 samples (see Figure 4d) although a reduction in the "update gain" $\alpha LA^2/2$ could stabilise the system.

Although the synchronous estimator with update delay can still implement useful cancelling notches, the equivalent transfer function in the passband is strongly influenced by the update delay. These passband effects are caused by the combination of the update delay and the heterodyning imposed upon the reference by the time variant weight trajectory. The transfer function does not change with each sample increase in delay Dl, but only moves to higher order, with the introduction of an extra pole pair, when the delay exceeds an integer multiple of 2. This is only true at the particular reference frequency chosen for the simulations reported in Figure 4 – generally the transfer function changes with each increase in compensated delay.

Increasing update delay is seen in Figure 4 to grossly change the response of the passband, introducing passband ripple and high Q resonances for larger delays. These passband effects can significantly reduce the subjective performance of a noise cancelling system – whilst the primary offending noise may have been successfully cancelled, any excitation of the system caused by a component of "d" at the frequency of the passband peaks will cause a potentially undesirable response at the output "e", when compared to the performance of the system with no compensated delay (Figure 4a).

PASSBAND EFFECTS IN SYNCHRONOUS CONTROLLERS

The description of the passband effects for a synchronous estimator with coefficient update delay, above, describes a case where the dynamics of a system appearing in the control loop of an adaptive filter have been perfectly compensated. This is never possible in real applications, where finite order compensator filter designs implemented in limited resolution processors leave some small error. The effects produced in the passband of the equivalent transfer function of synchronous narrowband estimators applied to more general control applications will be discussed in the same context of a synchronous system excited by the reference signal defined in equation 11.

The passband disturbance around the single cancelling notch produced when the reference signal defined in equation 11 is applied to a synchronous controller (equation 8) is strongly dependent upon the phase response of the compensated control loop filter $\Gamma(z)$. This dependence will be illustrated by using a first order filter:

$$\Gamma(z) = 1 + \beta z^{-1} \tag{14}$$

and varying the parameter β to move the zero from −1 to 1 in the z plane. The relationship between the phase response of the filter of equation 14 at the frequency of the reference and the parameter β is approximately linear, 1 + 45° when $\beta = -1$, -45° when $\beta = 0$).

The location of the poles of the equivalent transfer function:

$$\frac{E(z)}{D(z)} = \frac{1}{1 + \frac{\alpha LA^2}{2} \frac{(1 + \beta z^{-1})}{\sqrt{1 + \beta^2}} \left[\frac{1}{z^2 + 1} \right]} \tag{15}$$

was identified for various values of β. The change in gain of the filter at the reference frequency with β would cause a change in apparent update again, hence the introduction of the factor $1/\surd(1 + \beta^2)$.

A section of the root locus of equation 15 is plotted as Figure 5. The filter $\Gamma(z)$ is seen to displace the pole (pair) from the position approximately radially behind the cancelling zero expected for the estimator with no control loop filter (when $\beta = 0$) to a slightly different frequency. This gives the cancelling notch a marked asymmetry, as shown in the magnitude transfer function plots of Figure 6.

When the compensated control loop filter $\Gamma(z)$ has a phase lead at the frequency of the reference sinusoid, the magnitude equivalent transfer function has a peak in the passband immediately below the cancelling notch, whilst phase lag at the reference frequency introduces a peak above the frequency of the reference. This change in the notch shape is due to the combination of the weight modulation and the particular control loop filter $\Gamma(z)$ and can lead to the introduction of high Q effects into the passband which can degrade the subjective performance of active controllers in certain applications.

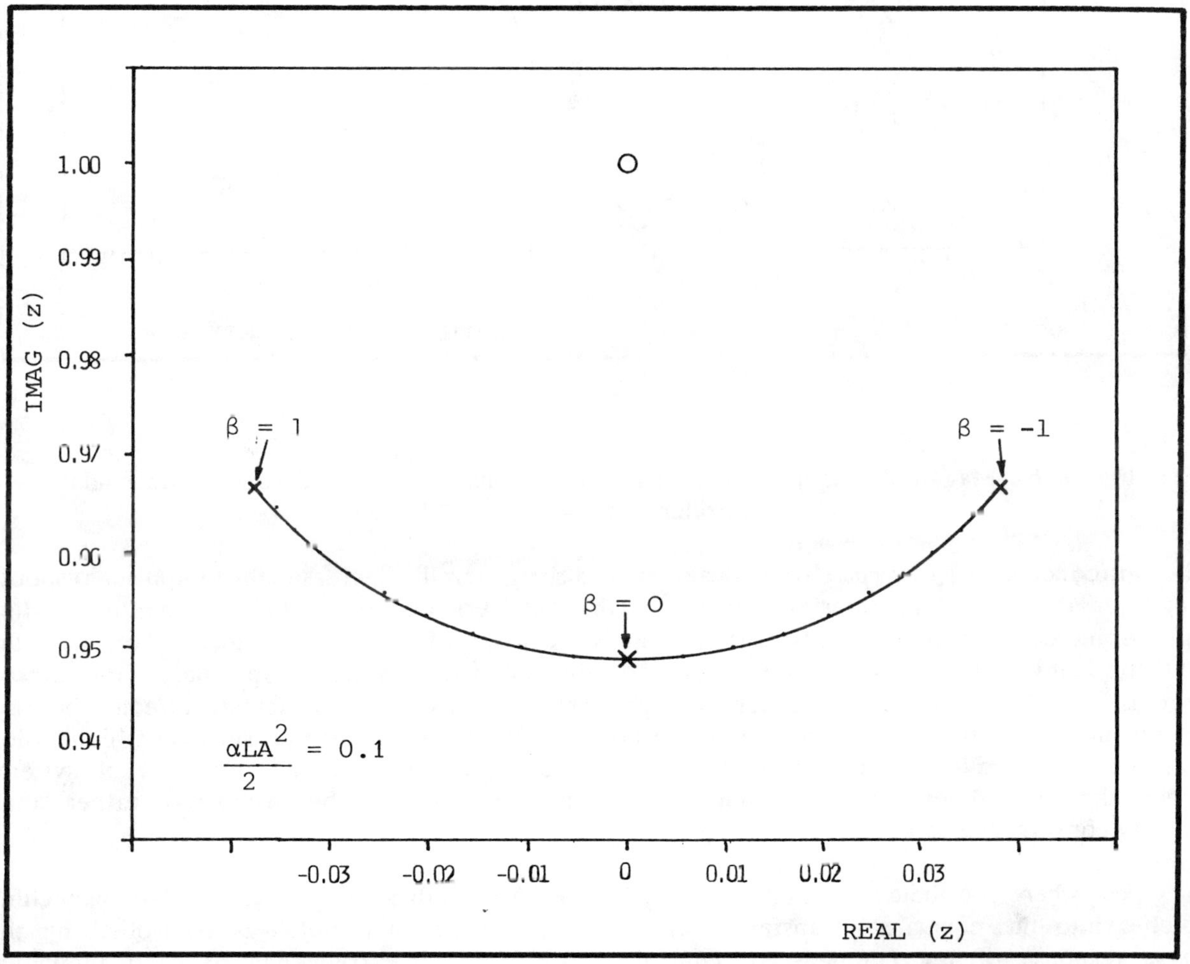

Figure 5 Section of the Root Locus of the Synchronous Sinusoidal Controller $C(z) = 1 + \beta z^{-1}$

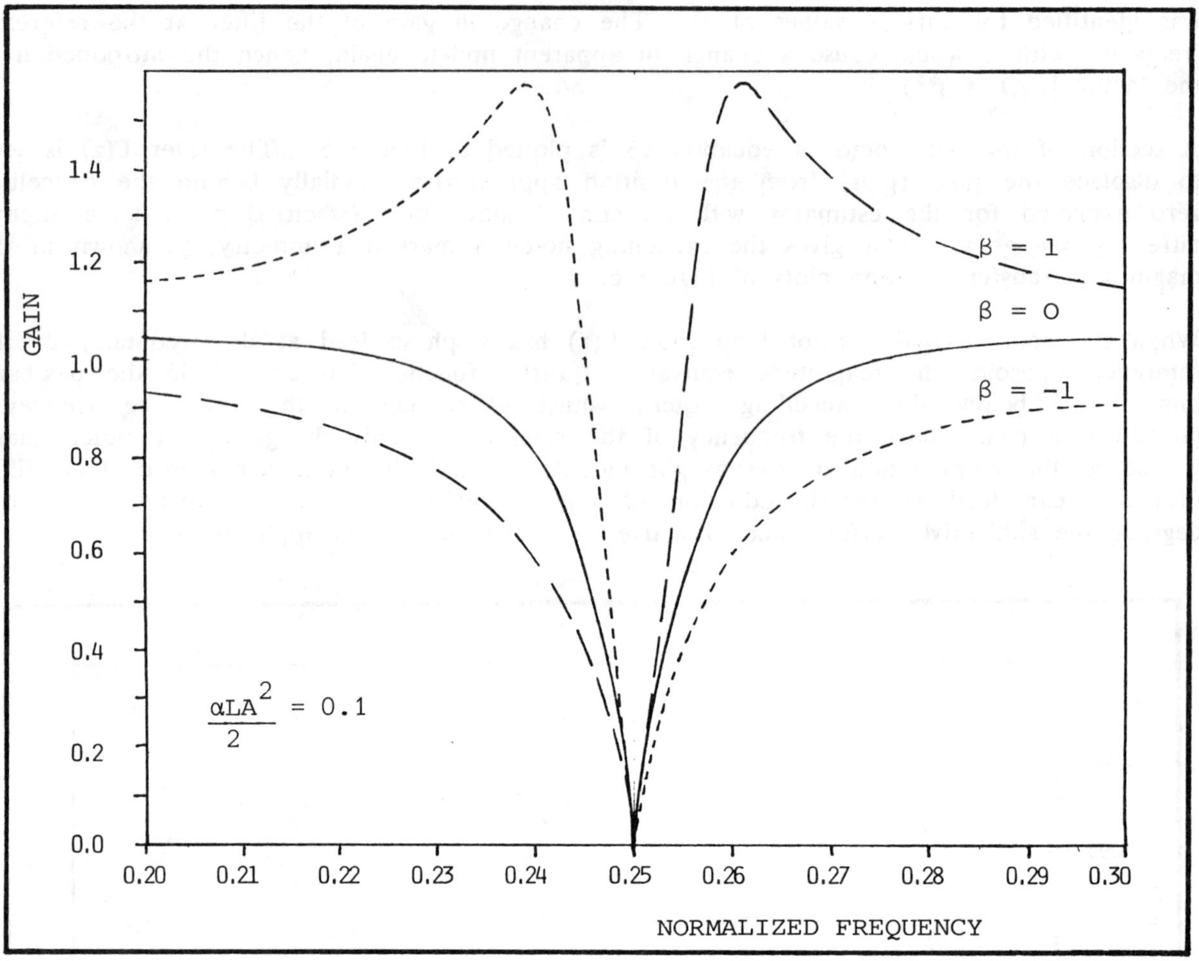

Figure 6 Magnitude Equivalent Transfer Functions of the Synchronous Sinusoidal Controller $C(z) = 1 + \beta z^{-1}$

The introduction of the passband peaks will also generally increase the impulse response length of the equivalent transfer function – the "convergence time" of the adaptive system will be increased. Experiments with the active control of sound in a duct have indicated that the convergence time is only weakly dependent upon control loop phase (within ±45 degrees at a reference frequency component) whereas the passband effects become significant for such control loop phase responses. This leads to the surprising conclusion that the most significant motivation for *accurate* compensation for the dynamics of systems under the control of adaptive filters is the optimization of the passband, rather than stability/convergence issues.

In cases where periodic of multi–sinusoidal references introduce a number of cancelling notches into the equivalent transfer function, the shape of each notch is controlled by the phase response of the compensated control loop $\Gamma(z)$ at the frequency of each individual notch. The notch asymmetry can be exploited as a diagnostic tool during system design, allowing the phase response of the compensating filters to be adjusted after a simple measurement of the magnitude equivalent transfer function.

CONCLUSIONS

The introduction of the linear filters to the control loop and reference path of narrowband adaptive estimators, as in the active control of sound, has been shown to introduce passband effects, in addition to the more widely understood consequence of influencing system stability. The passband ripple and high Q peaks could significantly degrade the subjective performance of a control system in certain applications. The passband phenomena have been shown to be particularly troublesome for "filtered-x" compensated systems, which are widely used in the control of non-minimum phase plants.

References

1. CLARKSON, P. M. and WHITE, P.R., 1987. "Simplified Analysis of the LMS Adaptive Filter Using a Transfer Function Approximation". IEEE Transactions on Acoustics, Speech and Signal Processing, ASSP-35(7): 987-993.

2. ELLIOTT, S. J. and DARLINGTON, P., 1985. "Adaptive Cancellation of Periodic, Synchronously Sampled Interference". IEEE Transactions on Acoustics, Speech and Signal Processing, ASSP-33(3): 715-717.

3. GLOVER, J. R. 1977 "Adaptive Noise Cancelling Applied to Sinusoidal Interferences". IEEE Transactions on Acoustics, Speech and Signal Processing, ASSP-25(6): 484-491.

4. DARLINGTON, P. 1987 "Applications of Adaptive Filters in Active Noise Control". PhD Thesis, I.S.V.R., University of Southampton, Southampton, UK.

5. MORGAN, D. R. 1980 "An Analysis of Multiple Correlation Cancellation Loops with a Filter in the Auxilliary Path". IEEE Transactions on Acoustics, Speech and Signal Processing, ASSP-28(4): 454.

6. WIDROW, B. and STERNS, S. D., 1985. Adaptive Signal Processing Englewood Cliffs, NJ: Prentice-Hall, Inc., pp.2288-294.

7. DARLINGTON, P. and ELLIOTT, S. J. 1988 "Synchronous Adaptive Filters with Delayed Coefficient Adaption" Proceedings, ICASSP '88 p.2586.

A METHOD TO IDENTIFY THE SECONDARY PATH IN ACTIVE NOISE CONTROL SYSTEMS

M. W. R. M. van Overbeek[1]

ABSTRACT

In active noise control the anti-noise signal, generated by the anti-noise source is computed with the help of adaptive filter-techniques. To update the filter coefficients we need knowledge of the parameters defining the secondary path or error path. This is the path from anti-noise signal output to residual sensor input, comprising not only the acoustical path, but also the hard-ware components (filters, transducers).

In most cases the parameters to be estimated are the coefficients (in frequency or time domain) of the transfer-function of the secondary path. There are several ways to estimate these coefficients, however most of these methods are (processor)time-consuming, so problems can occur if, due to a change in the environment, the parameters must be estimated during the generation of anti-noise.

In this paper a simple method to estimate the parameters, in other words to identify the secondary path, is presented. It can be used in active noise control systems for quasi-periodic noise with a slowly changing secondary path. With quasi-periodic noise we mean that the change in amplitude and phase is slow compared to the period. The method is based on a system-identification configuration, but it makes no use of additive noise.

Several algorithm-types which are based on the method have been developed and tested. There are algorithms for the (commercially available) single-input single-output active noise controller for periodical noise and for an experimental multiple-input multiple-output controller (both make computations in the frequency domain).

In addition algorithms have been developed and tested for a simple single-input single-output time-domain controller and for a controlled resonator (closed side branch of length 1/4 λ).

Experiments have been performed in the laboratory as well as in some industrial applications. Although simulations indicate that the method can work in some broadband applications too, the derivations will be made for quasi- periodic noise.

THE FILTER CONFIGURATION IN ANC

In ANC closed loop adaptation is used. The objective is to minimize the error signal e, being

[1]TNO Institute of Applied Physics, P.O. Box 155, 2600 AD Delft, The Netherlands; Laboratory of Seismics and Acoustics, Faculty of Applied Physics, Delft University of Technology, The Netherlands.

the difference between the primary signal (the noise) and the secondary signal (the anti-noise). In this paper we do not concern whether it is better to use (inverse) modelling, interference cancelling or what so ever to minimize the error signal.

Due to the (non-stationary) secondary path or transfer from anti-noise output to error input the basic closed loop configuration must be modified to describe the basic form of the ANC-problem. The secondary path h stands for everything between the (digital) computed controller output and the (digital) controller input. In figure 1a the modified configuration is shown (in a Single-Input Single-Output system). There are several mathematical forms (spectrum, IIR, FIR) in which the parameters of the secondary path can be expressed, this does not affect the configuration of figure 1a.

It is obvious that knowledge of the secondary path is essential because it affects the output signal, and so the adaptive process. This means that the controller needs an estimation of the secondary path within some region of convergence. The secondary path can be measured before operation but in most cases it must be adapted regularly due to the varying environment parameters. This can be done in a second adaptive loop, as is shown in figure 1b. The configuration illustrated in figure 1b is described by Widrow [1] in an application for adaptive inverse model control using the filtered-X-LMS algorithm.

In this paper a very simple method is described to adapt the secondary path. Due to its simplicity there are some restrictions, it cannot be applied to very rapid changing systems. However, it gives good results in different applications and has the great advantage that we do not need an initial guess.

THE TRANSFER ADAPTATION LOOP

The configuration in figure 1b shows the two adaptation loops, the first loop for cancelling the primary noise (the 'desired signal' p),and the second loop for cancelling the difference between real and estimated secondary path. There is a wide variety of possible control algorithms for the

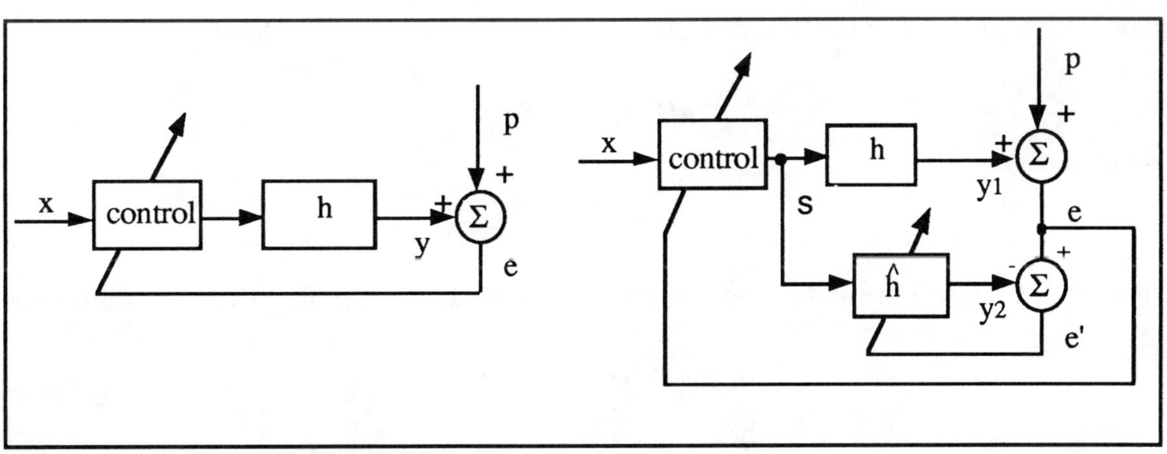

Figure 1a: Basic ANC configuration.

Figure 1b: Basic ANC configuration with second adaptive loop.

first adaptation loop. In ANC most algorithms till now are based on Wiener Filter Theory resulting in some form of the LMS algorithm. There is much literature about this subject. In this paper the LMS algorithm is used too, so the criterion is the squared instantaneous error $(e)^2$.

This paper is restricted to the second loop, the transfer adaptation loop, for convenience we will call this the secondary loop. As we can see in figure 1b the transfer adaptation loop is a system-identification configuration. The filter input-signal is s and the error signal e'. The primary noise, p, appears as additive noise. Due to the correlation of the filter-input and the primary signal, simulations with this configuration do not give satisfying results. There are other possibilities for the choice of the filter-input signal. Eriksson [2] suggests noise, not correlated with the primary noise signal p.

Our suggestion is originated in the basic idea to minimize the influence of the primary signal p in the loop. This can be done by subtracting the signals of two successive primary control loops. So, the filter input signal of the secondary-loop is found by subtracting two periods of input signal s, or, in the frequency-domain, by subtracting the fourier transforms of two successive blocks of s.

ADAPTATION IN THE FREQUENCY-DOMAIN

The implications of the method are probably best understood by giving the Wiener solution in the frequency domain for a SISO system. In the frequency domain for a SISO system the derivations are found very easily. Symbols in the frequency domain are in capitals. In this paragraph all the symbols are elements of a k-point Fourier transform of a time-signal, k being some power of 2. For convenience we leave the frequency dependence. We assume there is no additive noise.

WIENER SOLUTION, SISO

In figure 2 the configuration is illustrated.The filter-input of the secondary loop is

$$\Delta S_n = S_n - S_{n-1} \tag{1}$$

The 'desired' signal is

$$\Delta E_n = E_n - E_{n-1} \tag{2}$$

With the help of figure 2 the process equation of the secondary loop in the frequency domain is found:

$$E'_n = \Delta E_n - \hat{H}_n \Delta S_n \tag{3}$$

$\hat{H}_n$ is the filter coefficient to be adapted.

With help of the Wiener solution for the filter coefficient of the transfer adaptation loop, the

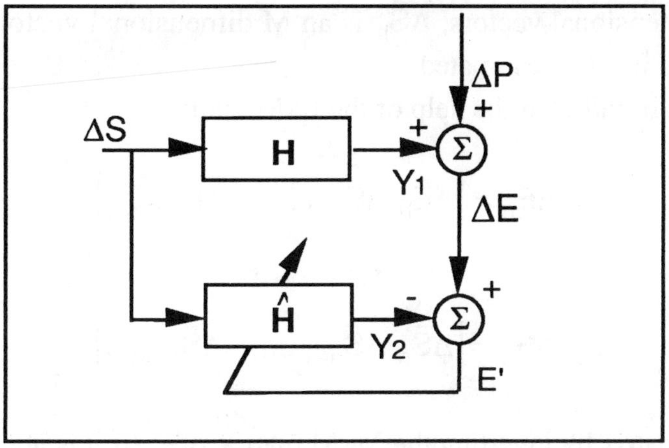

Figure 2 The adaptation loop of the secondary path

optimum estimation (* stands for optimum) of the transfer function of the secondary path using the configuration is:

$$\hat{H}_n^* = \frac{\Delta E_n}{\Delta S_n} \tag{4}$$

With the help of the frequency transforms of the primary signal and the transfer function we find the process equation of the primary loop

$$E_n = P_n + H_n S_n \tag{5}$$

and we easily see that if $\quad P_n = P_{n-1}$

and $\quad H_n = H_{n-1}$ (6)

while $\quad S_n \neq S_{n-1}$

the optimum estimation equals the instantaneous transfer function of the secondary path:

$$\hat{H}_n^* = H_n \tag{7}$$

WIENER SOLUTION, MIMO

We also can derive the Wiener solution for a multiple-input multiple-output system in the frequency domain.
Suppose there are M anti-noise signals and L error signals (L≥M).
The process equation of the secondary loop is

$$\vec{E}'_n = \Delta\vec{E}_n - \hat{H}_n \Delta\vec{S}_n \tag{8}$$

$\vec{E}'_n$ and $\Delta\vec{E}_n$ are L dimensional vectors, $\Delta\vec{S}_n$ is an M dimensional vector and $\hat{\mathbf{H}}_n$ is the LxM dimensional matrix that must be estimated.
The Wiener solution is found with the help of the LxM matrix

$$\Delta\mathbf{E}_n = \left[\Delta\vec{E}_n\ \Delta\vec{E}_{n-1} \ldots\ldots \Delta\vec{E}_{n-M-1}\right] \tag{9}$$

and the MxM matrix

$$\Delta\mathbf{S}_n = \left[\Delta\vec{S}_n\ \Delta\vec{S}_{n-1} \ldots\ldots \Delta\vec{S}_{n-M-1}\right] \tag{10}$$

we find the optimum matrix by inverting the MxM matrix $\Delta\mathbf{S}_n$ (if it is not singular)

$$\hat{\mathbf{H}}^*_n = \Delta\mathbf{E}_n[\Delta\mathbf{S}_n]^{-1} \tag{11}$$

The process equation of the primary loop is a L dimensional vector too:

$$\vec{E}_n = \vec{P}_n + \mathbf{H}_n\vec{S}_n \tag{12}$$

$\vec{E}_n$ and $\vec{P}_n$ are L dimensional vectors, $\vec{S}_n$ is a M dimensional vector and $\mathbf{H}_n$ is a LxM dimensional matrix. If the vector $\vec{P}_n$ and the matrix $\mathbf{H}_n$ are stationary during the M iterations and $\Delta\mathbf{S}_n$ is not singular, we will find

$$\hat{\mathbf{H}}^*_n = \mathbf{H}_n \tag{13}$$

The optimum estimation matrix equals the instantaneous transfer matrix.

APPLICATIONS

A SISO NEWTON-NEWTON SYSTEM

The Wiener solution will give the exact secondary path if the primary noise and the secondary path itself are stationary. This is not always true, and it can give rise to deviations in the estimated secondary path. Furthermore it will be clear that the method is sensitive to the signal to noise ratio in E. To test the value of the method in industrial applications, it is implemented in a SISO controller (based on a TMS c25). Both algorithms (for primary loop and secondary loop) operate in the frequency domain and are based on the *Newton* gradient search method. The primary loop is an inverse-modelling configuration, the filter input of this feed-forward loop is a signal synchronized with the primary source. We need the first and second derivative of the instantaneous squared error with respect to the adapted coefficient for the Newton algorithm.

Using the estimation of the transfer function of the secondary path the adaptive algorithm for the primary loop is (n is the iteration index, all capital symbols are in the frequency domain) :

$$S_{n+1} = S_n - \mu_1 \frac{E_n}{\hat{H}_n} \tag{14}$$

The adaptive algorithm for the secondary loop is

$$\hat{H}_{n+1} = \hat{H}_n - \mu_2 \frac{E'_n}{\Delta S_n} = (1-\mu_2)\hat{H}_n + \frac{\Delta E_n}{\Delta S_n} \tag{15}$$

and μ is the convergence rate.

The controller is tested in the laboratory in an acoustic one-dimensional system (a duct below the cut-on frequency). The fourier transforms are performed on 32 point data-blocks. The secondary path is measured for 11 frequencies with the help of pure sines. In the ANC experiment the primary source and the secondary path are kept stationary, the initial (complex) value $\hat{H}_o$ is taken (10.,0.). During ANC the primary source radiates a signal containing 4 frequency-components (S/N ratio > 40 dB). The result of the measurement with the pure sines is compared to the estimation made in the ANC configuration. In the figures 3a and 3b the amplitude and phase of the measurement (solid line) and of the estimation (dotted line) are given in one graph. The four frequency-components are marked with a small circle. The estimation is good for those elements, but poor for the other frequency-components. If the error signal E_n is near to the noise level the secondary loop will give bad results. This can be avoided by making the rate of convergence μ_2 equal to zero if the error signal is near the noise level. The noise level can be estimated during operation.

We measured the transfer function of the secondary path (amplitude and phase) as a function of

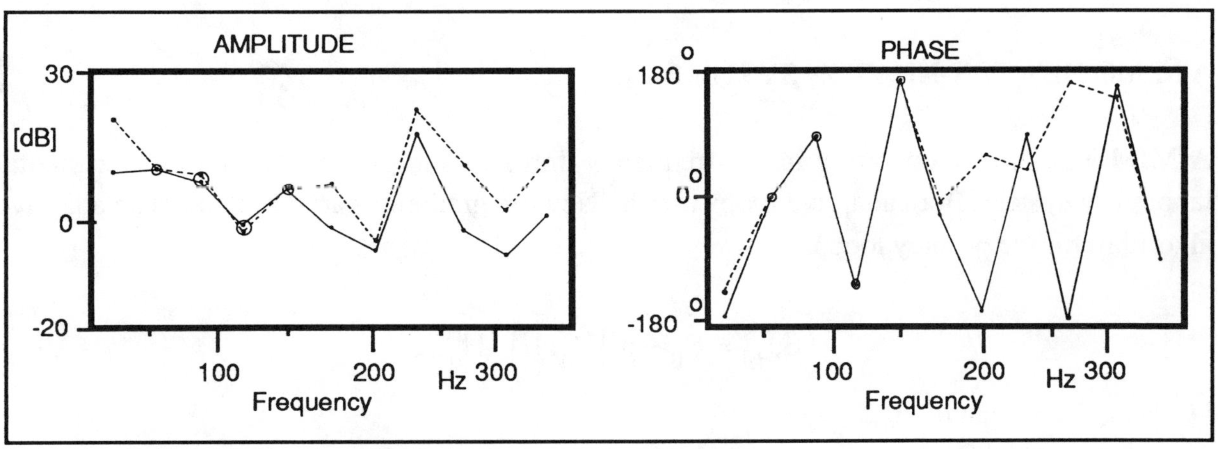

Figure 3a and 3b: Amplitude and phase of estimated and measured transfer function.

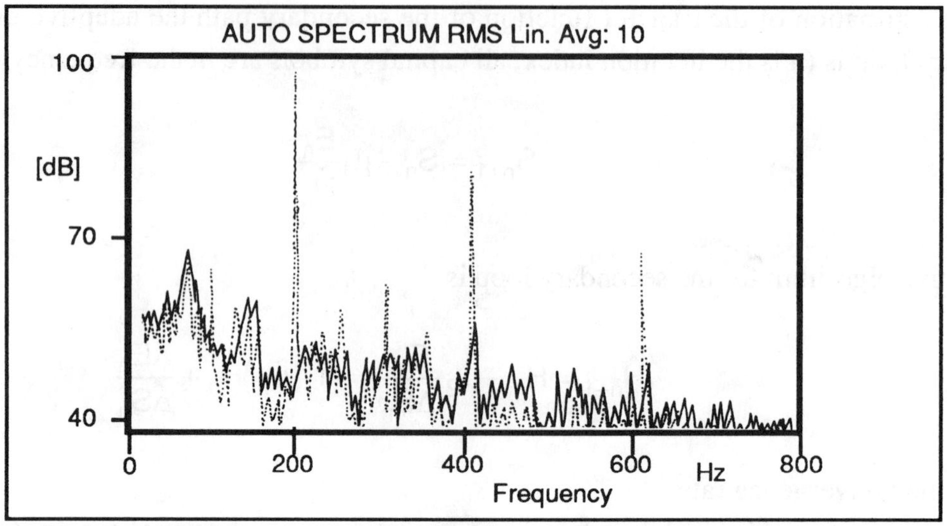

Figure 4 The sound pressure level with ANC (solid line) and without ANC (dotted line).

several physical parameters. Without adaptation of the estimated secondary path, there are maximum deviations of the physical parameters from the initial values that can be allowed to guarantee convergence of the primary algorithm. A change in the sound velocity of about 0.3 % for instance will cause divergence at certain frequencies if the estimation of the secondary path is not adapted.If the primary source and secondary path are not stationary, the Wiener solution will not give the secondary path. Adapting the estimation of the secondary path there is a maximum rate of change for each physical parameters (sound velocity for instance) we can allow. This maximum rate of change depends strongly on the acoustic environment. In the duct used in the experiment we can allow a continuous change in sound velocity of about 0.05% each iteration step (n) in a worst case situation.In an industrial application a complete ANC system based on the SISO controller has been installed on the outlet of a Roots Blower. The secondary path has not been measured before operation. The error signal E is measured in the outlet. In figure 4 the sound pressure level before operation of the ANC system (dotted line) and the sound pressure level during operation (solid line) is shown in one graph. The periodic noise is reduced to the noise level.

A MIMO NEWTON-NEWTON SYSTEM

A MIMO algorithm operating in the frequency domain is implemented in an experimental laboratory system. Both adaptive loops use the Newton gradient search method. The adaptive algorithm for the primary loop is

$$\vec{S}_{n+1} = \vec{S}_n - \mu_1 \left[\hat{H}_n^H \hat{H}_n\right]^{-1} \hat{H}_n^H \vec{E}_n \qquad 16)$$

$\hat{H}_n^H$ is the Hermitian of $\hat{H}_n$.

Since the matrix $\Delta \mathbf{S}_n$ (equation (10)) is a square matrix we find for the secondary loop

$$\hat{H}_{n+1} = (1-\mu_2)\hat{H}_n + \mu_2 \Delta E_n [\Delta S_n]^{-1} \quad (17)$$

In the algorithm the matrix $\Delta \mathbf{S}_n$ is tested for singularity and here too μ_2 is made zero if the error signals are near the noise level.

The experiments were performed in an enclosure. In an experiment the primary source and secondary source were kept stationary. Before the ANC experiment the transfer function-matrix of the secondary path is measured with the help of pure sines (2 sources , 4 sensors). P is the primary source, S1 and S2 are secondary sources, M are the 4 sensors (see figure 5). The frequency is 300 Hz. In the ANC experiment the initial values of the matrix $\hat{\mathbf{H}}_o$ were all zero except the elements $\hat{H}_o(1,1)$ and $\hat{H}_o(2,2)$ which were taken (10.,0.).The transfer function-matrix estimated with the ANC system is compared to the values measured with the pure sines. The estimation-matrix is found to be very close to the measured values, and certainly within the region of convergence.

The inverse matrix-operation is very sensitive to noise. Furthermore, the rate of change in the transfer-function that can be allowed in the MIMO application is much smaller than in the one-dimensional system. This means that the MIMO system will only work satisfactorily in very slowly changing environments. For instance in the laboratory it can be a great help since we do not need to measure the transfer function matrix before a measurement, and it will adapt to changes in the set-up during a measurement (filter-setting etc.).As a part of our acoustic research in enclosures a great deal of measurements have been performed with the method. In figure 5 the results are given of one of the measurements. The pressure level on an horizontal plane in an enclosure is measured on 100 points before and during operation of the ANC system. There are 4 sensors on a row and there are two anti-noise sources (S1 and S2) in two opposite corners. There is one primary noise source (P). The reduction on the line on which the sensors are placed is approximately 25 to 35 dB.

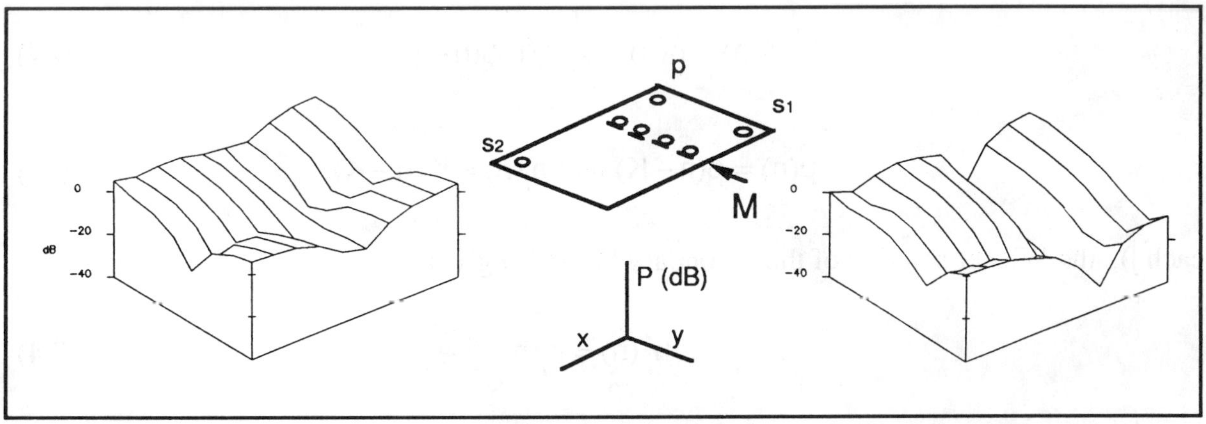

Figure 5 Pressure level measured in an horizontal plane with and without ANC.

A SISO TIME DOMAIN SYSTEM

In this paragraph all symbols are in the time-domain, and the iteration index is a sample number, and not a block number like in the frequency-domain. The filter input signal $x(n)$ is synchronized with the primary signal.
In the time-domain the filter-input of the secondary loop is

$$\Delta s(n) = s(n) - s(n-K) \tag{18}$$

and the desired signal is

$$\Delta e(n) = e(n) - e(n-K) \tag{19}$$

K is the number of samples in one period of the noise.
The estimation of the secondary path is modelled as a FIR filter of length J with coefficients $\hat{h}_j(n)$ The process equation of the secondary loop is

$$e'(n) = \Delta e(n) - \sum_{j=0}^{J} \hat{h}_j(n) \Delta s(n-j) \tag{20}$$

The input signal of the primary loop is $x(n)$, the filter-output is

$$s(n) = \sum_{i=0}^{I} w_i(n) x(n-i) \tag{21}$$

in which $w_i(n)$ are the filter coefficients of the primary filter.
The process equation of the primary loop now can be written as

$$e(n) = p(n) + \sum_{j=0}^{J} h_j(n) s(n-j) \tag{22}$$

If

$$p(n) = p(n-K) \text{ and } h_j(n) = h_j(n-K) \tag{23}$$

(each j), the Wiener solution of the secondary loop (20) gives

$$\hat{h}_j^*(n) = h_j(n) \tag{24}$$

The elements $\hat{h}_j(n)$ could be estimated with equation (20) if they are considered to be time-invariant during a number of samples N>J. However, we were looking for a very fast algorithm. This can be found using the filtered-x configuration ([1]). If there are only a few frequency-components the filtered-x algorithm is not as time-consuming as doing the convolutions in equation (20). Since $x(n)$ is periodic (K samples a period), equation (18) can be written as

$$\Delta s(n) = \sum_{i=0}^{I} \Delta w_i(n) x(n-i) \tag{25}$$

in which

$$\Delta w_i(n) = w_i(n) - w_i(n-K) \tag{26}$$

Considering $\Delta w_i(n)$ time-invariant equation (20) can be rearranged as

$$e'(n) = \Delta e(n) - \sum_{i=0}^{I} \Delta w_i(n) \sum_{j=0}^{J} \hat{h}_j(n) x(n-j-i) \tag{27}$$

And we introduce the 'filtered-x' coefficients

$$\hat{r}_{kni}(n) = \sum_{j=0}^{J} \hat{h}_j(n) x(n-j-i) \tag{28}$$

kni stands for $[(n\text{-}i) \text{ modulo } K]$
(if $n\text{-}i=\alpha K+\beta$; α,β integers and $\beta<K$; then $kni=\beta$)
Considering $w_i(n)$ time invariant equation (21) becomes

$$e(n) = p(n) + \sum_{i=0}^{I} w_i(n) r_{kni}(n) \tag{29}$$

It can be proven that we only need two filtered-x coefficients for each frequency component in the primary noise. If there are only a few frequency components, estimating the filtered-x coefficients $\hat{r}_{kni}(n)$ is much faster than estimating the filter coefficients $\hat{h}_j(n)$. In an experiment the algorithm is implemented in a micro-computer with a 8 bit AD-DA board. The estimation of the secondary path is based on a Newton gradient method with a fast convergence rate, the primary loop is based on a Steepest descent algorithm with a small convergence rate. The

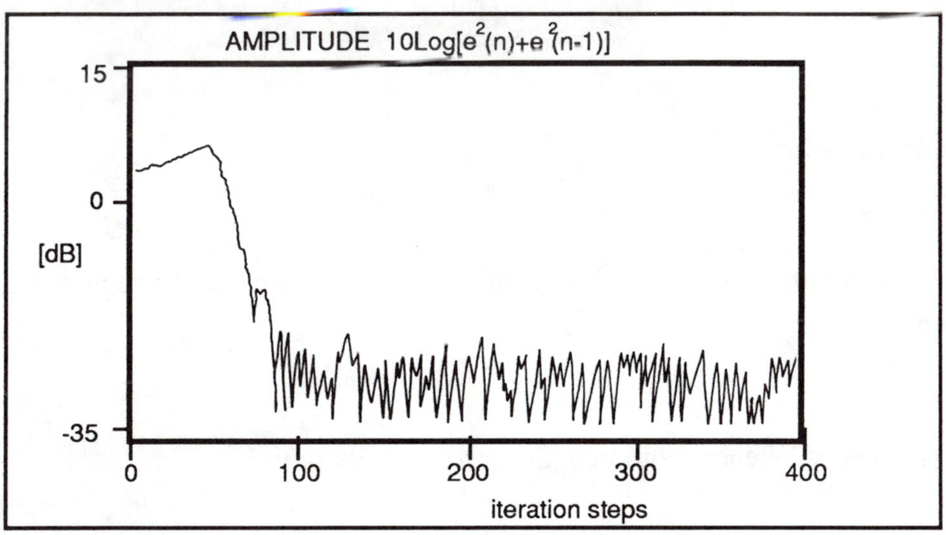

Figure 6 Amplitude of the error signal as a function of the iteration steps. Adaptation of the secondary loop starts after 40 samples.

estimated path is not updated each step, but every 40 steps. We tested the algorithm in an one dimensional acoustic system, the primary source having only one frequency component. The input signal x is synchronized with the primary source-signal, having 4 samples a period. In - figure 6 the amplitude of the error signal is plot as a function of the iteration step. The initial values of the filtered reference coefficients is set 1. The first estimation of the filtered-x coefficients is made after 40 iteration steps. The first 40 steps the error increases due to the (wrong) values 1. Near the noise level the convergence rate μ_2 is made zero.

RESONATOR

A last application we would like to mention is a tunable resonator, a side branch tuned to 1/4 of the wavelength λ. In some industrial applications a tunable resonator in combination with a conventional SISO ANC system can be a good alternative to a conventional ANC system with loudspeakers [3]. Derivations are in the frequency domain .

In the process equation of the primary loop here the error is the relation between the residuals measured in the duct and on the piston (see figure 7)

$$E_n = \left|\frac{E_d}{E_r}\right|_n = \left|\frac{P}{E_r}\right|_n + G_n(x_n) \tag{30}$$

P is the pressure in the duct caused by the primary source (the 'desired' signal) and G(x) is the function which gives the relation between the length x (the filter output) and the reduction by the side branch. Without concerning about this relation we know that E^2 has a global minimum within a certain area at about $x=x_0=1/4\ \lambda$ (figure 7). This area is found to be between x=0 and $x=1/2\ \lambda$. In a first experiment we linearized G(x)

$$E_n = \left|\frac{P}{E_r}\right|_n + \alpha_n x_n \tag{31}$$

The secondary loop now is

$$E'_n = \Delta E_n - \hat{\alpha}_n \Delta x_n \tag{32}$$

and $\hat{\alpha}_n$ is the estimation of α_n In the adaptive algorithm we used steepest descent in the primary loop

$$x_{n+1} = x_n - 2\mu E_n \hat{\alpha}_n \tag{33}$$

and the instantaneous Wiener solution in the secondary loop (μ_2=1)

$$\hat{\alpha}_n = \frac{\Delta E_n}{\Delta x_n} \tag{34}$$

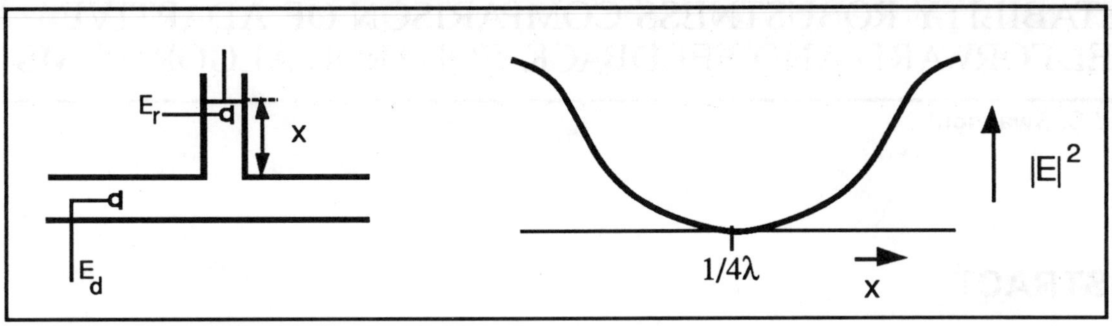

Figure 7 A tunable resonator and the relation between x and $|E|^2$.

To avoid oscillation around x_0, μ_2 is made zero if E is below a certain noise level.
In experiments we found with this very simple algorithm, using an μ-controller and an 8 bit AD-DA board reductions up to 30 dB, which we found a satisfactorily good result although of course much more clever choices for the function G(x) can be made.

CONCLUSIONS

A method to estimate on line the secondary path in ANC problems has been described using adaptive filter techniques. The method can be applied to several applications in frequency-domain as well as time domain algorithms. A number of experiments have been performed in which the algorithms are implemented on different hardware controllers. The systems are fully adaptive and do not need any initial knowledge of the secondary path.

ACKNOWLEDGEMENTS

The author gratefully acknowledges the contributions made to the research by J. van 't Hof and T.C. van den Dool of TPD/TNO Delft.

References

1. Widrow, Bernard and Samuel D.Stearns. 1985. Adaptive Signal Processing, Englewood Cliffs: Prentice-Hall, Inc., pp.288-300.

2.Eriksson, L.J. and M.C.Allie. February 1989. "Use of Random Noise for on-line Transducer Modelling in an Adaptive Active Attenuation System."Journal of the Acoustical Society of America, 85 (2):797-802.

3. van Overbeek, M.W.R.M. 1990. "Acoustical Elements in Active Noise Control." Proceedings of Internoise 90, 1263-1266.

A STABILITY ROBUSTNESS COMPARISON OF ADAPTIVE FEEDFORWARD AND FEEDBACK CONTROL ALGORITHMS

David C. Swanson[1]

ABSTRACT

The two basic forms of electronic control algorithms used in active sound and vibration control are examined to compare stability robustness for both the adaptive and non-adaptive cases. Active noise control is based on either a feedforward controller where a coherent reference noise signal is sensed *before* it propagates past the active control source, or feedback control where the active control source attempts to cancel the noise without the benefit of an "upstream" reference noise signal. Two feedforward systems are analyzed; one based on precalibrated feedback and error plant transfer functions, the other based on-line error plant estimation and pole-zero adaptive filtering. The two systems are fundamentally different both in operation and stability robustness. The analysis suggests that feedforward active control is generally more robust to instability than feedback control, particularly when the feedforward system has a reference noise signal isolated from the effects of the active control signal.

INTRODUCTION

This work compares the stability and robustness of the two basic forms of active noise controller: the feedforward controller and the feedback controller. The feedforward controller dates back to the original duct noise cancellation work of Lueg [1], where an "upstream" microphone senses the sound *before* it passes a loudspeaker "downstream" in the duct. The spacing between the microphone and loudspeaker is varied to allow precise phase cancellation at the speaker position for the frequency of interest. Modern adaptive feedforward systems rely on essentially the same principles of causality and high coherence between the upstream noise sensor and the wave to be canceled at the actuator. However, the modern adaptive feedforward controller is far more robust and effective at canceling noise in environments where the actuator corrupts the upstream noise sensor signal with unwanted feedback and where the physical parameters which dictate the required amplitude and phase for cancellation are time-varying.

[1]The Pennsylvania State University, The Applied Research Laboratory and The Graduate Program in Acoustics, Applied Science Building, University Park, PA 16802, U.S.A.

A different active control structure based on feedback was introduced by Olson [2]. Olson's basic non-adaptive controller is still in use today (in much improved versions) in the form of active hearing protectors [3]. Active noise cancellation using feedback offers the advantage of creating a null at the error microphone for all periodic noise sources without requiring correlated noise signals from reference noise sensors for each noise source. However, the stability requirements of a feedback cancellation system limit the amount of feedback gain which can be applied to cancel the noise, which also places a practical limit on the effectiveness of the active cancellation. Because of the time delay between the electrical input of the actuator and the electrical response of the error sensor in a feedback control system, only periodic noise signals can be canceled and the phase response of the controller must be very precisely matched to the physical system for stable cancellation. For non-adaptive feedback controllers using analog electronics, the phase matching problem is typically made easier by placing the error sensor very close to the actuator.

FEEDFORWARD ACTIVE NOISE CONTROL

Feedforward active control is defined as a controller where a reference signal, which is coherent with the noise to be canceled, is electrically transported to an advanced position in the propagation path where it can be re-introduced to the original wave to produce the desired response. Feedforward active noise cancellation is only possible for waves which propagate slower than the active cancellation system can respond. For waves described by second order differential equations, active "cancellation" of the wave is physically described by coupling the system to a parallel impedance of approximately zero. The wave energy not absorbed by the active cancellation source is reflected back towards the primary noise source due to an active impedance boundary condition often referred to as a "virtual earth" [4]. Feedforward control can be implemented in free space so long as the "upstream" reference noise sensor is observing waves in advance of their interaction with the active control source. Burgess [5], shows that purely periodic noise can be adaptively canceled by an appropriate noncausal phase adjustment when the "upstream" reference signal is not actually upstream of the active source. This is technically not feedforward active cancellation and the system will be ineffective for random or impulsive noise, if not unstable due to the lack of causality.

Feedforward control is typically employed for noise propagating in a duct or waveguide because active cancellation at an error sensor position stops all wave propagation past that point rather than simply creating an active node as would be the case in a 3-dimensional space. For simplicity, the discussion below will use a duct as an example physical system for the frequency range where the duct only supports plane waves propagating down its length. Figure 1 shows a general finite-length duct with an active source denoted by "S" located at the position x=A.

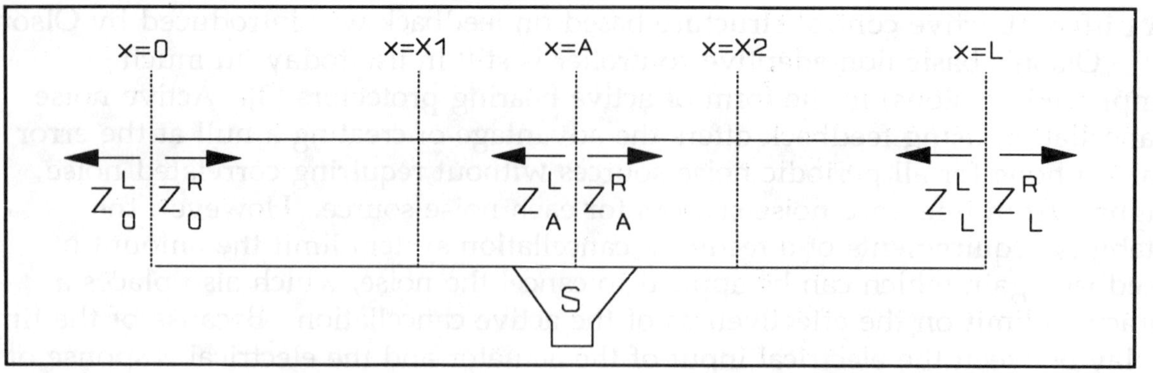

Figure 1 **A typical finite length duct showing directional acoustic impedance.**

The acoustic impedance is defined as the acoustic pressure divided by the volume velocity in the duct. Since velocity is a vector quantity, acoustic impedance is also a vector quantity and it can be seen that the radiation impedance of the source S is the parallel combination of the impedance in the right direction and the impedance in the left direction at x=A [6]. Note that this radiation impedance for the source is significantly more complicated than the assumption of that of a loudspeaker radiating into a room used for the duct analysis in [7].

$$Z_A^S = \frac{Z_A^R \cdot Z_A^L}{Z_A^R + Z_A^L} \tag{1}$$

The input impedance to any particular duct section is well known to be determined by the impedance at the end of the section and the length of duct in the section. For example, the input impedance to the duct at the left end in Figure 1 is given in equation (2) for a circular duct cross-section of radius r.

$$Z_0^R = \frac{\rho c}{\pi r^2} \cdot \frac{Z_L^R + j\frac{\rho c}{\pi r^2}\tan(kL)}{\frac{\rho c}{\pi r^2} + jZ_L^R\tan(kL)} \tag{2}$$

The input impedance function can be quite complicated with many peaks and valleys due to the duct system resonances and anti-resonances and the position of the active control source within the duct. The voltage-to-pressure transfer function for the control source will also be affected by the complex input impedance thereby causing particular source positions to be more effective for active noise cancellation at some frequencies than for other frequencies. If the active control source transfer function is zero at any frequency where active cancellation is attempted, an adaptive system will become unstable due to no controllability regardless of control system design. The duct impedances can be used to determine the reflections within the duct and the resulting acoustic pressure transfer functions between two arbitrary positions X1 and X2 in the duct in Figure 1 (with the active source S removed) as seen in equation (3) below.

$$\frac{p(\omega,X2)}{p(\omega,X1)} = \frac{\frac{\rho c}{\pi r^2}Z_L^R\cos(kX2) + \frac{\rho c}{\pi r^2}Z_L^R\tan(kL)\sin(kX2) + j\left(\frac{\rho c}{\pi r^2}\right)^2\tan(kL)\cos(kX2) - j\left(\frac{\rho c}{\pi r^2}\right)^2\sin(kX2)}{\frac{\rho c}{\pi r^2}Z_L^R\cos(kX1) + \frac{\rho c}{\pi r^2}Z_L^R\tan(kL)\sin(kX1) + j\left(\frac{\rho c}{\pi r^2}\right)^2\tan(kL)\cos(kX1) - j\left(\frac{\rho c}{\pi r^2}\right)^2\sin(kX1)} \tag{3}$$

The pressure transfer function between two arbitrary positions in a homogeneous duct section is a function of the input impedance to the duct at one end and the positions of the sensors within the duct, which is consistent with Munjal's analysis [8]. If a change in duct impedance due to a side branch or change in mechanical impedance of the wall exists between the pressure transfer function positions, the expression in equation (3) becomes more complicated, but still remains *passive*, or positive real. This means that there is no net increase in energy of waves propagating between the two positions; i.e., the wave energy either stays the same or decreases slightly due to absorption. As will be seen below, the positive real condition is a critical stability factor for stability when using pole-zero adaptive modeling in feedforward active noise control systems.

NON-ADAPTIVE FEEDFORWARD CANCELLATION

Consider the feedforward arrangement in Figure 2 where the reference signal n_t is fed through a control filter/amplifier A to produce the control signal u_t. The acoustic pressure transfer function between the active source S and the reference sensor is denoted as F(ω) (for feedback path), while the transfer function between the reference sensor and active source S is denoted as P(ω) (for plant). The complete error plant consists of the path through the source S and transfer function E(ω) between the source input and error sensor output. The transfer functions P(ω), F(ω), E(ω), S(ω), and A(ω), may be seen as having a defined magnitude and phase as a function of frequency to simplify the stability analysis.

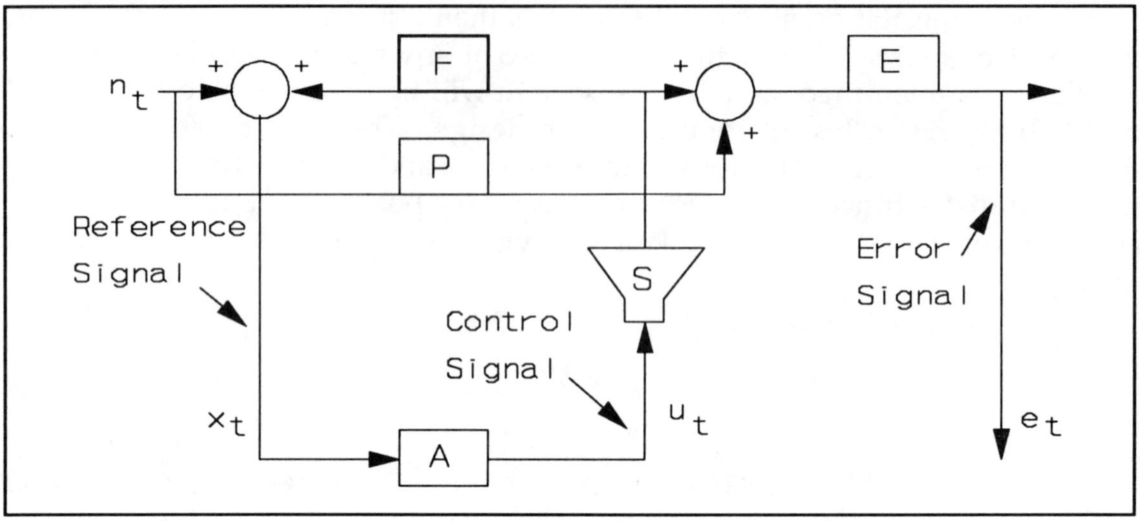

Figure 2 **The general feedforward controller used by Lueg and others showing F, P, and E as the existing acoustic plants, A as the controller, and S as the source.**

Lueg [1] employed the general system in Figure 2 using a simple amplifier for A and adjusting the spacing between the source S and reference sensor to control the cancellation of the propagating noise wave. For an infinite duct a spacing of approximately one half-wavelength would do the job of causing cancellation at the speaker, but also at the reference microphone. Consider the broadband response of the system at the error sensor position due to a reference noise impulse.

$$\frac{\mathcal{E}(\omega)}{N(\omega)} = P(\omega)E(\omega) + A(\omega)S(\omega)E(\omega)\cdot\sum_{n=0}^{\infty}[A(\omega)S(\omega)F(\omega)]^n \tag{4}$$

The system will be stable for any reference noise signal provided equation (4) converges, which will be the case if the loop gain $A(\omega)S(\omega)F(\omega)$ has magnitude less than unity for all frequencies. The system response can then be seen as

$$\frac{\mathcal{E}(\omega)}{N(\omega)} = P(\omega)E(\omega) + \frac{A(\omega)S(\omega)E(\omega)}{1 - A(\omega)S(\omega)F(\omega)} \tag{5}$$

Due to all the resonances of a finite duct system, the functions $P(\omega)$, $F(\omega)$, $E(\omega)$, and $S(\omega)$ are quite complicated but, Lueg's system could give quite reasonable cancellation for certain fixed frequencies provided the amplifier gain is sufficiently low for the loop gain where the magnitude of $A(\omega)S(\omega)F(\omega)$ is less than unity. With a stable amplifier gain, a change in noise frequency will not cause instability, but rather causes reduced cancellation or even amplification. Obviously, if the feedback path F is zero, the instability problem is nonexistent. Warnaka [9], reports much work on directional sensors and sources to reduce the feedback stability problem. Fundamentally, a feedforward active noise control system with isolated reference signal (zero feedback from the active source) is considered the most stable design possible.

ADAPTIVE FEEDFORWARD CONTROL WITH FEEDBACK COMPENSATION

Figure 3 shows the digital adaptive feedforward controller of Warnaka, et. al. [10], where the system *removes* the feedback from the reference sensor signal. Before operation, models for the feedback loop transfer function, $S(\omega)F(\omega)$, and the error signal transfer function, $S(\omega)E(\omega)$, are estimated. Removal of the acoustic feedback from the active source to the reference sensor adds a considerable margin of stability to the system so long as the feedback transfer function model is accurate. The error signal transfer function is necessary only for the adaptive LMS algorithm used to adjust the control filter $A(\omega)$ to give cancellation. A straightforward frequency domain analysis reveals the operation of the controller. The reference noise spectrum is

$$X(\omega) = N(\omega) + U(\omega)S(\omega)F(\omega) - U(\omega)D(\omega) \tag{6}$$

such that if $D(\omega)=S(\omega)F(\omega)$, the feedback is completely compensated and the control filter A can apply as much gain as needed without concern about instability because the feedback loop no longer exists. However, if the feedback transfer function changes due to temperature, flow, or source changes, the control loop gain again determines the system stability. The error signal spectrum is seen in equation (7) where $X(\omega)=N(\omega)$ due to precise feedback compensation.

$$\mathcal{E}(\omega) = N(\omega)P(\omega)E(\omega) + N(\omega)A(\omega)S(\omega)E(\omega) \tag{7}$$

Equation (7) shows that for cancellation where $E(\omega)\neq 0$, the control filter A equals

$$A(\omega) = \frac{-P(\omega)}{S(\omega)} \tag{8}$$

It can be seen from equations (7) and (8) that even with feedback compensation, the control system is unstable if there are frequency(s) $S(\omega)=0$ and ineffective where $E(\omega)=0$. A zero in the error transfer function causes *unobservable* control frequencies. A zero in the active source transfer function means there are *uncontrollable* frequencies. In general, an adaptive control system will be ineffective for unobservable frequencies and unstable for uncontrollable frequencies regardless of the adaptive algorithm or control architecture. Fortunately for a typical finite-length duct with some absorption in the duct radiation impedance, the transfer functions S and E will be small at some frequencies, but not exactly zero. However, some applications of feedforward control may have transfer function zeros (a good example is undamped active vibration control), or due to fixed-point digital signal processing, a transfer function may be small enough to be essentially zero to the signal processor, leading to instabilities.

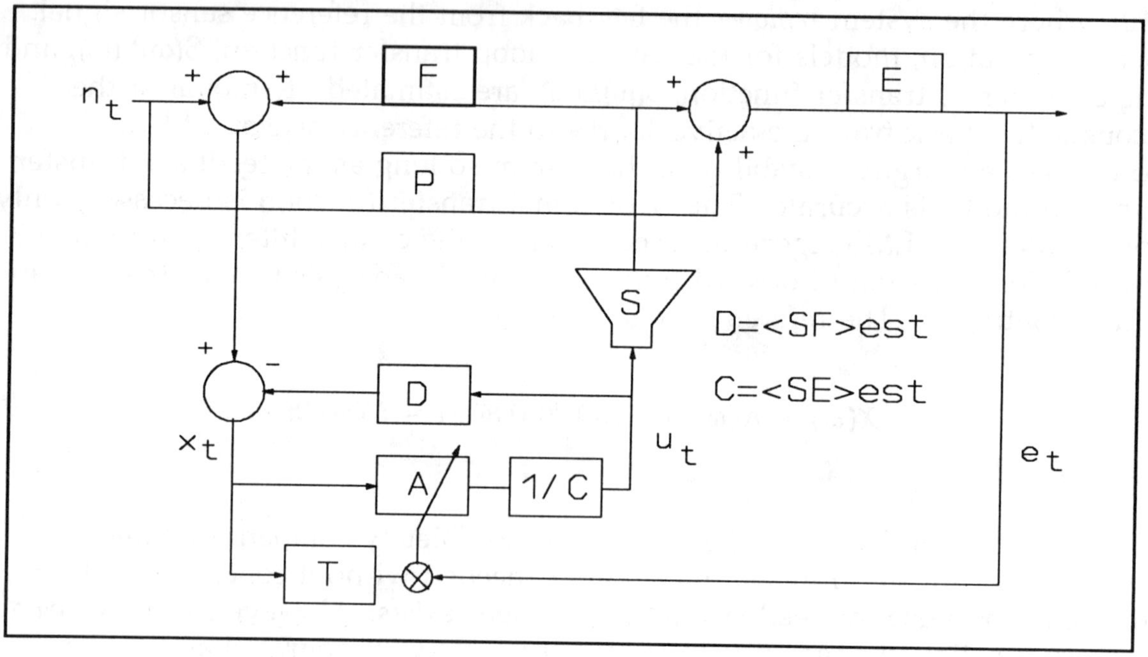

Figure 3 **Warnaka's adaptive feedforward controller with feedback compensation for robust stability.**

The error transfer function delay T is necessary to "time-align" the data with the error signal in the adaptive update. It should be noted that this is not the same as the "filtered-X" adaptive controller [11]. The inverse of the error plant, $C(\omega)=S(\omega)E(\omega)$, critically damps the control loop dynamics giving improved performance. For purely deterministic signals, the adaptive algorithm will manage to converge non-optimally even if estimate for $S(\omega)E(\omega)$ is slightly in error, but for random noise these errors are much more critical to stability.

POLE-ZERO CONTROL WITH ON-LINE MODELING

Figure 4 depicts the controller design of Eriksson [12] where the error plant is modeled as the active control system cancels the reference noise. The control approach taken allows the system to "track" changes in the error and feedback plant dynamically during cancellation operations. An analysis of the system stability reveals that it is fundamentally different from the Warnaka approach and must adhere to some previously unpublished constraints to assure convergence and control stability. As for the Warnaka design, the error transfer function must be both accurate *and current* for on-line stable adaptive control. As seen in Figure 4, injection of a known white noise sequence excites all the transfer function modes to give an accurate model, even if the active noise cancellation is for only a single frequency. The stability of the adaptive controller is defined by

$$\frac{U(\omega)}{X(\omega)} = \frac{A(\omega)}{1 - B(\omega)} \quad (9)$$

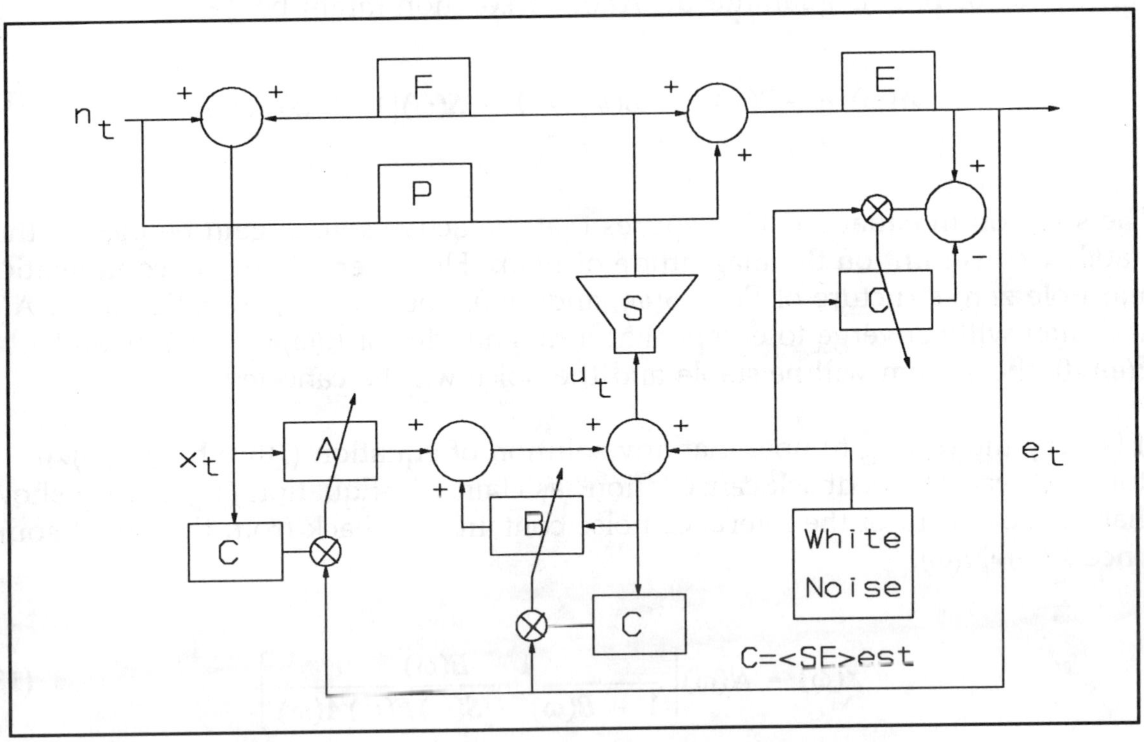

Figure 4 **Eriksson pole-zero adaptive noise cancellation system with On-line transfer function modeling.**

which clearly shows stability for all frequencies of system operation where the magnitude of B(ω)<1. This is straightforward to enforce using a single scale factor when B(ω) is an all-zero filter. Combining equations (7) and (9) one can write an expression for the entire control system transfer function with the error signal as output and the reference noise signal as input as seen in equation (10).

As for Warnaka's controller, the transfer function in equation (10) can be made zero independent of the reference noise signal provided the magnitude of E(ω) and S(ω) are both nonzero. However, the solution of two unknowns, A(ω), and

$$\frac{\mathscr{X}(\omega)}{N(\omega)} = P(\omega)E(\omega) + \frac{A(\omega)S(\omega)E(\omega)}{1 - B(\omega) - S(\omega)F(\omega)A(\omega)} \qquad (10)$$

B(ω), from one equation is not unique. One solution consistent with Munjal [7] is

$$A(\omega) = \frac{-P(\omega)}{S(\omega)}; \qquad B(\omega) = F(\omega)P(\omega) \qquad (11)$$

Since the duct transfer functions F(ω) and P(ω) are both passive, the required constraint for B(ω) is guaranteed. Another solution might be

$$A(\omega) = -P(\omega); \qquad B(\omega) = 1 - S(\omega)[1 - F(\omega)P(\omega)] \qquad (12)$$

The solution in equation (12) requires that the active source gain be part of the stability constraint on the magnitude of B(ω). However, given the complexities and pole-zero structure of P(ω), S(ω), and F(ω), one cannot generalize what A(ω) and B(ω) will converge to except that if magnitudes of $B(\omega)<1$, $S(\omega)>0$, and $E(\omega)>0$, the system will be stable and the noise will be canceled.

It is very interesting to note that any solution of equation (10) where $F(\omega)\neq 0$ does not lead to "feedback cancellation" as claimed. Equation (13) clearly shows that the spectrum of the reference noise contains feedback from the active source since $X(\omega)\neq N(\omega)$.

$$X(\omega) = N(\omega)\left[\frac{1 - B(\omega)}{1 - B(\omega) - S(\omega)F(\omega)A(\omega)}\right] \qquad (13)$$

It can be shown that for robust stability where the magnitude of $B(\omega)<1$, the magnitude of A(ω) should be less than the magnitude of $[1-B(\omega)]/S(\omega)F(\omega)$, to guarantee a stable reference signal. For finite impulse response, or all-zero, filters it is straightforward to implement this constraint through scaling. The solutions given in equations (11) and (12) insure stability of the system. The RLMS algorithm as depicted in Figure 4 is composed of two finite impulse response filters configured with feedback between them to give an infinite impulse response, or pole-zero adaptive filter. Therefore, the stability of the RLMS algorithm is controlled by scaling the magnitudes of each of the filter sections.

ACTIVE NOISE CANCELLATION USING FEEDBACK CONTROL

Feedback control is required for applications of active noise control where it is not possible (or practical), to use a coherent reference noise signal. Such applications include spatially-incoherent noise generated from turbulence, noise generated from many sources and paths, and induced resonance where no coherent reference signal is available. The most important characteristic about adaptive and non-adaptive feedback control is that its stability is *signal dependent.* All active noise cancellation systems using feedback control have a limitation due to the time delay from when a control signal is applied to the active source and when the response is detected at the error sensor. One effect of the delay is that only periodic, or sinusoidal, signals may be completely canceled. Another effect is that even the slightest change in delay causes a larger change in required controller phase (for cancellation), the higher in frequency one goes. Therefore, a delay change in an optimal feedback cancellation system will generally cause some high frequency(s) from the active source to change from high-gain negative feedback for cancellation to high-gain positive feedback for unstable oscillation. As will be seen in the analysis below, adaptive feedback is much less robust to instability than adaptive feedforward control.

NONADAPTIVE FEEDBACK-BASED ACTIVE CANCELLATION

Olson's feedback active cancellation system [2], simply used a carefully designed amplifier matched to the response of the microphone error sensor and loudspeaker active source. The design had the microphone very close to the loudspeaker in order to minimize the time delay and the system operated over a fixed frequency range of only a few hundred Hz. Olson's feedback amplifier design was so sensitive that after using the system for an extended time the heat from the vacuum tubes would alter the values of the capacitors, resistors, and inductors enough to give phase bias and sometimes instability [13]. While Olson's approach has its limitations, the basic design may be found in a variety of active hearing protector systems today [3]. Consider the feedback control system in Figure 5 below where G(ω) is the feedback control amplifier response, S(ω) is the active source response, and E(ω) is the error transfer function. Olson's system designed G(ω) such that

$$\angle G(\omega) = \angle \frac{-1}{S(\omega)E(\omega)}; \qquad |G(\omega)S(\omega)E(\omega)| < 1 \tag{14}$$

If the feedback loop has a response delay of τ seconds, the transfer function S(ω)E(ω) will have a linear phase of $-\omega\tau$. Carefully designing the control filter G(ω) such that its magnitude is slightly less than the inverse of the magnitude of S(ω)E(ω), and its phase is *exactly* $+\omega\tau$ will give stable cancellation of periodic noise. It can be seen that a slight change in delay in the feedback loop can cause a loss of cancellation or even amplification, but not instability so long as the loop gain G(ω)S(ω)E(ω) is less than unity. However, a stable loop gain also limits the maximum amount of cancellation to less than 3 dB. Noise signal changes will cause oscillations in the feedback loop, but a loop gain less than unity guarantees these oscillations will eventually be damped out.

$$\frac{\mathcal{E}(\omega)}{N(\omega)} = \frac{P(\omega)E(\omega)}{1 - G(\omega)S(\omega)E(\omega)} = \frac{P(\omega)E(\omega)}{2} \tag{15}$$

To obtain greater degrees of cancellation for periodic noise, the loop gain must be increased beyond unity which will be stable so long as cancellation is occurring. For low frequencies, where $|\omega\tau|<\pi$, it is fairly certain that stable negative feedback can be maintained. However at higher frequencies, the non-causal feedback control filter will almost instantly diverge into uncontrolled oscillations from any transient in the noise signal. High levels of feedback-type cancellation is considered impractical for all but totally-stationary periodic, or very low frequency noise. The above discussion illustrates the fundamental problem with using a feedback-type system for active noise cancellation purposes. As will be seen below, use of adaptive feedback control complicates the issue still further, but offers the only hope for high levels of cancellation in real-world active noise control applications.

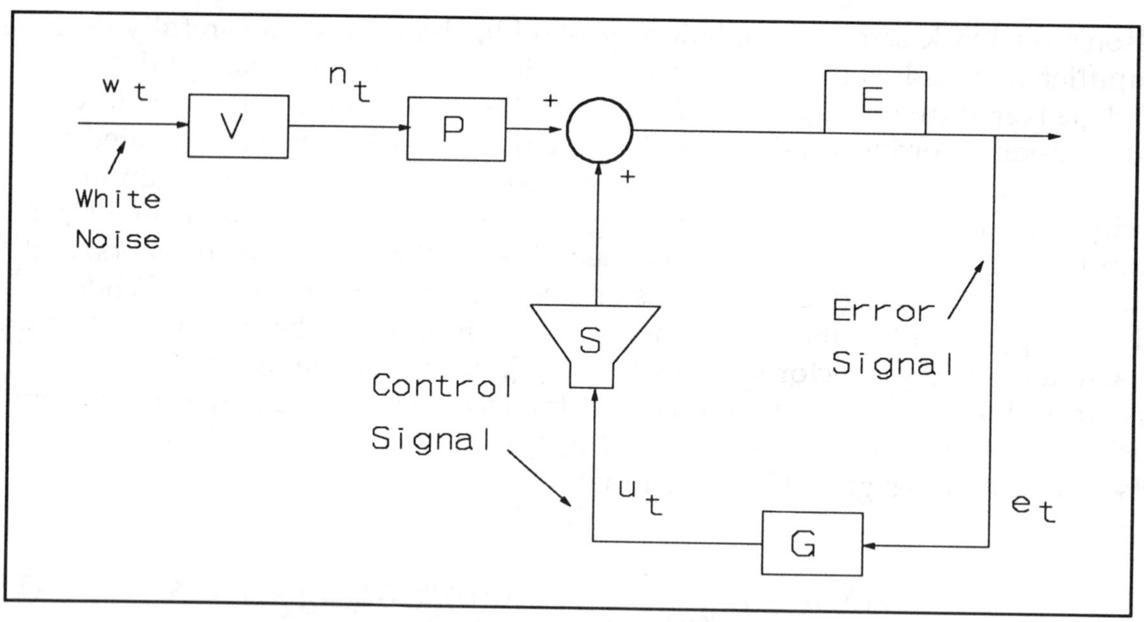

Figure 5 **A general feedback control system for active noise attenuation using the symbols of a feedforward system.**

ADAPTIVE FEEDBACK-BASED ACTIVE CANCELLATION

Adaptive feedback control systems perform two essential functions necessary to maintain stability during signal transients and/or error plant changes. First, the adaptive algorithm *models the error plant* transfer function S(ω)E(ω) including the response delay. Second, the control system *models the noise signal* as an output from a filter with white noise input. The noise signal model allows the noise waveform to be predicted in the future by the amount of the delay in the error plant. This allows the implementation of a causal control filter G(ω) and the elimination of "unpredictable" transients from passing around the feedback loop [14].

In practice, simultaneous modeling of the noise and error plant during cancellation is very difficult to achieve. Examples of adaptive feedback control in the literature describe very low-frequency industrial process control where the plant gain and delay are both known and stationary [15]. Consider the noise signal model for Figure 5 where V(ω) is a pole-zero filter providing an ARMA model for the noise N(ω).

$$N(\omega) = W(\omega)\,V(\omega) \tag{16}$$

If the error signal can be "whitened" to match W(ω), the removal of the spectral peaks of the noise (the periodic components) will give a considerable degree of active noise reduction. Since the ARMA noise is whitened by an auxiliary control signal, the adaptive control process is known as an ARMAX controller. The solution for G(ω) in equation (15) for whitening the error signal is

$$G(\omega) = \frac{1 - V(\omega)P(\omega)E(\omega)}{S(\omega)E(\omega)} \tag{17}$$

The additional term in equation (17), V(ω)P(ω)E(ω), causes the negative of the predicted noise waveform to be generated and prefiltered by the inverse of the error plant transfer function. As for all the above-mentioned adaptive systems, a zero in S(ω) or E(ω) causes the control filter, and hence, system to be unstable. Insertion of the control filter of equation (17) into equation (15) reveals that a zero in V(ω) or P(ω) will lead to instability in the closed loop control system. This condition is known as the *persistently exciting noise requirement* [15] and is typical of ARMAX systems. For active noise attenuation, the persistently exciting noise condition can be relaxed by modeling the noise as an AR process and not attempting to whiten the noise zeros, or dips in the noise spectrum below the whitened noise level [14].

CONCLUSIONS

The stability of two basic active noise control systems has been briefly examined in terms of sensitivity to noise signal changes as well as plant transfer function changes. The most stable and simple of the controllers is the feedforward controller where the reference noise is isolated from the active control source response. Given contamination of the reference noise by feedback from the active control source, it is not clear whether adaptive feedforward control with feedback compensation or on-line plant estimation offers the best measure of stability robustness. The on-line pole-zero modeling system will track plant transfer function changes but does carry additional adaptive stability constraints which must be enforced. The pole-zero system does not actually cancel feedback from the reference noise signal, but rather compensates the closed-loop system poles caused by feedback. Adaptive feedback control is the least stable system for active noise cancellation because of fundamental limitations of the feedback loop delay and the fact that high measures of cancellation are only possible with potentially unstable feedback loop gains.

REFERENCES

1 P. Lueg, "Process of Silencing Sound Oscillations," German Patent, DRP No. 655,508, submitted: Jan. 27, 1933.

2. H. F. Olson and E. G. May, "Electronic Sound Absorber," Journal of The Acoustical Society of America, **25**, 6, pp 1130-1136 (1953).

3. C. W. Nixon, R. L. McKinley, and J. W. Steuver, "Performance of Active Noise Reduction Headsets," Harry G. Armstrong Aerospace Medical Research Laboratory, Wright-Patterson AFB, Ohio 45433, U.S.A.

4. S. J. Elliott and P. A. Nelson, "Models for Describing Active Noise Control in Ducts," I.S.V.R. Report No. 127, Southhampton, April 1984.

5. J. Burgess, "Active Adaptive Sound Control in a Duct," Journal of The Acoustical Society of America, **70**, 3, pp. 715-726 (1981).

6. D. Swanson, "The Role of Impedance Coupling in Achieving Global Active Attenuation of Noise," Presented at The American Society of Mechanical Engineers Winter Annual Meeting, Chicago, IL, Nov. 27 - Dec. 2, 1988.

7. M.L. Munjal and L.J. Eriksson, "Analysis of a Linear One-Dimensional Active Noise Control System by Means of Block Diagrams and Transfer Functions," Journal of Sound and Vibration, **129**, 3, pp. 443-455.

8. M.L. Munjal and L.J. Eriksson, "An Analytical, One-Dimensional, Standing Wave Model of a Linear Active Noise Control System in a Duct," Journal of The Acoustical Society of America, **84**, 3, pp. 1086-1093 (1988).

9. G.E. Warnaka, "Active Attenuation of Noise - The State of the Art," Journal of Noise Control Engineering, Vol. 18, No. 3, pp. 100-110, (1982).

10. G.E. Warnaka, L.A. Poole, J. Tichy, "Active Acoustic Attenuator," US Patent No. 4,473,906, September, 1984.

11. B. Widrow and S. Stearns, Adaptive Signal Processing, (Englewood Cliffs, NJ: Prentice Hall, 1985).

12. L.J. Eriksson, "Active Sound Attenuation System With On-Line Adaptive Feedback Cancellation", US Patent No. 4,677,677, June 30, 1987.

13. RCA, "Study of Communication in High-Level Ambient Noise Fields - Final Report, Phase II: Electronic Noise Reduction," prepared for US Army Signal Corps Eng. Lab., Fort Monmouth, NJ, by Surface Comm. Eng., Camden, NJ, February, 1958.

14. D. Swanson, "A Computer Simulation of Active Attenuation of Computer Fan Noise," in Proceedings INTER-NOISE '87, The Pennsylvania State University, University Park, PA, June 9-12, 1987.

15. K.J. Astrom and B. Wittenmark, Adaptive Control, (New York: Addison-Wesley, 1989).

APPLICATION OF RECURSIVE ADAPTIVE ALGORITHMS FOR SYSTEM IDENTIFICATION AND VIBRATION CONTROL

R. Wimmel,[1] J. Melcher[1]

ABSTRACT

This paper describes advances in the identification and control of flexible structures by consequent application of recursive adaptive algorithms on controllers with load path-considering design. Control of large structures requires powerful algorithms that are able to influence certain modes of a system with high modal densities. The stability of the adaptation process must be maintained. This paper describes theoretical and experimental investigations within the framework of the DLR program ARES (Actively **Re**acting Flexible **S**tructures), showing the performance of the modal controllers that are based on recursive digital filters.

INTRODUCTION

In order to influence the dynamic behavior of a flexible structure with active control concepts, the structure should possess multifunctional members with load carrying, actuating and sensoring capabilities. A controller must calculate the actuator driving signals, considering both the feedback from the sensor signals and the external data that describe the desired behavior of the structure. For this purpose, adaptive digital filters have been proposed [1] *). The following chapters deal with the modal formulation of the structural dynamic behavior and the consequences for the controller design.

THE STRUCTURE GIVES THE PATTERN

Before designing a controller for application on adaptive structures, it is useful to look at the mathematical description of the dynamics of flexible structures. The transformation of the physical coordinates into generalized coordinates allows for decoupling of the underlying differential equation of motion if proportional damping of the structure is assumed, which leads to a modal description of the dynamic system. One result of this method is that the transfer characteristics can be formulated in an accelerance function $A_{k\ell}$, which assumes a finite number m of modal degrees and proportional damping of the structure:

$$A_{k\ell}(s) = \frac{\ddot{U}_k(s)}{F_\ell(s)} = \sum_{i=1}^{m} \frac{s^2}{K_i + D_i s + M_i s^2} \varphi_{ik} \varphi_{i\ell} \quad . \tag{1}$$

*) The cited paper describes the goals of the DLR program ARES, as well as the basic ideas of actively controlled structures.

[1]German Aerospace Research Establishment (DLR), Institute of Aeroelasticity, Structural Dynamics Division, Bunsenstr. 10, 3400 Goettingen, Germany.

$\ddot{U}_k(s)$ represents the Laplace transform of the k-th element of the acceleration vector $\ddot{u}(t)$, and F_ℓ represents the Laplace transform of the ℓ-th element of the force vector $f(t)$. K_i, D_i, M_i are the i-th elements of the corresponding modal matrices of stiffness, damping and mass, respectively. φ_{ik} and $\varphi_{i\ell}$ are elements of the i-th eigenvector and k and ℓ represent two measuring points. This equation combines physical values with modal quantities. Therefore, it is suitable for both system identification experiments and modeling through FEM.

If the goal is to actively control the structure, it is also advantageous to consider what kinds of internal forces must be affected to obtain a particular desired accelerance function given through modal parameters. For the evaluation of the force signals through a controller, the pattern is set by the modal description of the structure. However, there are strong requirements for a so-called modal controller, such as adaptivity, i.e. self-optimizing a defined objective, and real-time-ability, and in order to make use of the modal properties the controller design must be digital.

All of the above-mentioned requirements can be fulfilled by using digital filters. The following chapters will further discuss the use of digital filters as modal controllers.

THE MODAL FILTER -- A COPY OF THE STRUCTURE

A single mode of the accelerance function has the form

$$A_{ik\ell}(s) = \frac{s^2}{K_i + D_i s + M_i s^2} \varphi_{ik} \varphi_{i\ell} \quad . \tag{2}$$

[1] shows that such a second-order module can be approximated by second-order digital filters of the form

$$H_{ik\ell}(z) = \frac{a_{ik\ell}\left(1 - 2z^{-1} + z^{-2}\right)}{1 + b_{1,i} z^{-1} + b_{2,i} z^{-2}} \quad \forall\, i \in [1, m] \quad . \tag{3}$$

These second-order sections are taken to build up a complete model that considers all of the modal characteristics of the flexible structure:

$$H_{k\ell}(z) = \sum_{i=1}^{m} H_{ik\ell}(z) \quad . \tag{4}$$

This modal filter is shown in Figure 1. It is a copy of the modal representative of the structure and can be used to alter the structure's transfer characteristics.

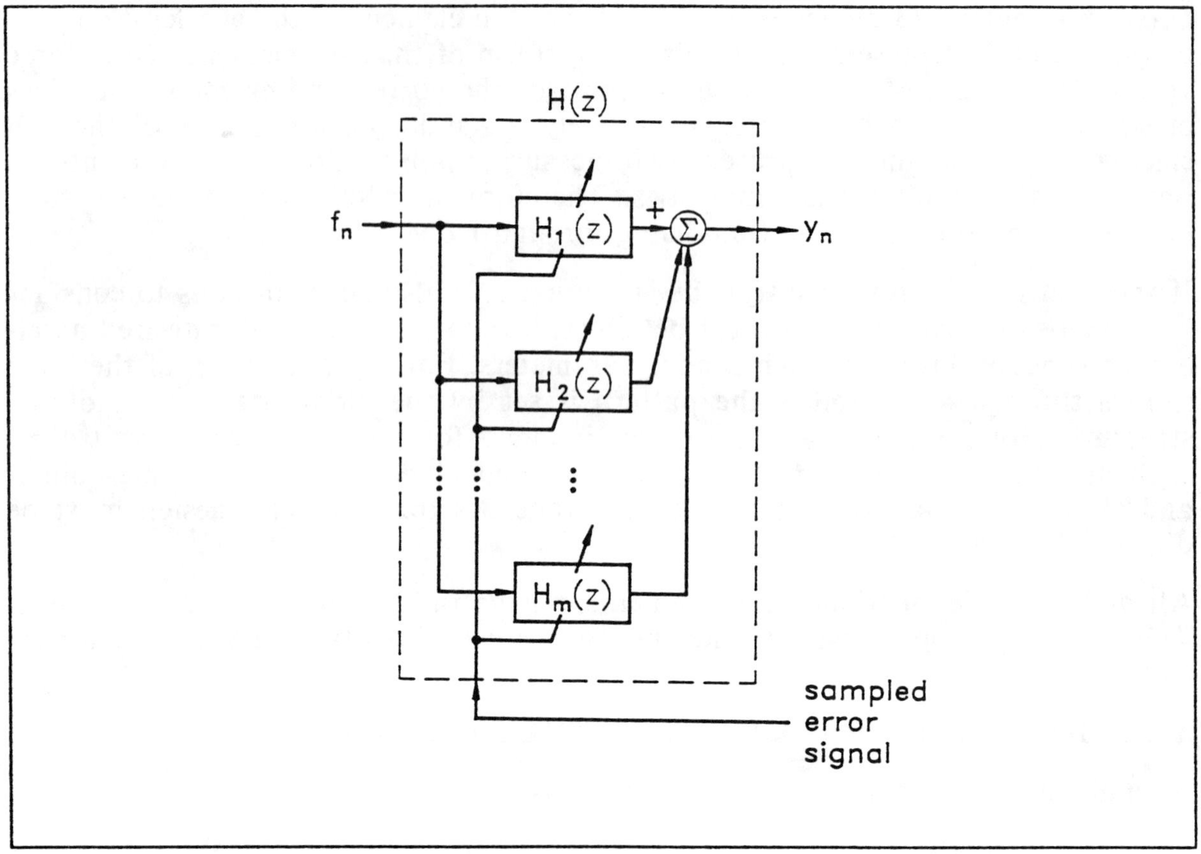

Figure 1. Parallel form of the adaptive modal filter

Altering the filter coefficients makes the filter applicable for the typical controller configuration, which is described in Figure 2.

Between the measuring points k and ℓ an accelerance function

$$A_{k\ell}(s) = \sum_{i=1}^{m} A_{ik\ell}(s) \tag{5}$$

exists, defined in the s-plane. In order to facilitate the notation, the indices $k\ell$ are excluded in the following. Assuming that these characteristics must be improved to accomplish a desired behavior $A_d(c)$, the filter $H(z)$ causes an actuator to exert internal forces so that the actual structural response matches the desired response. If the full accelerance matrix $\boldsymbol{A}(s)$ is to be altered, $H(z)$ must be replaced by the MIMO filter $\boldsymbol{H}(z)$. Due to changing external influences, the filter $H(z)$ must adapt its transfer function by minimizing the error signal

$$e(t) = \Delta\ddot{u}(t) = \ddot{u}_d(t) - \ddot{u}(t) \quad . \tag{6}$$

Suitable adaptation algorithms for this purpose are introduced in the next chapter.

ADAPTATION -- HOW TO GET AN OPTIMUM

Many existing algorithms use the mean square error as performance function. For the time step n it is given by

$$\psi(n) = E\left[e^2(n)\right] \tag{7}$$

In the pure system identification environment (Figure 5), which is part of the test philosophy used here, ψ is a $3m$-dimensional function of the a_i, $b_{1,i}$, $b_{2,i}$.

The full performance surface has at least one global minimum at which the transfer function $H(z)$ of the filter matches the accelerance function $A_d(s)$ of the structure being identified. The minimum is valued zero if the filter has as many as or more poles than the accelerance function $A_d(s)$.

Otherwise, $H(z)$ is undermodeled and not all of the modes of $A_d(s)$ can be identified. Local minima can exist because the performance surface is generally non-quadratic in regard to the denominator coefficients $b_{1,i}$ and $b_{2,i}$, but were not observed in the experiments.

An adaptation algorithm is used to find the global minimum. In the references used for this study (overview: [2,3]), several algorithms using gradient methods are described, all of which use the gradient of the instantaneous squared error itself

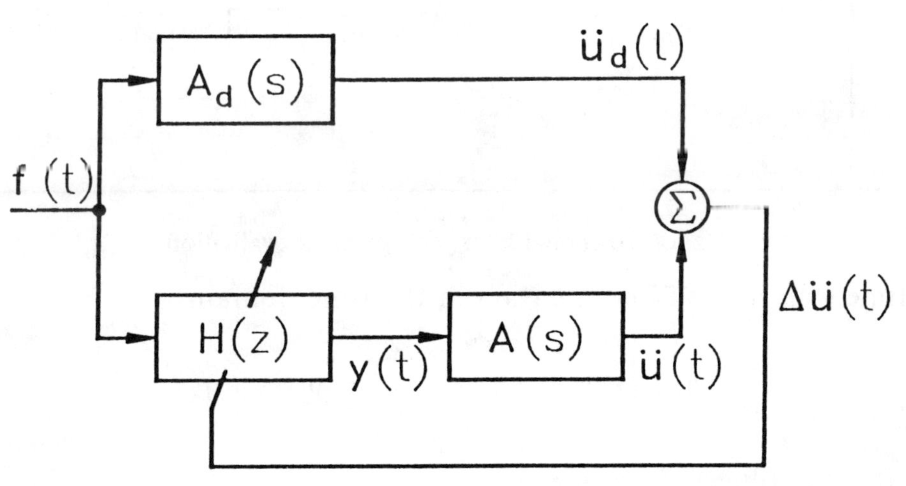

Figure 2. Typical controller configuration adapting desired system characteristics. $H(z)$: Digital time-varying transfer matrix of the adaptive filter system. $A(s)$: Behavior of the continuous uncontrolled structural system, e.g. given by the accelerance matrix. $A_d(s)$: Derived behavior of the structural system. The AD- and DA-converters are not explicitly carried out.

instead of the mean squared error. The filter described in Eq. (3) can be written in the form

$$H_i(z) = \frac{a_i \cdot C(z)}{1 + B_i(z)} \quad , \tag{8}$$

consisting of a transversal filter of the form

$$C(z) = 1 - 2z^{-1} + z^{-2} \tag{9}$$

and a recursive filter $a_i / (1 + B_i)$ (see Figure 3).

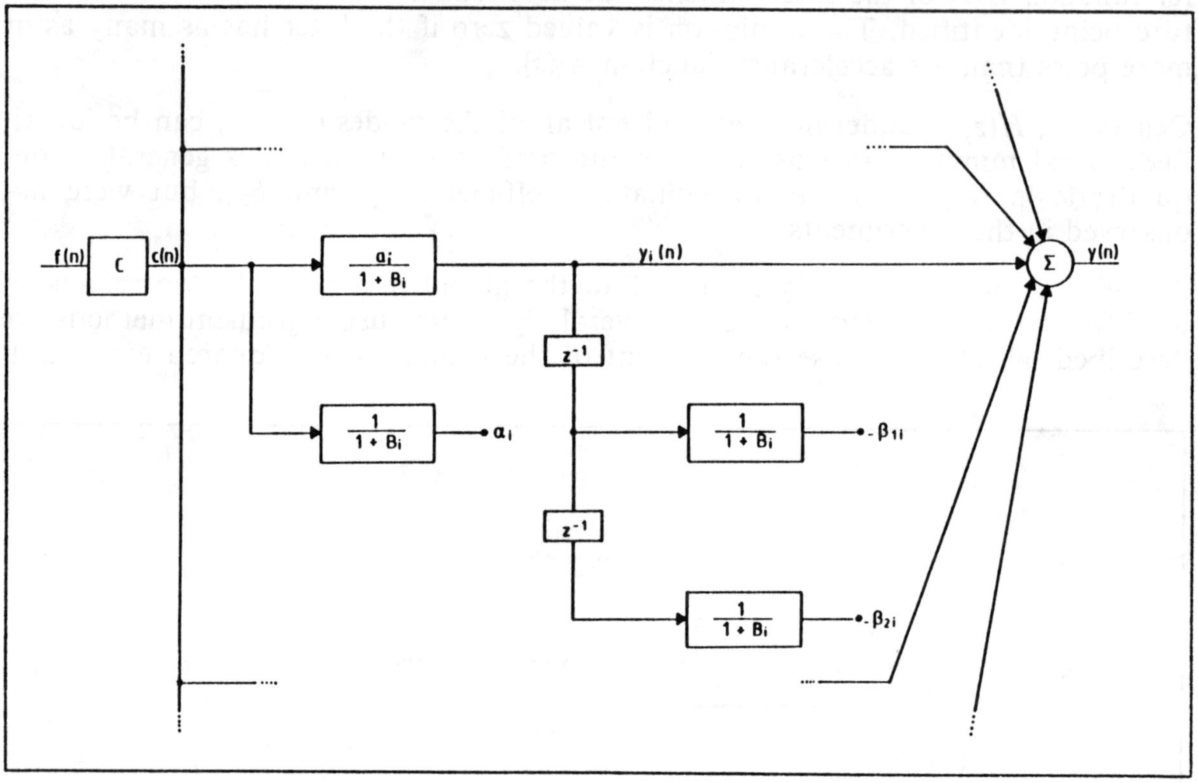

Figure 3. Subsystem $H_i(z)$ of the modal filter with gradient evaluation

For the time step n, the filter equation in the time domain is

$$y_i(n) = a_i \cdot c(n) + \sum_{j=1}^{2} b_{ji} y_i(n-j) \quad , \tag{10}$$

where $c(n)$ is the filtered force signal

$$c(n) = f(n) - 2f(n-1) + f(n-2) \quad . \tag{11}$$

The error signal is given by

$$e(n) = d(n) - \sum_{i=1}^{m} y_i(n) \quad . \tag{12}$$

In order to calculate the gradient of the squared error, the partial derivatives had to be determined:

$$\frac{\partial\, e^2(n)}{\partial\, a_i} = 2e(n)\,\frac{\partial\, e(n)}{\partial\, a_i}$$
$$\frac{\partial\, e^2}{\partial\, b_{ji}} = 2e(n)\,\frac{\partial\, e(n)}{\partial\, b_{ji}} \quad , j = 1,2 \quad . \tag{13}$$

Through Eqs. (10) and (12) we obtain:

$$\frac{\partial\, e(n)}{\partial\, a_i} = \frac{-\,\partial\, y_i(n)}{\partial\, a_i} = -\,c(n) + \sum_{r=1}^{2} b_r\,\frac{\partial\, y(n-r)}{\partial\, a}$$
$$\frac{\partial\, e(n)}{\partial\, b_{ji}} = \frac{-\,\partial\, y_i(n)}{\partial\, b_{ji}} = y_i(n-j) + \sum_{r=1}^{2} b_r\,\frac{\partial\, y(n-r)}{\partial\, a} \quad , j = 1,2 \quad . \tag{14}$$

Defining

$$\alpha_i(n) = \frac{\partial\, y_i(n)}{\partial\, a_i}$$
$$\beta_{ji}(n) = \frac{\partial\, y_i(n)}{\partial\, b_{ji}} \quad , j = 1,2 \tag{15}$$

yields

$$\alpha_i(n) = c(n) - \sum_{r=1}^{2} b_{ri}\,\alpha_i(n-r)$$
$$\beta_{ji}(n) = -\,y_i(n-j) - \sum_{r=1}^{2} b_{ri}\,\beta_{ji}(n-r) \quad , j = 1,2 \tag{16}$$

which basically is a filtering of $c(n)$ and $y_i(n-j)$ with the recursive filter $1\,/\,(1 + B_i)$. The adaptation formula is

$$\mathbf{w}_i(n+1) = \mathbf{w}_i(n) - \boldsymbol{\mu}_i\,\boldsymbol{\delta}_i\,(n) \quad , \tag{17}$$

where

$$\mathbf{w}_i(n) = \begin{pmatrix} a_i(n) \\ b_{1,i}(n) \\ b_{2,i}(n) \end{pmatrix} \tag{18}$$

is the coefficient vector at the time step n and $\boldsymbol{\mu}_i$ is the diagonal matrix of the convergence parameters. The gradient of the squared error δ_i can be expressed as

$$\boldsymbol{\delta}_i(n) = -\,2e(n)\cdot\begin{pmatrix} \alpha_i(n) \\ \beta_{1,i}(n) \\ \beta_{2,i}(n) \end{pmatrix} \quad . \tag{19}$$

This evaluation of the gradient was first proposed by White in 1975 [4]. In 1987, Hsia made a simplification of White's proposal, which reduces the computational costs [5]. Horvath described the parallel form with second order sections in 1976 [6], and Jenkins and Nayeri furthered this with examples and analysis in 1986 and

1988 [7,8]. In 1976, Feintuch suggested an algorithm that is a drastically simplified computation of the gradient [9]. Hsia's algorithm as well as the algorithm suggested by Feintuch is successful when applied to filters in the direct form. Whether they are able to adapt the coefficients of the parallel modal filter system described here is yet to be proven.

The z-transformation of Eq. (16) clearly shows the difference (see Figure 4) between the algorithms of White

$$\alpha_i^W(z) = \frac{C(z)}{1 + B_{i,n}(z)} \quad , \qquad \beta_{ji}^W(z) = \frac{Y(z) \cdot z^{-j}}{1 + B_{i,n}(z)} \tag{20}$$

the algorithms of Hsia, which used an approximation for the computation of $\beta_{ji}(z)$:

$$\beta_{ji}^H(z) = \frac{Y(z) \cdot z^{-j}}{1 + B_{i,n-j}(z)} \quad , \tag{21}$$

and of Feintuch, where the transform of input and output sequences are substituted for $\alpha_i(z)$ and $\beta_{ji}(z)$, respectively:

$$\alpha_i^F(z) = C(z) \qquad \beta_{ji}^F(z) = Y(z)\, z^{-j} \quad . \tag{22}$$

The update-formula (19) remains unchanged in all cases. The conclusion can be drawn that Feintuch's algorithm is not appropriate for use in parallel realizations, because the α_i of the different modes would be adjusted with the same value $c(n)$ for all modes at time step n. In the following chapter, examples of the performance of the modal filter will be given.

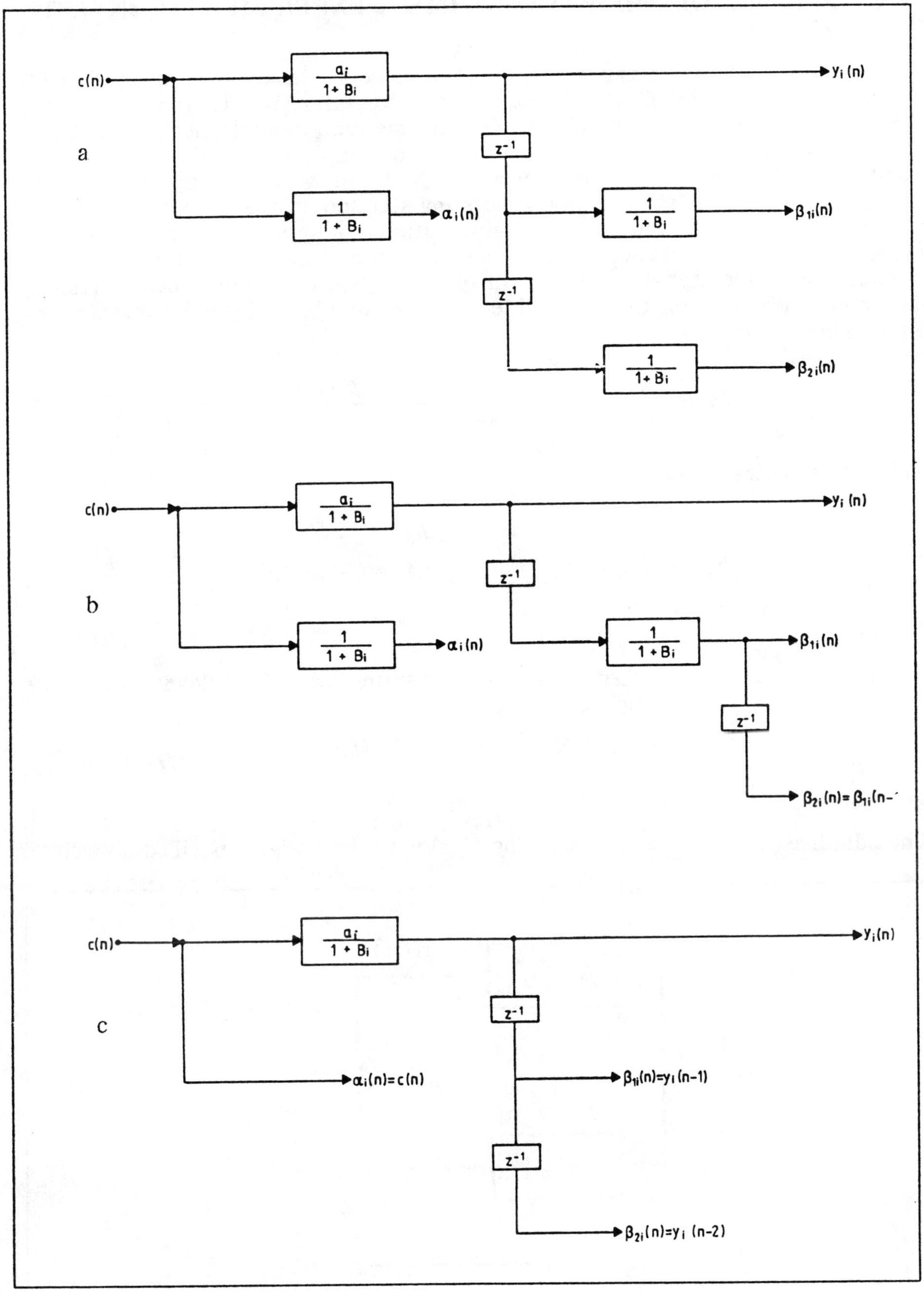

Figure 4. Comparison between the gradient based algorithms of a) White, b) Hsia and c) Feintuch

PERFORMANCE OF THE ALGORITHMS -- EXPERIMENTAL INVESTIGATIONS

The essential part of the experimental set-up used is the digital system TXP32, which contains the WE DSP32 Digital Signal Processor [10]. Its rapid hardware floating-point arithmetic unit fulfills the real-time requirement, meaning that Eq. (17) can be calculated completely between two sampling operations. All experiments described in this chapter were carried out with a sampling time of $\tau = 2ms$. From the adaptive signal processing standpoint, the simplest form of an adaptive controller is the system identification configuration (see Figure 5). Therefore, this configuration is part of the test philosophy to show the basic mechanisms of the modal controller. In order to guarantee reproducable results, the accelerance function to be identified $A_d(s)$ is simulated by the approximated digital two-section filter system

$$A_d(s) = \sum_{i=1}^{2} A_{di}(z) = \sum_{i=1}^{2} \frac{p_i\, c(z)}{1 + q_{1,i}z^{-1} + q_{2,i}z^{-2}} \tag{23}$$

with the numerical values:

$$\begin{aligned} p_1 &= 1.10 & p_2 &= 1.50 \\ q_{1,1} &= -1.71812 & q_{1,2} &= -1.07137 \\ q_{2,1} &= 0.89035 & q_{2,2} &= 0.63878 \end{aligned} \tag{24}$$

In this sense, $A_d(z)$ is equivalent to a structural system with the following values for the eigenfrequency and critical damping:

$$\begin{aligned} f_{01} &= 35\ Hz & f_{02} &= 75\ Hz \\ \vartheta_1 &= 14\% & \vartheta_2 &= 30\% \end{aligned} \tag{25}$$

The adaptive filter also has two sections $H_i(z)$ with the design described above.

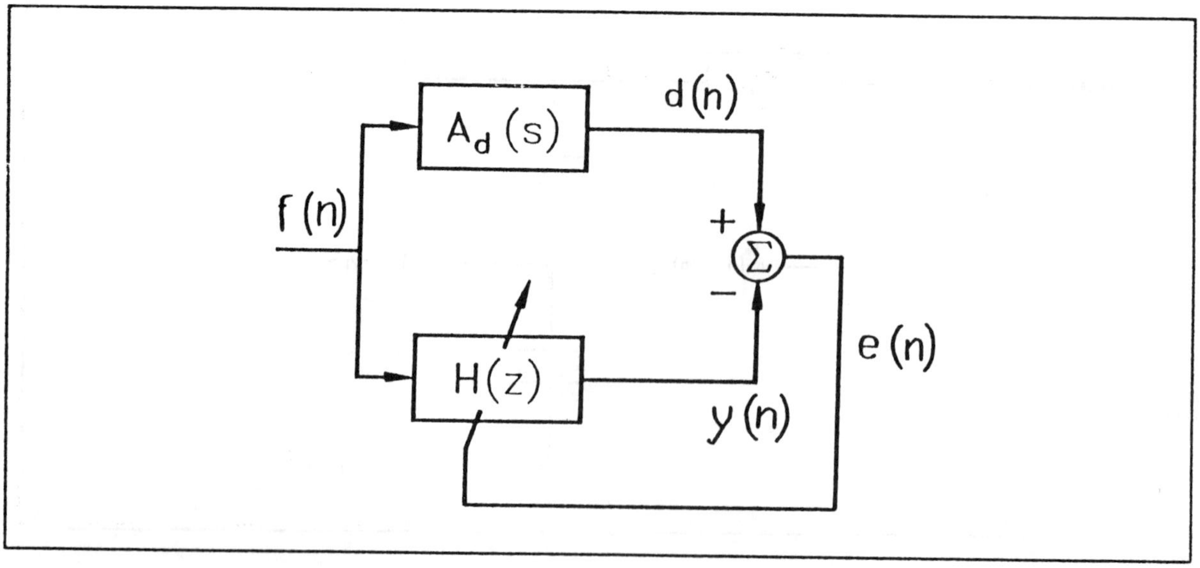

Figure 5. System identification configuration

The adaptation process can be represented clearly by using the filter-coefficient space, especially the $b_{1,i}$ / $b_{2,i}$-plane. The stability condition (poles inside the unit circle of the z-plane) demands that the pairs $b_{1,i}$ / $b_{2,i}$ lie inside the stability triangle, see Figure 6. The interesting lower frequencies are found in the left half-plane. Cases that are relevant to structural dynamics, having complex conjugate pairs of poles (underdamped), limit the $b_{1,i}$ / $b_{2,i}$-pairs into the area above the drawn-in parabola. In the following figures, only the upper left quadrant $b_{1,i}$ / $b_{2,i}$-plane is shown in order to enlarge the resolution.

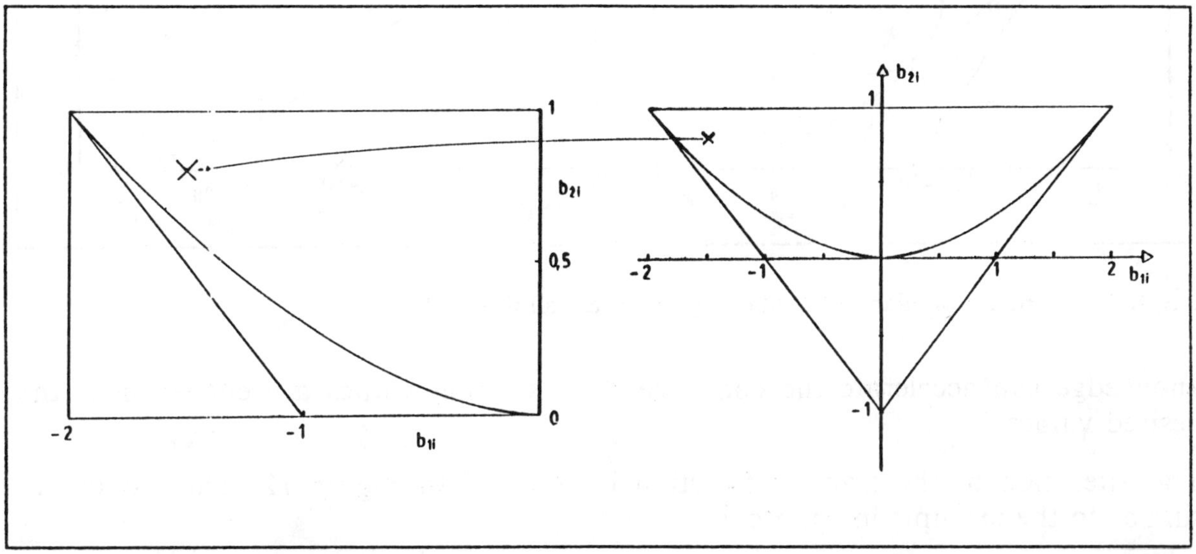

Figure 6. Stability triange interfilter coefficient space

The experimental investigations proved that the algorithms of White and Hsia converge when applied to the modal filter. A large variety of starting values can be chosen, which is important in system identification if little knowledge about the structure exists and coefficients must be set arbitrarily.

The elements of the convergence matrix μ_i have to be optimized for every new task in order to minimize convergence time. An overall value of 0.001 was suitable to track the coefficients. Figures 7-11 show five different cases for the starting values. The adaptation is accomplished by the algorithms of White and Hsia, respectively, which can be compared; thereby, two graphs are shown, both with two tracks which correspond to the b_1 / b_2-pairs of the two modes. The crosses mark the desired values.

Figure 7 illustrates a case where the starting values of all coefficients are different. The tracks may cross, but the corresponding coefficients are not equal at the same point in time.

In Figure 8 only the transversal coefficients a_i are different and convergence still happens.

If all coefficients of one mode match those of the other mode (see Figure 9), the adaptive process will end at a saddle point between two *global* minima which exist because the filter modes are equivalent. Eq. (16) shows that for this constellation the algorithm always evaluates the same gradient in both modes. If only the a_i slightly differ from each other ($a_1 = 0.5$; $a_2 = 0.51$), the adaptation process leads to optimal adjustment (see Figure 10). Figure 11 illustrates that a priori

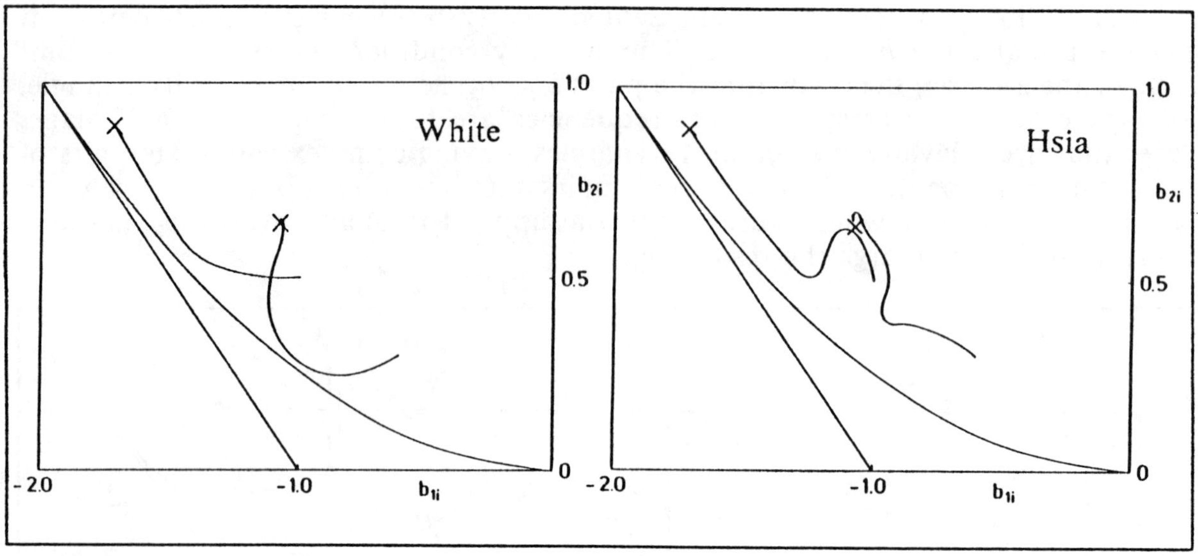

Figure 7. Process of adaptation: Starting values are all different

knowledge can accelerate the convergence if starting values are chosen near the desired values.

The alteration of the transfer function is depicted in Figure 12. The values are similar to the example in Figure 7.

Figure 13 refers to the problem of undermodeling. Several tracks with different starting values show that one filter mode can approximate two modes, but cannot identify them. The performance function has a minimum, which is found from all the starting points shown. This minimum is not zero, however, which means that the error will not fade away.

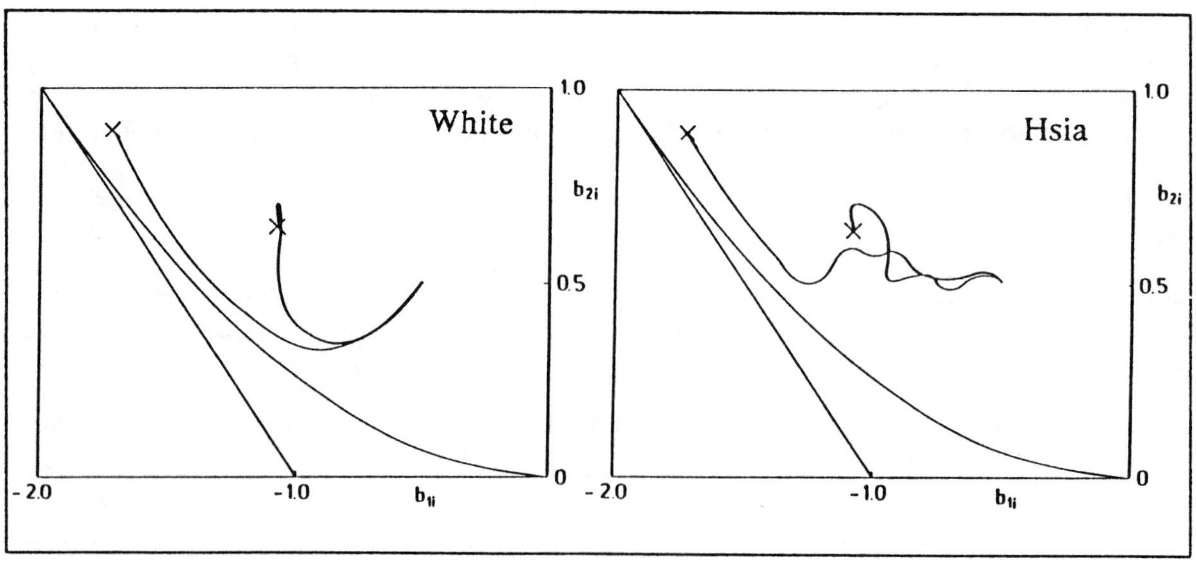

Figure 8. Process of adaptation: Starting values differ only at the a_i ($a_1 = 1.1$, $a_2 = 1.5$)

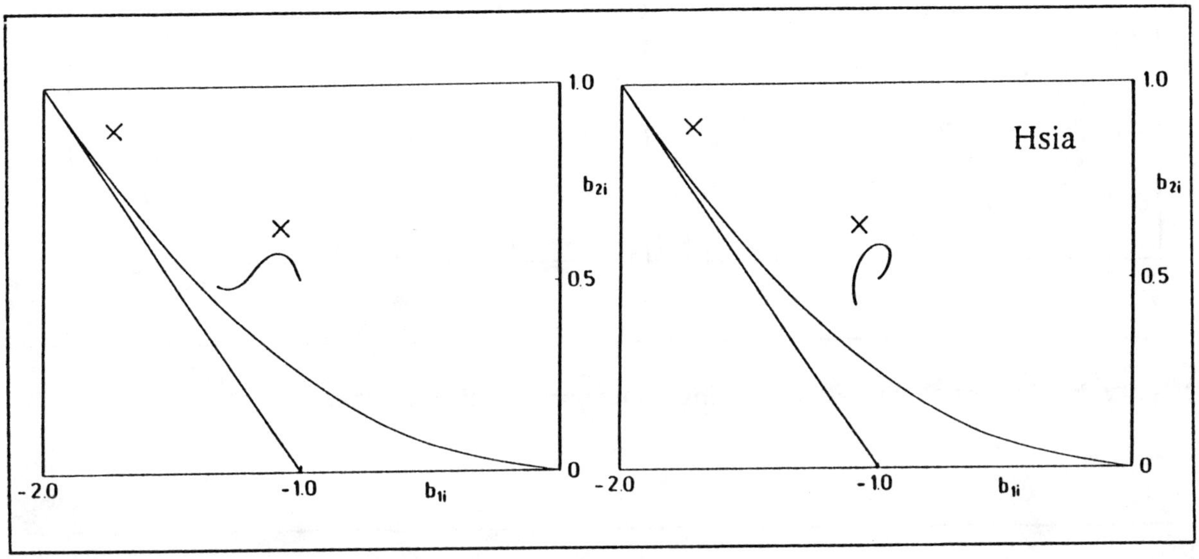

Figure 9. Process of adaptation: Starting values are all equal

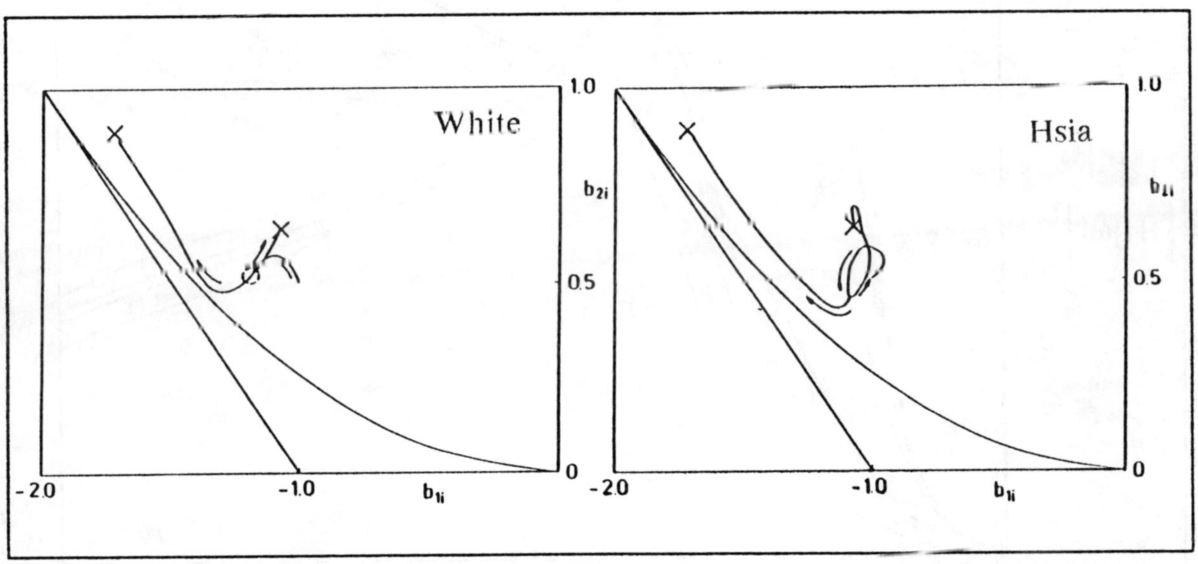

Figure 10. Process of adaptation: Starting values differ slightly at the a_i ($a_1 = 0.5$, $a_2 = 0.51$)

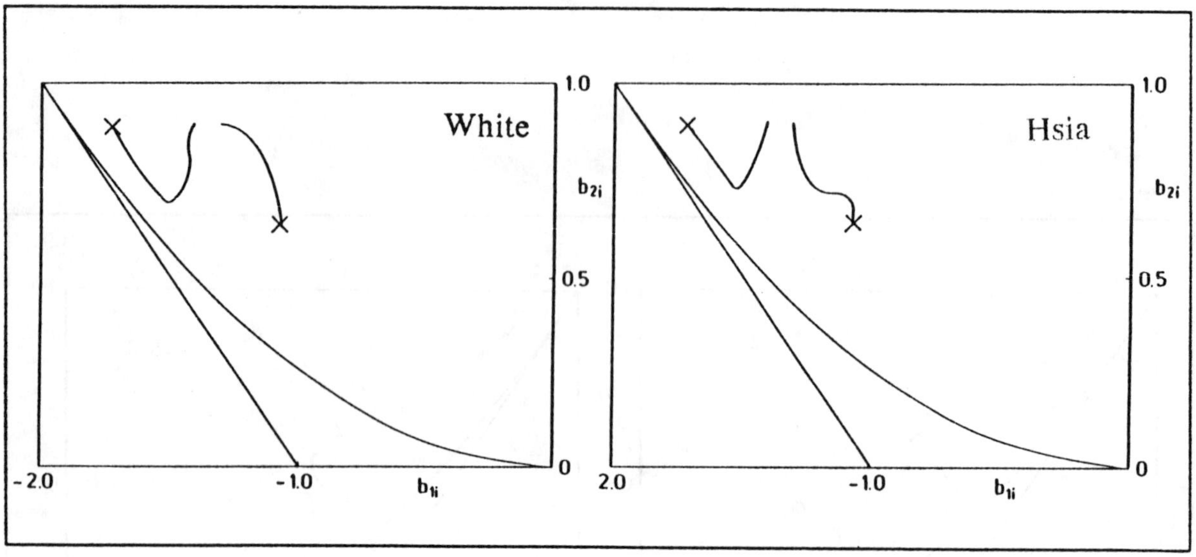

Figure 11. Process of adaptation: Starting values near the desired values

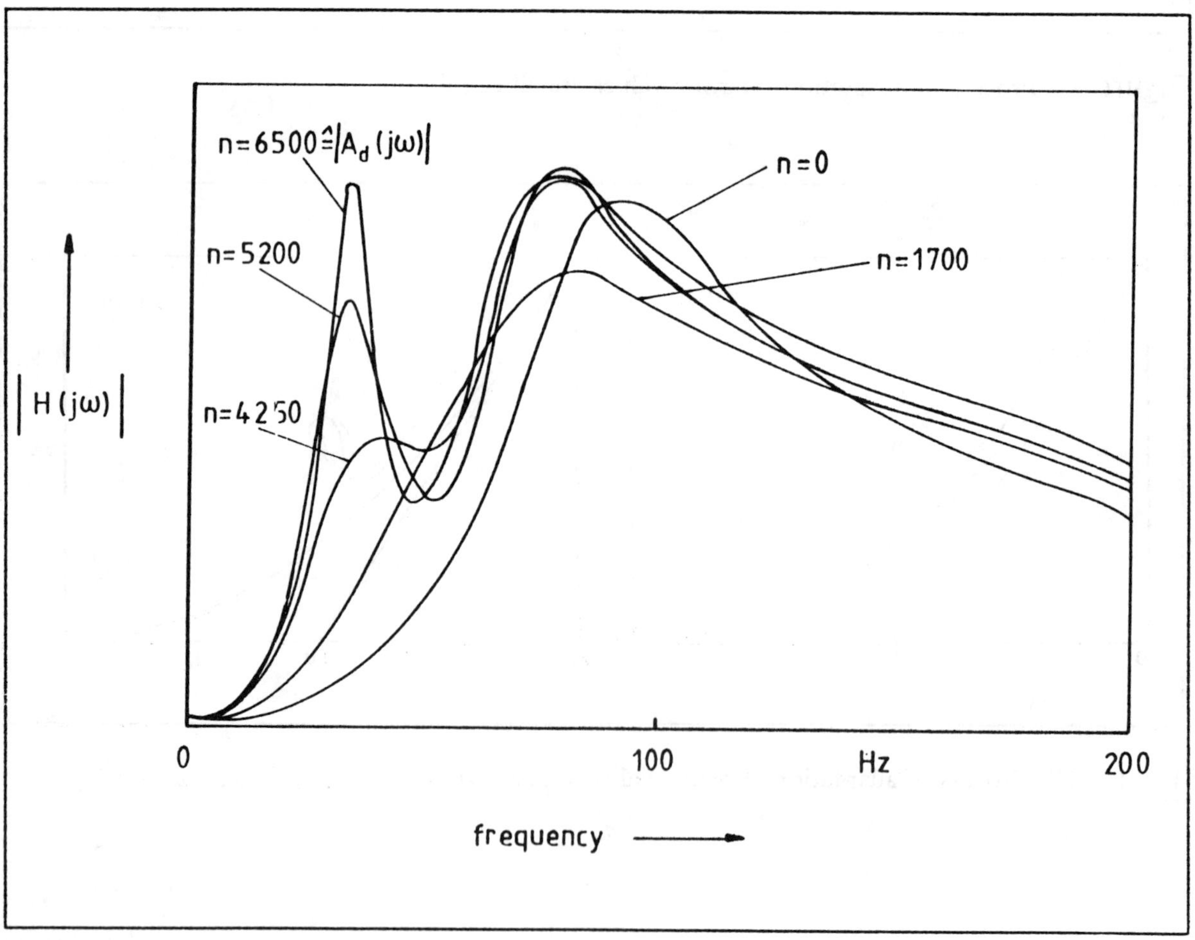

Figure 12. Adapted filter function $|H(jw)|$ at different points of time $n \cdot \tau$

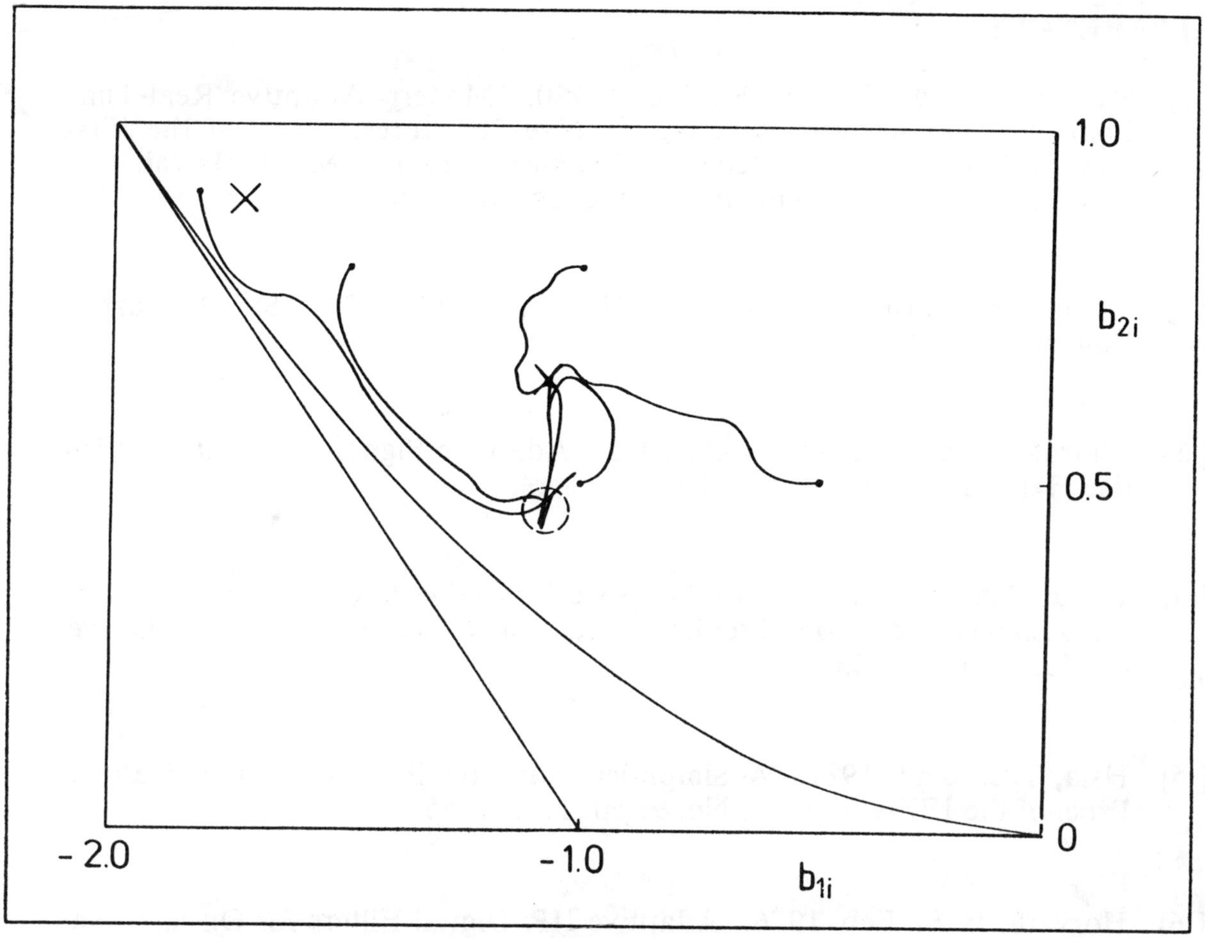

Figure 13. Undermodeling: $H(z)$ consists of only one filter section. The adaptation process converges, but the desired values cannot be reached.

CONCLUSION

This investigation shows that it is possible to identify or alter certain modes of a (still simulated) structural system. If the order of the adaptive system matches the number of modes of the structure, even modes with frequencies in close vicinity can be identified. Experiments in which the modal controller is applied to a truss-type structure are presently under preparation.

The investigated type of adaptive recursive filters can be called an interface between modal analysis and adaptive control. More of such interfaces to other disciplines are required if further advances are to be achieved in the interdisciplinary field of actively reacting flexible structures.

ACKNOWLEDGEMENTS

The authors wish to acknowledge Dr. E. Breitbach, head of the DLR program ARES, for many valuable discussions.

REFERENCES

[1] Melcher J. and Wimmel, R., Nov. 1990. "Modern Adaptive Real-Time Controllers for Actively Reacting Flexible Structures." Proc. of the First Joint U.S./Japan Conference on Adaptive Structures, Maui, Hawaii, will be published by Technomic Publishing Co., May 1991.

[2] Shynk, J.J., April 1989. "Adaptive IIR Filtering." IEEE ASSP Magazine, Volume 6, pp. 4-21.

[3] Widrow, B. and Stearns, S.D., 1985. "Adaptive Signal Processing." Prentice Hall, Inc., Englewood Cliffs, New Jersey.

[4] White, S.A., Nov. 1975. "An Adaptive Recursive Digital Filter." Proc. of 9th Asilomar Conf. on Circuits, Systems and Computers, Pacific Grove, California, pp. 21-25.

[5] Hsia, T.C., Sept. 1981. "A Simplified Adaptive Recursive Filter Design." Proc. of the IEEE, Vol. 69, No. 9, pp. 1153-1155.

[6] Horvath, Jr. S., Feb. 1976. "Adaptive IIR Digital Filters for On-line Time Domain Equalization and Linear Prediction." Paper presented at the IEEE Arden House Workshop on Digital Signal Processing, Harriman, NY.

[7] Jenkins, W.K. and Nayeri, M., 1986. "Adaptive Filters Realized with Second Order Sections." Int. Conf. on Acoust., Speech and Signal Proc., Tokyo, pp. 2103-2106.

[8] Nayeri, M. and Jenkins, W.K., June 1988. "Analysis of Alternate Realizations of Adaptive IIR-Filters." Proc. IEEE Int. Symp. Circuits Systems, Espoo, Finland, pp. 2157-2160.

[9] Feintuch, P.L., Nov. 1976. "An Adaptive Recursive LMS Filter." Proc. of the IEEE, pp. 1622-1624.

[10] Skalar Computer GmbH, 1988. "32 & TXA12 PC/AT Signal Processor and I/O Module." Skalar Computer GmbH, Göttingen, FRG.

DIGITAL SIGNAL PROCESSING METHODS FOR CONTROL OF VIBRATIONS IN STRUCTURES

W. H. Bennett[1]

ABSTRACT

In this paper we describe aspects of an approach to design of active control systems for vibration suppression and isolation in structures using both feedforward and feedback information processing. The computational approach is based on frequency response modeling of the elastic response of structures and exogenous disturbances representing noise sources. We will focus attention on a comprehensive approach to design based on Wiener-Hopf optimization which results in a frequency response model for a desired control action. We also report progress on a project to utilize digital signal processing methods to support both design computations and implementation of online control processing using high speed, special purpose DSP processors.

Introduction

An important issue in the design and implementation of active vibration control of structures is the integration of multiple measurements of structural flexure from a variety of spatially distributed sensors. In many applications the requirement for active control arises due to desire to isolate certain critical locations on the structure from vibrations emanating from a spatially noncollocated disturbance source. Often sensors can be placed at the principal disturbance source suggesting that an attractive solution for active control is to use feedforward methods to attempt to cancel the structural born vibration at the point where isolation is desired. However, feedforward processing alone is well known to be sensitive to modeling assumptions needed to predict the propagation of structural born vibration. Such models for flexible structures are sensitive to loads, structural configuration complexity and a variety of environmental conditions. The use of feedback control can offer robustness and reduced sensitivity to model uncertainty. In this paper we describe an approach to vibration control design which resolves tradeoffs in feedforward and feedback processing. The control design methods employ transfer function models for the structure born vibration response and sampled frequency response data for the design computations. Digital Signal Processing (DSP) techniques are described to support all aspects of the control design computations and implementation of the online control algorithm.

The approach described employs a comprehensive methodology for modeling of structure born vibrations based on irrational transfer function models, specification of disturbance sources via their Power Spectral Density (PSD), and resolution of control design tradeoffs for vibration isolation in terms of minimum energy spectral density performance measures. Frequency re-

[1]Techno-Sciences, Inc., 7833 Walker Drive, Suite 620, Greenbelt, MD 20770, U.S.A.

symbol	explanation
$j = \sqrt{-1}$	imaginary unit
$E\{\cdot\}$	expectation
$Tr\{A\}$	trace of matrix A
$A_*(s) = A^T(-s)$	transfer function of "causally reversed" system
$\mathbf{C}_+$	open right half complex s-plane, $\Re e\ s > 0$
$\mathbf{H}_\infty$	Hardy space of complex functions, analytic and essentially bounded in $\mathbf{C}_+$
$\mathbf{H}_2$	Hardy space of complex functions, $f(s)$, analytic in $\mathbf{C}_+$and such that; $\frac{1}{2\pi j}[\int_{-j\infty}^{j\infty} \|f(s)\|^2\, ds]^{1/2} < \infty$ for $s \in \mathbf{C}_+$.

Table 1: Notation

sponse methods for modeling vibrations in flexible structures has been described in detail in [1]. Wiener-Hopf methods are used to obtain the optimal, realizable controller employing both feedback and feedforward processing. Computational algorithms based on a sampled data approach to the frequency response computations are highlighted. DSP techniques for implementing the control laws are then described.

The paper reports progress on a project whose principal objective is to develop and demonstrate innovative implementations of active, realtime control laws for precision, wideband vibration control using advanced DSP technology. A workstation environment for modeling of structure born vibrations and design of active control based on sampled frequency response data together with a real time simulation facility for testing DSP based control laws is being developed and progress will be reported.

An Approach to Design of Integrated Multiloop Control for Vibration Suppression in Structures

A general architecture combining feedforward and feedback control is shown in block diagram form in Figure 1 where $P(s)$ is an $n \times m$ transfer function model of the structure response from active control to the point where vibration isolation is desired. (Notation used in the paper is summarized in Table 1.) Component models include $n \times n$ transfer functions: $F(s)$, feedback sensor measurement dynamics, $L(s)$, feedforward sensor dynamics, and $P_0(s)$, which models the response of the structure to a localized external, disturbance, d. Additional terms (viz. L_0, F_0) of compatible dimensions are included to model precision and random noise effects in the feedforward and feedback sensors. Focusing attention on vibration isolation in structures we make the additional assumption that the natural structure response is stable in the sense that $P, P_0 \in \mathbf{H}_\infty$.

Exogenous disturbances d, sensor noises n_ℓ, n_f, and desired set points, u, are modeled as zero mean, wide sense stationary random processes and are therefore characterized by their respective power spectral densities; $G_d(s)$, $G_{n_\ell}(s)$, $G_{n_f}(s)$, $G_u(s)$. Notice that no assumption is made about the form of the controller $C(s)$ except the obvious dimensions: $m \times n$.

With the architecture of the control loops defined it is now possible to address important, quantitative tradeoffs in the design of the controller based on an optimal control problem

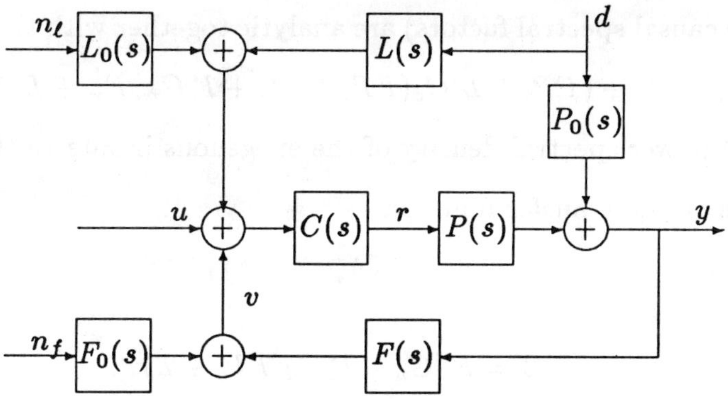

Figure 1: Multiloop Control Configuration for Wiener-Hopf Design

whose objective is to minimize *tracking error*, $e = u - y$, subject to a power-like constraint on the control, r. Defining the control objectives for tracking:

$$J_t = \frac{1}{2\pi i} E \left\{ Tr \int_{-i\infty}^{i\infty} e(s)e_*(s)ds \right\} \tag{1}$$

and control saturation

$$J_s = \frac{1}{2\pi i} E \left\{ Tr \int_{-i\infty}^{i\infty} r(s)r_*(s)(s)ds \right\}, \tag{2}$$

the control problem is to find stabilizing controller C which minimizes J_t subject to constraint on J_s. This is equivalent to minimizing the cost

$$J = J_t + \mu J_s \tag{3}$$

over all controllers which maintain closed loop stability, where $\mu > 0$ is a real parameter which plays the role of a Lagrange multiplier and effectively permits tradeoff between tracking and saturation.

The above Wiener-Hopf control problem is standard and a complete solution was first given by Youla et al [2]. The computational problem identified by Wiener-Hopf approach involves the computation of *causal solutions* to minimum mean square error problems. We indicate a causal projection of a transfer function via a (partial fraction) expansion $A(s) = \{A(s)\}_+ + \{A(s)\}_- + \{A(s)\}_\infty$ where $\{.\}_+$ (resp. $\{.\}_-$) is the part analytic in $\Re es > 0$–the causal part ($\Re es < 0$–the anti-causal part) and $\{.\}_\infty$ is the part associated with poles at infinity. The following procedure for its solution is specialized from Youla's method:

Wiener-Hopf Design Procedure

Step 1 Compute spectral factorizations[1]

$$\begin{aligned} [P_*P + \mu Q] &= \Lambda_*\Lambda & (4) \\ G_e &= \Omega\Omega_* & (5) \end{aligned}$$

[1]We remark that the computations of the spectral factors in (4) (resp. (5)) effectively replace the computational step of solving a Riccati matrix equation for the control (resp. filter) problem typically encountered in modern state-space methods for control design. For systems with distributed parameter models the Riccati equation is a PDE and therefore difficult to solve by standard methods.

where Ω, Λ (the causal spectral factors) are analytic together with their inverses in $\mathbf{C}_+$ and

$$G_e = G_u + (FP_o + L)G_d(FP_o + L)_* + F_oG_{n_f}F_{o*} + L_oG_{n_\ell}L_{o*}, \tag{6}$$

is the combined power spectral density of the exogenous inputs to the control system.

Step 2 Compute the causal transfer function

$$K = \Lambda^{-1}\{\Lambda_*^{-1}Z\Omega_*^{-1}\}_+\Omega^{-1} \tag{7}$$

where

$$Z = P_*[G_u + P_oG_d(FP_o + L)_*]. \tag{8}$$

Then the optimal controller has transfer function given by

$$C = K(I - PK)^{-1} \tag{9}$$

From the control architecture shown in Fig. 1 we can readily derive the following closed loop frequency responses:

$$y = PCS(u - F_on_f - L_on_\ell) + [P_o - PCS(FP_o + L)]d \tag{10}$$

$$r = CS[u - F_on_f - L_on_\ell - (FP_o + L)d] \tag{11}$$

$$e = (I - PCS)u + PCS(F_on_f + L_on_\ell) - [P_o - PCS(FP_o + L)]d \tag{12}$$

where the *system sensitivity operator*

$$S := [I + FPC]^{-1} = I - PK \tag{13}$$

is an important indicator of closed loop performance and robustness. Under the current assumptions the transfer function K satisfies $CS = K$ which illustrates its importance for closed loop control.

The basis for resolving the engineering tradeoff of closed loop control action in terms of the conflicting requirements for stability, disturbance rejection, command tracking, and saturation avoidance is the application of Wiener-Hopf method for the solution of the optimization problem outlined above. It can be easily shown that the optimal performance objectives J_t and J_s are readily computed from the frequency domain solution above from the following formulae [2]:

$$J_t = \frac{1}{2\pi j}\int_{-j\infty}^{j\infty} Tr\ \Phi_t(j\omega)d\omega, \tag{14}$$

$$J_s = \frac{1}{2\pi j}\int_{-j\infty}^{j\infty} Tr\ \Phi_s(j\omega)d\omega, \tag{15}$$

where closed loop tracking and saturation PSD's are:

$$\Phi_t = (I - PK)G_u(I - PK)_* + (PK)G_1(PK)_* + [P_o - PK(FP_o + L)]G_d[P_o - PK(FP_o + L)]_*, \tag{16}$$

$$\Phi_s = QKG_eK_*, \tag{17}$$

and

$$G_1 = F_oG_{n_f}F_{o*} + L_oG_{n_\ell}L_{o*}. \tag{18}$$

Thus the design method proceeds by resolving the tradeoff between tracking performance (requiring high loop gains) and saturation limitation of the critical plant modes (which will limit loop gains) by direct implementation of the composite cost relative scale factor μ in (3) which plays the role of a Lagrange multiplier.

Frequency Response Computations for Optimal Control.

Our approach to optimal control computation is motivated by distributed parameter models which arise in modeling vibrations in flexible structures [1] and uses standard methods of Digital Signal Processing (DSP): sampling and interpolation of the model frequency responses. The important computational problems in Wiener-Hopf design involve spectral factorization and the closely related problem of causal projection.

Recall that a transfer function $H(s) \in \mathbf{H}_2 \cap \mathbf{H}_\infty$ has a unique *spectral factorization* $H(s) = F(s)F_*(s)$ with $F \in \mathbf{H}_\infty$ if:

1. $\overline{H(s)} = H(\bar{s})$; i.e., $H(s)$ is the transform of a real-valued function $h(t)$.
2. $H(s) = H_*(s)$; i.e., $H(s)$ is "para-hermittian".
3. $H(s)$ is of normal rank; i.e., full rank almost everywhere in $\mathbf{C}$.
4. $H(i\omega)$ is positive, semi-definite and bounded for $\omega \in \Re$.

To see that causal projection is a closely related problem consider the following. If $H(s)$ is scalar-valued, then the function $\Phi(s) = \ln H(s)$ can be expanded as

$$\Phi(s) = \{\Phi(s)\}_+ + \{\Phi(s)\}_-, \tag{19}$$
$$= \ln F(s) + \ln F_*(s), \tag{20}$$

so that the causal, spectral factorization is related to causal projection via the logarithmic transformation; $F(s) = \exp\{\ln H(s)\}_+$.

Our goal is to obtain numerically stable approximations to both problems for transfer functions in $\mathbf{H}_2 \cap \mathbf{H}_\infty$. Considering first the problem of computing a causal projection, we motivate one class of algorithms from basic properties of the Hilbert transforms applied to the frequency response $\Phi(j\omega)$. Recall that the Hilbert transform of a time signal $f(t)$ is defined as a convolution; $\check{f}(t) = \int_{-\infty}^{\infty} \frac{f(\tau)}{\pi(t-\tau)}\, d\tau$ and it's Fourier transform has the property,

$$\check{f}(\omega) = \begin{cases} -jf(\omega), & \omega > 0 \\ jf(\omega), & \omega > 0 \end{cases}.$$

The inverse Fourier transform of $\check{\Phi}$ is $-j\text{sgn}(t)\phi(t)$ where $\phi(t)$ is the inverse Fourier transform of $\Phi(\omega)$. A consequence is that the casual projection can be obtained as

$$\{\Phi(\omega)\}_+ = \frac{1}{2}[\Phi(\omega) + j\check{\Phi}(\omega)].$$

In previous studies we reported computational algorithms for causal projection and scalar spectral factorization by numerical evaluation of the Hilbert transform integral [3]. A more efficient method for causal projection and spectral factorization was considered in [4] based on sampling and interpolation of the system frequency response. The algorithm developed in [4] employs results of Stenger [5] and is efficiently implemented using Fast Fourier Transform (FFT). Details of the algorithm used for the current studies and computer implementation are given in [4].

For the design of multiloop, vibration control we require the solution of two, matrix spectral factorization problems (analogous to the solution of control and filtering Riccati equations for time domain models). The computational approach we employ is based on a Newton-Raphson iteration for the matrix causal spectral factor which was first described in [6].

Frequency Sampling Filters for Real Time Control of Vibrations in Structures

A principal concern in implementation of high precision control laws for suppression and isolation of wide band vibrations in structures is the complexity of the control laws. One measure of controller complexity is the order or dimension of a state variable realization of the transfer function $C(s)$ which will be implemented for realtime control. However, this is not always meaningful for distributed parameter models of flexible structures. In [7] an example is given where a discrete time controller implementation is shown to be optimal for a simple, distributed parameter model of a flexible structure. The objective of the current research effort is to exploit the sampled data frequency response models obtained above as a specification of the desired controller frequency response and to obtain efficient, (time) sampled-data realization using DSP methods.

Digital computer implementations are primarily limited by computational speed and algorithm complexity effecting the ultimate obtainable sampling rate and considerations for reduced order realization of the controller may be required. However, the emergence of specialized computer hardware implemented in VLSI single chip circuits for digital signal processing opens new opportunities for realization of realtime control for flexible structures. We prefer to consider realization options for the optimal controller in discrete time for implemenation on a digital computer. Realization of the controller specified by its frequency samples can be obtained using a FIR digital filter implementation.

Given the specified frequency samples obtained for the optimal controller. $C_k = C(j\omega_k)$, at frequencies, $\omega_k = k\omega_o/N$, where ω_o is bandwidth and N the number of uniformly spaced frequency samples we describe the digital filter realization using z-transforms. With discrete time sampling rate $\omega_s > 2\omega_o$ the frequency samples correspond to interpolation points in the z-plane given by[2], $z_k = e^{jk2\pi/N_1}$, for $k = 0, \ldots, N-1$. The z transform which realizes the frequency sampling filter is

$$C(z) = \sum_{k=0}^{N-1} C_k F_k(z)$$

where the interpolating functions are,

$$F_k(z) = \frac{1 - z^{-N}}{N(1 - e^{jk2\pi/N} z^{-1})},$$

with $k = 0, \ldots, N-1$. A standard computation shows that the frequency sampling filter has transfer function

$$C(z) = \frac{1 - z^{-N}}{N} \sum_{k=0}^{N-1} \frac{C_k}{1 - e^{jk2\pi/N} z^{-1}} = \sum_{\ell=0}^{N-1} c_\ell z^{-\ell} \tag{21}$$

[2]Bandwidth and sampling requirements would typically require padding the sequence of frequency samples of length N with $N_1 - N$ zero values to avoid aliasing.

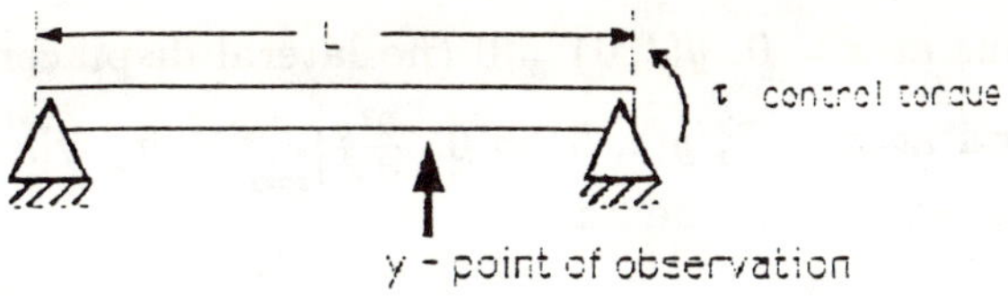

Figure 2: Pinned-Pinned Beam Control Problem.

where the coefficients,

$$c_\ell = \frac{1}{N}\sum_{k=0}^{N-1} C_k e^{j(2\pi/N)\ell k} \tag{22}$$

for $\ell = 0,\ldots,N-1$, are the *Inverse Discrete Fourier Transform (IDFT)* of the sequence $C_0,\ldots,C_{N-1}$. The final form in (21) shows that the realization is a FIR realization. Such realizations are nonrecursive and are efficiently implemented using high speed, single chip DSP processors which utilize highly pipelined architectures to achieve high throughput.

A workstation for design and testing of DSP methods in vibration control

Development and testing of FIR implementations of closed loop control for application to vibration control in structures will require extensive testing to validate the design approach. Currently available single chip, programmable digital signal processors (i.e., DSP chips) offer low cost, computational engines which are easily integrated with a personal computer for design and testing of DSP applications. A workstation is under development at TSI to test application of frequency sampling methods in modeling and design of DSP methods in control of vibrations in structures. The system configuration includes an Everex PC386/33Mhz. and two AT&T DSP32C processors with 1 Mbyte each of dedicated RAM. The DSP32C is single chip processor which offers 32 bit floating point arithmetic and can achieve computational rates of 25 MFLOPS. The system includes data acquisition system to support closed loop controlled tests of DSP processing for vibration control of laboratory experiments.

The software system, currenlty under development at TSI, includes three subsystems: 1) a design environment to support for sampled data frequency response modeling and control computations, 2) a design environment to support FIR filter software design and program development for the dedicated DSP-32C processors, and 3) a simulation environment to test FIR control laws in closed loop control of flexible structures. The simple vibration control example discussed in the following section includes data obtained from the system.

Vibration Isolation in a Simple Structure

In this example we consider a control system design problem for a relativly simple structure with well known transfer function. Consider a Bernoulli-Euler beam model with "pinned-pinned" boundary conditions as shown in Figure 2. The beam lateral deformation is given by $y(t,z)$ with $0 \le z \le L$ and has dynamics described by the PDE in dimensionless form:

$$\frac{\partial y^2}{\partial t^2} - 2\zeta\frac{\partial^3 y}{\partial t \partial z^2} + \frac{\partial^4 y}{\partial z^4} = 0. \tag{23}$$

with boundary conditions at $z = 0$, $y(t,0) = 0$ (no lateral displacement) and $\frac{\partial^2 y}{\partial z^2}\Big|_{z=0} = 0$, (no restraining moment), and at $z = L$, $y(t,L) = 0$, $\frac{\partial^2 y}{\partial z^2}\Big|_{z=L} = \tau$, where the control moment τ is applied at the right hand end of the beam.

The transfer function for beam control is

$$P(s,z) = L^2 \frac{\sin\lambda_1 \sinh\lambda_2 \frac{z}{L} - \sin\lambda_1 \frac{z}{L} \sinh\lambda_2}{(\lambda_1^2 + \lambda_2^2)\sin(\lambda_1)\sinh(\lambda_2)}, \tag{24}$$

or in a more useful form for numerical evaluation as

$$P(s,z) = \frac{L^2}{(\lambda_1^2 + \lambda_2^2)} \left(\frac{\sinh\lambda_2 \frac{z}{L}}{\sinh\lambda_2} - \frac{\sinh\lambda_1 \frac{z}{L}}{\sin\lambda_1} \right), \tag{25}$$

where

$$\lambda_1^2 = (-\zeta + i\sqrt{1-\zeta^2})sL^2, \tag{26}$$

$$\lambda_2^2 = (\zeta + i\sqrt{1-\zeta^2})sL^2, \tag{27}$$

L is the beam length, ζ is the damping factor, and $0 \leq z \leq L$ is the observation point on the beam. The transfer function model is meromorphic and stable ($P \in \mathbf{H}_\infty$) with uniform damping rate given by ζ.

In many applications of control of vibrations in structures frequency response models can be estimated using standard DSP method from measurements on the system. The class of simple beam models currently under consideration serve to focus attention on offline control design computational complexity and online processing complexity for prototype problems.

The control design problem we next examine is defined by a choice of the output point for regulation, $\frac{z}{L} = 0.7$, the length of the beam $L = 10.$, and the effective damping ratio, $\zeta = 0.01$. The beam frequency response P, shown in Figure 3, clearly involves complex multimodal resonant responses. The spectral properties of the exogenous inputs define the control problem. In this example, we focus attention on vibration isolation with a narrow band displacement disturbance d with PSD given by

$$G_d(s) = \frac{1}{s^4 + 2\omega_d^2(1-2\zeta_d^2)s^2 + \omega_d^4},$$

and we take $\omega_d = 1$, $\zeta_d = 0.1$. For the control architecture of Figure 1 we note that $P_o = 1$. The feedback and feedforward sensors are assumed to have negligible dynamics (i.e., $F_o = F = 1$ and $L_o = L = 1$). Assumptions on the shape of the PSD G_{n_f} (resp. G_{n_ℓ}) models precision of feedback (resp. feedforward) information and will effectively limit feedback control action in certain frequency bands. For the computational example we choose the feedback and feedforward PSD as rational functions as shown in Figure 4 which effectively models a feedback sensor with good low frequency precision and a feedforward sensor with good high frequency precision. The WH approach will then be used to assess the tradeoff in use of the available information for design of the controller C.

The vibration isolation problem considered here is the regulation of the beam displacement at the specified location and we therefore characterize the PSD of the set point as $G_u(s) = 0$. The PSD's for the exogenous inputs are shown in Figure 4.

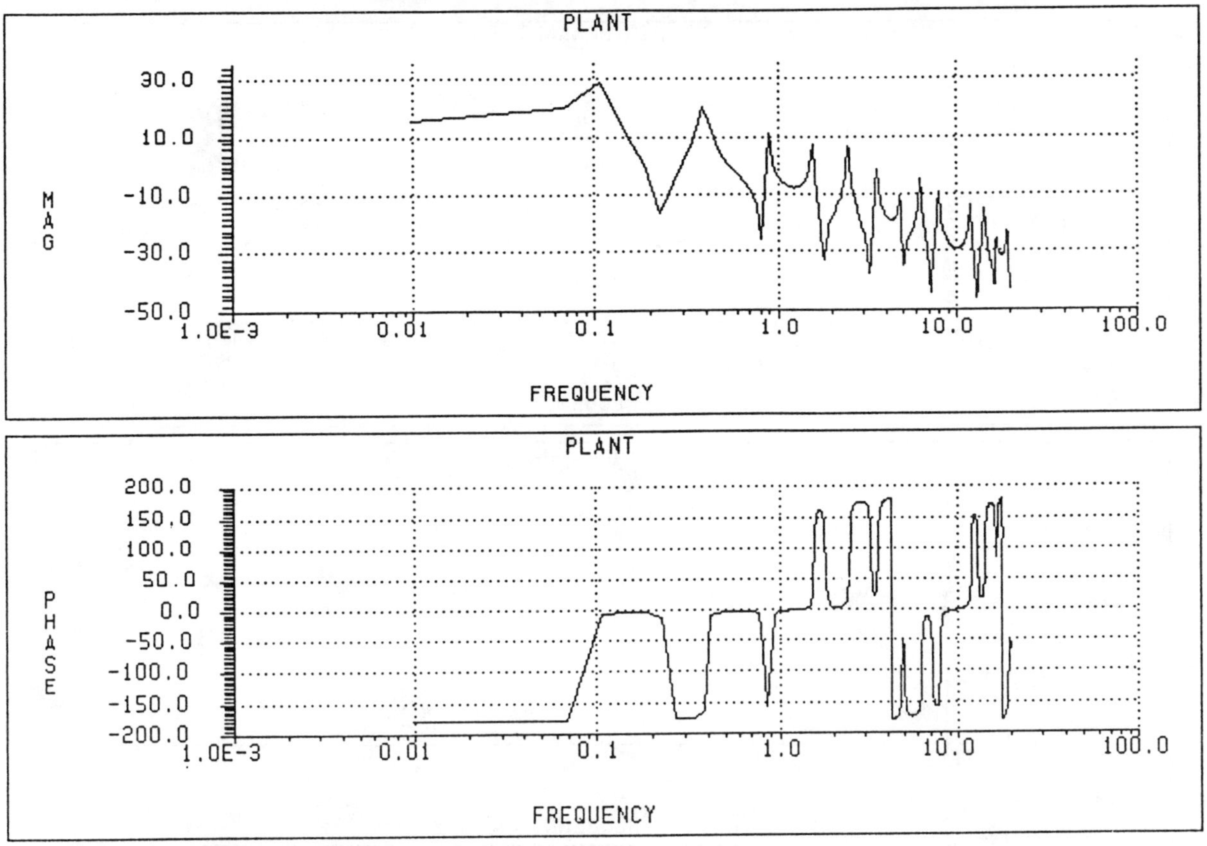

Figure 3: Frequency response for pinned-pinned beam control.

Control design proceeds by evaluation of the tradeoff between optimal tracking cost subject to a constraint on control torque saturation (i.e., we take $Q = 1$). Results shown are for the choice, $\mu = 0.1$.

The results shown here were obtained using 1024 point uniform frequency sampling of the modeled responses over the range $0.01 \leq \omega \leq 20.$. The required spectral factors, as obtained by frequency sampled computations, are shown in Figure 5. Spectral contribution to the cost components J_t, J_s can be determined from the frequency sampled PSD's Φ_t and Φ_s as shown in Figure 6. Finally, the optimal controller frequency response is obtained and displayed in Figure 7. This provides a specification for the online implementation which may be obtained using FIR realization which interpolates the given frequency samples.

Conclusions

The use of Wiener-Hopf optimization for design of control systems is not new. We have shown how sampled data methods for frequency response modeling and control design can be integrated to support computations for design of vibration isolation control which integrates both feedback and feedforward procesing of information from availble sensors. The emphasis in this work is on developing a workstation to support the frequency response tradeoff analysis suggested by Wiener-Hopf optimization for application to vibration control in flexible structures. The methods employed provide frequency response measures of known value in analysis and design of closed loop control systems. A direct approach to control implementation using FIR constructions to implement digital filters to interpolate the desired frequency response model

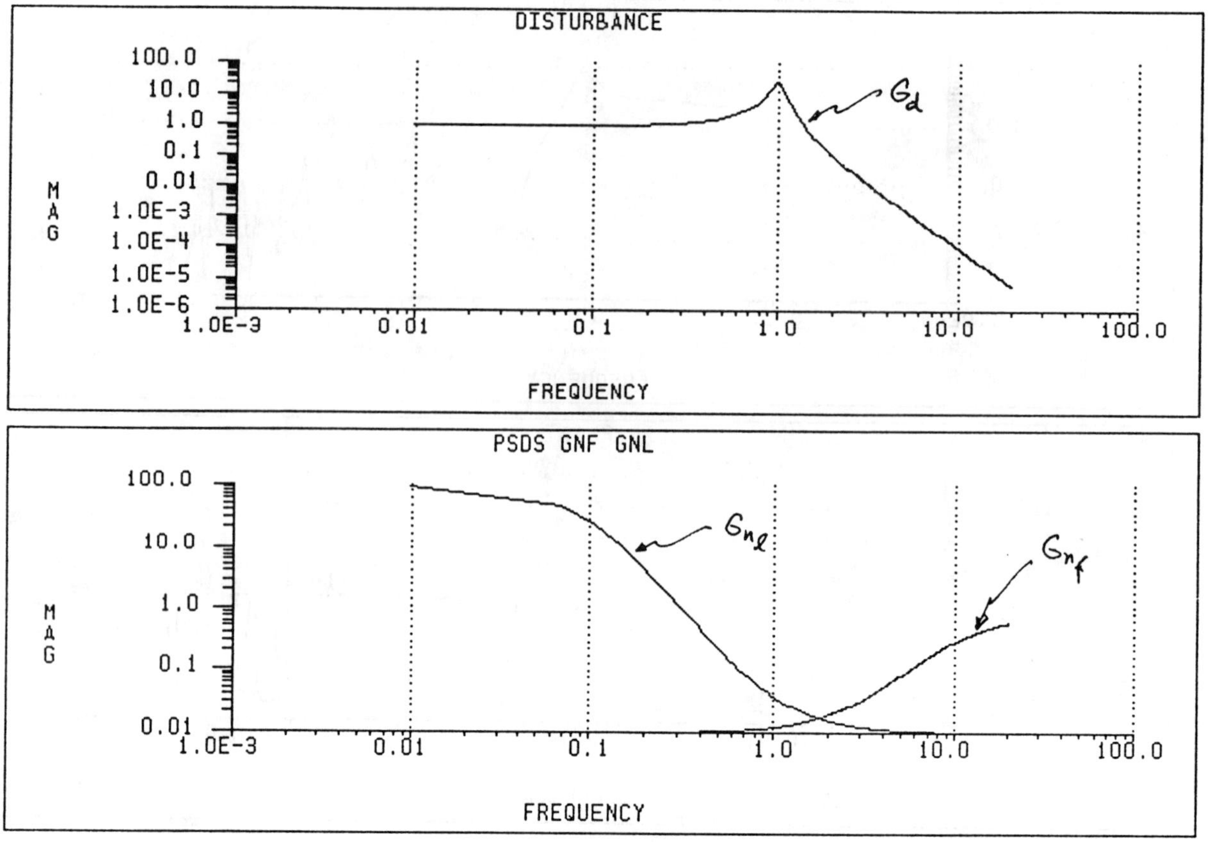

Figure 4: PSD for disturbance and sensor noise inputs for beam control.

of the controller is currenlty under development and test at TSI. The system utilizes a pair of AT&T DSP32C processors for high speed implementation of DSP operations.

Acknowledgement

The research project reported herein is supported by SDIO/SBIR and managed by Flight Dynamics Laboratory, Wright Research and Development Center, Wright-Patterson AFB, under contract F33615-90-C-3203.

References

[1] W. Bennett and H. Kwatny, "Continuum modeling of flexible structures with application to vibration control," *AIAA J.*, vol. 27, pp. 1264–1273, September 1989.

[2] D. Youla, H. Jabr, and J. J.J. Bongiorno, "Modern Wiener-Hopf design of optimal controllers–parts I and II," *IEEE Trans. on Auto. Cntrl.*, vol. AC-30, no. 7, pp. 3–13 and 319–338, 1976.

[3] W. Bennett and N. Barkakati, "FlexCAD: Prototype software for modeling and control of flexible structures," in *Proc. IEEE CACSD Symp.*, Sept. 1986.

[4] W. Bennett and I. Yan, "A computer algorithm for causal spectral factorization," in *Proc. 1988 NAECON*, May 1988.

[5] F. Stenger, "The approximate solution of Wiener-Hopf integral equations," *J. Math. Analysis and Applic.*, vol. 37, pp. 687–724, 1979.

[6] J. Davis and R. Dickinson, "Spectral factorization by optimal gain iteration," *J. Appl. Math.*, vol. 43, pp. 389–301, 1983.

[7] W. H. Bennett, "Computation and implementation of precision control for flexible structures," in *Advances in Computing and Control*, Springer-Verlag, 1988.

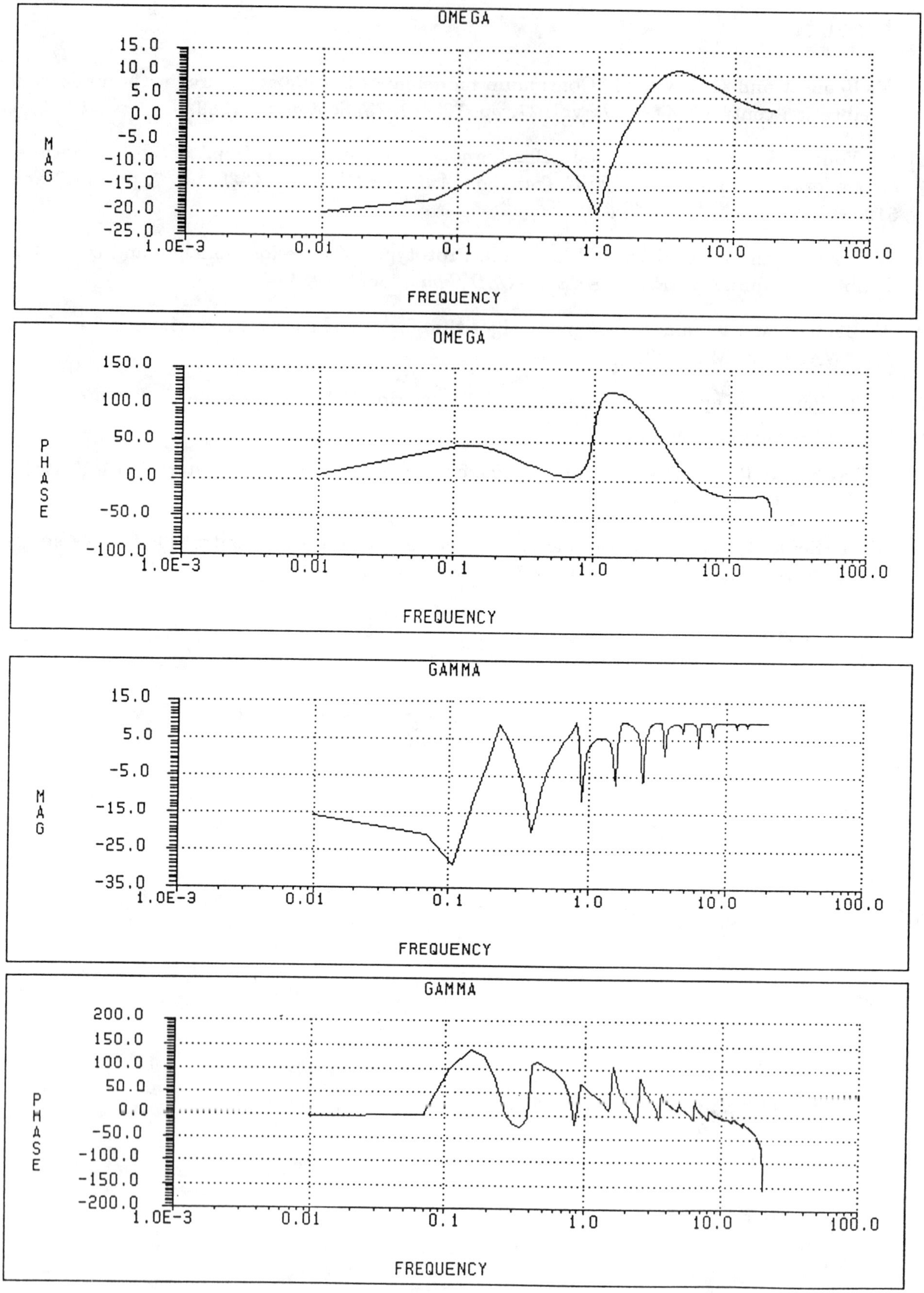

Figure 5: Spectral Factors for pinned-pinned beam control.

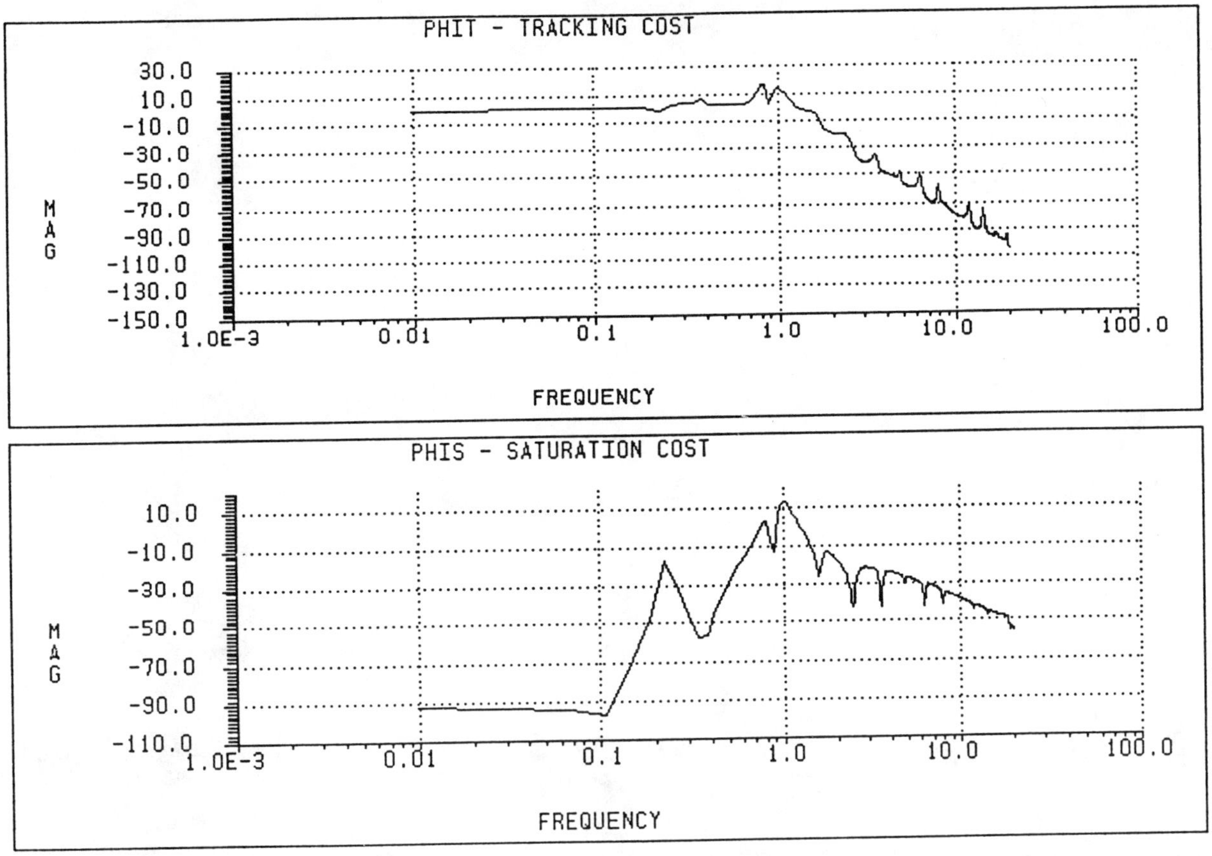

Figure 6: Tracking and Saturation PSD's.

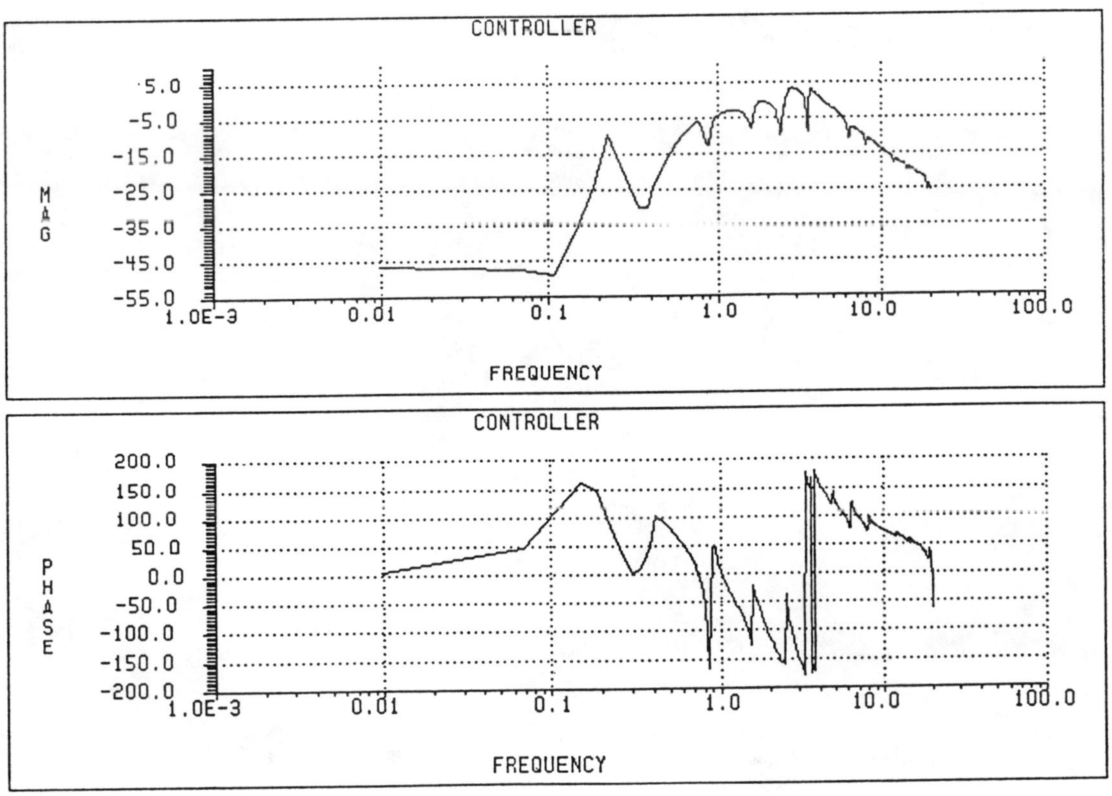

Figure 7: Optimal Controller Frequency Response

Flexible Structures II

NUMERICAL SIMULATION OF FLUTTER AND SUPPRESSION BY FEEDBACK CONTROL

B. Dong,[1] D. T. Mook[1]

ABSTRACT

A numerical procedure for simulating the response of an airfoil mounted on an elastic support in a freestream to the deflections of a trailing-edge flap is described. The airfoil, flowing air, and a feedback control system are treated as a single dynamic system. The general aerodynamic model describes attached flows. It is based on a continuous distribution of vorticity along the surface of the airfoil to represent the boundary layer and a system of densely packed discrete cores to represent the wake. It is shown that the present simulation accurately predicts the flutter speed of a flat plate at zero angle of attack. Then a flap is added to the trailing-edge and a control law and servo equation describing the response of the flap to a command are added to the simulation. It is shown that flutter can easily be controlled. Several control laws are considered and it is shown that using only the vertical velocity as the feedback gives excellent results. Finally, the response of the uncontrolled airfoil to a gust is simulated, and then it is shown that feedback control can significantly reduce the response. The present results demonstrate the feasibility of developing simulations of complete-aircraft systems.

1. Introduction

A classic problem in the study of flutter is the analysis of an airfoil on an elastic support. The first studies considered incompressible flows and linear springs. The equations of motion were linearized and the aerodynamic loads were calculated by Wagner's function as described in detail in the textbook by Fung [1]. Using this method Lee and LeBlanc [2] investigated the effect of nonlinear springs. About ten years ago, the analysis was extended to transonic flows [3-8]. Numerical methods were developed to integrate the nonlinear equations of motion of both the flowing fluid and the airfoil simultaneously. This work was the first that truly captured dynamic/aerodynamic interaction. Subsequently, Strganac and Mook [9,10] modelled the aerodynamic response of a finite-aspect-ratio wing mounted on an elastic support. The loads were predicted by the general unsteady vortex-lattice method for incompressible flows. In our recent work [11,12], a similar method was used for two dimensional problems. The aerodynamic model is a modified version of the vorticity-panel method developed by Kim and Mook [13] for incompressible flow. The aerodynamic loads, the motion of the airfoil, and the wake are obtained simultaneously.

[1]Department of Engineering Science and Mechanics, Virginia Polytechnic Institute and State University, Blacksburg, VA 24061, U.S.A.

The problem of controlling the motion of a wing is also of interest to researchers. Mracek [14] developed a numerical simulation of the control of wing rock by using a vortex-panel method to calculate the aerodynamic loads. Lee and Batina [15] did the same thing for supersonic flows by integrating the Euler equations. Matsuzaki and Miyazawa [16,17] did both analytical and experimental work on gust-load alleviation. In the present work, a control law is developed to suppress the two degree-of-freedom motion of the airfoil. The airfoil, the wake and the control system are considered as one system: the unsteady flow and aerodynamic loads are coupled with the motions of the airfoil and control flap at the trailing-edge. The problem is solved simultaneously by integrating the equations of motion of both the fluid and the airfoil and the servo equation governing the oscillation of the flap.

2. Description of the Method

A. Equations of the motion

The model is shown in Fig. 1: an airfoil with two degrees of freedom oscillates in plunge and pitch. The half chord length is b . The elastic axis is located at point E, and the mass center is located at point C. If both points E and C are on the chord line, then the elastic axis E is at a distance $a_h b$ from the midchord, and the mass center C is at a distance $x_\alpha b$ from the elastic axis; a_h and x_α are positive when measured toward the trailing-edge of the airfoil.

As shown in Fig. 1, a global coordinate system (X, Y) is set up with unit vectors $\vec{I}$ and $\vec{J}$. Point E has no motion in the X-direction, and the plunging displacement of point E is denoted by y, positive in the upward direction. The pitch angle oscillating around the static angle of attack α is denoted by θ, positive with the nose down.

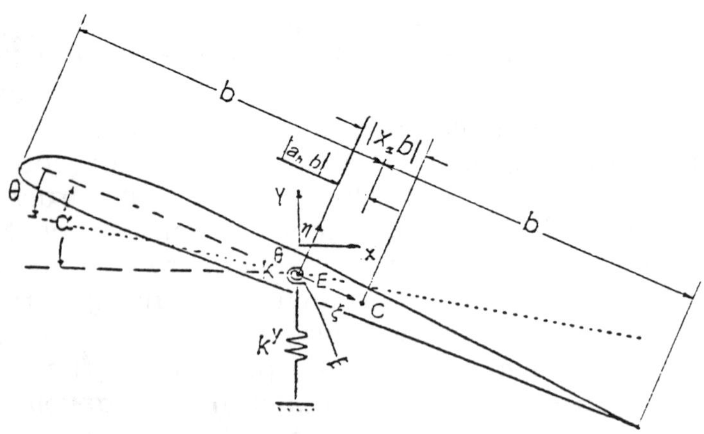

Figure 1. The model of the two-degree-of-freedom system.

For convenience, a local coordinate system (ξ, η) fixed on the airfoil is introduced with unit vectors $\vec{i}$ and $\vec{j}$. Its origin is located at point E, and the ξ-axis points toward the trailing-edge and is parallel to the chord line.

The equations of motion are

$$m\{\ddot{y} + [\xi_C \cos(\alpha - \theta) + \eta_C \sin(\alpha - \theta)]\ddot{\theta}\} = F_Y \\ - m\,[\xi_C \sin(\alpha - \theta) - \eta_C \cos(\alpha - \theta)]\dot{\theta}^2 \tag{1}$$

$$m\,[\xi_C \cos(\alpha - \theta) + \eta_C \sin(\alpha - \theta)]\ddot{y} + I_E\,\ddot{\theta} = F_\theta \tag{2}$$

where m is the mass of the airfoil per unit span, (ξ_C, η_C) are the coordinates of the mass center in the body-fixed system, I_E is the moment of inertia about point E, F_Y is the total force in the Y-direction on the airfoil, positive in the upward direction, and F_θ is the total moment acting about point E, positive in the counterclockwise direction.

Eq. (1) and Eq. (2) can be solved numerically when the loads are given. In the absence of structural damping, the loads are composed of three parts

$$\begin{Bmatrix} F_Y \\ F_\theta \end{Bmatrix} = \begin{Bmatrix} F_Y^g \\ F_\theta^g \end{Bmatrix} + \begin{Bmatrix} F_Y^e \\ F_\theta^e \end{Bmatrix} + \begin{Bmatrix} F_Y^a \\ F_\theta^a \end{Bmatrix} \tag{3}$$

where the superscript g represents gravitational forces, the superscript e represents the elastic forces and the superscript a represents the aerodynamic loads, which will be discussed later.

The gravitational forces are nonlinear and given by

$$\begin{Bmatrix} F_Y^g \\ F_\theta^g \end{Bmatrix} = -\begin{Bmatrix} mg \\ mg[\xi_C \cos(\alpha - \theta) + \eta_C \sin(\alpha - \theta)] \end{Bmatrix} \tag{4}$$

where g is the acceleration due to gravity

The elastic forces are taken in the form:

$$\begin{Bmatrix} F_Y^e \\ F_\theta^e \end{Bmatrix} = -\begin{Bmatrix} k_1^y\, y \\ k_1^\theta\, \theta \end{Bmatrix} + \dots \tag{5}$$

which can include nonlinear terms in y and θ .

B. Aerodynamic model

In the present model, it is assumed that the flow always remains attached. Then the boundary layer is approximated by an infinitesimally thin vortex sheet, and the wake is approximated by a system of discrete vortices. The flowfield is the superposition of the freestream and the flow induced by the vorticity and rotation of the airfoil [18]. For the numerical calculation, a panel method is used. The profile of the airfoil is approximated by a series of short straight segments that are connected at points, called nodes, lying on the actual surface. A linearly varying distribution of G, which represents the tangential component of the velocity of the fluid relative to the surface of the airfoil, is placed on each segment. The distribution is chosen in such a way that the corresponding velocity of fluid particles next to the surface is piecewise linear and continuous at all nodes. At each node G is unknown; G is obtained by imposing the no-penetration condition at the center of each segment and the conservation of the total vorticity in the flowfield. At the trailing-edge the surface velocities are finite and the pressures on the upper and lower surfaces are equal [19,21]. From this condition the vorticity shedding rate can be determined. The details of the present aerodynamic model were described in the earlier papers by Kim and Mook [13] and Mook and Dong [22].

For the 2-D problem considered here, the aerodynamic load can be derived from the Euler equation. On the surface of the airfoil, the Euler equation takes the form

$$-\frac{1}{\rho}\frac{\partial p}{\partial \ell} = \frac{\partial G}{\partial t} + G\frac{\partial G}{\partial \ell} + W_Y\ddot{y} + W_\theta\ddot{\theta} + W_q\dot{\theta}^2 \tag{6}$$

where ℓ is the distance along the surface, t is the time, p is the pressure, G is the tangential component of the velocity of the fluid relative to the surface, $W_Y = \vec{J}\bullet\vec{e}_l$, $W_\theta = (\vec{K}\times\vec{r})\bullet\vec{e}_l$, $W_q = -\vec{r}\bullet\vec{e}_l$, $\vec{e}_l$ is the unit vector tangent to the surface in the direction of increasing ℓ, $\vec{K} = \vec{I}\times\vec{J}$, and $\vec{r}$ is the position of a point on the surface relative to point E.

The aerodynamic load is obtained by integrating Eq.(6) and has the following form:

$$\begin{Bmatrix} F_Y^a \\ F_\theta^a \end{Bmatrix} = \ddot{y}\begin{Bmatrix} F_{YY}^a \\ F_{Y\theta}^a \end{Bmatrix} + \ddot{\theta}\begin{Bmatrix} F_{\theta Y}^a \\ F_{\theta\theta}^a \end{Bmatrix} + \dot{\theta}^2\begin{Bmatrix} F_{qY}^a \\ F_{q\theta}^a \end{Bmatrix} + \begin{Bmatrix} F_{0Y}^a \\ F_{0\theta}^a \end{Bmatrix} \tag{7}$$

If the airfoil is a flat plate, there is a singularity at the leading edge. so there is a suction force F_S there pointing away from the leading edge. This force can be obtained by the conformal-mapping method of potential theory:

$$F_S = \frac{1}{8}\left[2b\pi V_\infty \sin\alpha_a + \Gamma + 2\sum_{k=1}^{M}\frac{\Gamma_k(1+\xi_k)}{(1+\xi_k)^2+\eta_k^2}\right]^2 \tag{8}$$

where $2b$ is the chord of the flat airfoil, α_a is the instantaneous angle of attack with respect to the freestream, $\xi_k + i\eta_k = (z_k + \sqrt{z_k^2 - b^2})/b$ and $z_k = x_k + iy_k$ is the position of the kth vortex in the wake expressed in a complex plane.

With Eq. (3) and Eq. (7), Eq. (1) and Eq. (2) can be rewritten in matrix form:

$$\begin{bmatrix} m - F^a_{YY} & m(\xi_C \cos(\alpha - \theta) + \eta_C \sin(\alpha - \theta) - F^a_{\theta Y}) \\ m(\xi_C \cos(\alpha - \theta) + \eta_C \sin(\alpha - \theta) - F^a_{Y\theta}) & I_E - F^a_{\theta\theta} \end{bmatrix} \begin{Bmatrix} \ddot{y} \\ \ddot{\theta} \end{Bmatrix}$$

$$= \begin{Bmatrix} F^g_Y \\ F^g_\theta \end{Bmatrix} + \begin{Bmatrix} F^e_Y \\ F^e_\theta \end{Bmatrix} + \begin{Bmatrix} F^a_{0Y} \\ F^a_{0\theta} \end{Bmatrix} + \dot{\theta}^2 \begin{Bmatrix} F^a_{qY} + m(\eta_C \cos(\alpha - \theta) - \xi_C \sin(\alpha - \theta)) \\ F^a_{q\theta} \end{Bmatrix} \tag{9}$$

To solve Eq. (9), Hamming's fourth-order predictor-corrector method [23] is chosen. Eq. (9) can be rewritten as four first order ordinary differential equations as follows

$$\dot{y}_i = f_i(y_1, y_2, y_3, y_4) \qquad for \quad i = 1, 2, 3, 4 \tag{10}$$

where $y_1 = y$, $y_2 = \dot{y}$, $y_3 = \theta$, $y_4 = \dot{\theta}$. The method needs the information from the four previous steps to predict the solution of Eq. (10) at the current step. This predicted solution is used to evaluate the right-hand side of the equations. Then with the new right-hand side and the information from the four previous steps, the solution is corrected. The correction could, however, be repeated until the error is less than a desired bound. The vortices in the flowfield are convected at the speed of local fluid particles once at the beginning of each time step. Each time that the current aerodynamic loads are needed, the algorithm calls the subroutine that contains the aerodynamic model and the flowfield is recalculated.

C. Feedback control

A flap is added at the trailing-edge as shown in Fig. 2. The airfoil is approximated by a flat plate for computational convenience, and δ is the angle of the control flap relative to the airfoil.

Figure 2. The model of the control system.

The control flap oscillates in a tailored manner governed by a control law. In general, it is assumed that the angle of command δ_c is a function of the displacements and velocities of the airfoil

$$\delta_c = \delta_e + a_1(y - y_e) + a_2\dot{y} + a_3(\theta - \theta_e) + a_4\dot{\theta} \tag{11}$$

where the subscript *e* represents the equilibrium positions, which may or may not be stable.

For the problem of instant control, the flap angle satisfies

$$\delta = \delta_c \tag{12}$$

And for the delayed control, δ satisfies the following servo equation

$$\ddot{\delta} + c_1\dot{\delta} + c_2(\delta - \delta_c) = 0 \tag{13}$$

The larger c_2 is, the closer δ is to δ_c. The range over which δ varies is limited to δ_{max}. When Eq. (16) calls for a δ such that $|\delta| > \delta_{max}$, it is imposed that $|\delta| = \delta_{max}$ and $\dot{\delta} = 0$.

3. Examples

In this section, some numerical results are presented and compared to previous results when possible. A flat plate is used as the test model. Throughout this section, some of the parameters are as follows: $\alpha = 0°$, $b = 5$ in, $a_h = -0.15$, $x_\alpha = 0.25$, $r_\alpha^2 = I_E/mb^2 = 0.388$, $\mu = m/\pi\rho b^2 = 76$, $\omega_y = \sqrt{k_Y^l/m} = 55.9$ rad/sec, and $\omega_\theta = \sqrt{k_\theta^l/I_E} = 64.1$ rad/sec. Those parameters are taken from Fung's text book [1].

A. Simulation of flutter

To check the present method of flutter analysis of 2-D airfoils, the present results were compared with those given by Fung [1]. Fung considered a flat airfoil and solved the linear differential equations with the aerodynamic loads approximated by Wagner's function. The springs are linear, and there is no structural damping. In Fig. 3, the present results of y and θ are shown as functions time. The initial conditions for all three parts are $y(0) = 0$, $y(0) = 0$, $\theta(0) = -5°$ and $\theta(0) = 0$. In part (a), the freestream speed, V_∞, is 88 ft/sec. Both y and θ appear to be decaying with time to the values $y = y_e = 0.0093$ and $\theta = \theta_e = -0.93°$, which correspond to the static equilibrium position. The slow decay is due to the fact that V_∞ is near the flutter speed. In part (b), $V_\infty = 89$ft/sec, and y and θ appear to be neither growing nor decaying with time. Limit-cycle motions are possible for linear elastic restoring forces because the aerodynamic loads are inherently nonlinear. Both y and θ have the same frequency $\omega = 63$rad/sec, which satisfies the condition $\omega_y \le \omega_f \le \omega_\theta$, a characteristic of flutter motion. In part (c), $V_\infty = 90$ft/sec, and both y and θ are growing. Fung's analysis gives the flutter-onset speed as 90.1 ft/sec. The agreement of two sets of results is very

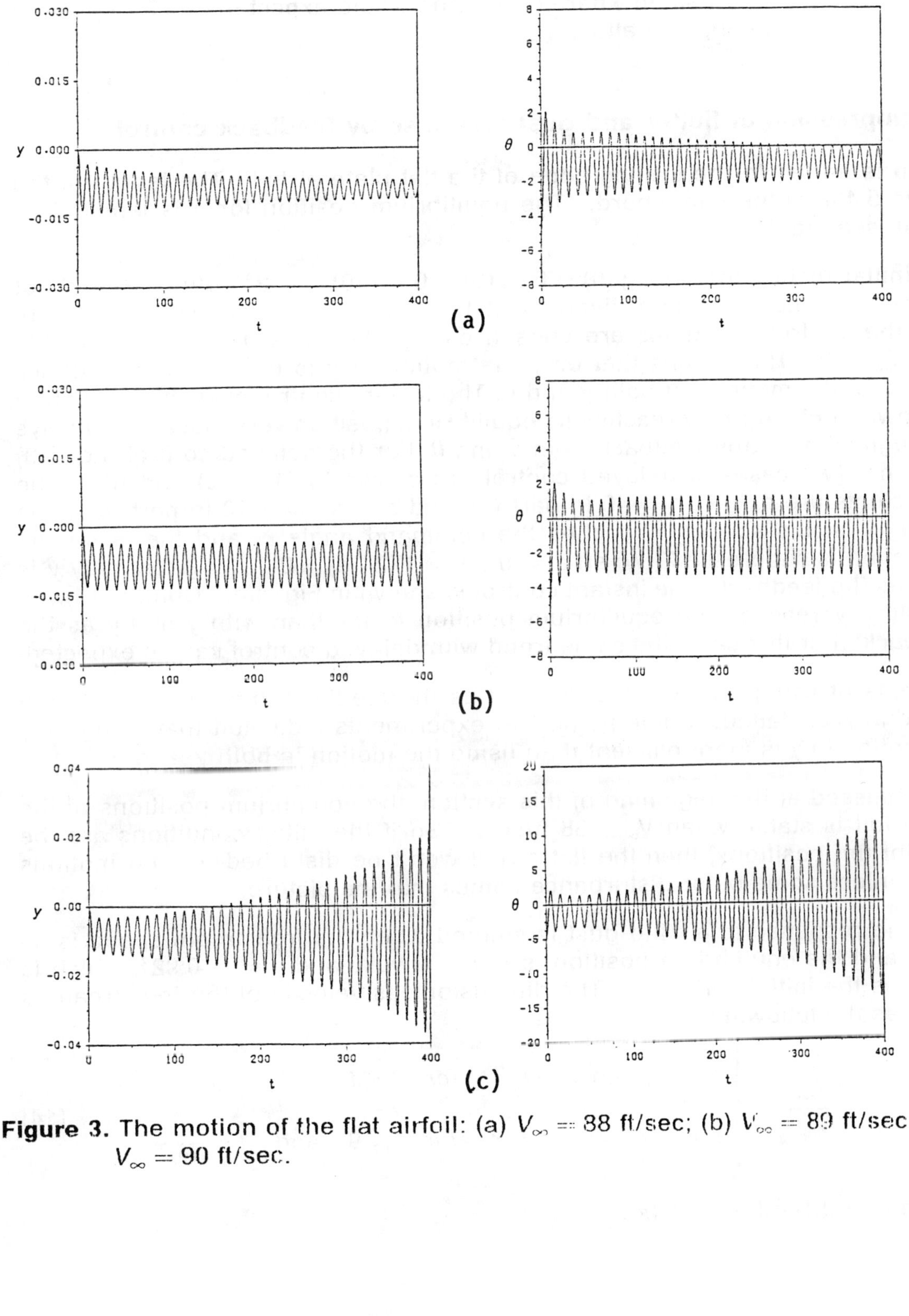

Figure 3. The motion of the flat airfoil: (a) $V_\infty = 88$ ft/sec; (b) $V_\infty = 89$ ft/sec; (c) $V_\infty = 90$ ft/sec.

close. The slight discrepancy between Fung's results and the present results is most likely due to the following: (1) Fung used Wagner's function to obtain the aerodynamic loads, and in the present analysis the general unsteady vorticity-panel method was used. (2) Fung considered oscillations around $\theta = 0°$, and in the present analysis the oscillations are approximately centered at an "equilibrium" position where $\theta \simeq 1°$. One would expect the flutter speed to decrease with the angle of attack.

B. Suppression of flutter and gust response by feedback control

A flap is added to the trailing-edge of the flat plate airfoil. The length of the flap is 5.4% of the total chord. The equilibrium position for δ is labelled as $\delta_e = 0$ (see Fig. 2).

The initial disturbance is: $y(0) = 0$, $\dot{y}(0) = 0$, $\theta(0) = -2°$, $\dot{\theta}(0) = 0$. Without any control, the oscillation diverges at $V_\infty = 90$ ft/sec as shown in Fig. 4(a). With the control, the gains are chosen as $a_1 = 114$, $a_2 = 172$, $a_3 = 0$, $a_4 = 0$. and $\delta_{max} = 5°$. This means that only the motion in y is used as the feedback instead of the motions in both y and θ. The instant control over the oscillation is shown in Fig. 4(b) : y reaches its equilibrium position very fast and θ decays although without any feedback from θ and $\dot{\theta}$. For the delayed control, Eq. (13) is used. Two cases of delayed control are shown in Fig. 4(c) and (d) : the coefficients are $c_1 = 4$, $c_2 = 40$ in part (c), and $c_1 = 4$, $c_2 = 10$ in part (d). The larger c_2 is, the sooner δ reaches the command angle δ_c and the faster the oscillation is suppressed. When $a_1 = 0$, $a_2 = 172$, $a_3 = 0$, $a_4 = 0$ (i.e. only y is used as the feedback), the instant control is shown in Fig. 4(e). Compared with Fig. 4(b), y reaches the equilibrium position faster than with y and $\dot{y}$ as the feedback. For this case, the same trend with delayed control can be expected.

Attempts at using only the motion in θ as the feedback (i.e., $a_1 = a_2 = 0$, but a_3 and $a_4 \neq 0$) failed. Some numerical experiments indicated that using only the motion in y is more efficient than using the motion in both y and θ.

As discussed at the beginning of this section, the equilibrium positions of the flat airfoil is stable when $V_\infty \leq 88$ ft/sec . So if the initial conditions are the equilibrium positions, then the flat airfoil won't be disturbed. i.e. no motions at all unless some other disturbance comes into the picture.

Now the response to a wind gust is studied when the freestream speed is 88 ft/sec and the equilibrium position is $y_e = -0.0096$ and $\theta_e = -0.92°$, which is taken as the initial condition. The dimensionless velocity of the freestream is given as the following

$$\frac{\vec{V}_\infty}{V_\infty} = \begin{cases} \vec{i} + (v_g \sin \omega_g t)\vec{j} & for \;\; 0 \leq t \leq \dfrac{2\pi}{\omega_g} \\ \vec{i} & for \;\; t < 0 \;\; \text{and} \;\; t > \dfrac{2\pi}{\omega_g} \end{cases} \tag{14}$$

where $v_g = 0.1$and $\omega_g = 0.4\pi$.

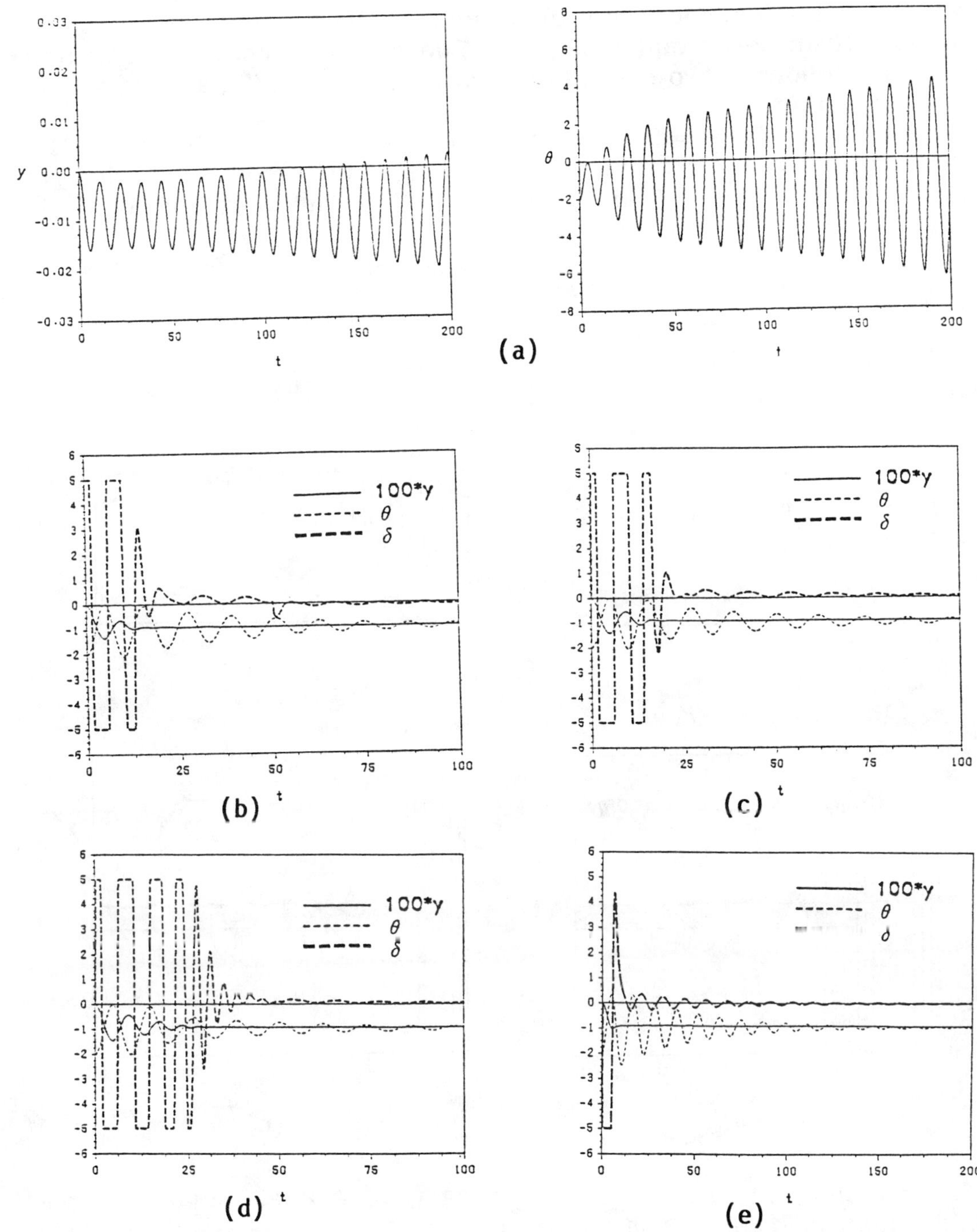

Figure 4. The motions of the airfoil with/without feedback control. $V_\infty = 90$ ft/sec and the initial condition is $y(0) = \dot{y}(0) = \dot{\theta}(0) = 0$, $\theta(0) = -2^\circ$. For the delayed control δ satisfies $\ddot{\delta} + c_1\dot{\delta} + c_2(\delta - \delta_c) = 0$ where $\delta_c = 114(y - y_e) + 172\dot{y}$ and $\delta_{max} = 5^\circ$. The motions (a) without control, (b) with instant control $\delta = \delta_c$, (c) with delayed control when $c_1 = 4$ and $c_2 = 40$; (d) with delayed control when $c_1 = 4$ and $c_2 = 10$, and (e) with instant control when $\delta = 172\dot{y}$.

The motions without any attempt to control the response to this wind gust are shown in Fig. 5(a). The oscillations decay gradually. The motions die out more rapidly when feedback control is applied to the flap by choosing $c_1 = 4$ and $c_2 = 10$ in servo equation (13). Two sets of gains are used and $\delta_{max} = 5°$ for the following two cases. First, $a_1 = 114$, $a_2 = 172$, $a_3 = 0$, $a_4 = 0$, and the motions are shown in Fig. 5(b) Second, $a_1 = 0$, $a_2 = 172$, $a_3 = 0$, $a_4 = 0$, and the motions are shown in Fig. 5(c). The motion in y dies out faster than the previous case, but the motion in θ dies out more slowly. For both cases, that the motions will die out more quickly if c_2 is larger, and the instant control is the most effective.

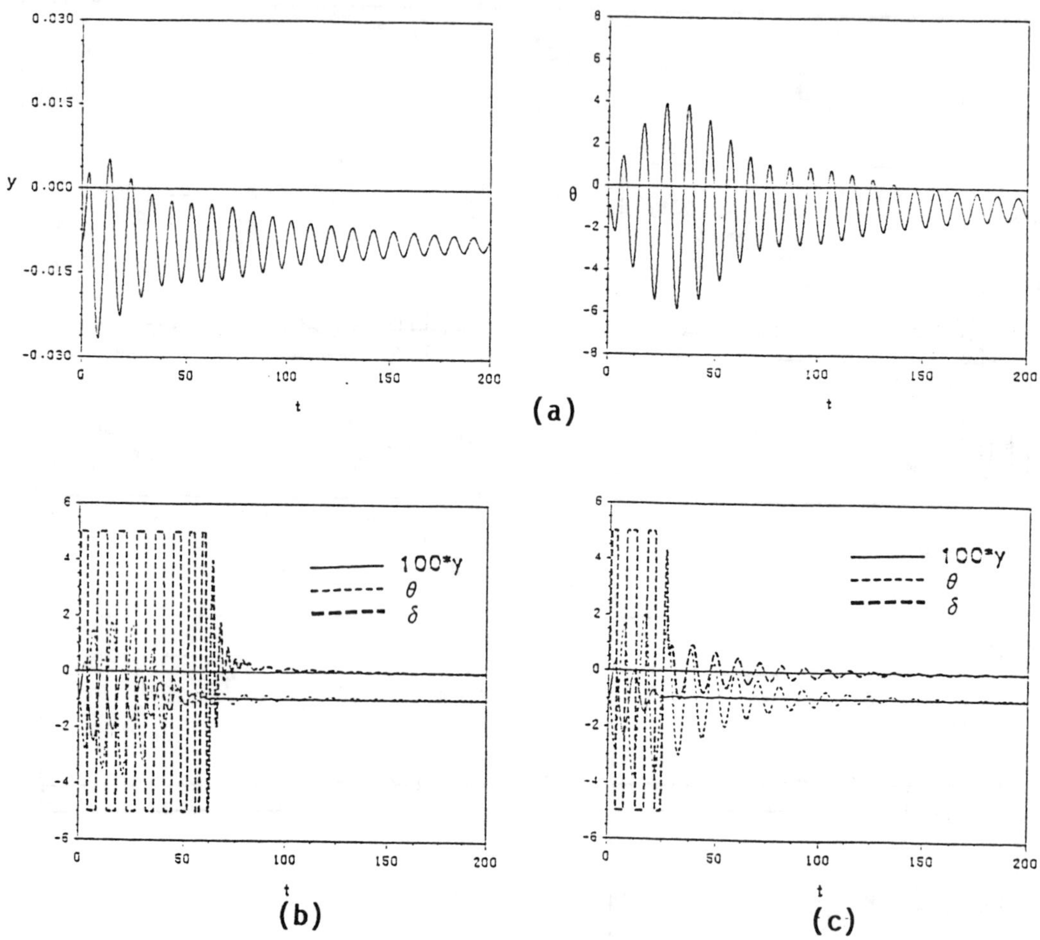

Figure 5. The response to a wind gust. $V_\infty = 88$ ft/sec and the initial condition is $y(0) = y_e$, $y(0) = 0$, $\theta(0) = \theta_e$, $\theta(0) = 0$. For the delayed control δ satisfies $\delta + 4\delta + 10(\delta - \delta_c) = 0$, and $\delta_{max} = 5°$. (a) the motions without control. (b) the motions with delayed control when $\delta_c = 114(y - y_e) + 172y$, and (c) the motions with delayed control when $\delta_c = 172y$.

4. Summary

A numerical model capable of simulating the arbitrary motion of a rigid 2-D airfoil mounted on a nonlinear elastic support was developed. The aerodynamic model for calculating loads acting on the airfoil is nonlinear, and the equations of motion are nonlinear. To suppress the response to some disturbances, a feedback control model was developed. The dynamic system, which includes airfoil, wake, elastic support and control devices, is considered integrally, i.e. the coupled aerodynamic -dynamic-control problems are solved simultaneously and interactively.

For the flat plate airfoil, the flutter speed obtained here is close to the result of Fung. Feedback control successfully suppressed the flutter and the response to a wind gust. Using the motion in y as the feedback appears to be more effective than using the motion in θ. Even with just y as the feedback, motions with two degrees of freedom can be suppressed. The present results demonstrate the feasibility of modelling complete-aircraft systems.

Acknowledgement

This work was supported by the Air Force Office of Scientific Research under Grant No. AFOSR-90-0032.

References

1. Fung, Y. C. 1969. **The Theory of Aeroelasticity,** Dover Publications, NY.

2. Lee, B. H. K., and P. LeBlanc, 1986. "Flutter Analysis of Two-Dimensional Airfoil with Cubic Non-Linear Restoring Force," Aeronautical Note, NAE-AN-36, NRC NO. 25438, N87-22628.

3. Rizetta, D. P., 1979. "Time-Dependent Responses of a Two-Dimensional Airfoil in Transonic Flow," AIAA Journal 17:26-32.

4. Rizetta, D. P., 1977. "Transonic Flutter Analysis of a Two-Dimensional Airfoil," AFFDL, TM-77-64-FBR.

5. Ballhaus, W. F. and P. M. Goorjian, 1978. "Computation of Unsteady Transonic Flows by the Indicial Method," AIAA Journal 16:117-124.

6. Yang, T. Y., P. Guruswamy, and A. G. Striz, 1979. "Aeroelastic Response Analysis of Two-Dimensional, Single and Two Degree of Freedom Airfoils in Low Frequency, Small Disturbance Unsteady Transonic Flow," AFFDL, TR-79-3077.

7. Yang, T. Y., P. Guruswamy, A. G. Striz, and J. J. Olsen, 1980. "Flutter Analysis of a NACA 64A006 Airfoil in Small Disturbance Transonic Flow," Journal of Aircraft 17:225-232.

8. Guruswamy, P. and T. Y. Yang, 1981. "Aeroelastic Time Response Analysis of Thin Airfoils by Transonic Code LTRAN2," Computers & Fluids 9:409-425.

9. Strganac, T. W. and D. T. Mook, 1987. " The Numerical Simulation of Subsonic Flutter," AIAA-87-1428.

10. Strganac, T. W. and D. T. Mook, 1990. "Numerical Model of Unsteady Subsonic Aeroelastic Behavior," AIAA Journal 28:903-909.

11. B. Dong and D. T. Mook, 1990. "Numerical Simulation of Unsteady Aeroelastic Behavior," Proceedings of the SECTAM XV 1990 Southeastern Conference on Theoretical and Applied Mechanics, Atlanta, GA, March 22-23.

12. B. Dong and D. T. Mook, 1990. "Numerical and Asymptotic Analysis of the Flutter Phenomenon for an Airfoil," Proceedings of the Third Conference on the Nonlinear Vibrations, Stability and Dynamics of Structures and Mechanism, Blacksburg, VA, Vol. XV, June 25-27, pp. 102-110.

13. Kim, M. J. and D. T. Mook, 1986. "Application of Continuous Vorticity Panels to General Unsteady Two-Dimensional Lifting Flows," Journal of Aircraft 23:464-471.

14. Mracek, C. C. P., 1988. "Unsteady Potential-Flow Solution Using Vortex Panels Coupled with Dynamics and Controls," Ph.D. Dissertation, Virginia Polytechnic Institute and State University.

15. Lee, E. M. and J. T. Batina, 1990. "Conical Euler Simulation and Active Suppression of Delta Wing Rocking Motion," NASA Technical Memorandum 102683.

16. Matsuzaki, Y. and T. Miyazawa, 1987. "Wind Tunnel Test and Analysis on Gust Load Alleviation of a Transport-Type Wing," AIAA-87-0781-CP.

17. Matsuzaki, Y. and T. Miyazawa, 1989. "Gust Load Alleviation of a Transport-Type Wing: Test and Analysis," Journal of Aircraft 26:322-327.

18. Wu, J. C. and N. L. Sankar, 1980. "Aerodynamic Force and Moment in Steady and Time-Dependent Viscous Flow," AIAA Paper No. 80-0011.

19. Giesing, J. P., 1968. "Nonlinear Two-Dimensional Unsteady Potential Flow with Lift," Journal of Aircraft, 5:135-143.

20. Maskell, E. C., 1971. "On the Katta-Joukowski Condition in Two-Dimensional Unsteady Flow," unpublished note, Royal Aircraft Establishment, Farnborough, England.

21. Poling, D. R. and D. P. Telionis, 1986. "The Response of Airfoils to Periodic Disturbances - The Unsteady Kutta Condition," AIAA Journal 24:193-199.

22. Mook, D. T. and B. Dong, 1990. "Application of Vortex Dynamics to Simulations of the Wake of Airfoils," Proceedings of the International Symposium on Nonsteady Fluid Dynamics, FED-Vol. 92, Toronto, Canada, June 4-7, pp. 435-448.

23. Carnahan, B., H. A. Luther, and J. O. Wilkes, 1969. **Applied Numerical Methods,** John Wiley & Sons Inc., NY.

COMPUTER CONTROL OF THE FIRST HARMONIC OF LOW FREQUENCY OSCILLATION IN MODEL OF END-OF-ARM TOOLING

Marek L. Szary, Ph.D.,[1] Kerwin Kassulker, M.S.[2]

ABSTRACT

This paper describes the use of computer and linear force actuator to control the first harmonic of low frequency oscillation. A single axis model of pneumatic robot with flexible cantilever beam mounted on end-of-the-arm is introduced. Two methodologies were decided upon for controlling vibrations in the tip of the cantilever: an open-loop system and a closed-loop system. Both methodologies make use of computer software for the control portion of the system, and both use DC-amplified signal to the linear force actuator. A testbed is constructed to verify the control theory applied. The testbed is built around an IBM PC/XT computer which controls the linear force actuator, monitors feedback from the linear hall effect sensor, and tests the limit switch which will trigger the experiment. The IBM PC performs its I/O operations to the testbed through a DAS-16 Data Acquisition and Control Board. The DAS-16 board has the necessary speed to sample the Hall sensor, and then send back a signal in 180 degrees out-of-phase. Therefore, the addition of a damping system to help decrease vibrations will thus help decrease the cycle time of the robot. The application of computer controlled active damping of the model of robot's arm with open and closed loop feedback control is discussed.

INTRODUCTION

Today's advanced manufacturing facilities have become highly computer-integrated, automated workplaces, capable of flexible manufacturing [1]. Robots are providing the necessary flexibility and repeated quality demanded of these modern facilities. However, a problem often encountered in implementing a robot lies in the development of the required end-of-arm tooling. During assembly, or pick-and-place operations, certain configurations of the robot arm may not be obtainable in order to perform the task. At this point, specialized end-of-arm tooling may have to be developed. A long flexible robot's arm with energized tools are especially prone to vibration.

Vibrations can also occur in end-of-arm tooling which could affect accuracy or possibly damage the product. Pneumatic robots which use "bang-bang control" [2] are most susceptible to these vibrations.

---

[1]Department of Technology, College of Engineering and Technology, Southern Illinois University at Carbondale, Carbondale, IL 62901-6603, U.S.A.

[2]Diversified Technical Systems, Inc., 2500 NE Broadway Street, Minneapolis, MN 55413, U.S.A.

When a pneumatic robot which uses "bang-bang control" reaches a fixed limit, there will be a force impulse induced in the system which will also transfer force to the object being moved. This type of force is sometimes referred to as "velocity shock" [5], and may harm fragile objects that are being moved. The force impulse will also cause vibrations to occur in the system, which may cause accuracy problems.

Vibration is not only a problem in pneumatic robots, but in other types of robots as well [4]. In different arm configurations and loads, a speed may be reached where the tip of the robot will vibrate. Vibrations may also occur in robots equipped with energized end-of-arm tooling (such as grinders, powered screwdrivers, etc.).

In robot arms that use a SCARA (Selective Compliance Assembly Robotic Arm), or Over Head Gantry configuration, there is an increased bending moment in the final link, which will make the end-of-arm tooling more susceptible to vibrations [2]. Once again, a damping system could help reduce these vibrations and improve the performance of the robot.

Although servo systems and programmable stops do exist for pneumatic robots, they tend to add cost and complexity to the system [3,4]. The servo systems that have been developed for pneumatic systems use "i" to "p" transducers that will transform a current input in proportion to a pressure output. A double acting cylinder supplied with controlled pressure on each side, and also a position sensor, will act as a servo system; but the response of these systems is poor, and few systems are in use today [3]. In order to minimize vibrations, the stiffness of the robot arm is increased, or often the speed of operation is reduced [2]. This will increase the mass of the robot, and possibly slow production.

TESTBED

The testbed was built around an IBM PC/XT which controls the linear force actuator, monitors feedback from the linear Hall effect sensor, and tests the limit switch which will trigger the experiment (Figure 1). The IBM PC performs its I/O operations to the testbed through a DAS-16 Data Acquisition and Control Board. The DAS-16 board has the necessary speed to sample the Hall sensor, and then send back a sign wave 180 degrees out-of-phase.

In order to insure a rigid test stand, a frame was built using 1-1/2" steel angle iron. The frame was then bolted to a 1/2" thick sheet of plywood, and clamped securely to a heavy table. This resulted in a test stand which is free from any undesired vibrations.

The air supply to the pneumatic cylinder is provided by a portable compressor set to 30 psi. The air line to the cylinder was clamped to make fine adjustments in air pres-

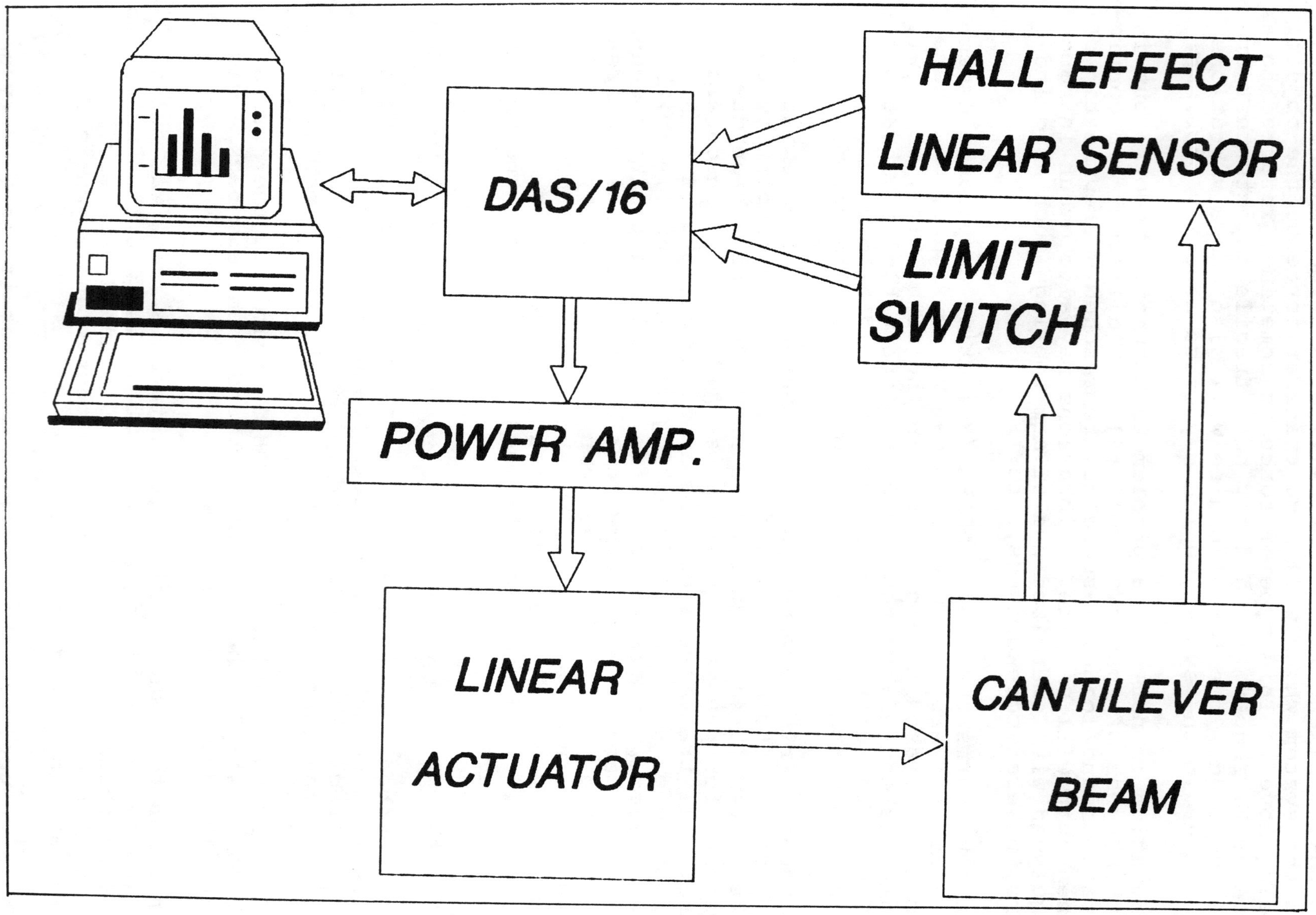

FIGURE 1-EXPERIMENTAL TESTBED LAYOUT

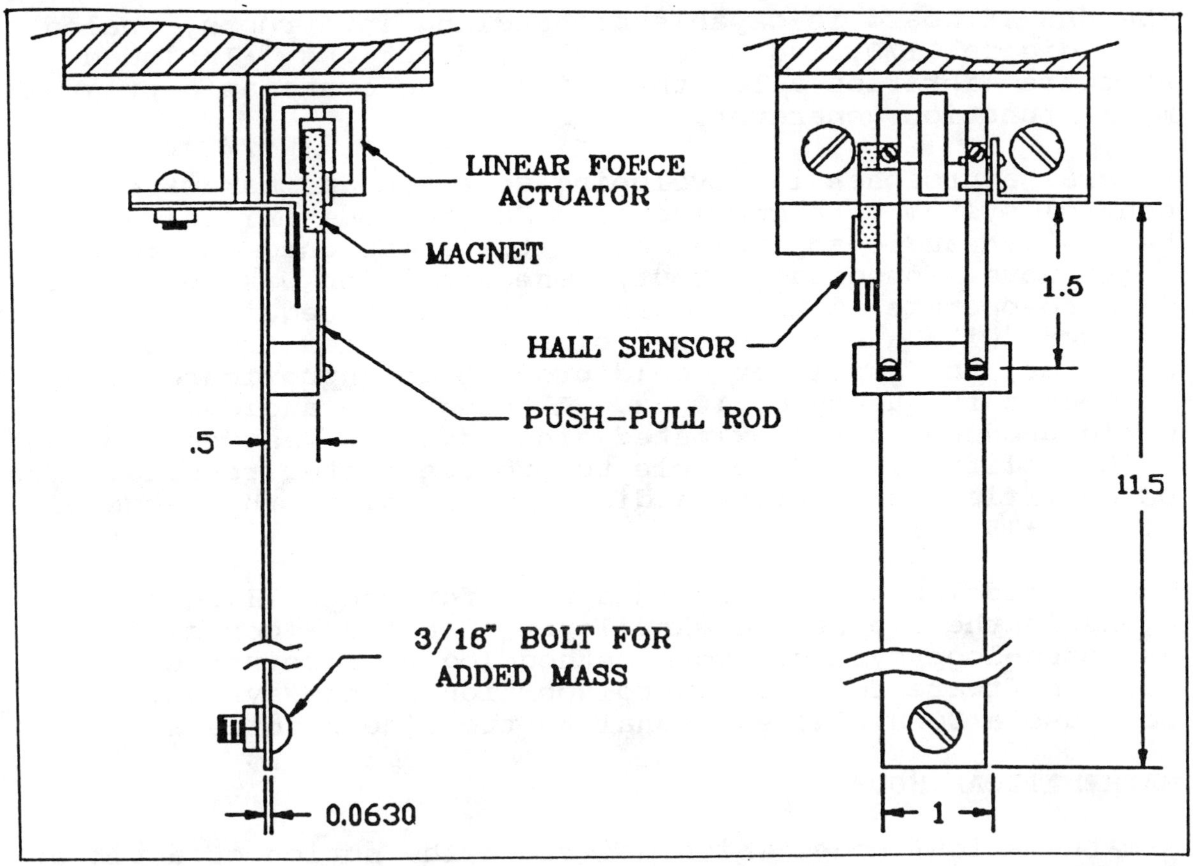

FIGURE 2-DRAWING OF CANTILEVER ASSEMBLY

sure. A twin-shaft, linear slide was mounted on the test stand to prevent side loading of the air cylinder and also to provide a mounting surface for the cantilever assembly.

Figure 2 shows a detailed drawing of the cantilever assembly. The linear force actuator was mounted near the top of the cantilever beam, because it provided a solid mounting surface. Two aluminum push-pull rods connect the force actuator coil to the beam. The length of the rods were determined by examining a deflection point on the beam, which would allow the force actuator coil to move its full stroke.

In order to provide the high current demands (2.3 amps at peak force) of the actuator's coil, an amplifier is required. During the initial setup of the testbed, a sinewave was inputted from a function generator and the measured output of the Hall sensor was displayed on an oscilloscope.

The DAS-16 board is capable of handling interrupts, analog I/O, digital I/O, and several other functions. Included software libraries allow the DAS-16 to operate as a programmable function generator.

Several approaches in developing software and hardware for a control system were considered. During some initial attempts, an audio amplifier circuit was designed to produce continuous sinusoidal output. The amplifier was not specified to operate at the low frequency range required 8-14 Hz; however, with a small enough signal outputted from the computer, the amplifier could produce an approximate sinewave at a frequency of 10 Hz. Although the audio amplifier could produce an approximated sine wave, it was decided that a DC amplifier would be able to produce better results. The DC amplifier can produce a discrete signal in the range of -15 to +15 volts.

Two methodologies were decided upon for controlling vibrations in the tip of the cantilever: an open-loop system and a closed-loop system. Both methodologies make use of computer software for the control portion of the system, and both use a DC-amplified signal to the linear force actuator.

MATHEMATICAL MODEL

A mathematical model which describes the motion of a vibrating spring mass system, with damping, can be described by equation (1). This model considers a damping force which is proportional to the velocity of the cantilever tip.

$$m \frac{d^2x}{dt^2} + c \frac{dx}{dt} + kx = 0 \qquad (1)$$

dividing through by m, gives:

$$\frac{d^2x}{dt^2} + \frac{c}{m} \frac{dx}{dt} + \frac{k}{m} x = 0 \qquad (2)$$

solving for d^2x/dt^2 equals:

$$\frac{d^2x}{dt^2} = - \left(\frac{c}{m} \frac{dx}{dt} + \frac{k}{m} x \right) \qquad (3)$$

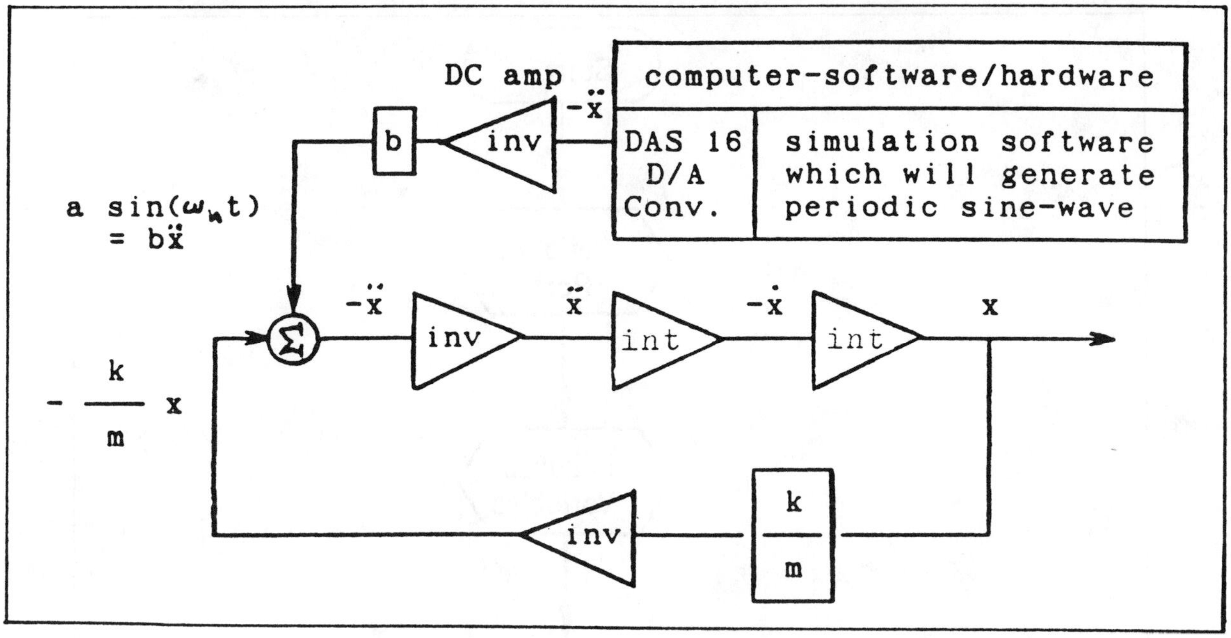

FIGURE 3-BLOCK DIAGRAM OF OPEN-LOOP CONTROL SYSTEM

METHODOLOGY

Based on the premise of available binary drivers for the DAS 16 I/O control card, an open-looped system was developed. Figure 3 shows a block diagram of the open-loop control system developed from existing drivers.

Figure 4 shows a flowchart of the open-loop control program. The program consists of routines which make calls to the binary drivers of the DAS 16 card. The program was difficult to modify for the testbed, because of the declarement "BYVAL". Individual FI/O statements could not be written in the program with this statement, and only a constant output of amplitude was obtainable.

The open-loop program operates in the following manner: A periodic waveform with a specified frequency and N number of cycles, will be outputted on interrupt to the DAS-16. Therefore, it is necessary to predetermine the frequency and number of damping cycles to apply.

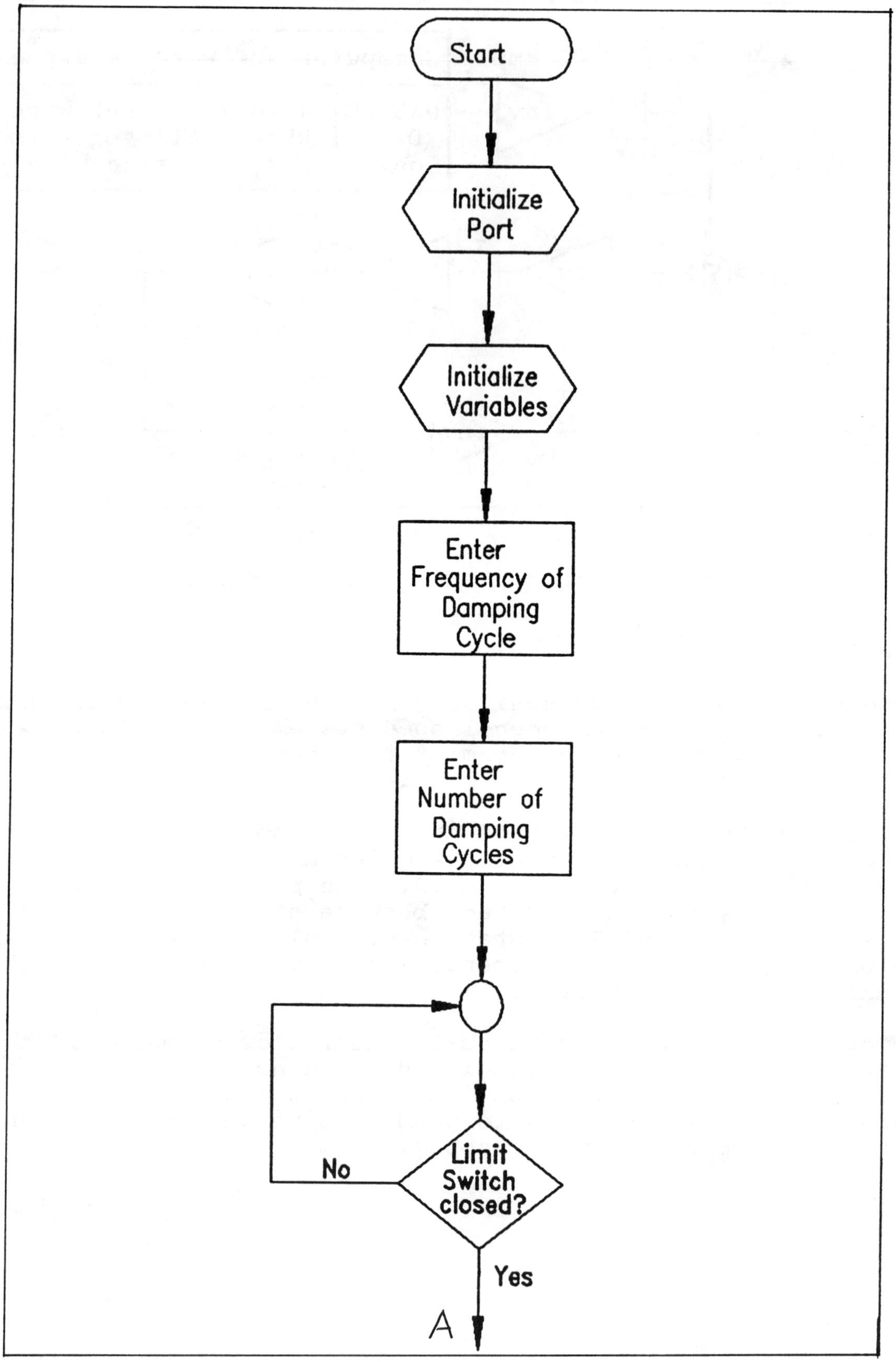

FIGURE 4-OPEN-LOOP CONTROL FLOWCHART

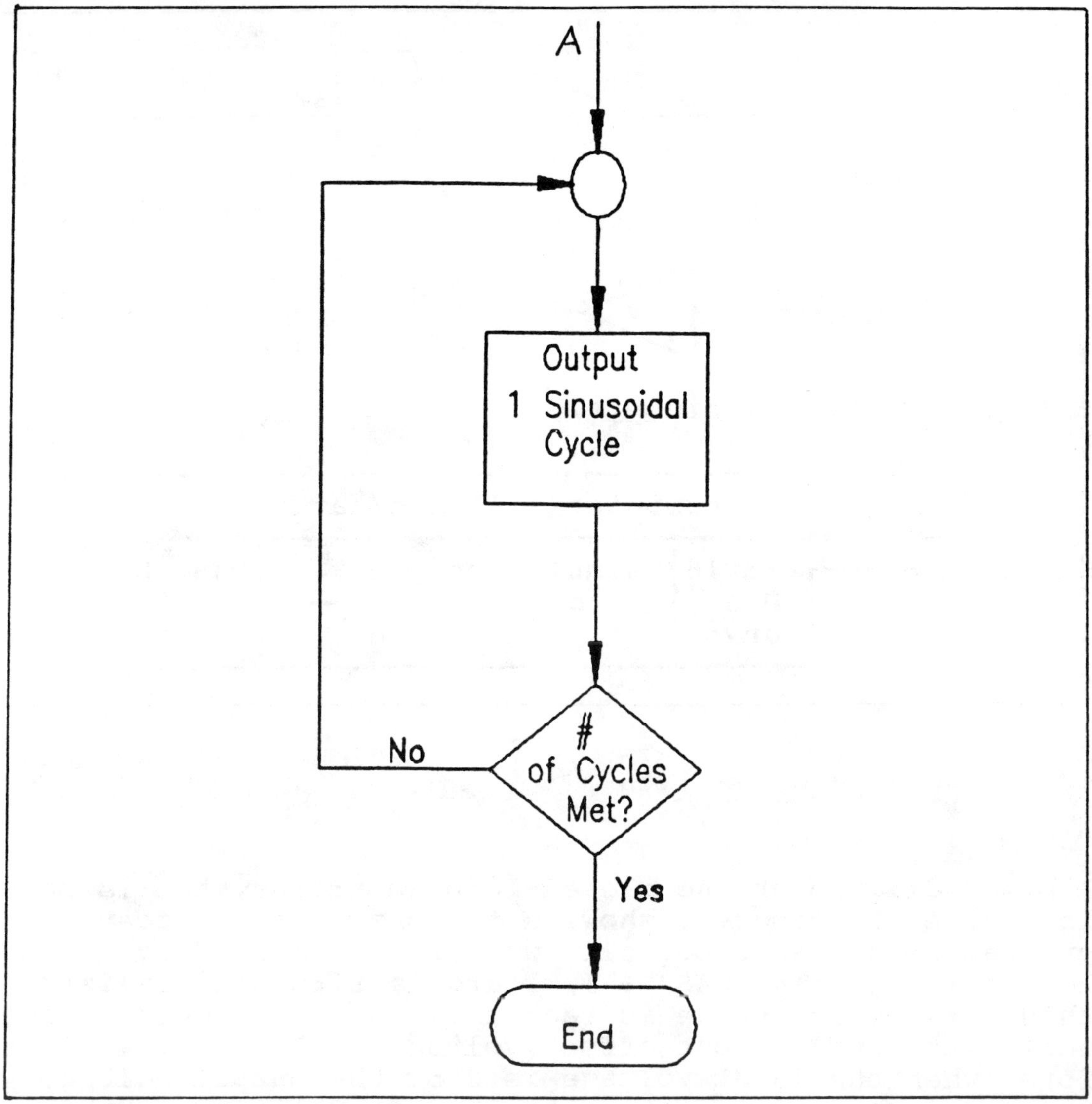

FIGURE 4-OPEN-LOOP CONTROL FLOWCHART

In order to determine the frequency of oscillation for a given mass, the linear actuator can drive the cantilever into a high amplitude resonance by inputting the proper frequency from a function generator. Once the frequency has been obtained, the number of damping cycles can then be determined experimentally.

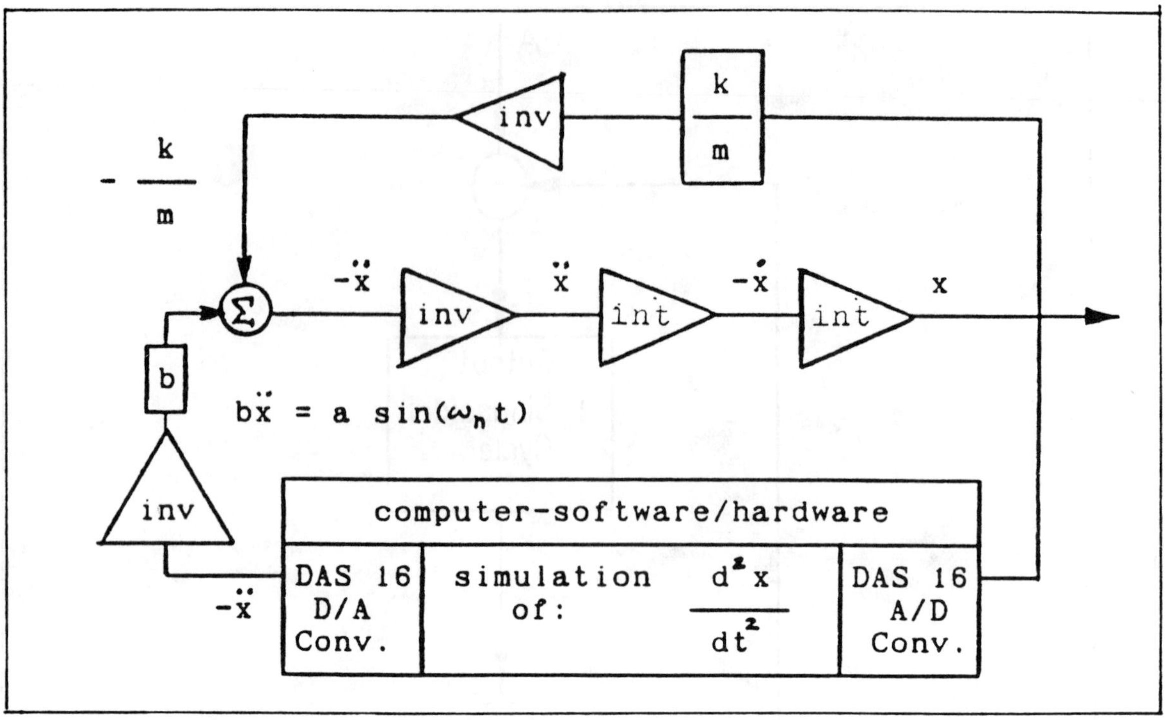

FIGURE 5-BLOCK DIAGRAM OF CLOSED-LOOP CONTROL SYSTEM

A block diagram of the closed-loop control system is shown in Figure 5. Figure 6 shows a flowchart of the control program which was especially written for providing closed-loop damping. The DAS/16 I/O card is first initialized, and then a reference value is read from the Hall sensor. The DC amplifier has an input offset voltage of 2.5 volts. Therefore, when the input voltage is 2.5, the output voltage will be zero. By sending output to the DC amplifier in the range of 0 to 5 volts, an output voltage of +/- 15 volts can be obtained.

Once started, the program will remain in a continuous loop. As long as the cantilever remains in its equilibrium position, there will be no output signal sent to the linear actuator. If the tip of the cantilever moves away from its reference (equilibrium) position, then a voltage will be outputted, which will create a force of opposite magnitude. The magnitude of force is determined by taking the difference between the present and reference voltage outputted from the Hall effect sensor.

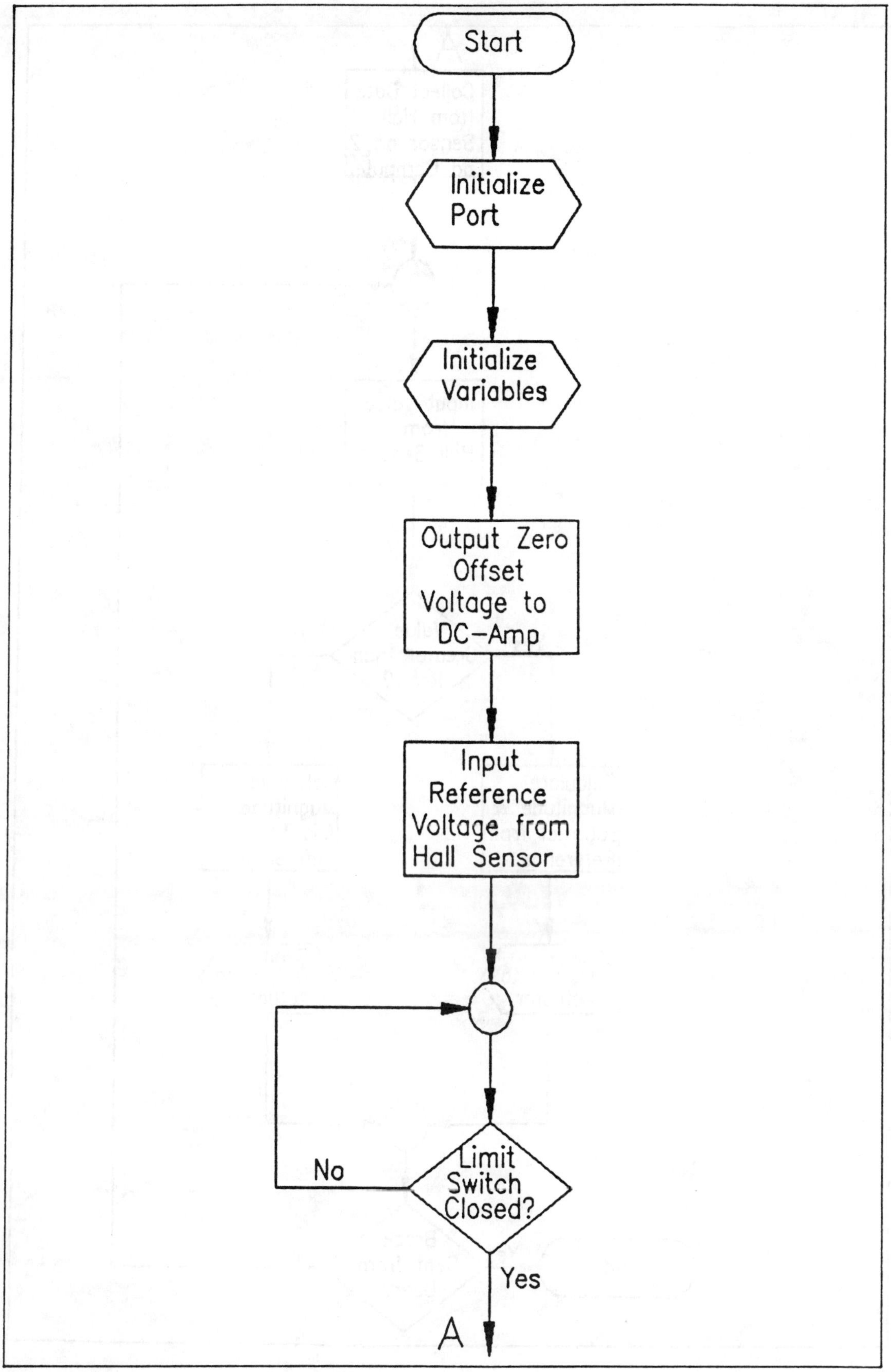

FIGURE 6-CLOSED-LOOP CONTROL FLOWCHART

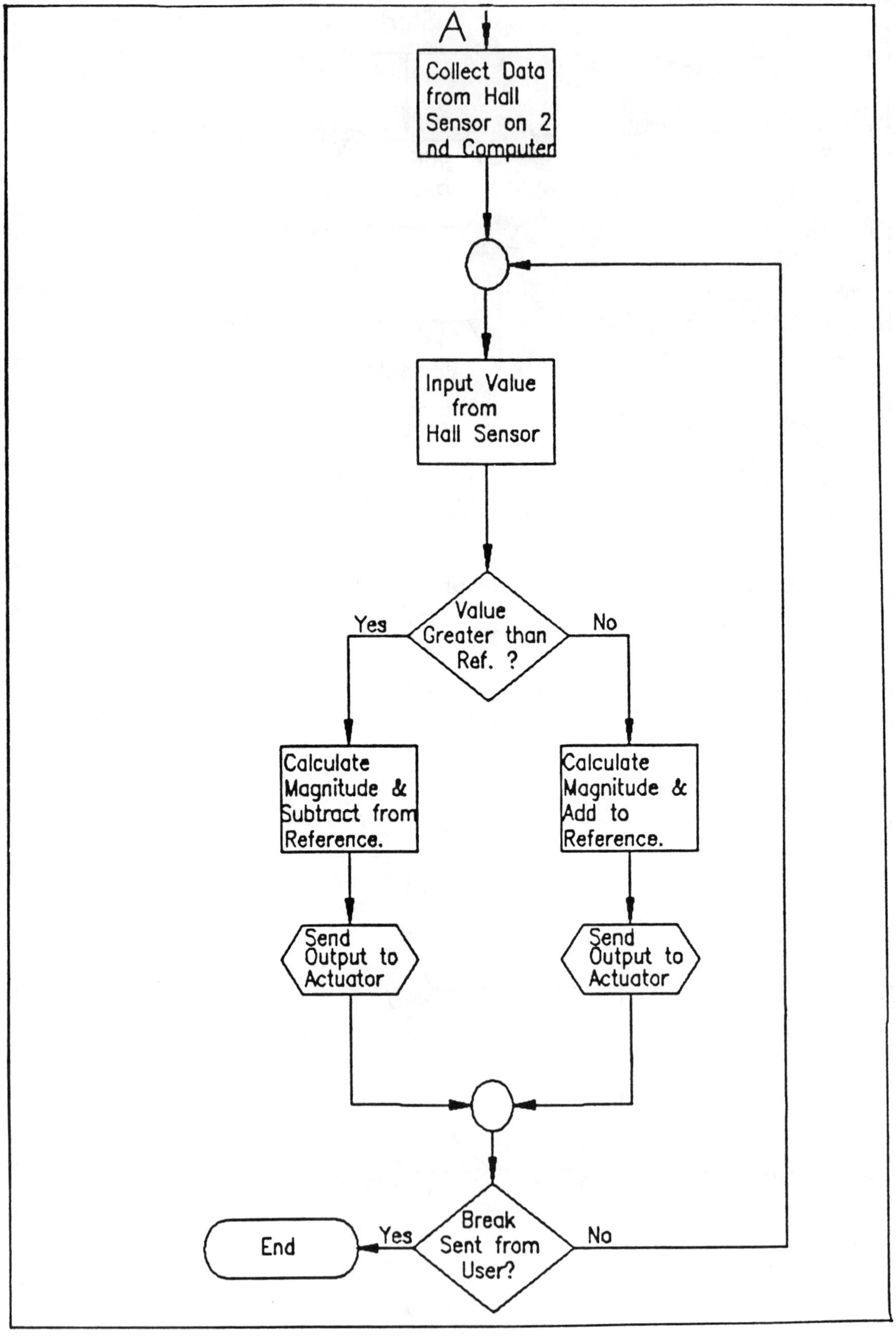

FIGURE 6-CLOSED-LOOP CONTROL FLOWCHART

LIMITATIONS OF THE SYSTEM

One of the obvious limitations of the system is the magnitude of the push-pull operation in the force actuator. The actuator's force output is a function of current, which is equal to 2.3 amps at its peak. Unfortunately, an amplifier with the required amperage of 2.3 amps was not obtainable for the testbed. The amplifier which was acquired only supplies 1 amp. Thus, the magnitude of force the actuator produces is only about 3.5 ounces, based on its specifications of 3.5 oz/amp.

The location of the actuator and sensor was chosen to provide unrestricted end-of-arm tooling and to also stay within the stroke limitations of the coil. This puts the actuator at a mechanical disadvantage and creates limited control over the extreme bending moments at the tip of the cantilever.

The linear Hall effect sensor did exhibit noise on initial impact which was caused by the vibration of the magnet. Therefore, unless a damping system is developed for the magnet, the first two to three cycles of oscillation in the cantilever tip could have inaccurate data feedback.

COLLECTION AND ANALYSIS OF THE DATA

Data specifying the movement of the force actuator coil is collected from a Linear Output Hall Effect Transducer. A magnet is mounted on a push-pull rod of the cantilever assembly, which will move with the actuator coil. Wang, Lu, and Hsu suggest that the location of the sensor has to be as close to the support of the arm as possible for a stable response [10]. The Hall sensor produces a linear differential output (sensitivity of 3.75 mV/gauss) which is a function of magnetic field intensity [15].

The collection of data will begin when the limit switch mounted on a test stand is triggered. When the pneumatic cylinder reaches its limit, triggering of the limit switch will occur. An interrupt is then sent to the DAS-16, which will begin collecting data from the Hall sensor. The data is stored in a file which can later be analyzed.

To verify damping in the cantilever beam, a digital timer will be used to measure settling time. The limit switch mounted on the test stand will activate the timer. Once the settling of the cantilever tip has been observed, a manual stop button will be pressed, and the time will be recorded. This will allow comparison of the undamped and damped response of the cantilever.

Data from the Hall effect sensor represents the movement of the force actuator coil. If the movement of the coil can be canceled, then you will have essentially canceled the movement of the cantilever tip as well.

The voltage output of the sensor is converted to binary with a range of 0 to 4095 bits (0 bits = 0 volts; 4095 bits = 10 volts). The Hall effect sensor will output a voltage proportional to its supply voltage. The supply voltage needs to be in a range of 8 to 16 volts for it to operate properly. With the cantilever at rest, the voltage from the Hall sensor is approximately three volts. The results for each experiment show that damping did occur without any loss in rate due to the increase in mass.

Table 1 shows the damping coefficients of both the undamped and damped conditions of the ten experiments. The damping coefficient was calculated by taking the peak magnitude of two cycles x_1 and x_2, and equating them to:

$$\frac{x_1}{x_2} = e^{\zeta \omega_n (\Delta t)} \tag{4}$$

taking the natural log of both sides, and then solving for gives:

$$\zeta \doteq \ln\left(\frac{x_1}{x_2}\right) \frac{1}{\omega_n (\Delta t)} \tag{5}$$

Figure 7 shows example of undamped and damped response from closed-loop control damping system.

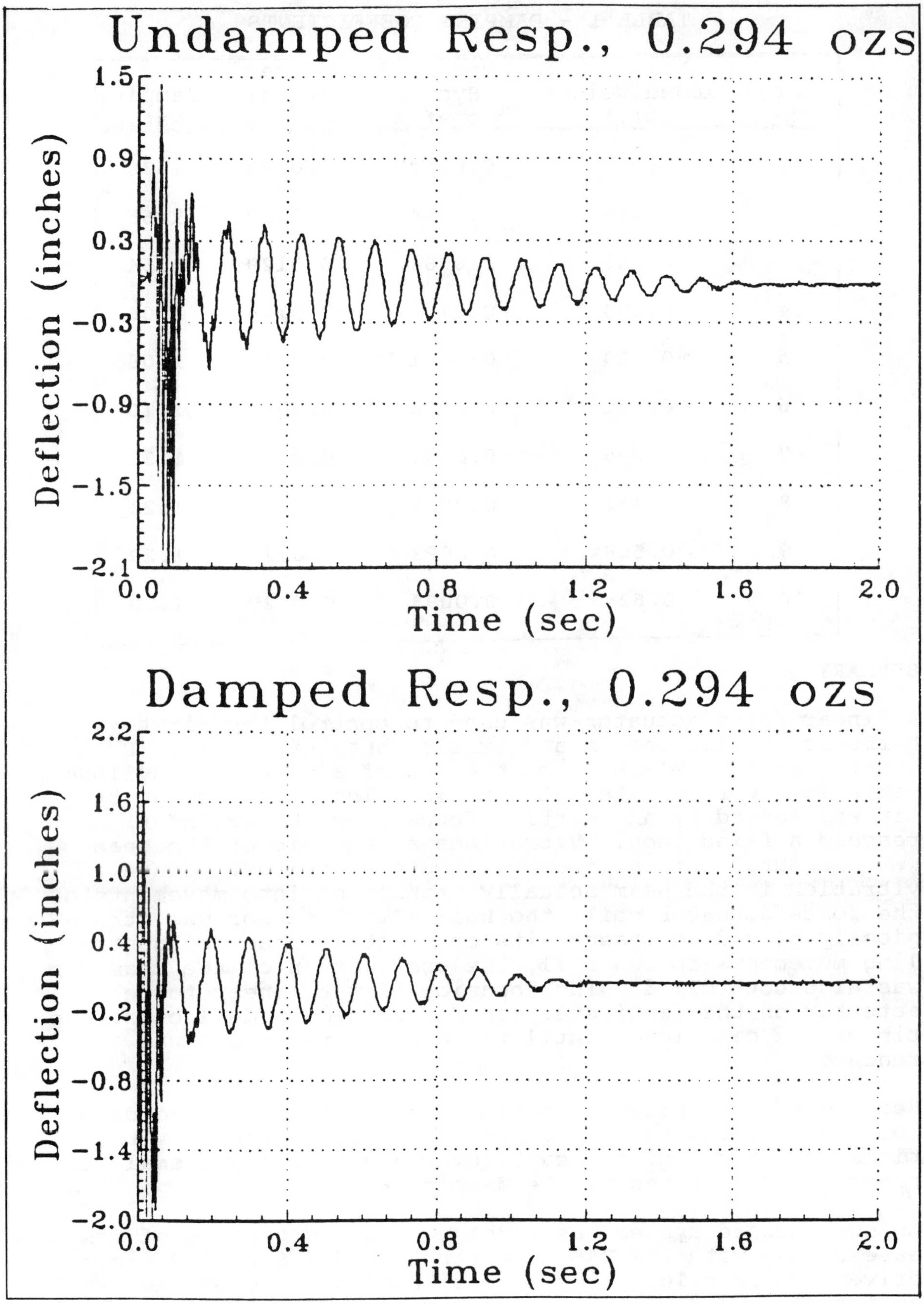

FIGURE 7-UNDAMPED AND DAMPED RESPONSE

TABLE 1 - DAMPING COEFFICIENTS

Ex No	Added Weight (oz.)	Undamped System coef's	Damped System coef's	Damping Ratio
1	0.000	0.00129	0.0178	1.38
2	0.256	0.0084	0.0139	1.65
3	0.259	0.0084	0.0129	1.54
4	0.294	0.0140	0.0148	1.06
5	0.323	0.0111	0.0120	1.08
6	0.361	0.0086	0.0095	1.10
7	0.396	0.0084	0.0111	1.32
8	0.432	0.0084	0.0102	1.21
9	0.508	0.0093	0.0120	1.29
10	0.624	0.0093	0.0129	1.39

SUMMARY

A linear force actuator was used to control the first harmonic oscillation in a flexible cantilever beam. By mounting a flexible beam on the end of a pneumatic cylinder, vibrations were created at the tip. The vibration at the tip was caused by an inertial force when the cylinder reached a fixed stop. Vibrations at the tip of the beam were measured with a linear Hall effect sensor. Since vibration in the beam actually translates into movements of the force actuator coil, the Hall effect sensor was strategically placed to measure its movements. Thus, by controlling movements in the coil, the oscillation of the beam tip was also controlled. The location of the linear force actuator on the cantilever was determined by deflecting the tip of the cantilever until the end of its stroke was reached.

Because of the necessary positioning of the force actuator coil and the limits of its force, the amplitude of vibrations in the tip of the cantilever had to be kept small in order to achieve recordable damping.

An open-looped system was first considered because of its ease of use with the DAS-16 control card. Supplied binary drivers would allow a periodic sinusoidal waveform to be outputted on a given interrupt signal to the DAS-16.

By triggering the start of the experiment with a limit switch mounted at the end of the pneumatic cylinder's stroke, a predetermined number of damping cycles was applied. The frequency of the applied damping was predetermined by using the linear actuator to drive the cantilever into resonance.

The open-loop system did not produce any recordable results because of a problem in the triggering of the damping output signal. During initial impact the tip of the cantilever was 90 degrees out-of-phase, which caused the output of the linear force actuator to be applied at the wrong interval of time.

The preferred methodology of control was applied using a closed-loop system. An initial voltage was read from the Hall sensor with the cantilever in its equilibrium state. This initial voltage was used as a reference to determine the magnitude of deflection in the cantilever tip. By being able to determine the magnitude of deflection, a damping force could be applied by simply outputting a force of opposite magnitude. The range of force outputted by the actuator was determined by assuming a peak deflection of the cantilever, and then by scaling the output to the DAS-16 so that it would remain within its 0 to 5 volt range (0 to 4095 bits).

Data collection for the closed-loop system involved the use of a second computer. By using a second computer data collection could be obtained much more accurately without interfering with the operation of the control program.

Damping coefficients were calculated for both undamped and damped cases of each experiment and are listed in Table 1. The damping ratio was computed for each of the experiments, which ranged form 1.06 to 1.65.

REFERENCES

[1] M.P. Grover, Automation, Production Systems, and Computer Integrated Manufacturing. Prentice-Hall, Inc., 1987.

[2] R.C. Dorf, Robotics and Automated Manufacturing, Reston Publishing Company, 1983.

[3] W.E. Snyder, Industrial Robots: Computer Interfacing and Control, Prentice-Hall, Industrial Robot Series, 1985.

[4] J.J. Craig, Introduction to Robotics, Mechanics, and Control, Addison Wesley, 1989.

[5] C.M. Harris, Shock Vibration Handbook, McGraw Hill, 1988.

[6] M.L. Szary, "Design and Laboratory Applications of Single and Differential Mechanical Vibration Exciter", December 1989, ASME Conference Proceedings.

[7] W.L. Hallauer Jr., and D.A. Wagie, "An Inexpensive Experiment for Undergraduates on Active Control of Structural Vibrations", 1989 ASEE Conference Proceedings.

[8] W.J. Wang, S.S. Lu, and C.F. Hsu, "Experiments on the Position Control of a One-Link Flexible Robot Arm", IEEE Transactions on Robotics and Automation, Volume 5, pp 373-377, 1989.

[9] H. Kanoh and H.G. Lee, "Vibration Control of One-Link Flexible Arm", Proceedings 24th IEEE CDC, Ft. Lauderdale, Florida, December 1985, pp 1172-1177.

[10] W.J. Wang, S.S. Lu, and C.F. Hsu, "Output Feedback Control of a Flexible Robot Arm", Proceedings 25th IEEE CDC, Athens, Greece, December 6-9, 1986.

[11] W.J. Book and B. Siciliano, "A Singular Perturbation Approach to Control of Lightweight Flexible Manipulators", The International Journal of Robotics Research, Volume 7, No. 4, p 79.

[12] D.X. Liao and C.K. Sung, "The Design of Flexible Robotic Manipulators with Optimal Arm Geometries Fabricated from Composite Laminates with Optimal Material Properties", The International Journal of Robotics Research, Volume 6, No. 3, p 116.

[13] G. Naganathan and A.H. Soni, "Coupling Effects of Kinematics and Flexibility in Manipulators", The International Journal of Robotics Research, Volume 6, No. 1, p 75.

[14] R.H. Cannon and E. Schmitz, "Precise Control of Flexible Manipulators", First International Symposium on Robotics Research, MIT Press, 1984.

[15] J.A. Allocca and Allen Stuart, Transducers Theory and Applications, Reston Publishing Company, 1984.

[16] K.W. Jeong and H.S. Cho, "Development of a Pneumatic Vibratory Wrist for Robotic Assembly", Robotics, Volume 7, 1989, pp 9-16.

SPATIAL SENSITIVITY OF ACTIVE STRUCTURAL VIBRATION CONTROL: A CASE STUDY

Robert Rost,[1] John Schultze[1]

ABSTRACT

This paper covers preliminary tests of the Digisonix dx-47 for active vibration control. Tests evaluated the performance of the controller in structural control situation with a quasi-ideal beam with cantilever boundary conditions. These tests consisted of a sine dwell source/driver signal set to resonances of the beam (as determined from initial experimentation combined with closed form analytical solutions). Different controller driven actuator and error transducer configurations were evaluated for attenuation effectiveness. Experimentation revealed optimal control with the error and controller actuator at the same position (without regard to the pair's position on the beam). Also experimentation showed that the controller gain can be as high as 90 decibels for some cases which can cause significant problems with actuator and system. Observation of mode shapes in the controlled and open-loop systems revealed local and global attenuation of vibration.

BACKGROUND

Adaptive control of systems has been applied to problems ranging from noise attenuation, to performance optimization, to 'intelligent' systems and currently to space structures. Adaptive control, as examined here, is a control scheme where through monitoring system variables, a plant model is updated and excitation signals are generated to drive the system to a minimum response at a point. (Some adaptive systems do not update plant model but drive system strictly through changes in the output signal bank based on this 'hard-wired' model, these can have acceptable performance in some cases.)

INTRODUCTION

Preliminary investigation of the dx-47 controller, research prototype, was performed to evaluate use of the controller in the structural vibration area. The controller presently is successfully used in industrial sound attenuation (such as air-duct silencing). The particular objective of these preliminary tests is to evaluate the present controller system/set-up (controller with accompanying hardware and software) for this application

[1]University of Cincinnati/Structural Dynamic Research Lab., Mechanical, Industrial, and Nuclear Engineering Department, Cincinnati, OH 45220, U.S.A.

and to acquire knowledge of the spatial characteristics of the structural vibration problem. This extension of the controller to the structural area requires the ability to solve the global attenuation problem. Where as in the acoustic plane wave/duct silencing area it is sufficient to drive a point to the desired noise level and all points down stream will be attenuated. In the structural situation creating a node at one point does not guarantee global attenuation, in fact it may cause other areas to increase level of vibration. This is covered in advanced text on modal analysis. Basically, if a point is pinned, points nearby will also be attenuated but those farther away will not be directly effected.

Therefore, the test were divided into two parts. The first set investigated the sensitivity of the configuration to transducer and excitor locations. This also allowed for evaluation of the controller's ability to capture torsional modes.

The second set of tests evaluated the type (global/local) and level of vibration attenuation of the pre-controlled (open-loop) and controlled (closed-loop) configurations.

A cantilever beam was chosen for preliminary testing because of its relatively simple dynamic analysis and dominant 1-dimensional motion. The length (42") chosen was arbitrary to the point that there where several modes with in the controller's prescribed frequency range (<800hz).

EXPERIMENTAL PROCEDURE

Two batteries of test were conducted. The objective of the first set was to determine the modes of vibration for which a satisfactory level of attenuation could be attained. (This level was set a 6 dB attenuation at the sensor location).

The second group of tests monitor the change in deformed shape of the beam while excited at a tonal frequency in open loop and closed loop configurations. The open loop configuration had the controller driven exciter attached but un-powered, while the beam was driven into one of its modes by a single sine wave. The closed loop set-up consisted of the controller powered and in a steady state controlling mode. Those modes, both sets, that where unable to be controlled after several restarts where recorded and not considered further in analysis.

The explicit procedure for the first set was as follows. First amplifiers and signal analyzers were powered up and allowed to reach steady state. The gains on the excitor signal amps were set at a prescribed level (11 dB). The signal analyzer was configured for 2-channel Power Spectral Density over appropriate frequency range. Set and engage signal generator at sine frequency and magnitude desired (1-800 hz .1V). Place transducers at desired locations. Record pre-control signal level from controller output and error transducer at driver frequency. Engage controller and wait for controller to capture mode. Record controlled signals. Set next frequency on signal generator and repeat procedure until completed.

The procedure utilized on the second battery was as follows. Set-up of HP-375 and HP-35650 systems including installation and calibration of all transducers. Set wave generator at frequency of desired mode. Average and record 50 pre-control response mode shapes and FRF's. Engage controller and allow to reach steady state control. Average and record 25 controlled response FRF's and modes shapes. Continue until complete with data acquisition, resetting controller before each attempt at control. Following data acquisition, records where converted for use in an in-house modal analysis code (UC MODAL) for parameter estimation and mode shape extraction. Quantitative analysis was performed through use of the Modal Analysis Criteria between controlled and pre-control modes. Qualitative mode-shape inspection also assisted in determining the nature of error reduction in the system.

EXPERIMENTAL EQUIPMENT

Controllability Battery

- (1) Digisonix dx-47 controller
- (1) H-P 35660 2-channel signal analyzer
- (2) MB Dynamics SS250 Amplifiers
- (2) PCB 480D06 Amplifiers
- (2) Ling Dynamic System V203 2 lb electro-mechanical shakers
- (1) PCB 303A03 accelerometer
- (1) PCB 208A02 force transducer

Mode Observance Battery

- (1) Digisonix dx-47 controller
- (1) H-P 375 work station (with UC MODAL research software package)
- (1) H-P 35650 main frame system and accompanying modules
- (1) PCB 483A02 rack amplifier, 6 channel (used with force transducers)
- (2) PCB 433A rack amplifier, 15 channel (used with Structcels)
- (2) Ling Dynamic System V203 2 lb electro-mechanical shakers
- (30) PCB 330A Structcel accelerometer
- (2) PCB 208A02 force transducer

EXPERIMENTAL RESULTS

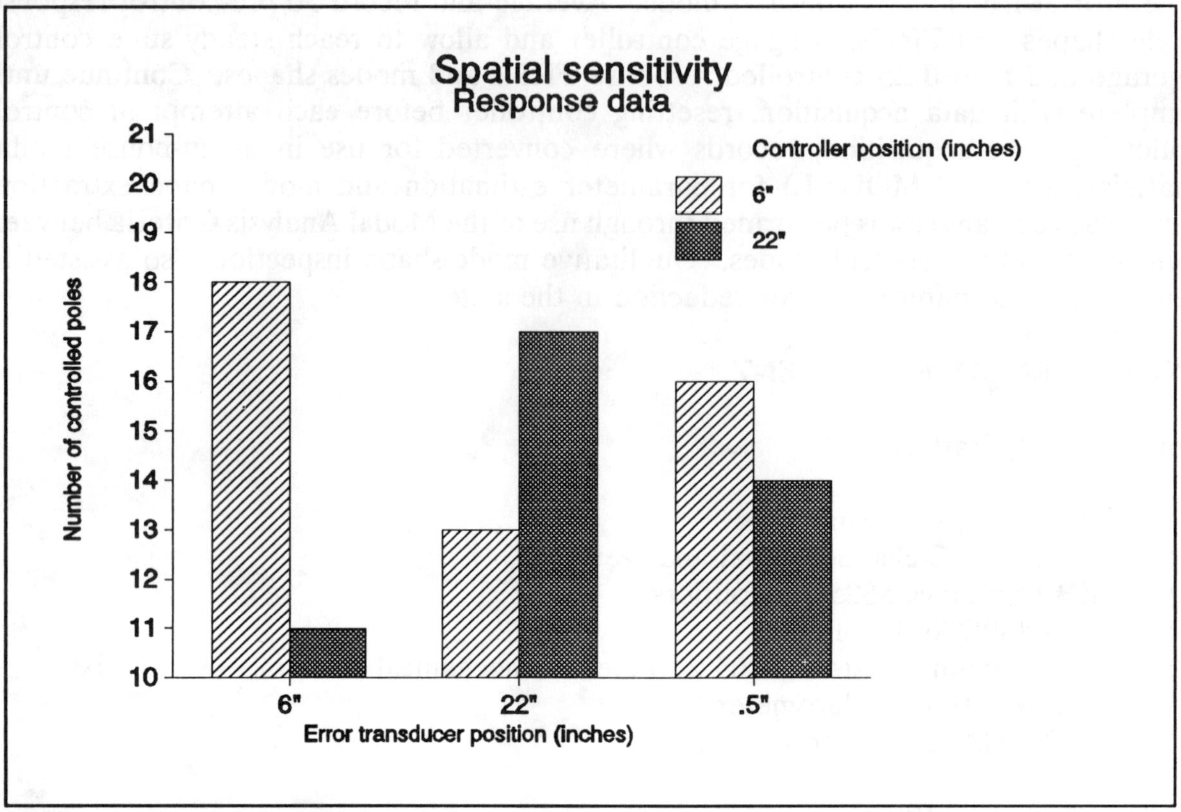

Figure 1. Spatial Sensitivity of Controller and Transducer Location

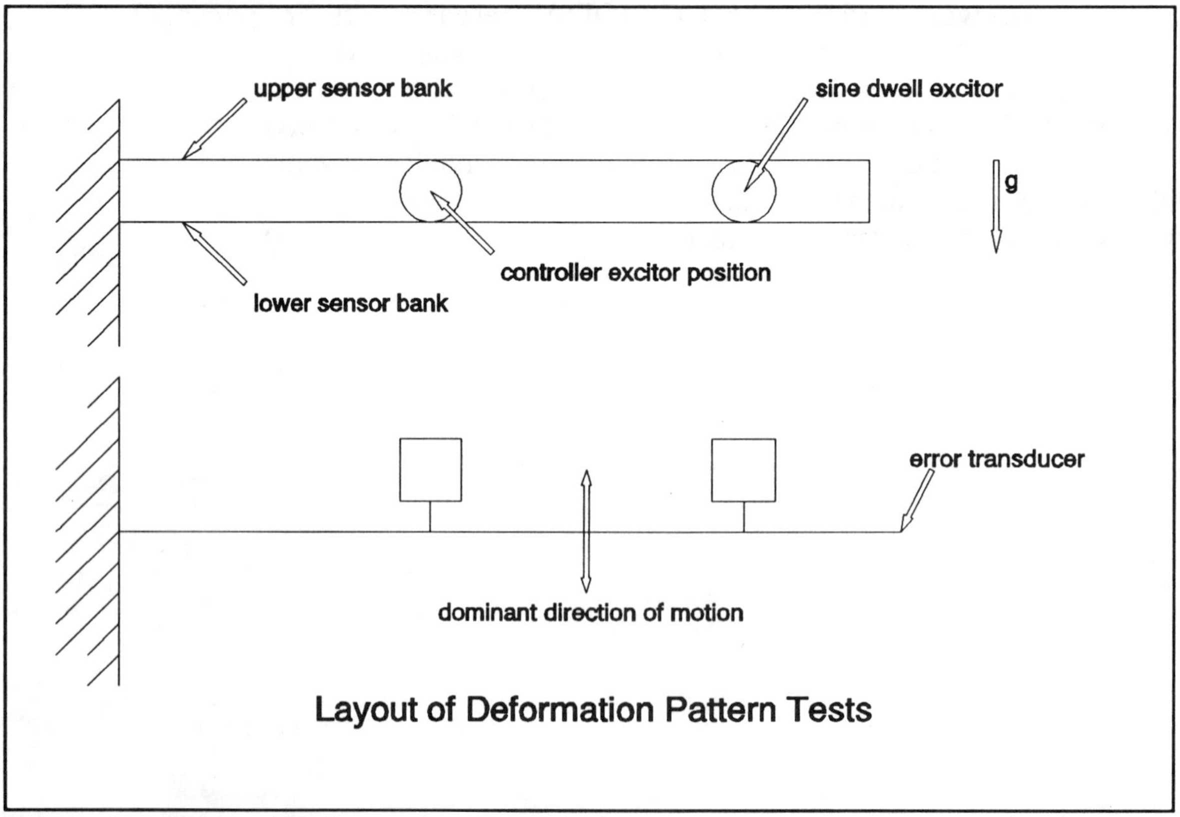

Figure 2. Layout of Deformation Pattern Tests

Table 1. Mode shape results

Freq Hz	Mode # open	Mode # closed	M.A.C.	Comments
124	5	22	.9727	global attenuation
180	6	23	.9889	" "
204	7	24	.0264	local attenuation
255	8	25	.9933	global attenuation
286	9	26	.4100	torsional modes, good local
294	10	27	.6478	local attenuation
329	11	28	.9613	local attenuation
349	12	29	.2107	very good local attenuation

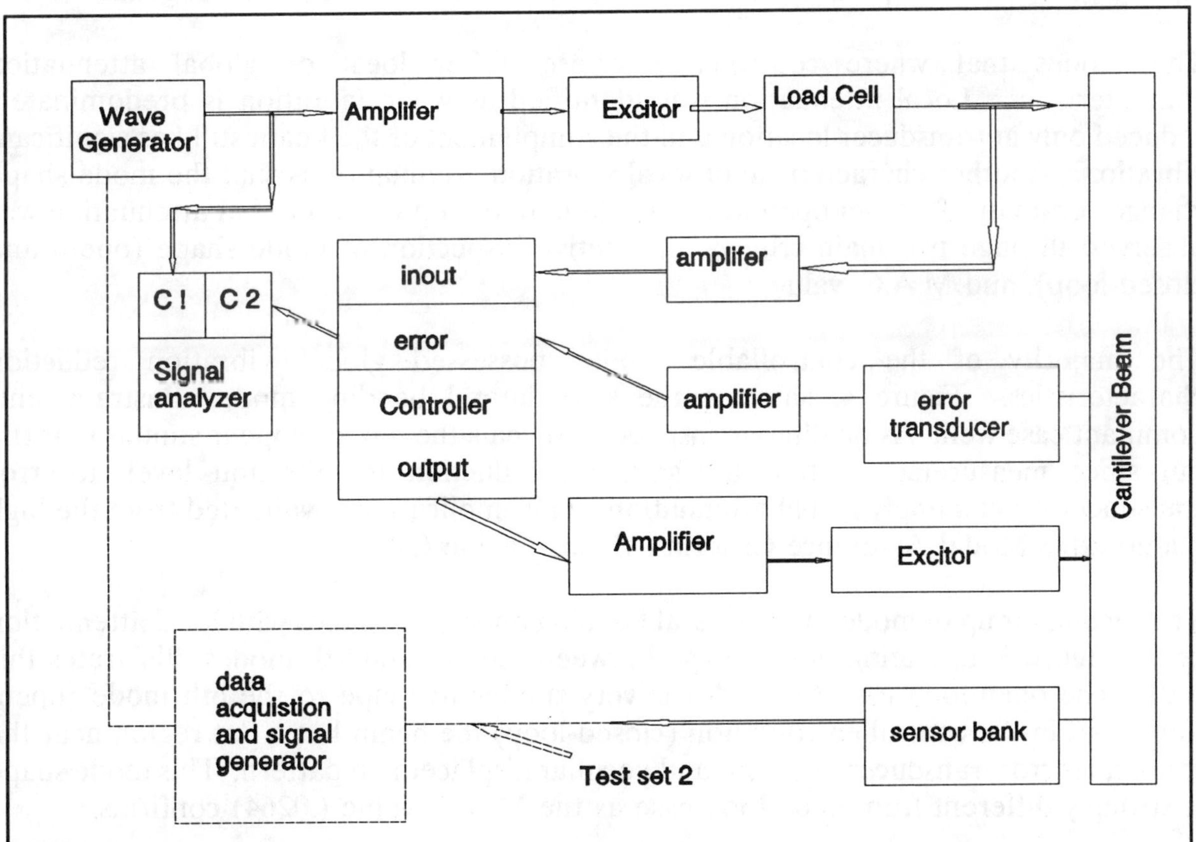

Figure 3. Schematic of Controller Configurations

DISCUSSION OF RESULTS

The most prominent characteristic of the first set of tests was that the ability to control modes of vibration was strongly linked to controller position and relative controller error signal location. When the error transducer was at the same location as the controller performance was optimum (see chart). Conversely, when the error transducer was located at the source of vibration, the performance was poor. The configurations performance showed marginal sensitivity to controller location. When the controller was at 6 inches from free-end (and vibration source at 22") performance was consistently increased by ten percent over the other configuration (controller at 22" and source at 6 ") regardless of error transducer location.

To interpret the second set of tests a basic explanation of modal analysis theory is required. When a structure is excited in to an operating mode shape, that shape is made up of contributions of the natural modes of vibration of the system. (The natural modes form a basis for the operating mode space.) Provided the boundary conditions remain relatively unchanged, a good estimate of these basis vector is available by exciting the structure with one excitor at a single resonant tone. Then by checking the orthogonality (normally weighted by the mass or stiffness matrix) of the operating vectors with the basis vectors relative contributions can be assessed. The orthogonality of these vectors can be estimated by the Modal Assurance Criteria (M.A.C.), a pseudo-dot product. A value of one indicates parallel vectors and a value of zero indicates orthogonal vectors.

The modes that where controllable where either local or global attenuation characteristics. Local attenuation was identified as when vibration is predominately reduced only at transducer location and the complement of the beam still has significant vibration. Another characteristic of local vibration attenuation is that the mode shape changes dramatically from open-loop case to closed-loop case. Global attenuation was observed through two main criteria, qualitative inspection of mode shape (open- and closed-loop), and M.A.C. value.

The majority of the controllable modes possessed global vibration reduction characteristics. Figure 4, showing the sixth lateral bending mode, illustrates this dominant case well. As qualitative inspection reveals the modes appear similar in form, but since measurements show a significant reduction in vibration level at error transducer we can imply global attenuation. This implication is validated from the high quantitative Modal Assurance Criteria (M.A.C.) value (.9889).

The second group of modes were lateral bending operating shapes with local attenuation only. Figure 5, operating mode shape between the 6th and 7th modes, illustrates this well. The open-loop case (left side) is very similar in shape to the 6th mode (open-loop), but in the controlled condition (closed-loop) the beam has a flat region near the free-end (error transducer location) and peculiar displacement pattern. This mode shape is strongly different from open loop case as the M.A.C. value (.0264) confirms.

The third main group of modes observed where torsional modes. (Note the positions for both the source and the control driven shakers was a poor location, beam centerline, for torsional excitation or suppression). In general, torsional modes had either local or intermittent controllability. A case with local attenuation can be seen in Figure 6, 3rd torsional mode shape. The mode shape again distorts greatly in the controlled case and has a relatively low M.A.C. value (.4100) indicating the presence of other signal content from the controller as it attempts to drive the local value acceleration to zero.

CONCLUSIONS

The adaptive acoustic controller (dx-47) performed very well in capturing (attenuating vibration) at the error transducer location. It was evident also that there is strong dependence on relative position controller to error transducer and weak dependence on absolute controller position. That is, if the controller and error are physically close, with out regard to position on beam, performance is optimum. The controller compensates for high impedance input location through increased gain.

While the error transducer was driven to acceptable levels in the preliminary tests, secondary tests showed attenuation to be local in nature for some cases and intermittent for others. For the cases of local attenuation the operating deformation shape was also changed dramatically, (low Modal Assurance Criteria values).

In summary the adaptive controller seems well suited to the structural vibration area, and future work will investigate how the spatial sensitivity could be reduced through a multiple input/multiple output algorithm (MIMO) which retains phase and relative magnitude information.

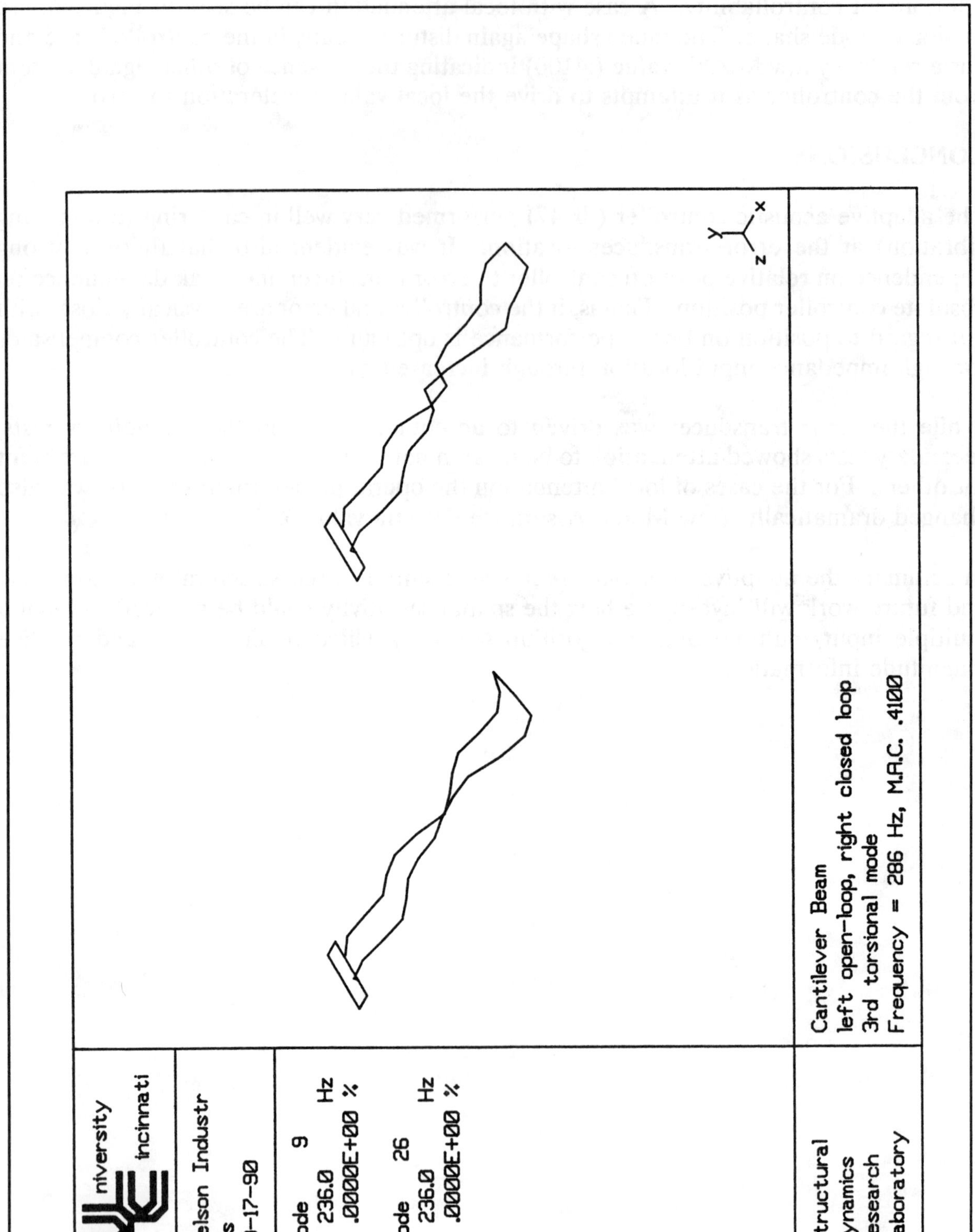

Figure 6. 3rd Torsional mode, Global attenuation

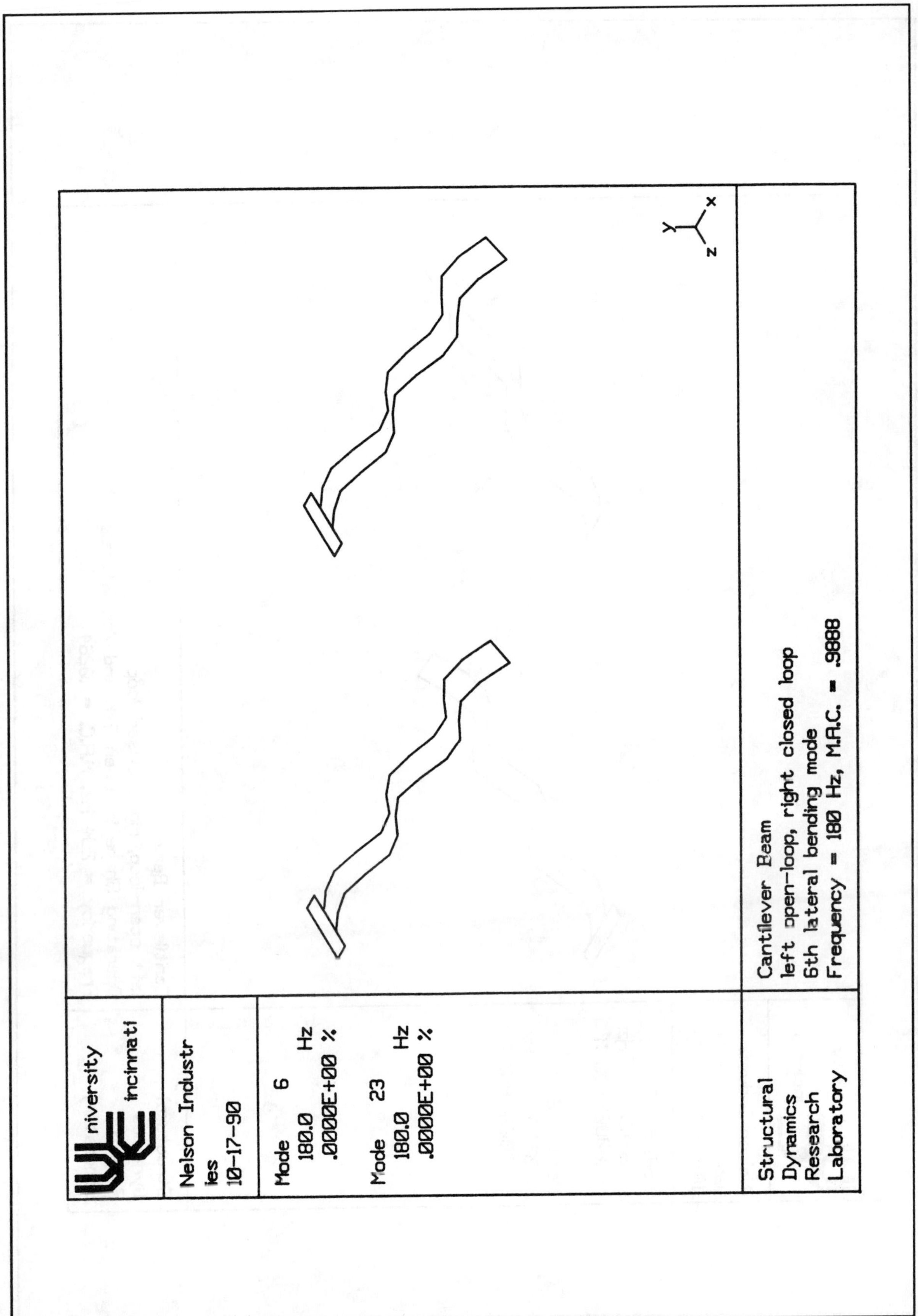

Figure 4. 6th Lateral Bending Mode, Global Attenuation

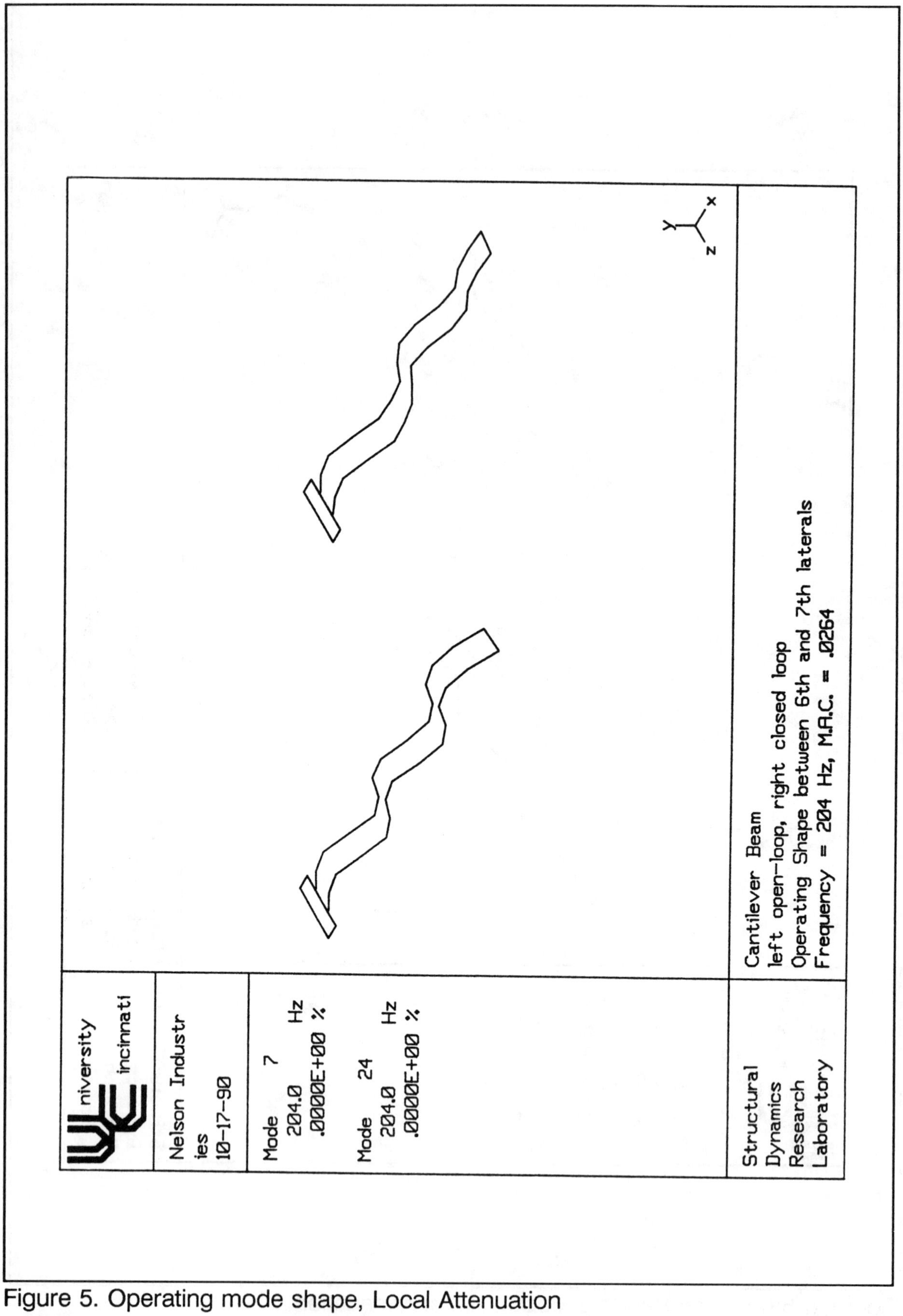

Figure 5. Operating mode shape, Local Attenuation

MODAL DOMAIN OPTICAL FIBER SENSORS FOR CONTROL OF ACOUSTIC RADIATION

Douglas K. Lindner,[1] William T. Baumann,[1] Fu-Sheng Ho,[1] Edward Bielecki[1]

ABSTRACT

Recently, it has been shown that modal domain optical fiber sensors can be used to sense vibrations in a flexible beam and that this signal can be used in a control system to damp vibrations in the beam. This sensor is novel in that it responds to strain in the beam over a significant gauge length, a so-called distributed-effect sensor. It has been shown that distributed-effect sensors can be configured to sense directly the desired structural measurable, such as modal amplitudes, according to a pre-specified weighting function; a capability unique to distributed-effect sensors. In this paper we will investigate the use of a weighted modal domain sensor in a control system to suppress the acoustic radiation from a vibrating beam.

We consider two different control algorithms. The first control system is based on modern linear control theory and yields a linear time invariant compensator. The design procedure, however, calls for incorporating a postfilter with the plant model to tailor the cost function to the requirement of suppressing the acoustic radiation. Here we will show that the complexity of this compensator can be significantly reduced by using a weighted distributed-effect sensor which can implement partial state feedback directly.

Secondly, we will consider an adaptive LMS control. We use the weighted modal domain sensor to provide modal amplitudes directly. The ability of this adaptive scheme to reduce the total radiated power using this novel sensor is compared to several conventional sensors and to the controller based on modern control theory.

1. INTRODUCTION

The suppression of vibration in flexible structures using active feedback control has received a great deal of attention in the last decade. More recently, these ideas have been applied to the suppression of acoustic radiation from elastic structures with success [1-4]. Current research has focused on the instrumentation required to implement these control laws, particularly, through the use of embedded sensors and/or actuators - so-called Smart Structures [5]. In this paper we concentrate on the impact this new generation of sensors will have on the design and implementation of control algorithms for the suppression of acoustic radiation for flexible structures.

One class of novel sensors which has shown potential for Smart Structure applications is optical fiber sensors. Recently, it has been shown [5] that modal domain optical fiber sensors can be used to sense vibrations in a flexible beam and that this signal can be used in a control system to damp vibrations in the beam. This sensor is revolutionary in that it

[1]Fiber & Electro-Optics Research Center, Bradley Department of Electrical Engineering, Virginia Tech, Blacksburg, VA 24061, U.S.A.

responds to strain in the beam over a significant gauge length, a so-called distributed-effect sensor. It has been shown [6] that distributed-effect sensors can be configured to sense directly the desired structural measurable, such as modal amplitudes, according to a pre-specified weighting function; a capability unique to distributed-effect sensors. Current research efforts are directed at extending this capability to optical fiber sensors. In this paper we will assume the existence of a weighted modal domain sensor and investigate the use of this sensor in the following control systems.

Here we will consider two different control algorithms. The first control system is based on modern linear control theory and yields a linear time invariant compensator [4]. The design procedure, however, calls for incorporating a postfilter with the plant model to tailor the cost function to the requirement of suppressing the acoustic radiation. This procedure typically leads to a compensator of rather high order. Here we show that the complexity of this compensator can be significantly reduced by using a weighted distributed-effect sensor. The basic idea is that the weighted sensor can implement a partial state feedback directly, thus reducing the order of the compensator. These results extend the preliminary results in this direction in [7].

Secondly, we will consider an adaptive LMS feedforward control [1]. Using the formulation of total radiated power in [4], we show how the sensor placement is related to the ability of the algorithm to suppress the radiated power. Then we show that using weighted modal domain sensor to provide modal amplitudes directly is the optimal sensor choice.

The results in this paper are illustrated using using a simulated clamped-clamped beam instrumented with a force actuator. The beam is driven with an external sinusoidal disturbance source. The beam is also instrumented with a strain gauge and a weighted and uniform modal domain optical fiber sensor. The ability of the two control algorithms to reduce the total radiated acoustic power will be compared for each of the sensors.

In Section 2 we describe the flexible beam model, the optical fiber sensors, and derive the cost function for the total radiated acoustic power. The LMS algorithm is studied in Section 3. Section 4 shows how distributed-effect sensors can reduce the complexity of standard linear time invariant controllers. Section 5 has the conclusions.

2. BACKGROUND

2.1 FLEXIBLE BEAM

We consider a uniform clamped-clamped beam. The parameters of this beam are shown in Table 2.1.1. This beam is modeled with an Euler-Bernoulli model

$$\begin{gathered} EI\frac{\partial^4 w(z,t)}{\partial z^4} + \rho_b\frac{\partial^2 w(z,t)}{\partial t^2} = \delta(z-.55)u(t) + \delta(z-0.36)d(t), \\ w(0,t) = w(L,t) = 0; \quad \frac{\partial w(0,t)}{\partial z} = \frac{\partial w(L,t)}{\partial z} = 0, \end{gathered} \qquad (2.1.1)$$

where $t \geq 0$ is the time variable, z, $0 \leq z \leq L$, is the space variable, w(z,t) is the displacement of the beam, u(t) is the control input force, and d(t) is the disturbance input force. Note that the actuator and the disturbance input are both assumed to be pure point force inputs. The nominal disturbance is assumed to be a sinusoid at 95.4 rad/sec.

The displacement is described in terms of the mode shapes, $\psi_n(z)$, and the modal amplitudes, $\eta_n(t)$, as

$$w(z,t) = \sum_{n=1}^{\infty} \psi_n(z)\, \eta_n(t). \tag{2.1.2}$$

For this simple structure the mode shapes are known [4]. For purposes of computation the infinite series in (2.1.2) is truncated to N modes. This approximation leads to a reduced order model (in state space)

$$\dot{x}_s = A_s x_s + B_s u(t) + L_s d(t). \tag{2.1.3}$$

The finite dimensional model also contained 1% damping in each mode.

In this paper we will consider this beam instrumented with a modal domain optical fiber sensor (discussed in detail in Section 2.3 below) and a strain gauge located at z=0.75.

2.2 ACOUSTIC RADIATED POWER

The total radiated power from the beam can be determined as follows [3,4]. The transfer function between the pressure, $p(t;R,\theta,\varphi)$, at the point (R,θ,φ) in the far field due to the velocity of the nth mode is given by

$$\frac{\widehat{p_n}(\omega;R,\theta,\phi)}{\widehat{\dot{\eta}_n}(\omega)} = \frac{\widehat{h_n}(\omega;\theta,\phi)e^{jkR}}{R} \tag{2.2.1}$$

where the carat denotes the Fourier Transform of the variable. The exponential factor accounts for the propagation delay and the R in the denominator accounts for the spherical spreading loss. This transfer function can be computed from the geometry and dynamics of the structure. The total far field pressure at a point is the sum of the contributions of each mode. Defining

$$H(\omega;\theta,\phi) = \left[\widehat{h_1}\ \widehat{h_2} \cdots \widehat{h_N}\right];\quad V(\omega) = \begin{bmatrix} \widehat{\dot{\eta}_1}(\omega) \\ \widehat{\dot{\eta}_2}(\omega) \\ \vdots \\ \widehat{\dot{\eta}_N}(\omega) \end{bmatrix}, \tag{2.2.2}$$

the total far field pressure at a point can be written as

$$\widehat{p}(\omega;R,\theta,\phi) = H(\omega;\theta,\phi)V(\omega)\frac{e^{jkR}}{R}. \tag{2.2.3}$$

The average power radiated in the direction (θ,ϕ) can be written in terms of the real pressure as

$$\lim_{T\to\infty} \frac{1}{\rho c T}\int_0^T p(t;R,\theta,\phi)^2\,dt \tag{2.2.4}$$

where ρc is the characteristic impedance. By considering the power spectral density of p and integrating over the a sphere of sufficiently large radius we find that the total radiated power, Π, in terms of the modal velocities is given by

$$\Pi = \frac{1}{2\pi}\int_{-\infty}^{\infty} \mathrm{tr}\{S_V(\omega)M(\omega)\}d\omega \tag{2.2.5}$$

where $S_V(\omega)$ is the power spectral density of $V(\omega)$ and

$$M(\omega) = \frac{1}{\rho c}\int_0^{2\pi}\int_0^{\pi} H^{T*}(\omega;\theta,\phi,)H(\omega;\theta,\phi,)\sin\theta\ d\theta\ d\phi\ . \tag{2.2.6}$$

We shall use the total radiated power to compare all of the control laws proposed below.

The matrix $M(\omega)$ can be used to identify what modes contribute to the total radiated power at a given frequency. Assume that the disturbance signal is such that the velocity components of the modes are given by the wide sense stationary signal

$$V(t) = V_0 e^{-j(\omega_0 t+\theta_s)} \tag{2.2.7}$$

where θ_s is uniformly distributed between 0 and 2π. The power spectral density of this process is given by

$$S_V(\omega) = 2\pi V_0 V_0^T \delta(\omega_0)\ . \tag{2.2.8}$$

Substituting (2.2.8) into (2.2.5) we find the total radiated power is

$$\Pi = V_0^T M(\omega_0) V_0\ . \tag{2.2.9}$$

From this quadratic form we see that the minimum and maximum radiated power per unit excitation is given by the singular values. The corresponding singular vectors identify the relative contributions of each of the modes. For the beam we consider here, excited at ω_0 = 95.4 rad/sec, we have

$$M(95.4) = \begin{bmatrix} 5.4906 & 0 & 2.4 \\ 0 & .00736 & 0 \\ 2.4 & 0 & 1.0491 \end{bmatrix} \times 10^{-2}\ . \tag{2.2.10}$$

This matrix has singular values

$$\Lambda = \begin{bmatrix} 0.065396 & 7.3\times 10^{-5} & 2.6998\times 10^{-8} \end{bmatrix} \tag{2.2.11}$$

corresponding to the singular vectors

$$\begin{bmatrix} .916 \\ 0 \\ .401 \end{bmatrix}; \begin{bmatrix} 0 \\ 1 \\ 0 \end{bmatrix}; \begin{bmatrix} .401 \\ 0 \\ -.916 \end{bmatrix}. \tag{2.2.12}$$

Hence, to minimize the radiated power for this beam when it is excited at this frequency, the control law should attempt to drive the structure such that the second mode is not excited and the first and third modes have a relative weighting of .401 and -.916; i.e. they are out of phase.

2.3 MODAL DOMAIN OPTICAL FIBER SENSORS

Recently, modal domain optical fiber sensors were used in a feedback control loop to damp vibrations in a flexible beam [5]. A detailed description of these sensors can be found there and in the references therein. Briefly, a modal domain optical fiber sensor consists of a section of optical fiber attached to the beam from $z=f^-$ to $z=f^+$, a coherent light source, and a photodetector. The geometry of the fiber and the wavelength of the source are chosen such that two electromagnetic modes, the LP_{00} and the LP_{01}, propagate in the fiber. The electric fields of these modes depend on the fiber geometry and the indices of refraction. These two modes interfere with each other to generate an intensity pattern at each cross section which varies as a function of the length of the fiber.

The photodetector monitors a point on the endface of the fiber and produces a current that is proportional to the intensity of the light at that point. It can be shown that the output of the photodetector has the form

$$y_{PD} = K_e\left[\bar{I}_0 + \bar{I}_1 \cos\Gamma\right] \tag{2.3.1}$$

where K_e is the gain of the electronics of the photodetector and all other quantities depend on the waveguide.

Since the fiber is attached to the beam, the strain at the surface of the beam is transferred to the fiber. When the fiber is strained the intensity of the light at the fiber endface changes in a predictable way. Recently, Sirkis and Haslach [8] gave a detailed description of this interaction. We summarize their results here. The direct effect of axial strain in the optical fiber is threefold. First, the length of the fiber will change. Secondly, the geometry of the fiber will change. Finally, the indices of refraction will change; the so-called photoelastic effect. As a result, all of the quantities in the brackets in (2.3.1) are known but very complicated functions of strain. It can be shown that the first order effects of strain are on the "phase" of the intensity, Γ in (2.3.1). Let $\varepsilon(z,t)$ be the axial strain in the fiber. Then the phase can be expanded in a Taylor series expansion around the total strain in the fiber

$$\Gamma(\varepsilon(z,t)) = \Gamma_0 + \Delta\tilde{\beta}\int_{f^-}^{f^+} \varepsilon(z,t)dz + \text{higher order terms.} \tag{2.3.2}$$

Drop the h.o.t. from (2.3.2) and substitute (2.3.2) into (2.3.1). Choose Γ_0 so that the cosine can be rewritten as a sine (This is called setting the Q-point of the sensor.) and filter the output of the photodetector to remove the dc bias. Finally, the output of the sensor is given by

$$y_{MD}(t) = K_{MD}\sin\left[\widetilde{\Delta\beta}\int_{f^-}^{f^+} \varepsilon(z,t)dz\right]. \tag{2.3.3}$$

Usually, the sensor is configured such that the sine in (2.3.3) can be replaced by its linear approximation. Making this assumption and assuming ideal bonding to the beam, the output of the modal domain sensor attached to the beam is given by

$$\begin{aligned} y_{MD}(t) &= K\int_{f^-}^{f^+} \frac{t_b}{2}\frac{\partial^2 w(z,t)}{\partial z^2}dz \\ &= \sum_{n=1}^{N}\left\{\int_{f^-}^{f^+} \frac{t_b}{2}\frac{\partial^2 \psi_n(z)}{\partial z^2}dz\right\}\eta_n(t) \\ &= \sum_{n=1}^{N} c_n\eta_n(t) = C_{MD}\eta(t). \end{aligned} \tag{2.3.4}$$

The fiber was assumed to be attached to the beam from $f^-=0.1$ to $f^+=0.7$. The asymmetrical configuration was chosen to reflect realistic structure placement.

In the model of the sensor above it was assumed that the local transfer of strain from the structure to the fiber was uniform along the beam. Suppose, however, that the transfer coefficient could be varied as a function of the length of the fiber according to the pre-specified function, say, W(z). Then the linear output of the fiber would be

$$y_{MDW}(t) = K\int_{f^-}^{f^+} W(z)\,\varepsilon(z,t)dz\,. \tag{2.3.5}$$

Chaing and Lee [6] have demonstrated a similar capacity for piezoelectric film.

The weighting function, W(z), now looks like a distributed filter gain from the system point of view. Suppose that we choose the weighting function as an arbitrary sum of mode shapes (2.1.3)

$$W(z) = \sum_{n=N_1}^{N_2} cw_n\psi_n''(z). \tag{2.3.6}$$

Substituting (2.3.6) into (2.3.4) we find that

$$\begin{aligned} y_{MDW}(t) &= \sum_{n=N_1}^{N_2} cw_n\eta_n(t), \\ &= C_{MDW}\eta(t)\,. \end{aligned} \tag{2.3.7}$$

In other words, by choosing W(z) as in (2.3.6), we can arbitrarily select the output matrix for the sensor model. Another interpretation of this result is to view (2.3.7) as N scalar sensors which we can combine in an arbitrary fashion. Thus, a weighted sensor acts as: 1) many scalar sensors, and 2) a signal processor. Below we speculate on the implications of this additional flexibility on the control system. Other ideas were suggested in [7].

3. ADAPTIVE CONTROL

3.1 THE LMS ALGORITHM

In this section we describe the implementation of the well known [1] LMS algorithm for the suppression of acoustic radiation using modal domain optical fiber sensors. Let

$$H_*(s) = C_*(sI - A_S)^{-1}B_s = \frac{Y_*(s)}{U(s)} \tag{3.1.1}$$

be the transfer function from the control input to the sensor output. Then the LMS controller generates the control signal u(t) using as inputs a measurement of the disturbance signal and the sensor output $y_*(t)$ in such a way as to drive the sensor output to zero. This controller is shown schematically in Figure 3.1.1.

The total radiated power can be related to the LMS algorithm. Assume that the disturbance is a pure sinusoid. Then the steady state amplitude of the modal velocities will be given by (2.2.7). Similarly, the steady state amplitudes of the modal displacements will be given by

$$P(t) = V_0\omega_0 e^{-j(\omega_0 t+\theta_s + \frac{\pi}{2})}. \tag{3.1.2}$$

Furthermore, we know that the output of the sensor is zero so that V_0 must satisfy

$$y_*(t) = C_*V_0 = 0. \tag{3.1.3}$$

It follows that the LMS algorithm will drive the vector of modal amplitudes into the nullspace of C_*. The worst case is determined by projecting the singular vectors of $M(\omega)$ onto the nullspace of C_*.

From the above analysis it also follows that the sensor placement can effect the performance of the algorithm. In particular, we should chose the sensor placement such that the singular vector, s_1, of $M(\omega)$ which corresponds to the largest singular value is orthogonal to the nullspace of C_*. It follows that

$$C_* = s_1^T. \tag{3.1.4}$$

With this selection of output matrix, the worst case radiated power will be given by the second largest singular value of $M(\omega)$.

3.2 WEIGHTED SENSORS

In general, (3.1.4) can't be satisfied with a point sensor. Using a distributed-effect sensor which admits a weighting, however, does give the flexibility to realize (3.1.4). Indeed, if s_1 contains zeros; that is, certain modes don't contribute to the far field power and should not be measured, then this modal filtering can also be achieved without the use of temporal filtering.

For the beam described above, we used a three mode model. For the same disturbance signal, we used three different sensor outputs, the strain gauge, (2.1.5), the uniform modal domain optical fiber sensor, (2.3.4), and a weighted modal domain optical fiber sensor, (2.3.7) where the weights for (2.3.6) were

$$\begin{bmatrix} c_1 \\ c_2 \\ c_3 \end{bmatrix} = \begin{bmatrix} .916 \\ 0 \\ .401 \end{bmatrix} . \tag{3.2.1}$$

See (2.2.12). The results are shown in Table 3.2.1 which shows that the LMS algorithm with the weighted modal domain sensor achieves the minimum radiated power because for this example there are only two modes which radiate.

The results in Table 3.2.1 illustrate that poor sensor placement can lead to an increase in the radiation power. On the other hand, weighted distributed-effect sensors allow for an optimal measurement.

4. CONTROL OF BROADBAND DISTURBANCES

4.1 RADIATION FILTERS

The expression for total radiated power (2.2.6) is expressed in terms of the modal velocities. Next we show that this expression can be rewritten in terms of auxiliary signals which can be incorporated directly into the control design process [4]. To that end, factor M(ω) as

$$M(\omega) = G^{T*}(\omega)G(\omega) . \tag{4.1.1}$$

Now the total radiated power (2.2.4,5) can be written as

$$\begin{aligned} P &= \frac{1}{2\pi}\int_{-\infty}^{\infty} \mathrm{tr}\{G(\omega)S_V(\omega)G^{T*}(\omega)\}d\omega \\ &= \lim_{T\to\infty} \frac{1}{T}\int_0^T z_f^T(t)z_f(t)\,dt \end{aligned} \tag{4.1.2}$$

where

$$\frac{z_f(s)}{V(s)} = G(s) \tag{4.1.3}$$

is a "radiation filter" (linear system). By including (4.1.3) with the model we can pose a standard optimization problem to minimize the cost function (4.1.2).

The beam in Section 2.1 is modeled in state space form (2.1.4) where the output matrix depends on the sensors used in the feedback loop. The disturbance input d(t) is modeled as a linear filter with a white noise input

$$\begin{aligned} \dot{s} &= A_D s + B_D v(t), \\ d(t) &= C_D s. \end{aligned} \tag{4.1.4}$$

Similarly, the radiation filter is modeled as

$$\begin{aligned} \dot{r} &= A_F r + B_F x_s(t), \\ z(t) &= C_F r + D_F x_s(t). \end{aligned} \tag{4.1.5}$$

Combining (2.1.4) and (4.1.4-5) the overall system can be written as

$$\begin{aligned} \dot{x} &= \begin{bmatrix} A_s & 0 & L_s C_D \\ B_F & A_F & 0 \\ 0 & 0 & A_D \end{bmatrix} x + \begin{bmatrix} B_s \\ 0 \\ 0 \end{bmatrix} u + \begin{bmatrix} 0 \\ 0 \\ B_D \end{bmatrix} v \\ &= Ax + Bu + Lv, \\ y &= [C_* \;\; 0 \;\; 0\,]x = Cx \\ z &= [D_F \; C_F \;\; 0\,]x\,. \end{aligned} \tag{4.1.6}$$

The controller that minimizes a cost function combining total radiated acoustic power and control effort

$$J = \lim_{T \to \infty} \int_0^T z_f^T(t) z_f(t) + \mu u^2(t)\, dt \tag{4.1.7}$$

is given by

$$\dot{\hat{x}} = A\hat{x} + Bu + K_o(y - C\hat{x}), \tag{4.1.8a}$$

$$u = -K_c \hat{x}. \tag{4.1.8b}$$

The feedback gain matrix K_c is chosen by minimizing the cost function (4.1.2). The observer gain matrix K_o is chosen based on the sensor noise characteristics or other performance criteria.

Note in particular that the order of the dynamic compensator in (4.1.8) is the same as the augmented plant. For realistic problems, it may not be possible to implement (4.1.8) because of limitations on the digital hardware. Next we propose an alternative to direct implementation of (4.1.8) using weighted optical fiber sensors.

4.2 FUNCTIONAL OBSERVERS

The dynamic compensator in (4.1.8a) is known as an observer because it reconstructs the state, x(t), of the plant (4.1.7). It is well known that the functional in (4.1.8b) can be reconstructed with a functional observer

$$\begin{aligned} \dot{x}_f &= F x_f + Hu + Gy, \\ u &= M x_f + Py. \end{aligned} \tag{4.2.1}$$

The order of the functional observer is the observability index of the system minus one.

For structure models (2.1.4) of large order with a small number of conventional sensors, the functional observer may not provide a dramatic reduction in order. If weighted distributed-effect sensors, such as weighted modal domain sensors, are available, we can assume that any combination of modal displacements can be measured. This assumption is equivalent to assuming that the output of the system is given by

$$y = [\,I \;\; 0\,]x_s = \eta\,. \tag{4.2.2}$$

For a modal system (2.1.4) with no augmentation for disturbances or radiation filtering, this output gives an observability index of 2 [7]. Hence, a first order functional observer of the form (4.2.1) can be used in the control loop to implement any desired state for any order model. The terms Gy and Py in (4.2.1) are simply weighted sums of modal amplitudes that can be provided by weighted distributed-effect sensors. For this case, each term can be realized by a single sensor and so only two sensors and a first order dynamic compensator are all that are needed for the implementation.

For the suppression of a single tone (pure sinusoid) disturbance, the system is augmented with the disturbance model and is of the form

$$\begin{bmatrix} \dot{x}_s \\ \dot{r} \end{bmatrix} = \begin{bmatrix} A_s & L_s C_D \\ 0 & A_D \end{bmatrix} \begin{bmatrix} x_s \\ r \end{bmatrix} + \begin{bmatrix} B_s \\ 0 \end{bmatrix} u \,. \qquad (4.2.3)$$

Under the assumption that all modal amplitudes are measurable, the observability index of the system is four. Thus, a functional observer of order three, requiring four weighted sensors, can be used to implement any state feedback control law, regardless of the order of the structural model. To minimize the acoustic radiation, the feedback would be chosen to optimally suppress those components of the state that radiate efficiently at the frequency of the disturbance, as discussed in Section 4.1. Note that in contrast to the LMS controller, no measurement of the disturbance is required and objectives other than disturbance rejection, such as increased modal damping, can be included in the feedback design. For the beam example above, a controller was designed to minimize the cost function

$$J = \int_0^\infty x_s^T C_*^T C_* x_s + 4 \times 10^{-5} u^2 \, dt \,. \qquad (4.2.4)$$

where C_* is given by (3.1.4). The total radiated power is shown in Table 3.2.1. This power could have been driven lower by decreasing the penalty on the control.

When a broadband disturbance is present, it is no longer possible to design the controllers as discussed above because the radiating components vary with frequency. It is necessary to introduce the radiation filters in (4.1.5-6). The states of these filters are unobservable from the sensors on the structure and, therefore, are not reconstructable by the compensator. The optimal feedback will be of the form

$$u(t) = K_{x_s} x_s + K_s s + K_r r \,. \qquad (4.2.5)$$

The compensator for this system would consist of a third order compensator for the first two terms in (4.2.4) as discussed above. The third term would be reconstructed by a compensator of the same order as the radiation filters. The performance of this type of controller is under investigation.

5. CONCLUSIONS

In this paper we have discussed the use of weighted modal domain optical fiber (distributed-effect) sensors for active suppression of acoustic radiation from flexible structures. It was shown that conventional sensors could lead to an increase of radiated power when the LMS algorithm was implemented to suppress a single disturbance source. A weighted distributed-effect sensor, however, was shown to give good results.

Secondly, it was shown that weighted distributed-effect sensors can be used to implement high order compensators with low order functional observers. The design and performance of these compensators for both narrow and wide band noise is under investigation.

6. ACKNOWLEDGMENT

This work was supported in part under DARPA/ONR Contract N00014-88-K-0721 and in part under NASA Grant NAG-1-1006.

7. REFERENCES

1. Elloitt, S. J., I. M. Stothers, and P. A. Nelson, 1987."A Multiple Error LMS Algorithm and Its Application to the Active Control of Sound and Vibration." IEEE Trans. on Acoustics, Speech, and Signal Processing, ASSP-35:1423-1434.

2. Fuller, C. R., and J. D. Jones, 1987. "Experiments on The Reduction of Propeller Induced Interior Noise By Active Control of Cylinder Vibration." Journal of Sound and Vibration, 112(2):389-395.

3. Baumann, W.T., W. R. Saunders, and Harry H. Robertshaw, 1990. "Active Suppression of Acoustic Radiation from Impulsively Excited Structures." 119th Meeting of the Acoustical Society of America.

4. Baumann, W. T., F. Ho, and H. H. Robertshaw, 1990. "Active Acoustic Control of Broadband Structural Disturbances." 120th Meeting of the Acoustical Society of America.

5. Cox, D. E. and D. K. Lindner, 1990. "Active Control For Vibration Suppression in a Flexible Beam Using a Modal Domain Optical Fiber Sensor." accepted for the ASME Journal of Vibration and Acoustics.

6. Chaing, Wie-Wen, and Chih-Kung Lee, 1989. "Critical Damping Control of a Flexible Slender Plate Using a Distributed Modal Actuator and Sensor." Proceedings of the 1989 American Control Conference, Pittsburgh, PA.

7. Lindner, D. K., K. M. Reichard, W. T. Baumann, and M. F. Barsky, 1990. "Measurement and Control of Flexible Structures Using Distributed Sensors." Proceedings of the 29th IEEE Conf. on Decision and Control, Honolulu, HI, pp. 2588-92.

8. Sirkis, J. S. and H. W. Haslach, Jr., 1990. "Complete Phase-Strain Model for Structurally Embedded Interferometric Optical Fiber Sensors." accepted for publication in the Journal of Intelligent Material Systems and Structures.

TABLE 2.1.1
BEAM PARAMETERS

Young's Modulus	E	$2.04 * 10^{11}$ Pa
Mass Density	r_b	0.491 kg/m
Length	L	1.0 m
Width	h_b	0.125 m
Thickness	t_b	0.5 mm

TABLE 3.2.1 COMPARISON OF CONTROLLERS	
The LMS Algorithm	
Sensor Type	Total Radiated Power
Open Loop	1.01×10^{-5}
Strain Gauge	1.6×10^{-5}
Modal Domain Sensor	8.5×10^{-5}
Weighted MD Sensor	2.301×10^{-8}
Modern Control	
Disturbance Type	Total Radiated Power
Pure Sinusoid	6.75×10^{-8}

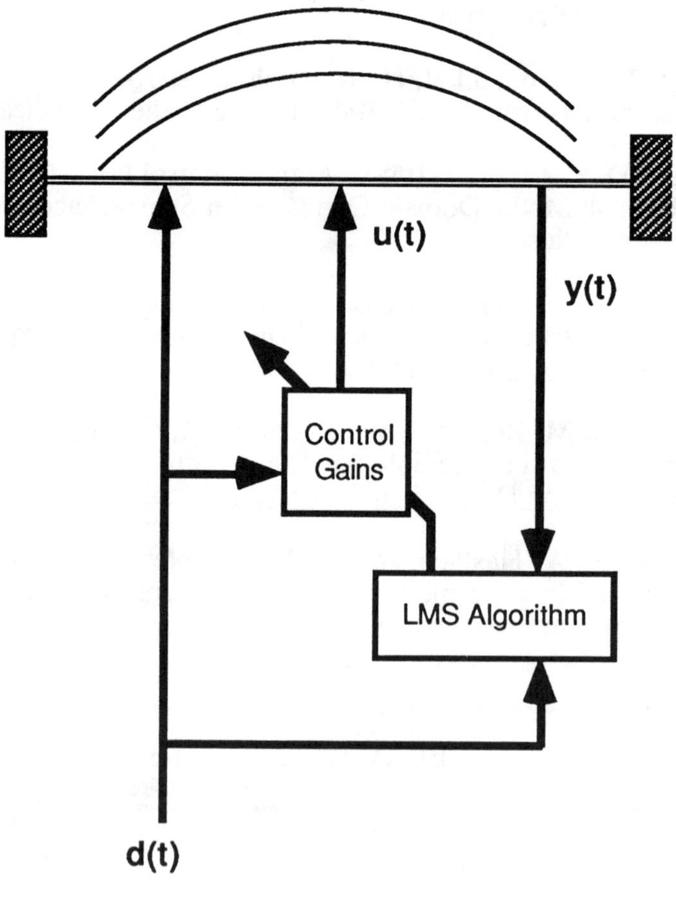

FIGURE 3.1.1. LMS ALGORITHM

EIGENPROPERTIES OF FEEDFORWARD CONTROLLED FLEXIBLE STRUCTURES

Ricardo A. Burdisso,[1] Chris R. Fuller[1]

ABSTRACT

Adaptive feedforward algorithms have been successfully applied in the active control of sound and vibration. However, the actual mechanisms of control inherent in the technique remain an area of much interest. The main objective of this research is to study the dynamic characteristics of a feedforward controlled elastic system. A entirely new mathematical approach to predict the dynamic characteristics of the controlled structure is presented. It is shown that the controlled system effectively behaves as having new eigenproperties. The controlled eigenvalues and eigenfunctions are function of the control force and error sensor locations, and independent of the input disturbance.

INTRODUCTION

Steady-state vibrations in elastic structures is a problems of particular relevance in engineering practice. Typical examples are vibrations induced by unbalanced rotors and reciprocating machines. Due to recent advances in high speed digital signal processing, these problems can be amenable to real time active control implementation. Adaptive feedforward algorithm have proved to be successful for application in which the excitation signal is steady state, i.e. pure tone, multiple frequencies and random. This control approach is based on the minimization of a quadratic cost function of a measurable state(s) of the system. The mean-square signal of the sum of a number of sensors on the structure is commonly used as the cost function.

Initial application of feedforward control was to one-dimensional acoustics fields as summarized in the review article by Warnaka [1]. More recently the technique has been extended to multi-dimensional acoustic fields [2] as well as control of structurally radiated noise [3]. For control of structural vibration, unlike feedback approaches, most feedforward algorithms have been applied to the active control of infinite or semi-infinite thin beams and in general rely on a traveling wave approach [4-9]. Recently, the behavior of a feedforward controlled simply-supported finite beam has been analytically investigated by Post [10] who parametrically characterized the behavior of the controlled system under harmonic, random and transient excitation.

Although much work has been carried out on demonstrating the potential of feedforward techniques, very little effort has been directed towards

[1]Virginia Polytechnic Institute and State University, Mechanical Engineering Department, Blacksburg, VA 24061, U.S.A.

understanding the behavior of the controlled system or the mechanisms of control. The traditional view is of "active cancellation" in which the noise excited modes of the uncontrolled structure are "canceled" by similar modes driven by an appropriately chosen control force(s). Similar observations were also held about the mechanisms of feedforward control applied to acoustic fields. Recent work has shown that real mechanisms for acoustic control for steady state disturbance is that, in general, the active source modifies the radiation impedance that the noise source "sees" thus leading to less power radiated [11]. Inherent in this generally accepted understanding is that the two cases of uncontrolled and controlled responses are, for harmonic steady state excitation, causally unrelated [12] and thus one can not relate (except mathematically) the behavior of the controlled system in terms of the behavior of the uncontrolled system. In other words, for steady state excitation the two systems have existed "ad infinitum" alongside each other and are not temporally connected.

The main aim of this paper is to investigate the dynamics of a single-input single-output (SISO) feedforward controlled finite system excited by a harmonic disturbance input, and it is based in the work of reference [13]. A entirely new mathematical formulation is presented that gives insight into the mechanisms of feedforward control under the constraints of a finite undamped system, in which there is theoretically no net power input. The analysis is valid for any modally responding finite system and is applied to the case of a simply supported finite beam for illustrative results. Finally, conclusions to the real behavior of the controlled system are presented and discussed.

EQUATIONS OF MOTION

The equations of motion for a linear discrete system subjected to harmonic excitation can be written as

$$\mathbf{M}\,\ddot{\underset{\sim}{X}} + \mathbf{C}\,\dot{\underset{\sim}{X}} + \mathbf{K}\,\underset{\sim}{X} = (\,\underset{\sim}{v}\,F + \underset{\sim}{u}\,U\,)\,e^{i\omega t} \tag{1}$$

where **M**, **C** and **K** are the mass, damping and stiffness matrices; $\underset{\sim}{X}$ is the displacement vector; F is the amplitude of the input load with spatial distribution defined by the vector $\underset{\sim}{v}$; and U is the amplitude of the control force load with spatial distribution defined by the vector $\underset{\sim}{u}$. The excitation frequency is ω, and i is the imaginary number.

The associated eigenvalue problem of Eq.(1)

$$\mathbf{K}\,\underset{\sim}{\phi}_n = \omega_n^2\,\mathbf{M}\,\underset{\sim}{\phi}_n \tag{2}$$

gives the mode shape vector $\underset{\sim}{\phi}_n$, and natural frequencies ω_n. The eigenvectors are assumed to be orthonormal. That is,

$$\underset{\sim}{\phi}_n^{\mathrm{T}}\,\mathbf{M}\,\underset{\sim}{\phi}_m = \delta_{mn} \tag{3}$$

where δ_{mn} is the Kronecker delta function, and the superscript (T) over a vector quantity represents its transpose.

By using the expansion theorem, the solution of Eq.(1) can be represented by the series expansion of the form

$$\underset{\sim}{X} = \sum_{n=1}^{N} \underset{\sim}{\phi}_n \, q_n(t) \tag{4}$$

where $q_n(t)$ is the nth modal coordinate, and N is the number of modes included in the analysis. The modal coordinate can be shown to be as follows

$$q_n(t) = H_n(\omega) \, \underset{\sim}{\phi}_n^T \, (\underset{\sim}{v} \, F + \underset{\sim}{u} \, U) \, e^{i\omega t} \tag{5}$$

where $H_n(\omega) = (\omega_n^2 - \omega^2 + i2\beta_n\omega_n\omega)^{-1}$ is the nth modal frequency response function (FRF); and β_n is the modal damping ratio. Replacing Eq.(5) into (4) gives

$$\underset{\sim}{X} = \sum_{n=1}^{N} \underset{\sim}{\phi}_n \, H_n(\omega) \, \underset{\sim}{\phi}_n^T \, (\underset{\sim}{v} \, F + \underset{\sim}{u} \, U) \, e^{i\omega t} \tag{6}$$

In feedforward control, the complex amplitude of the control force U seeks to minimize a quadratic function of a measurable response quantity such as the weighted mean-square of the system response. The selection of the cost function to be minimized dictates the controlled system response behavior. For the seek of clarity and without lost of generality, the cost function used here is the mean-square response at the eth degree of freedom

$$C(U) = X(e)X^*(e) \tag{7}$$

where $X(e)$ is the eth component of the response vector $\underset{\sim}{X}$, and is referred as the error sensor degree of freedom (DOF). The asterisk (*) over a variable denotes the complex conjugate. The optimum complex amplitude of the control force U is obtained by setting to zero the derivative of Eq.(7) with respect to the real and imaginary part of U. For a SISO system, the optimum control force amplitude U is obtained by setting $X(e)$ to zero. Thus,

$$U = - \frac{\sum_{n=1}^{N} H_n(\omega) \, \phi_n(e) \, \underset{\sim}{\phi}_n^T \, \underset{\sim}{v}}{\sum_{n=1}^{N} H_n(\omega) \, \phi_n(e) \, \underset{\sim}{\phi}_n^T \, \underset{\sim}{u}} \, F = - G(\omega) \, F \tag{8}$$

where $\phi_n(e)$ is the eth component of the nth mode shape vector. Equation (8) defines the controller $G(\omega)$ as the ratio of two transfer functions. The numerator is the transfer function between the input load and the error sensor DOF, where as the denominator is the transfer function between the control load and the error sensor DOF.

To stress the main purpose of this work, the response of an uncontrolled and controlled simply supported beam versus excitation frequency is shown in Figures 1(a) and 1(b), respectively. The properties of the system are given in the Numerical Example section. Control is achieved by a single point force and error sensor which measures displacement positioned as shown in the Figure. It is apparent from Figure 1 that the uncontrolled and controlled beam show peaks in their FRF at different frequencies and the response of the beam at these frequencies is markedly different. The

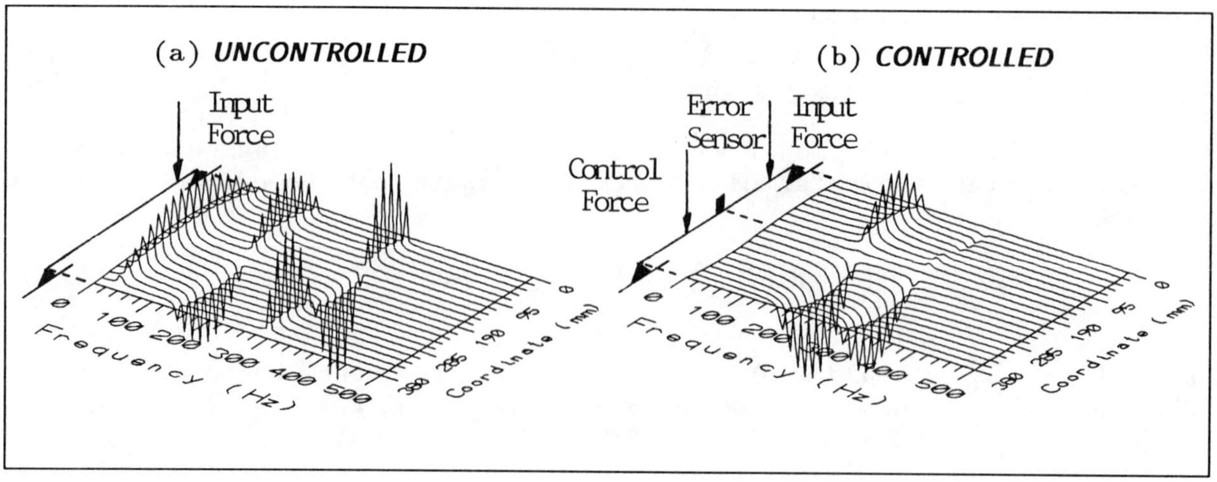

Figure 1: Beam frequency response curve.

thesis of the present work is that the controlled beam exhibits new eigenproperties to the input disturbance which account for its controlled behavior. To support this statement the dynamic response of the controlled system is further investigated in the following section.

DYNAMIC MODIFICATION DUE TO FEED-FORWARD CONTROL

For this analysis, we assume the system to be undamped although the introduction of damping is straight forward. The dynamic behavior of the controlled system can be achieved by introducing the controller $G(\omega)$ from Eq.(8) into the modal coordinate of Eq.(5). However, it is first convenient to rewrite Eq.(8) in a different form. Multiplying and dividing Eq.(8) by the product of the modal FRFs, $G(\omega)$ can be written as follows

$$G(\omega) = \frac{\sum_{n=1}^{N} \phi_n(e)\, \underset{\sim}{\phi}_n^T\, \underset{\sim}{v} \prod_{\substack{r=1 \\ r\neq n}}^{N} (\omega_r^2 - \omega^2)}{\sum_{m=1}^{N} \phi_m(e)\, \underset{\sim}{\phi}_n^T\, \underset{\sim}{u} \prod_{\substack{s=1 \\ s\neq m}}^{N} (\omega_s^2 - \omega^2)} = \frac{N(\omega)}{D(\omega)} \tag{9}$$

Thus, the controller $G(\omega)$ is the ratio of two polynomials of the same order in the variable ω^2. Solving for the (N−1) roots of these polynomials, Eq.(9) can be written as

$$G(\omega) = \kappa \prod_{l=1}^{N-1} (\rho_{cl}^2 - \omega^2) \Big/ \prod_{l=1}^{N-1} (\omega_{cl}^2 - \omega^2) \tag{10}$$

where

$$\kappa = \sum_{n=1}^{N} \phi_n(e)\, \underset{\sim}{\phi}_n^T\, \underset{\sim}{v} \Big/ \sum_{n=1}^{N} \phi_n(e)\, \underset{\sim}{\phi}_n^T\, \underset{\sim}{u} \tag{11}$$

The roots of the numerator and denominator in Eq.(10) are the zeros, ρ_{cl}, and poles, ω_{cl}, of the controller, respectively. By using a partial fraction expansion, Eq.(10) can be put in the following form

$$G(\omega) = \frac{\kappa}{N-1} \sum_{l=1}^{N-1} \frac{(A_l - \omega^2)}{(\omega_{cl}^2 - \omega^2)} = \frac{\kappa}{N-1} \sum_{l=1}^{N-1} (A_l - \omega^2)\, H_{cl}(\omega) \tag{12}$$

where the coefficients A_l of the partial fraction can be easily obtained. The controller written as in Eq.(12) is known as the parallel form realization. Each $H_{cl}(\omega)=(\omega_{cl}^2 - \omega^2)^{-1}$ is equivalent to the FRF of a single-undamped oscillator, where ω_{cl} is the associated natural frequency. These controller natural frequencies are in general different from the system frequencies ω_n, unless the error sensor is at a nodal point and/or the control force vector is orthogonal to the mode shape, $\boldsymbol{\phi}_n^T\, \mathbf{u} = 0$. This implies that the mode is unobservable and/or uncontrollable. We assume here that all modes are observable and controllable. The controller $G(\omega)$ can now be replaced into Eq.(8) and this into (5) to give

$$q_n(t) = \left[\boldsymbol{\phi}_n^T\, \mathbf{v}\, H_n(\omega) - \frac{\kappa}{N-1}\, \boldsymbol{\phi}_n^T\, \mathbf{u} \sum_{l=1}^{N-1} (A_l - \omega^2)\, H_{cl}(\omega)\, H_n(\omega) \right] F\, e^{i\omega t} \tag{13}$$

The product of the controller and system FRFs can be solved in partial fractions as follows

$$(A_l - \omega^2)\, H_{cl}(\omega)\, H_n(\omega) = C_{ln}\, H_{cl}(\omega) + D_{ln}\, H_n(\omega) \tag{14}$$

where the coefficients of the partial fraction are

$$C_{ln} = (\omega_{cl}^2 - A_l) \Big/ (\omega_{cl}^2 - \omega_n^2) \quad ; \quad D_{ln} = (A_l - \omega_n^2) \Big/ (\omega_{cl}^2 - \omega_n^2) \tag{15a,15b}$$

Substituting Eq.(14) into (13) and reordering terms gives

$$q_n(t) = \left\{ \left[\boldsymbol{\phi}_n^T\, \mathbf{v} - \frac{\kappa}{N-1}\, \boldsymbol{\phi}_n^T\, \mathbf{u} \sum_{l=1}^{N-1} D_{ln} \right] H_n(\omega) - \frac{\kappa}{N-1}\, \boldsymbol{\phi}_n^T\, \mathbf{u} \sum_{l=1}^{N-1} C_{ln}\, H_{cl}(\omega) \right\} F\, e^{i\omega t} \tag{16}$$

It is shown in reference [13] that the term in brackets vanishes. Then, replacing the modal coordinate $q_n(t)$ from Eq.(16) into Eq.(4) and interchanging the position of the summation terms, the response becomes

$$\mathbf{X} = \sum_{l=1}^{N-1} \left\{ - \frac{\kappa}{N-1} \sum_{n=1}^{N} \boldsymbol{\phi}_n^T\, \mathbf{u}\, C_{ln}\, \boldsymbol{\phi}_n \right\} H_{cl}(\omega)\, F\, e^{i\omega t} \tag{17}$$

The term in braces, in Eq.(17), is a vector multiplied by the FRF of the controller. Thus, we can rewrite Eq.(17) in the typical form found in modal analysis [14]. That is,

$$\mathbf{X} = \sum_{l=1}^{N-1} \frac{\boldsymbol{\phi}_{cl}^T\, \mathbf{v}}{m_{cl}}\, \boldsymbol{\phi}_{cl}\, H_{cl}(\omega)\, F\, e^{i\omega t} \tag{18}$$

where

$$\frac{\underset{\sim}{\phi}_{cl}^{T}\ \underset{\sim}{v}}{m_{cl}}\ \underset{\sim}{\phi}_{cl} = -\frac{\kappa}{N-1}\sum_{n=1}^{N}\underset{\sim}{\phi}_{n}^{T}\ \underset{\sim}{u}\ C_{ln}\ \underset{\sim}{\phi}_{n} \tag{19}$$

Equation (18) shows that the response of the controlled system is now expressed in terms of a new set of (N−1) eigenvectors ϕ_{cl} with associated natural frequencies ω_{cl}. The common assumption in feedforward control is that the minimization of the error sensor displacement is the result of change in the modal magnitude and phase [10]. However, Eq.(18) demonstrates that the controlled system effectively have new eigenproperties, similar to the shift in the system poles in feedback control.

Finally, we need to solve for the eigenvector ϕ_{cl} from Eq.(19). Thus, the eigenvector is normalized with respect the the mass matrix **M** as follows

$$\underset{\sim}{\phi}_{cl}^{T}\ \mathbf{M}\ \underset{\sim}{\phi}_{cl} = 1 \tag{20}$$

Introducing Eq.(19) into (20) and after some manipulation the controlled system eigenvector $\underset{\sim}{\phi}_{cl}$ is given as linear combination of the uncontrolled beam eigenvectors as follows [13]

$$\underset{\sim}{\phi}_{cl} = \sum_{n=1}^{N}\Gamma_{nl}\ \underset{\sim}{\phi}_{n} \tag{21}$$

where the expansion coefficients are given as

$$\Gamma_{nl} = \frac{\underset{\sim}{\phi}_{n}^{T}\ \underset{\sim}{u}}{(\omega_{cl}^{2}-\omega_{n}^{2})}\left\{\sum_{m=1}^{N}\left(\underset{\sim}{\phi}_{m}^{T}\ \underset{\sim}{u}\right)^{2} \Big/ \left(\omega_{cl}^{2}-\omega_{m}^{2}\right)^{2}\right\}^{-1/2} \tag{22}$$

It is noted that the expansion coefficients are computed in terms of the structure eigenproperties and the controlled eigenvalues. Therefore, the only computational effort required to obtain the controlled eigenproperties is the solution for the roots of $D(\omega)$ in Eq.(9) and in computing the coefficients Γ_{nl} in Eq.(22). None of the partial fraction coefficients from Eqs.(12) and (14) are needed.

The controlled eigenvalues ω_{cl}^{2} are the roots of the polynomial $D(\omega)$ in Eq.(9). Thus, this polynomial is the characteristic equation of the controlled system. Moreover, $D(\omega)$ is only a function of the control force distribution, location of the error sensor and the structure eigenproperties. Thus, the controlled eigenvalues and eigenfunctions depend only on these factors and are independent of the external loading. Once the control force and error sensor placement is specified, the frequency characteristics of the controlled system is fixed and the same for any coherent input disturbance. It is also important to mention that the applied control force reduces by one the dynamic degrees-of-freedom of the original structure.

It is also helpful to consider the above development in terms of other well known systems. As stated above, the new controlled eigenfunctions are a linear combination of the original eigenfunction. Thus a typical question might be "isn't then the structure still responding in terms of its old modes?". In reality, the uncontrolled modes have been conveniently used as a basis functions to derive the eigenfrequencies and eigenfunctions of the controlled system similar to approaches in analysis of modal modifications [15] and the behavior of composite plates [16] which also employ, for convenience, basis functions which are the eigenfunctions of the homogeneous system. The explicit solution for the system eigenvalues and their associated eigenfunctions determines the modal characteristics of the system independent of the form of the basis functions.

Finally, it should be noted that this system modal modification is for feedforward control where coherent information from the input disturbance is applied to the beam as a control action. To any other incoherent noise disturbance the system will exhibit its uncontrolled modal response. The implication of this are profound; a system can be made to appear to be totally different to a required noise disturbance (e.g. soft) while remaining unchanged to other input (e.g. stiff). This behavior should be contrasted to that of system with feedback control, in which the system characteristics generally change for all disturbance inputs.

NUMERICAL EXAMPLE

The applicability of the formulation developed here is demonstrated for a uniform simply supported beam of Figure 1. The beam has bending stiffness EI = 5.329 NT.m^2, mass per unit length m = 0.6265 NT.sec^2/m^2 and length L=0.38 m. The system is assumed to have 0.1% damping in each mode, and ten modes are included in the response analysis (N=10). The beam is excited by a concentrated force located at a=0.1L, and the control force is placed at b=0.65L to minimize the beam response at the error sensor located at c=0.45L. A finite element model with twenty beam elements is used to solve the eigenvalue problem. The natural frequencies are shown in Table I.

To illustrate the dynamic behavior of the beam before and after control, the response at the input force location was computed. Assuming an unit magnitude of the input force F, Figure 2 shows the amplitude and phase of the displacement at x=a as function of the excitation frequency ω. The dashed line is the response of the uncontrolled beam obtained by setting U to zero in Eq.(6). The continuous line corresponds to the controlled system calculated from Eqs.(6) and (8). The response of the uncontrolled beam shows resonance peaks at frequencies as given in Table II. On the other hand, when the control force U is applied, the response shows resonance behavior at excitation frequencies near 150, 270 and 630 Hz, respectively. The computed phase angle, in both the uncontrolled and

TABLE I – Simply Supported Beam Natural Frequencies.

Mode	1	2	3	4	5	6	7	8
ω_n [Hz]	31.7	126.9	285.5	507.7	793.2	1141.2	1553.3	2028.8

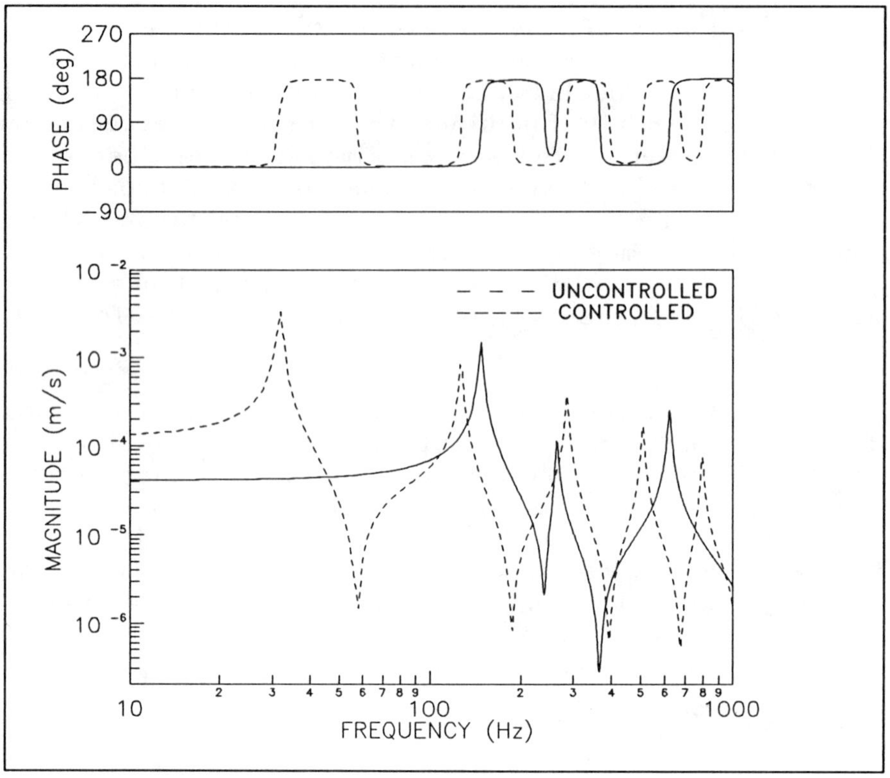

Figure 2: Displacement response at input force location.

controlled system, also shows the typical jump of 180° at the poles and zeros. Thus, it is clear from Figure 2 that the controlled system behaves as having new eigenproperties different from the original beam modes. The amplitude and phase of the controller $G(\omega)$, using Eq.(8), is computed and shown in Figure 3. Comparing Figures 2 and 3, a large control effort is required to minimize the response at the resonance frequencies of the controlled system. Thus, in practice the locations of the controlled beam eigenfrequencies are important because they indicate points of control instability for a particular arrangement.

The roots of the polynomial $D(\omega)$ in Eq.(9) were next computed by using the IMSL routine ZPOLR, and they are presented in Table II. As shown in the formulation, the controlled system have one less dynamic degree-of-freedom than the original beam. Since we have included ten uncontrolled modes in the analysis, there are nine controlled eigenvalues or roots. The last column in Table II lists the natural frequencies of the controlled system in Hz. The first three natural frequencies are the resonance condition of the controlled system as shown is Figure 2 (continuous line). The mode shape associated to the controlled natural frequencies were computed using Eq.(21). The first six mode shapes are shown in Figure 4a through 4f. The location of the input force, control force and error sensor are indicated in Figures 4a through 4f by the letters F, U and e, respectively. As can be seen, all the eigenfunctions have a nodal point at the error sensor location. Thus in line with the thesis of the paper, control is achieved by creating new modes which all have zero response at the error sensor location. This is in contrast to

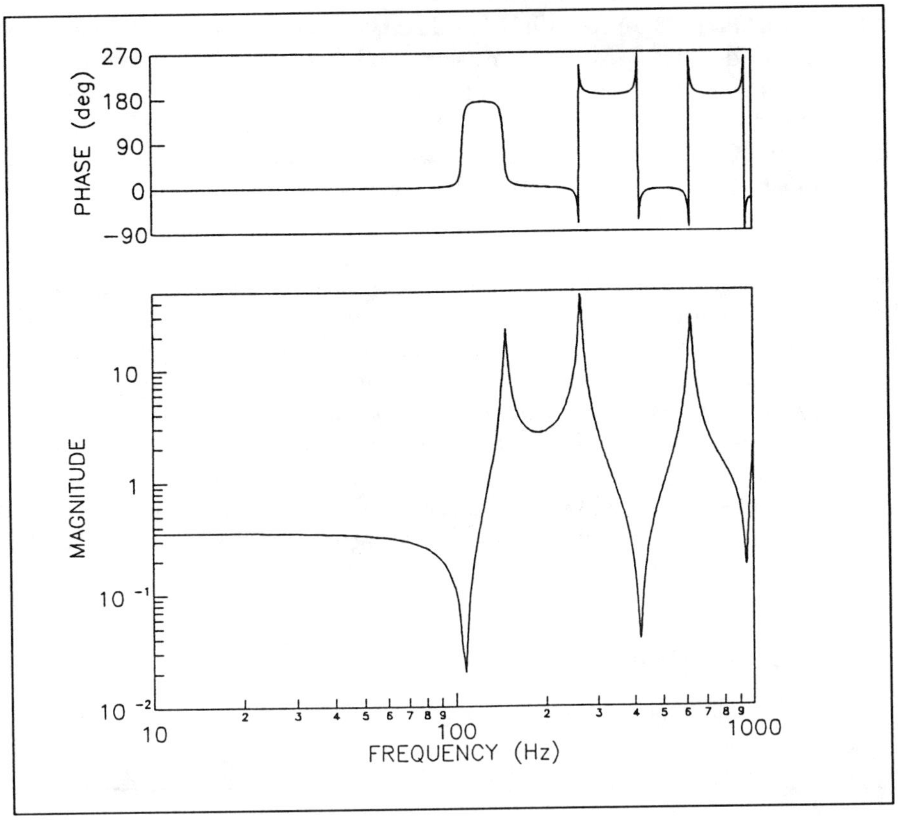

Figure 3: Magnitude and phase of controller $G(\omega)$.

the conventional understanding which explains the behavior in terms of the uncontrolled modes "phase canceling" the response to zero at that point.

Two of the controlled eigenvalues are seen to be negative, and their associated mode shapes are plotted in Figure 5. Again the input force, control force and error sensor location are indicated in the figure. These two modes can be seen to have a maximum amplitude at the driving

TABLE II – Eigenvalues and Natural Frequencies of Controlled Beam.

Mode l	Eigenvalue ω_{cl}^2 [Rad/s^2]	Nat. Freq. ω_{cl} [Hz]
1	0.894828E+06	150.55
2	0.286251E+07	269.27
3	0.157495E+08	631.61
4	0.435942E+08	1050.83
5	0.811189E+08	1433.44
6	0.220675E+09	2364.26
7	0.257342E+09	2553.14
8	-.338700E+08	i 926.24
9	-.367047E+09	i3049.16

point of the control force which decays away from that point. This phenomenon occurred in every numerical example investigated. This suggests that these modes are related to localize effects associated with the bending near field response to the control force. It is the opinion of the authors that these negative root are not spurious values due to the truncated number of modes included in the analysis. This behavior is a topic of future research.

Another important consideration pertains to the orthogonality of the controlled system mode shapes. As seen from Figures 4 and 5, the controlled system eigenvectors are not orthogonal. By definition, the mode shapes of a linear system are orthogonal with respect to the mass matrix. The beam-controller system is a linear system which could be represented by a new mass and stiffness matrices. Thus, the orthogonality property of the controlled system eigenvectors has to be satisfied with respect to the beam-controller system mass matrix; and therefore, the controlled modes do not need to be orthogonal with respect to the mass matrix **M**.

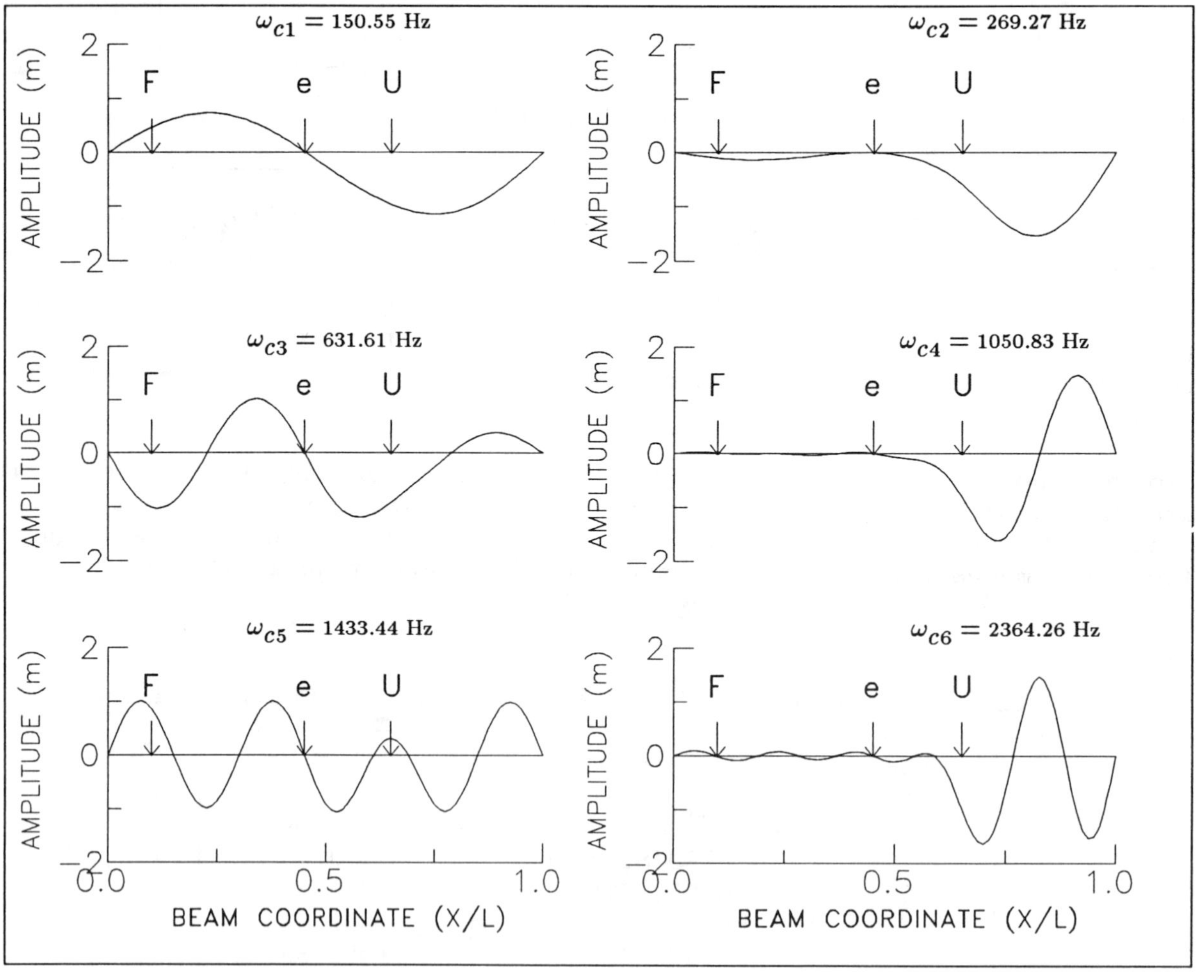

Figure 4: First six controlled system mode shapes.

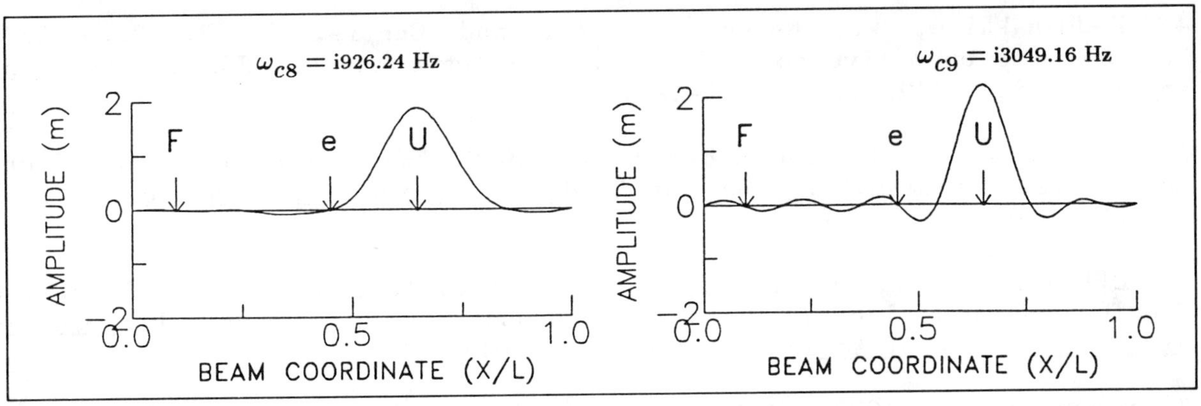

Figure 5: Controlled mode shapes corresponding to negative eigenvalues.

CONCLUSIONS

The dynamic behavior of a feedforward control system was investigated. The analysis presented here shows that feedforward controlled systems have effectively new eigenproperties. A completely new mathematical methodology was developed to compute the controlled system resonant frequencies and mode shapes. The controlled system eigenvalues are obtained by solving the roots of a polynomial, which is the characteristics equation of the controlled structure. The associated eigenvectors of the controlled structure are found as linear combination of the mode shapes of the original (uncontrolled) structure. These eigenproperties are functions of the applied control force and error sensor locations but independent of the input force. Control of a simply supported beam was investigated as a typical example to demonstrate the formulation. The formulation presented here is thus a entirely new method to predict the dynamics of feedforward control systems, and thus provides the basis for the design of such systems.

ACKNOWLEDGEMENTS

The authors gratefully acknowledge the support of this work by DARPA and the Office of Naval Research under grant ONR-00014-88-k-0721.

REFERENCES

1. Warnaka, G. E., 1982. "Active attenuation of noise: the state of the art." Noise Control Engineering, **18**(3), pp. 100–110.

2. Silcox, R. J., and Lester, H. C., 1989. "An evaluation of active noise control in a cylindrical shell." Journal of Vibration, Stress, and Reliability in Design, **111**, pp. 337–342.

3. Fuller, C. R., 1990. "Active control of sound transmission/radiation from elastic plates by vibration inputs: I. Analysis." Journal of Sound and Vibration, **136**(1), pp. 1–15.

4. Redman-While, W., Nelson, P. A., and Curtis, A. R. D., 1987. "Experiments on active control of flexural wave power." Journal of Sound and Vibration, **112**(1), pp. 181–187.

5. von Flotow, A. H., and Schafer, B., 1986. "Wave-absorbing controllers for a flexible beam." Journal of Guidance, Control and Dynamics, **9** (6), pp. 673–680.

6. Elliot, S. J., Stothers I. M., and Billet, L., 1990. "Adaptive feedforward control of flexural waves propagating in a beam." Proceedings of the Institute of Acoustics, **12**(1), pp. 613–622.

7. Scheuren, J., 1990. "Active attenuation of bending waves in beams." Proceedings of the Institute of Acoustics, **12**(1), pp. 623–629.

8. Gibbs, G. P., and Fuller, C. R., 1990. "Experiments on active control of vibrational power flow using piezoceramics actuators and sensors." AIAA Paper, 90-1132.

9. Fuller, C. R., Gibbs, G. P. and Silcox, R. J., 1990. "Simultaneous active control of flexural and extensional power flow in beams." Proceedings of International Congress on Recent Developments in Air and Structure-borne Sound and Vibration, Aurburn University, AL. March 6–8, pp. 657–662.

10. Post, J. T., 1989. "Active control of the forced and transient response of a finite beam." Master Thesis, The George Washington University, Washington, D.C..

11. Snyder, S. D., and Hansen, C. H., 1989. "Active noise control in ducts: Some physical insights." The Journal of the Acoustical Society of America, **86** (1), pp. 184–194.

12. Roebuck, I., 1990. "Energy considerations in active noise control." Proceedings of International Congress on Recent Developments in Air and Structure-borne Sound and Vibration, Aurburn University, AL. March 6–8, pp. 509–516.

13. Burdisso, R. A., and Fuller, C. R., 1991. "Theory of feedforward controlled system eigenproperties." Journal of Sound and Vibration, (Accepted for publication).

14. Clough, R. and Penzien, J., 1975. Dynamics of Structures. New York: McGraw Hill, 1975.

15. Formenti, D., and Welaratma, S., 1980. "Structural Dynamics Modification – An extension to Modal Analysis." SAE Paper 81-1043.

16. Whitney, J. M., 1987. Structural Analysis of Laminated Anisotropic Plates. Technomic Publishing Co., Lancaster, PA.

Active Isolation

THE APPLICATION OF ADAPTIVE FEEDFORWARD ALGORITHMS TO HELICOPTER VIBRATION CONTROL

Ian Lyndon[1]

ABSTRACT

This paper describes Active Control of Structural Response (ACSR), a vibration control developed system developed at Westland and successfully evaluated on a number of helicopters, including most recently, the EH101. The unique feature of this system is the actuation technology, whereby hydraulic struts are used to introduce vibratory forcing in the same dominant modes as are excited by the rotor blades, the source of the vibration problem. By superimposing this secondary vibration field on the primary rotor induced vibration, the net vibration response is considerably reduced.

A multivariable controller takes inputs from ten accelerometers and drives four actuators to minimise a performance index that comprises of the weighted sum of squares of vibration inputs and actuator demands. To achieve this the controller uses estimates of the frequency response functions between actuators and sensors at harmonics of the rotor frequency. These estimates are obtained by on line parameter estimation and thus previous knowledge of the structural dynamics is not required. This permits the system to compensate for in-flight changes in dynamics, due to fuel or payload for example.

The vibration problems experienced in helicopters are dominated by the structures response to rotor head forces that are harmonics of the rotor speed. Over short time intervals the rotor speed and magnitude of head force changes are small enough for the vibration to be considered as deterministic. If a feedforward rotor signal is used to maintain the phase relationship then an open loop scheme can be used with the output corrected at a relatively slow rate by feedback of frequency domain information. This control scheme is explained in more detail, the implementation is described and the results of flight trials are presented.

INTRODUCTION

Helicopter designers and operators alike have long regarded vibration as one of the prime technical challenges due to its undesirable effects on crew, passengers, reliability and

[1]Westland Helicopters Limited, Yeovil Somerset, England.

maintenance costs. The relative severity of the vibration environment on a helicopter in forward flight is a consequence of the forcing imbalance between the advancing and retreating sectors of the main rotor disk as it flies edgewise through the air. This generates a vibration response in the airframe at the blade passing frequency and its harmonics. In the past this vibration challenge has been met through the use of of a wide variety of passive devices which absorb vibration or isolate the fuselage from the source of vibration. More recently, with the rapid advancement of computer technology, attention has become focused on the development of more powerful active techniques for the alleviation of helicopter vibration.

Early research into active techniques concentrated upon Higher Harmonic Control (HHC), a technique aimed at reducing the vibratory forcing within the main rotor through the superposition of a controlled pitch excitation on the main rotor blades onto the primary uncontrolled vibration response of the rotor. The net effect of this superposition is to redistribute the vibratory airloads around the rotor azimuth. Although the majority of research and development has been conducted into the use of HHC for vibration control it also has application to actively improving rotor performance. The first studies into the potential of HHC were conducted in the early 1950's and subsequently a large number of analytical and experimental studies have been conducted, leading up to the the first flight evaluation of HHC on a Hughes OH-6A helicopter in 1984 (reference [1]). Subsequent flight trials (reference [2]) have emphasised the great potential of HHC for vibration reduction. The basis of HHC is that the secondary vibration inputs are composite waveforms comprising a number frequencies synchronised to higher harmonics of the main rotor rotational frequency. In this respect, the control approach adopted for HHC is based on a frequency domain transfer matrix characterisation of the vibration response (reference [3]), whereby optimal control inputs are based on the minimisation of the vibratory forcing at the rotor hub or more directly, the measured vibration response. The variation of the rotor dynamics with flight condition resulted in the development of an fully adaptive frequency domain control approach as detailed in reference [4], where Kalman filter estimators were incorporated within the controller to track the changes in rotor and airframe dynamics during flight and thus, maintain the optimal control inputs.

Whilst the HHC approach has shown great promise it has yet to be widely accepted as a practical proposition for production helicopters. In part, the reluctance of the industry to accept HHC is due to the relatively high power requirements and performance constraints at high forward speeds. It was for these reasons that Westland have pursued a an alternative approach, namely that of Active Control of Structural Response (ACSR). In common with HHC, the ACSR technique is based on a controller which schedules vibratory commands to actuators in such a manner as to reduce overall vibration. However, the principle differs in that the actuation devices are located within the structure and directly excite the structural response as opposed to indirectly controlling the structure through applying controlled forces to the rotor. Furthermore, unlike active isolation techniques ACSR does not attempt to isolate the structure from the source through direct control of key vibratory load paths but rather modifies the load path to control the vibration response at points remote from the actuator devices, as outlined in figure 1. This feature of the ACSR approach makes the technology makes the technology applicable to a wide variety of structural vibration problems, not least in the helicopter and aerospace fields. The ACSR technology was first proven through a comprehensive flight demonstration programme on a Westland 30 helicopter (reference [5]) in 1987, and subsequently ground vibration tests on both a Sikorsky S-76 (reference [6]) and Westland Lynx have expanded the technology base. Most recently, an ACSR system has been fitted to a number of the prototype Anglo-Italian EH101 helicopters and has now completed in excess of 90 hours of flight evaluation, showing significant reductions in vibration.

This paper outlines the ACSR control philosophy, describes the EH101 helicopter installation and presents a summary of the flight test results from this aircraft.

CONTROL PHILOSOPHY

The vibration spectra of helicopter structures are dominated by discrete frequency peaks related to the main rotor and tail rotor speeds. This means the problem of vibration control is well suited to frequency domain or synchronous type algorithms, where the periodic nature of the disturbance signal is used to simplify the control problem. In effect, the fundamental variable being controlled is converted from a time series vibration signal to a sum of frequency vectors. In doing this conversion the required bandwidth of the control is reduced by an order of magnitude. In fact, the rate of change of the frequency vectors in the helicopter case are slow enough for the dynamics to be ignored altogether and the problem treated as the steady state control of the individual frequency vectors.

Over a time interval of several vibration periods, variations in the rotor speed and vibratory forcing are small and the induced vibrations can be assumed to be deterministic rather than random. If a rotor position signal is used to maintain the phase relationship over the time interval, then an open loop control system can be used with the control corrected by the quasi-static feedback of the frequency domain information. The system can be described as feedforward in the time domain as a reference signal is continuously used to track frequency changes. In the frequency domain the system uses feedback of measured vibration but the update rate is relatively slow due to the delay imposed in accurately extracting the frequency measurements. In most helicopter flight conditions this is fast enough to track the changes in vibration forcing.

As the frequency domain loop is not dynamic, the system plant can be modelled by the frequency response transfer functions at the individual frequencies that are required to be controlled. Multiple frequencies are controlled using parallel sets of equations and summing the actuator demands. For each frequency, a linear model is used of the form:

$$\underline{Y} = [T]\ \underline{X} + \underline{B} \tag{1}$$

Here the vectors X and Y represent the actuator demands and the vibration measurements respectively. Currently, the system is configured to control up to four actuators with inputs from no more than ten accelerometers. The frequency response matrix [T] is a transfer function that includes the structures inertance, actuator dynamics and signal conditioning. The background vibration vector is B the result of rotor forcing alone, ie. the uncontrolled vibration. The above model is the global representation, which can alternatively be written in local-linear form as:

$$\underline{dY} = [T]\ \underline{dX} \tag{2}$$

where dY and dX are differential changes per control cycle of the Y and X vectors respectively.

Practically, accelerometers are light and easy to install so it is advantageous to use more sensors than actuators and minimise a cost function based on the sum of weighted sensor levels. Sensors are placed at positions where vibration reduction is most critical and in sufficient numbers to sense motion in all the dominant modes. The control emphasis can shifted to some extent by changing the individual sensor weightings. Actuator demands are included in the cost function with weightings adjusted to prevent saturation. The simplest form of optimisation is achieved by minimising a quadratic function of the form:

$$J = \sum_{i=1}^{M} W_i Y_i^2 + \sum_{j=1}^{N} A_j X_j^2 = \underline{Y}^T [W]\ \underline{Y} + \underline{X}^T [A]\ \underline{X} \tag{3}$$

where [W] and [A] are diagonal matrices of the respective output and input weightings.

As there are no dynamic terms in this optimisation, the solution is straight forward and can be calculated analytically provided the actuators and sensors are positioned to prevent ill-conditioning.

Two basic minimisation formulae have been used for generation of the optimal control inputs, these are as follows:

a. Global Deterministic

$$\underline{X}_{n+1} = - \{[T]^T [W][T]+[A]\}^{-1}_{n+1} \{[T]^T [W] \underline{B}\}_{n+1} \qquad (4)$$

b. Local Linear Deterministic

If the plant model is assumed be linear only around the operating point ie. local linear, then an iterative control update equation can be formulated that does not require an estimate of the background vibration vector B.

$$\underline{X}_{n+1} = \underline{X}_{n} - \{[T]^T [W][T]+[A]\}^{-1}_{n+1} \{[T]^T_{n+1} [W] \underline{Y}_{n} + [A] \underline{X}_{n}\} \qquad (5)$$

Matrix inversion required for equations (4) and (5) is not carried out explicitly. Far more efficiently, the problem is solved as a set of linear equations using the Cholesky decomposition.

The parameters [T] and B of the linear relationship (1) may be variable with flight condition and hence some form of parameter estimation is necessary to provide information to the controller. The on-line strategy uses a Recursive Least Squares (RLS) estimator with variable forgetting factor to track parameters (reference [7]). This technique operates recursively on information from the previous algorithm cycle to update the estimate and calculate a new optimal X input. Initial estimates for the parameters are obtained by open loop testing, ie. driving each actuator in turn and measuring the response.

The control actuators need a high force capability and open loop testing has indicated that significant non-linearity exists with respect to the forcing amplitude. However, this does not appear to be a problem. Whether this is because the algorithms are inherently robust or the parameter estimation eliminates the problem is currently under investigation.

The timing of algorithm operations is shown in figure 2, which indicates the cyclic nature of the operations and when various parameters become available. For the EH101 helicopter the cycle time is set to approximately 0.5 second. The major functional items are as follows:

(i) Transient Decay - this is a pure time delay during which the measurement transients are allowed to decay, prior to initiating the signal processing.

(ii) Signal Processing - the vibration measurements are extracted using a correlation method.

(iii) Parameter Estimation - this is a two stage operation. Firstly, the parameters not dependent on the vibration measurement are calculated. Finally, those parameters dependent on the measurement are calculated using a recursive least squares algorithm.

(iv) Output Minimisation - the updated parameter estimates are substituted in the minimisation formulae (eqn.4 or 5) and the optimal actuator demands for the next filter cycle are calculated.

Up to four frequencies can be controlled by running four sets of equations in parallel and then summing the demands before the A/D converter for the actuator drives. The frequencies can be non integer harmonics of up to two different rotor phase signals.

EH101 ACSR SYSTEM DESCRIPTION

The following sections provide an outline of the EH101 ACSR system and a summary of the flight test programme. A more comprehensive description of the EH101 system is given in reference [8].

The control is implemented digitally using Texas TMS32010 signal processors for the frequency analysis of the vibration data. These act as slave processors and are monitored and controlled by a Motorola MC68020 master processor that also implements the parameter estimation and minimisation. The control unit requires a synchronising signal from the main rotor for the sampling of both accelerometer inputs and actuator demand outputs. This is provided by a sensor in the main rotor gearbox that gives a clock signal of about 2 KHz. Operation of the control unit is governed in the first instance by the pilot via a master 'On/Off' switch on the Pilot Control Panel (PCP). In addition, the PCP incorporates a 'mode' switch, with indicator lamps, for the selection of the desired set of algorithm defaults and indication of the failure status of the ACSR system.

The ten ACSR accelerometers are located at strategic points within the aircraft, including the cockpit and cabin floor, the top structure and tail rotor gearbox.

A feasibility study based on forced response simulations using a finite element dynamics model of the EH101, identified the configuration of an ACSR actuator placed in parallel with each of the four main gearbox support struts as offering significant vibration reductions.

The ACSR main gearbox support actuator/strut configuration is shown in figure 3, the combined unit being an assembly of three components: the compliant ring element; the hydraulic actuator and the strut tube. In common with other ACSR helicopter applications, the actuator is mounted in parallel with the primary load bearing compliant element, and is required to input the controlling vibratory forces while accommodating the much higher displacements due to primary flight loads.

Design of the actuator and strut was carried out in an integrated process with Moog Controls Ltd. The actuator has a force capability of 30 kN at the 5R blade passing frequency of 17.5Hz. A dual-chamber single-piston actuator is controlled by a servo valve with pressure transducers for force feedback and an isolating solenoid valve. Moog have embodied novel sealing and surface finish techniques for long life under continuous high frequency operation. Hydraulic power consumption of the system using four actuators has been conservatively estimated at 5kW.

EH101 ACSR FLIGHT TRIALS

The ACSR system was assessed against the following flight envelope :

- o straight and level forward speeds in the range 40 to 160 kts
- o basic manoeuvres, including take-off and landing, max. power climb, banked turns, autorotation and transition-to-hover.

In addition, various aspects of ACSR system operation were investigated, including :

- o Optimisation of actuator and sensor weightings.
- o Degraded operation, three-actuator system.
- o Multi-frequency 5R and 10R operation.

The results for the steady state conditions were most impressive, and confirmed the potential indicated by the theoretical predictions. It can be seen from figure 4 that significant reductions in 5R vibration are achieved, with the Co-Pilot seat vertical vibration reduced from 0.38 'g' to 0.10 'g' at 40 kts and from 0.85 'g' to 0.07 'g' at 140kts, representing 74% and 92% reduction respectively. An examination of the vibration averages shown in figure 5 reveals substantial reductions for the ten ACSR control locations (between 59 and 88%), for the cabin and cockpit (between 61 and 75%) and for the aircraft as a whole, including engines and gearboxes (52 to 75%). Of most significance is the result for the cockpit and cabin, where average vibration is maintained well below the 0.15 'g' specification throughout the speed range tested, although at a few locations, the reduced vibration was still slightly above this level at the 40 knot condition.

Improvements in vibration reductions in the cockpit and cabin were achieved by adjusting the ACSR accelerometer control weightings, but these were marginal and led to slight degradations of the reduced levels elsewhere in the aircraft.

The actuator degraded operation tests were aimed at establishing the loss of vibration control following an actuator failure, but also to assess the potential of a three-actuator based system. This was carried out by weighting out the actuators in the ACSR control algorithm in turn. Results for straight and level flight for the least degrading case, removal of the aft starboard actuator, are given in figure 6. It is clear that the ACSR system performs equally well without this actuator, giving confidence in the validity of this configuration as the basis for an eventual production standard ACSR system for the EH101.

For the majority of the manoeuvre cases, the ACSR system maintained significantly reduced levels of 5R vibration, the benefits being most marked in take-off and power climb. However, the performance of ACSR in the transition-to-hover was not as impressive since the actuators were operating at full capacity due to the very high peak baseline vibration levels. Even so, useful reductions, in the order of 50%, were achieved, as figure 7 shows.

Multi-frequency operation at 5R plus 10R was assessed against the forward speed range. Though not presented here, the results were satisfactory, with the low baseline 10R vibration reduced by approximately 50% to levels probably at the noise threshold of the ACSR system, while the 5R performance was unchanged from the single frequency operation.

Further flight testing is planned, aimed at improving ACSR performance in both steady-state and manoeuvre conditions, particularly for the transition-to-hover. The former is to be achieved by investigating alternative locations for the ACSR control sensors, while the latter may require redesign of the integral actuator/strut to provide increased force capability.

Currently a retro-fit programme is underway to equip all the EH101 pre-production aircraft with ACSR and development activities are planned to further optimise ACSR performance. The demonstrated benefits of ACSR have led to the decision to include the system as part of the basic aircraft.

FUTURE DEVELOPMENTS

Although this control approach has proved very successful in that the steady state performance is close to the theoretical maximum, some improvement could be made in the transient performance. To achieve this the 'quasi-static' assumption must be dropped and a fully dynamic algorithm used.

The dynamics of the overall system are greatly influenced by the hydraulic actuators and they will always limit the performance, regardless of the data processing capability. If they are to be used to full capacity then the inherent non-linearity has to be considered in the control system design. The need to model the system dynamics is likely to make the implementation much more application specific. Another source of difficulty is that finite element models are not accurate enough to be used directly. System identification has to be used to obtain the model but there will always be uncertainty at high frequency. Computer simulations have shown the capability of time domain algorithms with perfect system models but robustness can only be demonstrated using real structures.

Despite the complexities of a fully dynamic algorithm approach, the rapid development of computational hardware has enabled high order multivariable control to be a feasable option for future developments of ACSR.

CONCLUSIONS

Westland have developed a technology using multiple actuators to introduce control forces into a structure to reduce the forced response, using information from sensors located throughout the structure.

The periodic nature of helicopter vibration makes the problem suitable for control using feedforward algorithms. A strategy of minimising a cost function in the frequency domain whilst maintaining phase accuracy by the feedforward of a rotor reference signal has proved to be very effective in flight trials.

REFERENCES

1. Wood,E.R., Powers,R.W., Hammond,C.E. and Cline,J.H., May 1983; "On Developing and Flight Testing a Higher Harmonic Control System."; 39th Annual Forum of the American Helicopter Society.

2. Achache,M. and Polychroniadis,M.; June 1988; "Higher Harmonic Control - Flight Tests of an Experimental System on SA349 Research Gazelle."; 42nd Annual Forum of the American Helicopter Society.

3. McCloud,J.L. and Kretz,M.; Feb 1974; "Multicyclic Jet-Flap Control for Alleviation of Helicopter Blade Stresses and Fuselage Vibration.";HASA SP352.

4. Molusis,J.A., Hammond,C.E. and Cline,J.H.; May 1981; "A Unified Approach to the Optimal Design of Adaptive and Gain Scheduled Controllers to Achieve Minimum Helicopter Rotor Vibration"; 37th Annual Forum of the American Helicopter Society.

5. Staple,A.E.; Sept 1989; "An Evaluation of Active Control of Structural Response as a Means of Reducing Helicopter Vibration."; 15th European Rotorcraft Forum.

6. Welsh,W.A., von Hardenberg,P.C., von Hardenberg,P.W. and Staple,A.E.; May 1990; "Test and Evaluation of Fuselage Vibration Utilising ACSR Optimised to ADS-27."; 48th Annual Forum of the American Helicopter Society.

7. Fortescue,T.R., Kershenbaum,L.S., and Ydstie,B.E.; Nov 1978; 'Implementation of Self Tuning Regulators with Variable Forgetting Factors.'; Automatica, Vol. 17 No. 6, pp 831-835, 1981.

8. Staple, A.E. and Wells, D.M ; Sept 1990; "The Development and Testing of an Active Control of Structural Response System for the EH101 Helicopter."; 16th European Rotorcraft Forum.

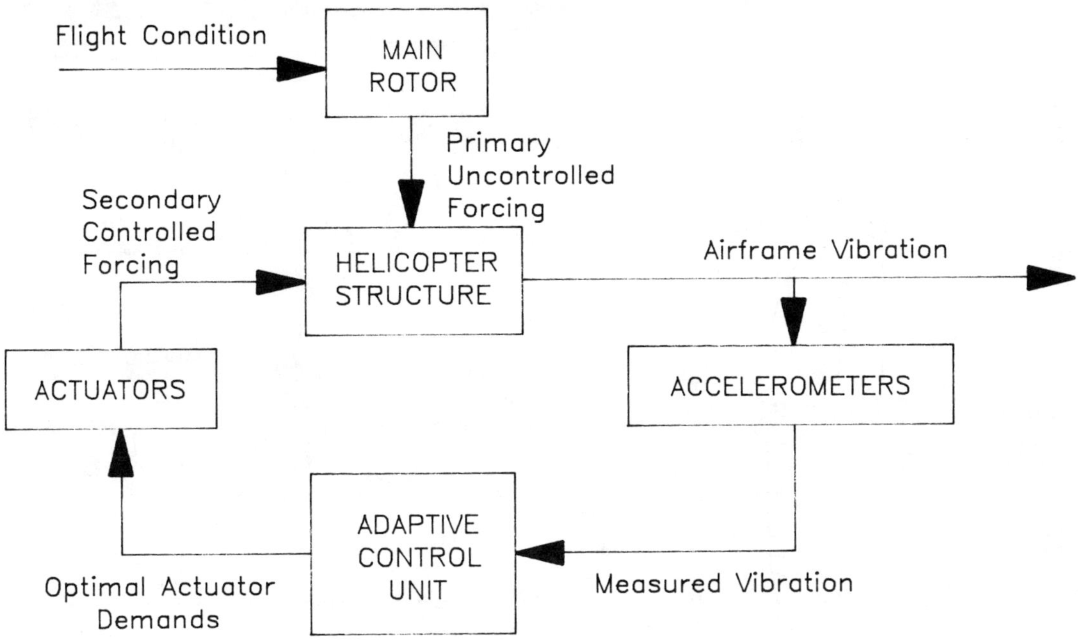

FIGURE 1 : ACSR OPERATIONAL PRINCIPLES

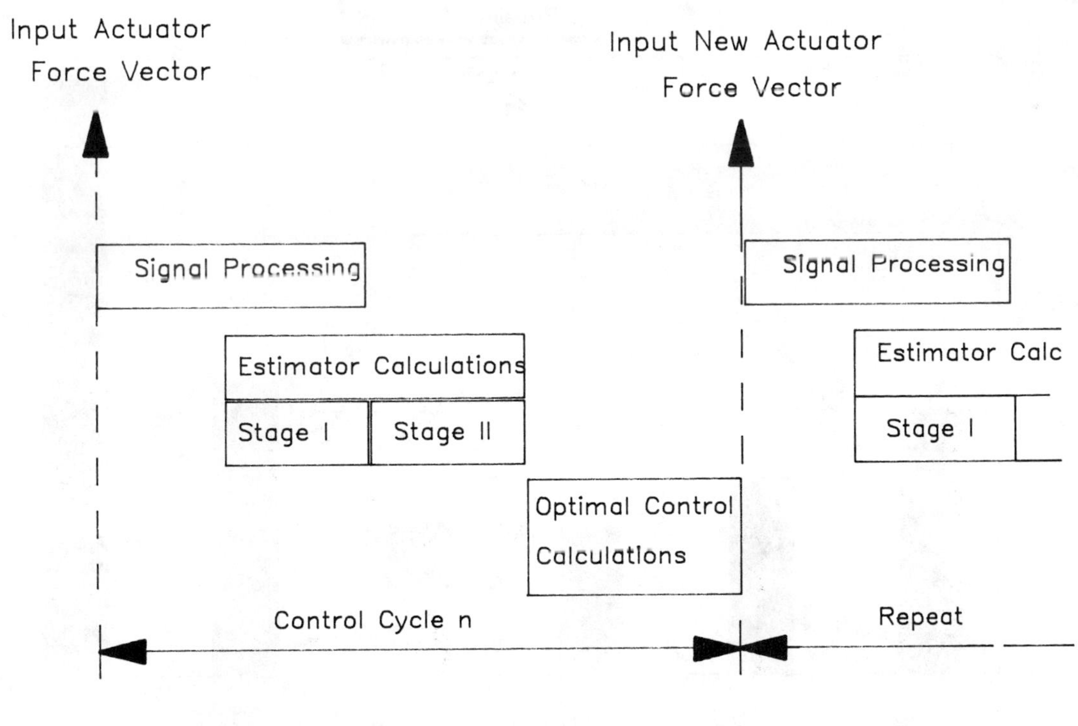

FIGURE 2 : CONTROL SYSTEM TIMING DIAGRAM

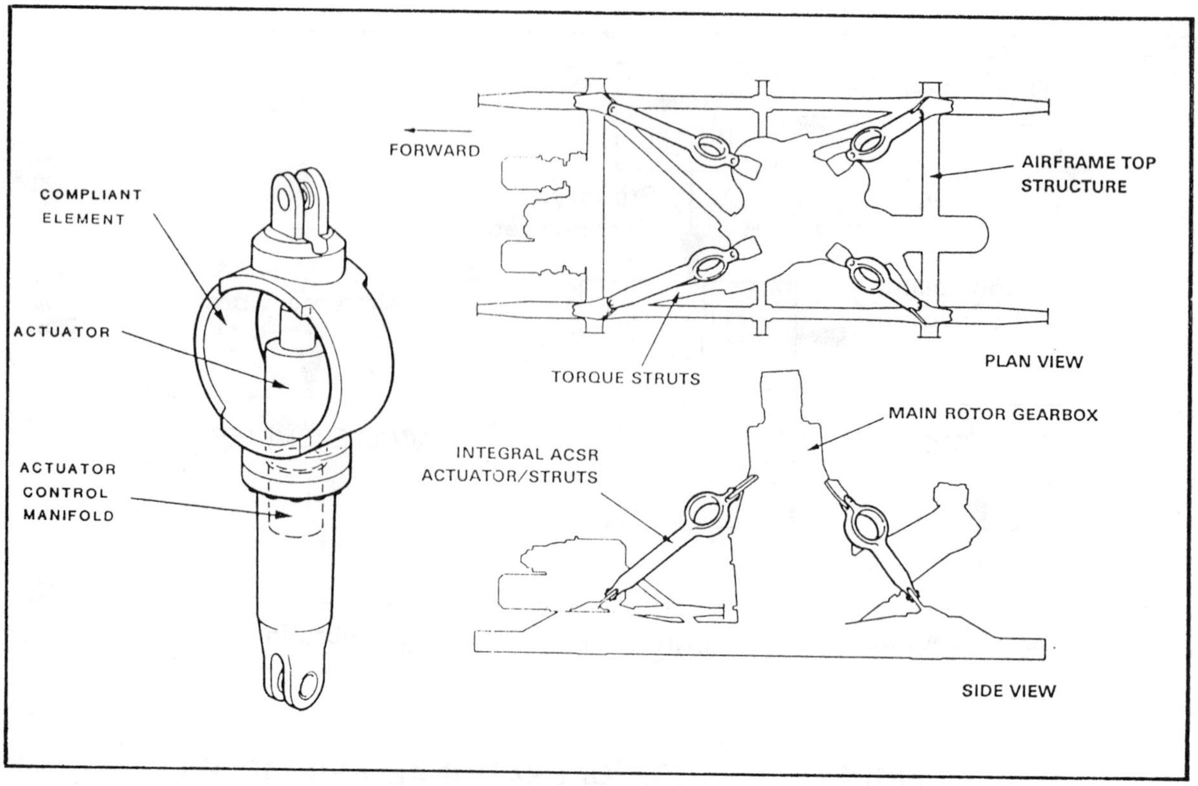

FIGURE 3 : EH101–ACSR INTEGRAL STRUT/ACTUATOR AND CONFIGURATION

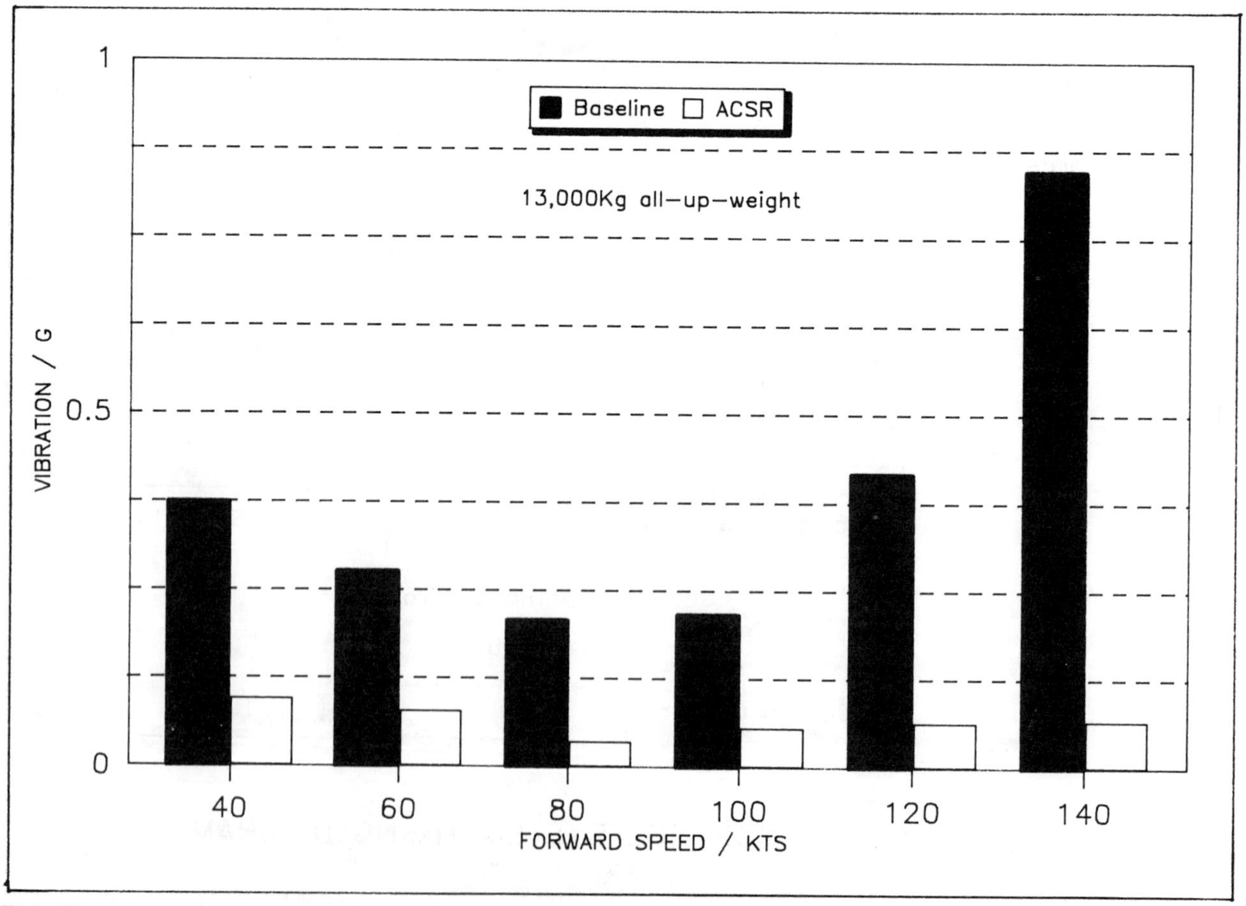

FIGURE 4 : EH101–ACSR FLIGHT TESTS, CO–PILOT VERTICAL VIBRATION

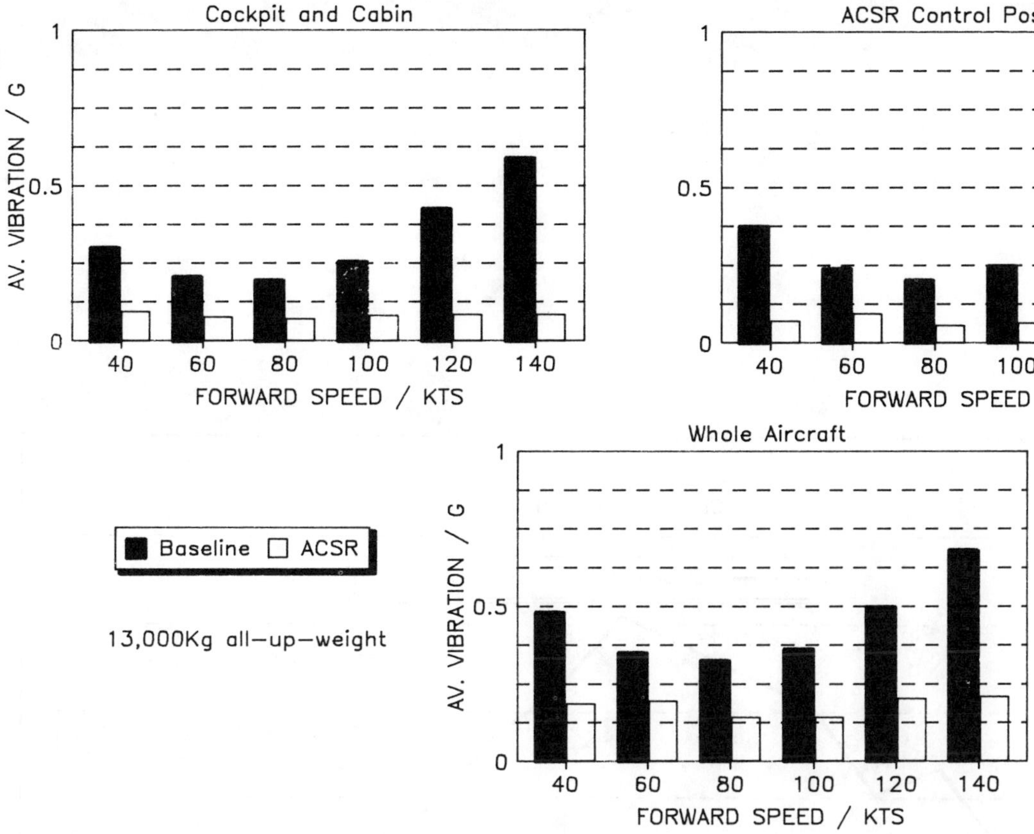

FIGURE 5 : EH101–ACSR FLIGHT TESTS – VIBRATION AVS VS FWD SPEED

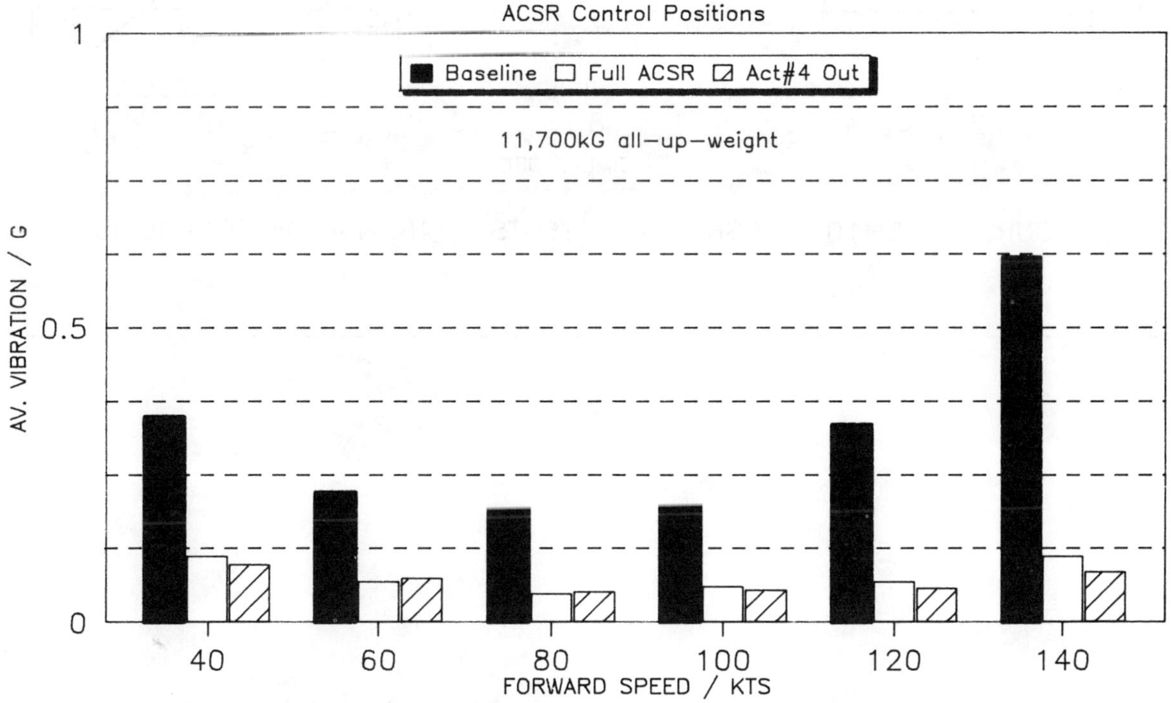

FIGURE 6 : EH101–ACSR FLIGHT TESTS – AV. VIBRATION VS FWD SPEED FOR OPTIMUM 3–ACTUATOR CONFIG (#4=AFT STBD)

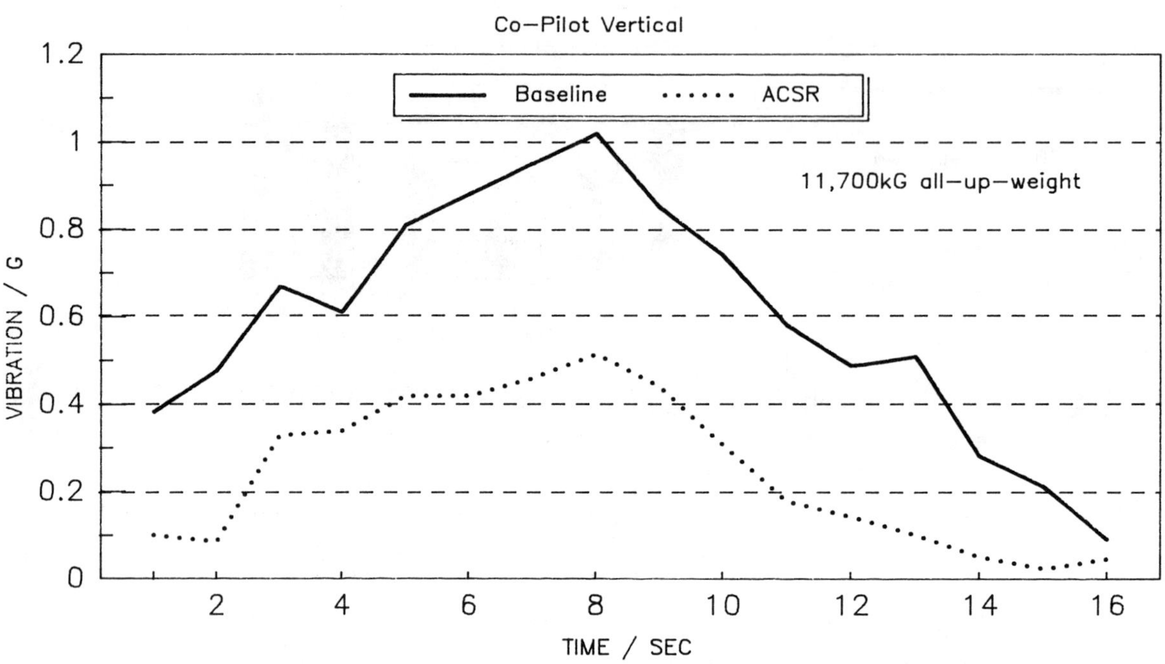

FIGURE 7 : EH101–ACSR FLIGHT TESTS – TRANSITION–TO–HOVER

STABILIZATION OF A DIGITALLY CONTROLLED ACTIVE-ISOLATION SYSTEM

S. E. Forsythe,[1] M. D. McCollum,[1] A. D. McCleary[1]

ABSTRACT

Digital control techniques have proved beneficial in many control applications because they are flexible and they lend themselves to adaptive implementations. However, in control systems that require feedback, rather than feedforward, as the control mechanism the advantages of digital control are offset by the difficulty of designing stable controllers. The time lag introduced by the sampling process itself is compounded by the time lag introduced by anti-aliasing filters. These lags, in conjunction with low sampling rates, make an otherwise stable system subject to instability if high open-loop gains are required. This paper will focus on a technique for adding an empirical model of the system under control to the control loop as a means of stabilizing the system.

INTRODUCTION

The objective of this effort is to create an active planar element that has one face held to zero motion in the presence of vibration at the other face. The active element consists of two flat piezoelectric elements bonded together: an actuator which is placed next to the vibrating face and a sensor which is used to measure motion of the other face. The actuator is used to counter motion of the vibrating surface and the sensor is used to measure residual motion of the controlled surface. The planar element has resonances but below the first resonance it can be assumed that the transfer of electrical and mechanical energy is primarily through thickness modes so that voltages applied and read are truly representative of the normal motion of the surface. An independent accelerometer is used to verify the performance of the surface element.

The strong coupling from the actuator to the sensor makes control of the system different from many active noise cancellation applications where the two are physically separated and there is little or no feedback from the compensator to the sensor.

This paper is confined to the control theory aspects of the system outlined above. The mechanical aspects of the problem are discussed in a companion paper [1].

This paper is organized into four sections: description of the control problem to be solved and a derivation of the control equation; details of the derivation of the model and control transfer functions; a presentation

[1]Naval Research Laboratory, Underwater Sound Reference Detachment, P.O. Box 568337, Orlando, FL 32856-8337, U.S.A.

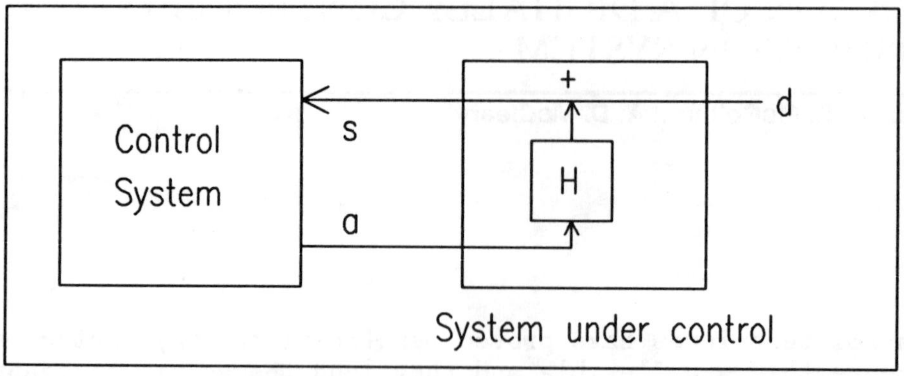

Figure 1 - overview of controller and controlled system

of experimental setup and actual real-time cancellation results; and a discussion of future work on this technique.

DESCRIPTION OF THE PROBLEM AND DERIVATION OF THE CONTROL EQUATION

Figure 1 is an illustration of the system we wish to control and the attached controller. At this point, the controller is specified as a black box so that we can focus on the mechanical/electrical system under control. A disturbance **d** is presented to the system. Without loss of generality, it can be assumed that a copy of **d** is available to the sensor input **s** of the control system (any system transfer function from **d** to **s** can be lumped with the source). Further, the control system can influence the behavior of the system at the summing point by applying an actuator control signal **a** to the system. The actuator's effect is felt at the summing point after passing through a loop transfer function, **H**. In our apparatus, the summing point represents the motion of the top surface. It is this motion and the corresponding signal **s** that we wish to drive to zero by application of the appropriate control signal **a**.

In terms of classical control theory, the controller in the loop can be represented by the transfer function **G** (which is to be determined) that converts the sensor's reading into a control output to the actuator. The equation relating **s** to **d** is then

$$s = \frac{d}{1 - GH}. \tag{1}$$

Note the sign. Both signals add at the junction. Classical control theory says that **s** can be driven toward zero over some frequency range by choosing **G** to make the product **GH** large over that range. However, this leads to the potential instability problems mentioned above.

Fig. 2 shows the "interior" of the controller: a model **M** of **H** is introduced in a parallel negative feedback loop implemented within the control system itself. This model **M** is designed to reproduce as closely as possible the response of **H**. Note that **M** can be implemented in any fashion (analog, digital, etc.).

The response of the compensated system at **s1** (the input to the control box, **G**) is

$$s1 = \frac{d}{1 + G(M - H)}. \tag{2}$$

Note the sign of **M** in the denominator; the output of **M** is subtracted at the summing point. When in the loop, the model reduces the effect of the actuator **a** on the input to **G**. In fact, if the model is perfect (**M=H**) then **s1 = d** and the input to **G** is only a replica of the original disturbance signal, uncontaminated by the effects of the actuator. This effectively opens the loop. Further, with the model in the loop, the new expression for **s** is

$$s = \frac{d(1 + GM)}{1 + G(M - H)}. \tag{3}$$

In the best possible case, if $M \equiv H$ and $GM \equiv -1$ ($G \equiv -1/M$), by construction, then **s** is identically zero independent of **d**, the disturbance. This is equivalent to perfect broadband cancellation of the disturbance at **s**.

In practice **M** cannot be set exactly equal to **H**, but it can be made arbitrarily close (see below for implementation). Making **M** close to **H** has the effect of increasing stability in the system by making the magnitude of the term **G(M-H)** in the denominator less than 1. This guarantees that the Nyquist criterion for stability will be satisfied.

Also, in practice, **G** cannot always be chosen to be exactly equal to **-1/M**. In some cases **M** may not be minimum phase, which would require that **G** be acausal. However, by constructing **G** to equal **-1/M** over a small range of frequencies it is possible to achieve very good cancellation over a small band of frequencies of interest at the expense of poor cancellation at all other frequencies.

The control system described so far is very general. Before proceeding to the implementation for digital control systems, it is worth reviewing exactly what has and has not been accomplished. The decomposition of the controller into **G** and **M** cannot improve the overall performance of the control system.

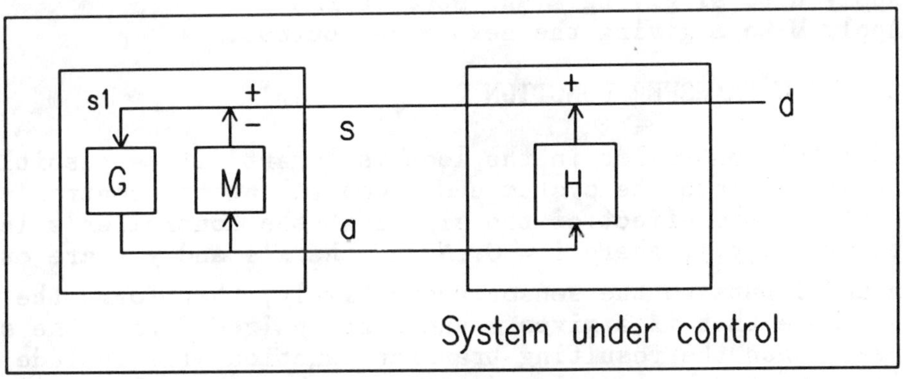

Figure 2 - control system decomposition

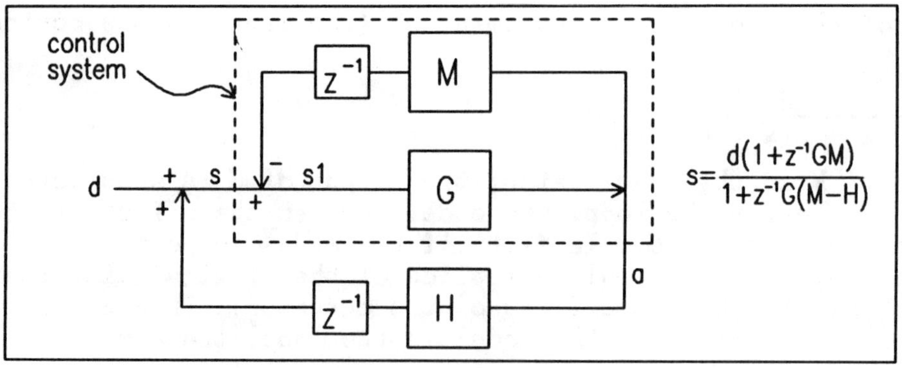

Figure 3 - digital control system with delays

That is, for a given **H**, noise level, signal ensemble, etc., the theoretical "best" controller performance remains the same when stated in terms of "greater than X dB cancellation of signals over a specified frequency range". What <u>can</u> be achieved, given the decomposition into **G** and **M**, is to determine **M** (the "modelling" aspect of the problem, which is easier than the "control" aspect) and, using **M**, then determine **G** using open-loop feedforward techniques. Future directions for adaptive implementation of the **M** and **G** functions will be discussed at the end of the paper.

DERIVATION OF THE MODEL AND CONTROL TRANSFER FUNCTIONS

We now proceed to show how the transfer functions **M** and **G** are derived for a digital control system. Figure 3 shows the system of Fig. 2 transformed into a more conventional closed-loop form. In addition, the **H** and **M** blocks of Fig. 2 have been decomposed further into new blocks concatenated with explicit sampling delays. The boundary of the control system is shown by a dotted line. For this discussion, the signal **s** is the digitized output of the control system's A/D converter (ADC) and signal **a** is the digital value to be sent to the system's D/A converter (DAC).

The program implemented in the control system is

1. Wait on clock for the next sample.
2. Read ADC and subtract the last model (**M**) output giving **s1**.
3. Apply **G** to **s1** giving **a** and send to DAC.
4. Apply **M** to **a** giving the next model output.

DERIVATION OF THE M TRANSFER FUNCTION

Because the digital controller in the loop is "smart" it is possible to inject a test signal into the system under control at the controller's output **a** and record the effect of the signal at the controller's input **s**. The set of pairs (x_i, y_i), where i = 0..N and the x's and y's are outputs to the actuator and inputs to the sensor respectively, then forms the basis for estimating **H**. Note that if a given **a** and **s** are paired during the same sample interval, then the resulting transfer function will include any sampling delays around the loop in addition to the external transfer function of the system under control.

The **M** transfer function is best realized in terms of an IIR (infinite impulse response) filter of the form

$$y_r = \sum_{k=0}^{m} b_k x_{r-k} + \sum_{k=1}^{n} a_k y_{r-k}. \tag{4}$$

This filter form includes both poles that will accurately model the mechanical or electrical response of **H** as well as zeroes that will model delays in **H**. Many techniques are available for estimating **H** based on input and output data [2].

We chose to drive the actuator with a succession of sinusoids and analyze the response at the sensor using the single-line DFT

$$H(z) = \sum_{k=0}^{N-1} h_k z^{-k}, \quad \text{where } z = e^{j\omega T}, \tag{5}$$

to determine the complex amplitude response at equally spaced frequencies below the Nyquist limit. Using this method it is possible to achieve a high signal-to-noise ratio in the estimates of **H** as a function of frequency. The actual system transfer function obtained this way is multiplied by **z** to remove a one sample delay; the resulting transfer function corresponds to **H** in Fig. 3 above. The estimate of **M** is then computed by using the least-squares fit of $\mathbf{H}(z_i)$ to an equation of the form

$$M(z) = \frac{\sum_{k=0}^{m} b_k z^{-k}}{1 - \sum_{k=1}^{n} a_k z^{-k}} \quad \text{or } M(z) = \frac{B(z)}{1 - A(z)}. \tag{6}$$

As a first approximation, the linear form of the equations

$$\left[\, 1 - A(z_i)\right] \bullet M(z_i) = B(z_i) \tag{7}$$

are fit setting $\mathbf{M}(z_i)$ equal to the measured $\mathbf{H}(z_i)$. The solution to this problem can then be used as a starting approximation for an iterative solution of Eq. (6) using the Levenburg-Marquadt or a similar technique.

Since Eq. (6) for $\mathbf{M}(z)$ (with the a_k and b_k coefficients determine above) is analytic and assumed stable, it can be evaluated at any value of z on the unit circle.

In practice, the least-squares solution to Eq. (7) was an adequate approximation of **H**.[1] Recalling Eq. (3), it is only necessary that $|\mathbf{M}-\mathbf{H}|$ be small for stability to be acceptable. For highly resonant systems, it may be necessary to fit Eq. (6) if noise in the system renders the first approximation unstable when implemented as an IIR filter.

DERIVATION OF THE G TRANSFER FUNCTION

Once **M** has been estimated, the construction of **G** such that $\mathbf{MG} \equiv -1$ over the frequency range of cancellation is theoretically straightforward.

First, it is necessary to choose a filter model for **G**. We chose the FIR (finite impulse response) filter model

$$G(z) = \sum_{k=0}^{m} g_k z^{-k} \tag{8}$$

because of its inherent stability. The construction of **G** then becomes a linear least-squares fit in the z domain using as "actual" values of $G(z_i)$

$$G(z_i) = -\frac{1}{M(z_i)} \tag{9}$$

at the appropriate z_i on the unit circle corresponding to the frequency range of interest.

It is important to use enough values z_i to overdetermine the problem. If, for example, the number of sample points on the unit circle is equal to the order of the FIR model, the solution will be exact for the frequencies specified, but may be poorly behaved even in the neighborhood of those frequencies. In practice, using roughly twice as many sample points as coefficients made the resulting FIR filters uniform in the region of interest.

EXPERIMENTAL INVESTIGATION

To investigate the above controller design, we first used a purely electrical system with no mechanical elements. Investigation of the design methodology on the mechanical system is still in progress. The electrical system consisted of

1. A summing op amp to implement the summing junction in Fig. 1.

2. An HP 35660A spectrum analyzer to generate the swept sinusoid disturbance **d** and to measure the signal **s** (the op amp's output) to verify the cancellation ability of the system.

3. A digital signal processor (ASPI's TMS32030-based product) and a 16 bit ADC and DAC module used to implement the digital control system. The sample rate of the control system was chosen as 125 kHz.

4. Software written by us to implement the **M** and **G** filters and to drive the output **a** during the determination of transfer function.
5. A cascaded low- and highpass filter pair (48 dB/octave) used to provide a realistic system transfer function **H**.

Consistent with the discussion above, the **H** of Fig. 2 consists of both the analog filters in the feedback loop and the digital delays in the ADC/DAC module. In fact, the time lag due to the I/O module alone is two sample times, arising from the buffered implementation of the ADC and DAC.

Figure 4 shows the amplitude response and group delay of the transfer function **H** from **a** to **s**.

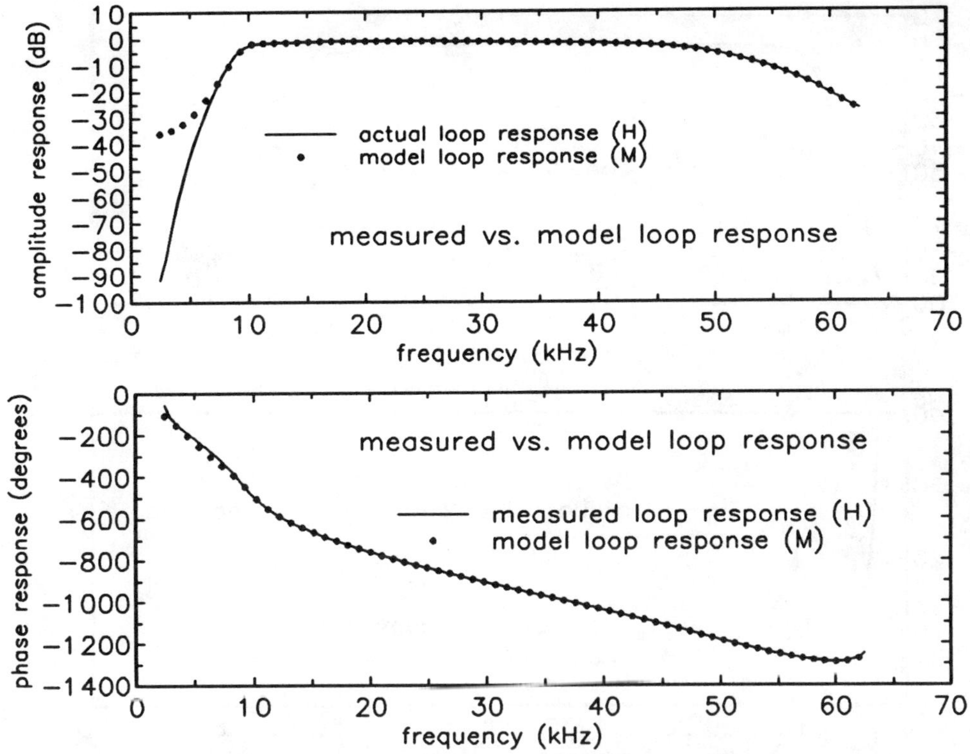

Figure 5 - actual vs. calculated response

DETERMINATION OF THE M TRANSFER FUNCTION

A filter order for **M** was chosen using m=9 and n=8 in Eq. (6) above. The comparison of **M** to the measured **H** is shown in Fig. 5. Note that the agreement of **M** and **H** is good except where |**H**| is small. The fit could be improved in this region by weighting the least-square design equations more heavily in the region below 10 kHz.

DETERMINATION OF THE G TRANSFER FUNCTION

A filter order of m=20 was chosen for the realization of **G** using Eq. (8). Forty equally spaced sample points, z_k, were chosen over each of the 4 kHz frequency ranges. A desired $G(z_k)$ was then calculated for each point using Eq. (9). These values were then fit to the FIR filter model in Eq. (8). As a check on stability, the term **G(M-H)** was plotted in the complex plane for all frequencies where **H** was measured. In all cases, the resulting path was entirely inside the unit circle with a maximum magnitude of |**G(M-H)**| equal to -6.8, -15, -18.5, and -20.3 dB (re unity) for the 10, 20, 30, and 40 kHz bands respectively. If the path had violated the Nyquist design criterion, it would have been neccssary to either:

> improve the fit of **M** to **H** at the problem frequencies to reduce the difference |**M-H**|;

or

> place additional constraints on **G** in the least-squares fit by requiring that it be small at critical frequencies.

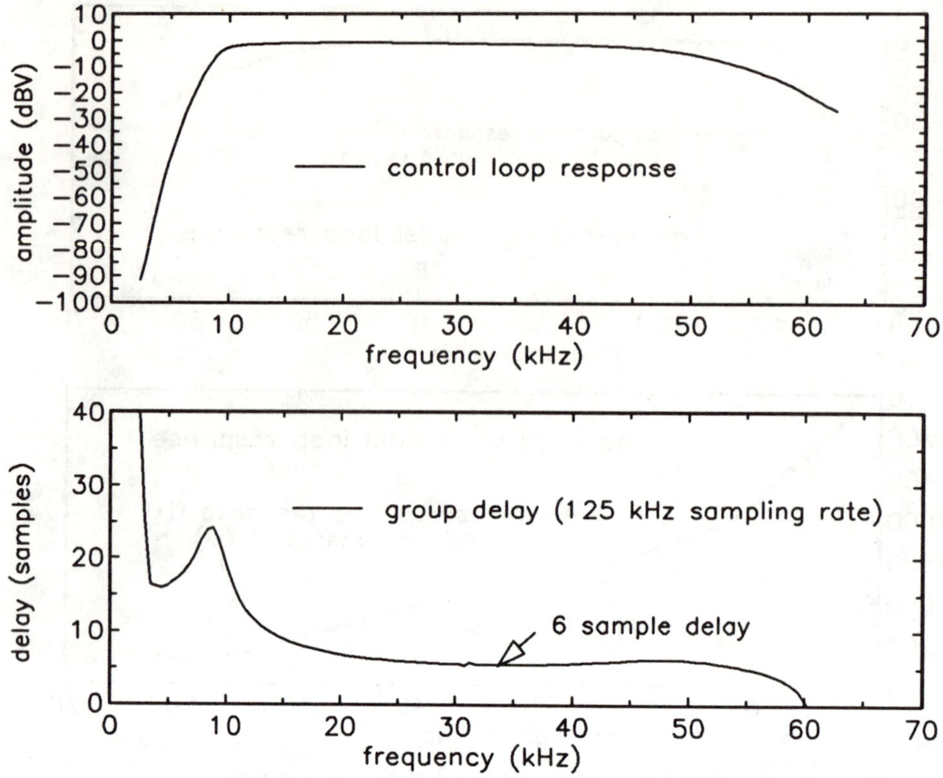

Figure 4 - measured control loop response

Note the large group delay of six samples or more in the system at virtually all frequencies. It is these large delays that make digital control difficult, since a delay of six samples produces a 180-degree phase shift every 10 kHz.

The design problem posed for this experiment was to construct four pairs of control functions **G** and **M** that would null the input **s** within a band 4 kHz wide starting at any one of four frequencies: 10, 20, 30, or 40 kHz.

Figure 6 shows the performance (20 log **s**/**d**) of the control systems designed for the specified frequencies. In all cases, the uncontrolled system has a flat magnitude response of unity (0 dB).

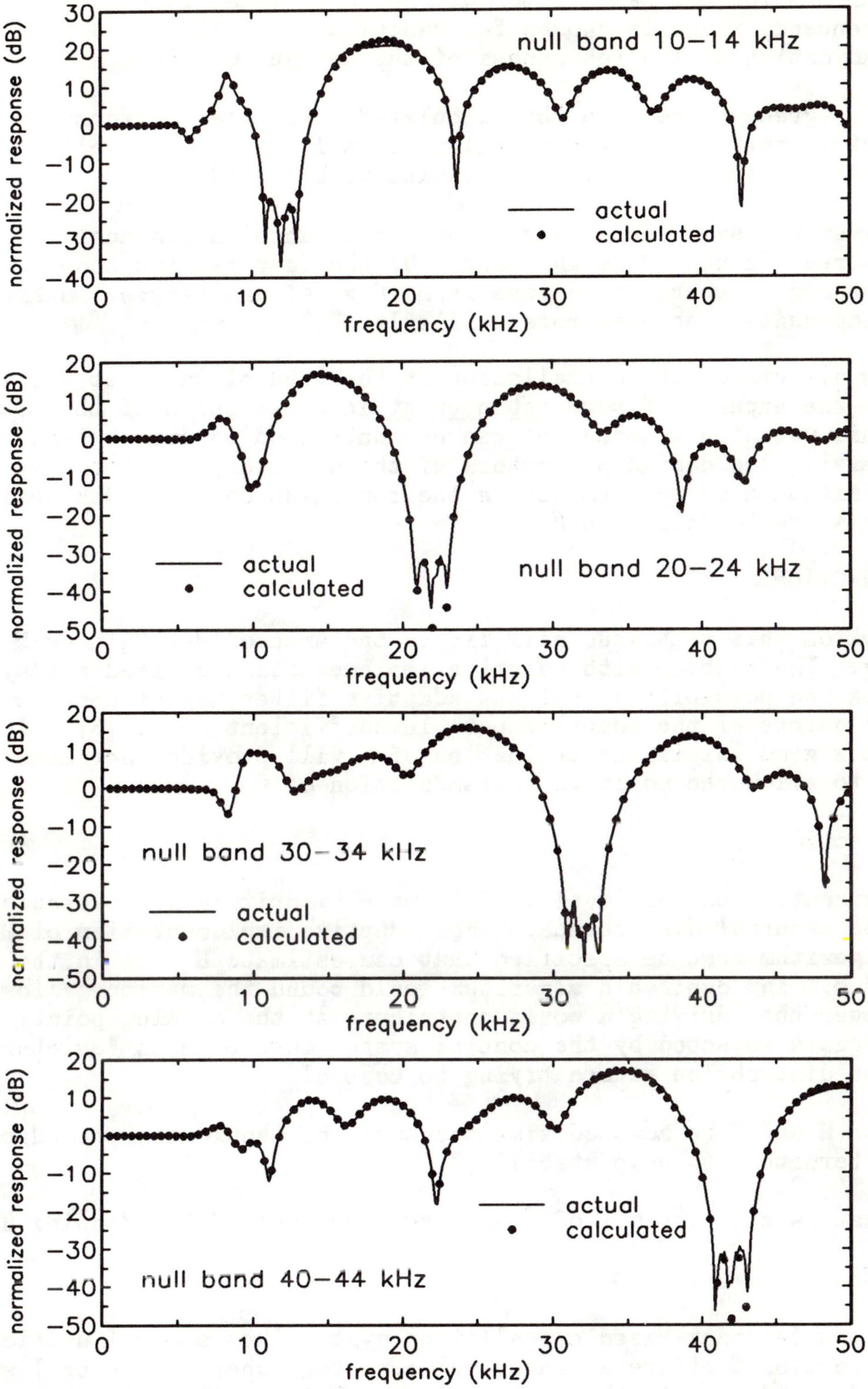

Figure 6 - performance of 10, 20, 30, and 40 kHz controllers

Four important qualitative results are evident from these plots:

1. Actual cancellation (roughly 30 dB in the center of the band) is uniform for the same filter design parameters, independent of which frequency range is chosen for cancellation. This is a good indication of the robustness of the design technique.

2. The agreement between the calculated controller performance and actual performance was excellent in all cases. This allows the designer to use simulation results with confidence.

3. Given the same design parameters, the cancellation achieved was poorest in the 10-14 kHz band. Of the four regions chosen, this is the region where the delays in **H** (Fig. 4) are largest, making the compensation problem more difficult.

4. In all cases, the cancellation in the band of interest was achieved at the expense of some <u>enhancement</u> in other parts of the spectrum. This unwanted enhancement can be controlled to some extent by varying the design parameters of the **G** filter, but the most serious limitation on performance is the cancellation bandwidth desired relative to delays in **H**.

FUTURE DIRECTIONS

Future work on this technique will lie in the area of determining **M** and **G** adaptively. The problem with adaptive implementation of feedback systems arises from the possibility that the adaptive filter may become unstable due to "noisy" nature of the adaptive path in coefficient space [3]. It is hoped that a good initial determination of **M** will provide increased stability to allow the adaptive implementation of **G**.

Future directions:

1. Currently, the estimate of **H** is made by driving **a** in absence of **d**, the external disturbance. For adaptive implementation of **M**, an algorithm must be specified that can estimate **H** even in the presence of **d**. Any desirable algorithm would bound the maximum allowable power that driving **a** would contribute at the summing point: test signals injected by the control system must be much "quieter" than the disturbance we are trying to control.

2. Can **M** and **G** be adapted simultaneously or should they be adapted alternately to help stability?

3. What is the best mix of computing time applied to adapting **M** and **G**?

CONCLUSIONS

Decomposing a feedback-based cancellation system into a model section **M** and a control section **G** offers a conceptual advantage when the control system to be designed has delays in the control loop. This conceptual advantage may result in real design advantages when the control system is to be implemented adaptively because of the increased decoupling of output to input offered by a good match of **M** to the actual control loop transfer function **H**.

References

1. M.D.McCollum, et al. 1991. "Electromechanical Modelling of an Active Isolation System" in Proceedings, Conference on Recent Advances in Active Control of Sound and Vibration, in preparation.

2. James H. McClellan. 1988. "Parametric Signal Modelling" in Advanced Topics in Signal Processing, Lim and Oppenheim, eds. Prentice Hall, pp. 1-58.

3. Widrow and Stearns. 1985. Adaptive Signal Processing, Prentice Hall, pp. 271-301.

4. Press, et al. 1986. Numerical Recipes, Cambridge University Press, pp. 52-60.

Footnotes

1. In solving all of the linear least-square systems described above, the Singular Value Decomposition (SVD) was used to overcome any near rank deficiencies in the L.S. design matrices [4]. In the least-squares formulation

$$A^tAx = A^tb,$$

the matrix A^tA was decomposed into $U\Sigma V$ and the inverse formed as

$$V^t\Sigma^{-1}U^t$$

with the elements of Σ^{-1} corresponding to $\sigma_i \leq 0.00001 \cdot \sigma_{max}$ in Σ set to 0.

ACTIVE VIBRATION CONTROL BY WAVE ISOLATION

Jiawei Lu,[1] Malcolm J. Crocker,[1] P. K. Raju[1]

ABSTRACT

Active vibration control of a flexible, finite length beam was studied both theoretically and experimentally using the wave isolation concept. In the study conducted, the beam was supported by two exciters. One exciter was located at one end of the beam and was used as a disturbing source. Another exciter was placed near the disturbing source and was used to prevent the vibration produced by the first exciter propagating to the other end of the beam. The theoretical model of the beam-exciter system was investigated and verified by experimental results. The experiment was first conducted using open-loop control for the vibration of the beam excited by a saw-tooth wave disturbing signal. The experiment was repeated using the LMS adaptive control. The vibration of the beam was greatly suppressed after only a few iterations of adaption.

1. INTRODUCTION

Modal control and wave control are the two main approaches used in the active control of the vibration of distributed dynamic systems. Modal control is used to suppress vibration in the modal space, while wave control is used to cancel or absorb waves in the real three-dimensional space. The wave control approach has been mainly studied in active noise control for acoustics problems.

Wave control approaches are used to block wave propagation or absorb wave energy in distributed dynamic systems. In structural vibration control, the junctions of the structures are usually the locations where the control is applied [1]. In principle, wave control requires the measurement of traveling waves in the propagation medium. To detect flexural traveling waves in a one-dimensional structure requires more efforts than to detect sound waves in a duct. In some applications, vibration minimisation at a single location results in wave cancellation in the system and the measurement of traveling waves is unnecessary. Wave isolation on a one-dimensional distributed system may be an example if the vibration response which indicates a control error can be measured at a single location somewhere in the system. This control error can be minimized by using the LMS (least-mean-square) adaptive control approach.

[1]Department of Mechanical Engineering, Auburn University, Auburn, AL 36849, U.S.A.

Control experiments have been conducted to investigate the effectiveness of wave isolation on a flexible beam, and are discussed in this paper. A theoretical model of the experimental beam system is also presented here.

2. EXPERIMENTAL SETUP

The schematic diagram of the control experiment is shown in Figure 1. A 3-feet long 2024 aluminum beam was supported by two B&K4809 electrodynamic exciters. The beam weighed 0.3 lb and had a cross section of 0.125×0.64 in^2. The material properties of the beam were $E=10.6\times10^6$ psi and $\rho=0.103$ lbm/in^3. The beam was fixed horizontally on the tables of the two exciters with two small aluminum blocks. Only transverse motion and rotation of the tables was possible in the vertical plane. Since the rotation stiffnesses of the tables were large, the rotation of the tables about the neutral axis of the beam was small. Rotation damping of the tables was also expected to exist. These rotation parameters were hard to evaluate. The parameters concerned with the transverse properties of the exciters could be determined experimentally.

The exciting and control signals were produced digitally by the microcomputer HP9817H, and then sent to the multi-programmer HP6944A. Digital signals were converted into analog signals via the buffer memories and D/A converters in the HP6944A. Two B&K2706 power amplifiers were used to amplify the analog signals and to drive the exciters. Two B&K4375 accelerometers were placed at the two ends of the beam to sense its vibration responses. The responses were first conditioned by Kistler 5004 charge amplifiers and then sent to the FFT analyzer B&K2032. The transfer function measured by the B&K2032 was sent to the microcomputer HP9817H and was used to modify the control signal until the error signal picked up at the free end of the beam was minimized.

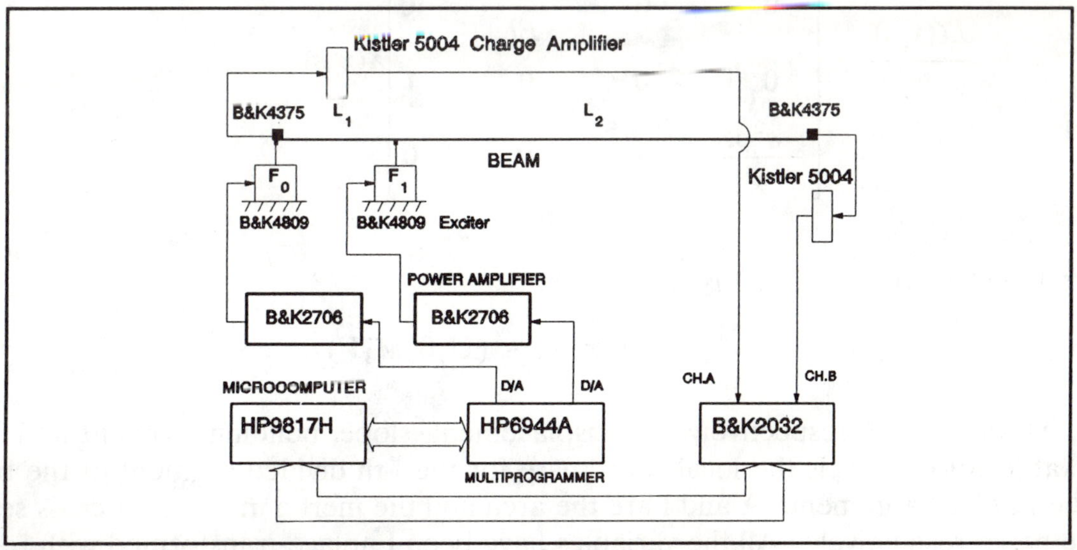

Figure 1. Schematic Diagram of the Control Experiment

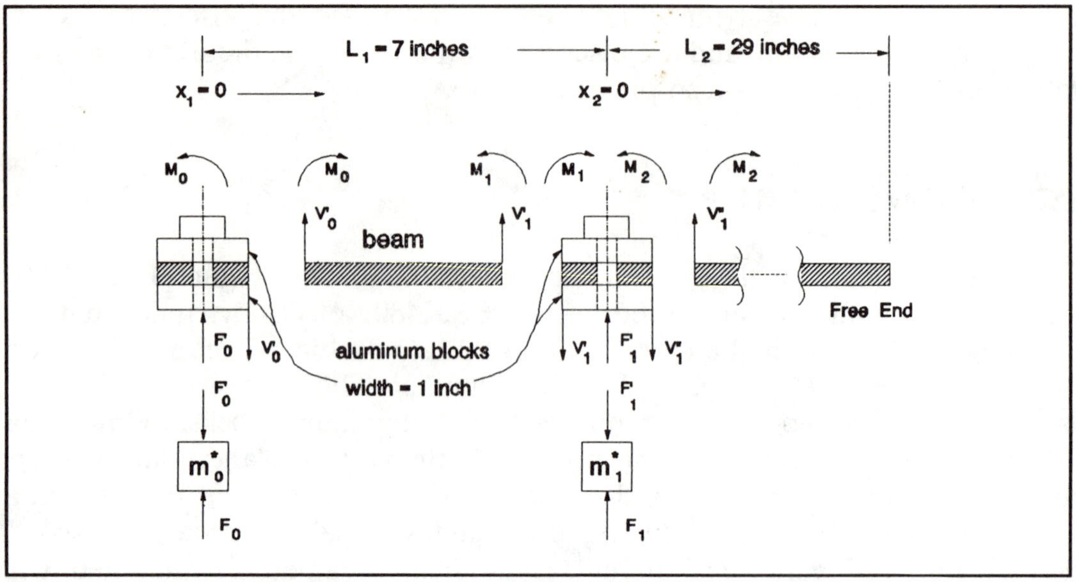

Figure 2. Free-Body Diagram of the Beam System

3. THEORETICAL MODEL

The free-body diagram of the beam-exciter system is shown in Figure 2. For the convenience of the theoretical analysis, the transverse properties of the two exciters are described by two apparent masses m_0^* and m_1^*. The experimental determination of m_0^* and m_1^* is discussed in Appendix A. The beam is divided into two segments with lengths L_i (i=1,2). For any arbitrary cross section along the i-th segment of the beam, the state equation of the bending motion can be described by

$$\frac{dX(x_i,s)}{dx_i} = \begin{bmatrix} 0 & 1 & 0 & 0 \\ 0 & 0 & \frac{1}{EI} & 0 \\ 0 & 0 & 0 & 1 \\ \frac{-s^2\rho A}{g} & 0 & 0 & 0 \end{bmatrix} X(x_i,s) \qquad (1)$$

where the state vector $\mathbf{X}(x_i,s)$ is

$$X(x_i,s) = (U,\theta,M,V)^T. \qquad (2)$$

U, θ, M and V are, respectively, the displacement, slope, bending moment and shear force at location x_i. x_i is the local coordinate for the i-th divided segment of the beam. s is the Laplace argument. A and I are the area and the inertia moment of cross section of the beam respectively. All the variables have been Laplace-transformed with respect to the time t and the assumed zero initial conditions. T represents the transpose of the matrix.

The boundary conditions at the two joints are

$$\begin{bmatrix} 0 & K_1 & -1 & 0.5 \\ m_0{}^*s^2 & 0 & 0 & 1 \end{bmatrix} X_1(0,s) = (0,F_0)^T, \tag{3}$$

$$\begin{bmatrix} 0 & K_2 & 1 & 0.5 \\ 0 & 0 & 0 & -1 \end{bmatrix} X_1(L_1,s) + \begin{bmatrix} 0 & 0 & -1 & 0.5 \\ m_1{}^*s^2 & 0 & 0 & 1 \end{bmatrix} X_2(0,s) = (0,F_1)^T, \tag{4}$$

$$\begin{bmatrix} 1 & 0 & 0 & 0 \\ 0 & 1 & 0 & 0 \end{bmatrix} X_1(L_1,s) + \begin{bmatrix} -1 & 0 & 0 & 0 \\ 0 & -1 & 0 & 0 \end{bmatrix} X_2(0,s) = (0,0)^T, \tag{5}$$

$$\begin{bmatrix} 0 & 0 & 1 & 0 \\ 0 & 0 & 0 & 1 \end{bmatrix} X_2(L_2,s) = (0,0)^T, \tag{6}$$

where K_1 and K_2 denote the rotation stiffnesses of the tables of the two exciters respectively. Equation (1) can also be written in a wave-mode form as

$$W(x_i,s) = \begin{bmatrix} e^{-ikx_i} & 0 & 0 & 0 \\ 0 & e^{-kx_i} & 0 & 0 \\ 0 & 0 & e^{ikx_i} & 0 \\ 0 & 0 & 0 & e^{kx_i} \end{bmatrix} \begin{bmatrix} A_p \\ A_n \\ B_p \\ B_n \end{bmatrix} \tag{7}$$

where

$$W(x_i,s) = (a_p(x_i), a_n(x_i), b_p(x_i), b_n(x_i))^T, \tag{8}$$

and a_p, b_p represent incident and reflected traveling waves; a_n, b_n are incident and reflected near-field waves. A_p, A_n, B_p and B_n are the incident and reflected wave amplitudes respectively. k is the wave number for the beam. The schematic diagram of thc wave propagation on the beam is shown in Figure 3.

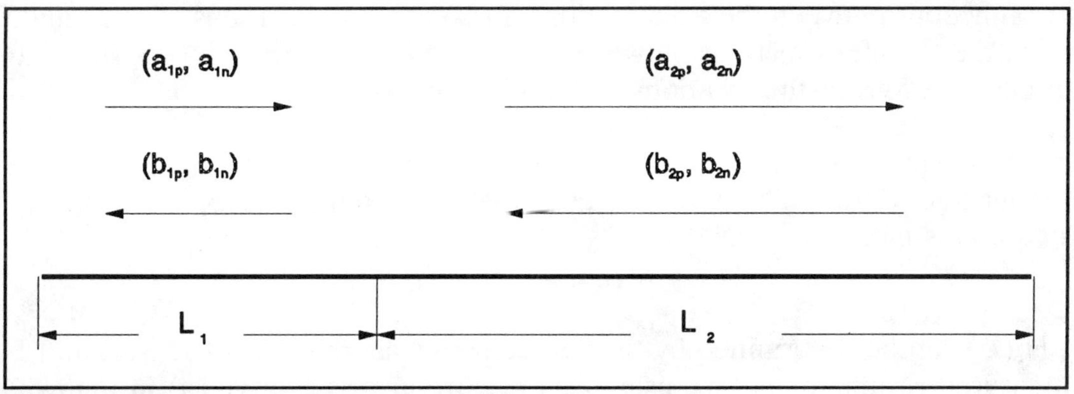

Figure 3. Schematic Diagram of Wave Propagation

Then, the boundary conditions can be rewritten as:

$$[A_{10}]\begin{bmatrix} a_{1p}(0) \\ a_{1n}(0) \end{bmatrix} + [B_{10}]\begin{bmatrix} b_{1p}(0) \\ b_{1n}(0) \end{bmatrix} = \begin{bmatrix} 0 \\ F_0 \end{bmatrix}, \tag{9}$$

$$[A_{11}]\begin{bmatrix} a_{1p}(L_1) \\ a_{1n}(L_1) \end{bmatrix} + [B_{11}]\begin{bmatrix} b_{1p}(L_1) \\ b_{1n}(L_1) \end{bmatrix} + [A_{20}]\begin{bmatrix} a_{2p}(0) \\ a_{2n}(0) \end{bmatrix} + [B_{20}]\begin{bmatrix} b_{2p}(0) \\ b_{2n}(0) \end{bmatrix} = \begin{bmatrix} 0 \\ F_1 \end{bmatrix}, \tag{10}$$

$$[A'_{11}]\begin{bmatrix} a_{1p}(L_1) \\ a_{1n}(L_1) \end{bmatrix} + [B'_{11}]\begin{bmatrix} b_{1p}(L_1) \\ b_{1n}(L_1) \end{bmatrix} + [A'_{20}]\begin{bmatrix} a_{2p}(0) \\ a_{2n}(0) \end{bmatrix} + [B'_{20}]\begin{bmatrix} b_{2p}(0) \\ b_{2n}(0) \end{bmatrix} = \begin{bmatrix} 0 \\ 0 \end{bmatrix}, \tag{11}$$

$$[A_{22}]\begin{bmatrix} a_{2p}(L_2) \\ a_{2n}(L_2) \end{bmatrix} + [B_{22}]\begin{bmatrix} b_{2p}(L_2) \\ b_{2n}(L_2) \end{bmatrix} = \begin{bmatrix} 0 \\ 0 \end{bmatrix}. \tag{12}$$

The A and B coefficient matrices are not given here since their derivations by transformation are trivial. A way of formulation suggested by B.R. Mace [2,3] can be applied here to assist in finding the solutions for a_p, a_n, b_p and b_n. Some theoretical results are presented in the following section.

4. OPEN-LOOP CONTROL TRANSFER FUNCTION $H_o(\omega)$

The control transfer function is defined as the transfer function between the disturbing signal $F_0(\omega)$ and the control signal $F_1(\omega)$. The open-loop control transfer function $H_o(\omega)$ can be determined experimentally by considering the minimisation of the vibration of the beam at the free end. The block diagram of the transfer functions for the beam system is shown in Figure 4, where it is assumed that the real experimental system is approximately linear so that the free-end response is the superposition of the effects of $F_0(\omega)$ and $F_1(\omega)$. $Y_i(\omega)$ (i=1,2,3) are the measured accelerations in the frequency domain at the locations x=0, L_1 and L_1+L_2, the free end of the beam. $H_{12}(\omega)$ and $H_{23}(\omega)$ are the transfer functions between the accelerations $Y_i(\omega)$ at these locations. $G_1(\omega)$ and $G_2(\omega)$ represent the transfer functions between the input voltage signals $F_0(\omega)$, $F_1(\omega)$ and the output accelerations at the locations x=0 and L_1 on the beam.

The block diagram in Figure 4 can be further simplified as given in Figure 5. By suppressing the vibration at the free end, the open-loop transfer function $H_o(\omega)$ is obtained and is given by

$$H_o(\omega) = -H_{n3}(\omega)/H_{c3}(\omega). \tag{13}$$

Thus, $H_o(\omega)$ can be determined by the respective measurements of $H_{n3}(\omega)$ and $H_{c3}(\omega)$. The measured results for $H_o(\omega)$, using Eq.(13), are given as curve (a) in Figures 6 and 7. The theoretical results for $H_o(\omega)$ are also plotted in the same figures.

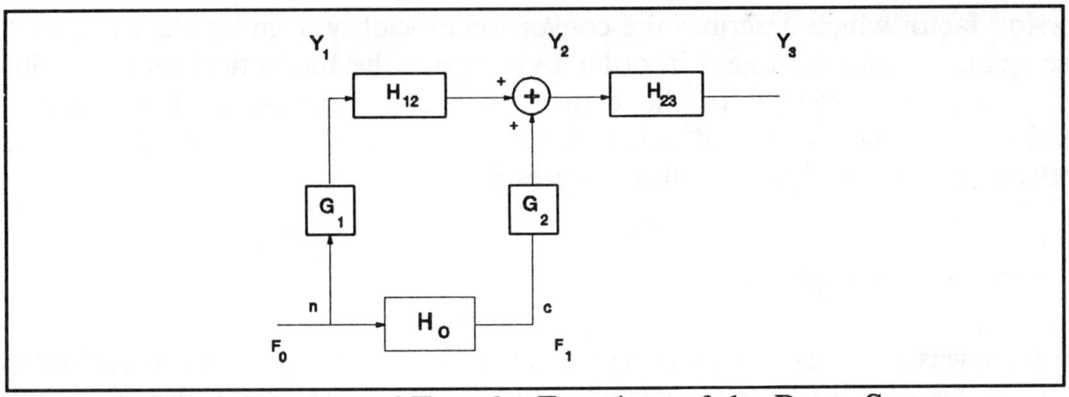

Figure 4. Block Diagram of Transfer Functions of the Beam System

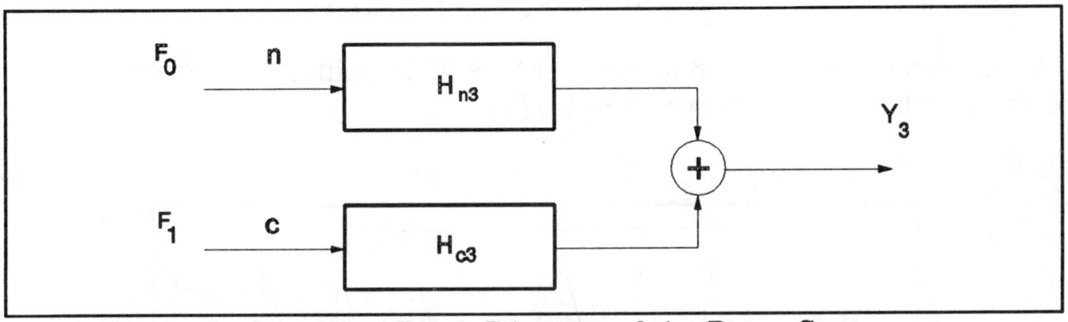

Figure 5. Simplified Block Diagram of the Beam System

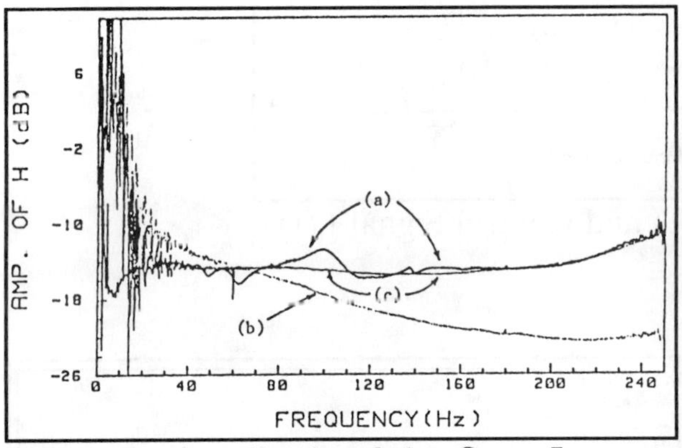

Figure 6. Amplitude of the Open-Loop Transfer Function $H_o(\omega)$

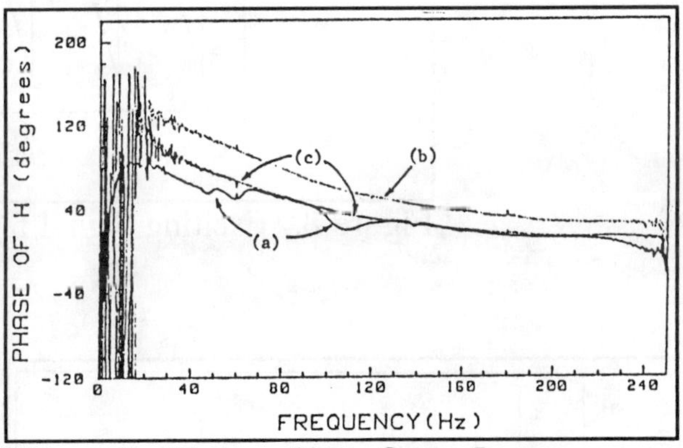

Figure 7. Phase of the Open-Loop Transfer Function $H_o(\omega)$

Two cases are considered in the theoretical results. The first case, given by curve (b), assumes that the stiffnesses K_1 and K_2 are infinite. In this case, the agreement between the experimental and theoretical results is not very good. The second case, given by curve (c), assumes that both K_1 and K_2 are 5×10^3 lb.in/rad. The theoretical results for this case agree very well with the experimental results. It is worth mentioning that the voltage inputs $F_0(\omega)$ and $F_1(\omega)$ were used in the measurements of $H_{n3}(\omega)$ and $H_{c3}(\omega)$, while the inputs $F_0(\omega)$ and $F_1(\omega)$ in force units were used for the theoretical results. A conversion factor which describes the conversion efficiency from an electric quantity to a force quantity must be found in order to compare the theoretical and experimental results. The measurement of the conversion factor denoted as T is discussed in Appendix A and the experimental results for the factor have been used to correct the theoretical results for $H_o(\omega)$ given in Figures 6 and 7 above.

5. CONTROL EXPERIMENTS

The control was first executed by using the open-loop transfer function $H_o(\omega)$ measured by Eq.(13). A 14 Hz saw-tooth periodic signal was chosen as the disturbance $F_0(\omega)$, whose time domain signal $f_0(t)$ is shown in Figure 8 as curve (a). The control signal was calculated by

$$f_1(t) = FFT^{-1}[F_0(\omega)H(\omega)], \tag{14}$$

which is also plotted in Figure 8 as curve (b). A 1024 point FFT program was used to realize the circular convolution given in Eq.(14).

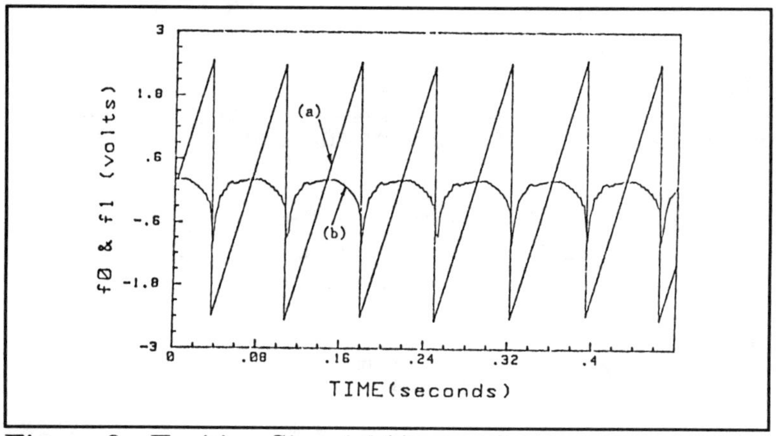

Figure 8. Exciting Signal $f_0(t)$ and Control Signal $f_1(t)$

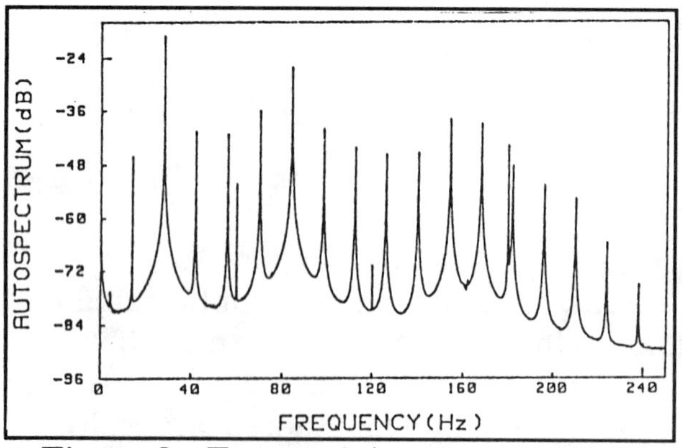

Figure 9. Free-end Acceleration before Control Force Applied

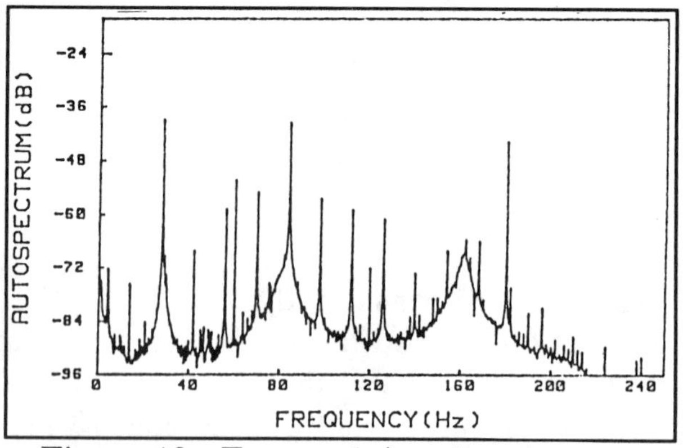

Figure 10. Free-end Acceleration after Control Force Applied

Figures 9 and 10 show the power spectral density of the free-end acceleration measured with and without the control force $f_1(t)$ applied. The corresponding vibration reductions obtained from Figures 9 and 10 are shown in the second column of Table I, where the reductions are shown only for the first eighteen harmonic components of 14 Hz. The reductions range from 13 to 31 dB. Better results can be obtained if a closed-loop control is used. This can be achieved using the LMS adaptive control approach. The block diagram for the closed-loop control system is shown in Figure 11. The free-end acceleration $Y_3(\omega)$ is used as an error response to modify the control transfer function $H(\omega)$. In practice, the real force $F_0(\omega)$ is not usually known, but the control force $F_1(\omega)$ applied is known. Hence a quantity, which represented the vibration contribution of $F_0(\omega)$ at the location x=0 on the beam, was measured instead of measuring $F_0(\omega)$. This quantity is denoted by $Y_1'(\omega)$, as shown in Figure 11, and is given by [4]

$$Y_1'(\omega) - Y_1(\omega) - H_{c1}(\omega)F_1(\omega). \qquad (15)$$

The second term on the right side of Eq.(15) represents the vibration contribution of $F_1(\omega)$ at the location x=0. The LMS adaptive algorithm for the control transfer function $H(\omega)$ is given by

$$H_{i+1}(\omega) - H_i(\omega) - H_i'(\omega)/H_{c3}(\omega), \qquad i-0,1,2,..... \qquad (16)$$

where $H_i'(\omega)$ is defined as

$$H_i'(\omega) - Y_{3,i}(\omega)/Y_{1,i}'(\omega). \qquad (17)$$

Here the subscript i is used to denote the number of the iterations.

The adaptive control experiment on the same beam model, shown in Figure 1, was conducted by using the algorithm discussed above. The disturbance $f_0(t)$ was chosen again to be a 14 Hz saw-tooth wave signal. The initial control transfer function $H_i(\omega)$ at i=0 was chosen to be zero. The vibration response at the free end of the beam with the control force $f_1(t)$ applied after the fifth iteration of Eq.(16) is shown in Figure 12.

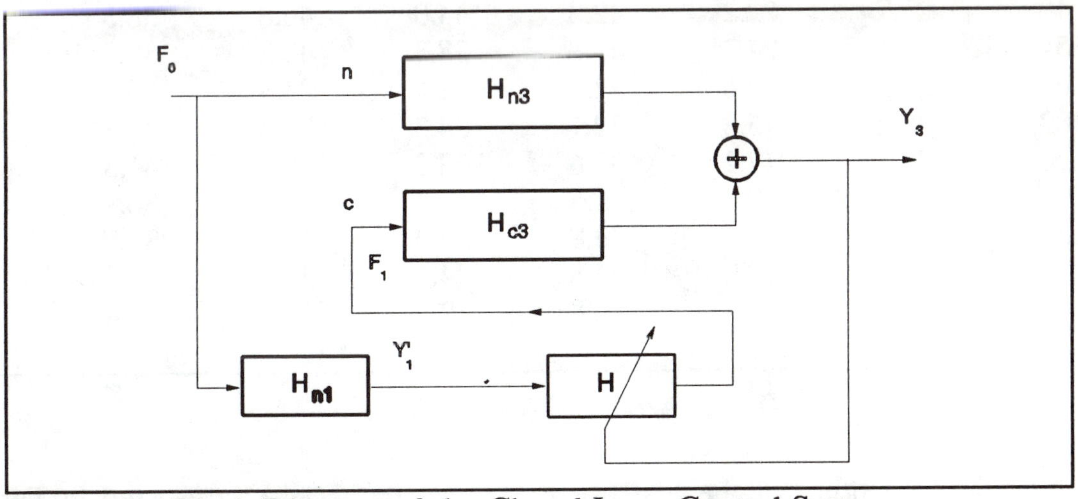

Figure 11. Block Diagram of the Closed-Loop Control System

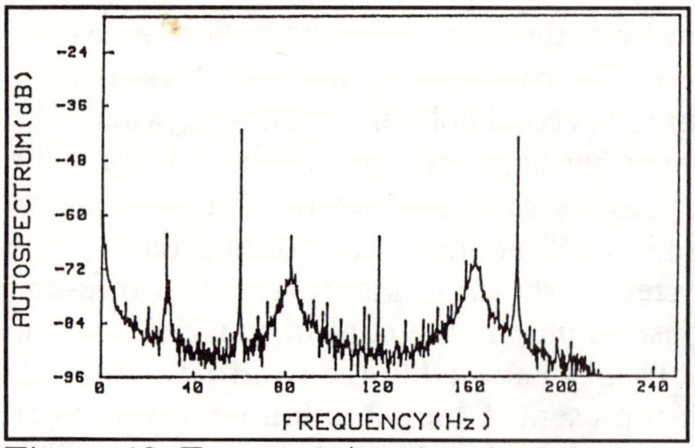

Figure 12. Free-end Acceleration after Control Force Applied for i=5

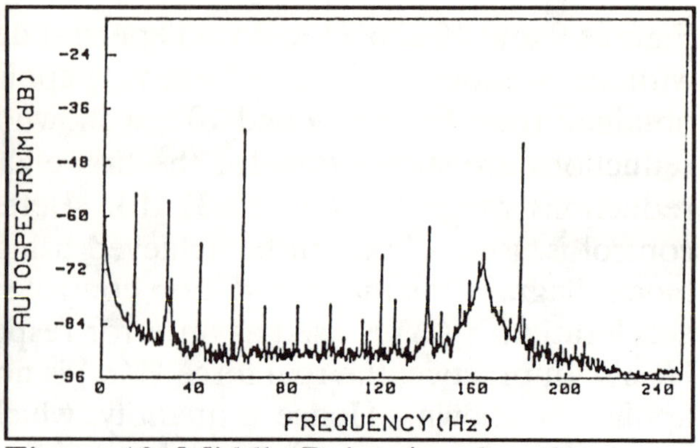

Figure 13. Middle-Point Acceleration after Control Force Applied for i=5

Table I. Vibration Reductions by Open-Loop Control (O.L.) and LMS Control

		Reductions(dB) for the Number of Iterations (i)				
	O.L.	LMS Adaptive Control				
Freq.(Hz)		i=1	i=2	i=3	i=4	i=5
14	29.4	25.6	34.7	34.4	35.8	38.5
28	19.9	20.0	35.2	40.8	43.0	47.0
42	27.6	21.9	41.3	46.0	50.0	48.4
56	17.4	19.3	41.3	40.0	49.1	45.6
70	19.0	31.1	50.5	44.8	42.7	57.5
84	13.5	20.3	38.0	46.3	46.1	46.7
98	16.3	15.2	29.9	41.3	42.7	40.3
112	14.7	18.1	34.8	39.0	46.8	45.1
126	15.2	26.6	42.5	44.2	46.3	45.2
140	27.7	22.2	42.8	44.0	46.8	38.3
154	30.1	24.7	43.8	38.3	42.2	38.3
168	26.9	12.6	24.8	32.2	34.3	46.8
182	28.2	15.8	32.1	34.3	30.8	35.8
196	28.1	17.2	35.6	32.3	38.2	41.5
210	31.5	19.4	31.5	35.4	39.4	36.9
224	23.8	20.6	33.8	32.3	33.0	36.7
238	17.5	18.0	24.8	24.1	25.9	29.2
252	16.2	7.9	13.8	12.6	12.9	12.4

The vibration reductions for the first five iterations (i=1,2,3,4,5) are listed in Table I. This shows that most frequency peaks corresponding to harmonic components disappear rapidly as the number of iterations is increased. The residual peaks at the frequencies

60 Hz, 120 Hz and 180 Hz in Figure 12 are caused by interference existing in the system due to an earth loop which is formed by the wiring connections between the instruments. The other three flatter peaks at the frequencies 28.75 Hz, 82.25 Hz and 161.5 Hz are the resonance peaks of the beam system, which are excited by residual noise. The vibration response at the middle-point (x_2=14.5 inches) of the 2nd segment of the beam was also measured and the result is shown in Figure 13. The experiment showed that the vibration on the 2nd segment of the beam was suppressed to a considerable degree, except for the range close to x_2=0, i.e. the location of the control force.

6. CONCLUSIONS

Active vibration control of a flexible beam using the wave isolation concept was investigated experimentally. The theoretical model of the beam-exciter system was developed and verified by the experimental results. The results show that the LMS/Newton adaptive control approach is effective and has the performance of fast convergence. The vibration of the beam can be greatly suppressed after a few iterations of adaption.

REFERENCES

1. A.H. von Flotow, 1986. "Traveling Wave Control for Large Spacecraft Structures," Journal of Sound and Vibration, Vol.**9**(4), 462-468.

2. B.R. Mace, 1984. "Wave Reflection and Transmission in Beams," Journal of Sound and Vibration, Vol.**97**(2), 237-246.

3. B.R. Mace, 1987. "Active Control of Flexural Vibrations," Journal of Sound and Vibration, Vol.**114**(2), 253-270.

4. J. Lu, M.J. Crocker and P.K. Raju, 1990. "Active Control of Random Vibration of a One-Dimensional Multiple Degree of Freedom Dynamic System," Proceedings of International Congress on Recent Developments in Air- and Structure-Borne Sound and Vibration, Vol.**1**, 237-242.

APPENDIX A

DETERMINATION OF APPARENT MASSES AND CONVERSION FACTORS

As shown in Figure 2, the transverse properties of the two B&K4809 electrodynamic exciters were described by two apparent masses m_0^* and m_1^*. This is convenient for analysis because m_0^* and m_1^* can be determined experimentally. Here the apparent mass is defined as the ratio of the driving force, which is generated with the driving coil, to the resulting acceleration of the exciter table. The only quantities which can be measured directly are the input voltage signal and the table acceleration. The schematic diagram of the measurement setup is shown in Figure 14. By Newton's law, we have

$$\mathbf{V}T = m^*\mathbf{a} + M\mathbf{a}, \tag{A1}$$

where **V** is the voltage signal applied to the exciters, and **a** is the measured acceleration of the mass M. The mass M is added deliberately. T and m^* are the conversion factor and the apparent mass respectively, which are two unknowns. The boldfaced letters denote those quantities which will be obtained directly from measurements. By changing the mass M, different values of acceleration **a** can be measured. T and m^* can be solved from two measurements with two different values of M.

If a transfer function $\mathbf{g}_i$, between the voltage $\mathbf{V}_i$ and the acceleration $\mathbf{a}_i$ at the mass M_i, is defined by

$$\mathbf{g}_i = \mathbf{a}_i / \mathbf{V}_i, \qquad i = 1,2, \tag{A2}$$

the apparent mass and the conversion factor of the exciter with the voltage signal $F_0(\omega)$ applied are given by (see Fig.2):

$$m_0^* = -\frac{M_1 \mathbf{g}_1 - M_2 \mathbf{g}_2}{\mathbf{g}_1 - \mathbf{g}_2}, \tag{A3}$$

and

$$T_0 = -\frac{M_1 - M_2}{\mathbf{g}_1 - \mathbf{g}_2} \mathbf{g}_1 \mathbf{g}_2. \tag{A4}$$

Equations (A3) and (A4) can also be used to solve for T_1 and m_1^* for the exciter with $F_1(\omega)$ applied. By letting M_1 be zero and M_2 be 0.132 lbm (60 gram), T_0 and m_0^* are calculated from Eqs.(A3) and (A4), and are shown in Figures 15 through 18. The values of T_1 and m_1^* are very close to those of T_0 and m_0^*, and are not shown here.

The transfer function $\mathbf{g}_i$ is frequency-dependent and may be insensitive to the increase of (M_2-M_1) in a certain frequency range so that the denominator $(\mathbf{g}_1-\mathbf{g}_2)$ in Eqs.(A3) and (A4) is close to zero in that frequency range. This effect has been observed in the experimental results for T_i and m_i^*. Drastic fluctuations in the results occur around 10 Hz as shown in Figures 15 through 18. The measurement accuracy of T_i and m_i^* deteriorates greatly within that frequency range.

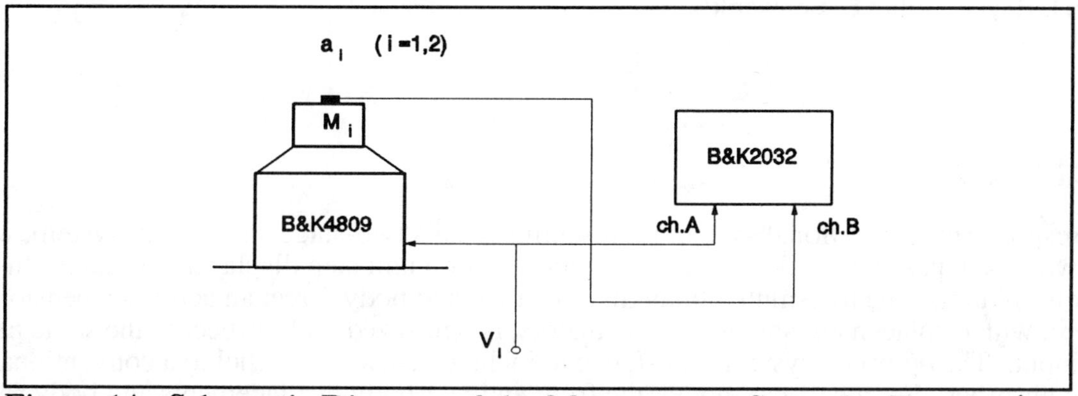

Figure 14. Schematic Diagram of the Measurement Setup for T_i and m_i^*

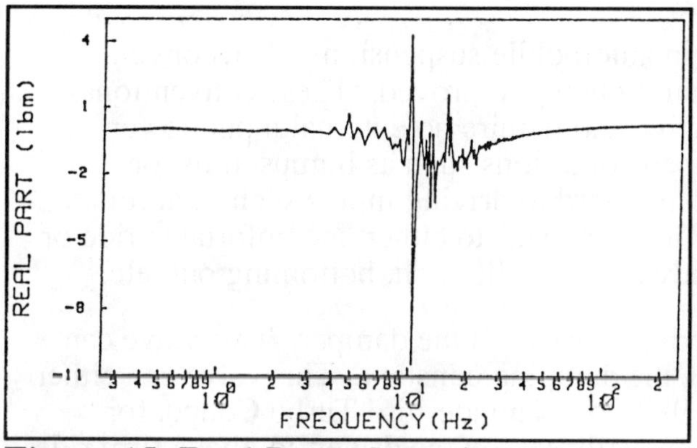

Figure 15. Real Part of the Apparent Mass m_0^*

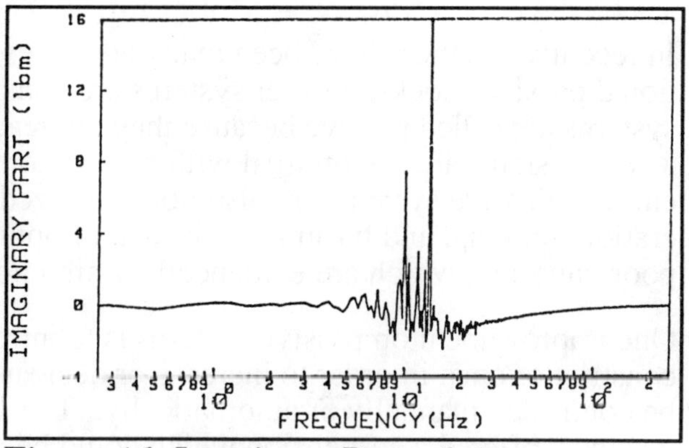

Figure 16. Imaginary Part of the Apparent Mass m_0^*

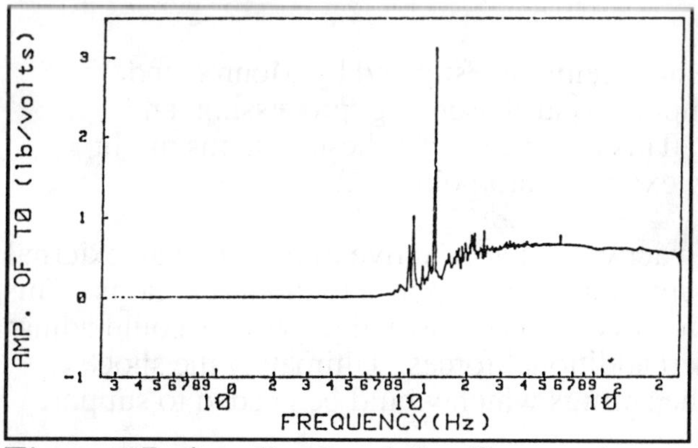

Figure 17. Amplitude of the Conversion Factor T_0

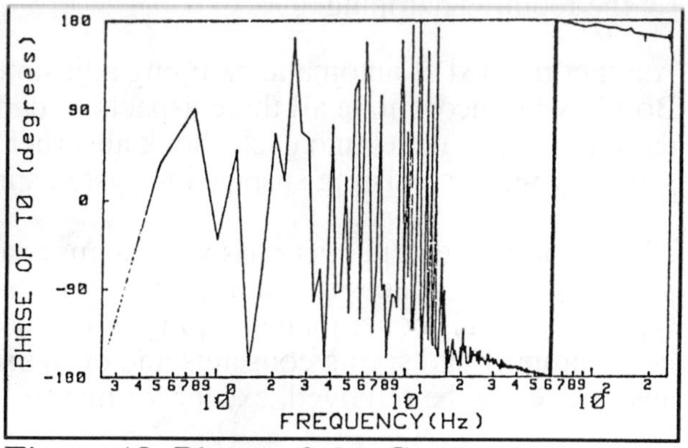

Figure 18. Phase of the Conversion Factor T_0 (Modulo 2π)

ACTIVE CONTROL OF AUTOMOBILE TWO-STAGE SUSPENSION SYSTEM – HALF CAR MODEL

Sabih I. Hayek,[1] Allan E. Schwenk[1]

ABSTRACT

The response of a vibrational system representing a car's two-stage suspension system, with a conventional passive shock absorber is simulated for a unit step displacement input. In order to minimize the transmitted body displacement and body force, an active suspension system, with displacement and velocity actuators, is optimized with respect to the same unit step input. The optimum system is designed to include actuators parallel to a conventional shock absorber. For a half-car model, the front and rear body displacements and body forces are simulated when the front and rear tires encounter unit step displacement inputs. Active control of the half car is optimized with independent displacement and velocity actuators at the front and rear suspensions.

INTRODUCTION

In recent years there have been many advances in automobile suspensions. The conventional passive shock-absorber systems are constantly being improved. These conventional systems are called passive because they are reactive and require no external input power. These systems are not optimal with respect to road conditions such as bumps, hills, or curves. Passive systems are also not optimized in regard to driving inputs such as acceleration, braking, and turning. This lack of optimization leads to either uncomfortable ride or poor handling, which are evidenced by high body forces, roll, pitch, bottoming out, etc.

One improvement on passive systems is to introduce a valve in the damper. This valve can constrict or open in order to increase or decrease the damping constant. This valve can either be controlled manually or automatically. The 1987 Ford Thunderbird Turbo Coupe, for instance, offered a suspension which could be manually set, by the driver, to give a firm ride or set on automatic [1]. A firm setting would have a high damping constant. The automatic selection would set the damping according to input conditions. This could be achieved with a series of accelerometers, to sense input, and a central microprocessor, to analyze the data and set the required damping.

Another method of automatic damping adjustment is being investigated by Monroe and Bosch, who incorporate all three aspects of damping control--sensing, processing, and damping adjustment--into each shock absorber. This is done so that these systems might someday replace existing suspension systems in existing automobiles [2].

Another improvement over passive systems is an active system. Active implies that an external power source produces hydraulic forces which are exerted in response to input functions. This type of system is better than damping-adjustment systems, because active systems could adjust both damping and spring constants and even exert additional forces. Ultimately the shock absorber could be removed, except for maybe the springs which would be needed to support

[1]Department of Engineering Science and Mechanics and the Applied Research Laboratory, 227 Hammond Building, Penn State University, University Park, PA 16802, U.S.A.

the dead weight [3]. Lotus and their parent company GM have been experimenting with these systems; however, active suspensions are presently limited by cost and computer power [4].

The computations that must be performed by an on-board computer limit the introduction of such systems because of the many different inputs that must be processed in each time step, and because of the necessity of small time steps. These computations have been investi-gated since the early 1970's. Thompson [5,6], for instance, has investigated a model representing one-fourth of a car that encounters a unit step input. Other authors have developed models for semi-active or active control of quarter-car suspension systems [7-9]. An active adaptive control of a two-stage mount system was developed recently by Sommerfeldt [10-11].

PASSIVE HALF CAR MODEL

The model for a passive half-car is shown in Fig. 1. In this model k_t is the spring stiffness constant of the tire, k_s is the spring constant of the shock absorber, and k_d is the damping constant of the shock absorber. The mass, m_1, is the mass of the front or rear tire, wheel, axle and suspension, and the mass, m_2, is roughly equal to one-half of the mass of the body of a car. These values are quoted from Thompson [5]. The displacement x_1 repre-sents the front tire's displacement; x_2 is the front end of the body's displacement (at a point directly above the front tire); x_3 is the rear tire's displacement; and x_4 represents the rear end of the body's displacement (at a point directly above the rear tire). The response of the car body is computed due to input displacements x_{01} and x_{02} imposed on the front and rear tires, respectively.

The differential equations of motion were kept to first order by introducing variables for the respective velocities. For a half car model, this results in eight equations for a four degree of freedom system:

$$\dot{x}_1 = x_5 \ , \quad \dot{x}_2 = x_6 \ , \quad \dot{x}_3 = x_7 \ , \quad \dot{x}_4 = x_8 \ ,$$
$$\dot{x}_5 = [k_t(x_{01}-x_1) - k_s(x_1-x_2) - k_d(x_5-x_6)]/m_1$$
$$\dot{x}_6 = [4k_s(x_1-x_2) + 4k_d(x_5-x_6) - 2k_s(x_3-x_4) - 2k_d(x_7-x_8)]/m_2$$
$$\dot{x}_7 = [k_t(x_{02}-x_3) - k_s(x_3-x_4) - k_d(x_7-x_8)]/m_1$$
$$\dot{x}_8 = [4k_s(x_3-x_4) + 4k_d(x_7-x_8) - 2k_s(x_1-x_2) - 2k_d(x_5-x_6)]/m_2 \qquad (1)$$

Making the change from absolute displacements to relative displacements:

$$\hat{x}_1 = x_1-x_{01} \ , \quad \hat{x}_2 = x_2-x_{01} \ , \quad \hat{x}_3 = x_3-x_{02} \ , \quad \hat{x}_4 = x_4-x_{02} \ ,$$
$$\hat{x}_5 = x_5 \ , \quad \hat{x}_6 = x_6 \ , \quad \hat{x}_7 = x_7 \ , \quad \hat{x}_8 = x_8, \qquad (2)$$

then equation (1) can be rewritten in the relative displacement form:

$$\dot{\hat{x}}_1 = \hat{x}_5 - \dot{x}_{01} \ , \quad \dot{\hat{x}}_2 = \hat{x}_6 - \dot{x}_{01} \ , \quad \dot{\hat{x}}_3 = \hat{x}_7 - \dot{x}_{02} \ , \quad \dot{\hat{x}}_4 = \hat{x}_8 - \dot{x}_{02} \ ,$$
$$\dot{\hat{x}}_5 = [-k_t\hat{x}_1 - k_s(\hat{x}_1-\hat{x}_2) - k_d(\hat{x}_5-\hat{x}_6)]/m_1$$
$$\dot{\hat{x}}_6 = [4k_s(\hat{x}_1-\hat{x}_2) + 4k_d(\hat{x}_5-\hat{x}_6) - 2k_s(\hat{x}_3-\hat{x}_4) - 2k_d(\hat{x}_7-\hat{x}_8)]/m_2$$
$$\dot{\hat{x}}_7 = [-k_t\hat{x}_3 - k_s(\hat{x}_3-\hat{x}_4) - k_d(\hat{x}_7-\hat{x}_8)]/m_1$$
$$\dot{\hat{x}}_8 = [4k_s(\hat{x}_3-\hat{x}_4) + 4k_d(\hat{x}_7-\hat{x}_8) - 2k_s(\hat{x}_1-\hat{x}_2) - 2k_d(\hat{x}_5-\hat{x}_6)]/m_2 \qquad (3)$$

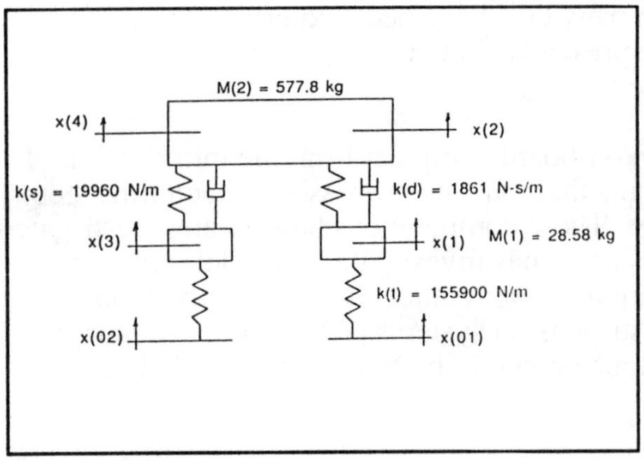

Fig. 1. Diagram of the Half Car Passive System

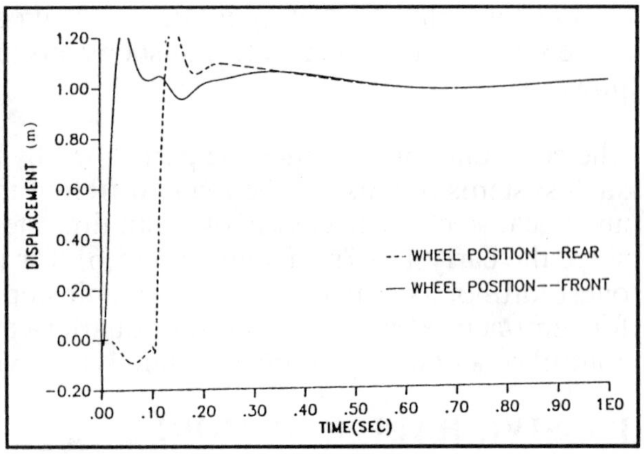

Fig. 2. Front and Rear Tire Displacements

THE RESPONSE TO THE INPUT FUNCTION

The input function used is the unit step function for both tires, separated by a time interval. This would be equivalent to a car's front tire hitting an obstacle first, and then, after some time interval, t_o, the rear tire would hit the same obstacle. For the front tire the initial relative displacements, namely $\hat{x}_1$ and $\hat{x}_2$, were set equal to -1, and then +1 was added to all of the calculated values of $\hat{x}_1$ and $\hat{x}_2$ in order to compute the absolute displacements x_1 and x_2. For the rear wheel this approach will not work. Instead of setting the initial conditions of $\hat{x}_3$ and $\hat{x}_4$, it was necessary to add the step function x_{02}. The input step function x_{02} remains 0 until time t_o, when it changes to 1. This t_o was set equal to 0.1 s. This represents a car with a 2.5 m long wheel base which is travelling at 25 m/s.

The response of the passive half car model can be seen in figures 2-4. In these figures, the response of the car is shown for a unit input of $x_{01} = x_{02} = 1.0$m. Due to the linearity of the system, the response of the car is proportional to any given input $x_{01} = x_{02}$ less than unity. In figure 2 the front and rear wheel displacements are plotted as functions of time. In figure 3 the front and rear body displacements are plotted versus time. The front of the body peaks at about 1.38 m and the rear body peaks at about 1.48 m. In figure 4 the forces exerted on the front end and the rear end of the body are plotted with respect to time. The force exerted on the front reaches a maximum value of about 315 kN, while the force exerted on the rear end peaks at a value of about 395 kN.

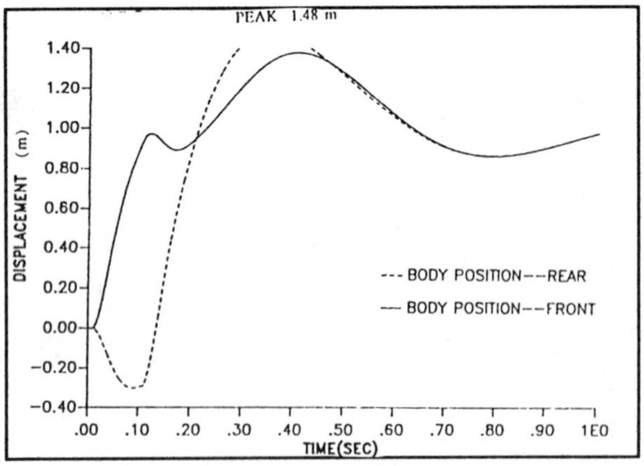

Fig. 3. Front and Rear Body Displacements

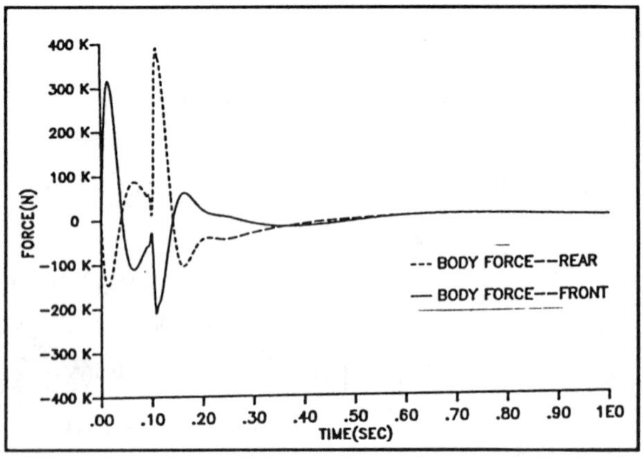

Fig. 4. Front and Rear Body Forces

ACTIVE HALF CAR MODEL

In an effort to minimize both the front and rear body displacements, the shock absorbers will be replaced by two actuator forces. This system is pictured in figure 5. The tire spring stiffness constant, k_t, and the tire masses, m_1, remain the same.

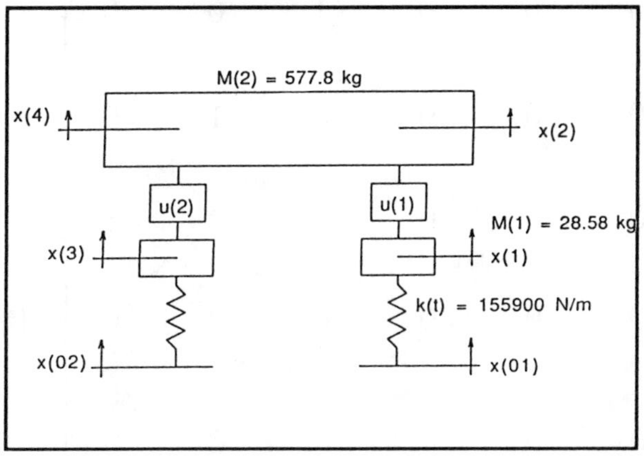

Fig. 5. Diagram of the Half Car Active System

The first order equations derived for the active half car are shown below.

$$\dot{x}_1 = x_5 \,, \quad \dot{x}_2 = x_6 \,, \quad \dot{x}_3 = x_7 \,, \quad \dot{x}_4 = x_8,$$
$$\dot{x}_5 = k_t(x_{01}-x_1)/m_1-u_1/m_1$$
$$\dot{x}_6 = 4u_1/m_2 - 2u_2/m_2$$
$$\dot{x}_7 = k_t(x_{02}-x_3)/m_1-u_2/m_1$$
$$\dot{x}_8 = - 2u_1/m_2 + 4u_2/m_2 \tag{4}$$

The active half car equations can also be written in relative form:

$$\dot{\hat{x}}_1 = \hat{x}_5 - \dot{x}_{01} \,, \quad \dot{\hat{x}}_2 = \hat{x}_6 - \dot{x}_{01} \,, \quad \dot{\hat{x}}_3 = \hat{x}_7 - \dot{x}_{02} \,, \quad \dot{\hat{x}}_4 = x_8 - \dot{x}_{02},$$
$$\dot{\hat{x}}_5 = -k_t\hat{x}_1/m_1-u_1/m_1$$
$$\dot{\hat{x}}_6 = 4u_1/m_2 - 2u_2/m_2$$
$$\dot{\hat{x}}_7 = -k_t\hat{x}_3/m_1-u_2/m_1$$
$$\dot{\hat{x}}_8 = - 2u_1/m_2 + 4u_2/m_2 \tag{5}$$

Again $\dot{x}_{01}$ and $\dot{x}_{02}$ are equal to 0; so, these equations can be written in matrix form:

$$[\dot{\hat{x}}] = [\mathbf{A}]\,[\hat{x}] + [\mathbf{b}]\,[\mathbf{u}] \tag{6}$$

where:

$$[A] = \begin{bmatrix} 0 & 0 & 0 & 0 & 1 & 0 & 0 & 0 \\ 0 & 0 & 0 & 0 & 0 & 1 & 0 & 0 \\ 0 & 0 & 0 & 0 & 0 & 0 & 1 & 0 \\ 0 & 0 & 0 & 0 & 0 & 0 & 0 & 1 \\ \frac{-k_t}{m_1} & 0 & 0 & 0 & 0 & 0 & 0 & 0 \\ 0 & 0 & 0 & 0 & 0 & 0 & 0 & 0 \\ 0 & 0 & \frac{-k_t}{m_1} & 0 & 0 & 0 & 0 & 0 \end{bmatrix} \qquad [\mathbf{b}] = \begin{bmatrix} 0 & 0 \\ 0 & 0 \\ 0 & 0 \\ 0 & 0 \\ \frac{-1}{m_1} & 0 \\ \frac{4}{m_2} & \frac{-2}{m_2} \\ 0 & \frac{-1}{m_1} \\ \frac{-2}{m_2} & \frac{4}{m_2} \end{bmatrix}$$

OPTIMIZATION

The performance index J is the integral of the weighted sum of squares of responses of the system:

$$J = \frac{1}{2}\int_0^{\infty} [\rho_1 u_1^2 + \rho_2 u_2^2 + q_1(x_{01} - x_1)^2 + q_2(x_1 - x_2)^2$$

$$+ q_3(x_{02} - x_3)^2 + q_4(x_3 - x_4)^2]\, dt \tag{7}$$

where ρ_1, ρ_2, q_1, q_2, q_3, and q_4 are weighting coefficients. One can pick one of these weighting coefficients, such as $q_4 = 1$, without loss of generality. This means that the optimization problem requires a five dimensional search of the remaining weighting coefficients. The performance index can be put into matrix form using relative displacements:

$$J = \frac{1}{2}\int_0^{\infty} \{[\mathbf{u}]^T [\rho] [\mathbf{u}] + [\hat{\mathbf{x}}]^T [\mathbf{Q}] [\hat{\mathbf{x}}]\}\, dt \tag{8}$$

The matrices in the performance index are:

$$[\mathbf{Q}] = \begin{vmatrix} q_1+q_2 & -q_1 & 0 & 0 & 0 & 0 & 0 & 0 \\ -q_1 & q_1 & 0 & 0 & 0 & 0 & 0 & 0 \\ 0 & 0 & q_3+q_4 & -q_3 & 0 & 0 & 0 & 0 \\ 0 & 0 & -q_3 & q_3 & 0 & 0 & 0 & 0 \\ 0 & 0 & 0 & 0 & 0 & 0 & 0 & 0 \\ 0 & 0 & 0 & 0 & 0 & 0 & 0 & 0 \\ 0 & 0 & 0 & 0 & 0 & 0 & 0 & 0 \\ 0 & 0 & 0 & 0 & 0 & 0 & 0 & 0 \end{vmatrix}$$

and:

$$[\mathbf{u}] = \begin{bmatrix} u_1 \\ u_2 \end{bmatrix} \qquad [\rho] = \begin{bmatrix} \rho_1 & 0 \\ 0 & \rho_2 \end{bmatrix} \tag{9}$$

The optimization problem is solved as a stochastic linear optimization problem [5], so that the optimum value of the input force [**u**] is given by:

$$[\mathbf{u}_{opt}] = [\mathbf{k}]^T[\hat{\mathbf{x}}] = -[\mathbf{b}]^T[\mathbf{P}][\hat{\mathbf{x}}][\rho]^{-1} \tag{10}$$

Where [**k**] is the [8 x 2] Kalman feedback gain matrix, and [P] is the [8 x 8] solution to the Riccati equation:

$$[\mathbf{P}][\mathbf{A}]+[\mathbf{A}]^T[\mathbf{P}] - [\mathbf{P}][\mathbf{b}][\mathbf{b}]^T[\mathbf{P}][\rho]^{-1}+[\mathbf{Q}]=0 \tag{11}$$

In this equation, [**A**] and [**b**] are obtained from the matrix representation of the system's differential equations (Eq. 6), and [ρ] and [**Q**] are from the matrix representation of the performance index (Eq. 9).

To simplify the search for an optimum solution of Equation (11) by reduction of the number of independent weighting coefficients, we elected to choose the same weighting coefficients for the front and rear suspensions, i.e., $q_2 = q_4 = 1.0$, $\rho_1 = \rho_2$ and $q_1 = q_3$. This reduces the number of weighting coefficients to two. To reduce the number to one, we elected to follow Thompson's [5] approach. To prevent the bottoming of the front and rear suspension systems or the loss of tires - road contacts for at least 99.7% of the time, the rms values of the relative front and rear wheel and tire deflections $\hat{x}_1$, $\hat{x}_2$, $\hat{x}_3$ and $\hat{x}_4$ are chosen to be less than one-third of the static dead-weight deflections of the wheels $x_w = 6.3$ cm and tires $x_t = 2.0$ cm. A convenient estimate of the weighting coefficients q_1 is

$$q_1 = q_3 = (x_w/x_t)^2 \cong 10 \tag{12}$$

Such an estimate was chosen by Thompson [5] based on the assumption that all the weighted integral squares of the six terms in the the performance index integral J is eq. (7) are approximately equal. Thus, the search for an optimum solution to eq. 11 is obtained for a one-dimensional search as a function of index $\rho = \rho_1 = \rho_2$.

Nine values of ρ were used, and the Riccati equation was solved by using the lqr function in the software package Control-C. The values of ρ and their corresponding values of [k] are shown in Table I for three values of ρ.

TABLE I - KALMAN FEEDBACK GAIN VALUES
FOR SELECTED ρ's

ρ	$.2\times10^{-9}$	1.0×10^{-9}	3.0×10^{-9}
k_{11}	172290	53240	22103
k_{21}	-11050	-5346	-2765
k_{12}	-70710	-31623	-18257
k_{22}	0	0	0
k_{13}	-11050	-5346	-2765
k_{23}	172290	53240	22103
k_{14}	0	0	0
k_{24}	-70710	-31623	-18257
k_{15}	2140	1221	762
k_{25}	160	43	15
k_{16}	-6440	-3582	-2587
k_{26}	-1600	-909	-672
k_{17}	160	43	15
k_{27}	2140	1221	762
k_{18}	-1600	-909	-672
k_{28}	-6440	-3582	-2587

For the half car model there are 16 values of k, i.e.,

$$[k]^T = \begin{matrix}[k_{11}\ k_{12}\ k_{13}\ k_{14}\ k_{15}\ k_{16}\ k_{17}\ k_{18}] \\ [k_{21}\ k_{22}\ k_{23}\ k_{24}\ k_{25}\ k_{26}\ k_{27}\ k_{28}]\end{matrix}$$

where one set of eight k's contribute to the evaluation of each u, with these eight weighting coefficients applied to the relative displacements. Changing back to absolute displacements gives:

$$u_{1opt} = k_{11}(x_1-x_{01}) + k_{12}(x_2-x_{01}) + k_{13}(x_3-x_{02}) + k_{14}(x_4-x_{02}) + k_{15}x_5 + k_{16}x_6 + k_{17}x_7 + k_{18}x_8 \quad (13)$$

A convenient algebraic rearrangement leads to:

$$u_{1opt} = k_{11}(x_1 - x_2) + k_{15}(x_5 - x_6) + (k_{11} + k_{12})(x_2 - x_{01}) + k_{13}(x_3 - x_{02})$$
$$+ k_{14}(x_4 - x_{02}) + (k_{15} + k_{16})x_6 + k_{17}x_7 + k_{18}x_8 \qquad (14)$$

From Eq. (14) one can see that u_{1opt} can be realized using, on the front tire, a passive shock absorber in parallel with several active elements. The passive element should have a spring with spring constant equal to k_{11} and a damper with damping constant equal to k_{15}. The remaining terms in u_{1opt} can be produced by the active elements, f_1:

$$f_1 = (k_{11}+k_{12})(x_2-x_{01}) + k_{13}(x_3-x_{02}) + k_{14}(x_4-x_{02}) + (k_{15}+k_{16})x_6$$
$$+ k_{17}x_7 + k_{18}x_8 \qquad (15)$$

The analysis used above also applies for u_{2opt}:

$$u_{2opt} = k_{21}(x_1-x_{01}) + k_{22}(x_2-x_{01}) + k_{23}(x_3-x_{02}) + k_{24}(x_4-x_{02})$$
$$+ k_{25}x_5 + k_{26}x_6 + k_{27}x_7 + k_{28}x_8 \qquad (16)$$

After algebraic rearrangement:

$$u_{2opt} = k_{23}(x_3-x_4) + k_{27}(x_7-x_8) + (k_{23}+k_{24})(x_4-x_{02}) + k_{21}(x_1-x_{01})$$
$$+ k_{22}(x_2-x_{01}) + (k_{27}+k_{28})x_8 + k_{25}x_5 + k_{26}x_6 \qquad (17)$$

In this case, u_{2opt} can be produced with a passive shock absorber and active elements in the rear suspension. The shock absorber should have a spring with spring constant equal to k_{23} and a damper with damping constant equal to k_{27}. The remaining terms in u_{2opt} will be produced by active elements f_2:

$$f_2 = (k_{23}+k_{24})(x_4-x_{02}) + k_{21}(x_1-x_{01}) + k_{22}(x_2-x_{01}) + (k_{27}+k_{28})x_8$$
$$+ k_{25}x_5 + k_{26}x_6 \qquad (18)$$

THE RESPONSE TO THE INPUT FUNCTION

The input function was the unit step function and was implemented in the same manner as in the passive half car. The response of various active systems can be seen in figures 6 through 14. For $\rho = 0.2 \times 10^{-9}$ the peak body positions, front and rear, are 1.07 m and 1.10 m, respectively; and the peak body forces, front and rear, are 411 kN and 550 kN, respectively. For $\rho = 1.0 \times 10^{-9}$ the peak body positions, front and rear, are 1.05 m and 1.08 m, respectively; and the body forces peak at 280 kN and 315 kN, front and rear. For $\rho = 3.0 \times 10^{-9}$ the maximum values of body position, front and rear, are 1.05 m and 1.06 m, respectively; and the maximum values of body force, front and rear, are 200 kN and 220 kN, respectively.

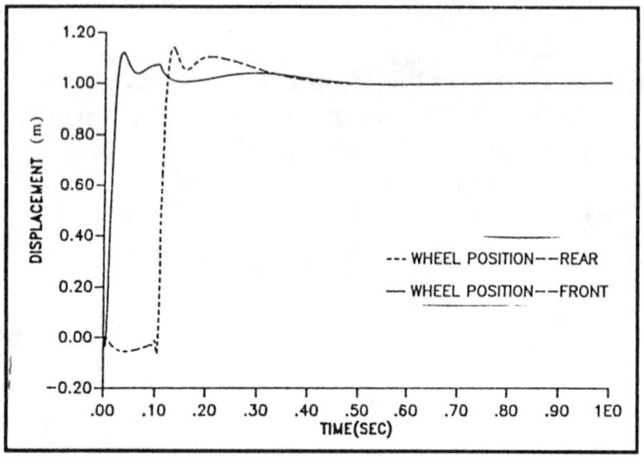

Fig. 6. Front and Rear Tire Displacements for Active Half Car $\rho = .2 \times 10^{-9}$

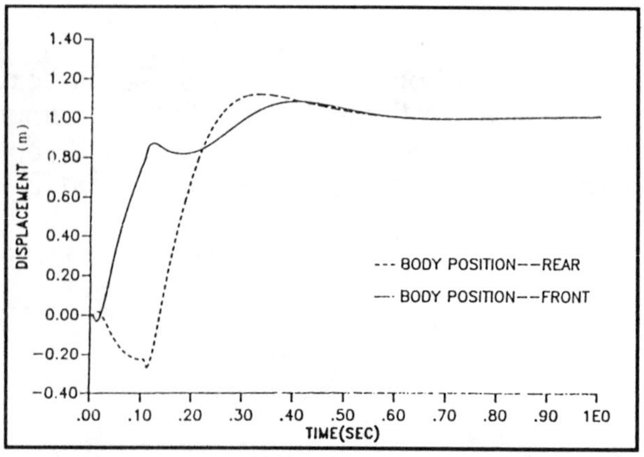

Fig. 7. Front and Rear Body Displacements for Active Half Car $\rho = .2 \times 10^{-9}$

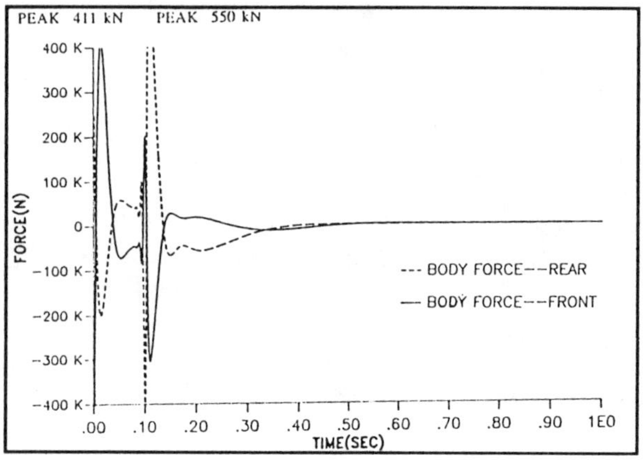

Fig. 8. Front and Rear Body Forces for Active Half Car $\rho = .2 \times 10^{-9}$

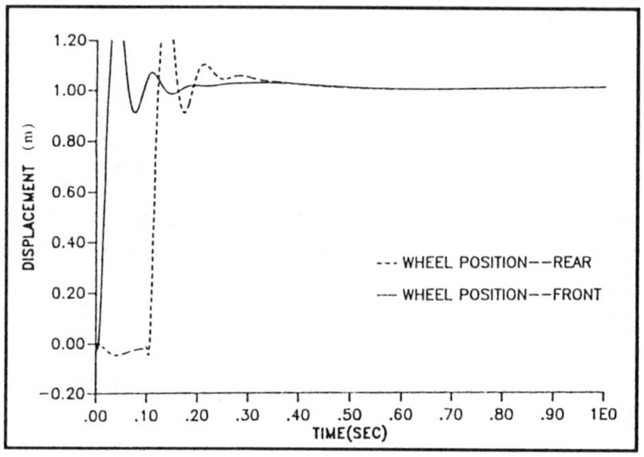

Fig. 9. Front and Rear Tire Displacements for Active Half Car $\rho = 1.0 \times 10^{-9}$

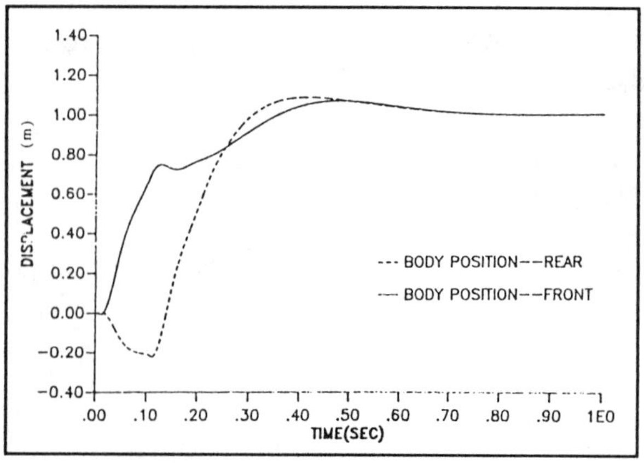

Fig. 10. Front and Rear Body Displacements for Active Half Car $\rho = 1.0 \times 10^{-9}$

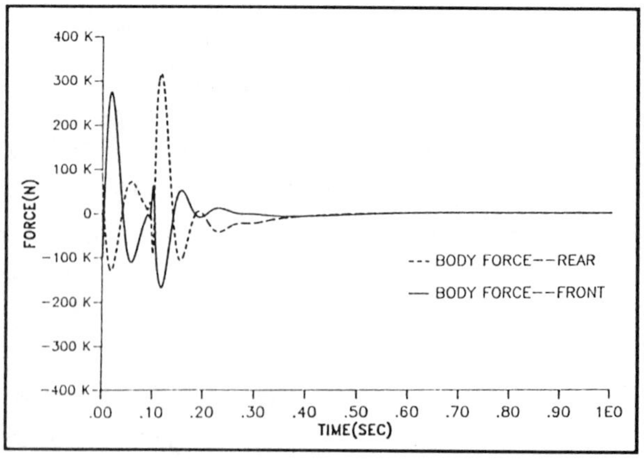

Fig. 11. Front and Rear Body Forces for Active Half Car $\rho = 1.0 \times 10^{-9}$

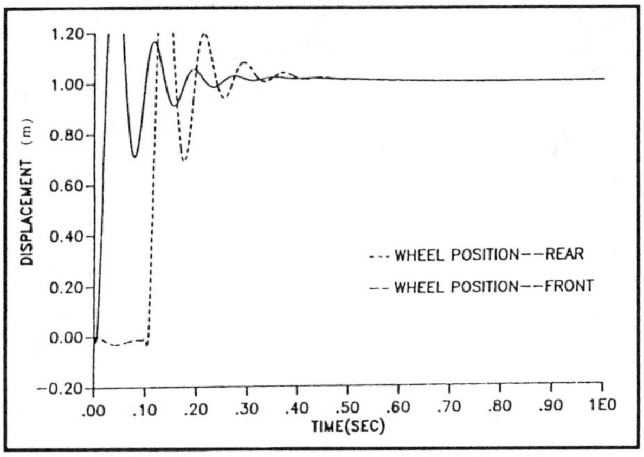

Fig. 12. Front and Rear Tire Displacements for Active Half Car $\rho = 3.0 \times 10^{-9}$

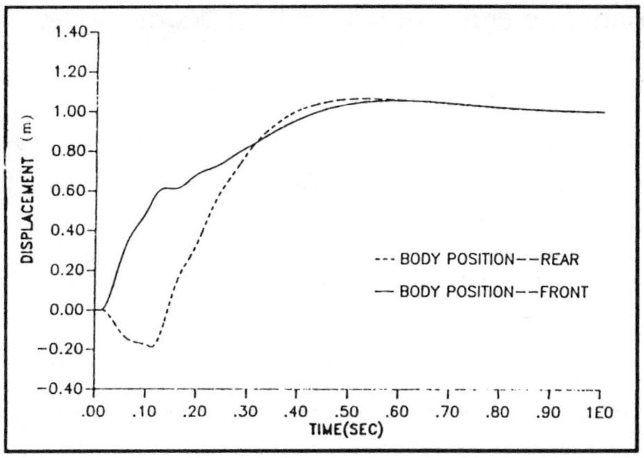

Fig. 13. Front and Rear Body Displacements for Active Half Car $\rho = 3.0 \times 10^{-9}$

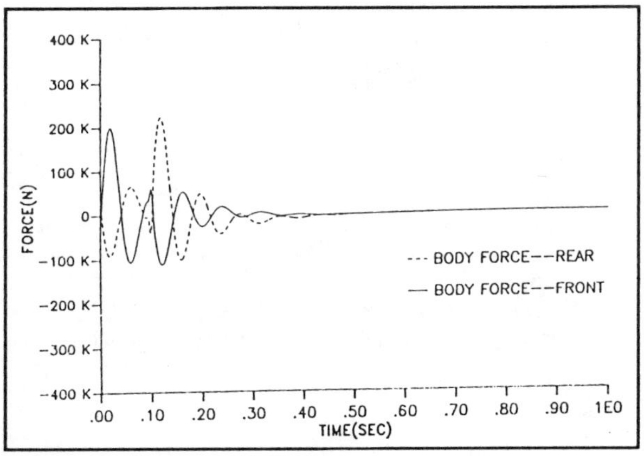

Fig. 14. Front and Rear Body Forces for Active Half Car $\rho = 3.0 \times 10^{-9}$

RESPONSE OF HALF-CAR WITH ACTIVE SYSTEM FAILURE

Having noticed above the variability in the spring and damping constant of the active system versus the passive system, it is interesting to consider what would happen if the active system would fail to function. This could be a result of a computer error, a mechan-ical malfunction in the hydraulics, or a sensor error. A series of simulations were conducted using the computed optimum values of [k], i.e. the spring constants equal to k_{11} and k_{23}, and the damping constants equal to k_{15} and k_{27}, but the active force suppressed. One can see in figures 15 and 16 that the body forces and the body displacements would be unacceptable in the case where $\rho = 0.2 \times 10^{-9}$. The maximum body displacements are, front and rear, 1.85 m and 2.38 m, respectively; and the maximum body forces, front and rear, are 497 kN (a negative value) and 726 kN, respectively. (Compare to passive response, fig's. 3 and 4) It is proposed that the spring constant be assigned the same value used in conventional passive systems. This would result in new active forces $\tilde{f}_1$ and $\tilde{f}_2$ dependent upon k_s and k_d.

spring constant = k_s = 19960 N/m

damping constant = k_d = 1861 Ns/m

$$\tilde{f}_1 = (k_{11}-k_s)(x_1-x_2) + (k_{15}-k_d)(x_5-x_6) + (k_{11}+k_{12})(x_2-x_{01}) + k_{13}(x_3-x_{02}) + k_{14}(x_4-x_{02}) + (k_{15}+k_{16})x_6 + k_{17}x_7 + k_{18}x_8 \quad (19)$$

$$\tilde{f}_2 = (k_{23}-k_s)(x_3-x_4) + (k_{27}-k_d)(x_7-x_8) + (k_{23}+k_{24})(x_4-x_{02}) + k_{21}(x_1-x_{01}) + k_{22}(x_2-x_{01}) + (k_{27}+k_{28})x_8 + k_{25}x_5 + k_{26}x_6 \quad (20)$$

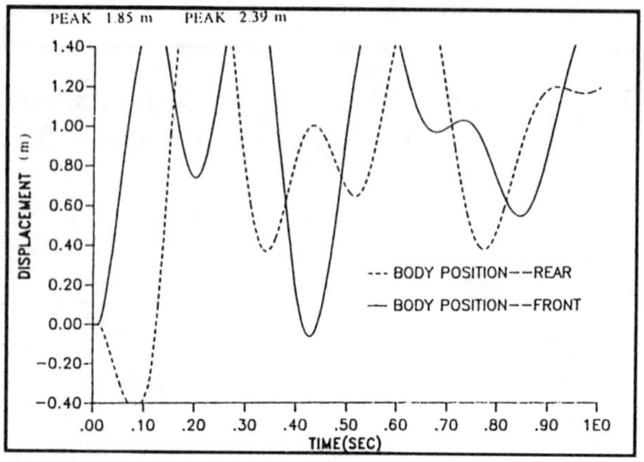

Fig. 15. Front and Rear Body Displacements Active Elements Suppressed $\rho = .2 \times 10^{-9}$

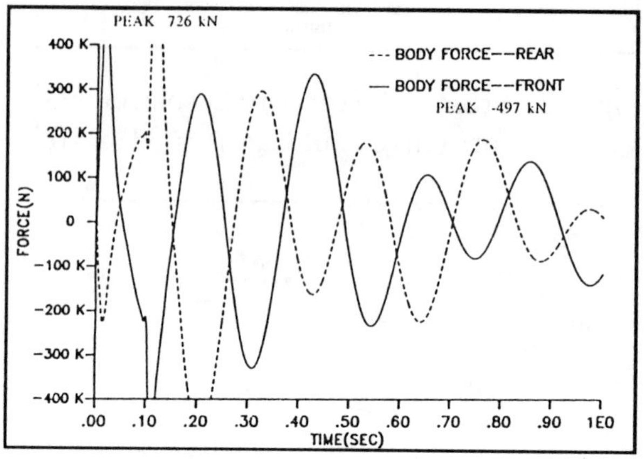

Fig. 16. Front and Rear Body Forces Active Elements Suppressed $\rho = .2 \times 10^{-9}$

CONCLUSIONS

There are three conclusions about the half-car active system: the active system is better than the passive system, $\rho = 3.0 \times 10^{-9}$ is the best of those investigated, and there is a better way to implement the active system in case of system failure. The active system, with a proper choice of ρ, shows a vast improvement in the indicators of ride comfort: peak body displacement and peak body force. The best choice of a controller is the one corresponding to $\rho = 3.0 \times 10^{-9}$, based on three criteria. First, the peak body position is the lowest. Second, the peak body force is the lowest of those simulated. Last, the cost function is the lowest, and this results in the least loss of power. Although the recovery time has increased with respect to the other ρ's it was felt that overall this is the best ρ. Finally, such active systems should be implemented in the manner that allows for conventional shock absorbers installed in parallel with actuators. This optimum system has a passive shock absorber with parameters equal to shock absorbers presently in use.

References

1. Soltis, M.W. Feb. 1987. "Programmed Suspension Provides Ride Control." Automotive Engineering, Society of Automotive Engineering, Inc., 95(2):122-132.

2. Birch, S. Sept. 1988. "Suspensions." Automotive Engineering, Society of Automotive Engineers, Inc., 96(9):73-76.

3. Barrie, C. Mar. 5, 1987. "Using the Right Formula for Suspension Control." The Enigneer, 264(6832):36.

4. McElroy, J. Sept. 1986. "GM's First Lotus Blossom: Active Suspension Corvette." Chilton's Automotive Industries, 166(9):108-110.

5. Thompson, A.G. 1976. "An Active Suspension with Optimal Linear State Feedback." J. Vehicle System Dynamics, 5:187-203.

6. Thompson, A.G. 1984. "Optimal and Suboptimal Linear Active Suspension for Road Vehicles." J. Vehicle System Dynamics, 13:61-72.

7. Margolis, D.L. 1982. "Semi-Active Heave and Pitch Control for Ground Vehicles." J. Vehicle System Dynamics, 11:31-42.

8. Karnopp, D. 1983. "Active Damping in Road Vehicle Suspension Systems." J. Vehicle System Dynamics, 12:291-316.

9. Wilson, D.A., R.S. Sharp and S.A. Hassan. 1986. "The Application of Linear Optimal Control Theory to the Design of Active Automotive Suspensions." J. Vehicle System Dynamics, 15:105-118.

10. Sommerfeldt, S.D. 1989. "Adaptive Vibration Control of Vibration Isolation Mounts, Using an LMS-Based Control Algorithm." Ph.D. Thesis, Penn State University, University Park, PA.

11. Sommerfeldt, S.D. and J. Tichy. December 4-6, 1989. "Adaptive Vibration Control Using an LMS-Based Control Algorithm." Proc. Internoise '89, held at Newport Beach, CA, USA, 513-518.

ELECTROMECHANICAL MODELLING OF AN ACTIVE ISOLATION SYSTEM

M. D. McCollum,[1] S. E. Forsythe,[1] A. D. McCleary[1]

ABSTRACT

The importance of detailed modelling of control element dynamics is demonstrated for the case of a distributed piezoelectric vibration isolator composed of a piezocomposite actuator/sensor combination mounted on a uniformly vibrating surface in air. Several system transfer functions were obtained during preliminary measurements, including the ratio of the acceleration of the top surface of the composite structure to the voltage out of the sensor element. It was discovered that this transfer function is different for the case of driving the base structure while leaving the actuator element open circuit, than for the case of driving the actuator while keeping the base rigid. The significance of this difference is that in the actual control situation, both the base structure and the actuator are driven; therefore, minimization of the sensor output does not necessarily imply minimization of the surface motion. The cause of and solution for this behavior is investigated using first a simple analytical model, and then a detailed finite element model. Results obtained using these models are compared with those obtained from experimental measurements.

INTRODUCTION

An accurate predictive model is an important tool for the designer of an active control system. Such a model can be used to determine optimum placement of actuators or to evaluate candidate materials without building and testing many different prototype systems. The amount of effort required to develop an adequate system model depends on both the nature of the control elements and the applicability of certain simplifying assumptions in the frequency range of interest. This paper will focus on the analysis of distributed piezoelectric actuators and sensors at frequencies below the first resonance of the composite isolator structure.

The usefulness of piezoelectric distributed sensors and actuators for active control/isolation applications has been demonstrated by a number of investigators. A variety of piezoelectric materials have been used, including various lead zirconate titanate (PZT) ceramics and piezoelectric polymers. Piezoceramics typically have higher electromechanical coupling coefficients than do piezopolymers, while the polymers are more rugged and offer the capability of bonding to curved surfaces.

[1]Naval Research Laboratory, Underwater Sound Reference Division, P.O. Box 568337, Orlando, FL 32856-8337, U.S.A.

Bailey and Hubbard [1] used a piezoelectric polymer known as polyvinylidene flouride (PVDF) bonded to one complete surface of a cantilever beam as an active vibration damper. The system is modelled as a two-layer Euler-Bernoulli beam. A voltage applied through the thickness of the PVDF results in a transverse strain in both layers, through the effect of the d_{33} piezoelectric constant. The net force produced in each layer acts as a moment about the neutral axis of the composite beam. The thickness strain in the PVDF layer resulting from the d_{33} constant has a negligible influence on the motion of the base beam because of the small inertia developed in the low density polymer.

Further work with piezoelectric actuators was performed by Crawley and deLuis [2] who used pieces of piezoceramic bonded symmetrically to both sides of a cantilever beam to excite flexure in the beam by driving the actuators out of phase. It was shown that for a one-dimensional beam, the transverse strain produced in the actuators can be modelled as a moment applied at the ends of the actuator. This work was extended by Dimitriadis, et al. [3] to two-dimensional "patches" of piezoceramic. This analysis shows that the transverse strains in the actuators result in moments applied around the perimeter of the actuator area. In both investigations, the influence of the d_{33} piezoelectric constant, and the resultant thickness strain, is neglected.

A rather different application of piezoelectric actuators was proposed by Tzou and Gadre [4]. They bonded Plexiglas plates to each side of a thin slab of PVDF, then added a heavy plate to the top, and finally mounted the composite structure on a shaker. The signal from an accelerometer placed on the top of the structure was filtered and applied across the thickness of the PVDF in order to null the vibration of the top surface. In the associated analytical work, the transverse strains developed in the piezopolymer are neglected because of the stiffness of the attached layers and the large ratio of area to thickness. The behavior of the piezoelectric layer is described by the coupling between the voltage applied through the thickness and the strain developed in the thickness dimension (the d_{33} constant). This one-dimensional model is shown to describe adequately the behavior of the system.

The objective of the work presented here is similar to that of Tzou and Gadre [4], that is, to create an effectively rigid surface over a vibrating base structure using active control techniques. The vibration isolation element is made up of an actuator, which is mounted on the vibrating surface, and a sensor, which is bonded to the top surface of the actuator. The material used for both elements is a piezoelectric composite comprising lead titanate particles in a neoprene matrix (tradename PZR [5]).

The sensor and actuator are each composed of two circular layers of PZR, poled in the thickness direction, which are electrically in parallel. Thin, stiff, elastic layers separate successive layers of piezoelectric material. A stiffening layer is also bonded to the top of the composite element. The voltage out of the sensor is a function of the area average of the thickness strain and the extensional strain in the sensor material. This voltage is filtered using an appropriate control algorithm, and then becomes an input to the actuator. The function of the control algorithm is to minimize the sensor voltage. A sketch of the control element configuration is shown in Figure 1 where V_s is the voltage out of the

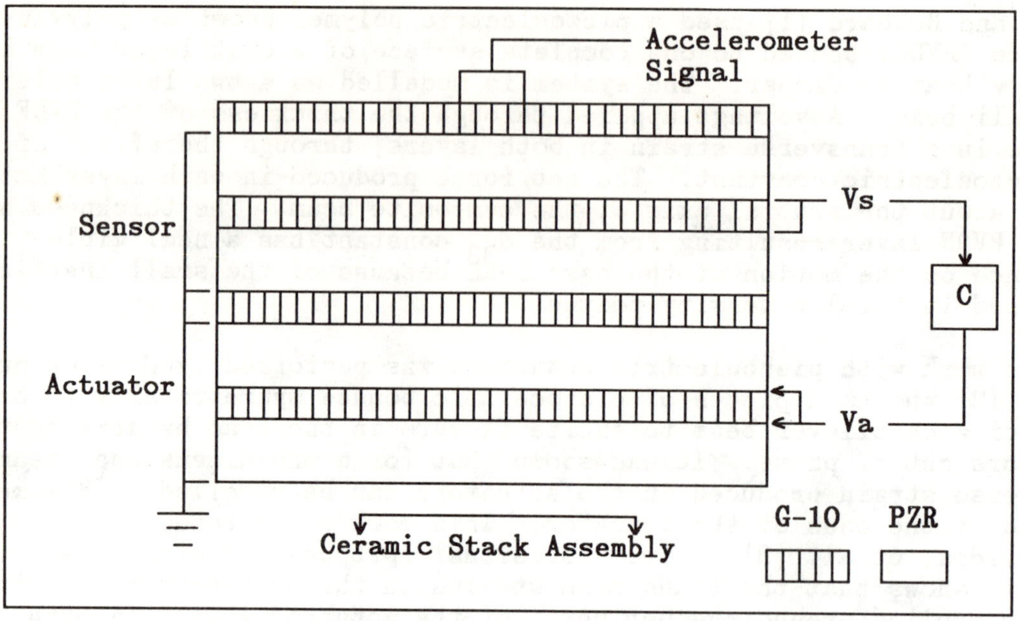

Figure 1. COMPOSITE ISOLATOR CONFIGURATION

sensor, V_a is the voltage into the actuator, and C is the controller. The development of the control algorithm is described in a companion paper [6].

Measurements of various transfer functions revealed that the ratio of the top surface acceleration, measured by an independent accelerometer, to the sensor output voltage is highly dependent on the drive conditions. This behavior was not predicted by a one-dimensional model of the isolator. Other possible explanations, including cross talk between the actuator and the sensor when driving the actuator, were eliminated. A more detailed model which includes extensional motion (through Poisson's ratio and the d_{31} constant) was developed using the finite element method. The transfer function predicted using this model demonstrated the same drive-dependent behavior as was found experimentally. In addition, the finite element method's capability of producing explicit graphical representations of the dynamic displacement field of the structure was useful in studying the behavior of the isolator structure.

ANALYSIS

In developing both the analytical and numerical models of the composite isolator, some basic assumptions are made about the nature of the materials. It is assumed that the elastic and piezoelectric properties of the materials are linear, lossless, and frequency independent over the frequency range of interest. In addition, the PZR layers are assumed to be perfectly bonded to the base plane and to the stiffening layers so that the effects of finite bonding layers are not considered.

The one-dimensional model of the composite isolator structure requires the further assumption that the effect of extensional motion of the piezoelectric layers on the thickness motion and on the sensor voltage is negligible. Also, the elastic stiffening layers are excluded from this model. The one-dimensional analysis is based on an equivalent circuit model of a single piezoelectric layer having only thickness motion [7] to represent each of the isolator components. Two such circuits are cascaded to form the composite isolator model. The impedance terms (electrical Z_b, mechanical Z_ϕ, and electromechanical Z_{em}) for each three-port network must be corrected to account for the fact that the actuator and sensor are each made up of two layers of PZR wired in parallel rather than a single layer of equivalent thickness. These corrections are as follows: $Z_b=Z_b'/4$, $Z_\phi=Z_\phi'$, and $Z_{em}=Z_{em}'/2$, where the prime indicates the impedance of a single layer of equivalent thickness. The system of equations is formed by writing an equation for each loop in the circuit in terms of the force at the base, the velocity of the actuator/sensor interface, the velocity of the top surface, the actuator current, and the sensor voltage. The base velocity and the actuator drive voltage are specified.

In developing the finite element model of the isolator, it is assumed that the vibration is axisymmetric so that it is necessary to model only a radial slice of the structure. This simplification reduces the number of degrees of freedom in the model from several hundred to about one hundred fifty, resulting in significantly shorter computational times. For a single isolator vibrating in air this is a reasonable assumption. The finite element code ATILA [8], which includes piezoelectricity, is used to develop the axisymmetric isolator model; the mesh is shown in Figure 2. The elements are eight-noded quadrilaterals based on a quadratic shape function. The boundary conditions are as follows: no displacements across the axis of symmetry, no extensional motion along the isolator/base interface, and uniform thickness displacement along the isolator/base interface. Results of the finite element and one-dimensional models will be presented later in the paper.

EXPERIMENTAL SET-UP

The actuator and sensor are each made up of two 0.125-inch-thick, 2.25-inch-diameter PZR discs which are poled in the thickness direction and electroded with thin copper foil and conductive adhesive (see Figure 1). The electroded PZR discs are separated by a 0.03-inch-thick disc of G-10 composite. The sensor is shielded from the actuator and the ceramic stack by copper foil. An accelerometer is mounted on the top surface of the isolator in order to provide an independent measurement of the surface to be controlled. An additional accelerometer is bonded to the base plate. The base is driven by a compliantly mounted piezoelectric ceramic stack having a fundamental resonance frequency near that of the isolator.

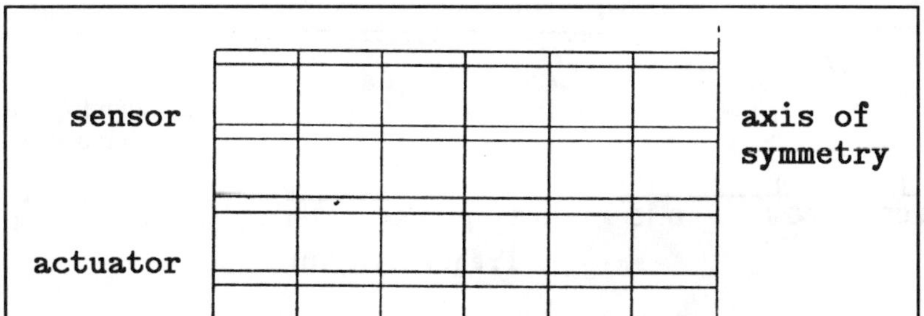

Figure 2. FINITE ELEMENT MESH

RESULTS AND DISCUSSION

Because of the frequency dependence of the ceramic stack compliance, measured accelerations and voltages depend on the response of not only the isolator but also the ceramic stack; therefore only ratios of measured quantities may be compared to model results. In this section, measurements of the acceleration to sensor voltage ratio are compared with results obtained from the one-dimensional and finite element models. Further model results for different boundary conditions and material properties are then presented in order to explain the cause of, as well as to determine a solution to, the problem of the drive-dependent transfer function.

In Figures 3(a), (b), and (c), the transfer function between the sensor voltage and the surface acceleration (obtained from measured data, the finite element model, and the one-dimensional model, respectively) is presented. The measured data demonstrates that this ratio is nearly constant with frequency when the base is excited; but when the actuator is driven, the ratio is approximately 10 dB down at the lowest frequencies and increases with frequency until the levels match those of the base-driven case. The transfer function obtained from the one-dimensional model is independent of drive conditions, indicating that the assumption of no extensional motion is not valid. Conversely, the results of the finite element model demonstrate the same drive dependence of the transfer function as was found experimentally. A comparison of Figures 3(a) and (b) shows that the model predicts approximately the same difference between the transfer functions as the measured results, although the absolute levels of the predicted results are between 1 and 3 dB lower than measured levels. This difference between predicted and measured levels is attributable to inaccurate material constants and the actual frequency dependence of the material properties in addition to experimental error.

In Figures 4 and 5, the harmonic displacement fields for the lowest ($0.05f_0$) and highest ($0.45f_0$) frequencies are shown for the two drive cases. Figures 4(a) and (b) demonstrate that the voltage applied across the actuator results in significant extensional displacement in the actuator, through the d_{31} piezoelectric coupling, and in the sensor, through the mechanical coupling of the two elements, at both the low and

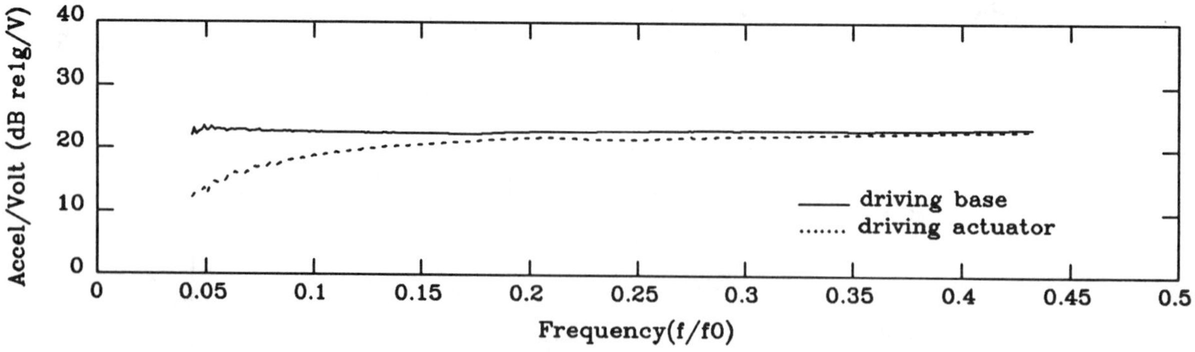

3(a) Measured Data (continued)

Figure 3. RATIO OF SURFACE ACCELERATION TO SENSOR VOLTAGE

(Figure 3. continues)

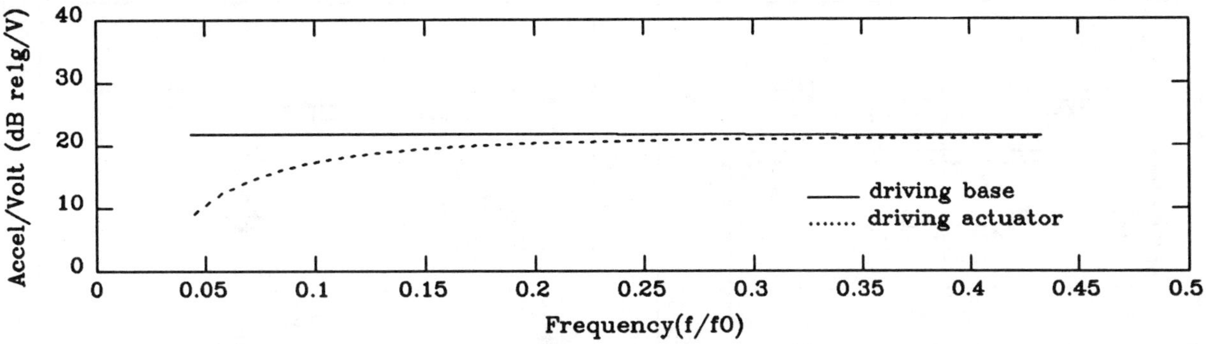

3(b) Finite Element Model

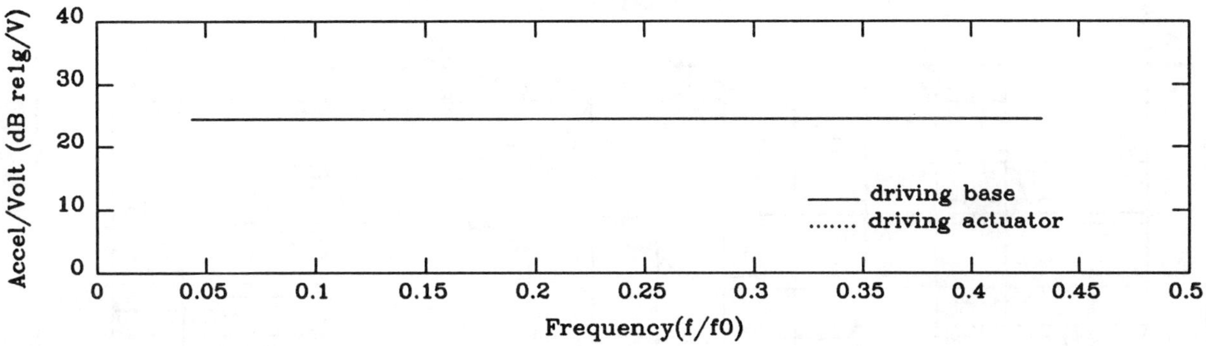

3(c) One-dimensional Model

high frequencies. Conversely, it is seen in Figure 5(a) that for the case of driving the base at low frequencies, the isolator moves as a rigid body with no extensional motion. At the high frequency, however, the base-driven structure demonstrates the Poisson effect and the displacement field becomes similar to that of the actuator-driven structure as shown in Figure 5(b). At frequencies for which the normalized displacement fields are similar for the two drive cases, the corresponding transfer function levels agree.

Results of the finite element and one-dimensional models are compared in Figures 6(a) and (b) for the surface acceleration and the sensor voltage, respectively, for the case of driving the actuator. The corresponding results for the case of driving the base are shown in Figures 7(a) and (b). These figures show that there is little difference between the surface accelerations predicted by the two models; however, there is a large difference between the results for the sensor voltage. The voltage predicted by the one-dimensional model is lower than that predicted by the finite element for both drive conditions. This is not surprising since the sensor voltage is a result of both the d_{33} and the d_{31} coupling of strain to electrical potential.

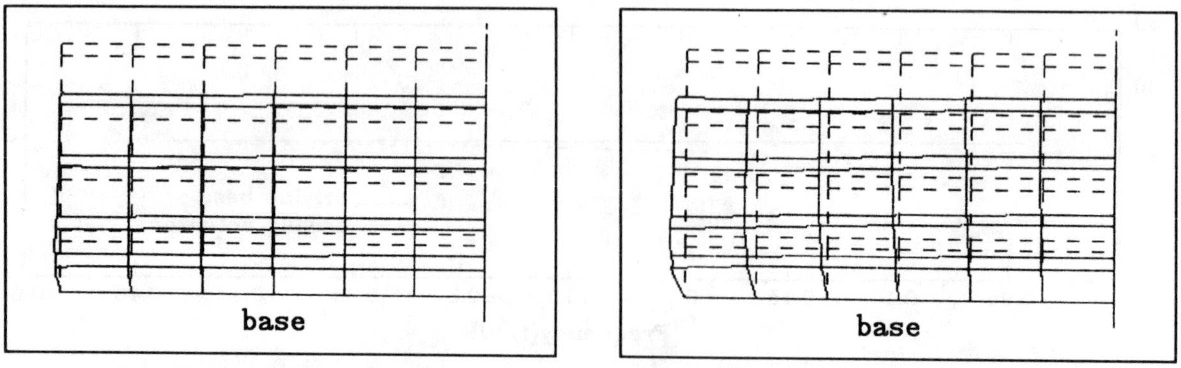

4(a) Frequency = $0.05f_0$ 4(b) Frequency = $0.45f_0$

Figure 4. NORMALIZED DISPLACEMENT FIELD FOR ACTUATOR-DRIVEN CASE
(___deformed, ---undeformed)

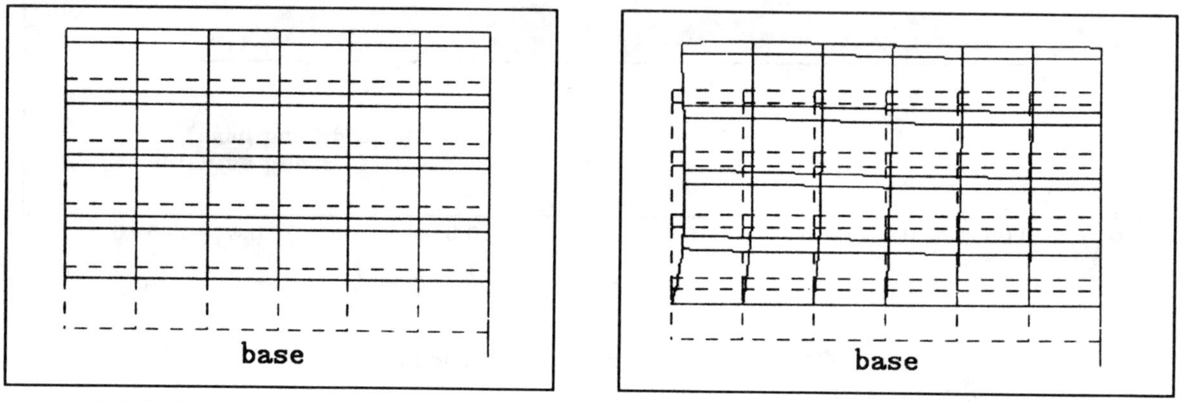

5(a) Frequency = $0.05f_0$ 5(b) Frequency = $0.45f_0$

Figure 5. NORMALIZED DISPLACEMENT FIELD FOR BASE-DRIVEN CASE
(___deformed, ---undeformed)

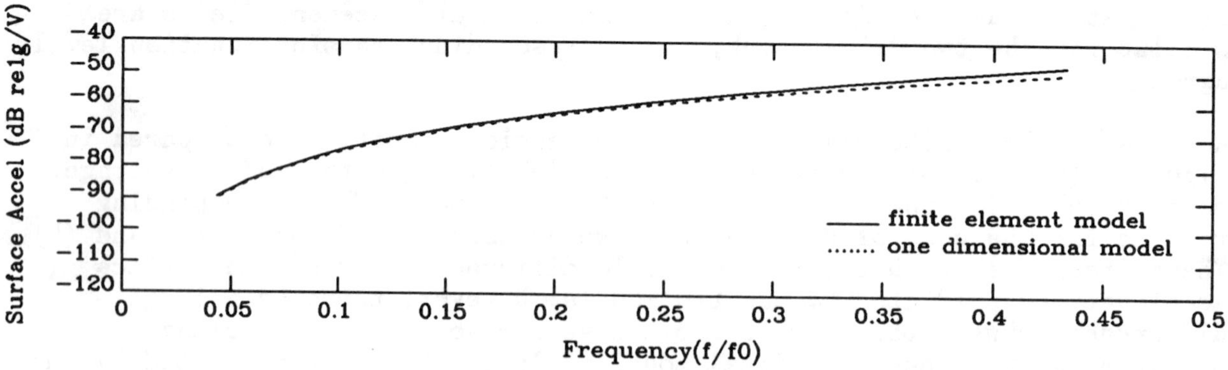

6(a) Surface Acceleration (continued)

Figure 6. MODEL RESULTS FOR ACTUATOR-DRIVEN CASE

(Figure 6. continues)

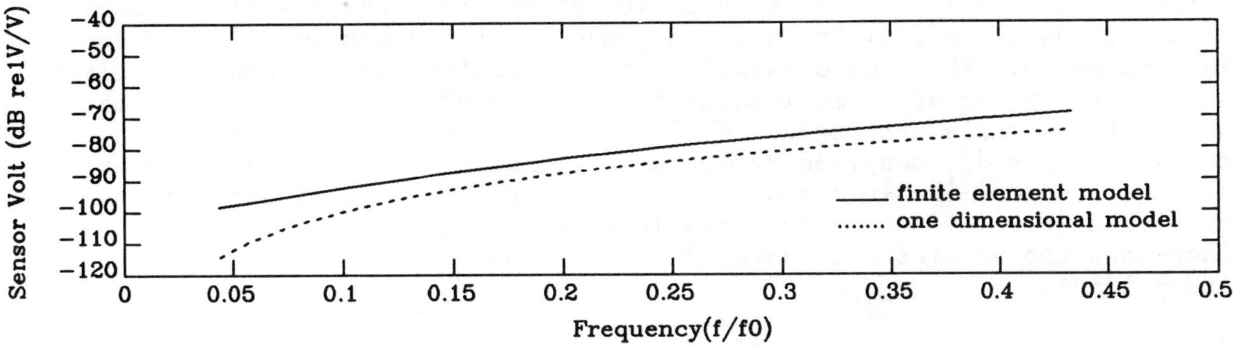

6(b) Sensor Voltage

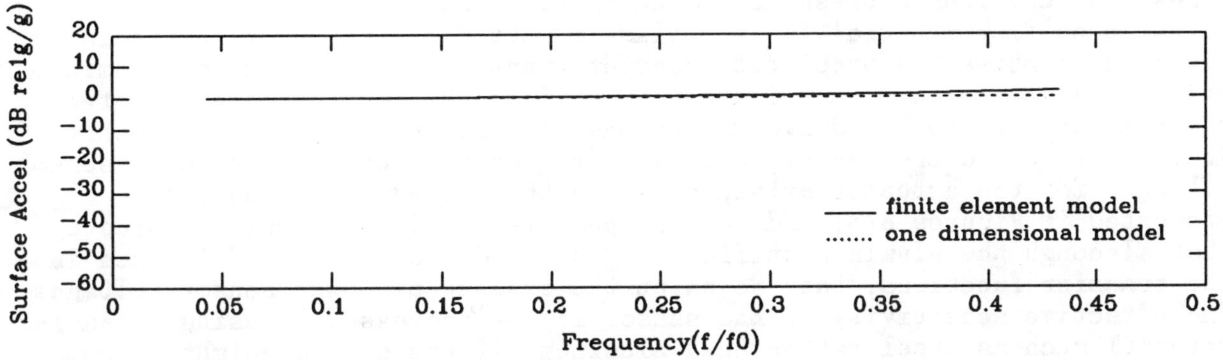

7(a) Surface Acceleration

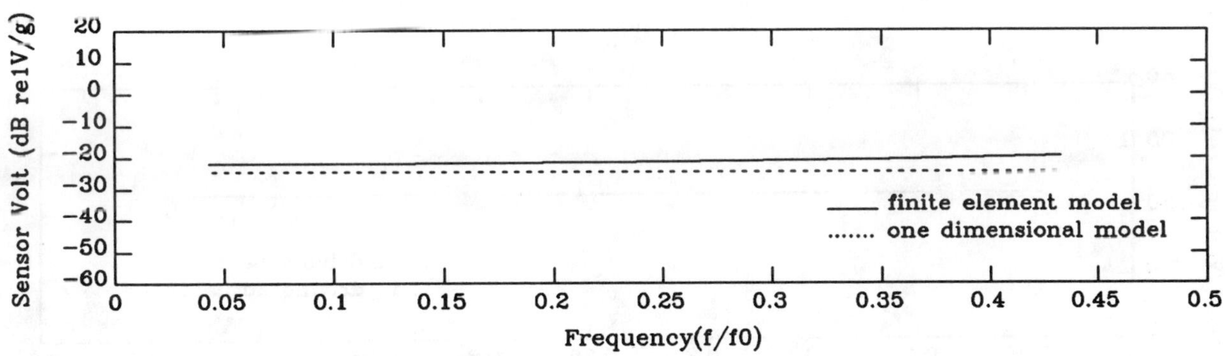

7(b) Sensor Voltage

Figure 7. MODEL RESULTS FOR BASE-DRIVEN CASE

Based on these results obtained from the baseline finite element model, several more computations were run with different material constants and/or boundary conditions. The two most interesting of these additional cases involve: 1) setting the d_{31} constant of the actuator material to zero and 2) specifying uniform displacements across the top surface of the actuator. In the first case, a voltage applied across the actuator drives only thickness strains in that element and the extensional motion of the sensor depends only on Poisson's ratio of the material so that the actuator-driven condition is similar to the base-driven condition. The results obtained from this model indicate no drive dependence of the acceleration to sensor voltage ratio. The second variation of the finite element model involves the specification of an additional boundary condition while leaving the material constants unchanged. With this model in the actuator-driven condition, the d_{31} coupling causes extensional strains in the actuator; however, because the in-plane mechanical coupling is eliminated, these strains do not lead to similar displacements in the sensor element. Therefore the predicted transfer functions for this model are drive independent.

The results of these model variations indicate two potential solutions to the problem of the drive dependence of the transfer function. However, since it is not possible to change the properties of the material itself, only one practical solution remains, that is, to eliminate the in-plane mechanical coupling between the actuator and sensor. One way to do this is to use a stiffer material for the elastic interface layers of the isolator. Figure 8(a) shows the predicted transfer function for the case of aluminum stiffening layers in place of the G-10 stiffeners. The difference between the results for the two drive conditions is less than 2 dB at the lowest frequencies and decreases rapidly with frequency. The corresponding sensor voltages for the actuator-driven case and the base-driven case are presented in Figures 8(b) and (c), respectively. These results indicate that although the aluminum stiffeners greatly decrease the differences in the transfer functions, they do so at the expense of lower sensor voltages. The effective sensitivity of the sensor may be increased by using a heavier material such as steel rather than aluminum, if the system weight is not constrained. In fact, the model indicates that steel stiffeners not only increase the sensor voltage output, but completely eliminate the differences in the transfer functions for the two drive conditions.

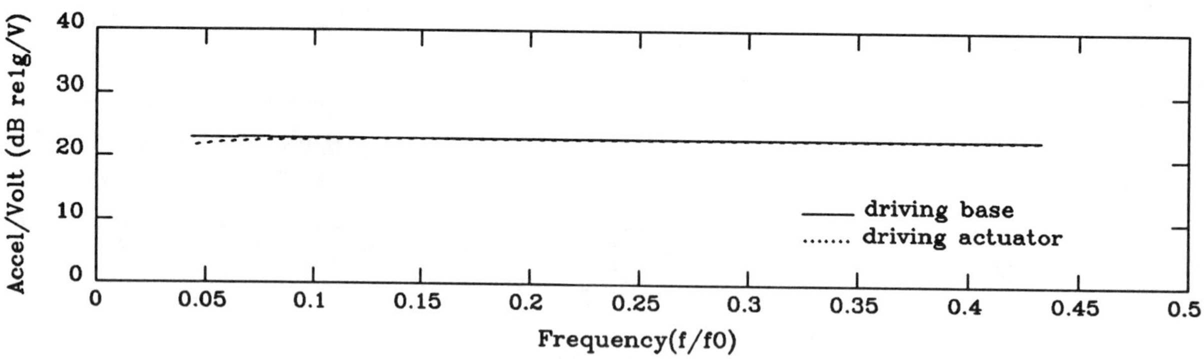

8(a) Ratio of Surface Acceleration to Sensor Voltage (continued)

Figure 8. COMPARISON OF G-10 AND ALUMINUM STIFFENING LAYERS

(Figure 8. continues)

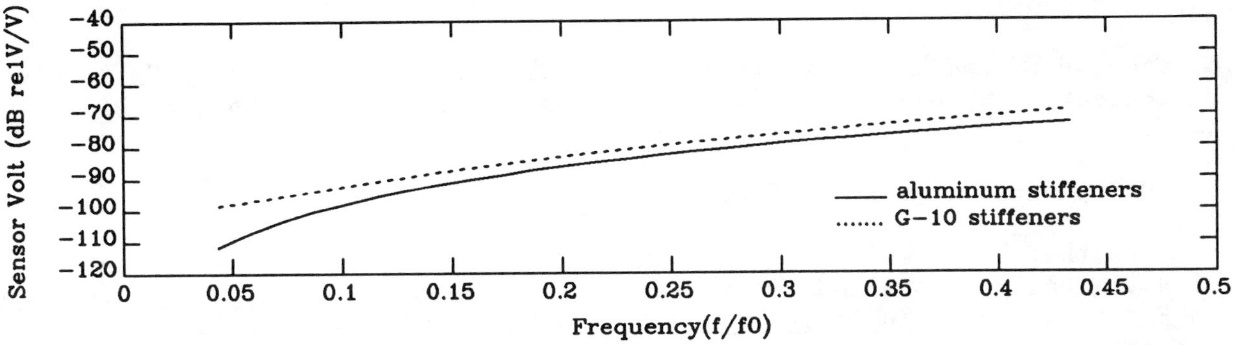

8(b) Sensor Voltage for Actuator-driven Case

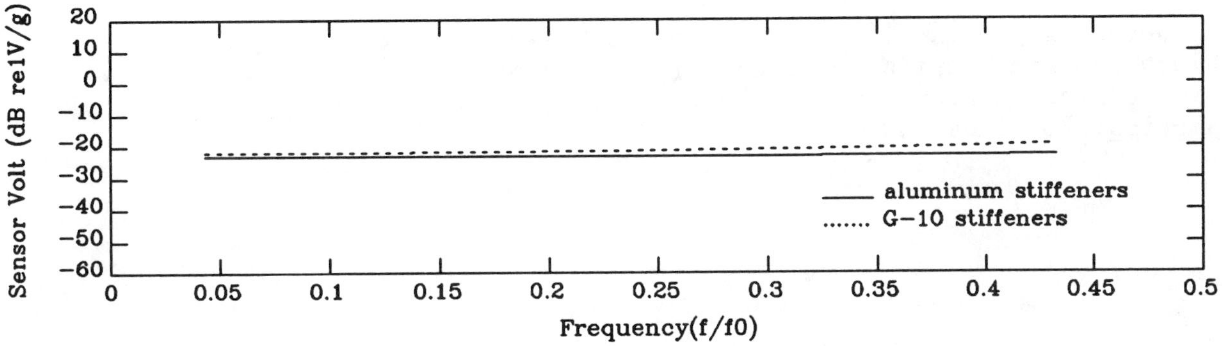

8(c) Sensor Voltage for Base-driven Case

SUMMARY

The work presented here demonstrates the importance of detailed modelling for a distributed isolation system using a piezoelectric composite material. A one-dimensional model fails to give an accurate prediction of the voltage output from the sensor element because of the absence of the d_{31} piezoelectric coupling and the Poisson effect in the model. The finite element model includes this coupling and gives good agreement with measured data. It has been shown that this thickness to extensional coupling is the cause of the dependence of the surface acceleration to sensor voltage ratio on the drive condition. A potential solution to this drive dependence lies in the use of stiffer, heavier elastic interface layers.

REFERENCES

1. Bailey, T. and J.E. Hubbard, 1985. "Distributed Piezoelectric-Polymer Active Vibration Control of a Cantilevered Beam." AIAA Journal of Guidance and Control, 6(5): 605-611.

2. Crawley, E.F. and J. deLuis, 1987. "Use of Piezoelectric Actuators as Elements of Intelligent Structures." AIAA JOURNAL, 25(10): 1373-1385.

3. Dimitriadis, E.K., C.R. Fuller, and C.A. Rogers, 1989. "Piezoelectric Actuators for Distributed Noise and Vibration Excitation of Thin Plates." (advance copy).

4. Tzou, H.S. and M. Gadre, 1988. "Active Vibration Isolation by Polymeric Piezoelectret with Variable Feedback Gains," AIAA Journal, 26(8): 1014-1017.

5. NGK Sparkplugs (U.S.A.),Inc.

6. Forsythe, S.E., M.D. McCollum, and A.D. McCleary, 1991. "Stabilization of a Digitally Controlled Active Isolation System." Proceedings, Conference on Recent Advances in Active Control of Sound and Vibration, in preparation.

7. Woollett, R.S. 1990. Sonar Transducer Fundamentals, New London,CT: Naval Underwater Systems Center.

8. Decarpigny, J.N., J.C. Debus, et al., 1985. "In Air Analysis of Piezoelectric Tonpilz Transducer in a Wide Frequency Band Using a Mixed Finite Element Plane Wave Method," Journal of the Acoustical Society of America, 78: 1499-1507.

Author Index